Stephen Hanessian, Simon Giroux,
and Bradley L. Merner

Design and Strategy in Organic Synthesis

Related Titles

Sierra, M. A., de la Torre, M.

More Dead Ends and Detours

En Route to Successful Total Synthesis

2013
SBN: 978-3-527-32976-2

Nicolaou, K. C., Chen, Jason S.

Classics in Total Synthesis III

Further Targets, Startegies, Methods

2011
ISBN-13: 978-3-527-32958-8

Wyatt, P., Warren, S.

Organic Synthesis

Strategy and Control

2007
ISBN: 978-0-471-92963-5

Warren, S., Wyatt, P.

Workbook for Organic Synthesis

Strategy and Control

2008
ISBN: 978-0-471-92964-2

Carreira, E. M., Kvaerno, L.

Classics in Stereoselective Synthesis

2008
ISBN: 978-3-527-32452-1

Hudlicky, T., Reed, J. W.

The Way of Synthesis

2007
ISBN-13: 978-3-527-32077-6

*Stephen Hanessian, Simon Giroux,
and Bradley L. Merner*

Design and Strategy in Organic Synthesis

From the *Chiron Approach* to Catalysis

WILEY-VCH

Verlag GmbH & Co. KGaA

The Authors

Prof. Dr. Stephen Hanessian
Department of Chemistry
University of Montreal
2900 Edouard Montpetit
Montreal, QC H3T 1J4
Canada

and

Departments of Pharmaceutical Sciences
and Chemistry
University of California, Irvine
Irvine, CA, 92697-2025
USA

Dr. Simon Giroux
Vertex Phamaceuticals
130, Wavely Street
Cambridge, MA 02139-4242
USA

Prof. Dr. Bradley L. Merner
Department of Chemistry
and Biochemistry
Auburn University
179 Chemistry Building
Auburn, AL 36849-5312
USA

Cover:
The completion of a total synthesis of a
complex organic molecule is often likened
to climbing a mountain and reaching the
summit. The cover depicts the structures
of nymania 1 (rubrin A) and bipleophylline
two molecules representing a high level
of structural complexity that have not yet
succumbed to total synthesis.

io 0699375 4

■ All books published by **Wiley-VCH** are
carefully produced. Nevertheless, authors,
editors, and publisher do not warrant the
information contained in these books,
including this book, to be free of errors.
Readers are advised to keep in mind that
statements, data, illustrations, procedural
details or other items may inadvertently be
inaccurate.

Library of Congress Card No.: applied for

**British Library Cataloguing-in-Publication
Data**
A catalogue record for this book is available
from the British Library.

**Bibliographic information published by the
Deutsche Nationalbibliothek**
The Deutsche Nationalbibliothek
lists this publication in the Deutsche
Nationalbibliografie; detailed bibliographic
data are available on the Internet at
<http://dnb.d-nb.de>.

© 2013 Wiley-VCH Verlag GmbH & Co.
KGaA, Boschstr. 12, 69469 Weinheim,
Germany

Composition Laserwords Private Ltd.,
Chennai
Printing and Binding betz-druck GmbH,
Darmstadt
Cover Design Adam-Design, Weinheim
Cover Art Benoît Deschênes Simard

Printed on acid-free paper
Printed in the Federal Republic of Germany

ISBN: 978-3-527-31964-0

To our families, children and grandchildren.

Contents

Foreword   *XVII*
Preface   *XIX*
Acknowledgement   *XXIII*
Abbreviations   *XXV*

1   **The Concept of Synthesis**   *1*
1.1   Organic Synthesis as a Central Science   *1*
1.2   Organic Chemistry and the Public   *3*
1.3   The "Small Molecules" of life   *6*
1.4   Nature's Rules   *9*
1.5   Organic Synthesis as a Mental and Visual Science   *11*
1.6   Art, Architecture, and Synthesis   *12*
1.7   Simplification of Complexity   *13*
1.8   Seeing Through the Mind's Eye   *15*
1.9   Beauty is in the Eye of the Beholder   *18*
            References   *19*

2   **The "Why" of Synthesis**   *25*
2.1   Nature the Provider, Healer, and Enticer   *25*
2.2   The Supply Problem   *26*
2.3   From Bench to Market   *28*
2.4   Thank you Nature!   *30*
2.5   Chasing Bugs with a Purpose   *32*
2.6   Structure-based Organic Synthesis   *33*
2.7   Almost There . . . or Just Arrived   *34*
2.8   The Futility of it All   *35*
2.9   Synthesis as a Seeker of the Truth   *36*
2.10   Nature as the Ultimate Synthesizer   *42*
2.11   A Brave New Chemical World   *43*
2.11.1   Beyond the molecule   *43*
2.11.2   Buckyballs and fullerenes   *45*
2.11.3   Dendrimers   *45*
2.11.4   Nanochemistry   *45*

2.11.5   Molecular machines   *46*
2.12   Exploring New Synthetic Methods   *47*
2.12.1   The Diels-Alder reaction   *48*
2.12.2   The direct aldol reaction   *54*
2.12.3   High impact catalytic oxidation and reduction reactions   *57*
2.12.4   High impact catalytic olefin-producing reactions   *59*
          References   *62*

3   **The "*What*" of Synthesis**   *73*
3.1   Periods, Trends, and Incentives   *73*
3.2   A Century of Synthesis   *74*
3.3   We the "Synthesis People"   *81*
3.4   Complex and Therapeutic too!   *82*
3.5   Peptidomimetics and Unnatural Compounds   *84*
3.6   Diversity Through Complexity   *89*
          References   *90*

4   **The "*How*" of Synthesis**   *97*
4.1   The Visual Dialogue   *97*
4.2   The Psychobiology of Synthesis Planning   *98*
4.2.1   "Psychosynthesis"   *101*
4.3   The Agony and Ecstasy of Synthesis   *102*
4.4   Rembrandt Meets Woodward   *104*
4.4.1   Cortisone   *104*
4.4.2   Strychnine   *108*
4.4.2.1   The Woodward Synthesis   *109*
4.4.2.2   The Overman synthesis   *111*
4.4.2.3   The Kuehne synthesis   *113*
4.4.2.4   The Bonjoch and Bosch synthesis   *114*
4.4.2.5   The Shibasaki synthesis   *116*
4.4.2.6   The Fukuyama synthesis   *118*
4.4.2.7   The Mori synthesis   *119*
4.4.2.8   The MacMillan synthesis   *121*
4.4.2.9   Strychnine syntheses: Synopsis   *122*
4.5   The Post Woodwardian Era   *123*
4.5.1   The convergent template-based approach   *123*
4.5.2   Chiral auxiliary approach   *125*
4.5.3   Substrate control approach in cycloadditions   *127*
4.5.4   Biomimetic cyclization approach   *129*
4.6   Catalysis and Chirality in Total Synthesis   *131*
4.6.1   Applications of asymmetric catalysis to drug discovery   *136*
          References   *139*

5   Sources of Enantiopure Compounds   *145*
5.1   Optical Resolution   *146*
5.2   Chemical Kinetic Resolution (KR)   *147*
5.2.1   Classical, natural, and parallel methods   *147*
5.2.2   Dynamic chemical kinetic resolution   *147*
5.3   Cell-free Enzyme-mediated Enantiopure Compounds   *149*
5.3.1   Hydrolases and ester formation   *149*
5.3.2   Nitrilases, amidases, and acylases   *152*
5.4   Cell-free Chemoenzymatic Methods   *154*
5.5   Metal-catalyzed Dynamic Kinetic Resolution (DKR)   *154*
5.6   Biocatalytic Methods for Enantiopure Compounds   *155*
5.6.1   Enzymatic reduction of ketones   *155*
5.6.2   Enzymatic hydroxylation and epoxidation   *156*
5.6.3   Enzymatic oxidation of alcohols   *157*
5.6.4   Enzymatic Baeyer-Villiger oxidation   *157*
5.7   Applications of Enzymatic and Chemoenzymatic Methods   *158*
5.8   Chemical Asymmetric Synthesis of Enantiopure Compounds   *160*
5.9   Enantiopure Compounds from Nature   *164*
         References   *165*

6   The *Chiron Approach*   *171*
6.1   Living Through a Total Synthesis   *171*
6.2   Principles of the *Chiron Approach*   *172*
6.2.1   Definition   *173*
6.2.2   The *Chiron Approach*   *175*
6.2.3   Two philosophies, one goal   *176*
6.2.4   There is more than meets the eye   *180*
6.2.5   The flipside of molecules   *183*
6.2.6   Common root, different MO: chirons and synthons   *184*
6.2.7   To chiron or not to chiron   *186*
6.3   Anatomy of a Synthesis   *186*
         References   *189*

7   Nature's Chirons   *193*
7.1 α-Amino Acids   *193*
7.2   Carbohydrates   *195*
7.3 α-Hydroxy Acids   *200*
7.4   Terpenes   *203*
7.5   Cyclitols   *206*
         References   *208*

8   From Target Molecule to Chiron   *213*
8.1   Where's Waldo?   *214*
8.2   Apparent Chirons   *217*
8.3   Partially Hidden Chirons   *220*

8.4   Hidden Chirons   *222*
8.5   Chirons as "Sacrificial Lambs"   *224*
8.6   Locating α-Amino Acid-type Substructures   *228*
8.6.1   Apparent amino acids   *229*
8.6.2   Partially hidden amino acids   *231*
8.6.3   Hidden amino acids   *232*
8.7   Locating Carbohydrate-type Substructures   *234*
8.7.1   Patterns and shapes   *235*
8.7.2   The "Rule of Five"   *236*
8.7.3   Apparent carbohydrates   *237*
8.7.4   Partially hidden carbohydrates   *239*
8.7.5   Hidden carbohydrates   *240*
8.8   Locating Hydroxy Acid-type Substructures   *243*
8.8.1   Apparent hydroxy acids   *243*
8.8.2   Partially hidden hydroxy acids   *245*
8.8.3   Hidden hydroxy acids   *248*
8.8.4   The Roche acid — a unique *C*-Methyl chiron   *251*
8.9   Locating Terpene-type Substructures   *254*
8.9.1   Apparent terpenes   *254*
8.9.2   Partially hidden terpenes   *258*
8.9.3   Hidden terpenes   *260*
8.9.3.1   The terpene route to taxol   *267*
8.10   Locating Carbocyclic-type Substructures   *270*
8.10.1   Apparent carbocycles   *271*
8.10.2   Partially hidden carbocycles   *272*
8.10.3   Hidden carbocycles   *275*
8.10.4   Quinic acid, cyclitols, and other carbocycles as chirons   *279*
8.11   Locating Chirons Derived from Lactones   *283*
8.11.1   Apparent lactones   *285*
8.11.2   Partially hidden lactones   *286*
8.11.3   Hidden lactones   *288*
8.11.4   The replicating lactone strategy   *292*
          References   *294*

9   **Applications of the *Chiron Approach***   *301*
9.1   Category I Target Molecules   *301*
9.1.1   Streptolic acid   *302*
9.1.2   *ent*-Gelsedine   *303*
9.1.3   Vincamine   *305*
9.1.4   Peribysin E   *307*
9.2   Category II Target Molecules   *308*
9.2.1   FK-506   *309*
9.2.2   Okadaic acid   *310*
9.2.3   Phorboxazole A   *312*
9.2.4   Brevetoxin B   *314*

9.3	Category III Target Molecules	*316*
9.3.1	Neocarzinostatin	*317*
9.3.2	Idiospermuline	*317*
9.4	Prelude to Total Synthesis of Category I Molecules	*320*
	References	*320*

10 **Total Synthesis from α-Amino Acid Precursors** *323*
10.1	Actinobolin	*323*
10.2	Aspochalasin B	*326*
10.3	Cephalotaxine	*329*
10.4	α-Kainic Acid (W. Oppolzer)	*332*
10.5	α-Kainic Acid (P. T. Gallagher)	*334*
10.6	Croomine	*336*
10.7	Biotin	*339*
10.8	Salinosporamide A	*342*
10.9	Thienamycin	*345*
10.10	FR901483	*348*
10.10.1	The Sorensen synthesis	*350*
10.11	Tuberostemonine	*353*
10.12	Phyllanthine	*358*
10.13	Oscillarin	*361*
10.14	*ent*-Cyclizidine	*364*
10.15	Pactamycin	*367*
10.16	Miscellanea	*371*
	References	*373*

11 **Total Synthesis from Carbohydrate Precursors** *377*
11.1	Ajmalicine	*377*
11.2	*ent*-Actinobolin	*381*
11.3	Trehazolin and Trehazolamine	*384*
11.4	Fomannosin	*390*
11.5	9a-Desmethoxy Mitomycin A	*394*
11.6	Saxitoxin and β-Saxitoxinol	*398*
11.6.1	Second generation synthesis	*400*
11.7	*ent*-Decarbamoyl Saxitoxin	*402*
11.8	Zaragozic acid A	*405*
11.9	Hemibrevetoxin B	*408*
11.10	Carbohydrates in Synthesis and in Biology	*416*
11.11	Miscellanea	*417*
	References	*422*

12 **Total Synthesis from Hydroxy Acids** *427*
12.1	Griseoviridin	*427*
12.2	Halicholactone	*431*
12.3	Brasilenyne	*435*

12.4   Octalactin A   *438*
12.5   (3Z)-Dactomelyne   *442*
12.6   UCS1025A   *446*
12.7   Jerangolid A   *449*
12.8   Miscellanea   *453*
          References   *456*

13   **Total Synthesis from Terpenes**   *459*
13.1   Picrotoxinin   *459*
13.2   Eucannabinolide   *463*
13.3   Trilobolide and Thapsivillosin F   *467*
13.4   Briarellin E and F   *472*
13.5   Samaderine Y   *477*
13.6   Ambiguine H and Hapalindole U   *481*
13.7   Platensimycin   *484*
13.7.1   The Nicolaou synthesis   *485*
13.7.2   The Ghosh synthesis   *488*
13.7.3   Nicolaou's two asymmetric syntheses   *491*
13.7.4   Yamamoto's organocatalytic asymmetric synthesis   *494*
13.7.5   Corey's catalytic enantioselective synthesis   *496*
13.7.6   Platensimycin and the mind's eye   *496*
13.8   Phomactin A   *500*
13.9   Pinnaic Acid   *503*
13.9.1   The Danishefsky and Zhao asymmetric syntheses   *507*
13.10   Fusicoauritone   *510*
13.11   Miscellanea   *514*
          References   *516*

14   **Total Synthesis from Carbocyclic Precursors**   *521*
14.1   Punctatin A   *521*
14.2   Acanthoic Acid   *524*
14.3   Stachybocin Spirolactam   *527*
14.4   Scabronine G   *529*
14.5   Chapecoderin A   *533*
14.6   Dragmacidin F   *533*
14.7   Reserpine   *538*
14.7.1   The Woodward synthesis   *539*
14.7.2   The Stork synthesis   *542*
14.7.3   The Hanessian synthesis   *545*
14.8   Fawcettimine   *548*
14.8.1   Toste's synthesis of fawcettimine   *548*
14.8.2   Heathcock's synthesis of (±)-fawcettimine   *551*
14.9   Tamiflu   *553*
14.9.1   The Fang and Wong synthesis   *553*
14.9.2   The Hudlicky and Banwell syntheses   *555*

14.9.3   The Shibasaki catalytic asymmetric Diels-Alder
           synthesis   *557*
14.9.4   Tamiflu synthesis in the age of catalysis: Synopsis   *558*
14.10   Miscellanea   *560*
           References   *563*

15   **Total Synthesis with Lactones as Precursors**   *567*
15.1   Megaphone   *567*
15.2   Dihydromevinolin   *569*
15.3   Mannostatin A   *572*
15.4   Furaquinocin C   *577*
15.5   Miscellanea   *577*
           References   *580*

16   **Single Target Molecule-oriented Synthesis**   *583*
16.1   Synchronicity   *583*
16.2   Joining Forces   *584*
16.3   Back-to-back Publishing   *586*
16.3.1   Veratramine (1967)   *587*
16.3.2   (±)-Lycopodine (1968)   *588*
16.3.3   Ionomycin (1990)   *589*
16.3.4   Vancomycin aglycone (1998)   *591*
16.4   Same Year Publications   *593*
16.4.1   (±)-Colchicine (1959)   *594*
16.4.2   (±)-Catharanthine (1970)   *595*
16.4.3   (±)-Cephalotaxine (1972)   *596*
16.4.4   Bleomycin A$_2$ (1982)   *597*
16.4.5   Kopsinine (1985)   *598*
16.4.6   Rapamycin (1993)   *600*
16.4.7   Phomoidrides (CP molecules) (2000)   *604*
16.4.7.1   The Nicolaou Synthesis   *604*
16.4.7.2   The Shair synthesis   *606*
16.4.7.3   The Fukuyama synthesis   *608*
16.4.7.4   The Danishefsky synthesis   *609*
16.4.7.5   What is in a drawing?   *611*
16.4.8   Borrelidin (2003–2004)   *612*
16.4.8.1   The Morken synthesis   *612*
16.4.8.2   The Hanessian synthesis   *614*
16.4.8.3   The Ōmura and Theodorakis syntheses   *615*
16.4.9   Amphidinolide E (2006)   *618*
16.5   Single Target Molecules with Special Relevance   *620*
16.6   Quinine   *621*
16.6.1   The Stork synthesis   *621*
16.6.2   Quinine: The Woodward and Doering formal vs total syntheses
           issue   *624*

16.6.3	Quinine: *Aprés* Woodward and Doering	*627*
16.6.3.1	The Uskoković synthesis	*627*
16.6.3.2	The Gates synthesis	*630*
16.6.3.3	The Taylor and Martin synthesis	*631*
16.6.4	Quinine: Total synthesis in the modern age of catalysis	*631*
16.6.4.1	The Jacobsen synthesis	*631*
16.6.4.2	The Kobayashi synthesis	*633*
16.6.4.3	The Williams and Krische syntheses of 7-hydroxyquinine	*635*
16.6.5	The total synthesis of quinine in the mind's eye	*637*
16.7	Lactacystin	*641*
16.7.1	The first Corey synthesis	*642*
16.7.2	The second Corey synthesis	*643*
16.7.3	The Baldwin synthesis	*645*
16.7.4	The Chida Synthesis	*647*
16.7.5	The Ōmura-Smith synthesis	*648*
16.7.6	The Panek synthesis	*650*
16.7.7	The Jacobsen synthesis	*651*
16.7.8	The Shibasaki synthesis	*654*
16.7.9	Lactacystin and omuralide: Alternative methods and synthetic approaches	*656*
16.7.10	The Kang approach	*656*
16.7.11	The Adams synthesis of omuralide	*657*
16.7.12	The Ohfune approach	*658*
16.7.13	The Pattenden approach	*659*
16.7.14	The Hatekayama approach	*659*
16.7.15	The Donohue synthesis of (±)-omuralide	*660*
16.7.16	The Wardrop approach	*661*
16.7.17	The Hayes synthesis of lactacystin	*662*
16.7.18	Total synthesis of lactacystin: Synopsis	*663*
16.8	Taxol	*665*
16.8.1	What mad pursuit	*666*
16.8.2	The Holton synthesis of taxol	*666*
16.8.3	The Nicolaou synthesis of taxol	*670*
16.8.4	The Danishefsky synthesis of taxol	*673*
16.8.5	The Wender synthesis of taxol	*676*
16.8.6	The Kuwajima synthesis of taxol	*679*
16.8.7	The Mukaiyama synthesis of taxol	*682*
16.8.8	The six total syntheses of taxol: The calm after the storm	*685*
16.8.9	Total syntheses of taxol in the mind's eye	*686*
	References	*690*
17	**Man, Machine, and Visual Imagery in Synthesis Planning**	*699*
17.1	The LHASA Program	*701*
17.2	SYNGEN	*702*
17.3	WODCA	*703*

17.4   The CHIRON Program   *704*
17.4.1   CASA (Computer-assisted stereochemical analysis)   *704*
17.4.2   CAPS (Computer-assisted precursor selection)   *705*
17.5   Computer-aided synthesis planning   *710*
            References   *711*

18   The Essence of Synthesis – A Retrospective   *713*
18.1   Lest we Forget   *714*
18.2   The Corey and Stork Schools   *714*
18.3   The Visual Dialogue with Molecules   *716*
18.4   Total Synthesis: From whence we came. . .   *717*
18.5   In Pursuit of the "Ideal Synthesis"   *722*
18.5.1   The problem with protecting groups – blessing or curse?   *724*
18.5.1.1   Protecting-group-free synthesis?   *725*
18.5.2   The "redox economy" problem   *725*
18.5.3   The "functional group adjustment" problem   *726*
18.5.4   "Chiral economy"   *727*
18.6   For the Love of Synthesis (*Synthephilia*)   *730*
18.6.1   Reaching the summit   *731*
18.7   Organic Synthesis: To where we are going   *732*
18.8   Synthesis at the Service of Humankind   *734*
18.9   From the *Chiron Approach* to Catalysis   *736*
18.9.1   The young, the brave, and the bold: Passing the baton   *740*
18.9.1.1   Himandrine   *740*
18.9.1.2   Palau'amine   *743*
18.9.1.3   Minfiensine   *745*
18.9.1.4   Maoecrystal Z   *747*
18.9.2   Parting thoughts   *749*
18.10   A Salute to the Vanguards of Synthesis   *749*
            References   *750*

Author Index [Natural product/Target]   *757*

Chiron/Starting Material to Natural Product/Target Index   *771*

Natural product/Target [Chiron]   *781*

Key (Named) Reactions Index   *791*

Foreword

Organic synthesis has been a major contributor to human welfare for millennia. From food, clothing, housing, communication, energy and medicine, our current world is made possible by organic synthesis. As fields allied to chemistry such as biology, medicine and engineering become more molecular, the science of organic synthesis is poised to have even broader impact in the future. Thus, the publication of *Design and Strategy in Organic Synthesis: From the Chiron Approach to Catalysis* comes at an opportune time. This superb book, whose central focus is the design and synthesis of enantiomerically enriched chiral organic molecules, uniquely interweaves a didactic exposition of many central aspects of this endeavor with considerations of the creative and human elements of the science of organic synthesis.

Readily available enantiomerically enriched molecules – of natural origin or ones now accessible by enantioselective synthesis – are the raw materials for building single enantiomers of chiral target molecules. Vast amounts of information on what these molecules are, how these starting molecules can be recognized in target molecules, and how their use in enantioselective synthesis can be orchestrated is found in this book. This discussion of the *chiron approach* extends and updates the senior author's previous book in this area. By its focus on enantioselective total syntheses of natural products that have been prepared in several different ways, *Design and Strategy in Organic Synthesis: From the Chiron Approach to Catalysis* also exposes the reader to the use of catalytic enantioselective reactions in complex molecule synthesis.

Students and practitioners of the science of chemical synthesis will find a vast wealth of welcomed information in this book. More than one hundred total syntheses of carbogenic and heterocyclic natural products are discussed in detail and illustrated in attractive schemes in which the mechanisms of key steps are highlighted. In cases where a number of routes to a target structure are presented, distinctive strategic elements of the various syntheses are compared and illustrated. A delightful feature of the book is the analysis of the thought process believed to have connected the target molecule's structure to the strategy developed for its synthesis.

Design and Strategy in Organic Synthesis: From the Chiron Approach to Catalysis informs and educates on a remarkable variety of levels: didactic, historical, philosophical, and human. We congratulate Stephen Hanessian, Simon Giroux and Bradley Merner for providing a unique book that teaches and vividly illustrates the power and vitality of organic synthesis.

Irvine, California and Princeton, New Jersey   *Larry E. Overman and*
January 4, 2013   *David W. C. MacMillan*

Preface

With merely a decade into the third millennium, the present-day level of achievements in organic synthesis in general, and natural products in particular, is the highest ever compared to as recently as a generation ago. Building on the monumental discoveries and the fundamentally important concepts that have been the foundations of the discipline of organic chemistry over the last century, we are in an unprecedented period of evolution in the science of chemical synthesis. Aided by powerful instrumental techniques and ingenious methods for stereocontrolled and site-selective bond forming reactions, the enabling aspect of organic synthesis has manifested itself in many areas that contribute directly to the well-being of humankind. Consider among other, the impact of life-sustaining drugs derived from purely synthetic efforts on one hand, and natural products on the other. In this context, we are in a position as synthetic chemists to engage in exciting research projects that aim at solving important unmet medical needs especially in conjunction with great advances in the biological sciences.

As synthetic chemists, our vocation is to "make molecules" either from basic components, or by chemical modification of existing ones. The rich armamentarium of available (and constantly evolving) synthetic methods, coupled with sophisticated techniques for separation, purification, and analysis, allow us, at least in principle, to propose strategies to access practically any target molecule of reasonable structural complexity. Once such an objective is defined, the synthetic chemist invariably begins, metaphorically speaking, "the long journey to the summit" (see book cover), by asking the following three questions: 1. How can I synthesize my molecule efficiently? 2. How can I make sure that I have come up with a viable strategy? How and where do I begin? Added to these questions is a long list of "wishes," not the least of which is the desire to innovate, to be competitive, and to contribute to coworker training, thus rendering service to humankind through chemistry.

Faced with a challenging target molecule to synthesize, our first contact is visual. What follows, once the adrenaline rush caused by its molecular complexity has subsided, is a subliminal interplay between the eye and the mind's eye, triggering a complex, yet quasi-instantaneous series of *visual relational* and *visual reflexive* chemical and mental thought processing events that are a part of the psychobiological basis of generating a synthesis plan. This heuristic aspect of design and strategy that starts with a visual dialogue between the molecule and the mind's eye, is a fundamentally human and personalized tenet of synthesis. It demonstrates the strength (and limitations) of our powers of perception and creativity when planning a synthetic strategy.

Much like the principle of Gestalt theory, as an entity, the target molecule is not equal to the sum of its parts. The visual continuum that links one or more chemical building blocks, initially conceived as suitable starting materials, will change in shape, form, and structure before being eventually embedded in the integral landscape of the intended target molecule. In the process, good times, and not-so-good times will be experienced. In this regard, Rudyard Kipling's famous quote: *"If you could meet with triumph and disaster, and treat those imposters just the same"* is *à propos*.

The basic objective of this book is to promulgate the discovery and utilization of *chi*ral synth*ons* (*chirons*) as starting materials in total synthesis. Over the years, this approach has had broad applications for the synthesis of virtually all classes of molecules possessing one or more stereogenic carbon atoms. The impact of catalysis and other methods of stereocontrolled synthesis as an important corollary to the *chiron approach* is also discussed. Chapter 1 delves in the psychobiology of synthesis planning and the way, as individuals, we "look" at molecules before engaging in the actual process of synthesis. Analogies of sculpting, painting, and architecture with synthesis are discussed. Thus, whereas the sculptor creates form by exclusion of matter, the painter, like the architect and the synthetic chemist, does so by filling in space. Chapters 2 and 3 deal with the *Why* and *What* of synthesis respectively, focusing primarily on synthetic and naturally derived molecules that are endowed with chirality and diverse functionality. The *How* of synthesis is discussed in Chapter 4, delineating the ways in which we analyze molecular complexity and how we conceive of strategies that rely primarily on *visual relational* and *visual reflexive* thinking modalities. Sources of enantiomerically pure or enriched small molecules are dealt with in Chapter 5. The principles of the *chiron approach*, and the synthetically useful features of Nature's chirons (amino acids, carbohydrates, hydroxy acids, terpenes, carbocycles, and lactones) are discussed in Chapters 6 and 7 respectively. In Chapter 8, the "rules" for discovering native and chemically modified chirons embedded in the intricate structures of representative natural products are delineated with examples showing *apparent, partially hidden,* and *hidden* motifs. Applications of the *chiron approach* to three general categories of molecules are shown in retrosynthetic terms in Chapter 9. Chapters 10–15 cover individual total syntheses of relevant natural products using amino acids, carbohydrates, hydroxy acids, terpenes, carbocycles, and lactones respectively as starting materials. With few exceptions, only syntheses of enantiopure molecules are addressed, showing their first published synthesis to provide a historical perspective. Whenever relevant, approaches relying on stoichiometric, or catalytic asymmetric methods are included, to show the evolution of different strategies toward one and the same target molecule. Chapter 16 deals with single target-molecule oriented syntheses, showing selected early examples of first-time syntheses of natural products published from more than one laboratory starting in 1967 to the present day. The impact of computer assisted synthesis, focusing primarily on the *Chiron Computer Program* is highlighted in Chapter 17. Finally, a retrospective analysis of the essence of synthesis as it has evolved over the last decades to reach our present-day levels of achievements is discussed in Chapter 18, ending with a salute to those vanguards of the field, who have laid the foundation for future generations of synthetic chemists to continue along the path of discovery and innovation.

An undertaking of this magnitude cannot be without its flaws and limitations. We know that even after multiple proof-readings and editing, there may still be errors in the structures, the text, and even the interpretations of results. The examples chosen to propagate the basic principle of the book are certainly not comprehensive. We have tried our best, within the realm of possibility, to be as accurate and fair in our citations. No doubt, many important contributions were unfortunately missed. To offset this oversight, we have made an effort to include selected references and syntheses published in 2011.

In this book we tried to portray natural product synthesis as a form of Dali-like art, where the eye and the mind's eye engage in the heuristic exercise of seeing through the intricate landscape of molecules, and coming up with ingenious strategies. Color coding of overlapping segments of target molecules with the starting chirons facilitates their visualization, especially when the relationships are not evident to the eye. The individual syntheses shown in Chapters 10–15 are organized in a pedagogically informative way, showing the starting chiron, details of reaction sequences, including conditions and yields. Each synthesis also features a Commentary section that highlights special aspects and relevant mechanisms for specific transformations. An index covering the natural products, and their respective starting materials (chirons), as well as a separate index showing the individual chirons and their applications are included at the end of the book. We are grateful to Professors Larry E. Overman and David W. C. MacMillan for the Foreword.

This book was written in the spirit of looking at synthesis as a form of creative art. It is our hope that its contents will stimulate young minds to appreciate the aesthetic beauty of organic molecules in general and natural products in particular, and to develop a sense of visual and mental awareness of how successful strategies are conceived before being rendered to practice in the laboratory.

Montréal
May 30, 2013

Stephen Hanessian
Simon Giroux
Bradley L. Merner

Acknowledgement

An undertaking of this magnitude cannot be completed without the substantial involvement of many individuals who have helped in a variety of capacities.

The initial version of the text was typed by a long-time past assistant Carol St. Vincent-Major, presently followed by Michèle Ammouche. Over the last six years, members of the Hanessian research group have read earlier versions of the Chapters and made valuable comments.

We are particularly grateful to Professors Larry Overman (University of California, Irvine) and David MacMillan (Princeton University) for writing the Foreword that succinctly captures the essence of the book.

Abbreviations

Ac	acetyl
acac	acetyl acetone
AIBN	2,2'-azobisisobutyronitrile
9-BBN	9-borabicyclo[3.3.1]nonane
BHT	butylated hydroxytoluene or (2,6-di-*tert*-butyl-*p*-cresol)
BINAP	2,2'-bis(diphenylphosphino)-1,1'-binaphthyl
BINAPO	2-diphenylphosphino-2'-diphenylphosphinyl-1,1'-binaphthalene
Bn	benzyl
Boc	*tert*-butyloxycarbonyl
brsm	based on recovered starting material
BTAF	benzyltrimethylammonium fluoride
BOP-Cl	bis(2-oxo-3-oxazolidinyl) phosphinic chloride
Bz	benzoyl
CAN	ceric ammonium nitrate
CBS	Corey-Bakshi-Shibata
CDI	1,1'-carbonyldiimidazole
cod	1,5-cyclooctadiene
Cp	cyclopentadienyl
m-CPBA	3-chloroperoxybenzoic acid
CSA	camphorsulfonic acid
DABCO	1,4-diazabicyclo[2.2.2]octane
dba	dibenzylideneacetone
DBAD	di-*tert*-butyl azodicarboxylate
DBU	1,8-diazabicyclo[5.4.0]undec-7-ene
DBM	3,4-dimethoxybenzyl
DCC	*N,N'*-dicyclohexylcarbodiimide
DDQ	2,3-dichloro-5,6-dicyano-1,4-hydroquinone
de	diastereomeric excess
DEAD	diethyl azodicarboxylate
DHP	dihydropyran
DHQD	dihydroquinidine
DIAD	diisropropyl azodicarboxylate
Dibal-H	diisobutylaluminium hydride

DIOP	*O*-isopropylidene-2,3-dihydroxy-1,4-bis(diphenylphosphino)-butane
DMAP	*N,N*-dimethyl-4-aminopyridine
DMBONPy	2-(3,4-dimethoxybenzyloxy)-3-nitropyridine
DMDO	dimethyldioxirane
DME	1,2-dimethoxyethane
DMF	*N,N*-dimethylformaide
DMP	1,1-dimethoxypropane
DMPU	*N,N'*-dimethylpropyleneurea
DMSO	dimethyl sulfoxide
DMTSF	dimethyl(methylthio)sulfonium tetrafluoroborate
DNA	deoxyribonucleic acid
DPPA	diphenylphosphoryl azide
dppb	1,4-bis(diphenylphosphino)butane
dppf	1,1'-bis(diphenylphosphino)ferrocene
dppp	1,3-bis(diphenylphosphino)propane
dr	diastereomeric ratio
EDCI	1-ethyl-3-(3-dimethylaminopropyl) carbodiimide
EE	ethoxyethyl
ee	enantiomeric excess
Fmoc	9-fluorenylmethoxycarbonyl
HMDS	bis(trimethylsilyl)amine or 1,1,1,3,3,3-hexamethyldisilazane
HMPA	hexamethylphosphoramide
HOMO	highest occupied molecular orbital
IBX	2-iodoxybenzoic acid
IC_{50}	50% inhibitory concentration
Ipc	isopinocampheyl
LDA	lithium diisopropylamide
LiDBB	lithium 4,4'-di-*tert*-butylbiphenylide
LUMO	lowest unoccupied molecular orbital
Mbs	*p*-methoxybenzenesulfonyl
MEM	methoxyethoxymethyl
Mes	mesityl or mesitoyl
MMTr	(*p*-methoxyphenyl)diphenylmethyl
MOM	methoxymethyl
Ms	methanesulfonyl
nbd	norbornadiene
NBS	*N*-bromosuccinimde
NIS	*N*-iodosuccinimide
NMM	*N*-methylmorpholine
NMO	*N*-methylmorpholine-N-oxide
NMR	nuclear magnetic resonance
PDC	pyridinium dichromate
PHAL	phthalazine
PhFl	9-phenylfluroenyl

Piv	pivaloyl
PMB	*p*-methoxybenzyl
PMS	(*p*-methylphenyl)methylsulfonyl
PNB	*p*-nitrobenzyl or *p*-nitrobenzene-sulfonyl
PP	pyrophosphate
PTSH	1-phenyltetrazole-5-thiol
PPTS	pyridinium *p*-toluenesulfonate
PS	polymer-supported
psi	pounds per square inch
pyr.	pyridine
Red-Al	sodium bis(2-methoxyethoxy) aluminum hydride
RNA	ribonucleic acid
SAD	Sharpless asymmetric dihydroxylation
SAE	Sharpless asymmetric epoxidation
SOMO	singly occupied molecular orbital
TBA	tribromoacetic acid or N-tetrabutylammonium
TBDPS	*tert*-butyldiphenylsilyl
TBHP	*tert*-butylhydroperoxide
TBPB	*tert*-butylperbenzoate
TBS	*tert*-butyldimethylsilyl
TDMPP	tris(2,6-dimethoxyphenyl)phosphine
TEMPO	(2,2,6,6-tetramethyl-piperidin-1-yl)oxyl
Teoc	2-(trimethylsilyl)ethoxycarbonyl
TES	triethylsilyl
Tf	trifluromethanesulfonyl
TFA	trifluoroacetic acid
TFAA	trifluoroacetic anhydride
THF	tetrahydrofuran
TIPS	triisopropylsilyl
TMEDA	*N,N,N'*, *N'*-tetramethylethylenediamine
TMP	tetramethylpiperidide
TMS	trimethylsilyl
TMSE	(trimethylsilylethyl)
tol.	toluene
TPAP	N-tetrapropylammonium perruthenate
TPS	triisopropylbenzenesulfonyl or triphenyl silyl
Ts	p-toluenesulfonyl

The domain in which chemical
synthesis exercises its creative
power is vaster than that of
nature itself.

Marcellin Berthelot

1
The Concept of Synthesis

1.1
Organic Synthesis as a Central Science

The profound impact of scientific discovery on the quality of life has been one of the major revelations of our time [1]. Advances in every field of scientific endeavor have contributed to the modern day comforts and amenities we enjoy today. Man has smashed the atom, raised a flag on the moon, eradicated certain epidemics, revolutionized communication and means of transportation, cracked the human genetic code, and invented "thinking machines"– all in the span of some fifty years [2].

Indeed, nowhere is the impact of scientific discovery more manifest than in the physical well-being of humankind, as it triumphantly marches through the inaugural years of the third millennium. Regrettably, the modern-day marvels of scientific discovery spanning the ever-intertwining realms of the physical, biological, and health sciences have little, if any, effect on many socio-economic and politically-motivated events, which beleaguer the human race across the globe today.

On a more optimistic note, and excluding man-made disasters, we do, in general, enjoy a longer span of productive life compared to a century ago, thanks to advances in medicine and chemistry [3]. Imagine for an instant the fate of a successfully transplanted heart into a human recipient. Adverse immune responses leading to rejection of the transplanted organ can only be counteracted by immunomodulatory agents of synthetic or natural origin [4]. There can be no more dramatic example of the impact of science on the well-being of man. Curiously, Greek and Chinese mythologies recount legends of successful transplantations in man [5]. In the "Legendary Exchange of Hearts," Pien Ch'iao the surgeon, is believed to have administered a mixture of strong narcotic drugs, then cut open the chest of two patients, removed and exchanged their hearts to restore equilibrium in their

Design and Strategy in Organic Synthesis: From the Chiron Approach to Catalysis, First Edition.
Stephen Hanessian, Simon Giroux, and Bradley L. Merner.
© 2013 Wiley-VCH Verlag GmbH & Co. KGaA. Published 2013 by Wiley-VCH Verlag GmbH & Co. KGaA.

"unbalanced energies." In the post-operative period Pien Ch'iao is also believed to have given the patients highly potent drugs, after which they recovered and returned home! Healing the ills of man is as old as man himself, whether in mythological times or present-day experience. Chemistry, as a discipline, particularly synthetic organic chemistry, has played a dominant role in this process [3]. Jons Jacob Berzelius, a Swedish chemist, defined the term "organic" as a substance produced by a living organism, plant or animal.

Before we proceed with the enterprise of organic synthesis as a subdiscipline of chemistry, it would be opportune to define the term that has evolved in its practice, and used by thousands of practitioners of the métier for nearly three hundred years, pre-dating ideas of molecular structure and atomic theory [6]. *Syn-thesis* from the Greek, means "putting or bringing together." It follows that *organic synthesis*, whether involving the study of a bond-forming process, or the elaboration of the entire carbon framework of a natural product (*i.e.*, total synthesis) for example, is an enabling science.

The first synthesis of a natural product is attributed to Friedrich Wöhler, who, in 1828, accidentally synthesized urea, a constituent of human and animal urine. Since then, the science of synthesis has evolved to a level of sophistication, where practically any organic molecule of reasonable structural and stereochemical complexity can now be made in the laboratory given the necessary time and means [7, 8]. In this regard, it is interesting to ponder how Wöhler would react today if he was reincarnated in this age of asymmetric synthesis, catalysis, supramolecular chemistry, nanotechnology, and man-made, life-saving drugs, not to mention the invaluable advances in instrumentation available to us.

We can also discuss the impact of achievements in total synthesis considering the analytical and preparative techniques available at a given time. How well are we really doing at present with respect to practicality and efficiency, when a target molecule of modest complexity is synthesized? Apart from absolute stereochemical control, how much better would a new synthesis of camphor be today over a century after its ingenious synthesis by Komppa, who relied on combustion analysis as a principal means of characterization? In the same context, how will our present achievements in synthesis be looked at a century from now? For that matter, will organic synthesis remotely resemble how we practice it today? Rather than speculate on the topic, it is *à propos* to quote the Greek historian Thucydides:

"How are we to divine the unseen future that lies hidden in the present"

The reality of today's concept of organic synthesis, whether it involves the discovery of a revolutionary chemotherapeutic agent in the laboratory, or the control of a vital physiological process in the body, is that it remains the very basis of life itself. The enterprise and *modus operandi* may change with time, but the ultimate aim of organic synthesis will be primarily associated with the wellness of humankind. Much like the symbolism portrayed by Leonardo da Vinci's Vitruvian Man [9], organic synthesis with its far-reaching potential, will be the vital force for propagating the health and welfare of the human race.

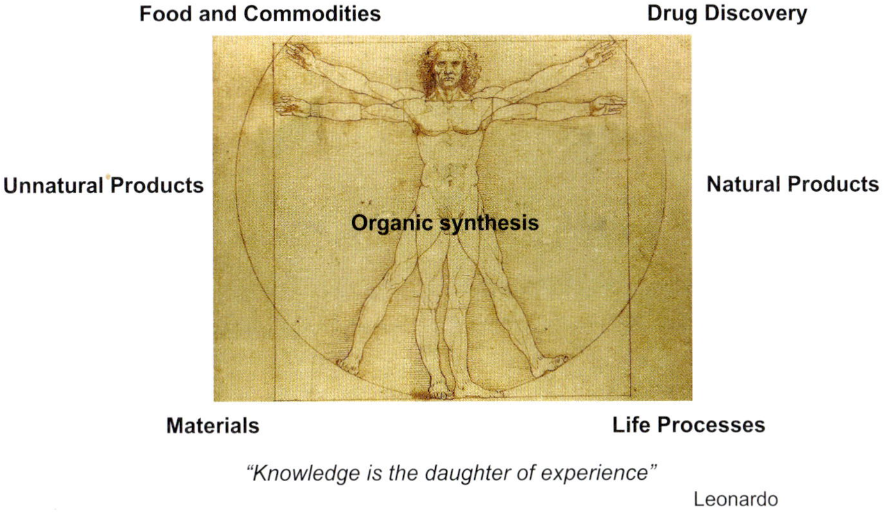

Figure 1.1 Organic synthesis as a central science adapted to Leonardo da Vinci's Vitruvian Man (Gallerie dell'academia, Venice, Italy).

1.2
Organic Chemistry and the Public

Aboard AC Flight 761: "Is that chemistry? I really had trouble with it in college," comments the passenger in seat 2K. Caught in the act, one's immediate reaction is to find a polite and civil way out of what may be a long conversation. After all, the intention was to correct the overdue galley proofs of a manuscript, and what better place than a seven hour transatlantic flight, even if the actual reading time would be much less due to in-flight services.

Although there are several escape tactics from such situations, the conversation eventually drifts to what it is a research scientist actually does as a synthetic organic chemist. People are fascinated by how drugs are made and how they ultimately reach the market. This sometimes carries the risk of having to justify the high cost of medicines, or being asked about generic versus brand names, and if there will be a cure for cancer and Alzheimer's disease soon.

Whether one is awakened from a structure-doodling interlude on the back of the boarding card, or interrupted from reading a manuscript, such encounters with fellow passengers can be mutually fulfilling. It is our social – and moral – obligation to inform the public of what we do "in the lab," of how, as mentors, we shape and mold the careers of young scientists, and what impact such work could have on the welfare of humankind.

Explaining what we do also reasserts our own convictions and commitments as research scientists, reaching out and contributing to the noble cause of coworker training, advancing the frontiers of science, and addressing fundamental questions relating to life processes.

To think that we do this by drawing lines between letters of the alphabet that become geometric shapes, which we then proceed to synthesize and study as chemical entities is truly amazing. What poets express in words, synthetic chemists express in chemical structures. Perhaps more than anything, this abstract process of transforming symbolic drawings into powders or liquid materials that can relieve a headache or become part of life's everyday conveniences, is what fascinates the public.

In these troubling environmental times, in which pollution has become headline news and a documentary on global warming can win an Academy Award, the same public reacts with concern and confusion. Unfortunately, the word "chemical" has become our environmental threat, from the air we breathe to the water that surrounds us, and to the agricultural products that we rely on for sustenance. Current efforts to promote green chemistry are certainly steps in the right direction and public awareness of such developments is crucial [10]. As a community, scientists must strive to inform the public at large of the immense positive impact of chemistry on our lives, especially in the area of beneficial therapy through many drugs developed during the past century [3].

Public awareness of the benefits of chemistry is a kind of education [11]. The man on the street relates to familiar names of compounds without knowing their chemical structures or historical origins. Aspirin and Tylenol are medicine cabinet and travel kit items. Their remarkably simple chemical structures do not do justice to the profound analgesic effects they have on people. Even more impressive is their relatively quick action, although the exact physiological mechanisms are still the subject of research and discussion. Consider the impact of such "small molecules" in the context of their benefit to man, alleviating among other, mild pain, with enormous savings in lost time from the workplace (Figure 1.2).

The man on the street also knows Viagra (the blue pill), the "accidental" and felicitous discovery for males experiencing erectile dysfunction [12]. For those enjoying the benefits of Viagra in their lives, little do they know that it was initially intended for cardiovascular use and that the remarkably *un*expected effect of the drug was manifested only during early phase clinical trials with male volunteers – much to their delight.

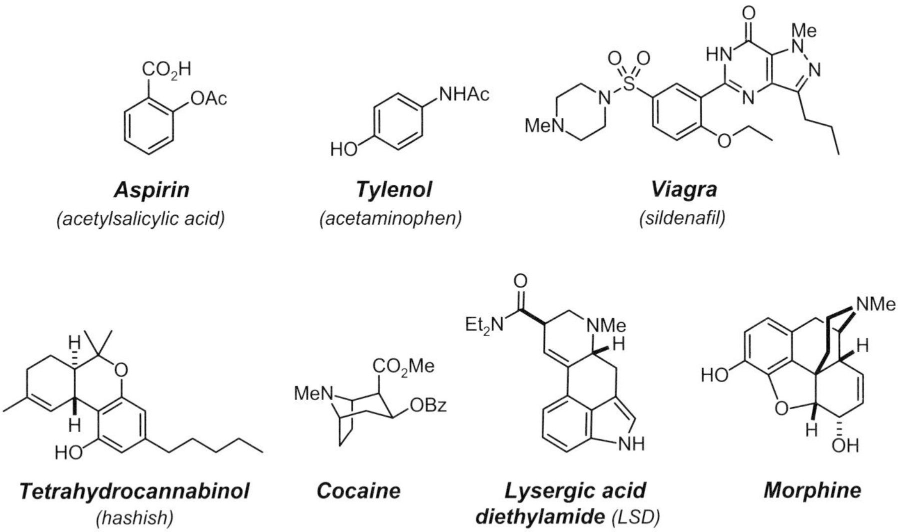

Figure 1.2 Familiar chemical compound names.

Serendipity has played a major role in drug discovery [13]. Four of the historically more relevant examples in Figure 1.2 are morphine, LSD, cocaine, and hashish, all familiar names to the public. While Tylenol, Aspirin, and Viagra are produced in bulk by total synthesis, morphine, cocaine, and hashish are extracted from natural plant sources. Morphine, known to man for many centuries, is one of the most potent pain-killers [14]. It is only used in a hospital setting for patients experiencing extreme pain. Morphine is also highly addictive. Ironically, with the good comes the bad, since diacetyl morphine (heroin) is one of the most addictive substances known. The monomethyl ether of morphine (codeine) on the other hand, is a constituent of a number of over-the-counter medications for alleviating the discomforts of coughs and colds.

Lysergic acid diethylamide (LSD), a potent hallucinogen was accidentally discovered as a result of a chance observation by Albert Hofmann in 1938, when he was exposed to this compound [15]. Ironically, he was looking for a drug to treat migraine headache. Like heroin, and morphine, LSD is a controlled substance of natural origin, and an illicit street drug akin to cocaine.

Perhaps the more controversial of the layman's natural products with hallucinogenic and related effects is tetrahydrocannabinol (hashish), which is smoked or inhaled rather than injected. It has been the subject of intensive research in the medical community, reportedly imparting beneficial effects to cancer patients in terms of their overall mental state. Non-addictive pharmacotherapy is high on the list of priorities in the pharmaceutical industry with obvious public and health agency approvals [16].

Returning to morphine (a constituent of opium), a passage from Homer in the Odyssey is attributed to Helen of Troy, who offered wine to Telemacchus, to which she had added *"a drug to quit all pain and strife, and to bring forgetfulness of every ill."*

1.3
The "Small Molecules" of life

Since ancient times, man has turned to nature for remedies to alleviate his ills. Quinine and morphine, known for centuries for their curative and analgesic powers are still in use today [14]. Of the so-called "small molecules" that are prescribed to the public for a variety of disease indications, a large proportion are natural products [17, 18], or products arising from their chemical modification. Many of these are life-saving or life-prolonging drugs. Joining this elite class of therapeutic compounds is a long list of synthetic, non-natural compounds, which, in many cases, are manufactured in ton quantities [19]. The molecular basis of drug action and the advent of genomics have greatly enhanced the drug discovery process [20]. Today, many drugs are the result of so-called "rational design" based on structural data gleaned from their biological targets such as proteins, and other natural macromolecules [21] (see also Chapter 2, section 2.6; Chapter 3, section 3.5.)

Heroic efforts in the pharmaceutical industry, often aided by elegant contributions to basic science from academic laboratories, have produced medicines to treat a large number of diseases and afflictions relating to man. Today, one is reasonably certain to receive medication for practically any disease indication, although some may be more effective than others. Cancer and cardiovascular disease are among the more serious medical conditions that affect a large segment of the population worldwide. Surgical intervention is often a necessary procedure, in which case anticancer or cardiovascular drugs are prescribed pre- or post-operatively. To illustrate the chemical nature of some anticancer, anticoagulant, and antihypercholesterolemic agents, two naturally derived small molecules, and two synthetic molecules exemplified by Taxol, Zocor, Gleevec, and Coumadin respectively are shown in Figure 1.3.

Synthetic compounds are also among the most widely prescribed medicines for bacterial infections (Cipro), hyperacidity (Nexium), and depression (Zoloft). Although the above mentioned three drugs combined, constitute several billion dollars of sales figures for their manufacturers [22], there are a number of alternative medicines also available from many pharmaceutical companies.

High profile diseases associated with the HIV virus (AIDS) for example, have commanded attention since the first incidence of the virus in man some 50 years ago. Two compounds, Viracept and Retrovir (AZT) representing different modes of action to attenuate viral proliferation are shown in Figure 1.3. The re-emergence of parasitic infections, such as malaria, has instigated much effort to optimize the activity of antimalarial agents such as the Artemether-Lumefantrine combination. Finally, controlled immunosuppression in conjunction with organ transplantation can be achieved with effective drugs such as Sandimmune, which was originally isolated from natural sources.

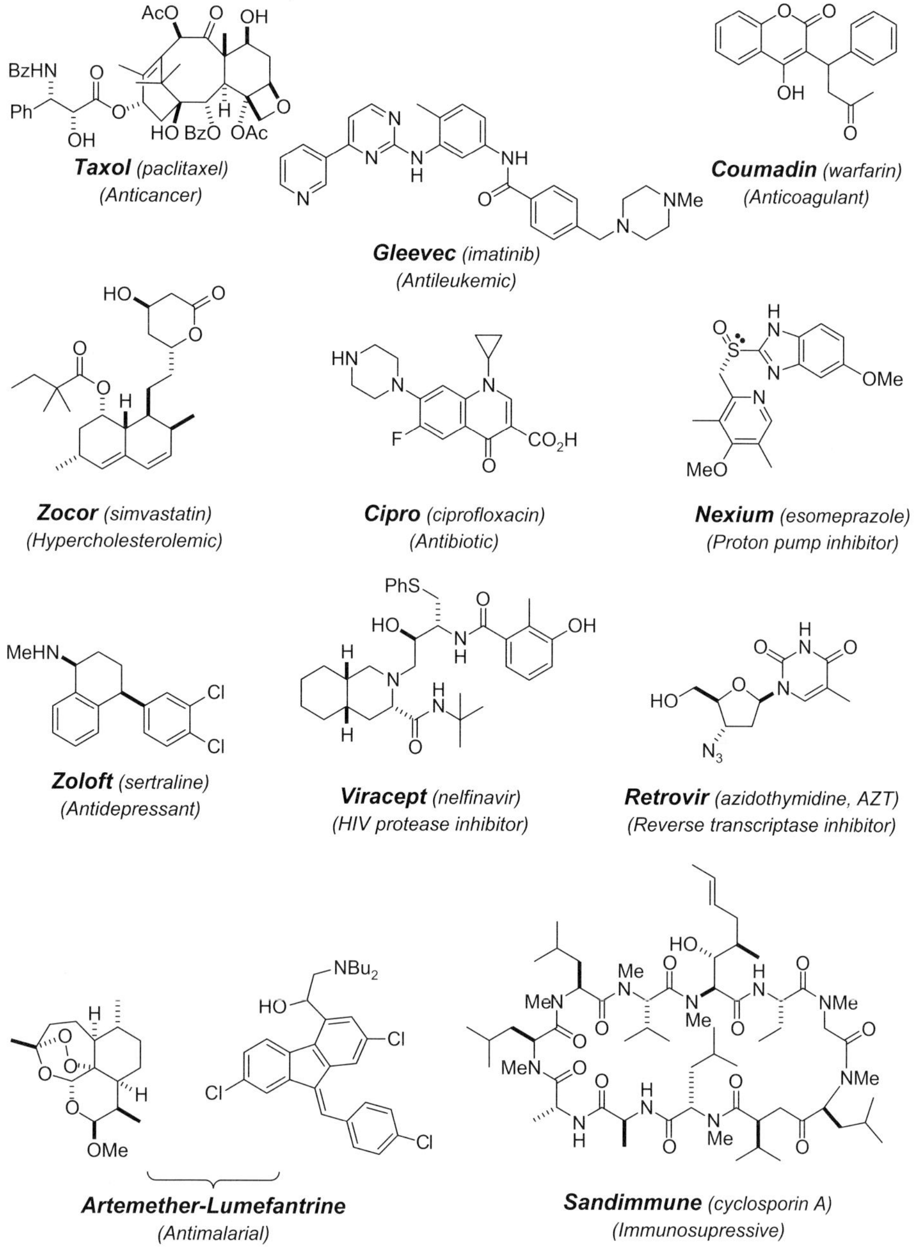

Taxol (paclitaxel)
(Anticancer)

Gleevec (imatinib)
(Antileukemic)

Coumadin (warfarin)
(Anticoagulant)

Zocor (simvastatin)
(Hypercholesterolemic)

Cipro (ciprofloxacin)
(Antibiotic)

Nexium (esomeprazole)
(Proton pump inhibitor)

Zoloft (sertraline)
(Antidepressant)

Viracept (nelfinavir)
(HIV protease inhibitor)

Retrovir (azidothymidine, AZT)
(Reverse transcriptase inhibitor)

Artemether-Lumefantrine
(Antimalarial)

Sandimmune (cyclosporin A)
(Immunosupressive)

Figure 1.3 The "small molecules" of life.

The naturally occurring and synthetic compounds shown in Figure 1.3 are chosen to represent different therapeutic areas. They have many variants that are marketed by different companies under different trade names.

The "small molecules" listed in Figure 1.3, representing a cross-section of medically relevant indications, serve to illustrate the very large diversity of architectures and levels of structural complexity. The modes of action for most of these compounds are well understood. The X-ray structural data of drug-enzyme complexes [21, 23] reveal binding interactions that can be used to develop future generations of optimized drugs exhibiting broader spectrum activities, higher potencies, and more favorably tolerated doses (see also Chapter 3, section 3.5). For example, protein-small molecule X-ray crystallography has been instrumental in the development of potent HIV protease inhibitors such as Viracept [24]. On the down side, the continued use of chemotherapeutic agents in the clinic will also cause resistance mechanisms to emerge, rendering their use precariously short-lived despite their potent curing abilities [25].

The discovery and development of useful medicines is fraught with obstacles and disappointments [26]. It is estimated that it takes seven to fifteen years to launch a drug from its inception at a cost of over one billion dollars [27]. Moreover, many initially successful programs may have to be halted or abandoned for a variety of reasons resulting in the loss of precious time and money (see also Chapter 2, section 2.8).

Drug-altering mechanisms, and related adverse pharmacological manifestations are the ultimate nemesis of any drug discovery program, particularly after unstinting efforts to successfully overcome the many hurdles on the way to the marketplace. The other potential pitfall is unforeseen or undetected toxicity (usually affecting the heart or liver) after a highly promising drug is launched, resulting in the loss of millions of dollars and long years devoted to its development [28]. Although drugs are launched after extensive critical scrutiny by clinicians to ensure their safety, and are eventually approved by the FDA and similar organizations, toxicity may manifest itself only when large numbers of the population have used a particular drug.

Chemistry and biology work hand in hand in the drug discovery process toward the development of small molecules as eventual medicines [29]. Depending on the particular indication, other therapeutic means are also available, like vaccines and gene therapy [30], in addition to emerging technologies such as antisense and other RNA-based methods [31]. Indeed some disease states may be better suited for non-small molecule treatment, relying on genetic engineering and protein therapeutics [32].

Exciting (and much debated) stem cell research is opening new opportunities in human cell biology and potential therapy [33]. Synthetic molecules are making inroads as activators and potentiators, ranging from self-renewal of embryonic stem cells to differentiation of adult ones through interaction with individual proteins [33, 34].

1.4
Nature's Rules

Appearances can be misleading. In front of a mirror we reflect a nearly perfect bilateral symmetry of ourselves. The left and right sides of our bodies appear to be symmetrical to our eyes. The same is true of the animal kingdom and in large parts of the insect and plant world. Suffice it to look at the majestic figure of a lion's face, the beauty of a monarch butterfly, or the exquisite petal arrangement of a daisy [35].

The left-right symmetry becomes asymmetrical in the vertebrate world once the positions of the visceral organs are considered [36]. Not only are such organs placed in different parts of our bodies, thus creating internal asymmetry, but they tend to loop to the right or the left depending on their development. Studies suggest that signalling molecules may be transiently expressed on one side of an emerging chick embryo leading to the beginning of morphological asymmetry.

A unique feature of life is the asymmetry of molecules that arise through biosynthesis [37]. Only L-amino acids are incorporated into proteins during translation. Some naturally occurring molecules incorporate D-amino acids, but only the so-called proteinogenic L-amino acids are involved in vital enzymatic processes. Our biological world is asymmetric, and it responds differently to left- or right-handed chirality in molecules relying on molecular recognition, complementarity, and mutual interactions. Natural products, synthetic molecules, and drugs in general, are recognized by macromolecules such as proteins, DNA, and RNA through non-covalent interactions involving H-bonding, charge-charge contact, and hydrophobic juxtaposition of specific residues. Such interactions result in highly productive contacts within the active sites of enzymes and complimentary recognition within critical domains of receptors for example, leading to specific biological responses. In the case of drugs, the anticipated favorable response may eventually lead to a useful therapeutic entity. A large number of drugs on the market exert their beneficial effect due to specific interactions with target proteins or other macromolecules [23]. Occasionally, racemic mixtures are used without any adverse effects. Such is the case for 3TC and amlodipine for example (Figure 1.4). However, the annals of medicinal chemistry contain numerous examples of dramatic differences in biological responses to a pair of enantiomers. Perhaps the most publicized case was thalidomide, where the (R)-enantiomer was a sedative, while the mirror image (S)-enantiomer had teratogenic properties. In fact, even the active enantiomer was racemized *in vivo*. This unforeseen and unfortunate outcome of marketing thalidomide as the racemate without knowledge of the biological effects in humans to each enantiomer individually, instigated the imposition of strict criteria of enantiomeric purity and its relationship to biological activities by the regulatory authorities such as the FDA. Today, the pharmacological effects of each enantiomer of a compound must be evaluated in animal studies for every investigational new drug application (INDA). Nature's way and the response to small molecule interactions can be manifested in bitter or sweet tastes, as experienced for (S)- and (R)-asparagines, as well as in more significant differences exhibited by the enantiomeric pairs of penicillamines or ethambutol for example (Figure 1.4).

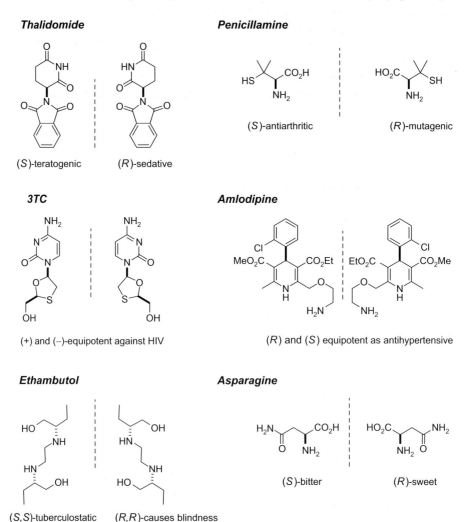

Thalidomide

(S)-teratogenic   (R)-sedative

Penicillamine

(S)-antiarthritic   (R)-mutagenic

3TC

(+) and (−)-equipotent against HIV

Amlodipine

(R) and (S) equipotent as antihypertensive

Ethambutol

(S,S)-tuberculostatic   (R,R)-causes blindness

Asparagine

(S)-bitter   (R)-sweet

Figure 1.4 Nature's rules.

It is therefore clear that organic synthesis of chiral, non-racemic molecules, especially those that may have therapeutic use in humans, must be planned so as to produce one or the other enantiomer [38]. This "natural" requirement has been one of the main reasons to explore asymmetric processes to construct molecules containing one or more stereogenic centers. Great strides toward this objective have been made in academia as well as in industry [39]. Of course, the relation of structure and absolute stereochemistry to function and property, manifests itself in many areas outside the domains of drug discovery and pharmaceuticals. The food, flavoring, and beverage industries rely heavily on receptor-mediated recognition in their marketing research endeavors [40]. The perfume industry's highly competitive edge relies on fragrances extracted and synthesized under watchful eyes and discriminating noses [41]. It is estimated that 25% of agrochemicals consist of chiral compounds [42].

Asymmetry goes hand in hand with life processes, and synthesis is the enabling enterprise that provides the distinction between "right" and "wrong" in molecular recognition as it pertains to the chirality of organic compounds and their natural receptors.

1.5
Organic Synthesis as a Mental and Visual Science

The adage *"The first taste is with the eyes"* is attributed to Sophocles. Whether it is a breathtaking sunset, a captivating landscape, a sumptuous dish, or a fascinating new chemical structure, our first contact (and impression) is visual (see also, Chapter 4, section 4.2).

In his De Anima, Aristotle had posited that *"thought without an image is not possible"* and *"the soul never thinks without a mental image"* [43]. Külpe, Ach, and Bühler, on the other hand, discussed *"knowledge without images,"* which others of the same school later qualified as *"thought without images."* J. B. Watson believes that mental activities consist of scenes that had been put into words long ago. In this context how are we to interpret Shakespeare: *"Shall I compare thee to a summer's day?"* Irrespective of different interpretations of the intricacies of mind and imagery, the psychobiological approach to thought and visualization is a topic of great interest to this day, especially with the advent of Gestalt psychology and experimental results on how mental images are produced and used [44]. The complex nature of the human thought process has been studied since man became aware of his consciousness and invented the alphabet to express himself in words. Vision developed before man practiced verbal language, and images remain as a natural part of our cognitive world. The mere mention of the name of a place or a molecule for that matter will instantly call up an image in our mind's eye even though our brain stores visual information, not words. Einstein said, *"I very rarely think in words at all."*

Consider the three images in Figure 1.5. At first sight, the Leaning Tower of Pisa may appear to be disorientating to our senses. However, the reality of the visual process makes us think in relational terms. The tower is leaning in relation to a perceived vertical axis as a plane of reference. There is a logic and knowledge-based imagery in this realization. The visual effect in Joan Miró's *"Seated Woman"* in relation to the title given by the artist is hard to reconcile. Only the artist sees a seated woman. The thought process is abstract and symbolic, rather than relational. The visual effect of Leonardo da Vinci's Mona Lisa elicits emotion rather than logic. It makes one think and feel, hence the suggestive connotation of a story as yet not told.

It was the impressionist artist Edgar Degas who said *"The artist does not draw what he sees, but what he must make others see."* In "Seated Woman," the abstract artist Miró deviates from this practice, and presents a challenge to the uninitiated viewer [45].

Visual, Relational

(Tower of Pisa)
Bonnano Pisano
Giovanni di Simone
(1174-1273)

Abstract, Symbolic

(Seated Woman)
Joan Miró
(1931)

Emotional, Suggestive

(Mona Lisa)
Leonardo da Vinci
(1503-1507)

Figure 1.5 Imagery and the thought process. ("Seated Woman" reproduced with permission by VG Bild-Kunst, Bonn 2011; copyright Successió Miró.)

When presented with a structure of a target molecule to synthesize, our first "taste" is with our eyes. What follows in the planning of a synthesis strategy is a complex series of logic and knowledge-based mental events that will undergo many iterations before settling on a preferred route to follow. Leonardo da Vinci wrote: *"The eye keeps in itself the image of luminous bodies for some time."* Metaphorically speaking, the same could be said of molecules.

1.6
Art, Architecture, and Synthesis

The steps involved in a total synthesis can be likened to the design and building of an edifice by an architect. Organic synthesis and architecture have similarities with art, since all three are visual forms of expression and creativity.

However, the rules are quite different in the world of an artist since the subject may be the product of the imagination and personal interpretation. The chemist's synthesis plan, much like the architect's blueprint, follows a given path until the target molecule (or the designed building) is completed. The architect must adhere to a building code, while the chemist must satisfy the rules of valence bonding and thermodynamics.

To illustrate the similarities and differences between art, architecture, and synthesis, the circular shape was chosen as a forum. Thus in "*Dance*," Henri Matisse portrays five human figures forming a circle that give the illusion of movement. Shapes, forms, and values are used to radiate dynamism. The process is visual and mental with an emotional component that the viewer actually feels.

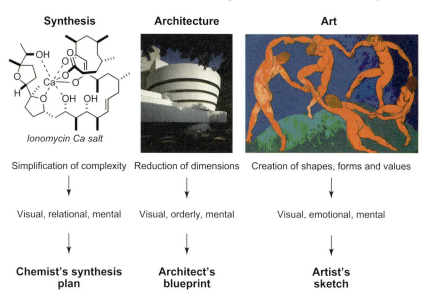

Synthesis	Architecture	Art
Ionomycin Ca salt		
Simplification of complexity	Reduction of dimensions	Creation of shapes, forms and values
↓	↓	↓
Visual, relational, mental	Visual, orderly, mental	Visual, emotional, mental
↓	↓	↓
Chemist's synthesis plan	**Architect's blueprint**	**Artist's sketch**

Figure 1.6 Art, architecture, and synthesis. Right to left: "La Danse (II)," (Henri Matisse, 1909, with permission by The State Hermitage Museum, St. Petersburg); Guggenheim Museum, New York (Frank Lloyd Wright, 1959, permission by The Solomon R. Guggenheim Foundation ("SRGF"), New York. FLWW-3. Photo: David Heald); ionomycin calcium salt.

In planning for the inverted concentric layers of the Guggenheim Museum, Frank Lloyd Wright had to follow a strict code while reducing macroscopic dimensions to the two-dimensional surface of a blueprint. The thought process is visual, mental, and highly orderly. The quasi-circular shape of ionomycin calcium salt does not have the aesthetic exterior quality of the Guggenheim museum, or the eye-pleasing expressionism of the Matisse masterpiece. However, the "beauty" of this natural product will become evident to the synthetic chemist, once it is viewed in a 3-dimensional form.

1.7
Simplification of Complexity

The synthetic chemist cannot consider laying down a synthesis plan for ionomycin, unless its complexity is simplified. The process is visual, relational, and mental.

All three pictures in Figure 1.6 are works of art from different perspectives. The ultimate objective of the synthetic chemist and the architect is the attainment of a pre-designated target molecule or building respectively. In both cases, a bottom-up approach is practiced where "building blocks" are eventually integrated into microscopic and macroscopic spaces respectively. No substitutes are allowed. On the other hand, the sculptor or artist may set out to paint or sculpt not knowing exactly what he or she will ultimately create. The sculptor creates form by exclusion of matter, while the painter (like the architect and synthetic chemist) does so by filling in space.

The complexity of ionomycin as a chemical entity as seen from its X-ray crystal structure [46], is simplified once the molecule is depicted in its stretched-out acyclic representation as shown in Figure 1.7 (for a discussion of syntheses, see Chapter 16, section 16.3.3). The same methyl and hydroxyl substituents shown in the simulated 3-dimensional portrayal in Figure 1.6, now adopt more visually familiar "up" and "down" orientations on a zigzag carbon backbone. Not only is relative stereochemistry simplified to the eye, but chemically feasible disconnections and assembly strategies can be envisaged by imaginary bond breaking and bond reforming operations on paper. At this point, the synthetic chemist and the architect will diverge in their thought process of execution. There may be *many* ways to synthesize ionomycin [47], but only one way to build the Guggenheim museum as we know it with a limited variety of approved building materials (Figure 1.6). It is said that creativity pertains to bringing forth something that was not there before. Therefore, in the strictest sense of definitions, the word creativity should apply to the synthesis of a *totally* new entity, and not to one of known structure.

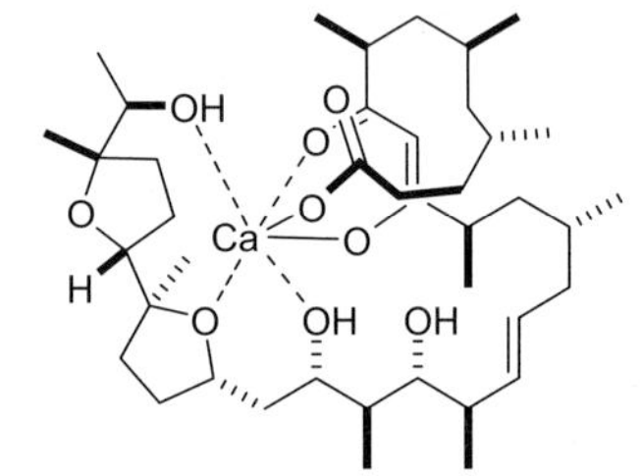

Ionomycin Ca salt

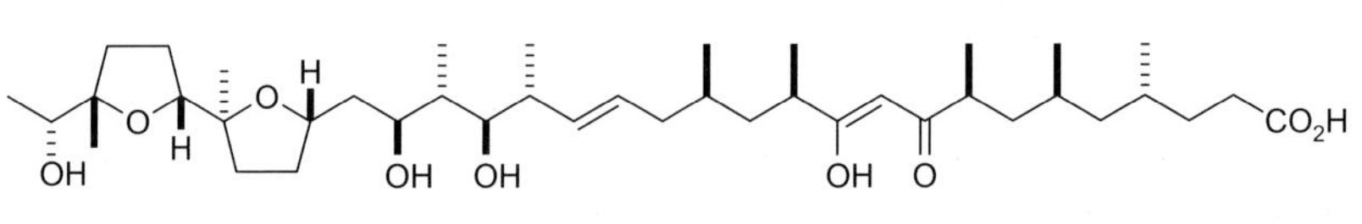

Ionomycin free acid

Figure 1.7 Simplification of complexity.

1.8
Seeing Through the Mind's Eye

The annals of organic synthesis contain a large number of total syntheses in which the structure of the target compound cannot be related to that of the starting material. Nowhere in the intricate skeletal framework of the target compound can one see any obvious remains of the smaller building block. This Gestalt loss of identity is particularly fascinating, even to the seasoned viewer, when functional groups are replaced by non-obvious ones in the eventual target compound, or when skeletal rearrangements are performed early in the synthesis. Many examples of non-obvious relationships resulting from drastic modifications of the initial starting material can be found in the literature, as will be demonstrated in Chapter 8 and other chapters of this book. The two examples, shown in Figures 1.8 A and B, dating back over twenty years ago, serve to illustrate the process of synthesis planning in the mind's eye.

The visual and mental process of simplifying complex structures of molecules to smaller elements was pioneered by E. J. Corey in the form of a retrosynthetic analysis [48]. Here, the proposed intermediates are drawn and separated by retrosynthetic arrows in the backward sense until a suitable building block or starting material is found. However, the path from initial viewing of the structure of a target molecule, to finding a given starting material to initiate the process of synthesis in the mind's eye is not simple (see Chapter 4, section 4.2).

For example, how and where did the carbon frameworks of L-glutamic acid and (+)-dibromocamphor end up in their respective target compounds, dihydromevinolin and ophiobolin C (Figure 1.8 A and B)? In the first case, a 5-carbon α-amino acid was used as an acyclic precursor to ring A of the dehydrodecalin portion of dihydromevinolin, which contains no nitrogen [49]. Formally, the amino group in L-glutamic acid became a hydroxyl group (with retention of configuration), and a C-methyl group was introduced at C-4. As seen in Figure 1.8 A, the 5-carbon framework of the amino acid was transformed to a butenolide-diene which would undergo an intramolecular Diels-Alder reaction to generate a tricyclic lactone with stereochemical control.

Synthesis strategies are carefully planned on paper from beginning to end, with alternative pathways kept in reserve. For every total synthesis, there may be one or more key reaction steps. In the case of dihydromevinolin, a key step was the intramolecular Diels-Alder reaction which led directly to the desired substitution pattern and absolute stereochemistry, in addition to the correct positioning of the double bond in the dehydrodecalin core. The initial visual connection with the target molecule in the mind's eye was the butenolide-diene intermediate with the appealing prospect of an intramolecular Diels-Alder reaction that would create the cyclohexene (ring B) system. Thus, a synthesis plan was instantly formulated in the mind's eye, based on a flash-card like imagery connection between the structure of the target molecule and the butenolide-diene intermediate. The stepwise relational thought process led to L-glutamic acid as a "hidden" precursor to the butenolide (see Chapter 15, section 15.2).

(The False Mirror)
R. Magritte
(1928)

A.

Dihydromevinolin

L-Glutamic acid

B.

Ophiobolin C

(+)-9,10-Dibromocamphor

Figure 1.8 Seeing through the mind's eye: from target molecule to starting material. (Reproduction of "The false mirror" with permission of VG Bild-Kunst, Bonn 2013; copyright MoMa, NYC.)

The synthesis of ophiobolin C from (+)-9,10-dibromocamphor by Kishi and coworkers [50] is an example of knowledge-based imagery. Inspection of the target structure reveals no direct visual progeny from camphor (Figure 1.8 B). However, prior knowledge that dibromocamphor can be converted to a vicinally substituted cyclopentane derivative containing a stereochemically distinct quaternary center bearing a C-methyl group, paved the way to a highly stereocontrolled synthesis of ophiobolin C (see Chapter 8, section 8.1). This choice was made with the realization that dibromocamphor would provide only ring C of the target molecule, while the remaining B and A rings must be generated in stepwise manner. As in many other syntheses, stereochemical and functional group adjustments were necessary. Thus, while the two stereogenic centers intentionally produced in the initial Grob-type fragmentation product of dibromocamphor were functionally useful, the exocyclic double bond had to be oxidatively cleaved, *en route* to an allylic alcohol derivative. In the end, all ten carbons atoms of (+)-9,10-dibromocamphor save one were used, and ultimately incorporated in ring C and part of ring B of the target compound. There is no doubt that seeing the cyclopentane ring C bearing the angular C-methyl group as part of the structure of ophiobolin C, and prior knowledge of the product of fragmentation from dibromocamphor, were strong visual signals (or stimuli) in Kishi's mind's eye leading to the chosen strategy. Once again, citing from Leonardo da Vinci would be appropriate: *"Knowledge is the daughter of experience."*

The two classical examples shown in Figures 1.8 A and B serve to illustrate different ways in which the mind's eye sees hidden precursors and conceives of enabling reactions when planning strategies for total synthesis. In each case, the image of the target molecule remains as a frozen optic array while the process of visual thinking simplifies complexity in an orderly and sequential manner. Unlike the exterior plan of a building, which must strictly adhere to the architect's specifications in the final blueprint during the construction phase, there can be many different plans and eventual ways to construct a target molecule. This template-fitting approach to total synthesis from suitable precursors or starting materials as illustrated above, can also be complemented by alternative strategies, wherein key bond constructs relying on asymmetric processes create useful intermediates *en route* to the intended target molecule. Therein lies the power of synthesis that allows for diversity of individual thought and freedom to innovate (see Chapter 4, section 4.2).

There is aesthetic beauty and intellectual satisfaction in the conception of a particular synthesis strategy, especially when the target molecule has unusual structural features. The further visually remote the starting material is from the target molecule, the more intriguing is the process of its evolution toward that target. The principles of Gestalt "Good Communication" apply equally well to the visual continuum from the beginning to the end of a logical synthesis plan. As an entity, the target molecule is different from the sum of its parts, but the visual chasm that separates the two gradually disappears as the synthesis plan takes shape, like a progressive epiphany of thoughts and images. The thought process that goes into a synthesis plan differs from one person to another and is poignantly articulated by E. J. Corey:

"Asking a chemist how he came upon precisely the starting materials and reactions that so elegantly led to the desired result would probably be as meaningless as asking Picasso why he painted as he did."

1.9
Beauty is in the Eye of the Beholder

To conclude this chapter, we illustrate the prowess of organic synthesis in the context of shapes and forms of selected topologically and functionally unique molecules [51]. The well-known hydrocarbon adamantane (from *adamas*, meaning diamond in Greek) was discovered by Landa and Machachaek in 1933 [52] who isolated it from petroleum (Figure 1.9). Viewed by the layman or a young child, other polyhedral hydrocarbons such as cubane (Necker Cube) [53], twistane [54], homopentaprismane [55], natone [56], corranulene [57], dodecahedrane [58] and C_{60} (Buckminster)fullerene [59] (see also Chapter 2, section 2.11.2) may appear to be eye-teasing geometrical shapes. Their three-dimensional kaleidoscopic forms become evident on paper only because of the wider lines portraying depth and perception in space. The trivial names given to them and reflected in their shapes, has instigated a curiosity-driven incentive to synthesize other "anes" such as housane (bicyclo[2.1.0]pentane), pagodane, basketane, etc. A beautiful example of insightful ingenuity is crafted in the synthesis of bullvalene, and the prediction of its electronic properties based on valence tautomerism [60]. The spatial disposition of a cyclopropane, and the three symmetrically spaced double bonds confined in a tricyclic structure, lead to rapid, degenerate Cope rearrangements, such that *all* the carbon and hydrogen atoms are equivalent within a given temperature range. Thus, bullvalene is a molecule that never rests!

The syntheses of these unique chemical entities constitute monumental achievements in the annals of organic synthesis of the twentieth century. The availability of such compounds through synthesis greatly enhanced the basic concepts of physical organic chemistry dealing with bond energies and strain. As a consequence, our understanding of the principles of reactivity and proximity effects were better defined.

The linear array of five interlocked rings, resulting from the self-assembly of a template-directed synthesis protocol by Stoddart and coworkers [61] was inspired by the symbol of the International Olympic Games, hence the name Olympiadane.

Finally, the natural product CP-263,114 (phomoidride B) [62] is included in this list of shapes and forms of architecturally unique molecules for its aesthetic beauty and its resemblance to a bird in flight, at least in the eyes of the authors of this book (see also Chapter 16, section 16.4.7).

We conclude our portrayal of the concept of synthesis and its many facets by quoting Proust:

"The real voyage of discovery consists not in seeking new landscapes, but in having new eyes"

Marcel Proust

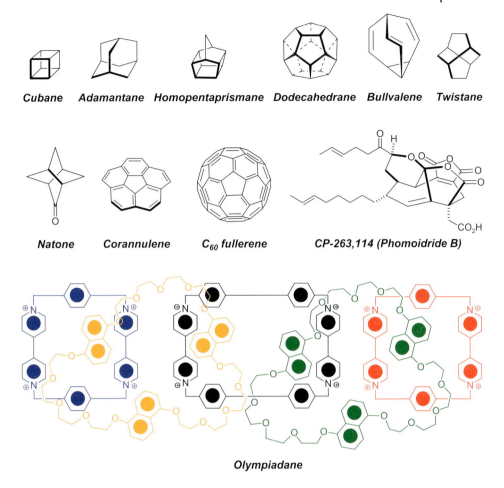

Figure 1.9 Shapes and forms of selected man-made molecules and the natural product phomoidride *B*.

References

1. For example, see: *The Immortal Cell: One Scientist's Quest to Solve the Mystery of Human Aging*, West, M. D. 2003, Doubleday, NY.

2. (a) *The Age of Intelligent Machines*, Kurzweil, R., 1990, MIT Press, Cambridge, MA; (b) *The Age of Spiritual Machines: When Computers Exceed Human Intelligence*, Kurzweil, R. 1999, MIT Press, Cambridge, MA; (c) *Explorations in Parallel Processing: a Handbook of Models, Programs, and Exercises*, Rumelhart, D. E.; McClelland, J. L.; Eds.; 1989, MIT Press, Cambridge, MA.

3. (a) *Molecules that Changed the World*, Nicolaou, K. C.; Montagnon, T. 2008, Wiley-VCH, Weinheim; (b) *Molecules and Medicine*, Corey, E. J.; Kürti, L.; Czakó, B. 2007, Wiley, NY.

4. (a) Morris, P. J. *N. Engl. J. Med.* 2004, **351**, 2678; (b) For a 50-year perspective, see: Sayegh, M. H.; Carpenter,

C. B. *N. Engl. J. Med.* 2004, **351**, 2761.

5. For a historical perspective, see: Ventura, H. O.; Muhammed, K. *Curr. Opin. Cardiology* 2001, **16**, 118.

6. Cornforth, J. W. *Aldrichim. Acta* 1994, **27**, 71.

7. For selected books, see: (a) *Enantioselective Chemical Synthesis: Methods, Logic and Practice*, Corey, E. J.; Kürti, L., 2010, Dallas, TX; (b) *Elements of Synthesis Planning*, Hoffmann, R. W., 2009, Springer, Heidelberg; (c) *Classics in Stereoselective Synthesis*, Carreira, E. M.; Kvaerno, L., 2009, Wiley-VCH, Weinheim; (d) *Strategies and Tactics in Organic Synthesis*, Harmata, M.; Ed.; 2008, vol. 1-7, Academic Press, NY; (e) *Organic Synthesis: Strategy and Control*, Wyatt, P.; Warren, S. 2007, J. Wiley, Chichester, UK; (f) *Organic Synthesis: State of the Art*, Taber, D. F. 2003-2005, 2005-2007, Wiley, Hoboken, NJ; (g) *The Way of Synthesis*, Hudlicky, T.; Reed, J. W. 2007, Wiley-VCH, Weinheim; (h) *Classics in Total Synthesis, Targets, Strategies, Methods*. Nicolaou, K. C.; Sorensen, E. J. 1996, Wiley-VCH, Weinheim; (i) *Classics in Total Synthesis II: More Targets, Strategies, Methods*. Nicolaou K. C.; Snyder, S. A. 2003, Wiley-VCH, Weinheim; (j) *Classics in Total Synthesis III: Further Targets, Strategies, Methods*. Nicolaou K. C.; Chen, J. S. 2011, Wiley-VCH, Weinheim; (k) *The Chemical Synthesis of Natural Products*, Hale, K. J. Ed.; 2000, Sheffield Academic Press, Sheffield, UK; (l) *Studies in Natural Products Chemistry*, Atta-ur-Rahman, Ed.; Elsevier, Amsterdam; (m) *The Logic of Chemical Synthesis*, Corey, E. J.; Cheng, X.-M., 1989, Wiley, NY.

8. For selected recent reviews, see: (a) Nicolaou, K. C.; Vourloumis, D.; Winssinger, N.; Baran, P. S. *Angew. Chem. Int. Ed.* 2000, **39**, 44; (b) Corey, E. J. *Angew. Chem. Int. Ed.* 1991, **30**, 455; see also: (c) Carroll, W. F. Jr.; Raber, D. J. *The Chemistry Enterprise in 2015*, Chemistry.org/chemistryenterprise2015.html; Baum, R. *Chem. Eng. News*, Jan. 30, 2006, p. 3.

9. *Art, Science, and Technology, The Genius of Leonardo*, Seeger, R. J.; Cohen, R. S.; Eds.; 1974, Reidel Publishing Company, Boston, USA; see also http://en.wikipedia.org/wiki/vitruvius

10. (a) *The Handbook of Green Chemistry*, Leitner, W.; Jessop, P. G.; Li, C.-J.; Wasserscheid, P.; Stark, A. Eds.; 2010, Wiley-VCH, Weinheim; (b) Green Chemistry in the Pharmaceutical Industry, Dunn, P.; Wells, A.; Williams, M. T. Eds.; 2010, Wiley-VCH, Weinheim; (c) *Green Catalysis*, Crabtree, R. H.; Anastas, P. T. Eds.; 2010, Wiley-VCH, Weinheim; (d) *Green Chemistry and Catalysis*, Sheldon, R. A.; Arends, S.; Hanefeld, U. 2007, Wiley-VCH, Weinheim; (e) *Green Chemistry-Thematic issue* Horváth, I. T.; Anastas, P. T. Guest editors, *Chem. Rev.* 2007, **107**, 2167. (f) *Green Chemistry*, Lancaster, M. Royal Society of Chemistry, 2002, London, UK; (g) Lipschutz, B. H.; Ghorai, S. *Aldrichim. Acta* 2008, **41**, 59.

11. For example, see: (a) *The Joy of Science*, R. H. Azan, 2009, The Teaching Company, Chantilly, VA; (b) *The Public Image of Chemistry*, Schummer, J.; Bensaude-Vincent, B.; Van Tiggelen, B.; Eds.; 2007, World Scientific Publishing Co. Hackensack, NJ; (c) Emsley, J., *Better Looking, Better Living, Better Loving – How Chemistry Can Help you Achieve Life's Goals*, 2007, Wiley-VCH, Weinheim; (d) Ball, P., *Elegant Solutions: Ten Beautiful Experiments in Chemistry*, 2005, Royal Society of Chemistry, Cambridge, UK; (e) Berson, J. A. *Chemical Discovery and the Logician's Program: A Problematic Pairing*, 2003, Wiley-VCH, Weinheim; (f) Whitesides, G. M. *Angew. Chem. Int. Ed.* 2004, **43**, 3632; see also: (g) Whitesides, G. M.; Deutch, J. *Nature* 2011, **469**, 21.

12. *Laughing Gas, Viagra, and Lipitor: The Human Stories Behind the Drugs we Use*, Li, J. J., 2006, Oxford University Press, UK.

13. (a) *Happy Accidents: Serendipity in Modern Medical Breakthroughs*, Meyers, M. A. 2007, Arcade Publishing, NY; (b) *Serendipity: Accidental Discoveries in Science*, Roberts, R. M., 1989, Wiley, NY.

14. For a historical perspective, see: (a) Hudlicky, T.; Butora, G.; Fearnley, S. P.; Gum, A. G.; Stabile, M. R. in *Studies in Natural Products Chemistry*, Atta-ur-Rahman; Ed.; 1996, vol. 18, p. 43, Elsevier Science, Amsterdam; (b) Blakemore, P. R.; White, J. D. *Chem. Commun.* 2002, 1159; (c) *Medicinal Plants of the World*, Van Wyk, B.-E.; Wink, M., 2004, Timber Press Portland, OR.

15. Hofmann, A. in *Discoveries in Biological Psychiatry*, Ayd, F. J. Jr.; Blackwell, B.; Eds.; 1970, Chapter 7, Lippincott, J. B. Company, Philadelphia, PA.

16. For medicinal applications of marijuana, see: Voth, E. A.; Schwartz, R. H. *Ann. Int. Medicine*, 1997, **126**, 791.

17. For example, see: (a) Cragg, G. M.; Grothaus, P. G.; Newman, D. J. *Chem. Rev.* 2009, **109**, 3012; (b) Newman, D. J.; Cragg, G. M. *J. Nat. Prod.* 2007, **70**, 461; (c) Baker, D. D.; Chu, M.; Oza, U.; Rajgarhia, V. *Nat. Prod. Rep.* 2007, **24**, 1225; (d) Butler, M. S. *Nat. Prod. Rep.* 2005, **22**, 162.

18. (a) *Medicinal Natural Products: A Biosynthetic Approach*, Dewick, P. M., 2009, Wiley, NY; (b) *Natural Compounds as Drugs*, 2008, Volumes I, II, Petersen, F.; Amstutz, R.; Eds.; Birkhäuser, Springer, NY; (c) *Natural Products-Drug Discovery and Therapeutic Medicine*, Zhang, L.; Demain, A. L.; Eds.; 2005, Humana Press, Totowa, NJ; (d) *The Role of Natural Products in Drug Discovery*, Mulzer, J.; Bohlmann, R.; Eds.; 2001, Springer-Verlag, Heidelberg; (e) *Drugs of Natural Origin – A Textbook of Pharmacognosy*, 5th edition, Samuelsson, G., 2004, CRC Press, Boca Raton, FL; (f) *Drugs from Natural Products: Pharmaceuticals and Agrochemicals*, Harvey, A. L., 1993, Taylor & Francis, NY.

19. *Pharmaceutical Manufacturing Encyclopaedia*, Sittig, M., 2008, 3rd edition, Noyes Publications, Norwich, NY.

20. (a) *Functional Informatics in Drug Discovery*, Ilyin, S.; Ed.; 2007, CRC Press, Boca Raton, FL; (b) *Target Validation in Drug Discovery*, Metcalf, B. W.; Dillon, S., 2006, Academic Press, NY; (c) *Genomics: Applications in Human Biology*, Primrose, S. B.; Twyman, R. M.; Eds.; 2004, Blackwell Publishing, UK; (d) *Chemogenomics in Drug Discovery: A Medicinal Chemistry Perspective*, Kubinyi, H.; Müller, G. 2004, Wiley-VCH, Weinhem; (e) *Modern Methods of Drug Discovery*, Hillisch, A.; Hilgenfeld, R. Eds.; 2003, Birkhäuser, Springer, NY; (f) *Integrated Drug Discovery Technologies*, Mei, H.-Y.; Czarnik, A. W. 2002, CRC Press, Boca Raton, FL; for a thematic issue on advances in genomics and proteomics, see: Yates J. R. III.; Osterman, A. L., Guest editors, *Chem. Rev.* 2007, **107**, 3363.

21. For example, see: (a) Abraham, D. J. in *Comprehensive Medicinal Chemistry II*, Triggle, D.; Taylor, J. Eds. 2006, Elsevier, Amsterdam; (b) *Structure-based Drug Discovery - An Overview*, Hubbard, R. E.; Ed.; 2006, RSC Publishing, London; (c) *Analogue-based Drug Discovery*, Fischer, J.; Ganellin, C. R. Eds. 2006, Wiley-VCH, Weinheim.

22. See: (a) http://en.wikipedia.org/wiki/medicinal_drug; (b) http://en.wikipedia.org/wiki/Pharmaceutical_company.

23. (a) *Enzymes and their Inhibition: Drug Development*, Smith, H. J.; Simons, C., 2005, CRC Press Boca Raton, FL; see also: (b) Tyndall, J. D. A.; Nall, T.; Fairlie, D. P. *Chem. Rev.* 2005, **105**, 973; (c) Babine, R. E.; Bender, S. L. *Chem. Rev.* 1997, **97**, 1359.

24. Kaldor, S. W.; Kalish, V. J.; Davies, J. F. II; Shetty, B. V.; Fritz, J. E.; Appelt, K.; Burgess, J. A.; Campanale, K. M.; Chirgadze, N. Y.; Clawson, D. K.; Dressman, B. A.; Hatch, S. D.; Khalil, D. A.; Kosa, M. B.; Lubbehusen, P. P.; Muesing, M. A.; Patick, A. K.; Reich, S. H.; Su, K. S.; Tatlock, J. H. *J. Med. Chem.* 1997, **40**, 3979.

25. *HIV Chemotherapy: A Critical Review*, Butera, S. T.; Ed.; 2005, Caister, Academic Press, Norfolk, UK.

26. (a) *Clinical Drug Trials and Tribulations*, Cato, A. E.; Sutton, L., 2002, Informa Healthcare, NY; (b) *Drugs – From Discovery to Approval*, Ng, R. 2004, Wiley-IEEE, NY.

27. For the cost of drug development, see: (a) Collier, R. *Canadian Medical Association Journal* 2009, **180**, 279, February 3;

(b) Mullin, R. *Chem. Eng. News* 2003, **81**, No. 50, December 15.

28. For example, see: www.fda.gov/cder/drug/infopage/vioxx/default-gov.

29. For exploring biology with small molecules, see (a) Wild, H.; Heimbach, D.; Huwe, D. *Angew. Chem. Int. Ed.* 2011, **50**, 7452; (b) Stockwell, B. R. *Nature* 2004, **432**, 846; (c) MacCoss, M.; Baillie, T. A. *Science* 2004, **303**, 1810; (d) Austin, C. P. *Curr. Opin. Chem. Biol.* 2003, **7**, 511.

30. *Gene Therapy (Health and Medical Issues Today)*, Kelly, E. B., 2007, Greenwood Press, Westport, CT; see also: *Vaccine Design: Innovative Approaches and Novel Strategies*, Rappuoli, R.; Bangoli, F. Eds.; 2011, Caister Academic Press, 2011, UK.

31. *Antisense Drug Technology: Principles, Strategies, and Applications*, Crooke, S. T. Ed.; 2007, CRC Press, Boca Raton, FL.

32. For a review on protein therapeutics, see: Leader, B.; Baca, Q. J.; Golan, D. E. *Nat. Rev. Drug Discovery* 2008, **7**, 21.

33. *Stem Cell and Gene-Based Therapy: Frontiers in Regenerative Medicine*, Battler, A.; Leor, J.; Eds.; 2005, Springer, Heidelberg.

34. For example, see: (a) Gonzalez, R.; Lee, J. W. Snyder, E. Y.; Schultz, P. G. *Angew. Chem. Int. Ed.* 2011, **50**, 3439; (b) Sakurada, K.; McDonald, F. M.; Shimada, F. *Angew. Chem. Int. Ed.* 2008, **47**, 5718; (c) Xu, Y.; Shi, Y.; Ding, S. *Nature* 2008, **453**, 338; (d) Emre, N.; Coleman, R.; Ding, S. *Curr. Opin, Chem. Biol.* 2007, **11**, 252; (e) Walsh, D. P.; Chang, Y.-T. *Chem. Rev.* 2006, **106**, 2476; (f) Ding, S.; Schultz, P. G. *Curr. Top. Med. Chem.* 2005, **5**, 383; (g) Everts, S. *Chem. Eng. News*, Jan. 15, 2007, p. 19.

35. *Visual Symmetry*, Hargittai, I.; Hargittai, M. World Scientific Publishing, Singapore, 2009.

36. Robertson, E. J. *Science* 1997, **275**, 1280.

37. (a) Bada, J. L. *Science* 1997, **275**, 942; (b) Cintas, P. *Angew. Chem. Int. Ed.* 2007, **46**, 4016.

38. *Chirality in Drug Research*, Francotte, E.; Linder, W.; Eds.; 2007, Wiley-VCH, Weinheim.

39. (a) *Asymmetric Synthesis: The Essentials*, Christmann, M.; Bräse, S., Eds.; 2007, Wiley-VCH, Weinheim; (b) *Asymmetric Synthesis with Chemical and Biological Methods*, Enders, D.; Jaeger, K.-E. Eds.; 2007, Wiley-VCH, Weinheim; (c) *Enantioselective Organocatalysis: Reactions and Experimental Procedures*, Dalko, P. I.; Ed.; 2007, Wiley-VCH, Weinheim; (d) *Asymmetric Organocatalysis: From Biomimetic Concepts to Applications in Asymmetric Synthesis*, Berkessel, A.; Gröger, H. 2005, Wiley-VCH, Weinheim; (e) *Asymmetric Catalysis on Industrial Scale: Challenges, Approaches and Solutions*, Blaser, H.-U.; Schmidt, E. 2004, Wiley-VCH, Weinheim; (f) *The Impact of Stereochemistry on Drug Development and Use*, Aboul-Enein, H. Y.; Wainer, I. G. 1997, Wiley, NY.

40. *Flavourings: Production, Composition, Applications, Regulations*, Ziegler, H.; Ed.; 2007, Wiley-VCH, Weinheim.

41. Brenna, E.; Fuganti, C.; Serra, S. *Tetrahedron: Asymmetry* 2003, **14**, 1.

42. *Chirality in Agrochemicals*, Kurihara, N.; Miyamoto, J. Eds.; 1998, Wiley, NY.

43. *A Commentary on Aristotle's De Anima*, Aquinas, T. 1999, Yale University Press, New Haven, CT.

44. (a) *The Psychology of Graphic Images: Seeing, Drawing, Communicating*, Massironi, M., 2002, Lawrence Erlbaum Associates; (b) *Perception and Cognition at Century's End: History, Philosophy, Theory*, Caterette, E. C.; Friedman, M. P.; Hochberg, J. Eds.; 1998, Academic, NY; (c) *Visual Intelligence*, Barry, A. M. S. 1997, State University of New York Press; (d) *Imagery, Creativity and Discovery: A Cognitive Perspective*, Roskos-Ewoldsen, B.; Intos-Peterson, M. J.; Anderson, R. E. 1993, Amsterdam; (e) *The Mind*, Reslak, R. M. 1988, Bantam Books, NY; (f) *Image and the Mind*, Kosslyn, S. M. 1980, Harvard University Press, Cambridge, MA; (g) *Seeing with the Mind's Eye: The History, Techniques and Uses of Visualization*, Samuels, M., 1975, Random House, NY; (h) *The Logic of Scientific Discovery*, Popper, K. 1959, Basic Books, NY.

45. *Art and Visual Perception – A Psychology of the Creative Eye*, Arnhim, R. 1974, University of California Press, Berkeley, CA.

46. Toeplitz, B. K.; Cohen, A. I.; Funke, P. T.; Parker, W. L.; Gougoutas, J. Z. *J. Am. Chem. Soc.* 1979, **101**, 3344.

47. (a) Hanessian, S.; Cooke, N. G.; DeHof, B.; Sakito, Y. *J. Am. Chem. Soc.* 1990, **112**, 5276; (b) Evans, D. A.; Dow, R. L.; Shih, T. L.; Takacs, J. M.; Zahler, R. *J. Am. Chem. Soc.* 1990, **112**, 5290; see also: Dow, R. L. Ph.D. dissertation, Harvard University, 1985; (c) Lautens, M.; Colucci, J. T.; Hiebert, S.; Smith, N. D.; Bouchain, G. *Org. Lett.* 2002, **4**, 1879; for a review, see: (d) Faul, M. M.; Huff, B. E. *Chem. Rev.* 2000, **100**, 2407. For a recent synthesis of ionomycin Ca complex see: (a) Gao, Z.; Li, Y.; Cooksey, J. P.; Snaddon, T. N.; Schunk, S.; Viseux, E. M. E.; McAteer, S. M.; Kocienski, P. J. *Angew. Chem. Int. Ed.* 2009, **48**, 5022.

48. Corey, E. J. *Pure Appl. Chem.* 1967, **14**, 19; see also ref. 7k and 8b.

49. Hanessian, S.; Roy, P. J.; Petrini, M.; Hodges, P. J.; Di Fabio, R.; Carganico, G. *J. Org. Chem.* 1990, **55**, 5766.

50. Rowley, M.; Tsukamoto, M.; Kishi, Y. *J. Am. Chem. Soc.* 1989, **111**, 2735.

51. (a) *Fascinating Molecules in Organic Chemistry*, Vögtle, F., 1992, Wiley, NY; (b) *Organic Chemistry: The Name Game: Modern Coined Terms and their Origins*, Nickon, A.; Silversmith, E. F. 1987, Pergamon, NY; (c) *Stimulating Concepts in Chemistry*, Vögtle, F.; Stoddart, J. F.; Shibasaki, M. 2000, Wiley-VCH, Weinheim; (d) *Reflections on Symmetry: In Chemistry and Elsewhere*, Heilbronner, E.; Dunitz, J. D. 1993, VCH Publishers, NY; see also: (e) Mann, S. *Angew. Chem. Int. Ed.* 2000, **39**, 3392; (f) *Organic Synthesis: The Science behind The Art*, Smit, W. A.; Bochkov, A. F.; Caple, R. The Royal Society, Cambridge, UK, Chapter 4, p. 301.

52. For a review, see: (a) Stetter, H. *Angew. Chem. Int. Ed.* 1962, **2**, 286; see also: (b) Landa, S.; Hala, S. *Coll. Czech. Chem. Commun.* 1959, **24**, 93.

53. Eaton, P. E.; Cole, T. W. Jr. *J. Am. Chem. Soc.* 1964, **86**, 3157.

54. (a) Whitlock, H. W. Jr. *J. Am. Chem. Soc.* 1962, **84**, 3412; (b) Gauthier, J.; Deslongchamps, P. *Can. J. Chem.* 1967, **45**, 297.

55. Marchand, A. P.; Chou, T.-C.; Ekstrand, J. D.; Van der Helm, D. *J. Org. Chem.* 1976, **41**, 1438.

56. Sauers, R. R.; Kelly, K. W.; Sickles, B. R. *J. Org. Chem.* 1972, **37**, 537.

57. (a) Barth, W. E.; Lawton, R. L. *J. Am. Chem. Soc.* 1966, **88**, 380; (b) Barth, W. E.; Lawton, R. G. *J. Am. Chem. Soc.* 1971, **93**, 1730.

58. Ternansky, R. J.; Balogh, D. W.; Paquette, L. A. *J. Am. Chem. Soc.* 1982, **104**, 4503. For a review, see: Paquette, L. A. *Chem. Rev.* 1989, **89**, 1051.

59. Scott, L. T.; Boorum, M. M.; McMahon, B. J.; Hagen, S.; Mack, J.; Blank, J.; Wegner, H.; de Meijere, A. *Science* 2002, **295**, 1500.

60. For a review, see: Ault, A. *J. Chem. Ed.* 2001, **78**, 924.

61. Amabilino, D. B.; Ashton, P. R.; Reder, A. S.; Spencer, N.; Stoddart, J. F. *Angew. Chem. Int. Ed.* 1994, **106**, 1316.

62. For reviews, see: (a) Speigel, D. A.; Njardarson, J. T.; McDonald, I. M.; Wood, J. L. *Chem. Rev.* 2003, **103**, 2691; (b) Nicolaou, K. C.; Baran, P. S. *Angew. Chem. Int. Ed.* 2002, **41**, 2678.

2
The "*Why*" of Synthesis

2.1
Nature the Provider, Healer, and Enticer

Throughout the millennia, Nature and man have lived in a symbiotic relationship. The plant world has been a primary source of sustenance since man learned how to sow the first seeds in the earth he stood on. Today, the agricultural industry is a gigantic enterprise, extending its know-how beyond the traditional crop protection techniques, to include genetically engineered products [1]. In spite of such advances, famine and hunger in the third world as a result of drought and poverty are disturbing realities. A major problem in supplying the agricultural needs of the world, other than climactic disasters, is pest control [2]. On-going extensive research efforts aim to develop effective and safe insecticides, fungicides, and herbicides especially in the industrial sector. The driving force behind these efforts comes primarily from synthetic chemistry [3]. Unlike medicines intended for human therapy, an agrochemical must be produced and marketed at low cost per acreage. Thus, the market challenges of synthetic chemistry in the agrochemical and pharmaceutical industries are very different, even though the biological premise of using chemical entities to interact with a target protein or other cellular components is common to both.

Man's reliance on Nature (especially the plant world) for cures and remedies to alleviate pains and ills is well-known. Quinine, extracted from the bark of the *cinchona* tree is still in use today [4]. Plants in particular have been a continuous and rich source of natural products with therapeutic value. As the means for scientific exploration developed, especially during the past century, harvesting natural molecules from aquatic sources, moulds and soil micro-organisms provided a rich diversity of fascinating and biologically active structures [5]. Time-honored discoveries such as penicillin [6], have paved the way to the development of a significant number of life-saving drugs. Today, the number of natural products of varied structural complexity increases steadily as new extracts from a variety of sources are analyzed for their constituents. A compelling impetus for such studies is to discover new natural products with therapeutic interest.

The discovery of totally new structural types as future antibiotics is rare. The historic example of course, is penicillin, with its unique bicyclic β-lactam structure.

Design and Strategy in Organic Synthesis: From the Chiron Approach to Catalysis, First Edition.
Stephen Hanessian, Simon Giroux, and Bradley L. Merner.
© 2013 Wiley-VCH Verlag GmbH & Co. KGaA. Published 2013 by Wiley-VCH Verlag GmbH & Co. KGaA.

Platensimycin [7] and platencin [8] discovered by Merck scientists, are intriguing examples of a structurally new class of antibiotic with a unique mode of action by blocking bacterial fatty acid biosynthesis (Figure 2.1) [9]. The potential for activity against multi-resistant micro-organisms by further exploration of these new offerings from Nature is an exciting prospect [10]. It remains to be seen if the pharmaceutical industry shares the same view [11]. Following the first total synthesis of racemic platensimycin by Nicolaou and coworkers in 2006 [12], several other notable contributions followed showing efficient enantioselective syntheses [13] (see Chapter 13, section 13.7). Total syntheses of platencin were disclosed in 2008 by Nicolaou [14], Rawal [15], and their respective coworkers.

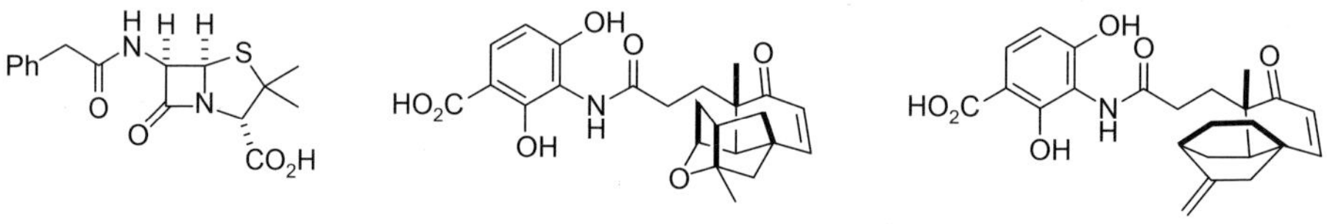

Penicillin G
(1928)

Platensimycin
(2006)

Platencin
(2007)

Figure 2.1 From petri dish (penicillin G) to superbug challengers (platensimycin and platencin).

In view of the novelty of their structures among natural products, platensimycin and platencin could, in principle, be considered as the penicillins of today with which they share the property of targeting the same bacterial foes, albeit using different modes of action. However, penicillin was a highly effective antibiotic in its time, and still dominates the market through its chemically modified variants and congeners. Whether platensimycin and platencin can live up to their superbug challenger reputation is at best, only a speculation at this time. A synthetic analogue, adamantaplatensimycin, retains substantial activity, but does not improve upon the natural product [16].

2.2
The Supply Problem

Sophisticated analytical techniques have greatly facilitated the rapid characterization and structure elucidation of new natural products often isolated in small amounts. In spite of enhanced modern isolation methods, including fermentation techniques, the quantities of new natural products ultimately produced and made available can be insufficient for extended biological tests beyond *in vitro* studies. This has instigated synthetic chemists in academia and in industry to develop innovative methods to address the supply problem of therapeutically important natural products such as taxol, discodermolide, and some members of the epothilone family for example (Figure 2.2). Although taxol is used in clinical practice as a potent anticancer agent, it is not manufactured by total synthesis (but rather by semi-synthesis). However, extensive efforts culminating with its total synthesis (see Chapters 8 and 16), as well as the study of analogues have contributed enormously to understanding the chemistry and biology of this remarkable drug [17]. The anticancer activity of the epothilones have also stimulated a large number of laboratories worldwide to explore methods of synthesis [18]. The relatively less complex nature of the epothilone structure compared to that of taxol, has led to the elaboration of practical syntheses that can be amenable to scale-up. The lactam analogue of epothilone B (Ixempra) was launched by Bristol Myers Squibb in 2007 for the treatment of advanced breast cancer (see also Chapter 3, section 3.5).

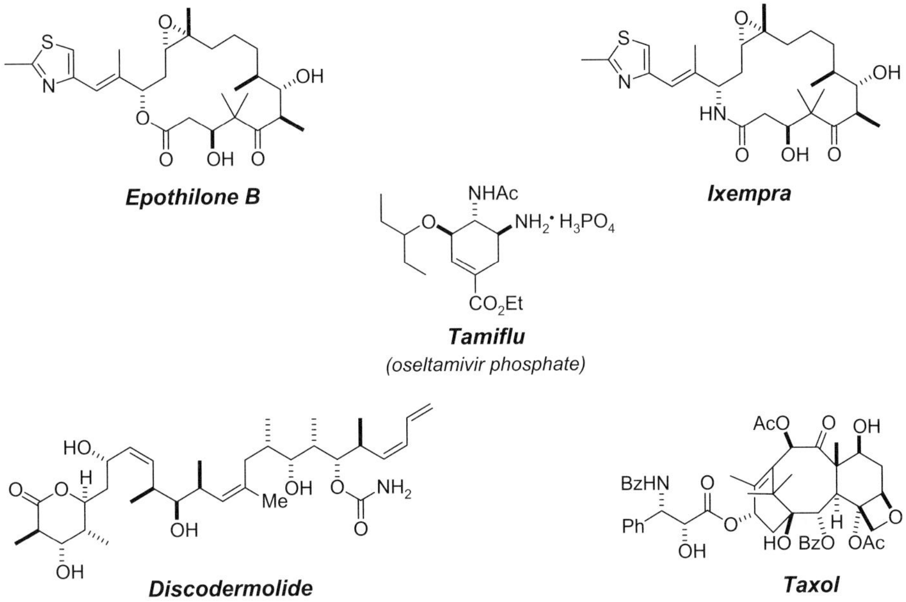

Epothilone B

Ixempra

Tamiflu
(oseltamivir phosphate)

Discodermolide

Taxol

Figure 2.2 Addressing the supply problem through synthesis.

The need to produce discodermolide by total synthesis in quantities sufficient for clinical studies culminated with a heroic effort within the Novartis process chemistry group in Basel, Switzerland [19]. Several total syntheses in academic laboratories [20], notably by Smith [21], Paterson [22] and their respective groups, laid the ground work with potential applications to a large scale synthesis of discodermolide. This serves to illustrate the power of multistep organic synthesis, even in an industrial context. Unfortunately, clinical trials with discodermolide as an anticancer agent were stopped due to toxicity (see section 2.8).

A particularly good example of the *"why"* in synthesis comes from efforts by several academic groups to explore alternative approaches to Tamiflu, a marketed antiviral agent against influenza [23]. The industrial synthesis utilizes shikimic and quinic acids as intermediates [24], which are isolated from the family of flowering plants *Illiciaceae*. Their limited supply has also led to the development of a recombinant microbial biocatalytic method, which in part, supplements the present requirements [25]. In view of the projected need for large quantities of Tamiflu in case of a flu epidemic, there has been an increasing demand to seek alternative routes [23]. The application of catalytic asymmetric methods for a practical synthesis of Tamiflu was reported in 2006 by Corey [26a], Shibasaki [26b,c,h], and their coworkers. A variety of other catalytic methods reported toward the total synthesis of Tamiflu by Fukuyama [26d], Trost [26e], Fang [26f], Hayashi [26g], Hudlicky [26i,j] and their respective groups, augur well for the development of a scalable and efficient synthesis (see Chapter 14, section 14.9).

2.3
From Bench to Market

The impressive variety of organic compounds for the treatment of diverse medical conditions in man is a great testament to the success of synthetic chemistry in the pharmaceutical industry. However, the road from bench to market is fraught with difficulties and obstacles (Chapter 1, section 1.3). Initially, these problems are exacerbated by the stringent criteria imposed by pharmacology, toxicology, and pharmacokinetics, among other factors. As a lead compound passes these tests and is declared as a pre-clinical candidate, the practicality of the synthesis becomes a primordial consideration. Invariably, the original synthesis developed during the discovery phase of the research has to be modified to satisfy cost effectiveness on a manufacturing scale. With very few exceptions involving natural products used directly from fermentation, the potential for large scale production is explored in a Process Chemistry Department within a pharmaceutical company [27]. The same applies to molecules produced by chemical synthesis in other related high-volume industries [28].

The challenges of developing practical and cost-effective syntheses are heightened when target molecules possess one or more stereogenic centers. The challenge is even greater when quaternary or multiple stereogenic centers are present. The compounds shown in Figure 2.3 represent a cross-section of such structurally demanding drugs manufactured by total synthesis, and marketed for a variety of applications (see also Chapter 18, Figure 18.3). Although resolution of racemic mixtures is still used industrially for the separation of the desired enantiomer whenever appropriate, asymmetric processes are becoming increasingly popular (see Chapter 5).

One of the important reasons to partake in the development of new synthetic methods, including *catalytic* asymmetric synthesis, is to find applications in drug discovery among other areas of research. Often, such methods are discovered in academic laboratories, and the technology is then successfully adapted to practical applications in the pharmaceutical sector [29] (see also, Chapter 4, section 4.6.1). The demands for practicality have also resulted in the invention of useful synthetic methods covered under proprietary patents.

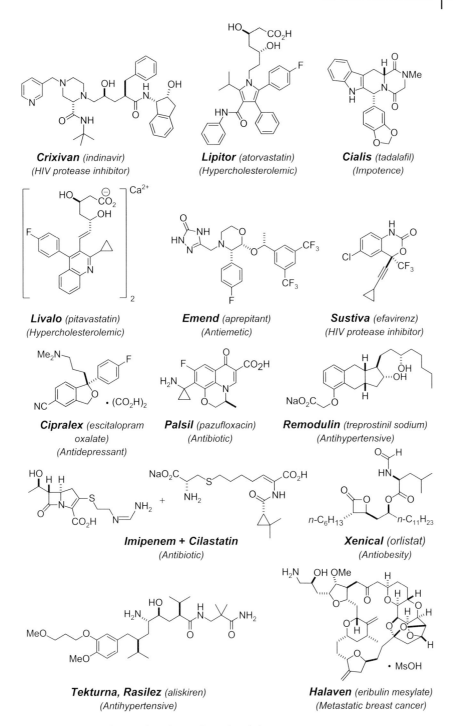

Figure 2.3 Selected examples of recently marketed drugs produced by total synthesis in an industrial laboratory.

2.4
Thank you Nature!

Some natural products have exhibited remarkable therapeutic properties from the time of their isolation and structure elucidation. Historic examples include the β-lactam, macrolide, aminoglycoside, and anthracycline antibiotics which have been in clinical use for many years as direct fermentation products. As the mode of action of many of these natural products became better understood, important insights into their interactions with biological receptors led to new generations of analogues. Thus, chemical modification at strategic sites in a given molecule could lead to congeners having a broader spectrum profile, higher potency, avoiding resistance mechanisms, improved water solubility, and exhibiting better pharmacological properties compared to the parent compound [30].

Two of the more interesting classes of marketed antibiotics that have benefited from chemical modification are the lincosaminides and the macrolides [31], whose mode of action at the bacterial ribosome level is well understood [32]. Dalacin (clindamycin) is simply the C-7 inverted chloro analogue of lincomycin in which a hydroxyl group was replaced by a chlorine atom (Figure 2.4). Biaxin (clarithromycin) is the 6-O-methyl ether of erythromycin, with an increased half-life in the blood stream (by avoiding intramolecular bis-ketal formation). Zithromax (azithromycin) is a ring-expanded aza-erythromycin analogue. The newest entry in the macrolide antibiotic arena is Ketek (telithromycin), a 3-keto macrolide also known as a ketolide. Although erythromycin was already marketed over 30 years ago, it is only recently that previously unimaginable modifications such as those found in the ketolide derivatives, were investigated based on reliable structural data gleaned from the X-ray co-crystal structures of subsites within the bacterial ribosome [33].

Elegant research on the mode of action of β-lactam antibiotics has led to a variety of synthetic penicillins and cephalosporins over the years [34]. It is ironic that decades after its development, amoxicillin (simply, a p-hydroxybenzyl penicillin) is now marketed in combination with the β-lactamase inhibitor, clavulanic acid, which is also a natural product, under the trade name of Augmentin. Certican (everolimus) is the (2-hydroxyethyl) ether of the parent sirolimus, an immunosuppressive drug [35]. Yasmin is a semi-synthetic oral contraceptive.

The aminoglycoside group of antibiotics have been in clinical use since the discovery of streptomycin over 50 years ago [36]. Amikacin, the N-1-(2S)-2-hydroxy-4-aminobutyryl analogue of kanaymcin A, and arbekacin, its 3′,4′-dideoxy analogue possess improved Gram-negative activity. The specific interaction of aminoglycosides with the A-site of the bacterial ribosomal 16S subunit as seen in a co-crystal structure complex has opened new vistas for further chemical modification in this potent class of bactericidal compounds [37]. Issues of nephrotoxicity, ototoxicity, bacterial resistance, and lack of oral activity remain unresolved [38].

Belotecan and idarubicin are added examples of how synthetic chemistry has been used to chemically modify traditionally well-known natural products to produce potent antitumor drugs [39].

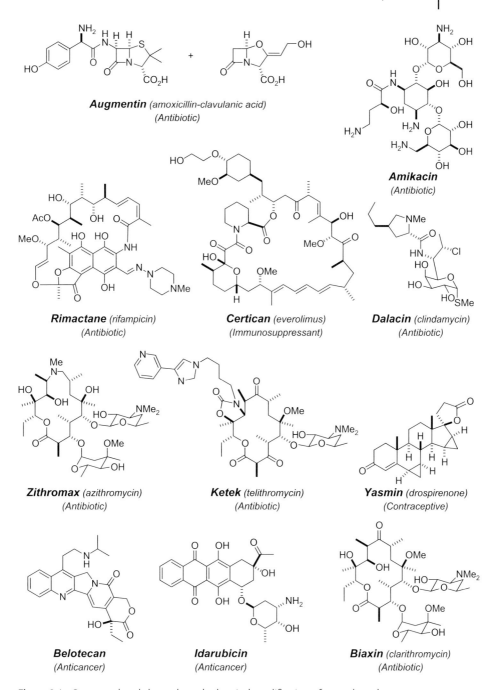

Figure 2.4 Some marketed drugs through chemical modification of natural products.

2.5
Chasing Bugs with a Purpose

Compared to other classes of chemotherapeutic agents, very few new antibacterial classes have entered the market since 1970. One reason for the decline in the discovery of new antibiotics is the decision by many pharmaceutical companies to reduce or terminate their research and development activities in this area. Unfortunately, the extensive use of existing antibiotics for decades has resulted in the emergence of varying degrees of resistance to nearly every class, thus increasing the risk of rendering them inefficient [32, 38].

The availability of most therapeutically relevant classes of antibiotics by fermentation, has turned the attention to their chemical modification through semi-synthesis. Indeed, a significant number of successfully marketed natural products are direct chemically modified analogues as discussed in the previous section. This approach has many advantages provided supply is not a problem. To overcome the onset of resistance, and to continue the quest for better potency,

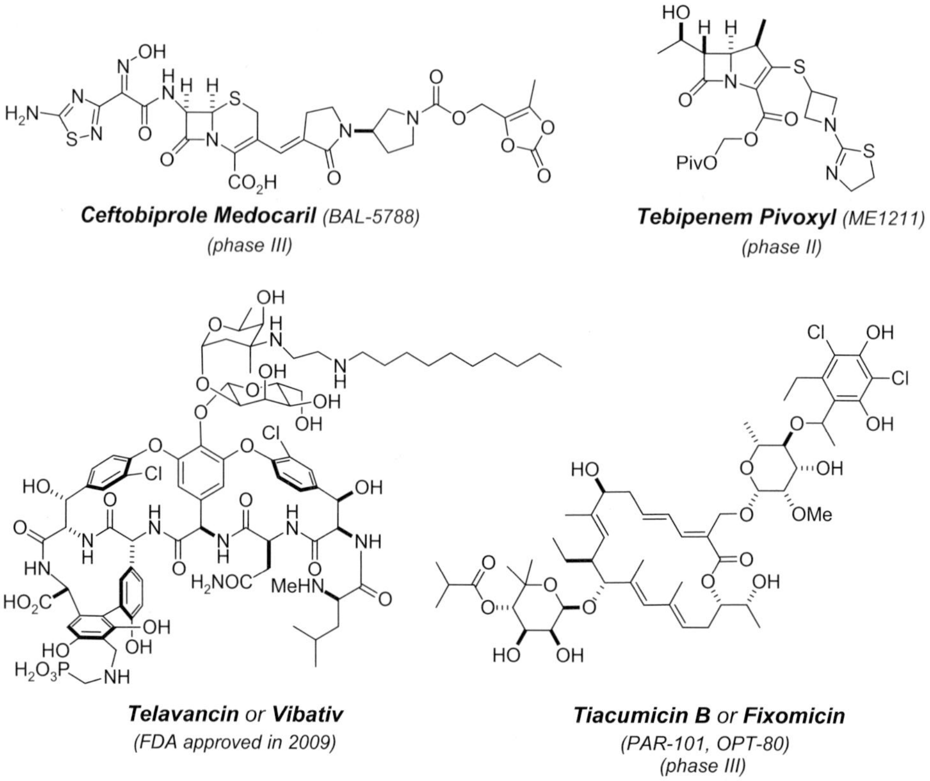

Ceftobiprole Medocaril (BAL-5788)
(phase III)

Tebipenem Pivoxyl (ME1211)
(phase II)

Telavancin or Vibativ
(FDA approved in 2009)

Tiacumicin B or Fixomicin
(PAR-101, OPT-80)
(phase III)

Figure 2.5 Chemically modified antibiotics undergoing late phase clinical trials or recently marketed.

known antibiotics are being chemically modified with increasing success in clinical trials. A few examples of chemically modified antibiotics that are currently undergoing late phase clinical trials or have been recently marketed are shown in Figure 2.5.

It is also an inescapable fact that even a simple modification which maintains, or better yet, improves activity of a known drug, can offer a major competitive patent advantage.

2.6
Structure-based Organic Synthesis

The advent of protein crystallography has revolutionized the drug discovery process during the past twenty years. The first successful examples of so-called structure-based design of enzyme inhibitors were demonstrated with the development of inhibitors of the angiotensin-converting enzyme (ACE) [40]. Information gleaned from the co-crystal structure determination of ACE with captopril proved to be invaluable for the development of other synthetic ACE inhibitors such as ramipril and enalapril, among others, for controlling hypertension. The impact of organic synthesis to develop peptidomimetics of the natural peptidic substrates that serve to inhibit harmful degradative enzymes has been strongly evident in the development of protease inhibitors associated with the HIV retrovirus [41].

In spite of the apparent difference in overall structures of the four HIV protease inhibitors shown in Figure 2.6, all have a hydroxyethylene or a hydroxyethylamino isostere subunit, which is crucial to their binding in the active site of the enzyme.

These relatively demanding structural entities, rich with stereochemistry, have instigated the development of elegant asymmetric processes for the synthesis of β-amino alcohols and related isosteric subunits [42] encompassed within specific entities. These methods have been very useful in the synthesis of many peptidomimetics, eventually culminating with clinical candidates and marketed drugs [43]. The practice of drug design based on peptidomimetic synthesis is prevalent in nearly all research endeavors that target soluble proteins. However, even when a potent inhibitor can be visualized as a complex with a protein target by X-ray crystallography, the prospects of developing a drug based on this information alone are not always encouraging.

The task is more challenging when the target proteins are membrane-bound and associated with GPCRs (G-protein coupled receptors) for example. Another major hurdle to overcome is in the area of synthetic CNS drugs, and their ability to penetrate the blood brain barrier. Non-selective action on receptor subtypes is often the cause of side-effects in animal studies and beyond [44].

Organic synthesis has also been instrumental in conjunction with other techniques for studying protein or macromolecule-small molecule interactions by exploiting the so-called "fragment-based drug discovery approach" [45] (see also Chapter 3, section 3.5).

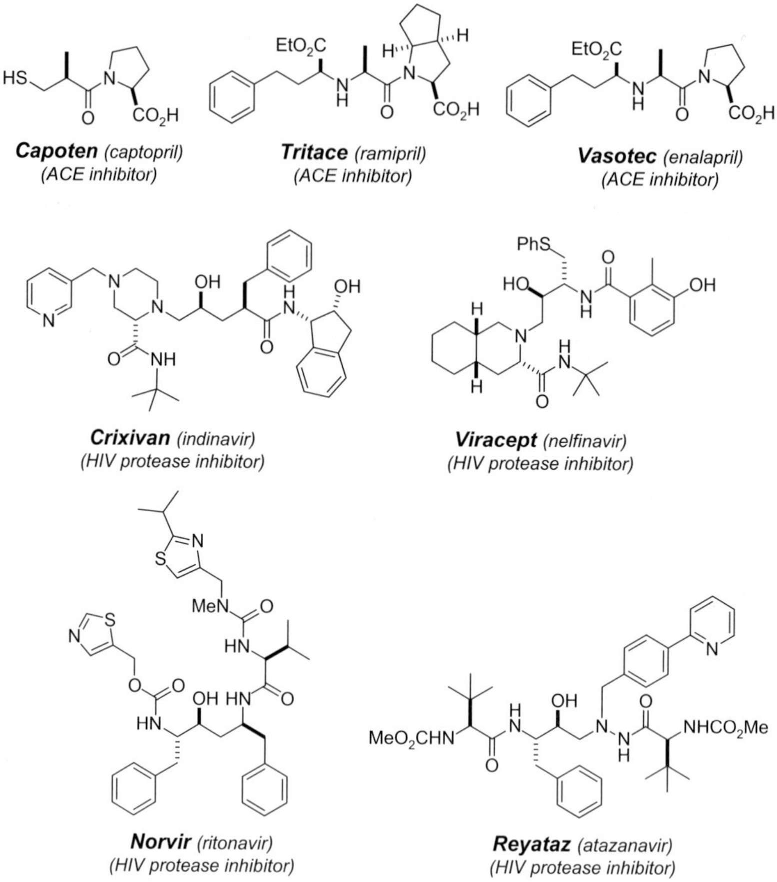

Figure 2.6 Some synthetic marketed drugs based on protein X-ray structures (structure-based design).

2.7
Almost There . . . or Just Arrived

For every class of promising therapeutic compounds, there are dozens of ongoing clinical trials at different stages and in different parts of the world. Those entities that make it to the late phases of such trials are potentially in an excellent position to be launched as products on the market, barring unexpected results. The compounds shown in Figure 2.7 represent a small selection of protease inhibitors that were undergoing late phase human trials in 2006 [43]. They were chosen as excellent examples of how organic synthesis has had an impact in their development, from the discovery stage to process chemistry and beyond within the pharmaceutical industry. Once again, synthetic methods that rely heavily on asymmetric bond-forming reactions were successfully used in each case.

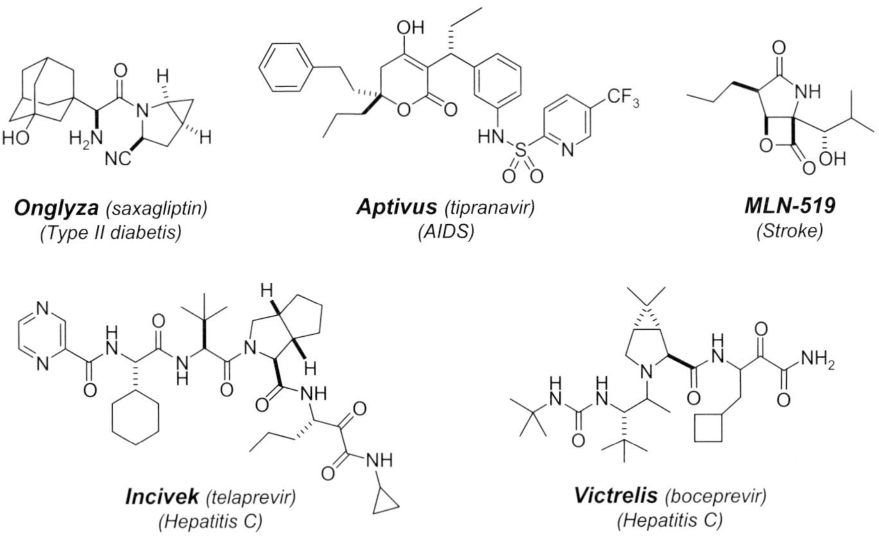

Onglyza *(saxagliptin)*
(Type II diabetis)

Aptivus *(tipranavir)*
(AIDS)

MLN-519
(Stroke)

Incivek *(telaprevir)*
(Hepatitis C)

Victrelis *(boceprevir)*
(Hepatitis C)

Figure 2.7 Selected synthetic protease inhibitors in late phase clinical trial studies or recently marketed for various therapeutic applications.

2.8
The Futility of it All

Organic synthesis is the engine that drives the drug discovery process forward. Its power and elegance is continually manifested in the structures of drug candidates that make it to the various stages of clinical trials and beyond. In the process, millions of dollars are spent in hope of a successful launch, offering remedies to many diseases. The same applies to the agrochemical, food, and related high volume industries.

However, the performance of a drug during human clinical trials is unpredictable in spite of safe animal studies. The development chemistry that went into the synthesis of discodermolide and BILN 2061, individually, was impressive and commensurate with the challenging structural and stereochemical demands of the respective chemical structures. Discodermolide succumbed to a total synthesis as a natural product anticancer agent on a kilogram scale [19]. BILN 2061 with an impressive activity against the hepatitis C virus was the result of extensive structure-based design and substantial chemical development [46]. Unfortunately, both of these highly promising candidates had to be withdrawn from their respective clinical trials due to adverse effects.

The anticoagulant ximelagatran (Exanta) [47], also conceived from knowledge of the mode of action of other thrombin inhibitors, was already launched and in clinical use for thromboembolism, before it was recalled and withdrawn from the marketplace after a short period of time.

It is important to point out that these "failed" drugs were the products of monumental synthetic efforts in the pharmaceutical industry. It is equally important that their rescue, if needed, will also rely on newer syntheses of entities in which the pharmacophores are preserved, but the possible toxophores are replaced by innocuous counterparts.

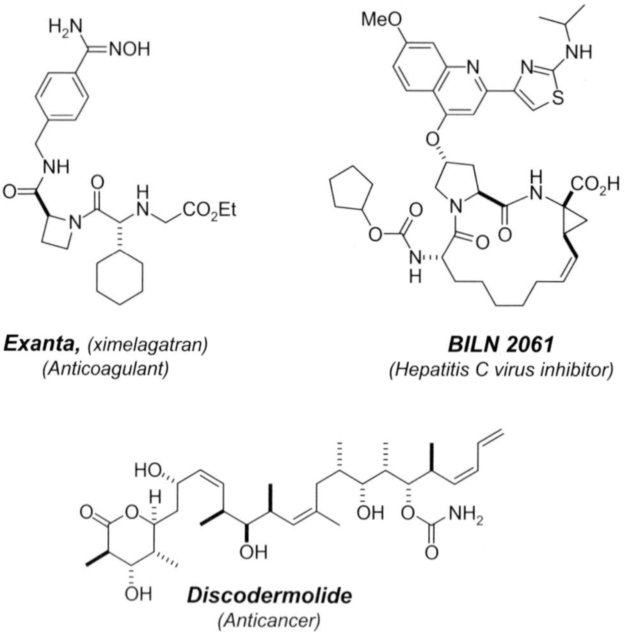

Exanta, *(ximelagatran)*
(Anticoagulant)

BILN 2061
(Hepatitis C virus inhibitor)

Discodermolide
(Anticancer)

Figure 2.8 Discontinued clinical studies of protease or tubulin polymerization inhibitors.

2.9
Synthesis as a Seeker of the Truth

Achieving the total synthesis of a complex natural product is a highly satisfactory feeling for a research team as a whole. Usually, the first indication of reaching the summit of the synthesis mountain, before raising the flag, is spectroscopic confirmation and other useful physical data. This can be easily achieved when such data is available from the literature or better yet, from a sample of the authentic material. Multistep total syntheses of complex natural products in academic laboratories rarely yield more than milligram quantities of the intended product. Oftentimes, purification results in even lesser amounts, which, thanks to high field NMR, advanced mass spectrometry techniques, and HPLC separation methods, may be enough to validate the structure and claim victory. A crystalline product can also be analyzed by single crystal X-ray techniques, which will definitively confirm its structure and the relative, if not absolute, stereochemistry. Quantities permitting, the characterization protocol can be completed with an optical rotation, an IR spectrum, and a fraction of a milligram to spare for a melting point (no joke). Time to hoist the victory flag and indulge in a good bottle of champagne!

In a sporting event, the match is not over until the final whistle is blown. There is a clear winner and loser. In synthesis, the ultimate "winner" is the team that completes the journey and conquers the intended target molecule in the most practical and innovative way. The good fortune to have been the first to complete the synthesis may also be a stimulus for striving to do better for those who follow with second and third generation syntheses. Sometimes, being first to synthesize a complex natural product may have unforeseen consequences. In one scenario, the originally assigned structure could be erroneous. Usually, this has a rewarding consequence when the correct structure is ultimately synthesized and the originally misassigned structure (constitutionally or stereochemically) is corrected [48]. This is a happy ending for total synthesis (see also Chapter 18, section 18.6).

The original and revised structures of the natural products shown in Figures 2.9 and 2.10 are classified according to two categories. In *structural* and *constitutional* revisions (Figure 2.9), segments of the original molecule had been incorrectly elucidated (isochizogamine, robustadial A, azadirachtin, sclerophytin A and hexa-cyclinol). Clearly, revisions of such intricacy in connectivity, especially in polycyclic compounds could only be validated through total synthesis.

In *functional* and *stereochemical* revisions, the originally assigned structures were constitutionally correct except for the incorrect placement of functional groups, double bond geometry or mistaken functionality (diazonamide A, oscillarin, aerug-inosin 205B, polyoxin A, alcyonin), or misassignment of absolute stereochemistry (azaspiracid-1, amphidinolide A). Corrections of originally misassigned structures are an important component of the practice of total synthesis and an excellent scientific justification for the *why* of synthesis [6, 7].

As a lead-in to the next section, we provide the present-day ultimate challenge for synthesis by depicting the originally proposed, and alternate, NMR suggested structures of the marine toxin, maitotoxin (Figure 2.11) [49].

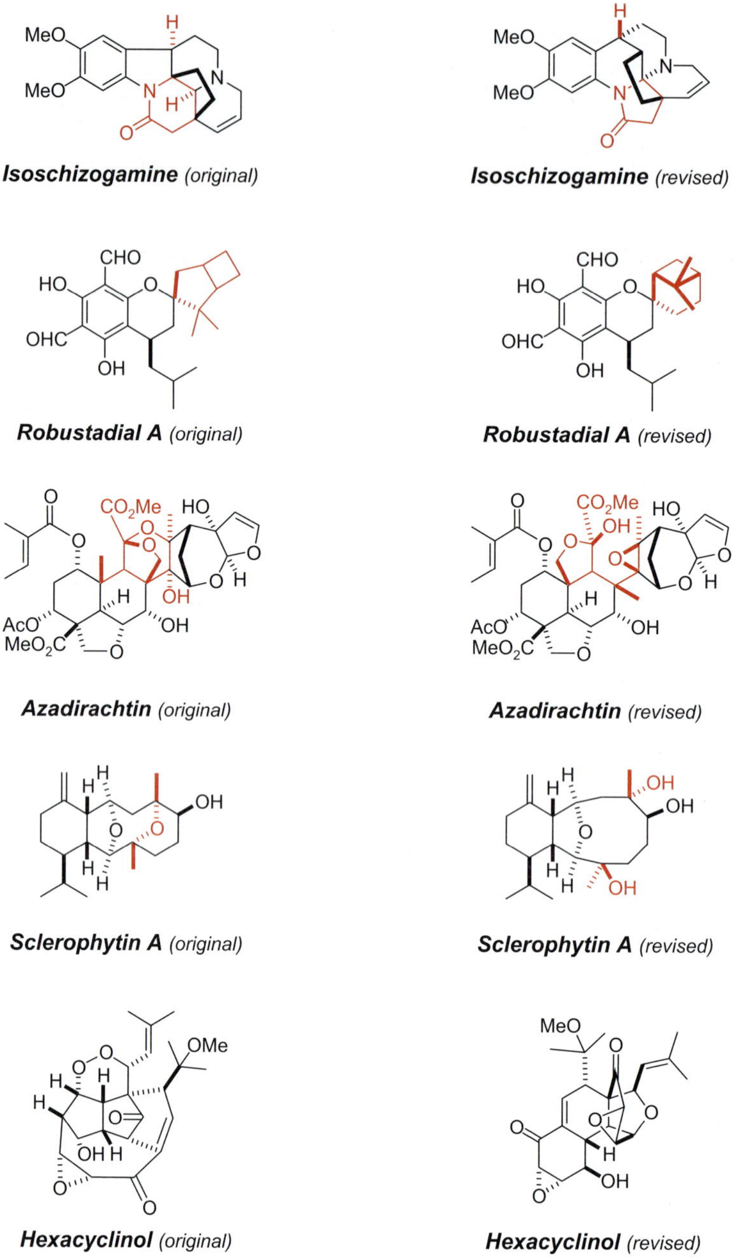

Isoschizogamine (original)

Isoschizogamine (revised)

Robustadial A (original)

Robustadial A (revised)

Azadirachtin (original)

Azadirachtin (revised)

Sclerophytin A (original)

Sclerophytin A (revised)

Hexacyclinol (original)

Hexacyclinol (revised)

Figure 2.9 Correction of selected misassigned structures. Structural and constitutional revisions [48].

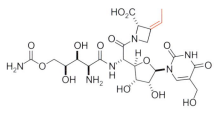

Diazonamide A *(original)*

Diazonamide A *(revised)*

Polyoxin A *(original)*

Polyoxin A *(revised)*

Azaspiracid-1 *(original)*

Azaspiracid-1 *(revised)*

Figure 2.10 Correction of selected misassigned structures. Functional group and stereochemical revisions [48].

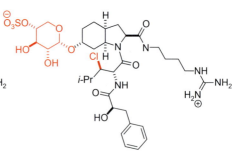

Amphidinolide A (original)   *Amphidinolide A* (revised)

Oscillarin (original)   *Oscillarin* (revised)

Aeruginosin 205B (original)   *Aeruginosin 205B* (revised)

Alcyonin (original)   *Alcyonin* (revised)

Figure 2.10   *(continued)*.

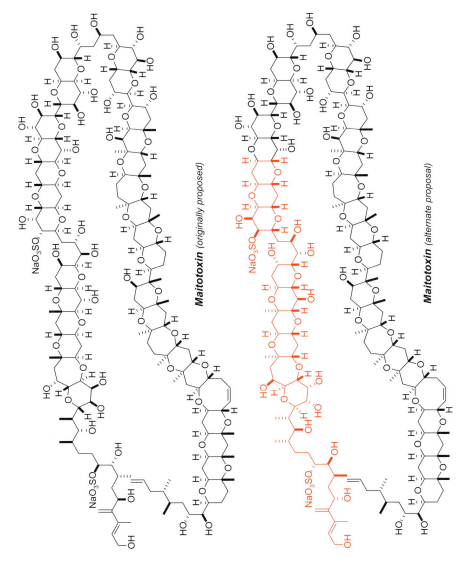

Figure 2.11 On the structure of maitotoxin. Top: originally proposed; Bottom: alternate proposal (red color).

2.10
Nature as the Ultimate Synthesizer

The way Nature builds its molecules is the ultimate challenge to synthetic chemists. The infinite variety of metabolites, some with great medicinal value, are true works of art. Remarkably, Nature produces these molecules from simple nutrients that repeat themselves in diverse classes of final natural products. In the words of Rustum Roy: "*Nature optimizes the system, never the components.*" In contrast, as synthetic chemists, we are continually trying to optimize the components, but to date we are incapable of optimizing the "system" as Nature does it.

Understanding the way biochemical pathways lead to natural products has been an area of extensive study over the years. Already more than thirty years ago the biosynthesis of macrolides and polyether antibiotics through the polyketide pathway delineated how arrays of carbon and oxygen atoms were incorporated into such complex molecules. These pioneering studies are summarized for monensin [50] and erythromycin A [51] in Figure 2.12 A. The availability of synthetic intermediates with and without heavy atom labels was crucial in these studies. The science of biosynthesis has been dramatically enhanced with the identification and cloning of genes responsible for the rate limiting steps of the enzymatic processes.

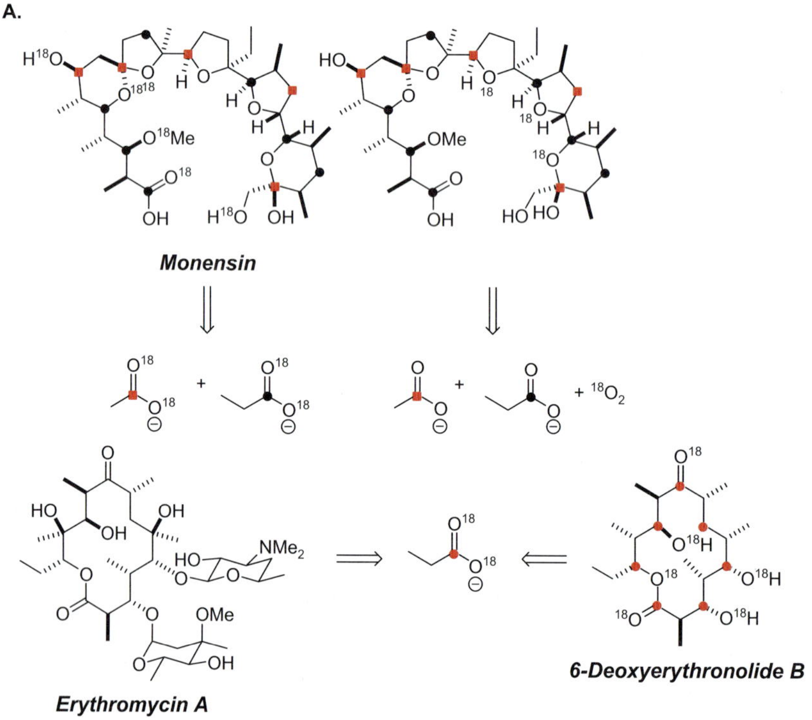

A.

Monensin

Erythromycin A

6-Deoxyerythronolide B

Figure 2.12 Nature's way to molecules. A. Validating biochemical pathways by labeling studies; B. Biogenetic synthesis of unnatural alkaloids.

B.

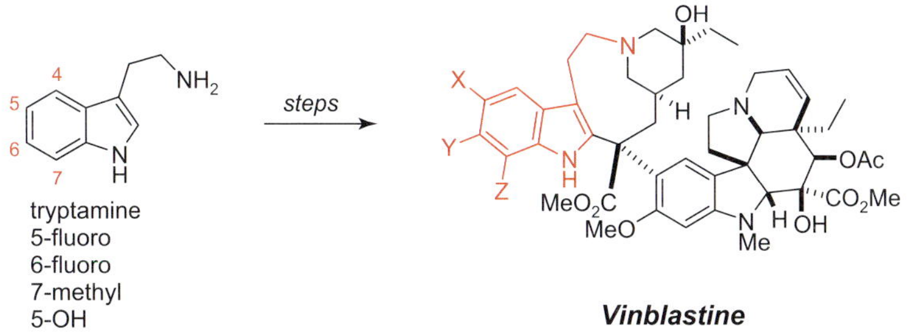

tryptamine
5-fluoro
6-fluoro
7-methyl
5-OH

Vinblastine

Figure 2.12 (*continued*).

Metabolic engineering, directed evolution of enzymes, and a host of enzyme-mediated functional group modifications are able to turn micro-organisms into factories of natural and unnatural small molecules [52] (see Chapter 5). Chemistry and biology are working hand in hand to produce precursors, intermediates and compounds of importance to the pharmaceutical, agrochemical, and other fine-chemical industries. Thus, in a sense, chemists are using Nature to voluntarily produce entirely new entities (see Chapter 5). For example, analogues of the antitumor agent vinblastine can now be prepared relying on a biogenetic incorporation of unnatural components (Figure 2.12 B) [53].

During a radio interview with the Canadian Broadcasting Corporation, one of the authors of this book was asked by the late Peter Gzowski if "a molecule was beautiful." Without hesitation the author replied: "Of course, it is a direct creation of Nature, or by man through synthesis."

2.11
A Brave New Chemical World

2.11.1
Beyond the molecule

Among the many innovations in organic synthesis was the development of supramolecular chemistry, for which the 1987 Nobel Prize in Chemistry was awarded to Cram, Lehn, and Pedersen. The ability of guest molecules to be recognized by host molecules has been the foundation of molecular recognition in chemistry [54]. Since the initial discovery of crown ethers and their ability to selectively form complexes with cationic species, many innovations have been introduced in this field. Cryptands and cavitands [55] are only two early examples of a large variety of innovative macrocyclic structures capable of highly selective recognition of specific guest molecules or charged entities (Figure 2.13 A).

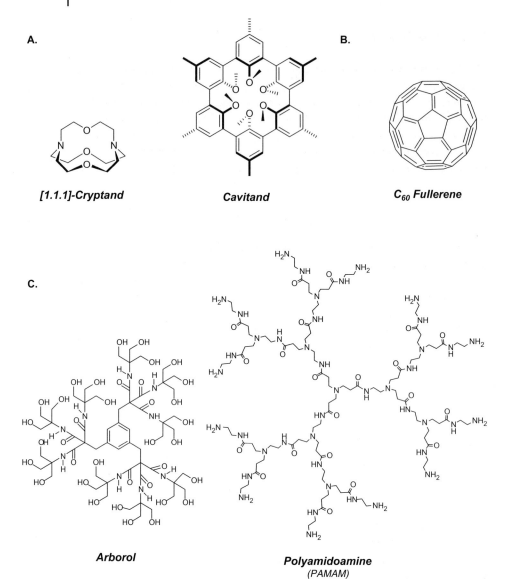

A.

[1.1.1]-Cryptand

Cavitand

B.

C$_{60}$ Fullerene

C.

Arborol

Polyamidoamine
(PAMAM)

Figure 2.13 Creating new entities. A. Cryptands and cavitands; B. C$_{60}$ Fullerene; C. Arborol and PAMAM (dendrimers).

2.11.2
Buckyballs and fullerenes

The discovery of C_{60} (Buckminster)fullerene as a new allotrope of carbon was also recognized as one of the great achievements in chemistry of the twentieth century, for which the Nobel prize in chemistry was awarded in 1996 to Curl, Kroto, and Smalley. Although traditional synthetic chemistry of this all-carbon compound was not instrumental in its discovery, a large body of work was generated by synthesizing functionalized derivatives and fragments of C_{60} and other fullerenes with applications ranging from materials science to medicine (Figure 2.13 B) [56].

2.11.3
Dendrimers

The bottom-up approach to synthesis on a macroscopic scale was adopted in the so-called dendrimeric (from *dendron* meaning a tree in Greek) compounds [57]. Here, in its simplest form, a central functionalized residue is used to branch out in a regular array producing dense architectures. Two original examples named Arborols (branched polyols), and PAMAM, (consisting of basic side-chains) have found extensive use in many industrial and biomedical applications with several newer-generation variants (Figure 2.13 C).

2.11.4
Nanochemistry

The enabling aspect of organic synthesis has produced a plethora of architecturally complex molecules. By and large, this can be related to a bottom-up approach in which basic building blocks are used to construct medium and relatively large-size molecules. However, synthetic chemistry can also be practiced in exciting new ways using a top-down paradigm. This concept, well-known in physics, was elegantly realized with the advent of nanotechnology. Nanoscale structures comprising 1–100 nm dimensions are now possible with organic and inorganic compounds [58]. Miniaturization is essential in the invention of new devices using single atoms or single molecules. Already, applications of nanowires in molecular electronics are making great progress in materials science. Nanochemistry will assume major relevance in medically-oriented applications [59] such as drug delivery [60], non-invasive therapy [61], and imaging [62] (Figure 2.14). For example, the antitumor compound camptothecin and related compounds have been covalently attached as an ester, through a linker, to an aminopolyvinyl alcohol polymer containing ultrasmall superparamagnetic iron oxide nanoparticles (USPIONS) [63], as a means to magnetically direct the drug conjugate to the site of a tumor by application of an external magnetic field [64]. Internalization and cleavage by intracellular esterases is expected to release the drug in the least invasive way while circulating in the blood stream (Figure 2.14 A). Methotrexate has been conjugated to a polymer carrier through a silicon-bearing tether attached to an amide linker. Proteases in the lysozome are intended to release the drug once into the cell [65].

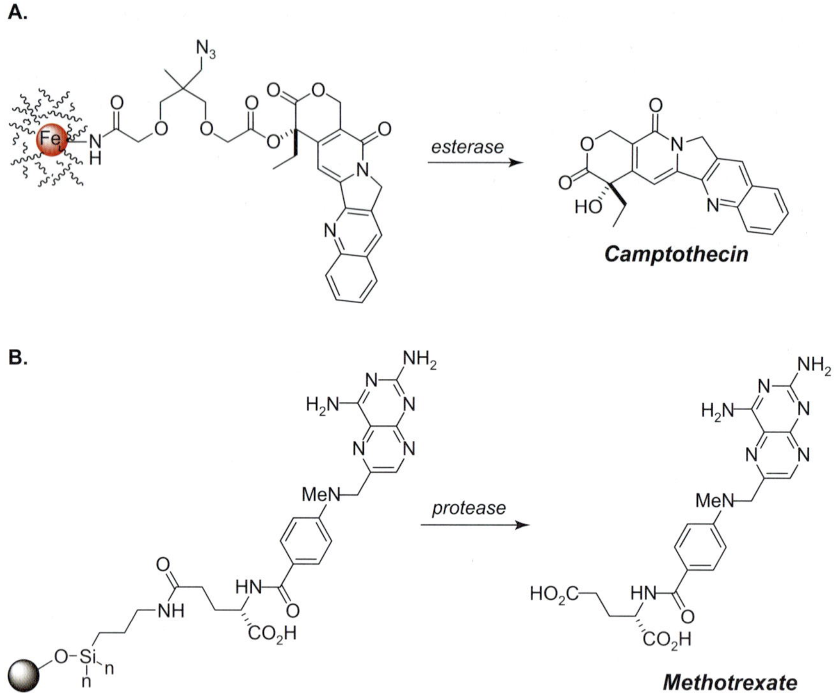

A.

Camptothecin

B.

Methotrexate

Figure 2.14 Drug delivery. A. Magnetically directed, esterase released delivery of a camptothecin – iron oxide/PVA nanoparticle conjugate; B. Protease mediated release of methotrexate nanoparticle conjugate in lysozomes.

2.11.5
Molecular machines

More futuristic applications of synthetic organic chemistry adopt a bottom-up approach where molecular components made by conventional organic synthesis are designed to perform machine-like movements under the influence of external sources of energy such as light and heat [66]. Molecular-level machines have already produced motorized nano cars [67] that actually move on a surface, a molecular elevator [68] that responds to a redox system, and molecular motors [69a–c] that convert chemical energy into rotary or linear motion. As well, molecules capable of performing logic at the molecular level are actively being pursued. Recent advances in this area include the synthesis of systems capable of performing numerical computations [69d–e].

2.12
Exploring New Synthetic Methods

One of the strongest reasons for the *"why"* of synthesis is the development of new and preparatively useful synthetic methods. Every creative endeavor utilizes means and ways to achieve its goals. The poet uses words, the painter applies colorful brush strokes, and the violinist varies the movement of his or her bow. Synthetic chemists rely on reagents and use them in reactions to prepare a given chemical entity. Just as there is no limit to the creativity of the poet and painter, and the virtuosity of the violinist, there is also a quasi-infinite number of chemical combinations of atoms and bonds to construct an unlimited array of organic compounds. For example, the familiar C_6H_6 molecular formula of benzene is predicted (via a computer) to give 217 possible structural combinations - both cyclic and acyclic!

Organic synthesis starts with the chemistry of carbon and its property to form bonds with other atoms based on their mutually compatible valence states. The tetrahedral nature of an sp^3 hybridized carbon atom is the basis of stereochemistry, and consequently, of life itself. We owe a lot to the early day alchemists and their followers into the latter half of the 19th century for laying the ground work of organic synthesis. Over time, organic synthetic methodology has evolved in a continuous mode offering methods of bond formation and breaking for practically any molecule or its substructure. Important insights into thermodynamic and kinetic processes have been gained from studies of physical organic chemistry in conjunction with organic synthesis. Today, synthetic methods for organic transformations are abundantly rich, and offer a large selection of reagents and procedures particularly for oxidation and reduction reactions. Carbon-carbon bonds can be formed in a myriad of ways, and the products can be used in equally abundant ways to build molecules of all shapes and complexities practically at will. Only time is the major intangible factor on the path to the intended target molecules (see also Chapter 18, sections 18.4–18.7).

Progress and efficiency in natural product or target-oriented synthesis are highly dependent on the methodologies available at the time. Occasionally, difficulties encountered in a particular synthetic step may incite the investigator to invent a new method or reaction in order to overcome a synthetic impasse. This may detract from the progress of the intended synthesis, but in the long run, the time spent is very worthwhile. In fact, historically, the great innovations and advances in synthetic methodology have come as a result of needs in natural product or target-oriented synthesis. Academic and industrial chemists have found many solutions to stubborn or recalcitrant chemical reactions. The mere inclusion of an additive, or changing the ligand in a catalytic reaction for example, can make a dramatic difference in yield and efficiency. The power of observation and the prepared mind, as aptly pronounced by Louis Pasteur, have contributed to many discoveries in science and in organic synthesis as well.

Just as every decade has ushered in new classes of compounds to be synthesized, so it has been for the development of synthetic methods to facilitate the task of reaching such objectives. We may recall the eras of sesquiterpenes (quadrone, coriolin, pentalenolactone), macrolides (erythromycin, carbomycin, methymicin), polyether antibiotics (lasalocid A, monensin), alkaloids (strychnine, morphine, ibogamine, reserpine), and the era of the highly complex natural products (taxol, phorbol, brevetoxin B, calicheamycin γ_1, vancomycin) [70].

Developing or creating new synthetic methods for a given transformation is a highly stimulating research activity, which evolves and improves with time [71]. As long as organic synthesis remains as a reagent and reaction-type of enterprise in the laboratory, there will always be a need for new synthetic methods. In the foreseeable near future, the majority of known organic chemical transformations will see their catalytic versions come to fruition. Absolute stereochemical control relying on asymmetric synthesis will be common place for a given bond forming reaction, including the preparation of quaternary stereogenic centers. However, small molecules abundantly provided by Nature such as amino acids, hydroxy acids, carbohydrates, and terpenes will still maintain their appeal as starting materials for organic synthesis. Advances in metal-catalyzed C–H bond activation and functionalization will open new possibilities in synthesis planning (see Chapter 18, section 18.6). Water may replace organic solvents in many reactions [72], and the word "chemical" will hopefully assume a more people-friendly and "greener" meaning.

There is good reason for such optimism if one surveys the evolution of the science of organic synthesis in the past one hundred or so years. To illustrate this evolution in synthesis, two of the most fundamental C–C bond forming reactions are briefly discussed in the following sections.

2.12.1
The Diels-Alder reaction

Arguably one of the most important contributions to organic synthesis, the Diels-Alder reaction, has greatly evolved from its original version first published in 1928 by Otto Diels and Kurt Alder [73] (Figure 2.15 A). Three decades passed before the importance of Lewis acid catalysis on enhancing *endo* selectivity of the Diels-Alder reaction was realized by Yates and Eaton [74]. Lewis acid catalysis has been used in many creative ways to achieve asymmetric versions [75] of the Diels-Alder reaction, first utilizing chiral auxiliaries, [76] and more recently with chiral reagents as catalysts [77] (Figure 2.15 B–E). An organocatalytic Diels-Alder cyclization was introduced in 2000 by MacMillan [78] using a chiral, non-racemic imidazolidinone catalyst (Figure 2.15 F).

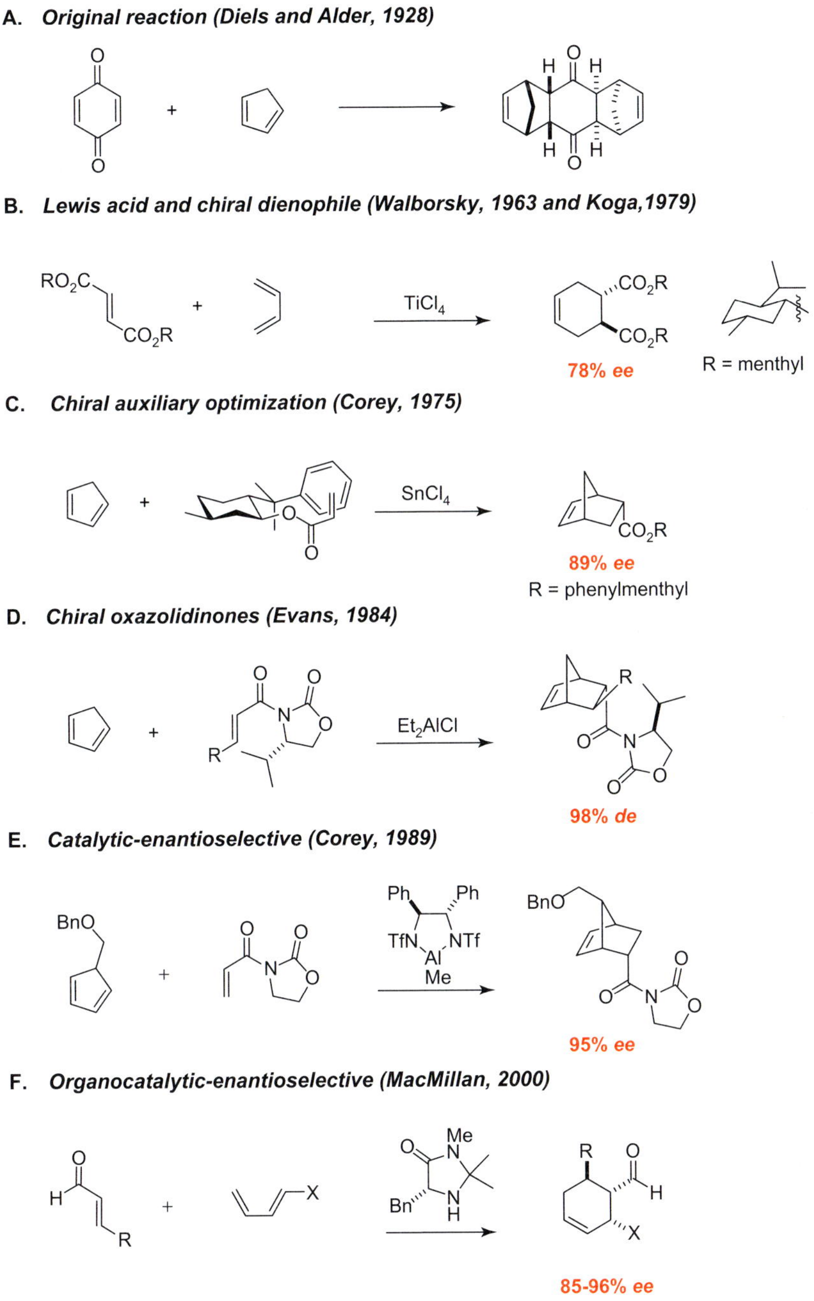
A. Original reaction (Diels and Alder, 1928)
B. Lewis acid and chiral dienophile (Walborsky, 1963 and Koga,1979)
RO2C
CO2R
TiCl4
CO2R
CO2R
78% ee
R = menthyl
C. Chiral auxiliary optimization (Corey, 1975)
SnCl4
CO2R
89% ee
R = phenylmenthyl
D. Chiral oxazolidinones (Evans, 1984)
Et2AlCl
98% de
E. Catalytic-enantioselective (Corey, 1989)
BnO
Ph
Ph
TfN
NTf
Al
Me
95% ee
F. Organocatalytic-enantioselective (MacMillan, 2000)
Me
Bn
85-96% ee

Figure 2.15 Evolution of the Diels-Alder reaction.

A number of outstanding examples of catalysis in the Diels-Alder reaction have been recorded in the literature, especially in conjunction with natural product synthesis (Figure 2.16) [79]. Thus, the synthesis of colombiasin A by Jacobsen and coworkers [80a] capitalized on the successful application of a hetero Diels-Alder reaction between ethyl vinyl ether and crotonaldehyde in the presence of a chiral chromium catalyst **A** to give an enantioenriched dihydropyran ethyl glycoside.

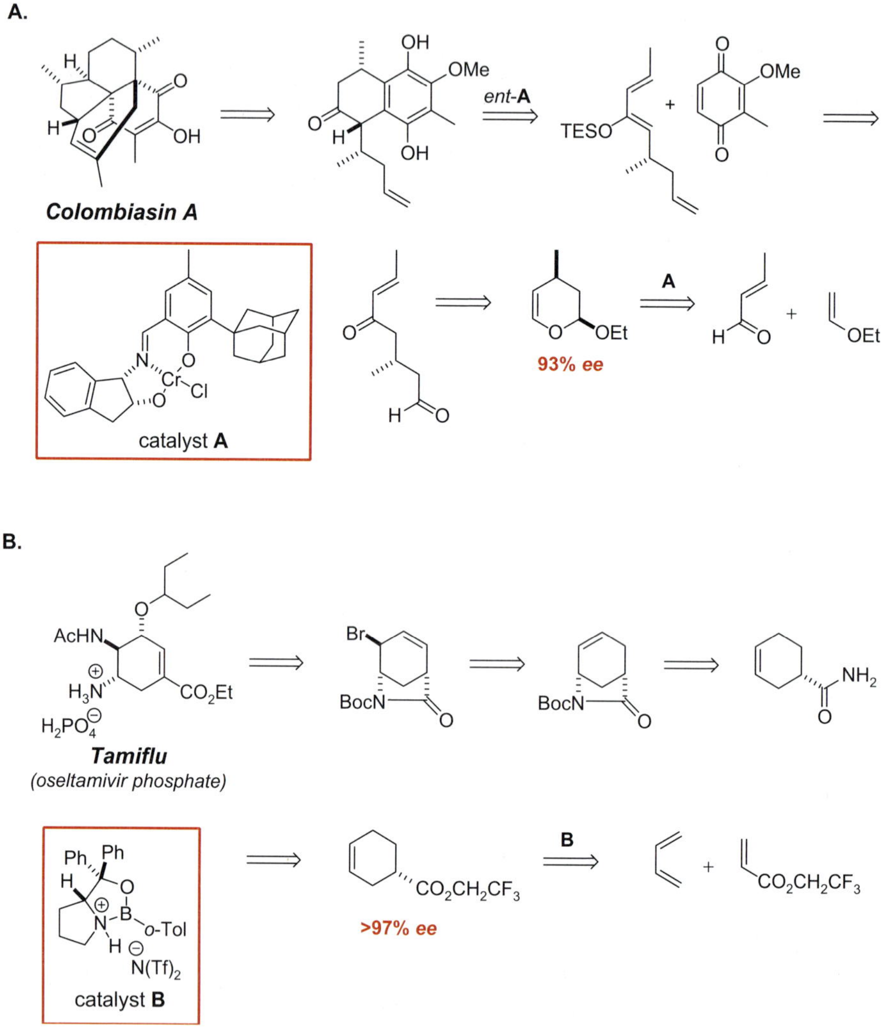

A.

Colombiasin A

catalyst **A**

93% ee

B.

Tamiflu
(oseltamivir phosphate)

catalyst **B**

>97% ee

Figure 2.16 Selected examples of catalytic asymmetric hetero Diels-Alder, and Diels-Alder reactions in target oriented syntheses.

A second Diels-Alder reaction starting with a methoxy quinone derivative, and using the enantiomer of the same catalyst, set the stage for the bicyclic precursor (Figure 2.16 A). Nicolaou and coworkers [81] used an (S)-BINOL/TiCl₂-mediated catalytic Diels-Alder reaction in their synthesis of colombiasin A. A substrate based approach starting with (−)-dihydrocarvone was adopted by Rychnovsky and Kim [82].

A practical application of the catalytic asymmetric Diels-Alder reaction was exemplified in the total synthesis of Tamiflu by Corey and coworkers [83] relying on a chiral, non-racemic oxazaborolidine catalyst (Figure 2.16 B, see also Chapter 14, section 14.9).

A much larger number of inter- and intramolecular Diels-Alder reactions have been recorded in conjunction with the total synthesis of polycyclic natural products [73d, 79]. In the vast majority of cases, enantio- and/or diastereoselectivity has been achieved relying on substrate-based induction with predisposed resident chirality. The selected examples shown in Figure 2.17 A–I, illustrate the diverse conditions that were used in critical intramolecular [84] Diels-Alder carbocyclization steps (see also Chapter 4, section 4.5.3).

Organocatalysis has also found interesting applications of the intramolecular Diels-Alder reaction in natural products synthesis [85]. A recent example involves the synthesis of amaminol B by Jacobs and Christmann [86] using MacMillan's imidazolidinone catalyst (Figure 2.17 H, see also Chapter 18, section 18.9.1).

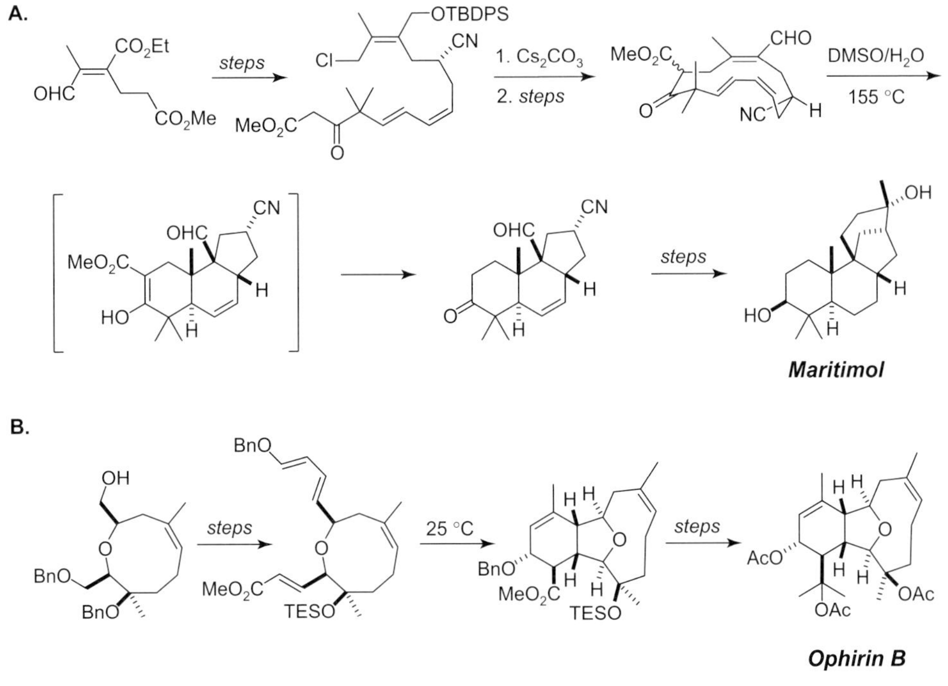

A.

Maritimol

B.

Ophirin B

Figure 2.17 Selected examples of intramolecular Diels-Alder reactions in natural product synthesis.

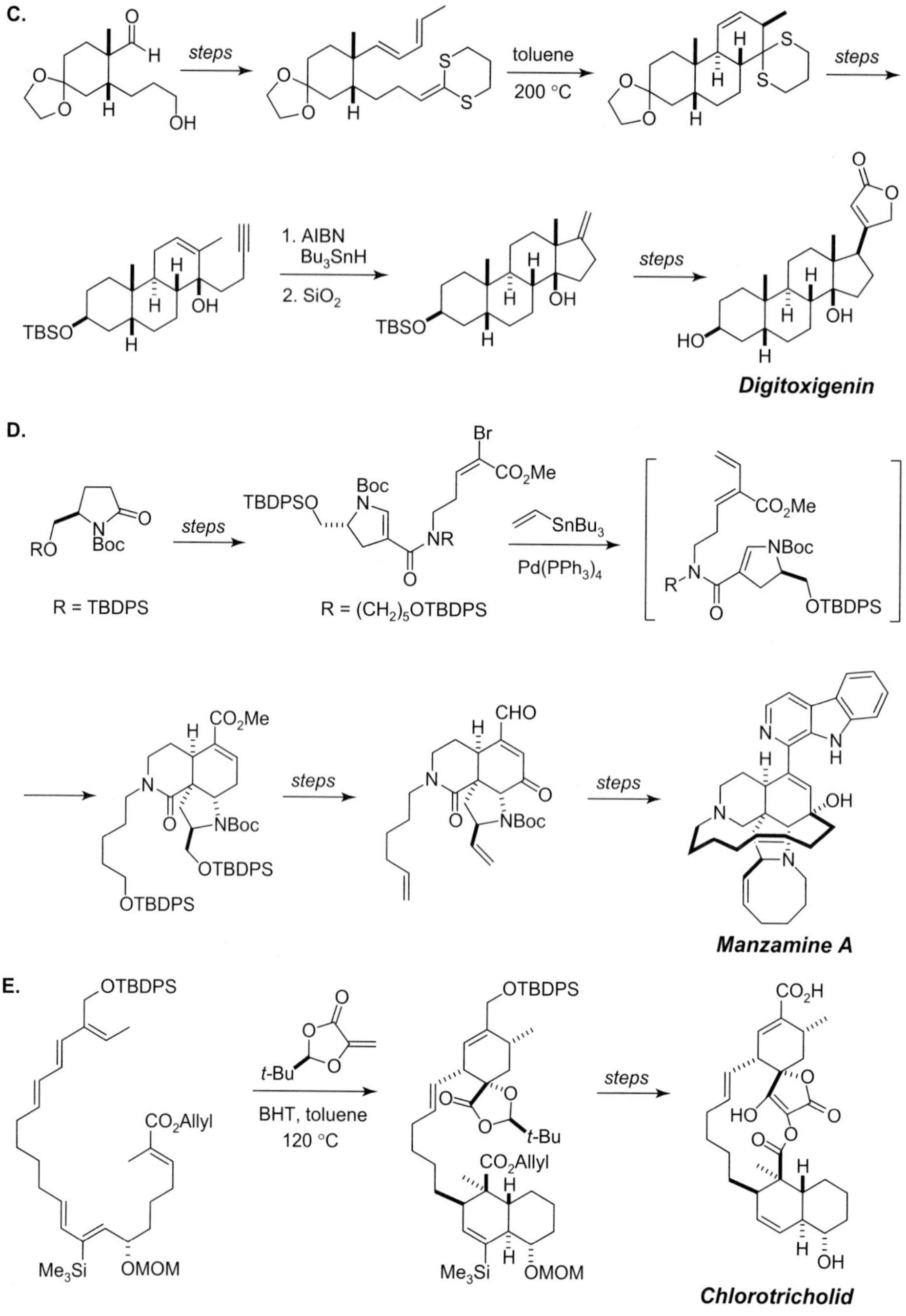

C.

Digitoxigenin

D.

Manzamine A

E.

Chlorotricholid

Figure 2.17 (*continued*).

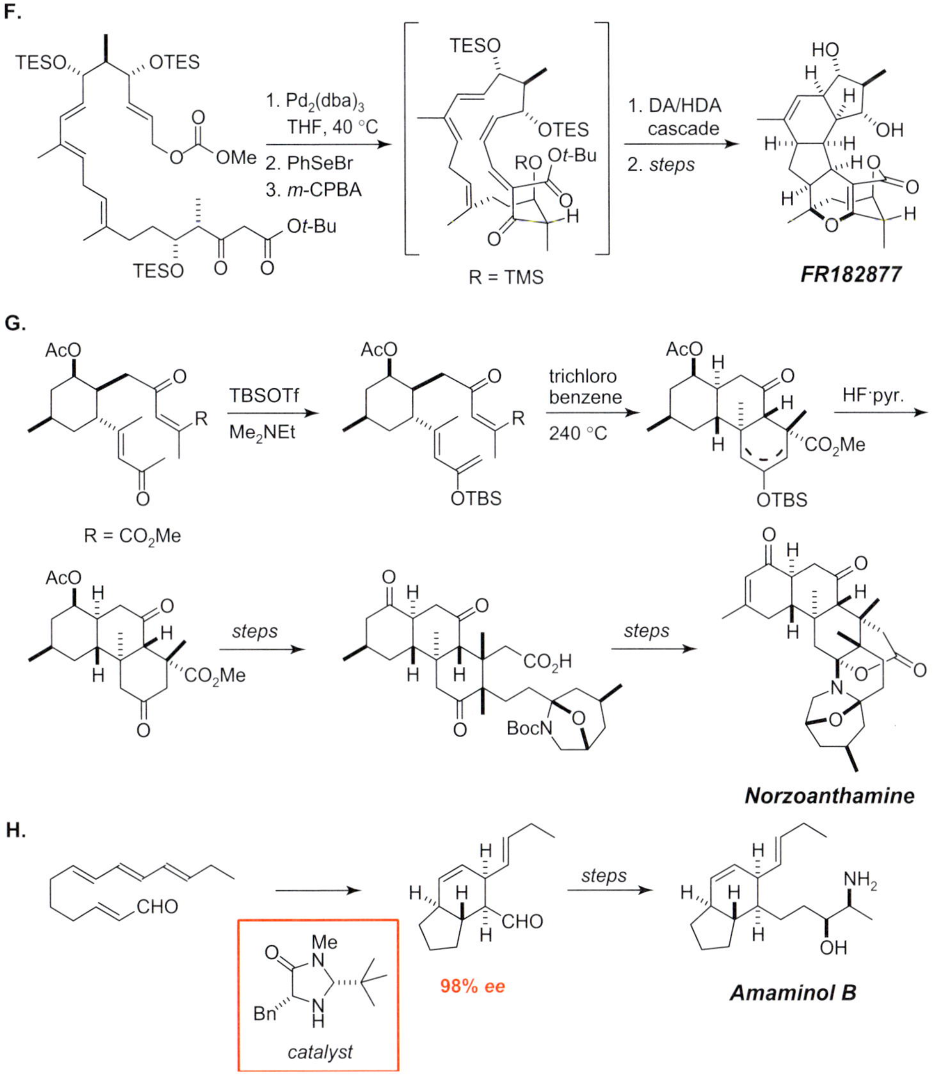

F.

G.

R = CO₂Me

Norzoanthamine

H.

catalyst

98% ee

Amaminol B

Figure 2.17 (*continued*).

I.

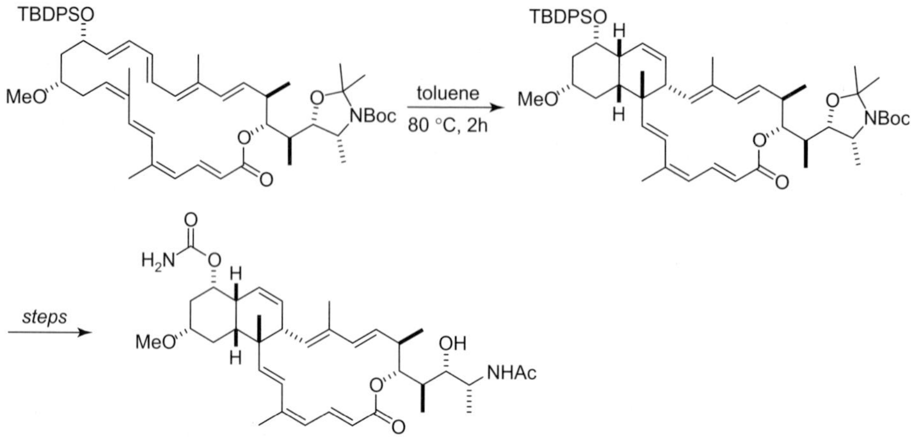

Superstolide A

Figure 2.17 (continued).

2.12.2
The direct aldol reaction

The condensation between two carbonyl compounds containing an enolizable hydrogen under acidic or basic conditions is one of the oldest known reactions in organic chemistry. The synthesis of β-hydroxybutanal from two molecules of acetaldehyde was published in 1872 by Wurtz [87], who coined the term "aldol" (Figure 2.18 A).

The advent of enolate chemistry almost half a century later offered new synthetic possibilities for the aldol reaction [88]. In order to avoid self-condensation and polymerisation products, many ingenious variants were introduced, including the pre-conversion of the ketone moiety to a silyl enol ether for example. This variant is known as the Mukaiyama cross-aldol reaction [89]. Other methods of carbonyl

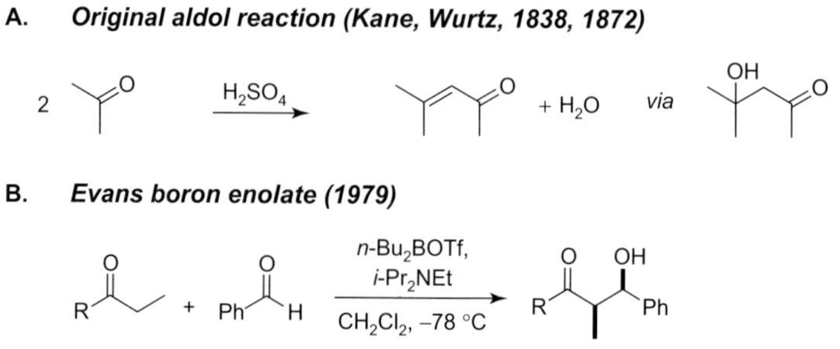

A. *Original aldol reaction (Kane, Wurtz, 1838, 1872)*

B. *Evans boron enolate (1979)*

Figure 2.18 Evolution of the direct aldol reaction.

C. *Shibasaki (1999) and Trost (2000) catalytic and enantioselective direct aldol reaction*

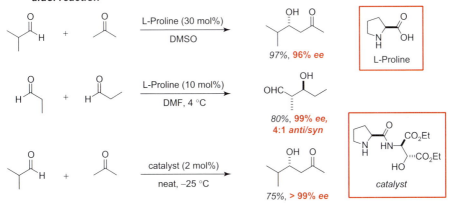

D. *List-Barbas (2000), MacMillan (2002) and Gong (2005) organocatalytic direct aldol reaction*

Figure 2.18 *(continued)*.

activation such as lithium, boron, and titanium enolates, allowed for controlled condensation reactions of ketones with aldehydes under stoichiometric conditions (Figure 2.18 B). Stereoregulation could be achieved in kinetic aldol condensations by controlling the geometry of the enolate to give a preponderance of *syn-* or *anti-β*-hydroxy ketones arising from a Zimmerman-Traxler transition state model as illustrated from the scholarly contributions of D. A. Evans (Figure 2.18 B) [90].

The catalytic version of a *direct* aldol condensation between a ketone and an aldehyde is a highly challenging goal. Progress in this direction was reported by Shibasaki [91] in the presence of an heterobimetallic catalyst derived from (*R*)- or (*S*)-BINOL, and by Trost [92] who used a zinc-coordinated C_2-symmetric chiral ligand (Figure 2.18 C).

Of particular importance is the more recent metal-free, organocatalytic, versions of aldol condensations mediated by proline and other nitrogen containing chiral compounds through the intermediacy of iminium salts and enamines (Figure 2.18 D) [93, 94] (see also Chapter 4, section 4.6, and Chapter 5, section 5.8).

The examples shown in Figure 2.18 A–D are representative of *direct* aldol reactions between ketones and aldehydes showing the evolution of the process since 1872. Clearly, the organocatalytic versions of this reaction have expanded the horizons of the cross-aldol reaction and revived the landmark Hajos-Parrish-Eder-Sauer-Wiechert (intramolecular aldol condensation) reaction reported in 1974! [95] (see Chapter 5, section 5.8). The first application of a proline catalyzed intermolecular aldol condensation in complex molecule synthesis can be found in Woodward's total synthesis of erythromycin A in 1981 [96]. A number of biologically active compounds have been synthesized from aldehyde or ketone precursors using organocatalysis as an enantioselective step in *direct* aldol reactions (Figure 2.19) [97].

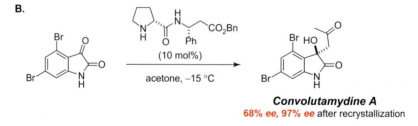

A.

99% ee

Trichostatin A

B.

(10 mol%)

acetone, −15 °C

Convolutamydine A
68% ee, 97% ee after recrystallization

Figure 2.19 Selected examples of organocatalytic aldol cross-coupling reactions (A and B), and an intramolecular aldehyde-ketone variant (C).

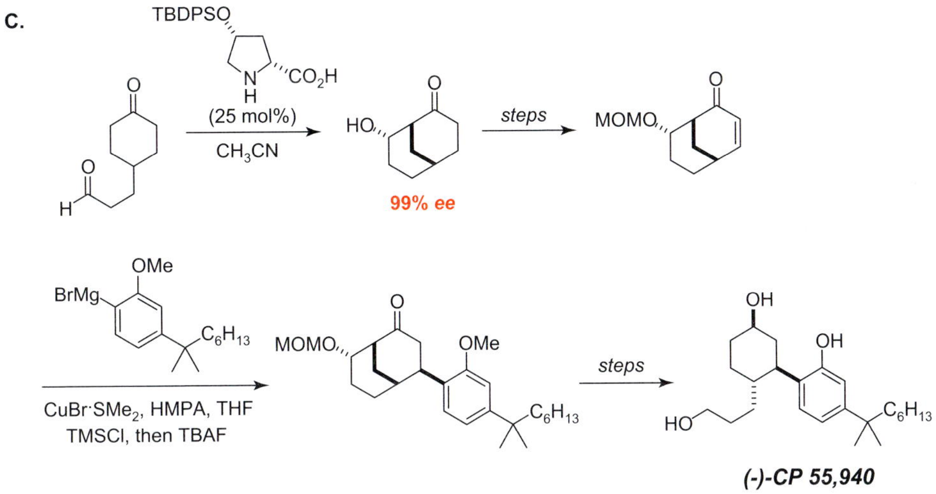

Figure 2.19 (continued).

2.12.3
High impact catalytic oxidation and reduction reactions

The timely discovery of the Sharpless asymmetric epoxidation reaction of allylic and homoallylic alcohols in 1982 heralded a new chapter in the annals of modern organic synthesis. (Figure 2.20 A) [98]. This remarkable reaction with enormous potential for industrial applications demonstrated the importance of Group IV elements in catalysis, particularly in conjunction with cleverly designed ligands. Soon after its discovery, the number of successfully completed total syntheses rose sharply due to the incorporation of a Sharpless asymmetric epoxidation step in the synthetic sequences. Ingenious methods of epoxidations of olefins that do not require an anchoring allylic alcohol have also been developed, notably by Jacobsen [99] and Shi [100] independently.

The Sharpless asymmetric dihydroxylation reaction provided another superb tool to introduce diol units with excellent enantioselectivities from achiral olefins (Figure 2.20 B) [101]. Like the tartrate esters used in asymmetric epoxidation reactions, the dihydroxylation catalysts were optimized based on C_2-symmetric ligands derived from quinine.

Catalytic asymmetric hydrogenation of ketones and double bonds was made possible thanks to the development of C_2-symmetric Ru-catalysts with appropriate ligands by Noyori and coworkers [102] (Figure 2.20 C). Enormously important applications of these catalytic reductions were extensively exploited in natural product synthesis, as well as in industrial processes, including the original Knowles asymmetric synthesis of L-DOPA, used for the treatment of Parkinson's disease.

Surely, it has not eluded the reader that the reactions shown in Figure 2.20 A−D are all well-earned Nobel-worthy contributions relating to catalysis (see also Chapter 18, section 18.4).

The conceptually simple design of a catalytic asymmetric hydride reagent is embodied in the prolinol-derived oxazaborolidine reagent developed by Corey, Bakshi, and Shibata (CBS) [103] (Figure 2.20 E). The CBS reagent is used extensively in the reduction of prochiral ketones.

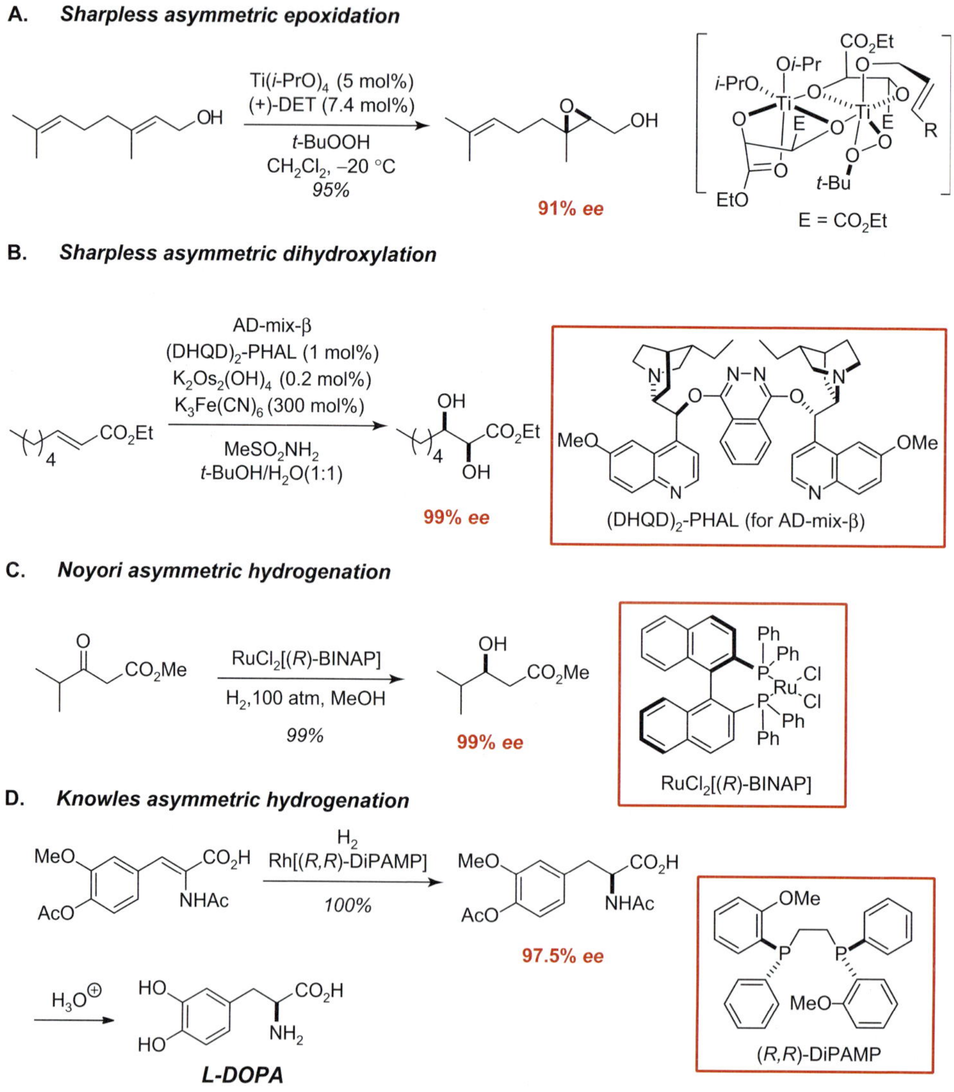

A. *Sharpless asymmetric epoxidation*

B. *Sharpless asymmetric dihydroxylation*

C. *Noyori asymmetric hydrogenation*

D. *Knowles asymmetric hydrogenation*

Figure 2.20 Examples of high-impact asymmetric catalytic oxidation and reduction reactions.

E. *Corey-Bakshi-Shibata asymmetric hydride reduction*

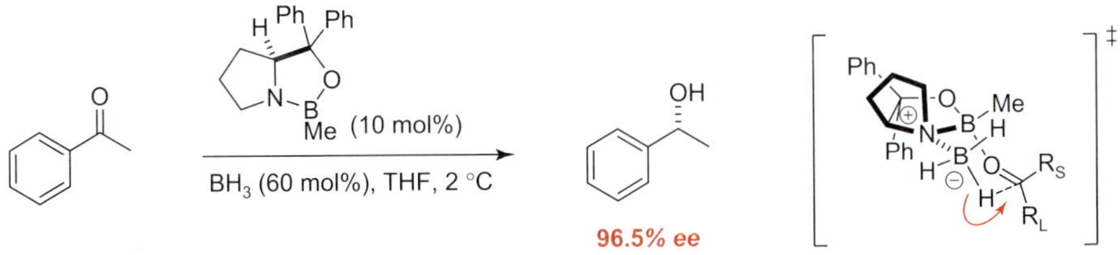

Figure 2.20 *(continued)*.

2.12.4
High impact catalytic olefin-producing reactions

Transition metal-catalyzed organic reactions are among the most useful for C–C bond formation. The ring-closing metathesis (RCM) reaction in particular, developed independently by Grubbs [104] and Schrock [105], is applicable to a remarkably large variety of olefin substrates. The efficiency of the carbenoid Ru-catalyst has been improved with different generations of catalysts as shown in Figure 2.21. The RCM reaction occupies a prominent position in natural product synthesis as illustrated by the synthesis of an intermediate to amphidinolide E in Figure 2.21 A [106]. The 2005 Nobel prize in chemistry was awarded to R. H. Grubbs, R. R. Schrock, and Y. Chauvin for their contributions to olefin metathesis reactions.

Cross-coupling of functionalized olefins with unfunctionalized counterparts is the basis of the Suzuki-Miyaura reaction [107] utilizing appropriate transition metal catalysts [108] as exemplified in the synthesis of epothilone A [109] (Figure 2.21 B). Equally useful palladium and other transition metal-catalyzed cross-couplings are possible in the Heck [110], Stille [111], Sonogashira [112], Negishi [113], Buchwald [114], and Hartwig [115] reactions among others. These methodologies have found extensive applications in the derivatization of aromatic and heteroaromatic compounds especially in the pharmaceutical industry. The 2010 Nobel Prize in chemistry was awarded to R. F. Heck, E.-i. Negishi, and A. Suzuki for their contributions to palladium-catalyzed cross-coupling reactions and its impact on organic synthesis.

The Heck, and Stille couplings of appropriately functionalized olefin partners have been admirably suited for fragment assembly in many complex natural product syntheses as illustrated by xestoquinone [116] and bafilomycin A$_1$ [117] (Figure 2.21 C and D).

Finally, the Pd-catalyzed Tsuji-Trost reaction [118] continues to enjoy extensive applications in C–C bond-forming protocols through the intermediacy of allylic esters and carbonates as reported in the synthesis of hamigeran B [119] (Figure 2.21 E). An excellent review covering Pd-catalyzed cross-coupling reactions in total synthesis has been reported by Nicolaou and coworkers [120a]. The application of transition metal-catalyzed functionalizations of C–H bonds offers exciting prospects for synthesis of complex organic molecules, including natural products [121] (see also Chapter 4, section 4.6, and Chapter 18, section 18.4).

A.   Olefin Metathesis

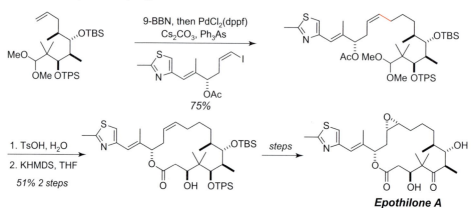

Figure 2.21 Application of current high-impact methods for olefin coupling based on transition metal catalysis to natural product synthesis.

D. *Stille coupling*

R = TBS

1. PdCl$_2$(dppf),
   AsPh$_3$, *i*-Pr$_2$NEt,
   THF/DMF (1:1)

2. *steps*

60% (for step 1)

Bafilomycin A$_1$

E. *Tsuji-Trost reaction*

LDA, Me$_3$SnCl, *t*-BuOH

[η^3-C$_3$H$_5$PdCl]$_2$, DME

83%, 95%ee

1. Me$_2$CuLi, Et$_2$O
2. LDA, PhNTf$_2$, THF

77% 2 steps

1. OsO$_4$, NMO, THF
   H$_2$O, then NaIO$_4$

2. ArLi, DME, −55 °C
3. Dess-Martin oxid.

75% 3 steps

steps

Hamigeran B

ArLi =

Figure 2.21 (*continued*).

References

1. *The Future of Genetically Modified Crops: Lessons from the Green Revolution*, Wu, F.; Butz, W. P., 2004, Rand Corporation Monographs.

2. *Pesticide Chemistry*, Ohkawa, H.; Miyagawa, H.; Lee, P. W., 2007, Wiley-VCH, Weinheim.

3. (a) *Modern Crop Protection Compounds*, Krämer, W.; Schirmer, U., Eds.; Wiley-VCH, Weinheim, 2007; (b) *Ullmann's Agrochemicals*, 2007, edited by Wiley-VCH, Weinheim; (c) *Chemistry of Crop Protection*, Voss, G.; Ramos, G., Eds.; 2002, Wiley-VCH, Weinheim.

4. *Medicinal Plants of the World*, VanWyk, B.-E.; Wink, M., 2004, Timber Press, Portland, OR.

5. (a) Fullbeck, M.; Michalsky, E.; Dunkel, M.; Preissner, R. *Nat. Prod. Rep.* 2006, **23**, 347; (b) *Review of Natural Products: the Most Complete Source of Natural Product Information*, DerMarderosian, A.; Beutler, J. A.; 2001, Lippincott, Williams and Wilkins, Philadelphia, PA; (c) Nicolaou, K. C.; Chen, J. S.; Edmonds, D. J.; Estrada, A. A. *Angew. Chem. Int. Ed.* 2009, **48**, 660.

6. (a) Bentley, R., *Perspectives in Biology and Medicine* 2005, **48**, 444; (b) *Penicillin: Triumph and Tragedy*, Bud, R., 2007, Oxford University Press, Oxford, UK; (c) *The Enchanted Ring – The Untold Story of Penicillin*, Sheehan, J. C., 1982, MIT Press, Cambridge, MA.

7. (a) Singh, S. B.; Jayasuriya, H.; Ondeyka, J. G.; Herath, K. B.; Zhang, C.; Zink, D. L.; Tsou, N. N.; Ball, R. G.; Basilio, A.; Genilloud, O.; Diez, M. T.; Vicente, F.; Pelaez, F.; Young, K.; Wang, J. *J. Am. Chem. Soc.* 2006, **128**, 15547; (b) Singh, S. B.; Jayasuriya, H.; Ondeyka, J. G.; Herath, K. B.; Zhang, C.; Zink, D. L.; Tsou, N. N.; Ball, R. G.; Basilio, A.; Genilloud, O.; Diez, M. T.; Vicente, F.; Pelaez, F.; Young, K.; Wang, J. *J. Am. Chem. Soc.* 2006, **128**, 11916.

8. Jayasuriya, H.; Herath, K. B.; Zhang, C.; Zink, D. L.; Basilio, A.; Genilloud, O.; Diez, M. T.; Vicente, F.; Gonzalez, I.; Salazar, O.; Pelaez, F.; Cummings, R.: Ha, S.; Wang, J.; Singh, S. B. *Angew. Chem. Int. Ed.* 2007, **46**, 4684.

9. Wang, J.; Soisson, S. M.; Young, K.; Shoop, W.; Kodali, S.; Galgoci, A.; Painter, R.; Parthasarathy, G.; Tang, Y. S.; Cummings, R.; Ha, S.; Dorso, K.; Motyl, M.; Jayasuriya, H.; Ondeyka, J.; Herath, K.; Zhang, C.; Hernandez, L.; Allocco, J.; Basilio, A.; Tormo, J. R.; Genilloud, O.; Vicente, F.; Pelaez, F.; Colwell, L.; Lee, S. H.; Michael, B.; Felcetto, T.; Gill, C.; Silver, L. L.; Hermes, J. D.; Bartizal, K.; Barrett, J.; Schmatz, D.; Becker, J. W.; Cully, D.; Singh, S. B. *Nature* 2006, **441**, 358.

10. (a) Fishbach, M. A.; Walsh, C. T. *Science* 2009, **325**, 1089; (b) Pearson, H. *Nature* 2002, **418**, 469; (c) Walsh, C. T. *Nat. Rev. Microbiol.* 2003, **1**, 65; (d) Singh, S. B.; Barrett, J. F. *Biochem. Pharmacol.* 2006, **71**, 1006; see also: *Emerging Trends in Antibacterial Discovery*, Miller, A. A.; Miller, D. F., 2011, ISBC Inc.. Portland, OR; Högberg, L. D.; Heddini, A.; Cars, O. *Trends in Pharm. Sci.* 2010, **31**, 509.

11. (a) Häbich, D.; von Nussbaum, J. F. *Chem. Med. Chem.* 2006, **1**, 951; (b) von Nussbaum, J.F.; Brands, M.; Hinzen, B.; Weigand, S.; Häbich, D. *Angew. Chem. Int. Ed.* 2006, **45**, 5072; see also: (c) Leeb, M. *Nature* 2004, **431**, 892; (d) Nathan, C. *Nature* 2004, **431**, 899; (e) Payne, D. J.; Gwynn, M. N. Holmes, D. J.; Pompliano, D. L. *Nat. Rev. Drug Discov.* 2007, **6**, 29; for recent reviews on drug discovery see: Bennani, Y. L. *Drug Discovery Today* June 2011; Swinney, D. C.; Anthony, J. *Nat. Rev. Drug Discov.* 2011, **10**, 507.

12. Nicolaou, K. C.; Li, A.; Edmonds, D. J. *Angew. Chem. Int. Ed.* 2006, **45**, 7086.

13. For a review, see Tiefenbacher, K.; Mulzer, J. *Angew. Chem. Int. Ed.* 2008, **47**, 2548.

14. (a) Nicolaou, K. C.; Toh, Q.-Y.; Chen, D. Y.-K. *J. Am. Chem. Soc.* 2008, **130**, 11292; (b) Nicolaou, K. C.; Tria, G. S.; Edmonds, D. J. *Angew. Chem. Int. Ed.* 2008, **47**, 1780; see also: (c) Tiefenbacher, K.; Mulzer, J. *J. Org.*

Chem. 2009, **74**, 2937; (d) Ghosh, A. K.; Xi, K. *Angew. Chem. Int. Ed.* 2009, **48**, 5372; (e) Matsuo, J.-I.; Takeuchi, K.; Ishibashi, H. *Org. Lett.* 2008, **10**, 4049.

15. Hayashida, J.; Rawal, V. H. *Angew. Chem. Int. Ed.* 2008, **47**, 4373.

16. Nicolaou, K. C.; Lister, T.; Denton, R. M.; Montero, A.; Edmonds, D. J. *Angew. Chem. Int. Ed.* 2007, **46**, 4712.

17. (a) Nicolaou, K. C.; Dai, W.-M.; Guy, R. K. *Angew. Chem. Int. Ed.* 1994, **33**, 15; see also: (b) *The Story of Taxol: Nature and Politics in the Pursuit of an Anti-Cancer Drug*, Goodman, J.; Walsh, V., 2001, Cambridge, University Press, UK.

18. For reviews, see: (a) Feyen, F.; Cachoux, F.; Gertsch, J.; Wartmann, M.; Altmann, K.-H. *Acc. Chem. Res.* 2008, **41**, 21; (b) Altmann, K.-H. *Curr. Pharm. Design* 2005, **11**, 1595; (c) Wartmann, M.; Altmann, K.-H. *in Curr. Med. Chem. – Anticancer Agents* 2002, **2**, 123; (d) Bollag, D. M. *Exp. Opin. Invest Drugs* 1997, **6**, 867.

19. Mickel, S. J.; Sedelmeier, G. H.; Niederer, D.; Daeffler, R.; Osmani, A.; Schreiner, K.; Seeger-Weibel, M.; Bérod, B.; Schaer, K.; Gamboni, R.; Chen, S.; Chen, W.; Jagoe, C. T.; Kinder, F. R. Jr.; Loo, M.; Prasad, K.; Repic, O.; Shieh, W.-C.; Wang, R.-M.; Waykole, L.; Xu, D. D.; Xue, S. *Org. Proc. Res. Dev.* 2004, **8**, 92, and following papers.

20. For reviews, see: (a) Smith, A. B. III; Freeze, B. S. *Tetrahedron* 2008, **64**, 261; (b) Kalesse, M. *Chem. Bio. Chem.* 2000, **1**, 171, and references cited therein.

21. Smith, A. B. III; Beauchamp, T. J.; LaMarche, M. J.; Kaufman, M. D.; Qiu, Y.; Arimoto, H.; Jones, D. R.; Kobayashi, K. *J. Am. Chem. Soc.* 2000, **122**, 8654.

22. Paterson, I.; Florence, G. J.; Gerlach, K.; Scott, J. P. *Angew. Chem. Int. Ed.* 2000, **39**, 377.

23. For reviews, see: (a) Magano, J.; *Chem. Rev.* 2009, **109**, 4439; (b) Shibasaki, M.; Kanai, M. *Eur. J. Org. Chem.* 2008, 1839; (c) Farina, V.; Brown, J. D. *Angew. Chem. Int. Ed.* 2006, **45**, 7330.

24. (a) Karpf, M.; Trussardi, R. *J. Org. Chem.* 2001, **66**, 2044; (b) Rohloff, J. C.; Kent, K. M.; Postich, M. J.; Becker, M. W.; Chapman, H. H.; Kelly, D. E.; Lew, W.; Louie, M. S.; McGee, L. R.; Prisbe, E. J.; Schultze, L. M.; Yu, R. H.; Zhang. L. *J. Org. Chem.* 1998, **63**, 4545.

25. Bertau, M. *Curr. Org. Chem.* 2002, **6**, 987.

26. (a) Yeung, Y.-Y.; Hong, S. Corey, E. J. *J. Am. Chem. Soc.* 2006, **128**, 6310; (b) Fukuta, Y.; Mita, T.; Fukuda, N.; Kanai, M.; Shibasaki, M. *J. Am. Chem. Soc.* 2006, **128**, 6312; (c) Mita, T.; Fukuda, N.; Roca, F. X.; Kanai, M.; Shibasaki, M. *Org. Lett.* 2007, **9**, 259; (d) Satoh, N.; Akiba, T.; Yokoshima, S.; Fukuyama, T. *Angew. Chem. Int. Ed.* 2007, **46**, 5734; (e) Trost, B. M.; Zhang, T. *Angew. Chem. Int. Ed.* 2008, **47**, 3759; (f) Shie, J. J.; Fang, J.-M.; Wong, C.-H. *Angew. Chem. Int. Ed.* 2008, **47**, 5788; (g) Ishikawa, H.; Suzuki, T.; Hayashi, Y. *Angew. Chem. Int. Ed.* 2009, **48**, 1304; (h) Yamatsugu, K.; Yin, L.; Kamijo, S.; Kimura, Y.; Kanai, M.; Shibasaki, M. *Angew. Chem. Int. Ed.* 2009, **48**, 1070; (i) Sullivan, B.; Carrera, I.; Drouin, M.; Hudlicky, T. *Angew. Chem. Int. Ed.* 2009, **48**, 4229; (j) Werner, L.; Machara, A. Hudlicky, T. *Adv. Synth. Catal.* 2010, **352**, 195; see also: reference [23].

27. (a) *From Bench to Market – The Evolution of Chemical Synthesis*, Cabri, W.; Di Fabio, R., 2000, Oxford University Press, Oxford, UK; (b) *Process Chemistry in the Pharmaceutical Industry*, Gadamasetti, K. G., Ed.; 1999, Vol. 1, Dekker, NY; (c) *Process Chemistry in the Pharmaceutical Industry*, vol. 2, Gadamasetti, K.; Braish, T., Eds.; 2007, CRC Press, Boca Raton, FL; (d) for a thematic issue, see: *Chem. Rev.* 2006, **106**, 2581, Lipton, M. F.; Barrett, A. G. M. Guest editors; see also: (e) Liu, K. K.-C.; Sakya, S. M.; O'Donell, C. J.; Flick, A. C.; Li, J. *Bioorg. Med. Chem.* 2011, **19**, 1136.

28. (a) *The Management of Chemical Process Development in the Pharmaceutical Industry*, Walker, D., 2008, Wiley, NY;

(b) *The Chemical Process Development in the Fine Chemical and Pharmaceutical Industry*, Rao, C. S., 2006, Asian Books Pvt. Ltd.

29. (a) *Asymmetric Catalysis on Industrial Scale: Challenges, Approaches and Solutions*, Blaser, H. U.; Schmidt, E., 2004, Wiley-VCH, Weinheim; (b) *Organometallics in Process Chemistry*, Larsen, R. D., Ed.; 2004, Springer, Heidelberg; (c) *Comprehensive Asymmetric Catalysis*, Jacobsen, E. N.; Pfaltz, A.; Yamamoto, H., Eds.; 1999, vol. I-III, Springer, Verlag, Berlin; see also: (d) Breuer, M.; Ditrich, K.; Habicher, T.; Hauer, B.; Kebeler, M.; Stürmer, R.; Zelinski, T. *Angew. Chem. Int. Ed.* 2004, **43**, 788.

30. For natural products and drug discovery, see: (a) Reymond, J.-L.; van Deursen, R.; Blum, L. C.; Ruddigkeit, L. *Med. Chem. Commun.* 2010, **1**, 30; (b) Gordaliza, M. *Clinical and Transl. Oncology* 2007, **9**, 767; (c) Baker, D. D.; Chu, M.; Oza, U.; Rajgarhia, V. *Nat. Prod. Rep.* 2007, **24**, 1225; (d) Gullo, V. R.; McAlpine, J.; Law, K. S.; Baker, D.; Petersen, F. *J. Industr. Microbiol. Biotech.* 2006, **33**, 523; (e) Butler, M. S.; Buss, A. D. *Biochem. Pharmacol.* 2006, **71**, 919; (f) *Natural Products: Drug Discovery and Therapeutic Medicine*, Zhang, L.; Demain, A. L. 2005, Humana Press, Totowa, NJ; (g) *Drug Discovery: A History*, Sneader, W., 2005, Wiley, NY; (h) Butler, M. S. *Nat. Prod. Rep.* 2005, **22**, 162; (i) Clardy, J.; Walsh, C. T. *Nature* 2004, **432**, 829; (j) *Drug Discovery from Nature*, Grabley, S.; Thiericke, R., Eds.; 1999, Springer, Heidelberg; (k) *Anticancer Agents from Natural Products*, Cragg, G. M.; Kingston, D. G. I.; Newman, D. J., Eds.; 2005, CRC Press, Boca Raton, FL.

31. *Macrolide Antibiotics*, Schönfeld, W.; Kirst, H. A., Eds.; 2002, Birkhäuser, Springer.

32. (a) *The Organic Chemistry of Drug Design and Drug Action*, 2nd; Silverman, R. B., Ed; 2004, Academic Press; (b) *Antibiotics: Actions, Origins, Resistance*, Walsh, C. T., 2003, ASM Press, Washington, DC; (c) *The Antimicrobial Drugs*, Scholar, E. M.; Pratt, W. B., 2000, Oxford University Press, UK; (d) *The Ribosome: Structure, Function, Antibiotics, and Cellular Interactions*, Garrett, R. A.; Douthwaite, S. R.; Liljas, A.; Matheson, A. T.; Moore, P. B.; Noller, H. F., Ed.; 2000, ASM Press, Washington, DC.

33. Pioletti, M.; Schlunzen, F.; Harms, J.; Zarivach, R.; Gluhmann, M.; Avila, H.; Bashan, A.; Bartels, H.; Auerbach, T.; Jacobi, C.; Hartsch, T.; Yonath, A. Franceschi, F. *EMBO* 2001, **20**, 1829.

34. (a) *Synthesis of Beta-lactam Antibiotics*, Bruggink, A. Springer, 2001; (b) *The Organic Chemistry of Beta-lactams*, Georg, G. I., Ed.; 1993, VCH, Weinheim.

35. (a) *Concepts in Immunology and Immunotherpeutics*, 4th Edition, Smith, B. T., 2008, Am. Soc. Health-System Pharmacists; (b) *Modern Immunosupressives*, Schuurman, H.-J.; Feutren, G.; Bach, J.-F., Eds.; 2001, Birkhäuser, Springer, Heidelberg.

36. *Aminoglycoside Antibiotics - From Chemical Biology to Drug Discovery*, Arya, D. P., 2007, Wiley-Interscience, NY.

37. (a) Lynch, S. R.; Gonzalez, R. L.; Puglisi, J. D. *Structure* 2003, **11**, 43; (b) Vicens, Q.; Westhof, E. *Chem. Biol.* 2002, **9**, 747; (c) Vicens, Q.; Westhof, E. *Structure* 2001, **9**, 647; (d) Carter, A. P.; Clemons, W. M.; Brodersen, D. E.; Morgan-Warren, R. J.; Wimberly, B. T.; Ramakrishnan, V. *Nature* 2000, **407**, 340.

38. (a) *Bacterial Resistance to Antimicrobials*, Wax, W. G.; Wax, R. G.; Lewis, K.; Salyers, A.; Taber, H., 2007, CRC Press, Boca Raton, FL; (b) Magnet, S.; Blanchard, J. S. *Chem. Rev.* 2005, **105**, 477.

39. *Cancer Drug Design and Discovery*, Neidle, S., 2007, Academic Press, NY; (b) *Medicinal Chemistry of Anticancer Drugs*, Avendaño, C.; Menéndez, J. C., 2008, Elsevier, Amsterdam.

40. (a) Ryan, M. J.; Sigmund, C. D. *Circ. Res.* 2004, **94**, 1; (b) Acharya, K. R.; Sturrock, E. D.; Riordan, J. F.; Ehlers, M. R. *Nat. Rev. Drug Discov.* 2003, **2**, 891.

41. De Clercq, E. *Nat. Drug Discovery* 2007, **6**, 1001; see also: *HIV Chemotherapy – A Critical Review*, Butera, S. T., Ed., 2005, Caister Academic Press, Norfolk, UK.

42. For example, see: (a) Cooper, J. B. *Curr. Drug Targets* 2002, **3**, 155; (b) Leung, D.; Abbenante, G.; Fairlie, D. P. *J. Med. Chem.* 2000, **43**, 305.

43. (a) Abbenante, G.; Fairlie, D. P. *Med. Chem.* 2005, **1**, 71; (b) Robertson, J. G. *Biochemistry* 2005, **44**, 5561.

44. For authoritative texts, see: (a) *The Practice of Medicinal Chemistry*, Wermuth, C. G., Ed.; 2008, 3rd Edition, Academic Press, MA; (b) *Drug-like Properties: Concepts, Structure-Design and Methods*, Kerns, E. H.; Di, L., 2008, Academic Press, NY; (c) *Foye's Principles of Medicinal Chemistry*, Lemmke, T. L.; Williams, D. A., Eds.; 2008, 6th Edition, Lippincott, Williams and Wilkins, Baltimore, MD; see also: (d) *Strategies for Organic Drug Synthesis and Design*, Lednicer, D, 2008, Wiley-VCH, Weinheim.

45. (a) Congreve, M.; Chessari, C.; Tisi, D.; Woodhead, A. J. *J. Med. Chem.* 2008, **51**, 3661; see also: (b) Schulz, M. N. Hubbard, R. E. *Curr. Opin. Phamacol.* 2009, **9**, 615; (c) *Fragment-Based Drug Discovery*, Zartler, E. R., Ed.; 2008, J. Wiley & Sons, Chichester, UK; (d) *The Art of Drug Synthesis*, Johnson, D. S.; Li, J. J., Eds; 2007, Wiley-Interscience, NY; (e) *Analogue-based Drug Discovery*, Fischer, J.; Ganellin, C. R., 2006, Wiley-VCH, Weinheim.

46. (a) Lamarre, D.; Anderson, P. C.; Bailey, M. D.; Beaulieu, P.; Bolger, G.; Bonneau, P.; Bos, M.; Cameron, D. R.; Cartier, M.; Cordingley, M. G.; Faucher, A.-M.; Goudreau, N.; Kawai, S. H.; Kukolj, G.; Lagace, L.; Laplante, S. R.; Narjes, H.; Poupart, M.-A.; Rancourt, J.; Sentjens, R. E.; St. George, R.; Simoneau, B.; Steinman, G.; Thibeault, D.; Tsantrizos, Y. S.; Weldon, S. M.; Young, C.-L.; Llinàs-Brunet, M. *Nature* 2003, **426**, 186; (b) Llinàs-Brunet, M.; Bailey, M. D.; Bolger, G.; Brochu, C.; Faucher, A.-M.; Ferland, J. M.; Garneau, M.; Ghiro, E.; Gorys, V.; Grand-Maître, C.; Halmos, T.; Lapeyre-Paquette, N.; Liard, F.; Poirer, M.; Rhéaume, M.; Tsantrizos, Y. S.; Lamarre, D. *J. Med. Chem.* 2004, **47**, 1605; for a large scale synthesis of BILN 2061 see: (c) Nicola, T.; Brenner, M.; Donsbach, K.; Kreye, P. *Org. Process Res. Dev.* 2005, **9**, 513; for optimization, see: (d) Shu, C.; Zeng, X.; Hao, M.-H.; Wei, X.; Yee, N. K.; Busacca, C. A.; Han, Z.; Farina, V.; Senanayake, C. H. *Org. Lett.* 2008, **10**, 1303.

47. (a) Gustafsson, D.; Bylund, R.; Antonsson, T.; Nilsson, I.; Nyström, J.-E.; Eriksson, U.; Bredberg, U.; Teger-Nilsson, A.-C. *Nat. Rev. Drug. Discovery* 2004, **3**, 649; (b) Petersen, P. *Curr. Pharm. Des.* 2005, **11**, 527; see also: (c) Food and Drug Administration (FDA) Cardiovascular and Renal Drugs Advisory Committee, September 10, 2004.

48. (a) Nicolaou, K. C.; Snyder, S. A. *Angew. Chem. Int. Ed.* 2005, **44**, 1012; for hexacyclinol see: (b) (i) La Clair, J. J. *Angew. Chem. Int. Ed.* 2006, **45**, 2769; (ii) Porco, J. A.; Su, X.; Lei, X.; Bardhan, S.; Rychnovsky, S. D. *Angew. Chem. Int. Ed.* 2006, **45**, 5790; (iii) Rychnovsky, S. D. *Org. Lett.* 2006, **8**, 2895; for polyoxin A, see: (c) Hanessian, S.; Fu, J.-M.; Tu, Y.; Isono, K. *Tetrahedron Lett.* 1993, **34**, 4153; for oscillarin, see: (d) Hanessian, S.; Tremblay, M.; Petersen, J. F. W. *J. Am. Chem. Soc.* 2004, **126**, 6064; for aeruginosin 205B see: (e) Hanessian, S.; Wang, X.; Ersmark, K.; Del Valle, J. R.; Kelgraf, E. *Org. Lett.* 2009, **11**, 4232; for alcyonin, see: (f) Corminboeuf, O.; Overman, L. E.; Pennigton, L. D. *Org. Lett.* 2003, **5**, 1543.

49. (a) Nicolaou, K. C.; Frederick, M. O. *Angew. Chem. Int. Ed.* 2007, **46**, 5278; (b) Nicolaou, K. C.; Frederick, M. O.; Aversa, R. J. *Angew. Chem. Int. Ed.* 2008, **47**, 7182; for recent reviews, see: (c) Inoue, M. *Chem. Rev.* 2005, **105**, 4379; (d) Sasaki, M.; Fuwa, H. *Synlett* 2004, 1851; for recent work towards the synthesis of maitotoxin, see: (e) Nicolaou, K. C.; Aversa, R. J.;

Jin, F.; Rivas, J. *J. Am. Chem. Soc.* 2010, **132**, 6855; (f) Nicolaou, K. C.; Gelin, C. F.; Seo, J.-H.; Huang, Z.; Umezawa, T. *J. Am. Chem. Soc.* 2010, **132**, 9900.

50. (a) Cane, D. E.; Liang, T.-C.; Hasler, H. *J. Am. Chem. Soc.* 1981, **103**, 5962; (b) Cane, D. E.; Liang, T.-C.; Hasler, H. *J. Am. Chem. Soc.* 1982, **104**, 7274; (c) for a thematic issue, see: *Chem. Rev.* 1997, **97**, 2463, Cane, D. E. Guest editor.

51. (a) Cane, D. E.; Hasler, H.; Liang, T.-C. *J. Am. Chem. Soc.* 1981, **103**, 5960; (b) Khosla, C.; Gokhale, R. S.; Jacobsen, J. R.; Cane, D. E. *Ann. Rev. Biochem.* 1999, **68**, 219.

52. Chartrain, M.; Salmon, P. M.; Robinson, D. K.; Buckland, B. C. *Curr. Opin. Biotechnology*, 2000, **11**, 209.

53. (a) McCoy, E.; O'Connor, S. E. *J. Am. Chem. Soc.* 2006, **128**, 14276; (b) *Medicinal Natural Products: A Biosynthetic Approach*, Dewick, P. M., 2002, Wiley & Sons, NY.

54. (a) *Supramolecular Chemistry: Scope and Perspectives*, Lehn, J.-M.; 1995, Wiley-VCH, Weinheim; (b) *Protein-ligand Interactions from Molecular Recognition to Drug Design*, Böhm, H.-J.; Schneider, G., 2003, Wiley-VCH, Weinheim.

55. Sliwa, W.; Peske, J. *Mini-Reviews in Organic Chemistry* 2007, **4**, 125.

56. (a) *Buckminsterfullerenes*, Billups, W. E. Ciufolini, M. A., Eds.; 1993, Wiley-VCH, NY; see also: (b) *Organic Synthesis: The Science behind the Art*, Smit, W. A; Bochkov, A. F.; Caple, R., 1998, Chapter 4, p. 301, The Royal Society, Cambridge, UK.

57. (a) *Dendrimers and Dendrons: Concepts, Syntheses, Applications*, Newkome, G. R.; Moorefield, C. N.; Vögtle, F., 2001, Wiley-VCH, Weiheim; (b) *Dendrimers and other Dendritic Polymers*, Fréchet, J. M. J.; Tomalia, D. A., Eds.; 2001, Wiley, NY; (c) Tomalia, D. A. *Aldrichim. Acta* 2004, **37**, 39.

58. (a) *The Chemistry of Nanomaterials*, Rao, C. N. R.; Müller, A.; Cheetham, A. K., Eds.; 2004, vol. 1,2, Wiley-VCH, Weinheim; (b) *Nanochemistry: A General Approach to Nanomaterials*, Ozin, G. A.; Arsenault, A. C., 2005, Royal Society of Chemistry, Cambridge, UK.

59. *Nano Biotechnology: Concepts, Applications and Perspectives*; Niemeyer, C. M.; Mirkin, C. A., Eds.; 2004, Wiley-VCH, Weinheim.

60. Rabinow, B. E. *Nat. Rev. Drug Discov.* 2004, **3**, 785.

61. Davis, M. E.; Chen, Z.; Shin, D. M. *Nat. Rev. Drug Discov.* 2008, **7**, 771.

62. Jun, Y.-W.; Lee, J.-H.; Cheon, J. *Angew. Chem. Int. Ed.* 2008, **47**, 5122.

63. (a) Gupta, A. K.; Gupta, M. *Biomaterials* 2005, **26**, 3995; (b) Ito, A.; Shinkai, M.; Honda, H.; Kobayashi, T. *J. Biosci. Bioengin.* 2005, **100**, 1.

64. (a) Hanessian, S.; Grzyb, J. A.; Cengelli, F.; Juillerat-Jeanneret, L. *Bioorg. Med. Chem.* 2008, **16**, 2921; (b) Cengelli, F.; Grzyb, J. A.; Montoro, A.; Hofmann, H.; Hanessian, S.; Juillerat-Jeanneret, L. *Chem. Med. Chem.* 2009, **4**, 988.

65. Kohler, N.; Sun, C.; Wang, J.; Zhang, M. *Langmuir* 2005, **21**, 8858.

66. (a) *Molecular Motors*, Schliwa, M., Ed.; 2003, Wiley-VCH, Weinheim; (b) *Molecular Devices and Machines – A Journey into the Nanoworld*, Balzani, V.; Credi, A.; Venturi M., 2003, Wiley-VCH, Weinheim; (c) Balzani, V.; Credi, A.; Raymo, F. M.; Stoddart, J. F. *Angew. Chem. Int. Ed.* 2000, **39**, 3348; see also: (d) Kinbara, K.; Aida, T. *Chem. Rev.* 2005, **105**, 1377.

67. (a) Vives, G.; Kang, J.; Kelly, K. F.; Tour, J. M. *Org. Lett.* 2009, **11**, 5602; (b) Tour, J. M. *J. Org. Chem.* 2007, **72**, 7477; (c) James, D. K.; Tour, J. M. *Aldrichim. Acta* 2006, **39**, 47; (d) Morin, J.-F.; Shirai, Y.; Tour, J. M. *Org. Lett.* 2006, **8**, 1713. (e) Kelly, T. R.; Cai, X.; Damkaci, F.; Panicker, S. B.; Tu, B.; Bushell, S. M.; Cornella, I. Piggott, M. J.; Salives, R.; Cavero, M.; Zhao, Y.; Jasmin, S. *J. Am. Chem. Soc.* 2007, **129**, 376; see also, (f) *Molecular Machines*, Kelly, T. R.. Ed.; 2006, Springer, Heidelberg.

68. Badjic, J. D.; Balzani, V.; Credi, A.; Silvi, S.; Stoddart, J. F. *Science* 2004, **303**, 1845.

69. (a) Klok, M.; Boyle, N.; Pryce, M. T.; Meetsma, A.; Browne, W. R.; Feringa, B. L. *J. Am. Chem. Soc.* 2008, **130**, 10484; (b) Browne, W. R.; Feringa, B. L. *Nature Nanotechnology* 2006, **1**, 25; (c) Feringa, B. L. *J. Org. Chem.* 2007, **72**, 6635; see also: *Molecular Switches*. Feringa, B. Ed.; 2001, Wiley-VCH, Weinheim. For the synthesis of a molecular half-subtractor see: (d) Coskun, A.; Deniz, E.; Akkaya, E. U. *Org. Lett.* 2005, **7**, 5187; for a molecular half-adder see: (e) Andréasson, J.; Straight, S. D.; Kodis, G.; Park, C.-D.; Hambourger, M.; Gervaldo, M.; Albinsson, B.; Moore, T. A.; Moore, A. L.; Gust, D. *J. Am. Chem. Soc.* 2006, **128**, 16259.

70. For example, see: Nicolaou, K. C.; Vourloumis, D.; Winssinger, N.; Baran, P. S. *Angew. Chem. Int. Ed.* 2000, **39**, 44.

71. For example, see: (a) *Encyclopaedia of Reagents for Organic Synthesis*, Crich, D. Ed.; 2010, Wiley-Interscience, NY; (b) *Comprehensive Organic Name Reactions*, Wang, Z., 2009, Wiley, NY; (c) *Name reactions: A Collection of Detailed Mechanisms and Synthetic Applications*, Li. J. J. 2009, 4th Edition, Springer Verlag, Berlin; (d) *Strategic Applications of Named Reactions in Organic Synthesis: Background and Detailed Mechanisms*, Kürti, L. Czakó, B. 2005, Elsevier Science, NY.

72. For example, see: (a) *Comprehensive Organic Reactions in Aqueous Media*, 2nd Edition, Li, C.-J.; Chan, T.-H., 2007, Wiley-VCH, Hoboken, NJ; (b) *Green Chemistry and Catalysis*, Sheldon, R. A.; Arends, I.; Hanefeld, U., 2007, Wiley-VCH, Weinheim; (c) *Organic Reactions in Water: Principles, Strategies, and Applications*, Lindström, U. M., Ed.; 2007, Wiley-Blackwell, Publishing, UK.

73. (a) Diels, O.; Alder, K. *Justus Liebigs Ann. Chem.* 1928, **460**, 98; see also: (b) Berson, J. A. *Tetrahedron* 1992, **48**, 3; (c) *The Diels-Alder Reaction: Selected Practical Methods*, Fringuelli, F.; Taticchi, A., 2002, Wiley-VCH, Weinheim; (d) *Intramolecular Diels-Alder and Alder-Ene Reactions*, Taber, D. F., 1984, Springer-Verlag,

NY; for reviews, see: (e) Corey, E. J. *Angew. Chem. Int. Ed.* 2002, **41**, 1650; (f) Takao, K.-I.; Munakata, R.; Tadano, K.-I. *Chem. Rev.* 2005, **105**, 4779; (g) Marsault, E.; Toró, A.; Nowak, P.; Deslongchamps, P. *Tetrahedron* 2001, **57**, 4243.

74. Yates, P.; Eaton, P. *J. Am. Chem. Soc.* 1960, **82**, 4436.

75. For an early example, see: (a) Hashimoto, S.-I.; Komeshima, N.; Koga, K. *Chem. Commun.* 1979, 437; see also: (b) Sauer, J.; Kredel, J. *Tetrahedron Lett.* 1966, **7**, 6359.

76. For example, see: (a) Evans, D. A.; Johnson, J. S. in *Comprehensive Asymmetric Catalysis*, vol. 3, Jacobsen, E. N.; Pfaltz, A.; Yamamoto, H., Eds.; 1999, p. 1177, Springer-Verlag, Berlin; (b) Corey, E. J.; Guzman-Perez, A. *Angew. Chem. Int. Ed.* 1998, **37**, 389; (c) Kagan, H. B; Riant, O. *Chem. Rev.* 1992, **92**, 1007; (d) Oppolzer, W. in *Comprehensive Organic Synthesis*, vol. 5, Trost, B. M.; Fleming, I.; Paquette, L. A., Eds.; 1991, Chapter 4, pp. 352, Pergamon; (e) Corey, E. J.; Imwinkelried, R.; Pikul, S.; Xiang, Y. B. *J. Am. Chem. Soc.* 1989, **111**, 5493; (f) Oppolzer, W.; *Angew. Chem. Int. Ed.* 1984, **96**, 840; (g) Evans, D. A.; Chapman, K. T.; Bisaha, J. *J. Am. Chem. Soc.* 1984, **106**, 4261; (h) Helmchen, G.; Schmierer, R. *Angew. Chem. Int. Ed.* 1981, **20**, 205; (i) Corey, E. J.; Ensley, H. E. *J. Am. Chem. Soc.* 1975, **97**, 6908; (j) Walborsky, H. M.; Barash, L.; Davis, T. C. *Tetrahedron* 1963, **19**, 2333.

77. For example, see: (a) Corey, E. J. *Angew. Chem. Int. Ed.* 2009, **48**, 2100; (b) Corey, E. J.; Loh, T.-P.; Roper, T. D.; Azimioara, M. D.; Noe, M. C. *J. Am. Chem. Soc.* 1992, **114**, 8290; (c) Corey, E. J.; Barnes-Seeman, D.; Lee, T. W. *Tetrahedron Lett.* 1997, **38**, 4351; (d) Corey, E. J. *Pure Appl. Chem.* 1990, **62**, 1209; for a review, see: (e) Dias, L. C. *J. Braz. Chem. Soc.* 1997, **8**, 289; see also ref. [73e].

78. Ahrendt, K. A.; Borths, C. J.; MacMillan, D. W. C. *J. Am. Chem. Soc.* 2000, **122**, 4243.

79. For selected reviews, see: (a) Juhl, M.; Tanner, D. *Chem. Soc. Rev.* 2009, **38**, 2983; (b) Nicolaou, K. C.; Snyder, S. A.; Montagnon, T.; Vassilikogiannakis, G. *Angew. Chem. Int. Ed.* 2002, **41**, 1668.

80. (a) Boezio, A. A.; Jarvo, E. R.; Lawrence, B. M.; Jacobsen, E. N. *Angew. Chem. Int. Ed.* 2005, **44**, 6046; for reviews, see: (b) Lin, L.; Liu, X.; Feng, X. *Synlett* 2007, 2147; (c) Jørgensen, K. A. *Angew. Chem. Int. Ed.* 2000, **39**, 3558.

81. Nicolaou, K. C.; Vassilikogiannakis, G.; Magerlein, W.; Kranich, R. *Angew. Chem. Int. Ed.* 2001, **40**, 2482; see also, Nicolaou, K. C. Vassilikogiannakis, G.; Mägerlein, W.; Kranich, R. *Chem. Eur. J.* 2001, **7**, 5359.

82. Kim, A. I.; Rychnovsky, S. D. *Angew. Chem. Int. Ed.* 2003, **42**, 1267; see also: Harrowven, D. C.; Pascoe, D. D.; Demurtas, D.; Bourne, H. O. *Angew. Chem. Int. Ed.* 2005, **44**, 1221.

83. Yeung, Y.-Y.; Hong, S.; Corey, E. J. *J. Am. Chem. Soc.* 2006, **128**, 6310; see also: ref. [23].

84. (a) Maritimol: Toró, A.; Nowak, P.; Deslongchamps, P. *J. Am. Chem. Soc.* 2000, **122**, 4526; (b) ophirin B: Crimmins, M. T.; Brown, B. H. *J. Am. Chem. Soc.* 2004, **126**, 10264; (c) digitoxigenin: Stork, G.; West, F.; Lee, H. Y.; Isaacs, R. C. A.; Manabe, S. *J. Am. Chem. Soc.* 1996, **118**, 10660; (d) manzamine A: Martin, S. F.; Humphrey, J. M.; Ali, A.; Hillier, M. C. *J. Am. Chem. Soc.* 1999, **121**, 866; see also: Humphrey, J. M.; Liao, Y.; Ali, A.; Rien, T.; Wong, Y.-L.; Chen, H.-J.; Courtney, A. K.; Martin, S. F. *J. Am. Chem. Soc.* 2002, **124**, 8584; (e) chlorothricolide: Roush, W. R.; Sciotti, R. J. *J. Am. Chem. Soc.* 1994, **116**, 6457; see also: (f) Roush, W. R.; Sciotti, R. J. *J. Am. Chem. Soc.* 1998, **120**, 7411; (g) FR182877: Vosburg, D. A.; Vanderwal, C. D.; Sorensen, E. J. *J. Am. Chem. Soc.* 2002, **124**, 4552; (h) norzoanthamine: Miyashita, M.; Sasaki, M.; Hattori, I.; Sakai, M.; Tanino, K. *Science* 2004, **305**, 495; (i) superstolide A: Tortosa, M.; Yakelis, N. A.; Roush, W. R. *J. Am. Chem. Soc.* 2008, **130**, 2722.

85. de Figueiredo, R. M.; Christmann, M. *Eur. J. Org. Chem.* 2007, 2575; see also: Marquéz-Lopez, E.; Herreva, R. P.; Christmann, M. *Nat. Prod. Rep.* 2010, **27**, 1138.

86. Jacobs, W. C.; Christmann, M. *Synlett* 2008, 247.

87. Wurtz, C.-A. *Bull. Soc. Chim. Fr.* 1872, **17**, 436.

88. For example, see: (a) *Modern Aldol Reactions*, vol. 1, 2. Mahrwald, R., Ed.; Wiley-VCH, Weinheim, 2004; see also: (b) Cowden, C. J.; Paterson, I. *Org. React.* 1997, **51**, 1; For recent reviews, see: (c) Trost, B. M.; Brindle, C. S. *Chem. Soc. Rev.* 2010, **39**, 1600. (d) Brodmann, T.; Lorenz, M.; Schäckel, R.; Simsek, S.; Kalesse, M. *Synlett* 2009, 174; (e) Palomo, C.; Oiarbide, M.; García, J. M. *Chem. Soc. Rev.* 2004, **33**, 65.

89. Mukaiyama, T. *Org. React.* 1982, **28**, 203.

90. (a) Evans, D. A.; Nelson, J. V.; Taber, T. R. in, *Topics in Stereochemistry*, Eliel, E.; Allinger, N., Eds.; 1982, vol. 13, p. 1, Wiley-Interscience, NY; (b) Evans, D. A. in, *Asymmetric Synthesis* Morrison, J. D., Ed.; 1984, vol. 3, p. 1 Academic, NY.

91. Yoshikawa, N.; Yamada, Y. M. A.; Das, J.; Sasai, H.; Shibasaki, M. *J. Am. Chem. Soc.* 1999, **121**, 4168.

92. Trost, B. M.; Ito, H. *J. Am. Chem. Soc.* 2000, **122**, 12003.

93. For selected recent reviews and monographs on organocatalysis, see: (a) Melchiorre, P.; Marigo, M.; Carlone, A.; Bartoli, G. *Angew. Chem. Int. Ed.* 2008, **47**, 6138; (b) Dondoni, A.; Massi, A. *Angew. Chem. Int. Ed.* 2008, **47**, 4638; (c) Barbas, C. F. III *Angew. Chem. Int. Ed.* 2008, **47**, 42; (d) Gaunt, M. J.; Johansson, C. C. C.; McNally, A.; Vo, N. T. *Drug Discov. Today* 2007, **12**, 8; (e) Erkkilä, A.; Majander, I.; Pihko, P. M. *Chem. Rev.* 2007, **107**, 5416; (f) *Enantioselective Organocatalysis: Reactions and Experimental Procedures*, Dalko, P. I. Ed.; 2007, Wiley-VCH, Weinheim; see also, Special Issues: (g) *Chem. Rev.* 2007, **107**, 5413,

List, B., Guest editor; (h) *Tetrahedron* 2006, **62**, 255, Kočovský, P.; Malkov, A. V., Guest editors; (i) Lelais, G.; MacMillan, D. W. C. *Aldrichim. Acta* 2006, **39**, 79; (j) *Asymmetric Organocatalysis: From Biomimetic Concepts to Applications in Asymmetric Synthesis*, Berkessel, A.; Gröger, H., 2005, Wiley-VCH, Weinheim.

94. (a) List, B.; Lerner, R. A.; Barbas, C. F. III *J. Am. Chem. Soc.* 2000, **122**, 2395; (b) Northrup, A. B.; MacMillan, D. W. C. *J. Am. Chem. Soc.* 2002, **124**, 6798; (c) Tang, Z.; Yang, Z.-H.; Chen, X.-H.; Cun, L.-F.; Mi, A.-Q.; Jiang, Y.-Z.; Gong, L.-Z. *J. Am. Chem. Soc.* 2005, **127**, 9285.

95. (a) Hajos, Z. G.; Parrish, D. R. *J. Org. Chem.* 1974, **39**, 1615; (b) Eder, U.; Sauer, G.; Wiechert, R. *Angew. Chem. Int. Ed.* 1971, **10**, 496; see also: (c) Allemann, C.; Gordillo, R.; Clemente, F. R.; Cheong, P. H.-Y.; Houk, K. N. *Acc. Chem. Res.* 2004 **37**, 558.

96. Woodward, R. B.; Logusch, E.; Nambiar, K. P.; Sakan, K.; Ward, D. E.; Au-Yeung, B.-W.; Balaram, P.; Browne, L. J.; Card, P. J.; Chen, C. H.; Chênevert, R. B.; Fliri, A.; Frobel, K.; Gais, H.-J.; Garratt, D. G.; Hayakawa, K.; Heggie, W.; Hesson, D. P.; Hoppe, D.; Hoppe, I.; Hyatt, J. A.; Ikeda, D.; Jacobi, P. A.; Kim, K. S.; Kobuke, Y.; Kojima, K.; Krowicki, K.; Lee, V. J.; Leutert, T.; Malchenko, S.; Martens, J.; Matthews, R. S.; Ong, B. S.; Press, J. B.; Rajan Babu, T. V.; Rousseau, G.; Sauter, H. M.; Suzuki, M.; Tatsuta, K.; Tolbert, L. M.; Truesdale, E. A.; Uchida, I.; Ueda, Y.; Uyehara, T.; Vasella, A. T.; Vladuchick, W. C.; Wade, P. A.; Williams, R. M.; Wong, H. N.-C. *J. Am. Chem. Soc.* 1981, **103**, 3210.

97. (a) Zhang, S.; Duan, W.; Wang, W. *Adv. Synth. Catal.* 2006, **348**, 1228; (b) Luppi, G.; Monari, M.; Correa, R. J.; Violante, F. de A.; Pinto, A. C.; Kaptein, B.; Broxterman, Q. B.; Garden, S. J.; Tomasini, C. *Tetrahedron* 2006, **62**, 12017; (c) Itagaki, N.; Sugahara, T.; Iwabuchi, Y. *Org. Lett.* 2005, **7**, 4181.

98. (a) Katsuki, T. in *Comprehensive Asymmetric Catalysis*, Jacobsen, E. N.; Pfaltz, A.; Yamamoto, H., Eds.; 1999, vol. II, p. 621, Springer-Verlag, Berlin; for selected reviews see: (b) Johnson, R. A.; Sharpless, K. B. in *Catalytic Asymmetric Synthesis*, Ojima, I., Ed.; 1993, p. 103, VCH, Weinheim; (c) Katsuki, T.; Sharpless, K. B. *J. Am. Chem. Soc.* 1980, **102**, 5974; (d) Hill, J. G.; Sharpless, K. B.; Exon, C. M.; Regenye, R. *Org. Synth.* 1984, **63**, 66.

99. Jacobsen, E. N.; Wu, M. H. in *Comprehensive Asymmetric Catalysis*, Jacobsen, E. N.; Pfaltz, A.; Yamamoto, H., Eds.; 1999, vol. II, p. 649, Springer-Verlag, Berlin.

100. Wang, Z.-X.; Tu, Y.; Frohn, M.; Zhang, J.-R., Shi, Y. *J. Am. Chem. Soc.* 1997, **119**, 11224.

101. (a) Jacobsen, E. N.; Markó, I.; Mungall, W. S.; Schröder, G.; Sharpless, K. B. *J. Am. Chem. Soc.* 1988, **110**, 1968; see also: (b) Johnson, R. A.; Sharpless, K. B. in *Catalytic Asymmetric Synthesis*, Ojima, I., Ed., 1993, p. 227, VCH, Weinheim; (c) Kolb, H. C.; Van Nieuwenhze, M. S.; Sharpless, K. B. *Chem. Rev.* 1994, **94**, 2483.

102. (a) *Modern Reduction Methods*, Andersson, P. G.; Munslow, I. J., 2008, Wiley-VCH, Weinheim; (b) Takaya, H.; Ohta, T.; Noyori, R. in *Catalytic Asymmetric Synthesis*, Ojima, I., Ed.; 1993, p. 1, Wiley-VCH, Weinheim.

103. (a) Corey, E. J.; Bakshi, R. K.; Shibata, S. *J. Am. Chem. Soc.* 1987, **109**, 5551 (b) Corey, E. J.; Helal, C. J. *Angew. Chem. Int. Ed.* 1998, **37**, 1986; (c) Xavier, L. C.; Mohan, J. J.; Mathre, D. J.; Thompson, A. S.; Carroll, J. D.; Corley, E. G.; Desmond, R. *Org. Synth.* 1996, **74**, 50; see also: (d) Itsuno, S.; Ito, K.; Hirao, A.; Nakahama, S. *J. Chem. Soc., Chem. Commun.* 1983, 469.

104. For reviews see: (a) Grubbs, R. H.; Chang, S. *Tetrahedron* 1998, **54**, 4413; (b) Grubbs, R. H. *Tetrahedron* 2004, **60**, 7117; see also: (c) *Handbook of Metathesis*, Grubbs, R. H., Ed; 2003, vol. 1-3, Wiley-VCH, Weinheim; see also: *Metathesis in Natural Product Synthesis: Strategies, Substrates and*

Catalysts, Cossy, J.; Arseniyadis, S.; Meyer, C., Eds.; 2010, Wiley-VCH, Weinheim.

105. For reviews see: (a) Schrock, R. R. *Tetrahedron* 1999, **55**, 8141; (b) Schrock, R. R. *J. Mol. Catal. A: Chemical* 2004, **213**, 21.

106. Va, P.; Roush, W. R. *J. Am. Chem. Soc.* 2006, **128**, 15960.

107. For reviews, see: (a) *Boronic Acids – Preparation, Applications in Organic Synthesis and Medicine*, Hall, D. G., Ed.; 2005, Wiley-VCH, Weinheim; (b) Bellina, F.; Carpita, A.; Rossi, R. *Synthesis* 2004, 2419; (c) Kotha, S.; Lahiri, K.; Kashinath, D. *Tetrahedron* 2002, **58**, 9633; (d) Chemler, S. R.; Trauner, D.; Danishefsky, S. J. *Angew. Chem. Int. Ed.* 2001, **40**, 4544; (e) Franzén, R. *Can. J. Chem.* 2000, **78**, 957; (f) Suzuki, A. *J. Organomet. Chem.* 1999, **576**, 147; (g) Miyaura, N.; Suzuki, A. *Chem. Rev.* 1995, **95**, 2457.

108. (a) *Transition Metal Reagents and Catalysts: Innovations in Organic Synthesis*, Tsuji, J., 2002, Wiley-Interscience, NY; (b) *Metal-catalyzed Cross-Coupling Reactions*, de Meijere, A.; Diederich, F., Eds.; 2004, vol. 1 and 2, Wiley-VCH, Weinheim.

109. Balog, A.; Meng, D.; Kamenecka, T.; Bertinato, P.; Su, D.-S.; Sorensen, E. J.; Danishefsky, S. J. *Angew. Chem. Int. Ed.* 1996, **35**, 2801.

110. (a) Heck, R. F.; Nolley, J. D. Jr. *J. Org. Chem.* 1972, **37**, 2320; (b) Dieck, H. A.; Heck, R. F. *J. Am. Chem. Soc.* 1974, **96**, 1133; (c) Heck, R. F. *Org. React.* 1982, **27**, 345; (d) de Meijere, A.; Meyer, F. E. *Angew. Chem. Int. Ed.* 1994, **33**, 2379; (e) Dounay, A. B. Overman, L. E. *Chem. Rev.* 2003, **103**, 2945; see also: (f) *The Mizoroki-Heck Reaction*, Oestreich, M. Ed.; 2009, Wiley, NY.

111. (a) Milstein, D. Stille, J. K. *J. Am. Chem. Soc.* 1978, **100**, 3636; (b) Farina, V.; Krishnamurthy, V.; Scott, W. J. *Org. React.* 1997, **50**, 1; (c) Duncton, M. A. J.; Pattenden, G. *J. Chem. Soc., Perkin Trans. 1* 1999, 1235; (d) Pattenden, G.; Sinclair,

D. J. *J. Organomet. Chem.* 2002, **653**, 261.

112. (a) Sonogashira, K.; Tohda, Y.; Hagihara, N. *Tetrahedron Lett.* 1975, **16**, 4467; (b) Sonogashira, K. *J. Organomet. Chem.* 2002, **653**, 46; (c) Tykwinski, R. R. *Angew. Chem. Int. Ed.* 2003, **42**, 1566; (d) Sonogashira, K. in, *Handbook of Organopalladium Chemistry for Organic Synthesis*, 2002, vol. 1, p. 493, Negishi, E.-i., Ed.; Wiley-Interscience, NY.

113. (a) Baba, S.; Negishi, E.-i. *J. Am. Chem. Soc.* 1976, **98**, 6729; (b) Knochel, P.; Singer, R. D. *Chem. Rev.* 1993, **93**, 2117; (c) Negishi, E.-i. *J. Organomet. Chem.* 2002, **653**, 34.

114. (a) Guram, A. S.; Buchwald, S. L. *J. Am. Chem. Soc.* 1994, **116**, 7901; (b) Wolfe, J. P.; Wagaw, S.; Marcoux, J.-F.; Buchwald, S. L. *Acc. Chem. Res.* 1998, **31**, 805.

115. (a) Paul, F.; Patt, J.; Hartwig, J. F. *J. Am. Chem. Soc.* 1994, **116**, 5969; (b) Hartwig, J. F. *Acc. Chem. Res.* 1998, **31**, 852.

116. Miyazaki, F.; Uotsu, K. Shibasaki, M. *Tetrahedron* 1998, **54**, 13073.

117. Hanessian, S.; Ma, J.; Wang, W.; Gai, Y. *J. Am. Chem. Soc.* 2001, **123**, 10200.

118. (a) Tsuji, J.; Takahashi, H.; Morikawa, M. *Tetrahedron Lett.* 1965, **6**, 4387; (b) Tsuji, J. in *Handbook of Organopalladium Chemistry for Organic Synthesis*, 2, 1669, 2002, Negishi, E.-i., Ed.; Wiley, NY; (c) Trost, B. M.; Van Vranken, D. L. *Chem. Rev.* 1996, **96**, 395; (d) Trost, B. M.; Fandrick, D. R. *Aldrichim. Acta* 2007, **40**, 59; (e) *Palladium Reagents and Catalysis: New Perspectives for the 21st Century*, Tsuji, J., 2004, Wiley, NY.

119. Trost, B. M. Pissot-Soldermann, C.; Chen, I.; Schroder, G. M. *J. Am. Chem. Soc.* 2004, **126**, 4480.

120. (a) Nicolaou, K. C.; Bulger, P. G. Jr.; Sarlah, D. *Angew. Chem. Int. Ed.* 2005, **44**, 4442; see also: (b) Seki, M. *Synthesis* 2006, 2975, and reference [108b].

121. For example see: (a) Newhouse, T.; Baran, P. S. *Angew. Chem. Int. Ed.* 2011, **50**, 3302; (b) Chen, M. S.;

White, M. C. *Science* 2007, **318**, 787; (c) Godula, K.; Sames, D. *Science* 2006, **312**, 67; (d) Hinman, A.; Du Bois, J. *J. Am. Chem. Soc.* 2003, **125**, 11510; (e) Chen, H.; Schlecht, S.; Semple, T. C. Hartwig, J. F. *Science* 2000, **287**, 1995; see also: (f) *Handbook of C-H Transformations*, Dyker, G. Ed.; 2005, Wiley-VCH, NY.

3
The "*What*" of Synthesis

3.1
Periods, Trends, and Incentives

Frequent comparisons between art and synthesis have been made over the years [1]. As previously alluded to in Chapters 1 and 2, the artist and synthetic chemist start their respective works with an image, either imaginary or real. Imagery preceded the written word and it could have been a complimentary form of the spoken expression from the time of primal man. Drawings in tombs and cave-dwellings are the roots of man's ability to translate the visual and mental world into a form of expression and communication.

The graphical expression of chemical events in the form of reagents and reactions relating to a synthesis came thousands of years after the cave man's drawings. Nevertheless, looking at the shape, form, and topology of organic molecules as targets for synthesis shows an exponential evolution towards molecular complexity in recent years. Historically, there have been periods of steady progress followed by periods of explosive productivity and ingenuity. As different art materials became available (also thanks to chemistry), artists ventured into new forms of expression. Analogously, as new enabling methods of bond formation were introduced on the palette of the synthetic chemist, more and more complex structures were tackled and successfully conquered [2]. The availability of sophisticated instrumentation for the separation and analysis of even minute quantities of advanced intermediates played a defining role in the successful completion of many total syntheses of natural and unnatural products. The history of total synthesis can be briefly described in terms of periods, trends, and incentives.

Design and Strategy in Organic Synthesis: From the Chiron Approach to Catalysis, First Edition.
Stephen Hanessian, Simon Giroux, and Bradley L. Merner.
© 2013 Wiley-VCH Verlag GmbH & Co. KGaA. Published 2013 by Wiley-VCH Verlag GmbH & Co. KGaA.

Period	Trends	Incentives
Pre 1950's	Isolation and structure elucidation of natural products	Structure validation; synthesis feasibility
1950's-1960's	Classical syntheses of specific classes of natural products	Target conquest, mechanistic insights, medicinal relevance; "art of synthesis"
1970's-1980's	Multistep, strategy and functional group based syntheses, asymmetric synthesis,	Stereochemical control, exploratory chemistry; the challenge of complex structures
1990's-present	Era of complex synthesis; emergence and applications of catalysis to asymmetric processes	Demonstration of synthetic prowess; extending the limits of feasibility; biological relevance
2013 and beyond	Natural products as biological tools; expanding the potential of catalysis in synthesis; green chemistry; chemical "omics," organic materials science and nanotechnology	Structure to function continuum; chemical basis to biology and medicine; organized macromolecular entities; advancing catalysis; global challenges, energy

3.2
A Century of Synthesis

Milestone achievements in the early part of the 20th century were focussed mainly on structure elucidation of organic natural products, and in establishing synthesis as a subdiscipline. Pertinent historic examples featuring the synthesis of (±)-camphor by Komppa and of (±)-equilenin by Bachmann are illustrative (Figure 3.1) [3].

The early part of the second half of the 20th century was dominated by basic concepts of bonding, mechanistic insights, and extending the art of total synthesis by a handful of research laboratories worldwide, exemplified by the legendary contributions of R. B. Woodward [4]. In spite of limited spectroscopic methods, this period of time demonstrated the remarkable power of organic synthesis in attaining different classes of natural products such as alkaloids, steroids, and various antibiotics. Concurrently, the revolutionary advances in combating bacteria with anti-infectives, and exercising means of birth control with steroids, were tremendous incentives as well as newsworthy achievements in organic synthesis [5].

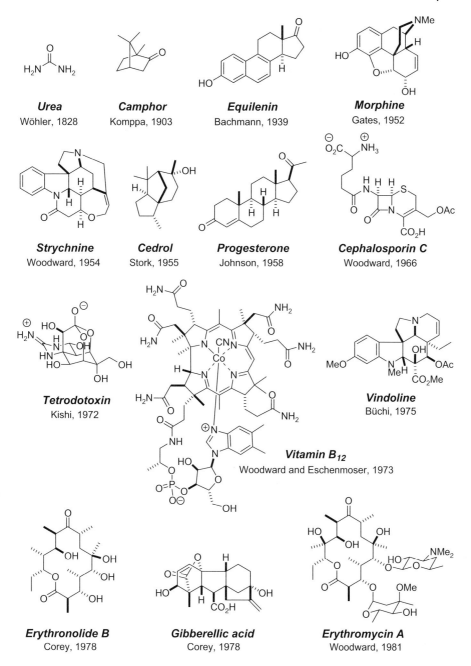

Figure 3.1 Selected syntheses of natural products (1828–1981).

The nineteen seventies and eighties witnessed remarkable displays of creativity in the development of stereocontrolled bond-forming reactions, and their application to natural product synthesis. Multistep syntheses with control of absolute stereochemistry became common place, as innovative methods for chemo-, stereo-, and regioselective chemical transformations emerged from the chemist's arsenal [6]. Organometallic chemistry gained importance especially in conjunction with Lewis acids and other carbonyl activation methods. Efforts concerned with total syntheses brought forth the Woodward-Hoffmann rules on the conservation of orbital symmetry [7], and Baldwin's rules for ring closure [8]. The variety and architectural complexity of a select group of natural products synthesized for the first time during the period 1828–1981 utilizing mostly UV and IR spectroscopy as a primary means of characterization is shown in Figure 3.1. The most functionally complex natural product of this period to be synthesized was vitamin B_{12}, a joint effort form the Woodward [9], and Eschenmoser groups [10], and a crowning achievement since Wöhler's synthesis of urea (see Chapter 16, section 16.2). The remaining examples are also among the masterpieces of this period, along with many more that can be found in the literature [2, 3].

Synthetic chemistry entered its golden period in the 1990's when complex natural product synthesis became a quasi-continuous euphoric activity in many laboratories worldwide. With the advent of high field NMR spectroscopy, powerful separation techniques, and a rich armamentarium of synthetic methods, the upper and lower limits of structural complexity of the molecule to be synthesized was self-imposed by the investigator. Thus, the height of the synthetic challenge bar could be raised or lowered at will depending on individual cases. For example, decades after the first landmark total synthesis of strychnine by Woodward [11], no less than 17 different syntheses were published between the period 1992–2011 (Figure 3.2) [12]. A discussion of some of the strategies used in the enantioselective total synthesis of strychnine is featured in Chapter 4 (section 4.4.2).

Perhaps the molecule that commanded the most notoriety and attention at the time of the public announcement of its synthesis by the Holton [13] and Nicolaou [14] groups independently was taxol – presently used for the treatment of ovarian cancer. There are now six total syntheses of taxol [13–15], each involving conceptually different strategies (see Chapter 16, section 16.8).

A well-designed strategy toward the synthesis of a high-profile molecule may provide access to valuable intermediates and flexibility to study structure-activity relationships (SAR), as well as to probe other pharmacologically relevant properties. Analogues that are not directly available from Nature may have improved activities as evidenced time and again in the field of antibiotics such as β-lactams and macrolides.

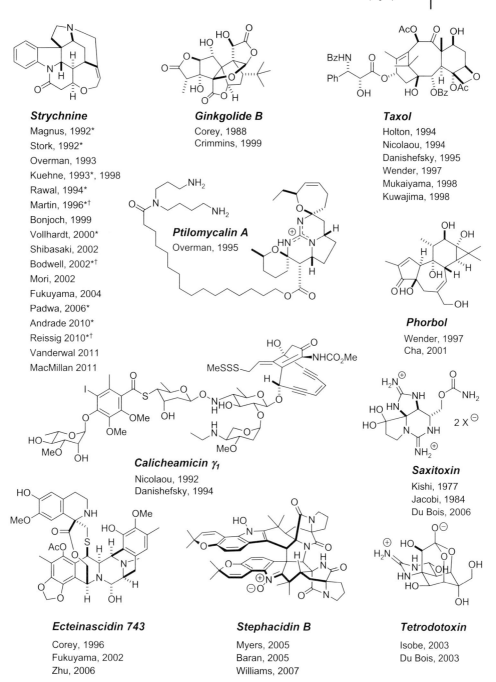

Strychnine

Magnus, 1992*
Stork, 1992*
Overman, 1993
Kuehne, 1993*, 1998
Rawal, 1994*
Martin, 1996*†
Bonjoch, 1999
Vollhardt, 2000*
Shibasaki, 2002
Bodwell, 2002*†
Mori, 2002
Fukuyama, 2004
Padwa, 2006*
Andrade 2010*
Reissig 2010*†
Vanderwal 2011
MacMillan 2011

Ginkgolide B

Corey, 1988
Crimmins, 1999

Taxol

Holton, 1994
Nicolaou, 1994
Danishefsky, 1995
Wender, 1997
Mukaiyama, 1998
Kuwajima, 1998

Ptilomycalin A

Overman, 1995

Phorbol

Wender, 1997
Cha, 2001

Calicheamicin γ₁

Nicolaou, 1992
Danishefsky, 1994

Saxitoxin

Kishi, 1977
Jacobi, 1984
Du Bois, 2006

Ecteinascidin 743

Corey, 1996
Fukuyama, 2002
Zhu, 2006

Stephacidin B

Myers, 2005
Baran, 2005
Williams, 2007

Tetrodotoxin

Isobe, 2003
Du Bois, 2003

Figure 3.2 Selected highlights in total synthesis of polyfunctional and architecturally complex natural products
(1977–2011); racemic = *; formal synthesis = †.

Tetrodotoxin [16], and saxitoxin [17], are some of the most potent toxins known to man. Phorbol ester is a powerful tumor promoter [18], yet its total synthesis has aided in understanding the origins of tumorigenesis by activation of oncogenic pathways. The unprecedented functionality and disposition of reactive groups in the calicheamicins has uncovered a unique mode of bioactivation [19], reminiscent of the aromatization of ene-diynes through diradical intermediates, as originally reported by Bergman [20]. The total synthesis of complex alkaloids such as ptilomycalin A [21], and ecteinascidin 743 [22] has further demonstrated the ability to predictably control the intricate juxtaposition of reactive functionalities.

Kishi's total synthesis of palytoxin, the largest man-made molecule of its time, was a *tour-de-force* in controlling the stereochemistry of 64 stereogenic centers, 8 olefinic bonds, and 41 hydroxyl groups, totalling 10^{21} possible isomers! [23] It ushered a new era of total synthesis, especially of highly oxygenated molecules (Figure 3.3). Stereochemical control in the elaboration of complex ionophores is exemplified by the total synthesis of X-206 by Evans [24], and azaspiracid-1 by Nicolaou [25] and Evans [26] independently (Figure 3.3). The power of asymmetric synthesis in the construction of highly oxygenated polyketides is illustrated in the case of macrocylic lactones such as cytovaricin [27], halichondrin B [28], and swinholide A [29]. Vancomycin, considered as the last resort antibiotic against certain bacterial infections, is another synthetic hurdle that was virtually conquered simultaneously by three independent research groups [30] (see Chapter 16, section 16.3.4). Complex fused oxacyclic marine natural products such as brevetoxin B [31] and gambierol [32] have pushed the frontiers of total synthesis even further. The complex macrocyclic peptide thiostrepton presents a variety of unusual amino acid building blocks [33]. The two decade long journey towards the total synthesis of azadiractin was successfully completed in 2007 by Ley and coworkers [34] (Figure 3.3).

The creative chemical thinking and synthetic prowess that is necessary in reaching the summit of many a natural product "mountain" is fully evident from the examples shown in Figures 3.2 and 3.3, in addition to many other elegant syntheses of related complex natural products discussed throughout this book. There seems to be no limit in total synthesis, although the perennial question, "*What* natural product (or compound) should be synthesized?" is ever present in our minds. Although there are numerous answers to this question, it is a common practice to choose target molecules which will enable the investigator to test the feasibility of synthetic methods developed in his or her own laboratory, especially for a critical step in the planned scheme of reactions.

Almost every class of therapeutically relevant natural product has been synthesized. Lessons learned from these syntheses have led to the design of unnatural products in an effort to improve biological activity or to validate hypotheses related to their mode of action.

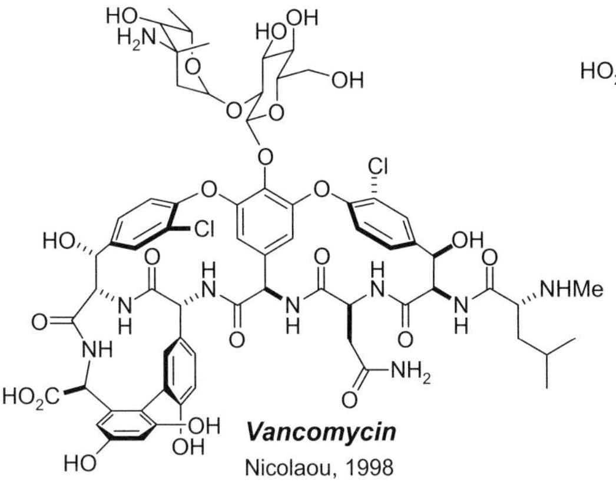

Vancomycin
Nicolaou, 1998
Evans (aglycone), 1998
Boger (aglycone), 1999

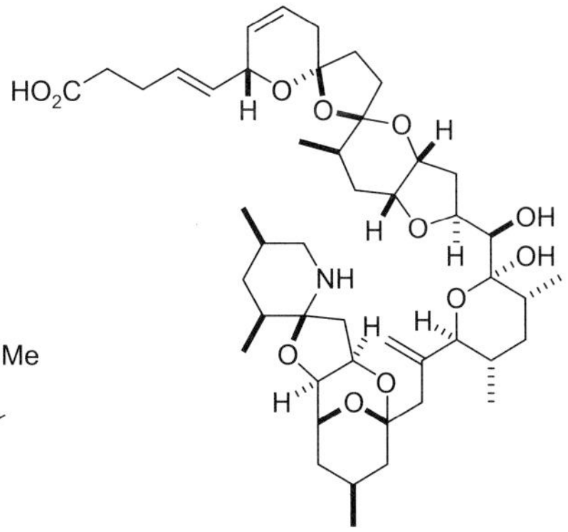

Azaspiracid-1
Nicolaou, 2005
Evans, 2007

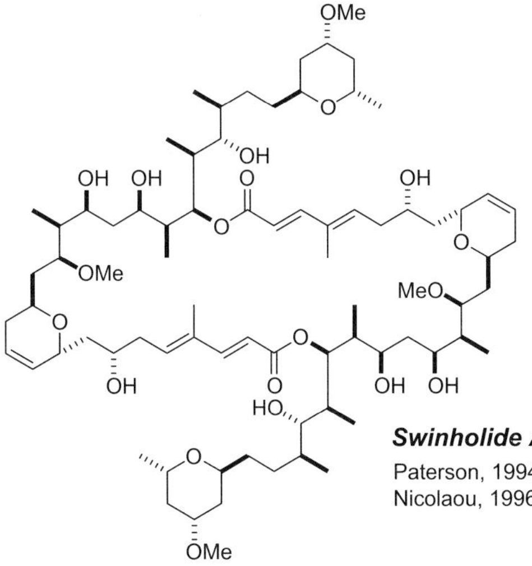

Swinholide A
Paterson, 1994
Nicolaou, 1996

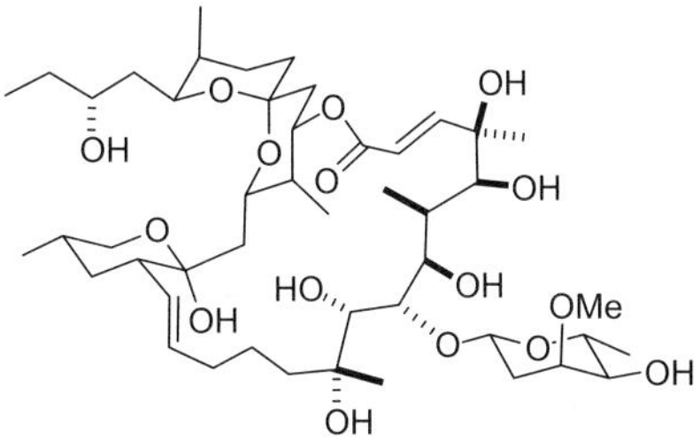

Cytovaricin
Evans, 1990

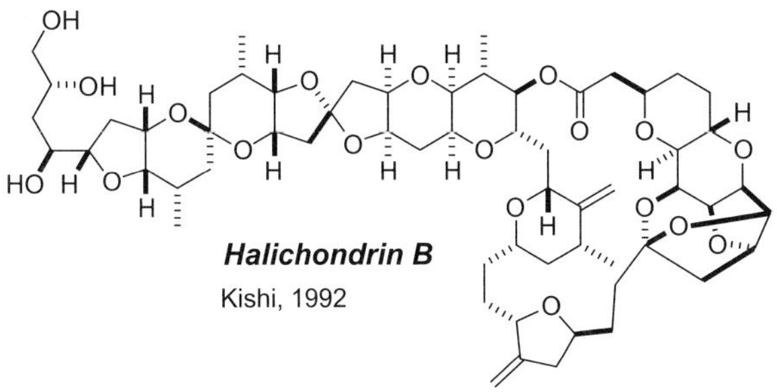

Halichondrin B
Kishi, 1992

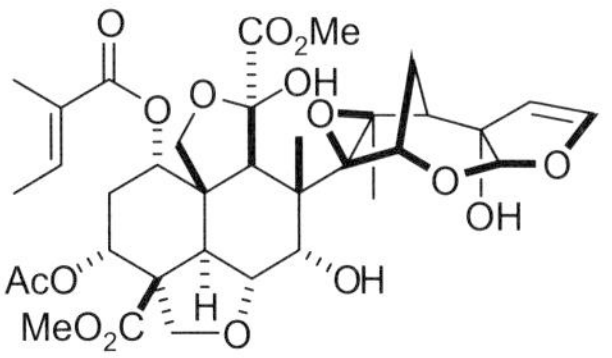

Azadirachtin
Ley, 2007

Figure 3.3 Selected highlights in multistep total synthesis of highly oxygenated and architecturally complex molecules (1989–2009).

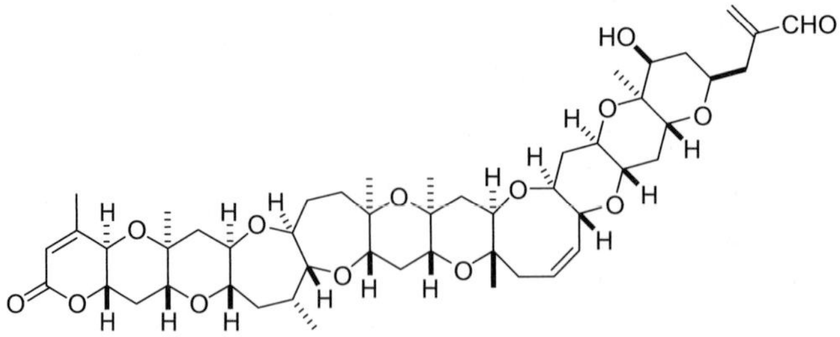

Brevetoxin B
Nicolaou, 1995
Nakata, 2004
Yamamoto, 2005

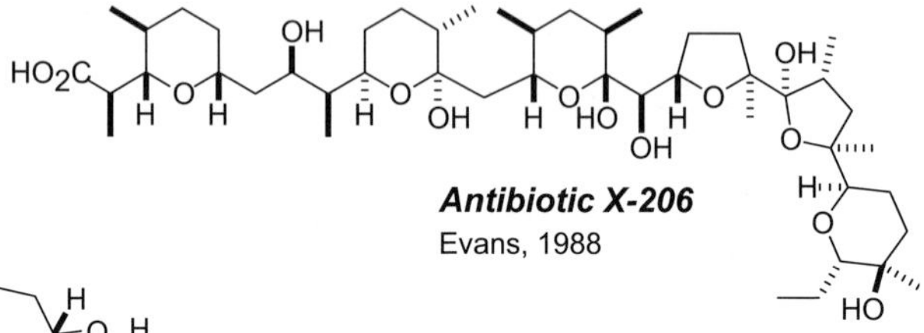

Antibiotic X-206
Evans, 1988

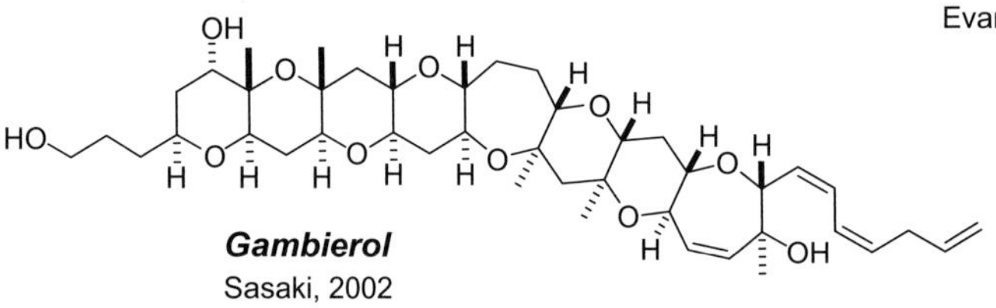

Gambierol
Sasaki, 2002
Yamamoto, 2003
Rainier, 2005
Mori, 2009

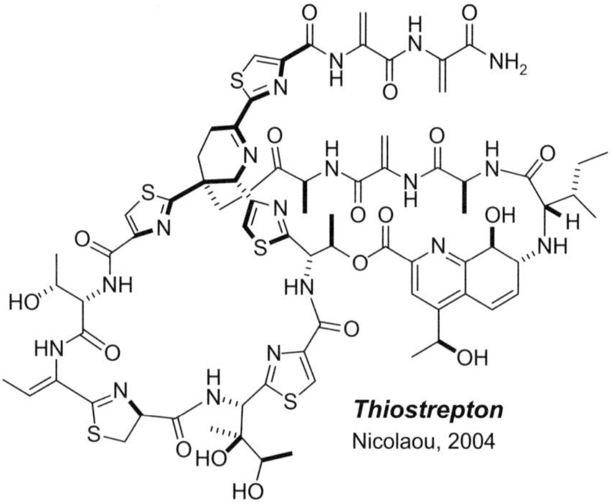

Thiostrepton
Nicolaou, 2004

Figure 3.3 *(continued)*.

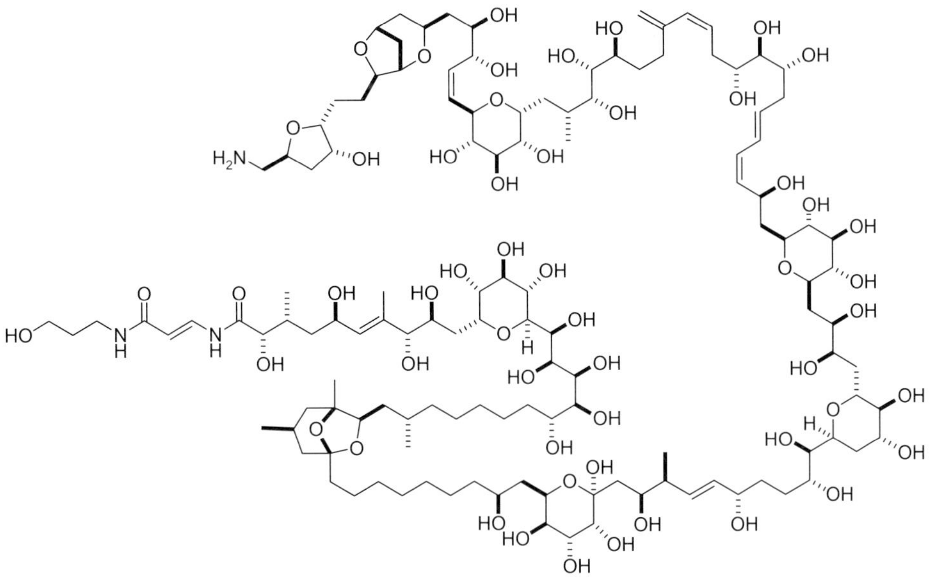

Palytoxin
Kishi, 1989 & 1994

Figure 3.3 (*continued*).

3.3
We the "Synthesis People"

Organic chemists are constant explorers. Some may even do it with a spirit of adventure. Although we are trained in specific areas during our schooling, we can also be highly adaptable to change. Some of this is self-imposed, but other factors may be responsible, not the least of which is research funding potential. Because we are well rooted in the fundamentals of organic chemistry, adapting to different areas and crossing over into less-charted subdisciplines are not considered as major obstacles. Great advances in the biological, materials, and polymer sciences, as well as nanotechnology, are due to a strong foundation in synthetic chemistry. Reluctance to cross boundaries is largely a matter of personal choice. It is interesting to reflect on the purely synthetic aspect of organic chemistry as it interfaces with other subdisciplines, such as the biological sciences. As a "people" we fall into the following categories:

- Methods people
- Methods and target people
- Target people
- Target and function people
- Function people

Within each of these categories are ample opportunities for innovation. Methodology encompasses all aspects of bond-forming processes, including catalysis, mechanism, and applications. Target-oriented synthetic chemists are primarily focussed on the design and execution of sophisticated, multistep syntheses of complex natural products, including unnatural compounds for medicinal, agricultural, and materials purposes. The emphasis is on attaining the intended target molecule.

Combining the best of the two worlds of methods development *and* applications to natural product synthesis brings in obvious rewards. The *"I did it my way"* (Sinatra) approach as opposed to the *"Just do it"* (Nike) approach, is intellectually satisfying, especially when new, in-house methodology is developed and its generality in total synthesis is validated [35].

Interfacing synthesis with biological function has its appeal in the study of structure and activity, thus probing the molecular basis of small molecule action. This approach to organic synthesis, often performed by learning biology through collaboration, is highly stimulating because it embodies the eternal promise of good health and wellness for humankind.

Some synthetic chemists have seen a different light at the end of the tunnel at some point in their careers, and made the conscious decision to focus primarily on the biological or materials science end of the structure to function continuum. This switch may come about during a planned sabbatical period, through fascination with the biological or macromolecular and "nano" worlds, or for reasons of survival in an ever-diminishing funding system.

A chemist will always remain a chemist at heart, especially in applying the analytical way of thinking and solving problems that are engrained in his or her formative years. No matter what "people" category we find ourselves in as synthetic chemists, the exhilaration of discovery is at the basis of our quest for excellence. In the process, generations of young investigators receive valuable training and mentorship in the enterprise of research, and will continue to do so for many generations to come [35].

3.4
Complex and Therapeutic too!

The *"What"* of total synthesis can manifest itself in other highly rewarding ways. The intellectually satisfying process of total synthesis, with its obvious coworker-training component, can in fact be applied to the synthesis of therapeutically relevant natural products. Synthetic chemists have always been on the lookout for challenging natural product structures. Occasionally, these are also developed as marketed drugs. With few exceptions, such compounds are produced by fermentation, although process chemists have made exemplary efforts in the high volume production of analogues. The interest of synthetic chemists in compounds of medical relevance became evident some 50 years ago, with the discovery of the potent pharmacological activities of steroids and prostaglandins. Combining relevance and synthetic challenge has been a compelling incentive for total synthesis in the academic community as evidenced by a sampling of marketed (or nearly-marketed) drugs shown in Figure 3.4 (see also Chapter 2, sections 2.7 and 2.8). Bryostatin 1 is included as an example of a natural product with potential therapeutic relevance (see Chapter 18, section 18.4 for bryostatin 16).

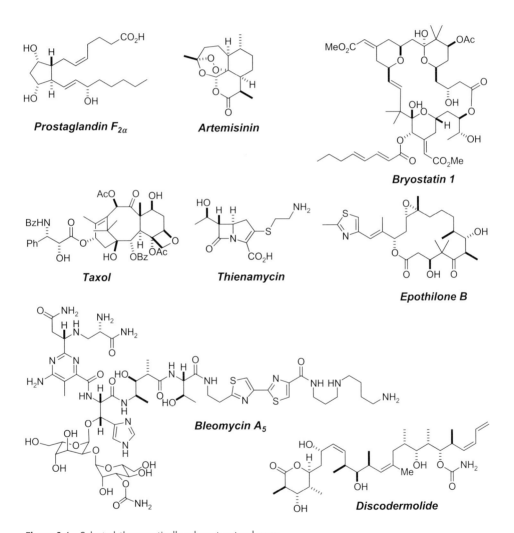

Figure 3.4 Selected therapeutically relevant natural compounds of interest to academic and industrial synthetic chemists.

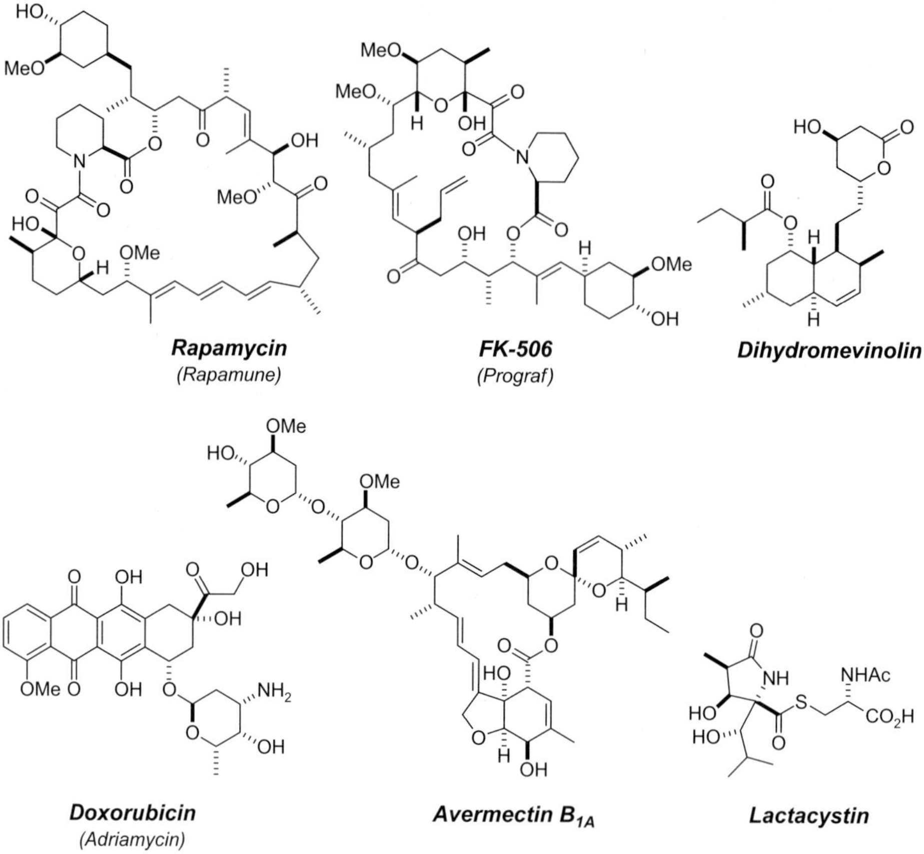

Rapamycin
(Rapamune)

FK-506
(Prograf)

Dihydromevinolin

Doxorubicin
(Adriamycin)

Avermectin B₁ₐ

Lactacystin

Figure 3.4 (continued).

In some instances, contributions to total synthesis of products ordinarily available from natural sources from industrial groups has brought added value to the existing methodology. This type of "friendly competition" has also fostered productive collaborations between academics and the pharmaceutical/agricultural industries [35]. The prostaglandins, anthacycline antibiotics, mevinic acids, epothilones, and discodermolide are illustrative of synthetic efforts covering nearly 40 years of research interfacing academia and industry.

3.5
Peptidomimetics and Unnatural Compounds

The "What" of synthesis extends to the design of enzyme inhibitors and small organic molecules that interact with receptors. Information obtained from X-ray crystal structures of enzymes involved in important *in vivo* processes has been invaluable for the design and synthesis of inhibitors [35, 36]. Frequently, such inhibitors are also natural products that can be co-crystallized with specific enzymes. Chemically modified or truncated variants of such natural products are excellent targets for synthesis under the structure- or fragment-based drug discovery paradigm [37]. NMR spectroscopy has also been instrumental in the design of potent inhibitors either directly, or by linking two entities to find dual binding sites in an enzyme active site [38]. Major advances in the development of zinc metaloproteases such as inhibitors of the angiotensin converting enzyme (ACE), for example, are the result of structure-based design refinements of the original co-crystal structure (see Chapter 2, section 2.6).

The natural substrates or cleavage products of certain aspartic, serine, and cysteine proteases are also known [39]. Based on this structural information, medicinal chemists have designed potent peptidomimetics that bind to these enzymes with high affinities. Such inhibitors have been developed for a host of physiologically harmful enzymes with excellent clinical results. The recently marketed 8-aryl carboxylic acid derivative Tekturna or Rasilez (aliskiren) [40] is a synthetic inhibitor of the aspartic protease renin. This enzyme is involved in a cascade of events that release vasoconstricting peptides in the blood stream leading to elevated blood pressure (hypertension) (Figure 3.5). The enzyme neuraminidase is responsible for the release and propagation of influenza virus from human epithelial cells. The synthetic compound A-315675 [41] has been shown to be an excellent inhibitor of neuraminidase and its interaction with the enzyme can actually be seen in an X-ray crystal structure of the complex. The development of antiviral agents BILN 2061 [42] and Incivek (telaprevir) [43] against the hepatitis C virus is the result of extensive structure-based design and optimization. The large number of marketed HIV protease inhibitors [44] is the direct result of structural information obtained from X-ray crystallography. Subtle structural modifications of approved drugs such as Prezista (darunavir) [45], may also result in an improved resistance profile.

Total synthesis of unnatural products exemplified by Tekturna (aliskiren) and its truncated congener (Figure 3.5) provides opportunities to develop alternative practical routes to medically relevant target compounds, or to find innovative methods that provide advanced intermediates [46]. The Jacobsen asymmetric synthesis of the aminoindanol component of Crixivan (indinavir) serves as a superb example of another academic contribution to an industrially and therapeutically relevant problem [47].

Alternative approaches to structure-based design and synthesis relying on biological data obtained from bioactive natural products involve structure modification or simplification. Bryologue [48], a truncated analogue of bryostatin, and ZK-EPO, an epothilone B [49] analogue, are two representative examples. The lactam analogue of epothilone B, Ixempra, was recently launched as an anticancer agent [50] (see also Chapter 2, section 2.2).

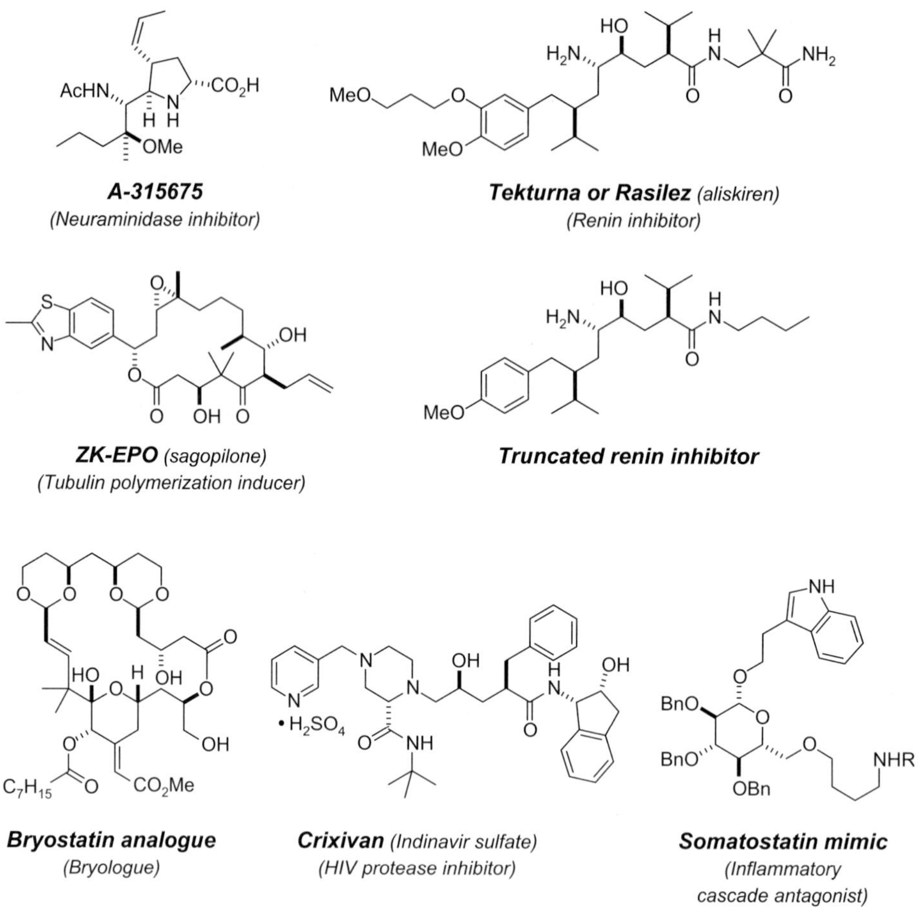

Figure 3.5 Synthesis of therapeutically relevant unnatural compounds.

The demonstration that an appropriately "decorated" hexopyranoside scaffold with pharmacophoric groups mimicking a segment of somatostatin, serves as an excellent example of structure simplification while maintaining biological activity [51]. The design and synthesis of potent thrombin inhibitors addresses an unmet medical need, and emphasizes academic contributions with added knowledge value to established entities such as Argatroban [52], and to inhibitors based on structural data [53].

Natural products have been particularly useful in the synthesis of analogues, hybrids, and truncated variants based on X-ray structural information in conjunction with biologically relevant enzymes [35]. For example, the aeruginosin family of natural products are a distinct group of some 20 members, isolated from geographically different aquatic regions. They exhibit *in vitro* activities against serine proteases such as thrombin, which is involved in the cascade of events leading to blood coagulation [54]. Dysinosin A [55], oscillarin [56], and chlorodysinosin A

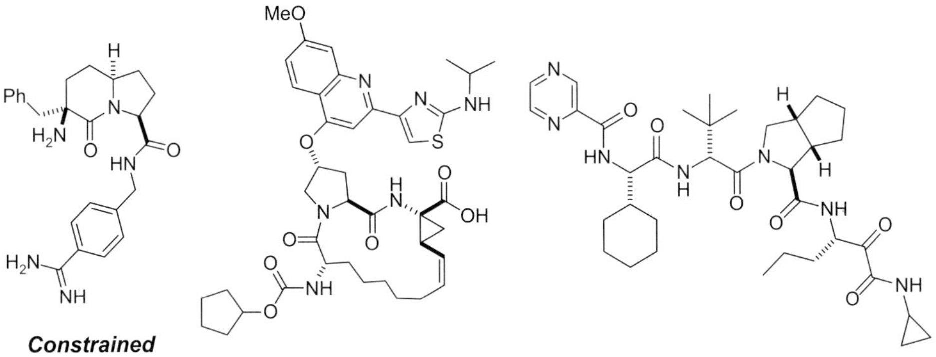

Constrained PPACK mimetic
(Thrombin inhibitor)

BILN 2061
(HCV protease inhibitor)

Incivek (telaprevir)
(HCV protease inhibitor)

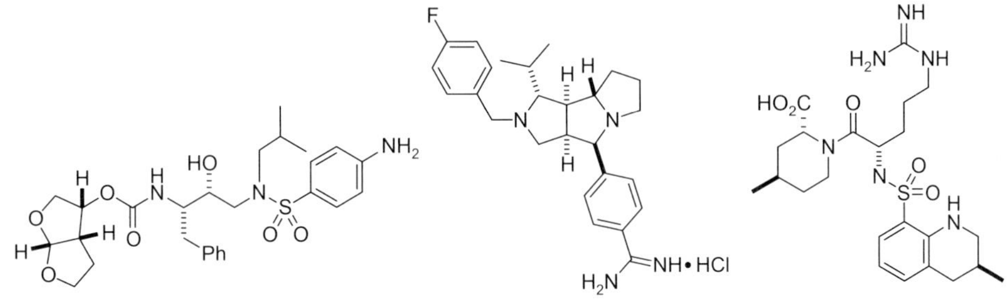

Prezista (darunavir)
(HIV protease inhibitor)

F-containing Thrombin inhibitor

Argatroban
(Thrombin inhibitor)

Figure 3.5 (continued).

[57] are among the most potent thrombin inhibitors known within the aeruginosin group (Figure 3.6). The dramatic influence of a chlorine atom in the D-leucine portion of chlorodysinosin A compared to dysinosin A (IC$_{50}$ = 5 nM versus IC$_{50}$ = 46 nM against thrombin), was rationalized on the basis of a more effective binding in the S$_3$ hydrophobic pocket of thrombin as concluded from X-ray co-crystal structure data [57]. Based on this observation, a number of truncated analogues were synthesized [58], including achiral molecules [59], resulting in excellent *in vitro* inhibitory activities. Dabigatran has recently received FDA approval as an anticoagulant drug, and can be used as an alternative to warfarin [60].

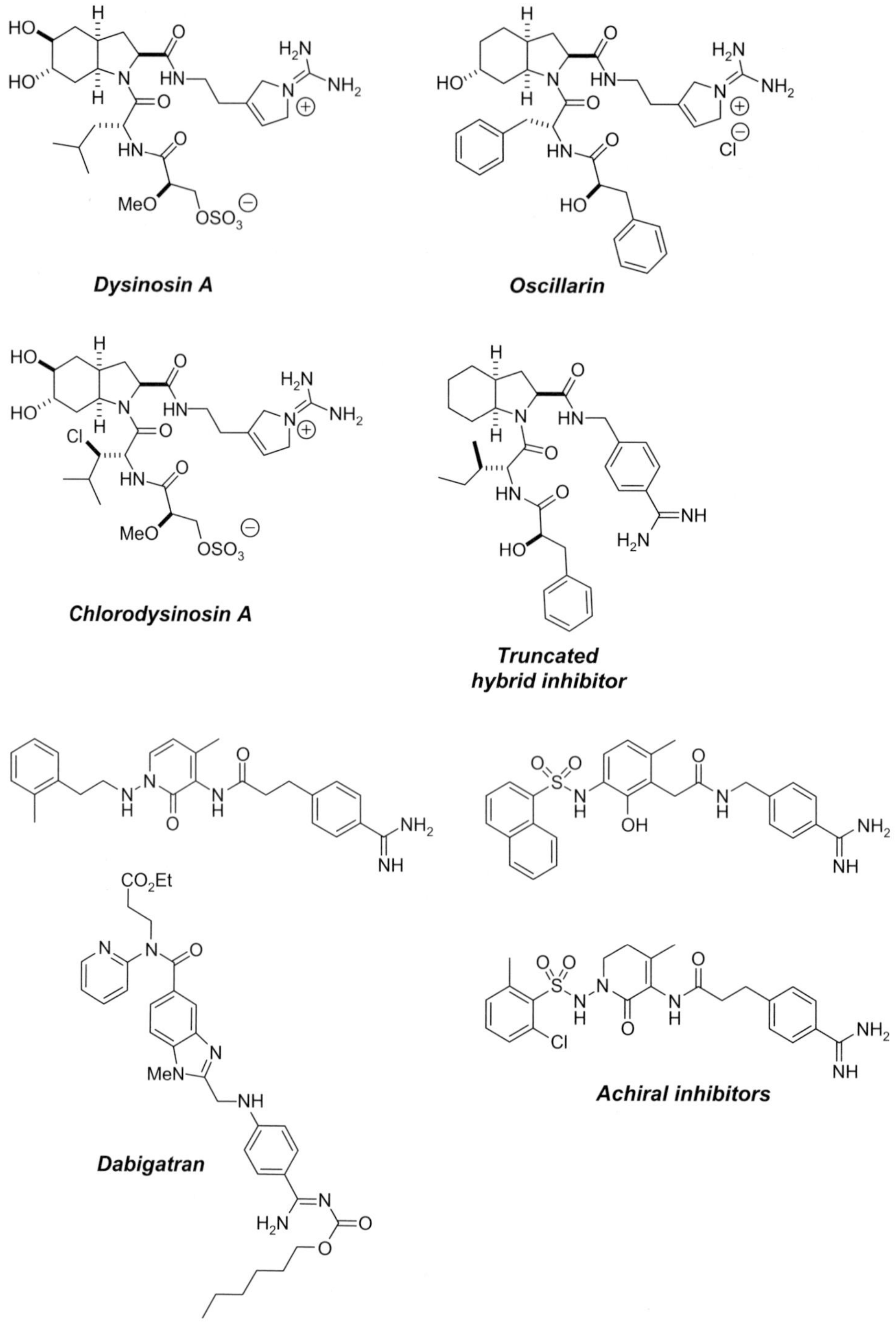

Figure 3.6 From natural products to analogues and achiral inhibitors of thrombin and related serine proteases.

Thus, the synthesis of surrogates and mimetics of natural products can be an exciting and rewarding area of research in the context of prototypical drug design.

3.6
Diversity Through Complexity

A conceptually and operationally different version of total synthesis is based on a diversity-oriented paradigm [61]. The notion of simplifying the structural complexity of a target molecule to the level of one or more building blocks is reversed by adopting a strategy in which a molecule can lend itself to a series of sequential reactions [62] creating a multitude of products. Ultimately, by adopting techniques of parallel or combinatorial synthesis [63], a large library of unnatural compounds can be produced with predetermined functional groups. With the advent of high-throughput biological screening, such libraries of compounds can provide "hits" that can be further developed into "leads." Testing libraries of synthetic compounds prepared in-house or acquired from outside sources has become a standard practice in the pharmaceutical industry [64]. The process is greatly facilitated with the availability of powerful instrumentation for automated separation and analysis [65]. Clearly, chance will play a critical role in obtaining hits, even in the most favorable cases where the diversity elements have been well-planned. Since chemical and biological spaces are vast, their exploitation with "smart" libraries of synthetic compounds provides a great incentive for innovation [66].

In Figure 3.7 are depicted six types (A–F) of basic core structures (scaffolds) representing millions of compounds resulting from automated parallel or combinatorial procedures utilizing solution or solid-phase techniques [67]. In spite of promising occasional good "hits" when confronting Nature's enzymes and receptors with such impressive numbers of compounds, the success rate of potential drugs remains dismally low. Clearly, much is to be learned with regard to the more efficient design of pharmacophoric elements that provide better drug-macromolecule interactions.

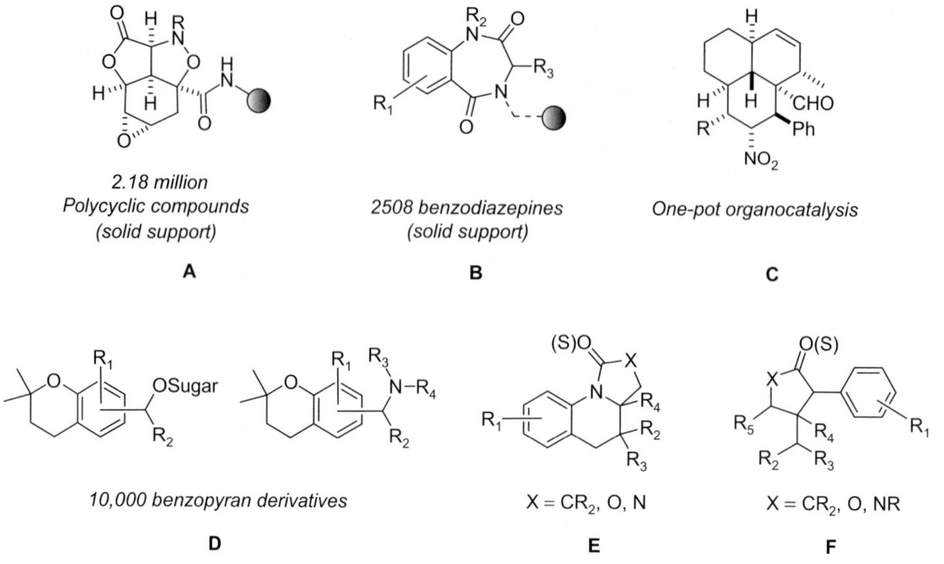

2.18 million
Polycyclic compounds
(solid support)

A

2508 benzodiazepines
(solid support)

B

One-pot organocatalysis

C

10,000 benzopyran derivatives

D

$X = CR_2$, O, N

E

$X = CR_2$, O, NR

F

Figure 3.7 The chemical "library" era.

The "*What*" domain of synthesis can also be extended beyond molecules of direct biological relevance. Thus, organic synthesis offers unique opportunities in new and exciting areas in materials science [68] and nanotechnology [69] to address global issues relating to energy sources among other needs (see also Chapter 2, section 2.10; Chapter 18, section 18.7) [70].

References

1. For example, see: (a) *The Art of Drug Synthesis*, Johnson, D. S.; Li, J. J. Eds.; 2007, Wiley-Interscience, NY; (b) *Art in Organic Synthesis*, Anand, N.; Bindra, J. S., Ranganathan, S., 1988, Wiley-Interscience, NY; see also: (c) *Organic Synthesis: The Science behind the Art*, Smit, W. A.; Bochkov, A. F.; Caple, R. 1998, Chapter 3, p. 232, The Royal Society, Cambridge, UK.

2. For selected books, see: (a) *Enantioselective Chemical Synthesis: Methods, Logic and Practice*, Corey, E. J.; Kürti, L., 2010, Dallas, TX; (b) *Strategies and Tactics in Organic Synthesis*, Harmata, M.; Ed.; 2008, vol. 1-7, Academic Press, NY; (c) *The Way of Synthesis*, Hudlicky, T.; Reed, J. W., 2007, Wiley-VCH,

Weinheim; (d) *Classics in Total Synthesis – Targets, Strategies, Methods*, Nicolaou, K. C.; Sorensen, E. J., 1996, Wiley-VCH, Weinheim; (e) *Classics in Total Synthesis II – More Targets, Strategies, Methods*, Nicolaou K. C.; Snyder, S. A., 2003, Wiley-VCH, Weinheim; (f) *Classics in Total Synthesis III: Further Targets, Strategies, Methods*. Nicolaou K. C.; Chen, J. S. 2011, Wiley-VCH, Weinheim; (g) *The Logic of Chemical Synthesis*, Corey, E. J.; Cheng, X.-M., 1989, Wiley-Interscience, NY.

3. For relevant examples, see: (a) Nicolaou, K. C.; Vourloumis, D.; Winssinger, N.; Baran, P. S. *Angew. Chem. Int. Ed.* 2000, **39**, 44; (b) Nicolaou, K. C.; Sorensen, E. J. Winssinger, N. *J. Chem. Ed.* 1998, **75**, 1225.

4. (a) Woodward, R. B. in *Perspectives in Organic Chemistry*, Todd, A. R. Ed.; 1956, p. 156, Interscience Publishers Inc., NY; for a retrospective on selected contributions of R. B. Woodward, see: (b) Kauffman, G. B. *Chem. Educator* 2004, **9**, 172; see also: (c) *Architect and Artist in the World of Molecules*, Benfey, O. T.; Morris, P. J. T. Eds.; 2001, Chemical Heritage Foundation, Philadelphia, PA.

5. For a historical perspective, see: (a) *Molecules that Changed the World*, Nicolaou, K. C.; Montagnon, T., 2008, Wiley-VCH, Weinheim; see also: (b) *Molecules and Medicine*, Corey, E. J.; Kürti, L.; Czakó, B., 2007, Wiley, NY.

6. For example, see: (a) *Encyclopaedia of Reagents for Organic Synthesis*, Crich, D. Ed.; 2010, Wiley-Interscience, NY; (b) *Comprehensive Organic Name Reactions*, Wang, Z. 2009, Wiley, NY; (c) *Name reactions: A collection of Detailed Mechanisms and Synthetic Applications*, Li. J. J. 2009 4th Edition, Springer Verlag, Berlin; (d) *Strategic Applications of Named Reactions in Organic Synthesis: Background and Detailed Mechanisms*, Kürti, L. Czakó, B. 2005, Elsevier Science, NY.

7. (a) *The Conservation of Orbital Symmetry*, Woodward, R. B.; Hoffmann, R., 1970, Academic Press, NY; see also: (b) Woodward, R. B.; Hoffmann, R. *J. Am. Chem. Soc.* 1965, **87**, 395; (c) Woodward, R. B.; Hoffmann, R. *Angew. Chem. Int. Ed.* 1969, **8**, 781.

8. Baldwin, J. E. *Chem. Comm.* 1976, 734.

9. (a) Woodward, R. B. *Pure Appl. Chem.* 1968, **17**, 519; (b) Woodward, R. B. *Pure Appl. Chem.* 1971, **25**, 283; (c) Woodward, R. B. *Pure Appl. Chem.* 1973, **33**, 145.

10. Eschenmoser, A.; Wintner, C. E. *Science* 1977, **196**, 1410.

11. (a) Woodward, R. B.; Sondheimer, F.; Taub, D. *J. Am. Chem. Soc.* 1951, **73**, 4057; (b) Woodward, R. B.; Sondheimer, F.; Taub, D.; Heusler, K.; McLamore, W. M. *J. Am. Chem. Soc.* 1952, **74**, 4223.

12. For a comprehensive review, see: Bonjoch, J.; Solé, D. *Chem. Rev.* 2000, **100**, 3455.

13. (a) Holton, R. A.; Somoza, C.; Kim, H.-B.; Liang, F.; Biediger, R. J.; Boatman, P. D.; Shindo, M.; Smith, C. C.; Kim, S.; Nadizadeh, H.; Suzuki, Y.; Tao, C.; Vu, P.; Tang, S.; Mucthi, K. K.; Gentile, L. N.; Liu, J. H. *J. Am. Chem. Soc.* 1994, **116**, 1597; (b) Holton, R. A.; Kim, H.-B.; Somoza, C.; Liang, F.; Biediger, R. J.; Boatman, P. D.; Shindo, M.; Smith, C. C.; Kim, S.; Nadizadeh, H.; Suzuki, Y.; Tao, C.; Vu, P.; Tang, S.; Zhang, P.; Murthi, K. K.; Gentile, L. N.; Liu, J. H. *J. Am. Chem. Soc.* 1994, **116**, 1599.

14. (a) Nicolaou, K. C.; Yang, Z.; Liu, J. J.; Ueno, H.; Nantermet, P. G.; Guy, R. K.; Claiborne, C. F.; Renaud, J.; Couladouros, E. A.; Paulvannan, K.; Sorensen, E. J. *Nature* 1994, **367**, 630; see also: (b) Nicolaou, K. C.; Dai, W.-M.; Guy, R. K. *Angew. Chem. Int. Ed.* 1994, **33**, 15.

15. (a) Danishefsky, S. J.; Masters, J. J.; Young, W. B.; Link, J. T.; Snyder, L. B.; Magee, T. V.; Jung, D. K.; Isaacs, R. C. A.; Bornmann, W. G.; Alaimo, C. A.; Coburn, C. A.; Di Grandi, M. J. *J. Am. Chem. Soc.* 1996, **118**, 2843; (b) Wender, P. A.; Badham, N. F.; Conway, S. P.; Floreancig, P. E.; Glass, T. E.; Houze, J. B.; Krauss, N. E.; Lee, D.; Marquess, D. G.; McGrane, P. L.; Meng, W.; Natchus, M. G.; Shuker, A. J.; Sutton, J. C.; Taylor, R. E. *J. Am. Chem. Soc.* 1997, **119**, 2757; (c) Morihira, K.; Hara, R.; Kawahara, S.; Nishimori, T.; Nakamura, N.; Kusama, H.; Kuwajima, I. *J. Am. Chem. Soc.* 1998, **120**, 12980; (d) Mukaiyama, T.; Shiina, I.; Iwadare, H.; Saitoh, M.; Nishimura, T.; Ohkawa, N.; Sakoh, H.; Nishimura, K.; Tani, Y.; Hasegawa, M.; Yamada, K.; Saitoh, K. *Chem. Eur. J.* 1999, **5**, 121.

16. (a) Kishi, Y.; Aratani, M.; Fukuyama, T.; Nakatsubo, F.; Goto, T.; Inoue, S.; Tanino, H.; Sugiura, S.; Kakoi, H. *J. Am. Chem. Soc.* 1972, **94**, 9217; (b) Ohyabu, N.; Nishikawa, T.; Isobe, M. *J. Am. Chem. Soc.* 2003, **125**, 8798; (c) Hinman, A.; Du Bois, J. *J. Am. Chem. Soc.* 2003, **125**, 11510.

17. (a) Tanino, H.; Nakata, T.; Kaneko, T.; Kishi, Y. *J. Am. Chem. Soc.* 1977,

99, 2818; (b) Jacobi, P. A.; Martinelli, M. J.; Polanc, S. *J. Am. Chem. Soc.* 1984, **106**, 5594; (c) Fleming, J. J.; Du Bois, J. *J. Am. Chem. Soc.* 2006, **128**, 3926.

18. (a) Wender, P. A.; Rice, K. D.; Schnute, M. E. *J. Am. Chem. Soc.* 1997, **119**, 7897; (b) Lee, K.; Cha, J. K. *J. Am. Chem. Soc.* 2001, **123**, 5590.

19. (a) Nicolaou, K. C.; Hummel, C. W.; Pitsinos, E. N.; Nakada, M.; Smith, A. L.; Shibayama, K.; Saimoto, H. *J. Am. Chem. Soc.* 1992, **114**, 10082; (b) Hitchcock, S. A.; Chu-Moyer, M. Y.; Boyer, S. H.; Olson, S. H.; Danishefsky, S. J. *J. Am. Chem. Soc.* 1995, **117**, 5750.

20. Bergman, R. G. *Acc. Chem. Res.* 1973, **6**, 25.

21. Overman, L. E.; Rabinowitz, M. H.; Renhowe, P. A. *J. Am. Chem. Soc.* 1995, **117**, 2657.

22. (a) Corey, E. J.; Gin, D. Y.; Kania, R. S. *J. Am. Chem. Soc.* 1996, **118**, 9202; (b) Endo, A.; Yanagisawa, A.; Abe, M.; Tohma, S.; Kan, T.; Fukuyama, T. *J. Am. Chem. Soc.* 2002, **124**, 6552; (c) Chen, J.; Chen, X.; Bois-Choussy, M.; Zhu, J. *J. Am. Chem. Soc.* 2006, **128**, 87.

23. (a) Armstrong, R. W.; Beau, J.-M.; Cheon, S. H.; Christ, W. J.; Fujioka, H.; Ham, W. H.; Hawkins, L. D.; Jin, H.; Kang, S. H.; Kishi, Y.; Martinelli, M. J.; McWhorter, W. W. Jr.; Mizuno, M.; Nakata, M.; Stutz, A. E.; Talamas, F. X.; Taniguchi, M.; Tino, J. A.; Ueda, K.; Uenishi, J.-I.; White, J. B.; Yonaga, M. *J. Am. Chem. Soc.* 1989, **111**, 7525; (b) Armstrong, R. W.; Beau, J.-M.; Cheon, S. H.; Christ, W. J.; Fujioka, H.; Ham, W.-H.; Hawkins, L. D.; Jin, H.; Kang, S. H.; Kishi, Y.; Martinelli, M. J.; McWhorter, W. W. Jr.; Mizuno, M.; Nakata, M.; Stutz, A. E.; Talamas, F. X.; Taniguchi, M.; Tino, J. A.; Ueda, K.; Uenishi, J.-I.; White, J. B.; Yonoga, M. *J. Am. Chem. Soc.* 1989, **111**, 7530.

24. Evans, D. A.; Bender, S. L.; Morris, J. *J. Am. Chem. Soc.* 1988, **110**, 2506.

25. (a) Nicolaou, K. C.; Li, Y.; Uesaka, N.; Koftis, T. V.; Vyskocil, S.; Ling, T.; Govindasamy, M.; Qian, W.; Bernal, F.; Chen, D. Y.-K. *Angew. Chem. Int. Ed.* 2003, **42**, 3643; (b) Nicolaou, K. C.; Chen, D. Y.-K.; Li, Y.; Qian, W.; Ling, T.; Vyskocil, S.; Koftis, T. V.; Govindasamy, M.; Uesaka, N. *Angew. Chem. Int. Ed.* 2003, **42**, 3649; (c) Nicolaou, K. C.; Vyskocil, S.; Koftis, T. V.; Yamada, Y. M. A.; Ling, T.; Chen. D. Y.-K.; Tang, W.; Petrovic, G.; Frederick, M. O.; Li, Y.; Satake, M. *Angew. Chem. Int. Ed.* 2004, **43**, 4312; (d) Nicolaou, K. C.; Koftis, T. V.; Vyskocil, S.; Petrovic, G.; Ling, T.; Yamada, Y. M. A.; Tang, W.; Frederick, M. O. *Angew. Chem. Int. Ed.* 2004, **43**, 4318; (e) Nicolaou, K. C.; Koftis, T. V.; Vyscocil, S.; Petrovic, G.; Tang, W.; Frederick, M. O.; Chen, D. Y.-K.; Li, Y.; Ling, T.; Yamada, Y. M. A. *J. Am. Chem. Soc.* 2006, **128**, 2859.

26. (a) Evans, D. A.; Kvaerno, L.; Mulder, J. A.; Raymer, B.; Dunn, T. B.; Beauchemin, A.; Olhava, E. J.; Juhl, M.; Kagechika, K. *Angew. Chem. Int. Ed.* 2007, **46**, 4693; (b) Evans, D. A.; Dunn, T. B.; Kvaerno, L.; Beauchemin, A.; Raymer, B.; Olhava, E. J.; Mulder, J. A.; Juhl, M.; Kagechika, K.; Favor, D. A. *Angew. Chem. Int. Ed.* 2007, **46**, 4698; (c) Evans, D. A.; Kvaerno, L.; Dunn, T. B.; Beauchemin, A.; Raymer, B.; Mulder, J. A.; Olhava, E. J.; Juhl, M.; Kagechika, K.; Favor, D. A. *J. Am. Chem. Soc.* 2008, **130**, 16295.

27. Evans, D. A.; Kaldor, S. W.; Jones, T. K.; Clardy, J.; Stout, T. J. *J. Am. Chem. Soc.* 1990, **112**, 7001.

28. Aicher, T. D.; Buszek, K. R.; Fang, F. G.; Forsyth, C. J.; Jung, S. H.; Kishi, Y.; Matelich, M. C.; Scola, P. M.; Spero, D. M.; Yoon, S. K. *J. Am. Chem. Soc.* 1992, **114**, 3162.

29. (a) Paterson, I.; Yeung, K.-S.; Ward, R. A.; Cumming, J. G.; Smith, J. D. *J. Am. Chem. Soc.* 1994, **116**, 9391; (b) Nicolaou, K. C.; Ajito, K.; Patron, A. P.; Khatuya, H.; Richter, P. K.; Bertinato, P. *J. Am. Chem. Soc.* 1996, **118**, 3059.

30. (a) Nicolaou, K. C.; Takayanagi, M.; Jain, N. F.; Natarajan, S.; Koumbis, A. E.; Bando, T.; Ramanjulu, J. M. *Angew. Chem. Int. Ed.* 1998, **37**, 2717; see also the two preceding papers: (b) Evans, D. A.; Dinsmore, C. J.; Watson, P. S.; Wood, M. R.; Richardson, T. I.; Trotter, B. W.; Katz, J. L. *Angew. Chem. Int. Ed.* 1998, **37**, 2704; (c) Boger, D. L.;

Miyazaki, S.; Kim, S. H.; Wu, J. H.; Loiseleur, O.; Castle, S. L. *J. Am. Chem. Soc.* 1999, **121**, 3226.

31. (a) Nicolaou, K. C.; Theodorakis, E. A.; Rutjes, F. P. J. T.; Tiebes, J.; Sato, M.; Untersteller, E.; Xiao, X.-Y. *J. Am. Chem. Soc.* 1995, **117**, 1171; (b) Nicolaou, K. C.; Rutjes, F. P. J. T.; Theodorakis, E. A.; Tiebes, J.; Sato, M.; Untersteller, E. *J. Am. Chem. Soc.* 1995, **117**, 1173; (c) Matsuo, G.; Kawamura, K.; Hori, N.; Matsukura, H.; Nakata, T. *J. Am. Chem. Soc.* 2004, **126**, 14374; (d) Kadota, I.; Takamura, H.; Nishii, H.; Yamamoto, Y. *J. Am. Chem. Soc.* 2005, **127**, 9246.

32. (a) Fuwa, H.; Sasaki, M.; Satake, M.; Tachibana, K. *Org. Lett.* 2002, **4**, 2981; (b) Kadota, I.; Takamura, H.; Sato, K.; Ohno, A.; Matsuda, K.; Yamamoto, Y. *J. Am. Chem. Soc.* 2003, **125**, 46; (c) Johnson, H. W. B.; Majumder, U.; Rainier, J. D. *J. Am. Chem. Soc.* 2005, **127**, 848; (d) for a recent review of marine natural products, see: Blunt, J. W.; Copp, B. R.; Hu, W.-P.; Munro, M. H. G.; Northcote, P. T.; Prinsep, M. R. *Nat. Prod. Rep.* 2008, **25**, 35.

33. (a) Nicolaou, K. C.; Safina, B. S.; Zak, M.; Estrada, A. A.; Lee, S. H. *Angew. Chem. Int. Ed.* 2004, **43**, 5087; (b) Nicolaou, K. C.; Zak, M.; Safina, B. S.; Lee, S. H.; Estrada, A. A. *Angew. Chem. Int. Ed.* 2004, **43**, 5092; (c) Nicolaou, K. C.; Safina, B. S.; Zak, M.; Lee, S. H.; Nevalainen, M.; Bella, M.; Estrada, A. A.; Funke, C.; Zécri, F. J.; Bulat, S. *J. Am. Chem. Soc.* 2005, **127**, 11159; (d) Nicolaou, K. C.; Zak, M.; Safina, B. S.; Estrada, A. A.; Lee, S. H.; Nevalainen, M. *J. Am. Chem. Soc.* 2005, **127**, 11176.

34. (a) Veitch, G. E.; Beckmann, E.; Burke, B. J.; Boyer, A.; Maslen, S. L.; Ley, S. V. *Angew. Chem. Int. Ed.* 2007, **46**, 7629; for a review of approaches to azadirachtin, see: (b) Jauch, J. *Angew. Chem. Int. Ed.* 2008, **47**, 34; for a retrospective, see: (c) Veitch, G. E.; Boyer, A.; Ley, S. V. *Angew. Chem. Int. Ed.* 2008, **47**, 9402.

35. For example, see: Hanessian, S. *Chem. Med. Chem.* 2006, **1**, 1300.

36. (a) Sharff, A.; Jhoti, H. *Curr. Opin. Chem. Biol.* 2003, **7**, 340; (b) Blundell, T. L.; Jhoti, H.; Abell, C. *Nat. Rev. Drug Discov.* 2002, **1**, 45; (c) Nienaber, V. L.; Richardson, P. L.; Klighofer, V.; Bouska, J. J.; Giranda, V. L.; Greer, J. *Nat. Biotechnol.* 2000, **18**, 1105.

37. For example, see: (a) Schulz, M. N.; Hubbard, R. E. *Curr. Opin. Pharmacol.* 2009, **9**, 615; (b) *Fragment-Based Drug Discovery*, Zartler, E. R.; Ed.; 2008, J. Wiley & Sons, Chichester, UK; (c) *Structure-based Discovery – An Overview*, Hubbard, R. F.; Ed.; 2006, RSC Publishing, London; (d) *Analogue-based Drug Discovery*, Fischer, J.; Ganellin, C. R.; Eds.; 2006, Wiley-VCH, Weinheim.

38. For example, see: (a) Pellecchia, M.; Sem, D. S.; Wüthrich, K. *Nat. Rev. Drug Discov.* 2002, **1**, 211; (b) Shuker, S. B.; Hajduk, P. J.; Meadows, R. P.; Fesik, S. W. *Science* 1996, **274**, 1531; (c) Fejzo, J.; Lepre, C. A.; Peng, J. W.; Bemis, G. W.; Ajay; Murcko, M. A.; Moore, J. M. *Chem. Biol.* 1999, **6**, 755.

39. For example, see: (a) *Enzymes and their Inhibition: Drug Development* Smith, H. J.; Simons, C.; Eds.; 2004, CRC Press, Boca Raton, FL; (b) Tyndall, J. D. A.; Nall, T.; Fairlie, D. P. *Chem. Rev.* 2005, **105**, 973; (c) Babine, R. E.; Bender, S. L. *Chem. Rev.* 1997, **97**, 1359.

40. For example, see: (a) Webb, R. L.; Schiering, N.; Sedrani, R.; Maibaum, J. *J. Med. Chem.* 2010, **53**, 7490; (b) Maibaum, J.; Feldman, D. L. *Ann. Rep. Med. Chem.* 2009, **44**, 105; (c) Jensen, C.; Herold, P.; Brunner, H. R. *Nat. Rev. Drug Discov.* 2008, **7**, 399; see also reference [46].

41. (a) Hanessian, S.; Bayrakdarian, M.; Luo, X. *J. Am. Chem. Soc.* 2002, **124**, 4716; (b) De Goey, D. A.; Chen, H.-J.; Flosi, W. J.; Grampovnik, D. J.; Yeung, C. M.; Klein, L. L.; Kempf, D. J. *J. Org. Chem.* 2002, **67**, 5445; (c) Barnes, D. M.; McLaughlin, M. A.; Oie, T.; Rasmussen, M. W.; Stewart, K. D.; Wittenberger, S. J. *Org. Lett.* 2002, **4**, 1427.

42. (a) Lamarre, D.; Anderson, P. C.; Bailey, M.; Beaulieu, P.; Bolger, G.; Bonneau,

P.; Bos, M.; Cameron, D. R.; Cartier, M.; Cordingley, M. G.; Faucher, A.-M.; Goudreau, N.; Kawai, S. H.; Kukolj, G.; Lagacé, L.; LaPlante, S. R.; Narjes, H.; Poupart, M.-A.; Rancourt, J.; Sentjens, R. E.; St. George, R.; Simoneau, B.; Steinman, G.; Thibeault, D.; Tsantrizos, Y. S.; Weldon, S. M.; Young, C.-L.; Llinàs-Brunet, M. *Nature* 2003, **426**, 186; (b) Llinàs-Brunet, M.; Bailey, M.; Bolger, G.; Brochu, C.; Faucher, A.-M.; Ferland, J. M.; Garneau, M.; Ghiro, E.; Gorys, V.; Grand-Maître, C.; Halmos, T.; Lapeyre-Paquette, N.; Liard, F.; Poirer, M.; Rhéaume, M.; Tsantrizos, Y. S.; Lamarre, D. *J. Med. Chem.* 2004, **47**, 1605; for a large scale synthesis of BILN-2061 see: (c) Nicola, T.; Brenner, M.; Donsbach, K.; Kreye, P. *Org. Process Res. Dev.* 2005, **9**, 513; for optimization, see: (d) Shu, C.; Zeng, X.; Hao, M.-H.; Wei, X.; Yee, N. K.; Busacca, C. A.; Han, Z.; Farina, V.; Senanayake, C. H. *Org. Lett.* 2008, **10**, 1303.

43. (a) Perni, R. B.; Chandorkar, G.; Chaturvedi, P.; Courtney, L. F.; Decker, C. J.; Gates, C. A.; Harbeson, S. L.; Kwong, A. D.; Lin, C.; Luong, Y.-P.; Markland, W.; Rao, B. G.; Thomson, J. A.; Tung, R. D. *Hepatology* 2003, **38**, (Suppl. I.), 624A; (b) Yip, Y.; Victor, F.; Lamar, J.; Johnson, R.; Wang, Q. M.; Glass, J. I.; Yumibe, N.; Wakulchik, M.; Munroe, J.; Chen, S.-H. *Bioorg. Med. Chem. Lett.* 2004, **14**, 5007.

44. For example, see: (a) De Clercq, E. *Nat. Rev. Drug Discov.* 2007, **6**, 1001; (b) *HIV Chemotherapy: A Critical Review*, Butera, S. T.; Ed.; 2005, Caister Academic Press, Norfold, UK.

45. Ghosh, A. K.; Chapsal, B. D.; Weber, I. T; Mitsuya H. *Acc. Chem. Res.* 2008, **41**, 78.

46. (a) Hanessian, S.; Claridge, S.; Johnstone, S. *J. Org. Chem.* 2002, **67**, 4261; (b) Hanessian, S.; Raghavan, S. *Bioorg. Med. Chem. Lett.* 1994, **4**, 1697; for a total synthesis of aliskiren see: (c) Hanessian, S.; Guesné, S.; Chenard, E. *Org. Lett.* 2010, **12**, 1816; for reviews, see ref. [40].

47. (a) Larrow, J. F.; Jacobsen, E. N. *J. Am. Chem. Soc.* 1994, **116**, 12129; (b) Maligres, P. E.; Upadhyay, V.;

Rossen, K.; Cianciosi, S. J.; Purick, R. M.; Eng, K. K.; Reamer, R. A.; Askin, D.; Volante, R. P.; Reider, P. J. *Tetrahedron Lett.* 1995, **36**, 2195.

48. (a) Wender, P. A.; Baryza, J. L; Bennett, C. E.; Bi, F. C.; Brenner, S. E.; Clarke, M. O.; Horan, J. C.; Kan, C.; Lacôte, E.; Lippa, B.; Nell, P. G.; Turner, T. M. *J. Am. Chem. Soc.* 2002, **124**, 13648; see also: (b) Wender, P. A.; Kee, J.-M.; Warrington, J. M. *Science* 2008, **320**, 649; (c) Wender, P. A.; D'Angelo, N.; Elitzin, V. I.; Ernst, M.; Jackson-Ugueto, E. E.; Kowalski, J. A.; McKendry, S.; Rehfeuter, M.; Sun, R.; Voigtlaender, D. *Org. Lett.* 2007, **9**, 1829; see also: (d) Trost, B. M.; Yang, H.; Thiel, O. R.; Frontier, A. J.; Brindle, C. S. *J. Am. Chem. Soc.* 2007, **129**, 2206; (e) Keck, G. E.; Poudel, Y. B.; Welch, D. S.; Kraft, M. B.; Truong, A. P.; Stephens, J. C.; Kedei, N.; Lewin, N. E.; Blumberg, P. M. *Org. Lett.* 2009, **11**, 593; for a review see: (f) Hale, K.; Hummersone, M. G.; Manaviazar, S.; Frigerio, M. *Nat. Prod. Rep.* 2002, **19**, 413.

49. Klar, U.; Buchmann, B.; Schwede, W.; Skuballa, W.; Hoffmann, J.; Lichtner, R. B. *Angew. Chem. Int. Ed.* 2006, **45**, 7942.

50. For recent reviews, see: (a) Borzilleri, R. M.; Vite, G. D. *Ann. Rep. Med. Chem.* 2009, **44**, 301; (b) Denduluri, N.; Swain, S. M. *Expert Opin. Investig. Drugs* 2008, **17**, 423.

51. (a) Hirschmann, R.; Nicolaou, K. C.; Pietranico, S.; Leahy, E. M.; Salvino, J.; Arison, B.; Cichy, M. A.; Spoors, P. G.; Shakespeare, W. C.; Sprengeler, P. A.; Hamley, P.; Smith, A. B. III, Reisine, T.; Raynor, K.; Maechler, L.; Donaldson, C.; Vale, W.; Freidinger, R. M.; Cascieri, M. R.; Strader, C. D. *J. Am. Chem. Soc.* 1993, **115**, 12550. (b) Hirschmann, R.; Nicolaou, K. C.; Pietranico, S.; Salvino, J.; Leahy, E. M.; Sprengeler, P. A.; Furst, G.; Smith, A. B. III, Strader, C. D.; Cascieri, M. A.; Candelore, M. R.; Donaldson, C.; Vale, W.; Maechler, L. *J. Am. Chem. Soc.* 1992, **114**, 9217.

52. For a synthesis see: Cossy, J.; Belotti, D. *Bioorg. Med. Chem. Lett.* 2001, **11**, 1989.

53. For an academic perspective of structure-based design, see: Zürcher,

M.; Diederich, F. *J. Org. Chem.* 2008, **73**, 4345.

54. For a review, see: (a) Ersmark, K.; Del Valle, J. R.; Hanessian, S. *Angew. Chem. Int. Ed.* 2008, **47**, 1202; for a recent total synthesis of microcin SF608 see: (b) Diethelm, S.; Schindler, C. S.; Carreira, E. M. *Org. Lett.* 2010, **12**, 3950.

55. Hanessian, S.; Margarita, R.; Hall, A.; Johnstone, S.; Tremblay, M.; Parlanti, L. *J. Am. Chem. Soc.* 2002, **124**, 13342.

56. Hanessian, S.; Tremblay, M.; Petersen, J. F. W. *J. Am. Chem. Soc.* 2004, **126**, 6064.

57. Hanessian, S.; Del Valle, J. R.; Xue, Y.; Blomberg, N. *J. Am. Chem. Soc.* 2006, **128**, 10491.

58. (a) Hanessian, S.; Ersmark, K.; Wang, X.; Del Valle, J. R.; Blomberg, N.; Xue, Y.; Fjellström, O. *Bioorg. Med. Chem. Lett.* 2007, **17**, 3480; (b) Hanessian, S.; Guillemette, S.; Ersmark, K. *Chimia* 2007, **61**, 361; see also: (c) Radau, G.; Gebel, J.; Rauh, D. *Archiv. Der Pharmazie* 2003, **336**, 372.

59. (a) Hanessian, S.; Therrien, E.; Van Otterlo, W. A. L.; Bayrakdarian, M.; Nilsson, I.; Fjellström, O.; Xue, Y. *Bioorg. Med. Chem. Lett.* 2006, **16**, 1032; (b) Hanessian, S.; Simard, D.; Bayrakdarian, M.; Therrien, E.; Nilsson, I.; Fjellström, O. *Bioorg. Med. Chem. Lett.* 2008, **18**, 1972; (c) Hanessian, S.; Therrien, E.; Zhang, J.; van Otterlo, W.; Xue, Y.; Gustafsson. D.; Nilsson, I.; Fjellström, O. *Bioorg. Med. Chem. Lett.* 2009, **19**, 5429.

60. For example, see: Hughes, B. *Nat. Rev. Drug. Discov.* 2010, **9**, 903.

61. For example, see: (a) Nielsen, T. E.; Schreiber, S. I.. *Angew. Chem. Int. Ed.* 2008, **47**, 48; (b) Burke, M. D.; Schreiber, S. L. *Angew. Chem. Int. Ed.* 2004, **43**, 46; (c) Schreiber, S. L. *Science* 2000, **287**, 1964; see also: (d) Walsh, D. P.; Chang, Y.-T. *Chem. Rev.* 2006, **106**, 2476.

62. For recent reviews, see: (a) Shindoh, N.; Takemoto, Y.; Takatsu, K. *Chem. Eur. J.* 2009, **15**, 12168; (b) Chapman, C. J.; Frost, C. G. *Synthesis* 2007, 1; (c) Nicolaou, K. C.; Edmonds, D. J.; Bulger, P. G. *Angew. Chem. Int. Ed.*

2006, **45**, 7134; (d) Pellissier, H. *Tetrahedron* 2006, **62**, 1619; (e) Pellissier, H. *Tetrahedron* 2006, **62**, 2143; (f) *Domino Reactions in Organic Synthesis*, Tietze, L. F.; Brasche, G.; Gericke, K. M., 2006, Wiley-VCH, Weinheim; (g) Guo, H.-C.; Ma, J.-A. *Angew. Chem. Int. Ed.* 2006, **45**, 354; (h) Wasilke, J.-C.; Obrey, S. J.; Baker, R. T; Bazan, G. C. *Chem. Rev.* 2005, **105**, 1001; (i) Ramón, D. J. *Angew. Chem. Int. Ed.* 2005, **44**, 1602; (j) *Multicomponent Reactions*, Zhu, J.; Benaymé, H.; Eds.; 2005, Wiley-VCH, Weinheim; (k) Tietze, L.-F. *Chem. Rev.* 1996, **96**, 115.

63. For example, see: (a) Morton, D.; Leach, S.; Cordier, C.; Warriner, S.; Nelson, A. *Angew. Chem. Int. Ed.* 2009, **48**, 104; see also: (b) *Combinatorial Synthesis of Natural Product-Based Libraries*; Boldi, A. M.; Ed.; 2006, CRC, Boca Raton, FL; (c) *Combinatorial Chemistry*, Bannwarth, W.; Hinzen, B.; Eds.; 2006, Wiley-VCH, Weinheim; (d) Ganesan, A. *Curr. Opin. Biotechnol.* 2004, **15**, 584; (e) Abreu, P. M.; Branco, P. S. *J. Braz. Chem. Soc.* 2003, **14**, 675; (f) Nielsen, J. *Curr. Opin. Chem. Biol.* 2002, **6**, 297; (g) *Combinatorial Chemistry*, Terrett, N. K., 1998, Oxford University Press, UK; (h) *Combinatorial Chemistry: Synthesis and Application*, Wilson, S. R.; Czarnik, A. W.; Eds.; 1997, Wiley-Interscience, NY.

64. (a) *High Throughput Screening in Drug Discovery*, Jörg, H.; Ed.; 2006, Wiley-VCH, Weinheim; (b) *High Throughput Screening*, Janzen, W. P.; Ed.; 2002, Humana Press, Totowa, NJ.

65. *Combinatorial Chemistry: Synthesis, Analysis Screening*, Jung, G.; Ed.; 1999, Wiley-VCH, Weinheim.

66. For example, see: (a) Kirkpatrick, P.; Ellis, C. *Nature* 2004, **432**, 823; (b) Lipinski, C.; Hopkins, A. *Nature* 2004, **432**, 85; (c) Dobson, C. M. *Nature* 2004, **432**, 824.

67. For example, see: (a) Tan, D. S.; Foley, M. A.; Stockwell, B. R.; Shair, M. D.; Schreiber, S. L. *J. Am. Chem. Soc.* 1999, **121**, 9073; (b) B. Boojamra, C. G.; Burow, K. M.; Thompson, L. A.; Ellman, J. A. *J. Org. Chem.* 1997, **62**, 1240; (c) Enders, D.; Hüttl, M. R. M.;

Runsink, J.; Raabe, G.; Wendt, B. *Angew. Chem. Int. Ed.* 2007, **46**, 467; (d) Enders, D.; Narine, A. A. *J. Org. Chem.* 2008, **73**, 7857;   (e) Nicolaou, K. C.; Pfefferkorn, J. A.; Mitchell, H. J.; Roecker, A. J.; Barluenga, S.; Cao, G.-Q.; Affleck, R. L.; Lillig, J. E. *J. Am. Chem. Soc.* 2000, **122**, 9954; (f) Nicoalou, K. C.; Zhong, Y.-L.; Baran, P. S. *Angew. Chem. Int. Ed.* 2000, **39**, 622;   (g) Nicoalou, K. C.; Zhong, Y.-L.; Baran, P. S. *Angew. Chem. Int. Ed.* 2000, **39**, 625.

68. For example, see: (a) *Hybrid Materials*, Kickelbick, G.; Ed.; 2007, Wiley-VCH, Weinheim;   (b) *Functional Organic Materials*, Müller, T. J. J. Bunz, H. F.; Eds.; 2007, Wiley-VCH, Weinheim.

69. For example, see: *Nanoparticle Assemblies and Superstructures*, Kotov, N. A.; Ed.; 2005, CRC Press, Boca Raton, FL.

70. For example, see: Whitesides, G. M.; Deutch., J. *Nature* 2011, **469**, 21; *Chem. Eng. News* 2010, 13.

4
The "*How*" of Synthesis

4.1
The Visual Dialogue

Imagine that you have just joined a first-rate organic synthesis research group for your Ph.D. or postdoctoral studies, and are anxiously waiting to know the structure of the natural product (or target molecule) you will be expected to synthesize. The first four structures shown in Figure 4.1 are revealed and the final onerous choice is left to you, with the comment: "I hope you are excited. Try and come up with some ideas over the weekend." There is no going back. The only way out is through.

   A different scenario presents itself the first day on the job for a process chemist with a seemingly less complex synthetic target, but with major conditions imposed for an industrially viable process. Once the adrenaline rush has subsided, a focused look at a given synthetic target, be it of academic or industrial interest, will elicit the usual questions:

1) How can I synthesize my target molecule efficiently?
2) How do I know that I have devised a viable route?
3) How and where do I begin?

   Conceptual elegance, innovative methodology, expediency, and economic considerations should also be inherent components of the best overall strategy. Naturally, different carbon frameworks and connectivities will conjure up a number of strategic disconnections in the mind's eye, strongly influenced by a host of direct and indirect factors. It will be immediately apparent to the viewer that four of the five natural products shown in Figure 4.1 possess quaternary stereogenic centers presenting different degrees of complexity [1]. After a brief visual tour of the remainder of the carbon framework of any one of these molecules, each with their own appended functionalities and heteroatom substituents, the eye will eventually settle on the "problem centers." In the process, many possible approaches to construct rings, chains, and appendages with required functionalities will flash before our eyes, and some will be subliminally registered in the mind as in Leonardo da Vinci's "luminous bodies" (see Chapter 1, section 1.5). Eventually, with pen and paper in hand, the rudiments of a strategy are sketched before the journey begins.

Design and Strategy in Organic Synthesis: From the Chiron Approach to Catalysis, First Edition.
Stephen Hanessian, Simon Giroux, and Bradley L. Merner.
© 2013 Wiley-VCH Verlag GmbH & Co. KGaA. Published 2013 by Wiley-VCH Verlag GmbH & Co. KGaA.

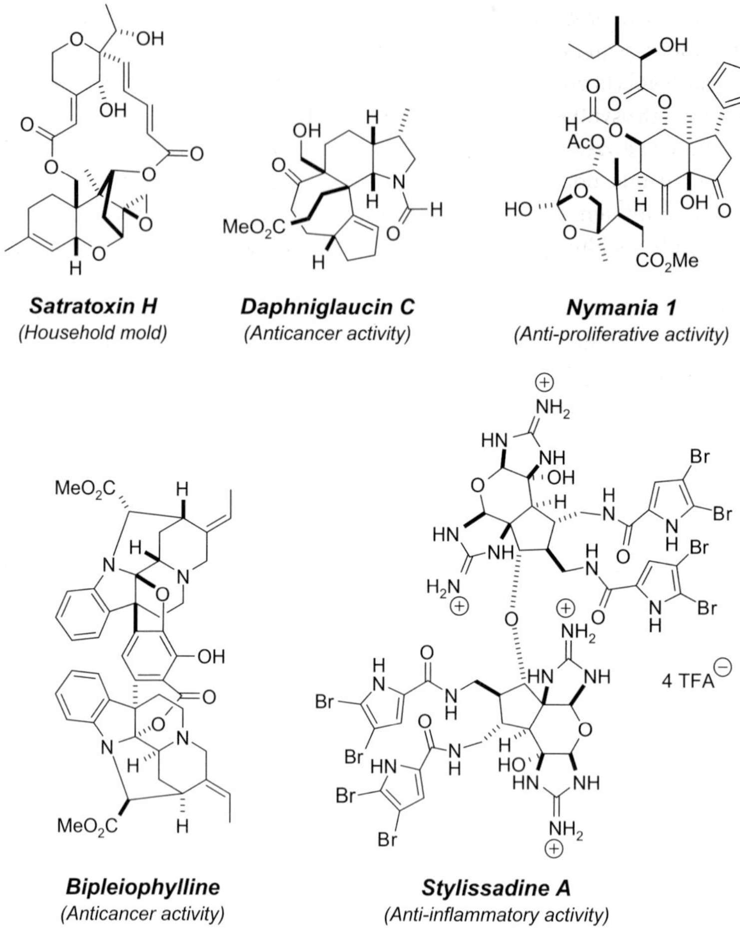

Satratoxin H
(Household mold)

Daphniglaucin C
(Anticancer activity)

Nymania 1
(Anti-proliferative activity)

Bipleiophylline
(Anticancer activity)

Stylissadine A
(Anti-inflammatory activity)

Figure 4.1 Natural products waiting to be synthesized.

4.2
The Psychobiology of Synthesis Planning

Our initial contact with a target molecule is visual. What follows is a fascinating series of mental processing events that engage the right and left hemispheres of the brain [2]. Extensive studies in the field of cognition and perception have validated the bilateral asymmetry of the human brain. The right hemisphere of the brain deals with non-verbal, synthetic, analogic, spatial, intuitive, and non-temporal processing. It accentuates awareness, relationships in space, and the way things come together to form wholes. These effects are manifested in part during the retrosynthetic analysis phase, where we attempt to simplify the complexity of target molecules in order to attain smaller substructures. The left hemisphere of the brain, on the other hand, is concerned with verbal, analytic, temporal, rational, numeric, and linear processing. Small bits of information represent the whole. Thoughts come in sequences, and logical conclusions emerge like a mathematical theorem. These effects are indoctrinated in our chemical thinking in the forward sense once the retrosynthetic disconnections have been found. Thus, whereas the left hemisphere analyzes over time, the right hemisphere synthesizes over space. These instantaneous non-voluntary chemical "seeing and thinking" processes have a profound influence in synthesis planning. The path from "thought at first sight" to a synthesis blueprint that is ready to be put to practice is illustrated in Figure 4.2.

Visual contact with the architecture of a target molecule will generate eye-to-mind communication resulting in an analysis of the carbon framework, discovering bond connectivities, decoding functional and stereochemical features as well as strategic bond-breaking possibilities *en route* to simplifying its complexity. What follows is a combination of *visual* and *relational* thinking events, which may lead to a smaller subunit that bears a skeletal and functional resemblance to a portion of the target molecule's framework. Alternatively, the visual contact may instantly evoke a *reflexive action*, relating a certain bond and atom connectivity pattern to a key reaction type. Eventually, following one or the other route, or a combination of both, will lead to a simplification of complexity to individual chemical entities and the generation of a synthesis blueprint. The plan is completed with the inclusion of reagents and conditions that convert simple starting materials to progressively

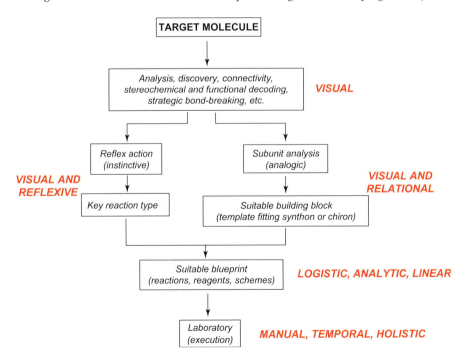

Figure 4.2 Visual "chemical thinking" in synthesis planning.

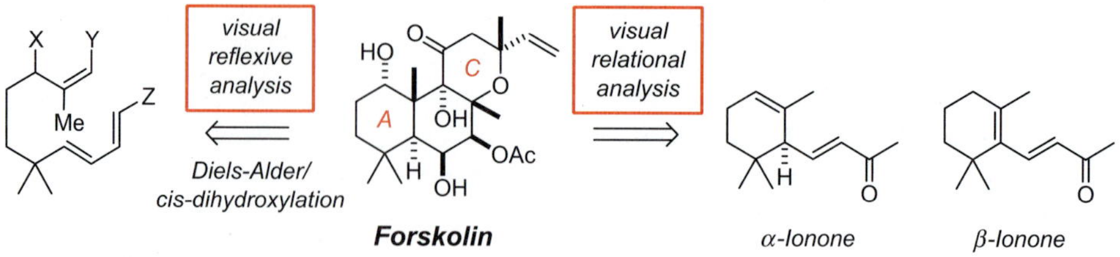

Figure 4.3 Template-fitting (right), and reaction-type based (left) synthesis planning.

more elaborate structures (intermediates or sub-targets), culminating with the drawing of a synthetic scheme. Following the proposed synthesis route in the laboratory and obtaining the target compound with high purity and acceptable yield completes the process. This heuristic analysis can be rendered into practice by considering synthetic approaches to forskolin as an example (Figure 4.3).

Visual analysis will produce a number of synthetic approaches. The *relational* aspect of the analysis may conjure the image of α- or β-ionone as suitable starting materials in the mind's eye, especially in view of the presence of the *gem*-dimethyl group. These starting materials offer skeletal and functional congruence with ring *A*, and a portion of ring *B* in the form of a suitably located enone motif on a carbocyclic template (template-fitting, Figure 4.3). The correct stereochemical overlap at the ring junction with α-ionone, and a methyl group that would eventually be located at the opposite ring junction are added assets. On the other hand, the unbiased eye relying on a logical, knowledge-based *reflexive* analysis may first make contact with the *cis*-diol in ring *B* of the target molecule, signalling an intramolecular Diels-Alder reaction to initially give a dehydrodecalin intermediate that can be dihydroxylated. In practice, both approaches and combinations thereof have been used in the numerous recorded syntheses of forskolin [3]. The first total synthesis of (±)-forskolin in a stereocontrolled manner by Corey and coworkers [4a] utilized α-ionone as a starting material. Conversion to the diene ester intermediate **A** set the stage for an intramolecular Diels-Alder reaction *en route* to the known lactone **B** (Ziegler intermediate [5], Figure 4.4 A). In the process, the existing chirality in α-ionone was sacrificed. In a second synthesis, Corey and coworkers obtained **A** in enantioenriched form by stereoselective reduction of the ketone precursor [4b]. A total synthesis by Ikegami and coworkers [6], published simultaneously with the Corey paper, relied on a different approach that capitalized directly on an intramolecular Diels-Alder reaction of intermediate **C** to construct a dehydrodecalin ring system, which eventually converged with the Ziegler intermediate (**B**) after functional group manipulation (Figure 4.4 B). In other approaches, α-ionone, hydroxy-β-ionone, and related naturally occurring compounds were used as starting materials to reach advanced intermediates (Figure 4.4 C) [3]. Thus, once the eye "sees" the synthesis through *relational* and *reflexive* imagery, the logical and knowledge-based thought process intervenes to evaluate feasibility.

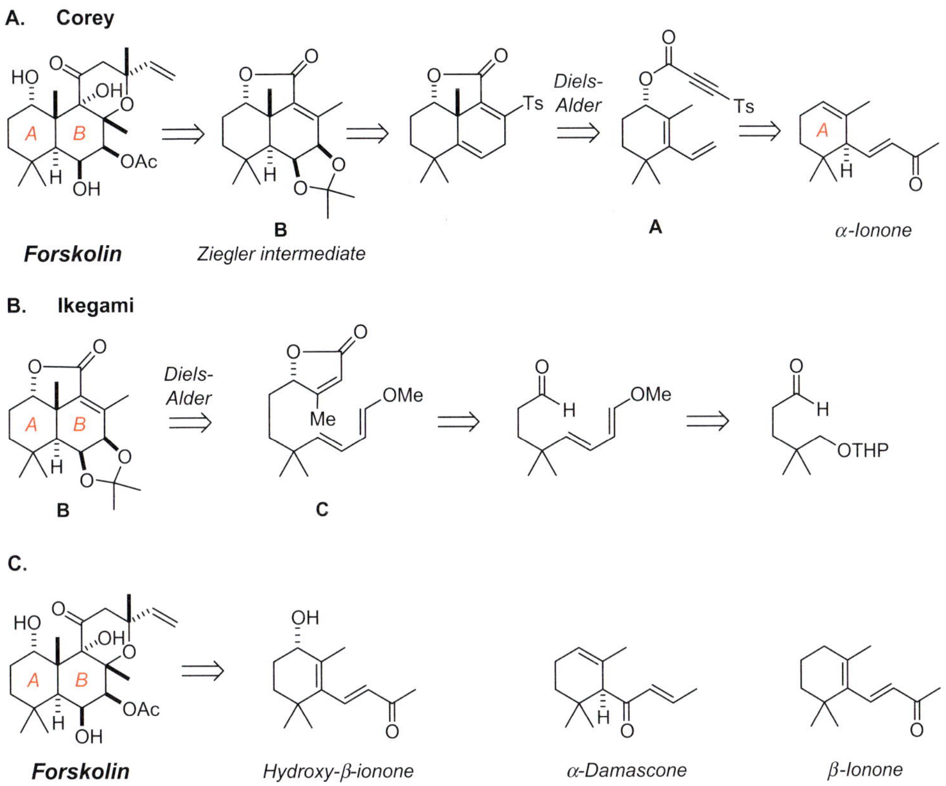

A.   Corey

Forskolin   *Ziegler intermediate*   **B**   **A**   *α-Ionone*

B.   Ikegami

B   **C**

C.

Forskolin   *Hydroxy-β-ionone*   *α-Damascone*   *β-Ionone*

Figure 4.4   The Corey (A), Ikegami (B), and β-ionone and related starting materials (C) approaches to the total synthesis of forskolin.

4.2.1
"Psychosynthesis"

As already demonstrated, and to be discussed further throughout this book, there are numerous approaches and strategies in the planning of a total synthesis. The realization that a particular strategy will lead to an advanced intermediate containing critical functionality and stereochemistry (the "a-ha factor"), has redeeming features in the conscious mind of the synthesis planner. When such a plan also embodies interesting bond-forming sequences demonstrating the application of methods developed or discovered in the home laboratory, the attractiveness level of the synthesis rises compared to a more traditional approach. In synthesis, one cannot separate the persona from the science and how certain ideas in the plan were conceived. Some synthesis strategies carry a subtle, yet unmistakable, signature of their respective planners. Others however, may bear the imprint of known literature methods where the innovative bond-forming step in a key reaction was first reported by others. Henri Poincaré had expressed the following words in his first memoir on the Fuchsian functions:

"The unconscious, or, as we say, the subliminal self plays an important role in mathematical creations. Usually, the subliminal self is considered as purely automatic ... but the subliminal self is in no way inferior to the conscious self; it is not purely automatic; it is capable of discernment." He continues by telling us that: *"only the interesting combinations of numbers break out of the subliminal self into the field of consciousness."* He then questions why some pass the threshold of the subliminal self while others remain below it.

Why do we think in such creatively different ways when we plan a synthesis? How are we influenced in our thought process, such that some of us see a molecule differently than others would, even though as chemists, we have acquired the same fundamental knowledge of the principles of bond-making and bond-breaking through our academic and professional lives?

Before we settle on a given strategy toward the synthesis of a target molecule, our conscious and subliminal self may have already rejected other possibilities, at least within the limits of visual perception. Fears, biases, and deliberate rejections put aside, we eventually test the plan on paper, especially with regard to how certain critical atoms would be introduced in a given intermediate. Once convinced, the potential success of a plan, the search for *individuation* is attained, and the exercise of "psychosynthesis" is completed. As a result, we unconsciously end up integrating our own egoistic and procreative instincts in the process of synthesis planning.

4.3
The Agony and Ecstasy of Synthesis

The master of organic synthesis of his time, R. B. Woodward, expressed his opinion regarding total synthesis as follows:

"Synthetic objectives are seldom if ever taken by chance. Synthesis must always be carried out by plan, and the synthetic frontier can be defined only in terms of the degree to which realistic planning is possible [7]."

As realistic as a plan can be on paper, it is rare that it can be fully executed without some change or modification. In the event, many unexpected and oftentimes surprising results can be the basis of important discoveries. These may even have a lasting impact on the way organic synthesis is done in the future. More often however, the progress of a total synthesis effort will be slowed down due to a number of unforeseen events although, in retrospect, some of these could have been avoided by more careful synthesis planning. Compatibility of protective groups to certain reaction conditions, and their eventual removal (deprotection) are problems that can be foreseen in an original plan by making judicious choices (see Chapter 18, section 18.5.1). Unreactivity and inefficient conversions can be the result of coordination with polar functional groups present in the substrate during organometallic reactions, or due to aggregation phenomena, or different kinetic acidities of hydrogen atoms. Although steric factors, solvent, and proximity effects are usually predictable, they are potential problems to consider. Unexpected rearrangements of the carbon skeleton are sometimes rationalized after the event, resulting in the loss of a precious advanced intermediate. However, such detours teach useful lessons to those involved, and often lay the foundation for the discovery of new science.

No matter how elegant and fail-safe the projected approach may be, there is invariably a price to pay when plans are turned into practice in the laboratory. Excluding the intangibles, the most frequent reasons for the length of synthetic sequences are the need to use protective groups, steps involving redox chemistry, functional group interconversions, and self-immolative transformations (see Chapter 18, sections 18.5.2 and 18.5.3). Performing one or more of these operations multiple times, especially in a synthesis of a complex molecule, will detract from the conceptual elegance of key steps that the chemist wants to showcase.

The climb to the summit is arduous and fraught with obstacles that necessitate frequent detours. Endurance and flexibility are indispensable criteria in order to complete a total synthesis. We may be attracted to, or even be fixated on, a particular synthetic route that may ultimately present an impasse (the Sinatra, *I did it my way*, approach, see Chapter 3, section 3.3). Eventually, it would be wise to detach ourselves from the fixation on the initial "a-ha factor," and move in a different direction. On the other hand, it is during times of duress and recalcitrant chemistry that synthetic organic chemists are at their best, for rare are the documented cases of abandoned syntheses.

There is no greater feeling than when the synthesis of a target molecule is successfully completed. Memories of popping champagne corks are forever etched in the minds of those who have experienced such achievements. Seemingly endless months, if not years, of long hours, and at times, frustrating experimental work in the laboratory, are soon forgotten at the sight of those few milligrams of pure synthetic material produced from one's own hands. In the process, much can be learned from successful as well as failed reactions. Turning failure into success, especially through logical and mechanistic reasoning, is a fulfilling and lasting experience. Here, the power of observation and creativity of the prepared mind are key elements. In many cases, new reactions, which often bear the name(s) of one or more of the investigators, have resulted from experiences in total synthesis [8].

Efficiency, practicality, and cost are important considerations in the industrial production of a chemical entity. Process chemists in the pharmaceutical, agrochemical, and fine chemicals sectors have excelled at changing a "discovery research" route to a drug candidate for example, to a viable, environmentally benign, and cost-effective alternative [9].

4.4
Rembrandt Meets Woodward

The 1965 Nobel Prize in chemistry was awarded to R. B. Woodward "for his outstanding achievements in the art of organic synthesis [10]." Like Rembrandt, who arguably was the father of classical painting in the Renaissance era, Woodward may be considered as the grandmaster of synthesis of his time, a legacy that is still honored to the present day. Just as Rembrandt's paintings hang in the museums of the world as masterpieces to be admired and to learn from, the annals of organic syntheses of the 1950's contain landmark contributions from the Woodward group that represent "gold standard" examples of achievement to emulate for present day and future synthetic chemists.

It can be argued that the ultimate success of a total synthesis is as good as the methods available at the time. In that regard, consider the synthetic methods and instrumental techniques that were available to chemists in the middle of the last century compared to fifty years later. Yet, the ingenuity of the synthetic chemist of the 1950's transcended all obstacles and adversities of the period. Armed with some twenty or so frequently used named reactions, common reagents, and instrumentation consisting mainly of infrared and UV spectrometers, pioneers such as Woodward and his contemporaries set and continuously raised the standards of excellence for generations to come (see also Chapters 16 and 18).

The key steps in the various syntheses of cortisone and strychnine discussed in the following two sections of this chapter were chosen to illustrate the evolution of the thought process and the impact of asymmetric technology in total synthesis since Woodward's original, groundbreaking work.

4.4.1
Cortisone

The publication of the first synthesis of cortisone by Woodward in 1951 drew immediate attention to the feasibility of synthesizing steroids of natural origin from petroleum products (Figure 4.5) [11]. The starting materials were 4-methoxytoluquinone **A** and butadiene **B**, which were engaged in a Diels-Alder reaction to give the *p*-quinone intermediate (±)-**C**. Further elaboration gave a didehydrodecalone intermediate (**D**). At first glance this may appear to be the precursor of the *AB* ring system in cortisone, but the synthesis plan actually considered it as rings *C* and *D*. C-Formylation and Michael addition reactions led to **E**, which upon treatment with NaOH furnished tricyclic trienone **F** via an intramolecular aldol/retro-aldol sequence. Further elaboration, that involved chemoselective dihydroxylation, stereoselective hydrogenation, and olefin isomerization, gave the keto acid **G**. Formation of the enol lactone **H**, then treatment with methylmagnesium bromide, and aldol cyclization gave **I**. Conversion to the dialdehyde **J**, cyclization under basic conditions, followed by oxidation, gave the new tetracyclic carboxylic acid **K**. A chemical resolution secured the optically pure acid, which was further elaborated through nitrile **L** to cortisone. It is remarkable how the tetracyclic ring

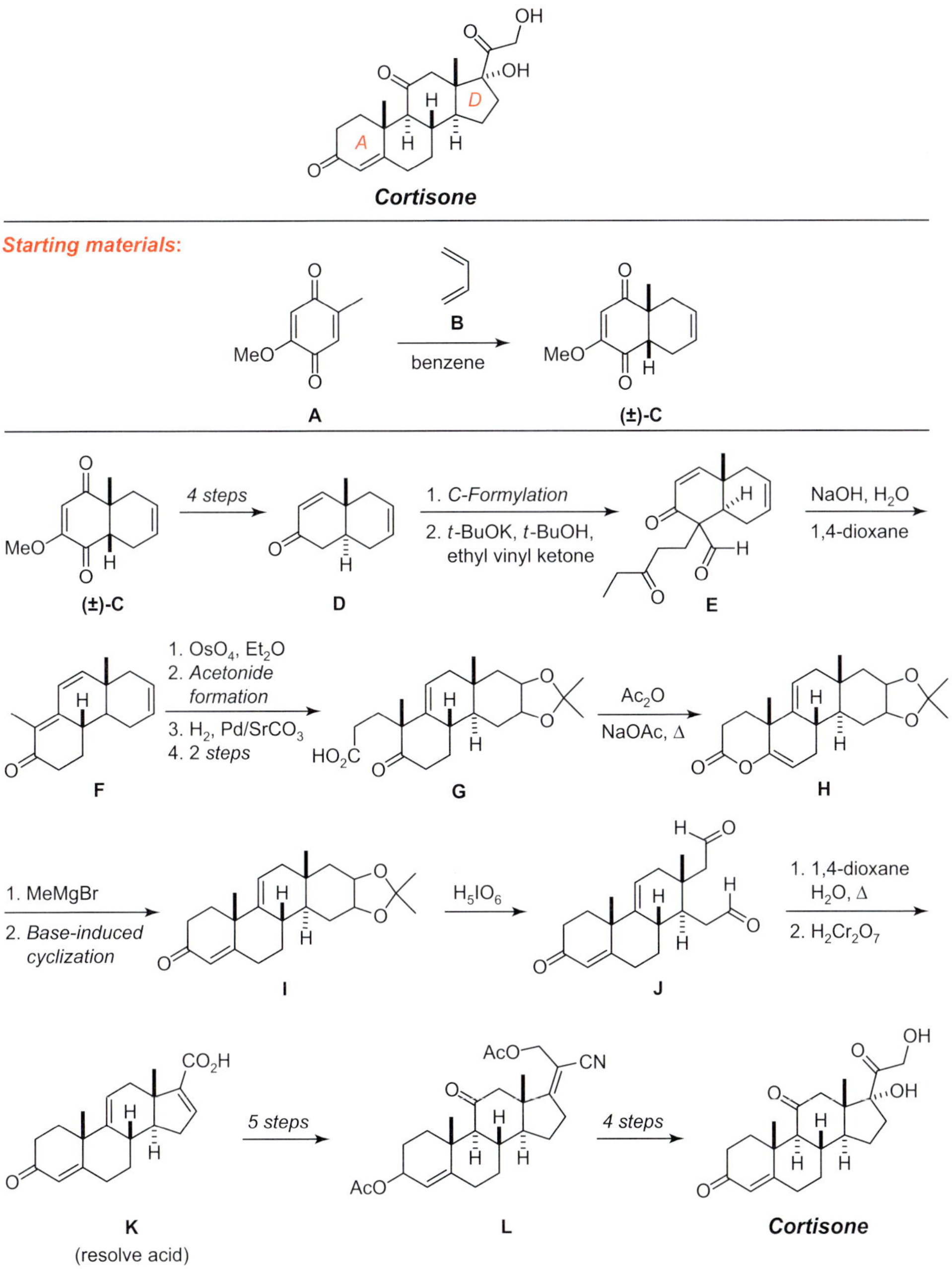

Cortisone

Starting materials:

Key reactions: *Intramolecular aldol/retro-aldol (**E** to **F**); intramolecular aldol/oxidation (**J** to **K**).*

Figure 4.5 Salient features of the 1951 total synthesis of cortisone by Woodward.

system in cortisone, with correctly placed functionality, was generated using simple, acid-base mediated condensations and some oxidative transformations with minimal use of protective groups. The total number of steps up to the resolved acid was under 25! The intermediates of this synthesis could also be used for the synthesis of other important steroids such as progesterone, testosterone, and cholesterol.

Four decades later, the first enantioselective total synthesis of cortisone was reported by Fukumoto and coworkers (Figure 4.6) [12]. The starting material was 3-(cyanoethyl)-4-bromoanisole (**A**), which was cyclized to a benzocyclobutane nitrile, then extended to the 3-carbon aldehyde **C**. Reaction with Eliel's chiral, non-racemic oxathiane anion **B** [13] (obtained from (+)-pulegone) afforded a mixture of diastereomeric ketones **D**, which was further transformed to a 2-propenyl tertiary alcohol **E**. An intramolecular thermal [4+2] cycloaddition reaction proceeded via an *o*-quinodimethane intermediate, which was generated upon an electrocyclic ring-opening reaction of the benzocyclobutane intermediate **E**. Reaction of this diene with the proximal dieneophile produced a single tricyclic product **F**, which was transformed to the enone **G**, corresponding to rings *B, C,* and *D* of cortisone. Oxidation at C-11 (steroid numbering) was achieved with selenium dioxide to give an enedione. The installation of an angular methyl group at the junction of rings *A* and *B* was achieved by dipolar cycloaddition of the 1,4-diketoenone prepared from **G** with diazomethane, and thermolysis of the resulting Δ^2-pyrazoline to give **H**. Ring *A* was then formed by introduction of the angular ketone appendage to give **I**, which was subjected to an intramolecular aldol condensation as in the Woodward synthesis to give **J**. A series of functional group manipulations included protection of the C-3 ketone as the ketal, reduction of the C-11 carbonyl, and introduction of the C-17 hydroxy ketone side-chain, followed by re-oxidation under Swern conditions to give **K**. Acid mediated deprotection of the MOM and ketal groups afforded cortisone.

Clearly the stereocontrolled [4+2] cycloaddition to generate the tricyclic precursor **F** with the correct absolute configuration of three stereogenic centers in a single synthetic operation is a highlight of the Fukumoto synthesis. However, the necessity to adjust a number of functional groups, the obligatory use protective groups *en route* to the construction of ring *A*, and the steps involved in the elaboration of the hydroxyketone side-chain, detract from an otherwise new approach to the precursor of rings *B, C,* and *D* of a steroidal skeleton.

The stepwise approach to the construction of the *A* and *B* rings in the Woodward synthesis of cortisone, and of ring *A* in the Fukumoto synthesis contrasts the biomimetic polyene cyclization strategy pioneered by W. S. Johnson in the late 1970's [14]. Here, an immediate precursor to 11-α-hydroxy progesterone was obtained "in one step" from a polyene intermediate by an acid-catalyzed biomimetic cyclization (see section 4.5.4, Figure 4.17 A).

With the powerful methodologies available today, the important steroid molecules of the Woodward era can no doubt be synthesized using a variety of asymmetric processes. However, the urgent supply needs of the corticosteroids at the middle of the last century have been supplanted by other natural products that command the attention of academic synthetic chemists. These are of a much

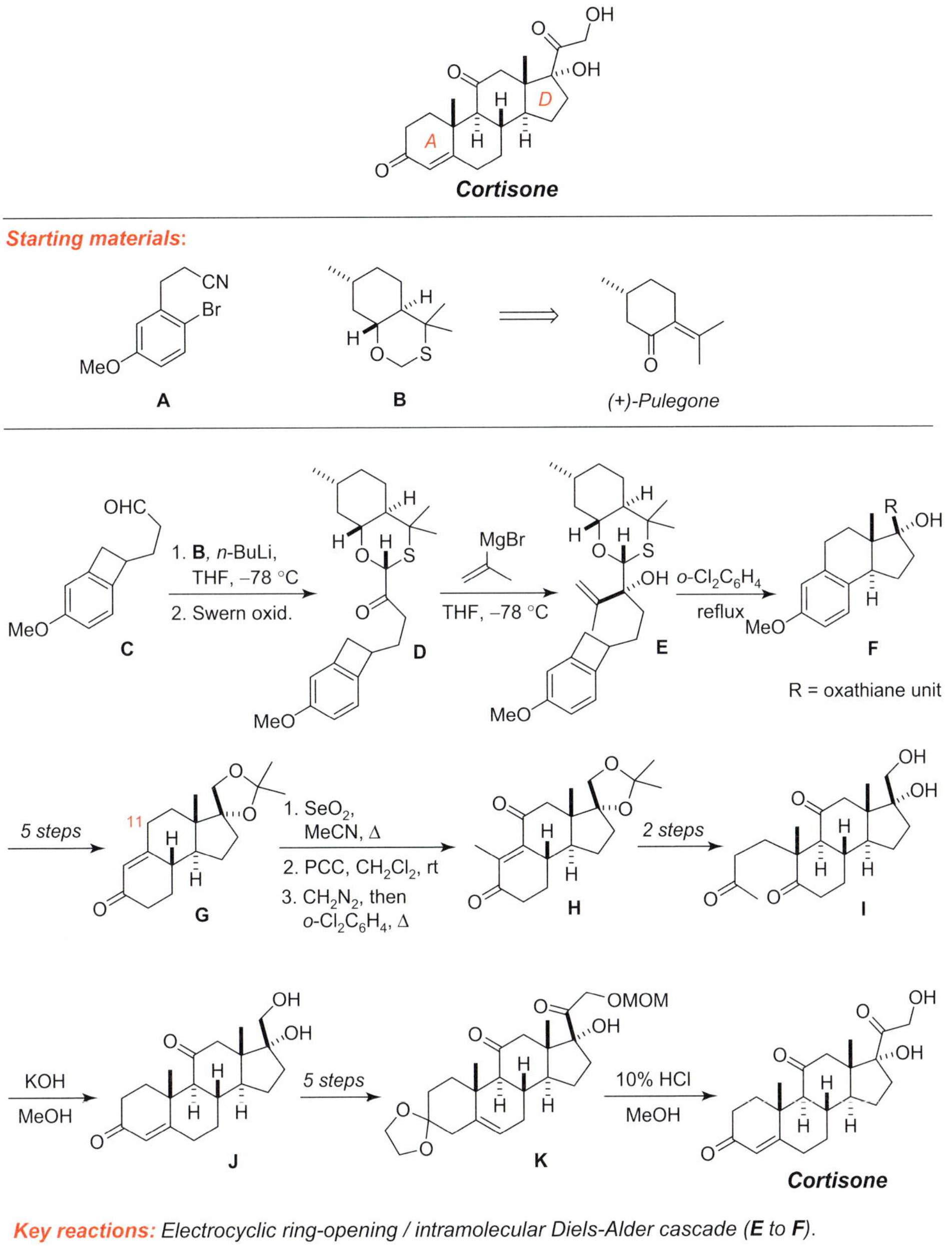
Cortisone
Starting materials:
A
B
(+)-Pulegone
C
1. B, n-BuLi, THF, –78 °C
2. Swern oxid.
D
MgBr
THF, –78 °C
E
o-Cl2C6H4
reflux
F
R = oxathiane unit
5 steps
G
1. SeO2, MeCN, Δ
2. PCC, CH2Cl2, rt
3. CH2N2, then o-Cl2C6H4, Δ
H
2 steps
I
KOH
MeOH
J
5 steps
K
10% HCl
MeOH
Cortisone
Key reactions: Electrocyclic ring-opening / intramolecular Diels-Alder cascade (E to F).

Figure 4.6 Salient features of the 1990 total synthesis of cortisone by Fukumoto.

higher level of structural and functional complexity, and many are endowed with potent therapeutic activities such as, for example, taxol, the epothilones, and related antitumor agents (see Chapters 3 and 16). Recent advances in olefin cascade reactions include the discovery by MacMillan and coworkers that distal aldehydes of polyolefin systems can initiate a series of 6-*endo*-trig cyclization reactions to furnish polycyclic hydrocarbon frameworks through the action of iminium ion catalysis [14c]. The general utility of this method is that it is amenable to the construction of tri-, tetra-, penta-, hexa- and even heptacyclic ring systems. Thus, in a single synthetic operation, up to five new C–C bonds and 11 stereogenic centers (five of which are quaternary) can be generated in high enantioselectivity.

4.4.2
Strychnine

The second Woodward synthesis to highlight in this chapter is that of strychnine, first isolated in the early eighteenth century from the seeds and bark of *Strychnos nux vomica* by Pelletier and Caventou [15]. Known for its lethal poisonous properties, strychnine has been the subject of extensive chemical and physiological studies for centuries [16].

The total synthesis of strychnine by Woodward and coworkers in 1954 stands to this day as a watershed accomplishment [17]. Even by contemporary standards, strychnine is considered as one of the most complex (for its size) natural product targets for synthesis. It comprises seven rings with six contiguous stereogenic carbon atoms, of which five are part of a central cyclohexane ring *E* (Figure 4.7). There are over fifteen recorded total syntheses of strychnine, in both racemic and enantioselctive forms, as well as a number of synthetic approaches (see Chapter 3, Figure 3.2). These have been surveyed in detail in excellent review articles [18]. Here, we show the key reactions and intermediates in eight asymmetric syntheses of strychnine, focussing on the evolution of the thought process and the conception of a strategy from different starting materials.

It is interesting to examine the three-dimensional structure, bond connectivities, and positions of the heteroatoms in the heptacyclic ring system of strychnine (Figure 4.7). While the hydrindole-like motif is apparent (*AB* ring system), not much more can be easily discerned with regard to a suitable starting material from which the target molecule could evolve. Today, the planning of most complex syntheses starts with a retrosynthetic analysis that eventually uncovers one or more convenient starting materials [19]. Key reactions in the forward sense will emerge from such an analysis, and the focus will remain on securing desired bond connectivities while ensuring absolute stereochemistry. The thought process may take us deeper into the retrosynthetic analysis while at the same time simplifying complexity (see Chapter 1, section 1.7). It is as a result of the emergence of a crucial intermediate in the mind's eye, that suitable starting materials are sought. In other words, the logic of a synthesis plan rarely starts from a given starting material and proceeds to build the molecule in a linear series of thoughts. Instead, each intermediate becomes a temporary target molecule which is related to its progenitor by chemical steps already performed, and to the next intermediate in the plan by steps yet to be realized. As in the labyrinth puzzles, or the snakes and ladders children's game, there are many "dead-ends" that require returning to the point of departure, and taking a different route [20]. We start by briefly analyzing eight asymmetric total syntheses of strychnine with the aforementioned mindset of relating a strategy to a given starting material.

4.4.2.1 **The Woodward synthesis** – The classical nature of the Woodward synthesis is evident from the recognition of the indole-like motif and the ingenious incorporation of an electron-rich *o*-dimethoxybenzene as an eventual source of rings *G* and *E* in strychnine (Figure 4.7) [17]. The 3-aminoethyl indole derivative **D**, readily available from **C**, in turn prepared from phenylhydrazine (**A**) and acetoveratrone (**B**), already contains 16 of the 21 carbon atoms of strychnine! Note that the oxidative cleavage of the *o*-dimethoxybenzene motif, followed by lactam formation, liberated an acetic acid appendage for a Dieckmann condensation (**F** to **G** to **H** to **I**). The enol form of the Dieckmann condensation product **I**, was deoxygenated via the corresponding vinylic benzylthio derivative to give an α,β-unsaturated ester. Catalytic hydrogenation led to the saturated *cis*-ester **J**. Treatment with KOH in aqueous methanol afforded the epimeric *trans*-acid **K**, which was identical to a product obtained by the controlled degradation of strychnine. Woodward then used intermediate **K** as an enantiopure relay compound to continue the synthesis, which was converted to ethyl ketone **N** in an intriguing sequence of reactions. Thus, treatment of **K** with acetic anhydride afforded a mixed anhydride which lost CO_2 through the intermediate formation of β-lactone **L**, to give the enol acetate **M**, then ketone **N**. Oxidation with SeO_2 gave the transient α-ketoaldehyde **O**, which was further elaborated through intermediates **P**, **Q**, and **R**, first to isostrychnine (**S**), then to strychnine. Woodward's strategy was conceptually elegant, while using "simple" reactions that capitalized on the inherent reactivity of existing functional groups (Figure 4.7).

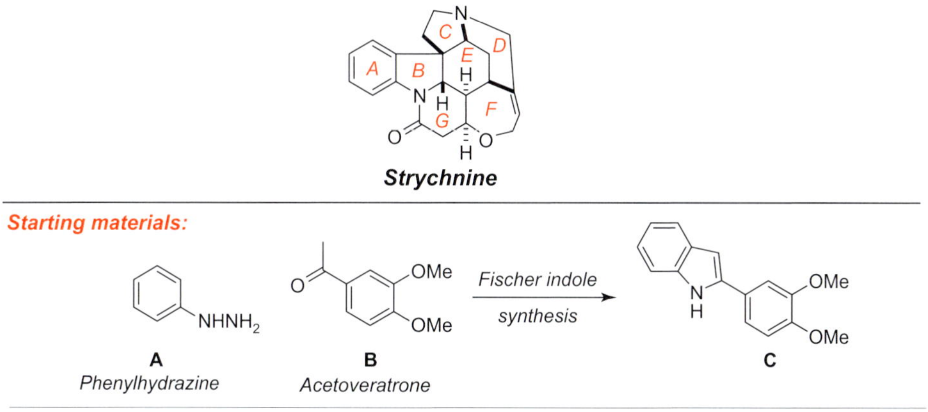

Strychnine

Starting materials:

A	**B**	**C**
Phenylhydrazine	*Acetoveratrone*	

Figure 4.7 Woodward's 1954 total synthesis of strychnine.

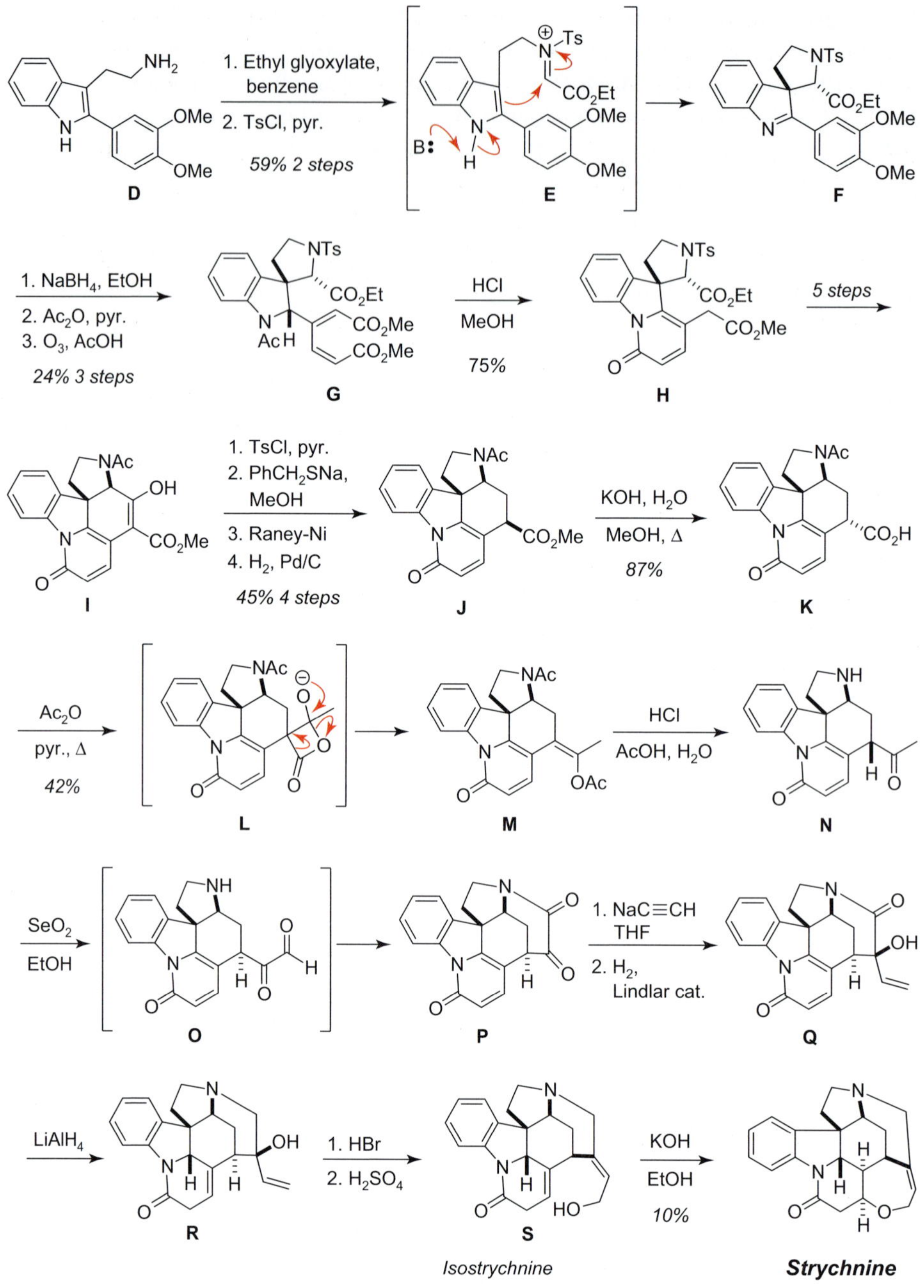

Key reactions: *Indole-iminium ion ring closure (D to F via E); oxidative aromatic cleavage (F to G); Dieckmann condensation/ketone transformation (H to N); selenium dioxide oxidation/cyclization cascade (N to P); stereoselective reduction (P to R); allylic alcohol isomerization (R to S).*

Figure 4.7 (*continued*).

4.4.2.2 **The Overman synthesis** – The first enantioselective total synthesis of strychnine by Overman and coworkers [21] demonstrates true ingenuity in design (Figure 4.8). Conceptually, it differs from the Woodward synthesis in the utilization of a chiral, non-racemic starting material, which is visually remote from any segment of the target molecule. The key step involved an aza-Cope/Mannich cyclization (**J** to **M**), which required the prior elaboration of appropriate precursors that would ensure access to intermediate **J**. Thus, the known esterase mediated desymmetrization of the *meso*-diacetate **B** led to enantiopure **C**. Two Pd-catalyzed C–C bond-forming reactions (**D** to **E**, and **G** to **H**) set the stage for intermediates **I** and **J**. The penultimate product was the well-known Wieland-Gumlich aldehyde (**O**), from which strychnine could be prepared in one synthetic operation. Overman targeted the Wieland-Gumlich aldehyde (**O**) and its precursors **M** and **N**, early in the synthesis plan by including an allylic alcohol in **F**. This should be compared with the Woodward strategy in which ring *G* was elaborated before *F*, leading to isostrychnine as a penultimate target. However, not until the aza-Cope/Mannich sequence was completed in the Overman synthesis (**J** to **K** to **L** to **M**), did the *C/D/E* ring system come into full view. It is also noteworthy that the conversion of the Wieland-Gumlich aldehyde to strychnine is much higher yielding than that of isostrychnine (*cf.* 65% to 10%) and that subsequent to their first synthesis, Overman and coworkers used essentially the same synthetic plan to prepare the unnatural isomer of strychnine.

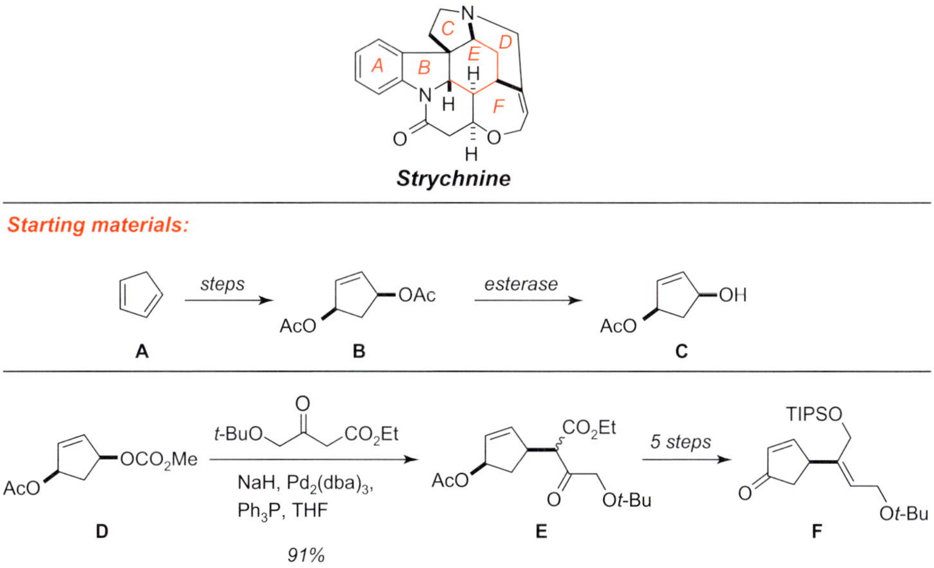

Strychnine

Starting materials:

91%

Figure 4.8 Overman's first enantioselective total synthesis of strychnine.

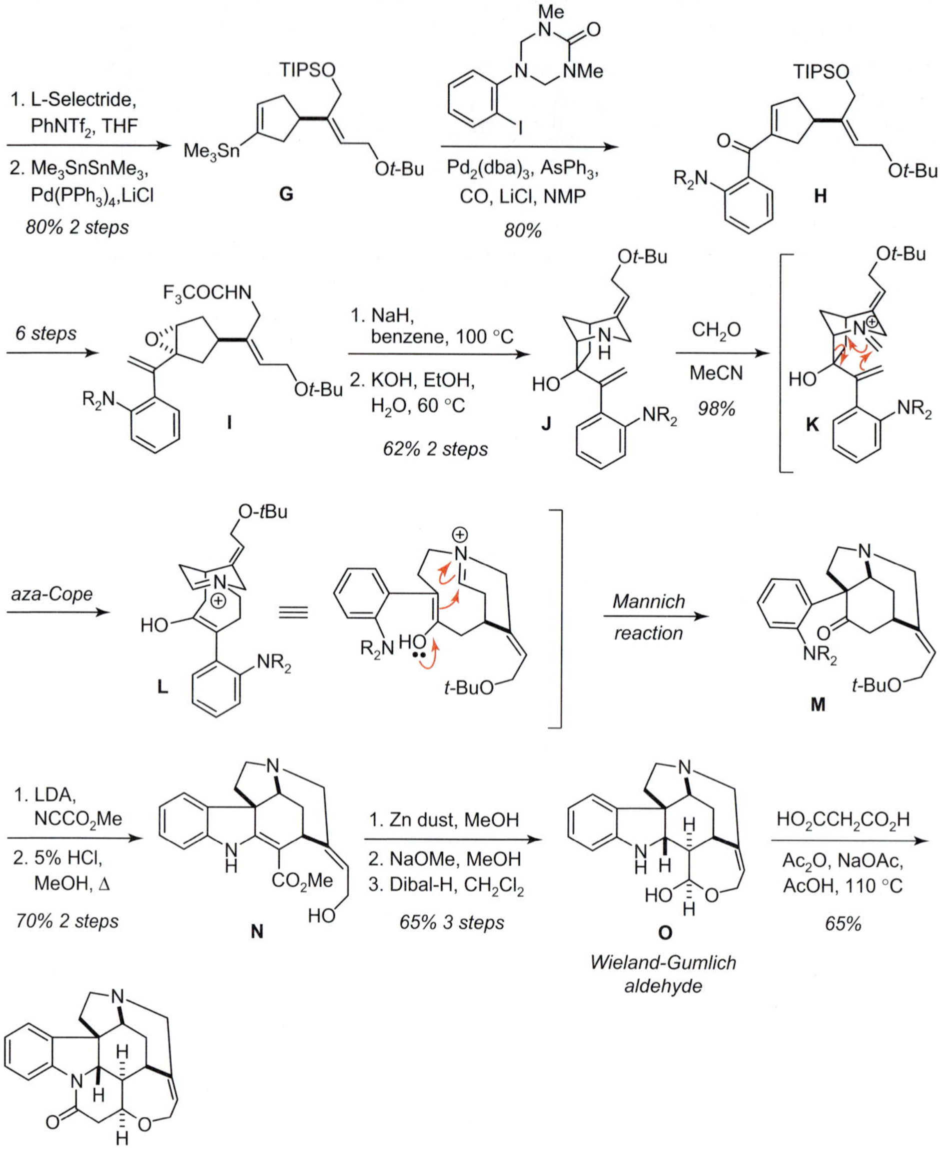

Strychnine

Key reactions: *Pd-catalyzed allylic enolate coupling (D to E); Pd-catalyzed carbonylative cross-coupling (G to H); intramolecular epoxide opening (I to J); aza-Cope/Mannich cyclization (J to M via K and L).*

Figure 4.8   (*continued*).

4.4.2.3 The Kuehne synthesis – In Kuehne's enantioselective total synthesis of strychnine, readily available L-tryptophan was used as starting material (Figure 4.9) [22]. A set of conceptually unique transformations furnished intermediate **B**, which was expected to provide access to rings *C* and *E*. In the event, the resident chirality in L-tryptophan ensured the correct stereochemistry in an elegant Mannich reaction/[3,3]-sigmatropic rearrangement cascade (**C** to **F**). Although the amino acid served its purpose admirably in the elaboration of the *C/E* ring system, a self-immolative process was part of the plan in the obligatory decyanation reaction (**G** to **H**), proceeding through the reductive elimination of an α-aminonitrile. Further elaboration of peripheral functionality proceeded through intermediates **I**, **J**, and **K**, which eventually led to the Wieland-Gumlich aldehyde

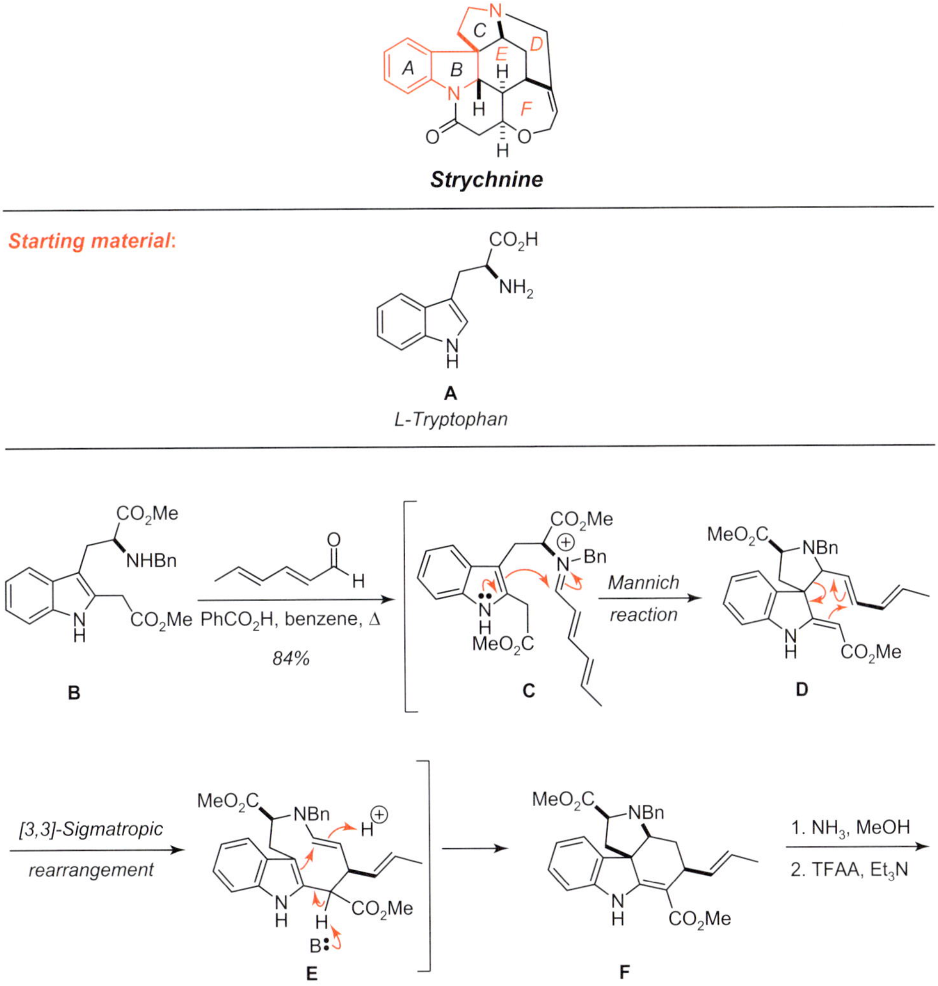

Strychnine

Starting material:

A
L-Tryptophan

Figure 4.9 Kuehne's total synthesis of strychnine from L-tryptophan.

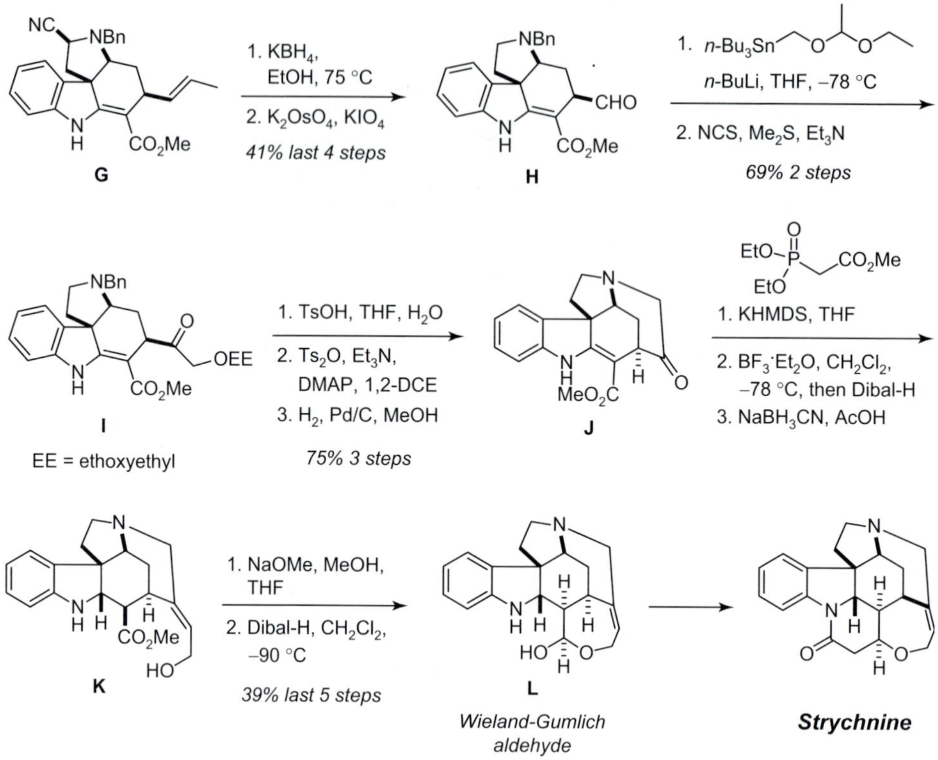

Key reactions: *Tandem Mannich reaction/[3,3]-sigmatropic rearrangement (**B** to **F** via **C**, **D**, **E**); α-aminonitrile reductive decyanation (**G** to **H**); Still-alkoxytin acetal extension/oxidation (**H** to **I**); intramolecular cyclization (**I** to **J**).*

Figure 4.9 *(continued).*

(**L**). It is noteworthy that the indole nitrogen remained unprotected throughout this relatively short synthesis of strychnine.

4.4.2.4 The Bonjoch and Bosch synthesis

Bonjoch and Bosch's total synthesis of strychnine presents a different approach to securing enantioselectivity (Figure 4.10) [23]. The key reaction was performed relatively early in the sequence and involved a double asymmetric reductive amination of the achiral 1,3-diketone aldehyde **D** with (*S*)-α-methylbenzylamine (**D** to **E**). A reductive Heck reaction of a vinyl iodide intermediate **F**, first utilized by Rawal [24] in his total synthesis of (±)-strychnine, eventually led to the familiar Wieland-Gumlich aldehyde (**I**). The visual connection of 1,3-cyclohexanedione (**A**) as starting material for ring *E*, can only be appreciated in the context of the doubly substituted diketone **C**, and the intention to utilize an *o*-nitrophenyl appendage as an indoline precursor much like in the Overman synthesis [21].

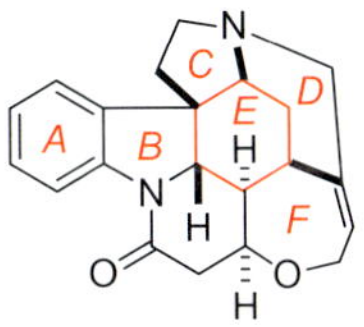

Strychnine

Starting materials:

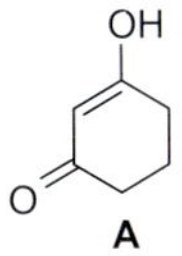

A

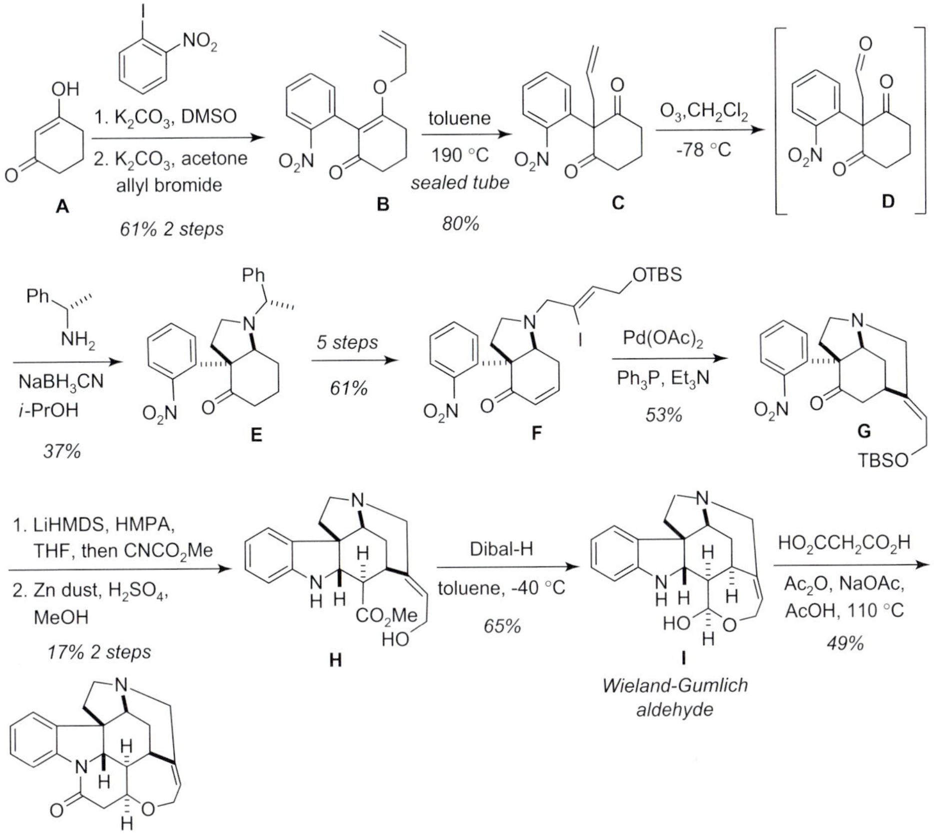

Strychnine

Key reactions: *Enolate aromatic substitution (**A** to **B**); double asymmetric reductive amination (**C** to **E**); intramolecular Pd-catalyzed reductive Heck cyclization (**F** to **G**).*

Figure 4.10 Bonjoch and Bosch's total synthesis of strychnine.

In each of the next three asymmetric syntheses of strychnine, catalysis was utilized in the first steps of the syntheses to generate enantioenriched starting materials.

4.4.2.5 The Shibasaki synthesis – Shibasaki and coworkers [25] started their synthesis with a highly efficient and scalable catalytic Michael addition of dimethyl malonate to 2-cyclohexenone (**A**) to give **B** (Figure 4.11). Eleven steps later, the enone intermediate **C** was further elaborated to **F**. A Stille coupling led to **G**, which was further transformed to the diethyl dithioacetal **H**. Tandem cyclization starting with 1,4-conjugate addition to form the piperidine ring, followed by reduction of the nitro group led to the tricyclic indole **I**. A key reaction to introduce the pyrrolidine subunit relied on the ring-closure of a thionium ion intermediate leading to **J**. Desulfurization of the ethylthio group, reduction, and *N*-acetylation, gave the pentacyclic ring system **K** similar to the Kuehne [22] and Bonjoch-Bosch [23] advanced intermediates. Base-catalyzed epimerization of the aldehyde group in **L** and cleavage of the acetyl group led to the Wieland-Gumlich aldehyde (**M**), and eventually strychnine. It should be noted that the stereogenic center created during the Michael addition (adduct **B**) was used to elaborate the remaining stereocenters through thermodynamic control of substituents and preferred topology. In the event, several steps were used to reach the crucial intermediate **H** with the inevitable use of orthogonal protective groups (Figure 4.11).

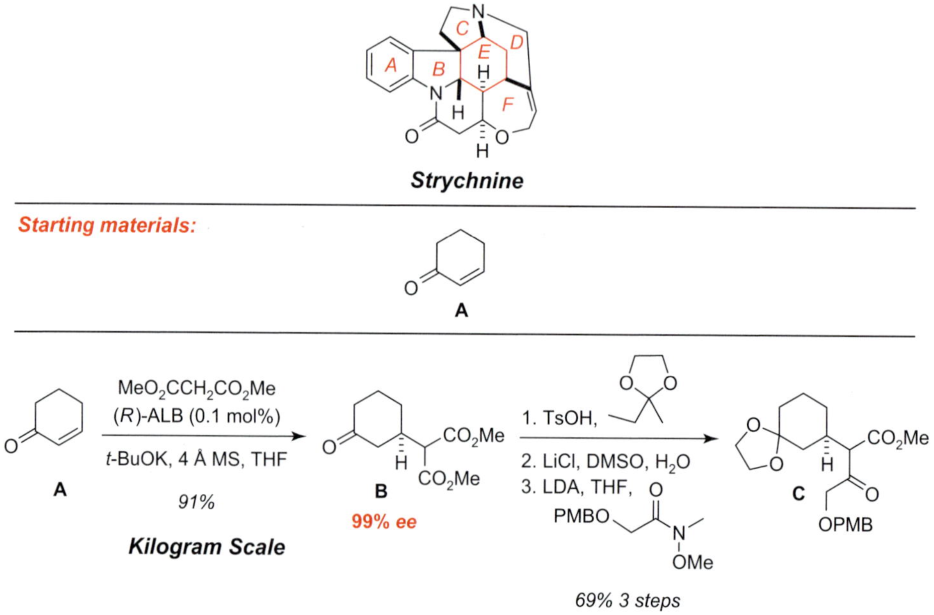

Strychnine

Starting materials:

A

Figure 4.11 Shibasaki's total synthesis of strychnine.

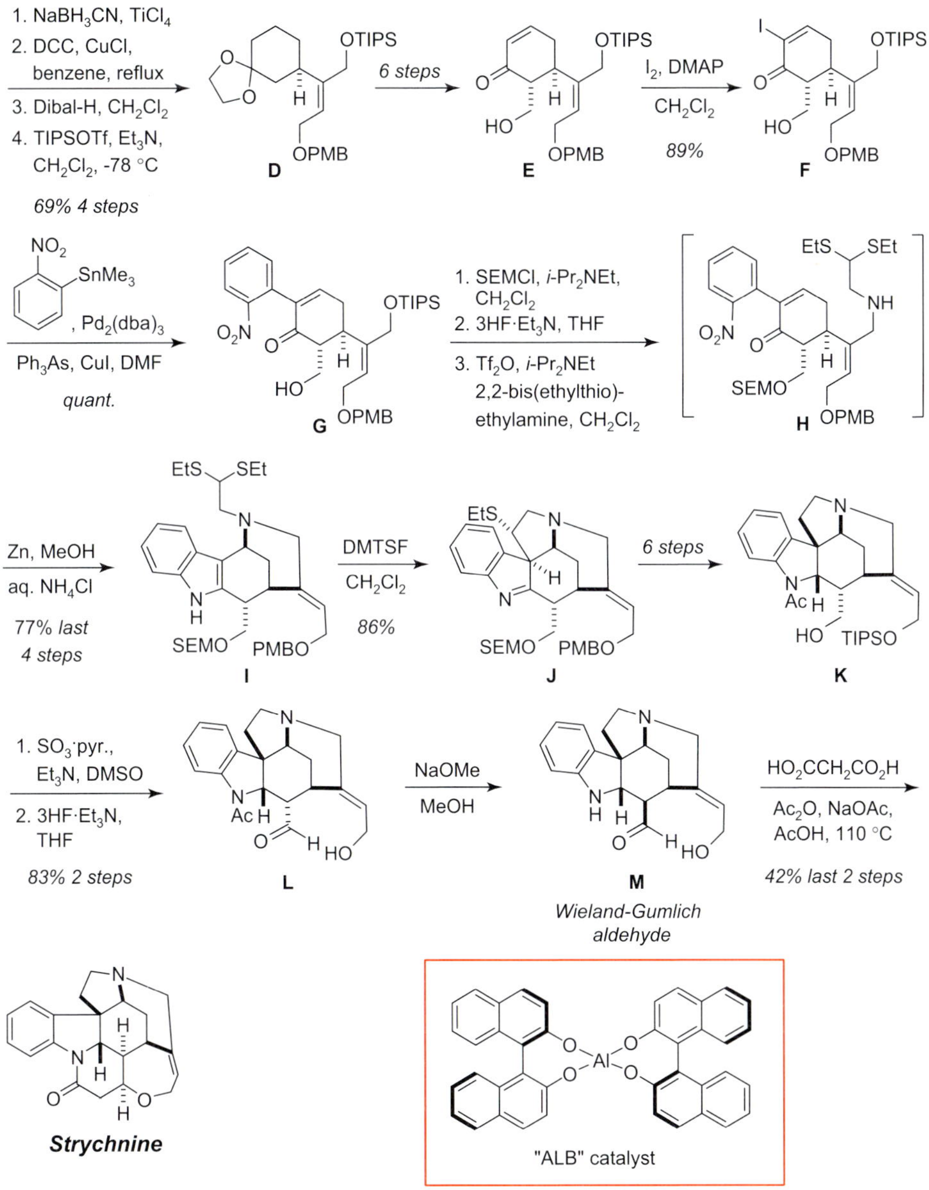

1. NaBH₃CN, TiCl₄
2. DCC, CuCl,
   benzene, reflux
3. Dibal-H, CH₂Cl₂
4. TIPSOTf, Et₃N,
   CH₂Cl₂, -78 °C

69% 4 steps

D → 6 steps → **E** → I₂, DMAP / CH₂Cl₂ 89% → **F**

NO₂ / SnMe₃ , Pd₂(dba)₃
Ph₃As, CuI, DMF
quant.

G →

1. SEMCl, *i*-Pr₂NEt, CH₂Cl₂
2. 3HF·Et₃N, THF
3. Tf₂O, *i*-Pr₂NEt
   2,2-bis(ethylthio)-
   ethylamine, CH₂Cl₂

→ **H**

Zn, MeOH
aq. NH₄Cl

77% last
4 steps

I → DMTSF / CH₂Cl₂ 86% → **J** → 6 steps → **K**

1. SO₃·pyr.,
   Et₃N, DMSO
2. 3HF·Et₃N,
   THF

83% 2 steps

L → NaOMe / MeOH → **M** → HO₂CCH₂CO₂H / Ac₂O, NaOAc, AcOH, 110 °C → 42% last 2 steps

M
*Wieland-Gumlich
aldehyde*

Strychnine

"ALB" catalyst

Key reactions: *Catalytic asymmetric Michael reaction (**A** to **B**); β-ketoester synthesis (**B** to **C**); introduction of double bond (**C** to **D**); Saegusa-Ito (silyl enol ether to enone) oxidation (**D** to **E**); Stille coupling (**F** to **G**); conjugate addition-cyclization (**H** to **I**); indole-thionium ion cyclization (**I** to **J**); DMTSF = dimethyl(methylthio)sulfonium tetrafluoroborate.*

Figure 4.11 (*continued*).

4.4.2.6 The Fukuyama synthesis – Fukuyama was the second to utilize enzymatic resolution to secure an enantiopure starting material early in the synthesis of strychnine (Figure 4.12) [26]. Epoxide **C**, readily obtained from benzoic acid (**A**) in seven steps, was subjected to a Pd-mediated coupling reaction with an intact 2,3-disubstituted indole moiety (**D** to **E**). The nine-membered azacyle **G** was obtained in a double Mitsunobu reaction with *p*-nitrobenzenesulfonamide (NsNH₂). The choice of the *N*-nosyl group was based on previous experience and compatibility with subsequent reactions. A self-immolative sequence was used to transform the original chiral cyclohexene unit in **F** and **G** to the aldehyde **I**. Intramolecular iminium ion cyclization was followed by a transannular cyclization through an indole nucleophilic attack with concomitant formation of the α,β-unsaturated ester **J**. Stereo- and chemoselective reductions of the ester functions individually, led to the Wieland-Gumlich aldehyde (**K**), which was in turn, converted to strychnine. As in the Woodward [17] and Kuehne [22] syntheses, Fukuyama recognized the "atom" value in starting with a 2,3-disubstituted indole. All the carbon atoms in the precursors **C** and **D**, except for one methoxycarbonyl group, remained embedded in the final target.

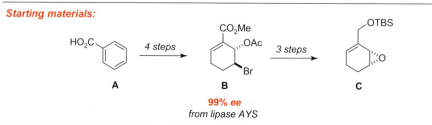

Figure 4.12 Fukuyama's total synthesis of strychnine.

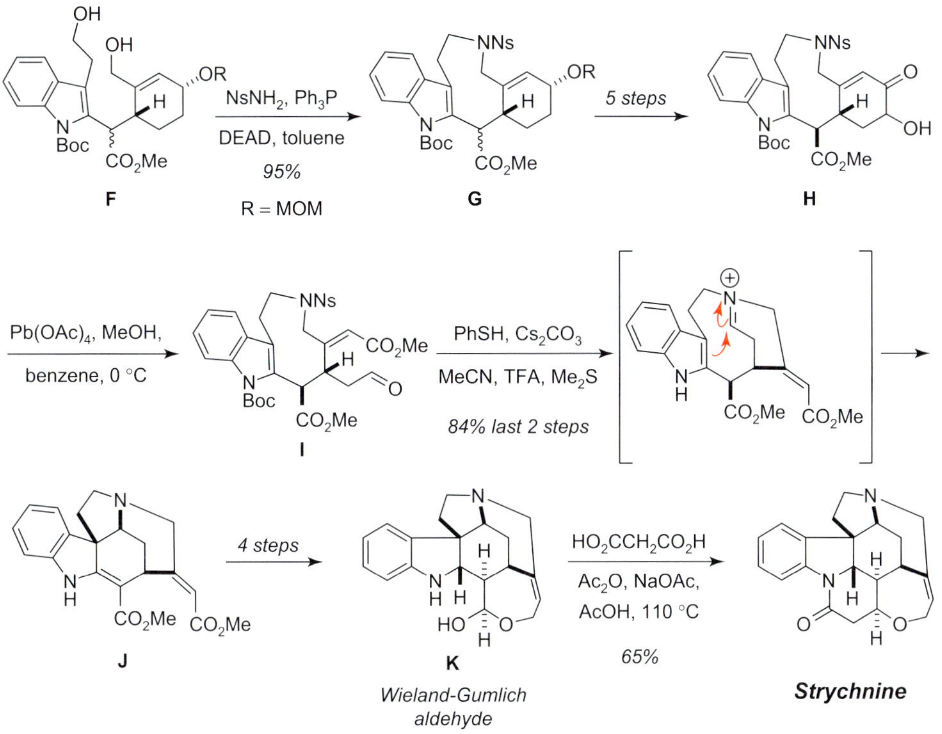

Key reactions: *Benzoic acid to chiral cyclohexene intermediate **B** (lipase resolution); Pd-mediated malonate coupling (**D+C** to **E**); double Mitsunobu N-nosyl amination (**F** to **G**); intramolecular iminium ion formation and base-catalyzed transannular cyclization (**I** to **J**).*

Figure 4.12 *(continued)*.

4.4.2.7 The Mori synthesis – Finally, the impact of organopalladium chemistry is further highlighted in Mori's total synthesis of strychnine (Figure 4.13) [27]. A tricyclic indoline **E** was obtained in enantioenriched form from an intramolecular Heck reaction (**C** to **D**). A second Pd-catalyzed activation and azacycle formation, followed by reductive elimination afforded **F**. Double bond manipulation and installation of ring G by a second Heck reaction afforded **I**, which was N-alkylated to give **J**, converging with the Woodward-like intermediate **K**. A total of four Pd-catalyzed reactions were used by Mori in the elaboration of the heptacyclic skeleton of strychnine.

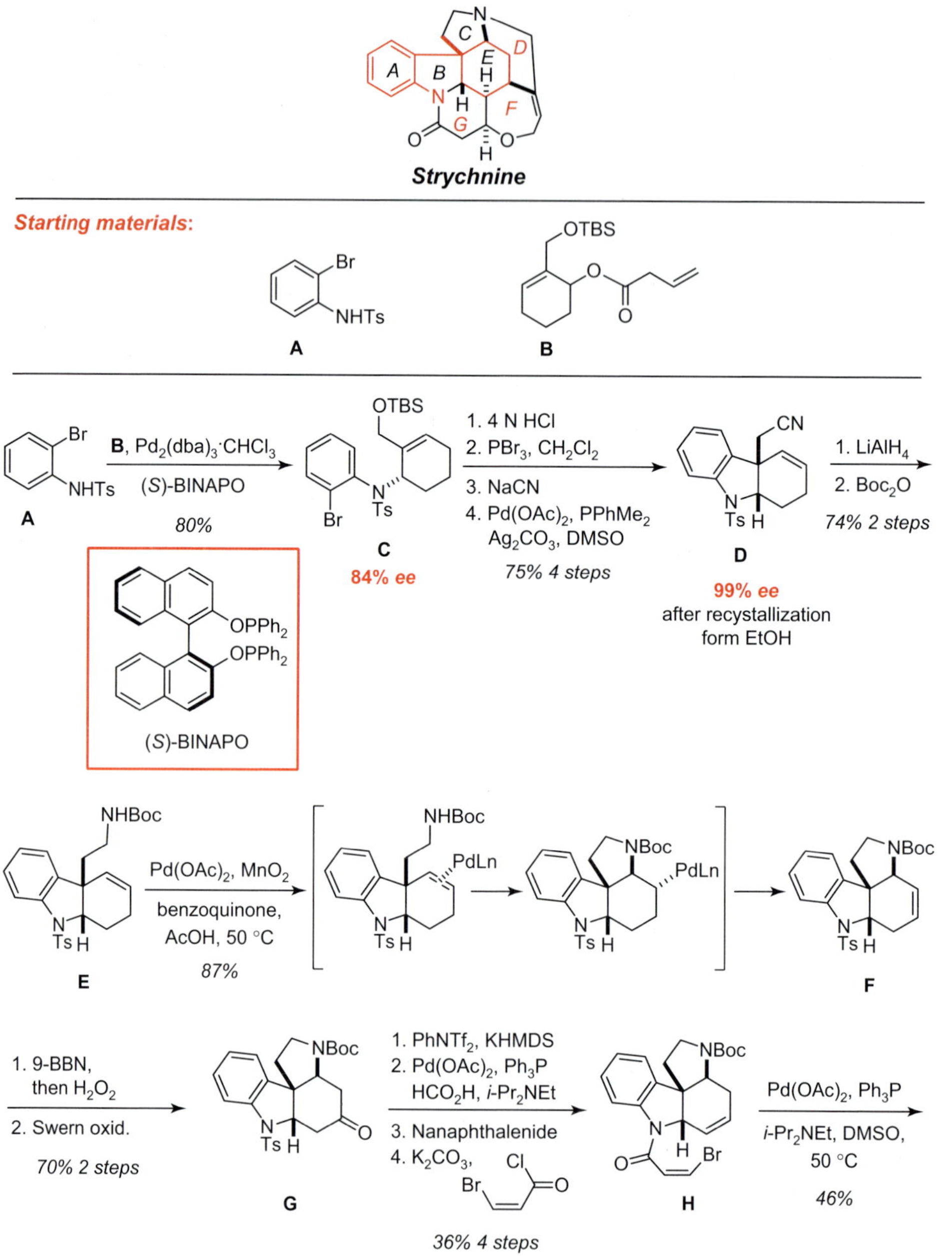

Strychnine

Starting materials:

A

B

Figure 4.13 Mori's total synthesis of strychnine.

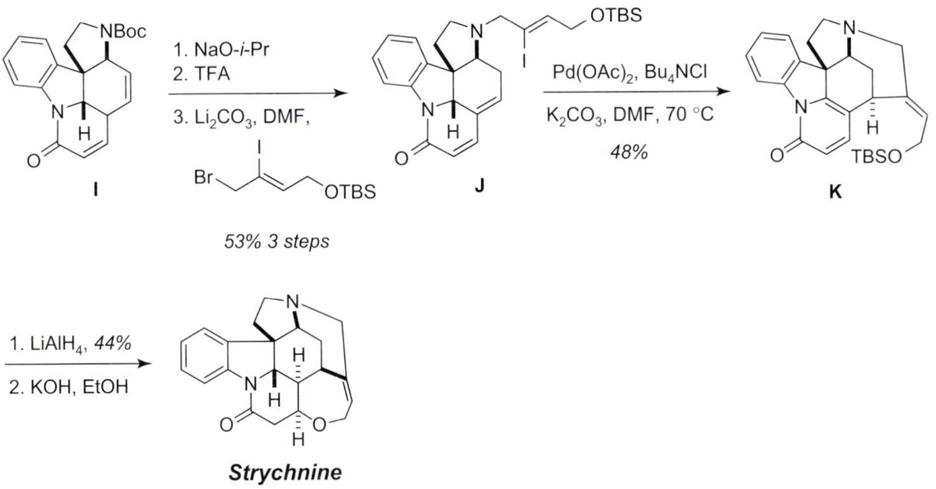

Key reactions: *Pd-catalyzed asymmetric allylic NHTs substitution (**A**+**B** to **C**); Pd-catalyzed intramolecular Heck cyclization (**C** to **D**); Pd-catalyzed olefin activation and pyrrolidine formation (**E** to **F**); Pd-catalyzed Heck cyclization (**H** to **I**); isomerization of double bond (**I** to **J**); Pd-catalyzed intramolecular Heck cyclization (**J** to **K**).*

Figure 4.13 *(continued).*

4.4.2.8 The MacMillan synthesis – The final synthesis to be discussed in this section showcases the power and importance of both organocatalytic and cascade reactions. In 2011, MacMillan and coworkers reported the shortest ever enantioselctive synthesis of strychnine [28]. Their synthesis commenced with β-carboline **A**, which was converted to 2,3-disubstituted indole **B** using a three step literature procedure (Figure 4.14). The 2-vinyl indole **B** served as the diene component in an organocatalytic Diels-Alder reaction with propynal (**C**) in the presence of the 1-naphthyl substituted imidazolidinone catalytic system **D**, which is the first step in a cascade process to tetracycle **E**. Upon [4+2] cycloaddition, the initially formed Diels-Alder adduct undergoes an elimination of methyl selenide and then succumbs to intramolecular pyrrolidine formation to give **E**. The productive organocascade addition-cyclization sequence is noteworthy in that it generates a functionalized spiroindoline system that was later used as a key intermediate in the synthesis of five other natural products. Completion of the synthesis of strychnine first involved conversion to a mixture of unsaturated esters **F**, *N*-alkylation, and ester reduction to furnish the bis-allylic alcohol **G**. An intramolecular Jeffery-Heck reaction/lactol formation sequence was followed by *N*-Boc cleavage to give the Wieland-Gumlich aldehyde (**H**), which was converted to strychnine in the familiar way.

At 12 steps, the MacMillian synthesis of strychnine is by far the shortest enantioselctive route to be reported. However, a much shorter route to (±)-strychnine (six steps from commercial material) was recently reported by Vanderwal and coworkers [18c], and the highest yielding synthesis of strychnine belongs to Rawal and his group [24a].

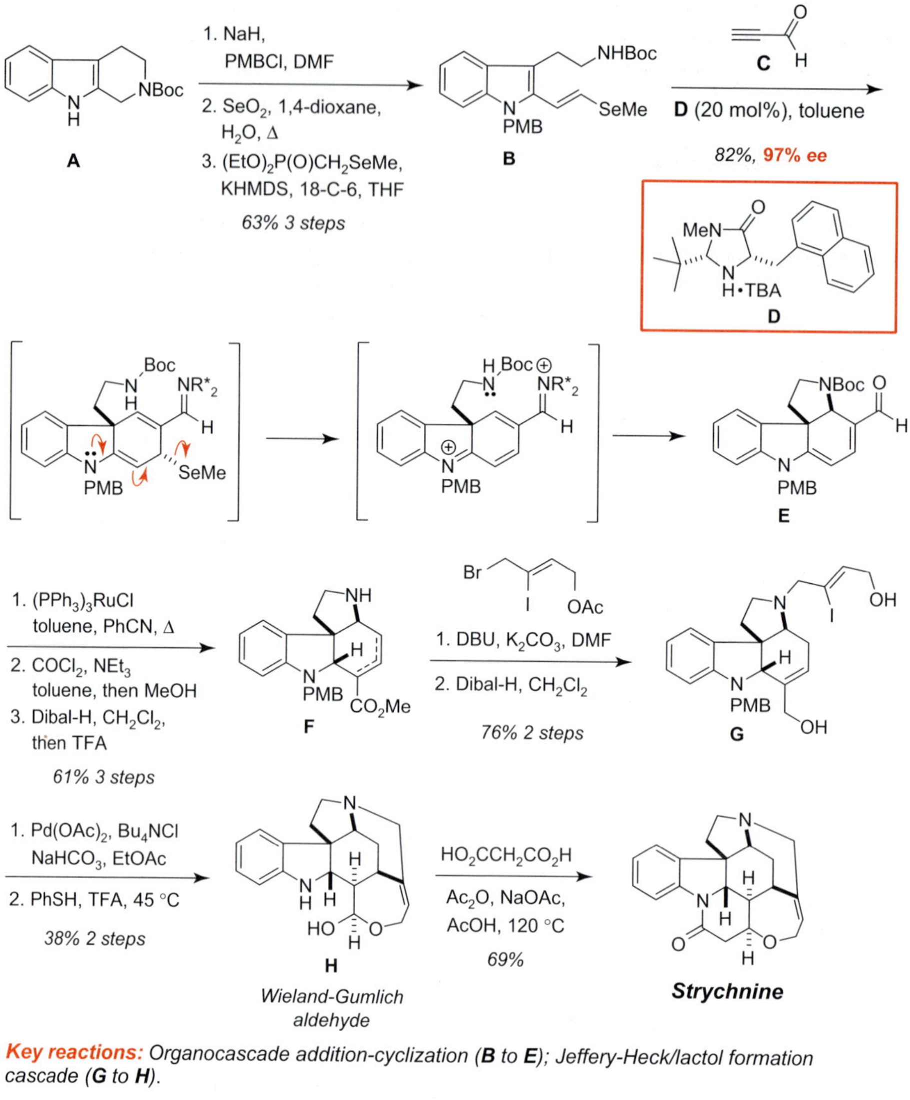

Key reactions: *Organocascade addition-cyclization (**B** to **E**); Jeffery-Heck/lactol formation cascade (**G** to **H**).*

Figure 4.14 MacMillan's synthesis of strychnine.

4.4.2.9 Strychnine syntheses: Synopsis – The above discussed eight total syntheses of strychnine are conceptually different in their approaches and, consequently, in the selection of starting materials. In the Woodward, Kuehne, Fukuyama, and MacMillan strategies, a visual connection from an indole derivative to product was evident. Remarkably, in the Woodward synthesis, only glyoxylic acid, acetic anhydride, and acetylene were used as sources of the five remaining carbons atoms to complete the carbogenic framework of strychnine starting from the Fischer indole intermediate (Figure 4.7). Of the eight total syntheses of strychnine discussed in this section, the most hidden visual disconnection to a starting material was conceived by Overman who chose to start with cyclopentadiene and proceeded to an enzymatic desymmetrization of a 1,4-diacetoxy cyclopentene (Figure 4.8).

4.5
The Post Woodwardian Era

As outlined in Chapter 3, different periods have promulgated trends and incentives in the total synthesis of natural products. With the advent of new instrumentation and powerful methods of stereocontrolled C–C bond-forming reactions, the complexity of molecular structures considered as targets for synthesis has increased accordingly. The rich legacy of total synthesis started by Woodward, Robinson, Stork, Corey, and their contemporaries, continued to flourish and expand through the end of the year 2000 and beyond (see also Chapter 18, sections 18.1 and 18.2). Emphasis on stereochemical control set new standards and guidelines in synthesis planning. Each class of natural product instigated the development of methods that were adapted to a particular type of carbon skeleton and stereochemical pattern. Stereodifferentiating reactions became accessible through *substrate* or *reagent controlled reactions* [29]. In the former, a resident chiral center would influence the stereochemical outcome of a particular reaction by processes involving proximity effects, chelation, and steric interactions among others. In the latter cases, chirality could be induced internally or externally in a particular bond-forming reaction through the use of, for example, a chiral auxiliary or a chiral catalyst respectively. Enantiofacial and diasterofacial selectivities can be controlled through metal-coordinated species, such as in enolate alkylations or aldol reactions. Normally, in such reactions chirality is introduced at one or at two vicinal carbon atoms at a time. Incorporation of chirality in a larger segment of the intended target molecule can be accomplished by choosing appropriate starting materials or chiral subunits as part of the reaction partner or the reagent. We illustrate three widely used strategies for the elaboration of natural products inspired by classical non-catalytic approaches.

4.5.1
The convergent template-based approach

The structure of the target molecule is analyzed to uncover one or more starting materials that can be elaborated to provide a suitable precursor for a segment of the intended target. Such precursors may, in turn, be prepared directly from chiral, non-racemic starting materials or by a given asymmetric process (see Chapter 6). The three classical examples chosen in Figure 4.14 A–C illustrate how one and the same precursor was used for different segments of erythromycin A, and erythronolide A and tylonolide hemiacetal, the aglycones of the corresponding macrolide antibiotics erythromycin A and tylosin respectively, at a time when stereocontrolled aldol condensations were not as well developed as they are today. Thus, Woodward and his 47 coworkers [30] conceived of a dithiadecalin template to converge with the C-3–C-8 and C-9–C-15 segments of erythromycin A. The

Raney-nickel desulfurization of the template would replace the carbon-sulfur bonds by hydrogen, thus generating the alternating *C*-methyl groups at C-4, C-6, C-8, C-10 and C-12 of the macrolide (Figure 4.15 A).

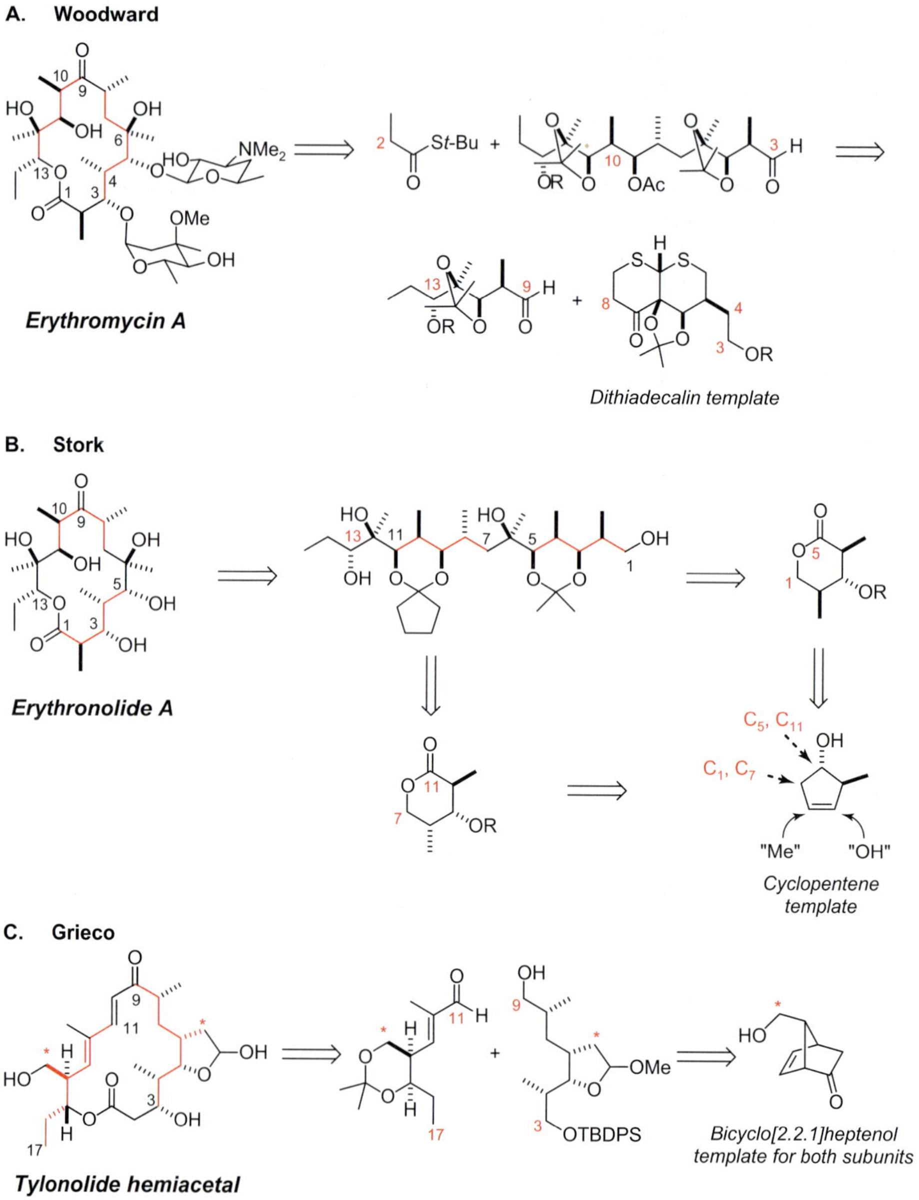
A. Woodward
Erythromycin A
St-Bu
Dithiadecalin template
B. Stork
Erythronolide A
Cyclopentene template
"Me"
"OH"
C. Grieco
Tylonolide hemiacetal
OTBDPS
Bicyclo[2.2.1]heptenol template for both subunits

Figure 4.15 Key strategic template-based approaches (common precursor or substrate control).

Stork and coworkers [31] used a chiral, non-racemic cyclopentene (prepared from an asymmetric hydroboration of 5-methyl-cyclopenta-1,3-diene), already harboring vicinal *C*-methyl and hydroxyl groups of known absolute configuration, as a template to elaborate the acyclic chain of erythronolide A *seco*-acid. Starting with a common cyclopentene intermediate, a series of reactions involving epoxidation, cuprate opening, and Baeyer-Villiger oxidation afforded δ-lactones that corresponded to the C-1–C-5 and C-7–C-11 segments respectively of erythronolide A (Figure 4.15 B) (see also, Chapter 8, section 8.10.3). An example using D-glucose as a common template in the construction of the erythronolide A *seco*-acid motif was originally conceived and reported by Hanessian [32].

A bicyclo[2.2.1]heptenol, obtained [33] in enantiopure form through resolution, was used as a precursor by Grieco and coworkers to eventually converge with C-3–C-9 and C-11–C-17 of tylonolide hemiacetal (Figure 4.15 C). The asterisk indicates the location of the original hydroxymethyl carbon atom of the bicyclic precursor (see also, Chapter 8, section 8.10.3).

The above three examples, dating back over three decades, demonstrate the astute visual analysis of the carbon framework of propionate-derived macrolides in relation to cleverly conceived methods of stereocontrolled reactions exploiting rigid cyclic templates as starting materials.

4.5.2
Chiral auxiliary approach

There exist powerful and general methods for the stepwise introduction of substituents on an acyclic framework that rely on the use of chiral auxiliaries. The well-known Evans (oxazolidinone) auxiliaries [34] can be prepared from enantiopure amino alcohols. The corresponding imides afford enolates, that when treated with appropriate electrophilic reagents, lead to the introduction of various carbon and heteroatom-type substituents with high diastereoselectivity. An early application is shown for the synthesis of ionomycin by Evans (Figure 4.16 A) [35]. Thus, a series of iterative diastereoselective enolate alkylations, led to the C-1–C-10 subunit of ionomycin. Ephedrine and pseudoephedrine (Myers auxiliaries) [36], as well as camphor sultams (Oppolzer auxiliaries) [37], are also popular methods for the highly stereoselective alkylation of enolates. The synthesis of the C-1–C-11 segment of borrelidin using the Myers auxiliary in an iterative fashion, was shown by Theodorakis and coworkers (Figure 4.16 B) [38]. Stereoselective α-alkylations of ketones has been achieved via hydrazones derived from D- or L-prolinol (Enders' SAMP/RAMP method) [39]. An example of the method is illustrated by Enders in the synthesis of pectinatone (Figure 4.16 C) [40]. A conjugate hydride ion addition to a camphor sultam intermediate was reported in the synthesis of kendomycin by the Mulzer group [41] (Figure 4.16 D). Each iteration in the Evans, Myers, and Enders auxiliary-based alkylations requires the cleavage of the chiral auxiliary, transformation to an iodide, and alkylation with the anion of the chiral auxiliary to extend the chain. These stoichiometric methods have been extensively used for the iterative synthesis of propionate and deoxypropionate triads that are commonly

A.   Evans enolate alkylation

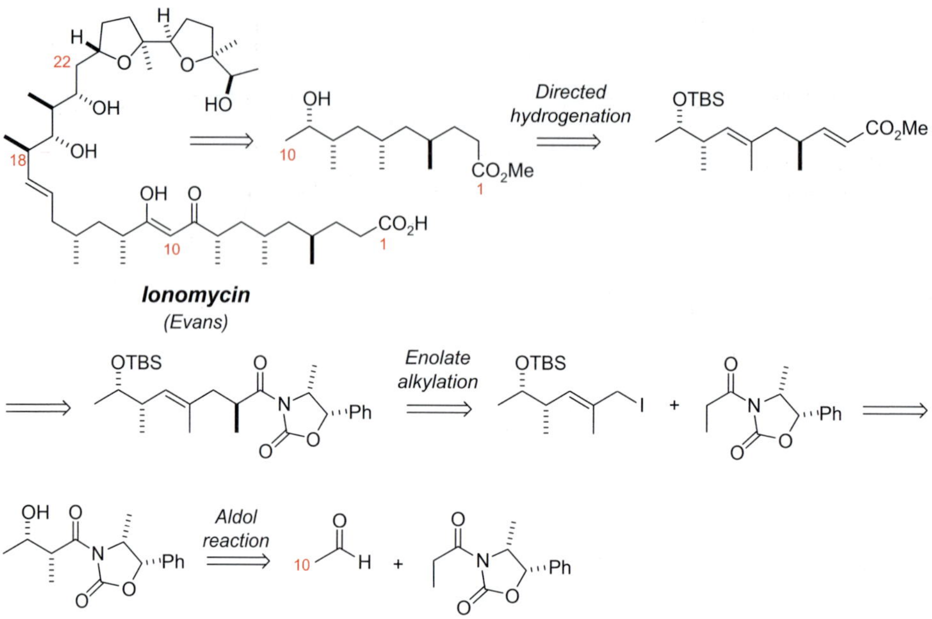

Ionomycin
(Evans)

B.   Myers enolate alkylation

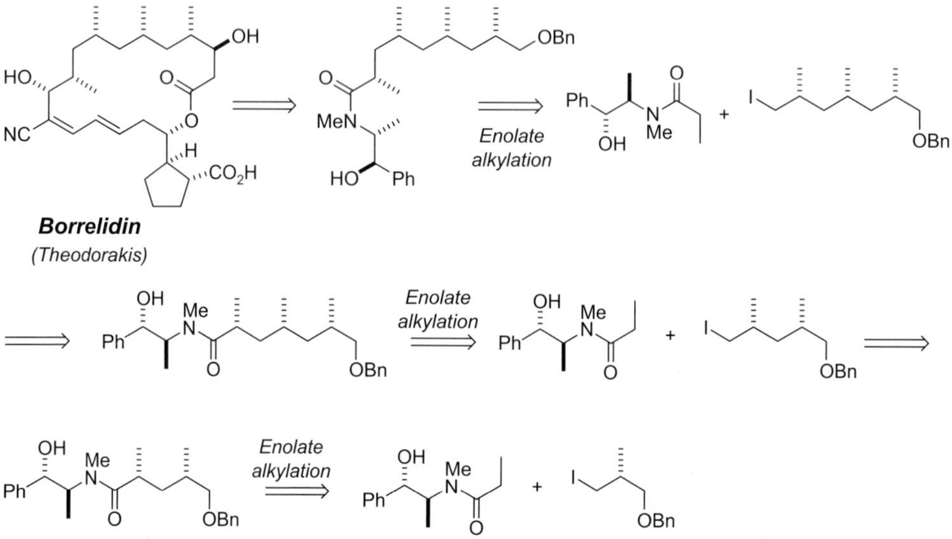

Borrelidin
(Theodorakis)

Figure 4.16 Key strategic asymmetric C–C bond-forming reactions that rely on chiral auxiliaries in natural product synthesis.

C. Enders SAMP/RAMP hydrazone anion alkylation

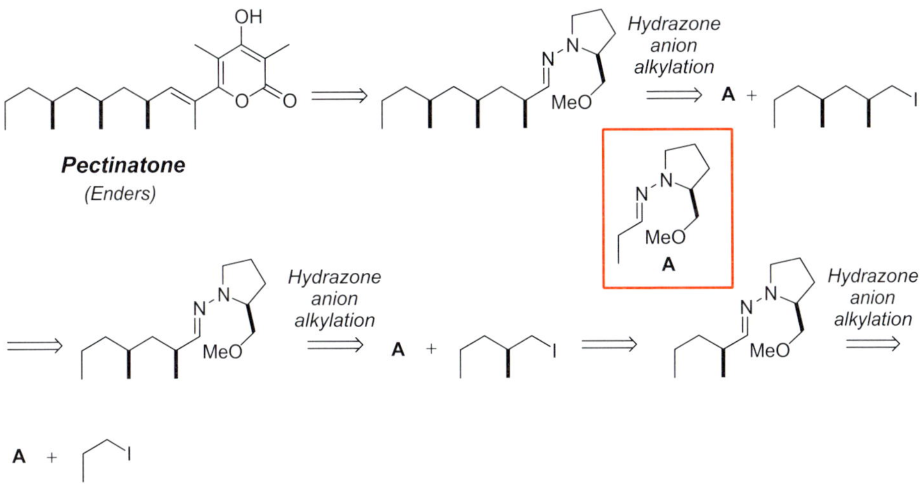

Pectinatone
(Enders)

D. Oppolzer sultam conjugate addition

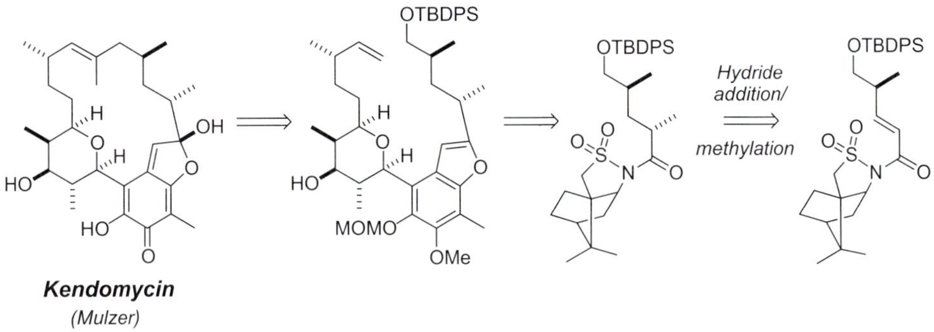

Kendomycin
(Mulzer)

Figure 4.16 (continued).

found in polyketide-derived natural products [42]. Many other pioneering and noteworthy contributions toward stereoselective C-alkylations and aldol reactions have enriched the repertoire of such stoichiometric asymmetric reactions [43].

4.5.3
Substrate control approach in cycloadditions

Pericyclic reactions, including the venerable Diels-Alder reaction, have featured prominently in many natural product synthesis, including asymmetric versions (see Chapter 2, section 2.11.1). Substrate-controlled hetero Diels-Alder (HDA) reactions, popularized by Danishefsky [44], have also been successfully applied to the synthesis of natural products containing tetrahydropyran motifs. For example, two hetero Diels-Alder cycloadditions were used by Danishefsky and coworkers [45] at different stages of the total synthesis of zincophorin (Figure 4.17 A).

An example of a diastereoselective intramolecular Diels-Alder reaction controlled by substrate conformation is shown in the case of the synthesis of abyssomycin C by Sorensen [46] (Figure 4.17 B). The resident C-methyl appendages in the chiral, non-racemic substrates exhibited a profound influence on the stereocontrolled cycloaddition reactions.

In the latest synthesis of yohimbine, Jacobsen and coworkers relied on a substrate-controlled intramolecular Diels-Alder reaction to establish four stereocenters simultaneously [47] (Figure 4.17 C).

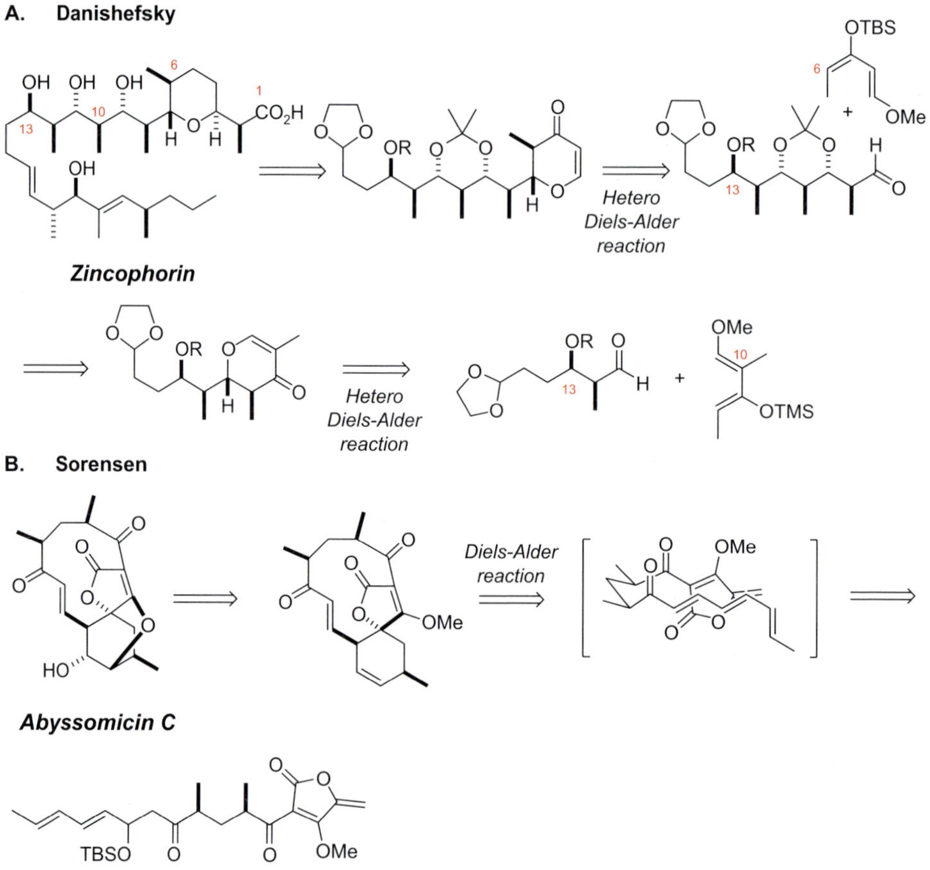

Figure 4.17 Substrate-controlled hetero Diels-Alder and Diels-Alder cycloaddition reactions in natural product synthesis.

C.   Jacobsen

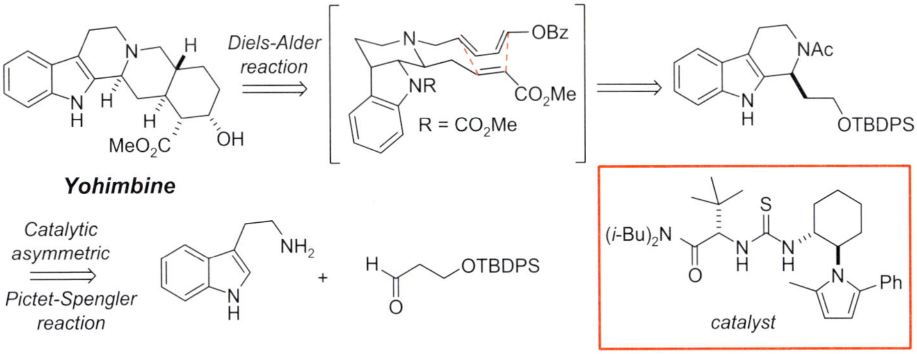

Figure 4.17   (*continued*).

4.5.4
Biomimetic cyclization approach

The preparative potential and utility of biomimetic cyclizations were elegantly demonstrated in the pioneering work of W. S. Johnson more than forty years ago [48]. The cationic polyene cascade reaction of strategically deployed double bonds in an acyclic hydrocarbon can be highly stereocontrolled, and leads directly to tetracyclic steroidal motifs as shown for the classical synthesis of progesterone (Figure 4.18 A) [14c, 48]. Polycyclic hydrocarbons, such as the endriadric acids, are other examples of accomplishments in total synthesis inspired by Nature's biosynthetic pathways. Sequential 8π and 6π electrocyclizations, followed by an intramolecular Diels-Alder reaction, allowed Nicolaou and coworkers to rapidly prepare the tetracyclic hydrocarbon acid (Figure 4.18 B) [49]. A masterful display of biomimetic synthesis is demonstrated in the beautifully crafted series of cascade reactions in the synthesis of methyl homosecodaphnyllate from simple starting materials by Heathcock and coworkers (Figure 4.18 C) [50].

The biomimetic synthesis approach shown in the above three examples is applicable to a large number of compounds belonging to one and the same family of natural products.

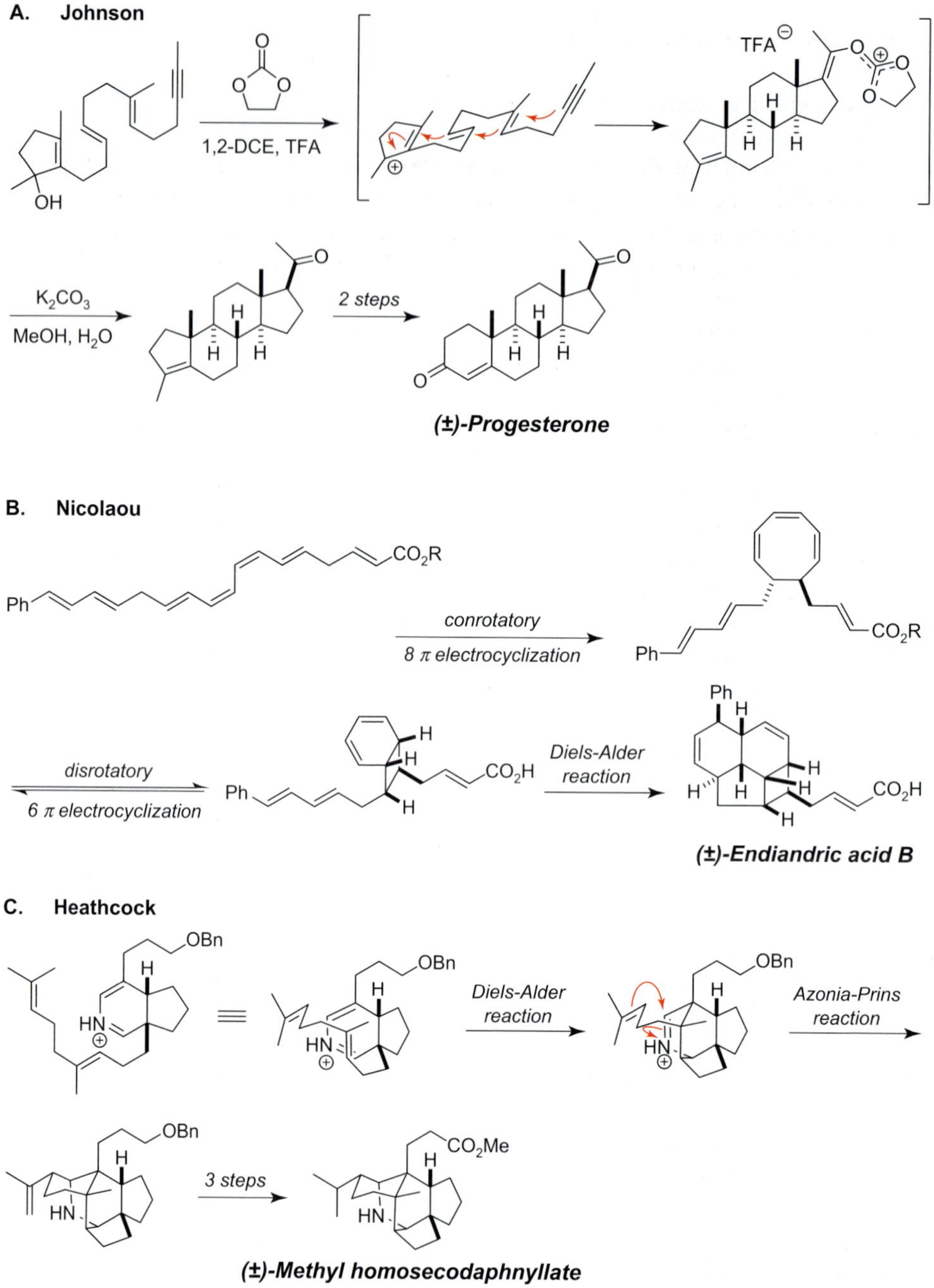

A. Johnson

B. Nicolaou

(±)-Endiandric acid B

C. Heathcock

(±)-Methyl homosecodaphnyllate

Figure 4.18 Selected examples of biomimetic cascade reactions.

4.6
Catalysis and Chirality in Total Synthesis

The development of catalytic methods for asymmetric synthesis has had an increasing impact on the strategic planning process in total synthesis of both natural and unnatural products in the last two decades [51]. The practicality of these methods can also be found in applications relating to the large scale synthesis of drug substances and polymers (see Chapter 2, section 2.3). The successful inclusion of one or more catalytic steps of bond formation, particularly involving stereochemical control, can be highly beneficial with regard to the efficiency and practicality of a synthetic sequence. Clearly, knowledge of the feasibility of a given catalytic asymmetric transformation will allow its judicious inclusion in the synthesis plan. In fact, strategic bond disconnections are often planned based on the desire to incorporate one or more catalytic steps in a given sequence *en route* to a target molecule.

Catalytic asymmetric reactions based on transition metals such as Pd, Rh, and Ru, in conjunction with appropriate chiral ligands, are among the most widely used in natural product synthesis. The design of the ligands and the successful development of a new catalytic system is a highly stimulating endeavor [51, 52]. However, subtle effects within the ligand structure, its coordinating capacity, and the nature of the solvent, added to electronic, geometric, and topological considerations, may have to be extensively studied before a catalytic reaction can be claimed to be of general utility.

Catalysis can also be achieved in metal-free systems, using organic acids and bases as sources of chirality in stereodifferentiating reactions (see Chapter 2, section 2.12, Chapter 5, section 5.8) [53]. Chiral Lewis acids [54] and bases [55] offer an exciting new forum to explore asymmetric C–C bond-forming reactions with high levels of selectivity.

The literature records numerous recent total syntheses in which critical C–C bonds are constructed through the application of catalytic reactions such as, ring-closing and cross-metathesis, Stille, Heck, Suzuki, Negishi, and Sonogashira couplings, among others (see Chapter 2, section 2.12.4). Pioneering methods for the catalytic enantioselective reduction of ketones, or the stereoselective reduction of substituted double bonds that generate single stereogenic centers are also well-known (see Chapter 2, section 2.12.3).

The impact of selected catalytic asymmetric C–C bond-forming reactions in natural product synthesis is illustrated by the three examples in Figure 4.19. In all cases, emphasis was placed on the application of methods that produce critical bonds with control of absolute stereochemistry, especially those involving quaternary centers and C–O bonds in the intended targets.

The synthesis of allocyathin B_2 by Trost and coworkers [56] involved an intermolecular Tsuji-Trost asymmetric allylic alkylation (AAA) reaction mediated by a C_2-symmetric bis-triphenylphosphine diamide ligand to provide a cyclopentanone intermediate containing a quaternary stereogenic center (Figure 4.19 A).

Catalytic asymmetric Mukaiyama-type aldol reactions in the presence of C_2-symmetric bis-oxazoline ligands and copper were used by Evans and coworkers [57] in the synthesis of a precursor to the tetrasubstituted tetrahydropyran segment of callipeltoside A (Figure 4.19 B) (see also, Chapter 8, section 8.8.4).

In the total synthesis of quadrigemine C, Overman and coworkers [58] demonstrated the power of catalytic asymmetric Heck reactions mediated by a BINAP ligand in the generation of quaternary stereogenic centers in complex natural product architectures (Figure 4.19 C).

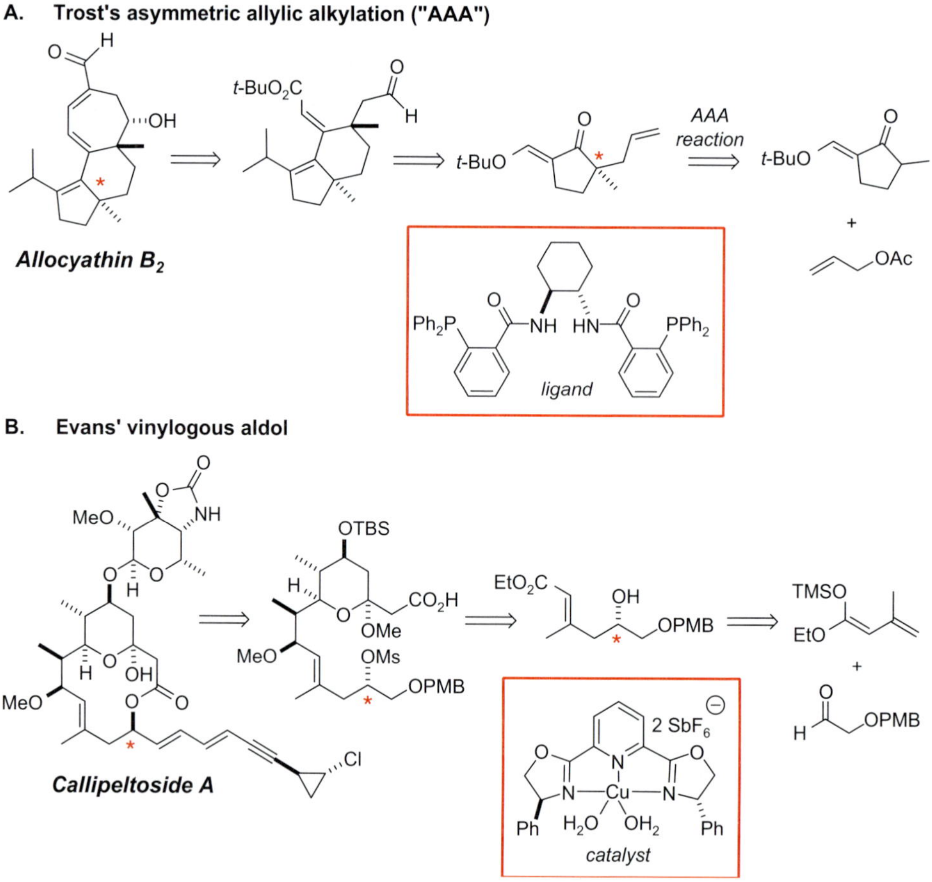

A.   Trost's asymmetric allylic alkylation ("AAA")

Allocyathin B$_2$

B.   Evans' vinylogous aldol

Callipeltoside A

Figure 4.19 Key strategic metal-based catalytic asymmetric C–C bond-forming reactions in natural product synthesis.

C. Overman's asymmetric Heck reaction

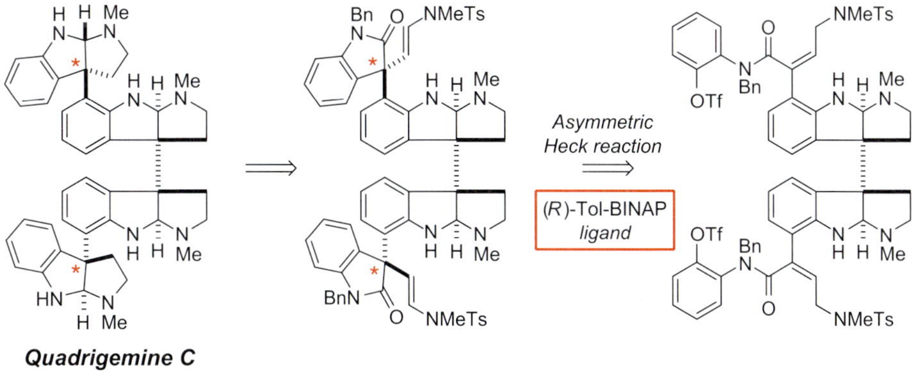

Quadrigemine C

Figure 4.19 (continued).

The discovery of the Sharpless asymmetric epoxidation reaction of allylic alcohols has had a profound impact on total synthesis for over three decades now (see Chapter 2, section 2.12.3). To illustrate the continuing relevance of this reaction in synthesis planning and execution, we show its use at three different stages in the total synthesis of the complex natural product methyl sarcophytoate by Nakata and coworkers [59] (Figure 4.20).

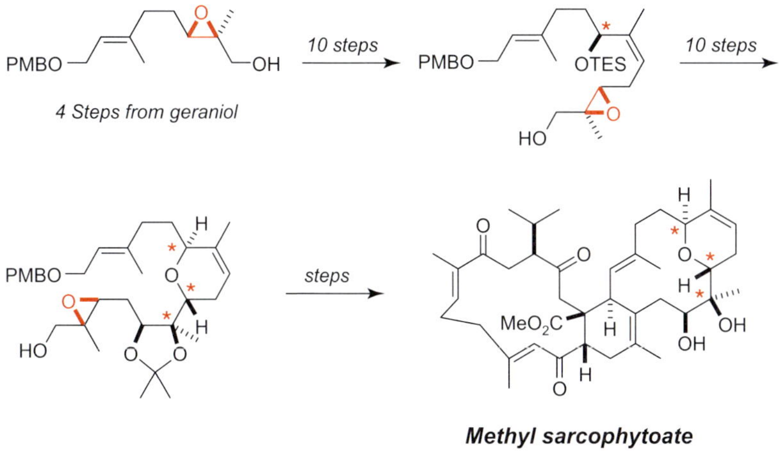

Methyl sarcophytoate

Figure 4.20 Impact of the Sharpless asymmetric epoxidation (SAE) reaction in natural product synthesis. Centers resulting from an SAE reaction are designated by an asterisk.

The broad utility of organocatalytic reactions was demonstrated by the MacMillan group in the total synthesis of flustramine B [60]. A crucial quaternary stereogenic center was established early in the synthetic sequence involving a sequential conjugate addition/cyclization cascade (Figure 4.21) (see also Chapter 5, section 5.8).

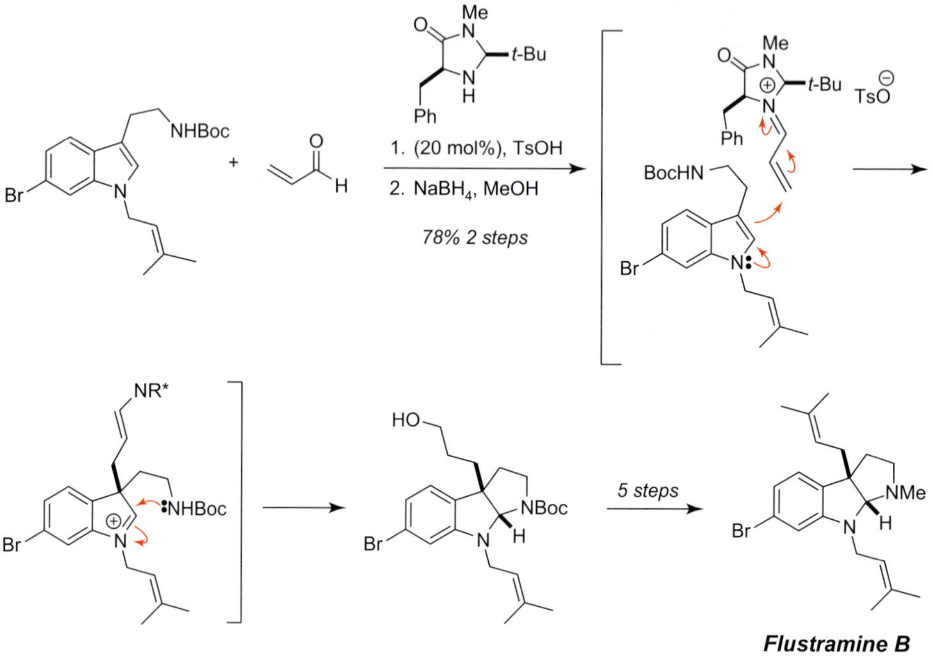

Figure 4.21 Total synthesis of flustramine B relying on an organocatalytic conjugate addition/cyclization cascade.

A selection of natural products syntheses utilizing various aspects of iminium and enamine catalysis is highlighted in Figure 4.22. Selected examples involving aldol reactions [61], Mannich reactions [62], α-hydroxylation of carbonyl compounds [63], vinylogous Michael additions [64], and intermolecular Diels-Alder reactions [65] are shown.

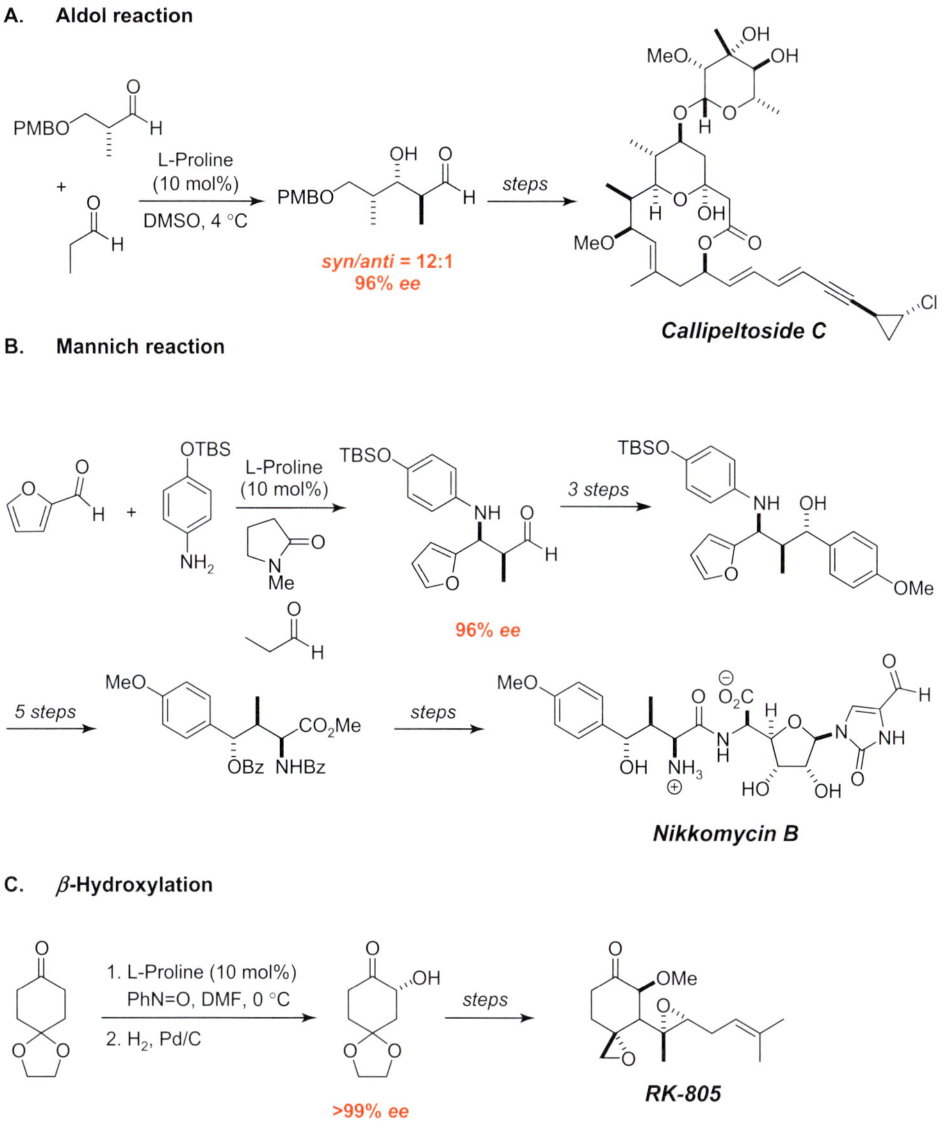

A. Aldol reaction

B. Mannich reaction

C. β-Hydroxylation

Figure 4.22 Selected organocatalytic reactions in natural product synthesis.

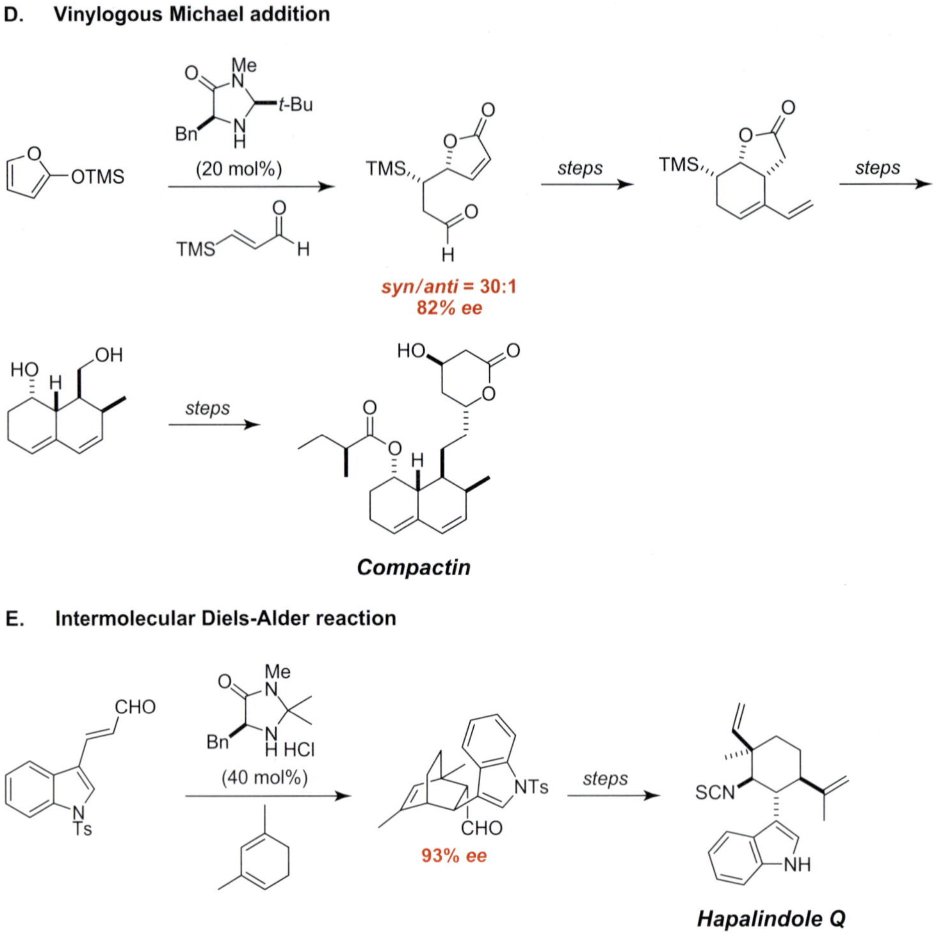
D. Vinylogous Michael addition
OTMS
Me
t-Bu
Bn
(20 mol%)
TMS
H
TMS,,
syn/anti = 30:1
82% ee
steps
TMS,,
steps
HO
OH
H
steps
HO
Compactin
E. Intermolecular Diels-Alder reaction
CHO
N
Ts
Me
Bn
H HCl
(40 mol%)
NTs
CHO
93% ee
steps
SCN
NH
Hapalindole Q

Figure 4.22 *(continued)*.

4.6.1
Applications of asymmetric catalysis to drug discovery

Pioneering efforts within the pharmaceutical industry have contributed to the discovery of many drug entities with enormous benefits to humankind. With increasing attention being paid to green chemistry and efficiency, many industrial processes have adopted catalysis as one or more key steps in the manufacture of drugs. However, the presence of multiple stereogenic centers in a molecule presents logistical and practical challenges. In Chapter 2, we highlighted a number of polyfunctional drug molecules, in which stereochemistry was a primordial concern in their synthesis.

Academic groups have excelled at the development of catalytic asymmetric reactions, especially in the context of natural product synthesis, over the last two decades. It is gratifying to see that the power of catalysis is also being exploited for the synthesis of drug substances. Salient features of metal-based and metal-free catalytic reactions are shown in Figure 4.23. Thus, Trost and Andersen [66] described the synthesis of a key fragment of the anti-HIV-1 protease inhibitor Aptivus using organopalladium chemistry (Figure 4.23 A). Shibasaki and coworkers [67] provided an alternative route to the influenza drug Tamiflu relying on an unusual organocatalyst derived from a sugar scaffold (Figure 4.23 B). This offered a different, albeit longer approach compared to Corey's synthesis [68] (see also Chapter 14, section 14.9) [69]. Ghosh and coworkers [70] relied on structure-based design to develop darunavir (Prezista), a new generation anti-HIV-1 drug. The hydroxyethylamine subunit was prepared from a Sharpless asymmetric epoxidation, followed by a regioselective hydroxy-assisted epoxide ring-opening with bis-azidoTi(OiPr)$_2$ (Figure 4.23 C). Oxazaborolidine-mediated Lewis acid catalysis was highly efficient in Diels-Alder reactions to produce steroidal motifs as shown by Corey and coworkers [71] in the synthesis of estrone and desogestrel (Figure 4.23 D).

A number of collaborations between academia and industry have resulted in marketed drugs or candidates for clinical studies [72]. It is hoped that this trend will continue in the future with the cooperation of expert groups that bridge the gap between synthetic chemistry and medicinally relevant biological applications [73] (see also, Chapter 18, section 18.7).

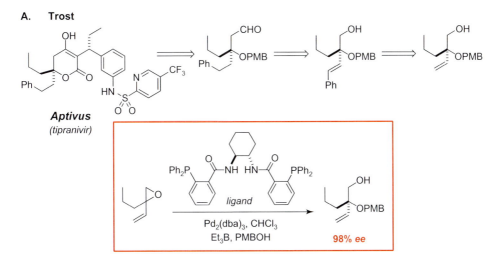

Figure 4.23 Catalysis in the total synthesis of medicines.

B.   Shibasaki

Tamiflu
(oseltamivir phosphate)

Asymmetric aziridine opening

91% ee

catalyst

C.   Ghosh

Prezista
(darunavir)

Sharpless asymmetric epoxidation

Figure 4.23   *(continued).*

D. Corey

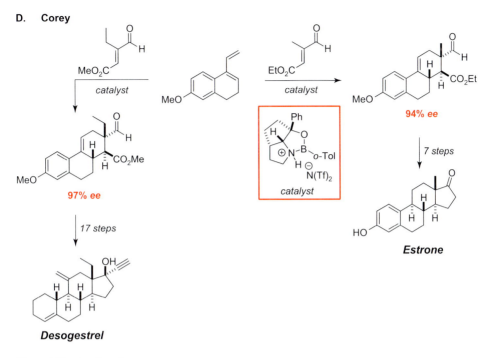

Figure 4.23 *(continued)*.

References

1. (a) Satratoxin H: Eppley, R. M.;
   Mazzola, E. P.; Highet, R. J.; Bailey,
   W. J. *J. Org. Chem.* 1977, **42**, 240;
   (b) daphniglaucin C: Morita, H.;
   Takatsu, H.; Shen, Y.-C.; Kobayashi,
   J. *Tetrahedron Lett.* 2004, **45**, 901;
   (c) nymania 1: Gunatilaka, A. A. L.;
   Bolzani, V.; Dagne, E.; Hofmann,
   G. A.; Johnson, R. K.; McCabe, F. L.;
   Mattern, M. R.; Kingston, D. G. I. *J.
   Nat. Prod.* 1998, **61**, 179; (d) stylissa-
   dine A: (i) Grube, A.; Kück, M. *Org.
   Lett.* 2006, **8**, 4675; (ii) Buchanan,
   M. S.; Carroll, A. R.; Addepalli, R.;
   Avery, V. M.; Hooper, J. N. A.; Quinn,
   R. J. *J. Org. Chem.* 2007, **72**, 2309;
   (e) bipleiophylline: Kam, T.-S.; Tan, S.-J.;
   Ng, S. W.; Komiyama, K. *Org. Lett.* 2008,
   10, 3749.

2. For example, see: (a) *Unleashing the
   Right Side of the Brain*, Williams, R. H.;
   Stockmyer, J., 1987, The Stephen Green
   Press Inc., Brattleboro, VT; (b) *Drawing
   on the Right Side of the Brain*, Edwards
   B., 1979, St-Martins Press, NY; see
   also: Hanessian, S.; Franco, J. Larouche,
   B. *Pure Appl. Chem.* 1990, **62**, 1887.

3. For reviews, see: (a) Bhat, S. V. *Progress
   Chem. Nat. Prod.* 1993, 621; (b)
   Colombo, M. I.; Zinczuk, J.; Rúveda,
   E. A. *Tetrahedron* 1992, **48**, 963.

4. (a) Corey, E. J.; Da Silva Jardine, P.;
   Rohloff, J. C. *J. Am. Chem. Soc.* 1988,
   110, 3672; (b) Corey, E. J.; DaSilva;
   Jardine, P.; Mohri, T. *Tetrahedron Lett.*
   1988, **29**, 6409; see also: (c) Nicolaou,
   K. C.; Li, W. S. *J. Chem. Soc., Chem.
   Commun.* 1985, 421; (d) Jenkins P. R.;
   Menear, K. A.; Barraclough, P.; Nobbs,
   M. S. *J. Chem. Soc., Chem. Commun.*
   1984, 1423.

5. Ziegler, F. E.; Jaynes, B. H.; Saindane, M. T. *J. Am. Chem. Soc.* 1987, **109**, 8115.

6. Hashimoto, S.-I.; Sakata, S.; Sonegawa, M.; Ikegami, S. *J. Am. Chem. Soc.* 1988, **110**, 3670.

7. Woodward, R. B. in *Perspectives in Organic Chemistry*, Todd, A. R.; Ed.; 1956, Interscience, Publishers Inc. NY.

8. For example, see: (a) *Strategic Applications of Named Reactions in Organic Synthesis: Background and Detailed Mechanisms*, Kürti, L.; Czakó, B., 2005, Elsevier, Academic Press, NY; see also: (b) *Comprehensive Organic Name Reactions*, Wand, Z. 2009, Wiley, NY; (c) *Name reactions: A collection of Detailed Mechanisms and Synthetic Applications*, Li. J. J. 2009, 4th Edition, Springer Verlag, Berlin.

9. (a) *From Bench to Market – The Evolution of Chemical Synthesis*, Cabri, W.; DiFabio, R., 2000, Oxford University Press, UK; (b) *Process Chemistry in the Pharmaceutical Industry*, Gadamasetti, K. G., Ed.; 1999, Vol. 1, Dekker, NY; (c) *Process Chemistry in the Pharmaceutical Industry*, vol. 2, Gadamasetti, K.; Braish, T.; Eds.; 2007, CRC Press, Boca Raton; FL; for a thematic issue, see: (d) *Chem. Rev.* 2006, **106**, 2581, Lipton, M. F.; Barrett, A. G. M. Guest editors.

10. (a) *Architect and Artist in the World of Molecules*, Benfey, O. T.; Morris, P. E.; Eds.; 2001, Chemical Heritage Foundation, Philadelphia, PA; see also: (b) *The Road to Stockholm*, Hargittai, I., 2002, Oxford University Press, UK; (c) Nobel Foundation. *Nobel Lectures in Chemistry*, 1999, World Scientific Publishing, Singapore.

11. (a) Woodward, R. B.; Sondheimer, F.; Taub, D. *J. Am. Chem. Soc.* 1951, **73**, 4057; (b) Woodward, R. B.; Sondheimer, F.; Taub, D.; Heusler, K.; McLamore, W. M. *J. Am. Chem. Soc.* 1952, **74**, 4223.

12. Nemoto, H.; Matsuhashi, N.; Imaizumi, M.; Nagai, M.; Fukumoto, K. *J. Org. Chem.* 1990, **55**, 5625.

13. Lynch, J. E.; Eliel, E. L. *J. Am. Chem. Soc.* 1984, **106**, 2943.

14. (a) Johnson, W. S.; Brinkmeyer, R. S.; Kapoor, V. M.; Yarnell, T. M. *J. Am. Chem. Soc.* 1977, **99**, 8341; (b) Johnson, W. S.; Frei, B.; Gopalan, A. S. *J. Org. Chem.* 1981, **46**, 1512; for a recent organocatalytic version of this reaction see: (c) Rendler, S.; MacMillan, D. W. C. *J. Am. Chem. Soc.* 2010, **132**, 5027.

15. Pelletier, P. J.; Caventou, J. B. *Ann. Chim. Phys.* 1818, **8**, 323.

16. *Bitter Nemesis – The Intimate History of Strychnine*, Buckingham, J. 2007, CRC Press, Boca Raton, FL.

17. Woodward, R. B.; Cava, M. P.; Ollis, W. D.; Hunger, A.; Daeniker, H. U.; Schenker, K. *J. Am. Chem. Soc.* 1954, **76**, 4749.

18. (a) Bonjoch, J.; Solé, D. *Chem. Rev.* 2000, **100**, 3455; (b) Beifuss, U. *Angew. Chem. Int. Ed.* 1994, **33**, 1144; see also: Chapter 3, section 3.2 for a complete list of authors who have prepared strychnine; for a recent synthesis of strychnine, see: (c) Martin, D. B. C.; Vanderwal, C. D. *Chem. Sci.* 2011, **2**, 649.

19. (a) *The Logic of Chemical Synthesis*, Corey, E. J.; Cheng, X.-M., 1989 Wiley, NY; see also: (b) Corey, E. J. *Angew. Chem. Int. Ed.* 1991, **30**, 455; (c) Corey, E. J. *Pure Appl. Chem.* 1967, **14**, 19.

20. *Dead-Ends and Detours*, Sierra, M. A.; de la Torre, M. C., 2004, Wiley-VCH, Weinheim.

21. (a) Knight, S. D.; Overman, L. E.; Pairaudeau, G. *J. Am. Chem. Soc.* 1993, **115**, 9293; see also: (b) Knight, S. D.; Overman, L. E.; Pairaudeau, G. *J. Am. Chem. Soc.* 1995, **117**, 5776; for a review of iminium ion cyclizations, see: (c) Royer, J.; Bonin, M.; Micouin, L. *Chem. Rev.* 2004, **104**, 2311.

22. (a) Kuehne, M. E.; Xu, F. *J. Org. Chem.* 1993, **58**, 7490; (b) Kuehne, M. E.; Xu, F. *J. Org. Chem.* 1998, **63**, 9427.

23. (a) Solé, D.; Bonjoch, J.; García-Rubio, S.; Peidró, E.; Bosch, J. *Angew. Chem. Int. Ed.* 1999, **38**, 395; (b) Solé, D.; Bonjoch, J.; García-Rubio, S.; Peidró, E.; Bosch, J. *Chem. Eur. J.* 2000, **6**, 655.

24. (a) Rawal, V. H.; Michoud, C. *Tetrahedron Lett.* 1991, **32**, 1695; (b) Rawal, V. H.; Iwasa, S. *J. Org. Chem.* 1994, **59**, 2685.

25. Ohshima, T.; Xu, Y.; Takita, R.; Shimizu, S.; Zhong, D.; Shibasaki, M. *J. Am. Chem. Soc.* 2002, **124**, 14546.

26. Kaburagi, Y.; Tokuyama, H.; Fukuyama, T. *J. Am. Chem. Soc.* 2004, **126**, 10246.

27. (a) Nakanishi, M.; Mori, M. *Angew. Chem. Int. Ed.* 2002, **41**, 1934; (b) Mori, M.; Nakanishi, M.; Kajishima, D.; Sato, Y. *J. Am. Chem. Soc.* 2003, **125**, 9801.

28. Jones, S. B.; Simmons, B.; Mastracchio, A.; MacMillan, D. W. C. *Nature* 2011, **475**, 183.

29. (a) Masamune, S.; Choy, W.; Petersen, J. S.; Rita, L. R. *Angew. Chem. Int. Ed.* 1985, **24**, 1; see also: (b) Hoveyda, A. H.; Evans, D. A.; Fu, G. C. *Chem. Rev.* 1993, **93**, 1307.

30. (a) Woodward, R. B.; Logusch, E.; Nambiar, K. P.; Sakan, K.; Ward, D. E.; Au-Yeung, B.-W.; Balaram, P.; Browne, L. J.; Card, P. J.; Chen, C. H.; Chênevert, R. B.; Fliri, A.; Frobel, K.; Gais, H. J.; Garratt, D. G.; Hayakawa, K.; Heggie, W.; Hoppe, D.; Hoppe, I.; Hyatt, J. A.; Ikeda, D.; Jacobi, P. A.; Kim, K. S.; Kobuke, Y.; Kojima, K.; Krowicki, K.; Lee, V. J.; Leutert, T.; Malchenko, S.; Martens, J.; Matthews, R. S.; Ong, B. S.; Press, J. B.; Rajan Babu, T. V.; Rousseau, G.; Sauter, H. M.; Suzuki, M.; Tatsuta, K.; Tolbert, L. M.; Truesdale, E. A.; Uchida, I.; Ueda, Y.; Uyehara, T.; Vasella, A. T.; Vladuchick, W. C.; Wade, P. A.; Williams, R. M.; Wong, H. N.-C. *J. Am. Chem. Soc.* 1981, **103**, 3210; (b) Woodward, R. B. *et. al. J. Am. Chem. Soc.* 1981, **103**, 3213; (c) Woodward, R. B. *et. al. J. Am. Chem. Soc.* 1981, **103**, 3215.

31. Stork, G.; Paterson, I.; Lee, F. K. C. *J. Am. Chem. Soc.* 1982, **104**, 4686; for the most recent synthesis of erythronolide A see: Muri, D.; Lohse-Fraefel, N.; Carreira, E. M. *Angew. Chem. Int. Ed.* 2005, **44**, 4036 and references cited therein.

32. (a) Hanessian, S.; Rancourt, G.; Guindon, Y. *Can. J. Chem.* 1978, **56**, 1843; (b) Hanessian, S.; Rancourt, G. *Can. J. Chem.* 1977, **55**, 1111; (c) Hanessian, S. *Acc. Chem. Res.* 1979, **12**, 159; see also: (d) Nakajima, N.; Hamada, T.; Tamaka, T.; Oikawa, Y.;

Yonemitsu, O. *J. Am. Chem. Soc.* 1986, **108**, 4645; (e) Sviridov, A. F.; Ermolenko, M. S.; Yashunsky, D. V.; Borodkin, V. S.; Kochetkov, N. K. *Tetrahedron Lett.* 1987, **28**, 3835, 3839.

33. Grieco, P. A.; Inanaga, J.; Lin, N.-H.; Yanami, T. *J. Am. Chem. Soc.* 1982, **104**, 5781.

34. For example, see: (a) Evans, D. A. *Aldrichimica Acta* 1982, **15**, 23; (b) Evans, D. A.; Helmchen, G.; Rüping, M. in, *Asymmetric Synthesis: the Essentials*, Christmann, M.; Bräse, S. Eds.; 2007, p. 3, Wiley-VCH, Weinheim; for reviews, see: (c) McManus, H. A.; Guiry, P. J. *Chem. Rev.* 2004, **104**, 4151; (d) Desimoni, G.; Faita, G.; Jørgensen, K. A. *Chem. Rev.* 2006, **106**, 3561.

35. Evans, D. A.; Dow, R. L.; Shih, T. L.; Takacs, J. M.; Zahler, R. *J. Am. Chem. Soc.* 1990, **112**, 5290.

36. (a) Myers, A. G.; Yang, B. H.; Chen, H.; Kopecky, D. J. *Synlett* 1997, 457; (b) Myers, A. G.; Yang, B. H.; Chen, H.; Gleason, J. L. *J. Am. Chem. Soc.* 1994, **116**, 9361.

37. (a) Oppolzer, W.; Moretti, R.; Bernardinelli, G. *Tetrahedron Lett.* 1986, **27**, 4713; (b) Oppolzer, W. *Tetrahedron* 1987, **43**, 1969.

38. Vong, B. G.; Kim, S. H.; Abraham, S.; Theodorakis, E. A. *Angew. Chem. Int. Ed.* 2004, **116**, 4037.

39. (a) Job, A.; Janeck, C. F.; Bettray, W.; Peters, R.; Enders, D. *Tetrahedron* 2002, **58**, 2253; see also: (b) Magauer, T.; Martin, H. J.; Mulzer, J. *Chem. Eur. J.* 2010, **16**, 507; for a review of other total syntheses, see: (c) Martin, H. J.; Magauer, T.; Mulzer, J. *Angew. Chem. Int. Ed.* 2010, **49**, 5614.

40. Birkbeck, A. A.; Enders, D. *Tetrahedron Lett.* 1998, **39**, 7823; for a review see: Enders, D.; Bettray, W. in, *Asymmetric Synthesis: the Essentials*, Christmann, M.; Bräse, S. Eds.; 2007, p. 23, Wiley-VCH, Weinheim;

41. Pichlmair, S.; Marques, M. M. B.; Green, M. P.; Martin, H. J.; Mulzer, J. *Org. Lett.* 2003, **5**, 4657.

42. (a) Hanessian, S.; Giroux, S.; Mascitti, V. *Synthesis* 2006, 1057; see also: (b) Hoffmann, R. W. *Angew. Chem. Int. Ed.* 2000, **39**, 2094; see also: (c)

Negishi, E.-i.; Liang, B.; Novak, T.; Tan, Z. in, *Asymmetric Synthesis: the Essentials*, M.; Bräse, S. Eds.; 2007, p. 133, Wiley-VCH, Weinheim.

43. For example, see: (a) *Chiral Auxiliaries and Ligands in Asymmetric Synthesis*, Seyden-Penne, J., 1995, Wiley-Interscience, NY; (b) *Asymmetric Synthetic Methodology*, Ager, D. J.; East, M. B., 1996, CRC Press, Boca Raton, FL; (c) *Principles of Asymmetric Synthesis* 2nd Edition, Gawley, R. E.; Aubé, J., 2012, Pergamon, NY; (d) Gnas, Y.; Glorius, F. *Synthesis* 2006, 1899; see also: (e) *Asymmetric Synthesis: the Essentials*, M.; Bräse, S. Eds.; 2007, p. 23, Wiley-VCH, Weinheim.

44. For example, see: (a) Danishefsky, S. J.; De Ninno, M. P. *Angew. Chem. Int. Ed.* 1987, **26**, 15; for reviews, see (b) Tietze, L.; Kettschau, G. in *Topics in Current Chem.* 1997, **189**, 1; (c) Lin, L.; Liu, X.; Feng, X. *Synlett* 2007, 2147.

45. Danishefsky, S. J.; Selnick, H. G.; DeNinno, M. P.; Zelle, R. E. *J. Am. Chem. Soc.* 1987, **109**, 1572.

46. (a) Zapf, C. W.; Harrison, B. A.; Drahl, C.; Sorensen, E. J. *Angew. Chem. Int. Ed.* 2005, **44**, 6533; see also: (b) Nicolaou, K. C.; Harrison, S. T.; Chen, J. S. *Synthesis* 2009, 33.

47. Mergott, D. J.; Zuend, S. J.; Jacobsen, E. N. *Org. Lett.* 2008, **10**, 745.

48. (a) Johnson, W. S.; Gravestock, M. B.; McCarry, B. E. *J. Am. Chem. Soc.* 1971, **93**, 4332; for a review, see: (b) Yoder, R. A.; Johnston, J. N. *Chem. Rev.* 2005, **105**, 4730.

49. (a) Nicolaou, K. C.; Petasis, N. A.; Zipkin, R. E.; Uenishi, J. *J. Am. Chem. Soc.* 1982, **104**, 5555; (b) Nicolaou, K. C.; Petasis, N. A.; Uenishi, J.; Zipkin, R. E. *J. Am. Chem. Soc.* 1982, **104**, 5557; (c) Nicolaou, K. C.; Zipkin, R. E.; Petasis, N. A. *J. Am. Chem. Soc.* 1982, **104**, 5558; (d) Nicolaou, K. C.; Petasis, N.; Zipkin, R. E. *J. Am. Chem. Soc.* 1982, **104**, 5560.

50. Ruggeri, R. B.; Hansen, M. M.; Heathcock, C. H. *J. Am. Chem. Soc.* 1988, **110**, 8734.

51. (a) *Fundamentals of Asymmetric Catalysis*, Walsh, P. J.; Kozlowski, M. C., 2008, University Science Books; Sausalito,

CA; (b) *Catalysis of Organic Reactions*, Schmidt, S. R.; Ed.; 2007, CRC Press, Boca Raton, FL; (c) *New Frontiers in Asymmetric Catalysis*, Mikami, K.; Lautens, M.; Eds.; 2007, Wiley-VCH, Weinheim; (d) *Catalysis from A to Z*, Cornils, B.; Herrmann, W. A.; Muhler, M.; Wong, C.-H.; Eds.; 2007, Wiley-VCH, Weinheim; (e) *Comprehensive Asymmetric Catalysis*, vol. I-III; Jacobsen, E. N.; Pfaltz, A.; Yamamoto, H.; Eds.; 2000, Springer-Verlag, Berlin; (f) *Catalytic Asymmetric Synthesis*, Ojima, I.; Ed.; 2000, Wiley-VCH, Weinheim; (g) *Principles and Applications of Asymmetric Synthesis*, Lin, G.-Q; Li, Y.-M; Chan, A. S. C., 2001, Wiley-Interscience, UK; see also: reference [42].

52. For example, see: (a) Hargaden, G. G.; Guiry, P. J. *Chem. Rev.* 2009, **109**, 2505; (b) Handy, S. T. *Curr. Org. Chem.* 2000, **4**, 363.

53. For example see: (a) *Enantioselective Organocatalysis: Reactions and Experimental Procedures*, Dalko, P. I. Ed.; 2007, Wiley-VCH, Weinheim; (b) *Asymmetric Organocatalysis: From Biomimetic Concepts to Applications in Asymmetric Synthesis*, Berkessel, A.; Gröger, H., 2005, Wiley-VCH, Weinheim; see also: (c) deFigueiredo, R. M.; Christmann, M. *Eur. J. Org. Chem.* 2007, 2575; (d) Marquéz-López, E.; Herrera, R. P.; Christmann, M. *Nat. Prod. Rep.* 2010, **27**, 1138.

54. *Lewis Acids in Organic Synthesis*, Yamamoto, H.; Ed.; 2000, Wiley-VCH, Weinheim; see also: Yamamoto, H.; Futatsugi, K. *Angew. Chem. Int. Ed.* 2005, **44**, 1924.

55. Denmark, S.; Beutner, G. L. *Angew. Chem. Int. Ed.* 2008, **47**, 1560.

56. Trost, B. M.; Dong, L.; Schroeder, G. M. *J. Am. Chem. Soc.* 2005, **127**, 2844.

57. (a) Evans, D. A.; Hu, E.; Burch, J. D.; Jaeschke, G. *J. Am. Chem. Soc.* 2002, **124**, 5654; see also: (b) Evans, D. A.; Burch, J. D. *Org. Lett.* 2001, **3**, 503.; (c) Evans, D. A.; Hu, E.; Tedrow, J. S. *Org. Lett.* 2001, **3**, 3133; for other syntheses, see Chapter 8, reference [54].

58. (a) Lebsack, A. D.; Link, J. T.; Overman, L. E.; Stearns, B. A. *J. Am. Chem. Soc.* 2002, **124**, 9008; for reviews, see:

(b) Steven, A.; Overman, L. E. *Angew. Chem. Int. Ed.* 2007, **46**, 5488; (c) Douglas, C. J.; Overman, L. E. *Proc. Natl. Acad. Sci. USA.* 2004, **101**, 5363.

59. Ichige, T.; Okano, Y.; Kanoh, N.; Nakata, M. *J. Am. Chem. Soc.* 2007, **129**, 9862.

60. Austin, J. F.; Kim, S.-G.; Sinz, C. J.; Xiao, W.-J.; MacMillan, D. W. C. *Proc. Natl. Acad. Sci. USA.* 2004, **101**, 5482.

61. Carpenter, J.; Northrup, A. B.; Chung, D.; Wiener, J. J. M.; Kim, S.-G.; MacMillan, D. W. C. *Angew. Chem. Int. Ed.* 2008, **47**, 3568.

62. Hayashi, Y.; Urushima, T.; Shin, M.; Shoji, M. *Tetrahedron* 2005, **61**, 11393.

63. Yamaguchi, J.; Toyoshima, M.; Shoji, M.; Kayeda, H.; Osada, H.; Hayashi, Y. *Angew. Chem. Int. Ed.* 2006, **45**, 789.

64. Robichaud, J.; Tremblay, F. *Org. Lett.* 2006, **8**, 597.

65. Kinsman, A. C.; Kerr, M. A. *J. Am. Chem. Soc.* 2003, **125**, 14120.

66. Trost, B. M.; Andersen, N. G. *J. Am. Chem. Soc.* 2002, **124**, 14320.

67. (a) Fukuta, Y.; Mita, T.; Fukuda, N.; Kanai, M.; Shibasaki, M. *J. Am. Chem. Soc.* 2006, **128**, 6313; see also: (b) Yamatsugu, K.; Yin, L.; Kamijo, S.; Kimura, Y.; Kanai, M.; Shibasaki, M. *Angew. Chem. Int. Ed.* 2009, **48**, 1070.

68. Yeung, Y.-Y.; Hong, S.; Corey, E. J. *J. Am. Chem. Soc.* 2006, **128**, 6310.

69. For reviews, see: (a) Magano, J. *Chem. Rev.* 2009, **109**, 4398; (b) Shibasaki, M.; Kanai, M. *Eur. J. Org. Chem.* 2008, 1839; (c) Farina, V.; Brown, J. D. *Angew. Chem. Int. Ed.* 2006, **45**, 7330.

70. (a) Ghosh, A. K.; Dawson, Z. L.; Mitsuya, H. *Bioorg. Med. Chem.* 2007, **15**, 7576; see also: (b) Ghosh, A. K.; Xu, C.-X.; Rao, K. V.; Baldridge, A.; Agniswamy, J.; Wang, Y.-F.; Weber, I. T.; Manabu, A.; Miguel, S. G. P.; Amano, M.; Mitsuya, H. *Chem. Med. Chem.* 2010, **5**, 1850; (c) Ghosh, A. K. *J. Org. Chem.* 2010, **75**, 7967.

71. Hu, Q.-Y.; Rege, P. D.; Corey, E. J. *J. Am. Chem. Soc.* 2004, **126**, 5984.

72. For example, see: (a) Baxendale, I. R.; Hayward, J. J.; Ley, S. V.; Tranner, G. K. *Chem. Med. Chem.* 2007, **2**, 788; (b) Gray, N. S. *Nature Chem. Biol.* 2006, **2**, 649; (c) Chin-Dusting, J.; Mizrahi, J.; Jennings, G.; Fitzgerald, D. *Nat. Rev. Drug Discov.* 2005, **4**, 891; (d) Abbenante, G.; Fairlie, D. P. *Med. Chem.* 2005, **1**, 71; see also: (e) Jarvis, L. M. in, *Chem. Eng. News* 2010, Nov. 8, p. 14.

73. For example, see: (a) Szpilman, A. M.; Carreira, E. M. *Angew. Chem. Int. Ed.* 2010, **49**, 9592; (b) Morton, D.; Leach, S.; Cordier, C.; Warriner, S.; Nelson, A. *Angew. Chem. Int. Ed.* 2009, **48**, 104; (c) Hanessian, S. *Chem. Med. Chem.* 2006, **1**, 1300.

5
Sources of Enantiopure Compounds

Pasteur's serendipitous resolution of enantiomeric tartaric acids through crystallization may have gone unnoticed were it not for his astute sense of observation [1]. Subsequent to his milestone discovery, he proffered the well-known dictum:

Dans les champs de l'observation, le hasard ne favorise que les esprits préparés (In the fields of observation, chance favors the prepared mind).

Chirality manifests itself directly or indirectly in virtually all life processes. In the domain of organic synthesis, the need to obtain enantiomerically pure compounds whenever applicable, is the accepted norm, especially for natural products or drug subtances [2]. Indeed, the stringent rules imposed by the Food and Drug Administration (FDA), in the United States, and related organizations around the world, regarding the approval of chiral drugs for clinical use has heightened the need to synthesize enantiopure compounds. Consequently, great advances have been made in conventional and non-conventional ways for securing enantiopurity in organic compounds [3]. Among the well-known practices toward this objective are optical resolution, microbial transformations (biocatalysis), enzymatic desymmetrization, asymmetric chemical synthesis, and the use of enantiopure natural substances such as α-amino acids, carbohydrates, hydroxy acids, terpenes and other readily available compounds (Table 5.1).

Table 5.1 General approaches to obtain enantiopure compounds.

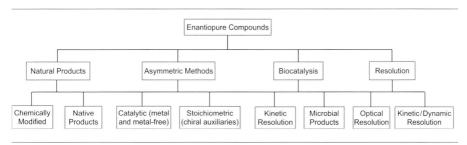

Recent advances in genetic engineering [4] have introduced sophisticated biotechnological methods involving biocatalysis, biotransformations and related

Design and Strategy in Organic Synthesis: From the Chiron Approach to Catalysis, First Edition.
Stephen Hanessian, Simon Giroux, and Bradley L. Merner.
© 2013 Wiley-VCH Verlag GmbH & Co. KGaA. Published 2013 by Wiley-VCH Verlag GmbH & Co. KGaA.

techniques to produce a variety of enantiopure or enantioenriched organic compounds. Chiral separation methods have enormously expedited the aquisition of both small and large quantities of enantiopure compounds [5].

Chemical and enzymatic methods of preparing enantiopure compounds will be briefly highlighted in this Chapter. Greater emphasis will be placed on functionalized compounds that can be used as starting materials for broad synthetic applications, especially toward the synthesis of natural products, chiral drug substances, and related entities.

5.1
Optical Resolution

Resolution of enantiomers from a racemic mixture is a time-honored method for the separation of enantiopure compounds [6]. Traditionally, carboxylic acids can be converted into salts using chiral, non-racemic amines exemplified by (R)- or (S)-α-methylbenzylamine, or by alkaloids such as brucine. Racemic amines can be resolved into the corresponding enantiopure (or enriched) ammonium carboxylate salts with chiral, non-racemic carboxylic acids such as, (R,R)- or (S,S)-tartaric acid, D- or L-malic acid, and enantiomers of camphorsulfonic acid. In all cases, crystallization techniques are used to separate pure diastereomeric salts, where one diastereomer selectively crystallizes and the other diastereomer remains in solution. More than one recrystallization may be necessary to achieve high enantiopurity after removal of the counter ion.

By its very nature, resolution also produces the unwanted isomer in equal amounts. This may be recycled by racemization and crystallization. In this regard, resolution may be very practical when the target compound contains a single stereogenic center. Many industrial processes still adopt optical resolution as a practical method to access a final enantiopure drug substance, or an early intermediate. The synthesis and resolution of the antihypertensive drug Vasotec (enalapril) is shown in Figure 5.1.

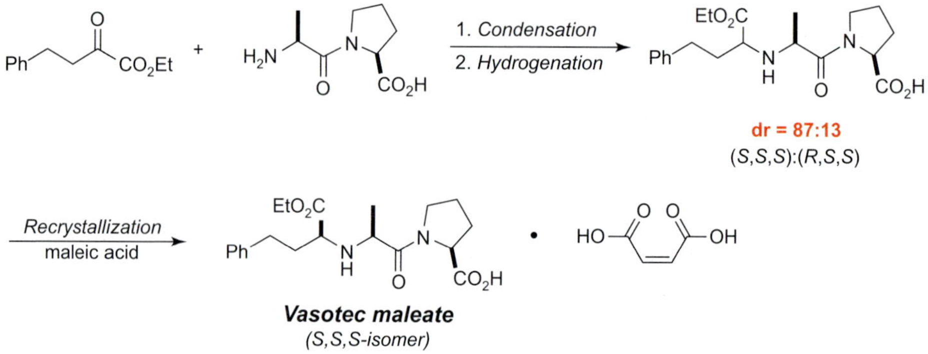

Figure 5.1 Synthesis and resolution of enantiopure Vasotec.

5.2
Chemical Kinetic Resolution (KR)

5.2.1
Classical, natural, and parallel methods

Enantiopure compounds can be obtained by kinetic resolution, a process in which one enantiomer in a racemic mixture reacts much faster than its optical antipode [6, 7]. In the classical example by Markwald and McKenzie [7c], esterification of racemic mandelic acid with (−)-menthol proceeded at a higher rate with the (R)-enantiomer when the reaction was stopped before being completed. In the process, the (S)-enantiomer remained unreacted and became enriched in the mixture. A more modern example of kinetic resolution relies on the differential reaction rates in the presence of a chiral catalyst, as in the epoxidation of a racemic allylic alcohol according to Sharpless (Figure 5.2 A) [8]. In mutual kinetic resolution (MKR) two sets of racemic compounds are allowed to react, and the resolution, through a kinetic pathway, is due to the mutual reaction of the racemic compounds. In parallel kinetic resolution (PKR), each enantiomer in a racemic mixture affords a pair of optically pure or enriched diastereomeric products (Figure 5.2 B) [9].

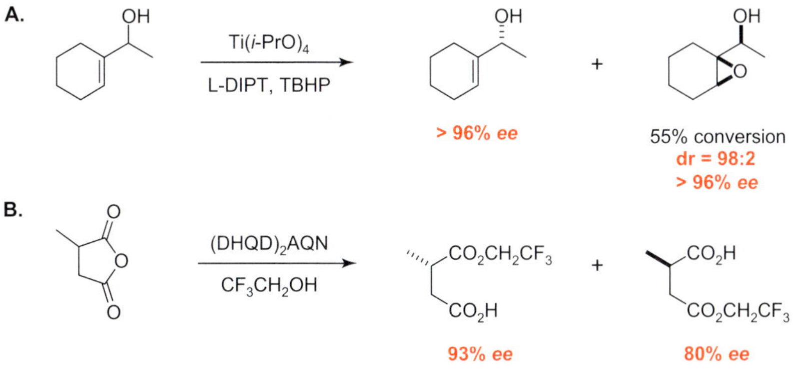

Figure 5.2   A. Chemical kinetic resolution; B. Parallel kinetic resolution.

5.2.2
Dynamic chemical kinetic resolution

The technique of dynamic chemical kinetic resolution is a useful way to continually recycle the unwanted isomer in a reaction medium containing two isomers that are in rapid equilibrium [10]. In this method, the rate of racemization of the unwanted isomer should exceed that of the desired isomer. It follows that the conversion of the unwanted isomer to the product should proceed at a much slower rate. In dynamic kinetic resolution it is possible to convert each enantiomer to product based on the Curtin-Hammett principle. Classical early examples from the Noyori [11a] and

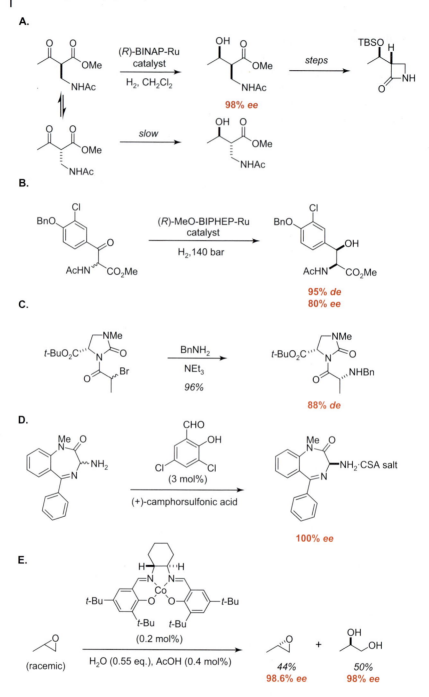

Figure 5.3 A. Noyori's dynamic kinetic resolution; B. Genêt's dynamic kinetic resolution; C and D. Crystallization-induced kinetic racemization/resolution method; E. Jacobsen's hydrolytic kinetic resolution (HKR) of terminal epoxides.

Genêt [12] groups are shown in Figures 5.3 A and B. Application to the synthesis of an azetidinone as a carbapenem intermediate shows the power of the method [11b].

Dynamic kinetic resolution is also possible in the S_N2 displacement of activated leaving groups in substrates harboring a chiral auxiliary or a proximal chiral motif. For example, treatment of a mixture of epimeric α-bromo amides [13a], sultams [14], or α-bromo ketones [13b] with benzylamine in the presence of triethylamine resulted in the formation of highly enriched diastereomeric α-benzylamino adducts, due to a faster reaction of one diastereomer and racemization of the slower reacting diastereomer (Figures 5.3 C).

A variant of chemical kinetic resolution relies on an *in situ* racemization/resolution by crystallization method [15]. A racemic mixture of compounds in which an amine is attached to a carbon atom that is prone to epimerization is treated with a catalytic amount of an aromatic aldehyde in the presence of the resolving chiral acid. The acidity of the α-proton in the intermediate Schiff base results in a spontaneous racemization of the non-crystalline amine salt, while the desired diastereomerically enriched (or pure) amine salt is continuously crystallizing out of solution. This crystallization-induced asymmetric transformation was successfully used in the synthesis of an enantiopure CCK antagonist [16] (Figure 5.3 D). A practical method to convert racemic, terminal epoxides to enantioenriched diols and recovered epoxides reported by Jacobsen and coworkers [17], relies on a hydrolytic kinetic resolution (HKR) in the presence of a chiral cobalt salen catalyst (Figure 5.3 E).

5.3
Cell-free Enzyme-mediated Enantiopure Compounds

5.3.1
Hydrolases and ester formation

Enzymes are capable of recognizing functional groups in substrates based on their spatial disposition, chirality, pro-chirality, and topology. The preparation of enantiopure compounds using enzymes is a well-established industrial process [18]. The procedures are simple, and involve adding an enzyme (or enzymes) to the substrates in aqueous solution or even in organic solvents [19].

The most widely used applications of enzymes are in the hydrolysis of esters or the acylation of alcohols [18, 20]. Hydrolases such as pig liver esterase (PLE) and lipases from different sources are capable of desymmetrizing prochiral *meso*-diesters in an enantiotopos-differentiating hydrolysis to give monoester monocarboxylic acids in moderate to excellent enantiomeric ratios (usually expressed as enantiomeric excess, *ee*) depending on the substrate (Figure 5.4 A) [21].

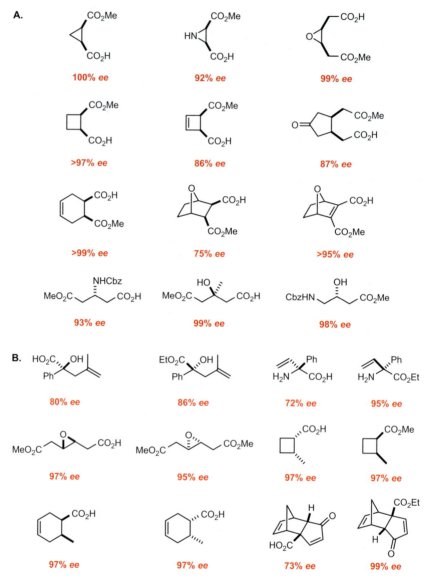

Figure 5.4 [20a,b]; [21] A. Enantiotopos-differentiated carboxylic acid and carboxylic acid monoesters (PLE, aqueous solution); B. Enantiomer-differentiated carboxylic acids and carboxylic acid ester pairs; C. Enantiotopos-differentiated monoacetate esters (lipases, in organic solvents with vinyl acetate); D. Enantiomer-differentiated primary and secondary acetate and alcohol pairs (lipases in organic solvents with vinyl acetate, ethyl acetate, etc).

C.

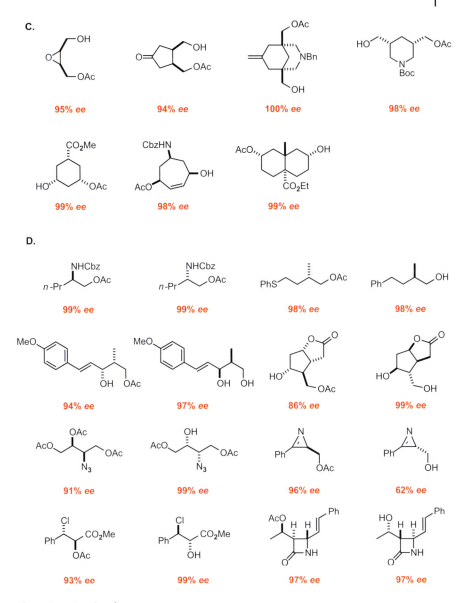

Figure 5.4 (continued).

Esterases are also efficient catalysts in the enantiomer-differentiating hydrolysis of racemic (carboxylic acid) esters and lactones. In ideal cases, the resulting enantiomeric carboxylic acid and corresponding (non-hydrolyzed) ester are obtained in high *ee*. (Figure 5.4 B).

Esterases, such as lipases, exhibit catalytic activity in organic solvents [19] or in biphasic systems. Selectivities in the acylation of enantiotopos or enantiomer-differentiating reactions of symmetrical prochiral diols and racemic non-symmetrical alcohols may vary with the nature of the solvent and the substrate. The acylating agents are usually alkenyl acetates, ethyl acetate, or acetic anhydride, which can also be used as solvents. Enantioselectivities in enzymatic hydrolysis of symmetrical diol diacetates, and the lipase-mediated monoacetylation of the corresponding diols are generally similar with regard to the chirality of the major enantiomer. Kinetic separation of enantiotopos-differentiated monoacetylations can be efficiently achieved with lipases (Figure 5.4 C). It is also possible to differentiate enantiomeric alcohols by acetylation with lipases in organic media where pairs of monoacetylation products and the non-acetylated enantiomeric alcohol (or diastereomer) can be isolated with excellent *ee* (or *de*) in the selected cases shown in Figure 5.4 D [19, 20]. Thus, it is possible to obtain a very large variety of enantiomerically pure (or enriched) alcohols, carboxylic acids, and esters using readily available esterases and lipases as ordinary laboratory reagents. A significant number of total syntheses have relied on an enzymatic reaction to secure enantiopure starting materials (as discussed in various sections of this book).

5.3.2
Nitrilases, amidases, and acylases

Enantiomerically pure or enriched α-hydroxy carboxylic acids and L-α-amino acids can be obtained from racemic cyanohydrins and nitriles by enantioselective enzymatic hydrolysis [22]. Nitrilases catalyze the hydrolysis of racemic α-hydroxynitriles (cyanohydrins) and α-aminonitriles to the corresponding α-hydroxy or α-amino carboxylic acids in high enantiopurity. A number of industrial preparations of carboxylic acids utilize nitrilases as immobilized enzymes (or as whole cell preparations) [23]. Dynamic resolution allows conversion of the remaining epimeric nitrile to the racemate, which undergoes catalysis, thus recycling the unconverted nitrile. Selected examples of enantiopure α-substituted carboxylic acids are shown in Figure 5.5 A.

Amidases can be used to prepare enantioenriched carboxylic acids and carboxamides from the racemic amides. D- and L-amino acid amidases have been isolated from various bacterial sources, although industrially, whole cells can also be used. Dynamic resolution through epimerization of a Schiff base provides efficient routes to D- and L-α-amino acids (see section 5.2.2). Thus, a racemic mixture of α-amino carboxamides will be hydrolyzed by an L-amidase to the enantiopure α-amino acid, while the D-amino component is racemized after treatment with an aromatic aldehyde at pH 12.5 [23]. Some examples of highly enantiopure products resulting from such processes are shown in Figure 5.5 B.

Thermophylic acylases hydrolyze racemic *N*-acyl α-amino acids to a mixture of the enantiopure α-amino acid and the corresponding unhydrolyzed *N*-acyl acid. This process is used on an industrial scale to produce many unnatural and highly functionalized α-amino acids, including α-alkyl amino acids (Figure 5.5 C) [24].

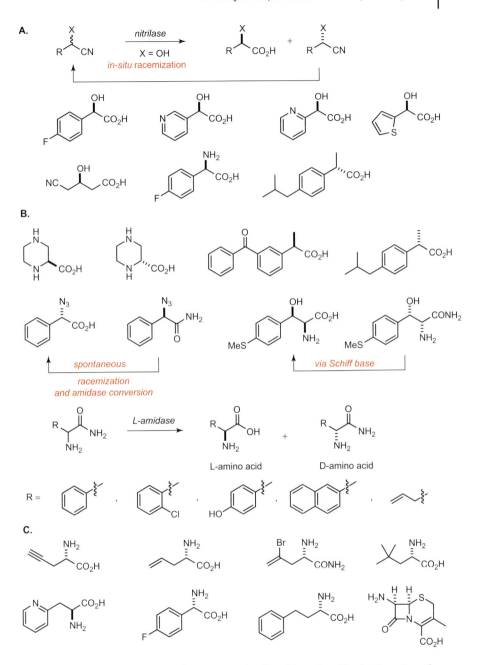

Figure 5.5 [23, 24] A. Formation of α-hydroxy and α-amino carboxylic acids from corresponding nitriles by nitrilase hydrolysis (including spontaneous *in situ* racemization and recycling); B. Formation of α-substituted carboxylic acids and amides by the action of D- and L-amidases (including spontaneous *in situ* racemization and recycling); C. Unnatural amino acids and carboxamides by the action of N-acyl acylases.

5.4
Cell-free Chemoenzymatic Methods

The combination of chemical methods of *in situ* racemization and an enzyme to provide a stereodifferentiating conversion of one enantiomer preferentially into an enantiopure product has had much success in asymmetric catalysis [25]. The Schiff base epimerization under mild conditions followed by the action of esterases or amidases, as demonstrated in the previous section, is an example of chemoenzymatic dynamic kinetic resolution.

5.5
Metal-catalyzed Dynamic Kinetic Resolution (DKR)

Enantiopure secondary alcohols bearing diverse functionality can be obtained by combining enzymatic kinetic resolution with a metal-catalyzed redox process. A bimetallic Ru-catalyst has been found to be compatible with the action of various lipases [25]. Thus, proton abstraction by the Ru-catalyst converts the unesterified alcohol to the corresponding ketone, which is reduced back to the racemic alcohol by the hydrido Ru-catalyst. Kinetic enzymatic esterification of one enantiomer enriches the mixture in that component while the unesterified alcohol is racemized by the metal-catalyzed redox process [26]. Primary amines are also substrates for metal-catalyzed dynamic kinetic resolution in the presence of ruthenium bimetallic catalysts and a lipase-catalyzed acyl transfer with isopropyl acetate [27]. Pertinent examples of preparatively useful enantioenriched compounds are shown in Figure 5.6 A.

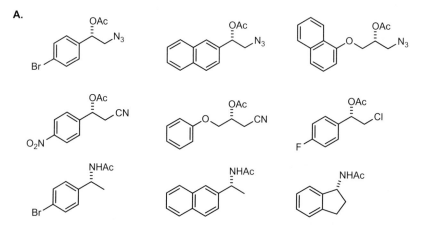

Figure 5.6 [25–28] A. Enantioenriched esters and amides from metal-catalyzed chemoenzymatic dynamic kinetic resolution; B. Pd-catalyzed chemoenzymatic dynamic kinetic resolution.

B.

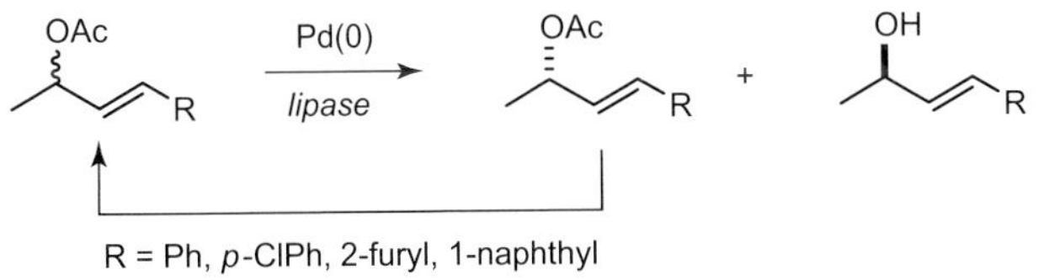

R = Ph, *p*-ClPh, 2-furyl, 1-naphthyl

Figure 5.6 (*continued*).

Lipases are also tolerant of neutral palladium catalysts for the enantioselective hydrolysis of racemic allylic acetates [28]. The unreacted ester is racemized through the well-known π-allyl complex [29], allowing the lipase to continue the catalytic cycle (Figure 5.6 B).

5.6
Biocatalytic Methods for Enantiopure Compounds

Nature is the ultimate synthetic chemist. It uses its catalysts in the most efficient and chemoselective way to provide a plethora of enantiomerically pure small molecules [18]. These are produced by highly stereoselective transformations catalyzed by a large variety of enzymes either as single entities, or in conjunction with the microorganisms from which they can be isolated. Remarkably efficient redox and bond-forming processes with control of absolute stereochemistry have been achieved in recent years, especially in industry. Large volume production of amino acids, terpenes, and related chiral, non-racemic compounds relies on biocatalytic methods that use immobilized enzymes, whole cell preparations, cultured cells, recombinant cells, microorganisms, yeast, and other natural sources. This subject has been discussed in many authoritative reviews and monographs [30].

5.6.1
Enzymatic reduction of ketones

The asymmetric reduction of ketones by biocatalysts relies on the enantioselective transfer of hydride from the coenzyme NADPH to the carbonyl group in an enzymatic reaction [31]. The oxidized cofactor is subsequently reduced by the same enzyme or by a two-enzyme system. Baker's yeast is the most common laboratory catalyst for the enantioselective reduction of prochiral ketones. It also catalyzes the reduction of azide and nitro groups to amines, and double bonds in α,β-unsaturated carbonyl compounds. Examples of highly selective reductions of ketones to enantioenriched alcohols using Baker's yeast are shown in Figure 5.7 [31].

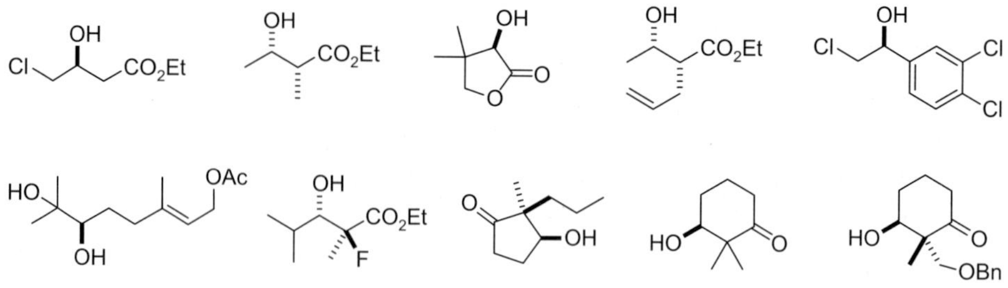

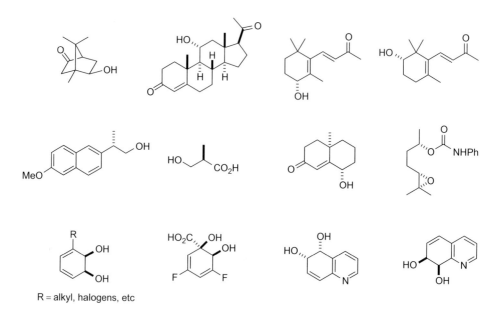

Figure 5.7 [30, 31] Enantioenriched alcohols from Baker's yeast reduction of ketones.

5.6.2
Enzymatic hydroxylation and epoxidation

Oxygenase enzymes are exquisite biocatalysts for the hydroxylation and epoxidation of a variety of organic compounds [32]. Many of these have been important starting materials for the synthesis of natural products and drug substances. Microbial monooxygenases have been known for many years for their ability to hydroxylate steroids and terpenoids. Selected examples of mono- and dihydroxylated enantioenriched compounds are shown in Figure 5.8. Cyclohexadiene-1,2-diols have been used as starting materials in a variety of total syntheses (see Chapter 14, section 14.9) [32].

R = alkyl, halogens, etc

Figure 5.8 [32] Enantioenriched alcohols, epoxides and diols from enzymatic oxidations.

5.6.3
Enzymatic oxidation of alcohols

Oxidation of alcohols by dehydrogenases such as horse liver alcohol dehydrogenase (HLADH) occurs in the presence of the $NADP^+/NADPH$ redox cofactors [33]. One of the most useful applications is in the preparation of lactones from *meso*-diols. Selected examples of enantiopure lactones are shown in Figure 5.9.

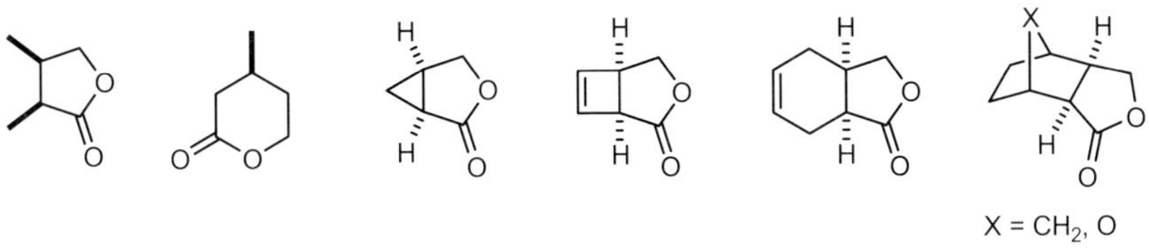

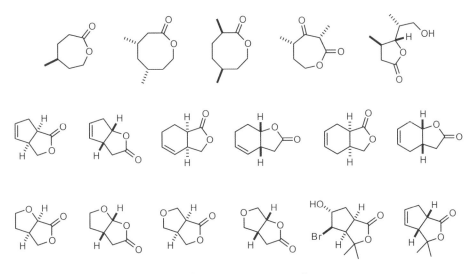

X = CH$_2$, O

Figure 5.9 [33] Enantioenriched lactones from diols oxidized with HLADH.

5.6.4
Enzymatic Baeyer-Villiger oxidation

Various microorganisms contain enzymes capable of oxidizing ketones to the corresponding lactones [34]. Interestingly, as in the chemical versions of the Baeyer-Villiger reaction, the product lactones result from the insertion of oxygen next to the more substituted carbon atom. Cyclohexanone monooxygenase converts various prochiral cyclic ketones to the corresponding enantiopure or highly enantioenriched lactones. Examples of such lactones that could be useful as starting materials in natural product synthesis are shown in Figure 5.10.

Figure 5.10 [34] Enantiopure lactones from enzymatic Baeyer-Villiger reactions.

5.7
Applications of Enzymatic and Chemoenzymatic Methods

Advances in biocatalysis technology and genetic engineering have allowed many high volume products to be manufactured using immobilized, partially purified enzymes, or whole cell preparations to achieve enantioselective transformations. Biocatalytic transformations for a given step in the synthesis of a drug substance for example, may present significant advantages compared to alternative chemical enantioselective methods. Immobilized enzymes allow easy separation of products although catalyst efficiency in a continuous process may be an issue. The topic has been extensively reviewed [30], and the examples shown below are only illustrative of the importance of chemoenzymatic methods in industrial processes (Figure 5.11) [2, 35].

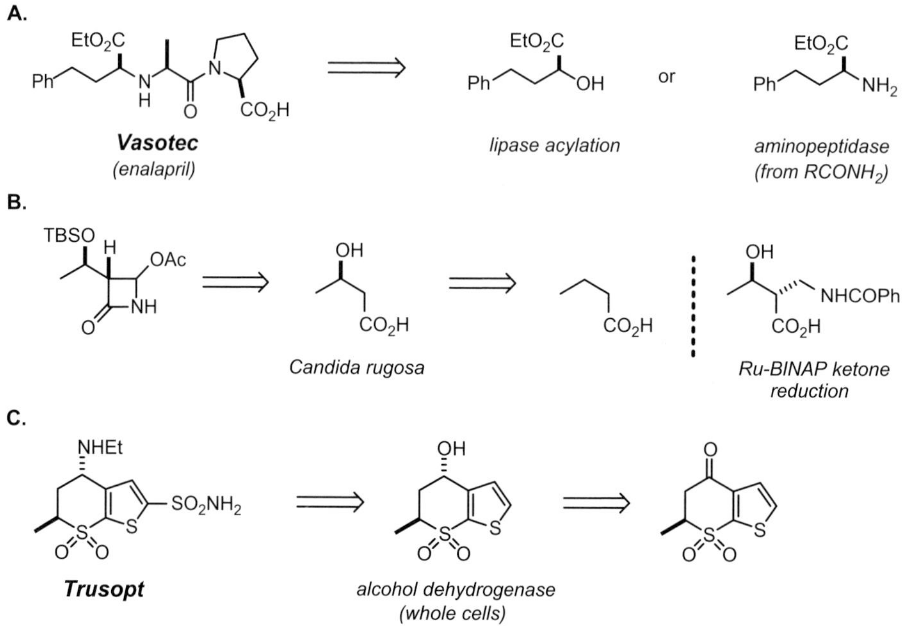

Figure 5.11 [2, 35] Uses of enzymatic and biocatalytic methods for the synthesis of selected active pharmaceutical compounds.

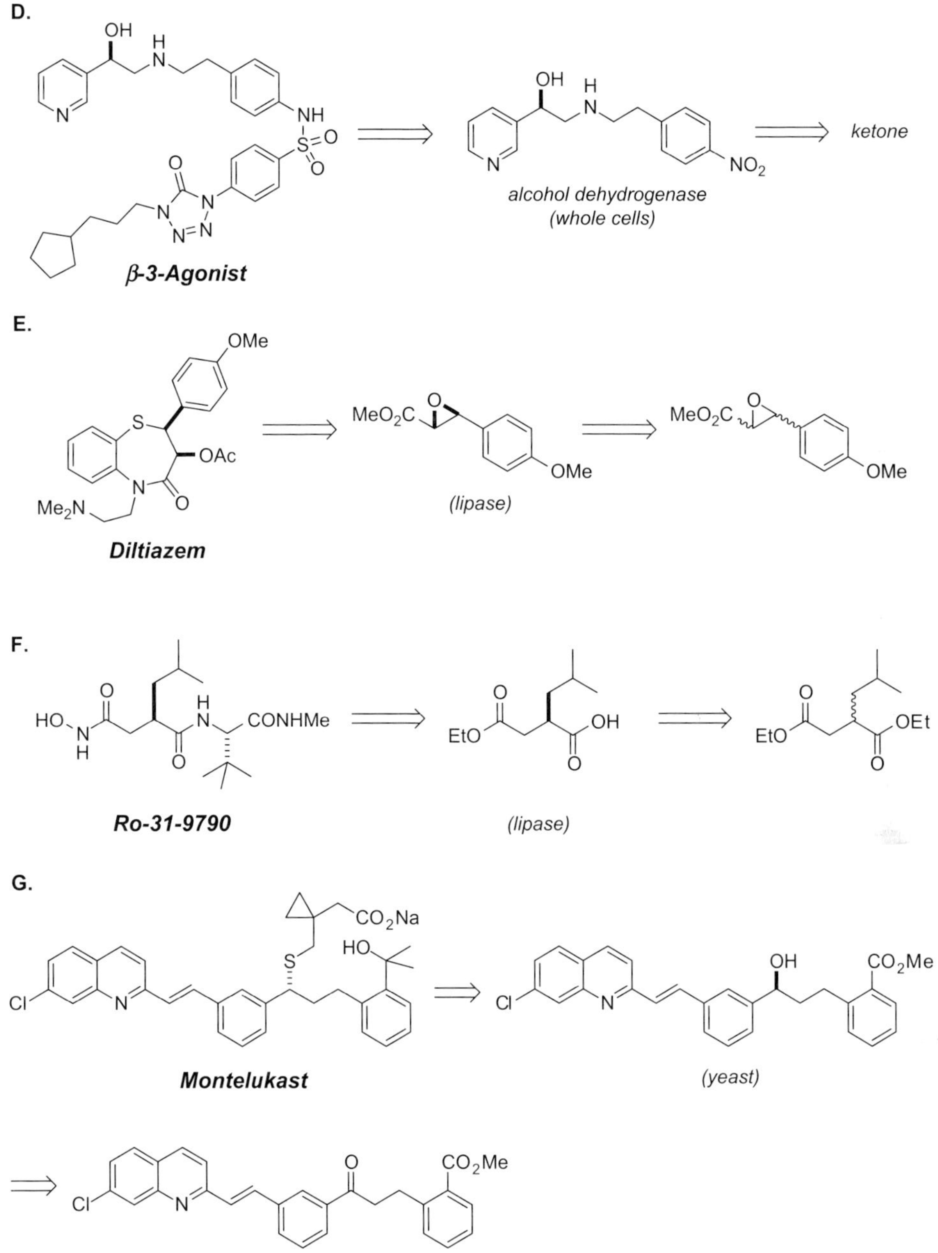

D.

β-3-Agonist

alcohol dehydrogenase
(whole cells)

ketone

E.

Diltiazem

(lipase)

F.

Ro-31-9790

(lipase)

G.

Montelukast

(yeast)

Figure 5.11 (continued).

5.8
Chemical Asymmetric Synthesis of Enantiopure Compounds

The enantioselective synthesis of natural products or other organic molecules containing one or more stereogenic carbon atoms requires a plan that includes an asymmetric bond-forming reaction at a predetermined stage. When considering enantiopure starting materials that contain one or more stereogenic carbon atoms with predisposed functionality, a common protocol is to use readily available chiral, non-racemic compounds that are available from amino acids, carbohydrates, hydroxy acids, and terpenes (the so called "chiral pool") [3]. Alternatively, starting materials harboring required functionality and stereochemistry can be accessed by *chemical asymmetric synthesis* [36]. A number of examples have already been highlighted in Chapters 2–4. We briefly summarize these methods below in the context of their utility in synthesis.

Stereodifferentiation can be achieved by the use of external chiral auxiliaries, chiral reagents, or through the influence of a proximal or nearby stereogenic center. Terms such as "substrate control" and "reagent control" have been used to describe asymmetric processes [37]. In general, stereogenic centers are formed at one or two carbon atoms at a time except in the Diels-Alder and related cycloaddition reactions where, in principle, up to four new stereogenic centers can be generated in a single step [38] (see Chapter 4, section 4.5.3 and Chapter 18, Figure 18.4). Many innovative contributions have been made in stoichiometric asymmetric synthesis using chiral auxiliaries in particular (see Chapters 2 and 4, sections 2.12.1 and 4.5.2) [36]. Important applications can be found in the aldol reaction [39], enolate alkylations, enolate α-substitutions, and 1,4-conjugate additions [40] to mention a few. Applications of asymmetric reactions to the synthesis of preclinical active pharmaceutical compounds have been reviewed [41].

The phenomenal progress made in *catalytic asymmetric reactions* [42] has changed the landscape of synthesis planning in recent years. An efficient chiral catalyst in a chemical reaction that generates one or more stereogenic centers is similar to an enzyme, except that it can produce either enantiomer at will, depending on the nature and availability of the chiral ligand. Catalytic asymmetric synthesis is also important on an industrial scale [41]. A large number of preparatively useful stoichiometric asymmetric reactions have their catalytic equivalents, affording products of high enantiopurity and quantity.

Catalytic asymmetric synthesis using transition metals in conjunction with chiral ligands has played a major role in natural product synthesis [43]. This has also instigated the development of architecturally and topologically unique organic ligands [44]. The area of asymmetric (homogenous) catalytic hydrogenation is an example where a multitude of phosphine ligands have been reported [42f]. Significant advances have been made in catalytic Diels-Alder, epoxidation, dihydroxylation, aldol, Michael, ene, allylic alkylation, cross-coupling, intramolecular Heck, and phase-transfer alkylation reactions [42] (see also Chapter 4, section 4.6).

Catalytic asymmetric reactions without the use of metals date back to the early part of the 20th century [45]. The term *organocatalysis* has been introduced to describe catalytic reactions in the absence of metals [46]. Organocatalysis may involve the intermediacy of covalent adducts or non-covalent interactions in the catalytic cycle. One of the earlier applications of asymmetric organocatalysis was the proline-catalyzed intramolecular aldol reaction of an achiral triketone to give an enantiomerically pure diketone after recrystallization (Hajos-Parrish-Eder-Sauer-Wiechert reaction, Figure 5.12) [47].

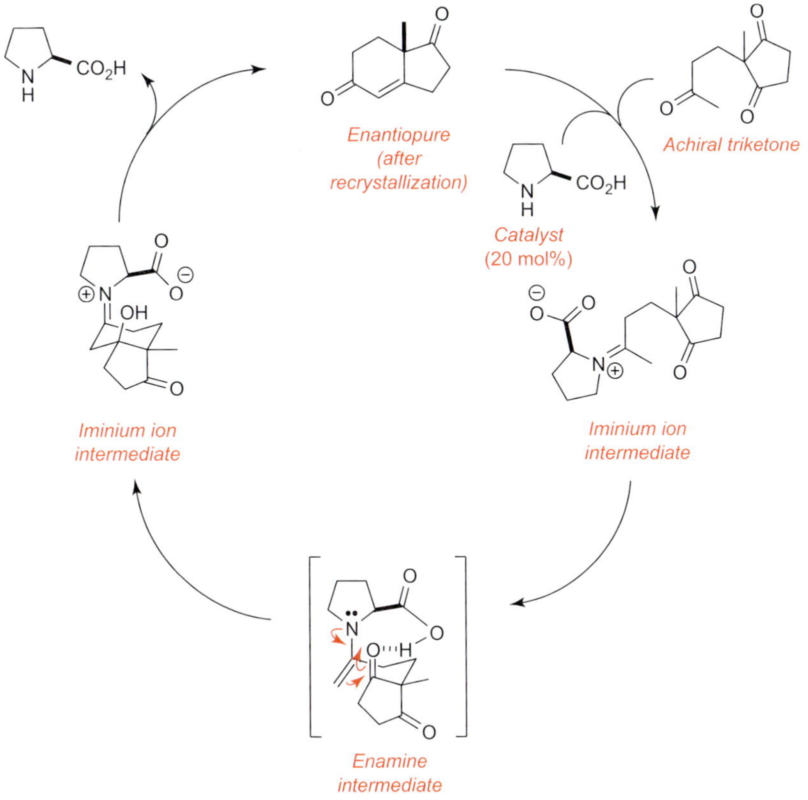

Figure 5.12   Hajos-Parrish-Eder-Sauer-Wiechert L-proline-catalyzed intramolecular aldol reaction.

Except for sporadic reports, this important proline enamine-mediated aldol reaction remained unexplored until it was revived in the laboratories of Barbas [48], and List [49], and further extended by others [46]. MacMillan [50] introduced iminium ion-mediated, and SOMO activated, organocatalytic reactions with many applications toward the synthesis of enantiopure and functionally useful compounds (Figure 5.13) (see also Chapters 2 and 4). The conceptually different generic activation modes in enamine and iminium-based catalytic reactions as summarized

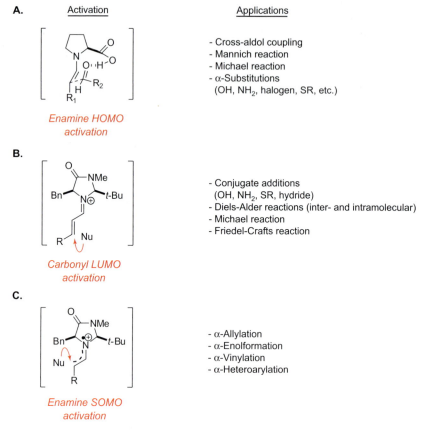

A. Activation

Applications

- Cross-aldol coupling
- Mannich reaction
- Michael reaction
- α-Substitutions
  (OH, NH$_2$, halogen, SR, etc.)

Enamine HOMO activation

B.

- Conjugate additions
  (OH, NH$_2$, SR, hydride)
- Diels-Alder reactions (inter- and intramolecular)
- Michael reaction
- Friedel-Crafts reaction

Carbonyl LUMO activation

C.

- α-Allylation
- α-Enolformation
- α-Vinylation
- α-Heteroarylation

Enamine SOMO activation

Figure 5.13 Generic modes of enamine and iminium activation in catalytic reactions according to MacMillan [50g].

by MacMillan [50g], are illustrated in Figure 5.13 A–C. Thus, by its very nature, enamine attack on a carbonyl partner involves HOMO (*h*ighest *o*ccupied *m*olecular *o*rbital) activation, whereas iminium catalysis relies on lowering the LUMO (*l*owest *u*noccupied *m*olecular *o*rbital) of the carbonyl component. SOMO lowering catalysis capitalizes on activation of a *s*ingly *o*ccupied *m*olecular *o*rbital generated in a redox reaction. In his insightful review, MacMillan [50g] delineates the many uses of his imidazolinone catalyst, as well as other modes of activation.

Quinine and cinchonidinium salts, as well as other chiral ammonium salts, are also excellent organocatalysts that act in a non-covalent manner through ionic interactions or H-bonding [51]. Many other elegant examples of organocatalytic reactions have been reported in recent years [46]. For the purposes of generating enantioenriched organic molecules with potential utility as functionalized starting materials, we list some examples in Figure 5.14 [52].

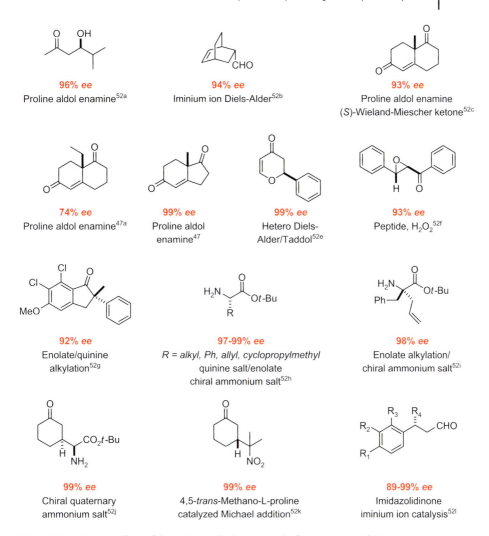

Figure 5.14 Functionally useful enantioenriched compounds from organocatalytic reactions.

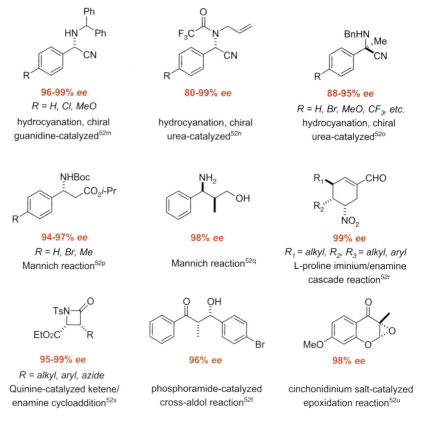

96-99% ee

R = H, Cl, MeO

hydrocyanation, chiral
guanidine-catalyzed[52m]

80-99% ee

hydrocyanation, chiral
urea-catalyzed[52n]

88-95% ee

R = H, Br, MeO, CF₃, etc.

hydrocyanation, chiral
urea-catalyzed[52o]

94-97% ee

R = H, Br, Me

Mannich reaction[52p]

98% ee

Mannich reaction[52q]

99% ee

R₁ = alkyl, R₂, R₃ = alkyl, aryl
L-proline iminium/enamine
cascade reaction[52r]

95-99% ee

R = alkyl, aryl, azide

Quinine-catalyzed ketene/
enamine cycloaddition[52s]

96% ee

phosphoramide-catalyzed
cross-aldol reaction[52t]

98% ee

cinchonidinium salt-catalyzed
epoxidation reaction[52u]

Figure 5.14 *(continued).*

5.9
Enantiopure Compounds from Nature

Nature's generosity in providing complex products is also manifested in the components it uses in its biosynthetic pathways. An abundant source of enantiopure small molecules that are ideal starting materials for the synthesis of natural products as well as active pharmaceuticals comes from natural α-amino acids, a variety of D- and L-sugars, a handful of chiral, non-racemic hydroxy and dihydroxy acids, and a larger number of terpenes. Considering their relative abundance and low cost, these four classes of molecules are Nature's gifts to the synthetic chemist. They are highly versatile starting materials that combine chirality with desired functionality depending on the intended target molecule (Figure 5.15 A). Nature also produces lactones, inositols [53], and a plethora of secondary metabolites. Depending on

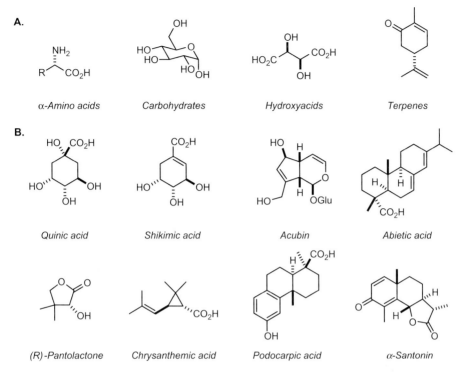

A.

α-Amino acids

Carbohydrates

Hydroxyacids

Terpenes

B.

Quinic acid

Shikimic acid

Acubin

Abietic acid

(R)-Pantolactone

Chrysanthemic acid

Podocarpic acid

α-Santonin

Figure 5.15 (A) Major classes of natural products as starting materials; (B) Miscellaneous commercially available heterocyclic and carbocyclic natural products as starting materials.

their availability, they also constitute useful starting materials for total synthesis (Figure 5.15 B).

References

1. Pasteur, L. "Sur les relations qui peuvent exister entre la forme cristallin, la composition chimique et le sens de la polarisation rotatoire," *Ann. Chim.* 1848, **24**, 442.

2. For relevant monographs, see: (a) *Chirality in Drug Research*, Francotte, E.; Lindner, W., Eds.; 2006, Wiley-VCH, Weinheim; (b) *Chirality in Drug Design and Synthesis*, Brown, C.; Ed.; 1997, Academic, NY; (c) *Chirality in Industry*, Collins, A. N.; Sheldrake, G.; Crosby, J.; Eds.; 1997, Wiley, Chichester; UK; *Chirotechnology*, Sheldon, R. A., 1993, Dekker, NY.

3. (a) *Handbook of Chiral Chemicals*, Ager, D.; Ed., 2006, CRC Press, Boca Raton, FL; (b) Blaser, H. U. *Chem. Rev.* 1992, **92**, 935; see also: (c) Seebach, D.; Kalinowski, H.-O. *Nachr. Chem. Tech. Lab.* 1976, **24**, 415; as cited in, Seebach, D.; Maestro, M. A.; Sefkow, M.; Adam, G.; Hintermann, S., Neidlein, A. *Liebigs. Ann. Chem.* 1994, 701.

4. For example, see: (a) Fischbach, M. A.; Walsh, C. T. *Science* 2006, **314**, 603; (b)

Fischbach, M. A.; Walsh, C. T. *Chem. Rev.* 2006, **106**, 3468 and references cited therein.

5. For example, see: (a) *Chiral Separations by Liquid Chromatography and Related Technologies* Aboul-Enein, H. Y.; Ali, I., 2003, Dekker, NY; *The Impact of Stereochemistry on Drug Development and Use*, Aboul-Enein, H.Y., 1997, Wiley-Interscience, NY.

6. For example, see: (a) *CRC Handbook of Optical Resolutions via Diastereomeric Salt Formation*, Kozma, D., 2002, CRC Press, Boca Raton, FL; (b) *Novel Optical Resolution Technologies* Sakai, K.; Hirayama, N.; Tamura, R.; Eds.; 2007, Springer, Heidelberg; (c) Jacques, J. in, *The Molecule and its Double*, 1993, McGraw-Hill Inc., NY; (d) Jacques, J.; Collet, A.; Wilen, S. H. *Enantiomers, Racemates and Resolutions*, 1981, Wiley, NY.

7. (a) Kagan, H. B.; Fiaud, J. D. *Top. Stereochem.* 1988, **18**, 249; (b) Ebbers, E. J.; Ariaans, G. J. A.; Houbiers, J. P. M.; Bruggink, A.; Zwanenburg, B. *Tetrahedron* 1997, **53**, 9417; For a classical and historic example, see: (c) Marckwald, W.; McKenzie, A. *Ber. Dtsch. Chem. Ges.* 1899, **32**, 2130.

8. Martin, V. S.; Woodward, S. S.; Katsuki, T.; Yamada, Y.; Ikeda, M.; Sharpless, K. B. *J. Am. Chem. Soc.* 1981, **103**, 6237.

9. Chen, Y.; Deng, L. *J. Am. Chem. Soc.*, 2001, **123**, 11302.

10. For relevant reviews, see: (a) Robinson, D. E. J. E.; Bull, S. D. *Tetrahedron: Asymmetry* 2003, **14**, 1407; (b) Pellissier, H. *Tetrahedron* 2003, **59**, 8291; (c) Huerta, F.; Minidis, A. B. E.; Bäckvall, J.-E. *Chem. Soc. Rev.* 2001, **30**, 321; (d) Azerad, R.; Buisson, D. *Curr. Opin. Biotech.* 2000, **11**, 565; (e) El Gihani, M. T.; Williams, J. M. J. *Curr. Opin. Chem. Biol.* 1999, **3**, 11; (f) Stecher, H.; Faber, K. *Synthesis* 1997, 1; (g) Caddick, S.; Jenkins, K. *Chem. Soc. Rev.* 1996, **25**, 447; (h) Ward, R. S. *Tetrahedron: Asymmetry* 1995, **6**, 1475.

11. (a) Noyori, R.; Ikeda, T.; Ohkuma, T.; Widhalm, M.; Kitamura, M.; Takaya, H.; Akutagawa, S.; Sayo, N.; Saito, T.; Taketomi, T.; Kumobayashi, H.

J. Am. Chem. Soc. 1989, **111**, 9134; (b) Mashima, K.; Matsumura, Y.; Kusano, K.; Kumobayashi, H.; Sayo, N.; Hori, Y.; Ishizaki, T.; Akutagawa, S.; Takaya, H. *J. Chem. Soc., Chem. Commun.* 1991, 609.

12. Girard, A.; Greck, C.; Ferrond, D.; Genêt, J.-P. *Tetrahedron Lett.* 1996, **37**, 7967.

13. (a) Nunami, K.; Kubota, H.; Kubo, A. *Tetrahedron Lett.* 1994, **35**, 8639; (b) Koh, K.; Ben, R. N.; Durst, T. *Tetrahedron Lett.* 1993, **34**, 4473.

14. Ward, R. S.; Pelter, A.; Goubet, D.; Pritchard, M. C. *Tetrahedron: Asymmetry* 1995, **6**, 469.

15. Tokunga, M.; Larrow, J. F.; Kakiuchi, F.; Jacobsen, E. N. *Science* 1997, **277**, 936; see also: (a) Annis, D. A.; Jacobsen, E. N. *J. Am. Chem. Soc.* 1999, **121**, 4147; (b) Schaus, S. E.; Brandes, B. D.; Larrow, J. F.; Tokunga, M.; Hansen, K. B.; Gould, A. E.; Furrow, M. E.; Jacobsen, E. N. *J. Am. Chem. Soc.* 2002, **124**, 1307.

16. Brands, K. M. J.; Davies, A. J. *Chem. Rev.* 2006, **106**, 2711.

17. (a) Reider, P. J.; Davies, P.; Hughes, D. L.; Grabowski, E. J. J. *J. Org. Chem.* 1987, **52**, 955; (b) Armstrong, J. D.; Eng, K. E.; Keller, J. L.; Purick, R. M.; Hartner, F. W.; Choi, W. B.; Askin, D.; Volante, R. P. *Tetrahedron Lett.* 1994, **35**, 3239.

18. For example see: (a) *Multi-Step Enzyme Catalysis*, Garcia-Junceda, E., Ed.; 2008, Wiley-VCH, Weinheim; (b) *Asymmetric Organic Synthesis with Enzymes*, Gotor, V.; Alfonso, I.; Garcia-Urdiales, E.; Eds, 2008, Wiley-VCH; (c) *Hydrolases in Organic Synthesis*, 2nd Edition., Bornscheuer, U. T.; Kazlauskas, R. J., 2006, Wiley-VCH, Weinheim; (d) Kula, M.-R. in, *Enzyme Catalysis in Organic Synthesis*, Drauz, K.; Waldmann, H.; Eds.; 2002, vol. 1, p. 1, Wiley-VCH, Weinheim; (e) Santaniello, E.; Ferraboschi, P.; Grisenti, P.; Manzocchi, A. *Chem. Rev.* 1992, **92**, 1071; see also: references [22–24].

19. For example see: (a) *Organic Synthesis with Enzymes in Non-Aqueous Media*, Carrea, G.; Riva, S., 2008, Wiley-VCH, Weinheim; (b) *Enzymes in Non-Aqueous Solvents. Methods and Protocols*, Vulfson,

E. N.; Halling, P. J.; Holland, H. C,
2001, Humana Press Springer, Heidelberg;   (c) *Enzymatic Reactions in
Organic Media*, Koskinen, A. M. P.;
Klibanov, A. M, Eds.; 1996, Blackie
Academic & Professional; NY;   (d)
Klibanov, A. M, *Nature* 2001, **409**, 241;
(e) Klibanov, A. M. *Acc. Chem. Res.* 1990,
23, 114; see also: references [22–24].

20. Gais, H. J., Theil, F. in, *Enzyme Catalysis in Organic Synthesis*, Drauz, K.;
Waldmann, H.; Eds.; 2002, vol. 2, p.
335, Wiley-VCH, Weinheim.

21. (a) Schoffers, E.; Golebiowski, A.;
Johnson, C. R. *Tetrahedron* 1996, **52**,
3769;   (b) Guanti, G.; Narisano, E.;
Podgorski, T.; Thea, S.; Williams, A.
Tetrahedron 1990, **46**, 7081;   (c) Ohno,
M.; Otsuka, *Org. React.* 1989, vol. **37**,
p. 1;   (d) Chen, C. S.; Sih, C. J. *Angew.
Chem. Int. Ed.* 1989, **28**, 695.

22. (a) Schulze, B. in, *Enzyme Catalysis in Organic Reactions*, Drauz, K.;
Waldmann, H.; Eds.; 2002, vol. **2**,.
p. 699, Wiley-VCH, Weinheim;   (b)
Wieser, M.; Nagasawa, T. *Stereoselective Biocatalysis*, Patel, R. N.; Ed.; 2000,
Chapter 17, p. 461, Dekker, NY;   (c)
Bunch, A. W. in, *Biotechnology*, Rehm,
H.-J.; Reed, G.; Puhler, A.; Stadler, P.;
Eds.; 1998, vol. **8a**, p. 277, Wiley-VCH,
Weinheim.

23. Schultze, B.; de Vroom, E. in, *Enzyme
Catalysis in Organic Reactions*, Drauz,
K.; Waldmann, H.; Eds.; 2002, vol. 2, p.
716, Wiley-VCH, Weinheim.

24. Bommarius, A. S. in, *Enzyme Catalysis in Organic Reactions*, Drauz, K.;
Waldmann, H.; Eds.; 2002, vol. **2**, p.
741, Wiley-VCH; Weinheim.

25. (a) Pàmies, O.; Bäckvall, J.-E. *Trends in
Biotechnology* 2004, **22**, 130;   (b) Pàmies,
O.; Bäckvall, J.-E. *Chem. Rev.* 2003, **103**,
3247; see also: ref. [10c].

26. For example, see: (a) Gladiali, S.;
Mestroni, G. in, *Transition Metals for
Organic Synthesis*, Beller, M.; Bolm, C.;
Ed.; 1998, p. 97, Wiley-VCH, Weinheim;
(b) Bäckvall, J.-E.; Chowdhury, R. L.;
Karlsson, U.; Wang, G. Z. in, *Perspectives
in Coordination Chemistry*, Williams,
A. F.; Floriani, C.; Merbach, A. E.; Eds.;
1992, p. 465, Verlag Helvetica Chimica Acta, Basel;   (c) Zassinovich, G.;
Mestroni, G.; Gladiali, S. *Chem. Rev.*
1992, **92**, 1051.

27. Paetzold, J.; Bäckvall, J.-E. *J. Am. Chem.
Soc.* 2005, **127**, 17620.

28. For example, see: (a) Choi, Y. K.; Suh,
J. H.; Lee, D.; Lim, I. T.; Jung, J. Y.;
Kim, M. J. *J. Org. Chem.* 1999, **64**,
8423;   (b) Allen, J. V.; Williams, J. M. J.
Tetrahedron Lett. 1996, **37**, 1859.

29. (a) Tsuji, J. in, *Palladium Reagents and
Catalysis: New Perspectives for the 21st
Century*, 2004, Wiley-Interscience, NY;
(b) Beletskaya, I. P.; Cheprakov, A. V. in,
*Handbook of Organopalladium Chemistry
for Organic Synthesis*, Negishi, E.-i.; de
Meijere, A.; Eds.; 2002, vol. **2**, p. 2957,
Wiley-Interscience, NY.

30. (a) *Biocatalysis in the Pharmaceutical and
Biotechnology Industries*, Patel, R. N.;
Ed.; 2007, CRC Press, Boca Raton, FL;
(b) Zaikov, G.E. in, Biocatalysis and
Biocatalytic Technologies, 2006;   (c)
Handbook of Industrial Biocatalysts, Hou,
C. T.; Ed.; 2005, CRC Press, Boca Raton, FL;   (d) Bommarius, A. S.; Riebel,
B. R. in, *Biocatalysis: Fundamentals and
Applications*, 2004, Wiley-VCH, Weinheim;   (e) Roberts, S. M.; Turner,
N. J.; Willetts, A. J.; Turner M. K.
in, *Introduction to Biocatalysis Using
Enzymes and Microorganisms*, 1995,
Cambridge University Press, Cambridge, UK;   (f) *Enzyme Technologies
for Pharmaceutcial and Biotechnological
Applications*, Kirst, H. A.; Yeh, W.-K.;
Zmijewski, M. J. Jr.; Eds.; 2001, M.
Dekker, NY;   (g) *Encyclopedia of Bioprocess Technology-Fermentation, Biocatalysis,
and Bioseparation*, Flickinger, M. C.;
Drew, S. W.; Eds.; 1999, vol. 1-5, John
Wiley & Sons, NY.

31. (a) Nakamura, K.; Matsuda, T. in,
Enzyme Catalysis in Organic Synthesis, Drauz, K.; Waldmann, H.; Eds.,
2002, vol. 3, p. 991, Wiley-VCH, Weinheim;   (b) Nakamura, K.; Yamanaka,
R.; Matsuda, T.; Harada, T. *Tetrahedron:
Asymmetry* 2003, **14**, 2659.

32. (a) Flitsch, S.; Grogan, G.; Ashcroft, D.
in, *Enzyme Catalysis in Organic Synthesis*,
Drauz, K.; Waldmann, H.; Eds.; 2002,
vol. 3, p. 1065, Wiley-VCH, Weinheim;
see also: (b) Hudlicky, T.; Gonzalez, D.;

Gibson, D. T. *Aldrichim. Acta* 1999, **32**, 35.

33. For example, see: (a) Schmid, A.; Hollmann, F.; Bühler, B. in, *Enzyme Catalysis in Organic Synthesis*, Drauz, K.; Waldmann, H.; Eds.; 2002, vol. **3**, p. 1108, Wiley-VCH, Weinheim; (b) Hummel, W. *Adv. Biochem. Eng. Biotech.* Scheper, T.; Ed.; 1997, vol. 58, p. 147, Springer, Heidelberg; (c) Jones J. B., Jakovac, I. J. *Can. J. Chem.* 1982, **60**, 19; (d) Eckstein, M.; Daussmann, T.; Kragl, *Biocatalysis and Biotransformation* 2004, **22**, 89.

34. (a) Flisch, S.; Grogan Ashcroft, D. in, *Enzyme Catalysis in Organic Reactions*, Drauz, K.; Waldmann; H.; Eds.; 2002, vol. **3**, p. 1202, Wiley-VCH, Weinheim; see also: *Asymmetric Synthesis: the Essentials*, M.; Bräse, S. Eds.; 2007, Wiley-VCH, Weinheim.

35. Liese, A. in, *Enzyme Catalysis in Organic Reactions*, Drauz, K.; Waldmann, H.; Eds.; 2002, vol. 3, p. 1419, Wiley-VCH, Weinheim.

36. *Asymmetric Synthesis with Chemical and Biological Methods*, Enders, D.; Jaeger, K.-E.; Eds.; 2007, Wiley-VCH, Weinheim.

37. (a) Masamune, S.; Choy, W.; Petersen, J. S.; Sita, L. R. *Angew. Chem. Int. Ed.* 1985, **24**, 1; see also: (b) Hoveyda, A. H.; Evans, D. A.; Fu, G. C. *Chem. Rev.* 1991, **91**, 1179.

38. For an authoritative review, see: Corey, E. J. *Angew. Chem. Int. Ed.* 2002, **41**, 1650; see also: Chapter 2, ref. 73.

39. For example, see: *Modern Aldol Reactions*; Mahrwald, R.; Ed.; 2004 vol. 1 and 2, Wiley-VCH, Weinheim.

40. For example, see: (a) López, F.; Feringa, B. in, *Asymmetric Synthesis: The Essentials*, Christmann, M.; Bräse, S., Eds.; 2007, p. 83, Wiley-VCH, Weinheim; (b) Krause, N.; Hoffmann-Röder, A. *Synthesis* 2001, 171; (c) Tomioka, K. in, *Modern Carbonyl Chemistry*, Otera, J. Ed,; 2000, p. 491, Wiley-VCH, Weinheim; (d) Perlmutter, P. *Conjugate Addition Reactions in Organic Synthesis*, 1992, Pergamon, Oxford; see also: (e) Hanessian, S.; Bennani, Y. L. *Chem. Rev.* 1997, **97**, 3161 and references cited therein.

41. For insightful reviews and monographs, see: (a) Farina, V.; Reeves, J. T.; Senanayake, C. H.; Song, J. J. *Chem. Rev.* 2006, **106**, 2734; (b) Federsel, H.-J. *Nat. Rev. Drug Discov.* 2005, **4**, 635; (c) *Asymmetric Catalysis on Industrial Scale*, Blaser, H. -U.; Schmidt, E.; Eds.; 2004, Wiley-VCH, Weinheim; (d) Kotha, S. *Tetrahedron* 1994, **50**, 3639; (e) Crosby, J. *Tetrahedron* 1991, **47**, 4789.

42. For excellent monographs, see: (a) *Fundamentals of Asymmetric Catalysis*, Walsh, P. J.; Kozlowski, M., 2008, University Science Books, Sausalito, CA; (b) *Catalysis of Organic Reactions*, Schmidt, S. R.; Ed.; 2007, CRC Press, Boca Raton, FL; (c) *New Frontiers in Asymmetric Catalysis*, Mikami, K.; Lautens, M.; Eds.; 2007, Wiley-VCH, Weinheim; (d) *Catalysis from A to Z*, Cornils, B.; Herrmann, W. A.; Muhler, M.; Wong, C.-H.; Eds.; 2007, Wiley-VCH, Weinheim; (e) *Asymmetric Synthesis: The Essentials*, Christmann, M.; Bräse, S.; Eds.; 2007, Wiley-VCH, Weinheim; (f) *Comprehensive Asymmetric Catalysis*, vol. I-III; Jacobsen, E. N.; Pfaltz, A.; Yamamoto, H.; Eds.; 2000, Springer-Verlag, Berlin; (g) *Catalytic Asymmetric Synthesis*, Ojima; Ed.; 2000, Wiley-VCH, Weinheim.

43. For a relevant review, see: Nicolaou, K. C.; Bulger, P. G. Jr.; Sarlah, D. *Angew. Chem. Int. Ed.* 2005, **44**, 4442.

44. For example, see: (a) *Chiral Reagents for Asymmetric Synthesis*, Paquette, L. A.; Ed.; 2003, Wiley, NY; (b) Handy, S. T. *Curr. Org. Chem.* 2000, **4**, 363.

45. Bredig, G.; Fiske, W. *Biochem. Z.* 1912, 7.

46. For an authoritative monograph, see: (a) Marquéz-López, E.; Herrera, R. P.; Christmann, M. *Nat. Prod. Rep.* 2010, **27**, 1138; (b) *Enantioselective Organocatalysis: Reactions and Experimental Procedures*, Dalko, P. I. Ed.; 2007, Wiley-VCH, Weinheim; (c) *Asymmetric Organocatalysis: From Biomimetic Concepts to Asymmetric Synthesis*, Berkessel, A.; Gröger, H, 2005, Wiley-VCH, Weinheim; for pertinent reviews and articles, see: (d) Dondoni, A.; Massi, A. *Angew. Chem. Int. Ed.* 2008, **47**, 4638; (e) Melchiorre, P.; Marigo,

M.; Carlone, A; Bartoli, G. *Angew. Chem. Int. Ed.* 2008, **47**, 6138; (f) Enders, D.; Niemeier, O.; Henseler, A. *Chem. Rev.* 2007, **107**, 5606; (g) Pellissier, H. *Tetrahedron* 2007, **63**, 9267; (h) Dalko, P. I.; Moisan, L. *Angew. Chem. Int. Ed.* 2004, **43**, 5138; Dalko, P. I.; Moisan, L. *Angew. Chem. Int. Ed.* 2001, **40**, 3726; (i) Jarvo, E. R.; Miller, S. J. *Tetrahedron* 2002, **58**, 2481. For a thematic issue on organocatalysis, see: (j) List, B., Guest Editor; *Chem. Rev.* 2007, **107**, 5413.

47. (a) Eder, U.; Sauer, G.; Wiechert, R. *Angew. Chem. Int. Ed.* 1971, **10**, 496; (b) Hajos, Z. G.; Parrish, D. R. *J. Org. Chem.* 1974, **39**, 1615; see also: (c) Hajos, Z. G.; Parrish, D. R. German Patent DE 2102623, 1971 (d) Hajos, Z. G.; Parrish, D. R. *Org. Synth.* 1985, **63**, 26; (e) Buchschacher, P.; Fürst, A. *Org. Synth.* 1986, **63**, 37; For selected recent articles on proline catalysis, see: (f) Kellogg, R. M. *Angew. Chem. Int. Ed.* 2007, **46**, 494; (g) Seebach, D.; Beck, A. K.; Badine, D. M.; Limbasch, M.; Eschenmoser, A.; Treasurywala, A. M.; Hobi, R. *Helv. Chim. Acta* 2007, **90**, 425; (h) Enders, D.; Grondal, C.; Hüttl, M. R. M. *Angew. Chem. Int. Ed.* 2007, **46**, 1570; (i) Jaroch, S.; Weinmann, H.; Zeitler, K. *Chem. Med. Chem.* 2007, **2**, 1261; (j) Klussmann, M.; White, A. J. P.; Iwamura, H.; Wells, D. H. Jr.; Armstrong, A.; Blackmond, D. G. *Angew. Chem. Int. Ed.* 2006, **45**, 7989; (k) Marigo, M.; Schulte, T.; Franzén, J.; Jørgensen, K. A. *J. Am. Chem. Soc.* 2005, **127**, 15710; (l) Córdova, A.; Ibrahem, I.; Casas, J.; Sundén, H.; Engquist, M.; Reyes, E. *Chem. Eur. J.* 2005, **11**, 4772; (m) Clemente, F.; Houk, K. N. *Angew. Chem. Int. Ed.* 2004, **43**, 5766.

48. For example, see: (a) Ramachary, D. B.; Barbas, C. F. III. *Chem. Eur. J.* 2004, **10**, 5325; (b) Ramachary, D. B.; Barbas, C. F. III. *J. Am. Chem. Soc.* 2001, **123**, 5260; (c) Barbas, C. F. III. *Angew. Chem. Int. Ed.* 2008, **47**, 42.

49. For example, see: (a) List, B. *Tetrahedron* 2002, **58**, 5573. (b) List, B. *Acc. Chem. Res.* 2004, **37**, 548; (c) Seayed, J.; List, B. *Org. Biomol. Chem.* 2005, **3**, 719; (d) Yang, J. W.; Hechavarria Fonseca, M. T.; List, B. *J. Am. Chem. Soc.* 2005, **127**,

15036; (e) List, B.; Yang, J. W. *Science* 2006, **313**, 584.

50. (a) Ahrendt, K. A.; Borths, C . J.; MacMillan, D. W. C. *J. Am. Chem. Soc.* 2000, **122**, 4243; (b) Huang, Y.; Walji, A. M.; Larsen, C. H.; MacMillan, D. W. C. *J. Am. Chem. Soc.* 2005, **127**, 15051; (c) Northrup, A. B.; MacMillan, D. W. C. *Science* 2004, **305**, 1752; (d) Beeson, T. D.; Mastracchio, A.; Hong, J.-B.; Ashton, K.; MacMillan, D. W. C. *Science* 2007, **316**, 582; see also: (e) Lelais, G.; MacMillan, D. W. C. *Aldrichimica Acta* 2006, **39**, 79; (f) Walji, A. M.; MacMillan, D. W. C. *Synlett* 2007, 1477; (g) MacMillan, D. W. C. *Nature* 2008, **455**, 304.

51. For excellent reviews, see: (a) O'Donnell, M. J. *Acc. Chem. Res.* 2004, **37**, 506; (b) Maruoka, K.; Ooi, T. *Chem. Rev.* 2003, **103**, 3013; (c) Kacprzak, K.; Gawrónski, J. *Synthesis* 2001, 961.

52. (a) List, B.; Lerner, R. A.; Barbas, C. F. III. *J. Am. Chem. Soc.* 2000, **122**, 2395; List, B.; Lerner, R. A.; Barbas, C. F. III. *Tetrahedron* 2002, **58**, 5573; (b) Ahrendt, K. A.; Borths, C. J.; MacMillan, D. W. C. *J. Am. Chem. Soc.* 2000, **122**, 4243; (c) (i) Gutzwiller, J.; Buchschacher, P. Furst, A. *Synthesis* 1997, 167 (ii) Harada, N.; Sugioka, T.;Uda.; Kuriki, T. *Synthesis* 1990, 53 (iii) Hajos, Z. G.; Parrish, D. R. *J. Org. Chem.* 1974, **39**, 1615 (d) Eder, U.; Sauer, G.; Wiechert, R. *Angew. Chem. Int. Ed.* 1971, **10**, 496; (e) Huang, Y.; Unni, A. K.; Thadani, A. N.; Rawal, V. H. *Nature* 2003, **424**, 146; (f) Juliá, S.; Guixer, J.; Masana, J.; Rocas, J.; Colonna, S.; Annuziata, R.; Molinari, H. *J. Chem. Soc., Perkin Trans. 1* 1982, 1317; (g) Hughes, D. L.; Dolling, U. H.; Ryan, K. M.; Schoenewaldt, E. F.; Grabowski, E. J. J. *J. Org. Chem.* 1987, **52**, 4745; (h) (i) Corey, E. J.; Xu, F.; Noe, M. C. *J. Am. Chem. Soc.* 1997, **119**, 12414; (ii) Lygo, B.; Crosby, J.; Lowdon, T. R.; Wainwright, P. G. *Tetrahedron Lett.* 1997, **38**, 2343; (iii) Ooi, T.; Kameda, M; Maruoka, K. *J. Am. Chem. Soc.* 1999, **121**, 6519; (i) Ooi, T.; Takeuchi, M.; Kameda, M.; Maruoka, K. *J. Am. Chem. Soc.* 2000, **122**, 5228; (j) Corey, E. J.; Noe, M. C.; Xu, F.

Tetrahedron Lett. 1998, **39**, 5347; (k) Hanessian, S.; Shao, Z.; Warrier, J. S. *Org. Lett.* 2006, **8**, 4787; (l) Austin, J. F.; MacMillan, D. W. C. *J. Am. Chem. Soc.* 2002, **124**, 1172; (m) Iyer, M. S.; Gigstad, K. M.; Namdev, N. D.; Lipton, M. *J. Am. Chem. Soc.* 1996, **118**, 4910; (n) (i) Sigman, M. S.; Vachal, P.; Jacobsen, E. N. *Angew. Chem. Int. Ed.* 2000, **39**, 1279; (ii) Sigman, M. S.; Jacobsen, E. N. *J. Am. Chem. Soc.* 1998, **120**, 4901; (iii) Vachal, P.; Jacobsen, E. N. *J. Am. Chem. Soc.* 2002, **124**, 10012; (o) Vachal, P.; Jacobsen, E. N. *Org. Lett.* 2000, **2**, 867; (p) Wenzel, A. G.; Jacobsen, E. N. *J. Am. Chem. Soc.* 2002, **124**, 12964; (q) (i) Córdova, A.; Barbas, C. F. III. *Tetrahedron Lett.* 2002, **43**, 7749; (ii) Hayashi, Y.; Tsuboi, W.; Ashimine, I.; Urishima, T.; Shoji, M.; Sakai, K. *Angew. Chem. Int. Ed.* 2003, **42**, 3677; (r) (i) Enders, D.; Hüttl, M. R. M.; Grondal, C.; Raabe, G.

Nature 2006, **441**, 861; (ii) Enders, D.; Grondal, C.; Hüttl, M. R. M. *Angew. Chem. Int. Ed.* 2007, **46**, 1570; (s) (i) Taggi, A. E.; Hafez, A. M.; Wack, H.; Brandon Young, B.; Drury, W. J. III; Lectka, T. *J. Am. Chem. Soc.* 2000, **122**, 7831; (ii) Taggi, A. E.; Hafez, A. M.; Wack, H.; Brandon Young, B.; Ferraris, D.; Lectka, T. *J. Am. Chem. Soc.* 2002, **1242**, 6626; (t) Denmark, S. E.; Ghosh, S. K. *Angew. Chem. Int. Ed.* 2001, **40**, 4759; (u) Adam, W.; Rao, P. B.; Degen, H.-G.; Levai, A.; Tamás Patonay, T.; Chantu R. Saha-Möller, C. R. *J. Org. Chem.* 2002, **67**, 259.

53. (a) *Cyclitols and Their Derivatives: A Handbook of Physical, Spectral, and Synthetic Data*, Hudlicky, T.; Cehulak, M., 1993, VCH, NY; see also: (b) Rassu, G.; Auzzas, L.; Pinna, L.; Battistini, L.; Curti, C. in, *Studies in Natural Products Chemistry*, Atta-ur- Rahman, Ed.; 2003, vol. 29, p. 449.

6
The *Chiron Approach*

6.1
Living Through a Total Synthesis

There are a number of reasons to choose a molecule as a target for organic synthesis. While the sheer challenge and intellectual curiosity aspects are always prime incentives, there are other factors that must also be considered, especially in the context of lasting value and contribution to science [1]. Great heights have been reached in the synthesis of complex natural products, as well as molecules of therapeutic relevance [2]. Every target molecule, no matter how simple, presents its challenges in terms of practicality and efficiency. The height of the synthetic complexity bar can be raised at will, depending on the structural and stereochemical intricacies of the target molecule. With the advent of newer and more powerful synthetic methods, enhanced by advanced separation and analytical techniques, the expectations for efficiency and creative design have also been heightened. More and more, reasons proffered for the synthesis of a given molecule and the chosen strategy, will be subject to scrutiny by peers and critics alike [1c]. Viewed in a positive context, such a process will also bring out the best in the those who practice the enterprise of synthesis. To put the process of target-oriented organic synthesis in perspective, and to portray it in human terms, let us consider "living" through its eight steps.

A.	Choice of target molecule	⇒	Relevance
B.	Perceptive powers	⇒	Heuristic analysis
C.	Emergence of a strategy	⇒	Individual prowess
D.	Generation of a syntheses plan	⇒	Attention to detail
E.	Execution	⇒	Efficiency, practicality
F.	Endurance	⇒	What is the "price" of synthesis?
G.	Contribution to science	⇒	New concepts, reactions, reagents
H.	Recognition	⇒	Rewards, legacy, fame and fortune (?)

Clearly, an element of relevance and timeliness must be associated with the choice of a target molecule, considering that such projects are time and resource dependent. Coworker training aside, there are many valid reasons to pick a particular target molecule for synthesis, ranging from its biological importance to its physicochemical properties in materials science for example.

Design and Strategy in Organic Synthesis: From the Chiron Approach to Catalysis, First Edition.
Stephen Hanessian, Simon Giroux, and Bradley L. Merner.
© 2013 Wiley-VCH Verlag GmbH & Co. KGaA. Published 2013 by Wiley-VCH Verlag GmbH & Co. KGaA.

As was amply discussed in the preceding chapters, visual perception and logic-based thinking will lead to a multitude of disconnective possibilities generating a number of solutions on paper. Each of these may also present a viable strategy and a synthesis plan to follow. Here is where individual prowess and ingenuity will be manifested, thus showing diversity of thought, and reflecting on preferences, biases, and even fears. The Sinatra (*"I did it my way"*) approach may become clearly evident relative to the Nike (*"just do it"*) type approach [1c].

Having finalized the synthesis blueprint, with well-chosen reagents, protective groups, and reaction conditions, it is time to render the paper synthesis to practice in the laboratory, where efficiency and practicality will be put to the test. Showing a strong preference for a given type of a reaction, which may be a planned highlight of the synthesis, could also be the Achilles heel of the entire operation. Alternate plan B routes are frequently called in to rescue the synthesis, sometimes at the expense of elegance or expediency.

It is well-known, that multi-step synthesis is labor intensive, and that the climb to the summit is arduous. Perseverance, resilience, and a "never say die" attitude are worthwhile traits to adopt when embarking on a synthesis journey, assuming that all other factors that keep a project alive are in place. The noble cause of synthesis can be best served when, in addition to valuable coworker training, there emerges an opportunity to contribute to the advancement of the sub-discipline in the form of new concepts, innovative reactions, or useful reagents. Finally, as deeply entrenched as scientists are in carrying out important research work, recognition in one form or another is always appreciated. In the final analysis, the legacy and the impact of the scientific contributions of a synthetic organic chemist will be measured not by the complexity of molecules that were synthesized, but by the quality of the science that was produced, and the benefits to the coworkers trained. In the words of Leonardo da Vinci:

Triste e quel diceplo che non avanza il suo maestro. (Sad is the case of the disciple who does not surpass his/her master.)

6.2
Principles of the *Chiron Approach*

The stereocontrolled synthesis of a chiral target compound as an enantiopure (or highly enantioenriched) entity is the standard of synthetic chemistry today. This practice has become a necessity in the pharmaceutical industry, particularly as chemistry and biology meet at the interface of stereochemistry and function.

Excluding microbiological, enzymatic, and related processes to obtain enantiopure starting materials, there are three main approaches to achieve the objective (see also Chapter 5). Of these, optical resolution of racemic mixtures, in its classical format, may be the most obvious and least logic-based approach. The two remaining important approaches rely on asymmetric processes, and on Nature's renewable reservoir of chiral, non-racemic small molecules whose inherent functionality and chirality can be fully exploited as building blocks for synthesis of enantiopure compounds. The recognition of these natural small molecules in the carbon framework of target molecules, and their judicious utilization in the synthesis proper, is the basis of the *chiron approach* [3].

6.2.1
Definition

The term *chiron* [4] is derived from *chi*ral synth*on*, referring to a chiral, non-racemic version of a synthon. Whereas synthons and retrons, originally proposed by E. J. Corey [5], can represent idealized fragments or actual chemical entities in an antithetic reasoning paradigm, chirons correspond to compounds originating from enantiopure amino acids, carbohydrates, hydroxy acids, and terpenes, as well as their chemical transformation products (see section 6.2.6). The term can be

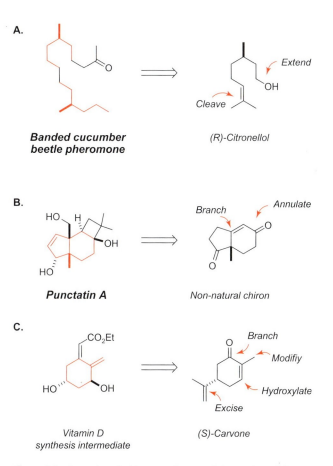

A.

Banded cucumber beetle pheromone

(R)-Citronellol

Extend

Cleave

B.

Punctatin A

Non-natural chiron

Branch

Annulate

C.

Vitamin D synthesis intermediate

(S)-Carvone

Branch

Modifiy

Hydroxylate

Excise

Figure 6.1 Examples of chirons and sites of chemical modification to provide advanced intermediates.

generalized to include any molecule that is easily obtained in an enantiomerically or diastereomerically pure form from natural or unnatural sources. A native chiron or its chemically modified version is most useful when it has a high level of skeletal, functional, and stereochemical overlap with a segment of the intended target molecule. Examples of obvious skeletal convergences of chirons are shown in Figure 6.1 (A–C) [6]. In each case, the native chiron was chemically modified as indicated with symbolic key words representing functional group transformations.

The heuristic nature of the approach is best demonstrated in total syntheses where the native chiron is extensively modified to fit the requisite substructure in the target molecule(s) [3]. In these cases, visual imagery is truly put to the test, because the modified chiron loses its original skeletal progeny compared to the native chiron from which it was made. The synthesis of the β-lactam antibiotic PS-5 from (1*R*,2*R*)-2-amino-4-cyclohexene-1-carboxylic acid is a pertinent example (Figure 6.2) [7]. On face value, the cyclic amino acid, which is readily available from *meso*-cyclohex-4-ene-1,2-dicarboxylic acid dimethyl ester by enzymatic desymmetrization, followed by epimerization, and a Curtius rearrangement, has no skeletal relationship with the target molecule. However, upon closer scrutiny, the embedded β-amino acid unit is a good match for the β-lactam unit in the target molecule. In practice, oxidative cleavage of the double bond generates an acyclic β-amino acid precursor that provides a perfect congruence with the "upper" framework of PS-5 with matching stereochemistry as a bonus. Several related examples are discussed in subsequent chapters of this book, and provide a *"the same and not the same"* [8] heuristic paradigm in synthesis planning (see also Chapter 17, section 17.4.2, Figure 17.11).

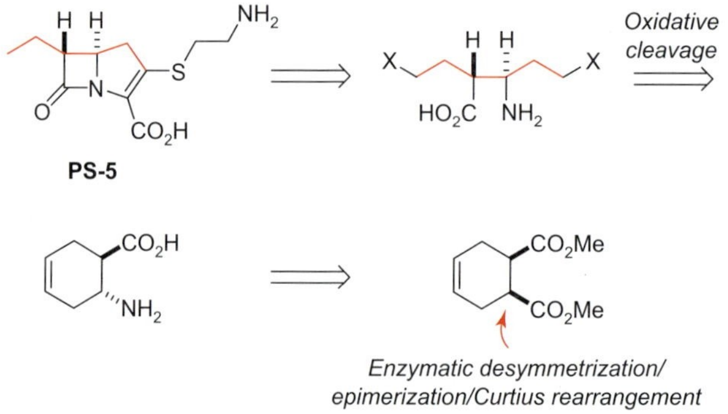

Figure 6.2 Utilization of cyclic chirons as their chemically modified acyclic equivalents.

6.2.2
The *Chiron Approach*

This approach relies on the visual recognition of chiral substructures in a target molecule, resulting from disconnections of strategic bonds in a retrosynthetic mode with minimal perturbation of stereogenic centers. Functional and stereochemical information present in the chiral substructures is then sought in the corresponding native chirons with directly overlapping carbon frameworks. Invariably, extension or truncation of such carbon frameworks will be required, in addition to functional group and stereochemical adjustments leading to modified chirons.

The *chiron approach* has been extensively used over the years, and it is applicable to virtually every class of molecule possessing one or more stereogenic carbon atoms. The approach has aesthetic appeal and offers a high level of predictability. Selected examples of total syntheses adopting the *chiron approach,* and spanning a 22-year period are shown in Figure 6.3 A–D [9]. The versatility of the approach is demonstrated in cases where multiple chirons were used as starting materials, then chemically manipulated so as to converge with different, non-overlapping sub-structures in the target molecule (see Chapters 8–16 and Chapter 18 for additional examples).

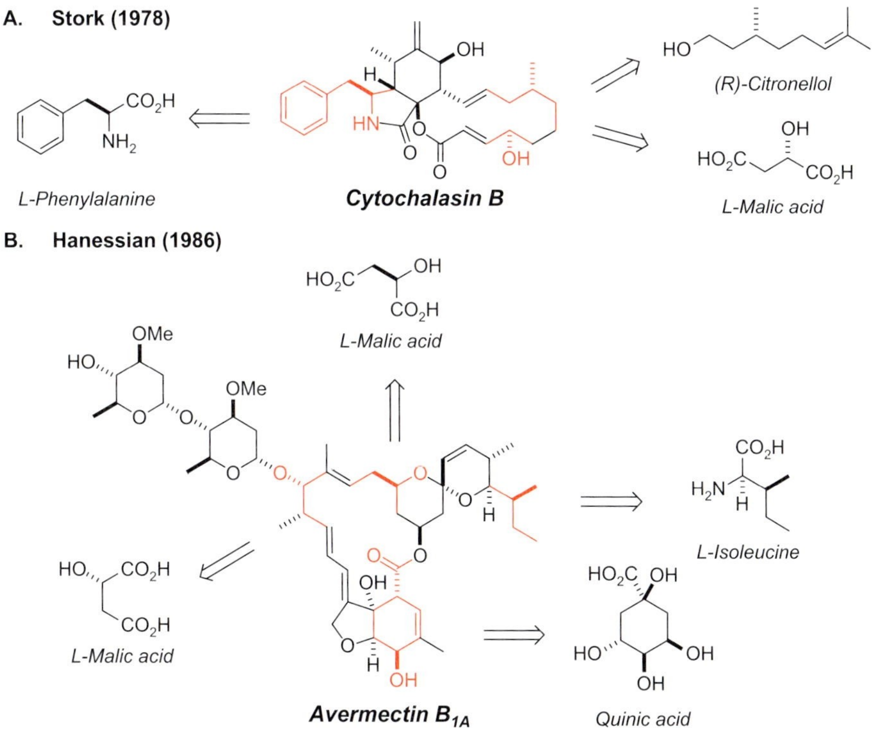

A. Stork (1978)

L-Phenylalanine   *Cytochalasin B*   L-Malic acid

(R)-Citronellol

B. Hanessian (1986)

L-Malic acid

L-Malic acid   *Avermectin B$_{1A}$*   Quinic acid

L-Isoleucine

Figure 6.3   Completed syntheses with native and chemically modified chirons as precursors for single target molecules.

C. Ireland (1996)

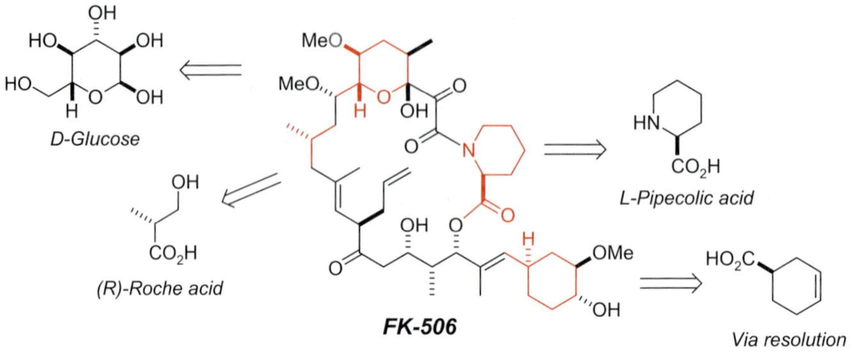

D-Glucose

(R)-Roche acid

FK-506

L-Pipecolic acid

Via resolution

D. Meyers (2000)

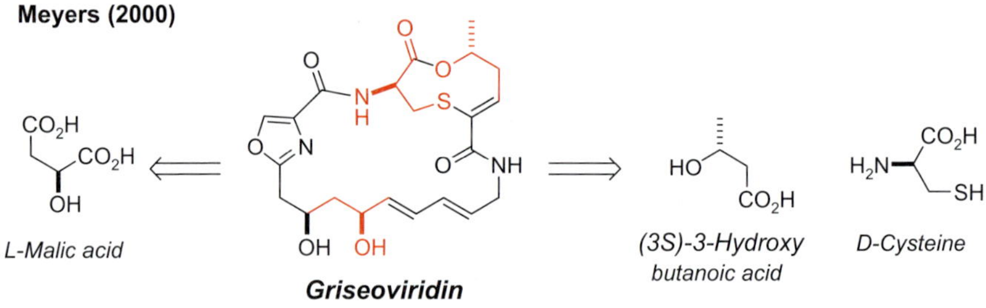

L-Malic acid

Griseoviridin

(3S)-3-Hydroxy butanoic acid

D-Cysteine

Figure 6.3 (*continued*).

In each of the above syntheses, there is a clear visual connection and an excellent overlap of carbon skeleton, functionality, and stereochemistry with the chosen chirons. The visual analysis has been rendered all the more evident in this book through the color coding of convergent substructures in the target molecules and the corresponding chirons. Although, mostly native chirons are shown for each synthesis in Figure 6.3, it is clear that further chemical modification was necessary in order to use them as functionally useful intermediates *en route* to the target molecules.

6.2.3
Two philosophies, one goal

As previously mentioned in section 6.2, another popular strategy to access enantiomerically pure (or enriched) compounds as starting materials is through asymmetric synthesis [10]. Here, in its most basic form, racemic or achiral entities may be engaged in asymmetric bond-forming steps relying on chiral auxiliaries or chiral catalysts to produce enantiopure (or enantioenriched) compounds as single enantiomers directly or as diastereomers in a given reaction. This strategy is generally applicable in cases where one or two stereogenic centers are generated per reaction. Many innovations have been introduced in asymmetric reactions, especially involving catalytic variants in recent years [10]. Whereas the *chiron approach* has the attributes of predictive power and general utility in total synthesis, the *asymmetric synthesis* approach offers intellectual stimulation and practical advantages in many cases, especially through ligand design in catalytic systems. Although the two approaches are philosophically different, they share the same objective of synthesizing segments of a given target molecule with a high level of stereochemical purity. Whenever applicable, chirons can also be the starting points as substrates for catalytic reactions *en route* to more elaborate intermediates, provided that the resident inherent chirality reinforces the stereodifferentiating event.

A large number of total syntheses of architecturally complex, polyfunctional natural products rely on Nature's chirons as starting materials (see Chapters 8–16, and Chapter 18 for examples). The total synthesis of ambruticin S, and a number of approaches to its "western" ring A tetrahydropyran unit, serves as a good example for comparing and contrasting the chiron and asymmetric synthesis approaches to one and the same target molecule (Figure 6.4 A–D; Figure 6.5 A–C).

In the first total synthesis of ambruticin S, Kende and coworkers [11a] used methyl α-D-glucopyranoside as a carbohydrate-based chiron. Considering the nature of ring A, a D-glucopyranose template was an obvious choice in view of the skeletal congruence and correct stereochemistry of substituents at C-2, C-3, and C-5 (D-glucose numbering). Deoxygenation at C-4 and chain extension at C-6 gave an advanced intermediate, which was functionalized as a *C*-glycoside encompassing a derivative of the substituted cyclopropane (ring B) (Figure 6.4 A). Hanessian and coworkers reported a similar approach to ring A [11b]. Martin, [11c] Lee [11d], and their respective groups utilized acyclic carbohydrate templates with a matching *trans*-diol unit and a fixed C-1 appendage as a hydroxymethyl group. Key reactions involved intramolecular stereocontrolled cyclizations to generate the eventual C-5 two-carbon carboxylic acid unit (tetrahydropyran numbering) (Figure 6.4 B and C). Finally, Donaldson and coworkers [11e] utilized L-arabinose as a chiron, which was chain-extended and subjected to cycloetherification *en route* to a methyl glycoside. *C*-allylation and functional group adjustment led to the intended advanced intermediate (Figure 6.4 D).

Each of the above syntheses of the ring A unit of ambruticin S was accomplished in 10–12 steps starting from readily available carbohydrates as native chirons. The main advantage of the carbohydrate-based *chiron approach* to ambruticin S is in securing the correct absolute stereochemistry of the *trans*-diol in ring A, as well as providing a good skeletal convergence.

The asymmetric synthesis approaches to ring A of ambruticin S have relied on stoichiometric and catalytic processes. Thus, Michelet [12a] used a hetero Diels-Alder reaction of an achiral α-keto-β,γ-unsaturated ester derivative and a vinyl ether containing a mandelate-derived appendage (Figure 6.5 A). An efficient hetero Diels-Alder reaction between a dienic enol ether and an aldehyde, catalyzed by a chromium-based chiral aminoindanol complex was devised by Liu and Jacobsen [12b] (Figure 6.5 B). The resulting endocyclic enol ether was subsequently transformed to the *trans*-diol required in ring A. Adjustment of oxidation states

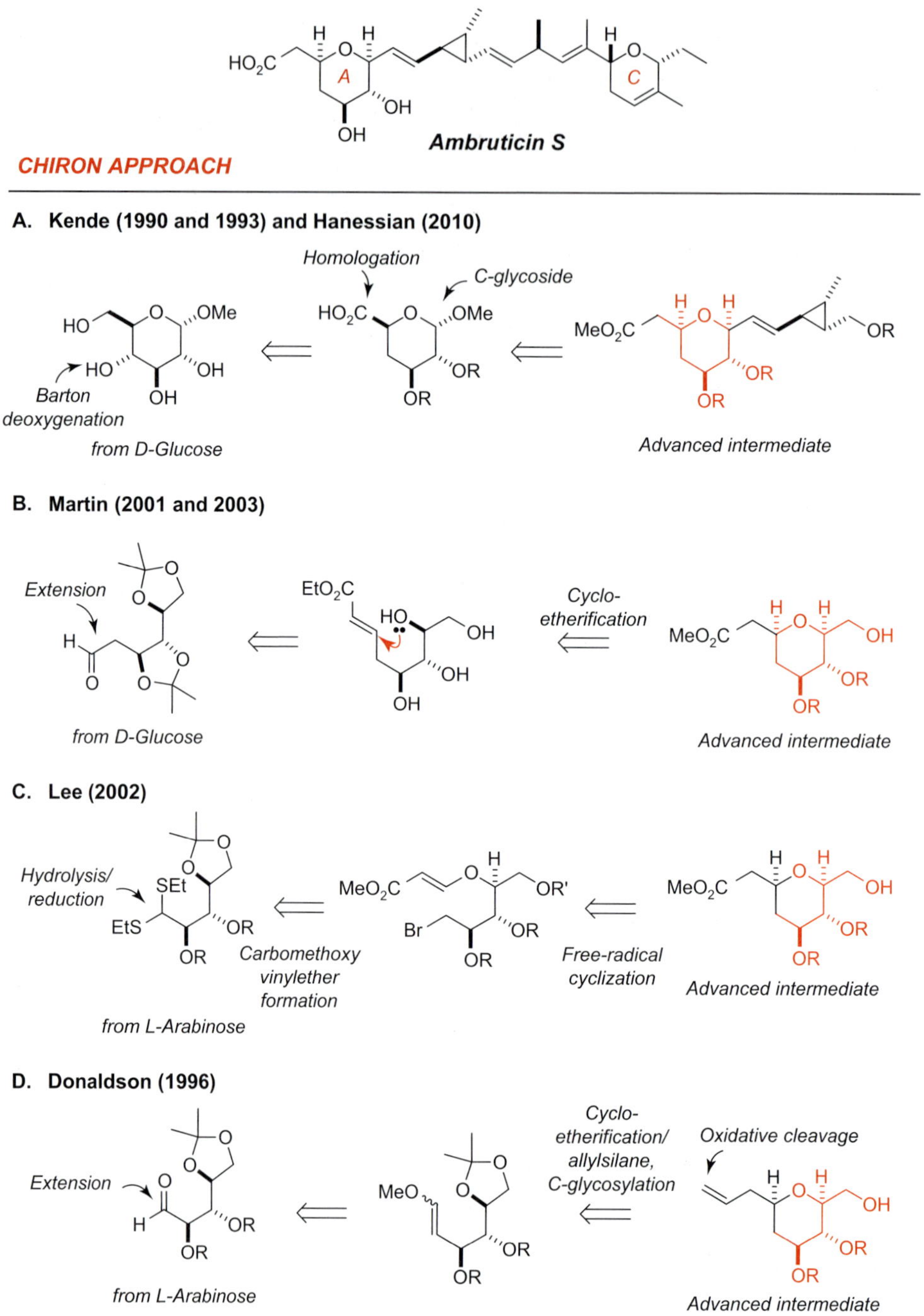

Ambruticin S

CHIRON APPROACH

A. Kende (1990 and 1993) and Hanessian (2010)

B. Martin (2001 and 2003)

C. Lee (2002)

D. Donaldson (1996)

Figure 6.4 Carbohydrate-derived chirons for ring A of ambruticin S.

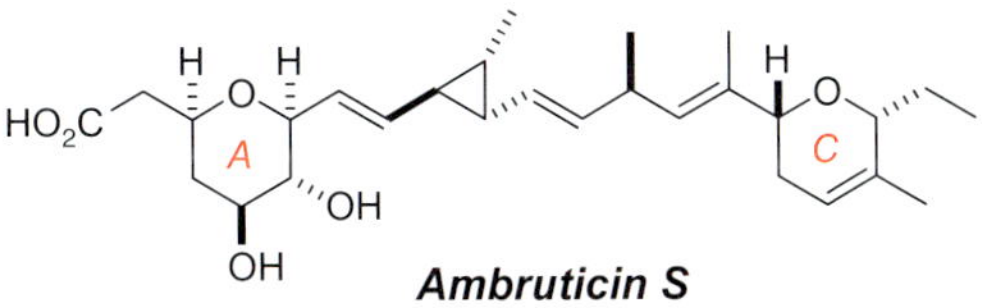

Ambruticin S

ASYMMETRIC SYNTHESIS APPROACH

A.   Michelet (1999): Auxiliary-mediated hetero Diels-Alder reaction

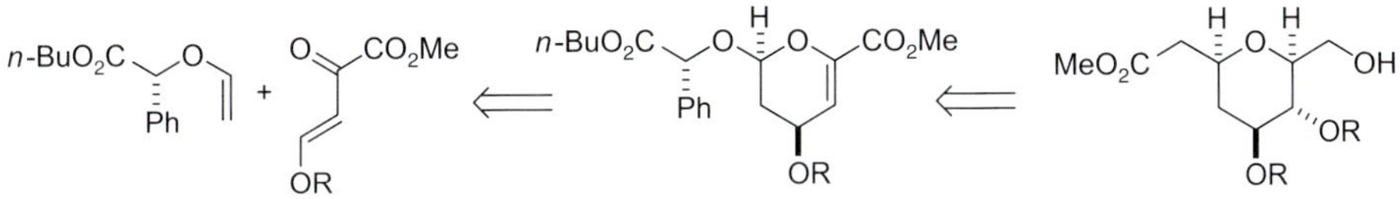

dr = 9:1

Advanced intermediate

B.   Jacobsen (2001): Catalyst-mediated hetero Diels-Alder reaction

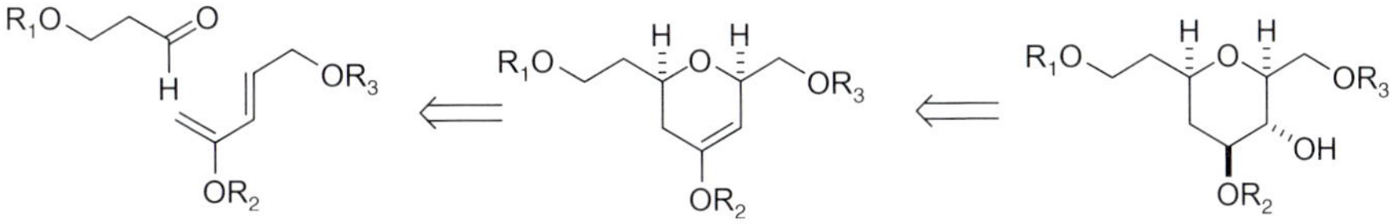

97% ee

Advanced intermediate

catalyst

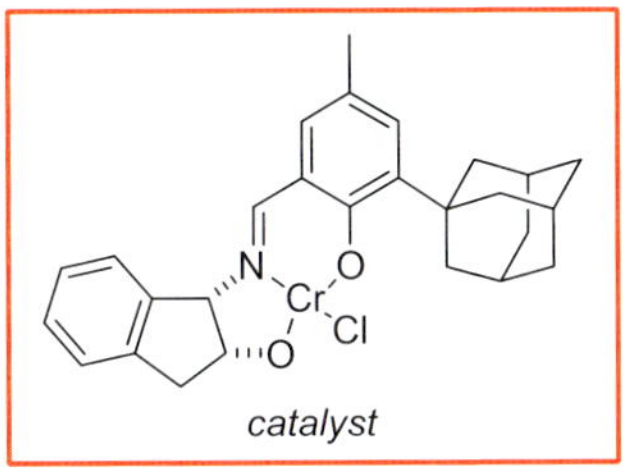

C.   Markó (1993): Silyl-modified Sakurai reaction

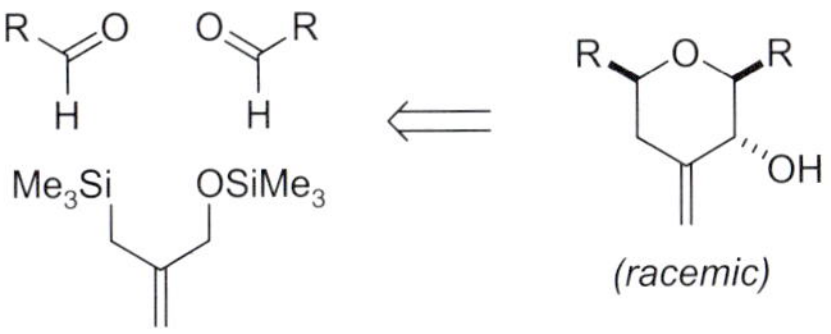

(racemic)

Figure 6.5   Asymmetric synthesis approaches to ring A of ambruticin S.

allowed further segment coupling *en route* to ambruticin S. Markó and his group [12c] developed a general synthesis of racemic tetrahydropyrans containing identical appendages at C-2 and C-6 (Figure 6.5 C). Hanessian and coworkers [12d] reported cationic palladium complexes, as well as Lewis acids, as efficient catalysts for the synthesis of dihydropyran subunits in connection with their work on the synthesis of jerangolid A, a naturally occurring congener of ambruticin S.

The asymmetric synthesis approach to ambruticin S, exemplified by the Jacobsen catalytic hetero Diels-Alder reaction, has the attribute of a one-step construction of functionalized tetrahydropyrans with the expected 2,6-disubstitution pattern. It should be noted that the construction of ring C in ambruticin S by Liu and Jacobsen also capitalized on a catalytic hetero Diels-Alder reaction with the enantiomeric aminoindanol as a ligand.

6.2.4
There is more than meets the eye

Wait a minute, I don't draw it that way is a comment we have said or heard more than once when discussing chemistry with colleagues or coworkers. Indeed, it can be argued that the way we draw a molecule may even dictate the manner in which we think when contemplating a strategy for its synthesis. Usually, we portray a key reaction in a visual representation that best approximates a plausible bonding event or a transition state model in a drawn structure. In other instances, target molecules may be drawn according to previously published formats, or changed for aesthetic reasons. The four independent total syntheses of calcimycin (A-23187) published during the period 1979–1989 serve as historical examples of the interplay between image and thought (or vice-versa) in synthesis planning. In the original Evans synthesis [13a], the focal point was to deploy stereochemically defined *C*-methyl groups on carbon chains that eventually underwent a stereoelectronically controlled spiroacetalization to give the 1,5-dioxaspiroacetal motif. Complexity was simplified by depicting the target ionophore as an acyclic α,ω-dihydroxy ketone precursor **A2**. Using hydrazone anion alkylation, chiron **A2** was built in a sequential two-directional manner from a common (2*S*)-2-methyl-3-hydroxy propionic acid ((*S*)-Roche acid) intermediate through its end-differentiated iodides **A3** and **A4** (Figure 6.6 A). Portrayal of an acyclic carbon chain as an α,ω-dihydroxy ketone leads the mind's eye to consider native or modified chirons as sources of *C*-methyl groups. Thus, the (*S*)-Roche acid was a logical and expedient choice by Evans, who exploited its "symmetry" in a most productive way.

Grieco [13b] utilized a different strategy in which an acyclic γ,δ-unsaturated ester **B1** harboring four *C*-methyl groups was derived from chiron **B2**. The latter was synthesized in a series of stereocontrolled steps that consisted of enolate alkylation of **B3**, Baeyer-Villiger oxidation (**B4** to **B3**), and oxidative ring contraction of a chiral bicyclo[2.2.1]heptenone bearing the "starter" methyl group as in **B5** (Figure 6.5 B). The drawing of calcimycin in a conformationally favorable chair-chair dioxaspiroacetal perspective presumably had little bearing, *per se*, on

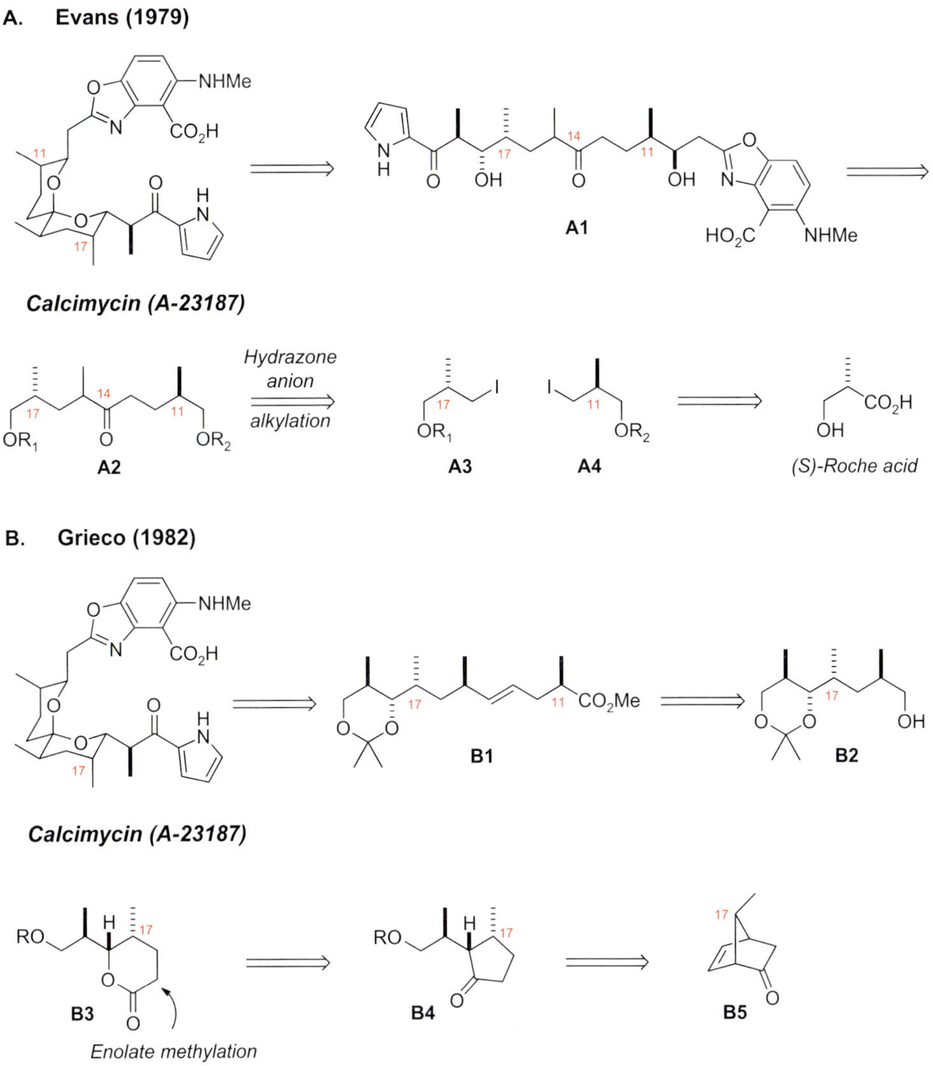

A. Evans (1979)

Calcimycin (A-23187)

B. Grieco (1982)

Calcimycin (A-23187)

Enolate methylation

Figure 6.6 *Chiron approach* strategies toward calcimycin.

the creative thought process leading to the synthesis plan. The remoteness of the Grieco bicyclic precursor (**B5**) from a calcimycin acyclic intermediate portrayed as **B1**, is an example of a very hidden chiron in the framework of a target molecule (see Chapter 8, section 8.4).

In Kishi's synthesis of calcimycin [13c], he capitalized on a carbocyclic starting chiron exemplified by (5*R*)-5-methyl-2-cyclohexenone (**C5**), which was used as a common intermediate to two fragments containing a *C*-methyl group (**C3** and **C4**) (Figure 6.6 C). Conceptually, the Kishi strategy relied on the intramolecular, base-catalyzed cyclization of an α,ω-hydroxy ketone **C1** to a hemiacetal **C2**, and

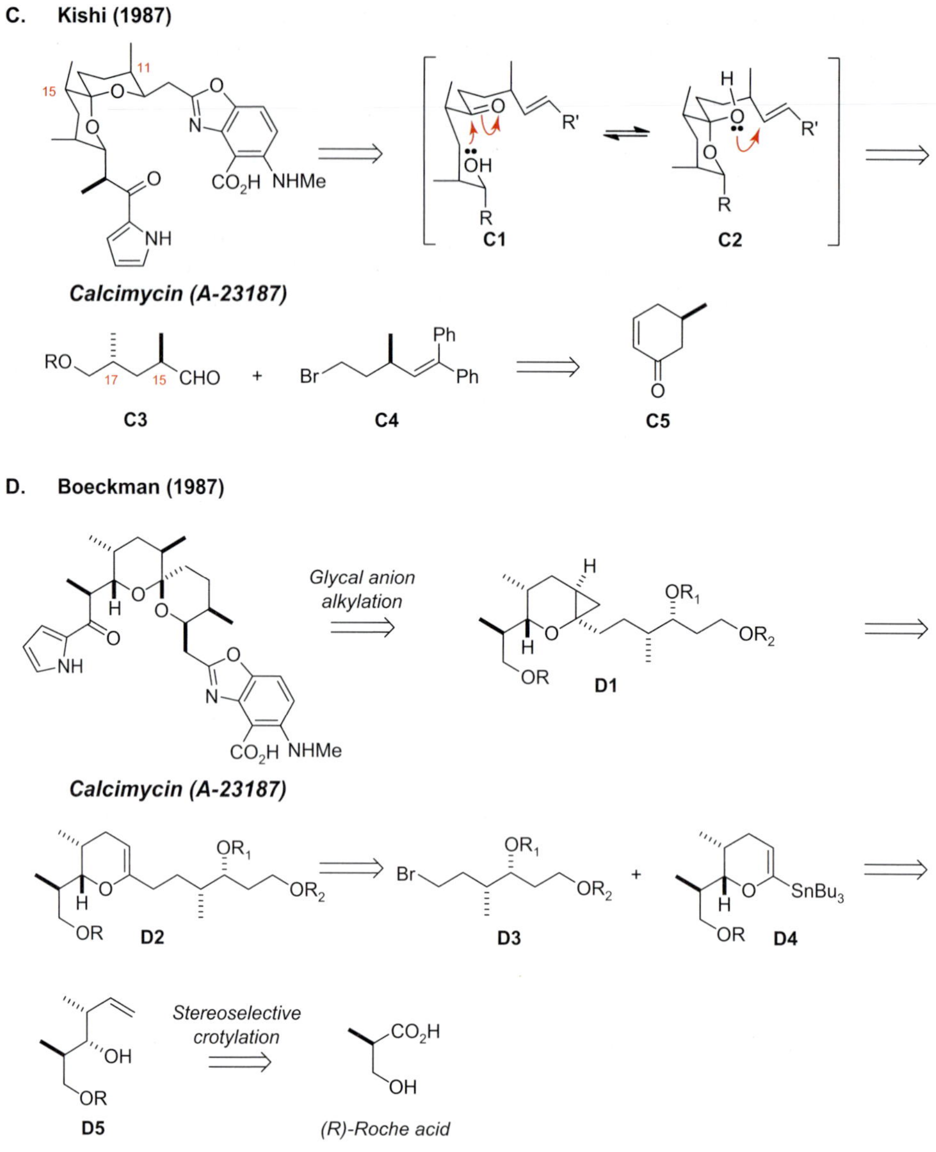

C. **Kishi (1987)**

Calcimycin (A-23187)

C3 + **C4** ⟹ **C5**

D. **Boeckman (1987)**

Glycal anion alkylation ⟹ **D1** ⟹

Calcimycin (A-23187)

D2 ⟹ **D3** + **D4** ⟹

D5 Stereoselective crotylation ⟹ (R)-Roche acid

Figure 6.6 (continued).

a subsequent Michael-type addition to an unsaturated heterocycle in a stereo-controlled manner leading to the desired dioxaspiroacetal diastereomer. For the purposes of clarity, if not for mechanistic insight, the structure of the precursor hydroxy ketone **C1** and the transient hemiacetal **C2** were drawn in chair-like, anomerically distinct, conformations. Only then would the reader appreciate the intended key step shown by the red arrows. Perhaps the same motivation also applied subliminally in the planning stage. In other words, what Kishi saw in the calcimycin structure as a key disconnect was the oxa-Michael intermediate **C2** and its precursor **C1**.

Finally, in the Boeckman [13d] synthesis of calcimycin, we see another aspect of imagery and the thought process (Figure 6.6 D). Contrary to the last three drawings of the target molecule, we now see a "flat" representation of the dioxaspiroacetal ring system. The (R)-Roche acid, was used as a "starter" chiron harboring a C-methyl group. This was chain-extended via a tri-n-butylcrotylstannyl-mediated reaction to **D5**, which was a precursor to the α-stannyl glycal intermediate **D4**. A key alkylation with **D3** gave a C-substituted glycal **D2**. Cyclopropanation afforded **D1** as a major isomer, which underwent an oxygen-assisted hydrolytic opening through an oxocarbenium ion to generate a C-methyl group in a stereocontrolled manner. This astute orchestration of the last C-methyl group was part of the original plan for which the α-lithio glycal alkylation reaction was a prerequisite. Starting with glycal alkylation, cyclopropanation, acid-mediated opening, and trapping of the incipient oxocarbenium ion to form the dioxaspiroacetal subunit, were part of the image-to-thought reflex action that would be best portrayed in the mind's eye as a "flat" perspective drawing.

6.2.5
The flipside of molecules

The manner in which molecules are drawn has other practical manifestations with regard to the type and order in which reaction sequences are performed. Consider the prospects of devising a synthesis of the bicyclic lactam shown in Figure 6.7 from the (5S)-5-hydroxy 2,3-unsaturated valerolactam as a designated precursor. Our right-brain processing will instinctively instruct us to draw the precursor as shown in expression **A**. Here, visual juxtaposition of the heterocyclic unit in the target and precursor in the mind's eye presents a natural bias, with the nitrogen atom and the carbonyl group occupying the same positions as in the target molecule. The chemical steps that need to be followed *en route* to the bicyclic lactam would require an activation of the allylic carbon atom to introduce carbon substitution with inversion of configuration, then finding a means to annulate at the C-6 position, perhaps through selective activation of an unsymmetrical cyclic imide.

Should the same starting material be drawn in a "flipped" orientation as in **B**, voluntarily, or otherwise, then a different set of reactions must be considered to give the intended bicyclic lactam. In this case, one could engage the same allylic hydroxyl group in a Johnson-Claisen reaction, thus transferring a 2-carbon acetic acid chain from "right to left," and simultaneously transposing the double bond to the correct position. Iminium ion chemistry can then be used to functionalize C-2, after which, appropriate chemistry can be devised for oxidation to the lactam.

Thus, if two investigators drew the *same* precursor differently, one as **A** and another as **B**, they would be doing different sets of reactions to reach the target. This aspect of image-driven chemistry planning is both curiously interesting and frightfully disturbing, since one strategy may lead to a better synthesis than the other. In the example shown, it would be the non-conventional flipped drawing **B** that would have a better chance of success for an enantioselective synthesis of the target molecule.

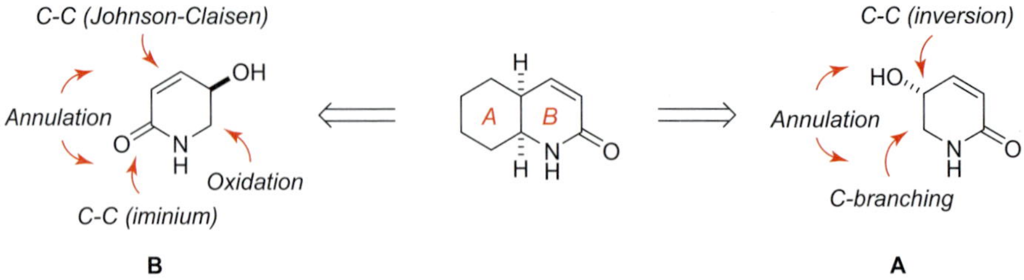

Figure 6.7 Access to the target molecule from one and the same precursor performing different sets of reactions.

6.2.6
Common root, different MO: chirons and synthons

The pioneering concepts of E. J. Corey regarding retrosynthetic thinking have been the logic-oriented basis of synthesis planning since its inception over 40 years ago [5]. In his originally formulated axiom, Corey defined synthons as: *"structural units within a molecule which are related to possible synthetic operations. A synthon may be almost as large as the molecule or as small as a single hydrogen. A functional group may be regarded as a synthon."* In considering the original synthon approach, disconnections are made at logical sites which facilitate bond formation in the forward sense based mostly on the ease of feasibility. It is generally the type of functionality present in the substructure of interest in the target molecule and chemical access to it, that dictates the strategy. For example, it may be convenient to break a bond adjacent to a stereogenic carbon bearing a hydroxyl group, simply because there is a reaction (aldol or ester enolate condensation) that reforms such a bond. This may indeed be a viable option since asymmetric versions of such condensations are now well-documented. The *chiron approach* on the other hand, will initially suggest a precursor in which the hydroxyl group is already present as in D- or L-malic acid for example.

In the total synthesis of the 42-membered macrocyclic lactone oasomycin A, Evans and coworkers [14] used sequential, auxiliary-mediated aldol reactions to generate subunits **A–D** containing *syn*- and *anti*-hydroxyl and methyl substitution (Figure 6.8). However, it was found practical to use D-malic acid as a native 4-carbon hydroxy acid chiron, that was chemically modified by a resident chirality-dependent C-allylation to give a *anti*-1,3-diol unit corresponding to C-22–C-28 of oasomycin A.

Another very useful reaction is the catalytic asymmetric dihydroxylation of olefins developed by Sharpless and coworkers [15]. This versatile synthesis of *cis*-diols with high diastereomeric control has been used as a critical step in many natural product syntheses. The total synthesis, and stereochemical revision of C-6, of amphidinolide

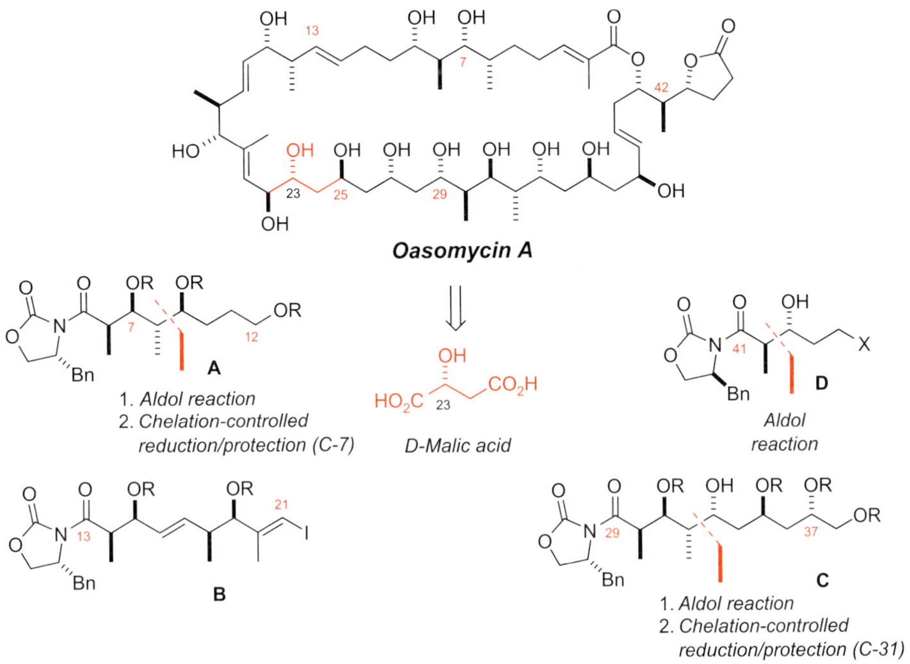

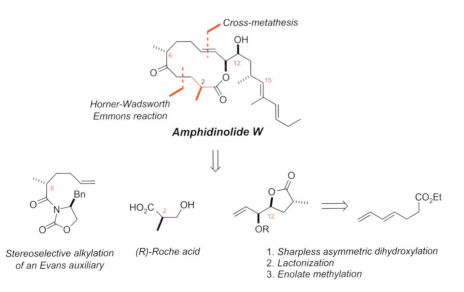

Figure 6.8 Evans' total synthesis of oasomycin A using an aldol strategy toward propionate-derived subunits. D-Malic acid as a 4-carbon chiron for C-22–C-25.

Figure 6.9 Ghosh's total synthesis and C-6 stereochemical revision of amphidinolide W.

W by Ghosh and coworkers [16] takes advantage of a stereoselective alkylation using an Evans auxiliary [17], a Sharpless asymmetric dihydroxylation, and the use of the (*R*)-Roche acid as a chiron at C-2 (Figure 6.9).

6.2.7
To chiron or not to chiron

The dictum *"If one thing is better than another, it is the way of Nature"* is attributed to Aristotle. Indeed, Nature's miraculous ways of healing have been recognized since the ascent of man. However, the utility of its chirons as building blocks must be viewed within the realms of practicality. The question may be asked: When is it advantageous to consider native or modified chirons as strategic components of a synthesis plan? From an operational perspective, the *chiron approach* can be likened to a block-by-block insertion of chiral entities into corresponding substructures in a target molecule. The alternative bond-by-bond construction mode would rely on asymmetric synthesis, involving its numerous stoichiometric and catalytic variants.

As previously stated, the *chiron approach* capitalizes on retrosynthetic analysis, while relying heavily on suitable, chiral, non-racemic starting materials with intact or modifiable functional groups that can be used, as such, or after modification. In the *chiron approach*, it is the type of chiral substructure in the target molecule, and its relationship to a chiral progenitor, that dictate the strategy of antithetic bond breaking. By locating such chiral substructures at the outset relying on visual analysis, and relating them to chirons from different classes, the scenario for synthesis is established early on with an element of predictability. The potential difficulty for a multi-synthon approach to complex natural products relying on known forward-sense reactions, can be simplified by a heuristic convergence of a set of chirons of predetermined structure, stereochemistry, and functionality, as illustrated in the classical examples shown in Figure 6.3. (see also Chapter 9, section 9.2). Thus, a chiron can be viewed as a chiral, non-racemic "piece" of the mosaic that eventually becomes an integral part of the target molecule. Although the term chiron was originally coined with respect to *chiral* synthons derived from Nature, it can also apply to any enantiomerically pure or enriched synthon prepared by purely synthetic methods. The possibility of using computer-aided recognition of chirons in target molecules is discussed in Chapter 17.

6.3
Anatomy of a Synthesis

Once a decision is made to adopt the *chiron approach* toward a given target molecule, a combination of visual and logic-based processes can be proposed to eventually generate a blueprint for the synthesis. To set the stage for a stepwise approach, the revised structure of the complex alkaloid palau'amine [18] was chosen as a target molecule (Figure 6.10. See also: Chapter 18, section 18.9.1.2).

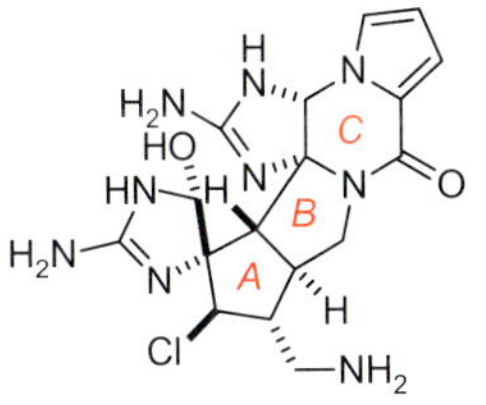

Palau'amine

TARGET $\Longrightarrow$ SUBSTRUCTURE $\Longrightarrow$ CHIRON $\Longrightarrow$ SOURCE

Phase 1: Analysis

1. Identify class
Alkaloid

$\downarrow$

2. Analyze framework
Heterocyclic, polycyclic, etc.

$\downarrow$

3. Analyze functionality
OH, NH_2, Cl, cyclic guanidine hemiaminal

$\downarrow$

4. Analyze connectivity
*Fused rings,
Vicinal, alternating, and
remote functional groups*

$\downarrow$

5. Identify problem centers
*Tertiary carbons.
Sensitive functionality, etc.*

$\downarrow$

6. Define absolute stereochemistry
Optional at this point

Phase 2: Discovery

7. Locate chiral substructures
*Allow maximum conservation of
asymmetric centers*

$\downarrow$

8. Disconnect at strategic bonds
*Minimum disruption of asymmetric
centers. Find sp^2 centers preferably*

$\downarrow$

9. Identify chiral substructures
*Substructures become primary goals.
Draw actual structures you would like to
have at this point in order to continue*

$\downarrow$

10. Search for appropriate chiron(s)
*Consider other sources such as
catalytic asymmetric synthesis,
enzymatic methods, and resolution*

Phase 3: Design

11. Blueprint
*Assembly of subunits.
Draw entire sequence with reagents.
Assess practicality, number of steps. etc.
Point of no return*

Phase 4: Execution

Figure 6.10 Anatomy of a synthesis based on the *chiron approach*, as well as asymmetric bond forming methods.

Phase 1 begins with *Analysis*, which encompasses six stages of visual dialogue with the target molecule. 1) A mental connection is immediately established by identifying the class of compound. In the case of palau'amine, the term *alkaloid* or *polycyclic nitrogen-containing compound* may be applicable; 2) One then proceeds to analyze the carbon framework of the molecule, which, in the present case, is an unusual hetero polycyclic entity; 3) Analysis of existing functionality and ring systems reveals isolated, peripheral, and interrelated types consisting of two cyclic guanidines, a primary aminomethyl group, and a secondary chloride on a cyclopentane fused to a pyrrolopiperazinone; 4) Connectivity reveals unusual vicinal amines as a bis-aminal and a spiro guanidine hemiaminal; 5) Problem centers are immediately apparent in the molecule, which is heavily functionalized with basic groups as well as a *trans*-fused azabicycle; 6) Defining absolute stereochemistry is optional at this point.

Having completed the *Analysis* phase, one proceeds to Phase 2, which deals with *Discovery*. 7) In order to find a suitable chiron, a reasonable level of functional and skeletal convergence must be satisfied. In the case of ring *A* of palau'amine, one can presumably start with a chiral, non-racemic functionalized cyclopentane derivative. Clearly, no obvious native chiron is evident, unless an acyclic precursor to the cyclopentane ring can be envisaged, or an alternative approach from a terpene can be devised. 8) Disconnection at strategic bonds to reveal intact chirons with minimum disruption of stereogenic centers may be difficult for palau'amine. Assuming that a suitably functionalized cyclopentane intermediate is in hand, the bond disconnections of choice would involve the middle ring *B* of the hexacyclic system. 9) The chiral substructure of interest in the case of palau'amine is the cyclopentane unit, which becomes the primary goal for synthesis, with appendages that would be used to construct the middle ring *B*. 10) An appropriate native chiron is searched for, and if found to be chemically manageable, it can be elaborated to a level of functionalization that is suitable for the continuation of the synthesis. Asymmetric methods of synthesis, including catalysis and enzymatic desymmetrization can also be considered. 11) With the information in hand, one proceeds to the *Design* phase and the generation of a plan for the synthesis starting with the chosen chiron(s), and defining reaction conditions, reagents, etc. This is a point of no return and feasibility is then tested in the *Execution* phase.

Various approaches to the synthesis of the chlorocyclopentane core of palau'amine and related alkaloids have been reported [19]. The first enantioselective total synthesis of palau'amine, which relied on a catalytic asymmetric Diels-Alder reaction, was accomplished by Baran and coworkers in 2011 [20] (for details, see Chapter 18, section 18.9.1.2).

Examples of potentially interesting modified chirons representing the *A* and *B* rings of palau'amine are shown in Figure 6.11.

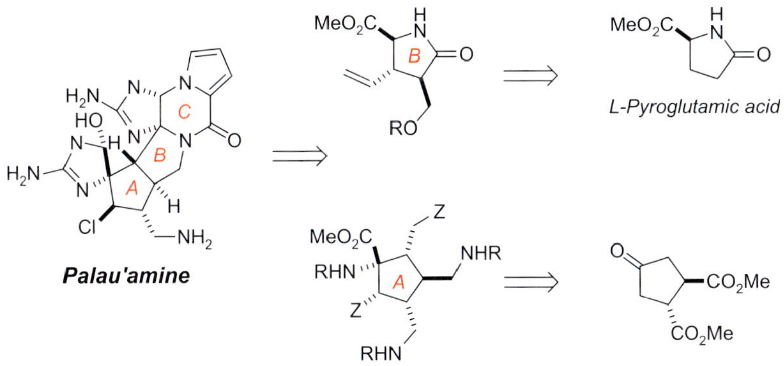

Figure 6.11 Palau'amine and possible chirons.

References

1. For example, see: (a) Hanessian, S. in, *Chemical Synthesis*, Chatgilialoglu, C.; Snieckus, V.; Eds.; Kluwer Academic Publishers, London, 1996; (b) Hanessian, S. *Pure Appl. Chem* 1993, **65**, 1189; (c) Hanessian, S. *Chem. Med. Chem* 2006, **1**, 1300; for some views and comments, see: (d) Sanderson, K. *Nature* 2007, **448**, 630; (e) Djerassi, C. *Science* 1999, **285**, 835; see also Chapter 18, section 18.7.

2. For example, see: (a) Nicolaou, K. C.; Vourloumis, D.; Winssinger, N.; Baran, P. S. *Angew. Chem. Int. Ed.* 2000, **39**, 44; (b) *Molecules that Changed the World*, Nicolaou, K. C.; Montagnon, T., 2008, Wiley-VCH, Weinheim; (c) *Molecules and Medicine*, Corey, E. J.; Czakó, B.; Kürti, L., 2007, Wiley, NY.

3. (a) Hanessian, S. *Aldrichimica Acta* 1989, **22**, 3; (b) *Total Synthesis of Natural Products: The Chiron Approach*, Hanessian, S., Pergamon, Oxford, UK, 1983; (c) Hanessian, S.; Franco, J.; Larouche, B. *Pure Appl. Chem.* 1990, **62**, 1887; see also reference 1.

4. According to Greek mythology, Chiron was a centaur (half man and half horse). Unlike other centaurs that were unruly with having fun and partying as their main goal in life, Chiron was wise and learned. He was a great healer, astrologer, and oracle, whose pupils came to him to fulfil their highest potential.

Chiron tutored Achilles, Jason, Hercules and Asclepius in music, morals, and medicine. He founded the Chironium, a temple for healing atop Mt. Pellion and he is considered as the father of ancient medicine. Chiron is also credited for the discovery of magnesium, hence the label, the Magnesian Centaur.

5. (a) Corey, E. J. *Pure Appl. Chem.* 1967, **14**, 19; (b) *The Logic of Chemical Synthesis*, Corey, E. J.; Cheng, X.-M., John Wiley & Sons, NY, 1989.

6. (a) Beetle sex pheromone: Mori, K.; Kishida, H. *Liebigs Ann. Chem.* 1988, 717; (b) punctatin A: Paquette, L. A.; Sugimura, T. *J. Am. Chem. Soc.* 1986, **108**, 3841. See also, Chapter 14, section 14.1; (c) vitamin D intermediate: Baggiolini, E. G.; Iacobelli, J. A.; Hennessy, B. M.; Batcho, A. D.; Sereno, J. F.; Uskoković, M. R. *J. Org. Chem.* 1986, **51**, 3098.

7. (a) Kobayashi, S.; Kamiyama, K.; Iimori, T.; Ohno, M. *Tetrahedron Lett.* 1984, **25**, 2557; (b) Okano, K.; Izawa, T.; Ohno, M. *Tetrahedron Lett.* 1983, **24**, 217.

8. *The Same and not the Same*, Hoffmann, R., Columbia University Press, NY, 1995.

9. (a) Cytochalasin B: Stork, G.; Nakahara, Y.; Nakahara, Y.; Greenlee, W. J. *J. Am. Chem. Soc.* 1978, **100**, 7775; (b) avermectin B$_{1A}$: (i) Hanessian, S.; Ugolini, A.; Dubé, D.; Hodges, P. J.; André, C.

J. Am. Chem. Soc. 1986, **108**, 2776; (ii) Hanessian, S.; Dubé, D.; Hodges, P. J. *J. Am. Chem. Soc.* 1987, **109**, 7063; (iii) Hanessian, S.; Ugolini, A.; Hodges, P. J.; Beaulieu, P.; Dubé, D.; André, C. *Pure Appl. Chem.* 1987, **59**, 29; (c) FK-506: (ii) Ireland, R. E.; Gleason, J. L.; Gegnas, L. D.; Highsmith, T. K. *J. Org. Chem.* 1996, **61**, 6856; see also: (ii) Jones, T. K.; Mills, S. G.; Reamer, R. A.; Askin, D.; Desmond, R.; Volante, R. P.; Shinkai, I. *J. Am. Chem. Soc.* 1989, **111**, 1157; for a recent review, see: (iii) Maddess, M. L.; Tackett, M. N.; Ley, S. V. *Prog. Drug Res.* 2008, **66**, 15; (d) Dvorak, C. A.; Schmitz, W. D.; Poon, D. J.; Pryde, D. C.; Lawson, J. P.; Ames, R. A.; Meyers, A. I. *Angew. Chem. Int. Ed.* 2000, **39**, 1664.

10. For excellent monographs, see: (a) *Fundamentals of Asymmetric Catalysis*, Walsh, P. J.; Kozlowski, M. C., 2008, University Science Books, Sausalito, CA; (b) *Catalysis of Organic Reactions*, Schmidt, S. R.; Ed.; 2007, CRC Press, Boca Raton, FL; (c) *New Frontiers in Asymmetric Catalysis*, Mikami, K.; Lautens, M.; Eds.; 2000, John Wiley & Sons, Hoboken, NJ; (d) *Catalysis from A to Z: A Concise Encyclopedia*, Cornils, B.; Herrmann, W. A.; Muhler, M.; Wong, C.-H.; Eds.; 2007, Wiley-VCH, Weinheim; (e) *Asymmetric Synthesis – The Essentials*, Christmann, M.; Bräse, S.; Ed.; 2007, 2nd Edition, Wiley-VCH, Weinheim; (f) *Comprehensive Asymmetric Catalysis*, vol. I-III; Jacobsen, E. N.; Pfaltz, A.; Yamamoto, H.; Eds.; 2000, Springer-Verlag, Berlin; (g) *Catalytic Asymmetric Synthesis*, Ojima, I.; Ed.; 2000, Wiley-VCH.

11. (a) (i) Kende, A. S.; Fujii, Y.; Mendoza, J. S. *J. Am. Chem. Soc.* 1990, **112**, 9645; (ii) Kende, A. S.; Mendoza, J. S.; Fujii, Y. *Tetrahedron* 1993, **49**, 8015; (b) Hanessian, S.; Focken, T.; Mi, X.; Oza, R.; Chen, B.; Ritson, D.; Beaudegnies, R. *J. Org. Chem.* 2010, **75**, 5601; (c) Berberich, S. M.; Cherney, R. J.; Colucci, J.; Courillon, C.; Geraci, L. S.; Kirkland, T. A.; Marx, M. A.; Schneider, M. F.; Martin, S. F. *Tetrahedron* 2003, **59**, 6819; (d) Kirkland, T. A.; Colucci, J.; Geraci, L. S.; Marx, M. A.; Schneider, M.; Kaelin, D. E.; Martin, S. F. *J. Am.*

Chem. Soc. 2001, **123**, 12432; (e) Lee, E.; Choi, J. C.; Kim, H.; Han, H. O.; Kim, Y. K.; Min, S. J.; Son, S. H.; Lim, S. M.; Jang, W. S. *Angew. Chem. Int. Ed.* 2002, **41**, 176; (f) Liu, L.; Donaldson, W. A. *Synlett* 1996, 103; (f) for a review on the synthesis of ambruticin S, see: Michelet, V.; Genêt, J.-P. *Curr. Org. Chem.* 2005, **9**, 405; for an organocatalytic approach to ring A, see: (g) Hanessian, S; Mi, X. *Synlett* 2010, 761.

12. (a) Michelet, V.; Adiey, K.; Bulic, B.; Genêt, J.-P.; Dujardin, G.; Rossignol, S.; Brown, E.; Toupet, L. *Eur. J. Org. Chem.* 1999, 2885; (b) Liu, P.; Jacobsen, E. N. *J. Am. Chem. Soc.* 2001, **123**, 10772; (c) Markó, I. E.; Bayston, D. J. *Tetrahedron Lett.* 1993, **34**, 6595; (d) Hanessian, S.; Focken, T.; Oza, R. *Org. Lett.* 2010, **12**, 3172.

13. (a) Evans, D. A.; Sacks, C. E.; Kleschick, W. A.; Taber, T. R. *J. Am. Chem. Soc.* 1979, **101**, 6789; (b) Martinez, G. R.; Grieco, P. A.; Williams, E.; Kanai, K.; Srinivasan, C. V. *J. Am. Chem. Soc.* 1982, **104**, 1436; (c) Negri, D. P.; Kishi, Y. *Tetrahedron Lett.* 1987, **28**, 1063; (d) Boeckman, R. K. Jr.; Charette, A. B.; Asberom, T.; Johnston, B. H. *J. Am. Chem. Soc.* 1987, **109**, 7553; for a review on ionophore antibiotics, see: (e) Faul, M. M.; Huff, B. E. *Chem. Rev.* 2000, **100**, 2407.

14. (a) Evans, D. A.; Nagorny, P.; McRae, K. J.; Reynolds, D. J.; Sonntag, L.-S.; Vounatsos, F.; Xu, R. *Angew. Chem. Int. Ed.* 2007, **46**, 537; (b) Evans, D. A.; Nagorny, P.; Reynolds, D. J.; McRae, K. J. *Angew. Chem. Int. Ed.* 2007, **46**, 541; (c) Evans, D. A.; Nagorny, P.; McRae, K. J.; Sonntag, L.-S.; Reynolds, D. J.; Vounatsos, F. *Angew. Chem. Int. Ed.* 2007, **46**, 545.

15. (a) Jacobsen, E. N.; Markó, I.; Mungall, W. S.; Schröder, G.; Sharpless, K. B. *J. Am. Chem. Soc.* 1988, **110**, 1968. For reviews see: (b) Johnson, R. A.; Sharpless, K. B. in, *Catalytic Asymmetric Synthesis*, Ojima, I.; Ed.; 1993, p. 227, VCH, Weinheim; (c) Kolb, H. C.; VanNieuwenhze, M. S.; Sharpless, K. B. *Chem. Rev.* 1994, **94**, 2483.

16. Ghosh, A. K.; Gong, G. *J. Org. Chem.* 2006, **71**, 1085.

17. Evans, D. A.; Dow, R. L.; Shih, T. L.; Takacs, J. M.; Zahler, R. *J. Am. Chem. Soc.* 1990, **112**, 5290.

18. (a) Köck, M.; Grube, A.; Seiple, I. B.; Baran, P. S. *Angew. Chem. Int. Ed.* 2007, **46**, 6586; (b) Grube, A.; Köck, M. *Angew. Chem. Int. Ed.* 2007, **46**, 2320; (c) Lanman, B. A.; Overman, L. E.; Paulini, R.; White, N. S. *J. Am. Chem. Soc.* 2007, **129**, 12896; (d) Buchanan, M. S.; Carroll, A. R.; Addepalli, R.; Avery, V. M.; Hooper, J. N. A.; Quinn, R. J. *J. Org. Chem.* 2007, **72**, 2309; for a related structure (carteramine A), see: (e) Kobayashi, H.; Kitamura, K.; Nagai, Y.; Nakuo, Y.; Fusetani, N.; Van Soest, R. W. M.; Matsunaga, S. *Tetrahedron Lett.* 2007, **48**, 2127.

19. For the synthesis of the chlorocyclopentane core subunits of palau'amine, massadine and axinellamine, see: (a) Yamaguchi, J.; Seiple, I. B.; Young, I. S.; O'Malley, D. P.; Maue, M.; Baran. P. S. *Angew. Chem. Int. Ed.* 2008, **47**, 3578; (b) Breder, A.; Chinigo, G. M.; Waltman, W. W.; Carreira, E. M. *Angew. Chem. Int. Ed.* 2008, **47**, 8514. (c) Bultman, M. S.; Ma, J.; Gin, D. Y. *Angew. Chem. Int. Ed.* 2008, **47**, 6821; (d) Hudon, J.; Cernak, T. A.; Ashenhurst, J. A.; Gleason, J. L. *Angew. Chem. Int. Ed.* 2008, **47**, 8889.

20. Seiple, I. B.; Su, S.; Young, I. S.; Nakamura, A.; Yamaguchi, J.; Jørgensen, L.; Rodriguez, R. A.; O'Malley, D. P.; Gaich, T.; Köck, M.; Baran, P. S. *J. Am. Chem. Soc.* 2011, DOI: 10.1021/ja2047232; for an initial communication of the synthesis of ($\pm$)-palau'amine, see: Seiple, I. B.; Su, S.; Young, I. S.; Lewis, C. A.; Yamaguchi, J.; Baran, P. S. *Angew. Chem. Int. Ed.* 2010, **49**, 1095.

7
Nature's Chirons

The four major classes of readily available naturally occurring chiral, non-racemic compounds comprise the α-amino acids, carbohydrates, α-hydroxy acids, and terpenes. Nature also produces a class of primary metabolites as well as chiral, hydroxylated six-membered carbocycles and carbocyclic acids. Within each class there exists a rich source of stereochemically and functionally diverse compounds, which can be exploited as starting materials in synthesis. We have defined these as chiral synthons or chirons, that can be used in their native or chemically modified forms. Native chirons have also found extensive use as chiral auxiliaries in stoichiometric and catalytic asymmetric reactions. In particular, they are used as ligands for metal-dependent catalysis or as chiral motifs in metal-free catalytic processes (organocatalysis) for a variety of C−C and related bond forming reactions.

In this chapter, we outline the pertinent characteristics of each of the above-mentioned classes of chirons, and discuss what they can offer as starting materials in the planning and execution of syntheses of target molecules that contain one or more stereogenic centers.

7.1
α-Amino Acids

The naturally occurring proteinogenic α-amino acids are abundantly available in their L-forms. Their enantiomers are also accessible by chemical and chemo-enzymatic methods in addition to resolution of racemates. A selection of representative L-amino acids is shown in Figure 7.1 [1]. α-Amino acids offer a plethora of possibilities as starting materials for organic synthesis. Their carbon frameworks correspond to acyclic, cyclic, aromatic, and heteroaromatic variants consisting of α-amino carboxylic acids, α-amino β-hydroxy (or thio) carboxylic acids, α-amino dicarboxylic acids, and α,ω-diamino carboxylic acids. α-Amino acids lend themselves to orthogonal protection with appropriate protecting groups, and the resulting products can be chemically (or enzymatically) modified to generate useful chirons. Depending on the nature of the substructure in a target molecule, it is possible to select the α-amino acid that best matches the corresponding carbon framework and chirality. The rich chemistry that

Design and Strategy in Organic Synthesis: From the Chiron Approach to Catalysis, First Edition.
Stephen Hanessian, Simon Giroux, and Bradley L. Merner.
© 2013 Wiley-VCH Verlag GmbH & Co. KGaA. Published 2013 by Wiley-VCH Verlag GmbH & Co. KGaA.

provides access to branching in different positions, coupled with the possibility of end-group differentiation in the case of dicarboxylic acids, offer attractive choices for potential uses as starting materials harboring functional groups besides the obvious amino group. Chemically modified amino acids are also excellent chiral auxiliaries for asymmetric processes [2]. When working with α-amino acids, it is imperative that their enantiomeric purity be maintained in the process of derivatization. Knowledge of protecting groups and their use in orthogonal protection/deprotection protocols is primordial [3]. Finally, it goes without saying that the presence of a nitrogen atom in the substructure of a target molecule may be conducive to consider an α-amino acid as an appropriate chiral precursor, especially when it can be suitably modified to achieve the desired level of functional, stereochemical, and skeletal convergence. In this regard, dicarboxylic amino acids, such as aspartic and glutamic acids, are particularly versatile as four and five carbon chirons respectively, especially since they can be differentially extended at both extremities. They are also amenable to enolate chemistry leading to α-carbon substitution with stereochemical control. A selection of chemically modified chirons readily available from α-amino acids are shown in Figure 7.2 [4]. α-Amino acids have been extensively used as starting materials in total synthesis and will be showcased in later chapters of this book (see Chapters 8–10, 16 and 18), [1, 5].

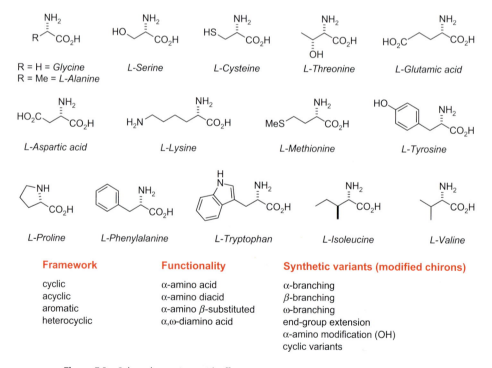

Figure 7.1 Selected α-amino acid offerings.

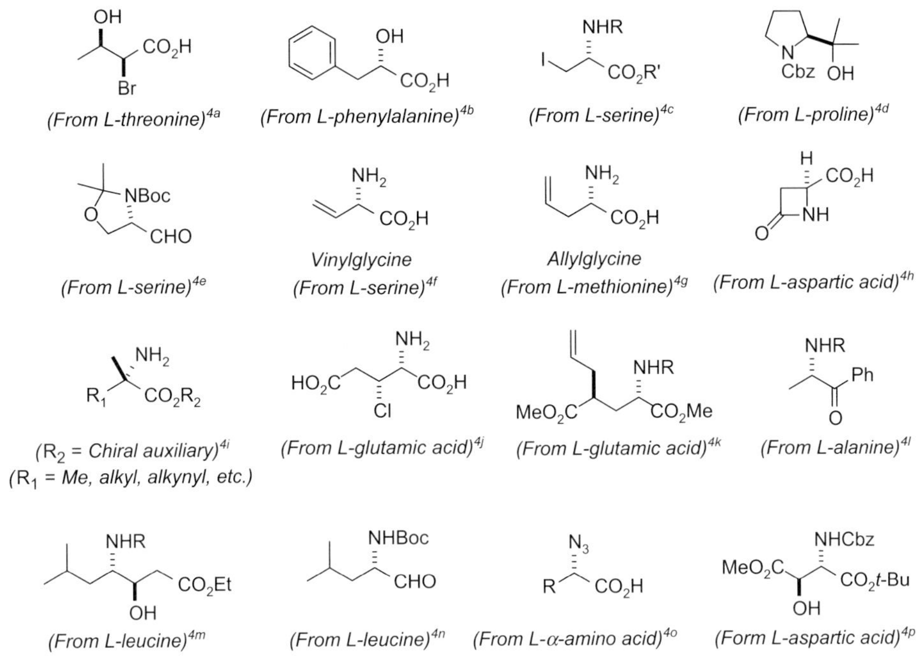

| (From L-threonine)[4a] | (From L-phenylalanine)[4b] | (From L-serine)[4c] | (From L-proline)[4d] |

| (From L-serine)[4e] | Vinylglycine (From L-serine)[4f] | Allylglycine (From L-methionine)[4g] | (From L-aspartic acid)[4h] |

| (R_2 = Chiral auxiliary)[4i] (R_1 = Me, alkyl, alkynyl, etc.) | (From L-glutamic acid)[4j] | (From L-glutamic acid)[4k] | (From L-alanine)[4l] |

| (From L-leucine)[4m] | (From L-leucine)[4n] | (From L-α-amino acid)[4o] | (Form L-aspartic acid)[4p] |

Figure 7.2 Selected chemically modified chirons from α-amino acids.

7.2
Carbohydrates

Carbohydrates constitute a rich reservoir of renewable carbon sources. They occur as monosaccharides, as disaccharides, and as components of polysaccharides from which they can be easily obtained by acidic hydrolysis [6]. Commercially available hexoses such as D-glucose, D-galactose, D-mannose, and pentoses such as D- and L-arabinose, D-xylose, and D-ribose are sold in bulk quantities.

Carbohydrates, exemplified by α-D-glucopyranose, are endowed with a number of attractive features that make them versatile precursors in synthesis [7]. In the pyranose form, α-D-glucose can be depicted in the so-called Haworth projection, a terpene-like projection, and in its $^4C_1$ chair-like conformation (Figure 7.3). The C-6 hydroxymethyl group and the all-equatorial orientation of the three hydroxyl groups at C-2–C-4, can be converted to a variety of O-protected derivatives. The anomeric hydroxyl group on the other hand, is subject to acetalization reactions leading to internal acetals and glycosides. 2-Amino-2-deoxy-D-glucopyranose (D-glucosamine) is a constituent of chitin, the hard outer shell carapace of marine crustaceans. It is abundantly available and inexpensive.

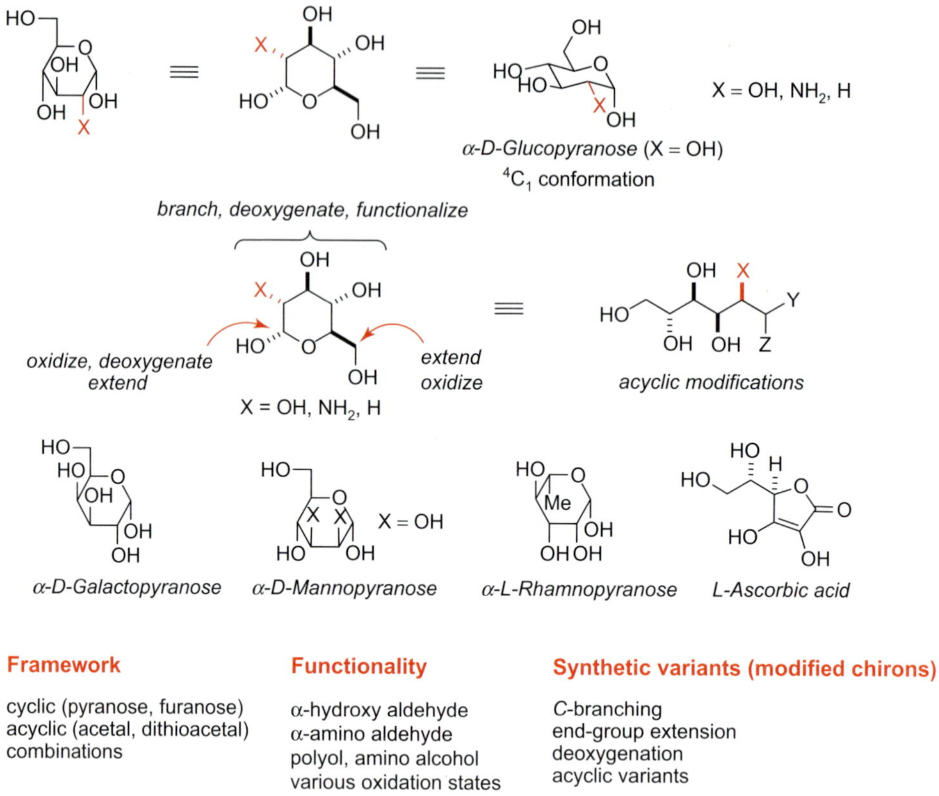

α-D-Glucopyranose (X = OH)
$^4C_1$ conformation

X = OH, NH$_2$, H

branch, deoxygenate, functionalize

oxidize, deoxygenate extend

extend oxidize

X = OH, NH$_2$, H

acyclic modifications

α-D-Galactopyranose α-D-Mannopyranose X = OH α-L-Rhamnopyranose   L-Ascorbic acid

Framework

cyclic (pyranose, furanose)
acyclic (acetal, dithioacetal)
combinations

Functionality

α-hydroxy aldehyde
α-amino aldehyde
polyol, amino alcohol
various oxidation states

Synthetic variants (modified chirons)

C-branching
end-group extension
deoxygenation
acyclic variants

Figure 7.3   Selected carbohydrate offerings.

The cyclic carbon framework of an anomerically protected hexopyranose, for example, can be subjected to many stereocontrolled and site-selective transformations. Other than extensions and oxidations at C-1 and C-6, the conformationally defined chair form lends itself to systematic manipulation of each hydroxyl group, often in an orthogonal protection-functionalization sequence. Thus, C-branching can be introduced via organometallic addition reactions to ketone intermediates (uloses) [8], and heteroatoms are readily inserted by S_N2-type reactions of O-sulfonates, or opening of epoxides [9]. Deoxy sugars are available through well-known protocols such as the Barton-McCombie free-radical deoxygenation [10], or reductive opening of epoxides for example (Figure 7.4). Steric, conformational, and stereoelectronic effects manifest themselves when working with carbohydrate derivatives. For example, 1,3-diaxial interactions may prevent the approach of a given nucleophile at C-3 of methyl α-D-glucopyranoside derivatives. Epoxides in alkyl hexopyranosides can be opened with nucleophiles according to the Fürst-Plattner rule.

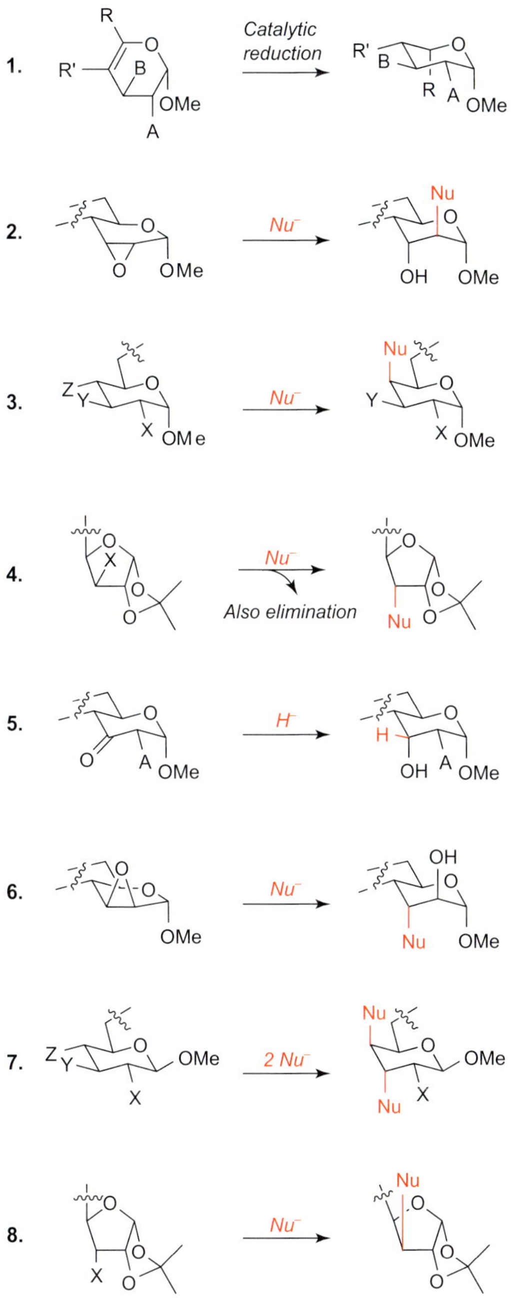

Figure 7.4 Branched and deoxygenated carbohydrates.

Although aldoses exist in the cyclic hemiacetal form, they can be easily converted to the corresponding polyols by reduction, or to the synthetically versatile acyclic dialkyl dithioacetals [11]. Aldoses are polyhydroxy aldehydes, masked as cyclic hemiacetals, which can be used as precursors to corresponding acyclic substructures in target molecules. Treatment of free sugars with nitromethane or Wittig-type reagents readily gives the corresponding unsaturated acyclic derivatives (Figure 7.5 A) [12].

Spatially proximal groups in carbohydrates are easily transformed into cyclic acetals by treatment with an aldehyde or a ketone in the presence of an acid catalyst [13]. It is of interest to point out that the stereoisomeric D-hexoses, such as D-glucose, D-galactose, and D-mannose each give different bis-acetonides (Figure 7.5 B). Knowledge of such derivatives can be very helpful in planning further chemical modifications in each series. For example, oxidation of each of these acetal derivatives will give a ketone, an aldehyde, and a lactone respectively! A variety of selectively functionalized derivatives can be obtained in relatively few steps from methyl D-hexopyranosides. Preferential O-protection can be achieved based on the relative spatial dispositions and acidities of hydroxyl groups (*eg.* silylation, and tosylation) [14]. Direct chlorination of methyl α-D-glucopyranoside with sulfuryl chloride affords the 4,6-dichloro derivative with inversion at C-4 [15]. Treatment of an O-benzylidene acetal with N-bromosuccinimide (NBS) gives the corresponding 6-bromo-4-benzoate ester [16] (Figure 7.5 C). Cyclic ethers (anhydro sugars) can be obtained from polyols or anomerically activated precursors (Figure 7.5 D). Carbohydrate lactones are easily obtained by direct oxidation as in the case of D-gluconolactone, or through degradative rearrangements (Figure 7.5 E) [17]. Unsaturation can be introduced in carbohydrate derivatives, and these can be utilized in a variety of applications (Figure 7.5 F). Finally, carbohydrates are excellent substrates for the synthesis of highly functionalized carbocycles, or as dienophiles in Diels-Alder reactions to give substituted cyclohexenes (Figure 7.5 G) [18].

Clearly, the most practical utilization of a carbohydrate molecule as a starting material is in those cases where tetrahydrofuran or tetrahydropyran-type structures are intended, especially if modifications of the hydroxyl groups are minimized. When considering carbohydrate precursors in total synthesis, extensive deoxygenation, protection/deprotection, and truncation should be avoided. Nevertheless, their abundance in a variety of chain lengths and stereochemical diversities offers attractive options in the synthesis of polyhydroxylated and related compounds (see Chapters 8, 9 and 11) [7].

A. Acyclic

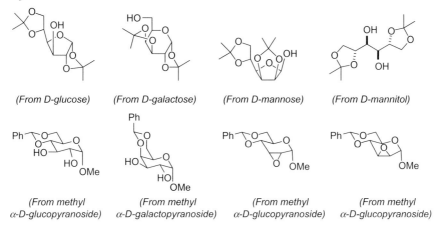

X = OH; *(From D-glucose)*   *D-Mannitol*   *(From D-ribose)*   *(From D-arabinose)*
X = NH₂; *(From 2-amino-*
2-deoxy-D-glucose)

B. Cyclic acetals

(From D-glucose)   *(From D-galactose)*   *(From D-mannose)*   *(From D-mannitol)*

(From methyl   *(From methyl*   *(From methyl*   *(From methyl*
α-D-glucopyranoside)   *α-D-galactopyranoside)*   *α-D-glucopyranoside)*   *α-D-glucopyranoside)*

C. Functionalized derivatives

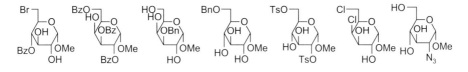

Figure 7.5 Selected chemically modified chirons from carbohydrates.

D. Anhydro sugars

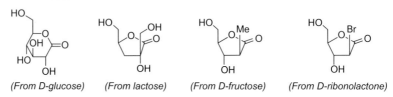

(From D-glucose)   (From D-mannitol)   (From 2-amino-2-deoxy-
                                                        D-glucose)

E. Aldonolactones

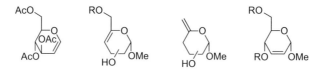

(From D-glucose)   (From lactose)   (From D-fructose)   (From D-ribonolactone)

F. Unsaturated derivatives

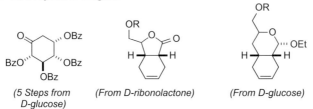

G. Carbocyclic analogues

(5 Steps from   (From D-ribonolactone)   (From D-glucose)
D-glucose)

Figure 7.5 (*continued*).

7.3
α-Hydroxy Acids

Although Nature produces only a handful of α-hydroxy acids, they occupy a
pre-eminent position in the annals of organic stereochemistry, as exemplified by
the legendary (R,R)- and (S,S)-tartaric acids (Figure 7.6) [19]. Pasteur's mechanical
separation of two forms of sodium ammonium tartrate in 1848, was the dawn of
chirality. Formally, the two enantiomers of tartaric acid are the "aric" acids of D-
and L-threose. The carbohydrate connection extends to D- and L-glyceraldehyde,
which can be built up through a cyanohydrin synthesis to diastereomeric tartaric

acids. The C_2-axis of symmetry of (R,R)- and (S,S)-tartaric acids makes them versatile starting materials for exploitation in total synthesis [19]. They have also been extensively used as chiral auxiliaries and as ligands in catalytic asymmetric reactions [2].

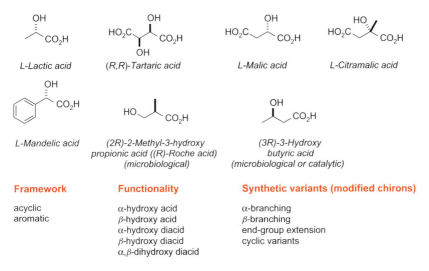

Framework	Functionality	Synthetic variants (modified chirons)
acyclic	α-hydroxy acid	α-branching
aromatic	β-hydroxy acid	β-branching
	α-hydroxy diacid	end-group extension
	β-hydroxy diacid	cyclic variants
	α,β-dihydroxy diacid	

Figure 7.6 Selected hydroxy acid offerings.

L-Lactic acid, or (2S)-hydroxypropanoic acid, occurs in sour milk and in human and animal muscle. It is produced from carbohydrates by fermentation, and it is available in bulk quantities at low cost. It is an ideal 3-carbon carboxylic acid chiron with a built-in C-methyl group, that can be related to its amino acid congener, L-alanine.

L-Malic acid, or (2S)-hydroxybutanedioic acid, is naturally found in various fruits. Its commercial production, at low cost on a ton-scale, is based on a biocatalytic fermentation process. The 4-carbon chiral dicarboxylic acid is the hydroxy analogue of L-aspartic acid from which it can be prepared by diazotization with retention of absolute configuration. D-Malic acid is readily available from the L-isomer by S_N2 inversion protocols. D- and L-Malic acids are amenable to a variety of chemical modifications including α- and β-branching of enolate dianions [20]. They have been extensively used as precursors to functionalized acyclic and cyclic intermediates in total synthesis (see Chapter 12).

L-Mandelic acid, or (2S)-hydroxy-2-phenylacetic acid (amygdalic acid), is a readily available and naturally occurring aromatic α-hydroxy carboxylic acid. As a hydroxy equivalent of L-phenylglycine, it is a versatile building block for aryl-terminated chiral subunits in synthetic targets. L-Mandelic acid, and its D-enantiomer, have been used as chiral auxiliaries in enolate alkylations and aldol reactions [2].

L-Citramalic acid, or (2S)-2-hydroxy-2-methylbutanedioic acid, the α-C-methyl analogue of L-malic acid, is produced by microbial transformation of mesaconic acid. It is also available by enolate alkylation of a dioxolanone derived from L-malic acid. L-Citramalic is acid an ideal precursor to acyclic or cyclic subunits bearing an α-C-methyl-α-hydroxy functionality of a predetermined chirality.

Acyclic hydroxy acids are available from microbiological oxidations or by catalytic asymmetric syntheses (see Chapter 5). Among these, (2R)-2-methyl-3-hydroxy-propionic acid ((R)-Roche acid), and (3R)-3-hydroxybutyric acid have found extensive applications as precursors to appropriate substructures in synthetic target molecules (see Chapter 12).

The acyclic framework of hydroxy acids lends itself to a variety of chemical trans-formations, taking full advantage of the existing chirality to generate branching sites using enolate chemistry. In the case of dicarboxylic acids, such as the tartaric and malic acids, it is possible to differentiate the end-groups, and to manipulate the hydroxyl group(s). Exploiting C_2-symmetry generates versatile 4-carbon interme-diates from (R,R)- and (S,S)-tartaric acids. The Roche ester in particular, presents different end-group oxidation states, which allows chemoselective functionaliza-tion, so that the chirality of the carbon bearing the 2-C-methyl group can be matched at will with corresponding subunits in target molecules. Selected modified chirons, readily available from the original hydroxy acids, are shown in Figure 7.7 [21].

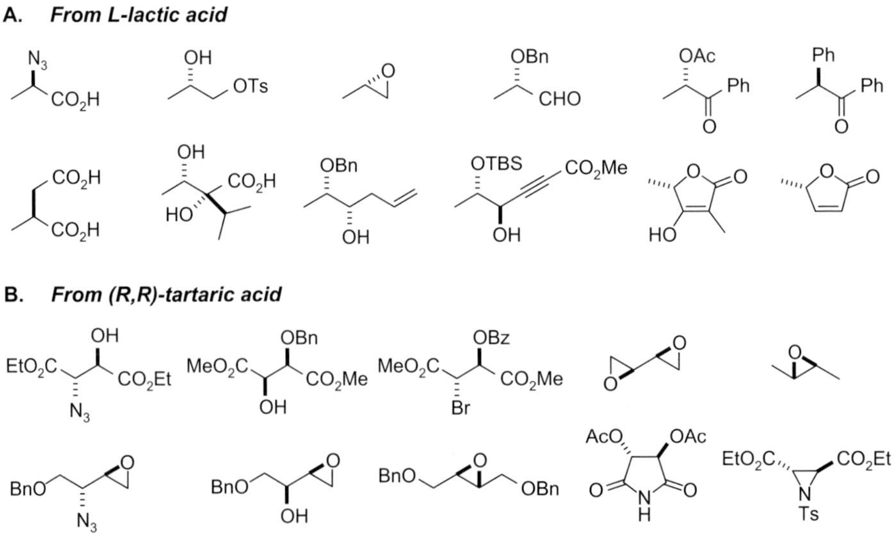

A.   From L-lactic acid

B.   From (R,R)-tartaric acid

Figure 7.7 Selected chemically modified chirons form hydroxy acids.

C.   From L-malic acid

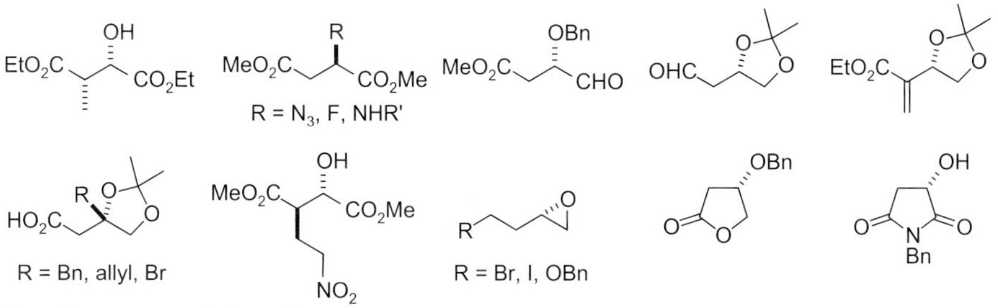

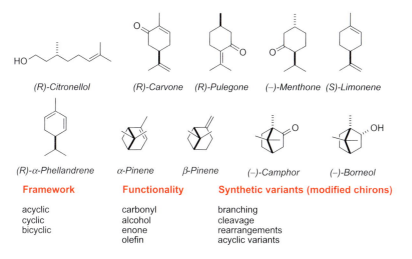

R = N₃, F, NHR'

R = Bn, allyl, Br

R = Br, I, OBn

Bn

D.   From L-mandelic acid

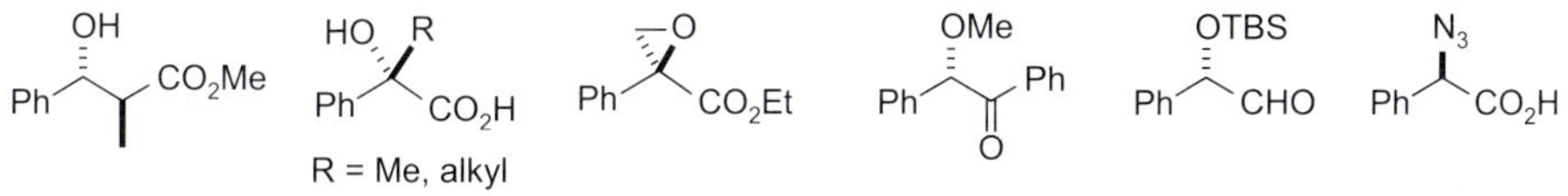

R = Me, alkyl

**E.   From (2R)-2-methyl-3-hydroxypropionic acid
and (3R)-3-hydroxybutyric acid**

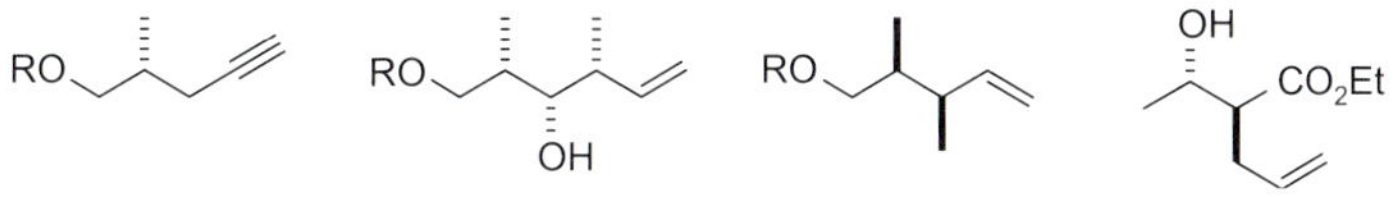

Figure 7.7 (*continued*).

7.4
Terpenes

Terpenes can be considered as Nature's chiral hydrocarbons. Except for
(*R*)-citronellol, naturally occurring chiral, non-racemic terpenes are monocyclic

(*R*)-Citronellol   (*R*)-Carvone   (*R*)-Pulegone   (–)-Menthone   (*S*)-Limonene

(*R*)-α-Phellandrene   α-Pinene   β-Pinene   (–)-Camphor   (–)-Borneol

Framework	**Functionality**	**Synthetic variants (modified chirons)**
acyclic	carbonyl	branching
cyclic	alcohol	cleavage
bicyclic	enone	rearrangements
	olefin	acyclic variants

Figure 7.8   Selected terpene offerings.

or bicyclic cycloalkanes with alkyl appendages on stereogenic carbon atoms. Some may contain a ketone or a hydroxyl group (Figure 7.8). Although they are commercially available, some terpenes are more accessible than others.

A. From (R)- and (S)-citronellol

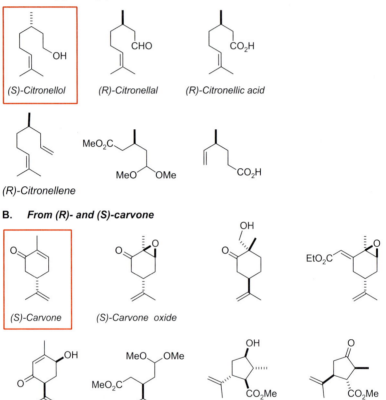

Figure 7.9 Selected chemically modified chirons from terpenes.

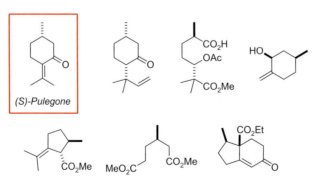

D. From (–)-menthone

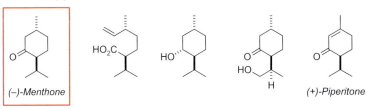

(–)-Menthone

(+)-Piperitone

E. From (R)- and (S)-limonene

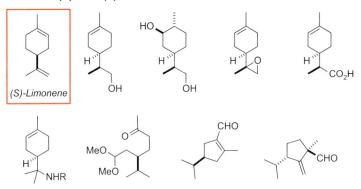

(S)-Limonene

F. From borneol and camphor

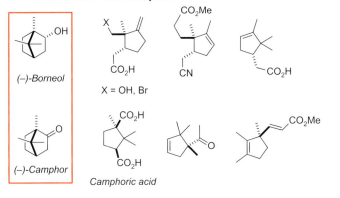

(–)-Borneol

X = OH, Br

(–)-Camphor

Camphoric acid

Figure 7.9 (*continued*).

Terpenes can be chemically manipulated through their existing olefinic, hydroxyl, and carbonyl groups to give a variety of synthetically useful building blocks [22]. Branching, extension, or degradative methods afford a host of modified acyclic and carbocyclic chirons (Figure 7.9) [23]. Of particular relevance are rearrangement reactions leading to vicinally substituted cyclopentanes, especially those containing quaternary centers.

The acyclic and carbocyclic terpenes are ideal precursors to corresponding substructures in natural products [23]. For example (*R*)-citronellol is a source of acyclic carbon chains with a configurationally fixed *C*-methyl group that can be matched in two spatial orientations depending on how the end-groups are differentiated (Figure 7.9 A). Monocyclic terpenes, such as (*R*)- and (*S*)-carvone and (*R*)- and (*S*)-pulegone, can converge with a matching cyclohexane fragment found in a target substructure. Their acyclic cleavage products, or densely functionalized cyclopentanes produced by rearrangements, serve as exquisite building blocks with pendant alkyl or alkenyl groups (Figure 7.9 B and C). Menthone and limonene are endowed with similar functionalizable groups (Figure 7.9 D and E). Finally, borneol and camphor are excellent sources of highly functionalized cyclopentanes (Figure 7.9 F) arising from stereocontrolled ring contractions (see Chapter 8, Figure 8.1). Camphorsulfonic acid, as the corresponding sultam, is a superb chiral auxiliary for asymmetric enolate alkylations and conjugate additions of enoylsultams [24].

A large number of natural products have been synthesized starting with natural and chemically modified terpenes as chirons (see Chapters 8, 9, 13 and 16).

7.5
Cyclitols

D-(−)-Quinic acid, or (1*S*,3*R*,4*S*,5*R*)-1,3,4,5-tetrahydroxycyclohexanecarboxylic acid, is a widely occurring plant metabolite (Figure 7.10). Biosynthetically, it is derived from D-glucose, and proceeding through several well-defined intermediates [25]. The same pathway also leads to (−)-shikimic acid, the dehydrated analogue of (−)-quinic acid (Figure 7.10). Quebrachitol is a chiral, non-racemic monomethyl ether cyclitol [26]. Although these cyclitols are commercially available, quinic acid is more accessible in moderately larger quantities.

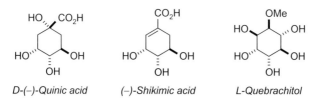

| *D-(−)-Quinic acid* | *(−)-Shikimic acid* | *L-Quebrachitol* |

Figure 7.10 Useful cyclitol chirons.

Quinic acid can be used as a starting material for carbocyclic subunits of target compounds in many creative ways [27]. The *syn/anti* disposition of the vicinal triol unit can be protected in orthogonal modes, especially if the carboxyl group is considered. The resulting lactone provides many opportunities for selective functionalization and voluntary manipulation of oxidation states. Introduction of carbon branching, heteroatom substitution, ring-expansion, and contraction methods offer many possibilities to give chemically modified quinic acids. These can then be used as starting materials to converge with carbocyclic substructures in target molecules. Selected examples of chemically modified chirons derived from quinic acid are shown in Figure 7.11 [27].

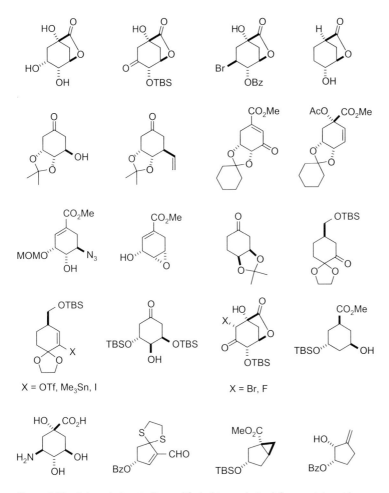

Figure 7.11 Selected chemically modified chirons derived from quinic acid.

Polyfunctional cycloalkanes are also known as "carbasugars." [18a–c]. Indeed many are derived from native sugar molecules by a variety of carbocyclization reactions that lead to 5–8 membered polyhydroxy cycloalkanes of known absolute configurations as shown in Figure 7.12 [18a]. Products derived from the microbiological dihydroxylation of aromatic compounds are a rich source of polyhydroxy cyclitols and their unsaturated precursors [28] (see Chapter 5, section 5.6.2).

Selected examples of the utilization of carbocycles in total synthesis are discussed in Chapter 14.

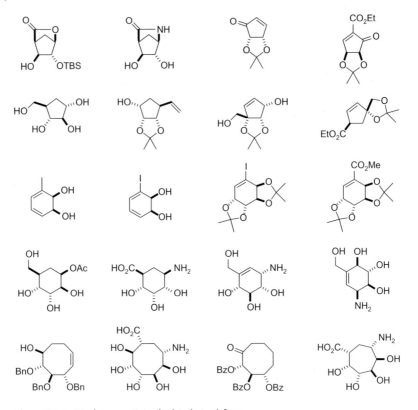

Figure 7.12 "Carbasugars" (cyclitols) derived from carbohydrates, by asymmetric synthesis from non-carbohydrates, and by microbiological oxidations.

References

1. For example, see: (a) *Synthesis of Optically Active α-Amino Acids*, Williams, R. M., 1989, Pergamon, NY; see also: (b) *Methods of Non α-Amino Acid Synthesis*, Smith, M. B., 1995, Dekker, NY.

2. For example, see: (a) *Chiral Auxiliaries and Ligands in Asymmetric Synthesis*, Seyden-Penne, J., 1995 Wiley-Interscience, NY; (b) *Asymmetric Synthetic Methodology*, Ager, D. J.; East, M. B., 1996, CRC Press, Boca Raton, FL; (c) *Principles of Asymmetric Synthesis*, Gawley, R. E.; Aubé, J., 1996, Pergamon, NY; (d) Gnas, Y.; Glorius, F. *Synthesis* 2006, 1899.

3. For example, see: (a) *Protecting Groups*, Kociénski, P. J., 1994, Thieme, Stuttgart; (b) *Protective Groups in Organic Synthesis*, 4th Edition, Wuts, P. G. M.; Greene, T. W.; Eds.; 2006, Wiley, NY.; see also: (c) Isidoro-Llobet, A.; Alvarez, M.; Alberico, F. *Chem. Rev.* 2009, **109**, 2455.

4. For example, see: (a) A. Shiozaki, M.; Ishida, N.; Hiraoka, T.; Yanagisawa, H. *Tetrahedron Lett.* 1981, **22**, 5205; (b) Cohen, S. G.; Weinstein, S. Y. *J. Am. Chem. Soc.* 1964, **86**, 5326; (c) Itaya, T.; Mizutani, A. *Tetrahedron Lett.* 1985, **26**, 347; (d) Overman, L. E.; Bell, K. L.; Ito, F. *J. Am. Chem. Soc.* 1984, **106**,

4192; (e) Garner aldehyde: Garner, P. *Tetrahedron Lett.* 1984, **25**, 5855; (f) Berkowitz, D. B.; Charette, B. D.; Karukurichi, K. R., McFadden, J. M. *Tetrahedron: Asymmetry* 2006, **17**, 869; (g) Walsh, C. *Tetrahedron* 1982, **38**, 871; (h) Salzmann, T. N.; Ratcliffe, R. W.; Christensen, B. G.; Bouffard, F. A. *J. Am. Chem. Soc.* 1980, **102**, 6161; (i) Seebach, D.; Aebi, J. D.; Gander-Coquoz, M.; Naef, R. *Helv. Chim. Acta,* 1987, **70**, 1194; (j) Silverman, R. B.; Holladay, M. W. *J. Am. Chem. Soc.* 1981, **103**, 7357; (k) Hanessian, S.; Schaum, R. *Tetrahedron Lett.* 1997, **38**, 163; (l) McClure, D. E.; Arison, B. H.; Jones, J. H.; Baldwin, J. J. *J. Org. Chem.* 1981, **46**, 2431; (m) Cooke, J. W. B.; Davies, S. G.; Naylor, A. *Tetrahedron* 1993, **49**, 7955; (n) Rich, D. H.; Sun, E. T.; Boparai, A. S. *J. Org. Chem.* 1978, **43**, 3624; (o) Goddard-Borger, E. D.; Stick, R. V. *Org. Lett.* 2007, **9**, 3797; (p) Hanessian, S.; Vanasse, B. *Can. J. Chem.* 1993, **71**, 1401.

5. For example, see: (a) *Amino Group Chemistry: From Synthesis to the Life Sciences*, Ricci, A.; Ed.; 2008, Wiley-VCH, Weinheim; (b) *Asymmetric Synthesis – Construction of Chiral Molecules Using Amino Acids*, Coppola, G. M.; Schuster, H. F., 1987, Wiley, NY; (c) Cintas, P. *Tetrahedron* 1991, **47**, 6079; (d) Kulkarni, Y. S. *Aldrichim. Acta* 1999, **32**, 18; (e) *Peptidomimetics Protocols*, Kazmierski, W. M.; Ed.; 1999, Humana Press, Totowa, NJ; (f) Martens, J. in, *Topics in Curr. Chem.* vol. 125, p. 167, Springer-Verlag, Heidelberg, 1984.

6. For recent books on methods, see: (a) *Best Synthetic Methods and Carbohydrates*, Osborn, H. M, Ed.; 2003, Academic Press, San Diego, CA; (b) *Carbohydrates: The Sweet Molecules of Life*, Stick, R. V. 2001, Academic Press, NY; (c) *Preparative Carbohydrate Chemistry*, Hanessian, S.; Ed.; 1997, Dekker, NY; (d) *Modern Methods in Carbohydrate Synthesis*, Khan, S. H.; O'Neill, R. A.; Eds.; 1996, Harwood Academic Publishers, Amsterdam; see also: (e) *Frontiers in Modern Carbohydrate Chemistry*,

Demchenko, A. V.; Ed.; 2007, American Chemical Society, Washington; (f) *Essentials of Carbohydrate Chemistry and Biochemistry*, Lindhorst, T. K., 2007, Wiley-VCH, Weinheim; (g) *Carbohydrate Chemistry and Biochemistry: Structure and Mechanism*, Sinnott, M. L., 2007, Royal Society of Chemistry, Cambridge, UK; (h) *The Organic Chemistry of Sugars*, Levy, D. E.; Fügedi, P.; Eds.; 2006, CRC/Taylor and Francis, Boca Raton, FL; (i) *Bioorganic Chemistry: Carbohydrates*, Hecht, S. M.; Ed.; 1999, Oxford University Press, Oxford, UK; (j) *Monosaccharides Sugars*, Györgydeák, Z; Pelyvás, I. F., 1999, Academic Press, San Diego, CA; (k) *Monosacharides: Their Chemistry and Their Roles in Natural Products*, Collins, P. M.; Ferrier, R. J., 1995, Wiley-Interscience, NY; see also: (l) *Adv. Carbohydr. Chem. Biochem*, Horton, D. Ed.; vol. 1-61; (m) *Carbohydrate Chemistry: Monosaccharides and Their Oligomers*, El Khadem, H., 1998, Academic Press, NY; (n) *Carbohydrate Mimics: Concepts and Methods*, Chapleur, Y. Ed.; 1998, Wiley-VCH, Weinheim.

7. For relevant monographs, see: (a) *Total Synthesis of Natural Products: The Chiron Approach*, Hanessian, S., 1983, Pergamon, Oxford, UK; for pertinent reviews, see: (b) Hudlicky, T.; Entwistle, D. A.; Pitzer, K. K.; Thorpe, A. J. *Chem. Rev.* 1996, **96**, 1195; (c) Ager, D. J.; East, M. B. *Tetrahedron* 1992, **48**, 2803; 1993, **49**, 5683; (d) Casiraghi, G.; Zanardi, F.; Rassu, G.; Spanu, P., *Chem. Rev.* 1995, **95**, 1677; (e) *Dictionary of Carbohydrates with CD-ROM*, Collins, P. M.; Ed.; CRC Press, Boca Raton, FL, 2006; (f) Hollingsworth, R. I.; Wang, G. *Chem. Rev.* 2000, **100**, 4267; (g) *Heterocycles from Carbohydrate Precursors. Topics in Heterocyclic Chemistry, 07*, El Ashry, E. S.; Ed.; Series Ed. Gupta, R. R., 2007, Springer Verlag, Berlin; (h) Hanessian, S. *Acc. Chem. Res.* 1979, **12**, 159; (i) Fraser-Reid, B. *Acc. Chem. Res.* 1975, **8**, 192; see also: (j) Rollin, P.; Klaffke, W. *J. Carbohydr. Chem.* 1991, **10**, 115; (k) *Carbohydrate Building Blocks*, Bols, M., 1996, Wiley, NY; (l) *Organic Synthesis with Carbohydrates*, Boons, G.-J.; Hale, K.–J., 2000, Blackwell Science,

Malden, MA; (m) Lichtenthaler, F. in *New Aspects of Organic Chemistry*, Yoshida, I. Z.; Shiba, T.; Ohshiro, Y.; Eds.; 1989, p. 351, Kodansha, Tokyo, and VCH, Weinheim; (n) Inch, T. D. *Tetrahedron* 1984, **40**, 3161; (o) Vasella, A. in, *Modern Synthetic Methods*, Scheffold, R.; Ed.; vol. 2, p. 173, Otto Salle Verlag, Frankfurt, 1980; (p) Chrétien, F.; Chapleur, Y. in, *The Organic Chemistry of Sugars*, Levy, D. E. Fügedi, D. Eds.; 2006, CRC Press.

8. Chapleur, Y.; Chrétien, F. in, *Preparative Carbohydrate Chemistry*, Hanessian, S.; Ed.; 1997, Chapter 10, p. 207, Dekker, NY.

9. For example, see: (a) Binkley, E. R.; Binkley, R. W. in, *Preparative Carbohydrate Chemistry*, Hanessian, S.; Ed.; 1997, Chapter 5, p. 87, Dekker, NY; (b) Vatèle, J.-M.; Hanessian, S. in, *Preparative Carbohydrate Chemistry*, Hanessian, S.; Ed.; 1997, Chapter 7, p. 127, Dekker, NY.

10. (a) Barton, D. H. R.; McCombie, S. W. *J. Chem. Soc., Perkin Trans 1* 1975, 1574; see also: (b) Barton, D. H. R.; Ferreira, J. A.; Jaszberenyi, J. C. in, *Preparative Carbohydrate Chemistry*, Hanessian, S.; Ed.; 1997, Chapter 8, p. 151, Dekker, NY.

11. Horton, D.; Norris, P. in, *Preparative Carbohydrate Chemistry*, Hanessian, S.; Ed.; 1997, Chapter 2, p. 35, Dekker, NY.

12. For a review of existing methods, see: Dondoni, A.; Marra, A. in, *Preparative Carbohydrate Chemistry*, Hanessian, S.; Ed.; 1997, Chapter 9, p. 173, Dekker, NY.

13. Calinaud, P.; Gelas, J. in, *Preparative Carbohydrate Chemistry*, Hanessian, S.; Ed.; 1997, Chapter 1, p. 3, Dekker, NY.

14. For example, see: Wang, C.-C.; Lee, J.-C.; Luo, S.-Y.; Kulkarni, S. S.; Huang, Y.-W.; Lee, C.-C.; Chang, K.-L.; Hung, S.-H. *Nature* 2007, **446**, 896.

15. Szarek, W. A.; Kong, X. in, *Preparative Carbohydrate Chemistry*, Hanessian, S.; Ed.; 1997, Chapter 6, p. 105, Dekker, NY.

16. (a) Hanessian, S. *Carbohydr. Res.* 1966, **2**, 86; (b) (i) Hanessian, S.; Plessas, N. R. *J. Org. Chem.* 1969, **34**, 1035; (ii) Hanessian, S.; Plessas, N. R. *J. Org.* *Chem.* 1969, **34**, 1045; (iii) Hanessian, S.; Plessas, N. R. *J. Org. Chem.* 1969, **34**, 1053; (c) Hanessian, S. *Org. Syn.* 1987, **65**, 243; (d) Failla, D. L.; Hullar, T. L.; Siskin, S. B. *J. Chem. Soc., Chem. Commun.* 1966, 716; (e) Hullar, T. L.; Siskin, S. B. *J. Org. Chem.* 1970, **35**, 225.

17. Monneret, C.; Florent, J.-C. *Synlett* 1994, 305.

18. (a) Arjoná, O. Gómez, A.-M.; López, J. C.; Plumet, J. *Chem. Rev.* 2007, **107**, 1919; (b) Ferrier, R. S. In, *Preparative Carbohydrate Chemistry*, Hanessian, S.; Ed.; 1997, Chapter 26, p. 569, Dekker, NY; (c) Rajan Babu, T. V. in, *Preparative Carbohydrate Chemistry*, Hanessian, S.; Ed.; 1997, Chapter 25, p. 545, Dekker, NY.

19. For example, see: (a) *α-Hydroxy Acids in Enantioselective Syntheses*, Coppola, G. M.; Schuster, H. F., 1997, VCH, Weinheim; (b) *Tartaric and Malic Acids in Synthesis*, Gawronski, J.; Gawronska, K., 1999, Wiley, NY; see also: Seebach, D. in, *Modern Synthetic Methods*, Scheffold, R.; Ed.; 1980, vol. 2, p. 91, Otto Salle Verlag, Frankfurt.

20. For example, see: (a) Seebach, D.; Wasmuth, D. *Helv. Chim. Acta* 1980, **63**, 197; (b) Fráter, G. *Helv. Chim. Acta* 1979, **62**, 2825; (c) Fráter, G.; Müller, U.; Gunther, W. *Tetrahedron* 1984, **40**, 1269; (d) Seebach, D.; Aebi, J.; Wasmuth, D. *Org. Syn.* 1985, **63**, 109.

21. For example, see ref. [19]: (a) lactic acid, Chapter 1, p. 1; (b) tartaric acids, Chapter 4, p. 313; (c) malic acid, Chapter 3, p. 167; (d) mandelic acid, Chapter 2, p. 137; (e) 3-hydroxybutyric acid: Müller, H. H.; Seebach, D. *Angew. Chem. Int. Ed.* 1993, **32**, 477.

22. (a) Money, T. *Nat. Prod. Rep.* 1985, **2**, 253; (b) Money, T. in, *Studies in Natural Products Chemistry*, Atta-ur-Rahman; Ed.; 1989, Vol. 4, p. 625; see also: (c) *Carbocycle Construction in Terpene Synthesis*, Ho, T.-L., 1988, VCH, Weinheim; (d) *Enantioselective Synthesis*, Ho, T.-L., 1992, Wiley-Interscience, NY; *Terpenes*, Breitmaier, E., 2006, Wiley-VCH, Weinheim.

23. (a) From citronellol, see: (i) Mori, K.; Kuwahara, S. *Tetrahedron* 1982, **38**, 521; (ii) Williams, D. R.; Barner, B. A.; Nishitani, K.; Phillips, J. G. *J. Am. Chem. Soc.* 1982, **104**, 4708; (b) from carvone: (i) Nishiyama, S.; Ikeda, Y.; Yoshida, S.-I.; Yamamura, S. *Tetrahedron Lett.* 1989, **30**, 105; (ii) Findlay, J. A.; Desai, D. N.; Lonergan, G. C.; White, P. S. *Can. J. Chem.* 1980, **58**, 2827; (iii) Baggiolini, E. G.; Iacobelli, J. A.; Hennessy, B. M.; Uskoković, M. R. *J. Am. Chem. Soc.* 1982, **104**, 2945; (iv) Castedo, L.; Mascareñas, J. L.; Mourino, A. *Tetrahedron Lett.* 1987, **28**, 2099; (v) Kitahara, T.; Kurata, H.; Matsuoka, T.; Mori, K. *Tetrahedron* 1985, **41**, 5475; (vi) Asaka, Y.; Kamikawa, T.; Kubota, T. *Tetrahedron* 1974, **30**, 3257; (vii) Kametani, T.; Suzuki, Y.; Ban, C.; Kanada, K.; Honda, T. *Heterocycles* 1987, **26**, 1789; (c) from pulegone: (i) Corey, E. J.; Pan, B. C.; Hua, D. H.; Deardorff, D. R. *J. Am. Chem. Soc.* 1982, **104**, 6816; (ii) White, J. D.; Avery, M. A.; Choudhry, S. C.; Dhingra, O. M.; Gray, B. D.; Kang, M. C.; Kuo, S. C.; Whittle, A. J. *J. Am. Chem. Soc.* 1989, **111**, 790; (iii) Wovkulich, P. M.; Tang, P. C.; Chadha, N. K.; Batcho, A. D.; Barrish, J. C.; Uskoković, M. R. *J. Am. Chem. Soc.* 1989, **111**, 2596; (iv) Wuest, J. D.; Madonik, A. M.; Gordon, D. C. *J. Org. Chem.* 1977, **42**, 2111; (v) Niwa, H.; Nisiwaki, M.; Tsukuda, I.; Ishigaki, T.; Ito, S.; Wakamatsu, K.; Mori, T.; Ikagawa, M.; Yamada, K. *J. Am. Chem. Soc.* 1990, **112**, 9001; (d) from menthone: (i) Davidson, B. S.; Plavcan, K. A.; Meinwald, J. *J. Org. Chem.* 1990, **55**, 3912; (ii) Schmid, G.; Hofheinz, W. *J. Am. Chem. Soc.* 1983, **105**, 624; (e) from limonene: (i) Broka, C. A.; Chan, S.; Peterson, B. *J. Org. Chem.* 1988, **53**, 1584; (ii) Kergomard, A.; Veschambre, H. *Tetrahedron Lett.* 1976, **17**, 4069; (iii) Meinwald, J.; Jones, T. H. *J. Am. Chem. Soc.* 1978, **100**, 1883; (iv) Darbre, T.; Nossbaumer, C.; Borschberg, H.-J. *Helv. Chim. Acta* 1984, **67**, 1040; (v) Heath, R. R.; Doolittle, R. E.; Sonnet, P. E.; Tumlinson, J. H.

J. Org. Chem. 1980, **45**, 2910; (vi) Lange, G. L.; Neidert, E. E.; Orrom, W. J.; Wallace, D. J. *Can. J. Chem.* 1978, **56**, 1628; (vii) Metha, G.; Krishnamurthy, N. *Tetrahedron Lett.* 1987, **28**, 5945; (f) from borneol and camphor: (i) Hutchinson, J. H.; Money, T. *Tetrahedron Lett.* 1985, **26**, 1819; (ii) Hutchinson, J. H.; Money, T. *J. Chem. Soc., Chem. Commun.* 1986, 288; (iii) Stevens, R. V.; Chang, J. H.; Lapalme, R.; Schow, S.; Schlageter, M. G.; Shapiro, R.; Weller, H. N. *J. Am. Chem. Soc.* 1983, **105**, 7719.

24. For a review, see: Oppolzer, W. *Tetrahedron*, 1987, **43**, 1969.

25. Campbell, M. M.; Sainsbury, M.; Searle, P. A. *Synthesis* 1993, 179.

26. For example, see: (a) Ogawa, S. *Trends Glycosci. Glycotechnol.* 2004, **16**, 33; (b) Suami, T.; Ogawa, S. *Adv. Carbohydr. Chem. Biochem.* 1990, **48**, 21; (c) Kobayashi, Y. in, *Glycoscience*, Fraser-Reid, B.; Tatsuta, K.; Thiem. J.; Eds.; 2001, p. 2595, Springer, Heidelberg; (d) *Carbohydrate Mimics*, Chapleur, Y.; Ed.; 1998, Wiley-VCH, Weinheim; *Cyclitols and Their Derivatives: A Handbook of Physical, Spectral, and Synthetic Data*, Hudlicky, T.; Cehulak, M., VCH, NY, 1993.

27. For reviews, see: (a) Barco, A; Benetti, S.; De Risi, C.; Marchetti, P.; Pollini, G.; Zanirato, V. *Tetrahedron: Asymmetry* 1997, **8**, 3515; (b) Jiang, S.; Singh, G. *Tetrahedron* 1998, **54**, 4697; (c) Rassu, G.; Auzzas, L.; Pinna, L.; Battistini, L.; Curti, C. in, *Studies in Natural products Chemistry*, Attu-ur-Rahman, Ed.; 2009, Vol. 29, p. 449

28. For example, see: (a) Hudlicky, T.; Reed, J. W. *Synlett* 2009, 685; (b) Johnson, R. A. *Org. React.* 2004, **63**, 117; (c) Banwell, M. G.; Edwards, A. J.; Harfoot, G. J.; Jolliffe, K. A.; McLeod, M. D.; McRae, K. J.; Stewart, S. G.; Vögtle, M. *Pure Appl. Chem.* 2003, **75**, 223; (d) Endoma, M. A; Bui, V. P.; Hansen, J.; Hudlicky, T. *Org. Process Res. Dev.* 2002, **6**, 525; (e) Boyd, D. R.; Sheldrake, G. N. *Nat. Prod. Rep.* 1998, **15**, 309.

8
From Target Molecule to Chiron

Coming face-to-face with the structure of a complex natural product or a therapeutically relevant molecule elicits many thoughts and emotions, especially when total synthesis becomes the ultimate objective. In fact, regardless of the structural and stereochemical complexity of a target molecule, the notion of having to devise a viable synthesis plan may bring about some excitement, mixed with trepidation, in the mind of the synthetic chemist. Questions like: "Is the plan a viable one, and will its execution *en route* to the final target compound proceed with minimal difficulties?" are not uncommon. As discussed earlier, the sequential thought process of bond disconnection in a retrosynthetic analysis will invariably generate several ideas, which will be carefully scrutinized by the investigator before settling on a given approach. The adrenaline rush that one feels when such an optimistic plan is scribed in the form of reagents and products on paper is all too familiar to synthetic chemists. Ultimately, whether it is a signature synthesis, admired for its conceptual elegance, or a more classical one utilizing well-established literature methods, there will be one or more starting materials to consider. In the case of molecules harboring single or multiple stereogenic carbon atoms with diverse functionality, chiral, non-racemic subunits must be generated as intermediates in part or as a whole. The availability of enantiopure chirons from a wide cross-section of natural sources, as articulated in Chapter 7, makes them obvious choices as starting materials that can be used as discrete entities, or in conjunction with stoichiometric or catalytic methods of asymmetric bond formation to generate appropriate synthetic intermediates.

In this chapter, we shall elaborate on how to locate native chirons in the structures of simple and complex organic molecules to be used as starting materials. Chirons may be *apparent, partially hidden*, or *hidden* within the carbon framework of intended target molecules [1]. Their recognition as direct or chemically modified precursors to chiral subunits in such molecules, and their elaboration into advanced intermediates, delineates the principles of the *chiron approach* to total synthesis. One must use this approach so as to derive full advantage of skeletal and stereochemical convergence with relevant segments of target molecules. When used judiciously, the *chiron approach* can be applied to virtually any organic molecule that harbors one or more stereogenic centers.

Design and Strategy in Organic Synthesis: From the Chiron Approach to Catalysis, First Edition.
Stephen Hanessian, Simon Giroux, and Bradley L. Merner.
© 2013 Wiley-VCH Verlag GmbH & Co. KGaA. Published 2013 by Wiley-VCH Verlag GmbH & Co. KGaA.

8.1
Where's Waldo?

The popular "Where's Waldo" books are veritable visual puzzles intended to find the young Waldo in a maze of other characters, akin to finding a needle in a haystack. Often, finding the residue of an appropriate native or modified chiron embedded in the structure of a target molecule upon visual analysis, is like locating Waldo in a kaleidoscope of similar figures. Three typical examples spanning a period of several decades are discussed below to show how chirons can be obvious, or not so, to the eye.

When considering two structurally and topologically different precursors to the California red scale pheromone, the notion of obvious and not so obvious starting points comes to the fore (Figure 8.1). Although, the target molecule is an acyclic hydrocarbon with two main-chain olefinic units, and a branched isopropenyl appendage, the choice of (S)-carvone as a precursor by Caine and Crews [2a] was not far fetched. Suffice it to focus on the isopropenyl group with its corresponding stereogenic center present in the target molecule, and to make a mental connection with (S)-carvone bearing the same appendage, albeit as a cyclic terpene structure. In practice, selective oxidative cleavage and appropriate extensions led to the natural pheromone. But what, if any, is the visual connection between the same pheromone and (+)-camphor? None whatsoever, unless prior knowledge of camphor chemistry by Hutchinson and Money [3] manifests itself in the mind's eye [4]. The readily available 9,10-dibromocamphor (A) is an exquisitely suitable substrate for a Grob-type fragmentation reaction under basic conditions (Figure 8.1). The resulting cyclopentene derivative B can now be manipulated to C, which undergoes a second Grob-type fragmentation to furnish the ester D, thereby unmasking a "hidden" isopropenyl group in A. These decades-old examples may have their limitations with respect to the number of synthetic operations compared to more direct syntheses relying on purely asymmetric methods. However, in both cases, the authors made very good use of the original terpene carbon frameworks.

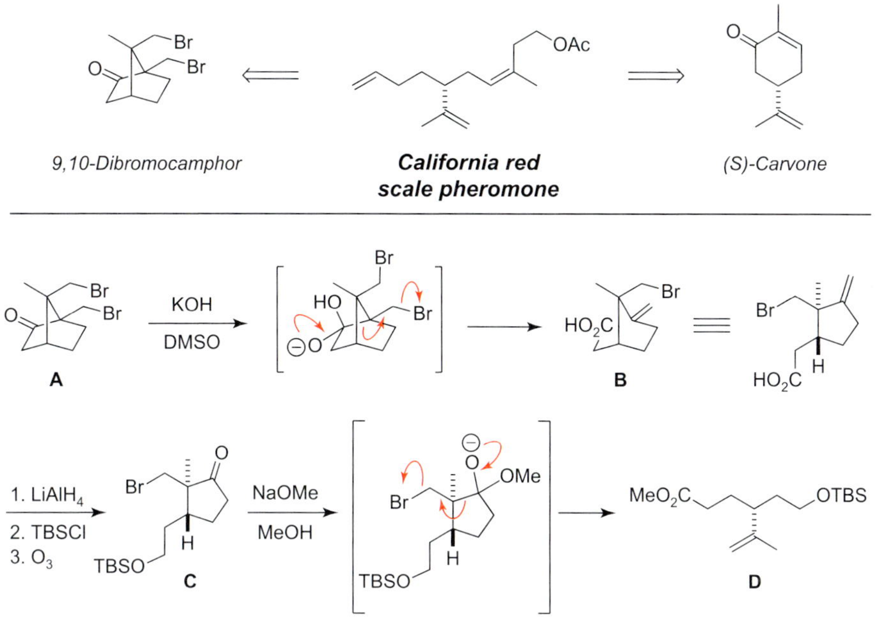

Figure 8.1 Obvious and not so obvious: Grob fragmentation route to a pheromone.

The synthesis of aristoteline from α-pinene by Gribble and Barden [5] is another interesting example of a not so obvious use of a structurally remote terpene as a chiron (Figure 8.2). 3-(2-Bromoethyl)indole derivative **A** was coupled with 8-amino menthene **B**, which is readily available from α-pinene, in an S_N2 reaction to afford **C**. Oxidation to the nitrone **D**, and heating in toluene triggered an intramolecular nitrone-olefin cycloaddition reaction to give the bridged adduct **E**. Reductive cleavage of the N–O bond revealed the aristoteline framework as in **F**, which upon dehydration led to hobartine (**G**). Further treatment with concentrated HCl furnished the target molecule. While a portion of carbocylic framework, the *gem*-dimethyl group, and a nitrogen atom in aristoteline can be traced back to chiron **B**, it took some astute retrosynthetic analysis to devise a highly stereocontrolled total synthesis.

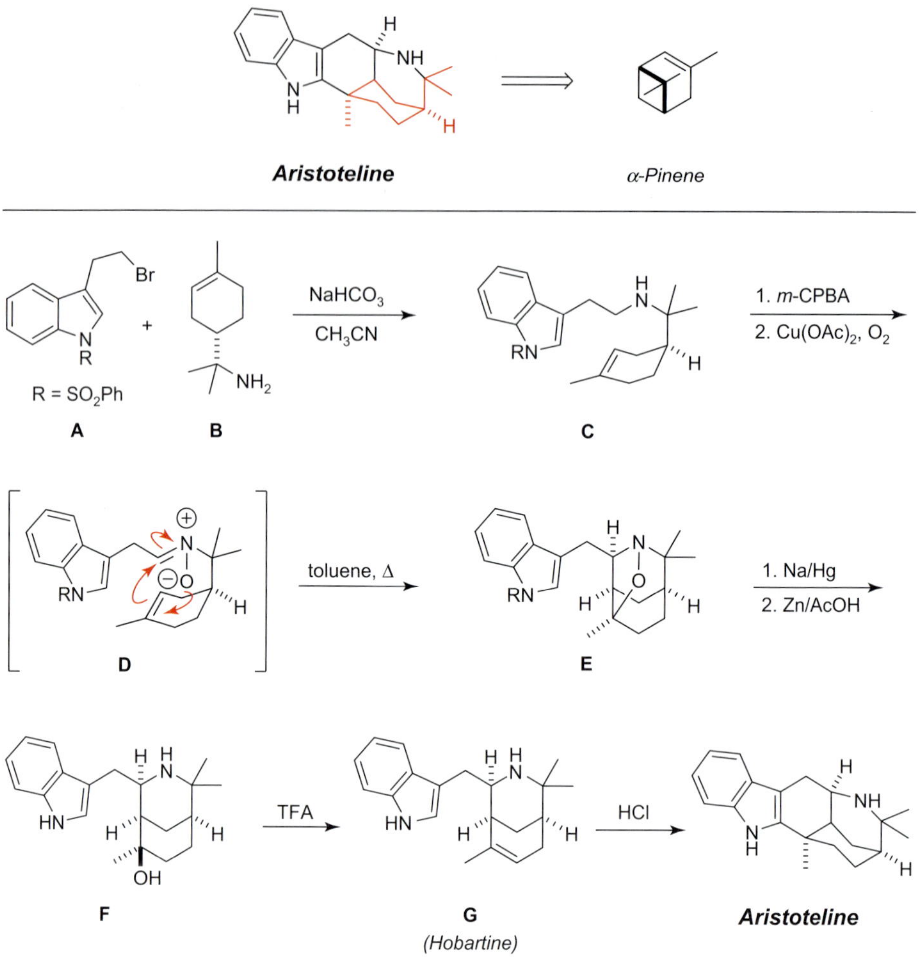

Aristoteline   α-Pinene

Figure 8.2 Obvious and not so obvious: intramolecular
nitrone-olefin cycloaddition route to aristoteline.

A third example of remote progeny within a subunit of a target molecule can
be found in the synthesis of virantmycin by Kogen and coworkers [6] (Figure 8.3).
The unnatural chiron (S)-indoline-2-carboxylic acid was converted to **A**, which was
subjected to a stereoselective Grignard reaction to give the branched tertiary alcohol
intermediate **B**. Treatment with tributylphosphine in CCl₄ triggered a remarkable
intramolecular ring expansion proceeding through the alkoxyphosphonium salt **C**,
and aziridinium ion **D**, before delivering the branched 3-chloro-tetrahydroquinoline
E. The successful realization of this plan is one of many manifestations of the power
of proximity effects in organic synthesis. A quote from the late R. B. Woodward is
à propos: "*Enforced propinquity often leads to greater intimacy.*"

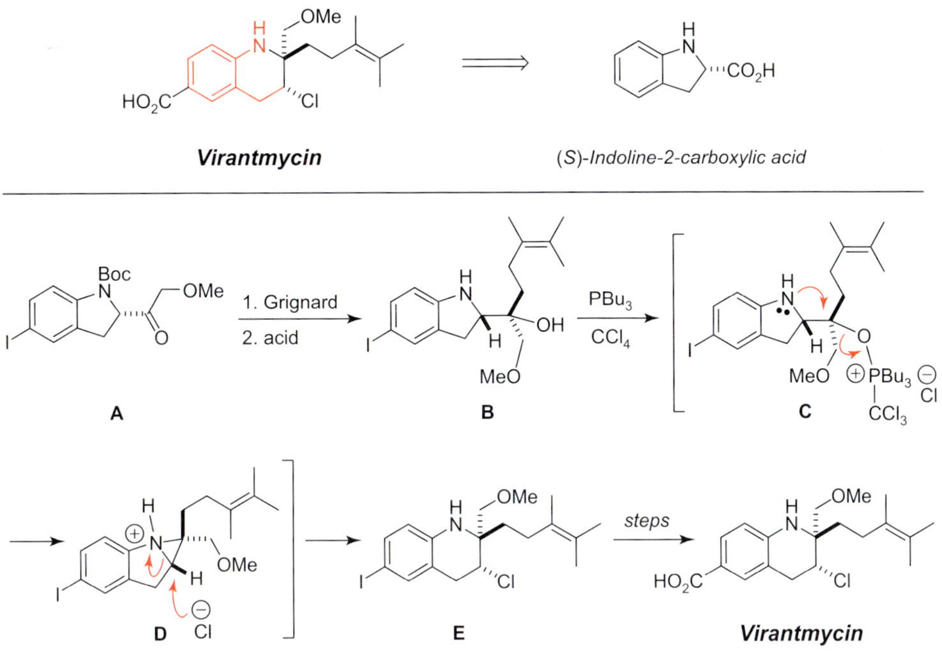

Virantmycin   (S)-Indoline-2-carboxylic acid

Figure 8.3 Obvious and not so obvious: alkoxyphosphonium-mediated ring expansion route to virantmycin.

8.2
Apparent Chirons

Chirons may be considered as being *apparent*, when a cursory inspection of the target molecule reveals an obvious connection with a native precursor available from the four naturally occurring classes consisting of: amino acids, carbohydrates, hydroxy acids, and terpenes. The examples shown in Figure 8.4 are representative of total syntheses spanning a 30 year period and covering a diverse set of natural products. The total synthesis of thienamycin from L-aspartic acid by Christensen and coworkers [7] at Merck Laboratories also demonstrated the utility of Rh-catalyzed carbenoid insertion reactions for this class of carbapenems (Figure 8.4 A, see details in Chapter 10, section 10.9).

In the first two syntheses of thromboxane B_2, Corey [8] and Hanessian [9] independently utilized the very apparent D-glucose framework as a starting chiron (Figure 8.4 B). In the synthesis of amphidinolide E [10] Roush took full advantage of the convergence of (S,S)-tartaric acid with a segment of the "left wing" diol of the natural product (Figure 8.4 C, see details in Chapter 16, section 16.4.9). Finally, the cyclohexane moiety in hapalindole G with the branching isopropenyl group inevitably led Fukuyama and Chen [11] to (R)-carvone as an obvious precursor (Figure 8.4 D).

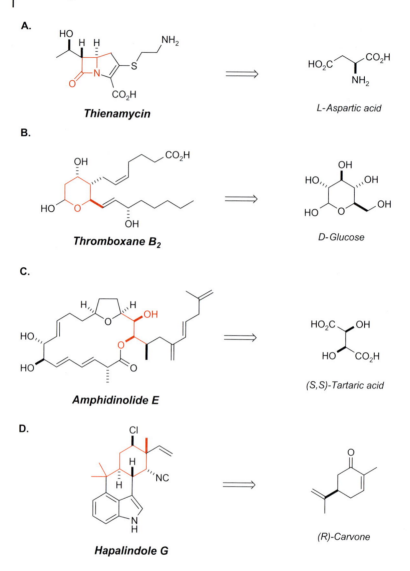

Figure 8.4 Apparent chirons.

In each of the above syntheses, the carbon skeleton and resident chirality of the native chiron was put to good use by recognizing the need for different levels of functional group modifications to converge with an intended subunit in the target molecule.

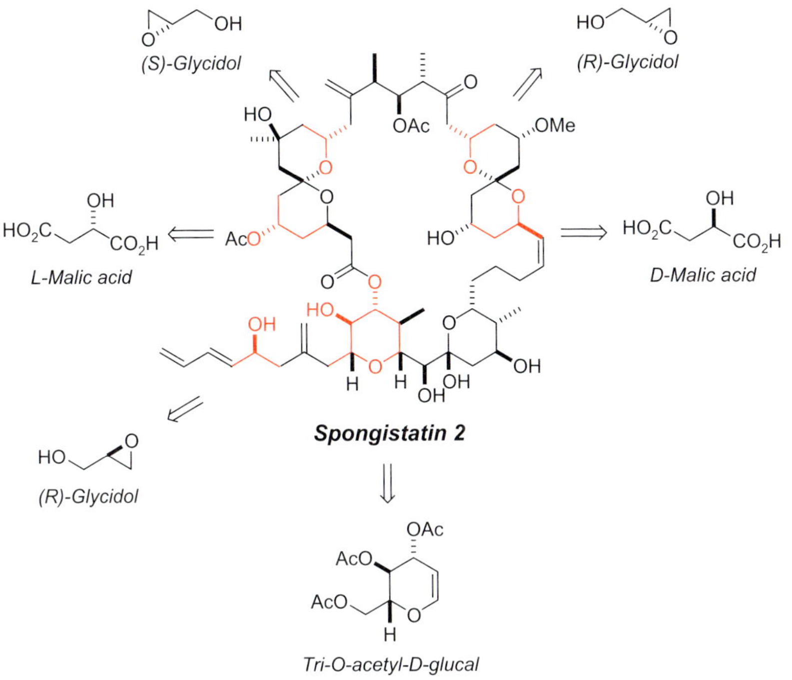

Figure 8.5 Multiple apparent chirons for spongistatin 2.

The versatility of the *chiron approach* can be appreciated in those cases where multiple apparent chirons can be used toward the synthesis of specific subunits in relatively large polyfunctional molecules (Figure 8.5). Heathcock and coworkers [12a–c] recognized the relationship between several chiral, oxygen-bearing subunits in spongistatin 2. Enantiomeric glycidols and malic acids, were used to provide 3- and 4-carbon building blocks that were amenable to end-group differentiation. The "southern" tetrahydropyran unit was elaborated from tri-*O*-acetyl-D-glucal, which is readily available from D-glucose.

Laulimalide is another popular molecule that has been used to showcase diverse strategies and synthetic prowess by several groups [13]. An assortment of chirons related to hydroxy acids, carbohydrates, and terpenes have been successfully used as precursors to corresponding segments of the macrolactone as seen in Figure 8.6 [14]. There are many similar prodigious uses of native chirons to provide functionally useful and stereochemically convergent synthetic intermediates to natural products of different levels of complexity. Some classical examples were already highlighted in Chapter 6 (see also Chapters 9–16).

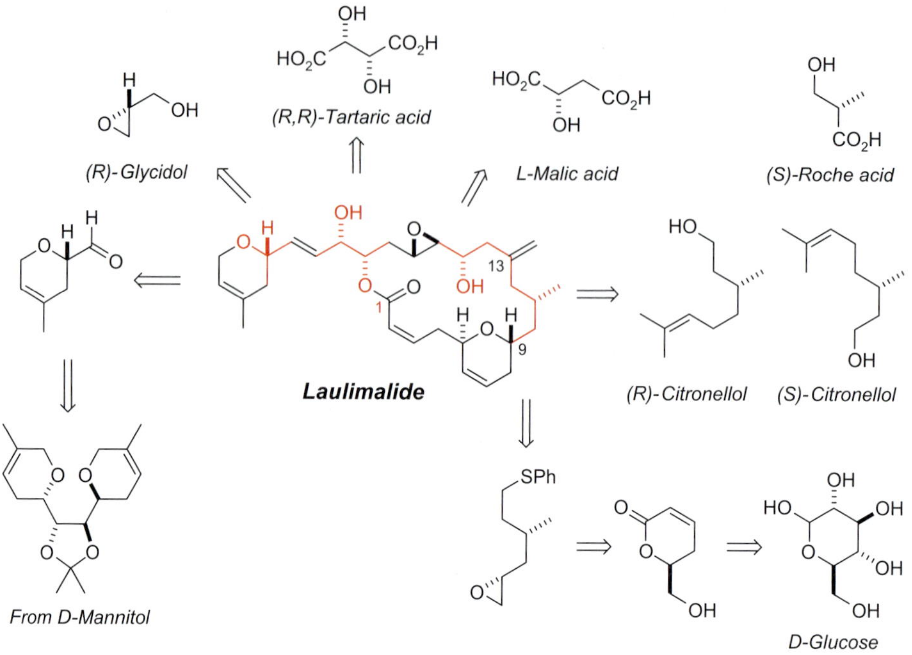

Figure 8.6   Multiple apparent chirons for laulimalide.

8.3
Partially Hidden Chirons

A chiron is considered as being *partially hidden* when only one or more functional groups or a ring-related feature can be traced back to a subunit in the target molecule. The chiron is considered to be partially hidden because it has been modified from its native form in order to accommodate the characteristics of the intended substructure. This practice has been extensively used over the years, as exemplified by the synthesis of the four target molecules shown in Figure 8.7. In tryptoquivaline, one recognizes the obvious challenge of elaborating a spirolactone and a highly functionalized indole core unit (Figure 8.7 A) [15]. Closer scrutiny reveals L-valine as a chiron for the isobutyl appendage bearing a secondary acetoxy group, with the knowledge that diazotization of the amino acid gives the corresponding α-hydroxy acid with retention of absolute configuration. However, focussing on the indole unit requires a deeper visual exploration of the chiral landscape of the molecule. This led to the unravelling of D-tryptophan as a matching chiron for the extended α-amino spirolactone segment of tryptoquivaline. The amino acid had to be extensively modified in order to introduce the remaining functionality, while providing stereochemical control at C-2 and C-3 in the dihydroindole ring.

The focal point in *endo*-brevicomin, at first glance, is the 2-methyl dioxolane unit with two contiguous stereogenic centers bearing the oxygen atoms that converge on the acetalic carbon atom (Figure 8.7 B) [16]. The masked vicinal diol can be related to D-erythronolactone, which also provides end-differentiated oxidation states for appropriate extensions, before reaching the dihydroxy methyl ketone stage, which undergoes a spontaneous stereoelectronically controlled cycloacetalization.

The macrocyclic dilactone portion of integerrimine contains a *C*-methyl group adjacent to a tertiary alcohol also bearing a methyl group. The (*R*)-Roche acid provided a viable starting point that was elaborated in both directions to build

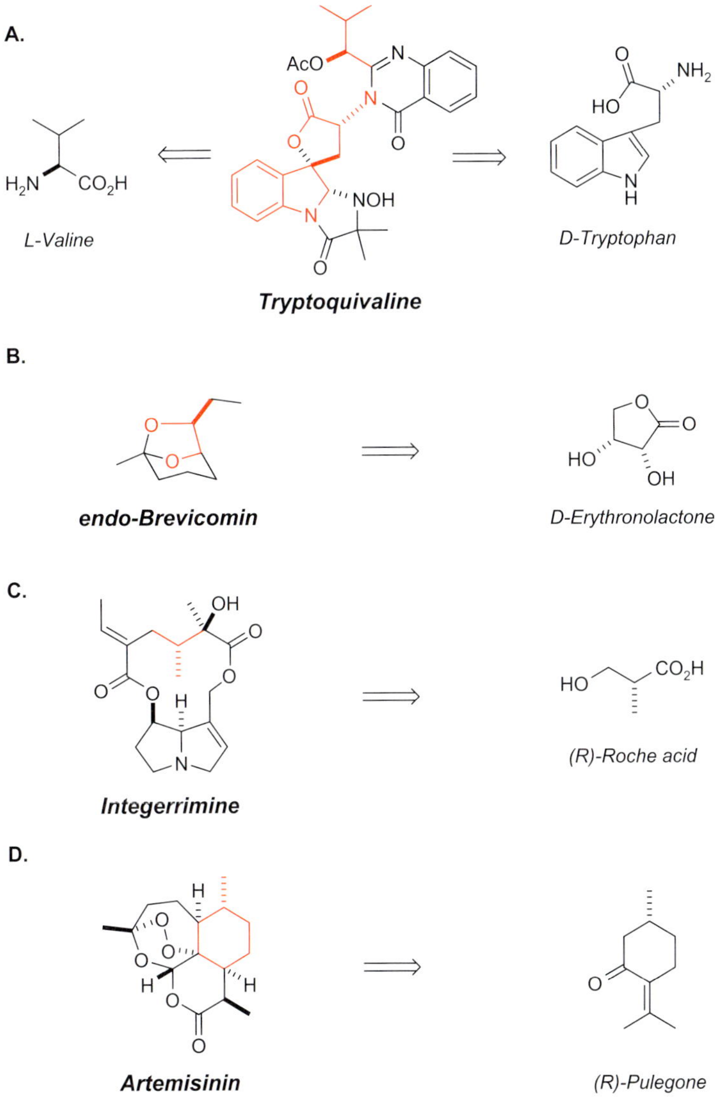

A.

L-Valine

Tryptoquivaline

D-Tryptophan

B.

endo-Brevicomin

D-Erythronolactone

C.

Integerrimine

(*R*)-Roche acid

D.

Artemisinin

(*R*)-Pulegone

Figure 8.7 Apparent and partially hidden chirons.

the trisubstituted dicarboxylic acid motif related to the macrocyclic portion of integerrimine (Figure 8.7 C) [17].

The condensed polycyclic ring system of artemisinin will sooner or later lead the wandering eye to the carbocyclic segment of the molecule (Figure 8.7 D) [18]. Closer inspection reveals a familiar substitution pattern, namely a methyl group at "12 o'clock," an oxygen atom at "8 o'clock" and a carbon substituent with a pendant C-methyl group at "6 o'clock." Instinctive reflex action cries for a terpene starting chiron, and in response the mind's eye offers (R)-pulegone. The matching C-methyl group and the prospects of introducing useful functionality via enolate and carbonyl activation chemistry suggested a plan with a viable solution for branching. The elaboration of the exocyclic gem-dimethyl olefin in (R)-pulegone, and completion of the synthesis presented additional challenges (see also section 8.9.2).

8.4
Hidden Chirons

The analogy with looking for a needle in a haystack (section 8.1) may be appropriate in those total syntheses where the native chiron has lost its progeny within the intricate network of the target molecule structure. Occasionally, atoms such as nitrogen and oxygen, and certain ring sizes may remind one of amino acids, hydroxy acids, or terpenes as starting materials for example. However, the visual path from substructure to a particular native chiron is at times tortuous. These also represent the most intellectually challenging syntheses when contemplating the use of the *chiron approach.*

Using (R)-pulegone as a starting material for ring A of epoxydictymene may have not been obvious where it not for the fact that Schreiber and coworkers had prior knowledge that the trisubstituted cyclopentane, with a matching stereochemistry, is readily available from the terpene in five short steps (Figure 8.8 A) [19]. In spite of the daunting structure of ecteinascidin 743, the presence of certain atoms conjure up thoughts about possible chirons *en route* to its total synthesis by the groups of Zhu [20a], Fukuyama [20b], and Corey [20c] (Figure 8.8 B). The thioether bridge is in fact part of an α-substituted L-cysteine, hence it is somewhat of an obvious choice. The second nitrogen atom in ring C may be related to a L-phenylalanine analogue. However, the branching pattern in ring G of the tetrahydroisoquinoline portion presented a challenge. Even though fitting L-serine at the junction of rings B and C was not so evident, it was successfully accomplished by exploiting its different oxidation states. The architecturally interesting *ent*-phomoidride B has been the target of elegant total syntheses by the Nicolaou [21a–c], Fukuyama [21d], Shair [21e], and Danishefsky [21f] groups (Figure 8.8 C, see details in Chapter 16, section 16.4.7). In more than one of these syntheses, a chiron derived from D-mannitol was used to elaborate the cyclic ether portion of *ent*-phomoidride B. Once elaborated, the original stereogenic center present in the starting chiron was destroyed. In contrast, the Shair synthesis made full use of D-glyceraldehyde acetonide, which is

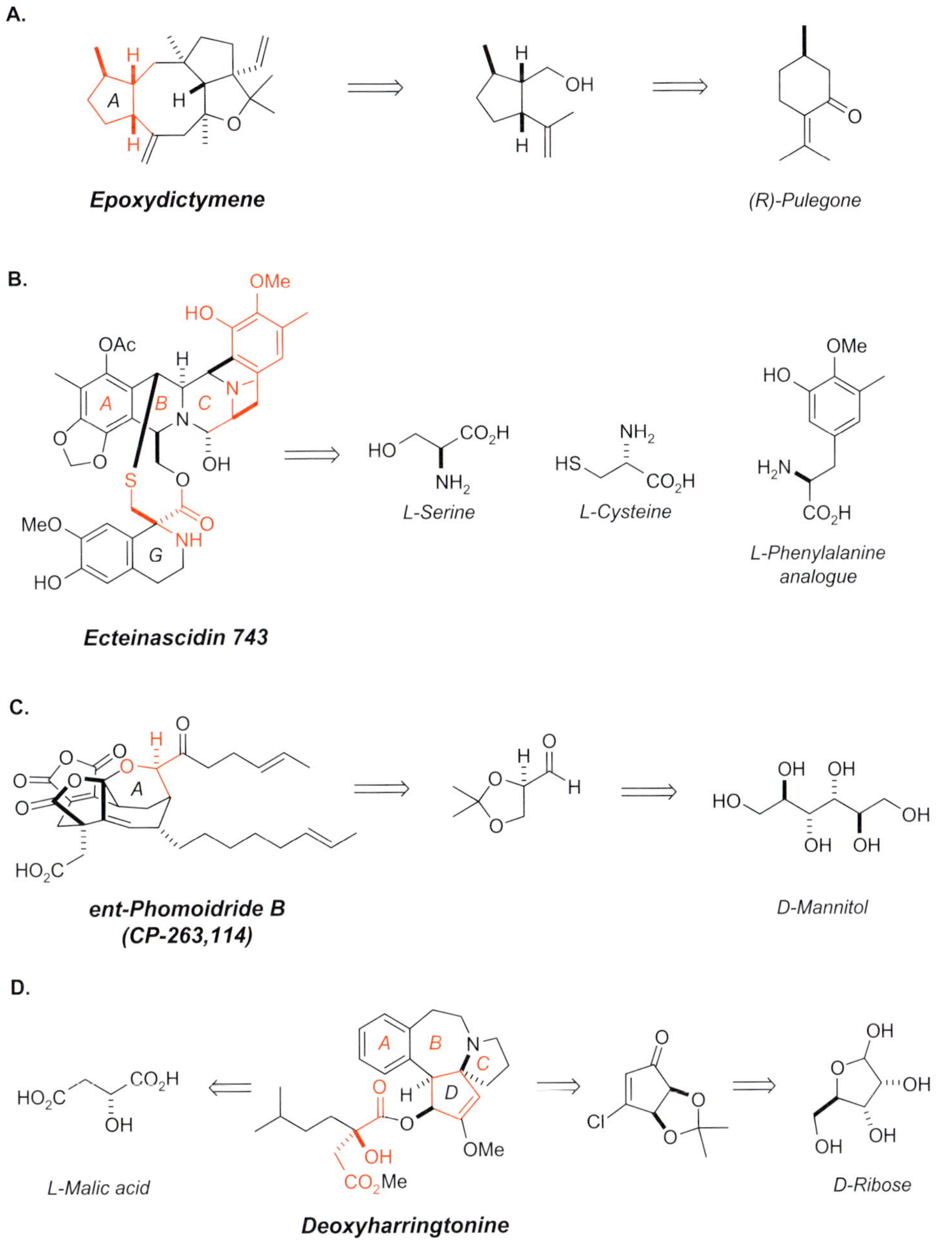

A.

Epoxydictymene

(R)-Pulegone

B.

Ecteinascidin 743

L-Serine

L-Cysteine

L-Phenylalanine
analogue

C.

**ent-Phomoidride B
(CP-263,114)**

D-Mannitol

D.

L-Malic acid

Deoxyharringtonine

D-Ribose

Figure 8.8   Hidden chirons.

readily available from D-mannitol. This 3-carbon chiron can be traced back to the oxacyclic ring *A* of the molecule, providing the ether oxygen atom.

Deoxyharringtonine harbors a secondary alcohol as part of a fused cyclopentene methyl enol ether as seen in ring *D* (Figure 8.8 D) [22]. A relationship to a carbohydrate precursor, if any, is very hidden since only one hydroxyl group is immediately present in the natural product. Other hydroxyl groups present in a potential carbohydrate precursor will have to be deoxygenated, if not extensively modified until the desired pattern and absolute stereochemistry is attained for convergence with ring *D*. Moreover, the carbocyclic nature of ring *D* does not lead one to consider a carbohydrate precursor at first glance. Yet, further analysis of the strategy required prior knowledge that five and six-membered aldoses can be converted to substituted cyclopentanes and cyclohexanes respectively with stereochemical control, as was cleverly exploited by Gin and coworkers [22]. Thus, a chlorocyclopentenone diol readily available from D-ribose was used as a precursor to ring *D*. The ester subunit containing a fully substituted stereogenic center was prepared from L-malic acid using appropriate enolate chemistry.

8.5
Chirons as "Sacrificial Lambs"

The age-old biblical custom of sacrificing a lamb on special family occasions is still practiced in many parts of the world. The lamb serves the purpose of feeding the guests and hosts alike in a festive atmosphere with musical accompaniment and savory odors emanating from a sizzling pit.

Chirons have been used as "sacrificial lambs" to satisfy the needs of a synthesis plan, but only after they have served their purpose of inducing chirality or securing absolute stereochemistry of critical stereogenic centers. The term "self-immolative process" was coined by Kurt Mislow, and will be used throught this book to describe syntheses were an original stereogenic center(s) of a chiron has been lost or removed once it has served its directing role. Four examples from the recent literature serve to illustrate the process in which only the carbon framework of the chiron, and occasionally a heteroatom, have remained in the structure of the intended target molecule once the synthesis was completed.

The complex polycyclic structure of biyouyanagin A presents a number of different synthetic strategies including the recognition that the hydrocarbon chain harboring the *C*-methyl group at C-24 can be derived from (*R*)-citronellal (Figure 8.9). Indeed, one of the reasons for its synthesis by the Nicolaou group [23] was to provide a full structure elucidation of the natural product and the establishment of the (*R*)-absolute configuration at C-24. The strategy involved an unprecedented [2+2] cycloaddition reaction to incorporate the central cyclobutane ring. Intermediate **A** (hyperlactone C), generated from propargylic alcohol **B** via a Pd-catalyzed cascade reaction, was chosen as a 2π component of the pivotal [2+2] cycloaddition reaction. L-Malic acid was converted to lactone **C** by a known procedure, thus incorporating the important quaternary stereogenic center at C-3.

Having served its stereodirecting purpose, the C-2 hydroxyl group in the original L-malic acid was oxidized to α-ketolactone **B**, then further elaborated to the spirocyclic intermediate **A**.

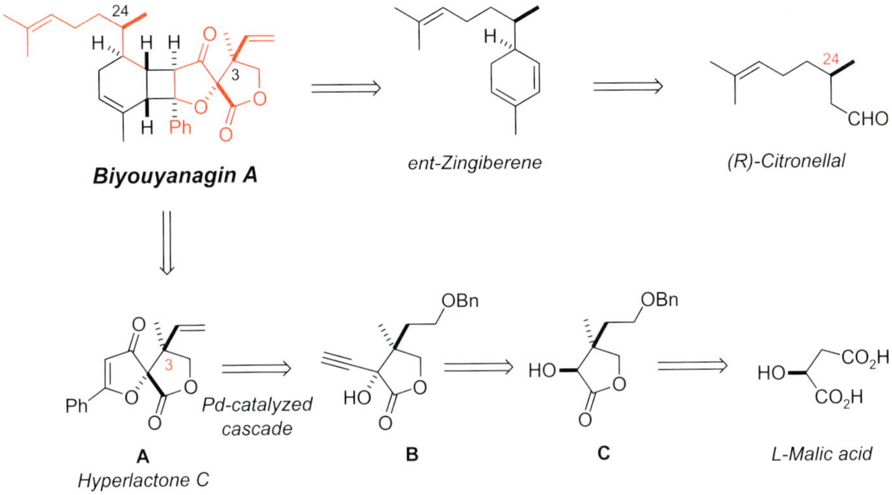

Biyouyanagin A

ent-Zingiberene

(*R*)-Citronellal

A
Hyperlactone C

Pd-catalyzed cascade

B

C

L-Malic acid

Figure 8.9 Self-immolative *chiron approach* to biyouyanagin A.

The second example involves the total synthesis of vindoline by Fukuyama and coworkers [24] from an enantiopure, lipase-derived actetate protected cyanohydrin (Figure 8.10). A dihydrofuran intermediate containing a primary aminomethyl group was first tethered to the indole as in **A**. Treatment of **A** with TFA gave hemiacetal **B**, which upon deprotection led to a ring-expanded iminium ion, followed by tautomerization to furnish the enamine intermediate **C**. The stage was now set for an intramolecular inverse electron demand Diels-Alder reaction between the enamine dienophile (piperidine ring) and indole based diene, to give the pentacyclic product **D** with high selectivity. The hydroxyl group in **D**, originating from the lipase derived cyanohydrin, was subjected to elimination to afford **E**, which was the converted to natural vindoline. The carbon and nitrogen atom(s) of the dihydrofuran precursor were all incorporated in the target molecule, but the original single stereogenic center that dictated the stereochemical outcome of the inverse electron demand Diels-Alder reaction was sacrificed in the end.

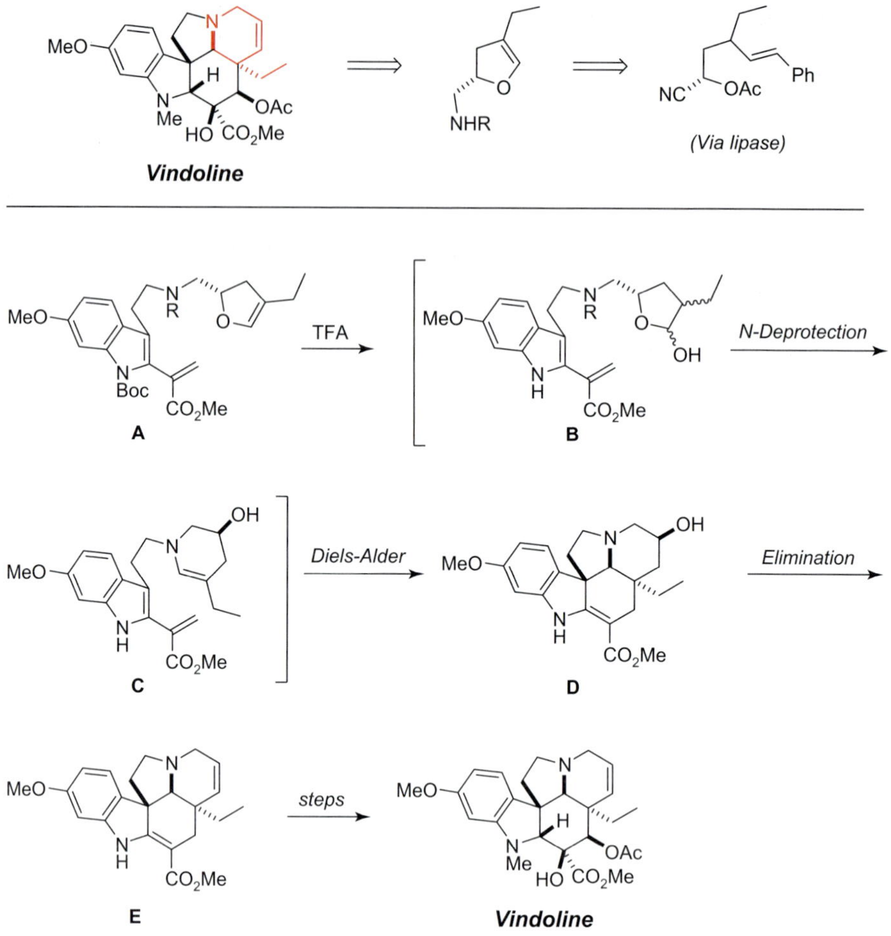

Figure 8.10 Self-immolative *chiron approach* to vindoline.

The final stages of the total synthesis of yatekamycin by Fukuyama and coworkers [25] were also planned based on a self-immolative strategy (Figure 8.11). The immediate precursor of the fused cyclopropane ring is a 3-hydroxypiperidine, which represents part of the central tetracyclic core yatekamycin. (*S*)-Epichlorohydrin was used to elaborate the central tricyclic unit of yatekamycin. Having served its purpose of incorporating the mesylate in **A**, the original stereogenic center was subsequently lost in a stereocontrolled spirocyclopropanation reaction to produce the fused cyclopropane ring in **B**.

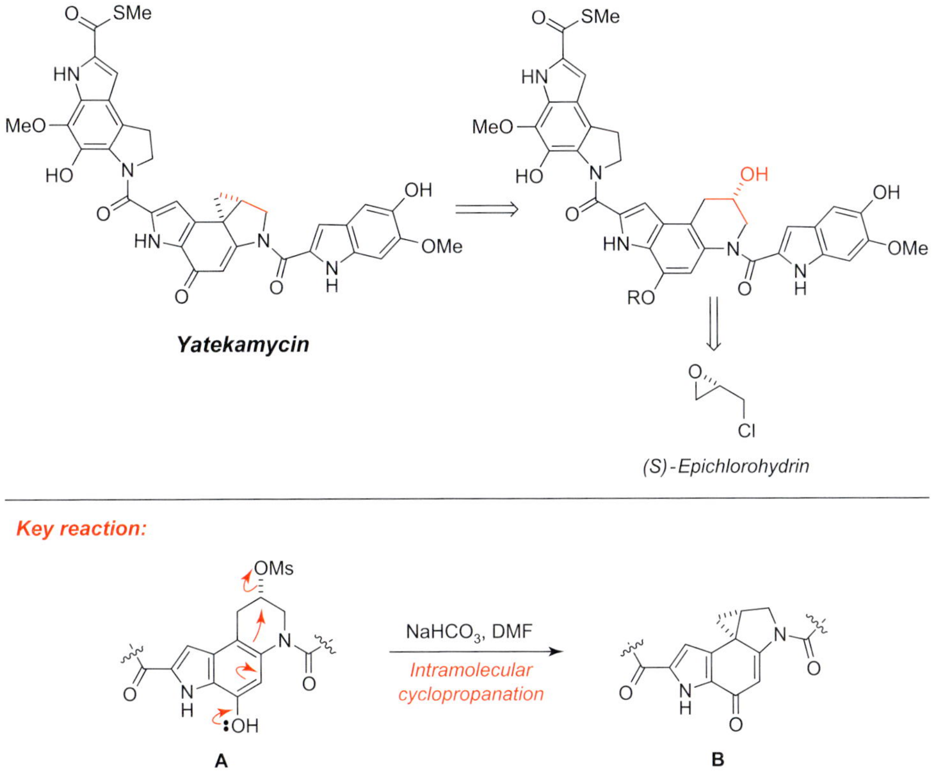

Yatekamycin

(S)-Epichlorohydrin

Key reaction:

NaHCO₃, DMF

Intramolecular cyclopropanation

A   **B**

Figure 8.11 Self-immolative *chiron approach* to yatekamycin.

Finally, disconnective analysis for the synthesis of ulapualide A by the Pattenden group [26] shows L-serine as the native chiron, which can be traced back to two of the three oxazole units (Figure 8.12). A series of highly stereocontrolled reactions involving cuprate additions to a γ-ureido-α,β-unsaturated ester (**C** to **B**), followed by oxazole formation (**B** to **A**) afforded a precursor to the C-1–C-12 segment bearing the C-9 *C*-methyl group in ulapualide A. Thus, L-serine was used to introduce the *C*-methyl group in a resident chirality-induced conjugate addition before sacrificing its stereogenic center and becoming part of the first achiral oxazole ring. L-Serine was also used in the Pattenden synthesis to build a second achiral oxazole unit of ulapualide A.

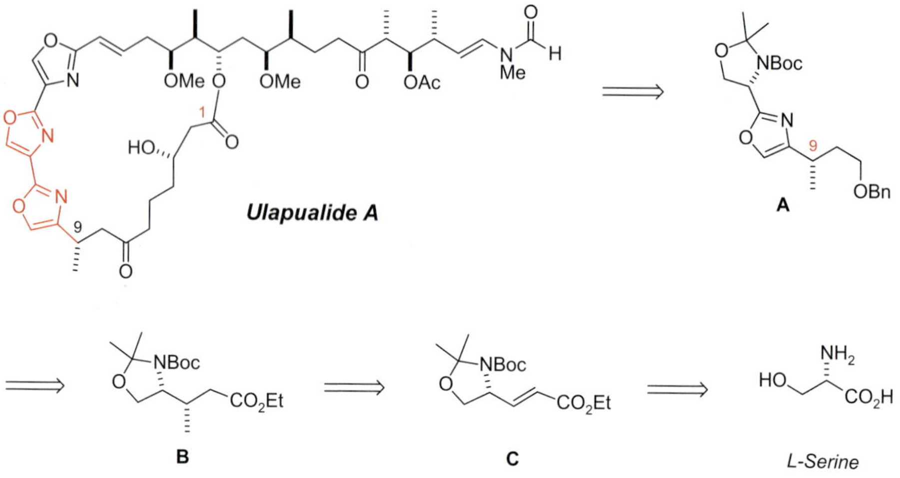

Figure 8.12 Self-immolative *chiron approach* to ulapualide A.

8.6
Locating α-Amino Acid-type Substructures

In order to appreciate the potential general utility of α-amino acids as chirons, they must first be located in their native or modified forms within substructures of target molecules. One begins by focussing on an amino group (or a nitrogen substituent) situated on a stereogenic carbon atom. A second reference point could be vicinal or distal sp^2 or sp^3-centers, as well as the presence of other substituents that can potentially be derived by chemical manipulation of an amino acid. Depending on the level of skeletal and functional complexity of the substructure, a specific amino acid-type that provides a good carbon framework and stereochemical overlap with the corresponding substructure in the target molecule may already be considered as a starting material. Existing or transformable functionality in an amino acid is an important consideration in order to fully exploit its utility as a suitable precursor. The use of α-amino acids (henceforth amino acids) as starting chirons in selected total syntheses is highlighted in Chapter 10.

8.6.1
Apparent amino acids

The target molecule may contain an amino group on a stereogenic carbon atom that matches the absolute configuration of a D- or L-amino acid. The following simple steps may be useful to consider in order to locate amino acid chirons in a given target substructure:

a. Determine the type of carbon framework required in the chiron to converge with the target substructure (acyclic, cyclic, or aromatic amino acids).
b. Search for α-, β-, or γ-substitution with respect to the position of the amino group in the target. For example, a 2-hydroxyethyl group may point to threonine or allothreonine as a potential precursor. A thioether will lead to cysteine. Aromatic or heterocyclic rings in α-, or β-positions with respect to the amino group, are easily related to corresponding amino acids.
c. Search for a bond-breaking site next to the amino group that could correspond to the carboxyl group of the original α-amino acid. Apply the same reasoning to end-differentiated amino acids such as serine, aspartic acid, and glutamic acid for proximal and distal bond-breaking possibilities.
d. A naturally occurring amino acid may have emerged at this point. Allow for further modifications such as carbon extensions on both sides of the amino group, utilizing existing or newly introduced functionality. Exploit enolate alkylation and carbocyclization reactions with appropriate appendages. Examples of the use of amino acids as apparent chirons are shown in Figure 8.13 [27].

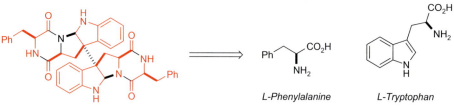

Figure 8.13 Apparent amino acid-type substructures.

8.6.2
Partially hidden amino acids

The substructure in the target molecule cannot be directly related to an amino acid because of a more complex, extended or modified architecture. The same guidelines suggested in section 8.6.1 can be applied except that a hydroxyl group or a halogen may replace the original amino group in the amino acid. In such cases, hydroxy acids and carbohydrates could also be used instead of an amino acid if it is more practical to do so. Examples of the utility of amino acids as partially hidden chirons are shown in Figure 8.14 [28].

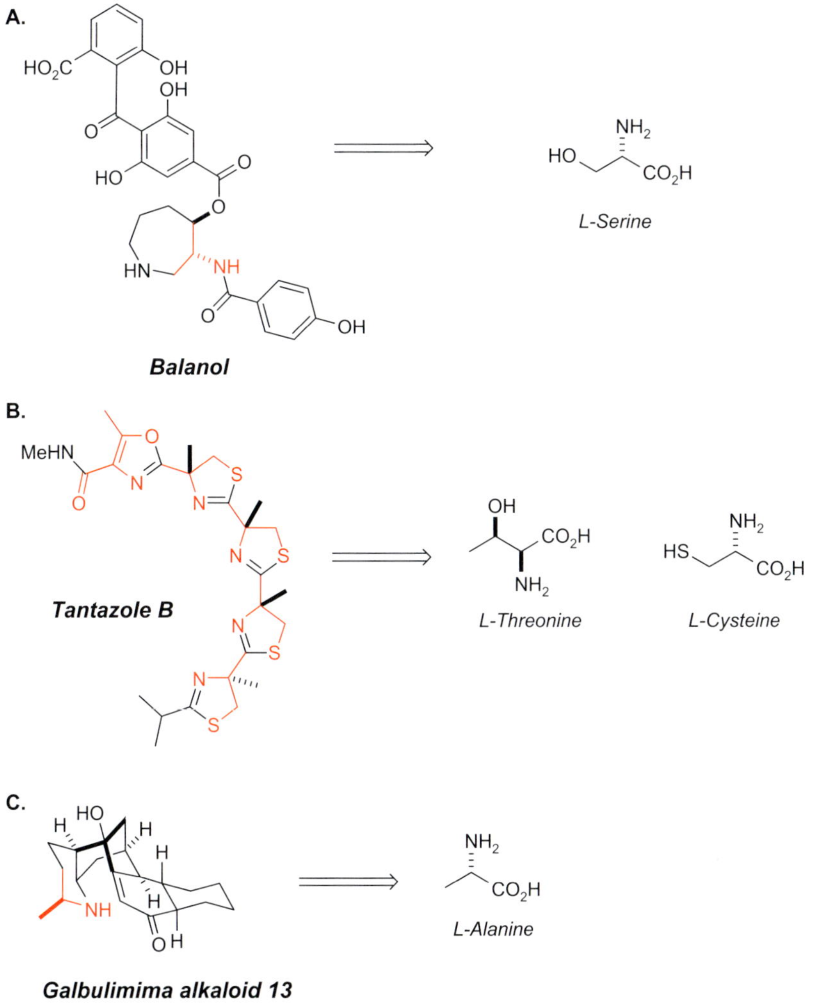
A.
HO2C
OH
OH
O
HO
O
O
HN
NH
O
OH
Balanol
NH2
HO
CO2H
L-Serine
B.
MeHN
O
N
O
S
N
N
S
Tantazole B
N
S
N
S
OH
CO2H
NH2
L-Threonine
NH2
HS
CO2H
L-Cysteine
C.
HO
H
H
H
H
NH
O
H
Galbulimima alkaloid 13
NH2
CO2H
L-Alanine

Figure 8.14 Partially hidden amino acids.

8.6.3
Hidden amino acids

The target molecule may or may not contain a nitrogen atom that can be traced to an amino acid. Even if present, the latter will have to be extensively modified from its native state so as to incorporate its carbon framework, in part or as a whole, in the target substructure while preserving the original stereogenic center.

a. Focus on a nitrogen atom, or a heteroatom, situated on a stereogenic center. Examine vicinal and neighboring functionality. Hydroxyl groups are easier to relate to amino acids as chirons compared to alkyl groups. Carbon substituents can be introduced from α-hydroxy esters through the corresponding sulfonate esters for example.

b. Search the chiral substructure for an embedded nitrogen atom, then assess the location of other functional groups.

c. Extend the analysis in both directions starting from the nitrogen atom, until a reasonable bond disconnection site is located on an sp^2 or sp^3-hybridized carbon atom.

d. Certain benchmarks can now be applied based on prior knowledge of specific amino acid transformations in order to relate a given substructure to a suitable hidden amino acid as a modified chiron.

In considering the total synthesis of perhydrohistrionicotoxin, Winkler and coworkers [29] recognized the hidden 5-carbon framework of L-glutamic acid, being aware that the stereochemistry of the amino acid was convergent with that of the azaspiro center in ring A (Figure 8.15 A). Intermediate **A** was readily prepared from L-pyroglutamic acid by conversion of the carboxylic acid group to a primary bromide, and then extension via a cuprate addition ($R_2 = C_5H_{11}$). The enamine formed with 1,3-cyclohexanedione and the corresponding γ-amino acid (after amide hydrolysis), followed by DCC mediated esterification gave **B**. Winkler's key reaction was predicated upon an alkoxide-induced transesterification/fragmentation of tetracyclic intermediate **D**, which was secured in a [2+2] cycloaddition from **C**. Reduction of the ketone with NaBH$_4$ was highly stereoselective to give the (R)-alcohol, which upon treatment with NaH brought about a transesterification with release of acetone and C−C bond fragmentation to afford lactone **E**. A small price to pay for this highly stereocontrolled synthesis was the obligatory removal of the ketone carbonyl group in **E**, which was done by conversion to the enol triflate and Pd-mediated hydrogenolysis. All five carbon atoms of L-glutamic acid were incorporated in the target structure. Furthermore, the resident chirality in **C** was used to full advantage in generating the remaing two stereogenic centers of the natural product.

An example of a very hidden and intentionally modified amino acid can be found in Mukaiyama's total synthesis of taxol (Figure 8.15 B) [30]. Rather than utilizing precursors with more carbogenic character to construct one or more of the four rings in taxol, a simple amino acid was used to converge with a 3-carbon portion of ring B that harbors the OBz group. It was then systematically modified to

A.

Perhydrohistrionicotoxin
$R_1 = C_4H_9; R_2 = C_5H_{11}$

L-Glutamic acid

A

$R = C_5H_{11}$

1. HCl,
   1,3-cyclohexanedione
2. DCC, MeOH

B

1. LDA, *t*-BuOAc
2. acetone, H⁺

C

hv

D

1. NaBH₄
2. NaH

E

steps

Perhydrohistrionicotoxin

B.

Taxol

L-Serine

A

(Via diazotization of L-serine)

steps

B

MeO₂C

1. LDA
2. PMBCl

C

anti-aldol

1. Dibal-H
2. MgBr₂

D

R = TBS

steps

E

1. SmI₂
2. Acetylation/
   elimination

F

Figure 8.15 Hidden amino acids.

ultimately give a functionalized eight-membered ring. Thus, L-serine was converted to L-glyceric acid ester **A**, which was transformed into L-glyceraldehyde derivative **B**. An *anti*-aldol directed condensation with 2-methylpropionic ester gave **C**, which was converted to the corresponding aldehyde, and then subjected to a Mukaiyama aldol condensation in the presence of $MgBr_2$ to give the all-*anti* adduct **D**. Further steps led to **E** in preparation for a SmI_2-induced Barbier-type reaction, eventually leading to the fully functionalized ring *B* precursor **F** of taxol. Thus, the Mukaiyama strategy relied heavily on utilizing substrate-based stereocontrol in the two aldol reactions, which took advantage of the resident chirality in the *anti*-aldol product (**B** to **C**). The choice of L-serine was no doubt due to the unavailability of **A** and **B** with L-configurations as starting materials in large quantites from natural sources (for details, see Chapter 16, section 16.8.7).

8.7
Locating Carbohydrate-type Substructures

Carbohydrates are obvious starting materials for oxacyclic compounds such as chiral, non-racemic tetrahydrofurans, tetrahydropyrans, and lactones as well as for acyclic polyhydroxylated compounds. In view of their polyfunctional nature, end-differentiated oxidation states, stereochemical diversity, and carbon chain lengths varying between 3 and 7 atoms, carbohydrates can be used as chirons for a large number of target compounds. The possibility of introducing other heteroatoms such as nitrogen, sulphur, and halogens by inversion or retention of configuration, as well as branching with carbon nucleophiles, greatly expands the utility of carbohydrates as versatile precursors in total synthesis. Chapter 11 showcases selected examples of total syntheses where carbohydrates have been used as starting chirons. When considering the use of carbohydrates, the following guidelines are useful:

a. Avoid extensive deoxygenations.
b. Avoid degradative procedures that conserve only one or two of the original stereogenic centers.
c. Avoid excessive stereochemical adjustment.
d. Avoid multiple protecting groups unless they are robust and can remain through the synthesis *en route* to advanced intermediates.
e. Make good use of the carbon framework and existing functionality.
f. Take advantage of conformational bias and stereoelectronic effects to maximize predictive outcomes in chemical transformations.

8.7.1
Patterns and shapes

Carbohydrates are among the most abundant natural product resources, especially as plant constituents. Monosaccharides such as D-glucose, D-galactose, and D-arabinose exist as cyclic hemiacetals that can be transformed into multiple oxidized or reduced products (see Chapter 7, section 7.2). The potential of carbohydrates in total synthesis can be expanded if they are viewed as virtual functionalized acyclic carbon compounds that can be made to adopt the contours of carbon frameworks of target molecule substructures. Thus, they can be chemically modified as conformationally defined alkyl hexopyranosides, for example, using well-known reactions to introduce diverse functionalities replacing one or more hydroxyl groups. Once the desired level of modification has been achieved, the resulting cyclic chiron can be converted to an acyclic derivative such as a dithioacetal, alditol, or aldonic acid for example. These can now be further used as acyclic precursors in a planned synthetic sequence toward the target structure or a segment thereof. In this regard, carbohydrates offer a great deal of versatility as precursors, provided that other methods of access are not vastly superior.

Stripped down frameworks corresponding to 3 to 6 carbon atoms that can be formally derived from carbohydrates are shown in Figure 8.16. The "shapes" correspond to contours of a variety of chiral, non-racemic substructures in target molecules. As such, these "shapes" do not reveal their carbohydrate ancestry and may not be first-choice chirons. However, the potential of introducing a variety of substituents through systematic modifications of cyclic carbohydrate precursors corresponding to some of these "shapes" prior to converting them into acyclic equivalents makes them useful chirons.

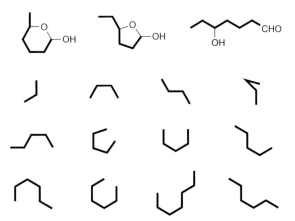

Figure 8.16 Locating carbohydrate-type substructures (patterns and shapes).

8.7.2
The "Rule of Five"

Locating a hidden carbohydrate-type substructure in a target molecule may be facilitated by considering the "rule of five" [1]. Basically, it involves selecting an sp, sp^2, or sp^3-hybridized carbon atom in a given target structure, then moving five bonds away and looking for a heteroatom, such as oxygen, regardless of what functionality lies in between. These two internal reference points correspond to the anomeric C-1 and the C-5 oxygen atoms in a hexopyranose structure. A "rule of four" can be considered for a pentose precursor. Once the two reference points are located and a cyclic carbohydrate derivative is envisaged as a precursor, then the intervening substituents in the target substructure are considered. The carbohydrate derivative can now be systematically manipulated to introduce the desired substituents that correspond to the target substructure. Due consideration should be given to the guidelines mentioned above, in order to assess the suitability of using a carbohydrate precursor *vis-à-vis* a chiron from another class or an alternative molecule prepared by an asymmetric process.

Application of the "rule of five" (and "four") to carbohydrate precursors is illustrated in Figure 8.17 (Note: target numbering corresponds to atom numbers in the carbohydrate starting material). As mentioned above, the necessity to introduce carbon substituents or deoxygenation may detract from the use of carbohydrate precursors. In this regard, the syntheses of thienamycin from a 3-amino-2,3-dideoxy-D-*xylo*-pyranose by Hanessian and coworkers [31], or the acyclic C-1−C-6 chain of neooxazolomycin from D-glucose by Kende and coworkers [32] may be only of pedagogical interest, especially considering that they were reported some 20 years ago. The amino acid substructure of calyculin by the Armstrong group [33a] illustrates the "rule of four." Thus, the 5-carbon framework of D-ribose harbors two of the required hydroxyl groups with stereochemical convergence within the target molecule substructure. The configuration of the C-4 hydroxyl group was inverted in the process of introducing the dimethylamino group.

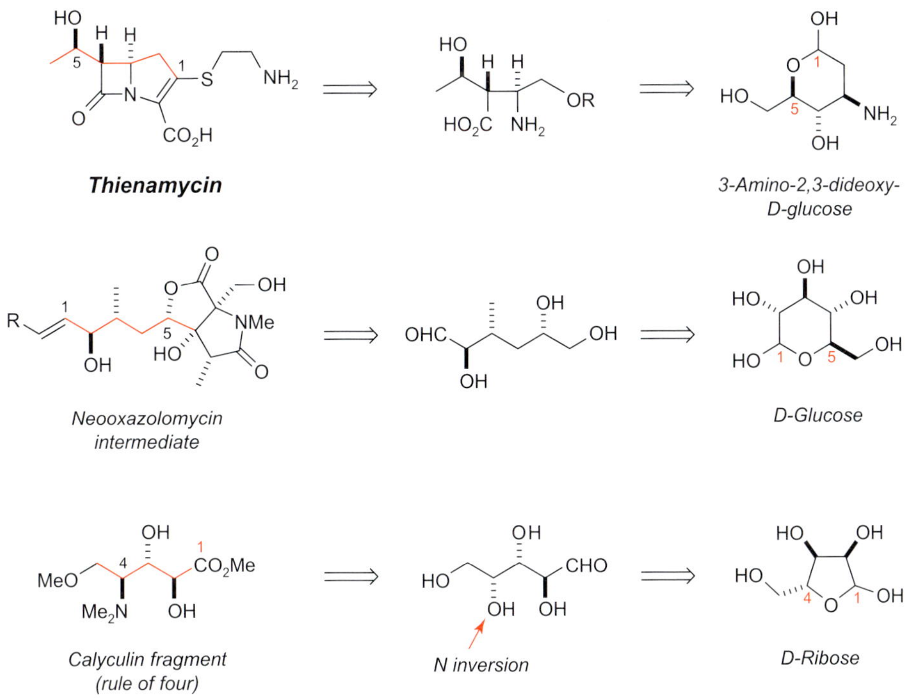

Thienamycin

3-Amino-2,3-dideoxy-
D-glucose

Neooxazolomycin
intermediate

D-Glucose

Calyculin fragment
(rule of four)

N inversion

D-Ribose

Figure 8.17 Rule of five and four (target numbering corresponds to atom numbers in the carbohydrate precursor).

8.7.3
Apparent carbohydrates

The carbohydrate-type motif may be quite evident in the target structure in the
form of a lactone, a cyclic ether, or as an acyclic chain with hydroxyl or related
substituents. The synthesis of thromboxane B_2 from D-glucose [8, 9] is a classic
example of framework and stereochemical convergence, especially with regard
to C-5, which is a "protected" hydroxyl group in a methyl D-glucopyranoside
framework (rule of five, Figure 8.18 A). Although the steps to the intended
target molecule involved deoxygenation at C-2, inversion at C-3, branching at
C-4, and extension at C-6, the intermediates were amenable to systematic and
stereocontrolled manipulation.

The synthesis of PA-48153 C (pironetin) from D-glucose also takes advantage
of the resident chirality at C-5, which is convergent with the target molecule
(Figure 8.18 B) [34]. Dehydration of the trans-diol and branching at C-4 of the
D-glucopyranose framework were required to reach a desired level of functional convergence within the lactone portion of the target molecule. Note that the (S)-Roche
acid was used to secure a 3-carbon subunit in the side-chain. The tetrahydrofuran
ring in laurallene has been synthesized from D-ribonolactone (Figure 8.18 C) [35].

The enantioselective synthesis of solandelactone E featured D-mannitol as a source of the C-6–C-8 segment (Figure 8.18 D) [36]. The vicinal diol subunit in *ent*-palmerolide A was readily constructed from D-arabitol (Figure 8.18 E) [37a].

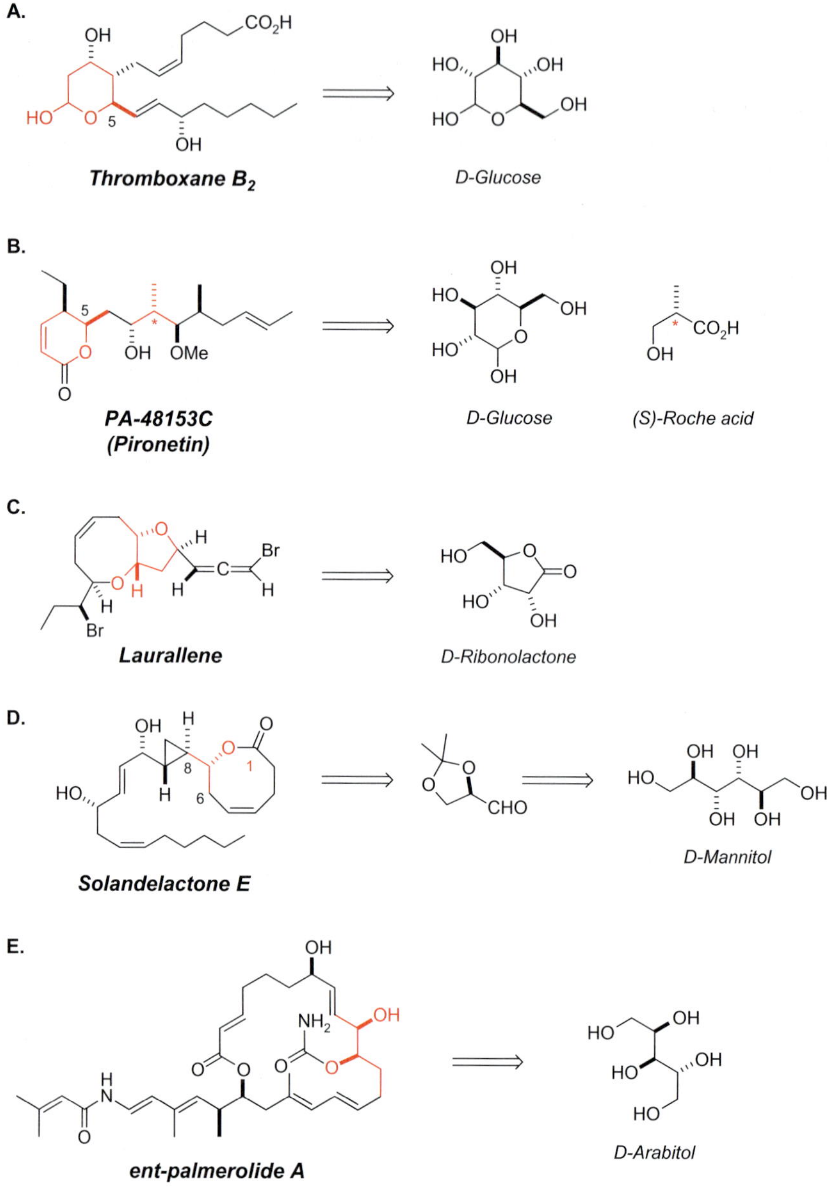

A.

Thromboxane B$_2$

D-Glucose

B.

PA-48153C
(Pironetin)

D-Glucose

(S)-Roche acid

C.

Laurallene

D-Ribonolactone

D.

Solandelactone E

D-Mannitol

E.

ent-palmerolide A

D-Arabitol

Figure 8.18 Apparent carbohydrate substructures.

8.7.4
Partially hidden carbohydrates

Here, the carbohydrate-type substructure may be part of an appendage or a segment in the carbon framework of the target molecule. Bond disconnections can be made to converge with a carbohydrate-derived chiron and the "rule of five" (or "four") may apply. The oxacyclic ring *C* in nanaomycin D has the features of a *C*-glycoside precursor when viewed with a carbohydrate perspective (Figure 8.19 A) [38]. Indeed, the ʟ-configuration of the stereogenic carbon next to the ring oxygen bearing a *C*-methyl group would suggest ʟ-rhamnose as precursor. However, all of the remaining hydroxyl groups had to be extensively modified by elimination at C-3–C-4, inversion at C-2, and branching at C-1 (carbohydrate numbering). The readily available intermediate **A** was efficiently transformed into **B** in a two-step sequence. Condensation of **B** with the lithium anion of sulfone **C** furnished the tricycle **D**, which was transformed to **E** via a stereoselective reduction and acid hydrolysis. Wittig reaction afforded the lactone **F** (and the ring-opened diastereomer), *en route* to nanaomycin D. The spiroacetal portion of augustamine originated form ᴅ-erythronolactone, which provided an end-differentiated 4-carbon chiron with matching stereochemistry [39]. Appropriate chain extensions at both extremities gave a functionalized intermediate **A**, which was subjected to a key

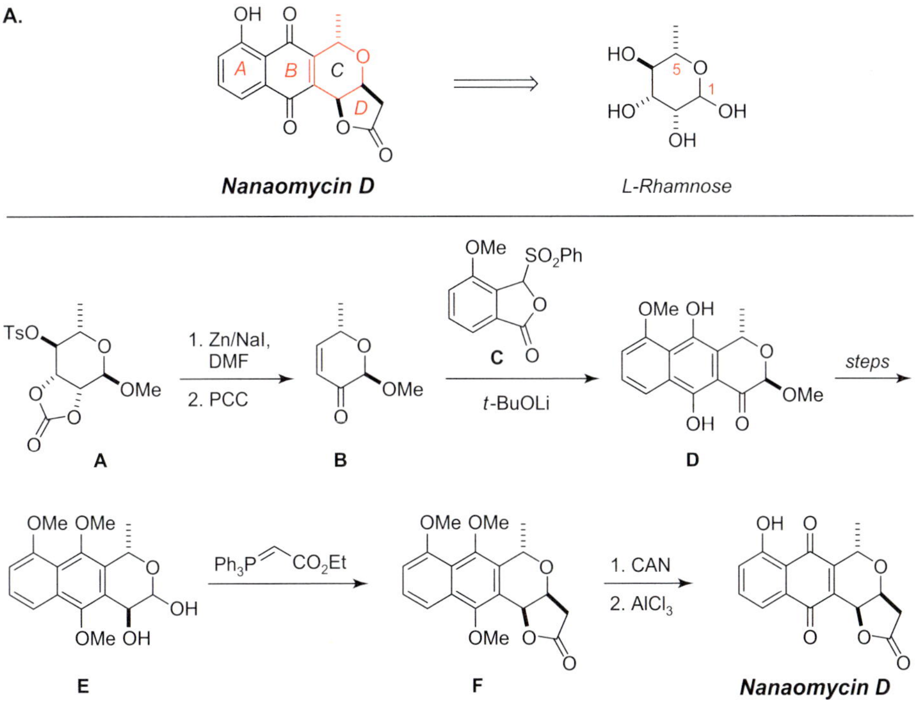

Figure 8.19 Partially hidden carbohydrate substructures.

B.

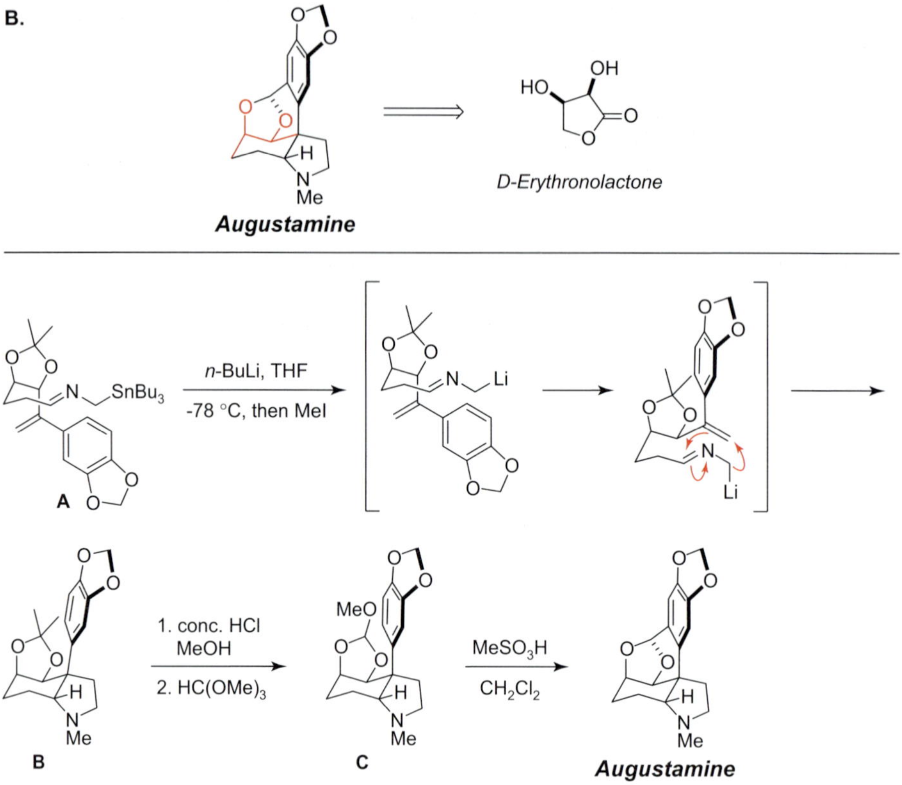

Augustamine

D-Erythronolactone

B

C

Augustamine

Figure 8.19 *(continued)*

2-azaallyl anion mediated annulation reaction to give **B** (Figure 8.19 B). Conversion of the acetal in **B** to an orthoester **C**, set the stage for an intramolecular Friedel-Crafts reaction to give augustamine.

8.7.5
Hidden carbohydrates

Examples of exploiting hidden carbohydrate-type substructures in synthesis are rare unless one or more hydroxyl groups can be located in the target structure. In the absence of a hydroxyl group (or minimal alcohol functional groups), it is not practical to consider a carbohydrate precursor in total synthesis. The "rule of five" helps to find internal reference points to correlate a particular substructure with a carbohydrate precursor that may need extensive functional group manipulation. An example which attracted attention to the use of carbohydrates as precursors in total synthesis over thirty years ago was the D-glucose approach to the macrolide aglycone erythronolide A (Figure 8.20) [40]. Applying the "rule of five" revealed the relationships with the hydroxyl group at C-5 and the lactone ring oxygen at C-13 respectively. Drawing the carbon framework with its substituents in a "pseudosugar" perspective **A**, reveals two methyl hexopyranoside precursors **B** and **C** that correspond to the C-1–C-6 and C-9–C-13 segments of erythronolide A respectively. In practice, systematic introduction of C-methyl groups by manipulating the hydroxyl groups in a predictable manner starting with a conformationally defined methyl α-ᴅ-glucopyranoside led to **B** and **C**. While the absolute configuration at C-5 of ᴅ-glucose corresponds to that at C-13 of the macrolide, that at C-6 corresponds to an "ʟ-sugar," hence the need to invert the configuration of the carbon atom bearing the hydroxymethyl group. Since these early studies, the field of macrolide synthesis has undergone a major revolution with regard to novel, stereocontrolled methods of elaborating polyketide substructures, notably using stereocontrolled aldol reactions [40c].

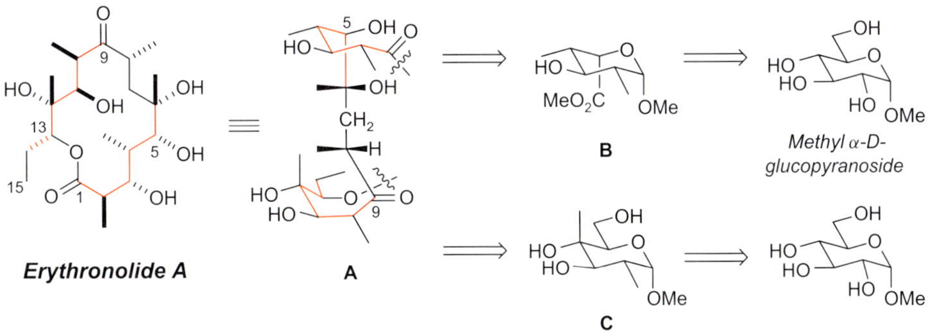

Figure 8.20   The ᴅ-glucose approach to erythronolide A.

Three examples of hidden carbohydrate-type substructures in target molecules are shown in Figure 8.21. Thus, the 4-carbon tricarboxylic acid motif in the spiroacetal subunit of zaragozic acid C was elaborated starting with ᴅ-erythrose [41]. In the process, only the 4-carbon framework and the original C-2 hydroxyl group remained intact (Figure 8.21 A, see details in Chapter 11, section 11.8). A segment comprising the ether oxygen in prelaureatin was prepared from ᴅ-galactose (Figure 8.21 B) [42].

The aminocyclitol ring C in lycoricidine originated from a synthetic cyclohexenone precursor, which in turn, was prepared from ᴅ-glucose (Figure 8.21 C) [43]. Intermediate **A** was subjected to dehydrohalogenation to give enol ether **B**. Intramolecular aldol cyclization mediated by Hg(OTFA)₂ produced **C**, which was β-eliminated, reduced, and O-protected to give **D**. Attachment of the 2-bromo-3,4-methylenedioxy carboxylic acid unit afforded **E**, which underwent an intramolecular Heck-type reaction to give precursor **F**. Further steps led to lycoricidine.

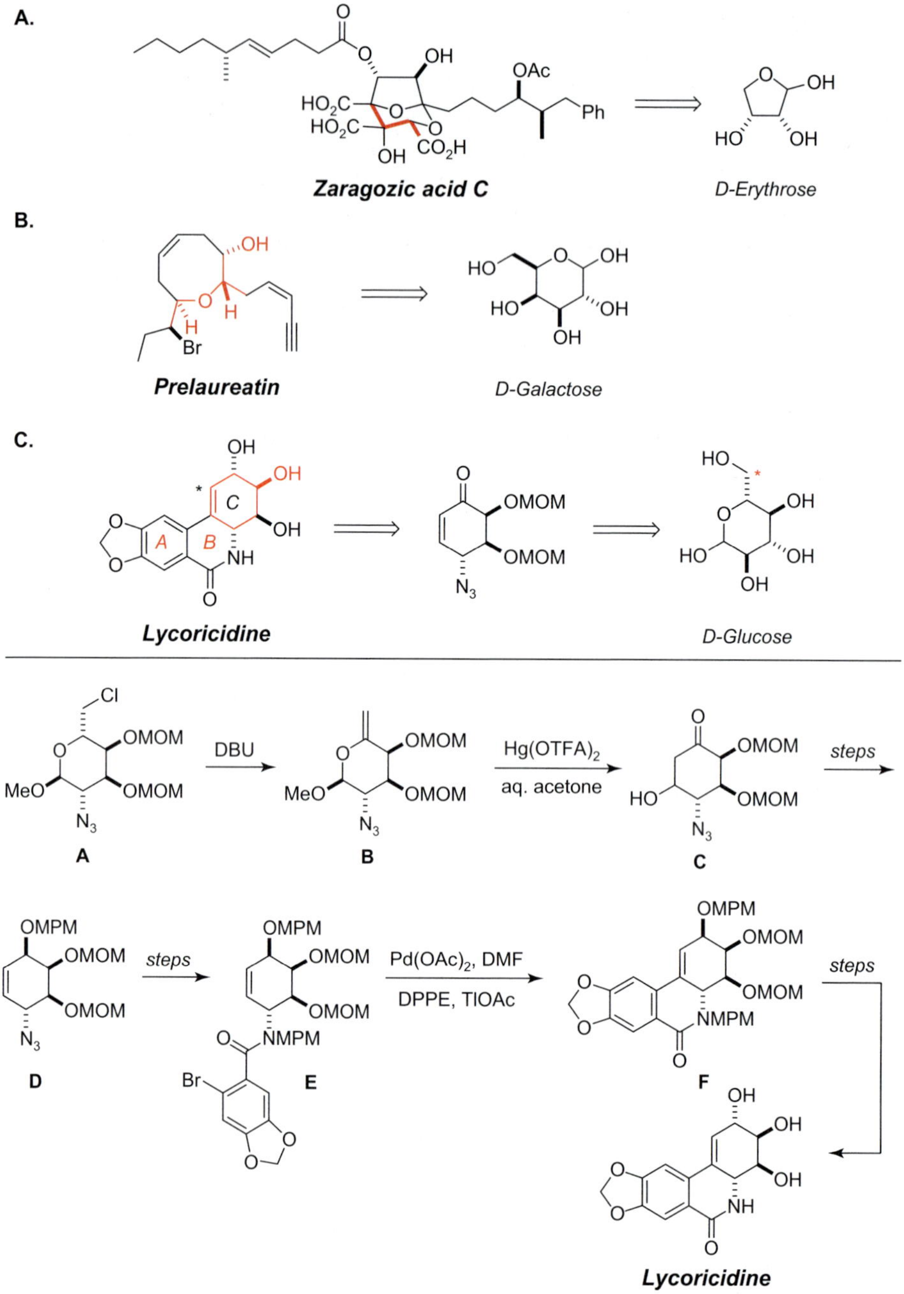

Figure 8.21 Hidden carbohydrate substructures.

8.8
Locating Hydroxy Acid-type Substructures

Hydroxy acids are ideal precursors to target substructures that contain one or two contiguous stereogenic centers, preferably with hydroxyl substituents. Mono-carboxylic acids, such as lactic and mandelic acids, offer excellent choices for substructures in which the terminal methyl or phenyl groups can be recognized in a target molecule. Displacement of the hydroxyl group by heteroatoms, and even carbon nucleophiles, provides entry into other α-substituted carboxylic acids. The commercially available (2R)-2-methyl-3-hydroxypropionic acid ((R)-Roche acid) offers an excellent end-differentiated hydroxy acid harboring a C-methyl group that can be incorporated in segments of target molecules with "up" or "down" orientations depending on which end of the acid is elaborated upon. Enolate chemistry using the dianion of (3R)-3-hydroxybutyric acid or D- or L-malic acids allows for the introduction of alkyl or heteroatom functionality with a high degree of stereocontrol, thereby expanding their utility as 4-carbon hydroxy acid chirons. Branching can also be introduced relying on enolate chemistry, thereby generating α-tertiary hydroxy carboxylic acids. L-Citramalic acid is a natural representative of this class of α-branched dicarboxylic acids. The C_2-symmetry in (R,R)- and (S,S)-tartaric acid provides ample opportunity to exploit them as versatile chirons in total synthesis. Their primordial importance as ligands in catalytic reactions is amply documented (see Chapter 7, section 7.3).

The presence of a hydroxyl group adjacent to an sp^2-hybridized carbon atom in a target molecule is a good starting point to locate a potential hydroxy acid precursor. Depending on the nature of the substitution and type of carbon skeleton, the choice of a particular hydroxy acid may become evident upon visual inspection of the intended target molecule. Chapter 12 will highlight the use of hydroxy acids as starting chirons in selected total syntheses.

8.8.1
Apparent hydroxy acids

An isolated hydroxyl group, a vicinal diol, or an oxygen atom on a stereogenic carbon may be easily seen in the target molecule, and related to a hydroxy acid precursor depending on the nature of neighboring functionality. When considering a hydroxy acid as a starting chiron for total synthesis, the following guidelines may be useful to consider:

a. Focus on a hydroxyl group or an oxygen atom on a stereogenic carbon in the target substructure. Do the same for a vicinal diol.
b. Consider the nature of the carbon skeleton to the "left and right" of the hydroxyl group, or its modified equivalent. An sp^2-hybridized carbon such as an olefin or a carbonyl group may coincide with the position of the original carboxyl group in the hydroxy acid.

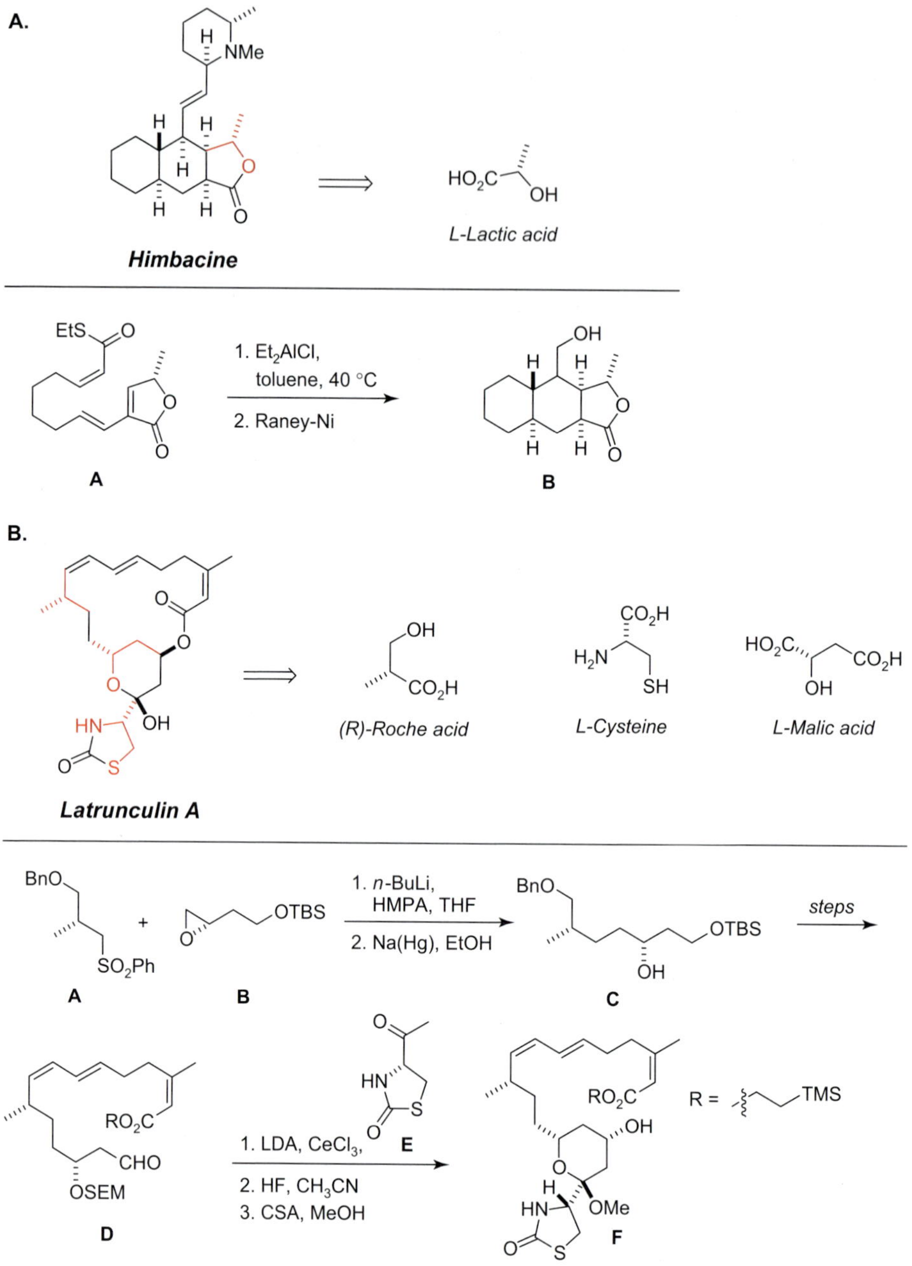
A.
H
NMe
H
H
H
O
H
H
O
Himbacine
HO2C
OH
L-Lactic acid
EtS
O
O
O
A
1. Et2AlCl,
toluene, 40 °C
2. Raney-Ni
OH
H
H
O
H
H
O
B
B.
O
O
O
HN
OH
O
S
Latrunculin A
OH
CO2H
(R)-Roche acid
CO2H
H2N
SH
L-Cysteine
HO2C
CO2H
OH
L-Malic acid
BnO
SO2Ph
A
+
OTBS
O
B
1. n-BuLi,
HMPA, THF
2. Na(Hg), EtOH
BnO
OTBS
OH
C
steps
RO2C
CHO
OSEM
D
O
HN
S
O
E
1. LDA, CeCl3,
2. HF, CH3CN
3. CSA, MeOH
RO2C
OH
O
H
HN
OMe
O
S
F
R =
TMS

Figure 8.22 Apparent hydroxy acid substructures.

c. Identify the suitability of a mono- or dicarboxylic hydroxy acid as a precursor in order to best match the configuration of a stereogenic center and skeletal convergence with a substructure in the target molecule.
d. Allow for α-branching using enolate chemistry.

The fused lactone ring in himbacine can be converged with L-lactic acid (Figure 8.22 A) [44]. Chemical modification led to a triene lactone **A**, which underwent an intramolecular Diels-Alder reaction (**A** to **B**) that established the desired stereochemistry at the ring junction in the tricyclic system. The five stereogenic centers in latrunculin A were derived from the (R)-Roche acid, L-malic acid, and L-cysteine (Figure 8.22 B) [45a]. Addition of the sulfone anion of **A** to the epoxide **B** led to **C** after desulfonylation with sodium amalgam. Extension using Wittig chemistry provided aldehyde intermediate **D**, which would undergo an aldol condensation with **E**, followed by acetal formation to give the advanced intermediate **F**.

8.8.2
Partially hidden hydroxy acids

In such cases, the hydroxy acid framework may be partially embedded in the target substructure, and the presence of other vicinal non-hydroxyl or non-oxygen-containing substituents may detract from locating it. The following guidelines may be useful:

a. Locate a hydroxyl group or an oxygen atom on a stereogenic carbon atom as part of a 3- or 4-carbon subunit in the target molecule.
b. Establish the nature of vicinal substituents and the location of potential extension sites that could correspond to the original carboxyl group of a hydroxy acid-type precursor.
c. Assess the extent and feasibility of chemical modifications to be done on the original hydroxy acid in order to converge it with the intended target substructure. Consider using enolate chemistry to introduce α-branching.
d. As always, consider practicality and merits compared to other types of precursors.

The C_2-symmetric structure of elaiolide, the aglycone of elaiophylin, contains several propionate-derived subunits, which were sequentially elaborated by exploiting highly diastereoselective auxiliary-mediated aldol condensations as reported by the Evans group (Figure 8.23 A) [46]. The C-14 *C*-ethyl group in the tetrahydropyran rings was derived from (3R)-3-hydroxybutyric acid ester **A** relying on a Fráter-Seebach [47] dianion α-ethylation to give the *anti*-product **B**. Transformation to the aldehyde, and substrate-controlled stannylcrotylation afforded the *syn*-alcohol **C**, which was in turn converted to ethyl ketone **D** in preparation for a "double" enolate aldol condensation with the 9,9′-dialdehyde **E**. Note that the carbonyl group in **D** would also be the site of a kinetically-controlled cyclic hemiacetal formation *en route* to the product. Although closer inspection of the hemiacetal subunits easily locates the ring oxygen, and the possible emergence of a hydroxy acid as a precursor, knowledge of the stereoselective alkylation of the dianion of (3R)-3-hydroxybutyric ester makes it a chiron of choice.

A.

Elaiolide

(3R)-Hydroxybutyric acid
(C15 and C15')

Figure 8.23 Partially hidden hydroxy acid substructures.

In their synthesis of rhizoxin, Ohno and coworkers [48] recognized the (*S*)-Roche ester (**A**) as a suitable precursor for elaboration at both of its extremities to produce intermediate **F** (Figure 8.23 B). Thus, the β-ketophosphonate **B** was converted to enone **C**, which was stereoselectively reduced relying on chelation and substrate-controlled delivery of hydride. The *O*-protected intermediate **D** was further elaborated to **E** and **F**. The lactone portion of rhizoxin was synthesized starting with a 3-substituted diethyl glutarate intermediate by sequential treatment with pig liver esterase and reduction of the desymmetrized monoester to the hydroxy acid. Further elaboration led to **G**, which underwent a highly stereocontrolled double regio- and stereoselective hydroboration to give the diol **H**, and then converted to sulfone **I**. The (*S*)-Roche ester (**A**) could have also been a good precursor to the 2-carbon sulfone appendage in **I**, considering the matching stereochemistry of the *C*-methyl group. This option was in fact considered by J. D. White and coworkers [49] in their studies toward the total synthesis of rhizoxin D, the bis-des-epoxy natural analogue of rhizoxin.

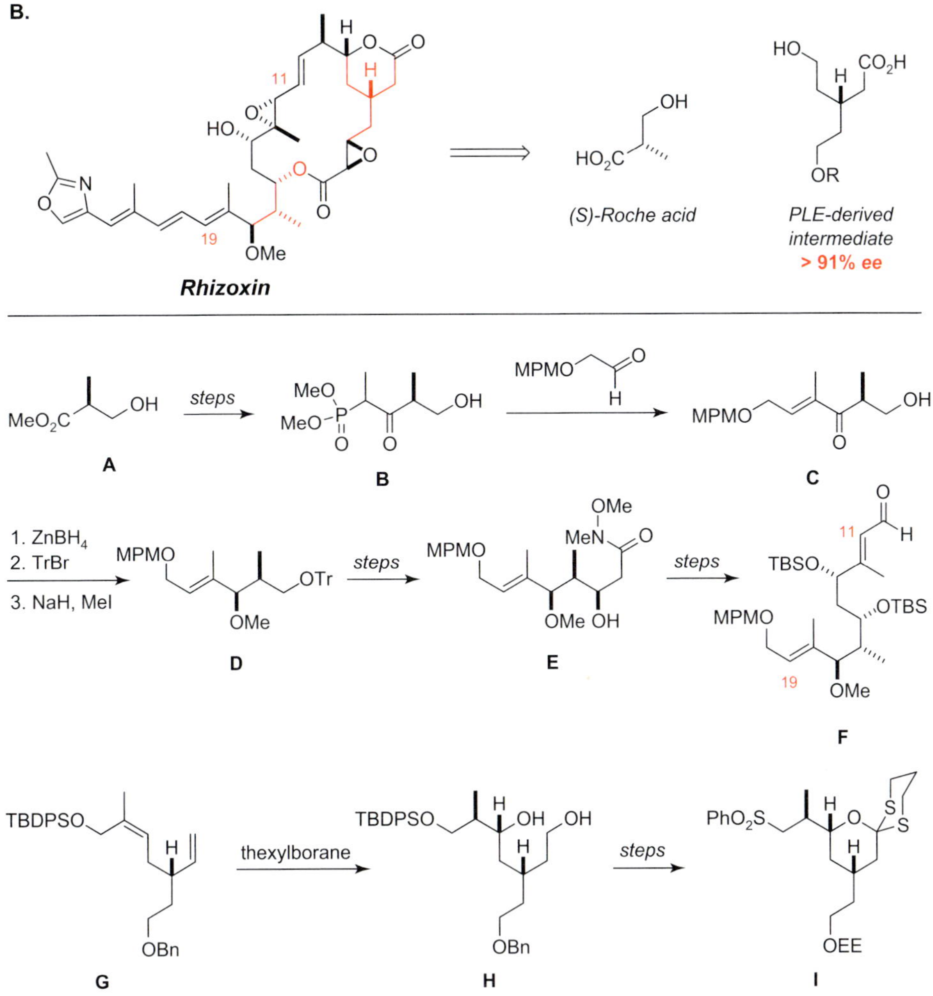

B.

Rhizoxin

(S)-Roche acid

PLE-derived
intermediate
> 91% ee

Figure 8.23 (*continued*)

8.8.3
Hidden hydroxy acids

The original hydroxy acid motif may be deeply imbedded or disguised as a surrogate in the substructure of the target molecule. It may also be transformed into an amino group, or even a carbon appendage such as an alkyl group. The search criteria outlined above for partially hidden hydroxy acids can be applied with greater latitude. An allylic alcohol derived from an α-hydroxy acid or aldehyde may be used in a Claisen or related [3,3]-sigmatropic rearrangement/reaction. Transfer of chirality to the distal carbon with loss of the original hydroxyl-bearing stereogenic center also generates an olefin.

A classical example of such an "acetate transfer" can be found in Boeckman's total synthesis of indanomycin (Figure 8.24 A) [50]. The *C*-methyl group in the side-chain of ring *C* can be easily related to the (*R*)-Roche acid. However, the ethyl group attached to ring *C* in the precursor to the penultimate intramolecular Diels-Alder cyclization was installed as part of a Johnson-Claisen rearrangement protocol from *cis*-allylic alcohol **A**. Treatment with trimethyl orthoacetate and heating in xylenes in the presence of propionic acid afforded the ester **B**, which was further extended to give **C**. Wittig coupling with ring *A* intermediate **D** gave **E**, which was subjected to an intramolecular Diels-Alder reaction to give **F**. This is another early example of a self-immolative process in which the original stereogenic carbon atom was transferred to give a *trans*-olefin in a [3,3]-sigmatropic rearrangement.

The two *C*-methyl groups in leinamycin can be traced to D-alanine and D- or L-lactic acid (Figure 8.24 B). In this respect, the chirons are actually of the apparent type. However, there is more than meets the eye in Fukuyama's plan for the substructures that harbor these *C*-methyl groups [51]. The C-17 methyl group originating from L-lactic acid by an S_N2 inversion of an alkynyl (5*S*)-hydroxyethyl thiazole **A** with azide ion to give **B**. Thus, L-lactic acid represents a hidden chiron in this case.

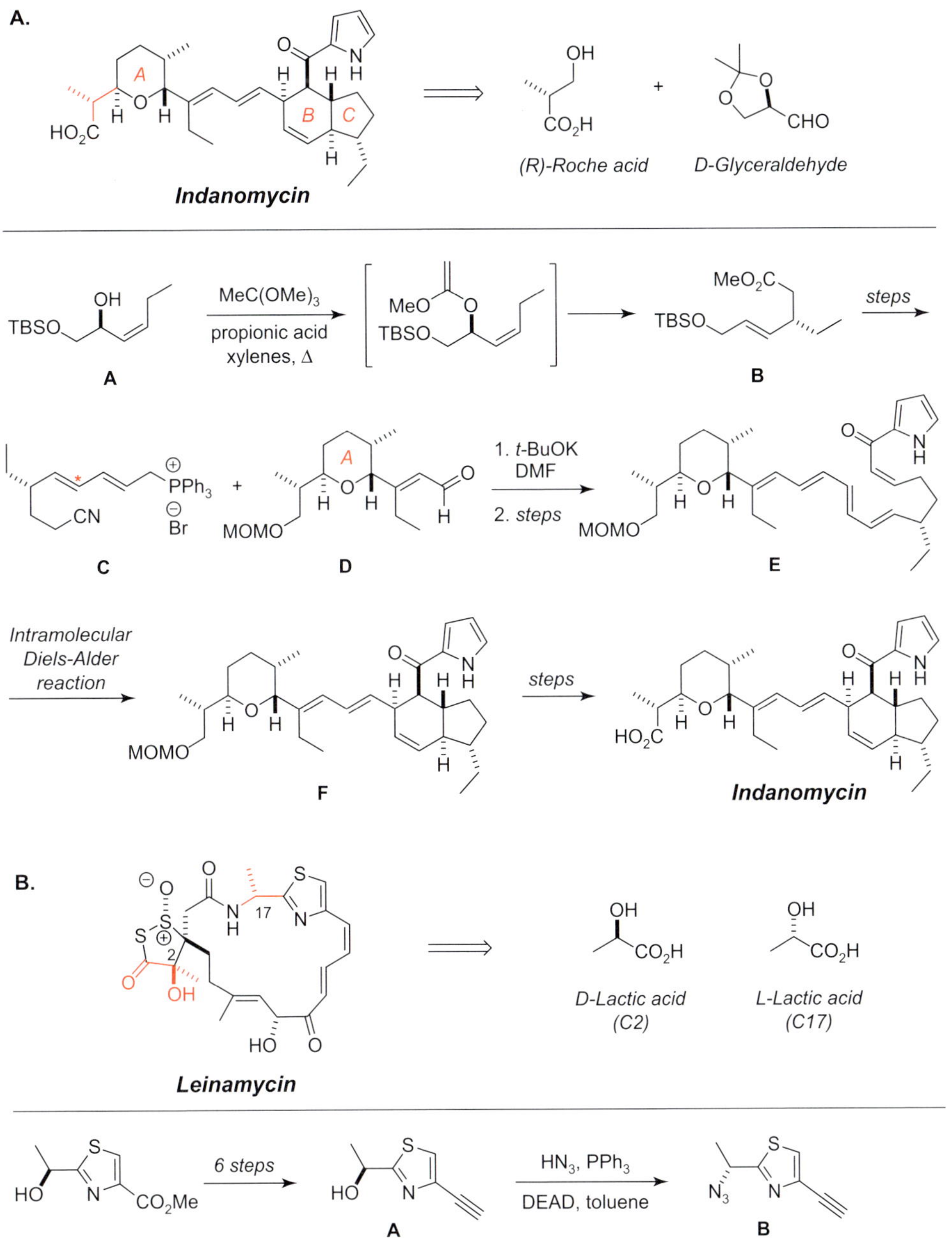

A.

Indanomycin

(R)-Roche acid   D-Glyceraldehyde

A

B

C   D   E

Intramolecular
Diels-Alder
reaction

F   *Indanomycin*

B.

Leinamycin

D-Lactic acid   L-Lactic acid
(C2)   (C17)

A   B

Figure 8.24 Hidden hydroxy acid substructures.

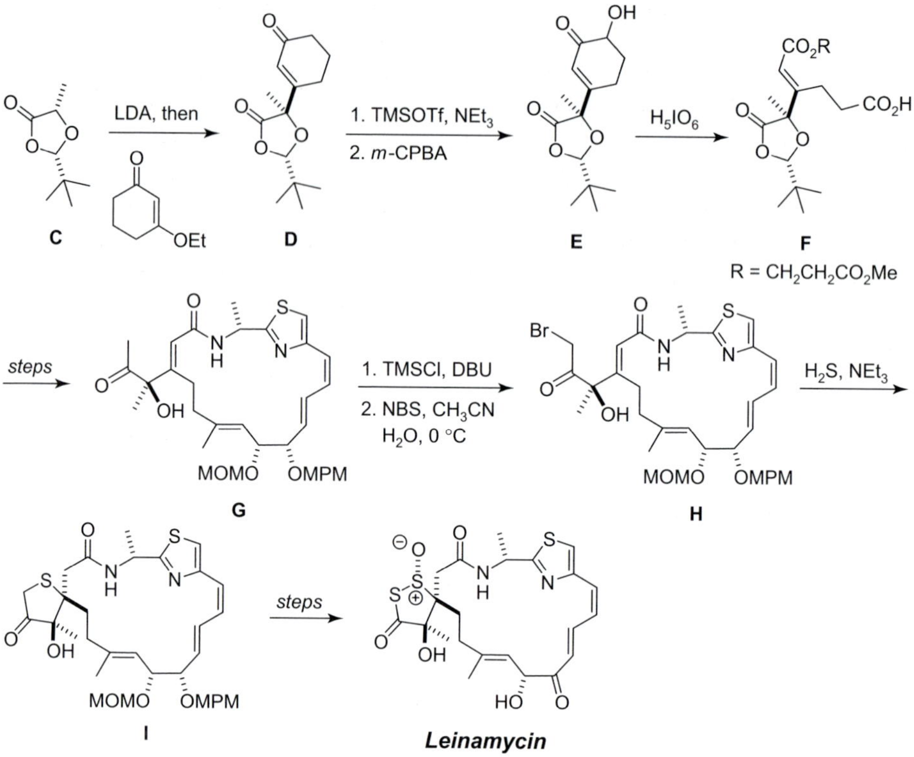

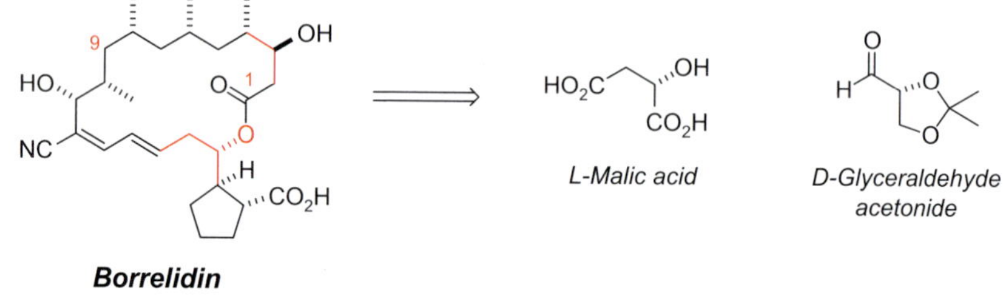

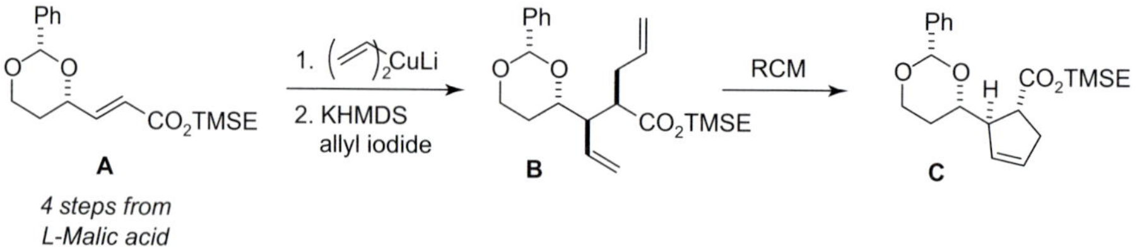

Figure 8.24 *(continued)*

The elaboration of the tertiary alcohol at C-2 in leinamycin involved an unusual strategy that also allowed the installation of an appropriate end-differentiated dicarboxylic acid as an intermediate. Thus, treatment of Seebach's dioxolanone (**C**) with LDA, and subsequent Michael addition to 3-ethoxy-2-cyclohexenone led to **D** as a single isomer. Having secured the correct stereochemistry of the tertiary hydroxyl and *C*-methyl substituent groups at C-2, the enone was carried through a number of steps, first giving α-hydroxy cyclohexenone **E**, which was subjected to oxidative cleavage to give diester **F**. Further steps afforded the macrocyclic lactam **G**. In the process, the original carboxyl group of D-lactic acid was converted to the corresponding methyl ketone. Bromoketone **H** was transformed to the 3-ketotetrahydrothiophene intermediate and further elaborated upon via an oximino ketone, to eventually introduce the unusual dithiolane oxide ring. The three carbon atoms of D-lactic acid, including the hydroxyl and *C*-methyl groups, remained in the C-1–C-2 framework of leinamycin. The original *C*-methyl group became part of the tertiary alcohol bearing stereocenter at C-2, while C-3 originated from the Michael addition reaction. In an independent approach to the macrocyclic lactam portion of leinamycin, Pattenden and Thom [52] utilized a thiazole intermediate derived from *N*-Boc-D-alanine.

In the Hanessian total synthesis of borrelidin [53a], the C-3 hydroxyl group was derived from L-malic acid, encompassing the C-1–C-4 substructure (Figure 8.24 C, see details in Chapter 16, section 16.4.8). The three alternating *syn*-related *C*-methyl groups in the C-1–C-9 substructure of borrelidin were elaborated starting with D-glyceraldehyde acetonide. The vicinally substituted cyclopentane carboxylic acid was prepared from L-malic acid, which was elaborated to **A**. A series of reactions involving substrate controlled conjugate addition and enolate alkylation gave **B**, which underwent ring-closing metathesis to the corresponding cyclopentene **C**. All the carbon atoms of L-malic acid were used to generate the required precursor. The hydroxyl group in L-malic acid became the macrolactone ester oxygen atom of borrelidin.

8.8.4
The Roche acid – a unique C-Methyl chiron

The availability of (2 *R*)-2-methyl-3-hydroxypropionic acid ((*R*)-Roche acid) by microbiological oxidation of 2-methylpropionic acid has greatly facilitated the elaboration of a multitude of substructures in target molecules bearing an isolated *C*-methyl group. Because of the possibility to perform orthogonally directed transformations with the (*R*)-Roche acid, its *C*-methyl group can be made to converge with either (*R*)- or (*S*)-configurations as required in the substructure. An early example of such a versatile use was demonstrated by Evans and coworkers in their synthesis of calcimycin (see Chapter 6, section 6.2.4). Since then, the (*R*)-Roche acid has been used as a starting chiron in numerous total syntheses, especially of complex natural products. In many of these, the "starter" *C*-methyl motif has played a major role in internal asymmetric induction to generate new stereogenic centers in subsequent reactions. Examples of syntheses where the (*R*)-Roche acid has served some "big clients" during the last decade are shown in Figure 8.25 [54] (see also Chapter 12).

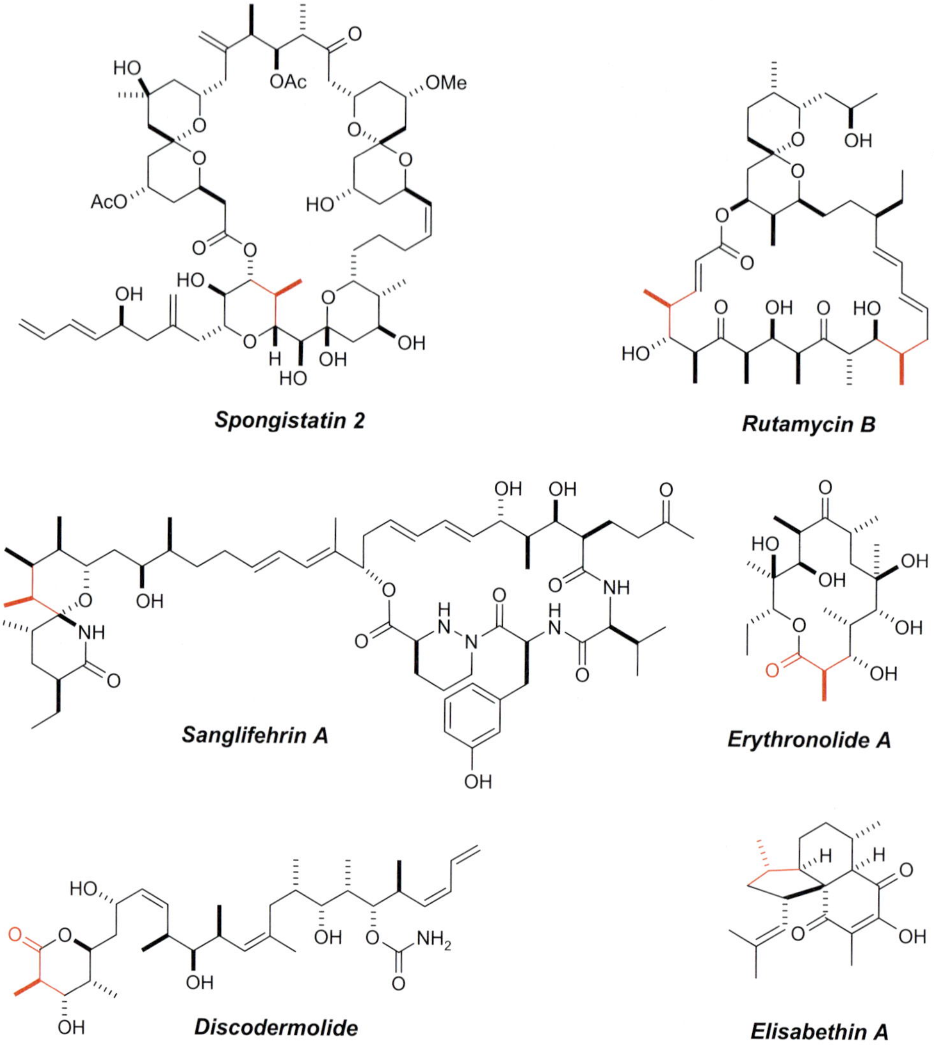

Spongistatin 2

Rutamycin B

Sanglifehrin A

Erythronolide A

Discodermolide

Elisabethin A

Figure 8.25 The Roche acid and its "big clients".

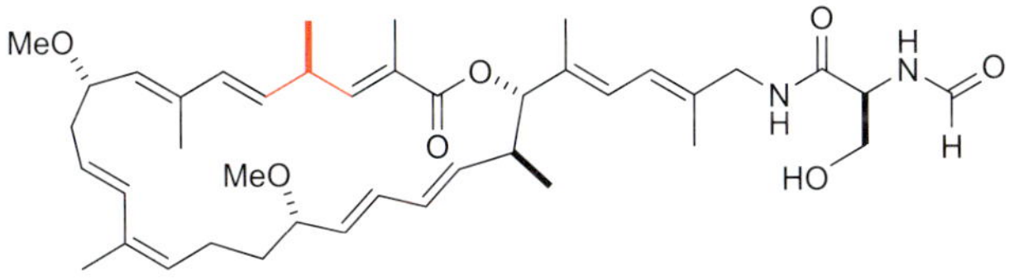

Iejimalide B

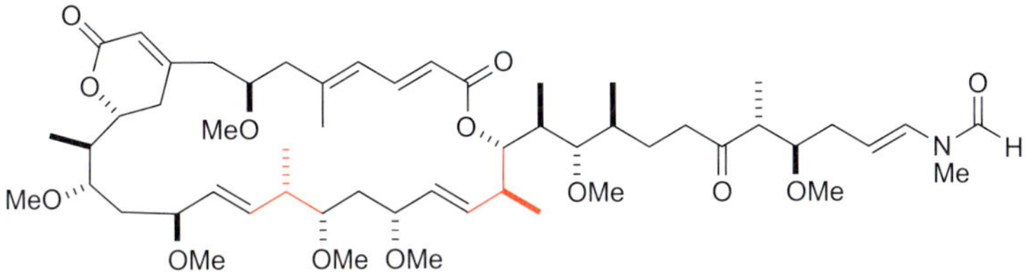

Reidispongiolide A

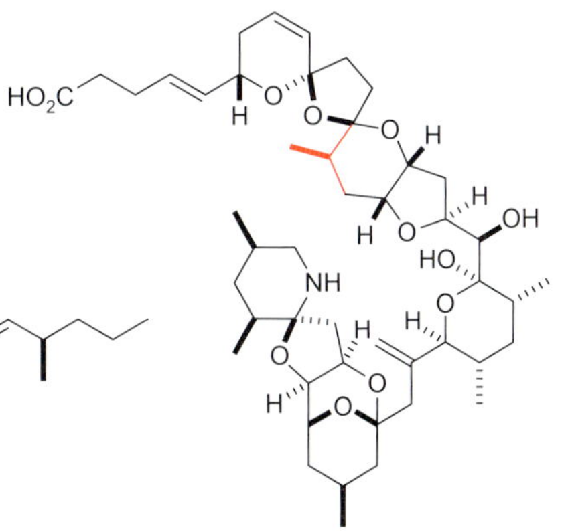

Zincophorin

Azaspiracid-1

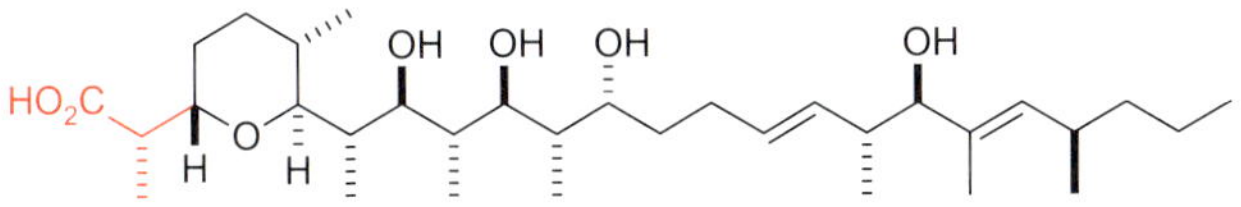

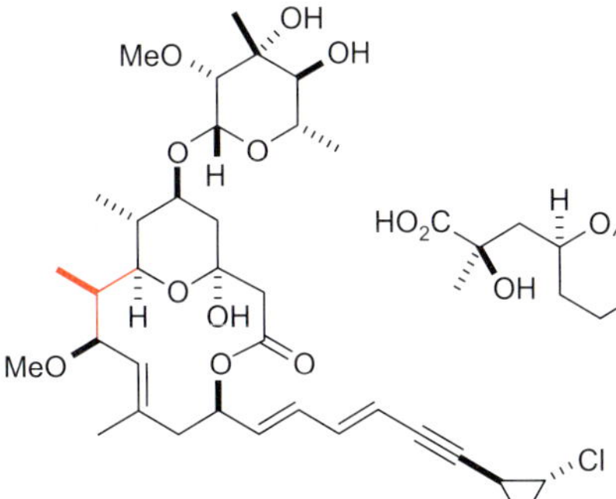

Callipeltoside C

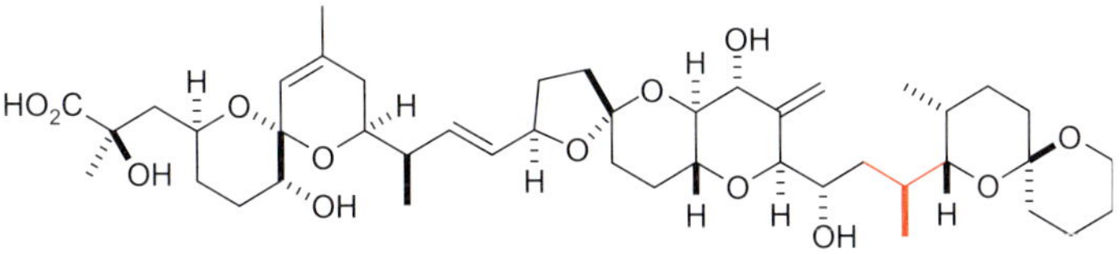

Okadaic acid

Figure 8.25   (continued)

8.9
Locating Terpene-type Substructures

Terpenes are among the most abundant sources of acyclic and carbocyclic compounds containing a single stereogenic center. It is therefore most appropriate to consider them as excellent chirons in the synthesis of the hydrocarbon frameworks of target molecules [55]. Based on the nature of the carbon framework, a direct convergence with an acyclic or cyclic terpene may be possible. In addition, the transformation products arising from cyclic terpenes in particular, can provide invaluable building blocks that encompass ring-contracted cyclopentanes with the added advantage of strategically situated substituents as illustrated in Chapter 7, section 7.4). A potential limitation in selecting terpenes as chirons could be their accessibility in both enantiomeric forms, and occasionally, their lack of availability in large quantities compared to other classes of compounds.

Although their carbogenic nature may suggest their use as starting materials to be limited toward the synthesis of hydrocarbon-based target molecules, they are also amenable to incorporating heteroatoms by chemical modification.

The rich history of terpenes is also associated with medicinally effective topical agents, such as menthol and camphor, since ancient times. Therefore, it is not surprising that already by the middle of the last century natural products chemists had turned to terpenes as preferred starting materials for the synthesis of certain classes of natural products. The need to use suitable protective groups, in view of the polyfunctional nature of other natural products such as amino acids, carbohydrates, and hydroxy acids, detracted from their widespread use as starting materials compared to terpenes up until a few decades ago. Selected examples showing the use of terpenes as starting chirons in total synthesis are presented in Chapter 13.

8.9.1
Apparent terpenes

A terpene-like substructure may be easily seen in the target molecule. Focus on a branch point that coincides with a *C*-methyl or a *C*-isopropyl group originating from a cyclic terpene such as (*R*)- or (*S*)-carvone for example. The nature of the carbon chain may determine if an acyclic terpene is appropriate, either directly after end-group or functional group adjustments, for example with (−)-citronellol, or indirectly, from a cyclic terpene such as carvone by oxidative ring-opening. The following guidelines may be useful:

a. An isopropyl group in the substructure of a target molecule may suggest (*R*)- or (*S*)-carvone, or a related terpene as a carbocyclic template.
b. A 1,4 ("para")-relationship of *C*-methyl and isopropyl groups in a ring substructure may further point to a starting material related to carvone.
c. Highly functionalized carbocycles may also be prepared from carvone as exemplified by the representative transformations shown in Figure 8.26 (see also, Chapter 7, section 7.4).

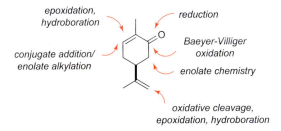

epoxidation, hydroboration

reduction

conjugate addition/ enolate alkylation

Baeyer-Villiger oxidation

enolate chemistry

oxidative cleavage, epoxidation, hydroboration

Figure 8.26 Selected chemical transformations of (S)-carvone.

The synthesis of the banded cucumber beetle pheromone from two molecules of (R)-citronellol was planned based on appropriate transformation products (Figure 8.27 A) [56]. Thus, (R)-citronellol was converted to the primary bromide, and the latter was extended by an acetoacetate ester condensation, hydrolysis, and ketalization sequence to afford **A**. Oxidative cleavage, reduction, and introduction of the phenylsulfone moiety by S_N2 displacement gave **B**. Intermediate **C**, also prepared from (R)-citronellol, was in turn converted to iodide **D**, following which, sulfone anion coupling and desulfonylation led to the pheromone. Thus, the presence of the C-methyl group(s) on a hydrocarbon-like framework in the target molecule clearly suggested (R)-citronellol as a logical starting chiron.

The carbocyclic ring substructure of sarcodictyin A (Figure 8.27 B) and the closely related eleutherobin, (a C-3 hydroxymethyl glycoside, see Figure 8.27 C for structure), led Nicolaou and coworkers [57a,c] to choose (S)-carvone as an apparent chiron, recognizing the "para" C-methyl/C-isopropyl substitution pattern. The challenge was to doubly functionalize the enone motif in (S)-carvone in order to elaborate the ten-membered macrocyclic motif. Thus, basic peroxide-mediated epoxidation of (S)-carvone led to an epoxyketone by an addition-elimination mechanism. Catalytic hydrogenation then gave **A**, which was subjected to a stereoselective aldol reaction affording the protected hydroxymethyl ketone **B**. Reduction of the ketone group, mesylation to **C**, and treatment with sodium naphthalenide gave allylic alcohol **D**. A Johnson-Claisen rearrangement in the presence of triethyl orthoacetate and propionic acid afforded intermediate **E**, in which the differentiated appendages were poised for further elaboration to the bicyclo[6.2.1]decane ring system of sarcodictyin A.

A.

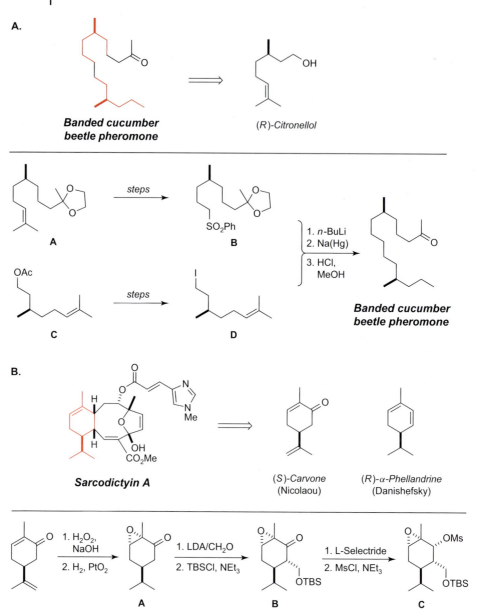

Figure 8.27 Apparent terpene-type substructures.

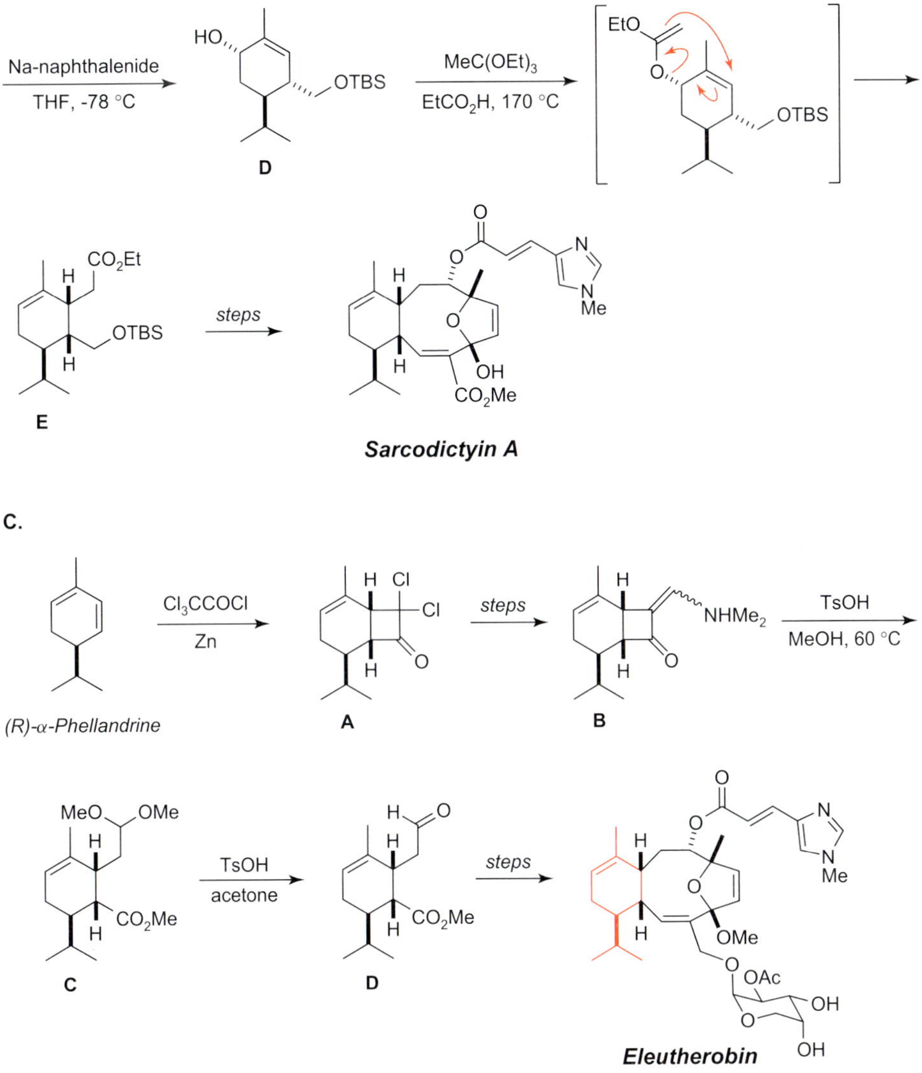

Figure 8.27 (*continued*)

A shorter route to a similar disubstituted intermediate was described by Danishefsky and coworkers [57b] in connection with their total synthesis of eleutherobin, which shares a common tricyclic core unit with sarcodictyin A (Figure 8.27 C). Thus, (*R*)-α-phellandrine was subjected to a regio- and stereoselective [2+2] cycloaddition to give bicyclic dichloroketone **A**. Reductive dehalogenation and conversion to α-dimethylaminomethylene ketone **B** through a Bredereck reaction, was followed by an acid-catalyzed ring-opening methanolysis to give **C**. Subsequent acetal hydrolysis furnished intermediate **D** as a precursor to the natural product.

8.9.2
Partially hidden terpenes

In these cases, a segment of the chiral substructure in the target molecule may be related to an acyclic or cyclic terpene due to the presence of a "methylene" chain, or a ring with a familiar *C*-methyl or *C*-isopropyl substituent. The target molecule structure may be bicyclic or polycyclic, while still revealing a terpene-like unit or appendage. The same guidelines can be followed as in the case of apparent terpenes, with allowance made for more elaborate modifications of the native chiron as shown in Figure 8.26.

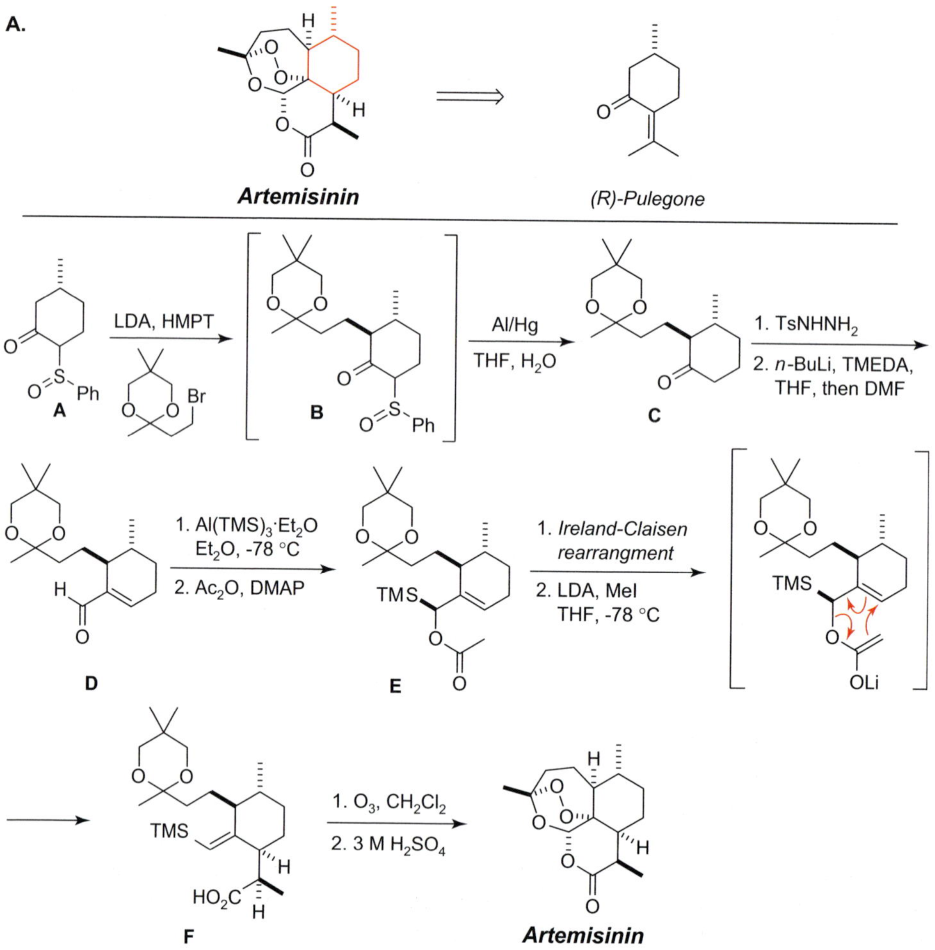

A.

Artemisinin (R)-Pulegone

Figure 8.28 Partially hidden terpene substructures.

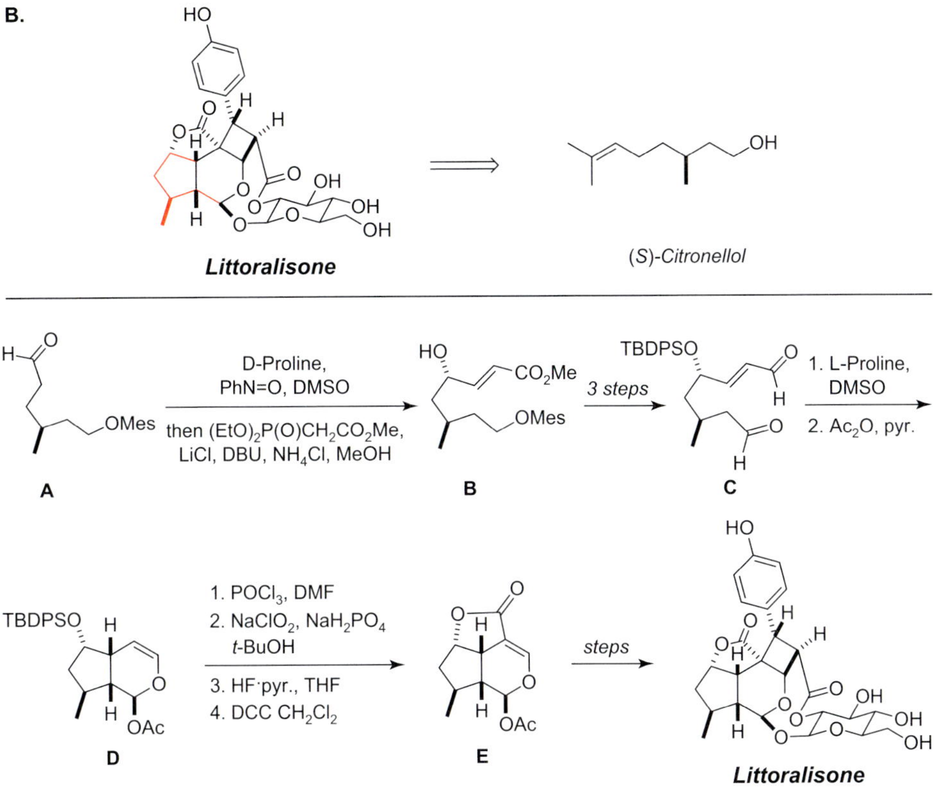

B.

Littoralisone

(S)-Citronellol

A

B

C

D

E

Littoralisone

Figure 8.28 *(continued)*

The tetrasubstituted cyclohexane substructure in artemisinin is reminiscent of a cyclic terpene-like entity, particularly in view of the 1,4-disposition of the *C*-methyl group and a *C*-isopropyl-like appendage (Figure 8.28 A). The chiron of choice for Avery and coworkers [18] was (*R*)-pulegone, which was modified by a series of oxidative transformations to the β-ketosulfoxide **A**. Dianion alkylation afforded **B**, which was treated with aluminum amalgam to give the key disubstituted ketone **C**. Installation of the α,β-unsaturated aldehyde led to **D**, which became a substrate for a Ireland-Claisen rearrangement starting with silane ester **E**. Alkylation of the product gave **F**, which was transformed to artemisinin under controlled oxidative conditions followed by hydrolysis.

The tetrasubstituted cyclopentane ring in the complex structure of littoralisone may conjure up a number of synthetic possibilities in the mind's eye. The MacMillan group [58] recognized the potential utility of (*S*)-citronellol to generate a formyl-enal intermediate (**C**) that would undergo an intramolecular organocatalytic Michael-type ring-closure to give bicyclic intermediate **D** (Figure 8.28 B). Citronellol was protected as the mesitoate ester (OMes) then oxidatively cleaved to give aldehyde **A**. Stereoselective α-hydroxylation was achieved by a D-proline catalyzed reaction in the presence of nitrosobenzene, and the resulting α-oxyaminophenyl

aldehyde product was directly subjected to a Horner-Wadsworth-Emmons olefination, which underwent N–O bond cleavage upon standing in methanol, to afford
B. Further elaboration gave the formyl-enal intermediate **C**, which would undergo
a contra-thermodynamic ring-closure reaction in the presence of ʟ-proline and
DMSO as solvent to give the *cis*-substituted bicyclic product **D** after acetylation.
Clearly, the choice of (*S*)-citronellol as an acyclic native chiron was driven by the
desire to explore proline-catalyzed reactions in two different contexts. Fortunately,
this choice was rewarded with a highly stereocontrolled *α*-hydroxylation, and by the
intramolecular ʟ-proline catalyzed ring-closure reaction.

8.9.3
Hidden terpenes

Relating a particular chiral substructure in a complex target molecule to a hidden
terpene-like motif may present a real challenge. Unlike other more functionalized
chirons, native terpenes offer mostly acyclic or carbocyclic motifs with characteristic
C-methyl and *C*-isopropyl appendages, and in some cases an oxygen atom in the
form of a carbonyl or a hydroxyl group. Chemical modification may totally alter
the original framework and functionality present in the starting terpene, which
offers greater versatility in relating it to a given substructure. This rich legacy
of terpene modification is the basis of many clever synthesis plans, where chiral
segments in a target molecule can be elaborated from chemically modified terpenes
as chirons. The internal reference points in such segments are most commonly
the familiar *C*-methyl and/or *C*-isopropyl groups. Extending the analysis of the
carbon framework on both sides of these appendages may reveal the congruence
of a hydrocarbon-like segment, which, in turn, can be related to an appropriate
acyclic or cyclic terpene. Cyclic terpenes are known to undergo fragmentation
reactions to give functionalized acyclic counterparts, or substituted cyclopentane
ring-contraction products (see Figure 8.1).

Beyond the visual recognition that terpenes could be possible starting chirons as
precursors to specific segments in target molecules, there are no specific guidelines
other than locating alkyl substituents such as a *C*-methyl or *C*-isopropyl group.
The choice of a terpene may become more attractive if, in addition to securing
stereochemical congruence with such *C*-alkyl groups for example, a good portion of
the carbon framework is also used to converge with the intended substructure. The
following two examples demonstrate the strategic use of terpenes as hidden chirons.
The literature is abound with many related examples, and later in Chapter 13 we
shall highlight selected total syntheses that commence with terpene precursors.

Analysis of the structure of the pseudoterane *ent*-kallolide B, shown in Figure 8.29
A, reveals two *C*-isopropenyl groups attached to the bridge of the cyclophane target.
The mind's eye immediately sees (*R*)- or (*S*)-carvone as a chiron, knowing that
in either case, the cyclic structure must be extensively modified and converted to
a usable acyclic equivalent, while maintaining the integrity of the *C*-isopropenyl
group. In their total synthesis of *ent*-kallolide B, Marshall and coworkers [59]
devised a plan in which (*S*)-perillyl alcohol was chosen as a chiron for the "eastern"

segment of the molecule. Thus, epoxide **A** was treated with periodic acid, and the resulting 6-formylhexanoic acid formed by an acid-catalyzed ring-opening, followed by diol cleavage, was esterified to give **B**. Subsequent transformations ensured the stereochemical and structural integrity of the C-isopropenyl group. Addition of the alkynyl stannane reagent derived from 1-bromo-2-butyne in the presence of SnCl$_2$, gave an allenyl carbinol, which was oxidized to the ketone **C**. Treatment with a catalytic quantity of AgNO$_3$ in acetone led to furan **D**, which was further elaborated to give the macrocyclic ether **E**. Marshall's plan to introduce the "western" C-isopropenyl group was based on the application of a diastereoselective intramolecular [2,3]-Wittig rearrangement. It was anticipated that the ring-contraction in **E** would proceed through a reactant-like transition state generated from the propargylic anion **F**. In the event, the major product **G**, now harboring the "western" C-isopropyl group was obtained with complete stereochemical control. It should be noted that the entire carbocyclic framework of (S)-perillyl alcohol remained in the "eastern" portion of the target molecule.

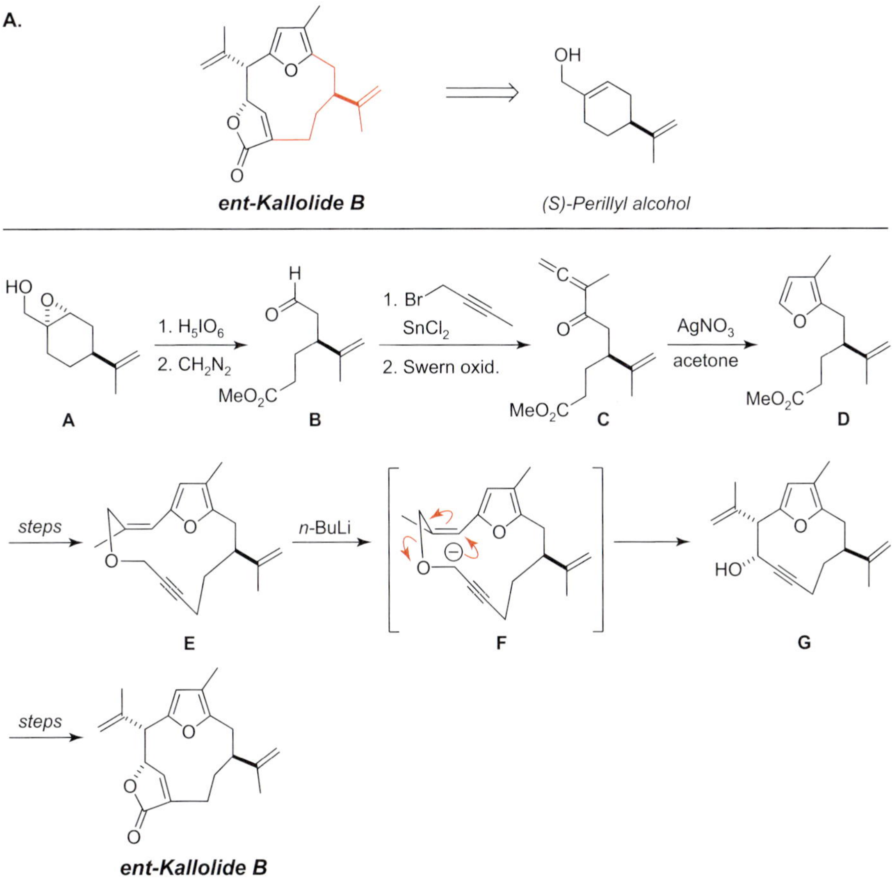

A.

ent-Kallolide B   (S)-Perillyl alcohol

Figure 8.29 Hidden terpene substructures.

B.

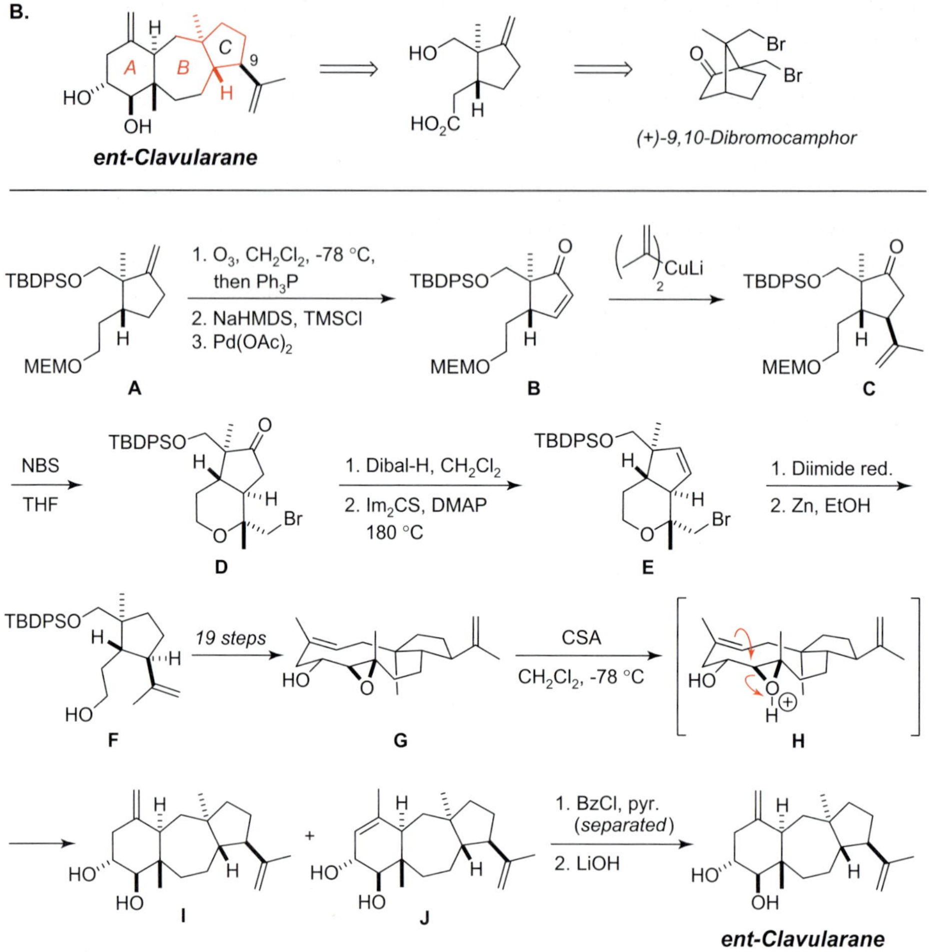

Figure 8.29 (continued)

The total synthesis of *ent*-clavularane by Williams and coworkers [60], also established the absolute configuration of the natural product (Figure 8.29 B). Analysis of the tricyclic framework reveals the presence of two angular *C*-methyl groups at quaternary ring junctions. The *C*-isopropenyl group at C-9 of ring *C*, and the carbotricyclic nature of the target molecule may, at first glance, lead to (*S*)-carvone as a possible starting material. However, this would require significant structural modification to produce an appropriately functionalized cyclopentane suitable for further elaboration. Aware that one of the fragmentation products of 9,10-dibromocamphor leads to an exquisitely substituted cyclopentane derivative (see section 8.1, Figure 8.1), Williams then devised a plan to utilize it toward building advanced intermediates that would eventually simulate a biomimetic cyclization to the intended target. Thus, chiron **A** was oxidatively cleaved to give the ketone, from which enone **B** was prepared by application of the Saegusa oxidation. 1,4-Conjugate addition of a 2-propenylcuprate gave **C**, thereby introducing the *C*-isopropenyl group "externally," and not, as first reflex action would seem to indicate, from (*S*)-carvone as a starting chiron. In order to "protect" the isopropenyl group from side-reactions during subsequent transformations of the ketone, Williams chose to generate the tetrahydropyran intermediate **D** in an intramolecular bromocycloetherification reaction (see Chapter 13, Figure 13.1 for a similar use of this reaction protocol). Deoxygenation of the ketone was accomplished by reduction to the alcohol, transformation to the thiocarbonylimidazole ester, and thermal *syn*-elimination to the olefin **E**. Reduction of the double bond with diimide, and restoration the *C*-isopropenyl group by a Zn-mediated ring-opening of the cyclic ether led to **F**. What followed was a systematic elaboration of this end-differentiated cyclopentane intermediate to reach the presumed biomimetic macrocyclic epoxide **G**. In a well-planned acid-catalyzed process, the epoxide ring was opened with concomitant transannular cyclization to give a mixture of alkenes, which were derivatized and separated to give *ent*-clavularane. It should be noted that all the carbon atoms of camphor were incorporated in rings *B* and *C* of *ent*-clavularane, while providing two contiguous stereogenic centers containing end-differentiated side-chains, and a precious angular *C*-methyl group. In the process, the exocyclic methylene group and its oxidized cyclopentanone product had to be subjected to functional group interconversions.

A selection of natural products in which terpenes were cleverly utilized in achieving total syntheses over a 20 year period are shown in Figures 8.30–8.34 [61] (see also Chapter 13).

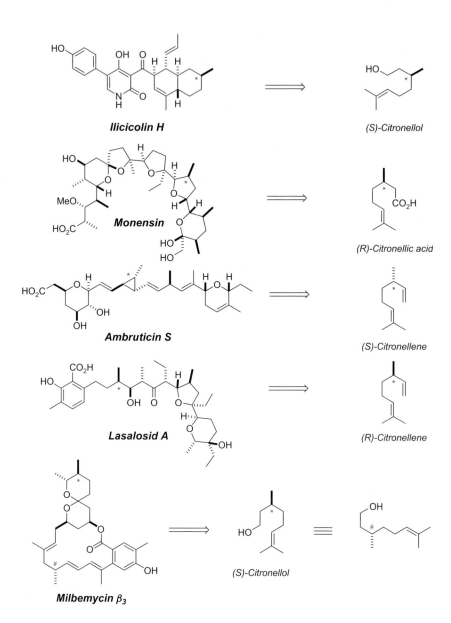

Figure 8.30 Setected total syntheses starting with (R)- and
(S)-citronellol and related chirons (1980–2007).

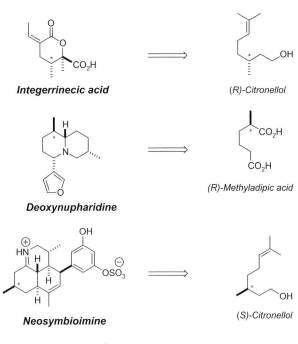

Integerrinecic acid ⟹ (R)-Citronellol

Deoxynupharidine ⟹ (R)-Methyladipic acid

Neosymbioimine ⟹ (S)-Citronellol

Figure 8.30 (*continued*)

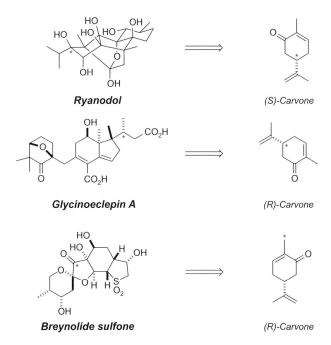

Ryanodol ⟹ (S)-Carvone

Glycinoeclepin A ⟹ (R)-Carvone

Breynolide sulfone ⟹ (R)-Carvone

Figure 8.31 Setected total syntheses starting with (*R*)- and (*S*)-carvone (1979–1989).

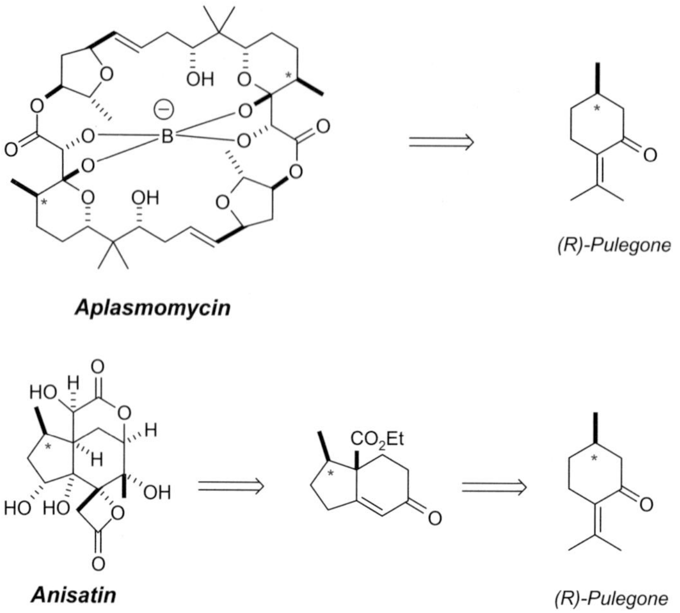

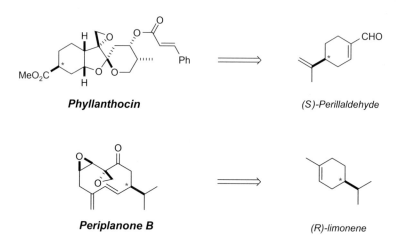

OH
OH
Aplasmomycin
(R)-Pulegone
CO2Et
Anisatin
(R)-Pulegone

Figure 8.32 Selected total syntheses starting with (*R*)-pulegone (1982–1994).

MeO2C
Ph
CHO
Phyllanthocin
(S)-Perillaldehyde
Periplanone B
(R)-limonene

Figure 8.33 Selected total syntheses starting from (*S*)-perillaldehyde and (*R*)-limonene (1982–1997).

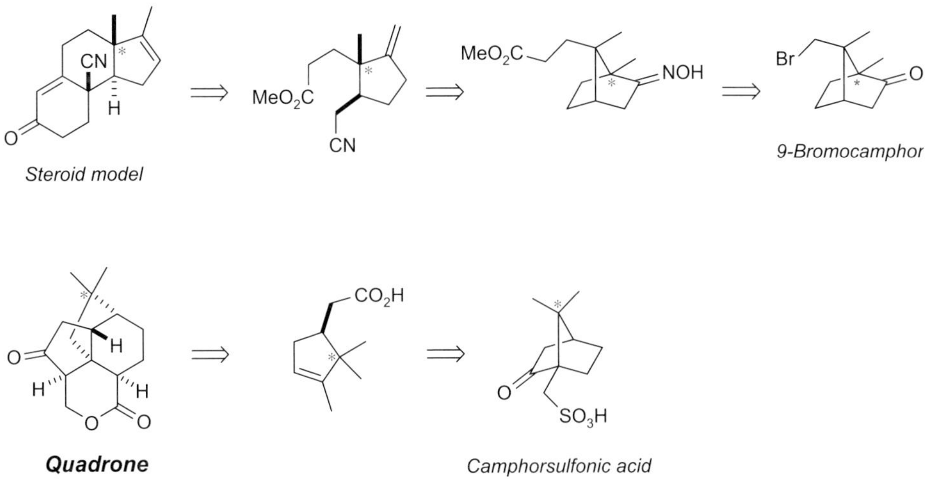

Steroid model

9-Bromocamphor

Quadrone

Camphorsulfonic acid

Figure 8.34 Selected total syntheses starting with camphor derivatives.

8.9.3.1 The terpene route to taxol – The total synthesis of taxol by the Holton [62] and Nicolaou [63] groups in 1994 was followed by four other elegant accomplishments toward this challenging synthetic target [64] (for details, see Chapter 16, sections 16.8.2–16.8.5). Two of the six total syntheses of taxol resorted to Nature's terpenes as starting materials and they are worthy of analysis. The visual relationship between rings A and B of taxol with the bicyclic terpenes (−)-camphor and α-pinene are difficult to appreciate at first glance. In hindsight, the daunting presence of multiple oxygen containing functionalities, and the tetracyclic nature of the taxol structure does reveal a glimpse of a terpene residue in the *gem*-dimethyl group common to rings A and B. For the aficionados of terpenes as chirons, it will be immediately apparent that a *gem*-dimethyl group is found in the readily available (+)- or (−)-camphor and α-pinene. In fact, α- or β-ionone (see Chapter 4, section 4.2 for structures) may also stir-up some interest as a precursor to ring A, although it may lack the desired oxygenation pattern to allow branching toward the construction of ring B (Figure 8.35).

The first example shown in Figure 8.35 is a contribution from the Paquette group, where (−)-camphor was used as a starting chiron in the synthesis of the related natural product taxusin [65]. The key feature of this terpene route to taxol involves a Wagner-Meerwein bridge migration reaction to furnish the AB ring system of the natural product (**D** to **C**). True to all other taxol routes (see Chapter 16, section 16.8), protecting group manipulations and redox chemistry of the highly oxygenated core were paramount to the completion of the total synthesis of taxusin.

Earlier work by the Holton group had shown that a tricyclic analogue, taxinine, could be synthesized as its enantiomer from camphoric acid, which is available from (+)-camphor.

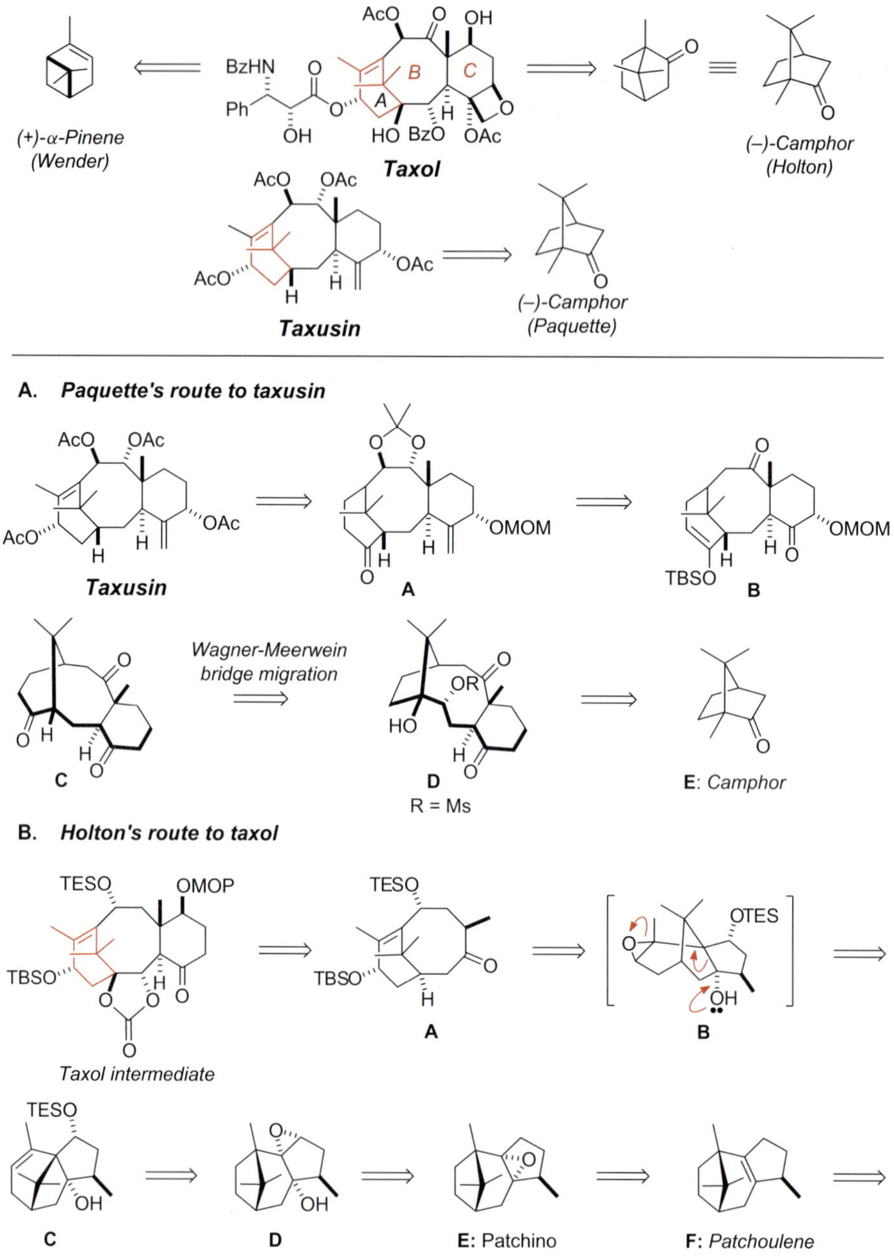
(+)-α-Pinene
(Wender)
BzHN
Ph
OH
AcO
O
OH
A
B
C
HO BzO
OAc
Taxol
(–)-Camphor
(Holton)
AcO
OAc
AcO
H
H
OAc
Taxusin
(–)-Camphor
(Paquette)
A. Paquette's route to taxusin
Taxusin
OMOM
A
TBSO
B
Wagner-Meerwein
bridge migration
C
OR
HO
D
R = Ms
E: Camphor
B. Holton's route to taxol
TESO
OMOP
TBSO
Taxol intermediate
A
OTES
OH
B
C
D
E: Patchino
F: Patchoulene

Figure 8.35 The camphor and pinene routes to taxol and taxusin.

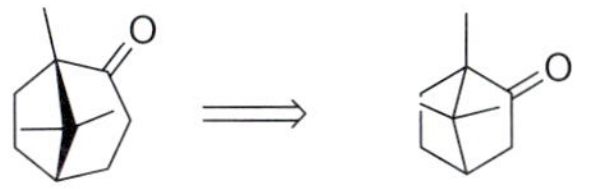

G: *Homocamphor*   *(−)-Camphor*

C. Wender's route to taxol

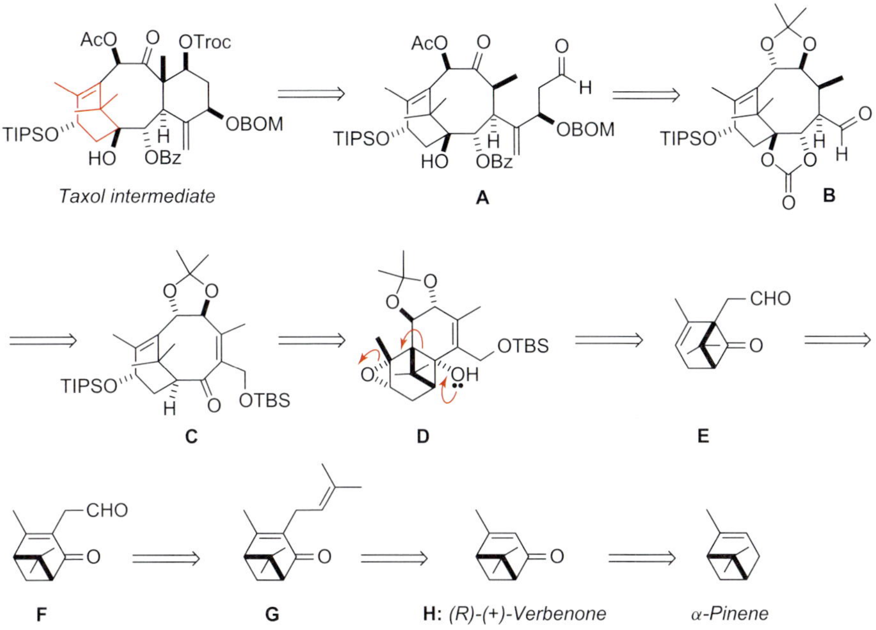

Taxol intermediate   **A**   **B**

C   **D**   **E**

F   **G**   **H:** *(R)-(+)-Verbenone*   *α-Pinene*

Figure 8.35   *(continued)*

A series of ring contractions and ring enlargements (camphor to **D**, and **C** to **B** respectively) led to a suitably substituted precursor **A** to ring *A* in taxol. The route that culminated with the total synthesis of taxol started with camphor (available in both enantiomeric forms), but capitalized on transformations reported earlier by Büchi and coworkers [66] who demonstrated that camphor could be transformed to patchino, the epoxide of β-patchoulene. Holton recognized the value of such intermediates early in his studies toward the elaboration of rings *A* and *B* of taxol. The unique transformations started from (−)-camphor and proceeded through two epoxy alcohol rearrangements (Figure 8.35 B). Thus, acid-catalyzed migration of the *gem*-dimethyl bridge in **D** led to **C** with introduction of the endocyclic olefin. Epoxidation of **C** enabled a Grob-type fragmentation reaction (as shown in **B**) that furnished an advanced precursor **A** containing the unique 6–8 bicyclic (*AB*) ring system of taxol. Intermediate **A** was then systematically manipulated *en route* to taxol (for details see Chapter 16, section 16.8.2).

Another elegant exploitation of Nature's terpenes, exemplified by α-pinene, was shown in Wender's [64d] total synthesis of taxol. Even the casual viewer will soon realize that the *gem*-dimethyl group in α-pinene must be "transposed" in order to consider any attempts at devising a synthetic route. This seemingly problematic step was astutely resolved by a photochemical reaction. Thus, air oxidation of α-pinene gave (+)-verbenone (**H**), which was α-prenylated to **G**, and the latter oxidized to the aldehyde **F**. Wender's experience in photochemical transformations, no doubt an incentive to explore the α-pinene approach, was rewarded with a highly efficient rearrangement of **F** to **E**. Elaboration to the bicyclic alcohol **D** set up a Grob-type fragmentation to construct the *AB* ring system of the natural product. The resulting eight-membered ring product **C** was taken through well-precedented steps to give **B**. Chain-extension to **A** and construction of the *C* ring via an intramolecular aldol reaction led to an advanced intermediate, which was further elaborated to taxol (for details, see Chapter 16, section 16.8.5).

These milestone achievements in the total synthesis of a therapeutically important antitumor agent such as taxol, support the notion of relying on naturally occurring terpenes and related chiral, non-racemic compounds as viable starting materials. Other approaches to build segments of the taxol skeleton from different enantiopure chirons are discussed in Chapter 16. However, the availability of such entities in quantities large enough to sustain a process to mass-produce a drug-like taxol presents major obstacles. The very reason to consider totally synthetic routes to taxol may be defeated if plant materials will have to be destroyed to secure the required raw products such as the correct isomers of camphor and pinene. Thus, the need for shorter, more economically acceptable and practical total syntheses of taxol and related life-saving drugs remain as major challenges in organic synthesis.

8.10
Locating Carbocyclic-type Substructures

Chiral, non-racemic cycloalkanes, cycloalkenes, and corresponding carbonyl derivatives can be versatile starting materials in total synthesis. They can be utilized directly as convergent substructures of a target molecule. Chemical modification of cyclic compounds by oxidative cleavage of double bonds, ring contraction, or other methods, can provide very useful acyclic equivalents as discussed for some cyclic terpenes. Chiral, non-racemic carbocycles are available from Nature directly, by transformation of terpenes through resolution, microbiological methods, or traditional asymmetric synthesis. A selection of natural and synthetic carbocyclic compounds can be found in Chapter 7 (Figures 7.8–7.12). In Chapter 14 we highlight the use of carbocycles as starting chirons in selected total syntheses.

8.10.1
Apparent carbocycles

Analysis of target molecule may reveal carbocyclic motifs that can be directly related to a chiral, non-racemic precursor. The congruence of other existing functional groups, as well as skeletal and stereochemical overlaps are good starting points in the planning stage.

The tricyclic structure of punctatin A does not reveal a particular terpene-like precursor (Figure 8.36 A). However, Paquette and coworkers [67] recognized the excellent convergence of rings *A* and *B* with the well-known Hajos-Parrish-Sauer-Eder-Wiechert ketone (see Chapter 5, section 5.8), which provides excellent skeletal and stereochemical overlap, especially considering the location of the angular *C*-methyl group, an enone function, and a second carbonyl that are well poised for appropriate elaboration toward the target molecule (for details see Chapter 14, section 14.1).

The tetrasubstituted *trans*-decalin ring in *B* of ilimaquinone can be immediately related to the familiar Wieland-Miescher ketone motif, easily prepared by catalysis with proline from an achiral triketone (Figure 8.36 B). Snapper and coworkers [68] sought to exploit such a chiron by choosing the 9-*C*-methyl analogue, which is available in enantioenriched form via an L-phenylalanine-mediated enantioselective Robinson annulation.

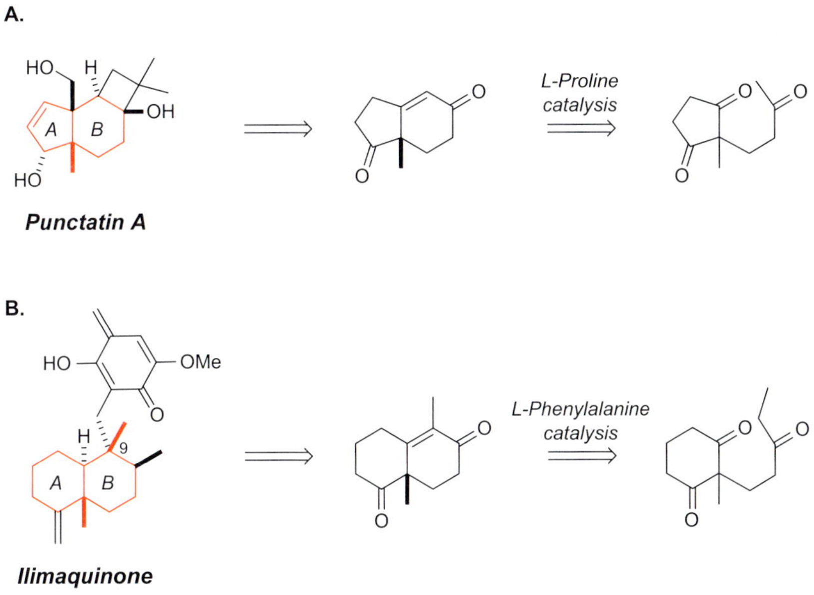

A.

Punctatin A

B.

Ilimaquinone

Figure 8.36 Apparent carbocyclic substructures.

8.10.2
Partially hidden carbocycles

The target substructure may contain a mono- or polycyclic unit, but a particular carbocyclic precursor may not be evident due to the presence of heteroatom substituents or carbon branching. As in the case of apparent carbocycles, one must focus on a ring system and "walk through" the substituents to find an anchoring site with adjacent workable functional groups. Retrosynthetic reasoning may relate a vicinal substitution pattern in the target molecule for example, to a double bond or a ketone in a carbocyclic precursor, provided the anchoring site corresponds to a stereogenic carbon atom or a heteroatom.

Analysis of phyllanthocin led Martin and coworkers [69a] to 3-cyclohexene (1R)-methanol as a potential starting material (Figure 8.37 A). In addition to the primary alcohol appendage, the double bond would be amenable to functionalization, leading to the desired cis-4,5-disubstitution pattern in the natural product. The C-methyl group in the spiropyran ring C could potentially be built from the (R)-Roche acid. In a different context, the C-methyl group at C-11 and vicinal hydroxyl groups at C-8 and C-7 of the pyran ring could also be introduced starting with (R,R)-tartaric acid. These visual connections were in fact put to practice by several groups during the period 1984–1991. Thus, Martin and coworkers [69a] started with readily available (R)-carboxylic acid A as a chiron, and systematically introduced functionality through iodolactonization to B, elimination to C, and a 1,3-dipolar cycloaddition to give cis-adduct D. With the required three cis-disposed substituents installed in D, a number of steps were performed exploiting the (R)-Roche acid as another chiron to prepare an acyclic appendage leading to E *en route* to the target molecule (Figure 8.37 A). Williams and coworkers had originally used racemic (4-methyl-cyclohex-3-enyl)methanol as a starting material in their synthesis of phyllanthocin, by exploiting the C_2-symmetric (R,R)-diethyl tartrate to elaborate the tetrahydropyran unit.

Starting with the enantiomeric lactone ent-C, Trost and coworkers [69b] obtained the allylic alcohol F, which was transformed into the acetal G in order to execute a highly selective ene-yne cycloreduction reaction to give the tricyclic motif H (Figure 8.37 A). Oxidative cleavage and spiroepoxidation, as in the Williams route [69c], led to phyllanthocin. The (R)-Roche acid was used as a versatile chiron in the respective Martin, Trost, and Burke [69d] total syntheses. The cyclohexenyl methanol chiron was also well utilized in the above syntheses, even though an obligatory deoxygenation was necessary in the Martin route (D to E), and a Mitsunubu inversion was required in the Trost protocol. Alternative approaches to rings A and C have been reported by Smith and coworkers [69e].

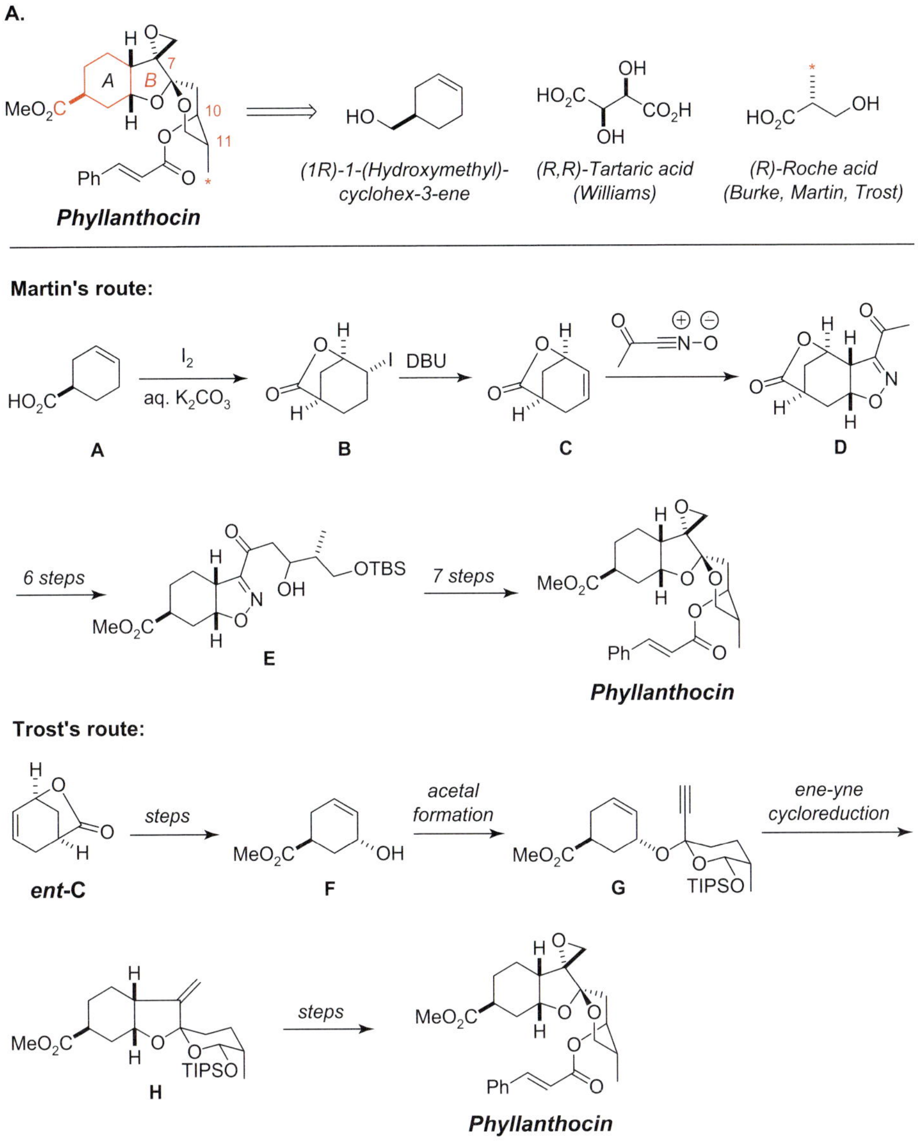
A.
Phyllanthocin
(1R)-1-(Hydroxymethyl)-cyclohex-3-ene
(R,R)-Tartaric acid (Williams)
(R)-Roche acid (Burke, Martin, Trost)
Martin's route:
I2
aq. K2CO3
DBU
6 steps
7 steps
Phyllanthocin
Trost's route:
ent-C
steps
acetal formation
ene-yne cycloreduction
steps
Phyllanthocin

Figure 8.37 Partially hidden carbocyclic substructures.

B.

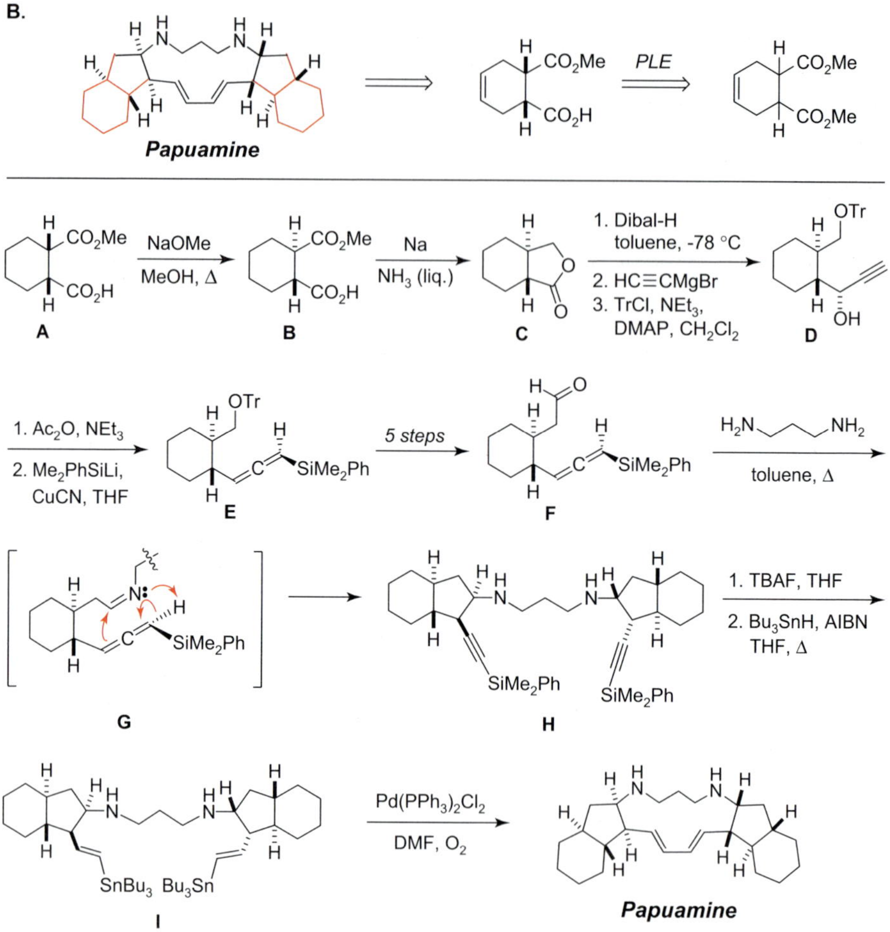

Figure 8.37 *(continued)*

The C_2-symmetric structure of papuamine led Weinreb and coworkers [70a] to a strategy that exploits a Pd-catalyzed intramolecular coupling of a bis-*E*-vinylstannane intermediate as an ultimate assembly step (Figure 8.37 B). The perhydroindane subunit was synthesized from the mono acid mono ester **A**, which is readily available in two steps and high enantiomeric purity from a PLE desymmetrization protocol. Prolonged treatment with NaOMe brought about a complete epimerization giving the mono ester **B**. Dissolving metal reduction led to lactone **C**, which was converted to the acetylenic alcohol **D** together with its epimer. Acetylation and addition of dimethylphenylsilyl lithiocuprate afforded the enantiopure allenylsilane **E**. Homologation of the primary alcohol gave aldehyde **F**, which, in the presence of 1,3-diamino propane, was converted to the C_2-symmetric imine intermediate **G**. This intermediate underwent an exquisitely aligned imino-ene reaction upon refluxing in toluene to give **H**.

Further elaboration by conversion to a bis-*E*-vinyl stannane and treatment with Pd(PPh$_3$)$_2$Cl$_2$ gave papuamine. The highlight of the synthesis rests in the novel imino-ene carbocyclization to produce the *cis*-substitution pattern in the cyclopentane ring portion of the perhydroindane.

8.10.3
Hidden carbocycles

When analyzing the carbon framework of a hydrocarbon-like chiral, non-racemic target molecule, it is normal to search for appropriate carbocyclic or acyclic precursors that provide good functional and stereochemical overlap. Because of their widespread occurrence, and our familiarity with the structures of the more common acyclic and carbocyclic terpenes, we may instinctively turn to them as possible chirons. As shown in the previous section, certain signature substituents and functional groups such as a *C*-methyl or *C*-isopropyl group may immediately lead to a terpene precursor upon analyzing the structure of the target molecule. Finding hidden carbocycles outside the realm of native cyclic terpenes may be quite difficult because the repertoire of commonly used chiral, non-racemic carbocycles is small, and not as easily visualized as segments of the target molecule.

The readily available naturally occurring cyclitols, and quinic acid, are excellent polyfunctional carbocycles that lend themselves to many chemical transformations leading to highly modified congeners. Unless one is familiar with these modified chirons, it is difficult to locate matching segments in target molecules. Often, it is the creativity of the synthetic chemists that leads them to consider hidden carbocycles or their acyclic modification products as chirons. Two classical examples of imaginative routes to macrolide antibiotic aglycones starting with carbocycles are shown in Figure 8.38.

The relationship of an achiral molecule such as 5-methyl-1,3-cyclopentadiene to the structure of erythronolide A is non-existent beyond the presence of a *C*-methyl group. However, a simple asymmetric hydroboration in the presence of (−)-diisopinocampheylborane and BH$_3$·THF according to Partridge and coworkers [71a] delivers the (+)-(1*S*,2*S*)-2-methyl-3-cyclopenten-1-ol in excellent enantiomeric purity. With this knowledge, Stork and coworkers [72] were able to devise a route to the C-1–C-15 chiral sequence of erythronolide A from a common intermediate (Figure 8.38 A). Hydroxyl-directed epoxidation of the precursor cyclopentenol gave the corresponding epoxide **A**. Oxidation and β-elimination, followed by protection of the resulting alcohol afforded enone **B**. A systematic substrate-controlled series of functional group modifications led to the cyclopentene enol ether **C**. Ozonolysis, followed by reduction of the resulting aldehyde, and acid-catalyzed lactonization gave **D** as the TBS ether. Further steps gave the acyclic ketone **E**, which corresponds to the C-1–C-5 portion of erythronolide A *seco*-alcohol. The C-7–C-11 segment of the acyclic target also utilized oxidation of intermediate **C**. Saegusa oxidation gave the enone **F**, which was further converted to the epimeric silyl enol ether **G**. Although the chirality at the *C*-methyl center was lost, it was nicely restored after hydrogenation to produce the *C*-methyl cyclopentenone **G** (a diastereomer of **C**!).

Ozonolysis of the silyl enol ether, followed by the same protocol used for the C-1–C-5 segment, led to lactone **H**. Further steps involved elaboration at both extremities to produce phenylsulfone **I** corresponding to the C-7–C-15 segment of erythronolide A. Julia coupling of the anion generated from **I** with methyl ketone **E**, eventually led to the intended *O*-protected acyclic congener of erythronolide A (**J**).

Like erythronolide A, the aglycone of the antibiotic tylosin (Figure 8.39 B) also harbors subunits originating from propionate and related biosynthetic pathways. While one could stretch one's imagination and recognize a *C*-methyl connection between Stork's achiral 5-methyl-1,3-cyclopentadiene approach to the propionate-derived pattern in erythronolide A, no such progeny can be found in the bicyclo[2.2.1]heptane derivative used by Grieco as a staring material for the synthesis of tylonolide hemiacetal. However, he had a far reaching plan to use topology and inherent chirality to his advantage (Figure 8.38 B) [73]. Two subunits derived from one and the same bicyclic precursor were envisaged, covering C-3–C-9 of the "eastern," and C-11–C-17 of the "western" segments of tylonolide hemiacetal. Intermediate **A**, available in large quantity and enantiomerically pure form was *C*-methylated to give **B**, and the latter subjected to a Baeyer-Villiger oxidation/transposition leading to bicyclic lactone **C**. Reduction of the lactone and the double bond, followed by protection as the TBDPS ether and Collins oxidation, afforded ketone **D**, harboring three contiguous stereogenic centers. Treatment with *m*-CPBA led to δ-lactone **E**, which was subjected to a stereocontrolled *C*-methylation to give **F**. Oxidation of the primary hydroxyl group followed by acetal formation with concomitant esterification led to **G**. Reduction of the ester to the alcohol, followed by Collins oxidation gave aldehyde **H** encompassing four contiguous stereogenic centers and the full complement of C-3–C-9 substituents present in tylonolide hemiacetal.

The synthesis of the C-11–C-17 segment started with the same bicyclic precursor having a shorter tether, which was subjected to a Dakin oxidation to give **I**. Further functional group adjustments led to cyclopentanol **J**, which was successively oxidized to the ketone, then to the corresponding lactone **K**. Methylation of the enolate, trapping with PhSeCl, and oxidative elimination gave **L**, which was reduced to the primary alcohol and converted to ketal **M**. Realizing the incorrect geometry of the double bond, a consequence of the *cis*oid orientation in the precursor lactone **L**, Grieco and coworkers resorted to a sulfenate-sulfoxide-sulfenate interconversion protocol, first reported by Evans and Andrews [74]. The resulting alcohol was oxidized to aldehyde **N**, which was extended to ene-yne intermediate **O**. Coupling of **H** and **O** through the corresponding lithium acetylide led to the C-3–C-17 segment, which was further elaborated to introduce the required 2-carbon appendage while creating the requisite stereochemistry at the C-3 alcohol.

The Stork [72] and Grieco [73] syntheses of their respective macrolides, incorporating the full complement of functional groups and correct stereochemistry, are classic examples of substrate-controlled planning of chemical transformations dating back several decades. In both cases, the entire carbon skeleton of the starting carbocyclic chirons was utilized in a visionary manner. These two syntheses constitute excellent pedagogic examples of voluntary manipulation of functional groups and chemical reactivity using steric and conformational

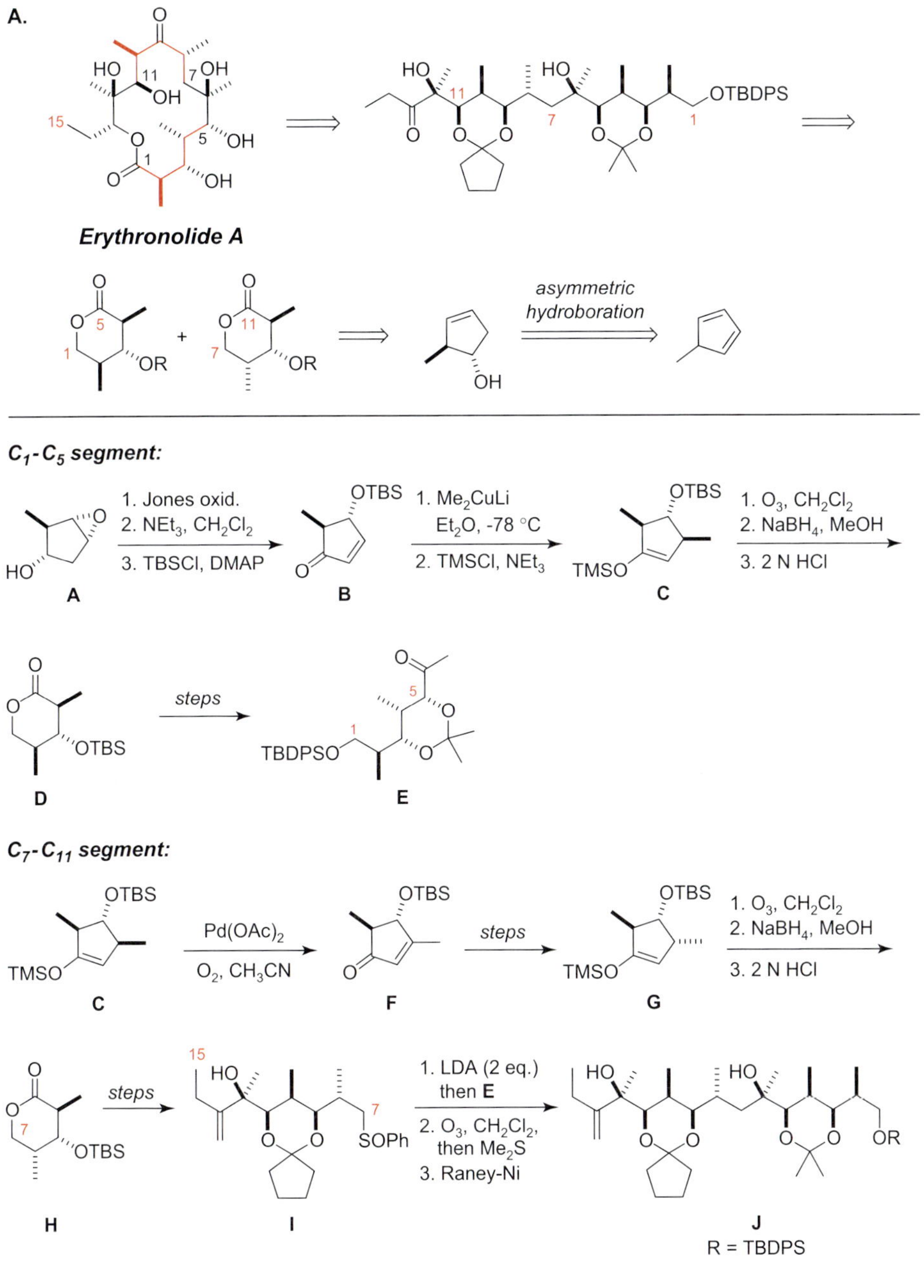

A.

Erythronolide A

asymmetric hydroboration

C_1-C_5 segment:

1. Jones oxid.
2. NEt$_3$, CH$_2$Cl$_2$
3. TBSCl, DMAP

1. Me$_2$CuLi
   Et$_2$O, -78 °C
2. TMSCl, NEt$_3$

1. O$_3$, CH$_2$Cl$_2$
2. NaBH$_4$, MeOH
3. 2 N HCl

A   B   C

steps

D   E

C_7-C_{11} segment:

Pd(OAc)$_2$
O$_2$, CH$_3$CN

steps

1. O$_3$, CH$_2$Cl$_2$
2. NaBH$_4$, MeOH
3. 2 N HCl

C   F   G

steps

1. LDA (2 eq.)
   then **E**
2. O$_3$, CH$_2$Cl$_2$,
   then Me$_2$S
3. Raney-Ni

H   I   J
R = TBDPS

Figure 8.38 Hidden carbocyclic substructures in acyclic segments of macrolides (vintage 1982).

B.

Tylonolide hemiacetal

C_3-C_9 segment:

C_{11}-C_{17} segment:

Figure 8.38 *(continued)*

bias advantageously. They also represent vintage 1982 syntheses using hidden precursors that start as carbocycles and end up as chiral acyclic segments. Today, advances in asymmetric methods in organic synthesis of complex, polyfunctional molecules such as macrolide antibiotics, may offer different options of controlling stereochemistry especially in acyclic motifs.

8.10.4
Quinic acid, cyclitols, and other carbocycles as chirons

The versatility of (−)-quinic acid as a polyfunctional carbocycle that lends itself to a large variety of chemical transformations is manifested in many total syntheses of natural products. The modified chirons available from (−)-quinic acid and a selected group of cyclitols were illustrated in Chapter 7 (section 7.5) [75]. Here we show some natural products and intermediates that were synthesized over a twelve year period from modified chirons originally prepared from (−)-quinic acid (Figure 8.39).

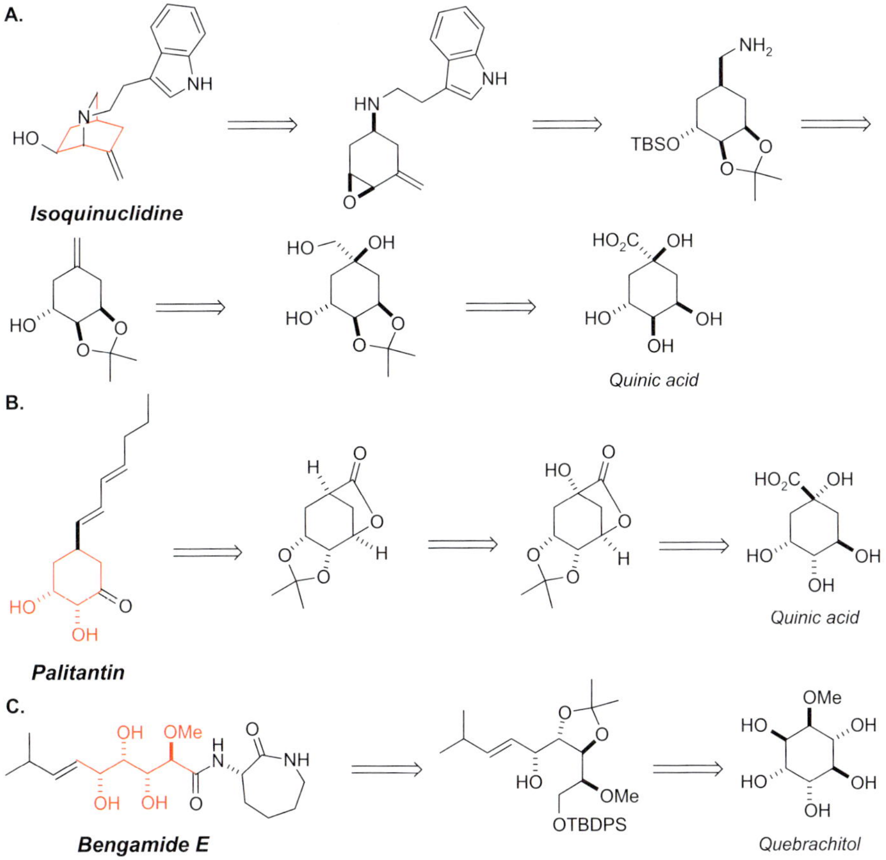

Figure 8.39 Selected natural products and advanced intermediates obtained from quinic acid and cyclitols (1986–1998).

D.

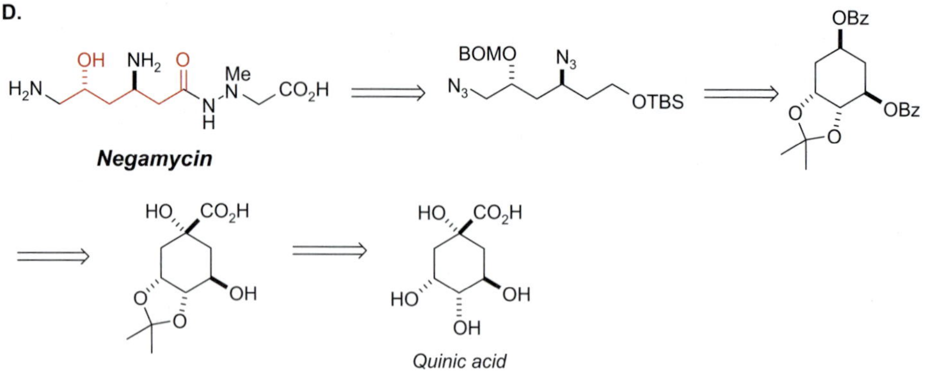

Negamycin

Quinic acid

E.

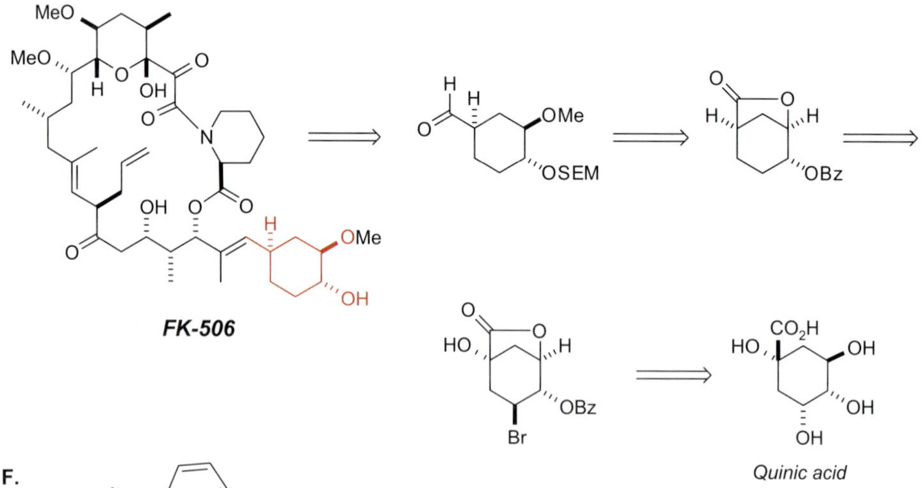

FK-506

Quinic acid

F.

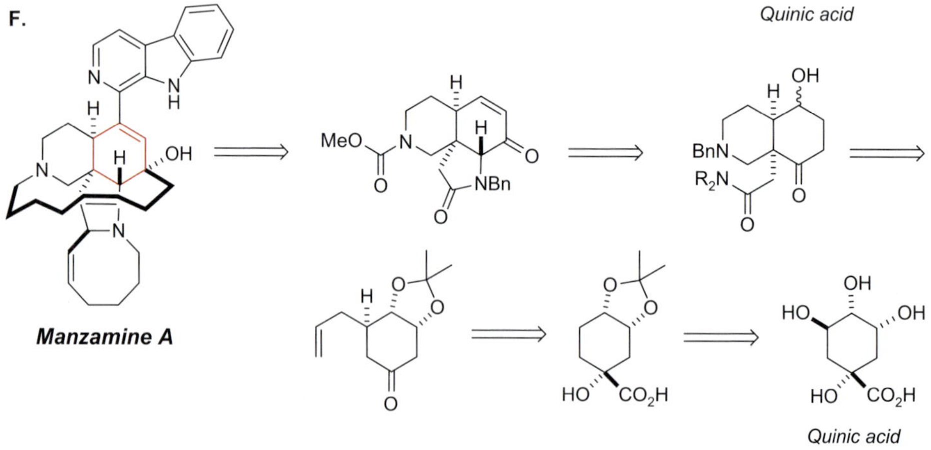

Manzamine A

Quinic acid

Figure 8.39 (continued)

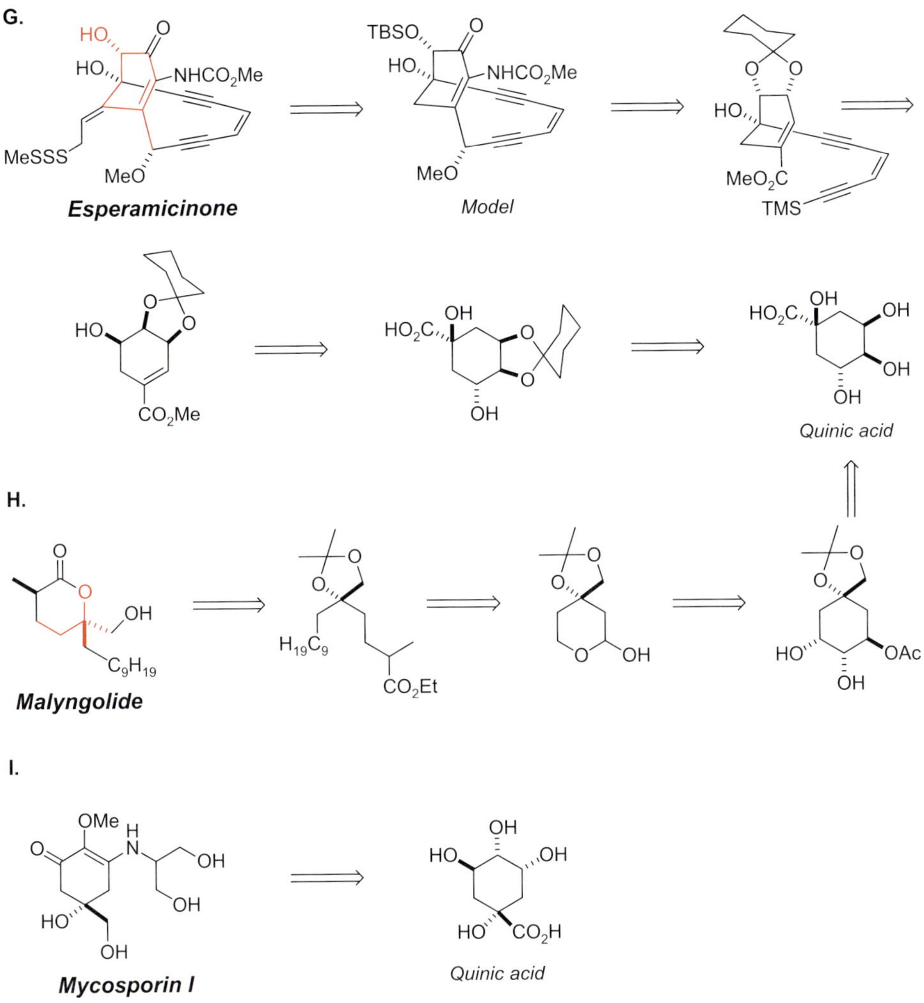

G.

Esperamicinone ⟹ Model ⟹

⟹ TMS

⟹

⟹ Quinic acid

H.

Malyngolide ⟹

⟹

⟹

I.

Mycosporin I ⟹ Quinic acid

Figure 8.39 *(continued)*

The readily available Wieland-Miescher ketone and its congeners have been extensively used as starting chirons for a number of natural product syntheses, particularly those containing a decalin-type substructure with an angular quaternary methyl group. Selected examples of such syntheses, as well as the utilization of other chiral, non-racemic carbocycles are shown in Figure 8.40 [76].

A.

Penitrem D   A   *Wieland-Miescher ketone*

B.

Nodulisporic acid A   A   *Wieland-Miescher ketone*

C.

Dysidiolide   A   B

D.

Norzoanthamine

Figure 8.40 Selected natural products and advanced intermediates obtained from carbocycles (1997–2007).

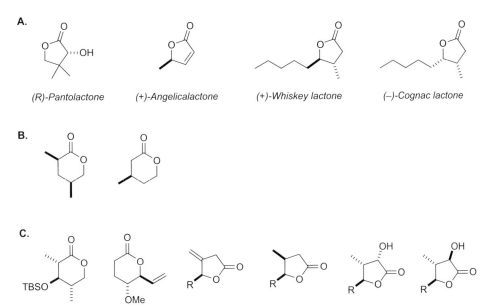

E.

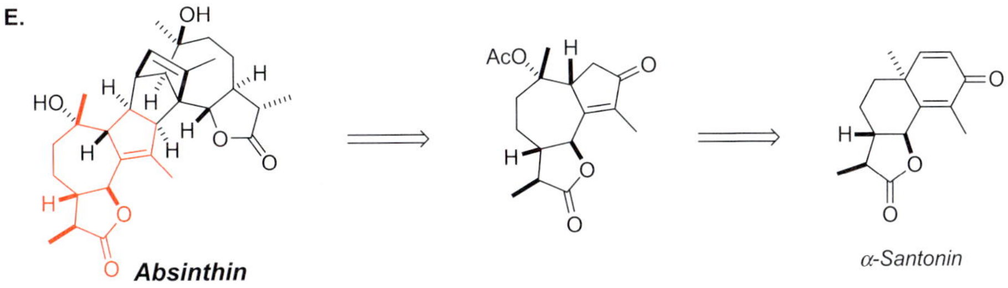

Absinthin

α-Santonin

Figure 8.40 *(continued)*

8.11
Locating Chirons Derived from Lactones

γ-Butyrolactones, butenolides, and their *C*-alkyl variants are ubiquitous in Nature, especially as components of plant natural products. The quercus lactones, (3*S*,4*R*)-3-methyl-4-butyl-γ-butyrolactone (whiskey lactone), and (3*S*,4*S*)-3-methyl-4-pentyl-γ-butyrolactone (cognac lactone) have been identified as key flavoring components in aged whiskey, cognac, wine, and brandy (Figure 8.41 A). In spite of the abundance of other substituted butyrolactones, 4-(hydroxymethyl)-γ-butyrolactones (5-(hydroxymethyl)-4,5-dihydro-2(3*H*)-furanones), and their 2,3-unsaturated congeners are only available in both enantiomeric form through synthesis. (*R*)-Pantolactone and (+)-angelicalactone are available in modest quantities from

A.

(R)-Pantolactone   *(+)-Angelicalactone*   *(+)-Whiskey lactone*   *(−)-Cognac lactone*

B.

C.

TBSO

OMe

Figure 8.41 A. Naturally occurring lactones; B. Lactones from enzymatic desymmetrization followed by chemical transformation; C. Lactones from asymmetric synthesis.

natural sources. Otherwise, a large variety of γ- and δ-lactones containing C-alkyl and heteroatom appendages can be obtained in enantiomerically pure form by synthetic methods relying on enzymatic desymmetrization, followed by chemical transformations (Figure 8.41 B) [77a]. Lactones can also be obtained from chiral epoxides, aldehydes, and by asymmetric synthesis (Figure 8.41 C) [77b].

The utility of chiral, non-racemic 4-(hydroxymethyl)-γ-butyrolactones and corresponding γ-butenolides as starting materials is manifested by the large number of applications found in natural product synthesis. (4R)- Or (4S)-4-(hydroxymethyl)-γ-butyrolactone [78] is available in large quantity through a three-step sequence from D- or L-glutamic acid respectively. Diazotization of L-glutamic acid leads to (S)-4-carboxy-γ-butyrolactone, which can be reduced to (4S)-4-hydroxymethyl-γ-butyrolactone (**A**) (Figure 8.42 A) [79]. The same lactone is also available from D-mannitol-bis-acetonide in a four-step sequence [80] through the intermediacy of butenolide **B**. Other methods are also available starting from chiral epoxides or by asymmetric processes.

Lactones **A** and **B** lend themselves to a variety of sterically controlled functionalizations, leading to the introduction of single or multiple substituents, including the generation of quaternary centers through enolate chemistry. Starting with O-protected butenolide **B**, an organocuprate addition can lead to a lactone such as **D**, and subsequent enolate alkylation can lead to **G** for example [81]. Lactone **C**, obtained from the Δ-^{3}pyrazoline leads to the diastereomeric reduction product **H**, which, in turn, can be alkylated to **I**. Direct C-methylation of O-protected **A** affords **F** as the major stereoisomer [82]. Reaction of the corresponding enolates with electrophiles allows for the introduction of a variety of other substituents including a hydroxyl group [81] (see section 8.11.4, Figure 8.46).

When analyzing target molecules for possible lactone precursors, one can apply the "rule of five" (or "four"), as with carbohydrates. Having correlated the position of the lactone carbonyl group with an sp^2-center in the target molecule for example, the intervening carbon chain can be further probed for existing branch points and heteroatom appendages such as a hydroxyl group that could correspond to the lactone ring oxygen. The requirement of (R)- or (S)-configurations of carbon atoms bearing substituents will dictate the choice of starting lactone. A bulky protecting group on the C-4 hydroxymethyl group in γ-butyrolactones and γ-butenolides, such as **A** and **B**, provides a high level of predictability in obtaining modified chirons (Figure 8.42). As always, alternative approaches relying on totally asymmetric reactions should also be contemplated.

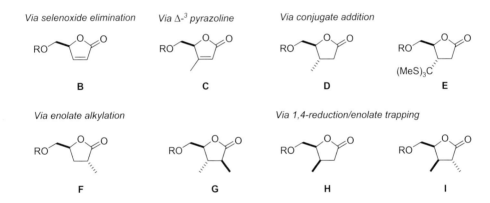

Figure 8.42 A. (4S)-4-(Hydroxymethyl)-γ-butyrolactone and B. (4R)-4-(Hydroxymethyl)-butenolides and their C-methyl chirons.

8.11.1
Apparent lactones

Analysis of the target molecule reveals a segment that can be directly related to a lactone in an acyclic or cyclic form. Characteristic groups, ring sizes, and substitution patterns may immediately reveal a lactone precursor.

Among the many research groups involved in the total syntheses of the therapeutically relevant epothilones, are those who considered the C-2–C-5 segment which harbors a *gem*-dimethyl group flanked by a carbonyl and an (S)-hydroxyl group (Figure 8.43 A). (*R*)-Pantolactone provides a direct overlap with the C-2–C-5 segment of epothilone D and it can be easily elaborated to intermediate **A** [83]. The 4-carbon framework of (*R*)-pantolactone including its inherent stereochemistry was incorporated in the intended segment in one such synthesis by Thomas and coworkers [83a].

The apparent lactone portion in isosteganone was synthesized by Brown and coworkers [84] from (4*S*)-4-(hydroxymethyl)-γ-butyrolactone by stereocontrolled enolate alkylation with a benzylic halide to give **A** (Figure 8.43 B). Subsequently, the original hydroxylmethyl group had to be excised to effect an intramolecular enolate ring-closure of **A**, leading eventually to the central eight-membered ring. Having served its directing role in the first enolate alkylation, the self-immolative process allowed Brown and coworkers to perform the second intramolecular enolate condensation in an *anti* fashion.

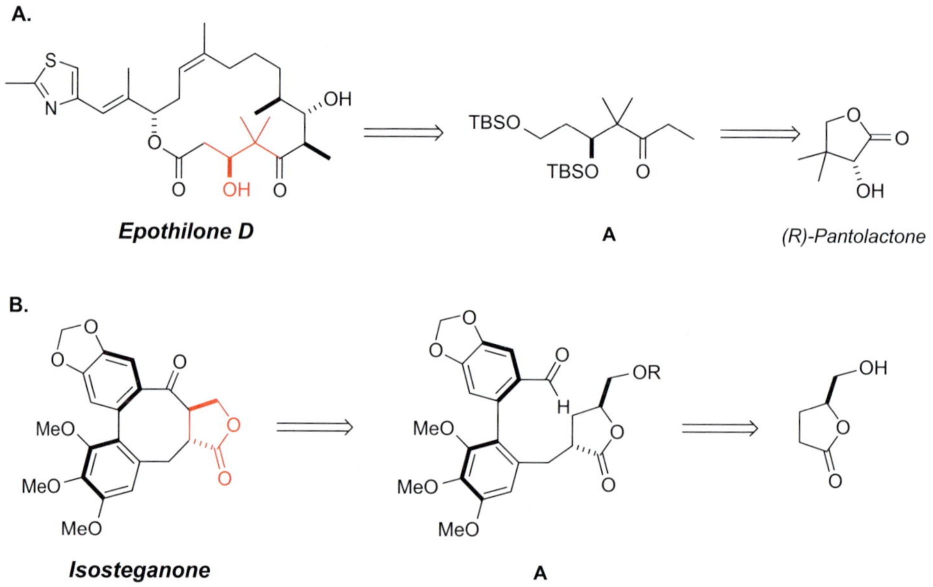

A.

Epothilone D　　　　　　　　　**A**　　　　　　　(*R*)-Pantolactone

B.

Isosteganone　　　　　　　　**A**

Figure 8.43 Apparent lactone substructures.

8.11.2
Partially hidden lactones

The original lactone may be extensively modified in the target molecule substructure, except for a remaining heteroatom, carbonyl group, or ring system. By scrutinizing adjacent functionality in the target substructure, one may be led to a substituted lactone precursor. In the synthesis of furaquinocin C by Smith and coworkers [85], a key intermediate was a functionalized ketomethyl lactone shown as the corresponding bis-TMS enol ether **A** (Figure 8.44 A, see also Chapter 15, section 15.4). (+)-β-Angelicalactone was utilized as a starting lactone, thereby ensuring the correct stereochemical outcome of a 1,4-conjugate addition in the generation of the geminal substitution at C-3 due to the influence of the resident chirality at C-4.

The presence of a butenolide ring in abyssomicin C could be misleading to the seeker of a lactone as a precursor (Figure 8.44 B). By focussing on the *meso*-2,4-dimethyl-1,5-diketone motif, Nicolaou and coworkers [86] envisaged an advanced intermediate **A** in which the ends of a potential 5-carbon precursor were differentiated. (2*R*,4*S*)-2,4-Dimethyl valerolactone (**B**) emerged as a suitable chiron, which could be readily prepared by enzymatic desymmetrization and further functionalization of *meso*-2,4-dimethylglutaric anhydride. The C-9 methoxy group in iejimalide B originated from the (2*S*)-2-hydroxy-γ-butyrolactone according to Fürstner's strategy and total synthesis of the natural product (Figure 8.44 C) [87].

The three examples of syntheses shown in Figure 8.45, take full advantage of the entire carbon skeleton and stereochemistry of the starting lactones.

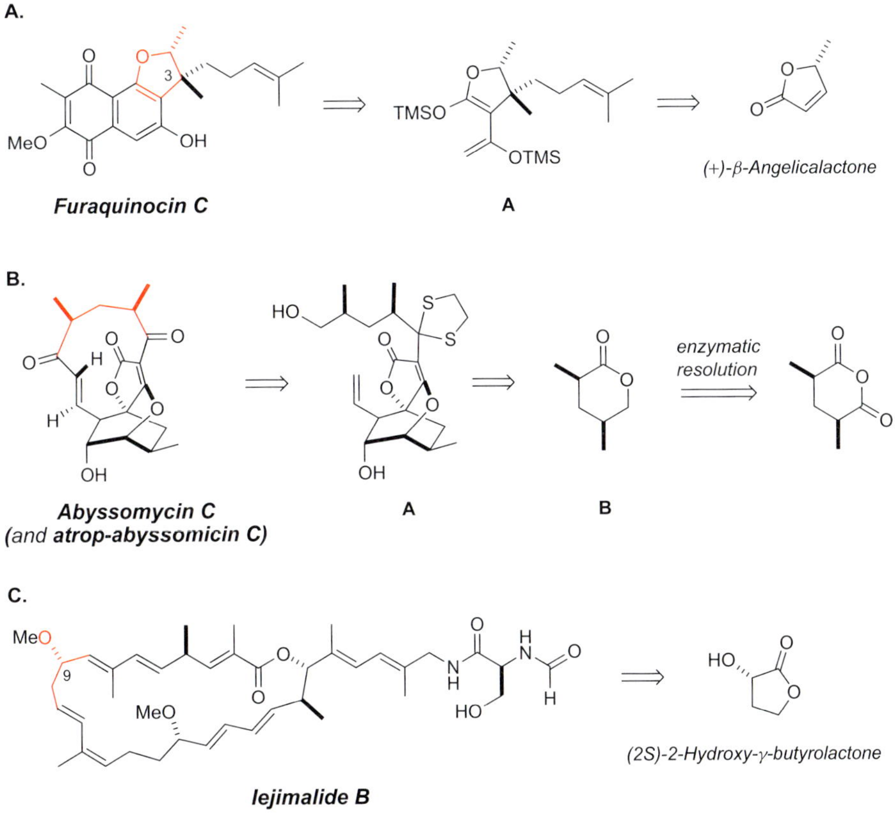

A.

Furaquinocin C　　　　　　　　**A**　　　　　　*(+)-β-Angelicalactone*

B.

enzymatic resolution

Abyssomicin C
(and **atrop-abyssomicin C**)　　　　**A**　　　**B**

C.

(2S)-2-Hydroxy-γ-butyrolactone

Iejimalide B

Figure 8.44 Partially hidden lactone substructures.

8.11.3
Hidden lactones

The original lactone may not be seen in the intricate structure of the target molecule, especially when no functional group correlations can be found between them. In some cases, the lactone may be used as a template to introduce desired branching patterns harboring intended stereochemical relationships, after which the original stereogenic center is destroyed. The examples shown in Figure 8.45 are genuinely crafted products of their times, when asymmetric (or catalytic asymmetric) methods were not as developed and widespread as they are today. Nevertheless, they illustrate how chiral, non-racemic lactone entities were predictably manipulated to generate advanced intermediates that were taken on to the target molecule.

Takano and coworkers [88] reported the synthesis of quebrachamine starting with a sterically biased lactone as the trityl ether (Figure 8.45 A). Enolate chemistry was used twice to introduce the C-2 carbon substituents. Following excision of the C-4 hydroxymethyl group, and condensation of a dialdehyde intermediate with tryptamine, the tetracyclic product **A** was obtained, then further elaborated to the target alkaloid. Thus, the original C-4 center in the starting lactone was used to direct the enolate alkylations, then "sacrificed" *en route* to **A**. Only four carbon atoms of the original lactones remained.

Stork and Rychnovsky [89] devised a total synthesis of (9*S*)-dihydroerythronolide A by coupling the Grignard reagent of sulfide **A** and ketone **B**, followed by further elaboration (Figure 8.45 B). Precursors **C** and **D** were fully substituted butyrolactones harboring the necessary complement of hydroxyl and *C*-methyl groups. Intermediate **B** had been previously prepared by Stork from lactone **E**, which originated from **F**. The synthesis of **C** involved an iterative homologation sequence starting with **D**.

The synthesis of an advanced C-3−C-15 intermediate to erythronolide A *seco*-acid by Chamberlin and coworkers [90] relied on lactones **G** and **I** representing C-3−C-7 and C-9−C-15 subunits respectively (Figure 8.45 C). These, in turn, were prepared from chiral, non-racemic allylic alcohol **H** and epoxide **J** utilizing methods of asymmetric synthesis.

In bourbonene, the original butenolide was incorporated in ring *C* after losing the hydroxymethyl group in a self-immolative process [91] (Figure 8.45 D). The 5-carbon framework of the butyrolactone chiron was incorporated in the acyclic hydroxyketone segment of hitachimycin by Smith and coworkers [92] (Figure 8.45 E). The polyol segment of amphotericin was elaborated by Hanessian and coworkers [93] using a stereocontrolled hydroxylation of the butyrolactone, and 1,4-conjugate additions to the butenolide. Segment coupling was achieved using an acetylene anion and an aldehyde prepared from a common 2-hydroxy lactone (Figure 8.45 F). Ziegler and coworkers [94] used a (3*S*)-3-methyl-*γ*-butyrolactone to construct a segment of ring *A* in calcimycin (Figure 8.45 F). However, the "extra" carboxyl appendage had to be excised.

A diene appendage obtained by enolate alkylation of (3*R*)-3-hydroxy-*γ*-butyrolactone was engaged in a biomimetic ring closure to generate the aplysistatin

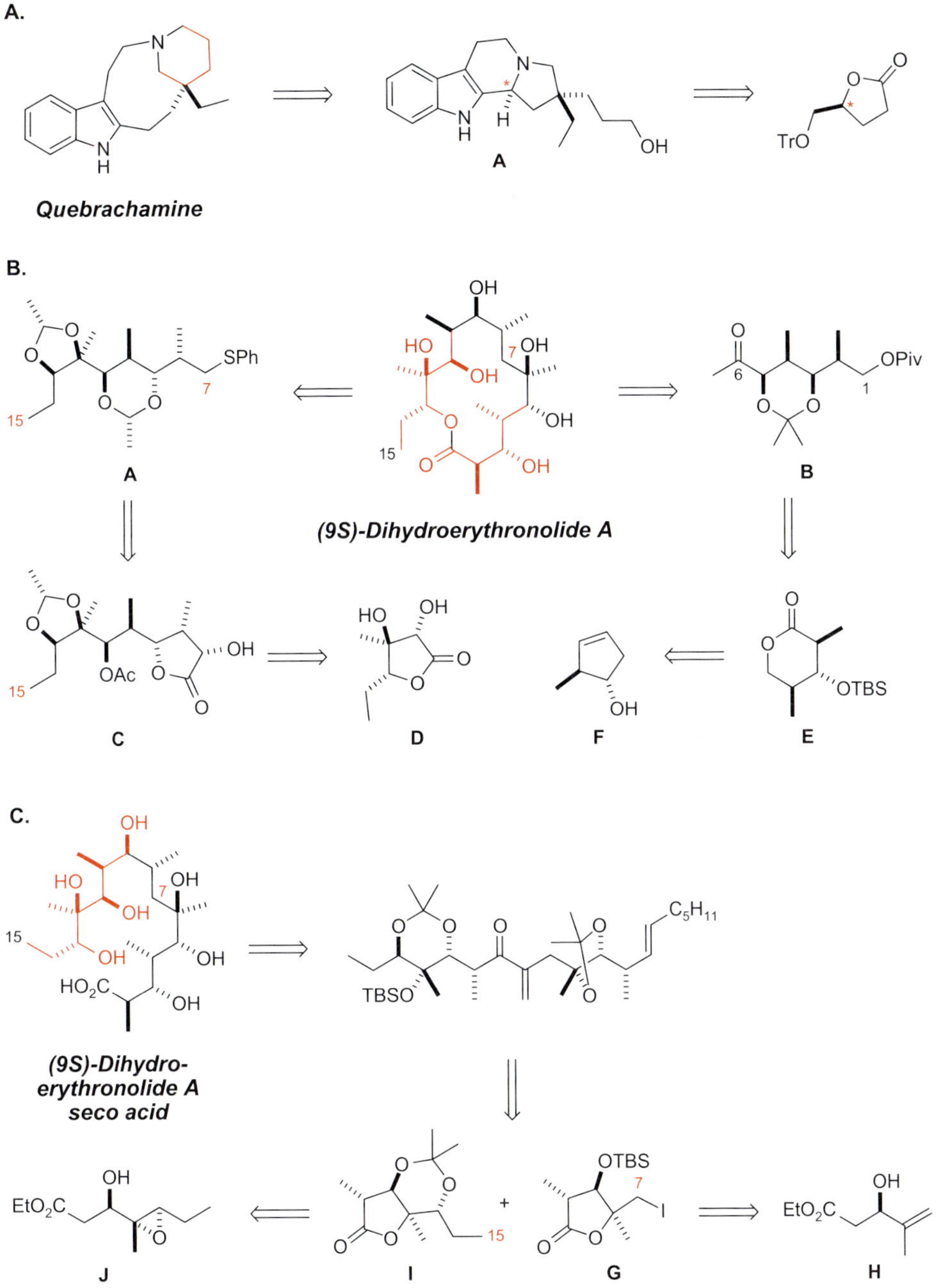

A.
Quebrachamine
A
TrO
OH
B.
(9S)-Dihydroerythronolide A
SPh
OPiv
OTBS
OAc
A
B
C
D
E
F
C.
(9S)-Dihydro-erythronolide A seco acid
HO_2C
TBSO
C_5H_{11}
OTBS
EtO_2C
J
I
G
H

Figure 8.45 Hidden and partially hidden lactone substructures.

D.

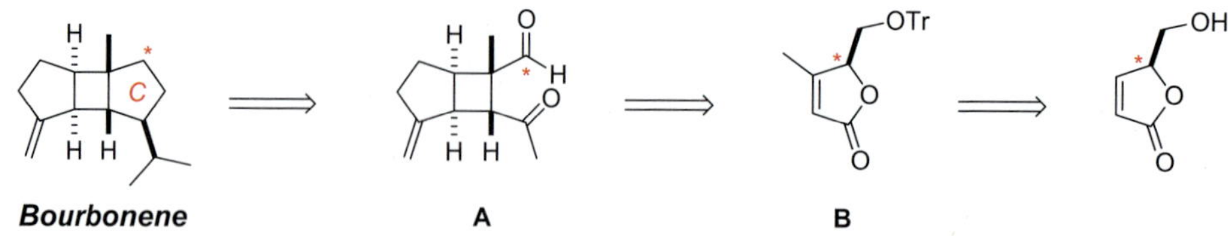
Bourbonene
A
B
OTr
OH

E.

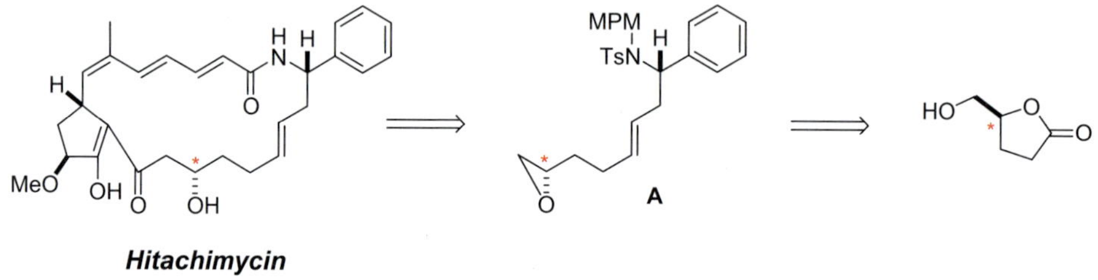
Hitachimycin
MPM
TsN
MeO
A

F.

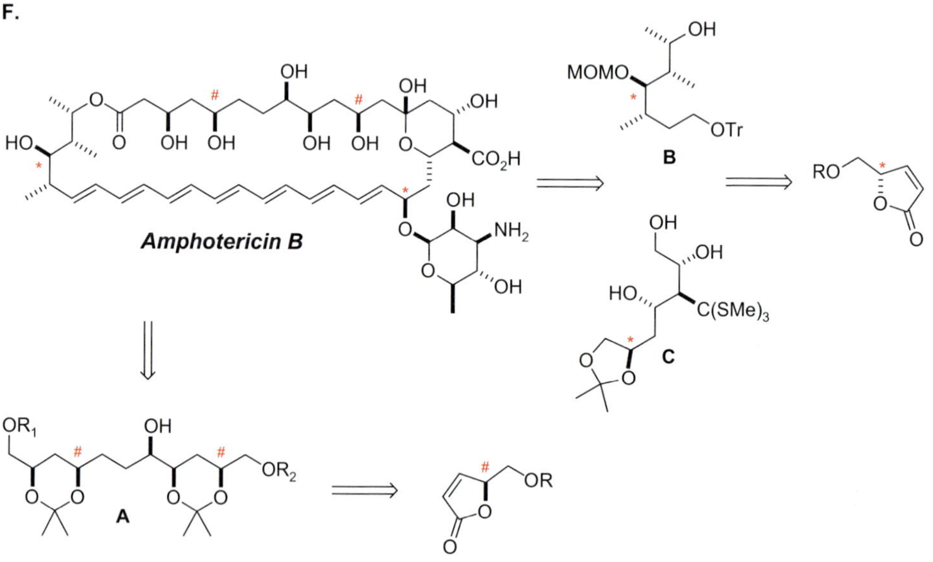
Amphotericin B
MOMO
OTr
B
C
C(SMe)3
RO
OR1
OR2
A
OR

Figure 8.45 (continued)

G.

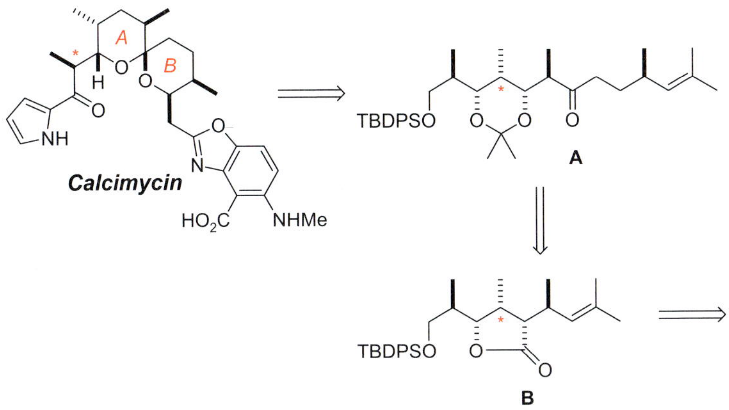

Calcimycin

H.

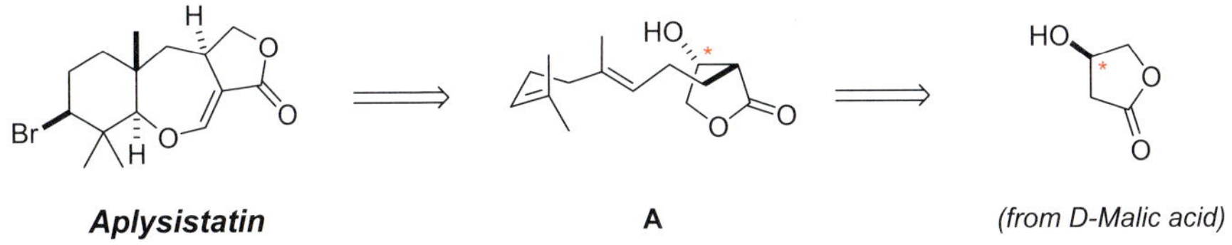

Aplysistatin   A   (from D-Malic acid)

I.

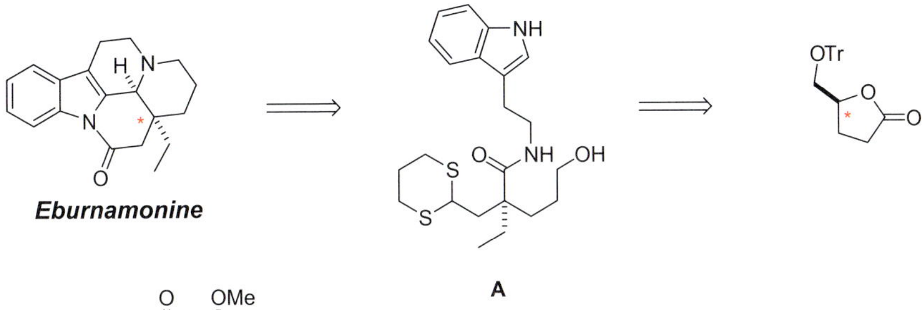

Eburnamonine

A

J.

Peloruside A   (S)-Pantolactone

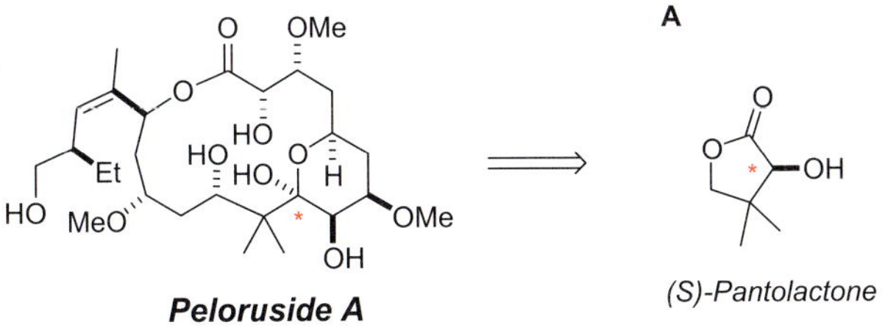

Figure 8.45 (continued)

framework by the Prestwich group [95] (Figure 8.45 H). A hidden lactone precursor was useful to elaborate the pentacyclic carbon framework of the alkaloid eburna-monine by Takano and coworkers [96] (Figure 8.45 I). Finally, (S)-pantolactone was used in the total synthesis of peloruside A by Evans and coworkers [97] (Figure 8.45 J). In their initial studies toward the total synthesis of bryostatin 16, Trost and Dong [98] used (R)-pantolactone to construct one of the tetrahydropyran rings bearing a *gem*-dimethyl group (see Chapter 18, section 18.4). Chapter 15 discusses the use of lactones as starting chirons in selected total syntheses.

8.11.4
The replicating lactone strategy

The versatility of butenolides as chiral templates for the systematic introduction of vicinal functionality in a highly stereocontrolled manner is shown for the iterative synthesis of polypropionate-derived subunits in Figure 8.46 A [81–84]. Thus, starting with **A**, a conjugate addition of lithium dimethylcuprate delivers a *C*-methyl group from the less hindered face of the double bond. Formation of the potassium enolate and treatment with MoOPH, introduces a hydroxyl group in an *anti*-mode as the major diastereomer **B**. A two-step sequence transforms lactone **B** to **C**, and then treatment with NaOMe funishes the epoxide **D**. A 2-carbon "acetate extension" provides the second butenolide template **E** in which the original C-4 substituent has now "grown" by the equivalent of a hydroxyethyl group. Application of the same protocol gives **F**, which is represented as its open-chain acid equivalent **G** harboring two full propionate subunits with an *anti/syn/anti/syn* orientation. Starting with the same first template **A**, but sulfonylating the secondary hydroxyl group as in **H** before borane reduction, then treatment with NaOMe, furnishes a terminal epoxide **I** (a diastereomer of **D**). Formation of the second butenolide template **J**, and iteration gives **K** (or **L**), now harboring an *anti/anti/anti/syn* orientation of propionate units. In principle, all the possible propionate triad orientations can be achieved depending on the stage at which new hydroxyl groups are inverted, or not, before each iteration [82].

A diastereomeric pair can be obtained by starting with an "inverted" C-3-methyl lactone **O** (Figure 8.46 B). Thus, treatment of the butenolide **A** with diazomethane gives a Δ^3-pyrazoline **M**, which upon heating in toluene gives butenolide **N**. Reduction in the presence of Rh/Al_2O_3 furnishes the diastereomeric *C*-methyl lactone **O**. Enolate hydroxylation and iteration as shown in Figure 8.46 A leads to another set of diastereomeric lactones (not shown).

The sterically biased lactone **O** can be *C*-methylated to **P**, and the latter hydroxylated to **Q** to generate a tertiary alcohol of high diastereomeric purity. Acyclic equivalents are represented by **R** and **S** respectively. Alkylation of the pyrazoline enolate from **M** gives **T**, which can be converted to the 2,3-disubstituted butenolide **U**, and finally, dihydroxylated to **V**. The acyclic equivalent is shown as **W**. Thus, starting from a single stereogenic center in butenolide **A**, one can generate stereochemically well-defined propionate and related subunits in an iterative and highly predictive manner [81, 82].

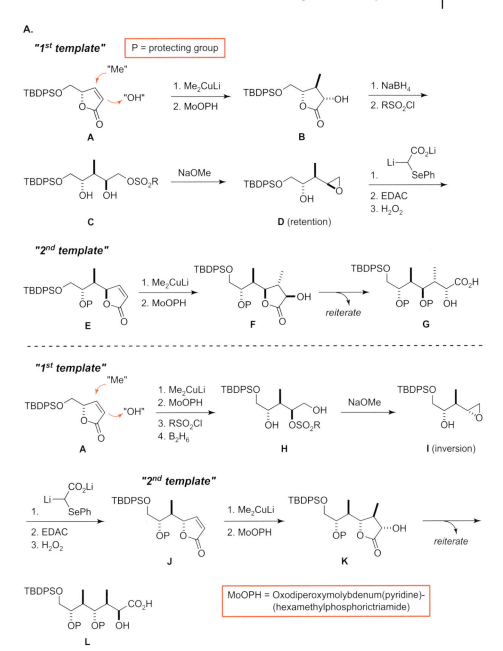

Figure 8.46 A. The "replicating lactone" route to polypropi-
onates; B. Vicinal C-methyl lactones and hydroxylactones.

B.

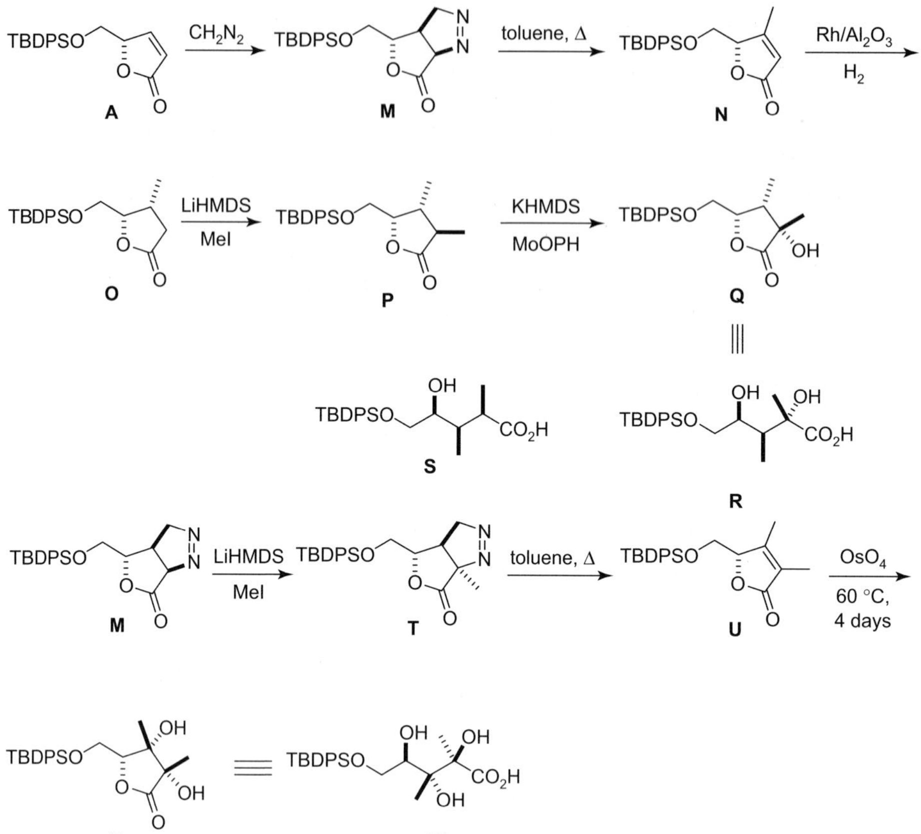

A M N O P Q R S M T U V W

Figure 8.46 (continued)

References

1. For the original definition, see: *Total Synthesis of Natural Products: The Chiron Approach*, Hanessian, S., Pergamon, Oxford, 1983.
2. For example, see: (a) Caine, D.; Crews, E. *Tetrahedron Lett.* 1984, **25**, 5359; (b) Roelofs, W. L.; Gieselmann, M. J.; Carde, A. M.; Tashiro, H.; Moreno, D. S.; Henrick, C. A.; Anderson, R. J. *Nature* 1977, **267**, 698; for a review, see: (c) Mori, K. *Tetrahedron* 1989, **45**, 3233; (d) Mori, K. in *The Total Synthesis of Natural Products*, ApSimon, J. Ed.; p. 1, Wiley & Sons, Inc. 1981.
3. Hutchinson, J. H.; Money, T. *Can. J. Chem.* 1985, **63**, 3182.
4. For a review see: (a) Money, T. *Nat. Prod. Rep.* 1985, **2**, 253; (b) see also reference [55].
5. (a) Gribble, G. W.; Barden, T. C. *J. Org. Chem.* 1985, **50**, 5900; (b) Burkard, S.;

Borschberg, H.-J. *Helv. Chim. Acta* 1989, **72**, 254.

6. Ori, M.; Toda, N.; Takami, K.; Tago, K.; Kogen, H. *Angew. Chem. Int. Ed.* 2003, **42**, 2540; see also: Steinhagen, H; Corey, E. J. *Org. Lett.* 1999, **1**, 843; Back, T. G.; Wulff, J. E. *Angew. Chem. Int. Ed.* 2004, **43**, 6493.

7. Thienamycin: Salzmann, T. N.; Ratcliffe, R. W.; Christensen, B. G.; Bouffard. F. A. *J. Am. Chem. Soc.* 1980, **102**, 6161.

8. Thromboxane B$_2$: Corey. E. J. Shibasaki, M.; Knolle, J. *Tetrahedron Lett.* 1997, **18**, 1625.

9. Thromboxane B$_2$: (a) Hanessian, S.; Lavallée, P. *Can. J. Chem.* 1977, **55**, 562; (b) Hanessian, S.; Lavallée, P. 1981, **59**, 870.

10. Amphidinolide P: Va, P.; Roush W. R. *J. Am. Chem. Soc.* 2006, **128**, 15960. For a review on the amphidinolides, see: Kobayashi, J.; Tsuda, M. *Nat. Prod. Rep.* 2004, **21**, 77.

11. Hapalindole G: (a) Fukuyama, T.; Chen. X. *J. Am. Chem. Soc.* 1994, **116**, 3125; see also hapaindole U: (b) Baran. P. S.; Richter, J. M. *J. Am. Chem. Soc.* 2004, **126**, 7450.

12. Spongistatin 2: (a) Heathcock, C. H.; McLaughlin, M.; Median, J.; Hubbs, J. L.; Wallace, G. A.; Scott, R. W; Claffey, M. M.; Hayes, C. J.; Ott, G. R. *J. Am. Chem. Soc.* 2003, **125**, 12844; (b) Wallace, G. A.; Scott, R. W.; Heathcock, C. H. *J. Org. Chem.* 2000, **65**, 4145; (c) Claffey, M. M.; Hayes, C. J.; Heathcock, C. H. *J. Org. Chem.* 1999, **64**, 8267; for other syntheses of spongistatins, see: (d) Smith, A. B. III; Zhu, W.; Shirakami, S.; Sfouggatakis, C.; Doughty, V. A.; Bennett, C. S.; Sakamoto, Y. *Org. Lett.* 2003, **5**, 761; (e) Crimmins, M. T.; Katz, J. D.; Washburn, D. G.; Allwein, S. P.; McAtee, L. F. *J. Am. Chem. Soc.* 2002, **124**, 5661.

13. (a) Ghosh, A. K.; Wang, Y. *J. Am. Chem. Soc.* 2000, **122**, 11027; (b) Enev, V. S.; Kaehlig, H.; Mulzer, J. *J. Am. Chem. Soc.* 2001, **123**, 10764; (c) Wender, P. A.; Hedge, S. G.; Hubbard, R. D.; Zhang, L. *J. Am. Chem. Soc.* 2002, **124**, 4956; (d) Crimmins, M. T.; Stanton, M. G.; Allwein, S. P. *J. Am. Chem. Soc.* 2002, **124**, 5958; (e) Nelson, S. G.; Cheung, W. G.; Kassick, A. J.; Hilfiker, M. A. *J. Am. Chem. Soc.* 2002, **124**, 13654.

14. For a review, see: Mulzer, J. Ohler, E. *Chem. Rev.* 2003, **103**, 3753.

15. (a) Nakagawa, M.; Ito, M.; Hasegawa, Y.; Akashi, S.; Taniguchi, M.; Hino, T. *Heterocycles* 1985, **23**, 224; (b) Nakagawa, M.; Ito, M.; Hasegawa, Y.; Akashi, S.; Hino, T. *Tetrahedron. Lett.* 1984, **25**, 3265.

16. Gypser, A.; Flasche, M.; Scharf, H.-D. *Liebigs. Ann. Chem.* 1994, 775.

17. White, J. D.; Amedio, J. C.; Gut, S.; Ohira, S.; Jayasinghe, L. R. *J. Org. Chem.* 1992, **57**, 2270.

18. Avery, M. A.; Chong, W. K. M.; Jennings-White, C. *J. Am. Chem. Soc.* 1992, **114**, 974.

19. (a) Jamison, T. F.; Shambayati, S.; Crowe, W. E.; Schreiber, S. L. *J. Am. Chem. Soc.* 1994, **116**, 5505; (b) Jamison, T. F.; Shambayati, S.; Crowe, W. E.; Schreiber, S. L. *J. Am. Chem. Soc.* 1997, **119**, 4353. See also: (c) Paquette, L. A.; Sun, L.-Q.; Friedrich, D.; Savage, P. B. *J. Am. Chem. Soc.* 1996, **118**, 9202.

20. (a) Chen, J.; Chen, X.; Bois-Choussy, M.; Zhu, J. *J. Am. Chem. Soc.* 2006, **128**, 87; (b) Endo, A.; Yanagisawa, A.; Abe, M.; Tohma, S.; Kan, T.; Fukuyama, T. *J. Am. Chem. Soc.* 2002, **124**, 6552; (c) Corey, E. J.; Gin, D. Y.; Kania, R. S. *J. Am. Chem. Soc.* 1996, **118**, 9202.

21. (a) Nicolaou, K. C.; Jung, J.; Yoon, W. H.; Fong, K. C.; Choi, H.-S.; He, Y.; Zhong, Y.-L.; Baran, P. S. *J. Am. Chem. Soc.* 2002, **124**, 2183; (b) Nicolaou, K. C.; Baran, P. S.; Zhong, Y.-L.; Fong, K. C.; Choi, H.-S. *J. Am. Chem. Soc.* 2002, **124**, 2190; (c) Nicolaou, K. C.; Zhong, Y.-L.; Baran, P. S.; Jung, J.; Choi, H.-S.; Yoon, W. H. *J. Am. Chem. Soc.* 2002, **124**, 2202; (d) Waizumi, N.; Itoh, T.; Fukuyama, T. *J. Am. Chem. Soc.* 2000, **122**, 7825; (e) Chen. C.; Layton, M. E.; Sheehan, S. M.; Shair, M. D. *J. Am. Chem. Soc.* 2000, **122**, 7424; (f) Tan, Q.; Danishefsky, S. J. *Angew. Chem. Int. Ed.* 2000, **39**, 4509; (g) for a

review see: Nicolaou, K. C.; Baran. P. S. *Angew. Chem. Int. Ed.* 2002, **41**, 2678.

22. Eckelbarger J. D.; Wilmot, J. T.; Gin, D. Y. *J. Am. Chem. Soc.* 2006, **128**, 10370.

23. (a) Nicolaou, K. C.; Wu, T. R.; Sarlah, D.; Shaw, D. M.; Rowliffe, E.; Burton D. R. *J. Am. Chem. Soc.* 2008, **130**, 11114; (b) Nicolaou, K. C.; Sarlah, D.; Shaw, D. M. *Angew. Chem. Int. Ed.* 2007, **46**, 4708; see also: (c) Du. C.; Li. L.; Li, Y.; Xie, Z. *Angew. Chem. Int. Ed.* 2009, **48**, 7853.

24. (a) Kobayashi, S.; Ueda, T.; Fukuyama, T. *Synlett*, 2000, 883; (b) Yokoshima, S.; Ueda, T.; Kobayashi, S.; Sato, A.; Kuboyama, T.; Tokuyama, H.; Fukuyama, T. *J. Am. Chem. Soc.* 2002, **124**, 2137.

25. (a) Okano, K.; Tokuyama, H.; Fukuyama, T. *J. Am. Chem. Soc.* 2006, **128**, 7136. See also: (b) Tichenor, M. S.; Boger, D. L. *Nat. Prod. Rep.* 2008, **25**, 220.

26. Pattenden, G.; Ashweek, N. J.; Baker-Glenn, C. A. G.; Walker, G. M.; Yee, J. G. K. *Angew. Chem. Int. Ed.* 2007, **46**, 4359.

27. Preussin: (a) Overhand, M.; Hecht, S. M. *J. Org. Chem.* 1994, **591**, 4721; pumiliotoxin 251D: (b) Overman, L. E.; Bell, K. L. *J. Am. Chem. Soc.* 1981, **103**, 1851; domoic acid: (c) Ohfune, Y.; Tomita, M. *J. Am. Chem. Soc.* 1982, **104**, 3511; sibirosamine: (d) Maurer, P. J. Knudsen, C. G.; Palkowitz, A. D.; Rapoport, H. *J. Org. Chem.* 1985, **50**, 325; WIN 6481: (e) Movassaghi, M.; Schmidt, M. A.; Ashenhurst, J. A. *Angew. Chem. Int. Ed.* 2008, **47**, 1485.

28. Balanol: (a) Nicolaou, K. C.; Bunnage, M. E.; Koide, K. *J. Am. Chem. Soc.* 1994, **116**, 8402; (b) Lampe, J. W.; Hughes, P. F.; Biggers, C. K.; Smith, S. H.; Hu, H. *J. Org. Chem.* 1994, **59**, 5147; tantazole B: (c) Parssons, R. L. Jr.; Heathcock, C. H. *Synlett* 1996, 1168; Galbulimina alkaloids: (d) Evans, D. A.; Adams, D. J. *J. Am. Chem. Soc.* 2007, **129**, 1048; (e) Movassaghi, M.; Hunt, D. K.; Tjandra, M. *J. Am. Chem. Soc.* 2006, **128**, 8126; for a review, see: Rinner, U.; Leutsch, C.; Aichinger, C. *Synthesis* 2010, 3763.

29. (a) Winkler, J. D.; Hershberger, P. M.; Springer, J. P. *Tetrahedron Lett.* 1986, **27**, 5177; (b) Winkler, J. D.; Hershberger, P. M. *J. Am. Chem. Soc.* 1989, **111**, 4852; (c) Stork, G.; Zhao, K. *J. Am. Chem. Soc.* 1990, **112**, 5875.

30. Mukayama, T. Shiina, I.; Iwadare, H.; Saitoh, M.; Nishimura, T.; Ohkawa, N.; Sakoh, H.; Nishimura, K.; Tani, Y.-i.; Hasegawa, M.; Yamada, K.; Saithoh, K. *Chem. Eur. J.* 1999, **5**, 121.

31. Hanessian, S.; Desilets, D.; Bennani, Y. L. *J. Org. Chem.* 1990, **55**, 3098.

32. Neooxazolomycin: Kende, A. S.; Kawamura, K.; DeVita, R. J. *J. Am. Chem. Soc.* 1990, **112**, 4070; for a review, see: Kang, S. H.; Kang, S. Y.; Lee, H.-S.; Buglass, A. J. *Chem. Rev.* 2005, **105**, 4537.

33. (a) Ogawa, A. K.; Armstrong, R. W. *J. Am. Chem. Soc.* 1998, **120**, 12435 and references cited therein; see also: (b) Smith, A. B. III.; Friestad, G. K.; Barbosa, J.; Bertounesque, E.; Hull, K. G.; Iwashima, M.; Qiu, Y.; Salvatore, B. A.; Grant Spoors, P.; Duan, J. J.-W. *J. Am. Chem. Soc.* 1999, **121**, 10468; 10478.

34. Yasui, K.; Tamura, Y.; Nakatani, T.; Kawada, K.; Ohtani, M. *J. Org. Chem.* 1995, **60**, 7567.

35. Saitoh, T.; Suzuki, T.; Sugimoto, M.; Hagiwara, H.; Hoshi, T. *Tetrahedron Lett.* 2003, **44**, 3175.

36. (a) Davoren, J. E.; Martin, S. F. *J. Am. Chem. Soc.* 2007, **129**, 510; (b) Davoren, J. E.; Harcken, C.; Martin, S. F. *J. Org. Chem.* 2008, **122**, 5473.

37. (a) Jiang, X.; Liu, B.; Lebreton, S.; De Brabander, J. K. *J. Am. Chem. Soc.* 2007, **129**, 6386; (b) Nicolaou, K. C.; Sun, Y.-P.; Guduru, R.; Banerji, B.; Chen, D. Y.-K. *J. Am. Chem. Soc.* 2008, **130**, 3633; See also: (c) Penner, M.; Rauniyar, V.; Kaspar, L. T.; Hall, D. G. *J. Am. Chem. Soc.* 2009, **131**, 14216.

38. Tatsuta, K.; Akimoto, K.; Annaka, M.; Ohno, Y.; Kinoshita, M. *Bull. Chem. Soc. Jpn.* 1985, **58**, 1699.

39. (a) Pearson, W. H.; Lovering, F. E. *J. Am. Chem. Soc.* 1995, **117**, 12336; (b) Pearson, W. H.; Lovering, F. E. *J. Org. Chem.* 1988, **63**, 3607.

40. (a) Hanessian, S.; Rancourt, G. *Can. J. Chem.* 1977, **55**, 1111; (b) Hanessian, S.; Rancourt, G.; Guindon, Y. *Can. J. Chem.* 1978, **56**, 1843; (c) Hanessian, S.; Rancourt, G. *Pure Appl. Chem.* 1977, **49**, 1201.

41. (a) Carreira, E. M.; DuBois, J. *J. Am. Chem. Soc.* 1994, **116**, 1082; (b) Carreira, E. M.; DuBois, J. *J. Am. Chem. Soc.* 1995, **117**, 8106; (c) Evans, D. A.; Barrow, J. C.; Leighton, J. L.; Robichaud, A. J.; Sefkow, M. *J. Am. Chem. Soc.* 1994, **116**, 12111; (d) Armstrong, A.; Barsanti, P. A.; Jones, L. H.; Ahmed, G. *J. Org. Chem.* 2000, **65**, 7020.

42. Crimmins, M. T.; Tabet, E. A. *J. Am. Chem. Soc.* 2000, **122**, 5473.

43. Chida, N.; Ohtsuka, M.; Ogawa, S. *J. Org. Chem.* 1993, **58**, 4441.

44. Hart, D. J.; Wu, W.-L.; Kozikowski, A. P. *J. Am. Chem. Soc.* 1995, **117**, 9369; (b) Chackalamannil, S.; Davies, R. J.; Asberom, T.; Doller, D.; Leone, D. *J. Am. Chem. Soc.* 1996, **118**, 9812.

45. (a) White, J. D.; Kawasaki, M. *J. Am. Chem. Soc.* 1990, **112**, 4991. See also: (b) Zibuck, R.; Liverton, N. J.; Smith, A. B. III. *J. Am. Chem. Soc.* 1986, **108**, 2451.

46. (a) Evans, D. A.; Fitch, D. M. *J. Org. Chem.* 1997, **62**, 454; (b) Seebach, D. Chow, H. F.; Jackson, R.; Lawson, K.; Sutter, M.; Thaisrivongs, S. Zimmermann, J. *J. Am. Chem. Soc.* 1985, **107**, 5292.

47. (a) Seebach, D.; Wasmuth, D. *Helv. Chim. Acta* 1980, **63**, 197; (b) Fráter, G.; Müller, U.; Gunther, W. *Tetrahedron* 1984, **40**, 1269.

48. (a) Nakada, M.; Kobayashi, S.; Iwasaki, S.; Ohno, M. *Tetrahedron Lett.* 1993, **34**, 1035; (b) Nakada, M.; Kobayashi, S.; Shibasaki, M.; Iwasaki, S.; Ohno, M. *Tetrahedron Lett.* 1993, **34**, 1039.

49. White, J. D.; Blakemore, P. R.; Green, N. J.; Hauser, E. B.; Holoboski, M. A.; Keown, L. E.; Nylund Koltz, C. S.; Phillips, B. W. *J. Org. Chem.* 2002, **67**, 7750.

50. Boeckman, R. K. Jr.; Enholm, E. J.; Demko, D. M.; Charette, A. B. *J. Org. Chem.* 1986, **51**, 4743; (b) Burke, S. D.; Piscopio, A. D.; Kort, M. E.; Matulenko, M. A.; Parker, M. H.; Armistead, D. M.; Shankaran, K. *J. Org. Chem.* 1994, **59**, 332; (c) Edwards, M. P.; Ley, S. V.; Lister, S. G.; Palmer, B. D.; Williams, D. J. *J. Org. Chem.* 1984, **49**, 3503.

51. Kanda, Y.; Fukuyama, T. *J. Am. Chem. Soc.* 1993, **115**, 8451.

52. Pattenden, G.; Thom, S. M. *Synlett* 1993, 215.

53. (a) Hanessian, S.; Yang, Y.; Giroux, S.; Mascitti, V.; Ma, J.; Raeppel, F. *J. Am. Chem. Soc.* 2003, **125**, 13784; for other syntheses of borrelidin see: (b) Duffey, M. O.; LeTiran, A.; Morken, J. P. *J. Am. Chem. Soc.* 2003, **125**, 1458; (c) Vong, B. G.; Kim, S. H.; Abraham, S.; Theodorakis, E. A. *Angew. Chem. Int. Ed.* 2004, **43**, 3947; (d) Togamitsu, T.; Takano, D.; Fukuda, T.; Otoguro, K.; Kuwajima, I.; Harigawa, Y.; Ōmura, S. *Org. Lett.* 2004, **6**, 1865.

54. Spongistatin 2: Smith, A. B. III; Doughty, V. A.; Sfouggatakis, C.; Bennett, C. S.; Koyanagi, J.; Takeuchi, M. *Org. Lett.* 2002, **4**, 783; rutamycin B: Panek, J. S.; Jain, N. F. *J. Org. Chem.* 2001, **66**, 2747; sanglifehrin A: Duan, M.; Paquette, L. A. *Angew. Chem. Int. Ed.* 2001, **40**, 3632; erythronolide A: Muri, D.; Lohse-Fraefel, N.; Carreira, E. M. *Angew. Chem. Int. Ed.* 2005, **44**, 4036; discodermolide: Smith, A. B. III; Freeze, B. S. *Tetrahedron* 2008, **64**, 261; elisabethin A: Heckrodt, T. J.; Mulzer, J. *J. Am. Chem. Soc.* 2003, **125**, 4680; iejemalide B: Fürstner, A.; Nevado, C.; Tremblay, M.; Chevrier, C.; Teply, F.; Aissa, C.; Waser, M. *Angew. Chem Int. Ed.* 2006, **45**, 5837; reidispongiolide A: Paterson, I.; Ashton, K.; Britton, R.; Cecere, G.; Chouraqui, G.; Florence, G. J.; Stafford, J. *Angew. Chem. Int. Ed.* 2007, **46**, 6167; callipeltoside C: Carpenter, J.; Northrup, A. B.; Chung, D.; Wiener, J. J. M.; Kim, S.-G.; MacMillan, D. W. C. *Angew. Chem. Int. Ed.* 2008, **47**, 3568; see also: Trost, B. M.; Gunzer, J. L.; Dirat, O.; Rhee, Y. H. *J. Am. Chem. Soc.* 2002, **124**, 10396; Evans, D. A.; Hu, E.; Burch, J. D.; Jaeschke, G. *J. Am. Chem. Soc.* 2002, **124**, 5654; Huang, H.; Panek, J. S. *Org. Lett.* 2004, **6**, 4383; okadaic

acid: Urbanek, R. A.; Sabes, S. F.;
Forsyth, C. J. *J. Am. Chem. Soc.* 1998,
120, 2523; azaspiracid: Evans, D. A.;
Kvaerno, L.; Dunn, T. B.; Beauchemin,
A.; Raymer, B.; Mulder, J. A.; Olhava,
E. J.; Juhl, M.; Kagechika, K.; Favor,
D. A. *J. Am. Chem. Soc.* 2008, **130**,
16295; zincophorin: Komatsu, K.;
Tanino, K.; Miyashita, M. *Angew. Chem.
Int. Ed.* 2004, **43**, 4341.

55. Money, T.; in *Studies in Natural Products
Chemistry*, Atta-ur-Rahman, Ed., 1989,
vol. 4, p. 625.

56. Mori, K.; Igarishi, Y. *Liebigs. Ann. Chem.*
1988, **717**.

57. Sarcodictyins: (a) Nicolaou, K. C.; Xu,
J. Y.; Kim, S.; Pfefferkorn, J.; Ohshima,
T.; Vourloumis, D.; Hosokawa, S. *J. Am.
Chem. Soc.* 1998, **120**, 8661; eleuther-
obin: (b) Chen, X.-T.; Bhattacharya,
S. K.; Zhou, B.; Gutteridge, C. E.;
Pettus, T. R. R.; Danishefsky, S. J.
J. Am. Chem. Soc. 1999, **121**, 6563;
(c) Nicolaou, K. C.; Ohshima, T.;
Hosokawa, S.; van Delft, F. L.;
Vourloumis, D.; Xu, J. Y.; Pfefferkorn,
J.; Kim, S. *J. Am. Chem. Soc.* 1998, **120**,
8674.

58. Mangion, I. K.; MacMillan, D. W. C. *J.
Am. Chem. Soc.* 2005, **127**, 3696.

59. (a) Marshall, J. A.; Wallace, E. M.;
Coan, P. S. *J. Org. Chem.* 1995, **60**,
796; (b) Marshall, J. A.; Bartley, G. S.;
Wallace, E. M. *J. Org. Chem.* 1996, **61**,
5729; (c) Marshall, J. M.; Liao, J. *J.
Org. Chem.* 1998, **63**, 5962. For reviews
on [2, 3]-Wittig rearrangements, see:
(d) Mikami, K.; Nakai, T. *Synthesis* 1991,
594; (e) Nakai, T.; Mikami, K. *Chem.
Rev.* 1986, **86**, 885.

60. Williams, D. R.; Coleman, P. J.; Henry,
S. S. *J. Am. Chem. Soc.* 1993, **115**,
11654.

61. Ilicicolin H: Williams, D. R.; Bremmer,
M. L.; Brown, D. L.; D'Antuono, J. *J.
Org. Chem.* 1985, **50**, 2807.; monensin:
Collum, D. B.; McDonald III, J. H.;
Still, W. C. *J. Am. Chem. Soc.* 1980,
102, 2118; ambruticin S: Kende, A. S.;
Mendoza, J. S.; Fujii, Y. *Tetrahedron*
1993, **89**, 8015; lasalosid A: Ireland,
R. E.; McGarvey, G. J.; Anderson,
R. C.; Badoud, R.; Fitzsimmons, B.;
Thaisrivongs, S. *J. Am. Chem. Soc.* 1980,

102, 6178; milbemycin β3: Williams,
D. R.; Barner, B. A.; Nishitani, K.;
Phillips, J. G. *J. Am. Chem. Soc.* 1982,
104, 4708; Integerrinecic acid: White,
J. D.; Jayasinghe, L. R. *Tetrahedron Lett.*
1988, **29**, 2139; deoxynupharidine:
Wong, C. F.; Auer, E.; LaLonde, R. T.
J. Org. Chem. 1970, **35**, 517; neosym-
bioimine: Varseev, G. N.; Maier, M. E.
Org. Lett. 2007, **9**, 1461; ryanodol:
Belanger, A.; Berney, D.; Borschber, H.;
Brousseau, R.; Doutheau, A.; Durand,
R.; Katayama, H.; Lapalme, R.; Leturc,
D.; Liao, C.; MacLachlan, F.; Maffrand,
J.; Maraza, F.; Martino, R.; Moreau, C.;
Saint-Laurent, L.; Saint-Onge, R.; Soucy,
P.; Ruest, L.; Deslongchamps, P. *Can. J.
Chem.* 1979, **57**, 3348; glycinoeclepin
A: Murai, A.; Tanimoto, N.; Sakamoto,
N.; Masamune, T. *J. Am. Chem. Soc.*
1988, **110**, 1985; breynolide sulfone:
Nishiyama, S.; Ikeda, Y.; Yoshida, S.;
Yamamura, S. *Tetrahedron Lett.* 1989,
30, 105; aplasmomycin: Corey, E. J.;
Pan, B. C.; Hua, D. H.; Deardorff,
D. R. *J. Am. Chem. Soc.* 1982, **104**,
6816; Anistatin: Niwa, H.; Nisiwaki,
M.; Tsukuda, I.; Ishigaki, T.; Ito, S.;
Wakamatsu, K.; Mori, T.; Ikagawa, M.;
Yamada. *J. Am. Chem. Soc.* 1990, **112**,
9001; phyllanthocin: McGuirk, P. R.;
Collum, D. B. *J. Am. Chem. Soc.* 1982,
104, 4496; periplanone B: Kitahara,
T.; Mori, M.; Mori, K. *Tetrahedron Lett.*
1987, **43**, 2689; steroid model: Stevens,
R. V.; Gaeta, F. C. A. *J. Am. Chem. Soc.*
1977, **99**, 6105; quadrone: Liu, H.-J.;
Llinas-Brunet, M. *Can. J. Chem.* 1988,
66, 528.

62. (a) Holton, R. A.; Somoza, C.; Kim,
H. B.; Liang, F.; Biediger, R. J.;
Boatman, P. D.; Shindo, M.; Smith,
C. C.; Kim, S.; Nadizadeh, H.; Suzukli,
Y.; Tao, C.; Vu, P.; Tang, S.; Zhang, P.;
Murthi, K. K.; Gentile, L. N.; Liu, J. H.
J. Am. Chem. Soc. 1994, **116**, 1597.
(b) Holton, R. A.; Somoza, C.; Kim,
H. B.; Liang, F.; Biediger, R. J.;
Boatman, P. D.; Shindo, M.; Smith,
C. C.; Kim, S.; Nadizadeh, H.; Suzukli,
Y.; Tao, C.; Vu, P.; Tang, S.; Zhang, P.;
Murthi, K. K.; Gentile, L. N.; Liu, J. H.
J. Am. Chem. Soc. 1994, **116**, 1599.

63. (a) Nicolaou, K. C.; Nantermet, P. G.; Ueno, H.; Guy, R. K.; Couladouros, E. A.; Sorensen, E. J. *J. Am. Chem. Soc.* 1995, **117**, 624; (b) Nicolaou, K. C.; Liu, J.-J.; Yang, Z.; Ueno, H.; Sorensen, E. J.; Claiborne, C. F.; Guy, R. K.; Hwang, C.-K.; Nakada, M.; Nantermet, P. G. *J. Am. Chem. Soc.* 1995, **117**, 634; (c) Nicolaou, K. C.; Yang, Z.; Liu, J.-J.; Nantermet, P. G.; Claiborne, C. F.; Renaud, J.; Guy, R. K.; Shibayama, K. *J. Am. Chem. Soc.* 1995, **117**, 645; (d) Nicolaou, K. C.; Ueno, H.; Liu, J.-J.; Nantermet, P. G.; Yang, Z.; Renaud, J.; Paulvannan, K.; Chadha, R. *J. Am. Chem. Soc.* 1995, **117**, 653.

64. (a) Danishefsky, S. J.; Masters, J. J.; Young, W. B,; Link, J. T.; Snyder, L. B.; Magee, T. V.; Jung, D. K.; Isaacs, R. C. A.; Bornmann, W. G.; Alaino, C. A.; Coburn, C. A.; DiGrandi, M. J. *J. Am. Chem. Soc.* 1996, **118**, 2843; (a) Wender, P. A.; Mucciaro, T. P. *J. Am. Chem. Soc.* 1992, **114**, 5878; (c) Wender, P. A.; Glass, T. E.; Krauss, N. E.; Mühlebach, M.; Peschke, B.; Rawlins, D. B. *J. Org. Chem.* 1996, **61**, 7662; (d) Wender, P. A.; Badham, N. F.; Conway, S. P.; Floreancig, P. E.; Glass, T. E.; Houze, J. B.; Krauss, N. E.; Lee, D.; Marquess, D. G.; McGrane, P. L.; Meng, W.; Natchus, M. G.; Shuker, A. J.; Sutton, J. C.; Taylor, R. E. *J. Am. Chem. Soc.* 1997, **119**, 2757; (e) Kusama, H.; Hara, R.; Kawahara, S.; Nishimori, T.; Kashima, H.; Nakamura, N.; Morihira, K.; Kuwajima, I. *J. Am. Chem. Soc.* 2000, **122**, 3811; (f) Morihira, K.; Hara, R.; Kawahara, S.; Nishimori, T.; Nakamura, N.; Kusama, H. Kuwajima, I. *J. Am. Chem. Soc.* 1998, **120**, 12980; (g) Mukaiyama, T.; Shiina, I.; Iwadare, H.; Saitoh, M.; Nishimura, T.; Ohkawa, N.; Sakoh, H.; Nishimura, K.; Tani, Y.; Hasegawa, M.; Yamada, K.; Saitoh, K. *Chem. Eur. J.* 1999, **5**, 121.

65. (a) Paquette, L. A.; Zhao, M.; *J. Am. Chem. Soc.* 1998, **120**, 5203; (b) Paqette, L. A.; Wang, H.-L.; Su, Z.; Zhao, M. *J. Am. Chem. Soc.* 1998, **120**, 5213.

66. Büchi, G.; MacLeod, W. D.; Padilla, O. J. *J. Am. Chem. Soc.* 1964, **86**, 4438.

67. (a) Paquette, L. A.; Sugimura, T. *J. Am. Chem. Soc.* 1986, **108**, 3841; (b) Sugimura, T.; Paquette, L. A. *J. Am. Chem. Soc.* 1987, **109**, 3017.

68. Bruner, S. D.; Radeke, H. S.; Tallarico, J. A.; Snapper, M. L. *J. Org. Chem.* 1995, **60**, 1114; see also: Toto, L.; Poupon, E.; Rueden, E. J.; Theodorakis, E. A. *Org. Lett.* 2002, **4**, 819; Poigny, S. Guyot, M.; Samadi, M. *J. Org. Chem.* 1998, **63**, 5890.

69. Phyllanthocin: (a) Martin, S. F.; Dappen, M. S.; Dupre, B.; Murphy, C. J. *J. Org. Chem.* 1987, **52**, 3706; (b) Trost, B. M.; Edstrom, E. D. *Angew. Chem. Int. Ed.* 1990, **29**, 520; (c) Williams, D. R.; Sit, S. Y. *J. Am. Chem. Soc.* 1984, **106**, 2949; (d) Burke, S. D.; Cobb, J. E.; Takeuchi, K. *J. Org. Chem.* 1985, **50**, 3420. 4496; (e) Smith, A. B. III.; Fukui, M. *J. Am. Chem. Soc.* 1987, **109**, 1269; (f) McGuirk, P. R.; Collum, D. B. *J. Am. Chem. Soc.* 1982, **104**, 1269.

70. (a) Borzelli, R. M.; Weinreb, S. M.; Parvez, M. *J. Am. Chem. Soc.* 1994, **116**, 9789. See also: (b) Barrett, A. G. M.; Boys, M. L; Boehm, T. L. *J. Org. Chem.* 1996, **61**, 685; (c) McDermott, T. S.; Mortlock, A. A.; Heathcock, C. H. *J. Org. Chem.* 1996, **61**, 700.

71. (a) Partridge, J. J.; Chadha, N. K.; Uskoković, M. R. *J. Am. Chem. Soc.* 1973, **95**, 532; (b) Partridge, J. J.; Chadha, N. K.; Uskoković, M. R. *Org. Synth.* 1985, **63**, 44.

72. Stork, G.; Paterson, I.; Lee, F. K. C. *J. Am. Chem. Soc.* 1982, **104**, 4686.

73. Grieco, P. A.; Inanagu, J.; Lin, N. H.; Yanami, T. *J. Am. Chem. Soc.* 1982, **104**, 5781.

74. Evans, D. A.; Andrews, G. C. *Acc. Chem. Res.* 1974, **7**, 147.

75. Barco, A.; Benetti, S.; De Risi, C.; Marchetti, P.; Pollini, G. P.; Zaniralo, V. *Tetrahedron: Asymmetry* 1997, **8**, 3515.

76. Isoquinuclidine: Trost, B. M.; Romero, A. G. *J. Org. Chem.* 1986, **51**, 233; palytatin: Hanessian, S.; Sakito, Y.; Dhanoa, D.; Baptistella, L. *Tetrahedron* 1989, **45**, 6623; bengamide E: Chida, N.; Tobe, T.; Ogawa, S. *Tetrahedron Lett.* 1991, **32**, 1063; negamycin: Maycock, C. D. *Tetrahedron Lett.* 1992, **33**, 4633; FK-506: White, J. D.; Toske, S. G.; Yakura, T. *Synlett* 1994, 591; manzamine A: Kamenecka, T. M.;

Overman, L. E. *Tetrahedron Lett.* 1994, **35**, 4279; esperamicinone: Ulibarri, G.; Nadler, W.; Skrysdstrup, T.; Audrain, H.; Chiaroni, A.; Riche, C.; Grierson, D. S. *J. Org. Chem.* 1995, **60**, 2753; malyngolide: Matsuo, K.; Matsumoto, T.; Nishiwaki, K. *Heterocycles* 1998, **48**, 1213; mycosporin 1: White, J. D.; Cammack, J. H.; Sakuma, K.; Rewcastle, G. W.; Widener, R. K. *J. Org. Chem.* 1995, **60**, 3600; penitrem D: Smith, A. B. III.; Kanoh, N.; Ishiyama, H.; Hartz, R. A. *J. Am. Chem. Soc.* 2000, **122**, 11254; nodulisporic acid A: Smith, A. B. III.; Kurti, L.; Davulcu, A. H.; Cho, Y. S. *Org. Process Res. Dev.* 2007, **11**, 19; dysidiolide: Corey, E. J.; Roberts, B. E. *J. Am. Chem. Soc.* 1997, **119**, 12425; nor-zoanthamine: Miyashita, M.; Sasaki, M.; Hattori, I.; Sakai, M.; Tanino, K. *Science* 2004, **305**, 495; absinthin: Zhang, W.; Luo, S.; Fang, F.; Chen, Q.; Hu, H.; Jia, X.; Zhai, H. *J. Am. Chem. Soc.* 2005, **127**, 18.

77. (a) Stork, G. Rychnovsky, S. D. *Pure Appl. Chem.* 1986, **58**, 767; (b) Stork, G. Rychnovsky, S. D. *Pure Appl. Chem.* 1987, **59**, 345; (c) Nakatsuka, M.; Ragan, J. A.; Sammakia, T.; Smith, D. B.; Uehling, D. E.; Schreiber, S. L. *J. Am. Chem. Soc.* 1990, **112**, 5583; (d) Beshore, D. C.; Smith, A. B. III. *J. Am. Chem. Soc.* 2007, **129**, 4148.

78. Gringore, O. H.; Rouessac, F. P. *Org. Synth.* 1985, **63**, 121.

79. (a) Hanessian, S.; Murray, P. J. *Tetrahedron* 1987, **43**, 5072; (b) Hanessian, S.; Murray, P. J.; Sahoo, S. P. *Tetrahedron Lett.* 1985, **26**, 5623; (c) Hanessian, S.; Murray, P. J.; Sahoo, S. P. *Tetrahedron Lett.* 1985, **26**, 5627.

80. (a) Häfele, B.; Jäger, V. *Liebigs Ann. Chem.* 1987, 85; (b) Takano, S.; Kurotaki, A.; Takahashi, M.; Ogasawara, K. *Synthesis* 1986, 403.

81. Hanessian, S. *Aldrichim. Acta*, 1989, **22**, 3.

82. Hanessian, S.; Murray, P. J. *J. Org. Chem.* 1987, **52**, 1170.

83. (a) Martin, N.; Thomas, E. J. *Tetrahedron Lett.* 2001, **42**, 8373. See also: (b) Klar, U.; Kuczynski, F.; Schwede, W.; Berger, M.; Skuballa, W.; Buchmann, B. *Synthesis* 2005, 301.

84. Robin, J. P.; Gringore, O.; Brown, E. *Tetrahedron Lett.* 1980, **21**, 2709.

85. Smith, A. B. III.; Sestelo, J. P.; Dormer, P. G. *J. Am. Chem. Soc.* 1995, **117**, 10755.

86. (a) Nicolaou, K. C.; Harrison, S. T. *Angew. Chem. Int. Ed.* 2006, **45**, 3256; (b) Nicolaou, K. C.; Harrison, S. T. *J. Am. Chem. Soc.* 2007, **129**, 429.

87. Fürstner, A.; Nevado, C.; Tremblay, M.; Chevrier, C.; Teply, F.; Aissa, C.; Waser, M. *Angew. Chem. Int. Ed.* 2006, **45**, 5837.

88. (a) Takano, S.; Hatakeyama, S.; Ogasawara, K. *J. Am. Chem. Soc.* 1976, **98**, 3022; (b) Takano, S.; Hirama, M.; Araki, T.; Ogasawara, K. *J. Am. Chem. Soc.* 1976, **98**, 7084. See also: (c) Malcolmson, S. J.; Meek, S. J.; Sattely, E. S.; Schrock, R. R.; Hoveyda, A. H. *Nature* 2008, **456**, 933.

89. (a) Stork, G.; Rychnovsky, S. D. *J. Am. Chem. Soc.* 1987, **109**, 1564; (b) Stork, G.; Rychnovsky, S. D. *J. Am. Chem. Soc.* 1987, **109**, 1565; See also: (c) Peng, Z.-H.; Woerpel, K. A.; *J. Am. Chem. Soc.* 2003, **125**, 6018.

90. Chamberlin, A. R.; Dezube, M.; Reich, S. H.; Sall, D. J. *J. Am. Chem. Soc.* 1989, **111**, 6247.

91. Tomioka, K.; Tanaka, M.; Koga, K. *Tetrahedron Lett.* 1982, **23**, 3401.

92. Smith, A. B. III.; Rano, T. A.; Chida, N.; Sulikowski, G. A.; Wood, J. L. *J. Am. Chem. Soc.* 1992, **114**, 8008.

93. (a) Hanessian, S.; Sahoo, S. P.; Murray, P. J. *Tetrahedron Lett.* 1985, **26**, 5631; (b) Hanessian, S.; Sahoo, S. P.; Botta, M. *Tetrahedron Lett.* 1987, **26**, 1143; (c) Hanessian, S.; Sahoo, S. P.; Botta, M. *Tetrahedron Lett.* 1987, **26**, 1147

94. Ziegler, F. E.; Cain, W. T. *J. Org. Chem.* 1989, **54**, 3347.

95. Shieh, H.-M.; Prestwich, G. D. *Tetrahedron Lett.* 1982, **23**, 4643.

96. Takano, S.; Yonaga, M.; Morimoto, M.; Ogasawara, K. *J. Chem. Soc., Perkin Trans. 1* 1985, 305.

97. Evans, D. A.; Welch, D. S.; Speed, A. W. H.; Moniz, G. A.; Reichelt, A.; Ho, S. *J. Am. Chem. Soc.* 2009, **131**, 3840.

98. Trost, B. M.; Dong, G. *Nature* 2008, **456**, 485.

9
Applications of the *Chiron Approach*

The principles of the *chiron approach* were delineated in Chapter 6 with many examples of the use of chiral, non-racemic starting materials that are readily available from natural sources. Other sources of enantiopure or enantioenriched compounds were also discussed in Chapter 5. Having access to such a diverse set of natural and unnatural compounds that incorporate functional, stereochemical, and topological features within their carbon framework, it is the investigator's prerogative to exploit their utility as starting materials for the elaboration of target molecules in the context of their total synthesis. We have seen that the extent of functional and stereochemical overlap of segments of target molecules with chirons in their native or chemically modified variants depends on their availability. The recognition of such chirons within the carbon framework of an intended structure is a prerequisite to their efficient utilization as versatile starting materials.

Before we discuss selected examples of the applications of the *chiron approach* to total synthesis utilizing amino acids, carbohydrates, hydroxy acids, terpenes, carbocycles, and lactones in the next six chapters, it is useful to divide target compounds into three categories. The terms "compound," "molecule," and "structure" shall be used interchangeably depending on the particular context.

For each synthesis, we shall emphasize the origin of the chiral starting materials and their elaboration into useful intermediates in a retrosynthetic manner, excluding simple operations such as minor functional group interconversions and protective group adjustments. Key reactions relating to the *chiron approach* and the general synthesis are indicated separately, as are relevant mechanisms and transformations for certain steps.

9.1
Category I Target Molecules

Examples in this category pertain to molecules that harbor two or more stereogenic centers within proximity to each other. They may be synthesized from an appropriate chiron that bears at least one functional group, or a carbon substituent on a stereogenic center.

Design and Strategy in Organic Synthesis: From the Chiron Approach to Catalysis, First Edition.
Stephen Hanessian, Simon Giroux, and Bradley L. Merner.
© 2013 Wiley-VCH Verlag GmbH & Co. KGaA. Published 2013 by Wiley-VCH Verlag GmbH & Co. KGaA.

This category of target molecules consists of compact acyclic or alicyclic structures with two or more stereogenic centers, monocyclic or polycyclic structures that may have appended or intervening acyclic segments, various heterocycles and related structures. The presence of a single functional group on a stereogenic carbon atom in the chosen chiron as a starting material, may strongly influence other bond-forming steps involved in the creation of new stereogenic carbons (internal induction). Alkaloids, steroids, triquinanes, various secondary metabolites, and fused heterocycles, are representative examples.

9.1.1
Streptolic acid

For the synthesis of streptolic acid, a subunit in the antibiotic streptolydigin, Ireland and Smith [1] utilized D-glucose as a starting material to converge with ring *A* (Figure 9.1). Appropriate elaboration of D-glucose led to intermediates **E** and **F**. Application of the Ireland-Claisen ester-enolate rearrangement [2] with transfer of chirality from C-3 to C-1 gave the corresponding branched propionate (**E** to **D**). Further elaboration to **C**, and Wittig extension led to the enone **B**. Introduction of the hydroxymethyl branch, followed by a Sharpless asymmetric epoxidation afforded the fully elaborated ring *A* (**B** to **A**). All six carbon atoms of D-glucose were utilized in the process, although none of the original stereogenic centers remained. The main focus in the synthesis was to exploit the D-glucopyranose scaffold, as its D-glucal template (**E**), in order to install the propionate branch at the anomeric position. Although full selectivity was not achieved at the off-template *C*-methyl group in the [3,3]-sigmatropic rearrangement (see relevant mechanism, Figure 9.1), the synthesis was successfully completed to yield the intended target compound.

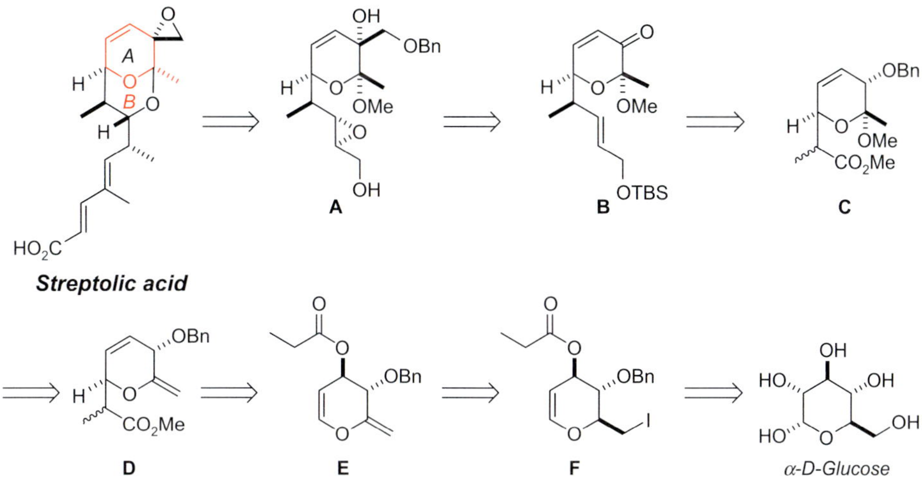

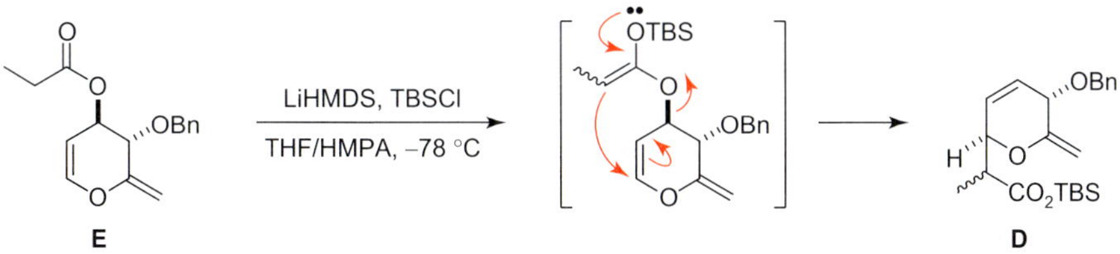

Figure 9.1 Retrosynthetic analysis of streptolic acid leading to D-glucose (Category 1).

9.1.2
ent-Gelsedine

The single source of chirality used in the elaboration of the bridged azacyclic ring system in *ent*-gelsedine was L-malic acid (Figure 9.2). Thus, Hiemstra and coworkers [3] were able to use proximity and constraint in the 4-hydroxylactam **G** (derived from L-malic acid), to secure the allene derivative **F**, which was subjected to an iodide-mediated azonia-allene cyclization in the presence of formic acid to provide the azabicyclo[4.2.1]nonane intermediate **E** (see relevant mechanism A, Figure 9.2). Palladium-catalyzed aminocarbonylation of **E** led to **D**, which underwent a Heck spirocyclization followed by oxidation of the alcohol and methylenation of the resulting ketone to give **C**. Hydroboration gave **B**, which was subjected to a Hg(II)-mediated cycloetherification and demercuration (see relevant mechanism B, Figure 9.2). Functional group manipulation relying on iminium ion functionalization then gave **A**, which is an immediate precursor of *ent*-gelsedine (see relevant transformation C, Figure 9.2). It should be noted that the original stereogenic center present in L-malic acid was sacrificed in a self-immolative process (**F** to **E** to **C**).

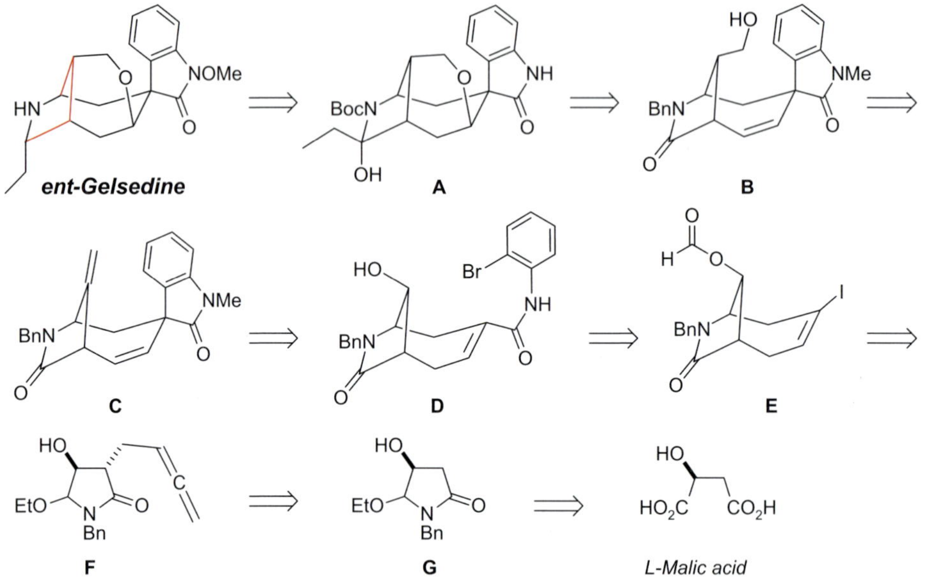

Key reactions: *Lactam dianion allenylmethylation (G to F); NaI/HCO₂H-promoted azonia-allene cyclization (F to E); Pd-catalyzed aminocarbonylation (E to D, after formate hydrolysis); Pd-mediated Heck spirocyclization (D to C); hydroboration (C to B); Hg(II)-mediated cycloetherification/reductive demercuration then oxidative de-N-methylation (B to A)*

Relevant transformations and mechanisms:

A.

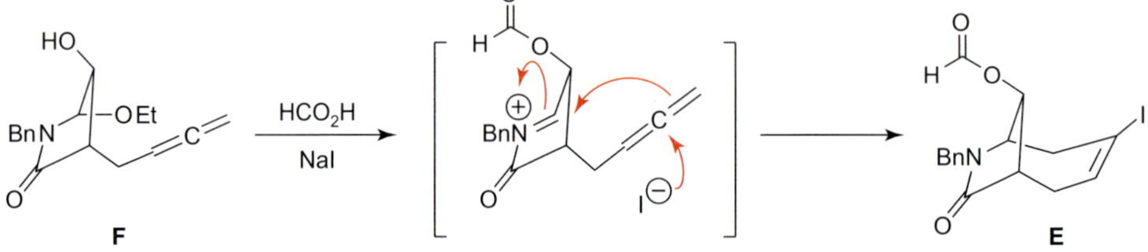

Figure 9.2 Retrosynthetic analysis of *ent*-gelsedine leading to L-malic acid (Category 1).

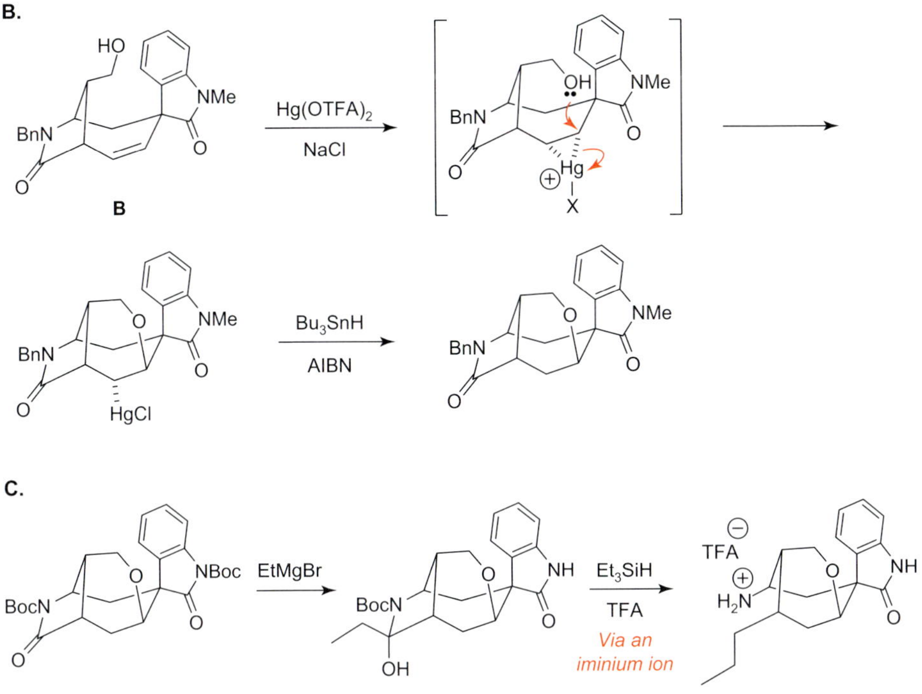

B.

C.

Figure 9.2 (continued).

9.1.3
Vincamine

The Rapoport synthesis of vincamine from ʟ-aspartic acid proceeded through an intramolecular enolate alkylation followed by a stereocontrolled enolate *C*-ethylation (**E** to **D**) (see relevant transformation A, Figure 9.3) [4]. Alkylation of **D** with tryptophyl bromide afforded the tethered intermediate **C**, which underwent a remarkable decarboxylative/iminium ion-mediated intramolecular cyclization to the tetracyclic intermediate **B** (see relevant mechanism B, Figure 9.3). Conversion

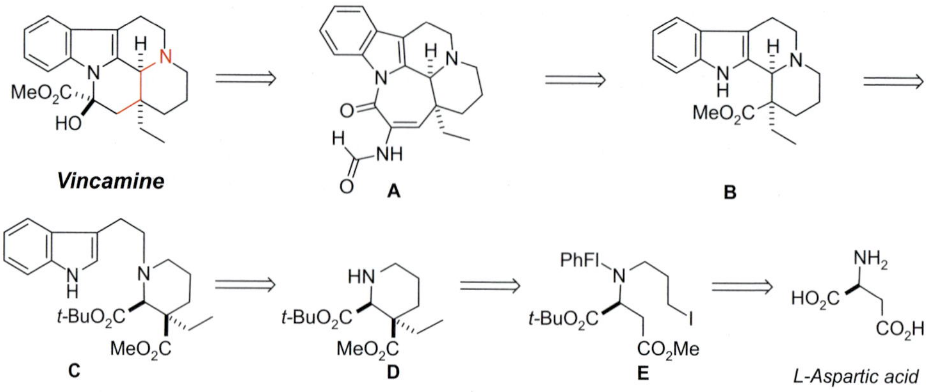

Key reactions: *Intramolecular enolate alkylation/C-ethylation (**E** to **D**); N-alkylation with tryptophyl bromide (**D** to **C**); decarbonylation/cyclization (free acid of **C** to **B**); methyl isocyanoacetate condensation/cyclization (**B** aldehyde to **A**); acid-catalyzed decarbonylation/cyclization (**A** to **target**); PhFl = 9-phenylfluorenyl*

Relevant transformations and mechanisms:

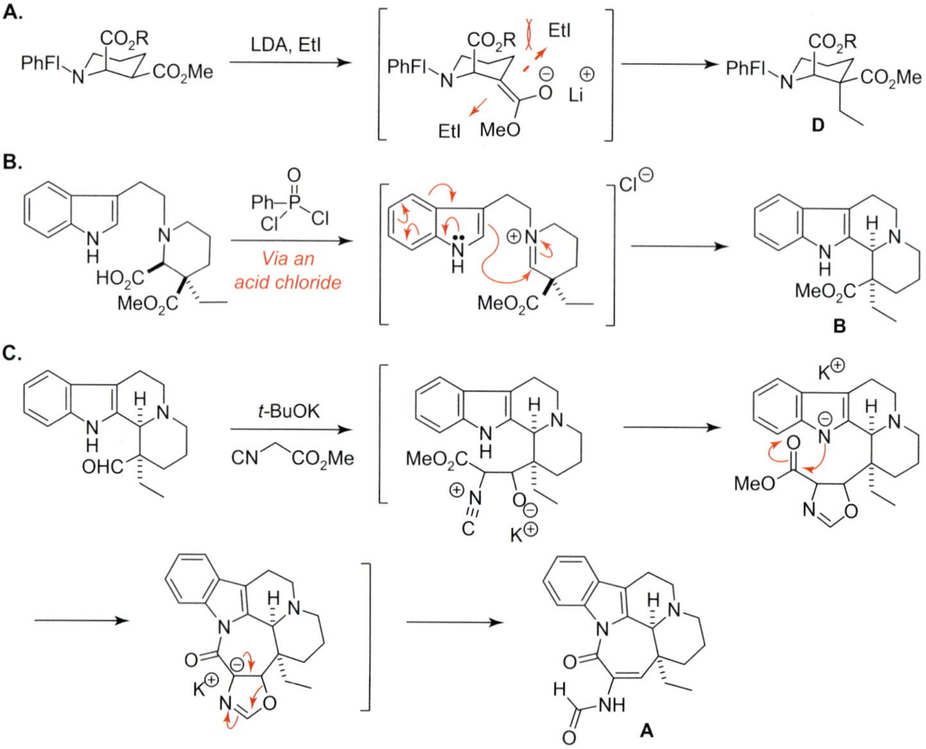

A.

B.

C.

Figure 9.3 Retrosynthetic analysis of vincamine leading to L-aspartic acid (Category 1).

to the aldehyde and condensation with methyl isocyanoacetate, following loss of ammonia from the transient *N*-formyl enaminolactam **A** under basic conditions, afforded vincamine (see relevant mechanism C, Figure 9.3). Thus, the single stereogenic center in L-aspartic acid was used to introduce a quaternary carbon and to control the stereochemistry of the indole-iminium ion cyclization (**E** to **D**, **C** to **B**). Three of the four carbon atoms of the starting α-amino acid were conserved in the structure of the target compound.

9.1.4
Peribysin E

The total synthesis of peribysin E by Danishefsky and coworkers [5] resulted in a revision of the previously assigned absolute configuration of the natural product (Figure 9.4). Considering the relative disposition of the two *C*-methyl groups in the target compound, an appropriate cyclohexane-type chiron was a reasonable starting point. Initially, the synthesis was completed starting with (*S*)-carvone, only to discover that it furnished *ent*-peribysin E. Hence, the synthesis was repeated with (*R*)-carvone. Diels-Alder reaction of (*R*)-carvone with the Danishefsky diene and further elaboration of the keto groups in **H**, chemoselectively, led to **G**, **F**, and then **E** respectively. Having served its stereodirecting role in the Diels-Alder cycloaddition, the isopropenyl group was oxidatively cleaved to the corresponding ketone **E**, then subjected to a Baeyer-Villiger oxidation with *m*-CPBA to give **D**. Transformation to the vinyl iodide **C**, Suzuki cross-coupling, followed by base-mediated epoxidation with H_2O_2, and stereoselective reduction led to epoxide **B**. A remarkably efficient stereocontrolled, $TiCl_4$-mediated ring contraction (semi-pinacol rearrangement) afforded **A**, which was transformed to peribysin E (see relevant mechanism, Figure 9.4).

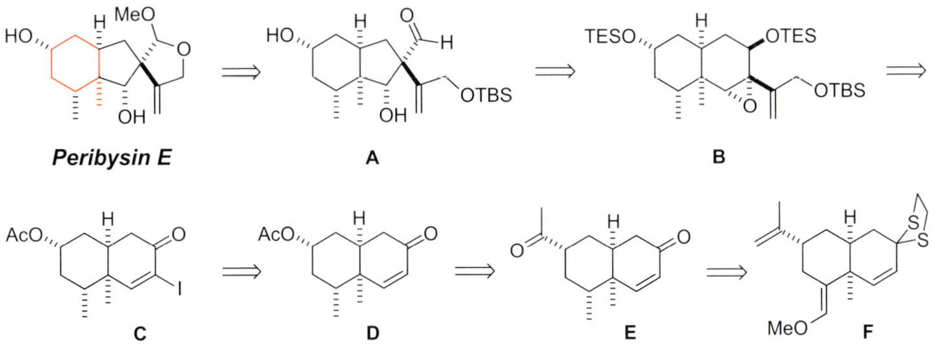

Figure 9.4 Retrosynthetic analysis of peribysin E leading to (*R*)-carvone (Category 1).

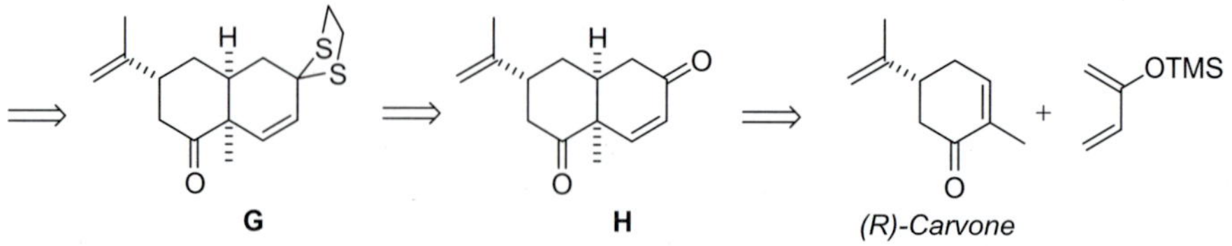

G   H   (R)-Carvone

Key reactions: *EtAlCl$_2$-catalyzed Diels-Alder cycloaddition ((R)-Carvone to **H**); chemoselective dithioketal formation (**H** to **G**); vinyl ether to aldehyde to C-methyl group (**F** to **E**); chemoselective Baeyer-Villiger oxidation (**E** to **D**); Suzuki cross-coupling/epoxidation (**C** to **B**); TiCl$_4$-mediated ring contraction (**B** to **A**); intramolecular methyl glycoside formation (**A** to **target**)*

Relevant mechanism:

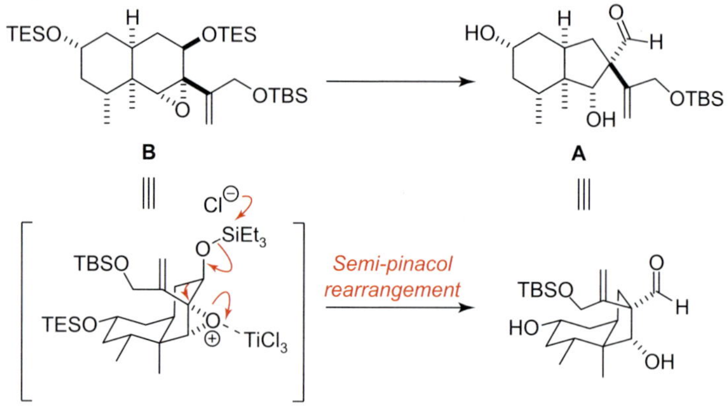

B   A

Semi-pinacol rearrangement

Figure 9.4 *(continued).*

9.2
Category II Target Molecules

Molecules belonging to this category are endowed with a more expansive structural landscape, and may contain multiple stereogenic centers. More than a single chiron may be required as starting materials. These must be eventually joined in practical ways, including the use of other chiral or achiral bridging units so as to provide the desired level of functional, stereochemical, and skeletal convergence with a given substructure in the target molecule.

This category of molecules consists among others, of macrocycles, linear polyethers, and extended polyketides harboring multiple stereogenic centers. Macrolides, ansa compounds, marine toxins, and polyether antibiotics are also representative classes.

9.2.1
FK-506

The total synthesis of the immunosuppressive compound FK-506 (Prograf) has been the subject of several elegant reports [6]. The approach adopted by Danishefsky and coworkers [7] in their formal synthesis capitalized heavily on the use of both apparent and hidden chirons (Figure 9.5).

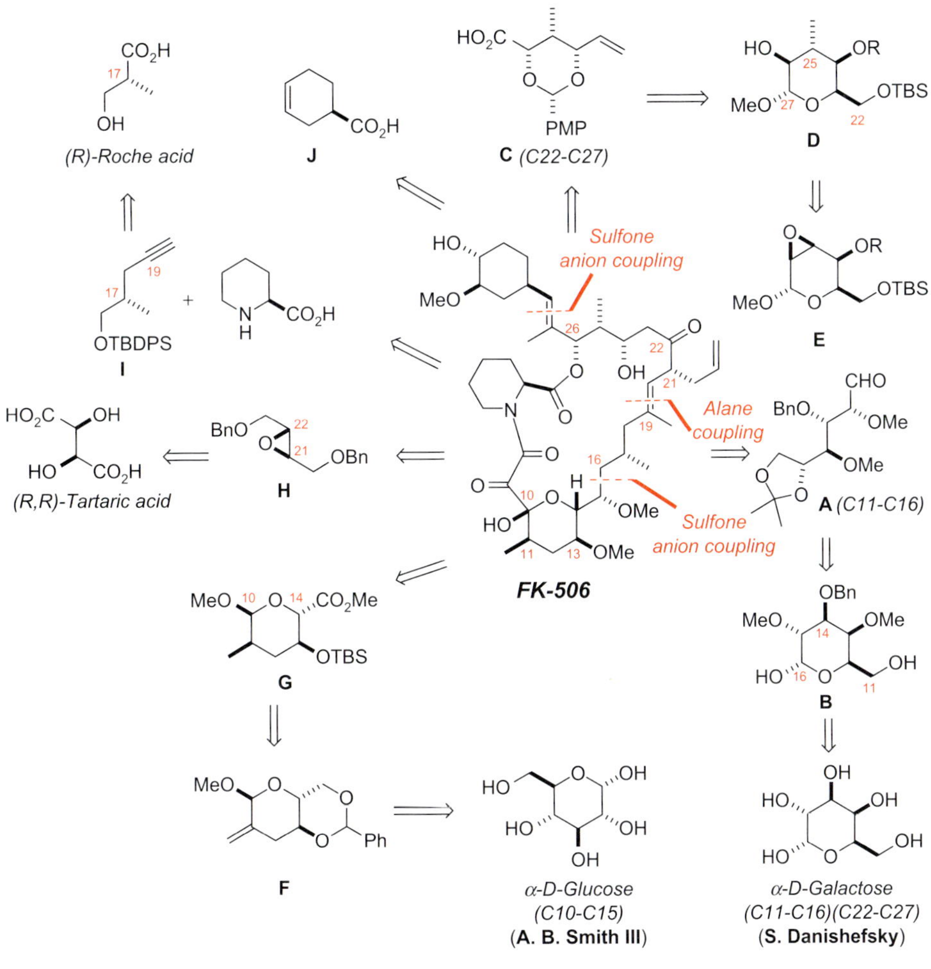

Key reactions: *Alane-mediated epoxide opening (C20-C21); Julia couplig (C15-C16; C27-C28); Negishi methylalumination (C19); Corey-Fuchs alkynylation (C19-C20); asymmetric Diels-Alder reaction using Oppolzer's sultam (for the synthesis of J)*

Figure 9.5 Elaboration of FK-506 from multiple chirons and key segment coupling methods (Category II).

Thus, D-galactose was chosen as a chiron for the C-11–C-16 and C-22–C-27 segments of FK-506. Systematic modification of existing hydroxyl groups by preferential *O*-methylation and conversion to acyclic analogue **A** was predictably executed from intermediate **B**. Introduction of a *C*-methyl group that corresponds to C-25 was done through the *trans*-diaxial opening of an *O*-protected epoxide **E** to afford **D**. This was transformed to the primary iodide, then subjected to a Zn-mediated fragmentation to give the carboxylic acid **C**. The pipecolic acid and cyclohexane subunits were prepared using well-established methods.

A. B. Smith III and coworkers [8] also used a carbohydrate approach to elaborate the hemiketal ring encompassing C-10–C-15 of FK-506 (Figure 9.5). Thus, methyl 4,6-*O*-benzylidene-α-D-glucopyranoside was transformed into the C-2-methylene precursor **F**, which was further elaborated to **G**. Regioselective opening of the C_2-symmetric epoxide **H**, readily available from (*R*,*R*)-tartaric acid with an alane precursor **I** prepared from the (*R*)-Roche acid, led to the C-16–C-23 segment of FK-506. The (1*R*)-3-cyclohexene-1-carboxylic acid (**J**) subunit was prepared by an asymmetric Diels-Alder reaction using Oppolzer's chiral auxiliary.

The combined Danishefsky [7] and Smith [8] approaches using carbohydrate precursors to elaborate acyclic and cyclic segments of FK-506 accounted for 10 of the 14 stereogenic carbon atoms in the macrocyclic lactam that contain an oxygen or a *C*-methyl function.

9.2.2
Okadaic acid

The extended structural landscape of okadaic acid lends itself to a variety of synthetic approaches as exemplified by the contributions of Isobe [9], Forsyth [10], and Ley [11]. Below we illustrate the most recent synthesis by Ley and coworkers [11]. A hetero Diels-Alder cyclization starting with an (*R*)-Roche aldehyde derivative according to Danishefsky, led to the 3-keto glycal **B**, corresponding to ring *B* in okadaic acid. After introduction of the branched *C*-methyl group, compound **A** was converted to the anomeric lithio derivative, and the latter was coupled to lactone **G** to give **H** (see relevant mechanism A, Figure 9.6). Further manipulation led to Julia-olefination precursor **I**. Reaction of the phosphonate anion derived from **C** with aldehyde **D** (prepared from D-mannose), followed by spirocyclization with release of the hydroxyl groups, provided the spiroacetal-dioxadecalin **E** encompassing rings *C*, *D*, and *E*. Finally, the (*S*)-Roche aldehyde was used, this time, in a Brown crotylation protocol [12] *en route* to spiroacetal intermediate **F** representing rings *F* and *G* (see relevant mechanism B, Figure 9.6). Segment coupling was accomplished by application of an alane-mediated *C*-glycosidation (rings *A* and *E*), and a Julia coupling (ring *B* to *C*). Thus, D-mannose, the (*R*)- and (*S*)-Roche acids, and (4*R*)-hydroxymethyl butyrolactone were efficiently utilized by Ley and coworkers [11] in elaborating rings *B*, *C*, *D* and *F* in okadaic acid. In each instance, the resident chirality in the original chiron was responsible for the generation of a second stereogenic center with the desired absolute configuration.

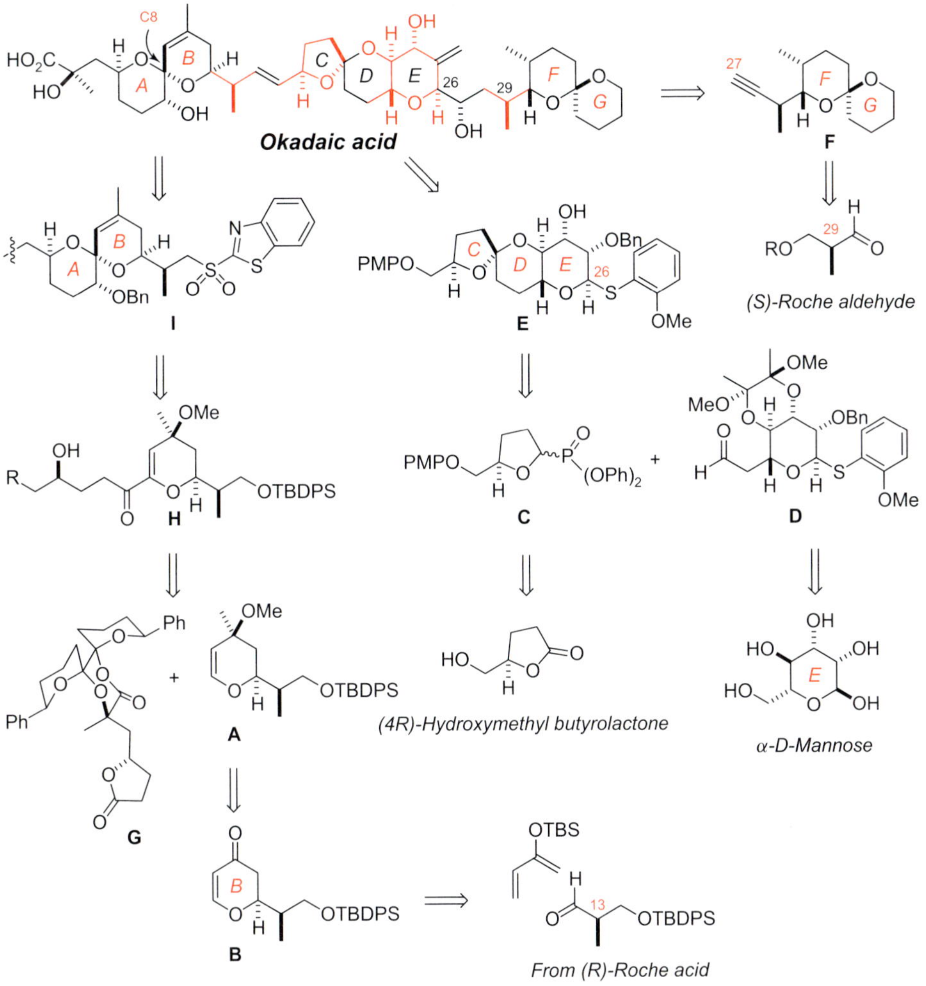

Key reactions: *Hetero Diels-Alder reaction (ring B); Horner-Wadsworth-Emmonsreaction/cyclization (ring C/D); Brown crotylation (ring F); C-glycoside formation (rings A and E, C8 and C26); Julia olefination; spiroacetalization (rings A/B, C/D, and F/G)*

Figure 9.6 Retrosynthetic analysis of okadaic acid leading to multiple chirons (Category II).

Relevant mechanisms and transformations:

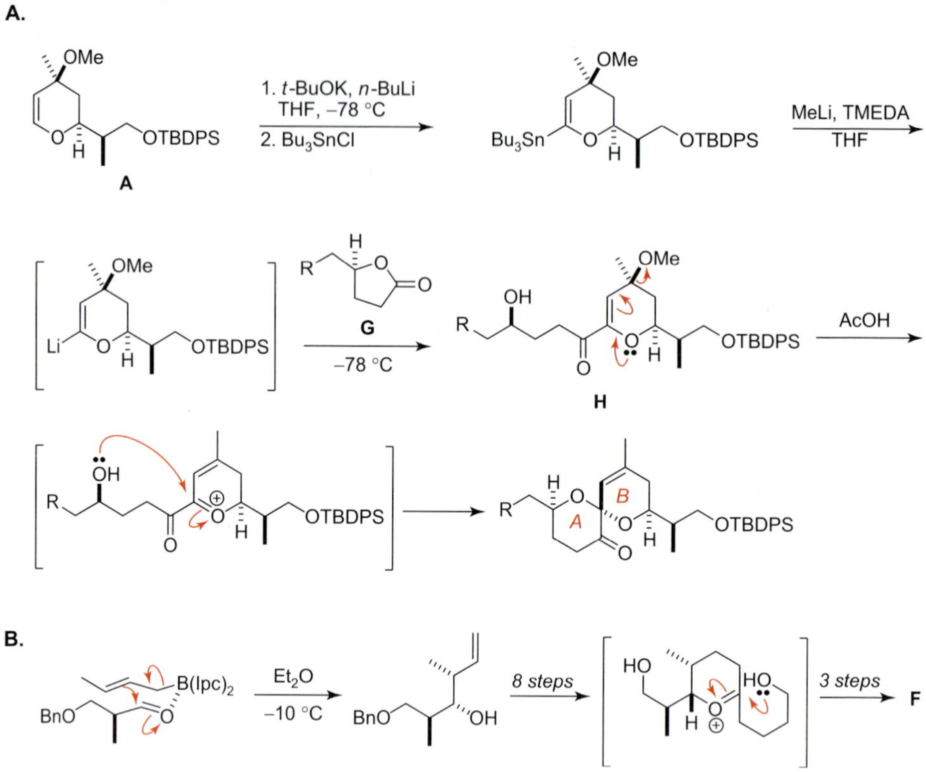

A.

B.

Figure 9.6 *(continued).*

9.2.3
Phorboxazole A

The first total synthesis of phorboxazole A was reported by Forsyth and coworkers [13]. Other syntheses by the groups of Evans [14], Williams [15], Smith [16], and Pattenden [17] have also been reported. A retrospective of these syntheses shows the various bond-breaking strategies in each case (Figure 9.7). Phorboxazole A is an ideal category II molecule due to its extended structure, especially when considering a *chiron approach* for its synthesis. Thus, Forsyth used a combination of chirons to converge with predetermined subunits in the target. D-Glyceraldehyde acetonide was a substrate for a hetero Diels-Alder reaction that would converge with ring B. The desired (*R*)-stereochemistry at C-15 was the result of a face-selective cycloaddition. The stereogenic center of the original chiron in the adduct was sacrificed in lieu of introducing a nitrogen (at C-16), and eventual ring closure to form the oxazole ring A. Forsyth used L-malic acid to correspond to C-5 of the second tetrahydropyran ring C. D-Malic acid was also used for the elaboration of C-37 in the cyclic ketal ring *F*. The stereogenic center corresponding to C-23 in ring *D* was generated from the (*S*)-Roche acid, while Garner's aldehyde was the 3-carbon source of C-29 in the oxazole ring *E* (with destruction of chirality). Thus, a combination of substrate and reagent-controlled reactions in conjunction with chiral, non-racemic starting materials, were used in the first synthesis of phorboxazole A by Forsyth. In the process, eight stereogenic centers originated directly or indirectly from the existing chirality of the starting materials used. The Evans [14] synthesis was a *tour-de-force* display of the power of aldol methodology developed in his laboratory over the years. (*R*)-Glycidol found its way as a 3-carbon chiron into the synthesis proper, converging with C-42–C-44. Williams [14] used L-malic acid for C-5 and also for C-11 (with inversion) of rings *C* and *B* respectively. (*R*)-Glycidol and D-glyceraldehyde acetonide were also used by Williams as precursors to C-42–C-44 and C-37–C-39 respectively.

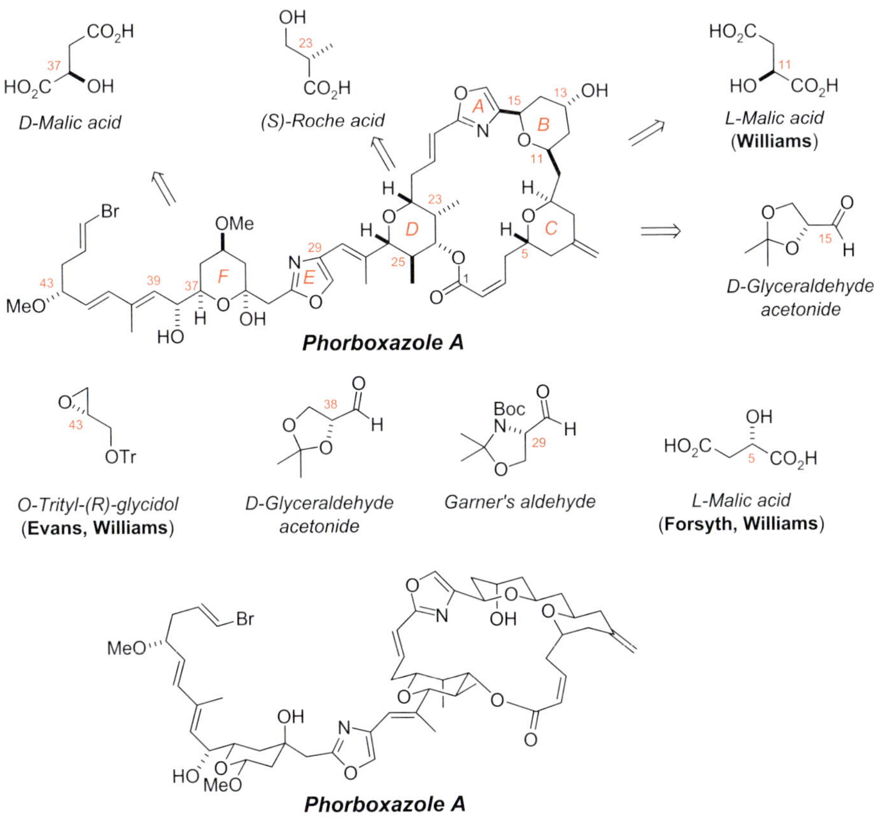

Key reactions: *Hetero Diels-Alder reaction (ring B); Nozaki-Hiyama-Kishi coupling (C38-C39); Still-Gennari olefination (C2-C3); Paterson anti-aldol (C24-C27); CBS reduction (C43); Michael-type addition (C22)*

Figure 9.7 Elaboration of phorboxazole A from multiple chirons (Category II).

9.2.4
Brevetoxin B

The daunting structure of brevetoxin B displays 10 or more "sugar"-like rings, arranged in an exquisitely alternating concatenation of fused oxacycles (Figure 9.8). Total syntheses have been reported by Nicolaou [18], Nakata [19], Yamamoto [20], and their respective groups. A key strategy utilized by Nicolaou and coworkers [18] in their total synthesis consisted in a series of hydroxy-mediated intramolecular cyclizations of activated epoxides and double bonds. Access to such intermediates started with 2-deoxy-D-ribose, on one hand, which provided the framework for rings *F* and *E*. On the other hand, the six-carbon framework and partial stereochemistry of D-mannose converged with ring *K*.

Thus, 2-deoxy-D-ribose was converted in a series of steps to intermediate **A**, which upon exposure to PPTS, underwent regioselective epoxide opening to afford intermediate **B** (ring *F*). Extension of the side-chain to the vinylic epoxide **C** and treatment with PPTS afforded the dioxabicyclic intermediate **D** (rings *F* and *G*), which was extended "westward" to **E** (rings *E*, *F*, and *G*), eventually joining ring *A*. The synthesis of the "eastern" ring *K* started with D-mannose, which was converted to the *C*-allyl glycoside **F**. Well-precedented steps led to the α,β-unsaturated ester **G**, which underwent base-catalyzed, stereocontrolled conjugate addition of the tertiary alcohol onto the α,β-unsaturated ester side-chain to afford **H** (rings *J* and *K*). Further extension to **I**, and cycloetherification gave **J** (rings *I*, *J*, and *K*). Segment couplings were accomplished through Wittig (rings *G* to *I*) and Nozaki-Hiyama-Kishi reactions [21] (rings *E* to *D* and *C* via **K** and **L**). Further elaboration led to the oxacyclic subunits **M**, **N**, and **O**.

It is not intended to elaborate on each and every step in this remarkable feat of total synthesis by Nicolaou and coworkers [18], but to validate the utility of carbohydrate starting materials and their versatility in substrate-based transformations with a very high degree of stereochemical control (see also Chapter 11).

Other examples of Category II molecules and the use of multiple chirons in their assembly can be found in Chapter 8 (Figures 8.5 and 8.6).

Figure 9.8 Elaboration of brevetoxin B from multiple chirons (Category II).

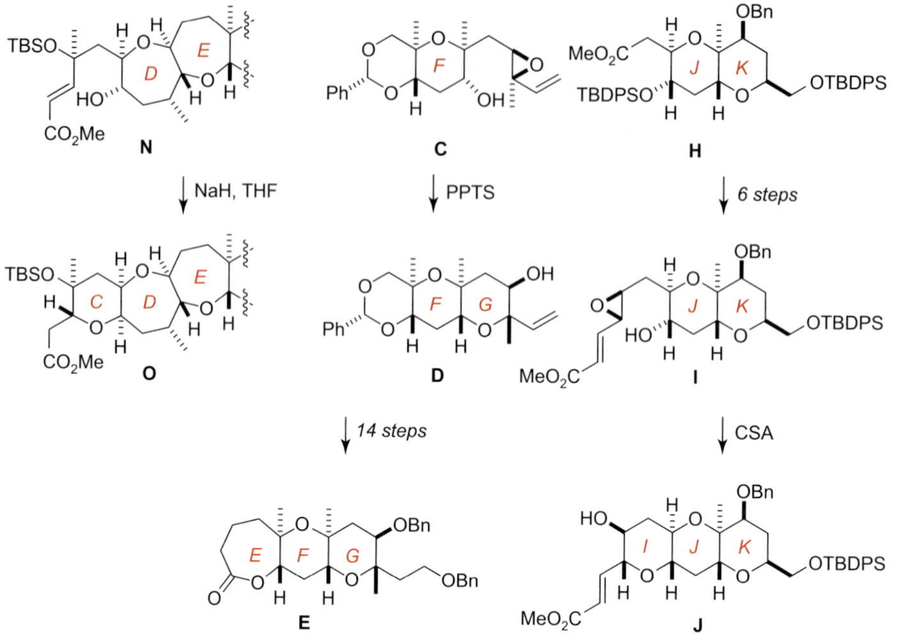

Key reactions: *Wittig extensions; Sharpless asymmetric epoxidation; Nozaki-Hiyama-Kishi coupling*

Relevant mechanism:

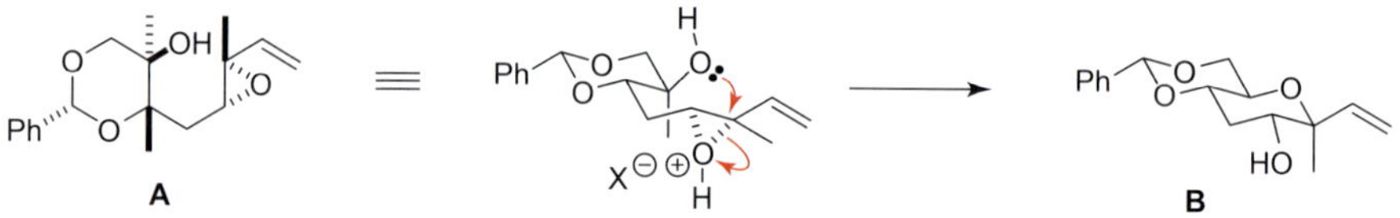

Figure 9.8 *(continued).*

9.3
Category III Target Molecules

Grouped in this category are molecules of varied levels of complexity, that may also contain stereochemical and functional features leading to the use of a single chiron for a relatively confined segment of their intricate structures. The remaining structural landscape in such molecules may or may not be "filled in" with groups of atoms or substructures that may take advantage of the existing chirality, or be independent from it. This category of molecules consists of fused or skipped aromatics and heteroaromatics with appended or embedded chiral subunits, complex alkaloids, extended polyunsaturated compounds, and highly functionalized metabolites.

9.3.1
Neocarzinostatin

In the total synthesis of neocarzinostatin, Myers and coworkers [22] started with
D-glyceraldehyde acetonide, which was further transformed to the chiral enediyne
segment **C** (Figure 9.9). A Sharpless asymmetric epoxidation mediated by the
stereochemically defined allylic alcohol group afforded **B**, which was further
functionalized to the cyclopentenol aldehyde **A**, *en route* to the target compound.
Installation of the remaining structural elements of the natural product was
completed without the aid of the existing stereogenic centers in **A**.

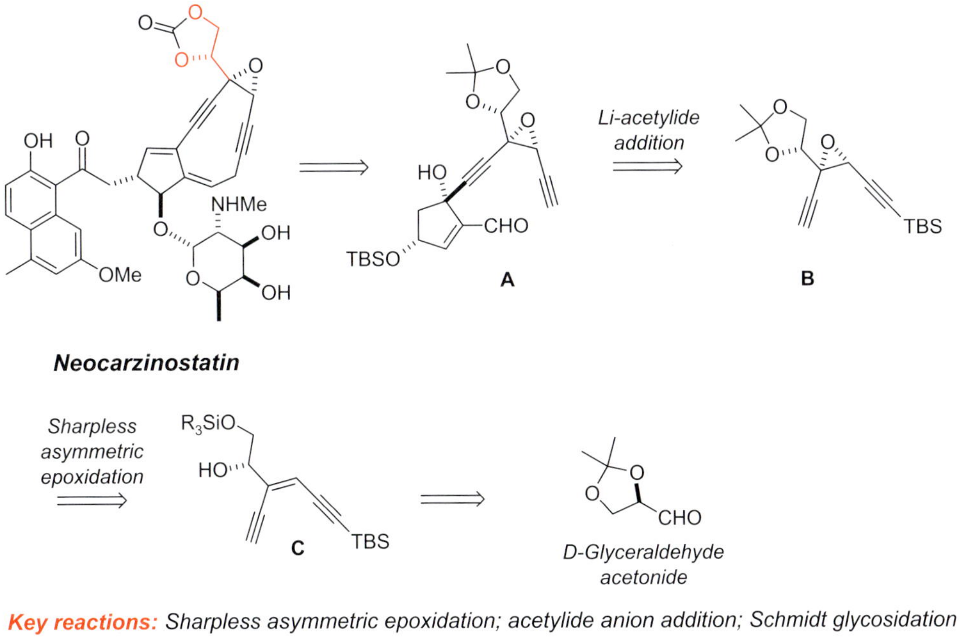

Neocarzinostatin

Key reactions: *Sharpless asymmetric epoxidation; acetylide anion addition; Schmidt glycosidation*

Figure 9.9 Retrosynthetic analysis of neocarzinostatin leading to D-glyceraldehyde
(Category III).

9.3.2
Idiospermuline

Members of the cyclotryptamine family of alkaloids are endowed with exquisitely
deployed 3a,3a'-bispyrrolidino[2,3-b]indolines as exemplified by idiospermuline
(Figure 9.10). The challenges present in the total synthesis of these fascinating
alkaloids is heightened by the presence of a pair of contiguous quaternary carbon
stereocenters bridging the tricyclic motifs [23]. Overman and coworkers [24] devised
a stereocontrolled synthesis of idiospermuline, and several of its congeners, initially
relying on the intrinsic stereochemistry of the C_2-symmetric bis-triflate **G**, readily

available from (R,R)-tartaric acid, as a "chiral" double-electrophile. Thus, a two-stage dialkylation of the dihydro isoindigo intermediate **F** provided the C_2-symmetric adduct **E** as a major diastereomer. Oxidative cleavage of the diol in **E**, followed by conversion to the bis-ethylamino tethers led to **D**. Thermal dehydration of the resulting hemiaminal, and reduction, led eventually to the bis-pyrrolindinoindoline **C** (R = H).

To facilitate *ortho*-lithiation, **C** was converted to the *N*-Boc carbamate to give the corresponding *ortho*-iodo intermediate (**C**, R = I). Stille coupling [25] was followed by conversion to the aryl triflate **B**, which underwent a Pd-catalyzed intramolecular asymmetric Heck cyclization [26] in the presence of (S)-BINAP to give the enamino oxindole **A**. Further steps, and ultimately ring-closure afforded idiospermuline. The effectiveness of the asymmetric intramolecular Heck reaction in generating the last quaternary carbon atom is one of the major highlights of this synthesis. Two carbon atoms have remained from the original (R,R)-tartaric acid, and they are encompassed in the outer pyrrolidine rings of intermediates, **A**, **B**, and **C**. This is another excellent example of a self-immolative process, using the C_2-symmetric diol **G** as the chiral anchor in the initial two-stage double alkylation process (**F** to **E**).

Other examples of Category III target molecules and self-immolative processes starting with a single chiron can be found in Chapter 8, section 8.5.

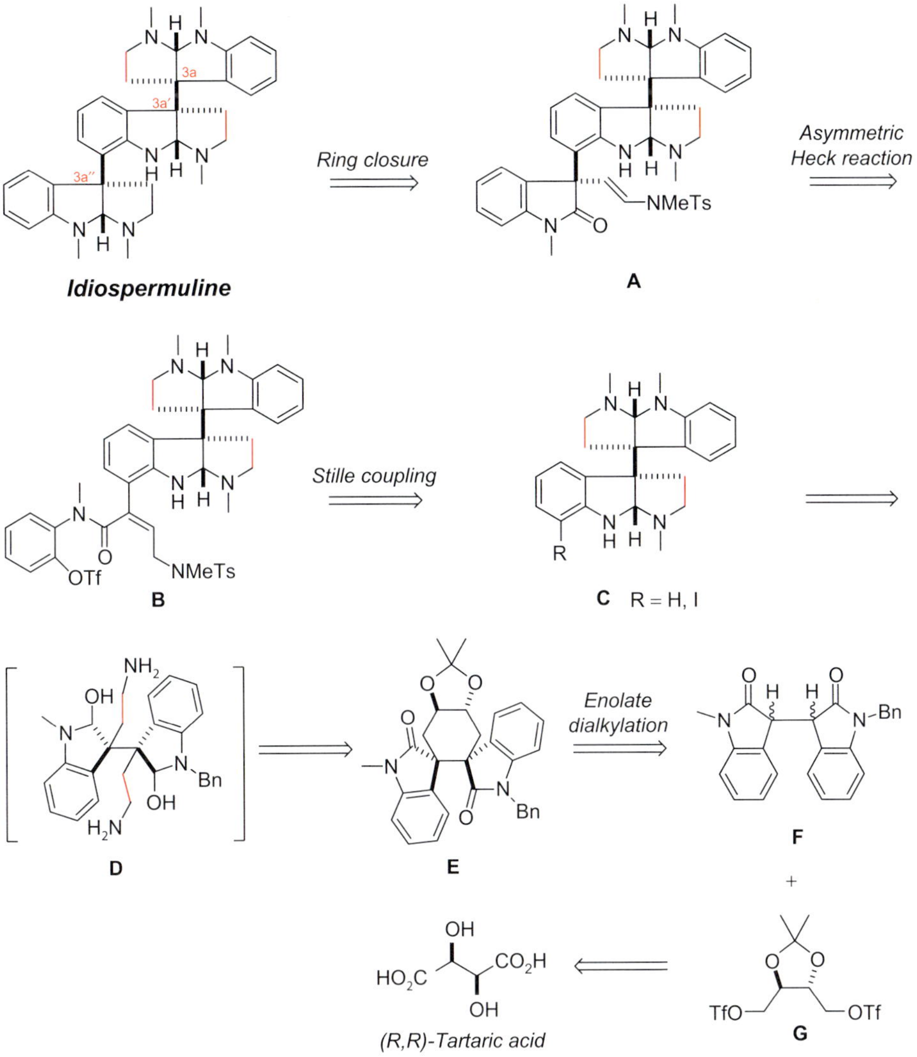

Key reactions: Stereocontrolled dialkylation (**F** to **E**); intramolecular hemiaminal formation (**D** to **C**); Stille coupling (**C** to **B**); intramolecular asymmetric Heck reaction (**B** to **A**)

Figure 9.10 Retrosynthetic analysis of idiospermuline leading to (*R,R*)-tartaric acid, and asymmetric Heck cyclization (Category III).

9.4
Prelude to Total Synthesis of Category I Molecules

As was discussed in section 9.1 of this Chapter, Category I molecules are those in which the chiron provides excellent structural and stereochemical convergence with a sizeable segment of the target molecule that may also include heteroatoms. An added criterion is for the resident stereogenic center(s) to play an important stereodirecting role in the elaboration of other stereogenic centers *en route* to the intended target molecule. In the next six Chapters, we will discuss the total syntheses of selected natural products from the four main sources of native chirons, as well as from readily available chiral, non-racemic carbocycles and lactones. In the majority of cases, the examples discussed represent the first reported syntheses, which were chosen for the interesting types of chemical transformations that were used in the elaboration of the target molecules. Thus, for each molecule, a disconnection to the precursor chiron will be shown by color coding to reveal apparent, partially hidden, or hidden relationships. Key intermediates, relevant mechanisms and transformations will also be highlighted. Pertinent reagents and reaction conditions will be shown in the Figures, but functional group transformations, and protection-deprotection operations may at times be designated by the word "steps" over the arrows. In some cases, independent syntheses from the same or different chirons will also be shown, to illustrate the diversity of approaches toward one and the same target molecule. Whenever applicable, alternative total syntheses involving catalysis as key asymmetric bond-forming steps will also be shown, while others will only be cited, but not discussed. Each synthesis will be followed by a "Commentary" section, in which aspects of the strategy and key reactions are discussed. The opinions expressed regarding the analyses of the strategies were done without consultation with the original authors, and may, in some instances, unintentionally misinterpret their reasons for certain choices in the formulation of their syntheses.

With a few exceptions, syntheses belonging to Category II and III molecules will not be discussed further since many were already highlighted here and in Chapter 8 in particular. Although chirons can be versatile and useful starting materials in the synthesis of Category II and III molecules, they usually address a smaller segment of the structural landscape, without necessarily playing a strategic stereodirecting role beyond their boundaries. Nevertheless, they constitute excellent sources of chirality in the synthesis of specific segments of target molecules with extended structural architectures (see also Chapters 11, 16 and 18).

References

1. Ireland, R. E.; Smith, M. G. *J. Am. Chem. Soc.* 1988, **110**, 854. For a later synthesis of streptolic acid see: Iwata; Y.; Maekawara, N.; Taninio, K.; Miyashita, M. *Angew. Chem. Int. Ed.* 2005, **44**, 1532.

2. Ireland, R. E.; Mueller, R. H.; Willard, A. K. *J. Am. Chem. Soc.* 1976, **98**, 2868.

3. Beyersbergen van Henegouwen, W. G.; Fieseler, R. M.; Rutjes, F. P. J. T.; Hiemstra, H. *Angew. Chem. Int. Ed.* 1999, **38**, 2214. See also:

Beyersbergen van Henegouwen, W. G.; Fieseler, R. M.; Rutjes, F. P. J. T.; Hiemstra, H. *J. Org. Chem.* 2000, **65**, 8317; Takayama, H.; Tominaga, Y.; Kitajima, M.; Aimi, N.; Sakai, S.-i. *J. Org. Chem.* 1994, **59**, 4381.

4. Gmeiner, P.; Feldman, P. L.; Chu-Moyer, M. Y.; Rapoport, H. *J. Org. Chem.* 1990, **55**, 3068. For selected earlier syntheses of vincamine see: Kuehne, M. E. *J. Am. Chem. Soc.* 1964, **86**, 2946; Hermann, J. L.; Cregge, R. J.; Richman, J. E.; Semmelhack, C. L. Schlessinger, R. H. *J. Am. Chem. Soc.* 1974, **96**, 3702; Oppolzer, W.; Hauth, H.; Pfäffli, P.; Wenger, R. *Helv. Chim. Acta* 1977, **60**, 1801; Rossey, G.; Wick, A.; Wenkert, E. *J. Org. Chem.* 1982, **47**, 4745; Desmaële, D.; Mekouar, K.; d'Angelo, J. *J. Org. Chem.* 1997, **62**, 3890.

5. Angeles, A. R.; Waters, S. P.; Danishefsky, S. J. *J. Am. Chem. Soc.* 2008, **130**, 13765.

6. (a) Norley, M. C. *Contemp. Org. Syn.* 1995, 345; (b) Maddess, M. L.; Tackett, M. N.; Ley, S. V. *Prog. Drug Res.* 2008, **66**, 15.

7. (a) Linde II, R. G.; Egbertson, M.; Coleman, R. S.; Jones, A. B.; Danishefsky, S. J. *J. Org. Chem.* 1990, **55**, 2771; (b) Villalobos, A.; Danishefsky, S. J. *J. Org. Chem.* 1990, **55**, 2776.

8. (a) Hale, K. J. in, *Organic Synthesis with Carbohydrates*, Boons, G.-J.; Hale, K. J., 2000, p. 292, Sheffield Academic Press, Sheffield, UK; see also: (b) Smith, A. B. III.; Hale, K. J. *Tetrahedron Lett.* 1989, **30**, 1037; (c) Smith, A. B. III.; Hale, K. J.; Laasko, L. M.; Chen, K.; Riera, A. *Tetrahedron Lett.* 1989, **30**, 6963; (d) Smith, A. B. III.; Chen, K.; Robinson, D. J.; Laasko, L. M.; Hale, K. J. *Tetrahedron Lett.* 1994, **35**, 4271.

9. (a) Isobe, M.; Ichikawa, Y.; Goto, T. *Tetrahedron Lett.* 1986, **27**, 963; (b) Isobe, M.; Ichikawa, Y.; Bai, D.-L.; Masaki, H. Goto, T. *Tetrahedron* 1987, **43**, 4767.

10. Forsyth, C. J.; Sabes, S. F.; Urbanek, R. A. *J. Am. Chem. Soc.* 1997, **119**, 8381.

11. Ley, S. V.; Humphries, A. C.; Eick, H.; Downham, R.; Ross, A. R.; Boyce, R. J.; Pavey, J. B. J.; Pietruszka, J. *J. Chem. Soc., Perkin Trans. 1* 1998, 3907.

12. Brown, H. C. Jadhav. P. K. *J. Am. Chem. Soc.* 1983, **105**, 2092.

13. Forsyth, C. J.; Ahmed, F.; Cink, R. D.; Lee, C. S. *J. Am. Chem. Soc.* 1998, **120**, 5597.

14. Evans, D. A.; Fitch, D. M.; Smith, T. E.; Cee, V. J. *J. Am. Chem. Soc.* 2000, **122**, 10033.

15. Williams, D. R.; Kiryanov, A. A.; Emde, U.; Clark, M. P.; Berliner, M. A.; Reeves, J. T. *Angew. Chem. Int. Ed.* 2003, **42**, 1258.

16. (a) Smith, A. B. III.; Verhoest, P. R.; Minbiole, K. P.; Schelhaas, M. *J. Am. Chem. Soc.* 2001, **123**, 4834; (b) Smith, A. B. III; Razler, T. M.; Ciavarri, J. P; Hirose, T.; Ishikawa, T.; Meis, R. M. *J. Org. Chem.* 2008, **73**, 1192.

17. (a) Gonzalez, M. A.; Pattenden, G. *Angew. Chem. Int. Ed.* 2003, **42**, 1255; see also: (b) Lucas, B. S.; Gopalsamuthiram, V.; Burke, S. D. *Angew. Chem. Int. Ed.* 2007, **46**, 769.

18. (a) Nicolaou, K. C.; Duggan, M. E.; Hwang, C.-K. *J. Am. Chem. Soc.* 1989, **111**, 6666; (b) Nicolaou, K. C.; Duggan, M. E.; Hwang, C.-K. *J. Am. Chem. Soc.* 1989, **111**, 6676; (c) Nicolaou, K. C.; Hwang, C.-K.; Duggan, M. E. *J. Am. Chem. Soc.* 1989, **111**, 6682; (d) Nicolaou, K. C.; Theodorakis, E. A.; Rutjes, F. P. J. T.; Tiebes, J.; Sato, M.; Untersteller, E.; Xiao, X.-Y. *J. Am. Chem. Soc.* 1995, **117**, 1171; (e) Nicolaou, K. C.; Rutjes, F. P. J. T.; Theodorakis, E. A.; Tiebes, J.; Sato, M.; Untersteller, E.; *J. Am. Chem. Soc.* 1995, **117**, 1173.

19. Matsuo, G.; Kawamura, K.; Hori, N.; Matsukura, H.; Nakata, T. *J. Am. Chem. Soc.* 2004, **126**, 14374.

20. Kadota, I.; Takamura, H.; Nishii, H., Yamamoto, Y. *J. Am. Chem. Soc.* 2005, **127**, 9246.

21. (a) Okude, Y.; Hirano, S.; Hiyama, T.; Nozaki, H. *J. Am. Chem. Soc.* 1977, **99**, 3179; (b) Jin, H.; Uenishi, J.; Christ, W. J.; Kishi, Y. *J. Am. Chem. Soc.* 1986, **108**, 5644.

22. Myers, A. G.; Liang, J.; Hammond, M.; Harrington, P. M.; Wu, Y.; Kuo, E. Y. *J. Am. Chem. Soc.* 1998, **120**, 5319.

23. Steven, A.; Overman, L. E. *Angew. Chem. Int. Ed.* 2007, **46**, 5488.

24. Overman, L. E.; Peterson, E. A. *Angew. Chem. Int. Ed.* 2003, **42**, 2525.

25. For a review, see: Farina, V.; Krishnamurthy, V.; Scott, W. J. *Org. React.* 1998, **50**, 1.

26. For a review, see: Dounay, A. B.; Overman, L. E. *Chem. Rev.* 2003, **103**, 2945.

10
Total Synthesis from α-Amino Acid Precursors

In this chapter we shall discuss the utilization of α-amino acids and their derivatives as precursors to the total synthesis of Category I-type natural products. What is intended is not a comprehensive compendium of total syntheses, but a selection that demonstrates the utility of α-amino acids in the first total synthesis of the intended targets. Obvious transformations are indicated by the word "steps" while key reactions and relevant mechanisms and transformations have been included with appropriate citations and commentary.

10.1
Actinobolin

The first asymmetric total synthesis of actinobolin was reported in 1984 by Ohno and coworkers [1, 2] (Figure 10.1). Visual analysis of the bicyclic lactone reveals the 4-carbon framework and absolute stereochemistry of an L-threonine-like motif.

The phenyloxazoline **A**, prepared in four steps from L-threonine, was O-tosylated then converted into the triphenylphosphonium salt **B** by heating to 130 °C, after which it was engaged in a *cis*-selective (*cis:trans* = 97:3) Wittig olefination with acrolein. Acid-catalyzed cleavage of the oxazoline gave the diene **C**, which was N-acylated to furnish **D**. A thermal intramolecular Diels-Alder reaction provided an excellent yield of the bicyclic lactam **E** with high diastereoselectivity (dr > 20:1). Exchange of the benzoate ester for a THP-ether followed by intramolecular iodolactonization led to the tricyclic intermediate **F**. Treatment with aqueous sodium hydroxide, followed by esterification, and subsequent treatment with TsOH gave the epoxide **G**, which was further elaborated to **H**, and then converted to the phenylseleno analogue **I** in three steps. Conversion to the N-(p-methylphenyl)methylsulfonyl (PMS) derivative followed by a methoxide-mediated ring expansion of the lactam unit in **I** led to bicyclic lactone **J**, which was further oxidized to the selenoxide, and then subjected to elimination to give the exocyclic methylene analogue **K**. Ozonolytic cleavage and deprotection afforded lactone **L**, which was coupled to Cbz-Ala, and the product hydrogenoylzed to give actinobolin.

Design and Strategy in Organic Synthesis: From the Chiron Approach to Catalysis, First Edition.
Stephen Hanessian, Simon Giroux, and Bradley L. Merner.
© 2013 Wiley-VCH Verlag GmbH & Co. KGaA. Published 2013 by Wiley-VCH Verlag GmbH & Co. KGaA.

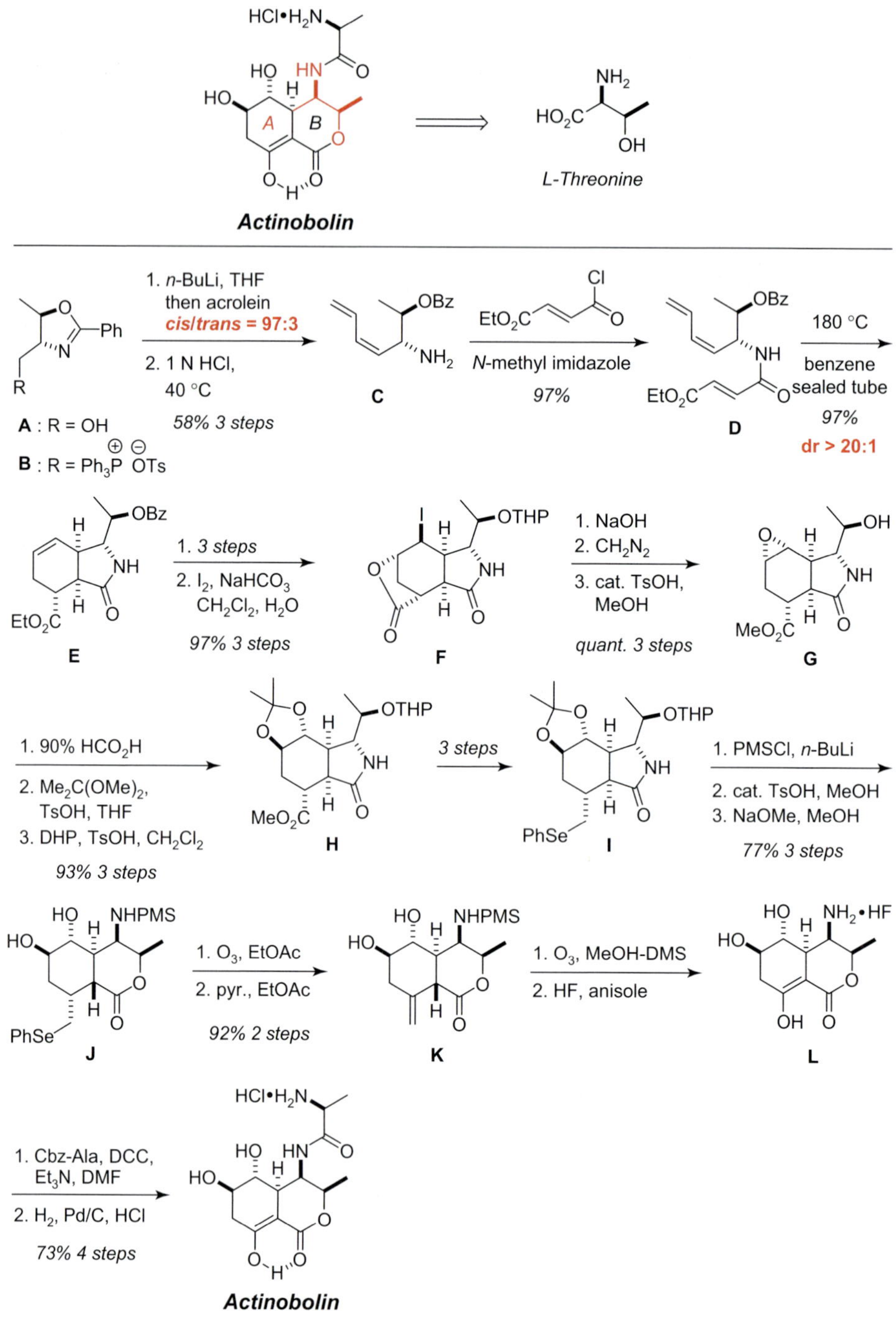

Figure 10.1 Total synthesis of actinobolin from L-threonine (apparent substructure).

Key reactions: *Intramolecular Diels-Alder reaction (**D** to **E**); double bond dihydroxylation via iodolactonization/epoxide opening (**E** to **H**); base-induced lactam to lactone ring expansion (**I** to **J**); PMS = (p-methylphenyl)methylsulfonyl*

Relevant transformation:

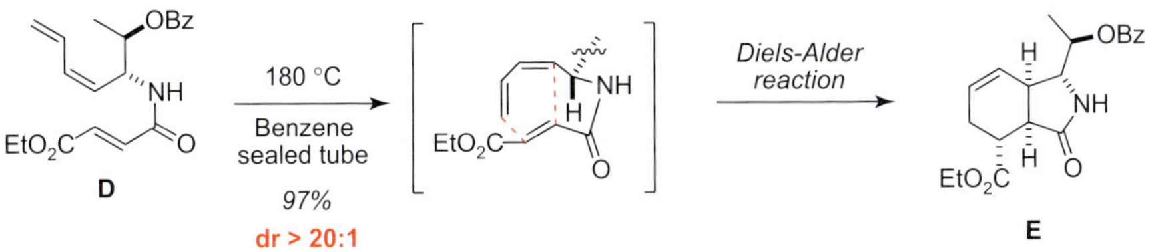

Figure 10.1 *(continued).*

Commentary: In Chapter 4 (section 4.2), we alluded to *relational* and *reflexive* visual analysis of target molecules. This is evident in the case of actinobolin which reveals an α-amino alcohol substructure in the lactone portion of ring *B*. A *relational* analysis leads to L-threonine as an apparent chiron of choice to start the synthesis of ring *B*. On the other hand, the presence of a *trans*-diol in ring *A* can be antithetically related to a carbocyclic olefin in the mind's eye. Epoxidation of such a virtual olefin, and regioselective ring-opening could, in principle, lead to the diol present in ring *A*. Consequently, a visual *reflexive* reaction-type based analysis, suggests an intramolecular Diels-Alder reaction as a key step to building the carbocyclic ring *A*.

The synthesis of the *cis*-diene intermediate **C** using the phosphorus ylide derived from **B** and acrolein is an interesting application of the Wittig reaction to an amino acid series. Notably, no β-elimination was observed in the formation of the phosphonium salt from the tosylate, at 130 °C, or during the subsequent olefination step. Curiously, a 20% yield of **C** was obtained when the iodide salt was used instead of the tosylate (*cf.* 72%).

To ensure the stereochemical outcome of the Diels-Alder reaction, the authors chose a *cis*-configured diene and an *trans*-configured ester dienophile component as seen in intermediate **D** (see relevant transformation, Figure 10.1). Indeed, this combination led to the adduct **E** in high diastereoselectivity (dr > 20:1). In spite of the biased topology of the bicyclic system, treatment of **E** with *m*-CPBA led to a 1:1 mixture of α- and β-epoxides. Indirect epoxidation was eventually achieved by an exchange of *O*-protecting groups, followed by intramolecular iodolactonization, and treatment with base. Acid-catalyzed cleavage of the resulting epoxide proceeded with high regioselectivity to afford the acetonide **H**. Although the stereochemistry of the six stereogenic carbon atoms in actinobolin were secured at this stage, there was a heavy penalty to pay in the obligatory excision of the extra methoxycarbonyl group. In fact, several steps were required to convert this ester to the phenylselenyl derivative **I**. However, prior to executing this sequence, a major functional group reorganization, which was no doubt factored in the original plan, was needed with regard to ring **B**. For this, Ohno transformed the lactam **I** into the corresponding (*p*-methylphenyl)methanesulfonamide, which underwent a ring expansion to the lactone **J** upon treatment of the corresponding triol with NaOMe. Oxidative elimination of the (phenylselenyl)methyl appendage led to the exocyclic olefin **K**, which was cleaved with ozone and the product treated with HF in anisole to give the amino lactone precursor **L** to actinobolin.

Clearly, the highly selective intramolecular Diels-Alder reaction was a key step that set the correct stereochemistry at the ring junction and placed the double bond in a strategic position for the installation of the *trans*-diol unit in the natural product. The presence of the ester group in the dienophile was deemed important for its activation. Although the disappointing results of the epoxidation of **E** could not be predicted *a priori*, the iodolactonization alternative provided the desired epoxide **G** in a five-step sequence. Thus, the pendant and superfluous ester group in **E** also served as a crucial participating group in the iodolactonization step.

10.2
Aspochalasin B

In the first total synthesis of aspochalasin B, Trost and coworkers [3] utilized L-leucine as an apparent chiron (Figure 10.2). The framework of the amino acid can be clearly seen in the lactam portion of the target molecule (visual *relational* analysis). The cyclohexene ring, on the other hand, can arise from an intermolecular [4+2] cycloaddition (visual *reflexive* analysis). However, only experimentation would reveal the challenges of regiochemistry, and the generation of a quaternary carbon atom with the expected absolute configuration in a highly substituted carbocyclic portion of the molecule. Nevertheless, internal induction from the L-leucine stereogenic center in the cycloaddition reaction would be expected to contribute to the observed stereochemistry at the ring junction. Reduction of *N*-Cbz-L-leucine methyl ester (**A**) to the aldehyde, and a Knoevenagel condensation reaction with dimethyl malonate gave diester **B** in 51% yield over two steps. A stereo- and regioselective Diels-Alder cyclization with diene **C** under thermal conditions led to the tricyclic lactone-lactam **D**, albeit in modest yield. Cleavage of the lactone and lactam rings under basic conditions, followed by sequential re-cyclization, acidification, and esterification gave the inverted ester, which was oxidized to the aldehyde **E**. Addition of 2-propenyl lithiocuprate, followed by treatment with methylphenylsulfone anion led to the β-ketosulfone **F**. Extension with ethyl 3,3-diethoxyacrylate under PPTS catalysis involved *in situ* formation of a mixed acetal, which underwent a [3,3]-sigmatropic rearrangement with double bond transposition to afford **G**. A series of steps led to the aldehyde **H**, which was extended with an ethoxyvinyl lithiocuprate reagent, and the resulting alcohol was converted to the methylcarbonate **I**. The key step involving a Pd(PPh$_3$)$_4$-catalyzed macrocyclization delivered the eleven-membered ring of the β-ketosulfone **J** as a single diastereomer. Chemoselective epoxidation enol ether double bond, and subsequent solvolysis furnished the desired α-hydroxy ketone **K**. Fluoride-catalyzed desulfonylation led to aspochalasin B.

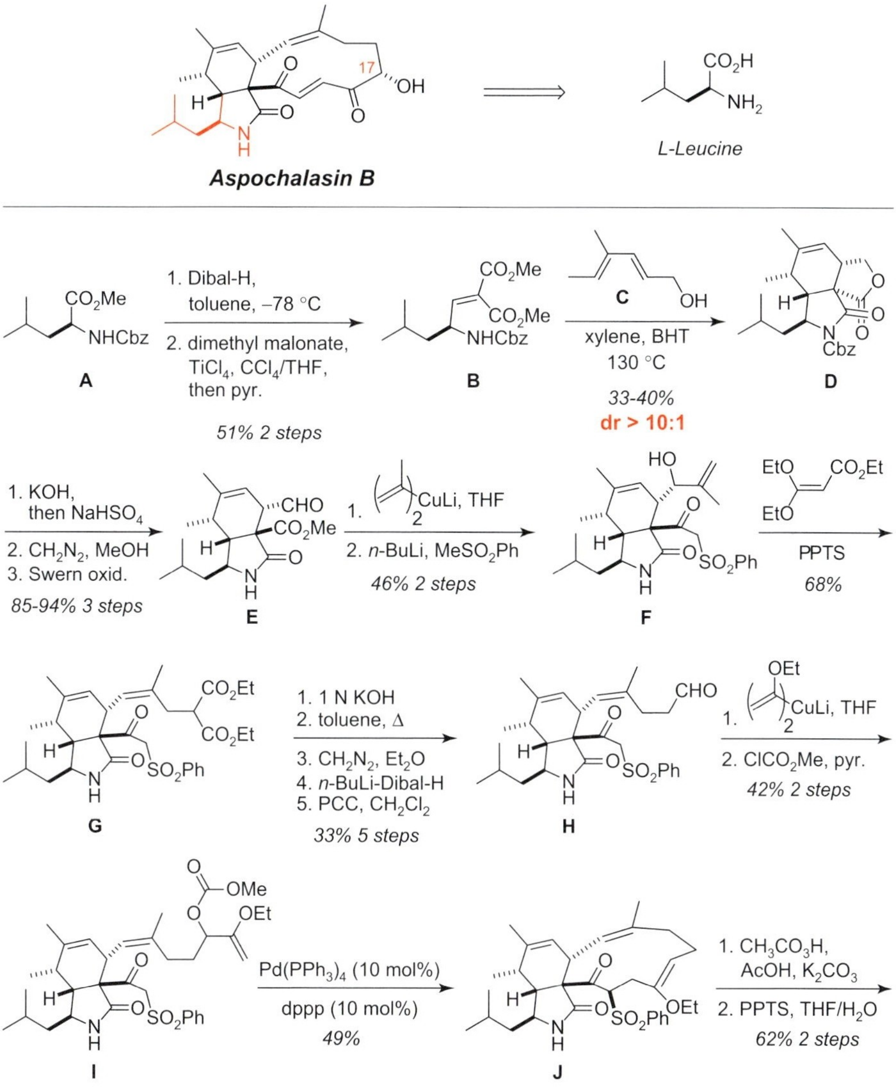

Figure 10.2 Total synthesis of aspochalasin B from L-leucine (apparent substructure).

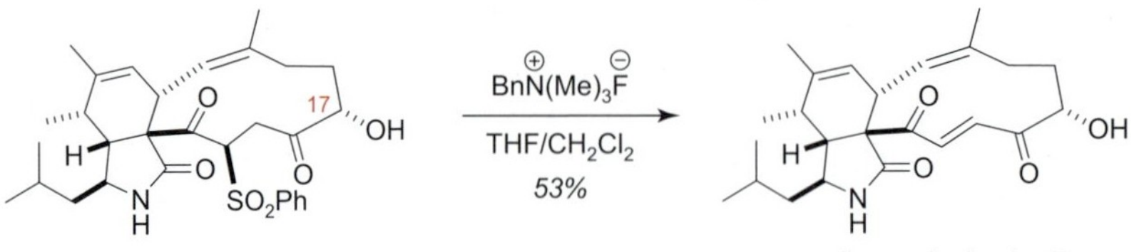

K → Aspochalasin B

Key reactions: Diels-Alder reaction (**B** to **D**); malonate extension via a [3,3]-sigmatropic rearrangement (**E** to **F** to **G**); Pd-catalyzed intramolecular macrocyclization (**I** to **J**); BHT = butylated hydroxytoluene (2,6-di-tert-butyl-p-cresol); dppp = 1,3-bis(diphenylphosphino)propane

Relevant transformations and mechanisms:

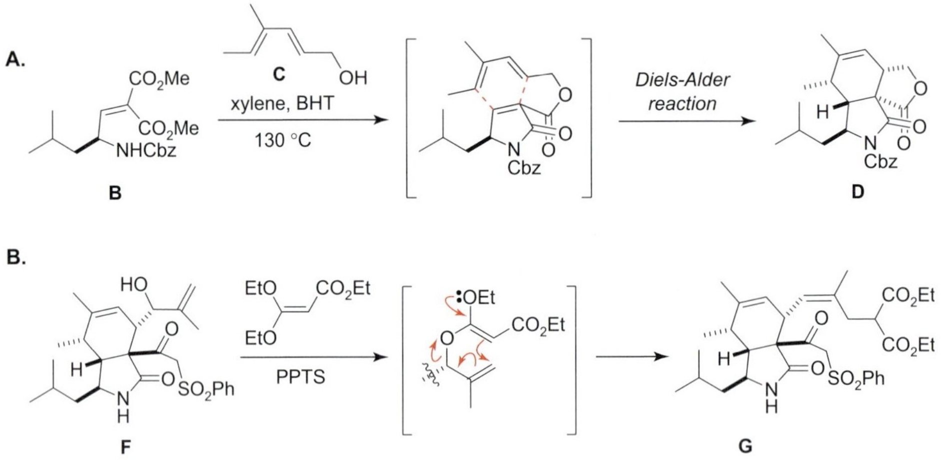

A.

B.

Figure 10.2 *(continued)*.

Commentary: Trost's synthesis strategy for aspochalasin B relied heavily on the desire to demonstrate the feasibility of the eleven-membered macrocycle formation under Pd(0) catalysis. Although such cyclizations had precedents with β-ketosulfones and aldehydes, application to polyfunctional natural products involving an allylic carbonate as part of an enol ether system, as in **I**, was a challenge. To reach this stage of the synthesis, the densely functionalized bicyclic lactam had to be constructed. L-Leucine was an obvious choice, since it offered good skeletal, functional, and stereochemical overlap with a portion of the lactam ring (Figure 10.2). Extension to the alkylidene malonate would ensure the correct electron demand of the Diels-Alder cycloaddition (**B** to **D**). Considering the reaction conditions, one would expect an *in situ* transesterification and decarbomethoxylation (see relevant transformation A, Figure 10.2). Subsequent intramolecular [4+2] cycloaddition would lead to **D** via an *endo*-transition state, possibly steered by the bulky β-isobutyl group of the lactam. However, the "inversion" of the carbomethoxy group at the ring junction with concomitant loss of the N-Cbz group must involve a base-catalyzed hydrolysis of the lactone and imide groups, followed by recyclization to the thermodynamically more stable β-carbomethoxy group, prior to a Swern oxidation to give the aldehyde **E**. The acid-catalyzed transposition of the allylic alcohol initiated by a Claisen-like [3,3]-sigmatropic rearrangement to the *E*-olefin was based on a protocol from Raucher, Chi, and Jones [4] (**F** to **G**) (see relevant mechanism B, Figure 10.2). The presence of the δ-hydroxy-α,β-unsaturated-1,4-diketone moiety in aspochalasin B, led Trost and coworkers to explore the reactivity of 2-alkoxyvinyl carbonates as sources for a π-allyl Pd-complex in the anticipated macrocyclization (**I** to **J**). In spite of the modest yield, the realization of this challenging reaction was a validation of the original plan. Of necessity, the allylic carbonate appendage had to incorporate the 2-ethoxy group in order to achieve the desired oxidation level in the natural product. Chemoselective epoxidation of the more electron rich ethoxy ether (compared to the two trisubstituted double bonds), ensured that solvolysis under PPTS-mediated conditions led to the desired stereochemistry at the C-17 hydroxyl group and correctly positioned the carbonyl group.

The establishment of crucial stereochemistry in the sequence **B** to **H** is an important feature of the synthesis, starting with the Diels-Alder cycloaddition step, where four contiguous stereogenic centers were generated including a quaternary branching point from L-leucine as a single chiral progenitor.

10.3
Cephalotaxine

In the first asymmetric synthesis of cephalotaxine, Mori and Isono [5] utilized D-proline as a precursor to the 1-azaspiro[4.4]nonane subunit of the fused benzazepine ring system (Figure 10.3). Visual analysis reveals a partially hidden proline as the source of the pyrrolidine subunit. An alternative consideration could involve an appropriately functionalized cyclopentanone as a precursor. However, the major challenge is the construction of the 1-azaspiro[4.4]nonane system with the correct stereochemistry at the spirocyclic carbon atom. In this regard, a proline-based approach would present a more attractive solution to the stereochemical issue with the advanced knowledge that α-alkylation of D-proline by Seebach's method [6] would secure the desired stereochemistry at the azaspiro carbon atom. Once the spirocyclic system is in place, the formation of the benzazepine ring would result with the anticipated stereochemical outcome in view of the unique topology of the tetracyclic ring system.

The lithium enolate of the bicyclic lactone **A**, which is readily prepared from D-proline [6], was alkylated with 1-(trimethylsilyl)allyl bromide to give the vinylsilane adduct **B**. Treatment with iodine monochloride and TFA according to Chan [7], followed by KF-mediated desilylation, afforded the *cis*-vinyl iodide which was converted to the corresponding *N*-Boc ester **C** in excellent overall yield. Installation of the 1-(3,4-dimethoxy)phenylethyl moiety as in **D**, followed by a two-step reduction/oxidation sequence led to aldehyde **E**. A key carbocyclization step involved a CsF-mediated formation of the spirocyclic allylic alcohol **F**. An intramolecular

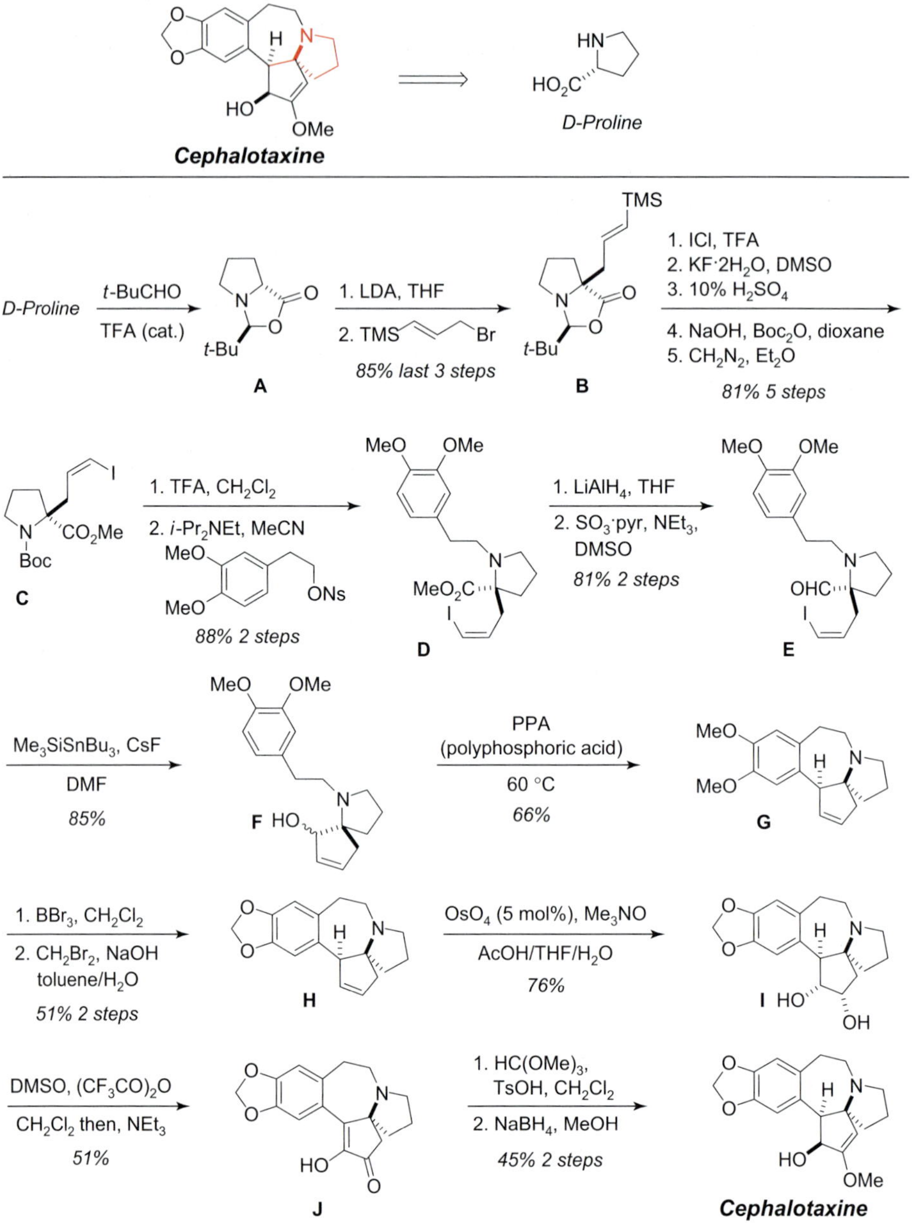

Figure 10.3 Total synthesis of cephalotaxine from D-proline (partially hidden substructure).

Key reactions: *Vinylsilane/vinyl iodide exchange (**B** to **C**); vinyl iodide stannylation and stannyl-anion-induced cyclization (**E** to **F**); intramolecular Friedel-Crafts-type cyclization (**F** to **G**)*

Relevant transformations and mechanisms:

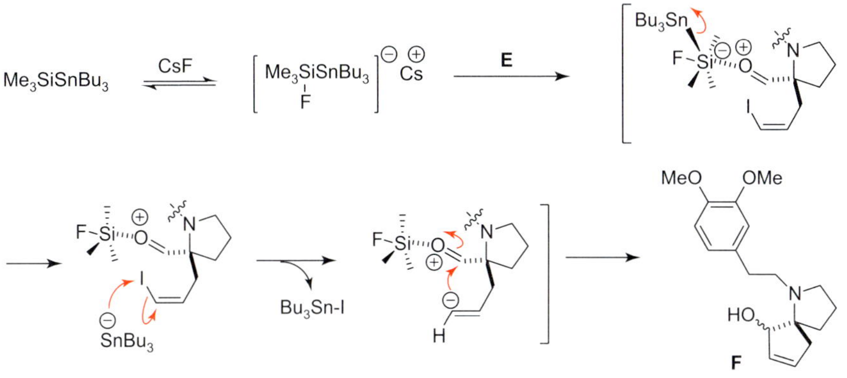

Figure 10.3 *(continued)*.

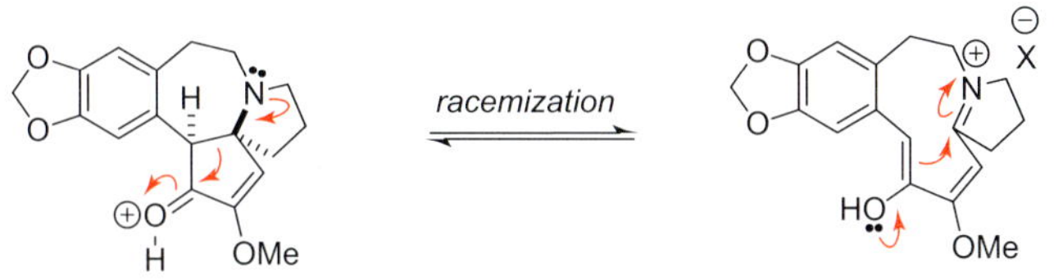

Friedel-Crafts-type reaction on the incipient allylic cation generated in the presence of polyphosphoric acid, led to the tetracyclic intermediate **G**. At this juncture, the *o*-dimethoxy groups were exchanged for the required 3,4-methylenedioxy group to give **H**. Dihydroxylation to **I** was followed by a Swern oxidation to give the corresponding hydroxyenone **J**. Treatment with trimethyl orthoformate in the presence of TsOH, followed by reduction led to cephalotaxine.

Commentary: The three key C–C bond forming steps in the Mori synthesis of cephalotaxine are the α-alkylation of the bicyclic proline lactone (**A** to **B**), the tin-mediated carbocyclization (**E** to **F**), and the Friedel-Crafts cyclization (**F** to **G**). Based on numerous precedents, the Seebach alkylation was successfully adapted to the vinylsilane containing electrophile (**A** to **B**). Securing this key intermediate, which generated the all-important spirocyclic juncture with the desired stereochemistry, was no doubt an essential go-ahead step in pursuing the subsequent vinyl stannane-mediated carbocyclization in the presence of CsF (**E** to **F**) (see relevant mechanism, Figure 10.3).

The authors had also attempted the conversion of **J** to cephalotaxinone under acidic conditions only to discover that the final product was racemic. They postulated two possible mechanisms that involved protonation of the ketone or the heterocyclic nitrogen atom (see relevant mechanism, Figure 10.3). In the first case, ring-opening would give a highly conjugated iminium ion, which would reclose to give racemic cephalotaxinone. Alternatively, abstraction of the benzylic hydrogen would proceed to a neutral 2-methoxy cyclopentadienone, which would revert to the ketone with a loss of stereochemical integrity. Milder conditions to form the enol ether obviated this end-game obstacle. Attempted Friedel-Crafts cyclization of the 3,4-methylenedioxy equivalent of intermediate **F** (**F** to **G**) was not successful, hence the initial use of the 3,4-dimethoxy variant, and the obligatory lengthening of the synthetic sequence by two steps (**G** to **H**).

10.4
α-Kainic Acid (W. Oppolzer)

The first enantioselective synthesis of α-kainic acid was reported in 1982 by Oppolzer and Thirring [8], thereby establishing its absolute configuration as well as those of the closely related congeners, α-allokainic acid, and domoic acid. A first glance at the structure of α-kainic acid reveals an apparent ʟ-proline substructure (Figure 10.4). However, Oppolzer and Thirring saw beyond that by focussing on the embedded α-amino dicarboxylic acid motif that harbors the C-3 branched acetic acid unit. Seen in this perspective, it becomes clear why ʟ-glutamic acid was chosen as a partially hidden chiron. The main challenge associated with the synthesis would involve the stereocontrolled construction of the trisubtituted pyrrolidine core with the correct stereochemistry at three contiguous stereogenic centers. ʟ-Glutamate monoethyl ester (**A**), was converted to the *N*-Boc derivative, then treated with borane in THF to selectively reduce the carboxylic acid to the primary alcohol, which was subsequently protected as the TBS ether **B**. Alkylation with 1-bromo-3-methyl-2-butene followed by enolate formation of the ester, quenching with phenylselenyl chloride, and oxidative elimination of the selenoxide gave the enoate **C**. The use of LiTMP was essential for enolate formation in this particular instance.

The critical ring closure was achieved by heating a toluene solution of **C** at 130 °C to furnish pyrrolidine **D** in 75% yield. Cleavage of the silyl ether was followed by a Jones oxidation to the carboxylic acid **E**. Saponification of the ethyl ester, followed by cleavage of the *N*-Boc group led to crystalline α-kainic acid in 5% overall yield from ʟ-glutamate monoethyl ester (**A**). The authors advanced the hypothesis that the biosynthesis of α-kainic might involve ʟ-glutamic acid and an isoprenoid unit as precursors, possibly proceeding by an intramolecular ene-type-reaction.

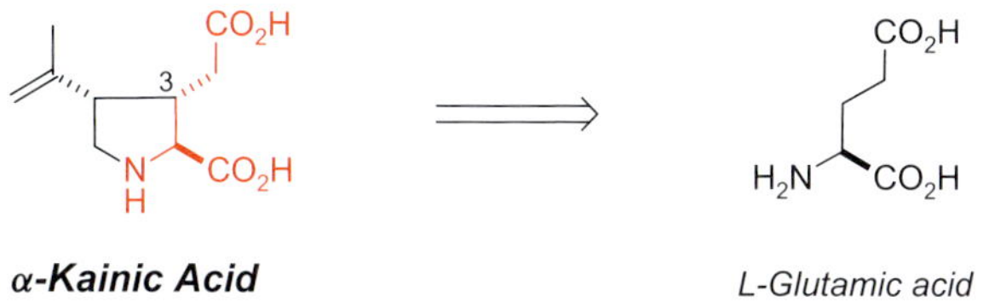

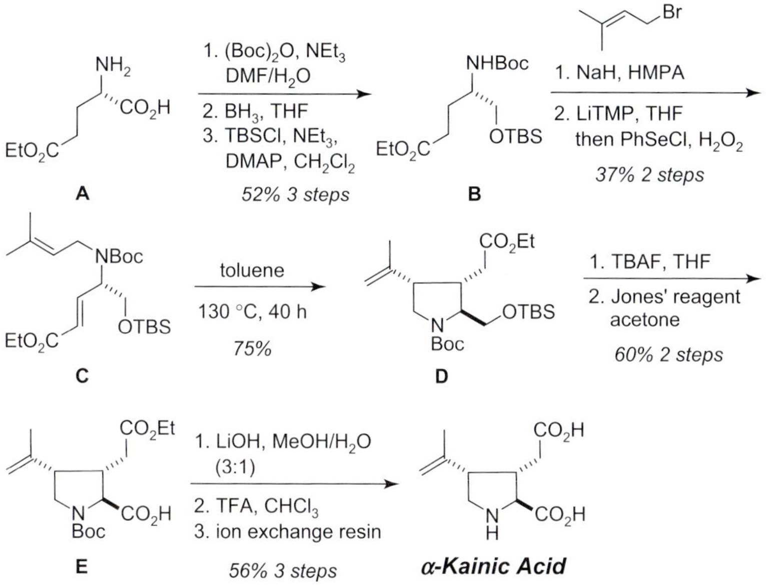

α-**Kainic Acid** L-Glutamic acid

Key reactions: *Intramolecular thermal type-I ene reaction (**C** to **D**);*
LiTMP = lithium tetramethylpiperidide

Relevant mechanism:

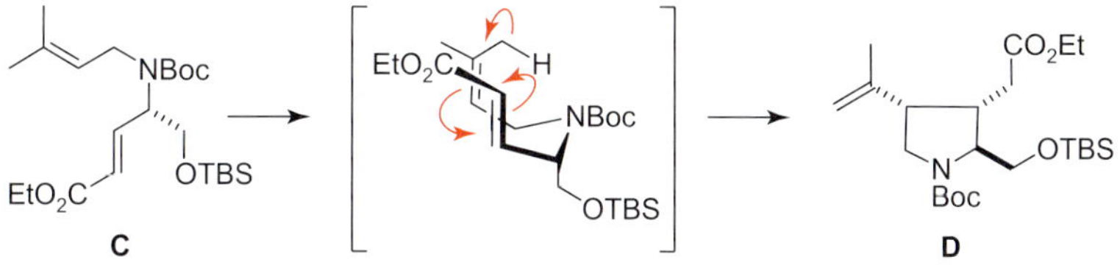

Figure 10.4 Total synthesis of α-kainic acid from L-glutamic acid (partially hidden substructure).

Commentary: The synthetic plan devised by Oppolzer and Thirring is a good example of the *relational* and *reflexive* aspects of the visual thought process. Clearly, L-glutamic acid fits in very nicely as a chiral precursor offering an α-amino acid starting material. However, in Oppolzer's mind's eye, it also provided an ideal chiral template to further manipulate as a potential substrate for an intramolecular type-I ene reaction (see relevant mechanism, Figure 10.4). The biomimetic relationship between α-kainic acid and L-glutamic acid was no doubt a strong impetus to pursue the synthesis. From a purely strategic standpoint, it is difficult, *a priori*, to know which thought process came first – the desire to build the pyrrolidine using an ene-reaction, or the realization that the carbon framework as well as absolute stereochemistry of L-glutamic acid would lend itself to manipulation toward to an α,β-unsaturated ester for the intended key reaction. The need to involve an α,β-unsaturated ester in the ene reaction necessitated the conversion of the α-carboxylic acid to an O-TBS protected primary alcohol to avoid conjugation, which would result in loss of stereochemical integrity. In fact, four of the ten steps in the synthesis necessitated such obligatory functional group minipulations. However, notwithstanding the planned detour, the bulky C-2 hydroxymethyl silyl ether may have played an important stereodirecting role in controlling the highly *cis*-selective type-I ene reaction. Thus, the original C-2 stereochemistry in L-glutamic acid was responsible for the generation of the C-3–C-4 stereogenic centers in the ene-reaction product **D**.

10.5
α-Kainic Acid (P. T. Gallagher)

A synthesis of α-kainic acid published in 1987 by Gallagher and coworkers [9] also involved the formation of a pyrrolidine ring by joining C-3 and C-4. However, a conceptually different strategy was implemented when compared to the Oppolzer and Thirring synthesis [8]. The potential for using an acyclic α-amino dicarboxylic acid was explored with a designated plan for the key carbocyclization reaction to involve a transannular Ireland-Claisen-type ring contraction (Figure 10.5). Thus, L-aspartic acid was converted into the silyl protected primary alcohol **A** (over five steps), and the latter was N-alkylated then converted to **B**. Treatment of **B** with 2-chloro-N-methylpyridinium iodide [10] gave the desired nine-membered azalactone **C**. Formation of the enolate at −100 °C, followed by trapping with TBSCl gave the corresponding cyclic ketene acetal, which underwent a smooth transannular Ireland-Claisen rearrangement, with ring contraction, in 55% yield to reveal the C-3–C-4 *cis*-substituted pyrrolidine motif **D**. With three contiguous stereogenic centers introduced, there remained only to make functional group adjustments to attain α-kainic acid. Thus, Arndt-Eistert homologation [11] of the carboxylic acid group in **D** gave the corresponding acetic acid methyl ester **E**. Desilylation and Jones oxidation led to the dicarboxylic acid, which was treated with TMSI and then KOH to remove the N-carbethoxy and methyl ester groups respectively. Purification by ion exchange chromatography as in the Oppolzer and Thirring synthesis [8] afforded crystalline α-kainic acid.

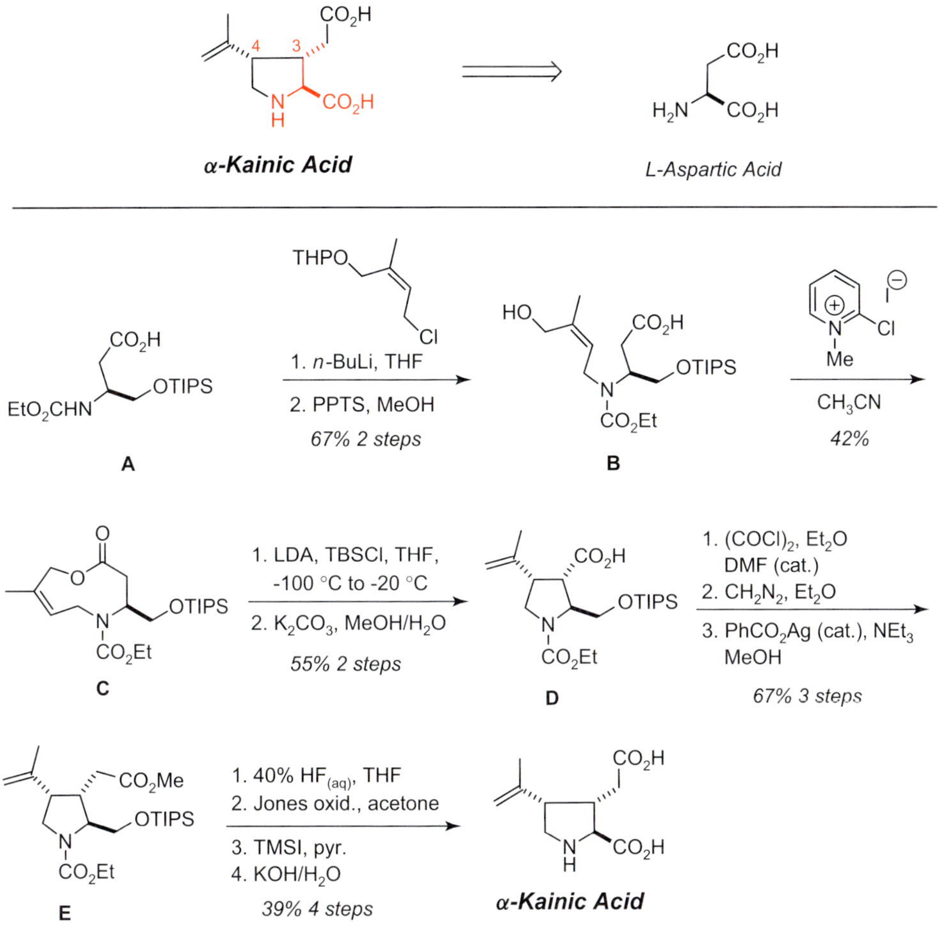

α-Kainic Acid

L-Aspartic Acid

Key reactions: *Transannular Ireland-Claisen-type ring contraction (**C** to **D**); Arndt-Eistert homologation (**D** to **E**)*

Relevant mechanism:

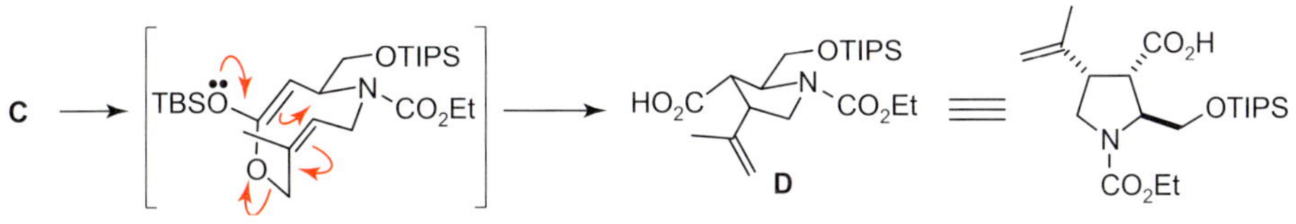

Figure 10.5 Total synthesis of α-kainic acid from L-aspartic acid (partially hidden substructure).

Commentary: The key step in the Gallagher synthesis of α-kainic acid revolved around the successful implementation of the transannular Ireland-Claisen rearrangement. A particularly interesting feature was the anticipated influence of the resident chirality at C-2, bearing a pseudoequatorial and bulky (CH_2OTIPS) group, on the stereochemical outcome of the rearrangement of the 4-aza-macrolactone intermediate **C** after conversion to the silyl ketene acetal (see relevant mechanism, Figure 10.5). The presence of the (Z)-double bond and the pendant side-chain would ensure the exclusive passage through a boat-like conformation approximating the transition state. Unlike L-glutamic acid that was used by Oppolzer and Thirring [8], L-aspartic acid fell short of having a "methylene" group representing the acetic acid appendage in α-kainic acid. However, Gallagher's plan was originally conceived based on a visual *reflexive* analysis, and the desire to create the C-3–C-4 bond relying on the above discussed transannular ring contraction. From that point on, there remained to see which amino acid would be most suitable as a chiral template to allow the implementation of the key reaction. The practical consequence of course, was that the N-butenyl side-chain had to be independently prepared. Furthermore, the nor-kainic acid precursor **D** required a three-step chain-homologation to the methyl ester **E**, and the α-carboxylic acid group had to be restored. Here too, the original stereogenic center (protected as the O-TIPS hydroxymethyl group) was a key stereodirecting group to generate the three contiguous centers in intermediate **D**.

10.6
Croomine

The first asymmetric synthesis of croomine was reported in 1996 by Martin and Barr [12]. The unique tetracyclic structure offers a number of possibilities for synthesis depending on which rings are first considered in a visual analysis (Figure 10.6). Of these, ring C may be considered as a focal substructure since it can be related to an L-pyroglutamic acid precursor. The C-methyl group would be a stereodirecting center in the elaboration of ring D, while the lactam carbon atom C-9a would constitute a second branching point to construct ring A. Martin and Barr addressed the elaboration of rings A and C first, followed by an intramolecular alkylation to introduce the seven-membered ring. The attachment of ring D would rely on a known decarboxylative substitution of an iminium ion intermediate [13].

A two-step sequence from methyl L-pyroglutamate gave intermediate **A**. Activation of the OMe group in **A** with TIPSOTf generated the corresponding iminium ion intermediate, which was coupled with O-TIPS alkoxyfuran **B** (prepared from 3-methyl-2-(5H)-furanone) in a vinylogous Mannich reaction [14] to give **C** as a crystalline *threo*-adduct.

Having secured the absolute configuration of **C** by X-ray crystallography, the N-Boc group was cleaved, and the α,β-unsaturated lactone was subjected to a diastereoselective reduction in the presence of the primary bromide group using 3 mol% Rh/C as a catalyst to give **D**. Cyclization in the presence of N-methylmorpholine, and acidic hydrolysis of the methyl ester led to tricyclic amino acid intermediate **E**. This was first treated with $POCl_3$, then with 3-methyl-2-siloxyfuran derivative **F** to give the *threo*-tetracycle **G** as the major product (dr = 2:1), which was then hydrogenated to give croomine.

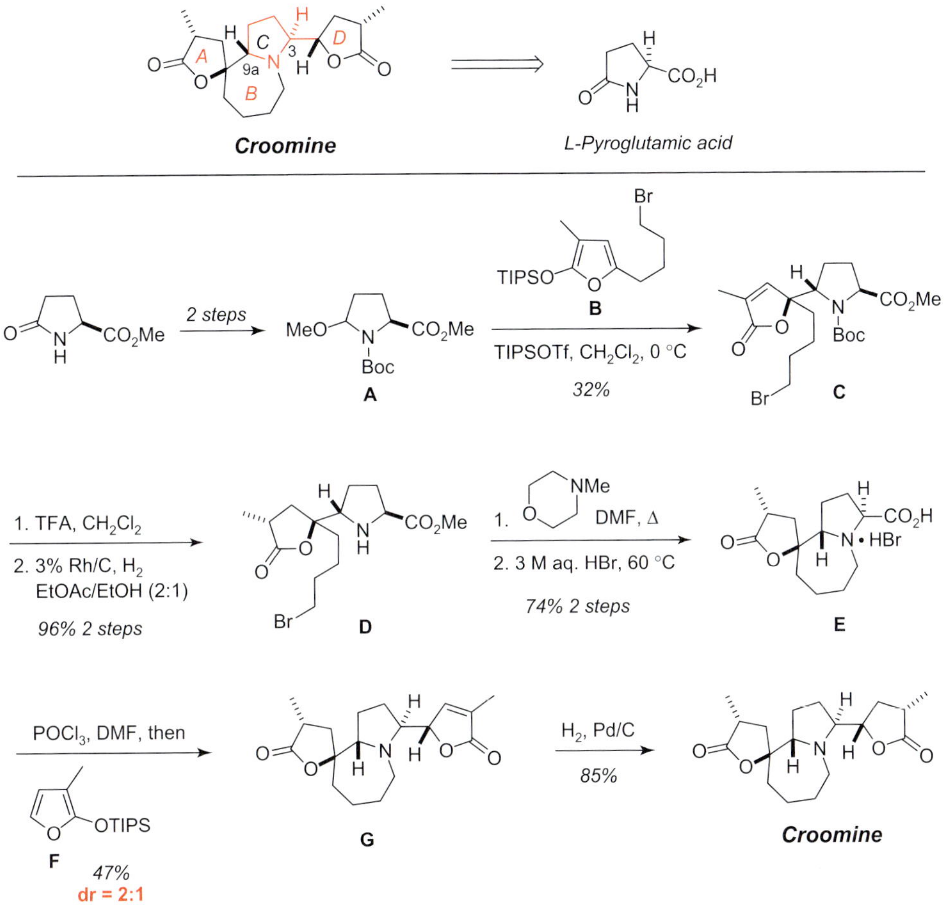

Croomine ⟹ L-Pyroglutamic acid

Key reactions: *Vinylogous Mannich reaction (**A** to **C**) and (**E** to **G**); decarboxylative iminium ion formation (**E** to **Ea**)*

Relevant mechanisms and transformations:

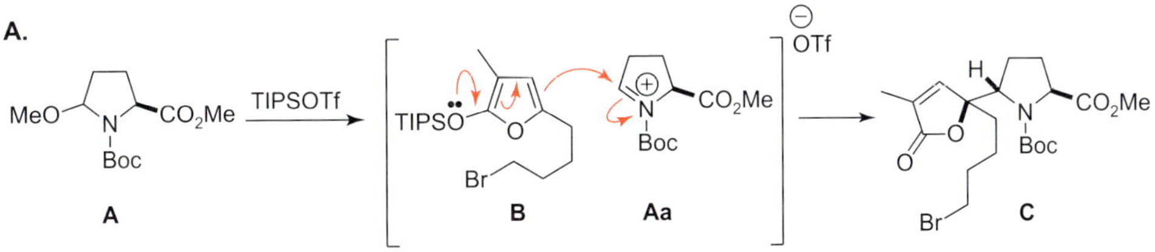

A.

Figure 10.6 Total synthesis of croomine from L-pyroglutamic acid (partially hidden substructure).

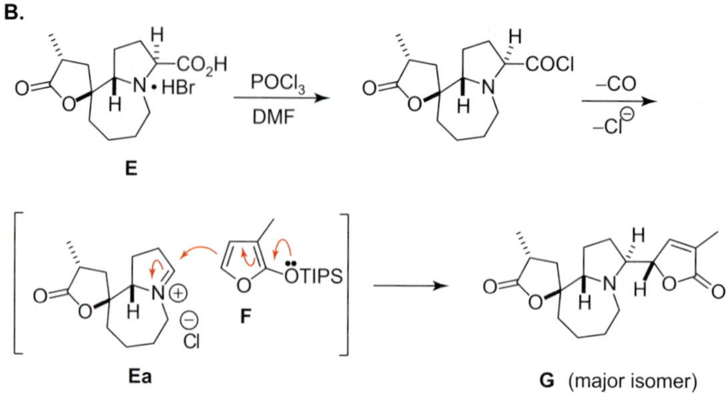

B.

E

Ea

G (major isomer)

Figure 10.6 (continued).

Commentary: Central to the Martin and Barr strategy was the successful application of vinylogous Mannich reactions at two stages in the synthesis with good diastereocontrol. In the first case, the nucleophilic siloxyfuran **B** was preferentially directed toward the iminium ion **Aa** from the *Re* face to give the *threo*-adduct **C** (see relevant mechanism A, Figure 10.6). The high selectivity at the iminium carbon atom can be rationalized based on the pseudoaxial orientation of the carbomethoxy group due to $A^{1,3}$-strain [15], hence its strong stereodirecting role. In spite of the modest yield, the successful application of the vinylogous Mannich reaction with the C-4 substituted siloxyfuran **B** is noteworthy. The selective reduction of the double bond in the presence of a Rh/C catalyst (**C** to **D**) was attributed to a possible directing effect by the basic nitrogen atom by virtue of its coordination to the catalyst.

The second vinylogous Mannich reaction was more challenging to plan because it was predicated upon the generation of an iminium ion intermediate from the decarbonylation of the acid chloride of **E**. Weinstein and Craig [13a] have postulated the loss of carbon monoxide and chloride ion in the thermal decomposition of acid chlorides of protonated tertiary amino acid salts. Original contributions from Rapoport [13c], Wasserman [13b], and their groups are also known. In the event, it was found that decarbonylation of the acid chloride of **E**, generated in the presence of POCl₃ in DMF, occurred at room temperature (see relevant mechanism B, Figure 10.6). The resulting iminium ion salt (**Ea**) was poised for the second vinylogous Mannich reaction, which gave the desired *threo*-adduct **G** resulting from a *Re* face attack, in addition to the *erythro*-diastereomer (2:1 respectively).

It should be noted that the entire carbon skeleton of the target molecule could be accounted for based on "building blocks" **A**, **B**, and **F**. Virtually complete stereochemical control at C-3 and modest diastereoselectivity at C-9a was provided by the original L-pyroglutamic acid motif in two vinylogous Mannich reactions. Having served its role as a stereodirecting group in the first iminium ion reaction with the siloxyfuran **B**, the carbomethoxy group was excised, to generate the second iminium ion **Ea**, now endowed with its own molecular topology, paving the way for a second stereocontrolled reaction to occur with the siloxyfuran **F**.

10.7
Biotin

The important vitamin biotin (vitamin H or B$_8$) has been the subject of extensive synthetic efforts over the years [16]. Analysis of its molecular structure reveals a partially hidden L-cysteine by virtue of the presence of an embedded α-aminoalkyl thioether substructure. In considering L-cysteine as a synthetic precursor, two additional stereogenic centers must be appended on the tetrahydrothiophene ring, consisting of an amino group at C-4 and an ω-pentanoic acid group at C-5. The highly stereocontrolled assembly of biotin can be appreciated in a cleverly planned synthesis by Baggiolini, Lee, and Uskoković [17] (Figure 10.7). Their synthesis

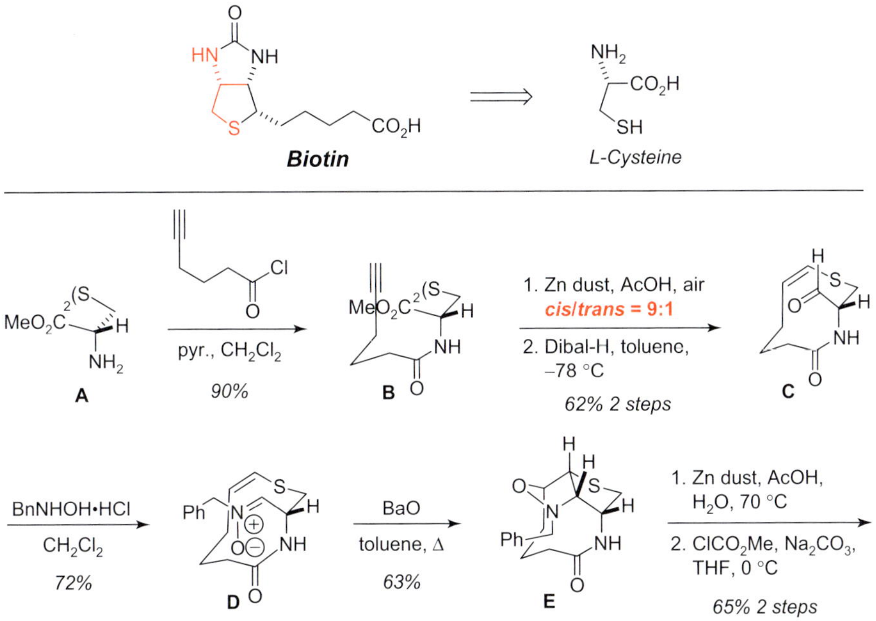

Figure 10.7 Total synthesis of biotin from L-cysteine (partially hidden substructure).

Key reactions: *Intramolecular nitrone-olefin cycloaddition (D to E); base-mediated macrolactam to cyclic urea ring contraction (F to G); episulfonium ion-mediated chlorination and reduction (G to H)*

Relevant mechanisms and transformations:

A.

B.

Figure 10.7 *(continued)*.

commenced with the *N*-acylation of L-cystine dimethyl ester (**A**) with 5-hexynoyl chloride to give **B**. In a crucial step that would generate a *cis*-thioenol ether contained within ten-membered ring, intermediate **B** was treated with Zn dust in acetic acid to reductively cleave the disulfide bond. Exposure to air led to the (*Z*)-thioenol ether as the major isomer, and subsequent reduction of the methyl ester with Dibal-H gave the corresponding aldehyde **C**. Treatment with *N*-benzylhydroxylamine afforded the nitrone **D**, setting the stage for the all-important intramolecular [3+2] nitrone-olefin cycloaddition. Upon refluxing a toluene solution of **D**, containing a small amount of barium oxide, exclusive formation of the tricyclic lactam **E** was observed. Cleavage of the isoxazolidine ring in **E** was effected with Zn in acetic acid, and the resulting *N*-benzylamino alcohol intermediate was converted to the *N*-methoxycarbonyl derivative **F**.

Unravelling of the tetrahydrothiophene ring was done by hydrolysis of the macrolactam portion of the bicyclic system in aqueous Ba(OH)$_2$ to give **G**. The unwanted hydroxyl group was removed by conversion to the chloride and reduction with NaBH$_4$ to furnish **H**. Treatment with aqueous HBr led to the cleavage of the *N*-benzyl and ester groups to give biotin. It is important to recall that L-cysteine is also the biosynthetic precursor of biotin.

Commentary: The Baggiolini, Lee, and Uskoković synthetic approach to biotin was the result of careful analysis of possible transition states in model [3+2] cycloaddition reactions of acyclic analogues of **B**. Their astute analysis of the prospects of an *intramolecular* cycloaddition (**D** to **E**) must have led to the realization that a *cis*-vicinal diamine motif would be formed from a *cis*-thioenol ether (see relevant mechanism A, Figure 10.7). However, reaching the crucial intermediate **D** required the installation of a "masked" 5-pentanoic acid precursor as in the alkynylamide **B**. Thioether formation as a ten-membered lactam would ensure the formation of the requisite *cis*-configured double bond as the preponderant isomer, setting the stage for the intramolecular cycloaddition to **E**, which was obtained as the major product. With the critical C–C bond being formed in a highly stereocontrolled manner, there remained to reveal the tetrahydrothiophene ring by hydrolysis of the lactam, thereby releasing the "masked" 5-pentanoic acid side-chain. The price to pay for the implementation of the stereocontrolled cycloaddition reaction, was the two steps required to deoxygenate the unwanted hydroxyl group in **G**. Treatment with thionyl chloride introduced the chloride with retention of configuration, most likely as a result of intramolecular participation through an episulfonium ion intermediate, followed by opening with chloride ion (*i.e.*, double inversion). Dechlorination with NaBH$_4$ in DMF at 80 °C may also proceed with anchimeric assistance through the same episulfonium ion leading to **H** (see relevant mechanism B, Figure 10.7).

Conceptually, the Baggiolini, Lee, and Uskoković synthesis of biotin is another elegant example of a heavily biased visual *reflexive* analysis with the desire to implement a specific reaction relying on macrocyclic constraint (**D** to **E**).

10.8
Salinosporamide A

The first total synthesis of the marine natural product salinosporamide A was reported in 2004 by Corey and coworkers [18] (Figure 10.8). Analysis of the fused β-lactone and γ-lactam moieties reveals several options for potential synthetic precursors, depending on which combinations of heteroatoms and carbon centers are considered. Corey and coworkers chose L-threonine as an appropriate starting amino acid precursor. Indeed, the entire carbon framework of L-threonine, including its absolute stereochemistry, can be seen in the β-lactone substructure. However, the establishment of two contiguous C–C appendages at the ring junction presents a veritable challenge in stereochemical control.

The methyl ester of L-threonine was converted over two steps to the N-p-methoxyphenyl oxazoline derivative **A** (Figure 10.8). Formation of the lithium enolate and quenching with chloromethyl benzyl ether gave **B** with the desired configuration at the tertiary center. Cleavage of the oxazoline ring with NaBH₃CN in acetic acid afforded **C**. A one-pot procedure, which first required temporary protection of the secondary alcohol in **C** as the TMS ether and subsequent N-acylation with acryloyl chloride, furnished the N-acryloyl derivative of **C**. Conversion to the ketone **D**, followed by an intramolecular Morita-Baylis-Hillman reaction in the presence of a catalytic amount of quinuclidine gave the lactam **E** as the major product (dr = 9:1). Protection of **E** as the (bromomethyl)dimethylsilyl ether facilitated separation of the diastereomers obtained in the previous step, and further enabled the free radical induced cyclization of **F** to give the oxasilane **G** with complete stereochemical control. The critical cyclohexenylation (**G** to **H**) was accomplished with high stereochemical control by treatment of the aldehyde generated from **G** with 2-(cyclohexenyl)zinc chloride to give **H**. A Tamao-Fleming oxidation [19] followed by cleavage of the N-PMB group furnished **I**. Hydrolysis of the ester group and treatment with BOP-Cl in pyridine led to the β-lactone **J**. Selective chlorination of the primary alcohol with Ph₃PCl₂ gave crystalline salinosporamide A.

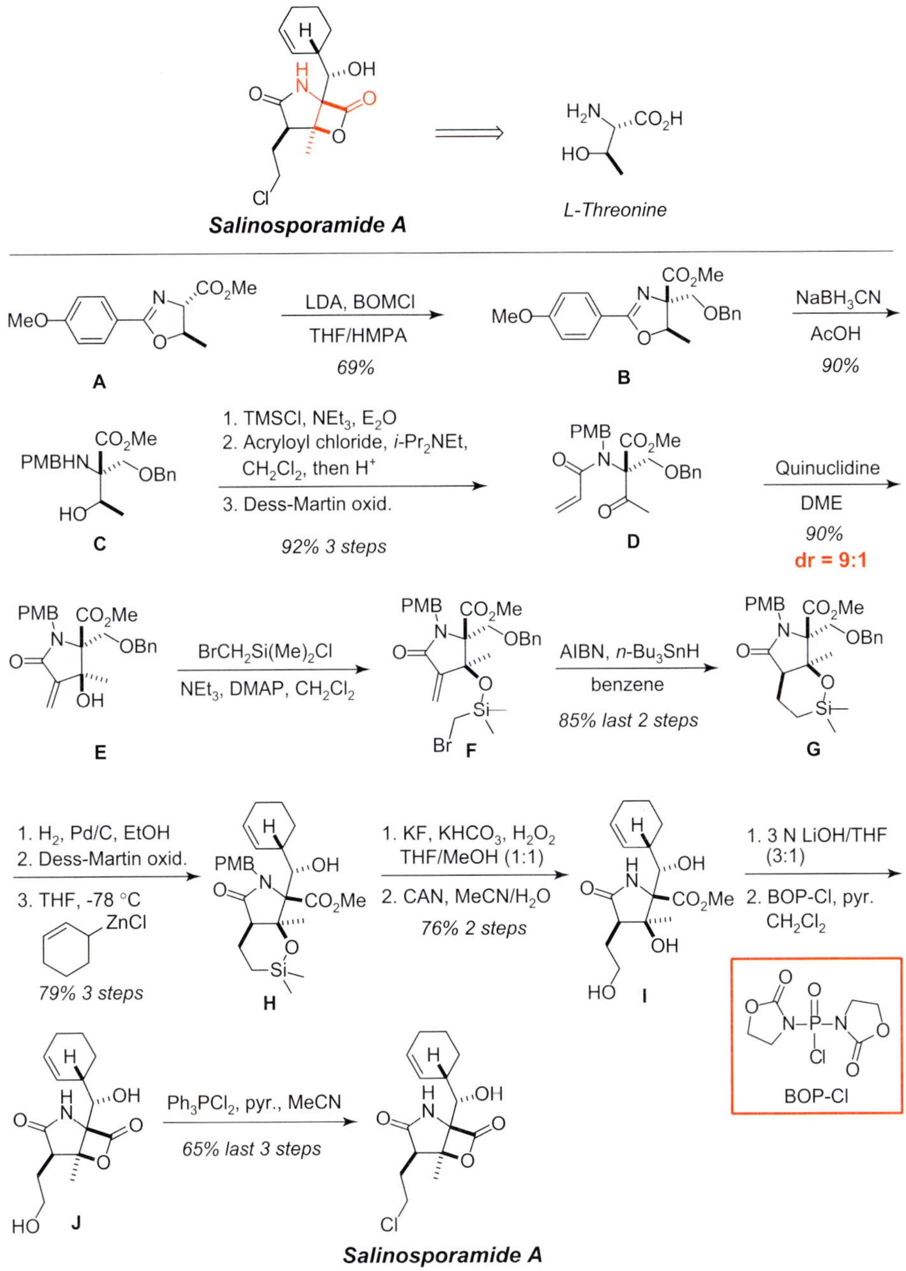

Figure 10.8 Total synthesis of salinosporamide A from L-threonine (partially hidden substructure).

Key reactions: *Enolate alkylation (**A** to **B**); Morita-Baylis-Hillman cyclization (**D** to **E**); free radical-mediated oxasilane cyclization (**F** to **G**); organozinc addition to aldehyde (**G** to **H**); Tamao-Fleming oxidation (**H** to **I**); β-lactone formation (**I** to **J**); BOP-Cl = bis(2-oxo-3-oxazolidinyl) phosphinic chloride; CAN = ceric ammonium nitrate.*

Relevant mechanisms and transformations:

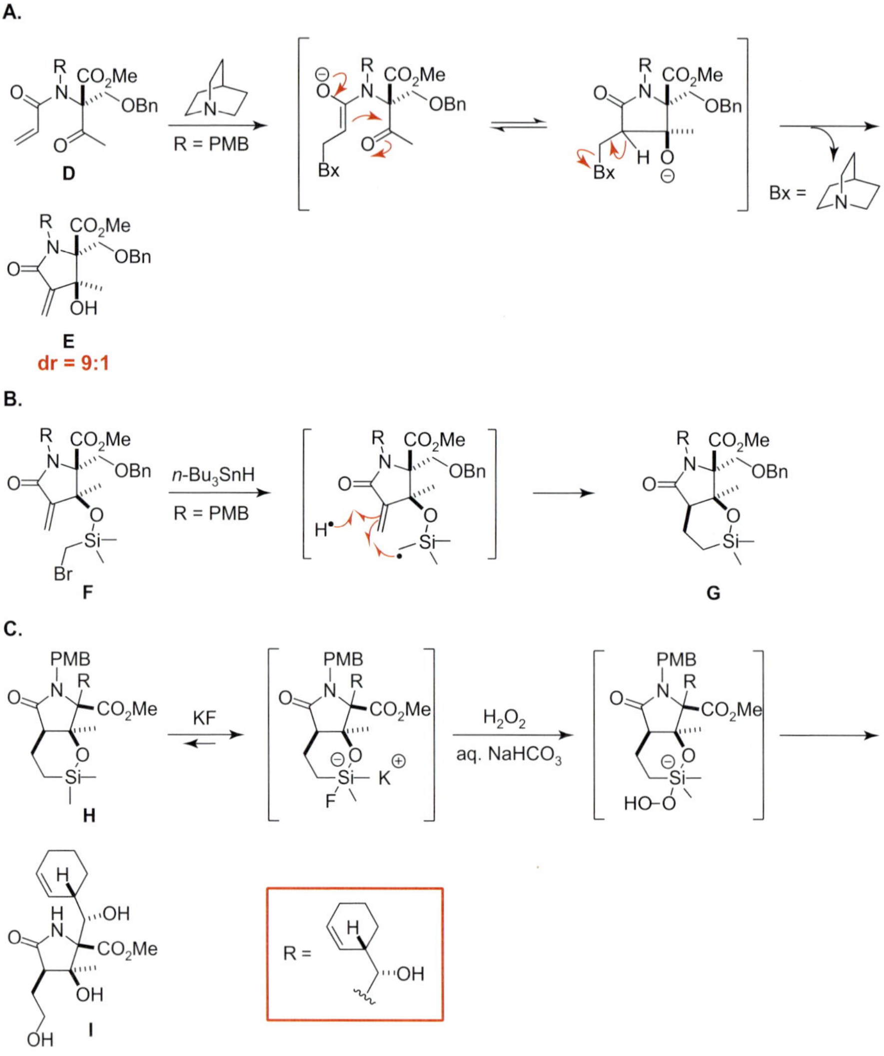

A.

E
dr = 9:1

B.

C.

Figure 10.8 *(continued)*.

Commentary: The incorporation of the first crucial tertiary center capitalized on the specific orientation of the *C*-methyl group in the lithium enolate generated from the oxazoline **A**. The intramolecular aldol reaction initiated by the conjugate addition of quinuclidine as a catalytic base (*i.e.*, Morita-Baylis-Hillman reaction sequence, **D** to **E**), was surprisingly face-selective (see relevant mechanism A, Figure 10.8). Quinuclidine was found to be the most efficient catalyst in this uniquely adapted cyclization. Furthermore, not only was the tertiary *C*-methyl group incorporated with the desired stereochemistry in the product **E**, but an exocyclic methylene group was generated, which was later converted to a hydroxyethyl group *en route* to **I**. First, a (bromomethyl)dimethylsilyl ether was installed on the tertiary alcohol as in **F**, then the latter was subjected to a radical-induced carbocyclization following a 6-*endo*-trig cyclization pathway, no doubt aided by the conjugated nature of the exocyclic methylene group to give **G** (see relevant mechanism B, Figure 10.8). The sequence was continued relying on a Tamao-Fleming oxidation, which proceeded through the intermediacy of a pentacoordinate fluorosilane followed by a peroxysilane. Fragmentation to the peroxysilane prior to the intramolecular migration of the C−Si bond ultimately furnished **I** (see relevant mechanism C, Figure 10.8).

Thus, the *C*-methyl group in the oxazoline derived from L-threonine may have rendered great service in the initial stereocontrolled enolate alkylation to give **B**. However, beyond this stage, the C−C bond forming events were controlled by preferential facial selectivities in the intramolecular Morita-Baylis-Hillman reaction (**D** to **E**), and the free-radical carbocyclization (**F** to **G**). The high degree of stereoselectivity in the organozinc addition to the aldehyde appears to be a felicitous event that only adds to the elegance of the overall plan.

10.9
Thienamycin

The discovery of the antibiotic thienamycin fostered a new era of β-lactam research [20]. Unlike the familiar penam nucleus in the penicillins, thienamycin is a carbapenem [21], with a (2*R*)-hydroxyethyl side-chain, and a configuration that is opposite at C-6 to that commonly found in the penicillins. The first enantioselective synthesis of thienamycin was reported in 1980, by Christensen, Saltzmann, and coworkers [22] (Figure 10.9). Numerous other approaches have been reported since then, including industrially scalable syntheses that are presently used in the manufacture of related congeners [23].

At first glance, L-threonine may be evident as a precursor, especially when focussing on the pendant (2*R*)-hydroxyethyl side-chain. With L-threonine as a native chiron, the C-2 amino group will have to be replaced by a C−C bond. The practical feasibility of such an approach has been validated in the synthesis of penems related to the original thienamycin structure [24]. Focussing on the nitrogen atom, and an alternative amino acid precursor, led the Merck scientists to explore L-aspartic acid as a starting material. Thus, the readily available intermediate **A** was transformed to the β-lactam by intramolecular cyclization via a magnesium amide salt (**A** to **B**). Conversion to the iodide **C** (over four steps) and treatment with TMS dithiane anion led to **D** in 70–80% yield. Enolate condensation with *N*-acetyl imidazole resulted in the transfer of an acetyl group with excellent diastereoselectivity to give **E**. Reduction of the carbonyl group with K-Selectride led to (*R*)-alcohol **F** as the major diastereomer. Cleavage of the dithiane unit, and oxidative workup afforded the acid **G**, which was in turn converted to β-ketoester **H**. Transformation to the diazo intermediate **I**, set the stage for a Rh(II)-catalyzed carbene insertion reaction [25] involving the β-lactam N–H bond giving bicyclic intermediate **J**. Further manipulation of the carbonyl group led to the protected precursor **K**, which was converted to thienamycin. It is noteworthy to mention that the entire sequence was executed without protection of the secondary hydroxyl group.

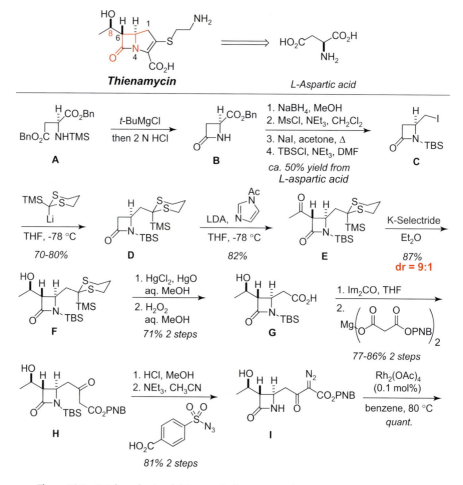

Figure 10.9 Total synthesis of thienamycin from L-aspartic acid (partially hidden substructure).

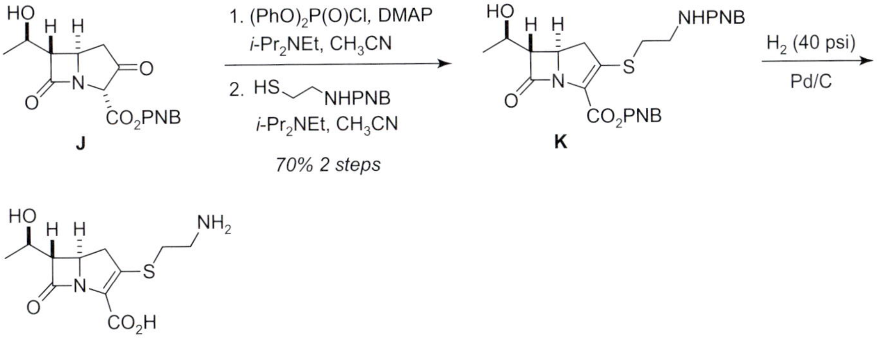

J → (via reagents) → **K** → **Thienamycin**

1. (PhO)₂P(O)Cl, DMAP
   i-Pr₂NEt, CH₃CN

2. HS⌒NHPNB
   i-Pr₂NEt, CH₃CN

70% 2 steps

H₂ (40 psi)
Pd/C

Thienamycin

Key reactions: Enolate acylation (**D** to **E**); malonate extension/decarboxylation (**G** to **H**); β-keto-diazoester Rh(II)-catalyzed carbene insertion reaction (**I** to **J**); PNB = p-nitrobenzyl

Relevant transformations:

A.

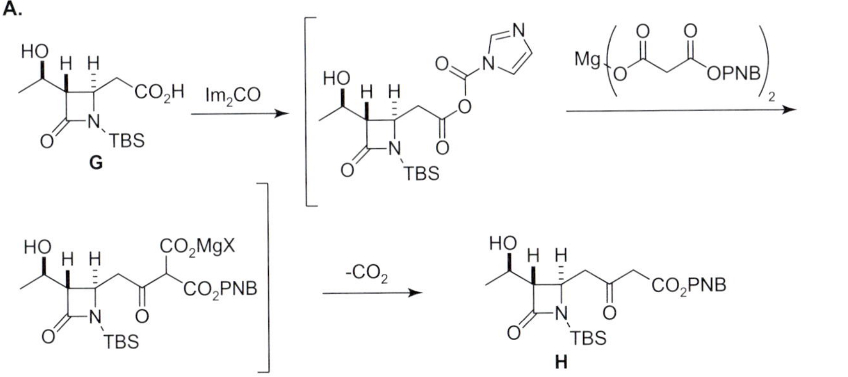

B.

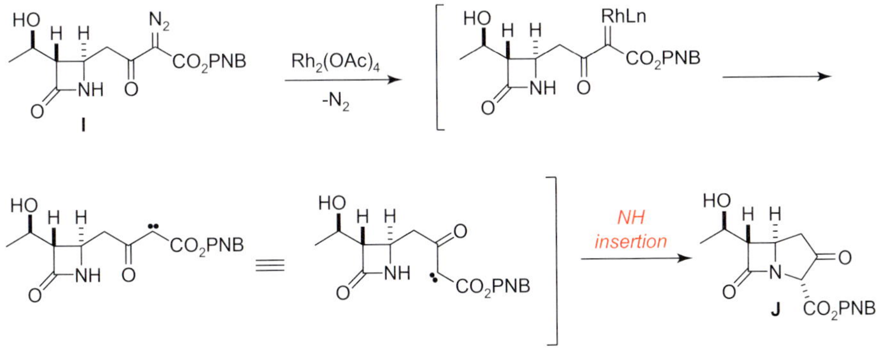

Figure 10.9 *(continued).*

Commentary: Internal induction played key roles in two successive steps of the synthesis of thienamycin. First, the acetyl transfer from *N*-acetylimidazole occurred with high selectivity with the electrophile approaching the β-lactam enolate from the least hindered side (**D** to **E**). The preponderance of the desired (R)-alcohol configuration during the reduction of the ketone **E** with K-Selectride could not have been predicted *a priori*. A second example of the utility of acylimidazoles as electrophilic reagents is seen in the transformation of **G** to the β-ketoester **H** via the acyl imidazole (mixed anhydride) intermediate (see relevant transformation A, Figure 10.9). The application of the Rh(II) acetate-catalyzed carbene insertion to form an oxo-carbapenam ring was unprecedented at the time of the synthesis, and the result of model studies carried out by Saltzmann and coworkers (see relevant transformation B, Figure 10.9).

10.10
FR901483

Three total syntheses of the potent immunosuppressive agent FR901483, starting with L-tyrosine as a common chiron, were independently reported by Snider [26] (1999), Sorensen [27] (2000), and Ciufolini [28] (2001). Other strategies adopting different approaches have also been reported [29]. The natural product contains a novel azaspiro[4.5]decane nucleus with an appended *p*-methoxybenzyl group in an equatorial orientation on the piperidine ring (Figure 10.10). Inspection of the molecular framework of FR901483 reveals an O-methyl-L-tyrosine subunit. Further scrutiny reveals the existence of a second, albeit concealed, L-tyrosine in the form of a reduced hydroxy cyclohexane (Figure 10.10 B and C). In fact, the prospect of a biosynthetic pathway arising from a tyrosinyl tyrosine subunit was recognized by the three groups, and exploited in different ways.

The three disconnective analyses by the Snider, Sorensen, and Ciufolini groups are shown in Figure 10.10. Although all three strategies relied on a key intramolecular aldol cyclization, the precursors to the respective intermediates were prepared based on conceptually different approaches. Snider [26] started with nitrone **D** prepared from the hydroxylamine **F** and the cyclohexanone ketal **E** (Figure 10.10 A). Dipolar [3+2] cycloaddition with ethyl acrylate gave the isoxazolidine **C** as a single isomer. Cleavage of the N–O bond and lactam formation afforded **B**, which was converted to ketoaldehyde **A**. A diastereoselective intramolecular aldol cyclocondensation, followed by stereochemical and functional group adjustments led to the intended target.

Sorensen's strategy [27] was guided by a possible biogenetic pathway utilizing two L-tyrosine moieties, involving a phenolic oxidation [29e] with concomitant formation of an azaspirocyclic system (Figure 10.10 B). Reductive coupling of the tyrosine-derived precursors **I** and **J** gave **H**, which was subjected to an oxidative spiroazacycle formation, eventually leading to ketoaldehyde **G**. Intramolecular aldol condensation according to Snider's protocol, followed by functional group modifications afforded FR901483.

Ciufolini [28] also relied on a tyrosinyl tyrosine precursor, which generated oxazoline **M**, in turn, readily available from **O** and **N** (Figure 10.10 C). Oxidative azaspirocyclization eventually afforded ketone **L**, which was converted to ketoalde-hyde **K** following Snider's protocol [26], *en route* to FR901483. Thus, the Ciufolini synthesis followed the same critical oxidative azaspirocyclization step as seen in the Sorensen synthesis, except for the utilization of the oxazoline as a nitrogen source.

We shall discuss the Sorensen synthesis in detail in the next section. However, we cannot leave this analysis without pointing out the three different perspective depictions of FR901483, by the three groups (Figure 10.10). Whether the conscious manner in which the molecules were drawn by each group has any bearing on the creative thought process involved in the respective synthesis plans is a matter of speculation.

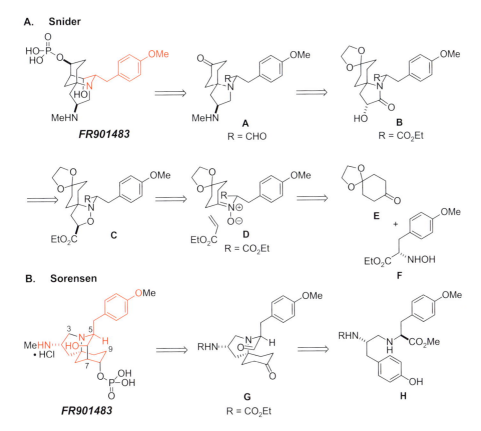

Figure 10.10 Retrosynthetic analyses of FR901483.

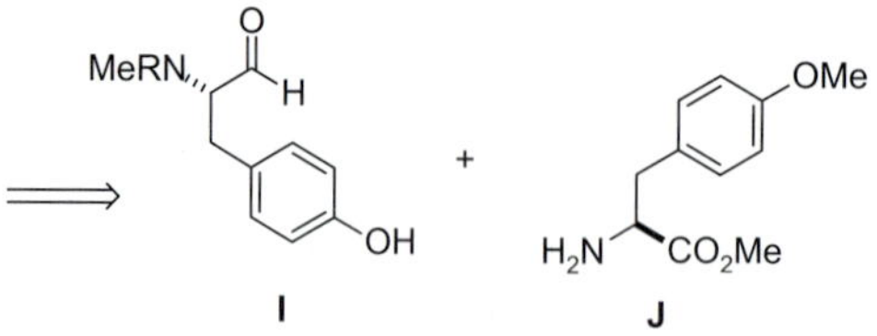

C.   Ciufolini

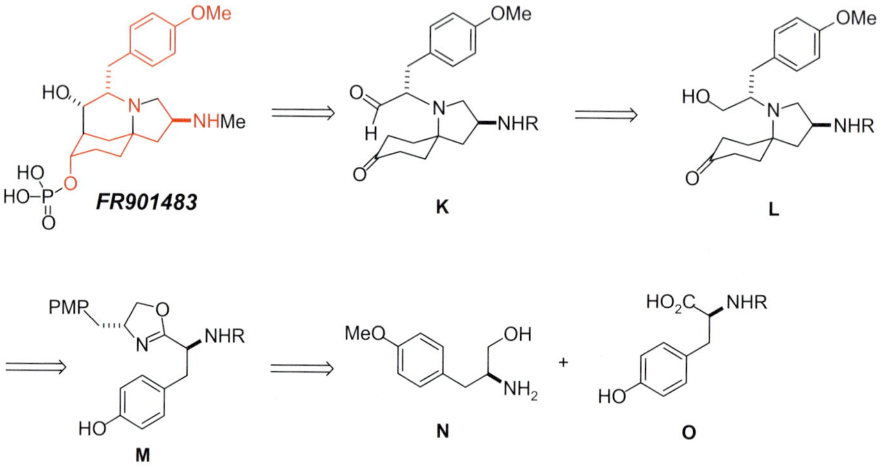

Figure 10.10   *(continued).*

10.10.1
The Sorensen synthesis

Coupling of the aldehyde intermediate **A** with the tyrosine methyl ester derivative **B** under conditions of reductive amination led to the amine **C** (Figure 10.11). Treatment with iodobenzene diacetate in hexafluoro-2-propanol as solvent effected a smooth azaspiroannulation to give **D** as a single isomer with recovery of some starting material (**C**). Exchanging the *N-p*-nitrobenzenesulfonyl group with *N*-Boc was accomplished in two steps, by first treatment with sodium thiophenoxide, followed by Boc$_2$O to give **E**. Reduction of the hexadienone to a mixture of saturated alcohols **F** in the presence of Raney-Ni, was followed by treatment with LiAlH$_4$ to give the corresponding diols. Swern oxidation of this mixture furnished the ketoaldehyde **G**. The crucial intramolecular aldol condensation was carried out in the presence of NaOMe in MeOH to give the tricyclic aldol product **H** albeit in modest yield (34%), following the same observation of Snider and coworkers [26]. Raney-Ni was the heterogeneous reducing agent of choice for the conversion of the ketone group in **H** to the equatorial alcohol **I**. An unprecedented Mitsunobu phosphorylation with inversion of configuration, followed by hydrogenolysis of the benzyl ester, and subsequent acid treatment, led to FR901483 as its HCl salt.

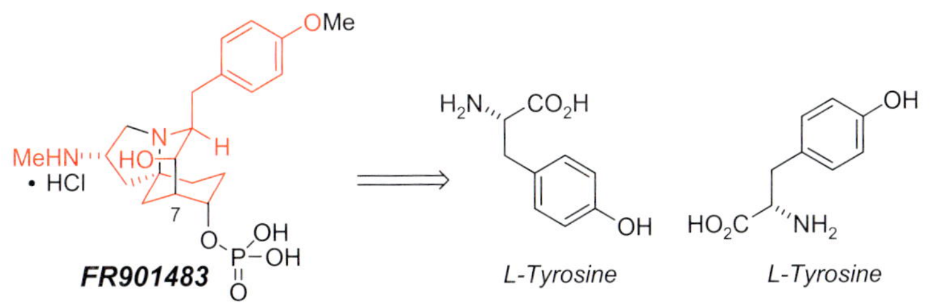

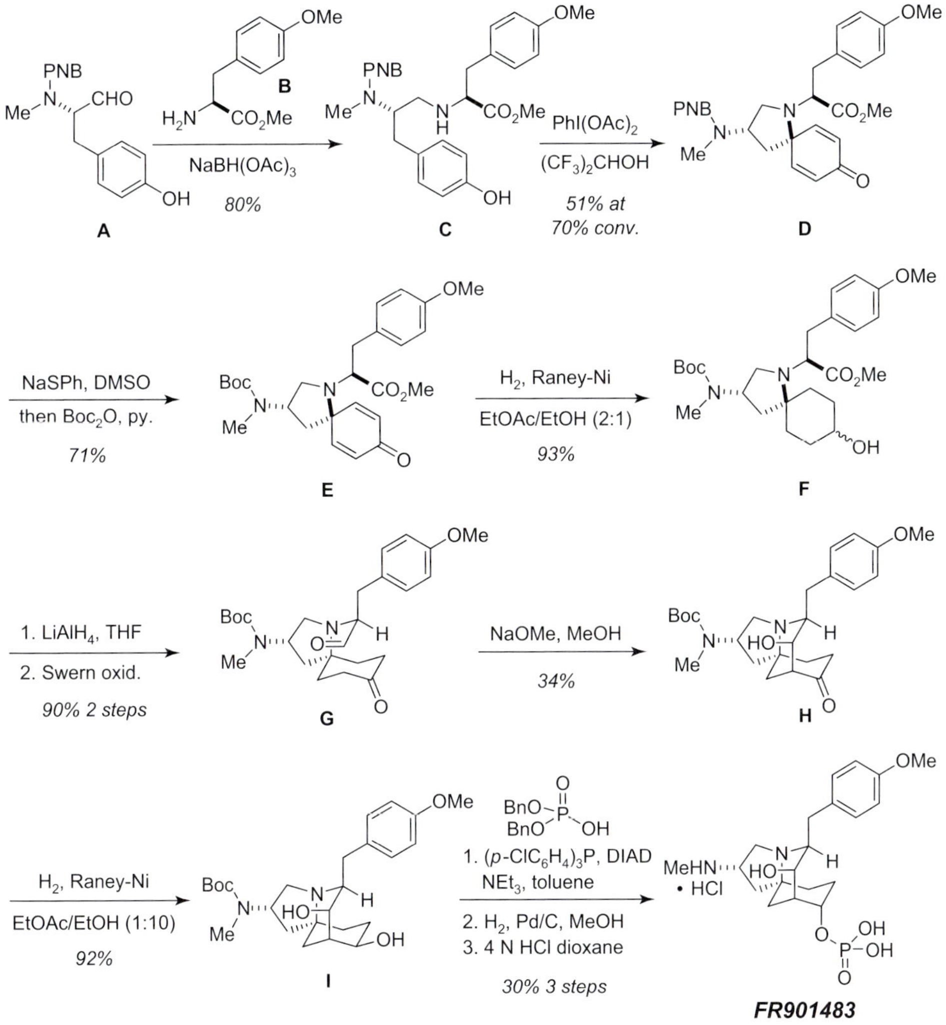

Figure 10.11 Sorensen's total synthesis of FR901483 from two L-tyrosines (one apparent, one hidden substructure).

Key reactions: *Oxidative spirocyclization (**C** to **D**); intramolecular aldol condensation (**G** to **H**); Mitsunobu phosphate synthesis (**I** to **target**); PNB = p-nitrobenzenesulfonyl; DIAD = diisopropyl azodicarboxylate*

Relevant mechanisms and transformations:

A.

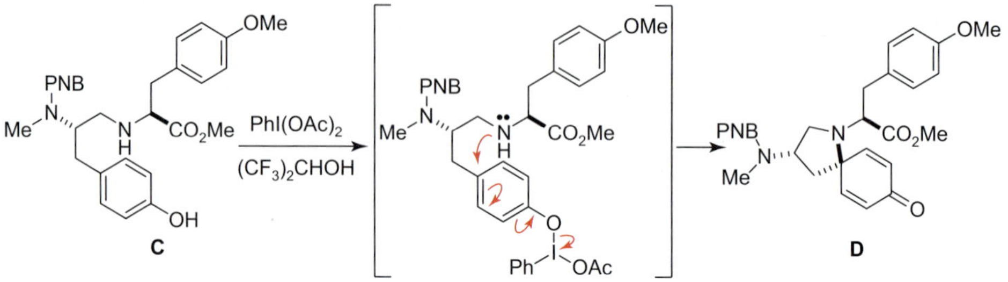

B.

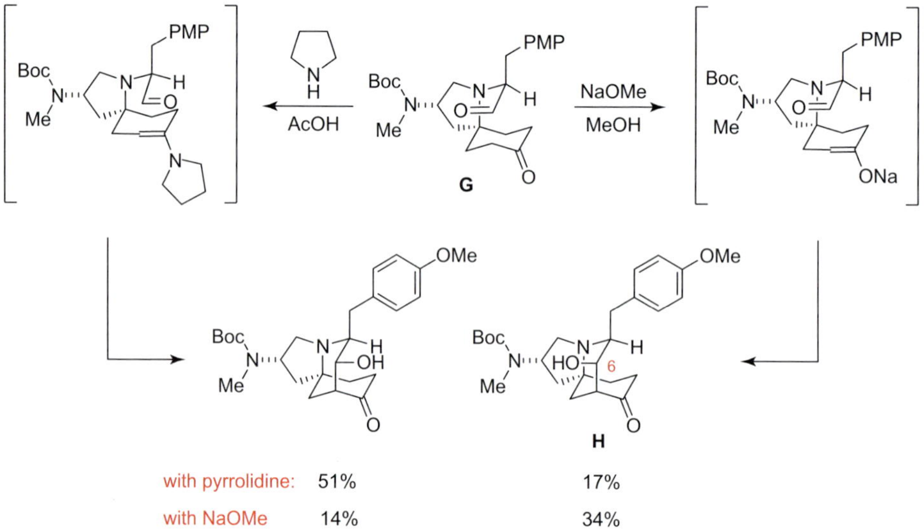

| | with pyrrolidine: | 51% | 17% |
| | with NaOMe | 14% | 34% |

Figure 10.11 *(continued).*

Commentary: Although the azaspiro cyclohexanol portion of FR901483 may be biogenetically related to a tyrosine precursor, visualizing such a connection and developing a synthetic strategy based on that notion is remarkably astute. As a *reflexive*, reaction-type thought process, combined with the desire to emulate a biomimetic Barton-type phenolic coupling [30], may have conjured the image of a cyclohexadienyl radical equivalent of the ring-modified tyrosine in Sorensen's mind's eye. Indeed, the successful **C** to **D** end-group differentiated azaspiroan-nulation was one of several key reactions used in this synthesis (see relevant mechanism A, Figure 10.11). This also led to a suitably disposed aldehyde group for the intramolecular aldol reaction (**G** to **H**). It is interesting to note that following the Snider protocol, the simplest of basic media, *i.e.*, NaOMe/MeOH, was sufficient to induce the cyclization which led to the axially disposed β-hydroxy ketone aldol product. The pyrrolidine enamine route in Sorensen's synthesis led to a 51% yield of the C-6-(*R*)-epimer of **H** (see relevant transformation B, Figure 10.11), which could not be put to good use toward the intended target. Presumably an *endo*-face inversion of the equatorial alcohol group, directly or indirectly, was not a facile process, considering the topology of the tricyclic ring system.

As remarked by Snider [26], the intramolecular aldol condensation of **G** can, in principle, lead to four products of which only **H** possesses the desired axial C-6 alcohol and equatorial *p*-methoxybenzyl group. Thus, regioselective enolization occurred more readily at C-7, possibly due to the stability of the formed product **H**. It should also be noted that enolization of the aldehyde under the reaction conditions did not affect the stereochemical outcome of the aldol reaction.

10.11
Tuberostemonine

The first total synthesis of tuberostemonine, was reported in 2002 by Wipf and collaborators [31]. The pentacyclic structure presents the viewer with a number of viable starting materials to consider. The mind's eye may well focus on the central *trans*-fused octahydroindole unit which contains seven of the ten stereogenic centers of the natural product (Figure 10.12). Viewed from an amino acid perspective reveals a 4,5-disubstituted bicyclic L-proline (ring *B*), or the more functionally useful L-pyroglutamic acid, which would allow the construction of the six-membered *trans*-fused ring. Wipf and coworkers utilized L-tyrosine as a hidden amino acid chiron in the elaboration of the carboxy octahydroindole unit corresponding to rings *A* and *B* of tuberostemonine. Although the connection between L-tyrosine and the octahydroindole unit is somewhat remote, a facile three-step reaction sequence, albeit in modest overall yield, provides rapid access to appropriately functionalized intermediates. Knowledge of this remarkable transformation from previous studies [32], allowed Wipf to formulate a viable strategy to elaborate the highly substituted, tricyclic *ABC* ring system in tuberostemonine with an excellent degree of stereocontrol.

Treatment of *N*-Cbz-L-tyrosine (**A**) with iodobenzene diacetate led to hexahy-droindole derivative **B** in a single step. Benzoylation and reduction afforded the allylic alcohol **C**. Reductive removal of the allylic benzoate group, while keep-ing the C–C double bond intact, was accomplished using a Tsuji deoxygenation [33]. Protection of the C-6 allylic alcohol and cleavage of the *N*-Cbz group was done under reductive conditions that did not affect the endocyclic double bond to give **D**. Introduction of the *N*-cinnamyl group, followed by de-*O*-silylation led to **E**. Oxidation to the ketone and a face selective *C*-allylation of the corresponding enolate afforded **F**. Construction of the azepane ring was achieved through a ring-closing metathesis reaction. Conjugate addition of benzenethiol led to the tricyclic intermediate **G** in excellent overall yield. Reduction of the double bond in the presence of Wilkinson's catalyst and β-elimination restored the 4,5-endocyclic double bond in intermediate **H**. Reduction of the C-6 carbonyl, silyl protection of the resulting alcohol, and conversion of the ester to the corresponding Weinreb amide gave **I** in 53% yield over three steps. Elaboration to the non-fused butyro-lactone ring was carried out in an another three-step operation. Thus, addition of the lithium anion of the bromo orthoester **J** to **I** gave the ketone **K**, which upon reduction with L-Selectride and acidification of the resulting alcohol, afforded the expected lactone appendage as in **L**. Next, a sequence of reactions was used to install the fused butyrolactone ring. An Eschenmoser-Claisen [34] rearrangement of the allylic alcohol system in ring *A* in the presence of dimethylacetamide dimethylacetal gave the *N,N*-dimethylamide **M** with double bond transposition. Treatment of **M** with PhSeCl gave the corresponding *epi*-selenenium ion, which was attacked intramoleculary by the adjacent *N,N*-dimethylamide carbonyl to give the β-oriented phenylseleno lactone. A free-radical mediated allylation was achieved with complete stereochemical control to give intermediate **N**. Transformation to the lithium enolate, and treatment with MeI afforded the *C*-methylated lactone, which was isomerized to the 2-propenyl isomer **O** in the presence of Grubbs' second generation catalyst and *N*-trityl allylamine according to Roy and coworkers [35]. A cross-metathesis reaction with the Grubbs-Hoveyda catalyst [36] in the presence of ethylene led to the *C*-vinyl compound which was hydrogenated to give tuberostemonine.

Commentary: Central to the formulation of this knowledge-based synthesis plan toward tuberostemonine, was the prior experience with the oxidative end-group differentiating sequence leading to **B** from *N*-Cbz-ʟ-tyrosine (**A** to **B**). Tsuji deoxy-genation of the allylic benzoate **C** led to the octahydroindole intermediate **D** for the construction of the azepane ring in **H**.

However, the planned ring-closing metathesis required activation of the C-7 position to enable the installation of a stereochemically defined *C*-allyl group. This was achieved by oxidation of **E** and an α-face selective allylation to give **F**. In order to selectively reduce the double bond resulting from the ring-closing metathesis, Wipf prepared the phenylthio ether **G**, then used Wilkinson's catalyst for the reduction, following which the enone system was restored upon β-elimination. Stereoselective reduction of the ketone in **H** was also anticipated due to the presence of the ring *C* azepane with a pseudoaxial substitution pattern at C-7. The elaboration of the non-fused butyrolactone by appending a masked Roche ester via an organolithium species derived from **I** is noteworthy (**I** + **J** to **K**). The allylic alcohol was now nicely poised to introduce the acetic acid chain at C-4 with control of stereochemistry in a [3,3]-sigmatropic rearrangement (**L** to **M**).

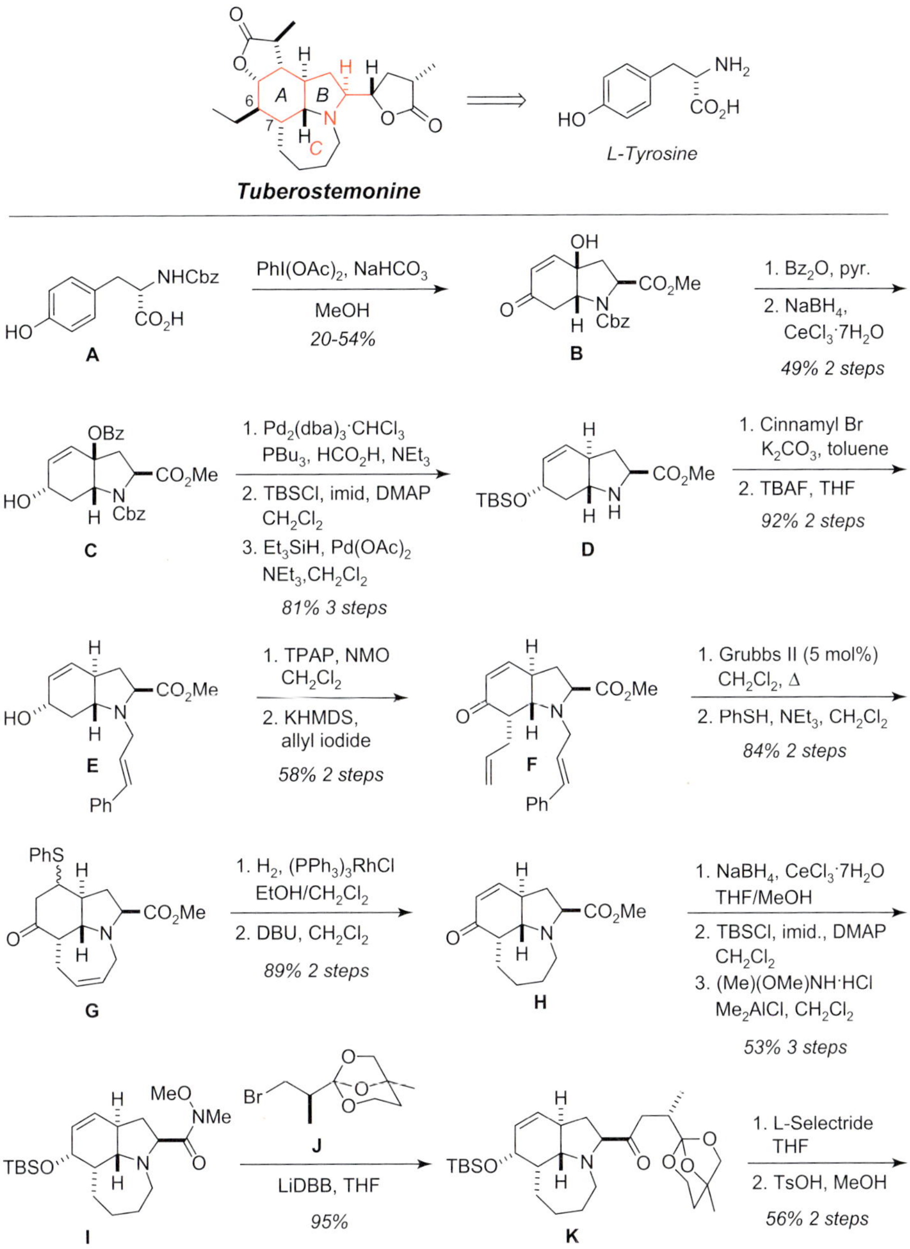

Figure 10.12 Total synthesis of tuberostemonine from L-tyrosine (hidden substructure).

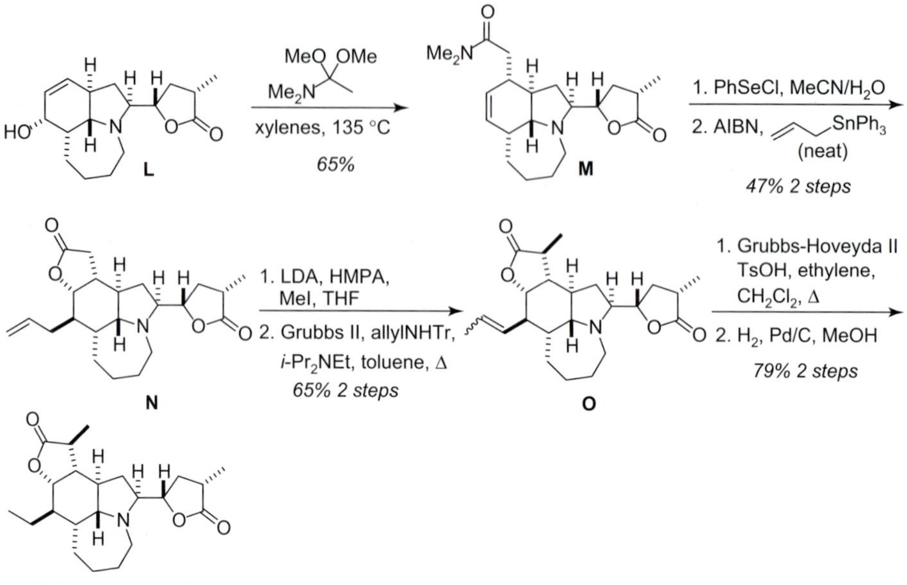

Tuberostemonine

Key reactions: *Oxidative cyclization (**A** to **B**); Pd-catalyzed allylic reduction (**C** to **D**); ring-closing metathesis (**F** to **G**); Weinreb amide alkylation/reduction/lactonization (**I** to **K** to **L**); Eschenmoser-Claisen rearrangement (**L** to **M**); phenylselenyl-mediated lactonization and radical allylation (**M** to **N**); double bond isomerization (**N** to **O**); LiDBB = lithium 4,4'-di-tert-butylbiphenylide*

Relevant transformations and mechanisms:

A.

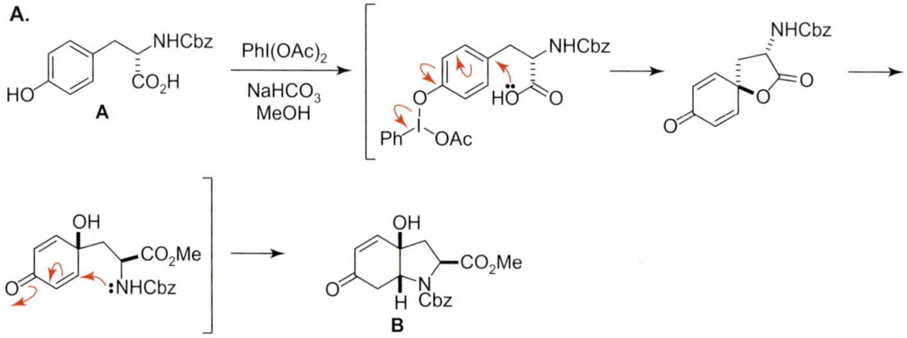

Figure 10.12 *(continued).*

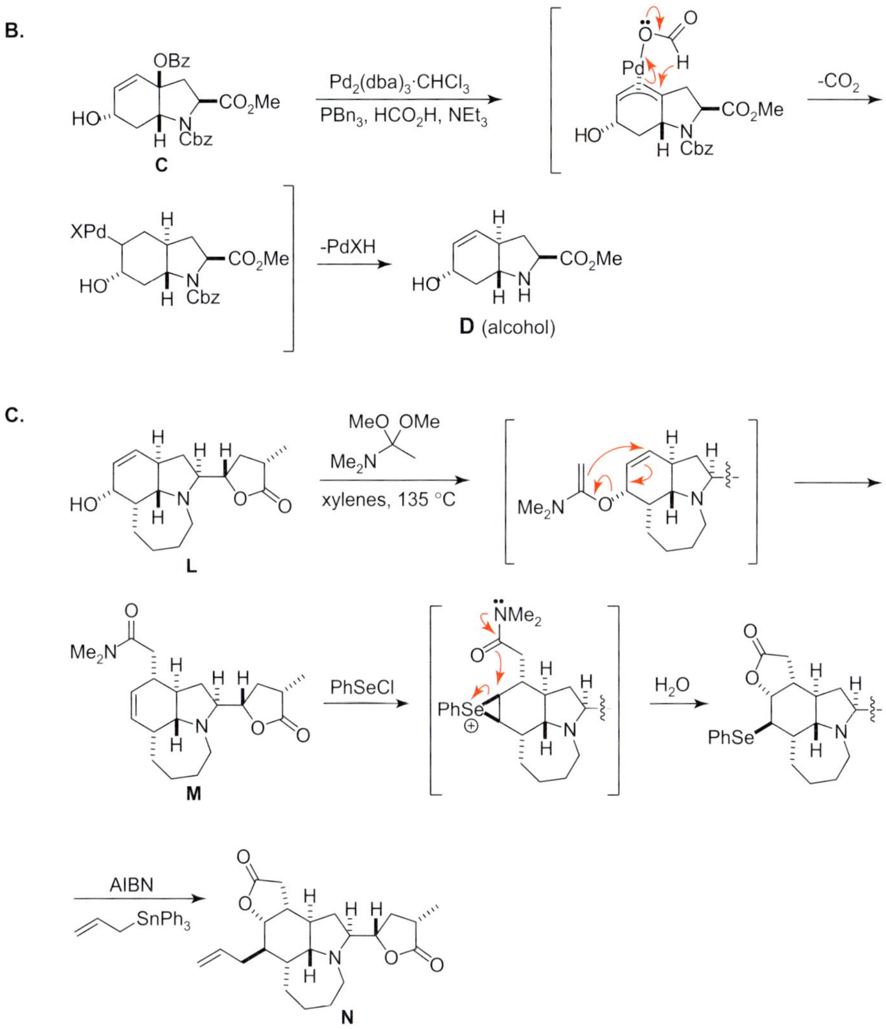

B.

C.

Figure 10.12 *(continued)*.

The six contiguous stereogenic centers in the octahydroindole unit of tuberostemonine were introduced with excellent stereochemical control starting with L-tyrosine. This was possible based on the availability of intermediate **D**, from which a logical sequence of reactions were executed, taking advantage of site-selective functionalizations. In the process, the Grubbs and Grubbs-Hoveyda second generation catalysts were utilized in three separate metathesis (both ring-closing and cross metathesis) reactions.

10.12
Phyllanthine

The first total synthesis of the alkaloid phyllanthine was reported in 2000 by Weinreb and coworkers [37]. A general strategy was developed to address the synthesis of other members of this class of alkaloids differing in the size of ring *A* and the configuration at C-2 (Figure 10.13). The intricate architecture of phyllanthine can be better appreciated when the piperidine ring is viewed in a chair conformation. An amino acid approach to the total synthesis could involve L-pyroglutamic acid or a suitably functionalized L-proline as viable starting chirons for ring *B*. Weinreb selected *trans*-4-hydroxy-L-proline, which would secure the chirality of C-7 as well as provide a functional group to elaborate ring *C*. The synthesis of phyllanthine required the development of an annulation strategy to elaborate ring *A* containing an axial C-4 methoxy group. The implementation of such a strategy was experimentally realized in a highly stereocontrolled manner.

Intermediate **A**, which is readily available from *trans*-4-hydroxy-L-proline in three steps, was chain-extended to the nitrile **B** after a reduction of the ester, followed by an oxidation/Wittig sequence. Reduction of the double bond, removal of the silyl protective group and oxidation, led to the 4-keto intermediate **C**. Exposure to SmI$_2$, followed by acidic hydrolysis led to the α-ketol **D**. Functional group modifications provided a suitably protected intermediate **E**, in which rings *B* and *C* with the correct configuration at C-7 (originally present in **A**), and the bridgehead hydroxyl at C-11, were secured. The construction of the six-membered ring *A* was achieved through a [4+2] cycloaddition reaction of the imine **F**, which was obtained in high yield by treatment of **E** with iodosobenzene [38]. Treatment of **F** with *trans*-3-methoxy-l-trimethylsiloxy-1,3-butadiene (Danishefsky's diene) in the presence of Yb(OTf)$_3$ under high pressure conditions, led to the cyclic vinylogous amide **G**. Functional group manipulation afforded the fully elaborated bridged bicyclic system **H**. The α-ketol **I** was converted to the α-phenylseleno enone **J** according to Sonoda [39], which was then deselenated in the presence of NaI [40]. Esterification of the bridgehead alcohol with diethyl phosphonoacetic acid led to **K**, which underwent intramolecular cyclization to give phyllanthine.

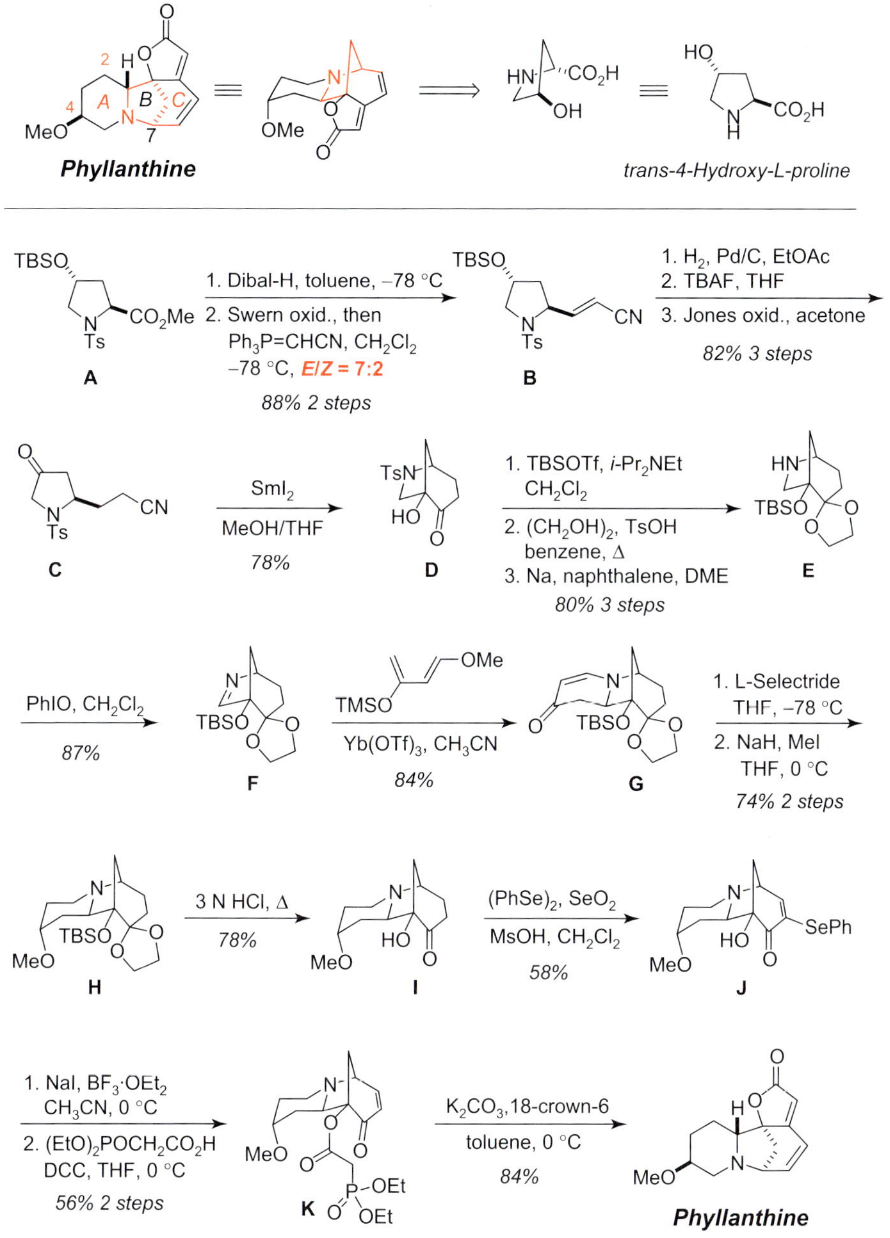

Phyllanthine

trans-4-Hydroxy-L-proline

Figure 10.13 Total synthesis of phyllanthine from
trans-4-hydroxy-L-proline (hidden substructure).

Key reactions: *Intramolecular SmI$_2$-mediated α-ketol formation (**C** to **D**);
iodosobenzene oxidation to imine (**E** to **F**); exo-hetero Diels-Alder reaction
(**F** to **G**); intramolecular β-ketophosphonate condensation (**K** to **target**).*

Relevant transformations and mechanisms:

A.

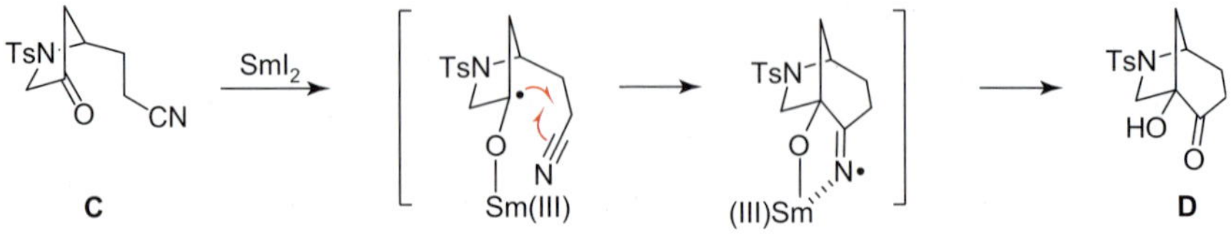

B.

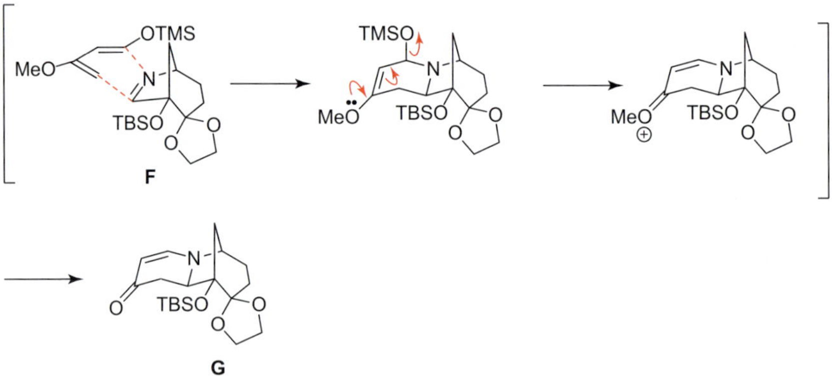

C.

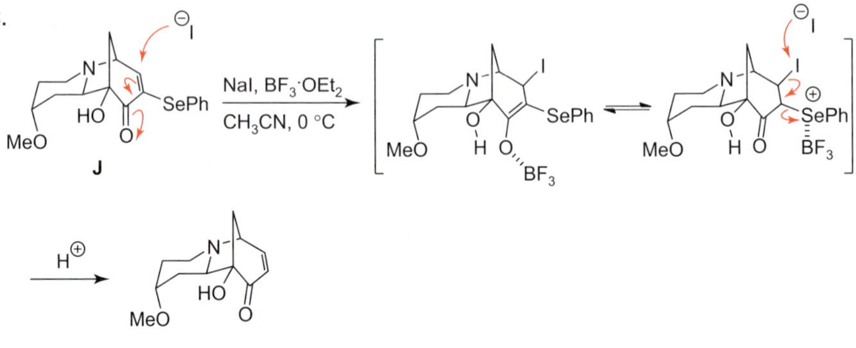

Figure 10.13 *(continued).*

Commentary: To reach intermediate **D** required a suitably branched azacyclic
precursor that would provide favorable spatial, functional, and topological features
for subsequent stereocontrolled transformations. The radical mediated SmI$_2$ ring
closure of a terminal nitrile with an endocyclic ketone (**C** to **D**), presented an
elegant solution to placing an α-ketol group in a strategically important position
in intermediate **D** (see relevant mechanism A, Figure 10.13). It becomes clear
why Weinreb chose the readily available *trans*-4-hydroxy-L-proline as the starting
chiron, realizing that the hydroxyl group was useful only as a site to create the

corresponding ketone as in **C**. The *reflexive* thought process that relates the *BC* aza-bicyclo[3.2.1] octane ring system to an appended ring *A* in phyllanthine led to the logical application of a hetero Diels-Alder reaction in the mind's eye (**E** to **G**) (see relevant mechanism B, Figure 10.13). Of necessity, this required the generation of an imine which was accomplished through an iodosobenzene mediated oxidation (**E** to **F**). The synthesis plan devised by Weinreb was highly dependent on the outcome of this cycloannulation. Danishefsky's diene was deemed to be a suitable reacting partner provided that the requisite regio- and stereochemical control was secured in the process. Indeed, an *exo*-selective transition state resulted in the formation of the intended adduct, placing the enone in an ideal position for further elaboration into the fully functionalized ring *A*, as exemplified by the adduct **G**. Formation of the enone relied on a selenium mediated process, avoiding an oxidative elimination (see possible mechanism C, Figure 10.13). From this central intermediate, Weinreb's synthesis plan would be assured a successful completion, especially capitalizing on an intramolecular phosphonate cyclocondensation as the last step.

10.13
Oscillarin

The first total synthesis of oscillarin, a member of the linear marine peptide family generally known as the aeruginosins [41], resulted in the correction of a misassigned structure for the 1-amidino-δ-3-pyrroline subunit (Figure 10.14) [42]. As a class of natural products, most of the 20-plus members of the aeruginosins family contain a common 1-aza[4.3.0]bicyclic core unit. In the case of oscillarin and aeruginosin 298A [43], for example, this bicyclic amino acid consists of a (2S,3aS,6R,7aS)-6-hydroxy octahydroindole-2-carboxylic acid (Choi subunit).

Visual analysis of the L-Choi subunit may reveal L-tyrosine as a hidden amino acid substructure, where the aromatic moiety would have to undergo stereoselective reduction with formation of a C-N bond to form the azabicycle. Indeed, Bonjoch [44] and Wipf [45] devised versatile routes to the L-Choi subunit of aeruginosin 298A from L-tyrosine in their total synthesis of the natural product.

The total synthesis and structural confirmation of oscillarin was reported in 2004 by Hanessian and coworkers [42] (Figure 10.14). A convenient starting material that would allow functionalization toward the intended L-Choi subunit could be L-pyroglutamic acid. However, the *cis*-ring junction and the relative stereochemical disposition of the carboxyl group necessitated starting with L-glutamic acid, rather than L-pyroglutamic acid. A similar strategy was employed in the total syntheses of two related aeruginosins, dysinosin A [46] and chlorodysinosin A [47] that also contain the novel 1-amidino-δ-3-pyrroline subunit.

Treatment of N-Boc dimethyl L-glutamate (**A**) with three equivalents of LiHMDS at −78 °C in THF, followed by addition of 3-butenyl triflate to the preformed dianion gave the *anti*-adduct **B** exclusively [48]. Cleavage of the N-Boc group, cyclization to the lactam, and reprotection gave **C**. Conversion of **C** to the acetate **D**, set the stage for an azonia-Prins [49] intramolecular cyclization *via* an iminium ion in the presence of SnBr₄ to give **E** as the major product. Displacement with Bu₄NOAc afforded **F**, in which all four stereocenters had the desired absolute stereochemistry. Cleavage of the *N*-Boc group and coupling with the *O*-MOM-protected D-phenyllactyl-D-phenylalanine **G** led to **H**. Attachment of the aminoethyl δ-3-pyrroline unit **I** necessitated the replacement of the C-6 acetate with a MOM group as in **J**. The choice of acid-labile protective groups facilitated the final one-step conversion of **J** to oscillarin.

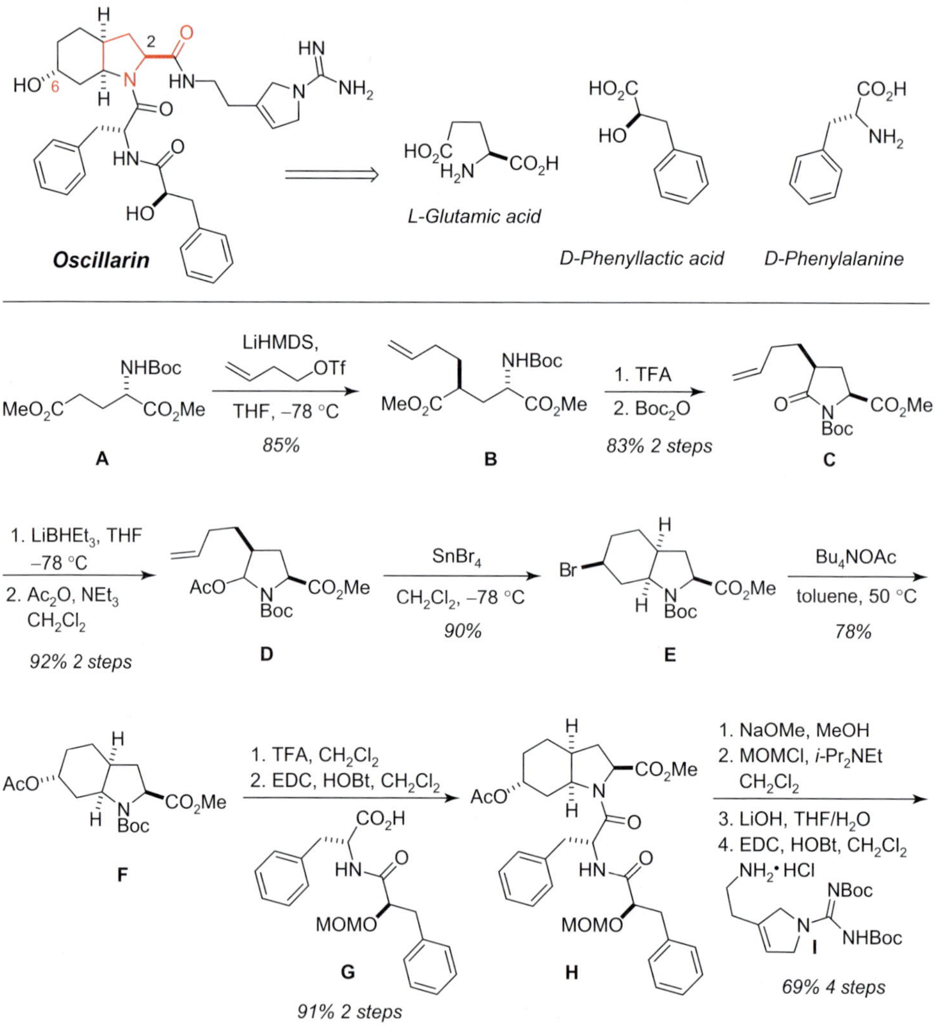

Figure 10.14 Total synthesis of oscillarin from L-glutamic acid (partially hidden substructure).

Key reactions: *Dianion alkylation (**A** to **B**); azonia-Prins intramolecular carbocyclization (**D** to **E**).*

Relevant mechanism:

antiperiplanar   *synclinal*   X = Cl, Br
attack   *attack*

Figure 10.14 *(continued)*.

Commentary: In the originally published structure of oscillarin, in the patent literature, an isomeric six-membered cyclic guanidine was proposed, which was subsequently changed to the 1-amidino-δ-3-pyrroline variant in a subsequent patent. The originally presumed oscillarin was also synthesized [46] and was found to be devoid of *in vitro* activity against thrombin. Thus, the total synthesis of oscillarin also allowed the revision of its structure in addition to establishing its absolute configuration (see Chapter 2, Figure 2.10).

The prospects of utilizing *N*-Boc methyl L-pyroglutamate as a starting chiron to construct the L-Choi subunit, based on a visual analysis, was alluded to. However, a C-4 alkylation of the corresponding enolate would have led to mixtures of 4-butenyl isomers with the *trans*-isomer predominating, in view of the strong *anti*-directing effect of the carbomethoxy group. It is well-known that a combination of $A^{1,2}$ and $A^{1,3}$-strain effects in related substrates results in a pseudoaxial orientation of the C-2 substituent, thus introducing a steric bias in favor of an *anti*-attack [47]. The "acyclic" route from *N*-Boc dimethyl L-glutamate (**A**) was therefore adopted to secure the desired C-2–C-4 *cis*-orientation in intermediate **C**.

There are numerous examples of *N*-tethered intramolecular carbocyclizations of iminium ions to the corresponding azabicycles. Surprisingly, there are much fewer examples of carbocyclizations of unactivated ω-olefinic carbon substituents attached as an integral part of the ring containing the incipient *N*-acyloxyiminium ion [47]. A plausible mechanism for the intramolecular azonia-Prins bromocyclization (**D** to **E**) could involve an antiperiplanar alignment of the pendant olefin *vis-à-vis* the iminium ion in a chair-like conformation, followed by cyclization to a cationic intermediate, which is subsequently trapped by a halide ion to give predominantly the C-6 equatorial isomer (see relevant mechanism, Figure 10.14). This stereoelectronically favored pathway would allow the maintenance of maximum orbital overlap of the alkenyl π-system with the developing lone pair on the nitrogen atom of the iminium ion intermediate. As such, a synclinal attack would be less favored. The bromide **E** has also been used in the total synthesis of chlorodysinosin A, in which the L-Choi subunit has an additional C-5 hydroxyl group [45].

10.14
ent-Cyclizidine

The first total synthesis of cyclizidine produced the enantiomer of the natural product due to a reliance on the initially reported crystal structure, which portrayed a relative configuration [50]. Hanessian and coworkers conceived of a synthetic approach starting with D-serine (Figure 10.15) [51]. Transformation of *N*-Boc-D-serine (**A**) to the corresponding Weinreb amide derivative **B** was accomplished in high yield. Formation of the methyl ketone **C**, followed by treatment with ethynyl-magnesium bromide afforded **D**, which was reduced with LiAlH$_4$ to give **E** in excellent overall yield. Asymmetric dihydroxylation and selective protection of the primary alcohol led to **F**, which was transformed to the cyclic carbamate **G** over two steps. Treatment with acid, and oxidation of the primary alcohol, followed by a Wittig reaction of the corresponding aldehyde furnished the terminal olefin **H**. Alkylation with 1-butenyl triflate gave **I**, which after cleavage of the TBDPS group and ring-closing metathesis led to **J**. Cleavage of the cyclic carbamate under basic conditions, followed by protection of the free amine as the Fmoc derivative afforded the diol **K** in excellent yield. Epoxidation with oxone in trifluoroacetone gave **L**, which was oxidized to the aldehyde, and the latter treated with propynylmagnesium bromide to give a mixture of diastereomeric alcohols **M** (dr = 3:2). Separation of the major isomer by chromatography and recycling the undesired isomer through an oxidation-reduction sequence augmented the diastereoselectivity to 8:1 in favor of the desired isomer **M**. Mesylation of the (*R*)-configured propargylic alcohol furnished **N**. Treatment with piperidine cleaved the Fmoc group and provided the indolizidine **O**. Formation of the vinyl stannane followed by treatment with iodine led to the vinyl iodide **P** in quantitative yield. A Suzuki-Miyaura coupling with the pinacol boronate ester **Q** led to the adduct **R**, which was deprotected in the presence of LiDBB to afford *ent*-cyclizidine. This product was identical in all respects to the natural product except for the opposite sign of its optical rotation. An X-ray crystal structure confirmed the assigned configuration.

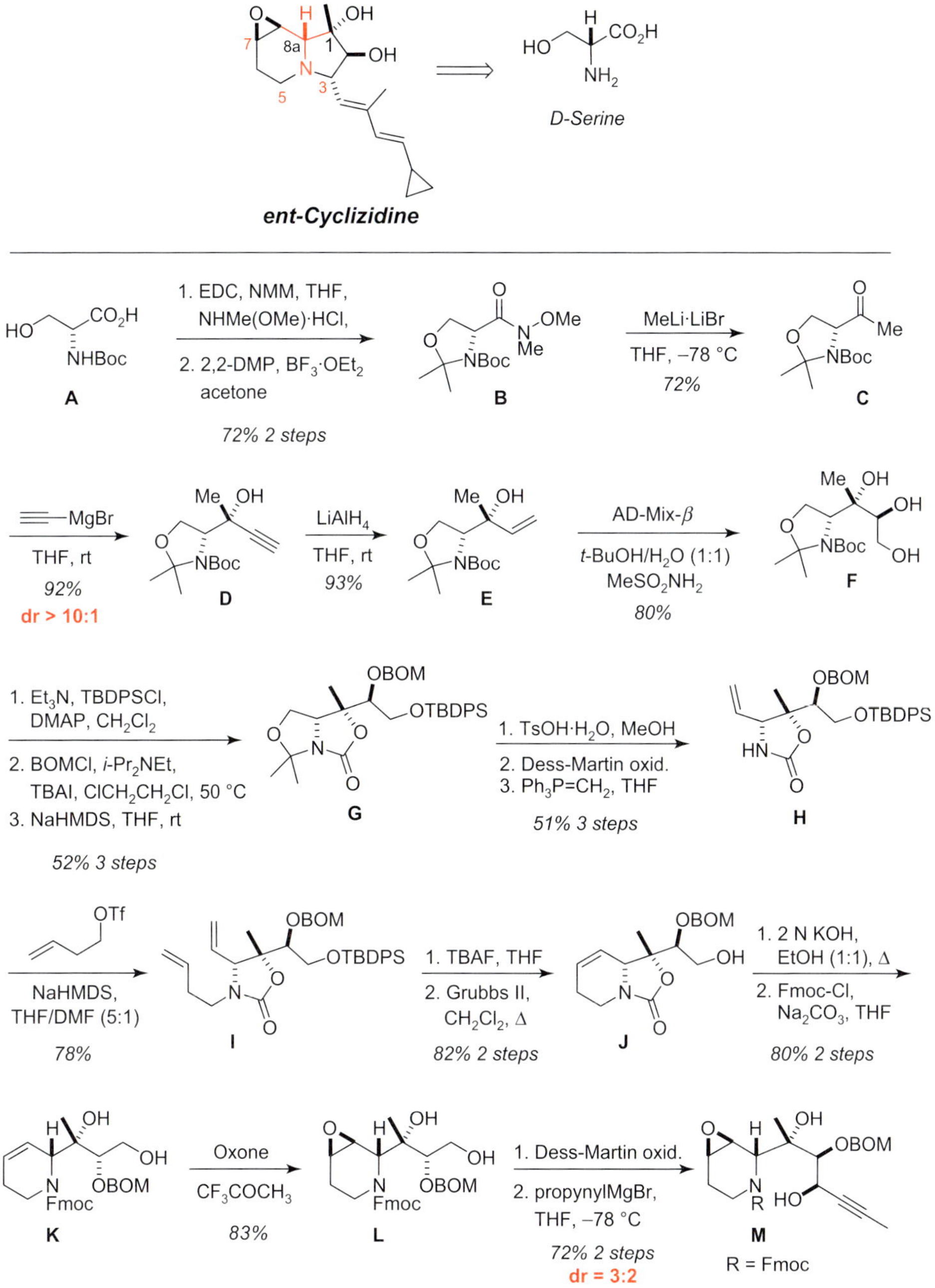

Figure 10.15 Synthesis of *ent*-cyclizidine from D-serine (partially hidden chiron).

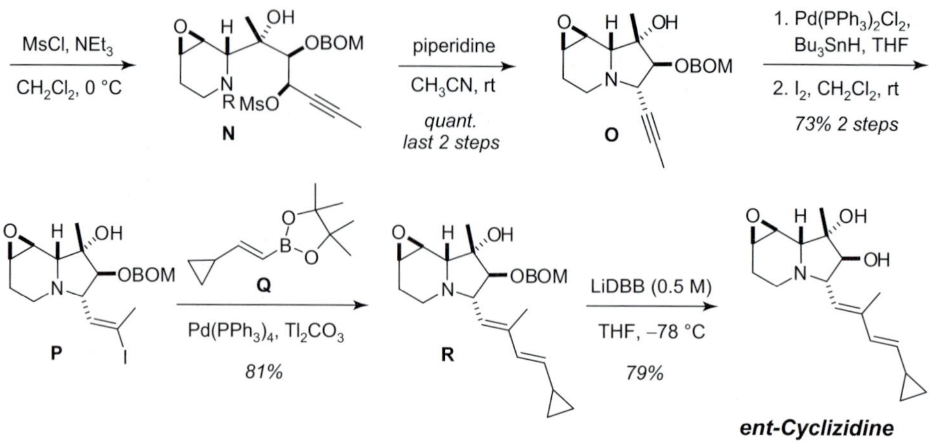

ent-Cyclizidine

Key Reactions: *Stereocontrolled acetylide addition (**C** to **D**); ring-closing metathesis (**H** to **I**); stereocontrolled epoxidation (**K** to **L**); intramolecular mesylate displacement (**N** to **O**); Suzuki-Miyaura coupling (**P** + **Q** to **R**).*

Figure 10.15 *(continued)*.

Commentary: Analysis of the structue of *ent*-cyclizidine reveals a number of challenges associated with its assembly. These are heightened by the presence of the C-7–C-8 epoxide as part of six contiguous stereogenic centers encompassing C-3–C-7 (cyclizidine numbering, Figure 10.15). Thus, elaboration of the highly functionalized carbon framework in a stereocontrolled manner would also require a judicious choice of reaction conditions, and a specific sequence of execution. Starting with D-serine provided a three-carbon building block that would accommodate the stereochemistry at C-8a and secure the position of the central nitrogen atom. A series of stereocontrolled branching and functionalization reactions led to an advanced acyclic precursor containing three contiguous stereogenic centers that were orthogonally protected as in **H**.

Relying on a ring-closing metathesis reaction ensured access to the unsaturated piperidine ring as the bicyclic carbamate **J**. Based on a careful choice of epoxidation conditions, it was possible to secure the correct stereochemistry for the epoxide **L**. Further branching led to a diastereomeric mixture of propargylic alcohols, from which an intramolecular S$_N$2 reaction was possible to give the indolizidine **O**. True to the original synthetic plan, the corresponding vinyl iodide **P** was engaged in a Suzuki-Miyaura coupling reaction with **Q** to give the diene **R**. Cleavage of the *O*-BOM group with LiDBB led to crystalline *ent*-cyclizidine.

10.15
Pactamycin

The first total synthesis of the highly substituted aminocyclopentitol, pactamycin, was reported by Hanessian and coworkers in 2011 [52] (Figure 10.16). In spite of its dense architecture and unique interaction with 30S ribosomal subunit, efforts toward a total synthesis have been sparse [53]. The synthesis commenced with the oxazoline derivative **A**, readily available from L-threonine. Treatment of the lithium enolate of **A** with *O*-TBDPS-2-(hydroxymethyl)acrolein, followed by protection of the resulting hydroxyl group led to **B** as a single isomer. A three-step conversion to the methyl ketone **C** was followed by ozonolytic cleavage of the exocyclic double bond, and a subsequent intramolecular Mukaiyama-type aldol condensation. Treatment of the resulting tertiary alcohol with trichloroacetyl chloride in pyridine resulted in sequential esterification and β-elimination to give the cyclopentenone intermediate **D** in excellent overall yield. Epoxidation under basic peroxide conditions led to **E**. Stereoselective reduction to **F** under Luche conditions, was followed by introduction of the azide group via S_N2 displacement of the triflate to give **G**. Cleavage of the TES ether, and oxidation to the ketone **H**, was followed by treatment with MeMgBr to give **I** as a single diastereomer. At this stage it was necessary to "invert" the epoxide function in **I**, which was done by treatment with $Zn(OTf)_2$ in AcOH to give **J**.

Manipulation of the diol unit via the triflate led to the desired epoxide **K** in excellent yield. The epoxide was then regio- and stereoselectively opened with *m*-(2-propenyl)aniline in the presence of $Yb(OTf)_3$ to give **L** in 81% yield. Conversion of the oxazoline to the amino ester **M** was achieved with dilute HCl, after which, the acetonide **N** was prepared in a two-step process in high overall yield. Treatment with diphosgene in the presence of activated charcoal, followed by exposure to dimethylamine (neat) led to the intended urea derivative **O** in 86% yield. Oxidative cleavage of the olefin and hydrolysis of the acetonide afforded the ketone **P**, which was esterified with (cyanomethyl)-2-hydroxy-6-methyl benzoate (**Q**) to give **R**. Finally, reduction of the azide group in the presence of Lindlar's catalyst gave pactamycin after chromatographic purification. Starting with intermediate **M**, it was also possible to synthesize the naturally occurring congener pactamycate and to confirm its structure and absolute configuration by X-ray analysis. The total synthesis of pactamycin was achieved in 3.0% overall yield over 29 linear steps starting from **A**.

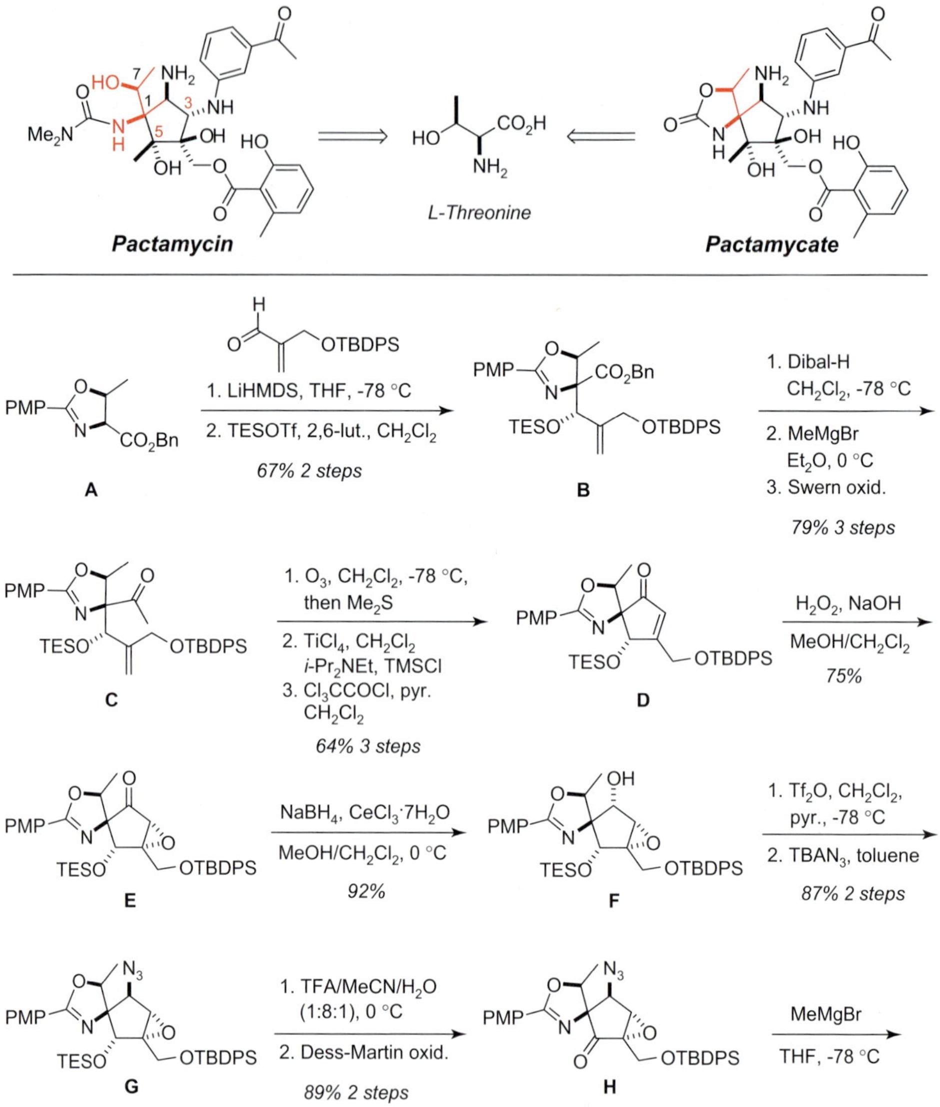

Figure 10.16 Total synthesis of pactamycin from L-threonine (partially hidden chiron).

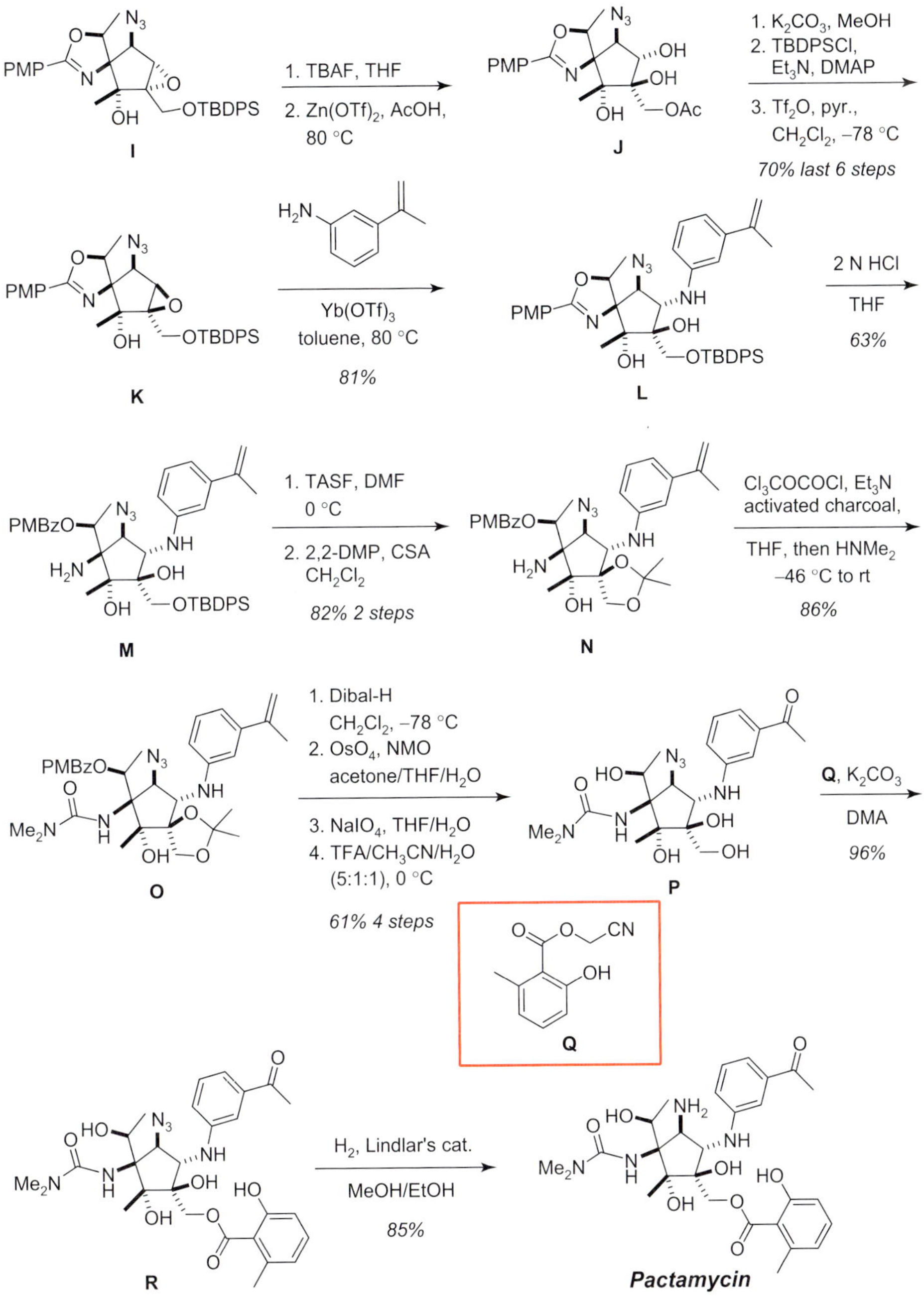

Figure 10.16 *(continued)*.

Key reactions: *Stereoselective aldol (**A** to **B**); epoxide inversion (**I** to **J** to **K**); Yb(TfO)₃-mediated epoxide opening; esterification (**P** to **R**)*

Relevant mechanisms and transformations:

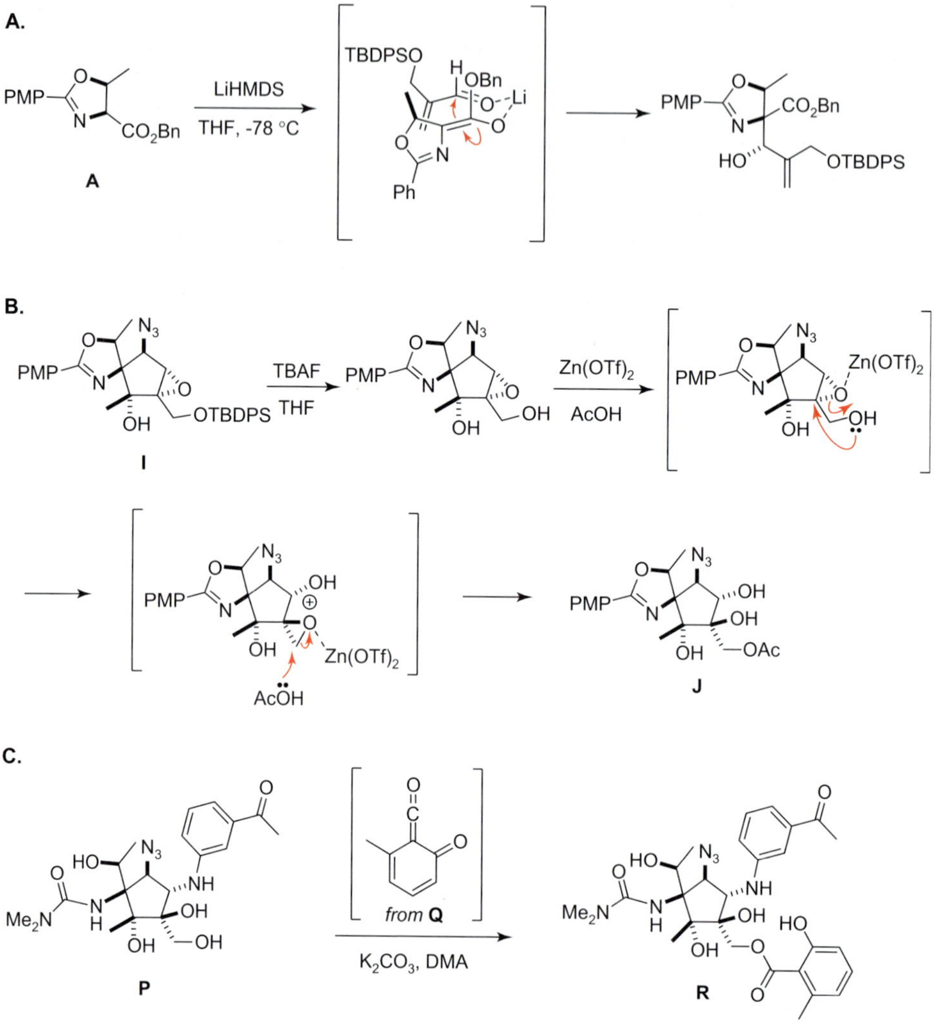

A.

B.

C.

Figure 10.16 *(continued).*

Commentary: Having recognized the partially hidden chiron, ʟ-threonine, as an integral part of pactamycin, the major challenge was to systematically elaborate the construction of the densely functionalized and uniquely substituted cyclopentane core. In considering various approaches that would benefit from the inherent chirality provided by ʟ-threonine, at least conserving the position and stereochemistry of the C-7 hydroxyl group, it was imperative to choose bond-forming sequences that would exploit the natural disposition of the amino alcohol function. Starting with the *p*-methoxyphenyloxazoline **A**, the authors were cognizant that the densely functionalized cyclopentane core harboring three contiguous tertiary centers would require a judicious choice of well-orchestrated bond-forming sequences. Passage from the oxazoline **A** to **B** was highly stereoselective (see relevant transformation A, Figure 10.16). Subsequent intamolecular aldol reaction led eventually to cyclopentenone **D**, while securing absolute stereochemistry from X-ray structure analysis of synthetic intermediates. A major question was concerned with the regio-and stereoselective introduction of branching corresponding to C-4 and C-5, which was achieved uneventfully to reach intermediate **I**. Introduction of the aniline moiety was envisaged to take place through opening of the C-3–C-4 epoxide with inversion at C-3. To achieve this, it was necessary to "invert" the alpha oriented epoxide in **I** to its diastereomer **K**. In the event, treatment of **I** (R = OH) with Zn(OTf)$_2$ in the presence of acetic acid led to the acetate ester **J**, presumably via a spiroepoxide intermediate (see relevant mechanism B, Figure 10.16). Conversion to the now "inverted" epoxide **K** (X-ray analysis), and an Yb(OTf)$_3$-mediated ring opening with the intended aniline, in which the double bond was a masked ketone, afforded **L**. Remarkably, only the desired regioselective epoxide opening with inversion took place. Systematic manipulations of functional groups eventually led to the *N,N*-dimethylurea **P**, which was esterified to give **R** (see relevant mechanism C, Figure 10.16), and converted to pactamycin.

The seemingly straightforward total synthesis of pactamycin (and pactamycate) by Hanessian and coworkers, underscores the importance (and frustrations) of many unwanted transformations due to the proximity of functional groups anchored on the cyclopentane core. These difficulties were eventually overcome by modifying the sequence of reactions and minimizing unexpected intramolecular reactions.

10.16
Miscellanea

A very large number of natural products and compounds containing one or more nitrogen atoms have been synthesized over the years. Selected examples where *α*-amino acids were used as hidden or partially hidden chirons are shown in Figure 10.17 [54] (see also Chapters 8 and 16).

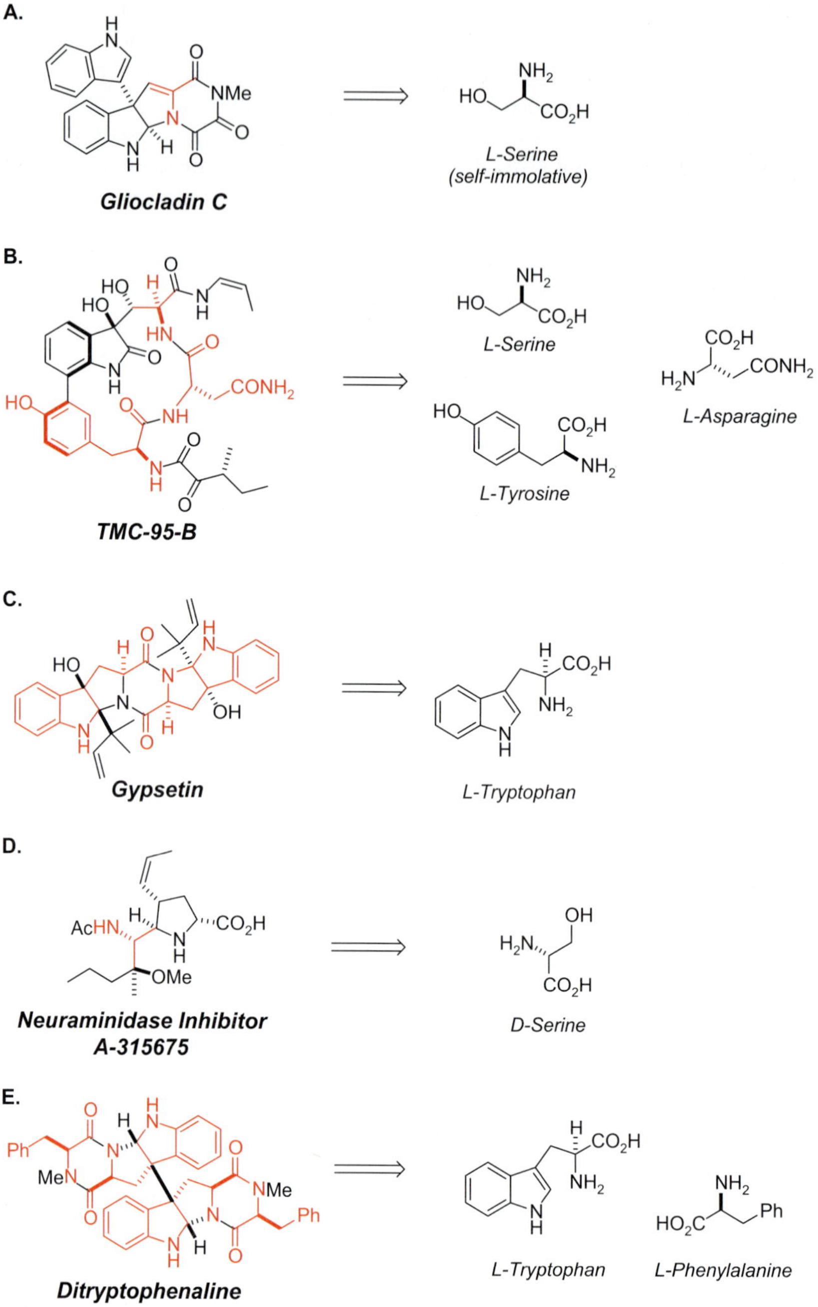

A.

Gliocladin C

L-Serine
(self-immolative)

B.

TMC-95-B

L-Serine

L-Tyrosine

L-Asparagine

C.

Gypsetin

L-Tryptophan

D.

**Neuraminidase Inhibitor
A-315675**

D-Serine

E.

Ditryptophenaline

L-Tryptophan

L-Phenylalanine

Figure 10.17 Miscellaneous examples of total syntheses using α-amino acids as partially hidden and hidden substructures.

References

1. (a) Yoshioka, M.; Nakai, H.; Ohno, M. *J. Am. Chem. Soc.* 1984, **106**, 1133; (b) Yoshioka, M.; Nakai, H.; Ohno, M. *Heterocycles* 1984, **31**, 151.
2. For reviews, see: Weinreb, S. M. in *Studies in Natural Products Chemistry*, Atta-ur-Rahman, Ed.; 2005; vol. 16, p. 3, Elsevier Science Publ. BV, Amsterdam; Fraser-Reid, B.; López, J. P. in *Recent Advances in the Chemical Synthesis of Antibiotics*, Lukacs, G.; Ohno, M.; Eds.; 1990, p. 285, Springer-Verlag, Heidelberg; for other syntheses, see: (a) Garigipati, R. S.; Tschaen, D. M.; Weinreb, S. M. *J. Am. Chem. Soc.* 1985, **107**, 7790; Garigipati, R. S.; Tschaen, D. M.; Weinreb, S. M. *J. Am. Chem. Soc.* 1990, **112**, 3475; (b) Kozikowski, A. P.; Konoike, T.; Nieduzak, T. R. *J. Chem. Soc., Chem. Commun.* 1986, 1350; (c) Kozikowski, A. P.; Nieduzak, T. R.; Konoike, T.; Springer, J. P. *J. Am. Chem. Soc.* 1987, **109**, 5167; (d) Askin, D.; Angst, C.; Danishefsky, S. J. *J. Org. Chem.* 1985, **50**, 5005; (e) Askin, D.; Angst, C.; Danishefsky, S. J. *J. Org. Chem.* 1987, **52**, 622. For syntheses from ᴅ-glucose, see Chapter 11, section 11.8
3. Trost, B. M.; Ohmari, M.; Boyd, S. A.; Okawara, H.; Brickner, S. J. *J. Am. Chem. Soc.* 1989, **111**, 8281.
4. Raucher, S.; Chi, K.-W.; Jones, D. S. *Tetrahedron Lett.* 1985, **26**, 6261.
5. Isono, N.; Mori, M. *J. Org. Chem.* 1995, **60**, 115; for a recent synthesis of (±)-cephalotaxine, see: Li, W.-D. Z.; Duo, W.-G.; Zhuang, C.-H. *Org. Lett.* 2011, **13**, 3538 and references cited therein; see also: Chapter 16, section 16.4.3.
6. Beck, A. K.; Blank, S.; Job, K.; Seebach, D.; Sommerfeld, T. *Org. Synth.* 1993, **72**, 62.
7. (a) Chan, T. H.; Koumaglo, K. *J. Organomet. Chem.* 1985, **285**, 109; (b) Miller, R. B.; McGarvey, G. *Synth. Commun.* 1978, **8**, 291.
8. Oppolzer, W.; Thirring, K. *J. Am. Chem. Soc.* 1982, **104**, 4978. For reviews on kainic acid, see: (a) Parsons, A. F. *Tetrahedron* 1996, **52**, 4149; (b) Moloney, M. G. *Nat. Prod. Rep.* 1998, **15**, 205; (c) Chamberlin, A. R.; Bridges, R. in *Drug Design for Neuroscience*, Kozikowski, A. P.; Ed.; 1993, p. 231, Raven Press, NY; for a summary of total syntheses, see: Sakaguchi, H.; Tokuyama, H.; Fukuyama, T. *Org. Lett.* 2007, **9**, 1635; for a recent total synthesis of kainic acid, see: Wei, G.; Chalker, J. M.; Cohen, T. *J. Org. Chem.* 2011, **76**, 7912 and references cited therein.
9. Cooper, J.; Knight, D. W.; Gallagher, P. T. *J. Chem. Soc., Chem. Commun.* 1987, 1220.
10. Mukaiyama, T.; Usui, M.; Saigo, K. *Chem. Lett.* 1976, **5**, 49.
11. Arndt, F.; Eistert, B. *Ber.* 1935, **68**, 200.
12. Martin, S. F.; Barr, K. J. *J. Am. Chem. Soc.* 1996, **118**, 3299; see also: Williams, D. R.; Brown, D. L.; Benbow, J. W. *J. Am. Chem. Soc.* 1989, **111**, 1923.
13. (a) Weinstein, B.; Craig, A. R. *J. Org. Chem.* 1976, **41**, 875; (b) Wasserman, H. H.; Tremper, A. W. *Tetrahedron Lett.* 1977, **18**, 1449; (c) Dean, R. T.; Padgett, H. C.; Rapoport, H. *J. Am. Chem. Soc.* 1976, **98**, 7448.
14. For a review, see: Casiraghi, G.; Zanardi, F.; Appendino, G.; Rassu, G. *Chem. Rev.* 2000, **100**, 1929.
15. Hanessian, S.; Tremblay, M.; Marzi, M.; Del Valle, J. R. *J. Org. Chem.* 2005, **70**, 5070.
16. For reviews, see: Lee, H. L.; Baggiolini, E. G.; Uskoković, M. R. *Tetrahedron* 1987, **43**, 4887; De Clercq, P. J. *Chem. Rev.* 1997, **97**, 1755.
17. Baggiolini, E. G.; Lee, H. L.; Pizzolato, G.; Uskoković, M. R. *J. Am. Chem. Soc.* 1982, **104**, 6460; for a recent synthesis of biotin, see: Dai, H.-F.; Chen, W.-X.; Zhao, L.; Xiong, F.; Sheng, H. *Adv. Synth. Catal.* 2008, **350**, 1635 and references cited therein.
18. Reddy, L. R.; Saravanan, P.; Corey, E. J. *J. Am. Chem. Soc.* 2004, **126**, 6230; see also: (a) Reddy, L. R.; Fournier, J.-F.; Reddy, B. V. S.; Corey, E. J. *Org. Lett.* 2005, **7**, 2699; (b) Endo, A.; Danishefsky, S. J. *J. Am. Chem. Soc.* 2005, **127**, 8298; (c) Mulholland, N. P.; Pattenden, G.; Walters, I. A. S. *Org. Biomol. Chem.* 2006, **4**, 2845; (d)

Ling, T.; Macherla, V. R.; Manam, R. R.; McArthur, K. A.; Potts, B. C. M. *Org. Lett.* 2007, **9**, 2289; (e) Takahashi, K.; Midori, M.; Kawano, K.; Ishihara, J.; Hatakeyama, S. *Angew. Chem. Int. Ed.* 2008, **47**, 6244; (f) Fukada, T.; Sugiyama, K.; Arima, S.; Harigaya, Y.; Nagamistsu, T.; Ōmura, S. *Org. Lett.* 2008, **10**, 4239. (g) Nguyen, H.; Ma, G.; Romo, D. *Chem. Commun.* 2010, **46**, 4803; (h) Satoh, N.; Yokoshima, S.; Fukuyama, T. *Org. Lett.* 2011, **13**, 3028 and references cited therein.

19. For a review, see: Jones, G. R.; Landais, Y. *Tetrahedron* 1996, **52**, 7599.

20. *The Organic Chemistry of β-Lactam Antibiotics*, Georg, G.; Ed.; 1993, VCH-Wiley, Weinheim.

21. For reviews on carbapenems, see: (a) Setti, E. L.; Micetich, R. G. *Curr. Med. Chem.* 1998, **5**, 101; (b) Mascaretti, O. A.; Boschetti, C. E.; Damelou, G. O.; Mata, E. G.; Roveri, O. *Curr. Med. Chem.* 1995, **1**, 441; (c) Kametani, T. *Heterocycles* 1982, **17**, 463.

22. Saltzmann, T. N.; Ratcliffe, R. W.; Christensen, B. G.; Bouffard, F. A. *J. Am. Chem. Soc.* 1980, **102**, 6161.

23. For selected contributions from the Merck group, see: (a) Reider, P. J.; Grabowski, E. J. J. *Tetrahedron Lett.* 1982, **23**, 2293; (b) Melillo, D. G.; Cvetovich, R. J.; Ryan, K. M.; Sletzinger, M. *J. Org. Chem.* 1986, **51**, 1498.

24. For examples, see: (a) Hanessian, S.; Bedeschi, A.; Battistini, C.; Mongelli, N. *J. Am. Chem. Soc.* 1985, **107**, 1438; (b) Shiozaki, M.; Ishida, N.; Hiraoka, T.; Yanagisawa, H. *Tetrahedron Lett.* 1981, **22**, 5205.

25. Ratcliffe, R. W.; Saltzmann, T. N.; Christensen, B. G. *Tetrahedron Lett.* 1980, **21**, 31.

26. Snider, B. B.; Lin, H. *J. Am. Chem. Soc.* 1999, **121**, 7778.

27. Scheffler, G.; Seike, H.; Sorensen, E. J. *Angew. Chem. Int. Ed.* 2000, **39**, 4593.

28. Gusmer, M.; Braun, N. A.; Bavoux, C.; Perrin, M.; Ciufolini, M. A. *J. Am. Chem. Soc.* 2001, **123**, 7534.

29. (a) Carson, C. A.; Kerr, M. A. *Org. Lett.* 2009, **11**, 777; (b) Kan, T.; Fujimoto, T.; Ieda, S.; Asoh, Y.; Kitaoka, H.; Fukuyama, T. *Org. Lett.* 2009, **6**, 2729;

(c) Maeng, J.-H.; Funk, R. L. *Org. Lett.* 2001, **3**, 1125; (d) Brummond, K. M.; Hong, S.-P. *J. Org. Chem.* 2005, **70**, 907, and references cited therein; (e) for a recent review on dearomatization strategies in total synthesis, see: Roche, S. P.; Porco, J. A. Jr. *Angew. Chem. Int. Ed.* 2011, **50**, 4068; for oxidative amidation of phenols, see: Ciufolini, M. A.; Braun, N. A.; Canesi, S.; Ousmer, M.; Chang, J.; Chai, D. *Synthesis* 2007, 3759.

30. (a) *Half a Century of Free Radical Chemistry*, Barton, D. H. R., Parekh, S. I.; Ed.; 1993, p. 7, Cambridge University Press, UK; (b) Barton, D. H. R.; Deflorin, A. M.; Edwards, O. E. *J. Chem. Soc.* 1956, 530; (c) Kametani, T.; Fukumoto, K. *Synthesis* 1972, 657.

31. (a) Wipf, P.; Rector, S. R.; Takahashi, M. *J. Am. Chem. Soc.* 2002, **124**, 14848; (b) Wipf, D.; Spencer, S. R. *J. Am. Chem. Soc.* 2005, **127**, 225.

32. (a) Wipf, P.; Kim, Y. *Tetrahedron Lett.* 1992, **33**, 5477; (b) Wipf, P.; Kim, Y.; Goldstein, D. M. *J. Am. Chem. Soc.* 1995, **117**, 11106.

33. For a review, see: Tsuji, J.; Mandai, T. *Synthesis* 1996, 1.

34. (a) Wick, A. E.; Felix, D.; Steen, K.; Eschenmoser, A. *Helv. Chim. Acta* 1964, **47**, 2425; (b) Wick, A. E.; Felix, D.; Gschwend-Steen, K.; Eschenmoser, A. *Helv. Chim. Acta* 1969, **52**, 1030.

35. Hu, Y.-J.; Dominique, R.; Das, S. K.; Roy, R. *Can. J. Chem.* 2000, **78**, 838.

36. (a) Kingsbury, J. S.; Harrity, J. P. A.; Bonitatebus, P. J.; Hoveyda, A. H. *J. Am. Chem. Soc.* 1999, **121**, 791; (b) Garber, S. B.; Kingsbury, J. S.; Gray, B. L.; Hoveyda, A. H. *J. Am. Chem. Soc.* 2000, **122**, 8168.

37. Han, G.; LaPorte, M. G.; Folmer, J. J.; Werner, K. M.; Weinreb, S. M. *Angew. Chem. Int. Ed.* 2000, **39**, 237.

38. (a) Mueller, P.; Gilabert, D. M. *Tetrahedron* 1988, **44**, 7171; (b) Larsen, J.; Jørgensen, K. A. *J. Chem. Soc., Perkin Trans. 2* 1992, 1213.

39. (a) Miyoshi, N.; Yamamoto, T.; Kambe, N.; Murai, N.; Sonoda, N. *Tetrahedron Lett.* 1982, **23**, 4813; see also: (b) Wantanabe, M.; Awen, B. Z.; Kato, M. *J. Org. Chem.* 1993, **58**, 3923.

40. For example see: Mandal, A. K.; Mahajan, S. W. *Tetrahedron* 1988, **44**, 2293.

41. For a recent review, see: Ersmark, K.; Del Valle, J. R.; Hanessian, S. *Angew. Chem. Int. Ed.* 2008, **47**, 1202.

42. Hanessian, S.; Tremblay, M.; Petersen, J. F. W. *J. Am. Chem. Soc.* 2004, **126**, 6064; see also: Hanessian, S.; Wang, X.; Ersmark, K.; Del Valle, J. R.; Klegraf, E. *Org. Lett.* 2009, **11**, 4232.

43. Steiner, J. L.; Murakami, M.; Tulinsky, A. *J. Am. Chem. Soc.* 1998, **120**, 597.

44. Valls, N.; López-Canet, M.; Vallribera, M.; Bonjoch, J. *J. Am. Chem. Soc.* 2000, **122**, 11248.

45. Wipf, P.; Methot, J.-L. *Org. Lett.* 2000, **2**, 4213.

46. Hanessian, S.; Margarita, R.; Hall, A.; Johnstone, S.; Tremblay, M.; Parlanti, L. *J. Am. Chem. Soc.* 2002, **124**, 13342.

47. Hanessian, S.; Del Valle, J. R.; Xue, Y.; Blomberg, N. *J. Am. Chem. Soc.* 2006, **128**, 10491.

48. Hanessian, S.; Margarita, R. *Tetrahedron Lett.* 1998, **39**, 5887.

49. For related cyclizations, see: Hanessian, S.; Tremblay, M.; Marzi, M.; Del Valle, J. R. *J. Org. Chem.* 2005, **70**, 5070.

50. Freer, A. A.; Gardner, D.; Greatbanks, D.; Poyser, P.; Sim, G. A. *J. Chem. Soc., Chem. Commun.* 1982, 1160.

51. Hanessian, S.; Soma, U.; Dorich, S.; Deschênes-Simard, B. *Org. Lett.* 2011, **13**, 1048.

52. Hanessian, S. Vakiti, R. R.; Dorich, S.; Banerjee, S.; Lecomte, F.; Del Valle, J. R.; Zhang, J.; Deschênes-Simard, B. *Angew. Chem. Int. Ed.* 2011, **50**, 3497.

53. (a) Knapp, S.; Yu, Y. *Org. Lett.* 2007, **9**, 1359; (b) Tsujimoto, T.; Nishikawa, T.; Urabe, D.; Isobe, M. *Synlett* 2005, 433.

54. Gliocladin C: Overman, L. A.; Shin, Y. *Org. Lett.* 2007, **9**, 339; TMC-95-B: Lin, S.; Danishefsky, S. J. *Angew. Chem. Int. Ed.* 2002, **41**, 512; gypsetin: Schkeryantz, J. M.; Woo, J. C. G.; Danishefsky, S. J. *J. Am. Chem. Soc.* 1995, **117**, 7025; A-315675: Hanessian, S.; Bayrakdarian, M.; Luo, X. *J. Am. Chem. Soc.* 2002, **124**, 4716; ditryptophenaline: Movassaghi, M.; Schmidt, M. A.; Ashenhurst, J. A. *Angew. Chem. Int. Ed.* 2008, **47**, 1485.

11
Total Synthesis from Carbohydrate Precursors

The stereochemically rich carbohydrate molecules are ideal starting materials for the synthesis of Category I-type natural products, especially if their carbon skeleton and existing hydroxyl groups are efficiently utilized [1]. The guidelines for such criteria were discussed in Chapter 8 in the context of locating apparent, partially hidden, and hidden carbohydrate-derived chirons in the structural framework of target compounds. In this chapter we show selected examples of the utilization of carbohydrates as precursors that provide good carbon skeletal and stereochemical convergence with Category I and Category II-type target molecules.

11.1
Ajmalicine

The heteroyohimbine alkaloid ajmalicine has been the subject of extensive synthetic studies over the years [2]. The total synthesis of ajmalicine from D-glucose was first reported in 1991 by Hanessian and Faucher [3] (Figure 11.1). The obvious inclusion of tryptamine in any synthesis plan towards ajmalicine presents the challenge of ensuring the correct stereochemistry at the ring junctions. A convergent approach starting with D-glucose relied on a systematic functionalization of the pyranose framework. Thus, tetra-O-acetyl-D-glucal (**A**), which is readily available from D-glucose in three steps, was converted to the *t*-butyl glycoside in the presence of BF$_3$·OEt$_2$ according to Ferrier [4]. The resulting ketone was treated with K$_2$CO$_3$ to effect β-elimination and the product thus formed was acetylated to give enone **B**. Conjugate addition of a mixed vinyl Grignard/cuprate reagent, and enolate trapping with methyl bromoacetate gave the *syn*-oriented adduct, which was converted to the *N*-tosylhydrazone derivative **C**. Treatment with sodium cyanoborohydride followed by NaOAc and NaOMe [5] completed the one-pot reductive deoxygenation of the keto group to give **D**. Chlorination of the primary alcohol, followed by radical-mediated conversion to the 6-deoxy derivative, and ozonolysis afforded the aldehyde **E**. Treatment with DBU resulted in a smooth epimerization to give the inverted aldehyde derivative **F**. With the required functionality in place, **F** was converted to the tryptamine analogue **G**, which underwent *in situ* lactam formation to give **H**. Hydrolysis of the *t*-butyl glycoside

Design and Strategy in Organic Synthesis: From the Chiron Approach to Catalysis, First Edition.
Stephen Hanessian, Simon Giroux, and Bradley L. Merner.
© 2013 Wiley-VCH Verlag GmbH & Co. KGaA. Published 2013 by Wiley-VCH Verlag GmbH & Co. KGaA.

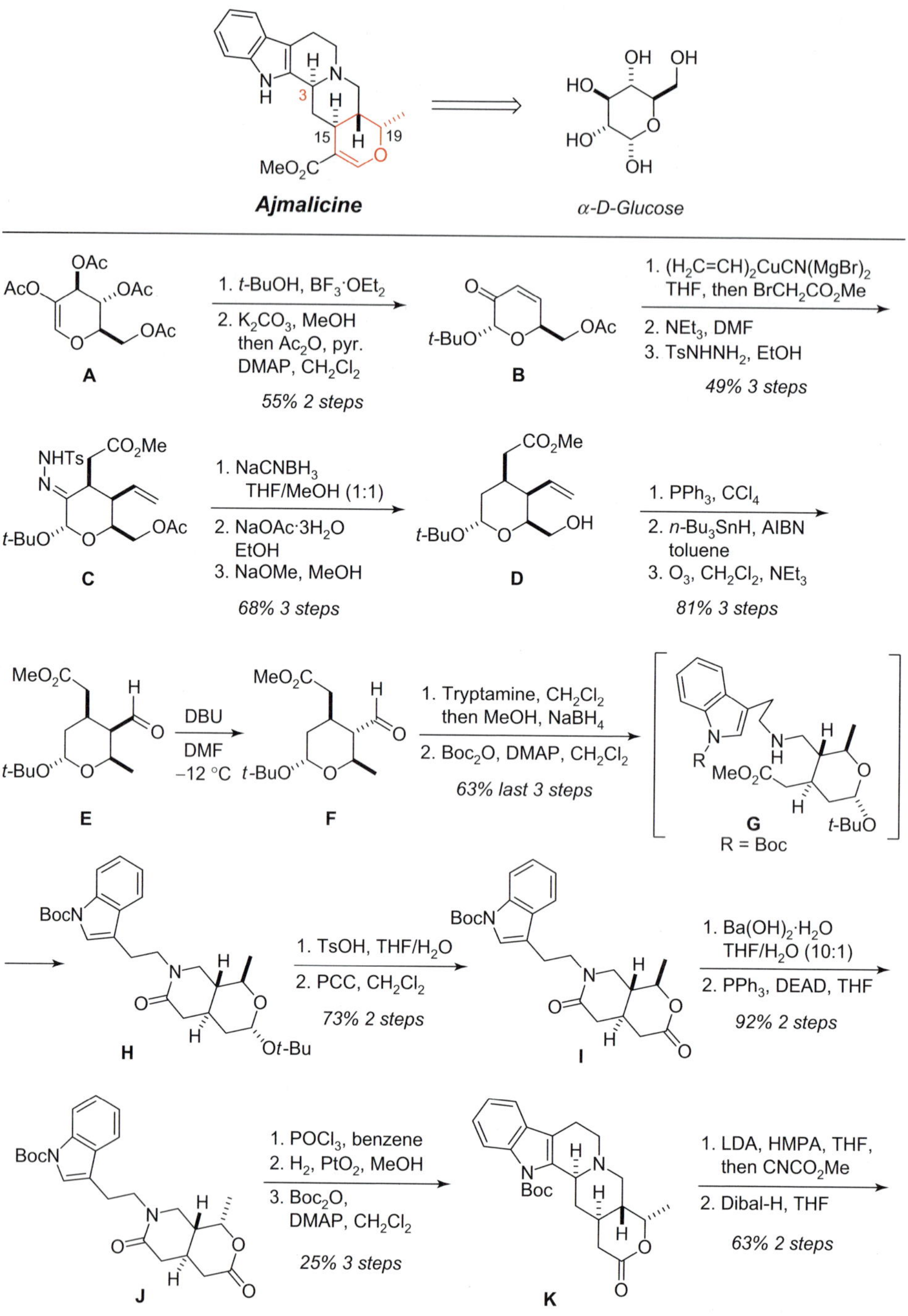

Ajmalicine

α-D-Glucose

Figure 11.1 Total synthesis of ajmalicine from D-glucose (partially hidden substructure).

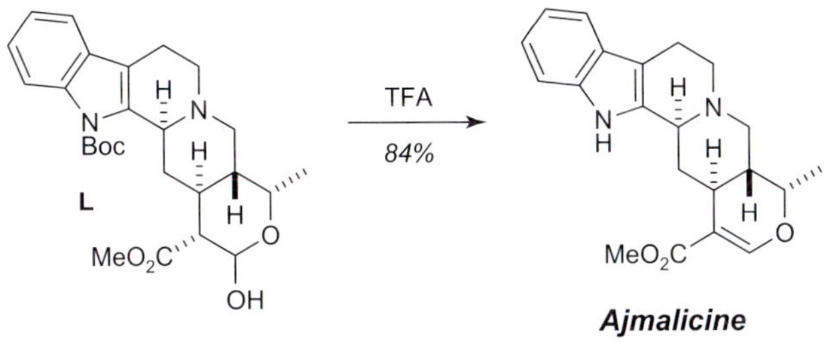

L

Ajmalicine

Key reactions: *Ferrier glycosidation (**A** to **B**); conjugate addition and enolate alkylation (**B** to **C**); DBU epimerization (**E** to **F**); intramolecular Mitsunobu lactonization/inversion (**I** to **J**); Bischler-Napieralski reaction (**J** to **K**).*

Relevant mechanisms and transformations:

A.

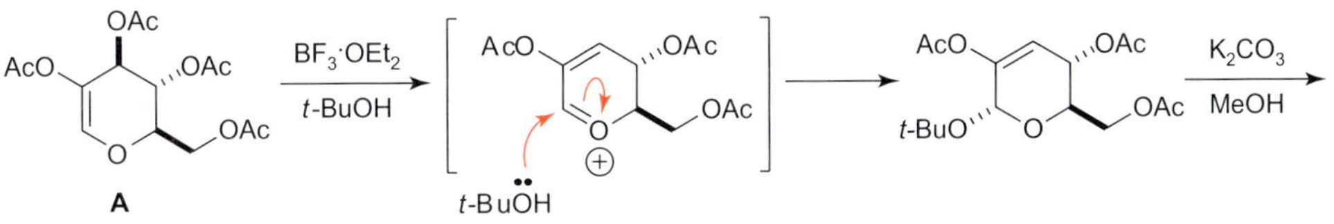

A

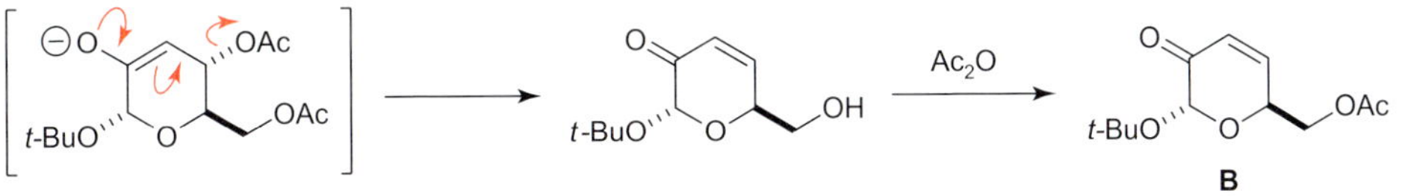

B

B.

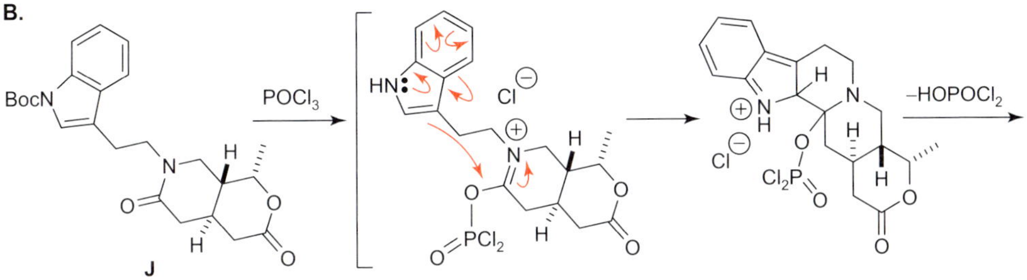

J

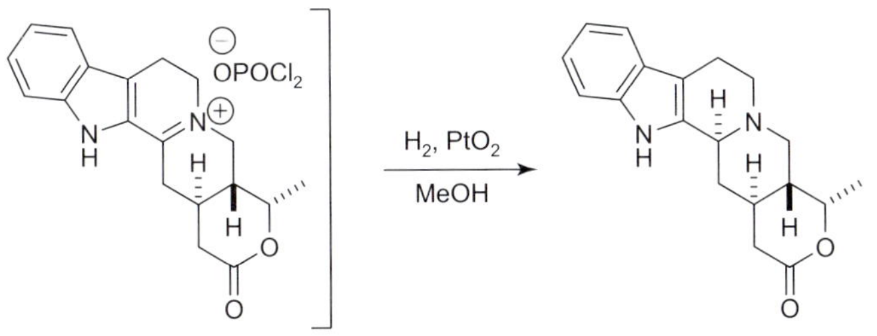

Figure 11.1 *(continued).*

was followed by oxidation to lactone **I**. Inversion of configuration was achieved by opening the lactone ring with Ba(OH)$_2$ and treatment of the resultant δ-hydroxy acid with Ph$_3$P and DEAD under Mitsunobu conditions to give the inverted lactone intermediate **J**. Cyclization under Bischler-Napieralski conditions [6], followed by catalytic hydrogenation of the iminium intermediate gave pentacyclic lactone **K**. Installation of the methoxycarbonyl group was accomplished by first generating the lithium enolate of **K**, followed by treatment with Mander's reagent [7]. Reduction to the hemiacetal **L** and β-elimination under acidic conditions gave ajmalicine. The same sequence was applied to intermediate **I**, which led to 19-*epi*-ajmalicine (mayumbine) [3].

Commentary: The visual relationship of the dihydropyran ring in ajmalicine to a carbohydrate precursor is not far fetched. Application of the "rule of 5" (Chapter 8, section 8.7.2), reveals the need to use an L-sugar to converge upon C-19 of ajmalicine. Embarking on an approach that commences from D-glucose was done with the full knowledge that extensive deoxygenation and branching had to take place. However, it was also expected that the steric and conformational bias embodied in the structure of enone **B** would secure the desired initial stereochemistry involving conjugate addition and enolate alkylation to give the α-keto precursor to intermediate **C**. Much of the internal control was provided initially with the choice of the *t*-butyl glycoside **B**, which is readily available through a Ferrier-type glycosidation of the glucal **A**. A stereoelectronically controlled attack of *t*-BuOH on the incipient oxocarbenium ion secured an axial orientation for the bulky aglycone (see relevant mechanisms A, Figure 11.1). Conjugate addition of the mixed vinyl Grignard cuprate leagent resulted in an axial attack from a trajectory that avoids interaction with the anomeric *t*-butyl group. The resulting enolate also favored an *anti*-attack of methyl bromoacetate leading to the *cis*-3,4-disubstituted ketone. Having served its purpose as a semi-rigid template for regio- and stereocontrolled C-functionalization, the 2-keto function in the product was then deoxygenated via the intermediacy of the corresponding tosylhydrazone derivative **C**. In order to construct the pentacyclic ring system, the stereochemistry at C-4 (carbohydrate numbering) of the intermediate aldehyde **E** had to be corrected. This was accomplished by treatment with DBU in DMF to effectively invert the aldehyde group and secure the desired 3,4-*trans*-stereochemistry. A final configurational adjustment was needed to eventually attain the correct stereochemistry at C-19 of ajmalicine. The predictable stereochemical outcome of the Bischler-Napieralski reaction, followed by catalytic hydrogenation was realized through the transformation of **J** to **K** (see relevant mechanism B, Figure 11.1).

The combination of tryptamine and D-glucose accounts for a large segment of the carbon framework of ajmalicine and 19-*epi*-ajmalicine. Although the native sugar chiron had to be extensively deoxygenated, it provided the conformational and steric bias to install the branch points needed to elaborate the pentacyclic intermediates with the correct stereochemistry. Nevertheless, the synthesis was accomplished in less than 25 steps from **A**.

11.2
ent-Actinobolin

A total synthesis of the antipode of natural actinobolin was completed by Chida and coworkers [8] starting with D-glucose (Figure 11.2) [9]. Analysis of the tetrahydroisochroman framework, harboring five contiguous stereogenic carbon atoms, reveals a carbohydrate-like substructure in the δ-lactone portion. In an earlier synthesis of the natural product, an unsaturated enone glycoside was utilized as a carbohydrate template for the lactone portion [9a]. The strategy envisaged by Chida and coworkers relied on a known Ferrier carbocyclization [10] that would convert a 5,6-unsaturated glycoside into a cyclohexanone derivative. Regioselective functionalization, and stereocontrolled annulations would eventually lead to *ent*-actinobolin.

Intermediate **A**, readily available from D-glucose in two steps, was treated with TsCl in pyridine, and the resulting ditosylate was subjected to reduction with LiAlH$_4$ according to Vis and Karrer [11]. The resulting 3-deoxy glycoside **B** was converted to benzyl ether **C** upon treatment with Dibal-H. Iodination at C-6 (carbohydrate numbering), and elimination gave the 5,6-unsaturated glycoside **D**. Exposure to 30 mol% mercuric trifluoroacetate, led to cyclohexanone derivative **E** via a Ferrier carbocyclization. Elimination of the corresponding mesylate gave the enone **F**. Sequential addition of vinyl cuprate and *O*-TES protected (*S*)-lactaldehyde in the presence of HMPA, led to *trans*-substituted adduct **G**. Stereoselective reduction of the carbonyl group (dr = 10:1), followed by chemoselective tosylation and protection of the remaining alcohol as the MOM ether gave **H**. Deprotection of the TES group with DDQ, and oxidative cleavage of the vinyl group gave lactone **I**. Displacement of the tosylate with sodium azide, and functional group manipulation led to **K**. Finally, reduction of the azide group and formation of the amide with Cbz-D-Ala, followed by hydrogenolysis afforded *ent*-actinobolin. A formal synthesis of the natural antipode was also executed by simply interchanging the *O*-benzyl and *O*-TBS groups in **F**. This was accomplished by *O*-benzylation of **B** and proceeding through the same steps shown in Figure 11.2.

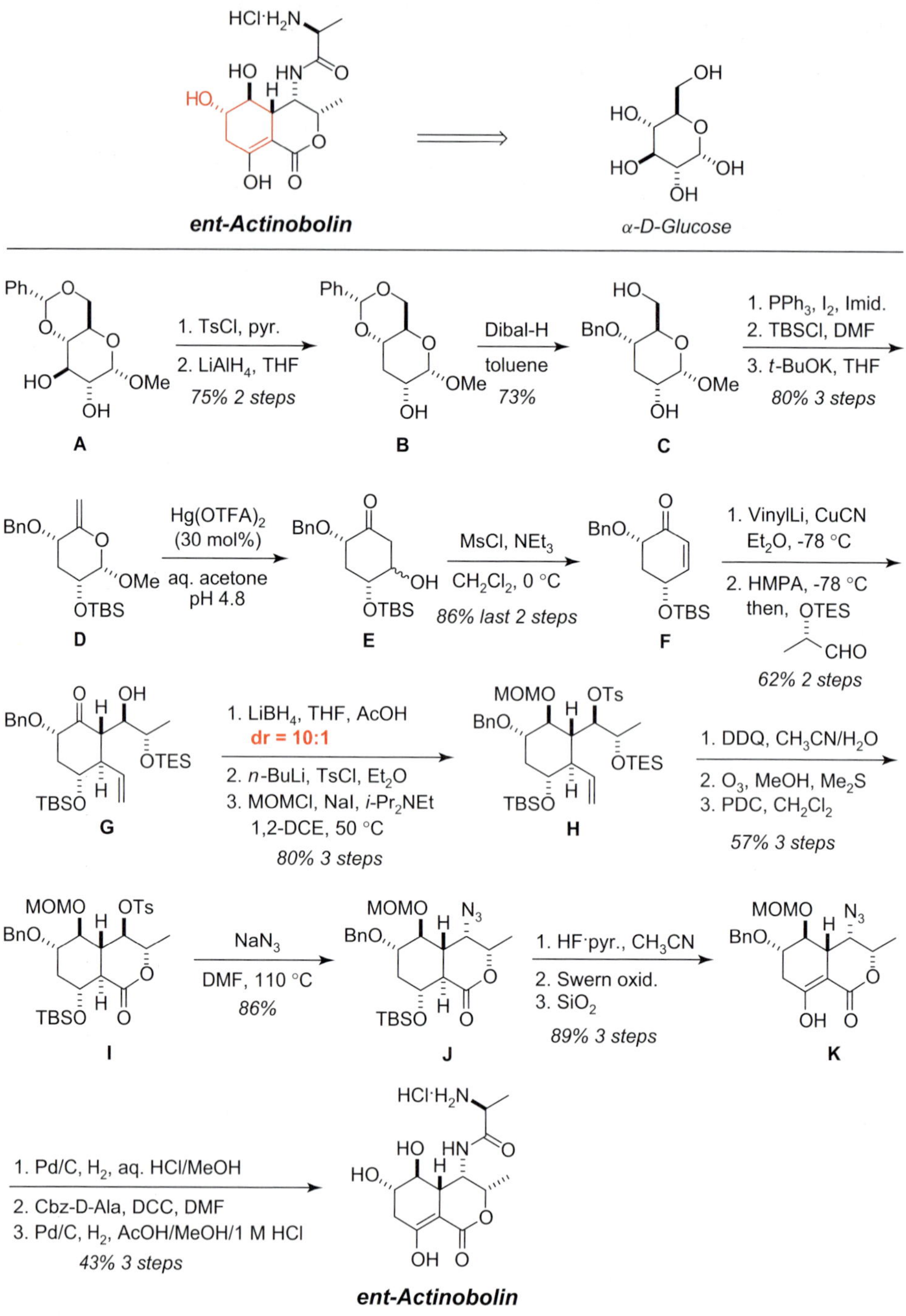

Figure 11.2 Total synthesis of *ent*-actinobolin from D-glucose (hidden substructure).

Key reactions: *Regioselective deoxygenation (**A** to **B**); reductive acetal cleavage (**B** to **C**); Ferrier carbocyclization (**D** to **E**); sequential cuprate addition and aldol reaction (**F** to **G**).*

Relevant mechanisms and transformations:

A.

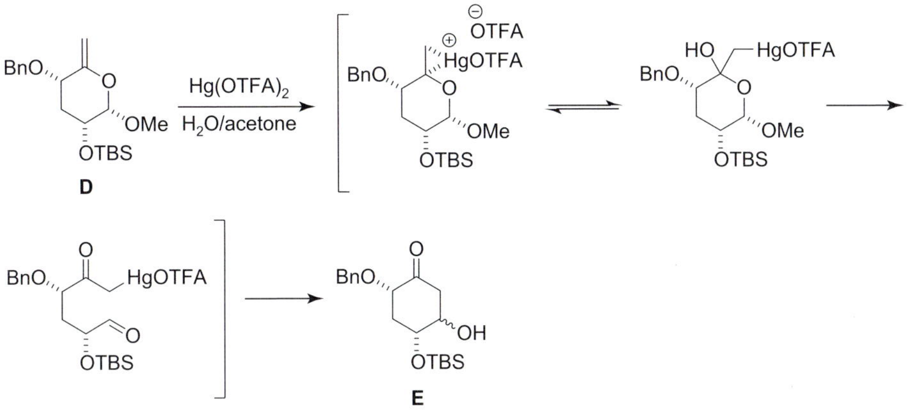

B.

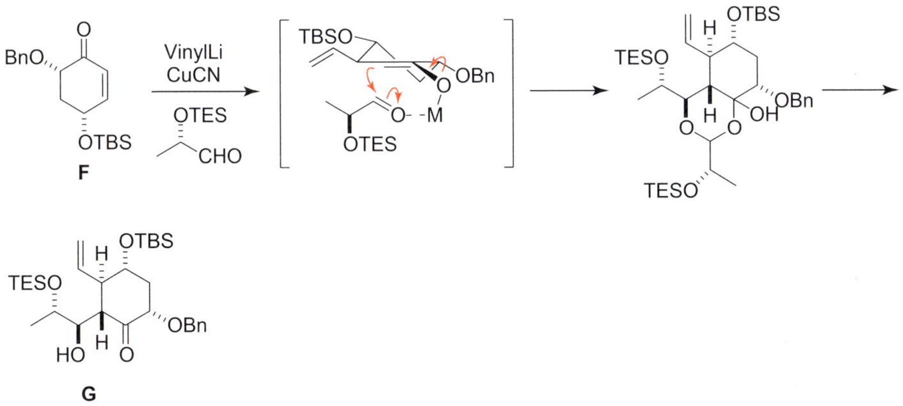

Figure 11.2 *(continued).*

Commentary: The utilization of an enone such as **F** to introduce vicinal functionality in a three-component coupling reaction is a critical design element in the overall strategy. Obtaining enone **F** from methyl α-D-glucopyranoside in nine steps and reasonable overall yield, adds value to the strategy since both stereogenic centers are important in order to secure the desired regio- and stereochemistry later in the synthesis. In this regard, the Ferrier carbocyclization is ideally suited to give the β-hydroxy ketone **E**, a precursor to the required enone **F** (see relevant mechanism A, Figure 11.2). Conjugate addition of the higher order vinyl cuprate gave, as expected, the *anti*-product with respect to the *O*-TBS group. The subsequent aldol reaction of the enolate with *O*-substituted L-lactaldehyde gave unexpected stereochemical outcomes depending on the nature of the *O*-protecting group. The expected product of double diastereoselection was indeed observed with an (*R*)-*O*-MPM ether. In fact, even racemic *O*-MPM-lactaldehyde afforded the expected diastereomer due to the faster reaction of the (*R*)-isomer relative to the (*S*)-isomer. This is an interesting case of kinetic resolution in an aldol reaction. However, a shorter route was devised using a common silyl ether protecting group. In the presence of HMPA, it was discovered that the *O*-TES-(*S*)-lactaldehyde was the preferred aldol partner (see relevant mechanism B, Figure 11.2).

The Chida strategy utilizing D-glucose as a starting material is operationally different compared to other related syntheses of actinobolin [9]. All six carbon atoms of D-glucose remain in the carbon framework of the target molecule as part of the dihydroxy carbocycle, which constitutes ring *A* of *ent*-actinobolin. However, only the original C-4 hydroxyl group in D-glucose was maintained as the C-6 hydroxyl group in the target. No fewer than six total syntheses of actinobolin have been published to date (see Chapter 10, section 10.1).

11.3
Trehazolin and Trehazolamine

The pseudodisaccharide trehazolin has been the subject of several synthetic studies, focusing primarily on the aminocyclopentane polyol core moiety [12] (Figure 11.3). The amino pentahydroxypentitol framework can be visually related to a polyol of matching absolute configuration with regard to the *anti,anti* disposition of the 3,4,5-triol system. Although structurally different, a suitable configurational match can be found in D-glucose [13, 14]. Thus, the main challenge for Giese and coworkers [14] was to devise a method to utilize the 6-carbon framework of a D-glucose derivative as a precursor to the branched 1-amino-2-(hydroxymethyl)-2,3,4,5-tetrahydroxy cyclopentane (**F**) with the required absolute configuration at the newly created stereogenic carbon atoms. Once assembled, trehazolamine was then converted to trehazolin following a literature precedent [13].

Intermediate **A**, readily available from methyl *α*-D-glucopyranoside [15], was treated with *O*-methylhydroxylamine for conversion to the acyclic methyl oxime ether, which was in turn oxidized with the Dess-Martin periodinane reagent to give keto-oxime ether **B** (Figure 11.3). Ring-closure was effected with SmI$_2$, and the resulting tertiary alcohol was acetylated to give **C**. This efficient carbocyclization was similar to reported examples by Marco-Contelles, Chiara, and coworkers [16] utilizing a perbenzylated variant of **B**. Although the required stereochemistry was secured at the tertiary alcohol site in **C**, an inversion of configuration of the carbon atom bearing the *N*-methoxyamine group was necessary. This was brought about by oxidation to the oxime ether **D** with Pb(OAc)$_4$ [17], followed by stereoselective reduction to **E**. Global deprotection under dissolving metal conditions afforded trehazolamine (**F**), which was converted to the glucosyl thiourea analogue **G**, and the latter cyclized to trehazolin.

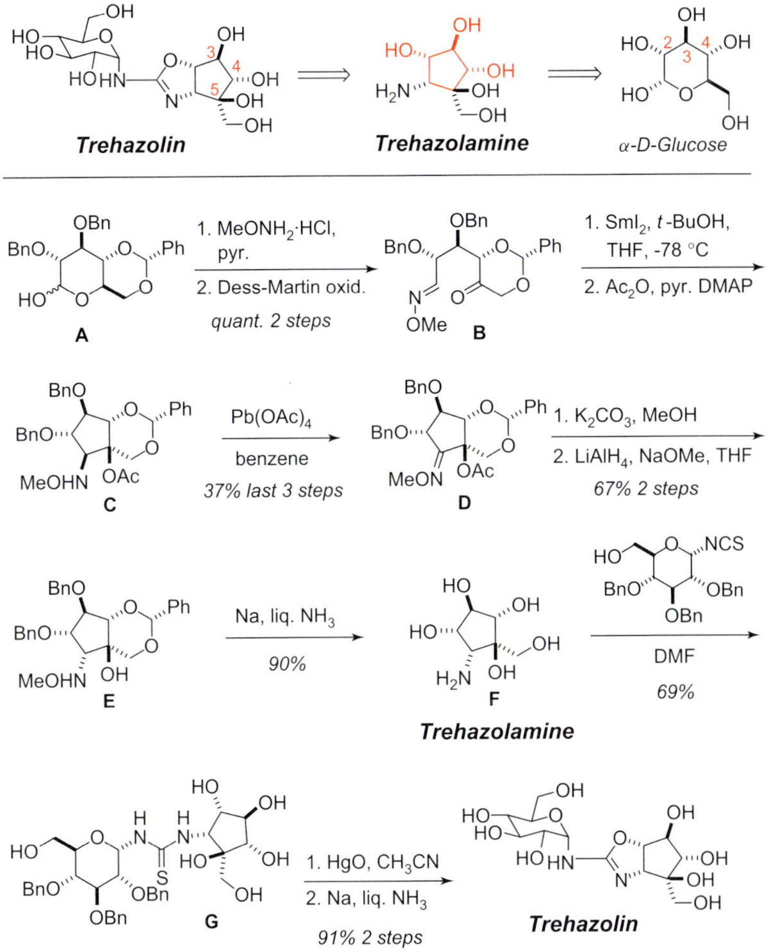

Key reactions: *Intramolecular SmI₂-mediated carbocyclization (**B** to **C**); N-methoxyamine oxidation to methyl oxime ether (**C** to **D**); stereoselective methyl oxime ether reduction (**D** to **E**).*

Relevant mechanisms and transformations:

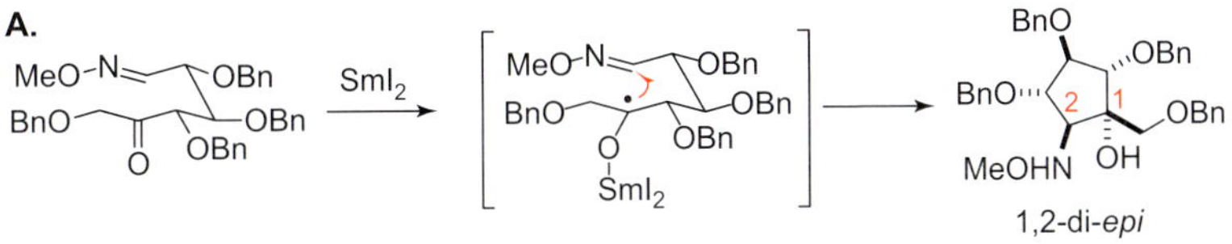

A.

Figure 11.3 Total synthesis of trehazolamine and trehazolin from D-glucose (hidden substructure).

B.

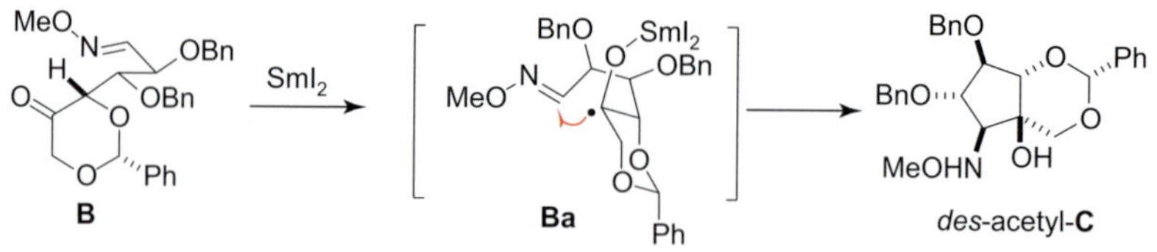

C.

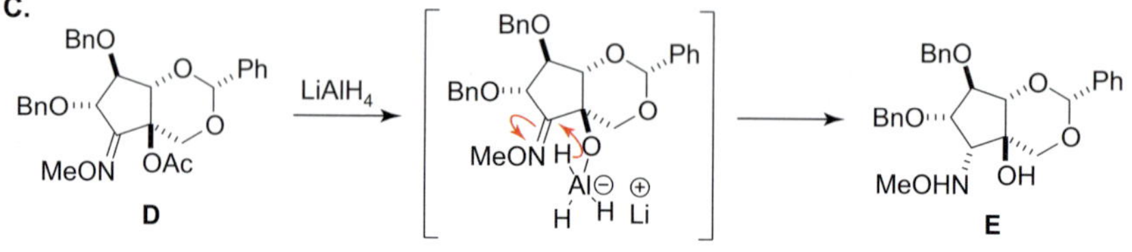

Figure 11.3 (continued).

Commentary: The relative stereochemistry of 3,4,5-triol system in trehazolamine and trehazolin is visually related to the 2,3,4-triol segment in D-glucose, especially when portrayed in a flat, terpene-like perspective (Figure 11.3). A practical method had to be devised to transform the oxacycle to a carbocycle. Fortunately, there are numerous examples of hydroxylated carbocycles originating from various carbohydrate precursors (see Chapter 7, section 7.12).

A SmI$_2$-induced carbocyclization of the acyclic tetra-O-benzyl methyl keto-oxime ether derived from 2,3,4,6-tetra-O-benzyl-D-glucose led to the incorrect stereochemistry of the tertiary hydroxyl group as shown by Marco-Contelles, Chiara and coworkers [16]. Giese surmised that introducing a 4,6-O-benzylidene acetal as in **B** would alter the orientation of the keto group, such that the intermediate ketyl radical anion **Ba** is suitably disposed for the desired carbocyclization (see relevant transformations A and B, Figure 11.3). A second electron transfer would lead to **C** after acetylation, with the desired stereochemistry at C-2. The conceptual design of a reactive intermediate allowed the carbocyclization to follow the anticipated stereochemical course, although a further stereochemical adjustment was needed at C-1 bearing the methoxyamino group in **C**. This was achieved by oxidation to the corresponding methyl oxime ether **D**, followed by stereoselective reduction. It is interesting to note that most oxidizing agents effected a cycloreversion to **B**. The stereoselective reduction of the methyl oxime ether **D** with LiAlH$_4$ in the presence of NaOMe was rationalized on the basis of an activated alkoxide, from which hydride could be delivered internally with complete selectivity to give the intended cyclopentylamine analogue **E** (see relevant mechanism C, Figure 11.3).

Key steps in alternative total syntheses of tetrahazolamine (and trehazolin) from carbohydrate precursors are highlighted in Figure 11.4. Thus, Shiozaki and coworkers [13a] utilized a 1,3-dipolar cycloaddition strategy to construct a trihydroxy bicyclic intermediate **B** from a suitably protected acyclic precursor **A**, which was prepared from D-glucose (Figure 11.4 A). This stereoselective synthesis also allowed the absolute configuration of trehazolin to be determined. Knapp and coworkers [13b] started with D-ribonolactone to obtain the known cyclopentene polyol **F**,

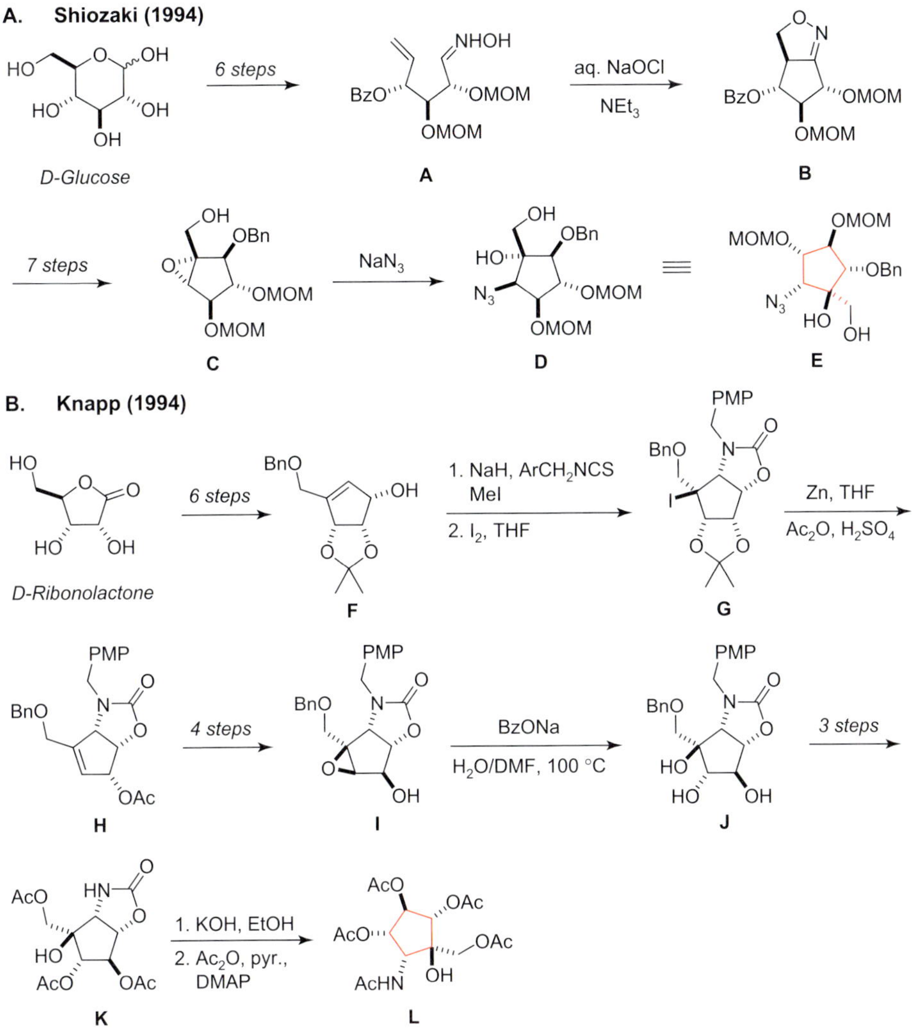

A. Shiozaki (1994)

B. Knapp (1994)

Figure 11.4 Total synthesis of trehazolamine and trehazolin from carbohydrate precursors.

which was elaborated through a number of steps to trehazolamine pentaacetate (Figure 11.4 B). Chiara, and coworkers [13e] utilized D-mannose as the carbohydrate source, and adopted a SmI$_2$-mediated protocol for carbocyclization following their previous studies [16] (Figure 11.4 C). They resorted to the cyclic acetal **M** to ensure proper alignment of the ketyl radical, as was originally conceived by Giese [14]. A number of steps led to intermediates **N** and **O**. Configurational inversion of the C-5 hydroxyl group was achieved through an intramolecular oxazoline formation via the triflate, which led to **P** *en route* to trehazolin.

Seepersaud and Al-Abed [13f] started with D-arabinose, which was transformed to diene **Q** as a mixture of two diastereomers. Ring-closing metathesis afforded

C. Chiara (1999)

D. Al-Abed (2001)

Figure 11.4 *(continued).*

cyclopentene polyol **R**, which was further manipulated through intermediates **S–U** to afford trehazolamine and trehazolin.

Finally, a synthesis of *N,O*-protected trehazolamine and trehazolin was achieved by Ledford and Carreira [12a] in 1995 starting with readily available (*R*)-epichlorohydrin (originally prepared from D-mannitol) and cyclopentadiene (Figure 11.5).

Trehazolin   (R)-Epichlorohydrin

Key reactions: *Regioselective epoxide opening with cyclopentadienyl-lithium (**A** to **B**); n-Bu₃SnH-mediated radical fragmentation of the cyclopropylmethyl bromide (**D** to **E**); Norrish type-II photochemical cleavage of the phenylketone (**G** to **H**); CpLi = cyclopentadienyllithium; DMDO = dimethyldioxirane*

Figure 11.5   Total synthesis of trehazolin from (R)-epichlorohydrin.

Relevant mechanisms and transformations:

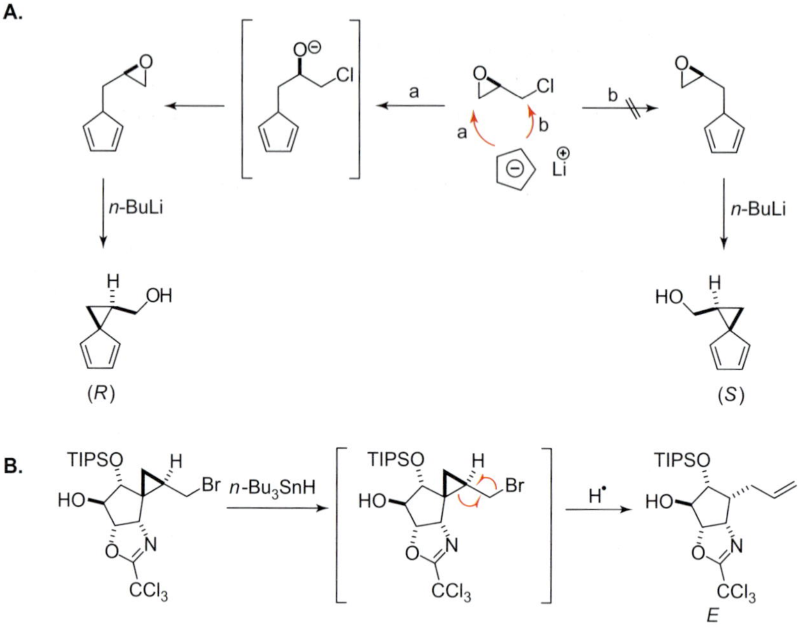

A.

B.

Figure 11.5 *(continued)*.

There are a number of interesting transformations in the Ledford and Carreira synthesis that are of general applicability to the synthesis of aminocyclopentanes. The unprecedented regioselective opening of (R)-epichlorohydrin with cyclopenta-dienyl lithium led to the (R)-alcohol corresponding to **B**. This has been rationalized on the basis of a faster relative rate of epoxide opening (path a), compared to S_N2 attack on the chloride (path b) (see relevant mechanisms A, Figure 11.5). Also noteworthy, is the regioselective radical induced fragmentation of the cyclo-propylcarbinyl bromide intermediate obtained from **D** to give the α-oriented *C*-allyl product **E** (see relevant mechanism B, Figure 11.5). A Norrish type-II cleavage of the phenyl ketone **G** led to the alkene **H**, which was further elaborated to trehazolin. It is of interest that the strategy used by Ledford and Carreira is self-immolative with respect to the (R)-epichlorhydrin, since only one carbon remains after the photochemical cleavage to the *exo*-methylene cyclopentane **H**.

11.4
Fomannosin

An enantioselective total synthesis of fomannosin by Paquette and coworkers [18] relied on a convergent approach that utilizes D-glucose as a starting material in a visually remote way (Figure 11.6). This phytopathogenic sesquiterpene lactone bears little, if any, skeletal or stereochemical relationship to D-glucose, even if one focuses attention on the bicyclic δ-lactone portion. The key reaction consisted of a zirconocene-mediated ring contraction of a *C*-vinyl furanoside, previously studied in the Paquette [19] and Taguchi/Hanzawa [20] laboratories.

D-Glucose was converted to the D-xylofuranose intermediate **A** utilizing known methods (Figure 11.6). Elimination of the tosylate followed by catalytic hydrogenation led to 3-deoxy derivative **B**, in which the configuration of the methyl ester substituent was inverted. Methanolysis and *O*-protection gave **C**, which upon treatment with LDA and formaldehyde gave a 1.9:1 ratio of the *C*-hydroxymethyl derivative **D** and its epimer.

Protection of the primary alcohol and manipulation of the ester led to the aldehyde **E**, which was further transformed to *C*-vinyl derivative **F**. Treatment with Cp_2ZrCl_2 in the presence of *n*-BuLi in THF afforded a 60% yield of the cyclobutanol, which was protected as the TBS ether to give **G**. The epimeric alcohol, isolated in 25% yield, could be oxidized to the ketone and recycled. Ozonolysis of the vinyl group followed by treatment with 5-lithio-4,4-dimethyl-1-pentene gave carbinol adduct **H**. Oxidation with PDC followed by a Petersen olefination led to diene **I**. Ring-closing metathesis in the presence of the Grubbs second generation catalyst, and *O*-deprotection gave the cyclopentene ring system in **J**. Esterification with ethylsulfanylcarbonyl acetic acid followed by oxidation, afforded ketoester **K**. The aldol product **L** was chemoselectively reduced in the presence of Pd/C and Et_3SiH, to give hydroxymethylene lactone **M**. Regioselective 1,4-reduction with $NaBH_4$ led to the α-hydroxymethyl lactone **N** as the major product. Dihydroxylation of the *O*-silyl protected **N**, followed by Swern oxidation gave α-ketol **O**, which was deoxygenated with SmI_2, then treated with TFA to produce diol **P** with the correct stereochemistry at C-9. β-Elimination of the cyclic sulfite ester was utilized to install the double bond of the lactone ring to give **Q**. Conversion to the triflate ester, elimination, and deprotection of the silyl ether completed the first synthesis of fomannosin. A total synthesis of racemic fomannosin had been accomplished by Semmelhack and coworkers in 1982 [21].

Commentary: No matter how closely one scrutinizes the carbon framework of fomannosin, it is highly unlikely that the visual thought process would lead to D-glucose as a chiral progenitor. However, prior knowledge within the Paquette group that *C*-vinyl furanosides can be contracted to vinyl cyclobutanols made the associative mental connection easier. The presence of a cyclobutene motif with a quaternary carbon atom at C-7, establishes a reflex-based connection reminiscent of *C*-vinyl-branched carbohydrate precursors, which are viable chirons for such carbocycles. Once this connection was made in Paquette's mind's eye, the assembly of fomannosin from D-glucose and subsequent transformations to construct the sesquiterpene framework followed a well-charted course.

The zirconocene mediated ring contraction is a fascinating reaction that could involve a transition state model which would result in a *cis*-disposition of the vinyl and hydroxyl groups in the major product as observed in **G** (see relevant mechanism, Figure 11.6). With the establishment of end-differentiated substituents at the quaternary center in **G**, the task of systematic manipulation relied on protective group

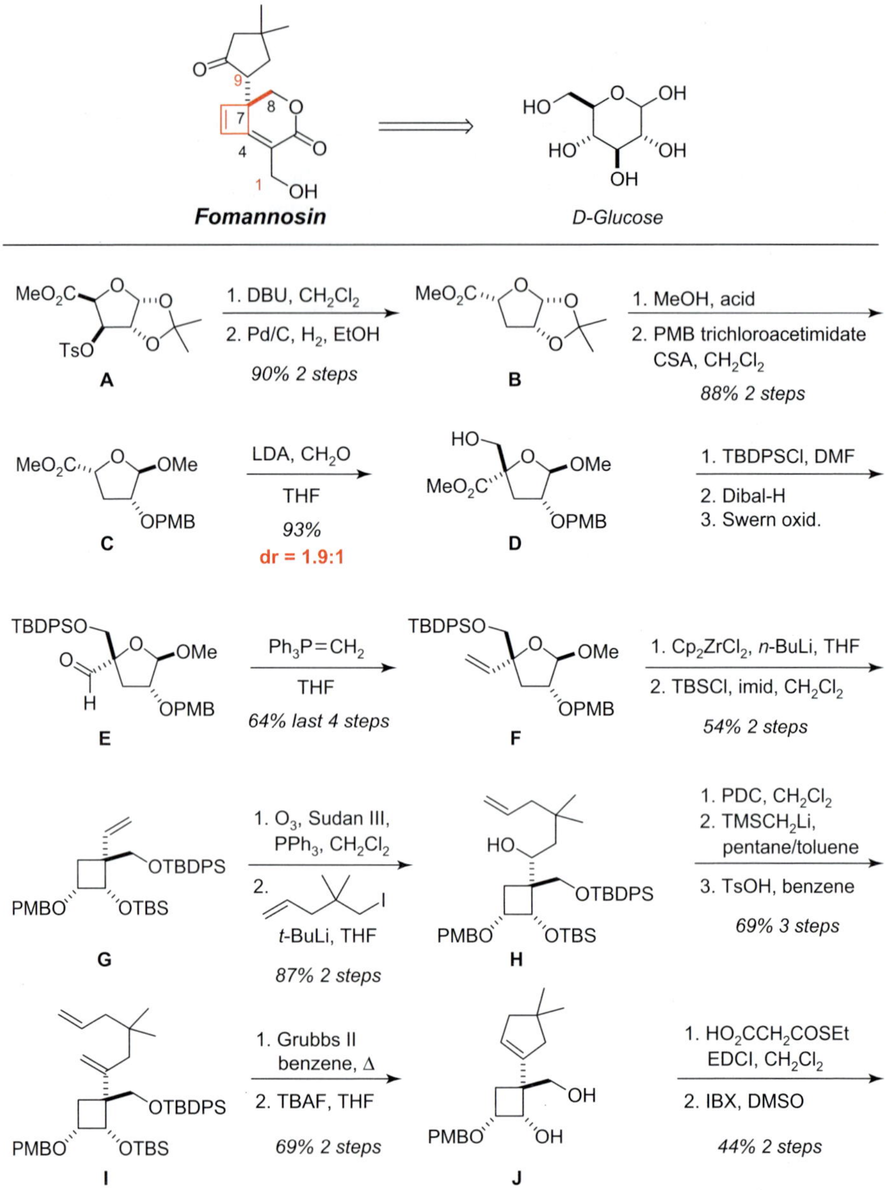

Figure 11.6 Total synthesis of fomannosin from D-glucose (hidden substructure).

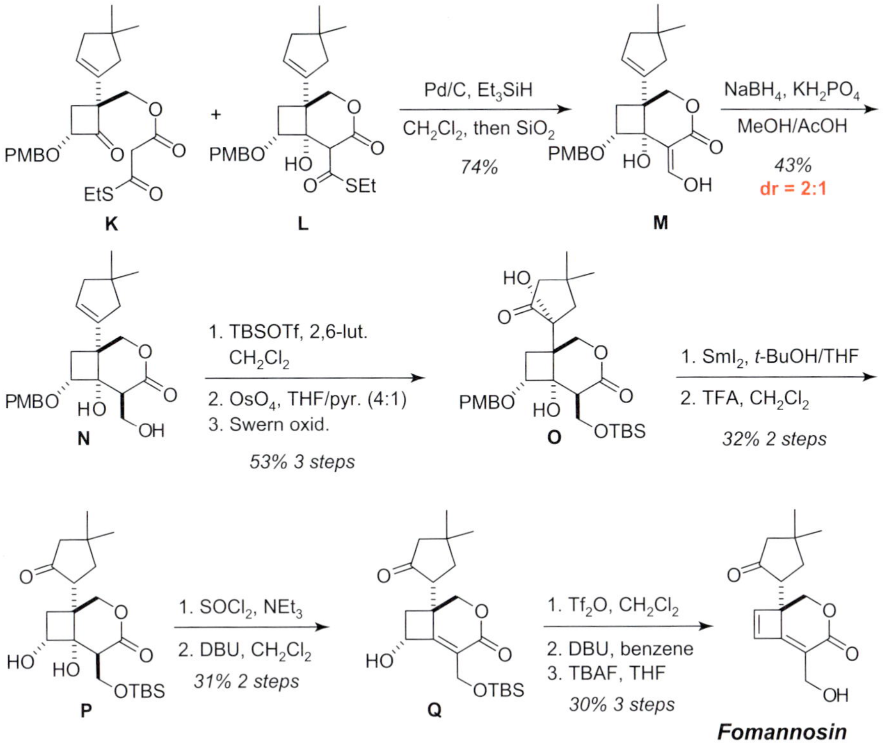

Key reactions: *Zirconocene-promoted ring contraction (F to G); ring-closing metathesis (I to J); Pd/C catalyzed silane reduction (L to M); chemoselective reduction (M to N); cyclic sulfite ester β-elimination (P to Q)*

Relevant mechanism:

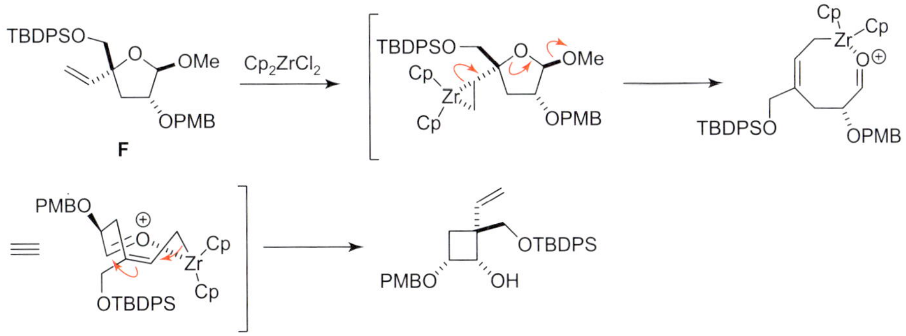

Figure 11.6 *(continued).*

compatibilities and projected chemoselective transformations. Construction of the cyclopentene by the venerable ring-closing metathesis reaction, and the intramolecular aldol reaction of intermediate **K** were well-planned sequences by Paquette and coworkers. Of particular note was the silane-mediated chemoselective reduction of thioester **L**, and the regioselective conjugate reduction of the α,β-unsaturated lactone **M** to give **N**. Functionalization of the cyclopentene adopting routine, well-planned oxidation-reduction sequences, resulted in intermediate **P**, which, by virtue of the positions of the remaining hydroxyl groups, was nicely poised to undergo two sequential β-eliminations *en route* to fomannosin.

11.5
9a-Desmethoxy Mitomycin A

The literature is abundant with numerous efforts toward the total synthesis of the mitomycins [22]. A strategically novel approach toward the mitosanes was reported by Coleman and coworkers [23]. The target molecule was 9a-desmethoxy mitomycin A, in which the angular methoxy group of the parent mitomycin A was missing (Figure 11.7). Analysis of the tricyclic quinonoid pyrrolidine led Coleman to utilize a trihydroxy pyrrolidine precursor to ring *C*, which would be prepared starting with D-ribose. The known sequence from D-ribose to intermediate **A** was executed in good overall yield [24]. Treatment of **A** with iodosobenzene and iodine, according to Suarez and coworkers, [24] led to pyrrolidine acetonide **B** in 72% yield. Introduction of an azide with inversion of configuration gave **C**, which was treated with the tributylstannyl allyl cinnamate **D** in the presence of $BF_3 \cdot OEt_2$ to give adduct **E** as the major isomer. The azide group was reduced in the presence of $SnCl_2$ and benzene thiol, and the resulting amine was protected as the *N*-Cbz derivative **F**. Intramolecular aziridine formation took place under Mitsunobu conditions to give **G**. An oxidation-reduction sequence converted the vinyl group to a primary alcohol, and changing the pyrrolidine *N*-protecting group to *N*-Cbz gave **H** in good overall yield. Subsequent manipulation of functional groups, adjustment of the oxidation state of the electron rich aromatic system, and an intramolecular Michael-type cyclization to form the fused pyrrolidine ring, was followed by air oxidation to give 9a-desmethoxy mitomycin A. Thus, a convergent and enantioselective synthesis of the target molecule was achieved in 20 synthetic operations starting with D-ribose. Slightly longer linear sequences to 9a-desmethoxy mitomycin A have been reported by the Danishefsky [25] and Ziegler [26] groups. The use of a chiral, non-racemic aziridine derivative by Ziegler led to 9a-desmethoxy mitomycin A (see below).

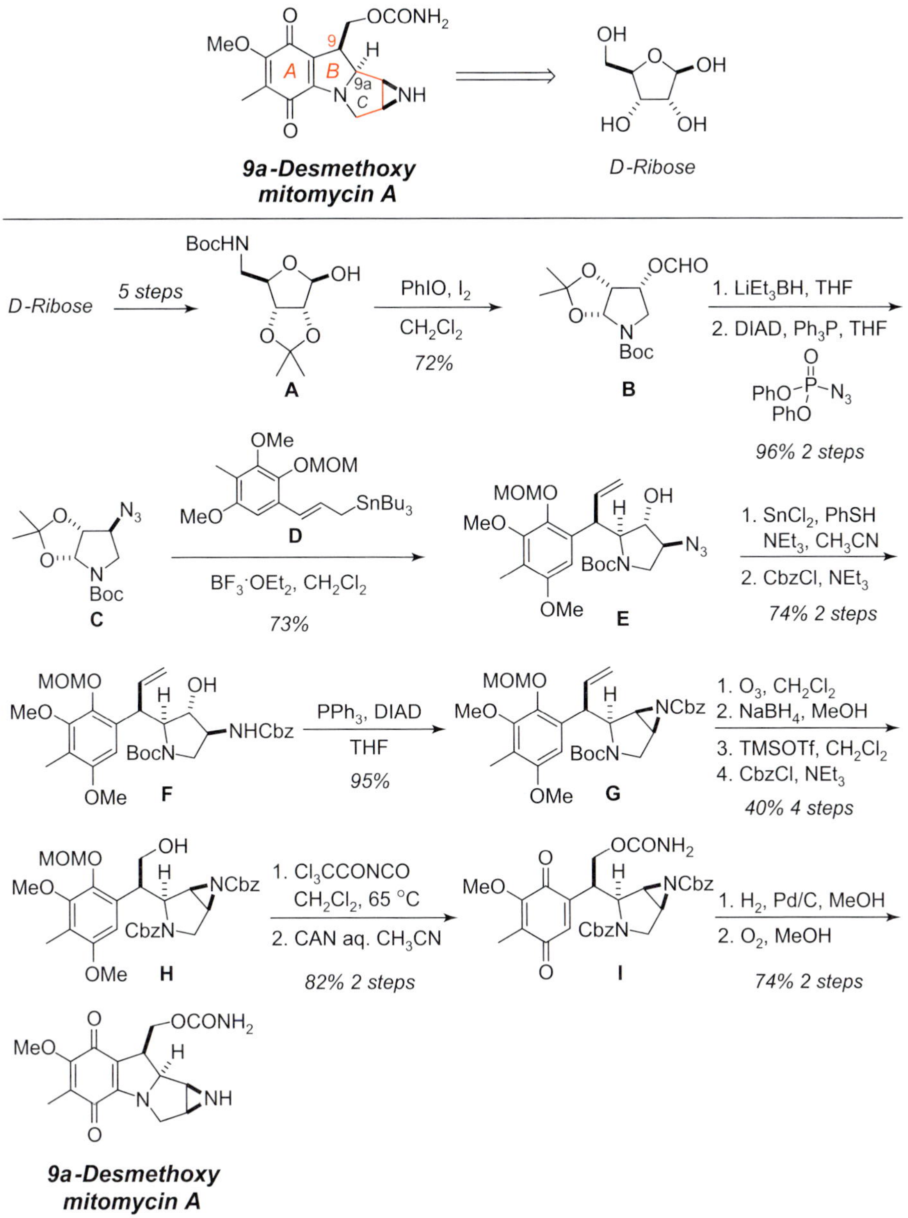

**9a-Desmethoxy
mitomycin A**

D-Ribose

**9a-Desmethoxy
mitomycin A**

Key reactions: *Furanose to pyrrolidine ring rearrangement (**A** to **B**); iminium ion alkylation (**C** to **E**); intramolecular aziridine formation (**F** to **G**)*

Figure 11.7 Total synthesis of 9a-desmethoxy mitomycin A from D-ribose (hidden subtructure).

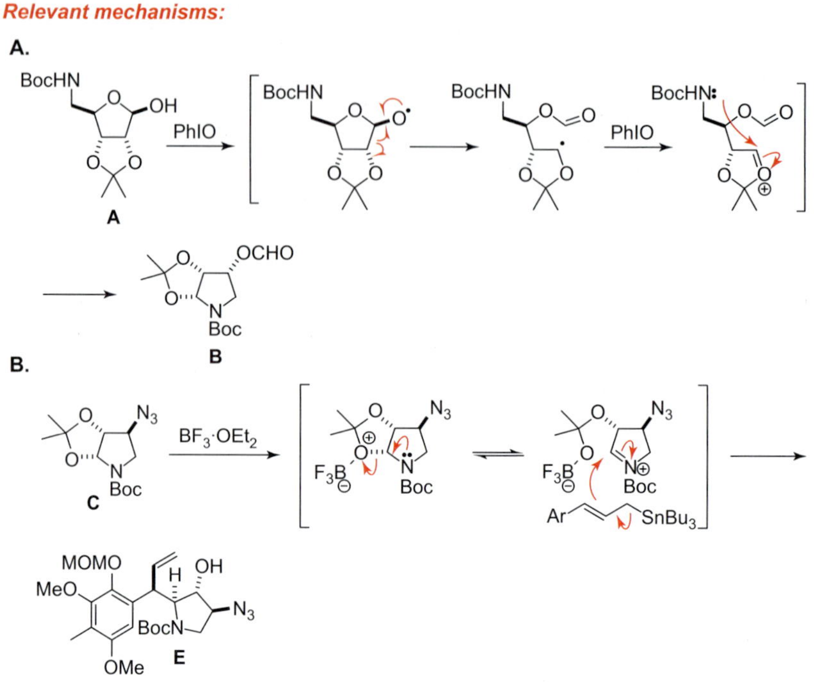

Relevant mechanisms:

A.

B.

Figure 11.7 *(continued).*

Commentary: Central to the strategy utilized by Coleman was the prospect of a stereoselective addition of a cinnamyl moiety onto an *N*-acyloxy iminium ion derived from an appropriately functionalized pyrrolidine intermediate (**C** to **E**). Having established this basic plan, it was opportune to apply the Suarez furanose-ring fragmentation [24] to **B**. In the event, an anomeric alkoxy radical underwent β-fragmentation to give a carbon-centered radical (see relevant mechanism A, Figure 11.7). A second electron transfer gave the oxocarbenium ion, which was intramolecularly attacked by the *N*-Boc group to give pyrrolidine **B**.

The "transfer" of the 2,3-*O*-isopropylidene group in **A** to the corresponding 2,3-acetal in the pyrrolidine **B** augured well for a BF$_3$·Et$_2$O-mediated activation with concomitant formation of an *N*-acyloxy iminium ion (see relevant mechanism B, Figure 11.7). The orientation of the BF$_3$-ligated hemiacetal appendage would be a good stereofacial directing group for the incoming cinnamylstannyl reagent. Indeed, formation of the major isomer **E** attests to the practical implementation of this notion.

It is of interest to present salient features of the Ziegler synthesis [26] of 9a-desmethoxy mitomycin A, since recourse was also made to a carbohydrate-derived progenitor to the pyrrolidine ring (Figure 11.8). Ziegler prepared an enantiopure 4-(hydroxymethyl)aziridine carboxylic acid ester **D** from methyl

(2R,3S)-4-hydroxy-2,3-epoxy-butyrate (**C**). Access to this chiron can be achieved by a number of methods including starting with D-*iso*-ascorbic acid (Figure 11.8 A).

Alkylation of the potassium salt of indole derivative **E** with triflate ester **F** gave **G**, which was decarboxylated to the bromo derivative **H** according to the Barton protocol [27]. Radical-induced cyclization of the aziridine gave the tricyclic precursor to 9a-desmethoxy mitomycin A.

9a-Desmethoxy mitomycin A

D-iso-Ascorbic acid

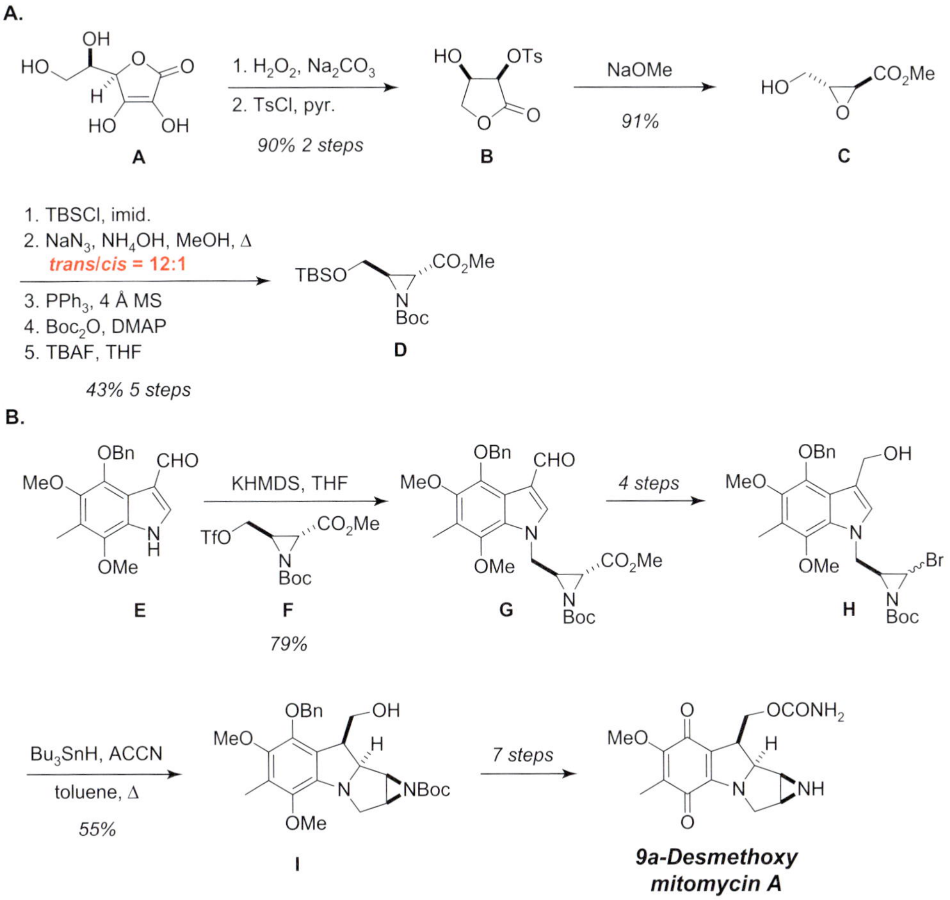

A.

B.

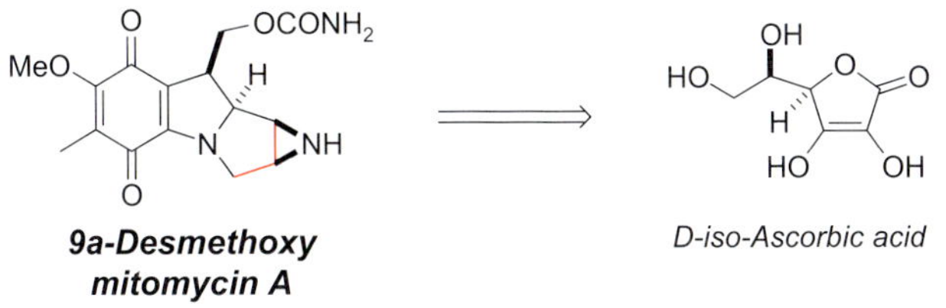

Figure 11.8 Total synthesis of 9a-desmethoxy mitomycin A from D-*iso*-ascorbic acid (hidden substructure).

11.6
Saxitoxin and β-Saxitoxinol

The first total synthesis of saxitoxin, a highly neurotoxic guanidinium alkaloid, was reported in 1977 by Kishi and coworkers [28]. A second total synthesis of the racemic toxin was communicated by Jacobi and coworkers in 1984 [29]. In considering a stereoselective synthesis of the natural (+)-isomer, Du Bois and coworkers [30] started with D-glycerol acetonide, which is readily available from D-mannitol (Figure 11.9). The plan called for the introduction of an amino group by inversion of configuration, which would ultimately correspond to C-6 in saxitoxin. Synthetic operations had to be devised to further elaborate the complex alkaloid skeleton with control of stereochemistry at C-4 and C-5.

Cyclic sulfamate *N,O*-acetal **A**, which is readily available from D-glycerol acetonide, was treated with an alkynyl organozinc reagent in the presence of $BF_3 \cdot OEt_2$ to give the oxathiazinane dioxide **B**. Catalytic reduction to the *cis*-olefin, followed by displacement of the tosylate group with azide and protection with *p*-methoxybenzyl chloride gave **C**. The *N*-Mbs (*p*-methoxybenzesulfonyl) methylthio amidine derivative was prepared from the amine, which was followed by displacement of the secondary triflate group by azide ion to give **D**. Cleavage of the *N*-PMB group and conversion to the Mbs guanidinyl derivative gave rise to **E**. Reduction of the azide group under Staudinger conditions led to the amine **F**. Treatment of **F** with $AgNO_3$ afforded the nine-membered cyclic guanidine derivative **G** in good overall yield for the two-step sequence. Introduction of the carbamate group and oxidation with catalytic $OsCl_3$ and oxone gave the ring-contracted cyclic guanidine **H**. Treatment of **H** with boron tris(trifluoroacetate) in TFA gave β-saxitoxinol (**I**), which, upon a Pfitzner-Moffatt oxidation, afforded saxitoxin.

Figure 11.9 Total synthesis of saxitoxin and β-saxitoxinol from D-mannitol (partially hidden substructure).

Key reactions: *Organozinc addition to N,O-acetal (**A** to **B**); intramolecular cyclic guanidine formation (**F** to **G**); ring contraction (**G** to **H**); Mbs = p-methoxybenzensulfonyl*

Relevant mechanisms and transformations:

A.

B.

C.

β-Saxitoxinol

Figure 11.9 *(continued).*

11.6.1

Second generation synthesis

Rather than starting from D-glycerol acetonide and introducing a nitrogen atom by inversion of configuration, Du Bois conceived a much shorter route to saxitoxin starting with *N*-Boc-L-serine methyl ester (**A**) (Figure 11.10) [31]. Thus, nitrone **B**, derived from *O*-TBDPS *N*-Boc serinal and obtained in five steps from L-serine, was converted to alkynyl hydroxylamine derivative **C** in 78% yield and 5:1 selectivity by following a literature precedent reported by Merino and coworkers [32]. Conversion to the *cis*-olefin via the *in situ* generation of diimide from tosyl hydrazine and reduction of the hydroxylamine gave **D**, which was transformed into bis-guanidine **E** and subsequently converted to the nine-membered cyclic guanidine **F** as shown in the first generation synthesis (Figure 11.9).

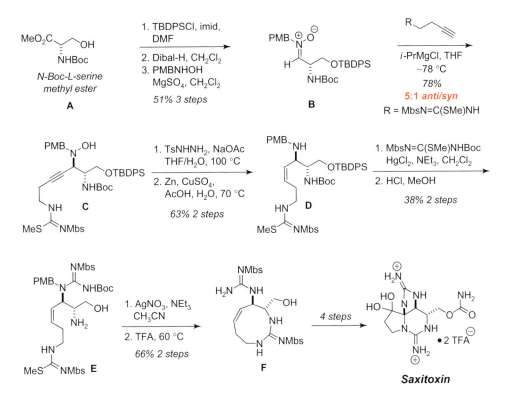

Figure 11.10 Second generation synthesis of saxitoxin from L-serine (partially hidden substructure).

Commentary: Visual analysis of the intricate substitution pattern present in the heterocyclic framework of saxitoxin revealed a vicinal diamino motif that demanded stereochemical consideration in the planning process. D-Glycerol acetonide offered a viable strategy that allowed the exploration of alkynylation of an iminium oxathiazinane dioxide as a novel precursor to vicinal amino alcohols (Figure 11.9, **A** to **B**). Extensive studies by Du Bois and coworkers demonstrated the utility of *N*,*O*-acetals such as **A** in stereoselective C–C bond forming reactions. The selectivity of acetylide addition to **A** (dr > 20:1) was rationalized on the basis of an intramolecular attack of a zinc-tethered alkoxide on an iminium ion (see relevant mechanism A, Figure 11.9).

The unprecedented formation of a nine-membered cyclic guanidine such as **G** was rationalized based on the intermediate formation of a carbodiimide, which underwent intramolecular cyclization by attack of the amino group (see relevant mechanism B, Figure 11.9). After extensive experimentation it was found that the reagent combination of $OsCl_3$ and oxone provided the optimal conditions to convert the endocyclic double bond in **G** to the ring-contracted bicyclic hemiaminal product **H** (see relevant mechanism C, Figure 11.9). Finally, the iminium ion resulting from the hemiaminal underwent intramolecular attack by the pendant C-6 guanidino group in the presence of boron tris(trifluoroacetate) to give β-saxitoxinol **I**.

Focussing on the inherent vicinal diamino motif in saxitoxin, especially considering the presence of the C-6 hydroxymethyl group, conjured the image of L-serine as a viable starting material. The visual connection of L-serine as a partially hidden chiron, in spite of the complex nature of the target molecule, was further solidified in Du Bois' mind's eye, with the knowledge of Merino's studies on the C-alkylation of nitrones [32]. A new plan for a second generation synthesis was thus formulated and successfully implemented (Figure 11.10). A total synthesis of saxitoxin starting from L-malic acid was recently reported by Nagasawa and coworkers [33b] (see Chapter 12, section 12.8)

11.7
ent-Decarbamoyl Saxitoxin

The basic strategy in the 1977 synthesis of racemic saxitoxin by Kishi and coworkers [28] involved the elaboration of the six-membered cyclic guanidine ring by a series of condensation reactions. A key construct was the single step, three-component cyclocondensation of a vinylogous carbamate prepared from a 3-oxo-thiopyrrolidinone with benzyloxyacetaldehyde in the presence of silicon tetraisothiocyanate (Figure 11.11). The resulting racemic bicyclic thiourea derivative was further elaborated to (±)-saxitoxin.

(±)-Saxitoxin

Figure 11.11 Three-component coupling strategy for construction of a bicyclic thiourea precursor to (±)-saxitoxin.

In revisiting the original synthetic plan for saxitoxin, Kishi utilized a chiral, non-racemic aldehyde for the three-component coupling instead of the achiral benzyloxyacetaldehyde [33a] (Figure 11.12). Using D-glyceraldehyde acetonide,

ent-Decarbamoyl saxitoxin

D-Glyceraldehyde acetonide

Key reactions: *Three-component cyclization (**A** to **B**); Curtius rearrangement (**E** to **F**)*

Figure 11.12 Total synthesis of *ent*-decarbamoyl saxitoxin from D-glyceraldehyde acetonide (partially hidden substructure).

Relevant mechanism:

Figure 11.12 *(continued).*

which is readily available from D-mannitol, and performing essentially the same cyclocondensation in the presence of silicon tetraisothiocyanate with the vinylogous carbamate derivative **A**, an enantiomerically pure isomer **B** was obtained in 80% yield. Subsequent steps to construct the cyclic thiourea, and further elaboration followed the originally reported route [28]. Thus, cleavage of the acetonide group and conversion of the cyclic thiourea into the ethylthio iminoether gave **C**. Oxidative cleavage of the diol to the aldehyde, and immediate reduction with NaBH$_4$ produced intermediate **D**. Protection as the benzyl ether and treatment of the resultant product with H$_2$S restored the thiourea motif in **E**. The carbomethoxy group was thus transformed to the hydrazide, and the latter was subjected to a Curtius rearrangement to eventually give the α,β-unsaturated urea **F**. Treatment with acid afforded the cyclic urea **G**, which was transformed to the cyclic guanidine **H**, and ultimately converted to *ent*-decarbamoyl saxitoxin.

Commentary: Kishi's observation that vinylogous urethane **A** exists in an intramolecularly H-bonded form, thereby conferring a degree of planarity, instigated the exploration of chiral ester auxiliaries as a means to control the stereochemistry at C-6. Failure to observe useful levels of asymmetric induction led to the choice of D-glyceraldehyde acetonide as a temporary chiral inducer. Indeed, the 3-component cyclocondensation of **A** with D-glyceraldehyde acetonide in the presence of silicon tetraisothiocyanate led to a mixture of diastereomers, from which the crystalline C-6 (*S*)-isomer **B** could be isolated in 72% yield. A plausible mechanism for this interesting cyclocondensation was proposed by Kishi (see relevant mechanism, Figure 11.12). The observed facial selectivity was explained by invoking a Felkin-Anh model.

It is of interest to note that having served its purpose as a temporary chirality inducer, the diol motif of the original chiron was sacrificed to what amounts to a self-immolative process. Thus, the C-6 hydroxymethyl group was originally C-2 in the D-glyceraldehyde acetonide. As a contrast, Du Bois used a related chiron as an *inherent* part of the molecule, in his first generation synthesis [31], while also inducing chirality in the intramolecular addition of an organometallic reagent to a vicinal iminium ion intermediate (1,2-induction) (Figure 11.9).

11.8
Zaragozic acid A

A group of squalene synthase inhibitors were reported independently under the names of zaragozic acids and squalestatins. The archetypical representative of this group is zaragozic acid A (squalestatin S1) (Figure 11.13) [34]. The total synthesis of zaragozic acid A from D-glucose was reported by Heathcock and coworkers [35], who exploited the sugar framework as well as resident chirality in an effective manner. Analysis of the dioxabicyclic core structure reveals a highly branched tetrahydrofuran diol that is encompassed within a [3.2.1]bicyclic ring system. The availability of zaragozic acid A allowed Heathcock to develop an efficient degradation sequence that produced the dimethyl acetal tri-*t*-butyl ester as a relay compound (Figure 11.13). Since this intermediate could be converted to the natural product by attaching the peripheral ester and disubstituted ω-phenylhexyl chain, the primary challenge became the synthesis of the relay compound.

Thus, D-glucose was transformed into 1,6-anhydro intermediate **A** according to a previous study by Heathcock [35]. Treatment of **A** with a (dimethylisopropoxysilyl)methylmagnesium chloride under Tamao conditions [36] gave a diol, which was isolated as TBDPS ether **B**. Acetolysis and cleavage of the ester groups afforded a triol, which was converted to the corresponding furanose and then oxidized to the γ-lactone **C**. Treatment with a 1-butenyl organocerium reagent, and exposure of the product to acid afforded the bicyclic acetal **D**, which is depicted in a perspective that resembles the core system of zaragozic acid A. Oxidation of the secondary alcohol in **D** gave the corresponding ketone, which was converted to the lithium enolate and condensed with paraformaldehyde to give hydroxymethyl adduct **E**, following silylation. Treatment of **E** with vinylmagnesium bromide in the presence of CeCl$_3$ led to branched alcohol **F** in excellent yield. With the carbon atoms corresponding to the intended carboxylic acid group in place, Heathcock proceeded to elaborate the alcohol groups. Thus, Dess-Martin oxidation, followed by a Pinnick oxidation, and esterification yielded bis-*t*-butyl ester **G**. Two successive selective ozonolyses then afforded aldehyde **H**, which was converted to the third *t*-butyl ester as in **I**. Oxidation of the primary alcohol to the aldehyde, conversion to the dimethyl acetal, and hydrogenolysis gave the relay compound, which was identical in all respects to that obtained by degradation of zaragozic acid A.

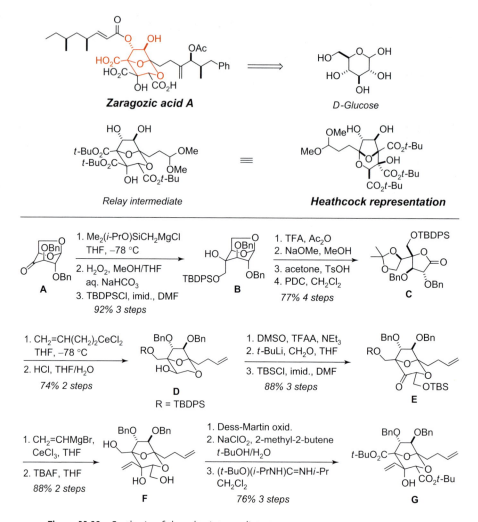

Figure 11.13 Synthesis of the relay intermediate to zaragozic acid A from D-glucose (hidden substructure).

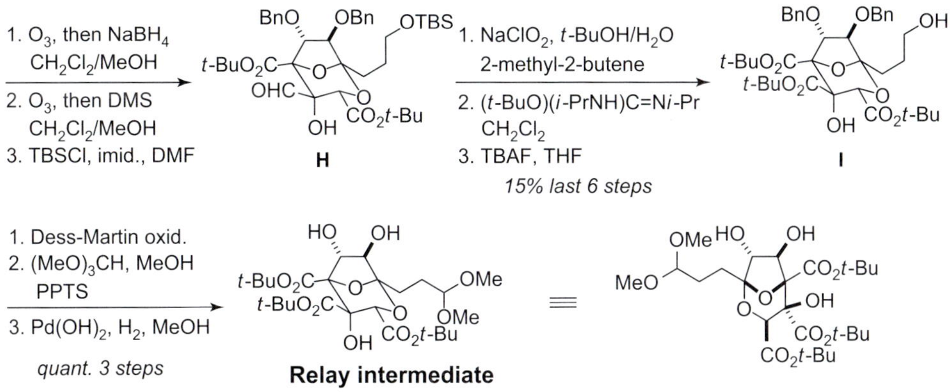

1. O₃, then NaBH₄ CH₂Cl₂/MeOH
2. O₃, then DMS CH₂Cl₂/MeOH
3. TBSCl, imid., DMF

→ **H**

1. NaClO₂, t-BuOH/H₂O 2-methyl-2-butene
2. (t-BuO)(i-PrNH)C=Ni-Pr CH₂Cl₂
3. TBAF, THF
15% last 6 steps

→ **I**

1. Dess-Martin oxid.
2. (MeO)₃CH, MeOH PPTS
3. Pd(OH)₂, H₂, MeOH
quant. 3 steps

Relay intermediate ≡

Key reactions: *Organocerium additions (C to D, E to F)*

Relevant transformations:

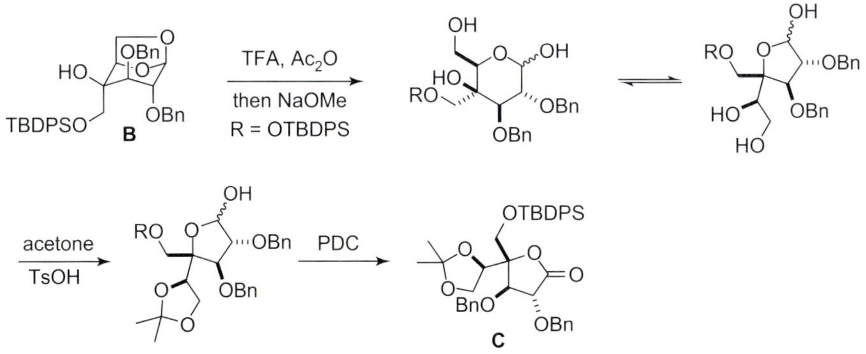

B — TFA, Ac₂O then NaOMe R = OTBDPS →

acetone TsOH → **C** PDC →

Figure 11.13 *(continued).*

Commentary: The various groups that have described total syntheses of the core bicyclic tricarboxylic acid structure of the zaragozic acids have employed conceptually different strategies of C–C bond formation [34]. Access to the bicyclic core of zaragozic acid from different chirons is illustrated in Figure 11.14. Of these, D-glucose provides the longest carbon chain as well as securing the C-6–C-7 diol system with the required stereochemistry. In each of these syntheses, carbonyl branching was a critical reaction as illustrated in the Heathcock synthesis (Figure 11.13). Once the protected C-4 hydroxymethyl lactone **C** was prepared from the precursor ketone, it was necessary to resort to a transmetallated organocerium reagent to minimize β-elimination (**C** to **D**). The topology of the dioxabicyclic acetal (**D**) ensured the correct stereochemical outcome of the enolate hydroxymethylation reaction. Further branching of the appropriately positioned ketone group with a vinyl cerium reagent also provided excellent facial selectivity, as observed independently by Carreira and Du Bois [37] in a related reaction with a lithium acetylide.

It should be noted that a key structural requirement in the Heathcock strategy was predicated upon the pyranose to furanose rearrangement (**B** to **C**, then oxidation) (see relevant transformations, Figure 11.13). Once in hand, the C-4 branched lactone **C** would serve as an ideal template to introduce the required branching regio- and stereoselectively.

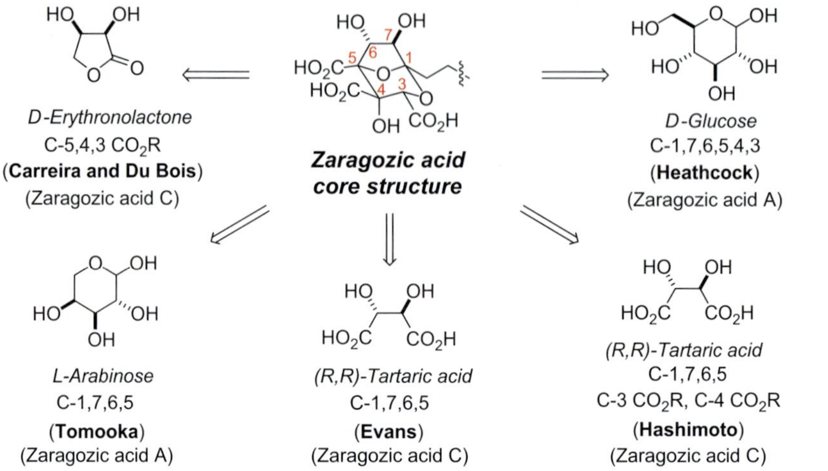

Figure 11.14 Access to zaragozic acid bicyclic core structure from different chirons.

The synthesis of zaragozic acid A from L-arabinose by Tomooka [34], and of zaragozic acid C from D-erythronolactone by Carreira and DuBois relied on different C–C bond forming methods. The Carreira and Du Bois [37] synthesis showed the feasibility of alkyne branching at C-4, followed by oxidative cleavage. An alternative strategy utilizing asymmetric Sharpless dihydroxylation by Nicolaou and coworkers [35a,b] provided the first asymmetric synthesis of zaragozic acid A in 1994. Zaragozic acids C and D share the same core bicyclic tri-carboxylic acid, but differ in the nature of the ester and pendant C-1 alkyl chain.

(*R,R*)-Tartaric acid was exploited as a chiron, in conjunction with other methods of asymmetric synthesis in the independent syntheses of zaragozic acid C by the Evans [38] and Hashimoto [39] groups. A synthesis of zaragozic acid C by Armstrong and coworkers [40] relied on a Sharpless asymmetric, double dihydroxylation of a prochiral diene.

11.9
Hemibrevetoxin B

The linear *trans*-fused polyoxacyclic neurotoxin hemibrevetoxin B belongs to an architecturally complex family of fused polyether marine natural products (Figure 11.15) [41]. The main challenge in considering synthetic strategies toward the total synthesis of hemibrevetoxin B and related compounds resides in the control of stereochemistry at the multitude of stereogenic centers involving C–O and C–C bonds. The presence of fused oxacycles has instigated various groups to consider carbohydrates as viable starting materials for the synthesis of hemi-brevetoxin B and related congeners. Indeed, in the first total synthesis, Nicolaou and coworkers [42] utilized D-mannose as a starting chiron to elaborate ring A. Yamamoto and Kadota [43] also recognized D-mannose as a suitable chiron for ring A. In a formal synthesis, Mori and coworkers [44] utilized D-glucose and L-glyceraldehyde (obtained from L-ascorbic acid), as sources for rings A and B respectively (Figure 11.15).

Figure 11.15 Total and formal syntheses of hemibrevetoxin B from carbohydrate precursors (apparent and partially hidden substructures).

Analysis of the structure of hemibrevetoxin B puts it in a Category II or III-type molecule (Chapter 9). Thus, the utilization of a single carbohydrate precursor such as D-mannose provides a carbon skeletal and partial stereochemical overlap with only one of the four oxacyclic units of hemibrevetoxin B. This necessitates the introduction of sequential appendages, which will have to be functionalized in a stepwise fashion in order to complete the structural landscape. The Nicolaou [42] and Yamamoto [43] linear syntheses are examples of such peripheral overlaps with the chosen carbohydrate precursor.

In 2003 Holton and coworkers [45] reported a total synthesis in which D-arabinose and D-glucose were utilized as viable carbohydrate precursors. The Holton strategy was based on a convergent approach where the "eastern" and "western" subunits representing rings A and D respectively were separately elaborated, then united in the assembly stage, and further manipulated *en route* to the target (Figure 11.16).

The C-2 branched intermediate **A** was prepared from D-arabinose according to an earlier report by Ireland [46]. Formation of the bicyclic lactone **B**, C-allylation to **C**, and elaboration of the terminal olefin by standard procedures led to the iodide **D**

Figure 11.16 Synthesis of hemibrevetoxin rings *A* and *D* subunit precursors I and II from D-glucose and D-arabinose (apparent and partially hidden substructures).

(subunit I) in 62% overall yield (Figure 11.16). The synthesis of subunit II (**J**) commenced from tri-*O*-acetyl D-glucal (**E**), which was converted to *C*-glycoside **F** using a Lewis acid mediated allylstannylation reaction. Hydroxy directed epoxidation of the endocyclic double bond led to **G**, which was converted to the *O*-protected diol **H**. Introduction of the methylacetylene appendage involved an S_N2-type chain-extension of a triflate ester to give **I**. Silylcupration and iodination in hexafluoro-2-propanol gave **J** (subunit II) (Figure 11.16).

Key reactions: *Regioselective alkylation (A to B); C-glycosidation with allyltin reagent (E to F); vinyl iodide synthesis (I to J); DMPU = Dimethylpropyleneurea*

Figure 11.16 *(continued).*

The coupling of the subunits was envisioned to take place by joining the iodozinc reagent prepared from subunit I with the vinyl iodide subunit II, following a catalytic Negishi coupling protocol [47a] in the presence of Pd(dppf)Cl₂ [47b] (Figure 11.17). In the event, a 76% isolated yield of **K** was obtained. Cleavage of the lactone ring in **K**, followed by iodolactonization proceeded with high diastereoselectivity to give **L** in excellent yield. The bis-silyl ether was subjected to methanolysis, which afforded the epoxy ester **M** in 95% yield. Introduction of a *Z*-ethylidene chain in a four-step sequence, and treatment of the epoxy olefin **N** with *N*-(phenylseleno)phthalimide in hexafluoro-2-propanol gave **O** as a single isomer in 83% yield. Oxidative elimination followed by *O*-debenzylation led to hemiacetal **P**. Petersen olefination followed by treatment with TBAF generated the diene intermediate **Q**. As a result, a *C*-vinyl group was introduced at the C-1 carbon of the original D-arabinose. Ring-closing metathesis using the Grubbs II catalyst afforded the seven-membered oxepane ring displayed in **R**, thus completing the 4-ring polyether motif (Figure 11.17).

The elaboration of the C-20–C-25 diene appendage in hemibrevetoxin B necessitated a 3-step sequence in which the original C-5 of D-arabinose was sacrificed

Figure 11.17 Total synthesis of hemibrevetoxin B - assembly of subunits.

Key reactions: *Negishi coupling (**D+J** to **K**); epoxyester formation (**L** to **M**); cis-olefination (**M** to **N**); selenoetherification (**N** to **O**); olefination (**O** to **P**); selective O-tosylation/elimination (**U** to **target**)*

Relevant mechanism:

Figure 11.17 *(continued)*.

to give the aldehyde **S**. Olefination with the *trans-γ*-silylallylboronate reagent led to the intended diene **T**. Final deprotection of the MOM ether led to the diol **U**. Monotosylation was effected via the intermediacy of a dibutyltin acetal, which was followed by Swern oxidation and *β*-elimination. Removal of the MOM group afforded hemibrevetoxin B in 69% yield for the last sequence of steps.

Commentary: In Chapter 4 we discussed the eye-to-brain signalling of messages when looking at the structure of a target molecule in the context of total synthesis. The visual *relational* analysis of hemibrevetoxin B may conjure up a carbohydrate-based imagery, especially for rings *A* and *B*. Indeed, in four of the synthetic approaches to hemibrevetoxin B, the individual groups utilized carbohydrate precursors (Figure 11.15).

A number of conceptually and practically interesting aspects of the Holton synthesis are noteworthy. Central to the strategy was the desire to adopt a biomimetic cascade epoxyalcohol cyclization as a key step (Figure 11.17, **N** to **O**). To reach the required intermediates, Holton prepared subunits I and II (Figure 11.16) with predisposed functionalities. The choice of a *cis*-propenyl group in intermediate **N** was based on the expectation that the facial selectivity in the cascade electrophilic attack on a suitably activated olefin intermediate would be controlled by minimization of $A^{1,3}$-strain (Figure 11.17). In the event, activation of the olefin in **N** as the *epi*-phenylseleninium intermediate led to the attack of the C-7 hydroxyl group on the C-10–C-11 epoxide, which after formation of the C-10 alcohol, underwent face-selective ring opening to give **O**, now displaying rings *B* and *C*, as a single isomer in 83% yield (see relevant mechanism, Figure 11.17). The successful application of this tandem epoxy alcohol cyclization was attributed in large measure to the effect of hexafluoro-2-propanol as a highly polar solvent.

The manner in which the two oxepan subunits were built is also of interest. Whereas ring *C* was the result of a highly stereocontrolled cascade cycloetherification sequence (Figure 11.17, **N** to **O**), ring *D* was formed by a ring-closing metathesis of a diene intermediate (Figure 11.17, **Q** to **R**). However, in order to introduce the second *C*-vinyl group, considerable skeletal reorganization had to be imposed on the original D-arabinose subunit (Figure 11.17, **P** to **Q**). Thus, the terminal olefin resulting from the selenoxide elimination from **O** was transformed to the lactol **P** by removing the *O*-benzyl groups with Li/NH₃. Treatment with Me₃SiCH₂MgCl followed by base, generated acyclic vinyl triol **Q** in excellent yield. Following the ring-closing metathesis reaction, the diol unit was cleaved to the aldehyde, and the latter was extended to the aldehyde precursor **S** for the introduction of the diene (Figure 11.17, **Q** to **T**).

In tracing the fate of D-arabinose throughout the process, only the C-1–C-4 portion, and the C-3 hydroxyl group were retained, which now corresponded to the C-18–C-20 segment of hemibrevetoxin B. Thus, whereas the progeny of ring *A* can be traced back to D-glucose (apparent substructure), that of ring *D* can only be partially related to D-arabinose (hidden substructure) (Figure 11.16).

Finally, the Holton synthesis of hemibrevetoxin B shows the power of a combination of visual *relational* and *reflexive* analysis in the planning stage. Once the cascade epoxy alcohol-olefin activation idea was conceived (Figure 11.17, **N** to **O**), the potential of utilizing the vinyl appendage resulting from the oxidative elimination of the phenylselenoxide intermediate **O** created the need to generate the second metathesis partner as depicted in **Q**. In selecting carbohydrate precursors, Holton also astutely planned the systematic modification of D-arabinose and D-glucose to give subunits I and II (Figure 11.16), which would be coupled using a Negishi coupling reaction to assemble the bridged precursor to the desired intermediate **N**. The generation of a regio- and stereochemically well-defined epoxide **M** from **L**, is another demonstration of sound planning and efficient execution.

According to Holton and coworkers [45], this convergent synthesis was accomplished in 39 steps and an overall yield of 4% in the longest linear sequence from tri-*O*-acetyl D-glucal (**E**). Comparisons of overall yields with the Nicolaou [42], Yamamoto [43] and Nakata [48] syntheses, all of which required an additional five or more steps, shows considerable improvement in the Holton synthesis with regard to efficiency.

An alternative visually *reflexive* analysis of hemibrevetoxin B, may lead one to consider a cycloetherification-type approach, where enantiomerically pure or enriched epoxy alcohols or hydroxy olefins can engage in intramolecular reactions.

Hemibrevetoxin B

Figure 11.18 Sharpless asymmetric epoxidation (SAE) strategy for tetrahydropyran and oxepan ring formation.

Nakata and coworkers [48] adopted such a strategy starting with a *trans,trans*-2,6-octadiene diol and elaborating the six-membered oxacyclic ring *D* from a chiral, non-racemic epoxy alcohol prepared by a Sharpless asymmetric epoxidation reaction (Figure 11.18 A). The oxepan motif **F** was obtained by a rearrangement/ring-expansion protocol from **E**. Presumably, the solvolytic ring expansion involves an antiperiplanar alignment of the mesylate, although a kinetically controlled solvolysis may also be operative (Figure 11.18 B).

Other approaches to different subunits of hemibrevetoxin B have relied on the exploitation of a desymmetrizing centrosymmetric intermediate [49], or a hetero Diels-Alder methodology [50]. An overview of related approaches is highlighted in Nakata's excellent review [41a].

11.10
Carbohydrates in Synthesis and in Biology

As a class of abundant natural compounds, carbohydrates were quickly recognized as a potential source of chiral, non-racemic starting materials in natural product synthesis [1]. They drew particular attention, over three decades ago, when the synthesis of relatively complex natural products such as prostaglandins [51], ionophores [52], and macrolides [53] among others were reported. Interest in carbohydrate-based chirons has remained strong over the years, although the emergence of powerful methodologies, especially using catalytic reactions to prepare chiral, non-racemic alcohols, polyols, and related polyfunctional compounds, has expanded the choice of strategies in total synthesis (see Chapter 4). As discussed in Chapters 6 and 8, the use of carbohydrates as starting materials in synthesis of non-carbohydrate molecules should be judiciously planned so as to take full advantage of what they offer in terms of rich stereochemistry, and a choice of chain lengths, spanning three to seven carbon atoms.

The practice of carbohydrate-based synthesis has also instigated the development of highly versatile methodologies of wider applicability in organic synthesis. A noteworthy example is the Ireland ester enolate version of the Claisen rearrangement [54], which was audaciously applied to the total synthesis of lasalocid A [52] in 1983, and explored for related polyether ionophores [55] (Figure 11.19). Carbohydrates have also been an excellent testing ground for the development of new protecting groups [56], as well as for S_N2-type reactions with novel leaving groups. The utility of carbohydrates as chiral auxiliaries and unique ligands in catalysis is also well-documented [57].

The exciting developments in the field of glycobiology [58], offering new vistas for carbohydrates in molecular recognition for important physiological processes, is an area of constant development. Carbohydrates, and their derivatives, are being studied for their potential as medicines for diabetes, thrombosis, Gaucher's disease among others, as well as synthetic vaccines [59]. Large volume consumables such as cellulose, starch, and related polysaccharides continue to be mainstays of important high-volume industrial activities [60].

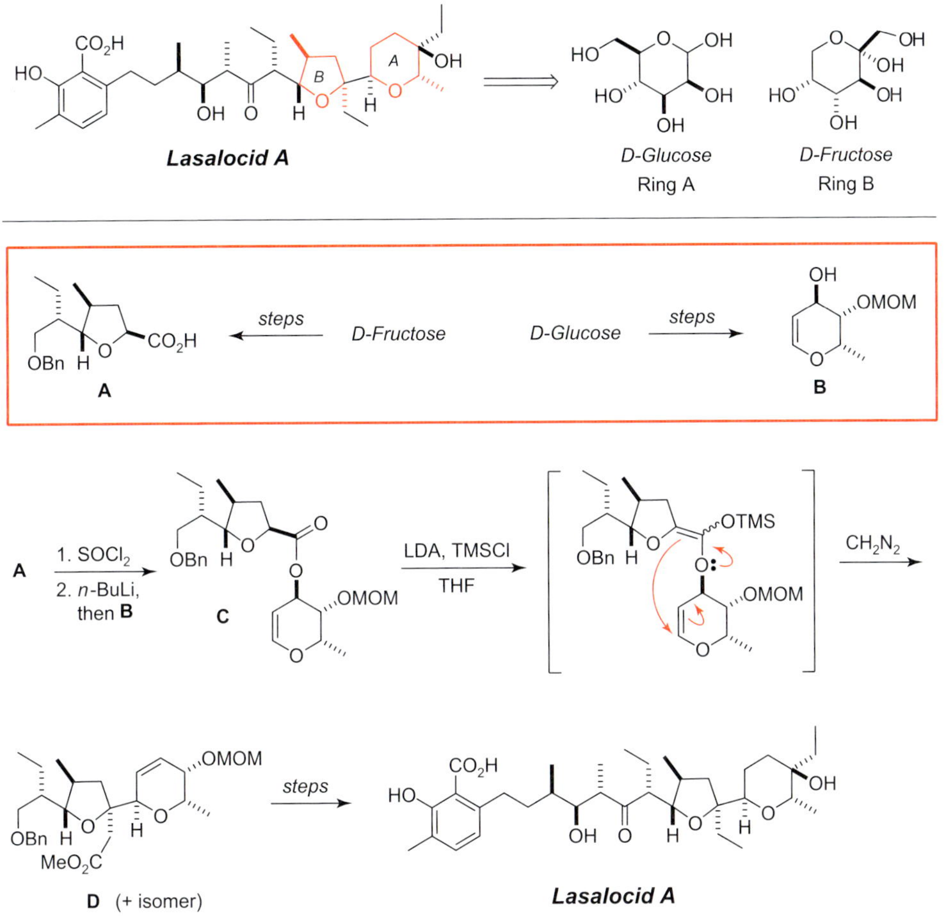

Figure 11.19 Total synthesis of lasalocid A from D-fructose and D-glucose. Implementation of the Ireland-Claisen rearrangement (apparent carbohydrate substructure).

11.11
Miscellanea

There are numerous molecules with diverse structures that have been synthesized starting from carbohydrates. A number of these were described in Chapter 8 in the context of apparent, partially hidden, and hidden substructures corresponding to carbohydrate motifs. In this chapter, we highlighted selected syntheses of Category I and II-type molecules in which the starting carbohydrate molecular framework was conveniently utilized in the elaboration of the target molecules.

Additional examples of hidden carbohydrate substructures can be identified in a large diversity of compounds as shown in Figure 11.20. Lycoricidine [61], FR65814 [62], and tetracycline [63] were prepared from functionalized carbocycles, which were readily available from the corresponding D-glucose and D-glucosamine chirons by application of the Ferrier aldose-to-carbocycle rearrangement [10]. The β-lactam intermediate used in the synthesis of tetracycline was prepared from D-glucosamine [64], albeit in a multi-step process that does not equal the efficiency of commercial processes. The complex alkaloid tetrodotoxin has been synthesized starting with either D-*iso*-ascorbic acid [65], or D-glucose [66]. The eight-membered cyclic ethers laurallene [67] and prelaureantin [68] were synthesized from D-ribonolactone and D-galactose respectively. Additionally, a recent synthesis of Tamiflu started with D-xylose [69]. (see also Chapter 14, section 14.9). Macrolides such as pikronolide [53a] and erythronolide B [53b] have been elaborated using D-glucose as a precursor. Polyol segments of the macrocyclic lactones migrastatin [70], *ent*-palmerolide A [71], and myxovirescin [72] were prepared from carbohydrate precursors. The macrolide lankacidin C was elaborated from D-arabinose and L-aspartic acid [73]. The total synthesis of pinnatoxin A was accomplished using D-ribose for rings *F* and *E* [74]. The enantiomeric forms of echinosporin can be prepared from D-galactose or L-ascorbic acid [75]. The cedranoid sesquiterpene α-pipitzol was synthesized starting with D-glucose [76]. The first total synthesis of polygalolides A and B was accomplished starting with D-arabinose [77]. The above are some representative examples of hidden and partially hidden carbohydrate-based substructures requiring considerable prior manipulation to provide partial skeletal and stereochemical convergence with the intended structures [78].

We close this chapter with some examples of Category II and III-type complex polyoxygenated natural products whose total syntheses have been achieved, in part, due to the opportune utilization of carbohydrate-based chirons for some of their subunits (Figure 11.21). Various monosaccharides were useful chirons in the

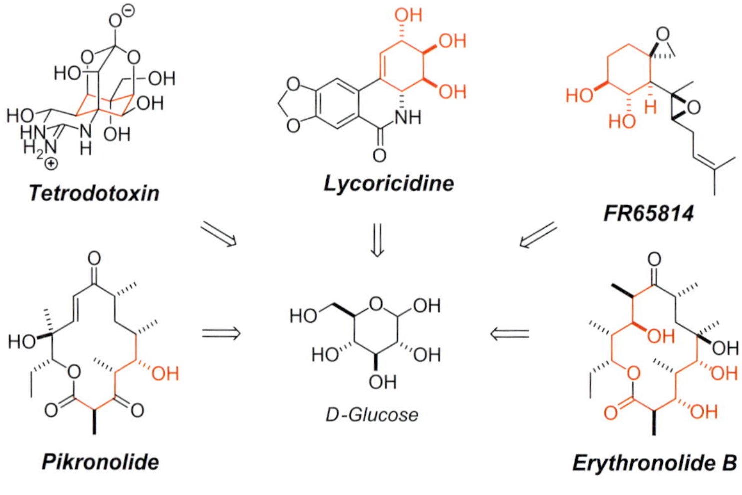

Figure 11.20 Miscellaneous examples of total syntheses using carbohydrate precursors (hidden and partially hidden substructures).

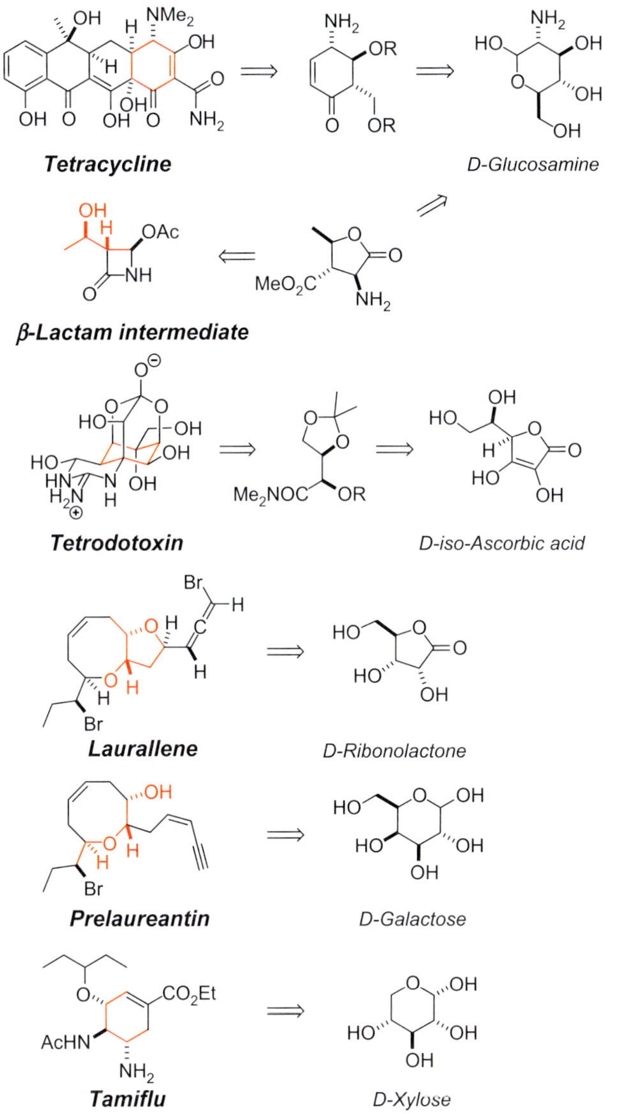

Tetracycline

D-Glucosamine

β-Lactam intermediate

Tetrodotoxin

D-iso-Ascorbic acid

Laurallene

D-Ribonolactone

Prelaureantin

D-Galactose

Tamiflu

D-Xylose

Figure 11.20 *(continued)*.

total syntheses of amphotericin B [79], swinholide A [80], brevetoxin B [81], and everninomicin [82], all from the Nicolaou laboratory. One or more oxacycles in the complex polyether marine natural products gambierol [83], ciguatoxin [84], and gymnocin A [85] have been synthesized form carbohydrate precursors. The total syntheses and approaches to other members of this fascinating family of polycyclic fused natural products have been reviewed [41].

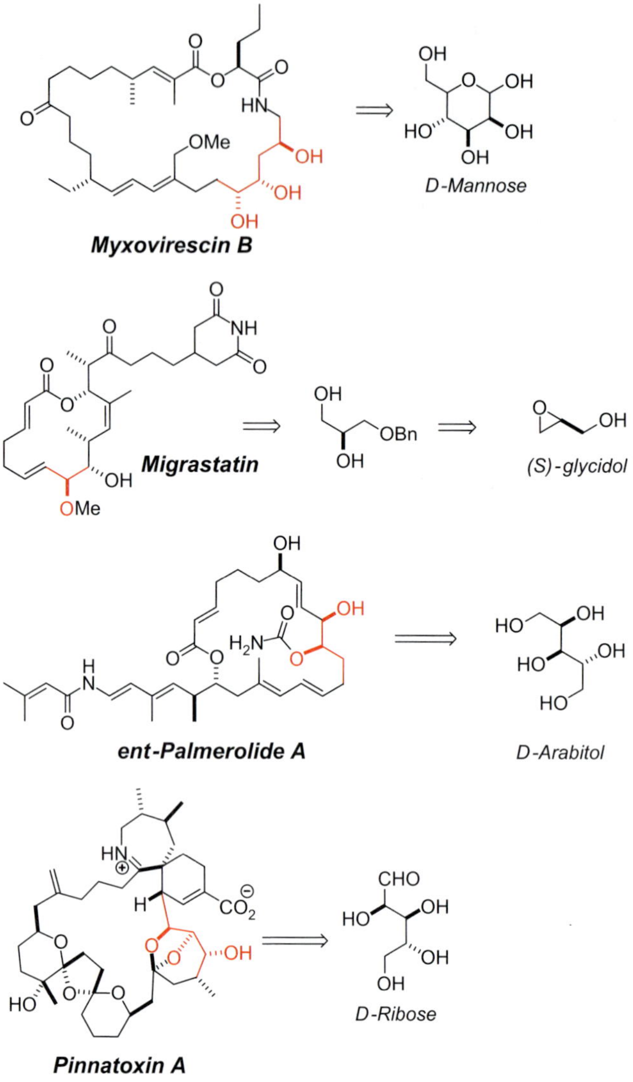

Myxovirescin B

D-Mannose

Migrastatin

(S)-glycidol

ent-Palmerolide A

D-Arabitol

Pinnatoxin A

D-Ribose

Figure 11.20 *(continued)*.

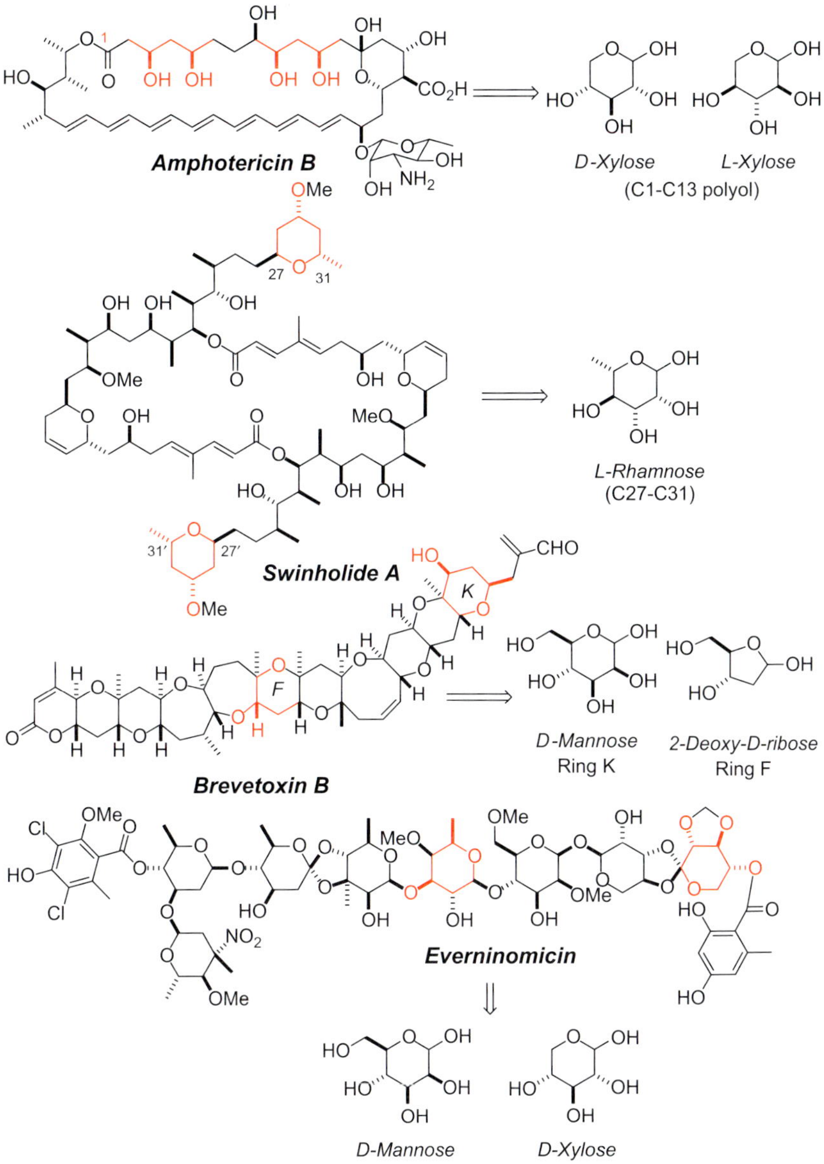

Figure 11.21 A selection of complex polyoxygenated natural products and their carbohydrate precursors.

References

1. For relevant books, see: (a) Hanessian, S. *Total Synthesis of Natural Products: The Chiron Approach*, 1983, Pergamon, Oxford; (b) Boons, G.-J.; Hale, K. J. *Organic Synthesis with Carbohydrates*, 2000, Sheffield Academic Press, Sheffield, U.K.; (c) Bols, M. *Carbohydrate Building Blocks*, 1996, Wiley-Inserscience, NY; (d) El Ashry, E. S.H.; El Nemr, A. *Nitrogen Heterocycles from Carbohydrates*, 2005, Blackwell Publishing, Oxford, UK; see also Chapter 7, references 6 and 7; (e) Levy, D. E.; Fügedi, P. *The Organic Chemistry of Sugars*, Eds.; CRC Press, 2006.

2. For reviews, see: (a) Ban, Y.; Murakami, Y.; Iwasawa, Y.; Tsuchiya, M.; Takano, N. *Med. Res. Rev.* 1988, **8**, 231; (b) Stöckigt, J. in *Indole and Biogenetically Related Alkaloids*, Phillipson, I. D.; Zenk, M. H.; Eds. 1980, Academic Press, NY.; for selected total syntheses, see: (±)-ajmalicine, and related alkaoids: (c) van Tamelen, E. E.; Placeway, C. J. *Am. Chem. Soc.* 1961, **83**, 2594; (d) Gutzwiller, J.; Pizzolato, G.; Uskoković, M. R. *J. Am. Chem. Soc.* 1971, **93**, 5907; (e) Martin, S. F.; Benage, B.; Hunter, J. E. *J. Am. Chem. Soc.* 1988, **110**, 5925; (f) Wenkert, E.; Chang, C.-J.; Chawla, H. P. S. Cochran, D. W.; Hagaman, E. W.; King, J. C.; Orito, K. *J. Am. Chem. Soc.* 1976, **98**, 3645. For enantiopure ajmalicine, see: (g) Uskoković, M. R.; Lewis, R. L.; Partridge, J. J.; Despreaux, C. W.; Preuss, D. L. *J. Am. Chem. Soc.* 1979, **101**, 6742; (h) Massiot, G.; Mulamba, T. *J. Chem. Soc., Chem. Comm.* 1984, 715; (i) Hatakeyama, S.; Saijo, K.; Takano, S. *Tetrahedron Lett.* 1985, **26**, 865; (j) Takano, S.; Satoh, S.; Ogasawara, K. *J. Chem. Soc., Chem. Comm.* 1988, 59; see also: Martin, S. F. in *Strategies and Tactics in Organic Synthesis*, Lindberg, T. Ed.; 1989, vol. 2, p. 291, Academic Press, San Diego, CA. and references cited therein. For an enantioselective and efficient synthesis of ajmalicine, see: (k) Lögers, M.; Overman, L. E.; Welmaker, G. S. *J. Am. Chem. Soc.* 1995, **117**, 9139.

3. Hanessian, S.; Faucher, A.-M. *J. Org. Chem.* 1991, **56**, 2947.

4. Ferrier, R. J.; Prasad, N.; Sankey, G. H. *J. Chem. Soc. C* 1969, 587.

5. (a) Kabalka, G. W.; Yang, D. T. C.; Baker, J. D., Jr. *J. Org. Chem.* 1976, **41**, 574; (b) Hutchins, R. O.; Milewski, C. A.; Maryanoff, B. E. *J. Am. Chem. Soc.* 1973, **95**, 3662.

6. For a review, see: Rozwadowska, M. D. *Heterocycles* 1994, **39**, 903.

7. (a) Mander, L. N.; Sethi, S. P. *Tetrahedron Lett.* 1983, **24**, 5725; (b) Mander, L. N.; Sethi, S. P. *Aldrichim. Acta* 1987, **20**, 53.

8. Imuta, S.; Tanimoto, H.; Momose, M. K.; Chida, N. *Tetrahedron* 2006, **62**, 6926.

9. For other syntheses from D-glucose, see: (a) Ward, D. E.; Gai, Y.; Kaller, B. F. *J. Org. Chem.* 1996, **61**, 5498; (b) Rahman, M. A.; Fraser-Reid, B. *J. Am. Chem. Soc.* 1985, **107**, 5576; for a review, see: (c) Fraser-Reid, B.; López, J. P. in *Recent Advances in the Chemical Synthesis of Antibiotics*, Lukacs, G.; Ohno, M. Eds.; 1990, p. 285, Springer-Verlag, Heidelberg; see also Chapter 10 reference 2.

10. (a) Ferrier, R .J.; Middleton, S. *Chem. Rev.* 1993, **93**, 2779; (b) Ferrier, R. J. *Top Curr. Chem.* 2001, **215**, 277; (c) Ferrier, R. J. in *Preparative Carbohydrate Chemistry*, 1997, p. 569, Hanessian, S. Ed.; Dekker, N.Y.

11. Vis, E.; Karrer, P. *Helv. Chim. Acta* 1954, **37**, 378.

12. For a review, see: Berecibar, A.; Grandjean, C.; Siriwardena, A. *Chem. Rev.* 1999, **99**, 779; for total syntheses, see: (a) Ledford, B. E.; Carreira, E. M. *J. Am. Chem. Soc.* 1995, **117**, 11811; (b) Li, J.; Lang, F.; Ganem, B. *J. Org. Chem.* 1998, **63**, 3403; (c) Ogawa, S.; Uchida, C.; Yuming, Y. *J. Chem. Soc., Chem. Commun.* 1992, **886**; see also: (d) Crimmins, M. T.; Tabet, E. A. *J. Org. Chem.* 2001, **66**, 4012; (e) Feng, X.; Duesler, E. N.; Mariano, P. S. *J. Org. Chem.* 2005, **70**, 5618; (f) Ogawa, S.; Uchida, C. *Chem. Lett.* 1993, 173; (g) Uchida, C.; Yamagishi, T.; Ogawa, S. *J.*

Chem. Soc., Perkin Trans 1, 1994, 58; for the total synthesis of (-)-allosamizoline, see: Donohue, T. J.; Rosa, C. P. *Org. Lett.* 2007, **9**, 5509, and references cited therein.

13. For syntheses starting with carbohydrates, see: (a) Kobayashi, Y.; Miyazaki, H.; Shiozaki, M. *J. Org. Chem.* 1994, **59**, 813;   (b) Knapp, S.; Purandare, A.; Rupitz, K.; Withers, S. G. *J. Am. Chem. Soc.* 1994, **116**, 7461;   (c) Bobo, S.; Storch de Garcia, I.; Chiara, J. L. *Synlett* 1999, 1551;   (d) Storch de Garcia, I.; Dietrich, H.; Bobo, S.; Chiara, J. L. *J. Org. Chem.* 1998, **63**, 5883; (e) Storch de Garcia, I.; Bobo, S.; Martin-Ortega, M. D.; Chiara, J. L. *Org. Lett.* 1999, **1**, 1705;   (f) Seepersaud, M.; Al-Abed, Y. *Tetrahedron Lett.* 2001, **42**, 1471.

14. Boiron, A.; Zillig, P.; Faber, D.; Giese, B. *J. Org. Chem.* 1998, **63**, 5877.

15. Qiao, L.; Vederas, J. C. *J. Org. Chem.* 1993, **58**, 3480.

16. Marco-Contelles, J.; Gallego, P. Rodríguez-Fernández, M.; Khiar, N.; Destabel, C.; Bernabé, M.; Martinéz-Grau, A.; Chiara, J. L. *J. Org. Chem.* 1997, **62**, 7397.

17. (a) Norman, R. O. C.; Purchase, R.; Thomas, C. B. *J. Chem. Soc., Perkin Trans 1* 1972, 1701;   (b) Weiss, R. H.; Furfine, E.; Hausleden, E.; Dixon, D. W. *J. Org. Chem.* 1984, **49**, 4969.

18. Paquette, L. A.; Peng, X.; Yang, J. *Angew. Chem. Int. Ed.* 2007, **46**, 7817.

19. (a) Paquette, L. A.; Kang, H. J. *Tetrahedron* 2004, **60**, 1353;   (b) Paquette, L. A. Cunière, N. *Org. Lett.* 2002, **4**, 1927; (c) Paquette, L. A. *J. Organomet. Chem.* 2006, **691**, 2083.

20. For a review, see: Hanzawa, Y.; Ito, H.; Taguchi, T. *Synlett* 1995, 299.

21. Semmelhack, M. F.; Tomoda, S.; Nagaoka, H.; Boettger, S. D.; Hurst, K. M. *J. Am. Chem. Soc.* 1982, **104**, 747.

22. For selected reviews, see: (a) Coleman, R. S. *Curr. Opin. Drug Discovery Dev.* 2001, **4**, 435;   (b) Danishefsky, S.; Schekeryantz, J. M. *Synlett* 1995, 475;   (c) Fukuyama, T.; Yang, L. in *Studies in Natural Products Chemistry,* Atta-ur-Rahman, Ed.; 1993, p. 433, Elsevier, NY;   (d) Kishi, Y. *J. Nat. Prod.* 1979, **42**, 549.

23. Coleman, R. S.; Felpin, F.-X.; Chen, W. *J. Org. Chem.* 2004, **69**, 7309.

24. (a) Francisco, C. G.; Freire, R.; González, C. C.; Suárez, E. *Tetrahedron: Asymmetry* 1997, **8**, 1971;   (b) de Armas, P.; Francisco, C. G.; Suárez, E. *J. Am. Chem. Soc.* 1993, **115**, 8865.

25. Danishefsky, S.J.; Berman, E. M.; Ciufolini, M.; Etheredge, S. J.; Segmuller, B. E. *J. Am. Chem. Soc.* 1985, **107**, 3891.

26. Ziegler, F. E.; Berlin, M. Y. *Tetrahedron Lett.* 1998, **39**, 2455;   see also: Ziegler, F. E.; Belema, M. *J. Org. Chem.* 1994, **59**, 7962 and refernces cited therein.

27. Barton, D. H. R.; Samadi, M. *Tetrahedron* 1992, **48**, 7083.

28. Tanino, H.; Nakata, T.; Kaneko, T.; Kishi, Y. *J. Am. Chem. Soc.* 1977, **99**, 2818;   for a review, see: Shimizu, Y. in *Progress in the Chemistry of Organic Natural Products*, Herz, W.; Grisebach, H.; Kirby, G. W. Eds.; 1984, p. 235, Springer-Verlag, NY.

29. Jacobi, P. A.; Martinelli, M. J.; Polanc, S. *J. Am. Chem. Soc.* 1984, **106**, 5594.

30. Fleming, J. J.; Du Bois, J. *J. Am. Chem. Soc.* 2006, **128**, 3926;   see also: Fleming, J. J.; Fiori, K. W.; Du Bois, J. *J. Am. Chem. Soc.* 2003, **125**, 2028;   for a review, see: Davies, H. M. L.; Long, M. S. *Angew. Chem. Int. Ed.* 2005, **44**, 3518.

31. Fleming, J. J.; McReynolds, M. D.; Du Bois, J. *J. Am. Chem. Soc.* 2007, **129**, 9964.

32. (a) Merino, P.; Franco, S.; Merchan, F. L.; Tejero, T. *Synlett* 2000, 442;   (b) Merino, P.; Lanaspa, A.; Merchan, F. L.; Tejero, T. *Tetrahedron: Asymmetry* 1998, **9**, 629;   see also: Hanessian, S.; Bayrakdarian, M.; Luo, X. *J. Am. Chem. Soc.* 2002, **124**, 4716.

33. (a) Hong, C. Y.; Kishi, Y. *J. Am. Chem. Soc.* 1992, **114**, 7001; for a synthesis of decarbamoyl and *ent*-decarbamoyl saxitoxin as well as saxitoxin, see: (b) Iwamoto, O.; Shinohara, R.; Nagasawa, K., *Chem. Asian J.* 2009, **4**, 277;   see also: Sawayama, Y.; Nishikawa, T. *Angew. Chem. Int. Ed.* 2011, **50**, 7176.

34. For reviews, see: (a) Armstrong, A.; Blench, T. J. *Tetrahedron* 2002, **58**, 9321; (b) Jotterland, N.; Vogel, P. *Curr. Org. Chem.* 2001, **5**, 637; (c) Koert, U. *Angew. Chem. Int. Ed.* 1995, **34**, 773; (d) Bergstrom, J. D.; Dufresne, C.; Bills, G. F.; Nallin-Omstead, M.; Byrne, K. *Ann. Rev. Microbiol.* 1995, **49**, 607. For total syntheses, see: (e) Nicolaou, K. C.; Nadin, A.; Leresche, J. E.; Yue, E. W.; La Greca, S. *Angew. Chem. Int. Ed.* 1994, **33**, 2190; (f) Nicolaou, K. C.; Yue, E. W.; La Greca, S.; Nadin, A.; Yang, Z.; Leresche, J. E.; Tsuri, T.; Naniwa, Y.; de Riccardis, F. *Chem. Eur. J.* 1995, **1**, 467; (g) Tomooka, K.; Kikuchi, M.; Igawa, K.; Suzuki, M.; Keong, P.-H.; Nakai, T. *Angew. Chem. Int. Ed.* 2000, **39**, 4502; (h) Nicewicz, D. A.; Statterfield, A. D.; Schmitt, D. C.; Johnson, J. S. *J. Am. Chem. Soc.* 2008, **130**, 17281.

35. (a) Caron, S.; Stoermer, D.; Mapp, A. K.; Heathcock, C. H. *J. Org. Chem.* 1996, **61**, 9126; (b) Stoermer, D.; Caron, S.; Heathcock, C. H. *J. Org. Chem.* 1996, **61**, 9115.

36. Tamao, K.; Ishida, N. *Tetrahedron Lett.* 1984, **25**, 4245.

37. (a) Carreira, E. M.; Du Bois, J. *J. Am. Chem. Soc.* 1994, **116**, 10825; (b) Carreira, E. M.; Du Bois, J. *J. Am. Chem. Soc.* 1995, **117**, 8106.

38. Evans, D. A.; Barrow, J. C.; Leighton, J. L.; Robichaud, A. J.; Sefkow, M. *J. Am. Chem. Soc.* 1994, **116**, 12111.

39. Sato, H.; Nakamura, S.; Watanabe, N.; Hashimoto, S. *Synlett* 1997, 451.

40. (a) Armstrong, A.; Jones, L. H.; Barsanti, P. A. *Tetrahedron Lett.* 1998, **39**, 3337; (b) Armstrong, A.; Barsanti, P. A.; Jones, L. H.; Ahmed, G. *J. Org. Chem.* 2000, **65**, 7020.

41. For recent reviews, see: (a) Morris, J. C.; Phillips, A. J. *Nat. Prod. Rep.* 2010, **27**, 1180; (b) Nakata, T. *Chem. Rev.* 2005, **105**, 4314; (c) Inoue, M. *Chem. Rev.* 2005, **105**, 4379; (d) Sasaki, M.; Fuwa, H. *Synlett* 2004, 1851; (e) Nicolaou, K. C. *Angew. Chem. Int. Ed.* 1996, **35**, 589; (f) Alvarez, E.; Candenas, M.-L.; Pérez, R.; Ravelo, J. L.; Marin, J. D. *Chem. Rev.* 1995, **95**, 1953; see also: *Tetrahedron* 2002, **58**, 1779, thematic issue, Hirama, M.; Rainier, J. D. Guest editors.

42. Nicolaou, K. C.; Reddy, K. R.; Skokotas, G.; Sato, F.; Xiao, X.-Y.; Huang, C.-K. *J. Am. Chem. Soc.* 1993, **115**, 3558.

43. (a) Kadota, I.; Yamamoto, Y. *J. Org. Chem.* 1998, **63**, 6597; (b) Kadota, I.; Takamura, H.; Nishii, H.; Yamamoto, Y. *J. Am. Chem. Soc.* 2005, **127**, 9246.

44. Mori, Y.; Yaegashi, K.; Furukawa, H. *J. Org. Chem.* 1998, **63**, 6200.

45. Zakarian, A.; Batch, A.; Holton, R. A. *J. Am. Chem. Soc.* 2003, **125**, 7822.

46. Ireland, R. E.; Courtney, L.; Fitzsimmons, B. J. *J. Org. Chem.* 1983, **48**, 5186.

47. (a) Negishi, E.; Okukado, N.; King, A. O.; Van Horn, D. E.; Spiegel, B. I. *J. Am. Chem. Soc.* 1978, **100**, 2554; (b) Hayashi, T.; Konishi, M.; Kobori, Y.; Kumada, M.; Higuchi, T.; Hirotsu, K. *J. Am. Chem. Soc.* 1984, **106**, 158;

48. (a) Nakata, T.; Nomura, S.; Matsukwa, H.; Morimoto, M. *Tetrahedron Lett.* 1996, **37**, 217; (b) Morimoto, M.; Matsukura, H.; Nakata, T. *Tetrahedron Lett.* 1996, **37**, 6365.

49. Holland, J. M.; Lewis, M.; Nelson, A. *Angew. Chem. Int. Ed.* 2001, **40**, 4082.

50. Rainier, J. D.; Allwein, S. P.; Cox, J. M. *J. Org. Chem.* 2001, **66**, 1380.

51. (a) Stork, G.; Takahashi, T. *J. Am. Chem. Soc.* 1977, **99**, 1275; (b) Stork, G.; Takahashi, T.; Kawamoto, I.; Suzuki, T. *J. Am. Chem. Soc.* 1978, **100**, 8272.

52. Ireland, R. E.; Anderson, R. C.; Badoud, R.; Fitzsimmons, B. J.; McGarvey, G. J.; Thaisrivongs, S.; Wilcox, C. S. *J. Am. Chem. Soc.* 1983, **105**, 1988.

53. (a) Nakajima, N.; Hamada, T.; Tamaka, T.; Oikawa, Y.; Yonemitsu, O. *J. Am. Chem. Soc.* 1986, **108**, 4645; (b) Sviridov, A. F.; Ermolenko, M. S.; Yashunsky, D. V.; Borodkin, V. S.; Kochetkov, N. K. *Tetrahedron Lett.* 1987, **28**, 3835, 3839; (c) Hanessian, S.; Rancourt, G.; Guindon, Y. *Can. J. Chem.* 1978, **56**, 1843.

54. Ireland, R. E.; Mueller, R. H.; Willard, A. K. *J. Am. Chem. Soc.* 1976, **98**, 2868; for a review see Werschkun, B.; Thiem, J. in *Topics in Current Chemistry*, 2001, vol. 215, p. 293, Springer, Heidelberg.

55. (a) Ireland, R. E.; Meissner, R. S.; Rizzacasa, M. A. *J. Am. Chem. Soc.* 1993, **115**, 7166; for reviews, see: (b) Faul, M. M.; Huff, B. E. *Chem. Rev.* 2000, **100**, 2407; (c) Chai, Y.; Hong, S.-P.; Lindsay, H. A.; McFarland, C.; McIntosh, M. C. *Tetrahedron* 2002, **58**, 2905; see also: *The Claisen Rearrangement*, Hiersemann, M.; Nubbemeyer, U. Eds.; 2007, Wiley-VCH, Weinheim.

56. For example, see; (a) Vatèle, J.-M.; Hanessian, S. in *Preparative Carbohydrate Chemistry*, Hanessian, S. Ed.; 1997, p. 127, Dekker, NY; see also: (b) Hanessian, S.; Kagotani, M. Komaglou, K. *Heterocycles* 1989, **28**, 1115; (c) Hanessian, S.; Vatèle, J.-M. *Tetrahedron Lett.* 1981, **22**, 3579.

57. For example, see: Hultin, P.G.; Earle, M. A.; Sudharshan, M. *Tetrahedron* 1997, **53**, 14823.

58. For example, see: (a) *Chemical Glycobiology*, Chen, X.; Halcomb, R.; Wang, P. G., Eds.; 2008, Oxford University Press, NY; (b) *Glycobiology*, Sanson, C.; Markman, O. Eds.; 2007, Scion Publishing Ltd, Bloxham, U.K.; see also: (c) Maryanoff, B. E. *J. Med. Chem.* 2009, **52**, 3432; (d) Dube, D. H.; Bertozzi, C. R. *Nat. Rev. Drug Discov.*, 2005, **4**, 477; (e) Shriver, Z.; Raguram, S.; Sasisekharan, R. *Nat. Rev. Drug Discov.* 2004, **3**, 863; (f) Davis, A. P.; Wareham, R. S. *Angew. Chem. Int. Ed.* 1999, **38**, 2979; (g) Dwek, R. A. *Chem. Rev.* 1996, **96**, 683; for carbohydrate-based vaccines, see: (h) Roy, R. *Drug Discovery Today: Technologies* 2004, **1**, 327; (i) Verez-Bencous, V.; Roy, R. *et. al. Science* 2004, **305**, 522.

59. For example, see: (a) *Iminosugars – From synthesis to therapeutic applications*, Compain, P.; Martin, O. R. Eds.; 2007, J. Wiley & Sons, NY; (b) *Carbohydrate-based Drug Discovery*, vol. 1, 2, Wong, C.-H. Ed.; 2003, Wiley-VCH, Weinheim; (c) Meutermans, W.; Le, G. T.; Becker, B. *Chem. Med. Chem.* 2006, **1**, 1164; (d) Schweizer, F. *Angew. Chem. Int. Ed.* 2002, **41**, 230; (e) Gruner, S. A. W.; Locardi, E.; Lohof, E.; Kessler, H. *Chem. Rev.* 2002, **102**, 491.

60. For example, see: *Renewable Bioresources*, Stevens, C.; Verhe, R., 2004, Wiley, NY; *Cellulose Derivatives: Modification, Characterization, and Nanostructures*, Heinze, T.; Glasser, W. Eds.; 1998, ACS Symposium Series, No. 688, Washington, DC; see also: Réczey, K.; Lázló, E.; Holló, J. *Starch* 2006, **38**, 306.

61. Chida, N.; Ohtsuka, M.; Ogawa, S. *J. Org. Chem.* 1993, **58**, 4441.

62. Amano, S.; Ogawa, N.; Ohtsuka, M.; Ogawa, S.; Chida, N. *Chem. Commun.* 1998, 1263.

63. Tatsuta, K.; Yoshimoto, T.; Gunji, H.; Okado, Y.; Takahashi, M. *Chem. Lett.* 2000, 646.

64. Tatsuta, K.; Takahashi, M.; Tanaka, N.; Chikauchi, K. *J. Antibiot.* 2000, **53**, 1231.

65. Hinman, A.; Du Bois, J. *J. Am. Chem. Soc.* 2003, **125**, 11510.; for a review, see: Kang, S. H.; Kang, S. Y.; Lee, H.-S.; Buglass, A. J. *Chem. Rev.* 2005, **105**, 4537.

66. Sato, K.-i.; Akai, S.; Shoji, H.; Sugita, N.; Yoshida, S.; Nagai, Y.; Suzuki, K.; Nakamura, Y.; Kajihara, Y.; Funabashi, M.; Yoshimura, J. *J. Org. Chem.* 2008, **73**, 1234.

67. Saitoh, T.; Suzuki, T.; Sugimoto, M.; Hagiwara, H.; Hoshi, T. *Tetrahedron Lett.* 2003, **44**, 3175.

68. Fujiwara, K.; Souma, S.-i.; Mishima, H.; Murai, A. *Synlett* 2002, 1493.

69. Shie, J.-J.; Fang, J.-M.; Wang, S.-Y.; Tsai, K.-C.; Cheng, Y.-S.E.; Yang, A.-S.; Hsiao, S.-C.; Su, C.-Y.; Wong, C.-H. *J. Am. Chem. Soc.* 2007, **129**, 11892; for a recent review, see: Magano, J. J. *Chem. Rev.* 2009, **109**, 4396.

70. Gaul, G.; Najardarson, J. T.; Shan, D.; Dorn, D. C.; Wu, K.-D.; Tong, W. P.; Huang, X.-Y.; Moore, M. A. S.; Danishefsky, S. J. *J. Am. Chem. Soc.* 2004, **126**, 11326.

71. (a) Jiang, X.; Liu, B.; Lebreton, S.; DeBrabander, J. K. *J. Am. Chem. Soc.* 2007, **129**, 6386; (b) Nicolaou, K. C.; Guduru, R.; Sun, Y.-P.; Banerji, B.; Chen, D. Y.-K. *Angew. Chem. Int. Ed.* 2007, **46**, 5896; (c) Nicolaou, K. C.; Sun, Y.-P.; Gudura, R.; Banerji, B.; Chen, D. Y.-K. *J. Am. Chem. Soc.* 2008, **130**, 3633; see also Penner, M.;

Rauniyar, V.; Kaspar, L. T.; Hall, D. G. *J. Am. Chem. Soc.* 2009, **131**, 14216.

72. (a) Williams, D. R.; McGill, J. M. *J. Org. Chem.* 1999, **55**, 3457; (b) Seebach, D.; Maestro, M. A.; Sefkow, M.; Adams, G.; Hintermann, S.; Neidlein, A. *Liebigs Ann. Chem.* 1994, **701**.

73. Kende, A. S.; Liu, K.; Kaldor, I.; Dorey, G.; Koch, K. *J. Am. Chem. Soc.* 1995, **117**, 8258.

74. (a) Stivala, C. E.; Zakarian, A. *J. Am. Chem. Soc.* 2008, **130**, 3774; (b) McCauley, J. A.; Nagasawa, K.; Lander, P. A.; Mischke, S. G.; Semones, M. A.; Kishi, Y. *J. Am. Chem. Soc.* 1998, **120**, 7647; (c) Nakamura, S.; Kikuchi, F.; Hashimoto, S. *Angew. Chem. Int. Ed.* 2008, **47**, 7091; for a review, see: Clive, D. L. J.; Yu, M.; Wang, J.; Yeh, V. S. C.; Kang, S. *Chem. Rev.* 2005, **105**, 4483.

75. Smith, A. B. III; Sulikowski, G.A.; Sulikowski, M. M.; Fujimoto, K. *J. Am. Chem. Soc.* 1992, **114**, 2567.

76. Pak, H.; Canalda, I. I.; Fraser-Reid, B. *J. Org. Chem.* 1990, **55**, 3009.

77. Nakamura, S.; Sugano, Y.; Kikuchi, F.; Hashimoto, S. *Angew. Chem. Int. Ed.* 2006, **45**, 6532.

78. For a review, see: Nicolaou, K. C.; Mitchell, H. J. *Angew. Chem. Int. Ed.* 2001, **40**, 1577.

79. (a) Nicolaou, K. C.; Daines, R. A.; Uenishi, D. J.; Li, W. S.; Papahatjis, D.; Chakraborty, T. K. *J. Am. Chem. Soc.* 1987, **109**, 2205; (b) Nicolaou, K. C.; Daines, R. A.; Chakraborty, T. K.; Ogawa, Y. *J. Am. Chem. Soc.* 1987, **109**, 2821; (c) for a review, see: Ceveghetti, D. M.; Carreira, E. M. *Synthesis* 2006, 914.

80. (a) Nicolaou, K. C.; Ajito, K.; Patron, A. P.; Khatuya, H.; Richter, P. K.; Bertinato, P. *J. Am. Chem. Soc.* 1996, **118**, 3059; (b) Paterson, I.; Yeung, K.-S.; Ward, R. A.; Cumming, J. G.; Smith, J. D. *J. Am. Chem. Soc.* 1994, **116**, 9391.

81. (a) Nicolaou, K. C.; Rutjes, F. P. J. T.; Theodorakis, E. A.; Tiebes, J.; Sato, Untersteller, E. *J. Am. Chem. Soc.* 1995, **117**, 1173. For reviews on polycyclic ethers, see: (b) Inoue, M. *Chem. Rev.* 2005, **105**, 4379; (c) Sasaki, M.; Fuwa, H. *Synlett*, 2004, 1851, see also: reference 41.

82. (a) Nicolaou, K. C.; Mitchell, H. J.; Suzuki, H.; Rodriguez, R. M.; Baudoin, O.; Fylaktakidou, K. C. *Angew. Chem. Int. Ed.* 1999, **38**, 3334; (b) Nicolaou, K. C.; Rodriguez, R. M.; Fylaktakidou, K. C.; Suzuki, H.; Mitchell, H. J. *Angew. Chem. Int. Ed.* 1999, **38**, 3340; (c) Nicolaou, K. C.; Mitchell, H. J.; Rodriguez, R. M.; Fylaktikidou, K. C.; Suzuki, H. *Angew. Chem. Int. Ed.* 1998, **38**, 3345.

83. (a) Fuwa, H.; Kainuma, N.; Tachibana, K.; Sasaki, M. *J. Am. Chem. Soc.* 2002, **124**, 14983; (b) Kadota, I. J.; Takamura, K.; Sato, K.; Ohno, A.; Matsuda, K.; Yamamoto, Y. *J. Am. Chem. Soc.* 2003, **125**, 46; (c) Johnson, H. W. B.; Majumder, U.; Rainer, J. D. *J. Am. Chem. Soc.* 2005, **127**, 848; (d) Furuta, H.; Hasegawa, Y.; Mori, Y. *Org. Lett.* 2009, **11**, 4382.

84. (a) Hirama, M.; Oishi, T.; Uehara, H.; Inoue, M.; Maruyama, M.; Oguri, H.; Satake, M. *Science* 2001, **294**, 1904; (b) Hamajima, A.; Isobe, M. *Angew. Chem. Int. Ed.* 2009, **48**, 2941; (c) Inoue, M.; Miyazaki, K.; Ishiara, Y.; Tatami, A.; Ohnuma, Y.; Kawada, Y.; Komano, K.; Yamashita, S.; Lee, N.; Hirama, M. *J. Am. Chem. Soc.* 2006, **128**, 9352; for a recent review, see: Isobe, M.; Hamajima, A. *Nat. Prod. Rev.* 2010, **27**, 1204.

85. Tsukano, C.; Sasaki, M. *J. Am. Chem. Soc.* 2003, **125**, 14294.

12
Total Synthesis from Hydroxy Acids

Hydroxy acids, such as L-lactic acid, L-malic acid and the enantiomeric tartaric acids, are versatile 3- and 4-carbon chirons that have found extensive utility in natural product synthesis [1]. Added to this select list are their enantiomeric counterparts as well as the highly popular (2R)-3-hydroxy-2-methyl propionic acid (i.e., the Roche acid). In this chapter, the utility of hydroxy acids as starting materials for Category I-type molecules will be shown with representative total syntheses.

12.1
Griseoviridin

The first total synthesis of griseoviridin was reported in 2000 by Meyers and coworkers [2]. Analysis of the macrocyclic structure reveals a nine-membered ene-thialactone incorporated in a larger macrolactam (Figure 12.1). An asymmetric Noyori reduction [3] of ethyl acetoacetate (A), followed by protection of the hydroxyl group, and a reduction-oxidation sequence gave aldehyde **B**. Extension to ketoester **C** was achieved by condensation with the lithium enolate of allyl acetate, followed by a Dess-Martin oxidation. Treatment of the sodium enolate of **C** with the S-phthalimido-D-cysteine derivative **D** (available from D-cysteine), delivered thioether **E** in excellent yield. Reduction of the ketone group, followed by mesylation and β-elimination gave ene-thioether **F**. Lactone formation was achieved by sequential cleavage of the silyl ether and t-butyl ester, inversion of configuration under Mitsunobu conditions, followed by deprotection of the Troc carbamate to give **G** (subunit I).

The oxazole-containing subunit II was prepared starting with L-malic acid, which was encompassed in the C-15–C-18 segment of griseoviridin (Figure 12.1). A series of known steps led to the Weinreb amide derivative **H**, which was treated with allylmagnesium bromide to give the corresponding ketone. A highly stereoselective reduction was achieved under chelation control with LiAlH$_4$ and LiI to give the 1,3-syn-diol, which was protected as the TBS ether **I**. Ozonolytic cleavage of the terminal double bond, followed by conversion to the acid, then amide formation

Design and Strategy in Organic Synthesis: From the Chiron Approach to Catalysis, First Edition.
Stephen Hanessian, Simon Giroux, and Bradley L. Merner.
© 2013 Wiley-VCH Verlag GmbH & Co. KGaA. Published 2013 by Wiley-VCH Verlag GmbH & Co. KGaA.

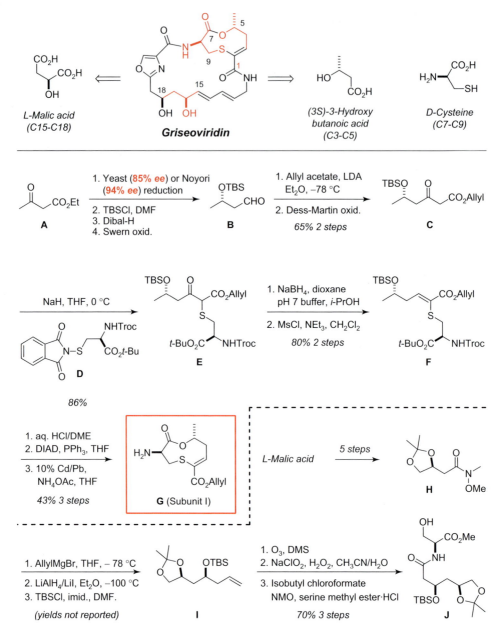

Figure 12.1 Total synthesis of griseoviridin from L-malic acid, (3S)-3-hydroxy butanoic acid and D-cysteine (apparent substructures).

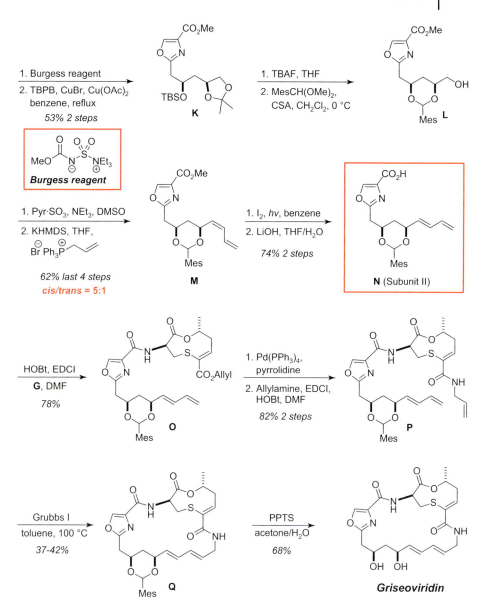

Figure 12.1 *(continued).*

Key reactions: *Enolate thioetherification (**D** to **E**); Mitsunobu lactonization (**F** to **G**); acetal exchange (**K** to **L**); olefin isomerization (**M** to **N**); ring-closing metathesis (**P** to **Q**), TBPB = t-butylperbenzoate; MesCH(OMe)$_2$ = 2,4,6-trimethylbenzaldehyde dimethyl acetal*

Relevant mechanisms and transformations:

A.

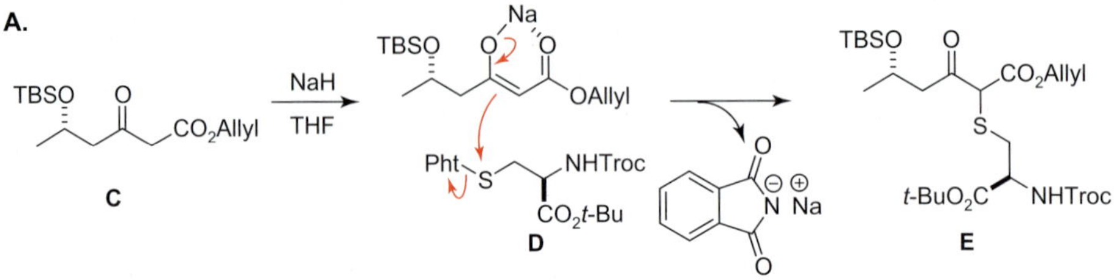

B.

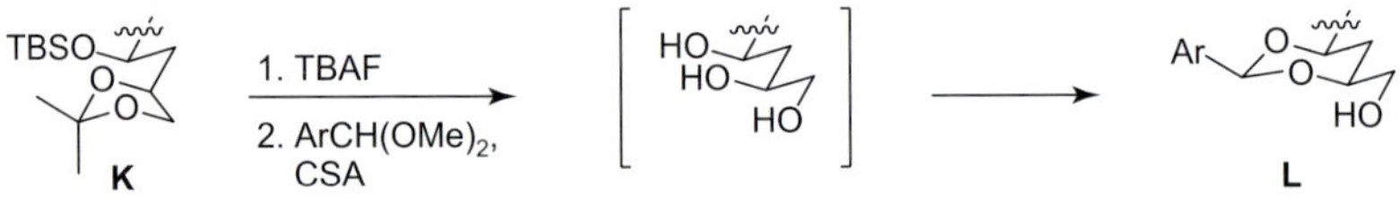

Figure 12.1 *(continued).*

with L-serine methyl ester gave **J**. Treatment with the Burgess reagent [4] effected cyclization to the dihydrooxazole, which was further oxidized to **K**.

At this juncture, it was necessary to change the 2,2-dimethyl-1,3-dioxolane protecting group in order to obtain the primary alcohol as in **L**. This was achieved by cleavage of the TBS ether, followed by transacetalization with 2,4,6-trimethylbenzaldehyde dimethyl acetal. Oxidation to the aldehyde and treatment with allyl triphenylphosphorane afforded *cis*-diene **M** as the major isomer. Photoisomerization in the presence of iodine, followed by hydrolysis of the methyl ester gave **N** (subunit II). Coupling of the two subunits, mediated by EDCI and HOBT, led to amide **O** in 78% yield. Cleavage of the allyl ester, followed by amide formation with allylamine led to **P**, which was subjected to ring-closing metathesis in the presence of Grubbs' first generation catalyst to give the *trans,trans*-macrocyclic diene **Q**. Finally, treatment with PPTS in acetone removed the acetal group to give griseoviridin.

Commentary: The presence of an ene-thioether linkage in the nine-membered thialactone subunit of griseoviridin presents a challenge, in spite of the apparent relationship with the amino acid D-cysteine. In order to secure the vinylic thioether linkage, Meyers and coworkers' strategy capitalized on the displacement of an S-phthalimido group with a soft enolate nucleophile (**C** + **D** to **E**; see relevant mechanism A, Figure 12.1). The presence of the keto group ensured that β-elimination would occur leading to vinyl thioether **F**. It was necessary to invert the configuration of the hydroxyl group in **F**, since this was originally obtained by a yeast mediated, or Noyori asymmetric reduction. This was accomplished by an intramolecular Mitsunobu reaction to obtain lactone **G**.

The acetal exchange from **K** to **L** was another operationally interesting strategy, which liberated the primary hydroxyl group for a Wittig extension. Presumably, the driving force in this equilibrating acetal exchange reaction is the all-equatorial disposition of the 2-mesityl-1,3-dioxane motif (see relevant mechanism B, Figure 12.1). Rather than deal with the 5:1 (*cis:trans*) mixture of olefins that resulted from the Wittig reaction to give **M**, it was decided to effect double bond isomerisation with iodine under photochemical conditions, which furnished the *trans*-diene **P**. A final challenge entailed the successful application a ring-closing metathesis reaction using Grubbs' first-generation catalyst to give a the macrolactam **Q**. This was achieved in *ca.* 40% yield under high dilution conditions and heating in toluene at 100 °C. The judicious choice of orthogonal S-, O-, and N-protective groups was key to ensure the compatibility of functional groups in a variety of chemical transformations.

12.2
Halicholactone

The first total synthesis of halicholactone was reported in 1995 by Wills and coworkers [5] (Figure 12.2). Analysis of the carbon framework of halicholactone reveals the presence of three hydroxyl groups at C-8, C-12, and C-15. The four carbon atoms of L-malic acid were seen as viable source of corresponding substructure in the target molecule that would provide the required stereochemistry at C-8. The C-15 stereogenic center of the natural product came from commercially available (3R)-1-octyn-3-ol.

L-Malic acid was converted to (3R)-3-hydroxy-butyrolactone PMB ether (**A**), which was reduced with Dibal-H to the corresponding hemiacetal, and the latter extended to the *cis*-olefin **B**. Esterification, oxidation to the aldehyde, and a Horner-Wadsworth-Emmons olefination, according to the Masamune and Roush protocol [6], gave *trans*-ester **C**.

Application of the Corey-Chaykovsky cyclopropanation reaction [7] employing trimethylsulfoxonium iodide in DMSO, led to the cyclopropane intermediate **D** as the major diastereomer (dr = 5:2). Conversion to the alcohol, and separation of isomers was achieved after hydrolysis of the methyl ester to give **E**. Yamaguchi lactonization [8] with 2,4,6-trichlorobenzoyl chloride, followed by ester hydrolysis

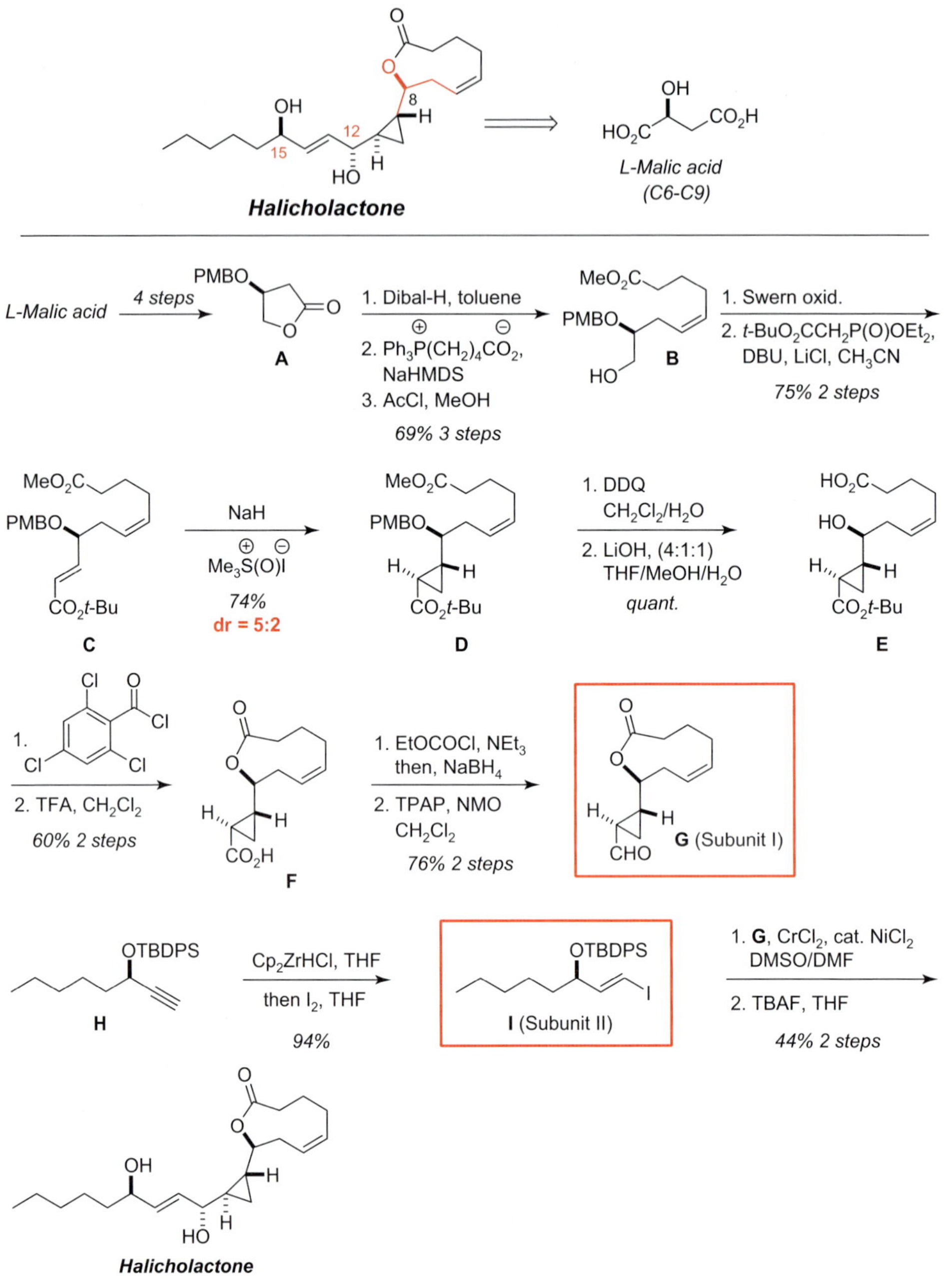

Halicholactone

Key reactions: *DMSO ylide cyclopropanation (**C** to **D**); Yamaguchi lactonization (**E** to **F**); hydrozirconation/iodination (**H** to **I**); Nozaki-Hiyama-Kishi coupling (**I** to TBS-halicholactone)*

Figure 12.2 Total synthesis of halicholactone from L-malic acid (apparent substructures).

Relevant mechanism:

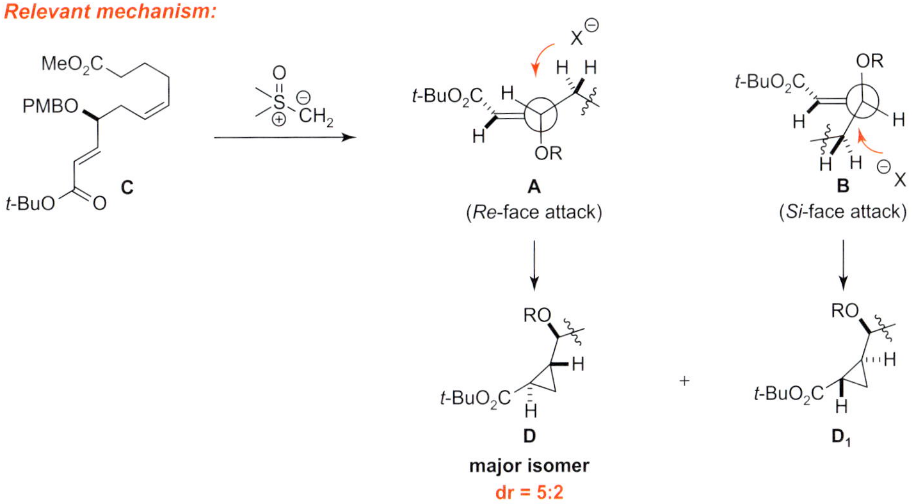

Figure 12.2 *(continued).*

led to the nine-membered lactone **F**. Conversion to the alcohol through reduction of the mixed anhydride, followed by oxidation with TPAP gave **G** (subunit I). This sequence, starting with L-malic acid, comprised 16 steps and proceeded in 7.5% overall yield.

(3*R*)-1-Octyn-1-ol was protected as the TBDPS ether **H**, which was subjected to a hydrozirconation/iodination reaction, according to Schwartz [9], to give *trans*-iodo olefin **I** (subunit II). Coupling of the subunits I and II was achieved following a method developed by Nozaki [10], Hiyama, and Kishi [11] to give a 2:1 mixture of epimeric alcohols at C-12. Chromatographic separation and treatment with TBAF led to halicholactone. Thus, a convergent synthesis comprising 20 steps from readily available starting materials was realized in 3% overall yield. As a result, the original assignment of absolute configuration by Yamada and Clardy [12] was also confirmed.

Commentary: The utility L-malic acid as 4-carbon precursor to a designated subunit in the target molecule is evident. The stereochemistry at C-8, originating from L-malic acid, was also critical in the elaboration of the *trans*-disubstituted cyclopropane (**C** to **D**, see relevant mechanism, Figure 12.2). Facile lactonization was achieved by virtue of the *cis*-configured olefin in **E** (**E** to **F**). Application of the Schwartz hydrozirconation/iodination reaction provided the desired *trans*-iodo olefin for the planned Nozaki-Hiyama-Kishi coupling of subunits I and II. This reaction was no doubt part of the synthesis plan that would be implemented in the penultimate step. Vinyl iodide **I** (subunit II) would be the cross-coupling partner with the cyclopropane aldehyde **G** (subunit I), delivering the *trans*-olefin as the major isomer.

Two alternative total syntheses of halicholactone have been reported, each adopting a different strategy to construct a common penultimate precursor. Takemoto

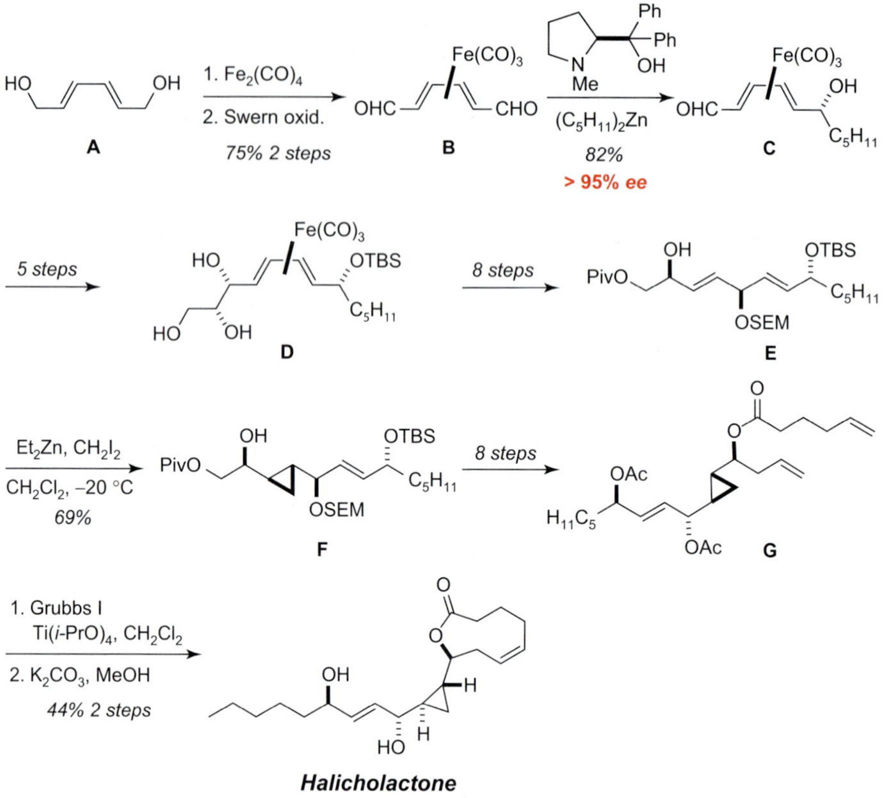

Figure 12.3 Total synthesis of halicholactone using iron complexes and a chiral ligand-mediated organozinc reaction.

and Tanaka [13] started with the Fe(CO)$_3$ *meso*-2,4-hexadien-1,6-diol complex **A**, which was oxidized to **B**, and then engaged in a highly stereoselective addition of dipentylzinc in the presence of an *N*-methyl prolinol ligand to give **C** (Figure 12.3). Further steps led to **D** and **E**. A hydroxyl-assisted cyclopropanation gave **F**, which was converted to **G** over eight steps. Ring-closing metathesis followed by ester hydrolysis afforded halicholactone.

In another synthesis, Kitahara and coworkers [14] relied on baker's yeast as a key step for the reduction of a cyclohexanone intermediate **A** (Figure 12.4). This highly selective biotransformation afforded **B**, which was converted to the tosylate **C**, and the latter subjected to strong base followed by TIPSOTf to give **D**. Ozonolytic cleavage and Wittig olefination gave *cis*-cyclopropane carboxylic acid **E**, which was eventually transformed to *trans*-aldehyde intermediate **F** (subunit I).

Commercially available acetylenic alcohol **G** was subjected to a hydrozirconation/iodination reaction as reported by Wills [5]. Completion of the synthesis relied on methodology described in the two previous syntheses as illustrated in Figures 12.2 and 12.3.

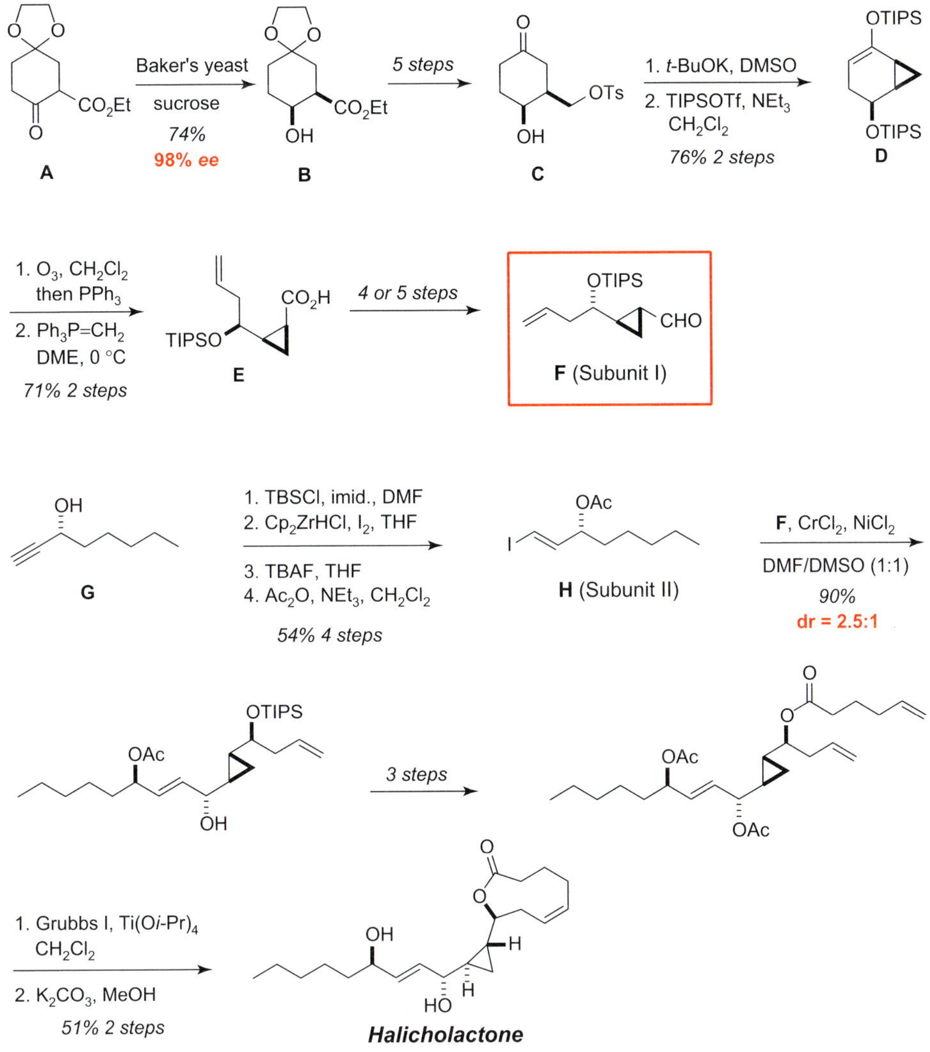

Figure 12.4 Total synthesis of halicholactone using a bio-transformation approach (partially hidden and hidden sub-structures).

12.3
Brasilenyne

Denmark and coworkers [15] reported the first total synthesis of brasilenyne, a novel nine-membered cyclic ether, in 2004 (Figure 12.5). Considering the stereochemistry of the carbon atom bearing the ether oxygen, L-malic acid offered a viable chiron that could be extended at either extremity. Major challenges resided in introducing the remaining stereogenic centers bearing both chloro and *C*-methyl groups.

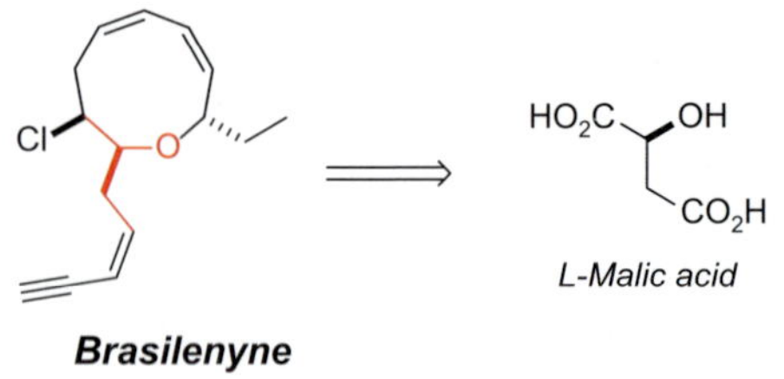

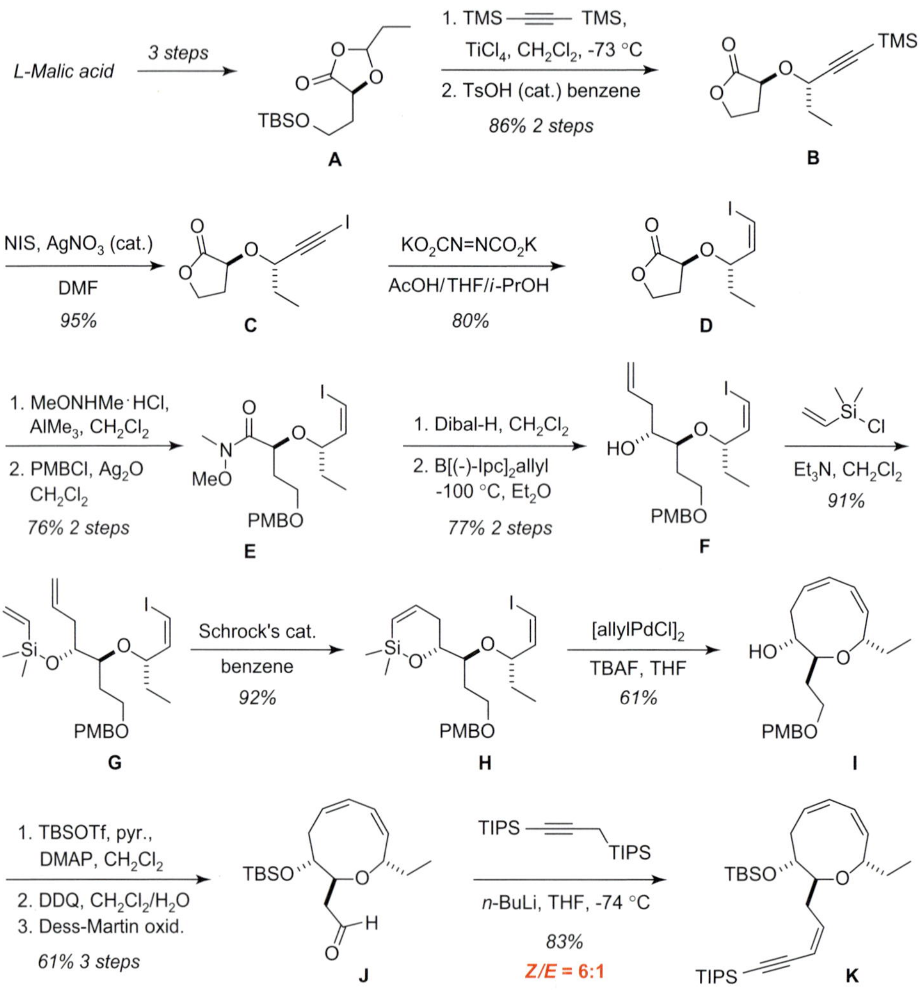

Brasilenyne

L-Malic acid

Figure 12.5 Total synthesis of brasilenyne from L-malic acid (partially hidden substructure).

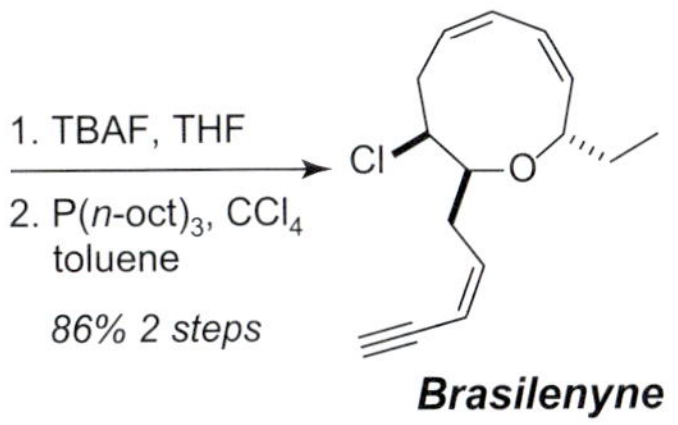

1. TBAF, THF
2. P(n-oct)3, CCl4
toluene
86% 2 steps
Cl
O
Brasilenyne

Key reactions: *Lewis acid-mediated acetal opening (**A** to **B**); Brown allylation (**E** to **F**); Si-tethered RCM reaction (**G** to **H**); Si-directed intramolecular cross-coupling (**H** to **I**)*

Relevant mechanism and transformation:

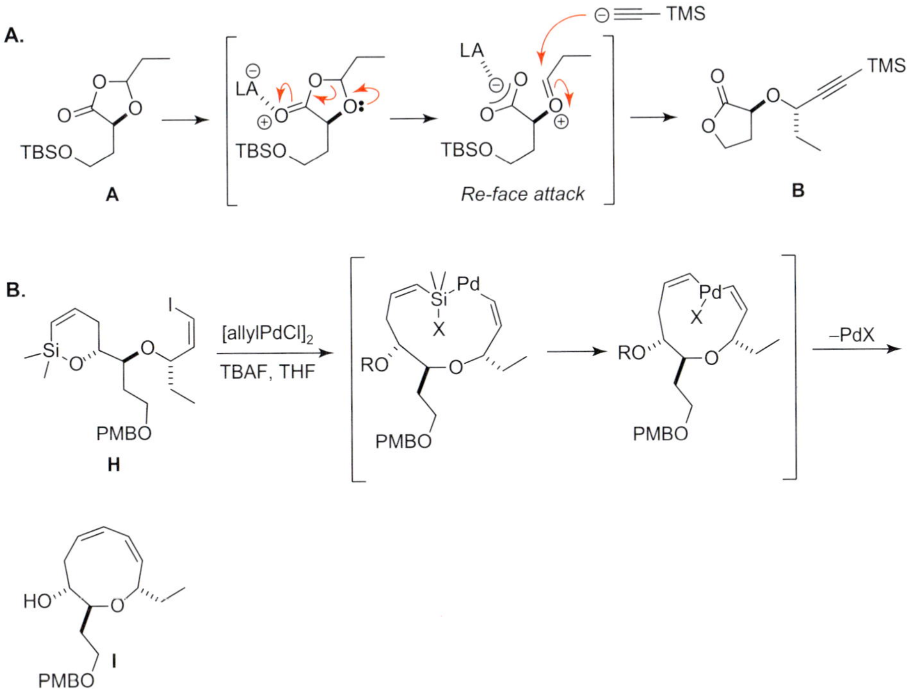

A.
TMS
LA
TBSO
A
Re-face attack
B
B.
[allylPdCl]2
TBAF, THF
Pd
Si
X
RO
PMBO
H
–PdX
HO
I

Figure 12.5 *(continued).*

L-Malic acid was converted to lactone acetal **A** in three steps. Lewis acid mediated ring-opening in the presence of bis(trimethylsilyl)acetylene, led to ether **B** in excellent yield. Iodination gave **C**, which was reduced with diimide to give the *cis*-vinyl iodide **D**. Conversion to the Weinreb amide, and protection as the PMB ether afforded **E**. Selective reduction to the aldehyde with Dibal-H, followed by a Brown allylation [16] gave alcohol **F** as the major diastereomer. Formation of the vinyldimethylsilyl ether **G** was followed by a ring-closing metathesis reaction using the Schrock catalyst [17] to give **H**. Treatment of **H** with 7.5 mol% [allylPdCl]$_2$, in the presence of TBAF as an activator, led to the formation of the nine-membered ring diene **I** in 61% yield. Manipulation of protecting groups and oxidation gave aldehyde **J**, which was subjected to a Petersen olefination reaction to give the ene-yne side-chain with a preponderance of the desired *cis*-isomer as in **K**. Finally, removal of the TBS group and treatment of the alcohol with tri-octylphosphine in CCl$_4$ and toluene effected a clean S$_N$2 (inversion) reaction to give brasilenyne. The authors completed the synthesis in 20 linear steps with a 5.1% overall yield from L-malic acid.

Commentary: Although there are a number of ways to plan the synthesis of brasilenyne, particularly in view of previous syntheses of analogous natural products, one of the major challenges associated with its construction is in the introduction of the *s-cis*-diene system present in the nine-membered cyclic ether. The strategy depended on two crucial back-to-back reactions involving ring-closing metathesis and silicon-directed cross-coupling reactions (**G** to **I**). To set the stage for this innovative approach, Denmark and coworkers devised an indirect etherification reaction of a 2-hydroxy-γ-butyrolactone. In initial studies, they realized that the desired ether derivative **B** could not be obtained by displacement of a propargylic triflate with 2-hydroxy-γ-butyrolactone. Reversing the order by nucleophilic displacement of a triflate ester of the lactone also failed. A solution to this problem was found in the diastereoselective ring-opening of 1,3-dioxolanone **A** in the presence of TiCl$_4$ (see relevant mechanism A, Figure 12.5). The product **B** was expected to arise from a favored oxocarbenium ion conformer and a *Re*-face attack by the TMS-acetylide anion. The exclusive formation of a single diastereomer in high yield reflects the importance of the stereogenic center bearing the original hydroxyl group of L-malic acid.

Noteworthy reactions leading to the *cis*-vinyl iodide **G** involved a diimide reduction of the acetylenic iodide **C**, and a reagent controlled allylboration to **F**. The sequential RCM and silicon-directed cross-coupling reactions had been previously applied to model vinylsilanes [18], showing that organosilanes could be effective donors in Pd-catalyzed reactions. In the case of intermediate **H**, the cross-coupling effectively generated a Pd-tethered oxa-cyclononadiene ring system in 61% yield (see relevant transformation B, Figure 12.5). It should be noted that the hydroxyl group released in the cross-coupling reaction was converted to the chloride in high yield without competitive dehydration.

12.4
Octalactin A

Subsequent to their first total synthesis of octalactin A, Buszek and coworkers [19] reported a more practical route in 2002 (Figure 12.6). The presence of *C*-methyl and secondary hydroxyl groups on the acyclic and lactone segments of octalactin A offers a number of possibilities to apply the *chiron approach*. Indeed, (*R*)- and (*S*)-Roche acids were recognized as being suitable starting materials for the C-7–C-9 and C-3–C-5 segments of the target molecule respectively. Protection of the (*R*)-Roche acid methyl ester as the TBDPS ether and reduction of the carboxylic acid group led to **A**, which was oxidized to the aldehyde, then chain-extended to allylic alcohol **B** (subunit I), and its epimer (dr = 1:1). Treatment of the (*S*)-Roche acid methyl ester under the same conditions gave **C** (*ent*-**B**). Protection of the hydroxyl group in **B** as the PMB ether, hydroboration with 9-BBN, followed by protection of the primary

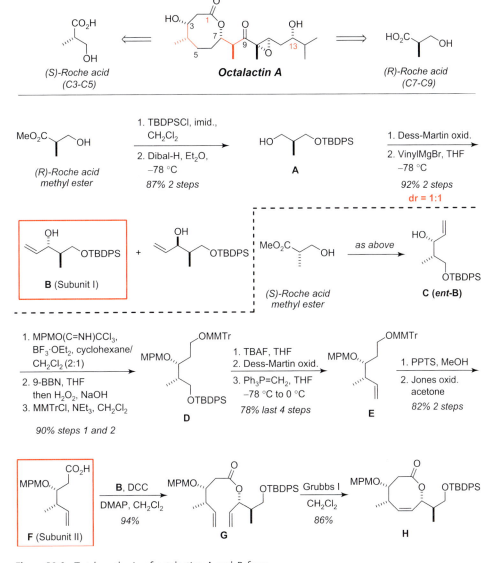

Figure 12.6 Total synthesis of octalactins A and B from (*R*)- and (*S*)-Roche acids (apparent substructures).

Key reactions: *Eight-membered lactone formation via RCM (**G** to **H**); Nozaki-Hiyama-Kishi coupling (**I** and **J** to **K**) reagent dependent allylic alcohol epoxidation (**syn** and **anti-K** to* **octalactin A** *precursor); MMTrCl = (p-methoxyphenyl)diphenylmethyl chloride*

Relevant transformations:

Figure 12.6 *(continued).*

alcohol as the (*p*-methoxyphenyl)diphenylmethyl ether afforded **D**. Cleavage of the TBDPS group, oxidation of the alcohol to the aldehyde, and Wittig olefination gave **E**, which was converted to carboxylic acid **F** (subunit II). Linking the two subunits as the ester led to **G**, which was followed by a ring-closing metathesis reaction with Grubbs' first generation catalyst to give the lactone **H**. Attention was then directed to the elaboration of the acyclic segment toward the octalactins. Thus, conversion of **H** to the aldehyde **I**, followed by cross-coupling with vinyl iodide **J** according to the Nozaki-Hiyama-Kishi [10, 11] protocol gave **K** and its C-9 epimer in a 1.5:1 ratio. Oxidation of *syn*-**K** (or its C-9 epimer) to the ketone, followed by deprotection led to octalactin B.

The *syn*-isomer **K** was epoxidized in the presence of *t*-BuOOH, VO(acac)$_2$, and the resulting epoxide was oxidized then deprotected to give octalactin A. Diastereomeric epoxides were obtained from the C-9 epimer (*anti*-**K**), depending on the nature of the oxidizing reagent (see relevant transformations, Figure 12.6). The total synthesis of octalactins A and B also established their absolute configuration.

Commentary: The first synthesis of octalactins A and B in 1994 [19a] relied on a segment coupling strategy via acetylide anion methodology, followed by lactone formation to the eight-membered ring. The advent of the Grubbs ring-closing metathesis reaction greatly facilitated the implementation of a more efficient total synthesis as highlighted above. A noteworthy feature of the Buszek synthesis is the utilization of the two antipodes of the Roche acid in order to converge with two 3-carbon segments in octalactins A and B. The successful application of a ring-closing metathesis reaction to produce the eight-membered oxocene is noteworthy, since related cyclizations from acyclic precursors have been, at times, problematic [20]. Total syntheses of octalactins A and B have also been reported by Shiina [21], Holmes [22a], and Toste [22b]. The total synthesis of the *ent*-octalactin A by McWilliams and Clardy [23] in 1994 started with (*R*)-citronellic acid as the source of the C-4 *C*-methyl group and (2*S*)-2-hydroxy-3-methylbutanoic acid as a building block for the isopropyl side-chain and the C-13 hydroxyl group (Figure 12.7).

(*R*)-Citronellic acid   ***ent*-Octalactin A**   (2*S*)-2-Hydroxy-3-methyl-butanoic acid

Figure 12.7 Total synthesis of *ent*-octalactin A from (*R*)-citronellic acid and (2*S*)-2-hydroxy-3-methylbutanoic acid.

12.5
(3Z)-Dactomelyne

The pyranopyranyl core motif in the (3Z)- and (3E)-dactomelynes has been the subject of synthetic interest [24] (Figure 12.8). Total syntheses of both isomers were first reported in 1995 by Lee and coworkers [25]. The tetrasubstituted *cis*-fused dioxadecalin motif presents a challenge in introducing a bromine and a chlorine substituent individually, adjacent to alkyl and alkenyl groups. Analysis of the dioxadecalin core motif reveals a relationship with (S,S)-tartaric acid as a viable precursor.

The synthesis of (3Z)-dactomelyne started with (S,S)-diethyl tartrate (**A**), which was converted to the benzylidene acetal **B** in a four-step sequence. Displacement of the triflate group in **B** with (trichloromethyl)lithium, prepared from $CHCl_3$ and LDA, led to **C** in 60% yield. Selective hydrogenolysis of the benzyl ether, and addition of the resulting alcohol to ethyl propiolate gave the *trans*-vinyl ether derivative **D**. Treatment of **D** with tricyclohexyltin hydride in the presence of AIBN under high dilution conditions afforded the *gem*-dichloride **E**. The reagent combination of tris(trimethylsilyl)silane and triethylborane gave the monochloro product **F** as the major isomer (dr = 13:1) in excellent yield. Reduction of the ester group, protection of the primary alcohol, and selective Lewis acid-mediated reductive opening of the acetal function furnished **G** in excellent overall yield. Conversion to the nitrile, followed by hydride reduction afforded amine **H**. Treatment with cupric bromide and *t*-butyl nitrite gave the *gem*-dibromide derivative **I**. Cleavage of the benzyl ether with BCl_3, followed by addition of ethyl acrylate led to the vinyl ether **J** in high yield. Intramolecular radical cyclization in the presence of tributyltin hydride and AIBN afforded bicyclic product **K**. There remained to modify the side-chains and to introduce appropriate appendages. Conversion of the ester group in **K** to the alcohol, formation of the iodide, and reduction with Super-hydride gave the *C*-ethyl substituted product **L**. Oxidation to the aldehyde **M**, and treatment with lithium 1,3-bis(tri-isopropylsilyl)propyne anion in a Petersen-type olefination, followed by deprotection, gave (3Z)-dactomelyne and (3E)-dactomelyne as the major and minor isomers respectively (*cis/trans* = 10:1).

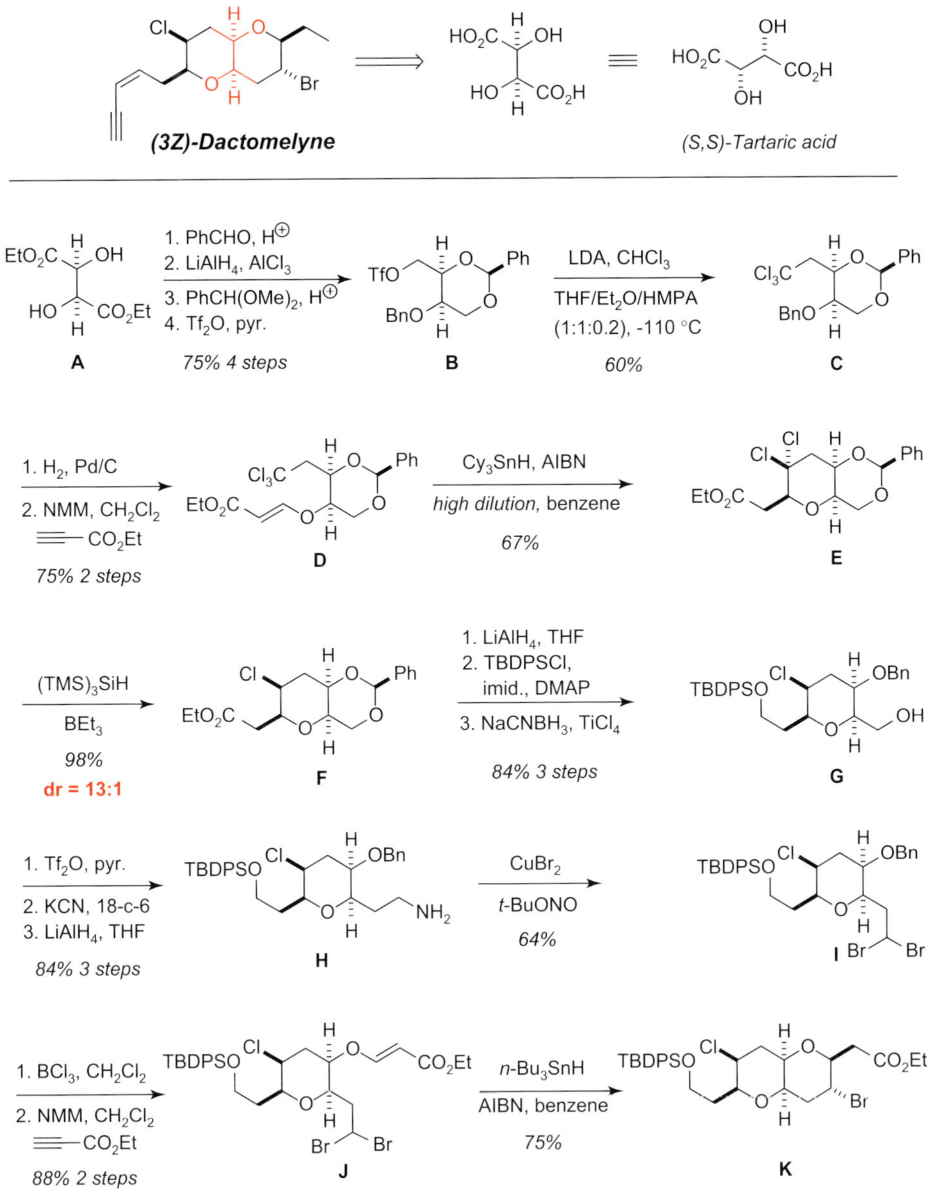

Figure 12.8 Total synthesis of (3Z)-dactomelyne from (S,S)-tartaric acid (apparent substructure).

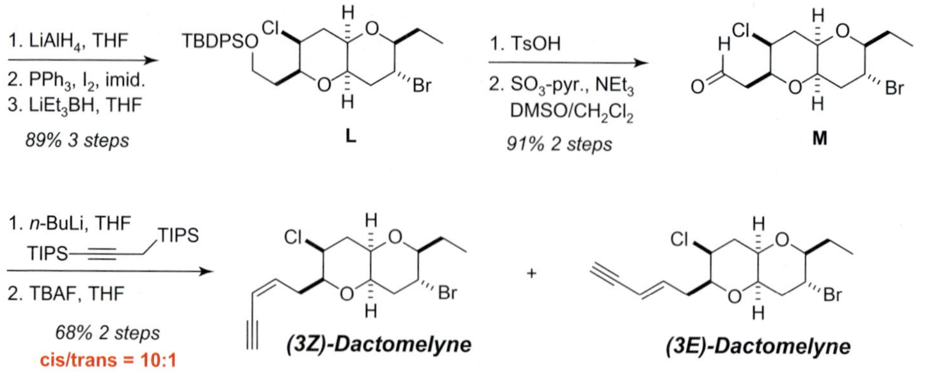
1. LiAlH4, THF
2. PPh3, I2, imid.
3. LiEt3BH, THF
89% 3 steps
TBDPSO
Cl
H
O
Br
L
1. TsOH
2. SO3-pyr., NEt3
DMSO/CH2Cl2
91% 2 steps
M
1. n-BuLi, THF
TIPS
TIPS
2. TBAF, THF
68% 2 steps
cis/trans = 10:1
(3Z)-Dactomelyne
+
(3E)-Dactomelyne

Key reactions: *Regioselective acetal formation (**A** to **B**); (trichloromethyl)lithium anion displacement (**B** to **C**); radical cyclization (**D** to **E**; **J** to **K**); regioselective acetal cleavage (**F** to **G**); ene-yne extension (**M** to **target**); NMM = N-methylmorpholine*

Relevant mechanisms:

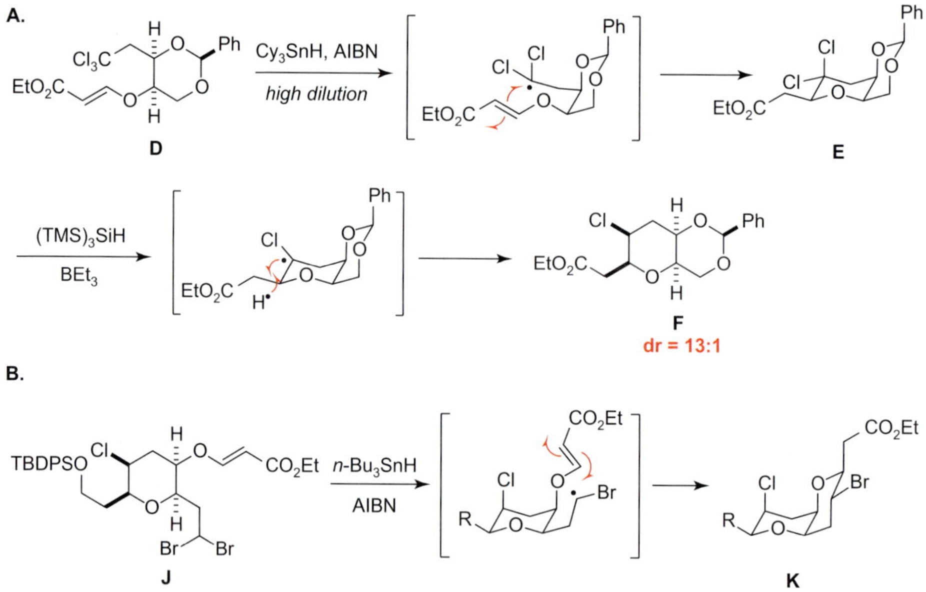
A.
Cl3C
EtO2C
Ph
D
Cy3SnH, AIBN
high dilution
Cl
E
(TMS)3SiH
BEt3
H•
F
dr = 13:1
B.
TBDPSO
CO2Et
n-Bu3SnH
AIBN
Br
J
R
K

Figure 12.8 *(continued).*

Commentary: It is interesting to reflect on the evolution of the thought process that led to the synthesis of the dactomelynes by Lee and coworkers. Previous studies in the group had shown that β-alkoxyacrylates were good radical acceptors in the formation of *cis*-2,5-disubstituted tetrahydrofurans and *cis*-2,6-disubstituted tetrahydropyrans [26]. The presence of bromine and chlorine substituents on the dioxadecalin core motif were strong incentives to explore a free-radical approach to cyclization. However, in order to secure the intact monochloride and mono-bromide groups, a strategy had to be devised for selective free-radical induced dehalogenation from polyhalo precursors, while ensuring stereochemical integrity in the formation of the tetrasubstituted tetrahydropyran intermediates. Once the plan to use free-radical cyclizations was in place, the search for a suitable source for the core "vicinal diol" precursor became apparent. Thus, (S,S)-tartaric acid emerged as a logical choice.

Introduction of a trichloromethyl group, by an S_N2-type reaction, set up the requisite reaction partners for radical cyclization in intermediate **D**. A stereoelectronically preferred *trans* conformation of the vinyl ether/ester appendage, and a chair-like conformation for the radical addition led to the first tetrahydropyran ring (**D** to **E**; see relevant mechanism A, Figure 12.8). Generation of the monochloromethyl radical and trapping with hydride radical led to **F**, containing an axially oriented chloro substituent in a *cis*-relationship with the adjacent ester side-chain. The stereochemical outcome of this reaction is of interest in view of the fact that it creates a 1,3-diaxial non-bonded interaction with the vicinal acetal ether oxygen. Thus, an early transition-state, known to be operative in such radical reactions, may also have a stereoelectronic component favoring an equatorial attack of a hydride radical on the chloromethinyl radical. The situation is different in the cyclization of the α-bromomethinyl radical generated from **J**. In this case it is the carbon radical, rather than hydride radical, that is the "nucleophilic" component in the chair-like transition-state of cyclization, leading to **K** with an equatorial bromo substituent (see relevant mechanism B, Figure 12.8).

It is not unusual to conceive of a free-radical-based strategy in conjunction with the cyclization of ω-alkoxy-α,β-unsaturated esters to oxacycles, especially, when both halogen atoms are the same. However, to extend the strategy to *different* halogens, as was done by the Lee group, requires prior knowledge.

Generally, when contemplating the formation of secondary halides, it is customary to do S_N2 displacement reactions of sulfonates with appropriate sources of halide ions. Burton and coworkers [27] adopted such a method to introduce two bromine substituents in elatenyne, a dibromo congener of (3Z)-dactomelyne (Figure 12.9). A two-directional strategy was conceived whereby *t*-butyl ester **A** was converted to the C_2-symmetric dilactone **B** using a sequential Sharpless dihydroxylation, bis-lactonization protocol. Subsequent ring expansion under acidic conditions, followed by elimination led to C_2-symmetric pyrano[3,2-b]pyran **C**. Epoxidation and C-allylation by stereo- and regioselective ring-opening gave **D**, still maintaining its exquisite symmetry element. Chain-elongation and a double S_N2 displacement of the bis-triflate with tetrabutylammonium bromide gave the dibromo product, which was further elaborated to elatenyne. Related monochloro analogues were also prepared using this approach.

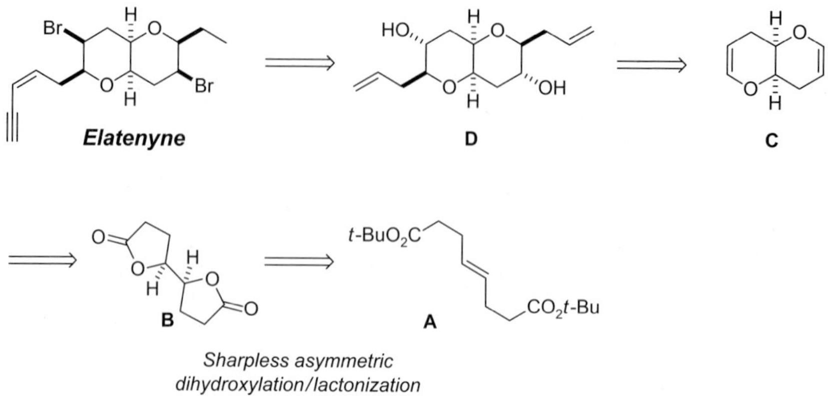

Figure 12.9 Total synthesis of elatenyne using a two-directional strategy.

12.6
UCS1025A

The total synthesis of the antiproliferative pyrrolizidinone alkaloid UCS1025A was reported in 2005 by Lambert and Danishefsky [28] (Figure 12.10). Analysis of this most unusual combination of pyrrolizidinone ketone and dehydrodecalin hydrocarbon subunits, reveals little if any progeny to a hydroxy acid. Although some skeletal relationship can be found to D- or L-pyroglutamic acid corresponding to the lactam portion of the pyrrolizidinone, Lambert and Danishefsky chose to use (R,R)-tartaric acid in an indirect way to converge with the lactam ring.

(R,R)-Tartaric acid was converted to known anhydride **A**, and the latter was condensed with amino acid methyl ester **B** to afford the C_2-symmetric imide derivative **C**. Treatment with TBS triflate effected intramolecular enolate cyclization to give bicyclic hemiaminal **D** as the major isomer. Removal of the TMS ether groups and treatment with triflic anhydride led to enol triflate **E**. Palladium-mediated reductive addition of n-Bu$_3$SnH resulted in the replacement of the triflate group by hydride to give **F**. Cleavage of the ester group and subsequent iodolactonization afforded tricyclic iodide **G** in high yield. A Reformatsky-type union of **G** with aldehyde **H**, obtained separately by an organocatalytic carbocyclization according to MacMillan [29], gave the β-hydroxy ketone **I** with complete stereoselectivity. Treatment with TBAF and oxidation under Dess-Martin conditions afforded UCS1025A.

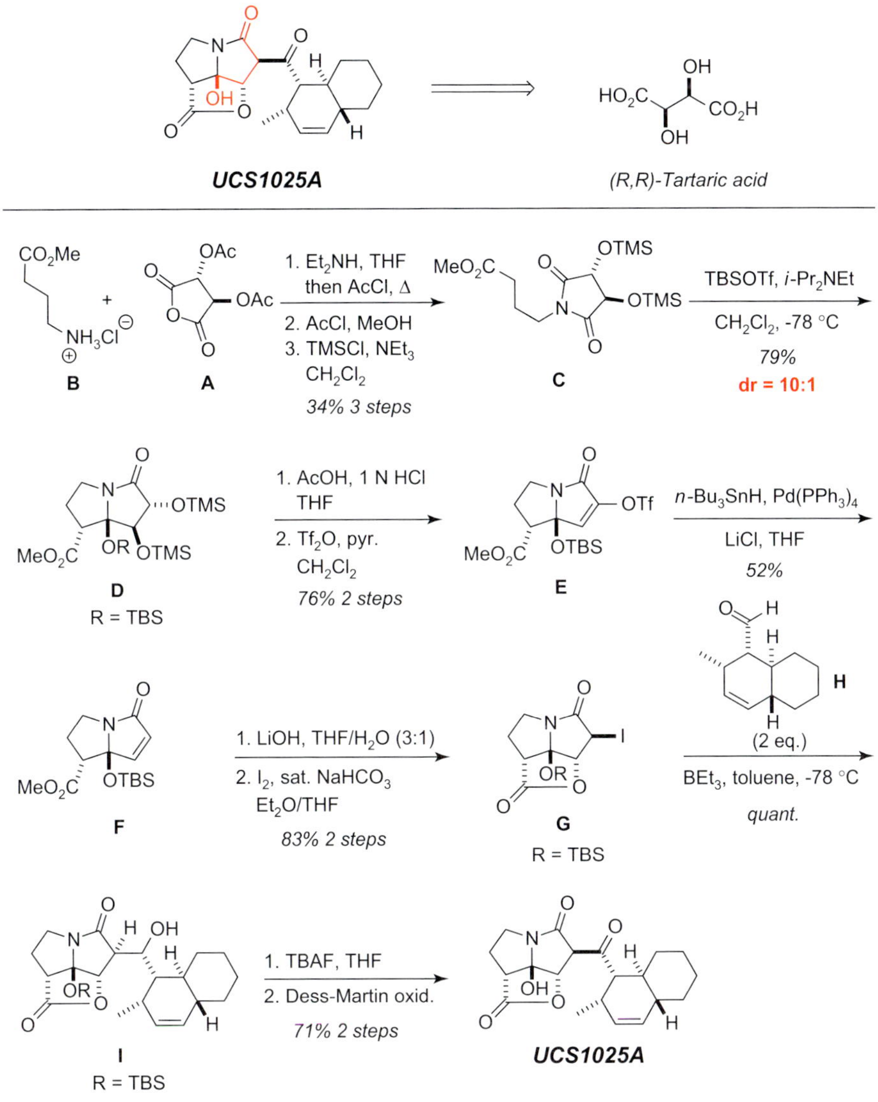

UCS1025A   (R,R)-Tartaric acid

Key reactions: *Intramolecular enolate cyclization (**C** to **D**); iodolactonization (**F** to **G**); boron-Reformatsky-type coupling (**G** + **H** to **I**)*

Figure 12.10 Total synthesis of UCS1025A from (R,R)-tartaric acid (hidden substructure).

Relevant mechanisms:

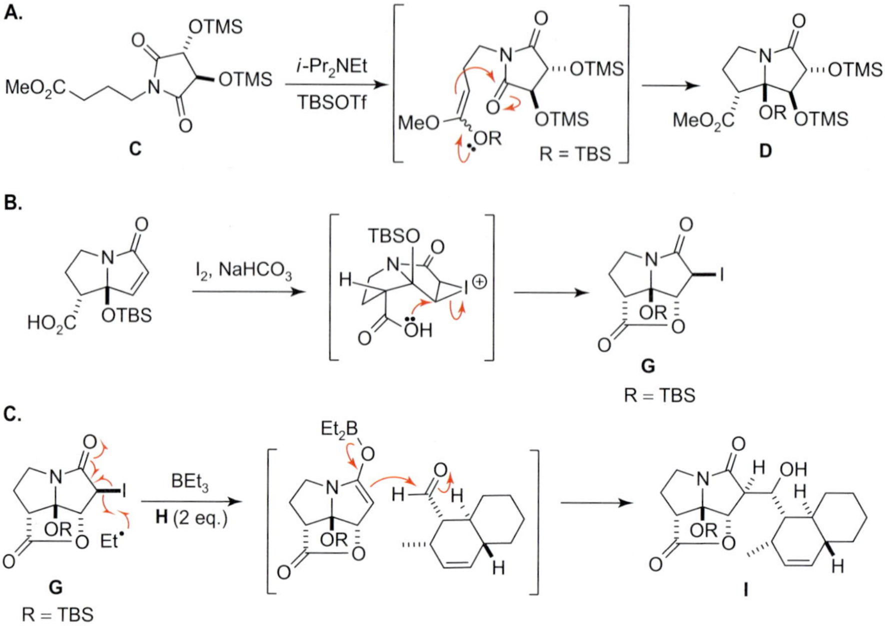

A.

B.

C.

Figure 12.10 *(continued).*

Commentary: The logical disconnection of the target ketone to a bicyclic lactam enolate and a dehydrodecalin carboxaldehyde is obvious. However, Lambert and Danishefsky had initially attempted such an aldol-type condensation with model bicyclic substituted pyrrolizidinones without success, which was rationalized on the basis of steric factors. An altogether different strategy was eventually adopted in the application of a boron enolate, Reformatsky-type reaction (**G** + **H** to **I**), with excellent results. In order to reach this crucial coupling step, the pyrrolizidinone had to be appropriately functionalized to provide enolate precursor **G**. To do so, the C_2-symmetry of imide **C** derived from (*R*,*R*)-tartaric acid was nicely exploited in the intramolecular Lewis acid-mediated enolate cyclization to give hemiaminal **D** as the major isomer (see relevant mechanism, A, Figure 12.10). Having fulfilled their stereodirecting role, the two hydroxyl groups protected as TMS ethers, were eliminated in a three-step sequence to eventually give the unsaturated lactam **F**. An iodolactonization was used to introduce the desired lactone and the iodo group to give **G** in a highly efficient manner, without affecting the *O*-protected hemiaminal (see relevant mechanism B, Figure 12.10). The conversion of iodoketone **G** to **I** in the presence of triethylborane has been shown to proceed through a radical-initiated formation of a boron enolate [30], which results in a remarkably diastereoselective Reformatsky-type reaction (see relevant mechanism C, Figure 12.10). In the process, two vicinal stereogenic centers were formed with virtually complete stereochemical control and without β-elimination. The four carbon atoms of the starting (R,R)-tartaric acids were embedded in the lactam portion of the pyrrolizidinone subunit of UCS1025A. In the process, the original stereogenic centers were lost to the benefit of well-planned stereo- and regioselective bond forming steps.

The impact of organocatalytic methods in the preparation of carbocyclic compounds was convincingly shown in the one-step synthesis of aldehyde **H** from achiral triene aldehyde **A** (Figure 12.11). MacMillan's imidazolidinone catalyst **B** (20 mol%), was highly efficient in promoting an intramolecular Diels-Alder cyclization to give aldehyde **H** directly in 71% yield, with high enantio- and diastereoselectivity.

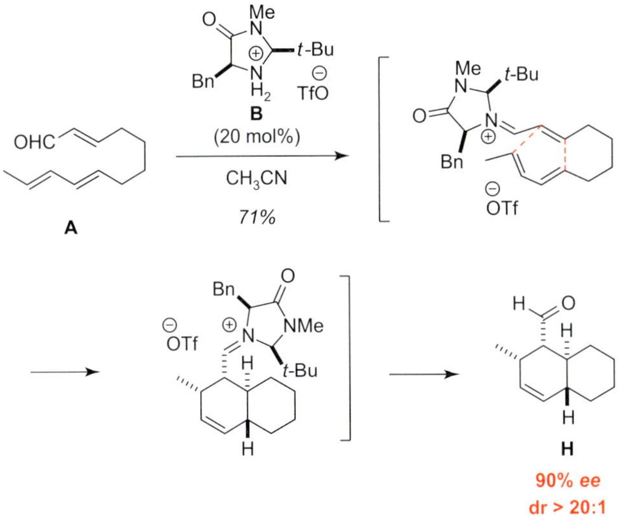

Figure 12.11 Organocatalytic intramolecular Diels-Alder reaction in the synthesis of a key bicyclic aldehyde intermediate in total synthesis of UCS1025A.

12.7
Jerangolid A

The first total synthesis of the antifungal polyketide jerangolid A was reported by Hanessian and coworkers in 2010 [31]. Recognizing the existence of the isolated C-methyl group flanked by two unsaturated chains, it was evident that the (R)-Roche acid could be used as an anchoring site for a convergent assembly of subunits. The utilization of (S)-glycidol as a common chiron to construct the lactone (ring A) and the dihydropyran (ring B) rings provided a secure basis to explore the stereocontrolled total synthesis of jerangolid A (Figure 12.12).

Treatment of **A** with the lithium anion of ethylpropiolate gave **B**, which was cyclized to the lactone **C** in the presence of NaOMe. Iodination to **D**, followed by formation of the Grignard reagent, and treatment with freshly cracked formalde-hyde led to the hydroxylmethyl derivative **E**. Protection as the TBS ether, cleavage of the PMB ether, and oxidation gave the aldehyde **F** (subunit I) in good overall yield. *O*-Benzyl-(*R*)-glycidol (**G**) was treated with the lithium anion generated from the dithiane **H** to give **I**. Protection of the resulting alcohol as the TBS ether and cleavage of the dithiane, according to Barton, afforded the ketone **J**. Reduction under Luche conditions, then deprotection led to a 3.6:1 mixture of diastereomeric alcohols **K**. In the presence of 5 mol% Pd(CH$_3$CN)$_2$(BF$_4$)$_2$, cyclization took place to give the *syn*-dihydropyran **L** as the only detectable isomer in 72% yield. The cyclization could also be done in the presence of BF$_3$·Et$_2$O (dr = 20:1, 82% yield). Cleavage of the benzyl ether, followed by a three-step sequence led to **M** (subunit II) in good overall yield.

The iodide **N**, readily available from the (*R*)-Roche acid, was treated with the sodium anion of *N*,*N*-dimethyl-2-oxo-1,3-diazaphospholidine (**O**) to give the phosphonamide **P**. Formation of the lithium anion, and reaction with ketone **M** led to **Q** as a 6:1 *E*/*Z* mixture of olefins in 58% yield. Conversion to the phenyltetrazole sulfone **R**, then coupling with **F** employing the Julia-Kocienski protocol led, after mild deprotection of the TBS ether group, to jerangolid A. The synthesis was accomplished in 17 linear steps and 6% overall yield, starting from glycidol **G**. The total synthesis of jerangolid D, with a *C*-methyl group replacing the hydroxymethyl group of jerangolid A, was reported by Markó and coworkers in 2007 [32].

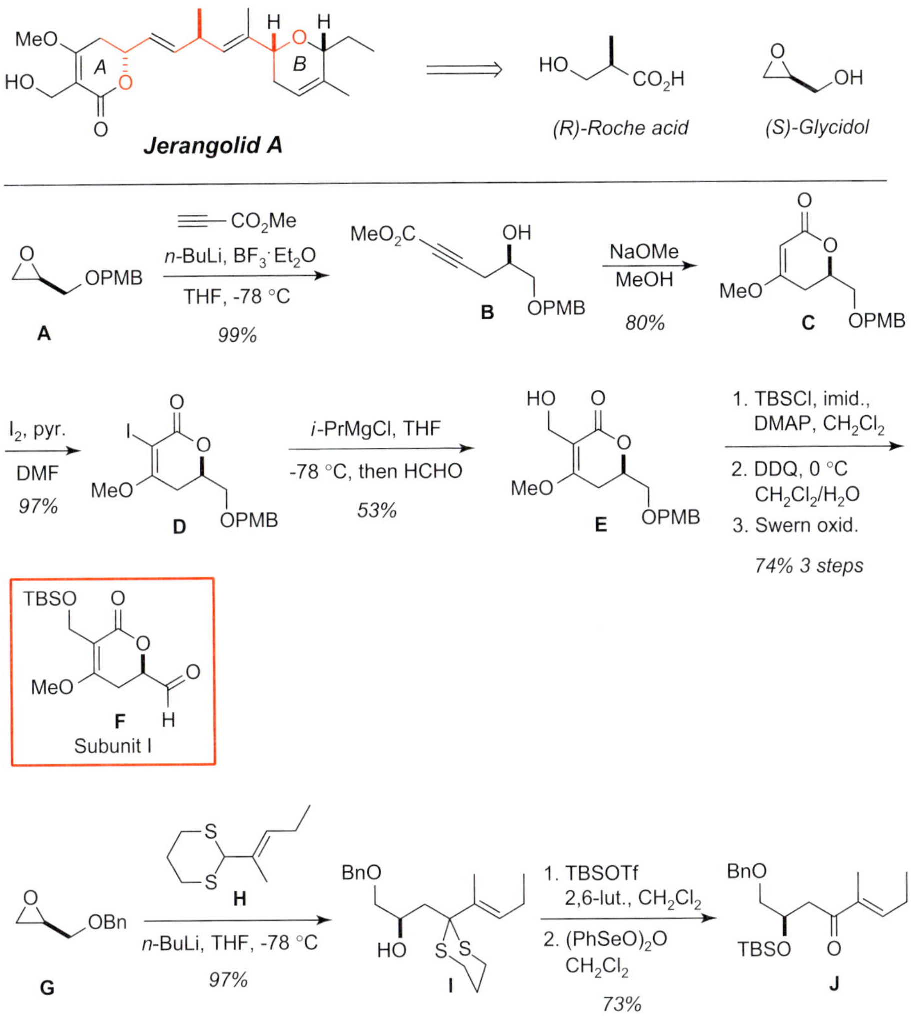

Figure 12.12 Total synthesis of jerangolid A from (S)-glycidol and the (R)-Roche acid (apparent substructures).

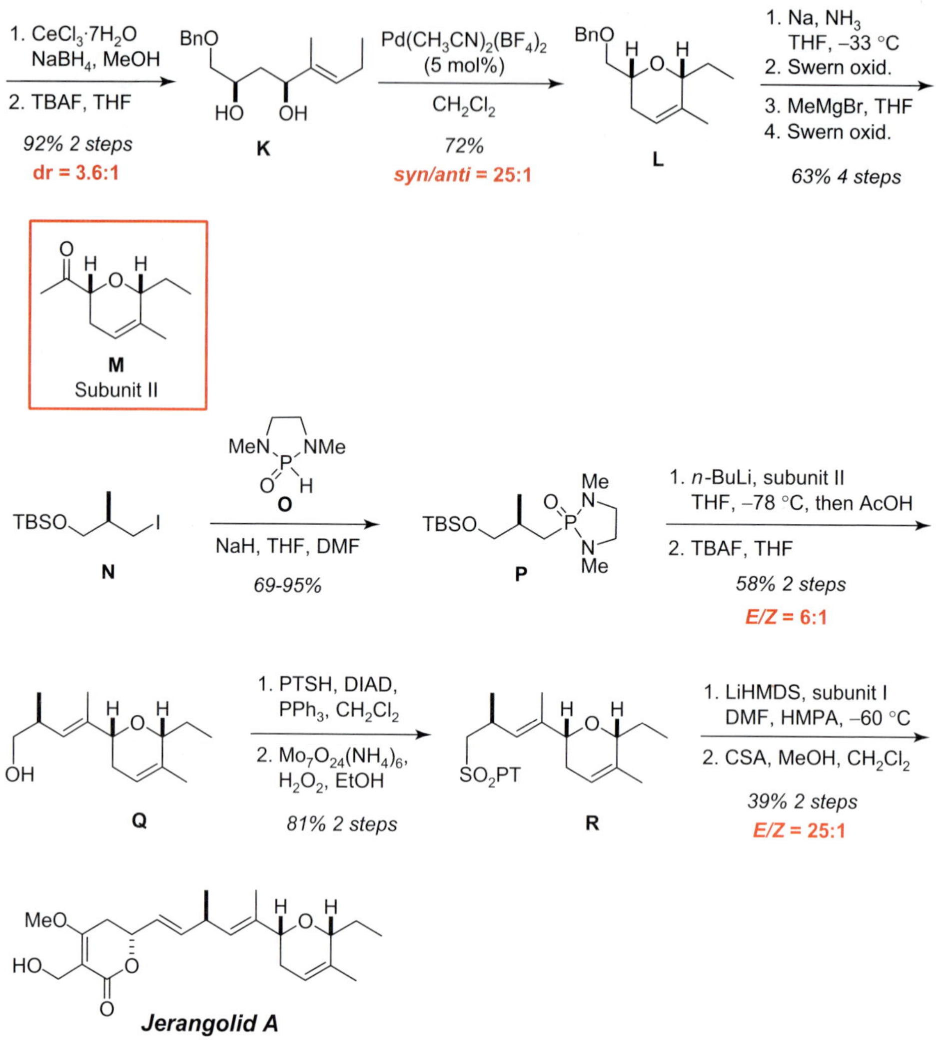

1. CeCl$_3$·7H$_2$O
   NaBH$_4$, MeOH

2. TBAF, THF

92% 2 steps
dr = 3.6:1

K

Pd(CH$_3$CN)$_2$(BF$_4$)$_2$
(5 mol%)

CH$_2$Cl$_2$

72%
***syn/anti* = 25:1**

L

1. Na, NH$_3$
   THF, −33 °C
2. Swern oxid.
3. MeMgBr, THF
4. Swern oxid.

63% 4 steps

M
Subunit II

N

MeN$\diagdown$P$\diagup$NMe
       ‖
       O H

O

NaH, THF, DMF

69-95%

P

1. *n*-BuLi, subunit II
   THF, −78 °C, then AcOH

2. TBAF, THF

58% 2 steps
E/Z = 6:1

Q

1. PTSH, DIAD,
   PPh$_3$, CH$_2$Cl$_2$

2. Mo$_7$O$_{24}$(NH$_4$)$_6$,
   H$_2$O$_2$, EtOH

81% 2 steps

R

1. LiHMDS, subunit I
   DMF, HMPA, −60 °C

2. CSA, MeOH, CH$_2$Cl$_2$

39% 2 steps
E/Z = 25:1

Jerangolid A

Key reactions: *Methoxide-mediated conjugate addition, then lactonization (**B** to **C**); vinyl iodide/Mg exchange, then hydroxymethylation (**D** to **E**); cationic Pd-cycloetherification (**K** to **L**); phosphonamide anion olefination (**P** to **Q**); PTSH = 1-phenyltetrazole-5-thiol*

Figure 12.12 *(continued)*.

Commentary: Having recognized the potential utility of (*S*)-glycidol and the (*R*)-Roche acid as apparent chirons to elaborate various subunits in jerangolid A, there remained to determine the order in which they would be assembled, particularly in presence of a sensitive unsaturated lactone. Although the elaboration of this subunit from glycidol derivative **A** proved uneventful, the aldehyde **F** was found to be unstable. The initial plan to use Pd(CH$_3$CN)$_2$Cl$_2$ as a catalyst in the

6-*endo-trig* ring closure of epimeric pairs of allylic alcohols corresponding to **K**, according to Uenishi's protocol [33], afforded mixtures of *syn-* and *anti*-dihydropyrans regardless of the stereochemistry of the diol in substrate **K**. However, a 3.6:1 diastereomeric mixture of allylic alcohols **K** in the presence of a cationic palladium catalyst $(Pd(CH_3CN)_2(BF_4)_2)$ led to *syn*-diastereomer **L** (ring B). In fact, 10 mol% of $BF_3 \cdot Et_2O$ alone was an effective catalyst for the same ring closure.

Apparently the pseudoaxial disposition of the terminal ethyl group in the *trans*-isomer is energetically less favored compared to that in the *syn*-isomer, even though the latter experiences $A^{1,2}$-strain. The stereocontrolled formation of the *trans*-trisubstituted olefin **Q** utilizing a phosphonamide anion **P** and ketone **M**, is noteworthy in spite of the modest yield.

12.8
Miscellanea

The utility of naturally occurring 3- and 4-carbon hydroxy acids in total synthesis has also been discussed in Chapters 6 and 8. Their short carbon chain-length is ideally suited to be used as chirons in total synthesis of Category I-type molecules. The proximity of hydroxyl and carbonyl groups can be exploited in orthogonal ways. Hydroxy acids derived from Nature provide only partial substructure convergence in segments of Category II and III-type target molecules because of their larger and more expansive structural landscapes. Nevertheless, chirons derived from lactic, malic, tartaric, and the versatile Roche acids have been valuable synthetic building blocks in a large number of natural products harboring one or more hydroxyl or *C*-methyl groups respectively. A selection of total syntheses based on hydroxy acids as chirons are shown in Figure 12.13 [34]. Examples of the utilization of the enantiomeric Roche acids in natural product synthesis were shown in Chapter 8 (Figure 8.25). Tartaric acids have been used as versatile chiral auxiliaries in the form of modified 1,3-dioxolanes (TADDOLs) [35].

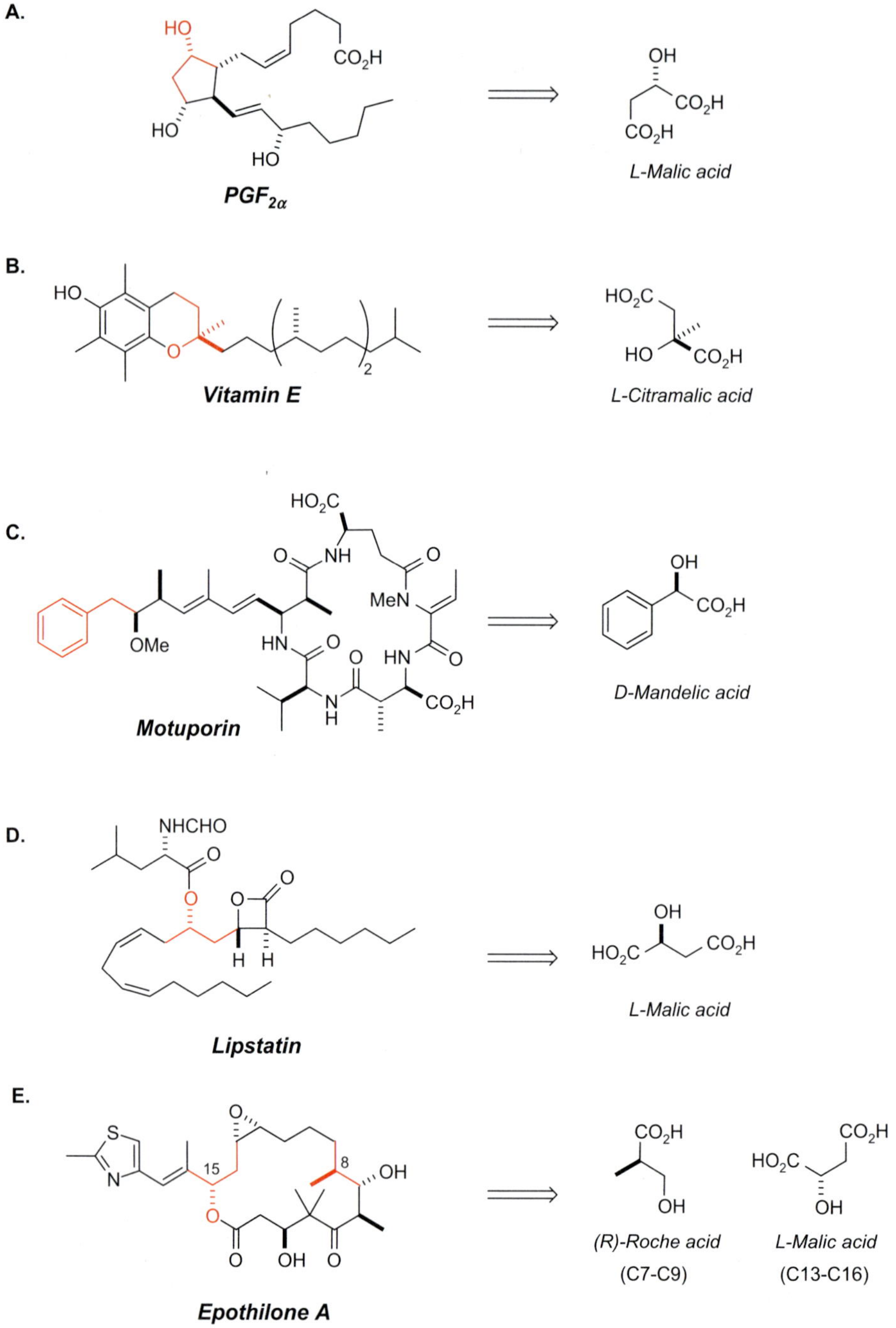

A.

PGF$_{2\alpha}$

L-Malic acid

B.

Vitamin E

L-Citramalic acid

C.

Motuporin

D-Mandelic acid

D.

Lipstatin

L-Malic acid

E.

Epothilone A

(R)-Roche acid
(C7-C9)

L-Malic acid
(C13-C16)

Figure 12.13 A selection of natural products and their hydroxy acid precursors (apparent and partially hidden substructures).

F.

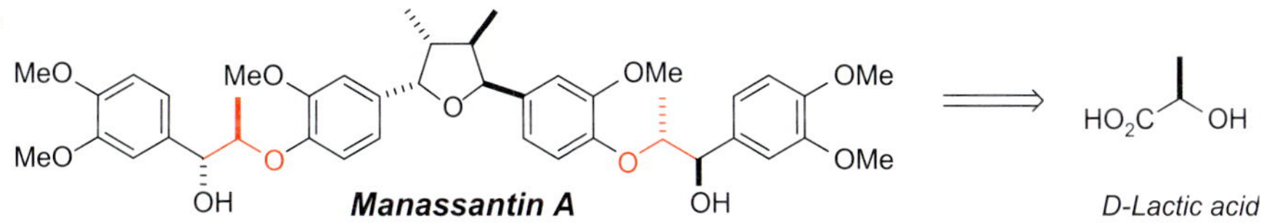

Manassantin A

D-Lactic acid

G.

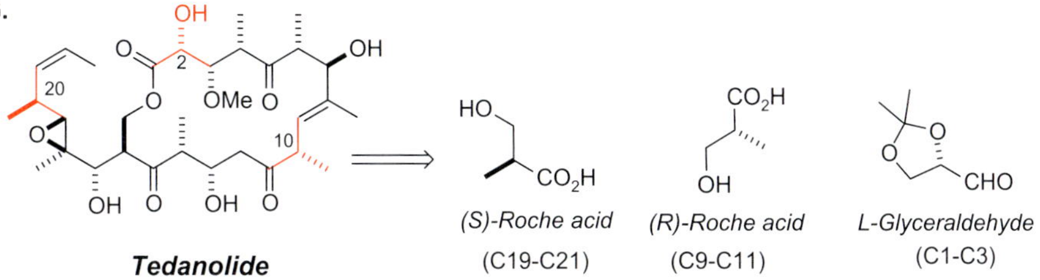

Tedanolide

(S)-Roche acid
(C19-C21)

(R)-Roche acid
(C9-C11)

L-Glyceraldehyde
(C1-C3)

H.

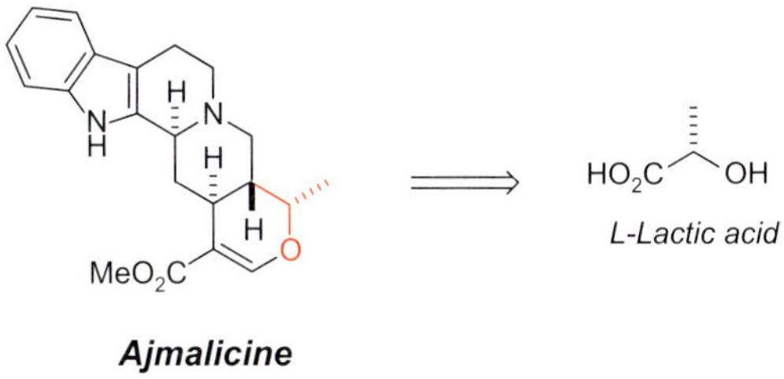

Ajmalicine

L-Lactic acid

I.

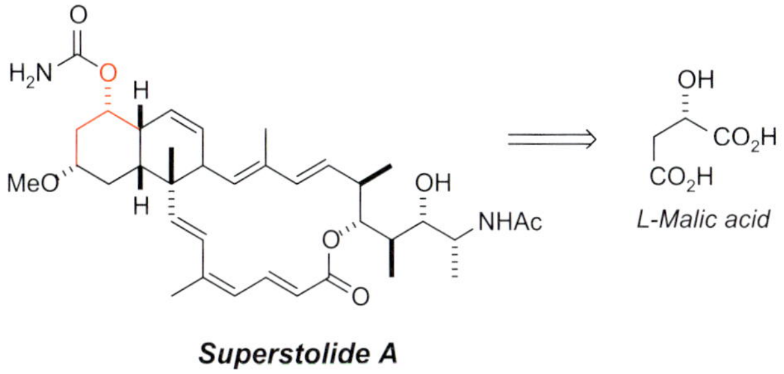

Superstolide A

L-Malic acid

J.

Saxitoxin

· 2 TFA

L-Malic acid

**Decarbamoyl
saxitoxin**

2 Cl⁻

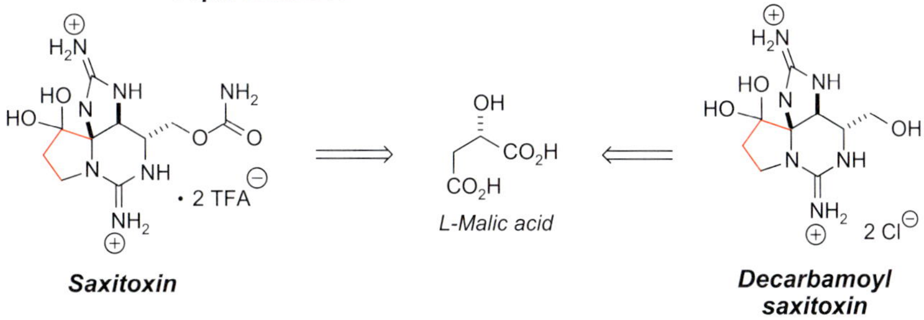

Figure 12.13 *(continued)*.

References

1. For example, see: (a) *α-Hydroxy Acids in Enantioselective Synthesis*, Coppola, G. M.; Schuster, H. F., VCH, Weinheim, 1997; (b) *Tartaric and Malic Acids in Synthesis*, Gawronski, J.; Gawronska, K., Wiley, N.Y., 1999; see also: (c) Seebach, D. in *Modern Synthetic Methods*, Scheffold, R.; Ed.; vol. 2, p. 91, Otto Salle Verlag, Frankfurt, 1980.

2. Dvorak, C. A.; Schmitz, W. D.; Poon, D. J.; Pryde, D. C.; Lawson, J. P.; Ames, R. A.; Meyers, A. I. *Angew. Chem. Int. Ed.* 2000, **39**, 1664.

3. (a) Noyori, R.; Ohkuma, T.; Kitamura, M.; Takaya, H.; Sayo, N.; Kumobayashi, H.; Akutagawa, S. *J. Am. Chem. Soc.* 1987, **109**, 5856; (b) Taber, D. F.; Silverberg, L. J. *Tetrahedron Lett.* 1991, **32**, 4227; see also: (c) Yang, H. W.; Romo, D. *J. Org. Chem.* 1998, **63**, 1344.

4. Burgess, E. M.; Penton, H. R.; Taylor, E. A.; Williams, W. M. *Org. Syn.*, Wiley, NY, 1988; Coll. Vol. VI, p. 788.

5. (a) Critcher, D. J.; Connolly, S.; Wills, M. *J. Chem. Soc., Chem. Commun.* 1995, 139; (b) Critcher, D. J.; Connolly, S.; Wills, M. *Tetrahedron Lett.* 1995, **36**, 3763; (c) Critcher, D. J.; Connolly, S.; Wills, M. *J. Org. Chem.* 1997, **62**, 6638; for a review, see: (d) Shiina, I. *Chem. Rev.* 2007, **107**, 239.

6. Blanchette, M. A.; Choy, W.; Davies, J. T.; Essenfeld, A. P.; Masamune, S.; Roush, W. R.; Sasaki, T. *Tetrahedron Lett.* 1984, **25**, 2183.

7. (a) Corey, E. J.; Chaykovsky, M. *J. Am. Chem. Soc.* 1962, **84**, 867; (b) Corey, E. J.; Chaykovsky, M. *J. Am. Chem. Soc.* 1964, **86**, 1640.

8. Inanaga, J.; Hirata, K.; Saeki, H.; Katsuki, T.; Yamaguchi, M. *Bull. Chem. Soc. Jpn.* 1979, **52**, 1989.

9. (a) Hart, D. W.; Blackburn, T. F.; Schwartz, J. *J. Am. Chem. Soc.* 1975, **97**, 679; (b) Schwartz, J.; Labinger, J. A. *Angew. Chem. Int. Ed.* 1976, **15**, 333.

10. Takai, K.; Tagashira, M.; Kuroda, T.; Utimoto, K.; Nozaki, H. *J. Am. Chem. Soc.* 1986, **108**, 6048.

11. Jin, H.; Uenishi, J.; Christ, W. S.; Kishi, Y. *J. Am. Chem. Soc.* 1986, **108**, 5644.

12. (a) Niwa, H.; Wakamatsu, K.; Yamada, K. *Tetrahedron Lett.* 1989, **30**, 4543; (b) Kigoshi, H.; Niwa, H.; Yamada, K.; Stout, T. J.; Clardy, J. *Tetrahedron Lett.* 1991, **32**, 2427.

13. (a) Takemoto, Y.; Baba, Y.; Saha, G.; Nakao, S.; Iwata, C.; Tanaka, T.; Ibuka, T. *Tetrahedron Lett.* 2000, **41**, 3653; (b) Baba, Y.; Saha, G.; Nakao, S.; Iwata, C.; Tanaka, T.; Ibuka, T.; Ohishi, H.; Takemoto, Y. *J. Org. Chem.* 2001, **66**, 81.

14. Takahashi, T.; Watanabe, H.; Kitahara, T. *Heterocycles* 2002, **58**, 99.

15. Denmark, S. E.; Yang, S.-M. *J. Am. Chem. Soc.* 2004, **126**, 12432.

16. (a) Brown, H. C.; Bhat, K. S.; Randad, R. S. *J. Org. Chem.* 1989, **54**, 1570; (b) Racherla, U. S.; Brown, H. C. *J. Org. Chem.* 1991, **56**, 401.

17. Fox, H. H.; Yap, K. B.; Robbins, J.; Cai, S.; Schrock, R. R. *Inorg. Chem.* 1992, **31**, 2287.

18. (a) Denmark, S. E.; Yang, S.-M. *J. Am. Chem. Soc.* 2002, **124**, 2102; (b) Itami, K.; Nokami, T.; Ishimura, Y.; Mitsudo, K.; Kamei, T.; Yoshida, J.-I. *J. Am. Chem. Soc.* 2001, **123**, 11577; for a recent review, see: Denmark, S. E.; Liu, J. H.-C. *Angew. Chem. Int. Ed.* 2010, **49**, 2978.

19. (a) Buszek, K. R.; Sato, N.; Jeong, Y. *J. Am. Chem. Soc.* 1994, **116**, 5511; (b) Buszek, K. R.; Sato, N.; Jeong, Y. *Tetrahedron Lett.* 2002, **43**, 181; for reviews, see: (c) Dieters, A.; Martin, S. F. *Chem. Rev.* 2004, **104**, 2199; (d) Shiina, I. *Chem. Rev.* 2007, **107**, 239; see also reference [5].

20. For example, see: (a) Illuminati, G.; Mandolini, L. *Acc. Chem. Res.* 1981, **14**, 95; (b) Miller, S. J.; Kim, S.-H.; Chen, Z.-R.; Grubbs, R. H. *J. Am. Chem. Soc.* 1995, **117**, 2108; (c) Tarling, C. A.; Holmes, A. B.; Markwell, R. E.; Pearson, N. D. *J. Chem. Soc., Perkin Trans. I* 1999, 1695; see also: (d) Hanessian, S. Guesné, S.; Chénard, E. *Org. Lett.* 2010, **12**, 1816.

21. Shiina, I.; Hashizume, M.; Yamai, Y.; Oshiumi, H.; Shimazaki, T.; Takasuma, Y.; Ibuka, R. *Chem. Eur. J.* 2005, **11**, 6601; see also ref. [19].

22. (a) O'Sullivan, P. T.; Buhr, W.; Fuhry, M. A. M.; Harrison, J. R.; Davies, J. E.; Feeder, N.; Marshall-Burton, J. W.; Holmes, A. B. *J. Am. Chem. Soc.* 2004, **126**, 2194; (b) Radosevich, A.; Chan, V. S.; Shih, H.-W.; Toste, F. D. *Angew. Chem. Int. Ed.* 2008, **47**, 3755.

23. McWilliams, J. C.; Clardy, J. *J. Am. Chem. Soc.* 1994, **116**, 8378.

24. For example, see: Kozikowski, A. P.; Lee, J. *J. Org. Chem.* 1990, **55**, 863.

25. Lee, E.; Park, C. M.; Yun, J. S. *J. Am. Chem. Soc.* 1995, **117**, 8017.

26. For example, see: (a) Lee, E.; Tae, J. S.; Lee, C.; Park, C. M. *Tetrahedron Lett.* 1993, **34**, 4831; (b) Lee, E.; Park, C. M. *J. Chem. Soc., Chem. Commun.* 1994, 293; see also: (c) Hanessian, S.; Dhanoa, D. S.; Beaulieu, P. L. *Can. J. Chem.* 1987, **65**, 1859.

27. Sheldrake, H. M.; Jamieson, C.; Burton, J. W. *Angew. Chem. Int. Ed.* 2006, **45**, 7199.

28. Lambert, T. H.; Danishefsky, S. J. *J. Am. Chem. Soc.* 2006, **128**, 426; for synthesis of (±)-USC1025A, see: Hoye, T. R.; Dvornikvos, V. *J. Am. Chem. Soc.* 2006, **128**, 2550.

29. Wilson, R. M.; Jen, W. S.; MacMillan, D. W. C. *J. Am. Chem. Soc.* 2005, **127**, 11616.

30. Nozaki, K.; Oshima, K.; Utimoto, K. *Bull. Chem. Soc. Jpn.* 1991, **64**, 403.

31. Hanessian, S.; Focken, T.; Oza, R. *Org. Lett.* 2010, **12**, 3172.

32. Pospišil, J.; Markó, I. E. *J. Am. Chem. Soc.* 2007, **129**, 3516.

33. Kawai, N.; Hande, S. M.; Uenishi, J. *Tetrahedron* 2007, **63**, 9049.

34. $PGF_{2\alpha}$: Paul, K. G.; Johnson, F. *J. Am. Chem. Soc.* 1976, **98**, 1285.; vitamin E: Barner, R.; Schmid, M. *Helv. Chim. Acta.* 1979, **62**, 2384.; motuporin: Valentekovich, R. J.; Schreiber, S. L. *J. Am. Chem. Soc.* 1995, **117**, 9069.; lipstatin: Pommier, A.; Pons, J.-M.; Kocienski, P. J. *J. Org. Chem.* 1995, **60**, 7334.; epothilone A: Mulzer, J. Mantoulidis, A.; Ohler, E. *Tetrahedron Lett.* 1997, **38**, 7725.; Mulzer, J. Mantoulidis, A.; Ohler, E. *J. Org. Chem.* 2000, **65**, 7456.; manassantin A: Hanessian, S.; Reddy, G. J.; Chahal, N. *Org. Lett.* 2006, **8**, 5477.; tedanolide: Smith, A. B. III.; Lee, D. *J. Am. Chem. Soc.* 2007, **129**, 10957.; ajmalicine: Hatakeyama, S.; Saijo, K.; Takano, S. *Tetrahedron Lett.* 1985, **26**, 865.; superstolide: Tortosa, M.; Yakelis, N. A.; Roush, W. R. *J. Am. Chem. Soc.* 2008, **130**, 2722; saxitoxin and decarbamoyl saxitoxin: Iwamoto, O.; Shinohara, R.; Nagasawa, K. *Chem. Asian J.* 2009, **4**, 277.

35. Seebach, D.; Beck, A. K.; Heckel, A. *Angew. Chem. Int. Ed.* 2001, **40**, 92; see also Pellissier, H. *Tetrahedron* 2008, **64**, 10279.

13
Total Synthesis from Terpenes

Naturally occurring terpenes are extremely useful acyclic and carbocyclic building blocks for synthesis [1]. Even more useful are the multitude of chemically transformed terpenes, providing access to chiral, non-racemic acyclic and carbocyclic molecules containing one or more stereogenic centers with appended functionality (see Chapter 7, section 7.5). Not all naturally occuring terpenes are equally abundant in both enantiomeric forms. Nevertheless, native terpenes and their modified chirons have been extensively used in natural product synthesis as highlighted in this chapter, and elsewhere in this book (see Chapters 8 and 16). The examples chosen herein represent interesting applications in the synthesis of Category I-type molecules.

13.1
Picrotoxinin

The first total synthesis of picrotoxinin was reported by Corey and Pierce [2] in 1979 (Figure 13.1). In spite of the complexity of the heavily functionalized pentacyclic structure of picrotoxinin, the combination of the isopropenyl group and the angular methyl group on the six-membered ring provided a visual connection with (R)-carvone as a carbocyclic starting material.

Treatment of N,N-dimethylhydrazone A with LDA in THF and HMPA, followed by addition of 3-bromopropionaldehyde dimethyl acetal, gave a mixture of B and its epimer (dr = 3:2) in 85% yield (Figure 13.1) [3]. Hydrolysis of the dimethylhydrazone to the ketone, and acid-catalyzed aldol carbocyclization with the carboxaldehyde tether furnished the desired alcohol with modest diastereoselectivity (dr = 3:2). Benzoylation led to C after purification by preparative HPLC, following which, treatment with lithium acetylide produced a single tertiary alcohol in 99% yield. Bromoetherification in the presence of NBS gave cyclic ether D in quantitative yield. Hydroboration of the triple bond with dicyclohexylborane, followed by oxidation of the intermediate borane, led to an unstable aldehyde that was converted to dithiolane intermediate E in good overall yield. A series of oxidative reactions [4] led first to diketone F, which, upon generation of the aldehyde from its dithiolane precursor, underwent intramolecular aldol cyclization to the hydrindene

Design and Strategy in Organic Synthesis: From the Chiron Approach to Catalysis, First Edition.
Stephen Hanessian, Simon Giroux, and Bradley L. Merner.
© 2013 Wiley-VCH Verlag GmbH & Co. KGaA. Published 2013 by Wiley-VCH Verlag GmbH & Co. KGaA.

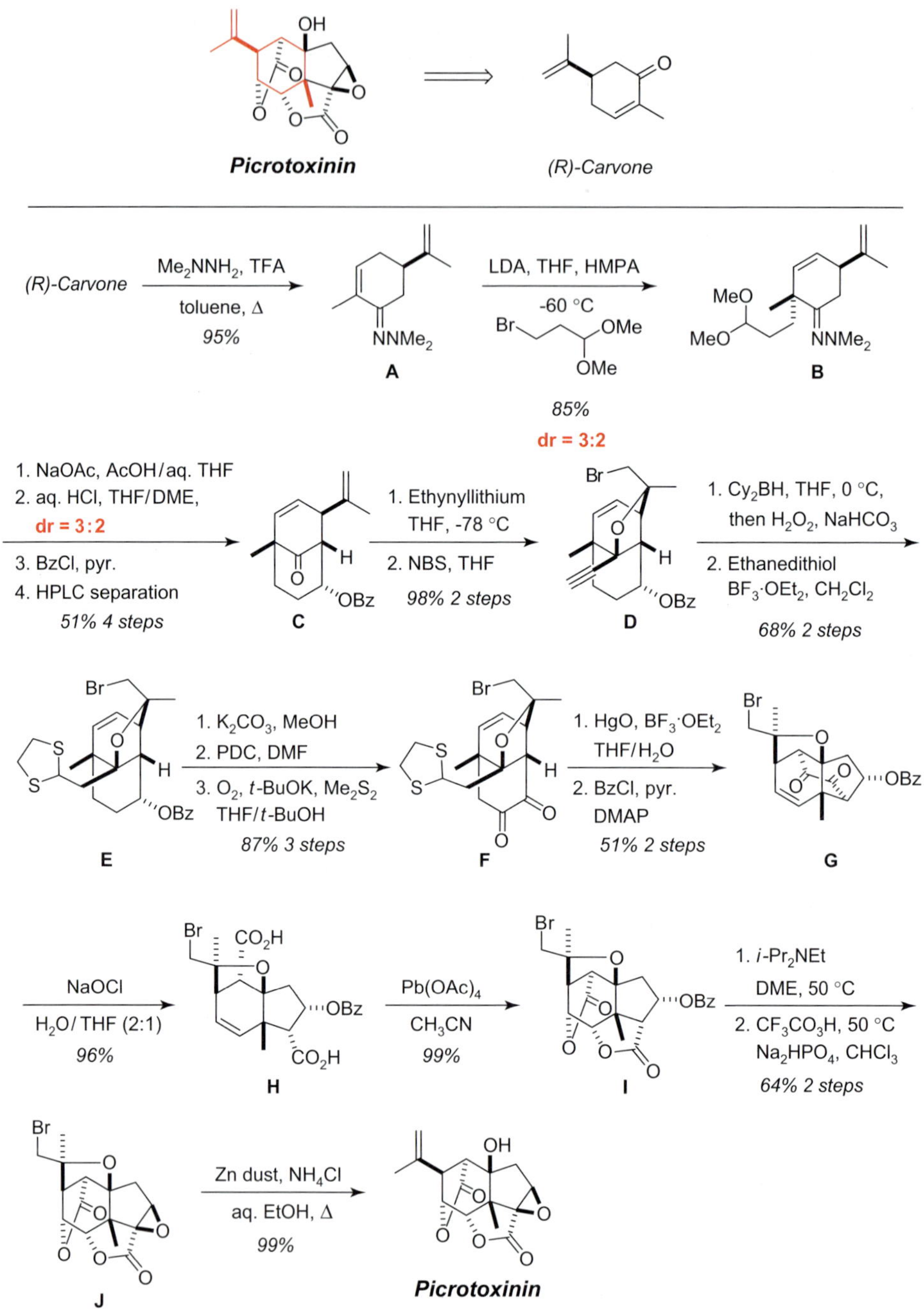

Figure 13.1 Total synthesis of picrotoxinin from (*R*)-carvone (apparent substructure).

Key reactions: *γ-Extended hydrazone enolate alkylation (**A** to **B**); bromocycloetherification (**C** to **D**); enolate oxidation (**E** to **F**); intramolecular aldol cyclization (**F** to **G** via aldehyde); oxidative cleavage to diacid (**G** to **H**); bis-lactonization (**H** to **I**); elimination/epoxidation (**I** to **J**); restoration of isopropenyl group (**I** to **target**)*

Relevant mechanisms and transformations:

Figure 13.1 *(continued)*.

diketone. Benzoylation afforded **G**, which was correlated with a degradation product obtained from a sample of authentic natural product. Oxidative cleavage to dicarboxylic acid **H**, followed by bis-lactonization in the presence of lead tetraacetate gave **I** in quantitative yield [5]. Treatment with *N,N*-diisopropylethylamine (Hünig's base) in DME at 50 °C effected the desired *β*-elimination to the olefin, which was epoxidized in the presence of excess peroxytrifluoroacetic acid to give the epoxide **J**. The identity of this penultimate intermediate was verified by chemical conversion of picrotoxinin to the same bromide in the presence of NBS. Finally, treatment of **J** with zinc dust and NH_4Cl in ethanol afforded synthetic picrotoxinin.

Commentary: The total synthesis of picrotoxinin starting with (*R*)-carvone is an exemplary testament to a visual *relational* analysis. The complex architecture of five contiguous interconnected rings including two *γ*-lactones, and the presence of eight contiguous stereogenic centers can be momentarily "simplified" by recognizing the isopropenyl and *C*-methyl groups as the peripheral appendages of (*R*)-carvone. However, methodology had to be devised to systematically introduce carbon and oxygen functionality in a stereochemically controlled manner. A key transformation was the intermolecular *α*-alkylation of the *γ*-extended enolate of *N,N*-dimethylhydrazone **A** (see relevant mechanism A, Figure 13.1). Once the bicyclo[3.3.1]nonanone structure was attained, as in **C**, a systematic formation of C–C and C–O bonds would follow, controlled by the topology of the bicyclic system, and the equatorial orientation of the isopropenyl group. Thus, with the highly stereocontrolled installation of the acetylenic appendage, *all* the required carbon atoms in picrotoxinin were in place. The plan to install an epoxide function in an advanced intermediate was impeded by the presence of a competing electron rich double bond in the isopropenyl side-chain. Thus, Corey and Pierce used a bromocycloetherification reaction to effectively "protect" the exocyclic double bond as in intermediate **D**, with the full knowledge that the restoration of the isopropenyl appendage would be required at a later stage. The acetylenic group provided the functionality necessary to generate an aldehyde that would undergo a Lewis acid-mediated intramolecular aldol carbocyclization to produce the hydrindene diketone, which was isolated as the benzoate ester **G** (see relevant mechanism B, Figure 13.1). Aided by the compact topology of the intermediates and the anticipated proximity of functional groups, it was possible to arrive at the dicarboxylic acid intermediate **H**. However, major difficulties in lactone formation led to the application of a seldom used method based on a lead tetraacetate oxidative bis-lactonization, which worked remarkably well to give **I**. It was at this juncture that the stereoselectivity of the intramolecular aldol carbocyclization (**F** to **G**), would be appreciated, since it allowed for a smooth *β*-elimination to the endocyclic olefin, which was converted to a single epoxide **J**. Finally, treatment of **J** with zinc dust effected electron transfer and fragmentation, restoring the isopropenyl group and the tertiary alcohol as the last step in the synthesis (see relevant mechanism C, Figure 13.1). It is remarkable that the entire (*R*)-carvone skeleton was embedded in the target structure, and that the remaining carbon atoms were provided by 3-bromopropionaldehyde dimethyl acetal (**A** to **B**) and acetylide anion (**C** to **D**).

The author's concluding sentence that "picrotoxinin now joins the list of long-known, but fiercely defiant naturally occurring substances which have been produced by total synthesis" was very much *à propos* considering the year of publication (1979).

13.2
Eucannabinolide

The first total synthesis of the germacranolide eucannabinolide was reported by Still and coworkers in 1983 [6]. The choice of (*S*)-carvone as a starting material was based on prior results obtained from the same laboratory while exploring anionic oxy-Cope mediated ring expansions.

Stereoselective reduction of (*S*)-carvone with LiAlH$_4$, was followed by a hydroxyl-directed epoxidation, and conversion to BOM ether **A** in 70% overall yield (Figure 13.2). Epoxide opening with PhSeK, oxidation to the selenoxide, and elimination gave the tertiary alcohol, which underwent an oxidative [3,3]-sigmatropic rearrangement upon treatment with CrO$_3$ to afford enone **B**. Addition of the 1-lithiated 3-cyclobutenone dimethyl acetal at −70 °C gave a single diastereomer **C**. The anionic oxy-Cope ring expansion [7] was carried out in the presence of KHMDS in DME at 85 °C to afford a 1:1 mixture of C-7 epimers, which was cleanly equilibrated in the presence of K$_2$CO$_3$ in MeOH to afford **D** (15:1 *S:R*, at C-7). Exposure to acid, followed by Baeyer-Villiger oxidation of the resulting cyclobutanone gave lactone **E**. Stereoselective reduction to the alcohol, and treatment with K$_2$CO$_3$ in MeOH effected lactone transposition leading to **F** in good overall yield over four steps. Oxidation to the ketone and equilibrating via the enolate with DBU in THF gave a 7:1 mixture of the C-7 epimerized lactone **G**. Stereoselective reduction with NaBH$_4$ followed by chemoselective cleavage of the BOM ether group led to **H**. Protection of the alcohol as the TMS ether and α-hydroxymethylation of the lactone enolate with formaldehyde was followed by mesylation and β-elimination to give the α-methylene lactone **I**. Deprotection of the two silyl ethers, followed by selective acetylation at C-3 gave **J**, which was esterified with dihydroxytiglic acid acetonide and then subjected to gentle acidic hydrolysis to give eucannabinolide.

Commentary: On face value, the visual relationship between the ten-membered germacranolide core structure in eucannabinolide and (*S*)-carvone is virtually non-existent, even to the trained synthetic chemist's eye (Figure 13.2). Thus, a visual *relational* analysis without prior knowledge or experience is hard to reconcile with the choice of a (*S*)-carvone as a starting material for synthesis. However, a visual *reflexive* analysis in Still's mind's eye was reminiscent of previous results in his laboratory that had demonstrated the utility of an anionic oxy-Cope rearrangement in the facile ring expansion of appropriately functionalized divinyl tertiary alcohols derived from isopiperitenone to the germacranone skeleton (Figure 13.3) [7b,c].

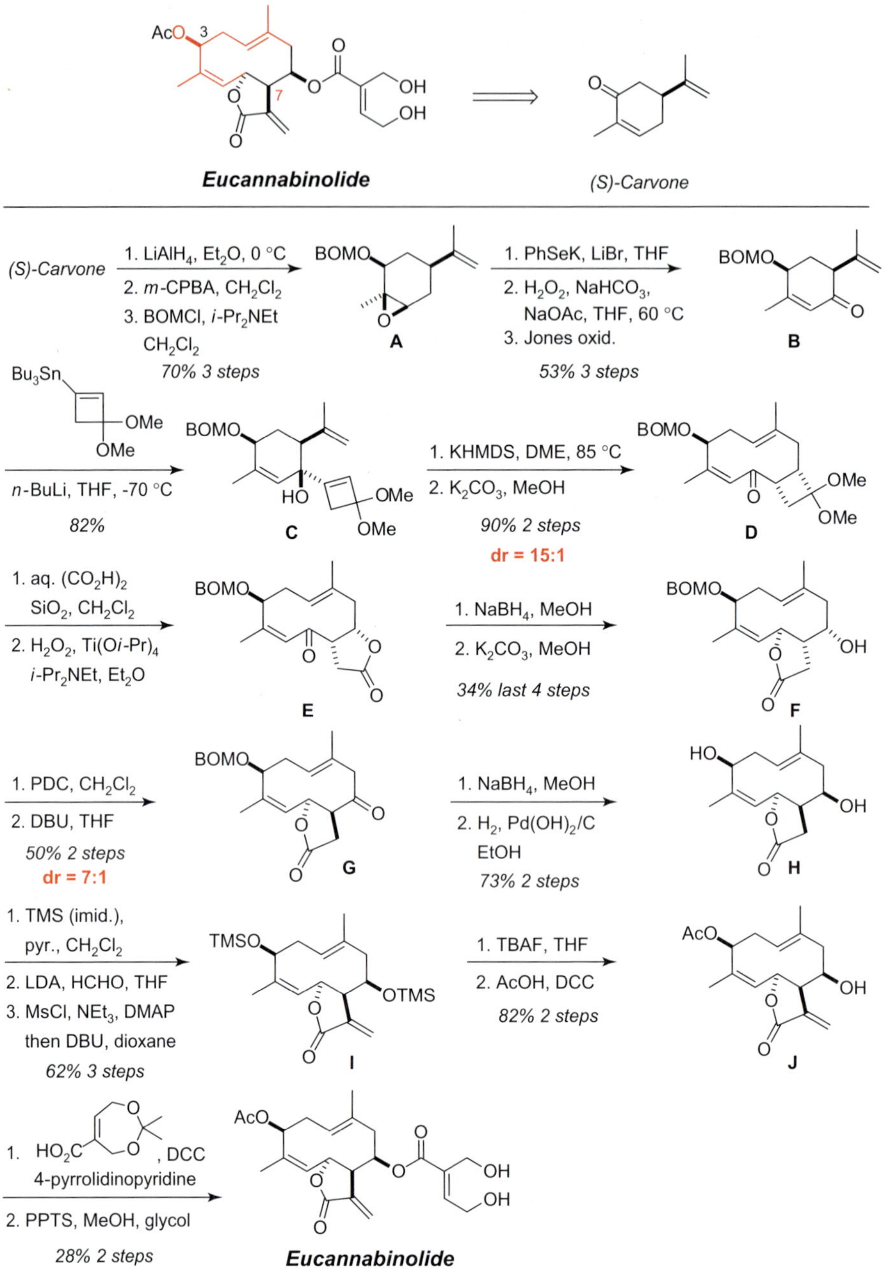

Figure 13.2 Total synthesis of eucannabinolide from (*S*)-carvone (hidden substructure).

Key reactions: *Enone formation (**A** to **B**); anionic oxy-Cope rearrangement (**C** to **D**); Baeyer-Villiger oxidation (**D** to **E**); Lactone transposition (**E** to **F**); lactone epimerization (**F** to **G**); selective acetylation (**I** to **J**)*

Relevant mechanisms and transformations:

A.

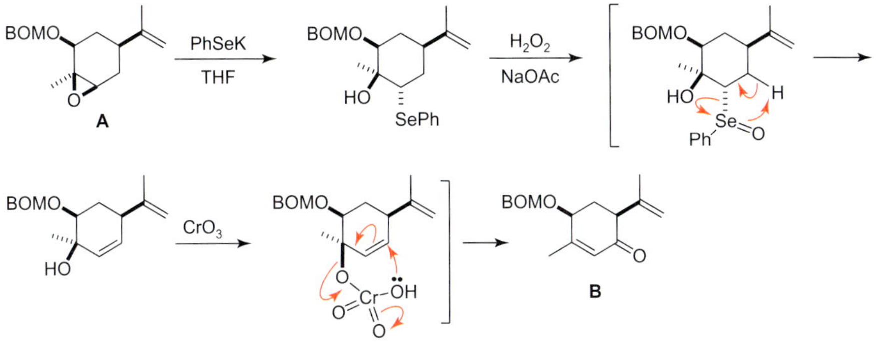

B.

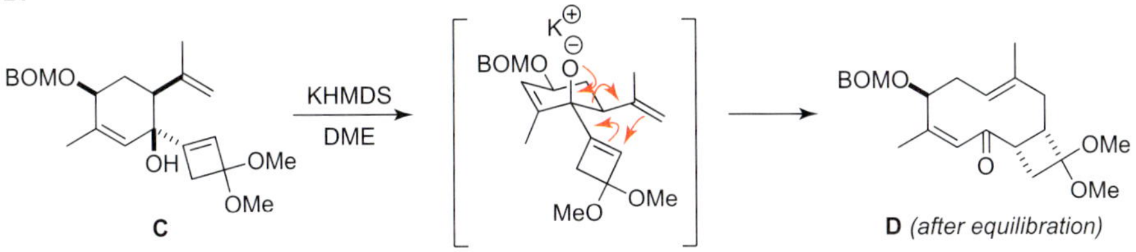

C.

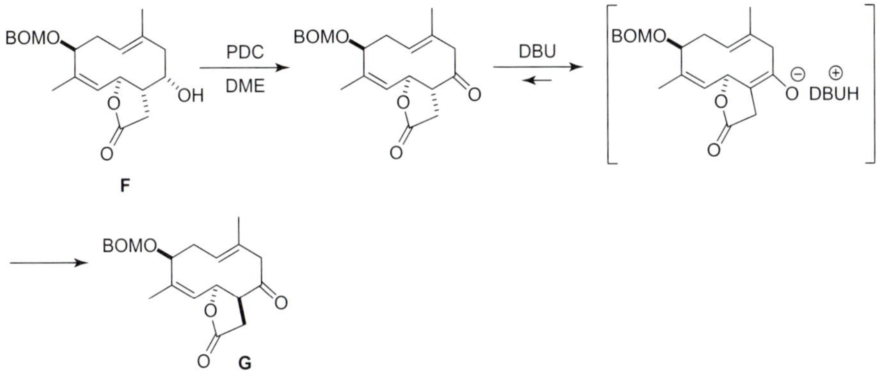

Figure 13.2 *(continued).*

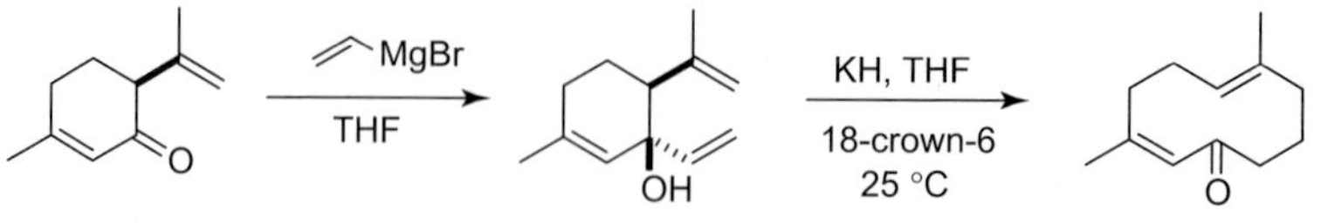

Figure 13.3 Anionic oxy-Cope-mediated ring expansion.

Armed with this knowledge, it became clear that a pathway from (S)-carvone to the functionalized germacrane could be a viable strategy toward eucannabinolide. To this end, a "tailor-made" cyclobutenyl lithium precursor reagent had to be prepared (Figure 13.2). The desire to utilize (S)-carvone (as in the previous model studies) necessitated the introduction of a ketone function in order to apply the anionic oxy-Cope rearrangement. This was done through judicious manipulation of the epoxide in **A** (see relevant mechanism A, Figure 13.2). The subsequent sequences involving the stereoselective introduction of the masked cyclobutenone motif, the anionic oxy-Cope ring expansion, and lactone transposition were done with remarkable substrate and thermodynamic control.

The spatial orientation of the isopropenyl and *O*-BOM groups ensured an α-face attack by the organometallic reagent leading to the tertiary alcohol **C** in excellent yield. The anionic oxy-Cope ring expansion was realized with high efficiency, and the product **D** was obtained with high diastereoselectivity (dr = 15:1) (see relevant mechanism B, Figure 13.2). It was assumed that the anionic oxy-Cope rearrangement proceeded via a chair-like transition state. Kinetic protonation led initially to a 1:1 mixture of *cis*- and *trans*-C-6–C-7 diastereomers, which could be equilibrated upon stirring in powdered K_2CO_3 in favor of the desired C-7 (S)-configured lactone **G**.

With the critical ring-expansion accomplished, there remained only to adjust oxidation states and to effect appropriate stereochemical alterations. However, the clever utilization of the fused cyclobutanone precursor **C** in the anionic oxy-Cope ring expansion to **D**, also ensured the regioselective Baeyer-Villiger oxidation to lactone **E**. Guided by MM2 calculations [8], it was anticipated that the transposed C-6/C-7 lactone **F** would be more stable than the C-7/C-8 lactone **E** by about 1 kcal/mol. Indeed, base-induced equilibration confirmed the thermodynamic preference for **F**. Oxidation to the ketone and treatment with DBU led to a 7:1 ratio of the C-7-epimerized lactone **G** as the major product (see relevant transformation C, Figure 13.2).

The total synthesis of eucannabinolide encompassed less than 25 steps and was characterized by elegant applications of fundamental concepts in organic synthesis. The utilization of (S)-carvone and tributyl(3,3-dimethoxy-cyclobut-1-enyl)stannane accounted for all the skeletal carbon atoms of the target molecule except for the α-methylene group, which originated from formaldehyde.

13.3
Trilobolide and Thapsivillosin F

The first total syntheses and stereochemical confirmation of guaianolides belonging to the "thapsigargins", exemplified by trilobolide and thapsivollosin F, were reported by Ley and coworkers [9] (Figure 13.4).

The known (S)-carvone epoxide **A** was converted to the chloroketone **B** by treatment with LiCl in THF containing TFA, followed by protection as the THP ether [10]. Upon reaction with NaOMe in MeOH at 0 °C, smooth ring contraction took place (Farvoskii rearrangement) to give the tetrasubstituted cyclopentane methyl ester **C** in excellent overall yield after protective group manipulation. Reduction of the ester group to the alcohol, protection as the PMB ether, and oxidative cleavage of the double bond in the isopropenyl appendage gave the ketone **D**. A series of sequential branching reactions first gave the tertiary alcohol corresponding to **E** as the major isomer, followed by conversion to an aldehyde and reaction with ethoxyvinyllithium. O-Silylation ultimately gave the diene **F** in excellent overall yield. Ring-closing metathesis was achieved in the presence of 2.5 mol% of the Grubbs second generation catalyst to give bicyclic enol ether **G**. Osmylation according to the Sharpless protocol in the absence of a chiral ligand led predominantly to **H** as the major isomer.

Esterification of the hydroxyl group with 2-(diethoxyphosphoryl)propionic acid, followed by treatment with NaH in refluxing THF effected ring-closure to give butenolide **I**. Difficulties in the direct dihydroxylation of the double bond in **I**, required an alternate approach. Thus, reduction to the diol, and preferential acetylation followed by protection with a MOM group gave **J**. Sharpless asymmetric dihydroxylation in the presence of quinuclidine as a basic ligand gave **K**, which was selectively deprotected, oxidized with TPAP, and then converted to the acetal **L**. Adjustment of oxidation states and enol ether formation led to **M**, which was epoxidized and then opened to give **N**. Treatment with PhSeBr initiated a catalytic reaction that resulted in an unexpected deoxygenation of the C-2 stereogenic center and ultimately led to enone **O**. There remained only to reduce the ketone to the alcohol affording the desired isomer in a 4:1 ratio, and to effect appropriate esterifications [11] and deprotections to give intermediates **P** and **Q**. Variations in the anhydride used afforded the intended natural products trilobolide and thapsivillosin F.

Commentary: Once again, knowledge-based analysis of the tricyclic guaianolide-type structure of trilobolide and thapsivillosin F led Ley and coworkers to (S)-carvone as a starting terpene. The literature has documented examples of ring-contracted cyclopentane analogues derived from (R)- and (S)-carvone (see Chapter 7, section 7.9). In considering intermediate **C** as a precursor to the cyclopentane ring of trilobolide, it was evident that extensive redox chemistry and configurational adjustments would be needed. The highly successful use of ring-closing metathesis

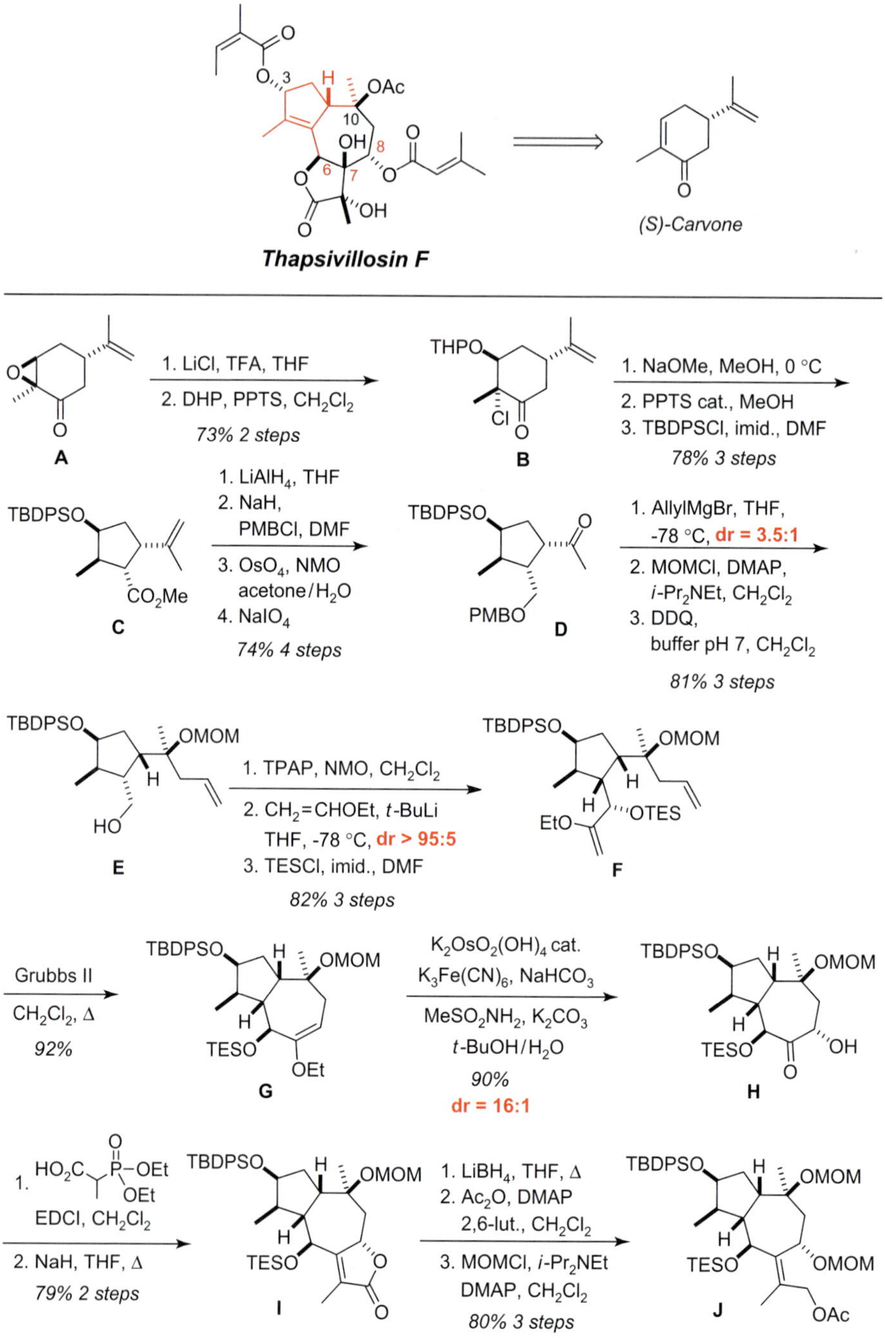

Thapsivillosin F

(S)-Carvone

Figure 13.4 Total synthesis of trilobolide and thapsivillosin F from (S)-carvone (hidden substructure).

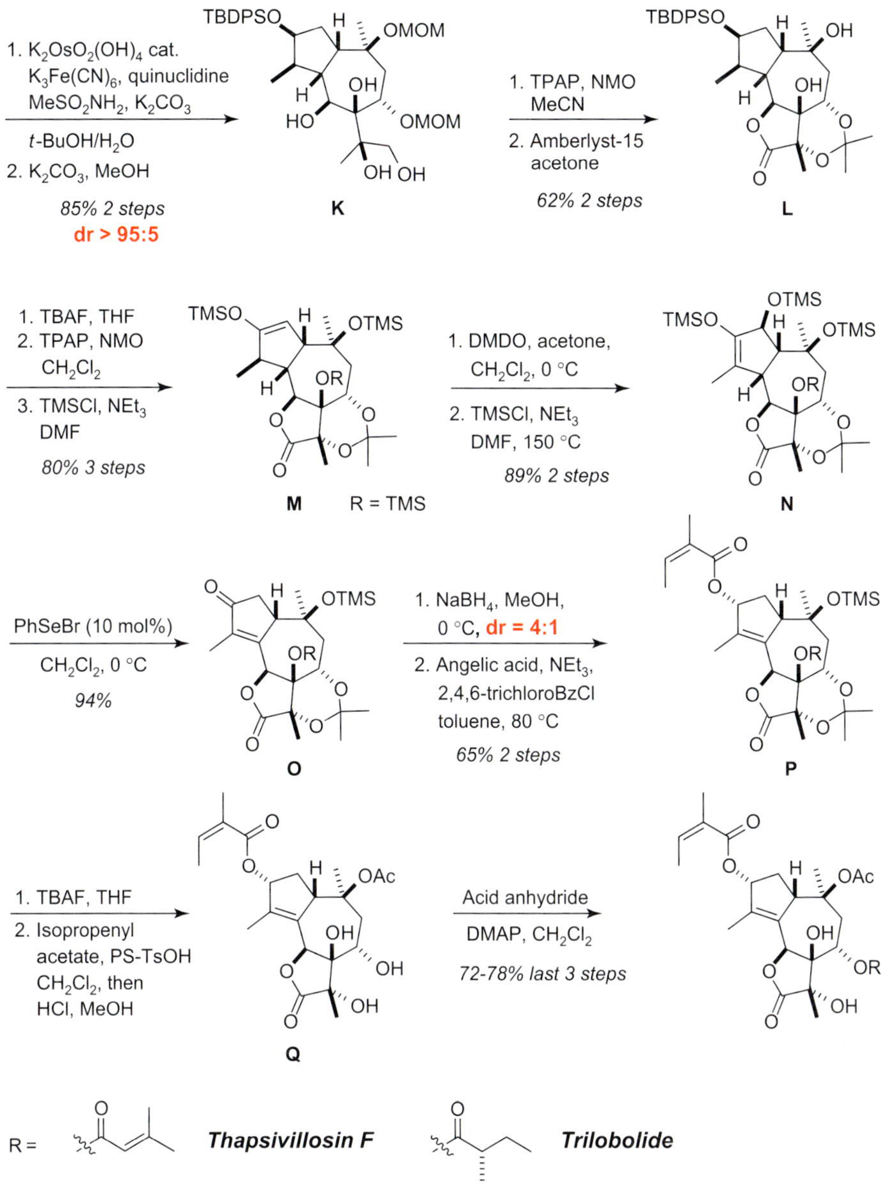

Key reactions: *Stereoselective chlorination (**A** to **B**); Favorskii rearrangement (**B** to **C**); ring-closing metathesis (**F** to **G**); Sharpless asymmetric dihydroxylation (**G** to **H** and **J** to **K**); DMDO = dimethyldioxirane; PS-TsOH = polymer-supported tosic acid*

Figure 13.4 *(continued).*

Relevant mechanisms and transformations:

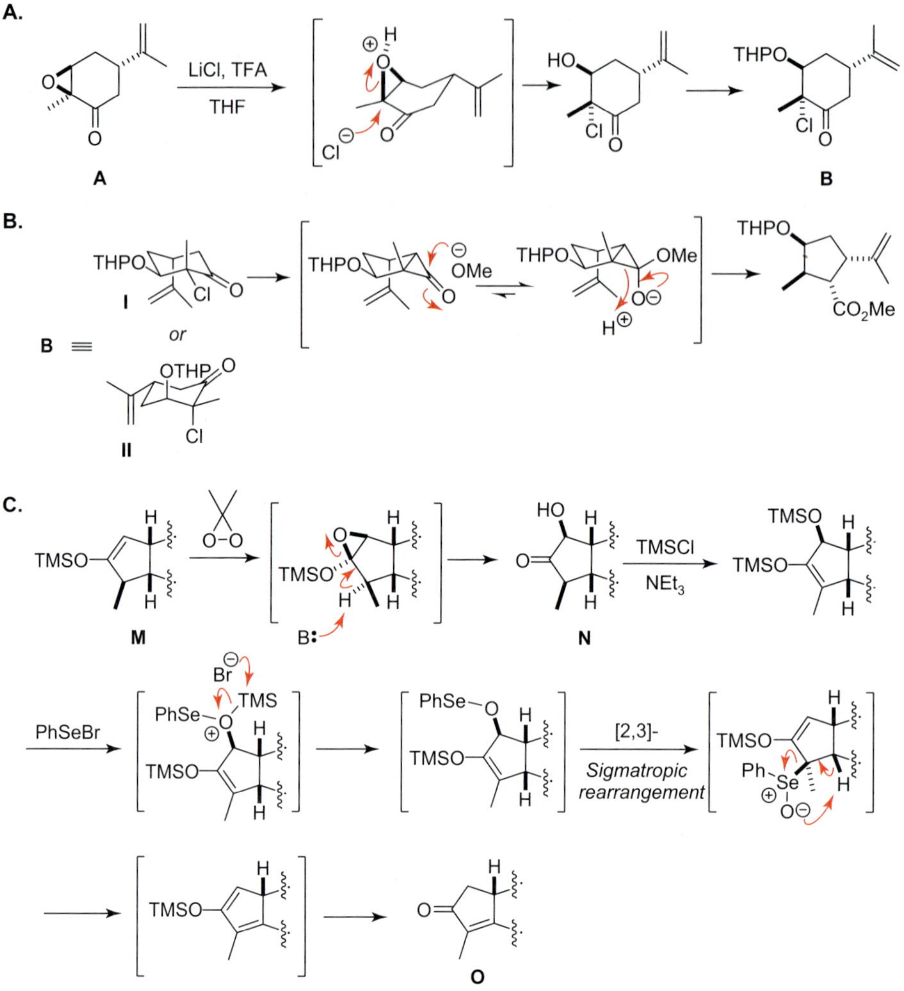

A.

B.

C.

Figure 13.4 *(continued).*

reactions provided the impetus to reach a bicyclic olefin such as **G**, provided that the necessary functionality was secured in order to allow a forward path to the natural product.

The synthesis of **C** was based on an acid-mediated ring-opening of α-epoxy ketone **A** in the presence of chloride ion (see relevant mechanism A, Figure 13.4). The resulting α-oriented chloride appears to have been formed by an S_N2-like modality in view of its *syn*-orientation *vis-à-vis* the bulky isopropenyl group, although a transient carbocation intermediate may also be operative. An axially oriented α-chloroketone would be stereoelectronically favored, however, the existence of other conformations cannot be ruled out. The facile Favorskii ring contraction of **B** to **C** at 0 °C with NaOMe in MeOH may be a reflection of a favorable bond alignment in the transition state via intermediates I or II (see relevant mechanism B, Figure 13.4). In order to introduce the appropriate olefinic tethers prior to the ring-closing metathesis to give **G**, two crucial stereochemical issues had to be secured. First, a Felkin-Anh approach of allylmagnesium bromide gave the tertiary alcohol at C-10 (thapsivillosin F numbering) in **E** as the major isomer, then a stereochemically controlled alkylation of the aldehyde derived from **E**, led to the required C-6 stereochemistry in **F** after *O*-silylation. These well-intended plans were in fact realized with great success, thus ensuring the stereochemical integrity of two critical alcohol groups. The topology of the bicyclic ring system and the orientation of substituents were determining factors in the stereoselective dihydroxylation of intermediates **G** and **J** individually. Thus, the stereochemistry at C-7, C-8, and the off-template tertiary alcohol were secured. Having achieved a significant milestone in the elaboration of the tricyclic ring systems, with six of the total seven stereogenic centers under control, there remained to invert C-3 and introduce a C-4–C-5 double bond. A single crystal X-ray analysis of advanced intermediates related to **O** provided the needed assurance to continue along the charted path *en route* to the intended targets. Considering the structure of intermediate **L**, it was not obvious how one would introduce unsaturation at C-4–C-5 while inverting the configuration at C-3. Ley and coworkers converted intermediate **L** to the TMS-enol ether **M**, which underwent a stereocontrolled epoxidation in the presence of DMDO to afford an α-hydroxy ketone (see relevant mechanism C, Figure 13.4). Protection of the secondary alcohol and formation of the silyl enol ether, upon treatment with TMSCl in the presence of triethylamine, gave **N**. Transforming **N** to enone **O** in the presence of 10 mol% PhSeBr was an unexpected result, which has been proposed by Ley and coworkers to involve a [2,3]-sigmatropic rearrangement and selenoxide elimination [9b].

The "hidden" (S)-carvone-based strategy allowed Ley and coworkers to develop a highly stereocontrolled route to the generic guaianolide framework while securing the stereochemistry at six stereogenic centers. Except for the exocyclic methylene group in the isopropenyl appendage of (S)-carvone, all other carbon atoms are retained in the carbon framework of the target compounds.

13.4
Briarellin E and F

The total syntheses of briarellin E (and its oxo-congener, briarellin F) by Over-man and coworkers [12] in 2003 also allowed the assignment of their absolute configurations (Figure 13.5). This family of cembranoid diterpenes comprises a number of biogenetically related cladiellin members such as sclerophytin and asbestinin-1.

Hydroboration of (S)-carvone afforded the hydroxymethyl analogue **A** as a mixture of diastereomers at the C-methyl branch site. Oxidation to the carboxylic acid followed by lactone formation led to **B** [13] in 75% overall yield from (S)-carvone. Formation of the silyl ketene acetal and protonolysis, followed by reduction with LiAlH$_4$, gave **C** as a pure crystalline compound after several recrystallizations. Protection as the TIPS ether, oxidation to the ketone **D**, and conversion to the kinetic enol triflate, was followed by a palladium-catalyzed coupling with (Me$_3$Sn)$_2$ and in situ iodination to give **E** [14]. Treatment of the cyclohexadienyl iodide with t-BuLi led to the dienyllithium species, which was coupled with aldehyde **F**, followed by cleavage of the isopropylidene acetal using PPTS/MeOH to give the diol **G** in good yield. The critical oxonia Prins-pinacol condensation reaction [15] was effected by sequential treatment with TsOH, followed by exposure to 10 mol% of SnCl$_4$ to give C-10 formyl tetrahydroisobenzofuran **I** as a single isomer in 84% yield over two steps. Photolytic deformylation [16], followed by selective cleavage of the TBDPS ether and protodesilylation under basic conditions afforded the alcohol **J**. Chemo-and stereoselective epoxidation with t-BuOOH in the presence of (t-BuO)$_3$Al [17], followed by acetylation gave **K** as the major isomer. Acetate-assisted opening of the epoxide in the presence of TFA afforded the alcohol, which was isolated as the diacetate **L**. Epoxidation of the C-11−C-12 double bond was accomplished with m-CPBA under buffered conditions, and the product was treated with TBAF to liberate the primary alcohol. Treatment with Tf$_2$O in the presence of 2,6-lutidine triggered cyclization to give the tricyclic epoxide **M**. Regio- and stereoselective hydration of the epoxide with aqueous acid gave the diol **N**, which was subjected to a four-step sequence to give intermediate **O** in excellent overall yield. Transformation of the propynyl side-chain to the 2-iodo-2-propenyl counterpart was accomplished by regioselective stannylalumination-protonolysis in the presence of Bu$_3$SnAlEt$_2$ and CuCN [18], followed by iododestannylation, affording intermediate **P** in 66% yield over two steps. The primary acetate group was selectively removed by treatment with di-t-butylchlorotin hydroxide [19], and the resulting alcohol was oxidized to the aldehyde. Application of the Nozaki-Hiyama-Kishi reaction [20] to intermediate **Q** afforded briarellin E as a single stereoisomer. Finally, Dess-Martin oxidation of briarellin E yielded briarellin F.

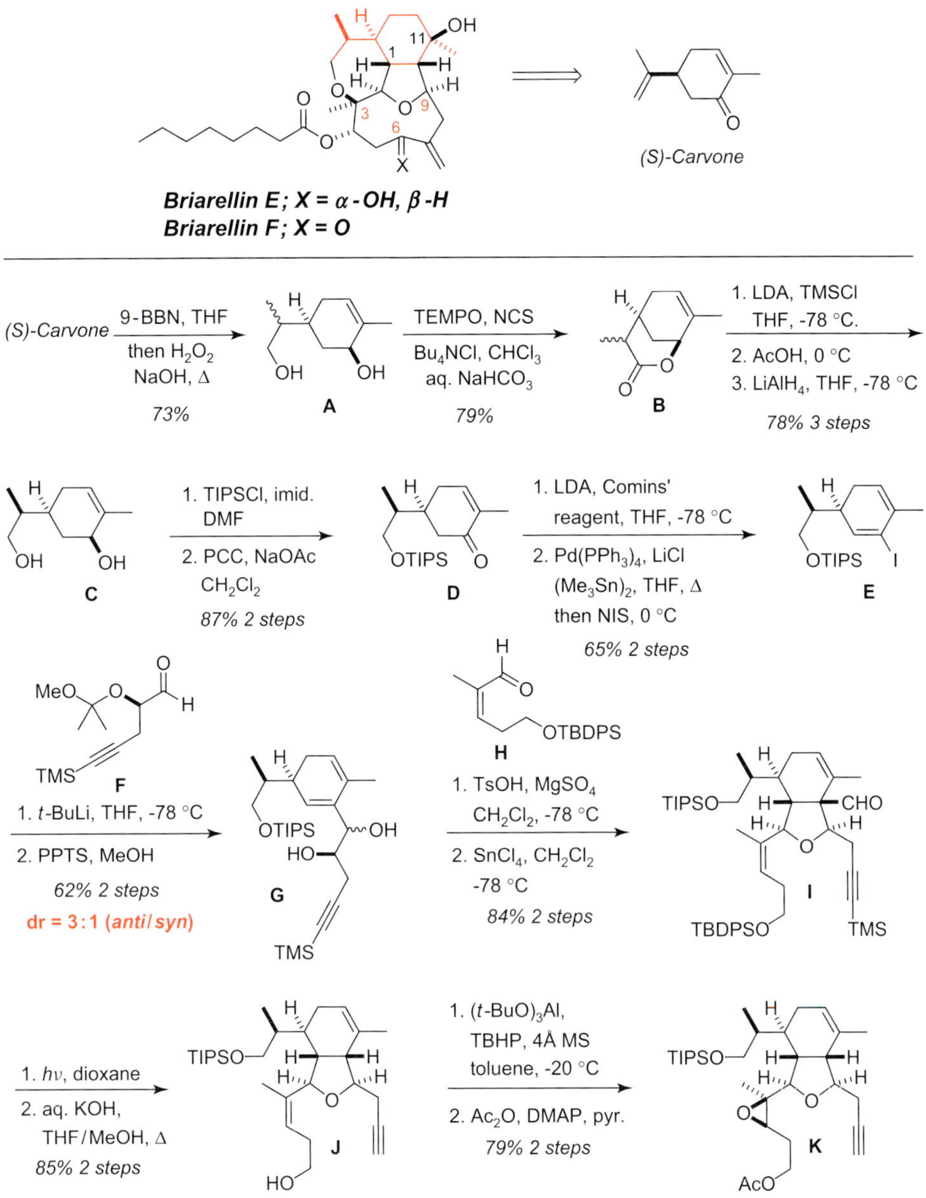

Briarellin E; X = α-OH, β-H
Briarellin F; X = O

(S)-Carvone

Figure 13.5 Total synthesis of briarellins E and F from (S)-carvone (hidden substructure).

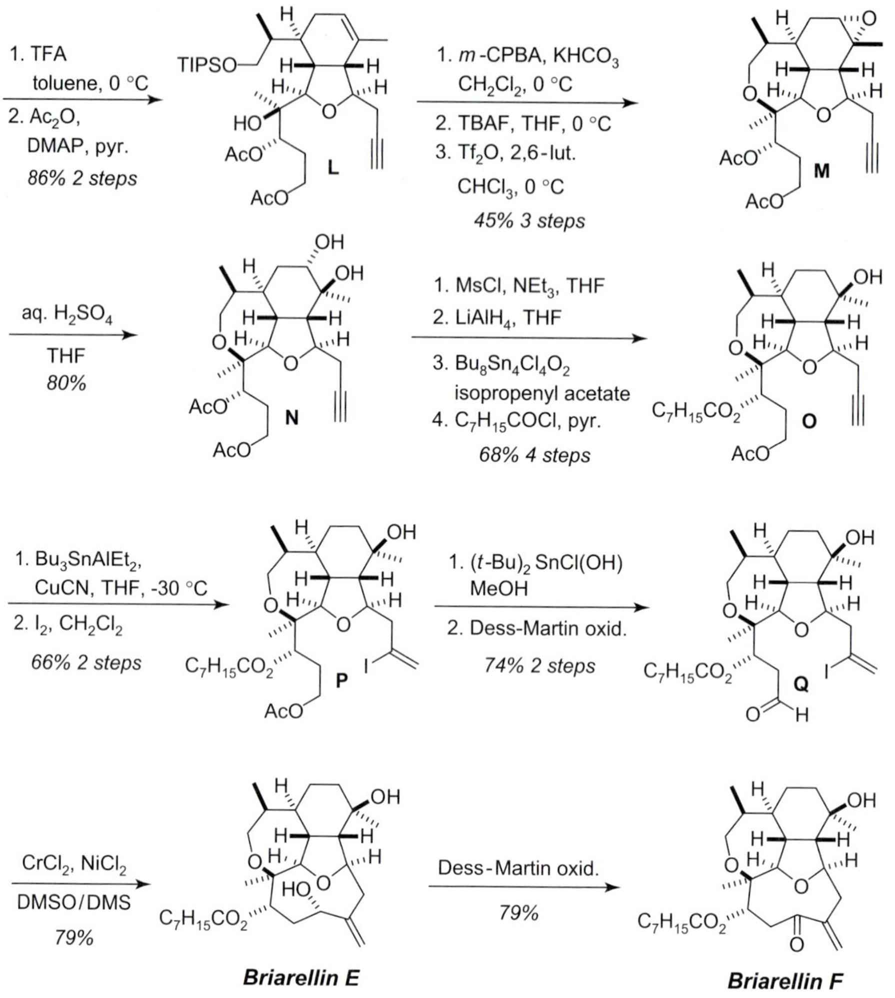

1. TFA
   toluene, 0 °C
2. Ac₂O,
   DMAP, pyr.
86% 2 steps

→ **L**

1. *m*-CPBA, KHCO₃
   CH₂Cl₂, 0 °C
2. TBAF, THF, 0 °C
3. Tf₂O, 2,6-lut.
   CHCl₃, 0 °C
45% 3 steps

→ **M**

aq. H₂SO₄
THF
80%

→ **N**

1. MsCl, NEt₃, THF
2. LiAlH₄, THF
3. Bu₈Sn₄Cl₄O₂
   isopropenyl acetate
4. C₇H₁₅COCl, pyr.
68% 4 steps

→ **O**

1. Bu₃SnAlEt₂,
   CuCN, THF, -30 °C
2. I₂, CH₂Cl₂
66% 2 steps

→ **P**

1. (*t*-Bu)₂ SnCl(OH)
   MeOH
2. Dess-Martin oxid.
74% 2 steps

→ **Q**

CrCl₂, NiCl₂
DMSO/DMS
79%

→ **Briarellin E**

Dess-Martin oxid.
79%

→ **Briarellin F**

Key reactions: *Regioselective hydroboration (**carvone** to **A**); enol triflate iodination (**D** to **E**); dienyl lithium-aldehyde coupling (**E** to **G**); oxonia Prins-pinacol reaction (**G** + **H** to **I**); photochemical deformylation (**I** to **J**); epoxide opening (**K** to **L**); cycloetherification (**L** to **M**); reductive cleavage of mesylate (**N** to **O**); stannylalumination-protonolysis, then iododestannylation (**O** to **P**); Nozaki-Hiyama-Kishi coupling (**Q** to **target**).*

Figure 13.5 *(continued).*

Relevant mechanisms and transformations:

A.   favored pathway

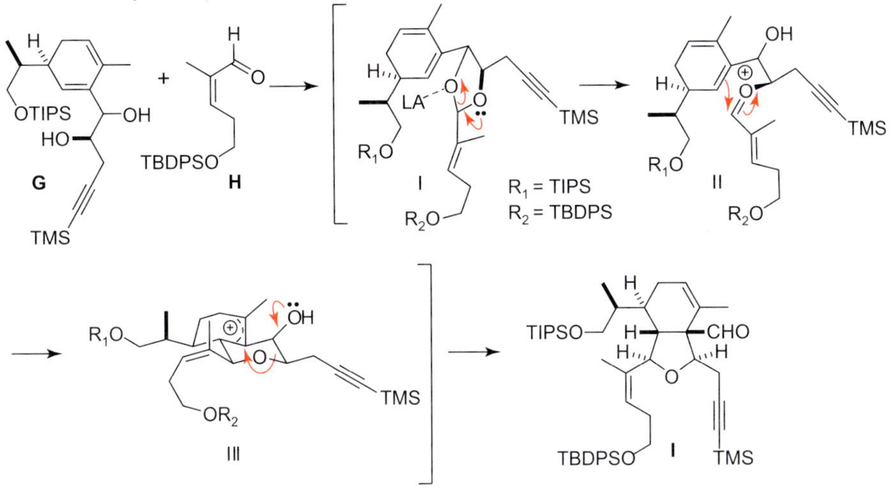

B.   disfavored pathway

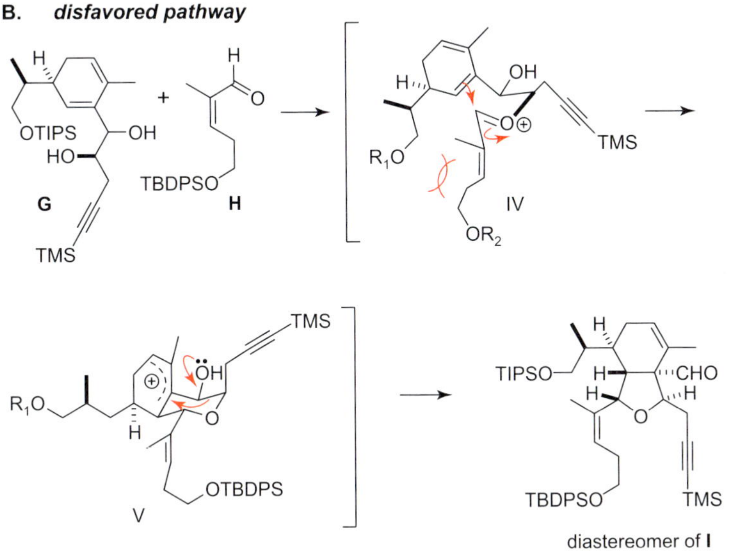

Figure 13.5 *(continued)*.

Commentary: Extensive studies toward the synthesis of a variety of stereochemically complex oxacyclic natural products have been reported by Overman and
coworkers [15]. The focal point in many of these syntheses was the application of

Lewis acid-mediated rearrangements of allylic diols and aldehydes to substituted tetrahydrofurans, in what amounts to an oxonia-Prins-pinacol rearrangement. Thus, the presence of the tetrasubstituted tetrahydrofuran in the briarellins was a visual stimulus to test stereocontrolled face-selective cyclizations of oxocarbenium ion intermediates in highly functionalized natural products such as the cladiellin diterpenes. Visual *reflexive* analysis of the tetracyclic architecture of briarellins E and F with an exocyclic double bond and the C-6—C-7 allylic alcohol motif, may have suggested an intramolecular Nozaki-Hiyama-Kishi reaction as a penultimate step, with the expectation of stereochemical control at the C-6 allylic alcohol center. With such a plan in mind, a visual *relational* analysis of the briarellins led Overman to consider (S)-carvone as a starting material. This chiron would provide the backbone for the oxonia-Prins-pinacol cyclization to construct the tetrasubstituted tetrahydrofuran ring with appended branches at C-2 and C-9. The choice of (S)-carvone, which was also successfully used in the synthesis of other members of the cladiellin diterpene family, was strengthened by the knowledge that the isopropenyl group could be hydroborated to the primary alcohol corresponding to the oxepan ring oxygen in briarellins E and F.

Elaboration of (S)-carvone into the carbocyclic precursor **C** proceeded uneventfully. In order to append the diol side-chain necessary for the implementation of the oxonia-Prins-pinacol rearrangement, a cyclohexadienyllithium reagent derived from vinyl iodide **E** had to be prepared. Cognizant that the stereochemistry of the allylic alcohol center in the diol **G** was not important, coupling of the lithium reagent with aldehyde **F** (prepared from (S)-glycidol), was followed by cleavage of the acetal protecting group. The crucial oxonia-Prins-pinacol rearrangement was initiated by the formation of the alkylidene acetal with aldehyde **H**, containing a Z-double bond (see relevant mechanism A, Figure 13.5). Treatment with $SnCl_4$ generated one of two possible oxocarbenium ions (II), which underwent cyclization via III, followed by ring-contraction to eventually give intermediate **I** in excellent yield. Clearly, this process would have to occur faster than Z to E isomerization of the conjugated oxocarbenium ion. The alternative chair-like cyclization would be subject to 1,3-diaxial interactions as shown in hypothetical intermediates IV and V (see relevant mechanism B, Figure 13.5).

The consequence of this highly stereocontrolled construction of the functionalized hexahydroisobenzofuran core subunit **I** was the residual formyl group at C-10, which was conveniently replaced by a hydrogen atom without stereochemical erosion based on a well-established photochemical reaction. Subsequent steps proceeded through a series of chemoselective and stereocontrolled reactions to introduce two tertiary alcohol groups at C-3 and C-11. Organotin chemistry was used for selective acetylation (**N** to **O**), and deacetylation (**P** to **Q**). Finally, it is remarkable that the Nozaki-Hiyama-Kishi intramolecular cyclization led to a single isomer corresponding to briarellin E.

The synthesis of these highly oxygenated diterpenes by Overman and coworkers demonstrates the power of functional group interconversions and compatibilities, in addition to efficient design and execution. Briarellin E was obtained in 28 steps comprising the longest linear sequence from intermediate **B**, which was in turn obtained from (*S*)-carvone in only two steps (Figure 13.5).

13.5
Samaderine Y

The first total synthesis of samaderine Y, a member of the quassinoid group of pentacyclic terpenoids was reported in 2005 by Shing and coworkers [21] (Figure 13.6). Extensive efforts toward the development of stereocontrolled methods for the synthesis of polycyclic natural products in this series culminated with the synthesis of samaderine Y, possessing a higher level of oxygenated substituents compared to the well-known archetypical quassin [22]. As in the case of quassin, Shing used (*S*)-carvone as a starting material to eventually overlap with the heavily substituted cyclohexane diol ring.

Formation of the lithium enolate of (*S*)-carvone and treatment with formaldehyde, followed by cyclic acetal formation provided **A** in two high-yielding steps. Allylic oxidation with CrO_3 and 3,5-dimethylpyrazole [23], followed by regio- and stereoselective reduction under Luche conditions [24], then *O*-silylation, gave **B**. Stereoselective epoxidation of the endocyclic enone double bond was acheived with alkaline hydroperoxide to give the α-epoxide, following which the ketone was reduced under Luche conditions to afford **C**. Upon treatment with TFA and 2,2-dimethoxypropane, migration of the acetonide group took place with opening of the epoxide at the tertiary carbon site to give the cyclic ether **D**. Protection of the resulting alcohol, cleavage of the acetonide, and oxidation with TPAP gave the ketoaldehyde **E**.

The construction of the *AB* ring system started with the reaction of **E** and 3-methyl pentadienylmagnesium bromide, which led to the rearranged 1,4-diene **F** as a single isomer. A cation and crown-ether accelerated [1,3]-sigmatropic rearrangement [25] was achieved under mild conditions, after which the product was acetylated to give **G** as a single diastereomer in excellent overall yield.

A thermal intramolecular Diels-Alder reaction afforded a mixture of *cis/trans*-isomers that were separated at a later stage. The mixture was converted to the triflate and then treated with *n*-Bu$_4$NOAc to give **H** via an S_N2 protocol. The *cis*-fused diastereomer underwent β-elimination instead of substitution, and could thus be separated. Formation of the enolate from the acetate ester of **H** was accompanied by an intramolecular aldol condensation with the C-14 ketone to give **I**. Treatment with $SOCl_2$ in pyridine followed by β-elimination to the α,β-unsaturated lactone, conjugate reduction with nickel boride to the lactol, and Fischer glycosidation gave **J** in good overall yield. Allylic oxidation of **J** to the enone, followed by α-keto acetoxylation [26] gave **K**. At this juncture, it was necessary to invert the configuration of the C-1 hydroxyl group, which was done in a three-step process relying on an oxidation-reduction protocol to give **L**. Finally, global deprotection and oxidation of the C-16 lactol to the lactone delivered samaderine Y.

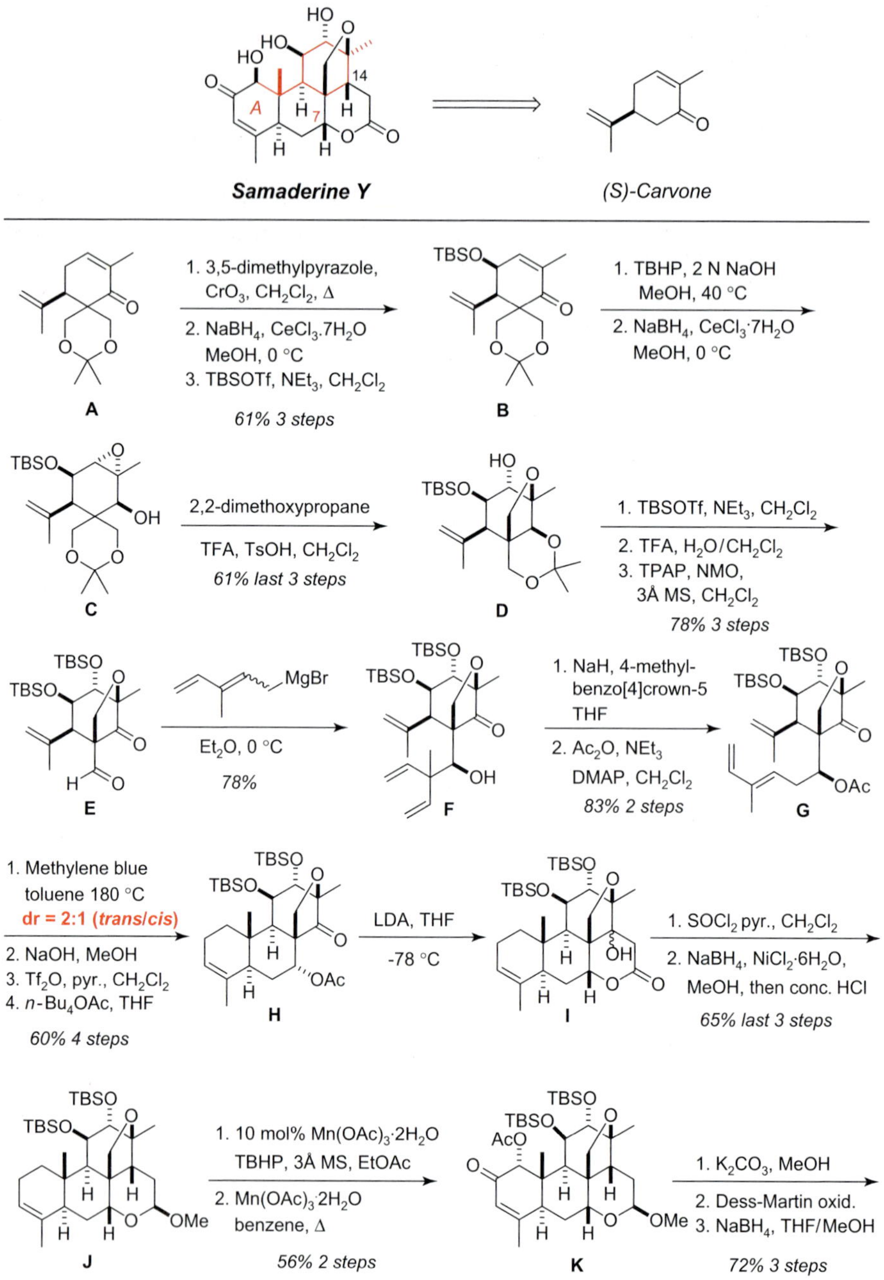

Samaderine Y ⟹ (S)-Carvone

Figure 13.6 Total synthesis of samaderine Y from (S)-carvone (partially hidden substructure).

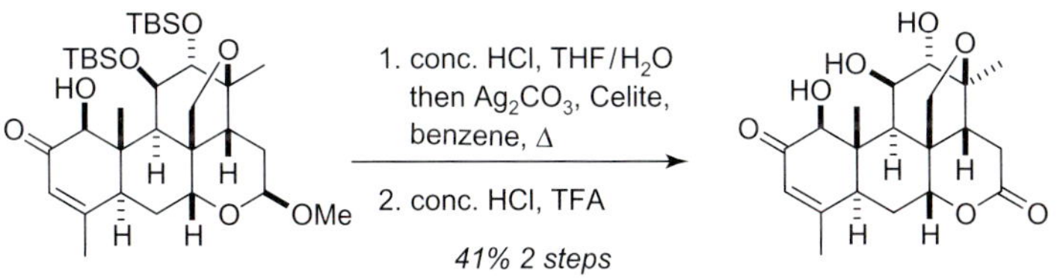

L **Samaderine Y**

Key reactions: *Cr-mediated allylic oxidation (**A** to **B**); [1,3]-sigmatropic rearrangement (**F** to **G**); Diels-Alder reaction (**G** to **H**); intramolecular aldol reaction (**H** to **I**); Mn-mediated allylic oxidation (**J** to **K**); TBHP = t-butylhydroperoxide*

Relevant mechanisms:

A.

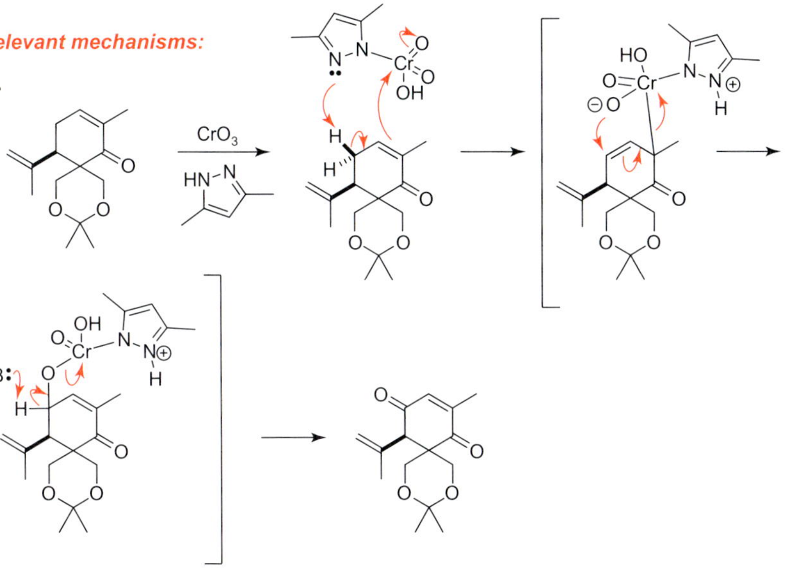

B.

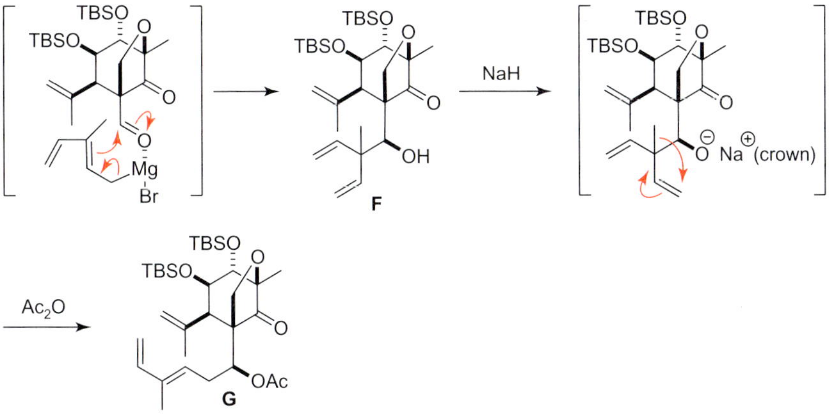

Figure 13.6 *(continued).*

Commentary: It is not unusual to consider an intramolecular Diels-Alder reaction as one of the key steps in the elaboration of polycyclic natural products such as those found in the quassin family. In fact, analysis of the *trans*-fused rings, *A* and *B*, in the decalin substructure with a C-3—C-4 double bond in samaderine Y may lead one to think of such a visual *reflexive* reaction-type strategy. Shing and coworkers had previously employed such a strategy in their total synthesis of quassin [22]. Furthermore, they had shown the utility of (S)-carvone as a versatile starting material by taking advantage of the available functionality. An added challenge in the case of samaderine Y was the presence of several peripheral oxygen bearing stereogenic centers. To consider (S)-carvone as a progenitor of the highly substituted cyclohexane diol ring system would require considerable chemical modification. To this end, the double aldol product **A** would admirably serve a dual purpose in a series of highly regio- and stereocontrolled reactions.

Allylic oxidation of **A**, previously applied to steroidal substrates, was successfully used in the synthesis of **B** (see relevant mechanism A, Figure 13.6). Although the exact mechanism of this oxidation has not been definitively established [23], an ene-like process has been suggested based on a CrO_3-3,5-dimethylpyrazole complex. Alternatively, a radical cation mechanism can also be considered. The Luche reduction was remarkably selective in giving the β-alcohol, no doubt due to a sterically less impeded attack as evidenced in products **B** and **C**. Formation of the thermodynamically more stable acetonide **D** allowed a smooth, acid-catalyzed intramolecular epoxide opening to take place.

An intriguing chemoselective addition of the 3-methylpentadienyl Grignard reagent to ketoaldehyde **E** deserves comment. Cognizant that the ultimate objective was to prepare a *trans*-diene as in **G**, the authors chose a totally different route to achieve it. First, the Grignard product was not the expected 1-(3-methylpentadienyl) adduct of the aldehyde. Rather, addition took place via a chelated six-membered intermediate to give the 1,4-dienyl homoallylic alcohol **F** (see relevant mechanism B, Figure 13.6). Relying on known cationic, and crown ether accelerated [1,3]-sigmatropic rearrangements [25], Shing and coworkers attempted the same with a highly functionalized substrate **F**. The transposition worked remarkably well, without erosion of stereochemistry to give the "planned" diene **G**, ready for the Diels-Alder reaction. Prior to the construction of the eventual lactone ring, an inversion of configuration at C-7 was required (**G** to **H**). An intramolecular aldol reaction using the acetate enolate led to lactone **I**. Subsequent conjugate reduction of an α,β-unsaturated lactone under nickel boride conditions would also favor a β-hydride approach (**I** to **J**).

The synthesis of samaderine Y was achieved in 21 steps from (S)-carvone with an average yield of 81% for each step. A number of regio- and stereoselective reactions were realized utilizing the original (S)-carvone template. The two hydroxymethyl branches in intermediate **A** provided a practical approach to secure the C-8—C-13 anhydro bridge early in the synthesis. The isopropenyl group of (S)-carvone served as the internal dienophile for the crucial intramolecular Diels-Alder reaction, unfortunately leading to only modest stereoselectivity. Nevertheless, the synthesis of samaderine Y demonstrates the importance of chemoselective transformations in the elaboration of densely functionalized polycyclic compounds.

13.6
Ambiguine H and Hapalindole U

Indole alkaloids are some of Nature's most commonly occurring nitrogen-containing metabolites [27]. A select group of biogenetically related polycyclic indole alkaloids belong to the Stigonemataceae family, and comprise isonitrile-containing indoles such as ambiguine H and its prenylated congener hapalindole U. The total synthesis of these two alkaloids in enantiomerically pure form has been reported by Baran and coworkers [28] (Figure 13.7).

Starting with (S)-carvone, a four-step procedure led to the *C*-vinyl analogue **A**, which was subjected to a radical-mediated coupling reaction with an intermediate radical partner derived from 4-bromoindole (**B**). In the event, treatment of a mixture of **A** and 4-bromoindole (**B**) with LiHMDS in THF, followed by oxidation with Cu(II)-2-ethylhexanoate afforded the coupling product **C** as a single isomer. A Heck-type cyclization in the presence of Hermann's palladacycle catalyst [29] gave the tetracyclic indole **D**. A three-step process involving reductive amination under microwave conditions, *N*-formylation, and dehydration led to hapalindole U in enantiomerically pure form. Subjecting the latter to *t*-BuOCl, followed by addition of prenyl-9-BBN [30] produced the crystalline adduct **E**. Irradiation in benzene containing triethylamine gave ambiguine H in 63% yield based on recovered starting material.

Commentary: Having gone through the individual steps of the syntheses of ambiguine H and hapalindole U in Figure 13.7, it is easy to understand why Baran and coworkers entitled their paper "Total synthesis of marine natural products without using protecting groups." Alas, not all complex natural products will succumb to such a utopian strategy, unless a certain mindset is adopted wherever such opportunities present themselves (see Chapter 18, section 18.5.1.1). In spite of their architectural complexity, these fused polycyclic indole alkaloids are conducive to the invention of new coupling methods. Analysis of the connectivity of the polycyclic targets, especially in relation to the 3,6-disubstitution pattern of the indole nucleus, relates to a knowledge-based design element. Thus, radical partners consisting of the carvone-derived chiron **A** and 4-bromoindole radical **B** were united starting from the corresponding "anionic" precursors through oxidation with a Cu(II) carboxylate (see relevant mechanism A, Figure 13.7).

Not only was this hetero-coupling experimentally realized, a single stereoisomer was produced (as in **C**) most likely because of the sterically controlling isopropenyl group. With the bromoindole intermediate in hand, Baran and coworkers had several options to effect cyclization including a free-radical induced ring-closure. They studied a number of Pd-catalyzed reductive cyclizations with the favorably situated bromine on the indole and the exocyclic double bond on the chiron derived from (S)-carvone with modest success. A preparatively useful carbocyclization was

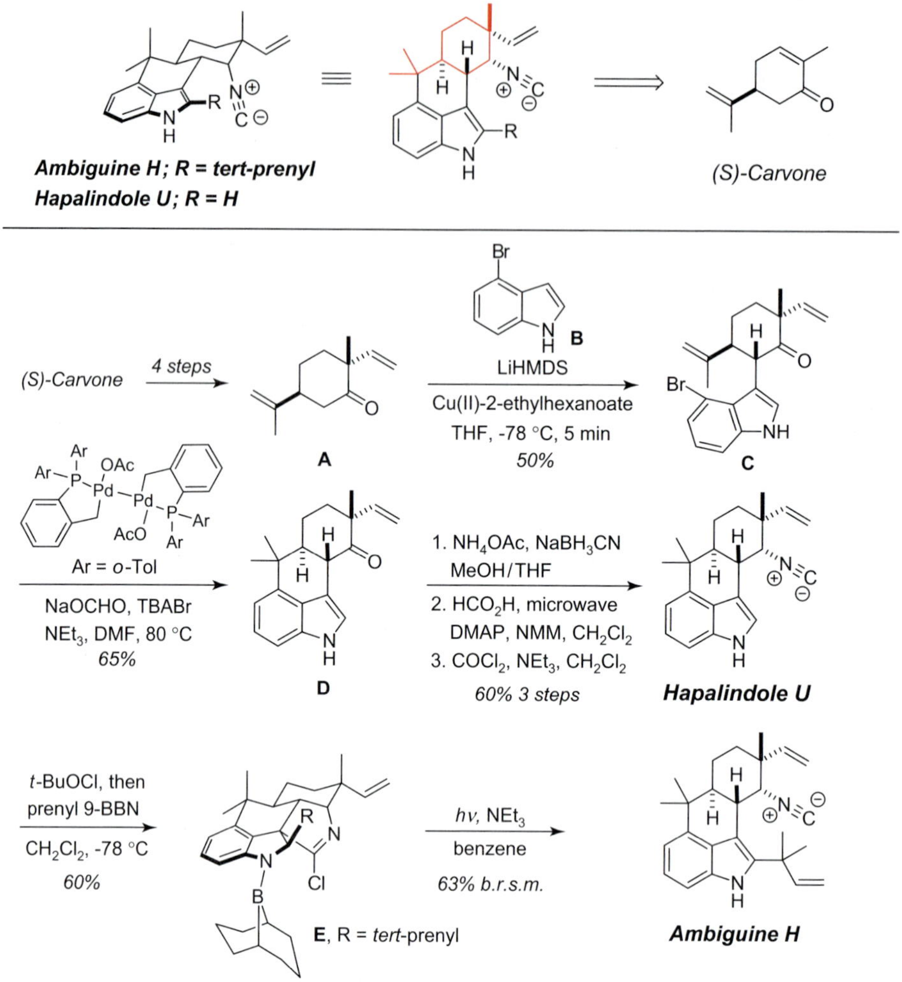

Ambiguine H; R = tert-prenyl
Hapalindole U; R = H

(S)-Carvone

Key reactions: *Radical coupling (**A** + **B** to **C**); Heck reaction (**C** to **D**); reductive amination-formylation-dehydration (**D** to **hapalindole U**); isonitrile to α-chloroimine (**hapalindole U** to **E**); radical-induced fragmentation (**E** to **ambiguine H**); TBABr = tetrabutylammonium bromide*

Figure 13.7 Total synthesis of ambiguine H and hapalindole U from (S)-carvone (partially hidden substructure).

Relevant mechanisms and transformations:

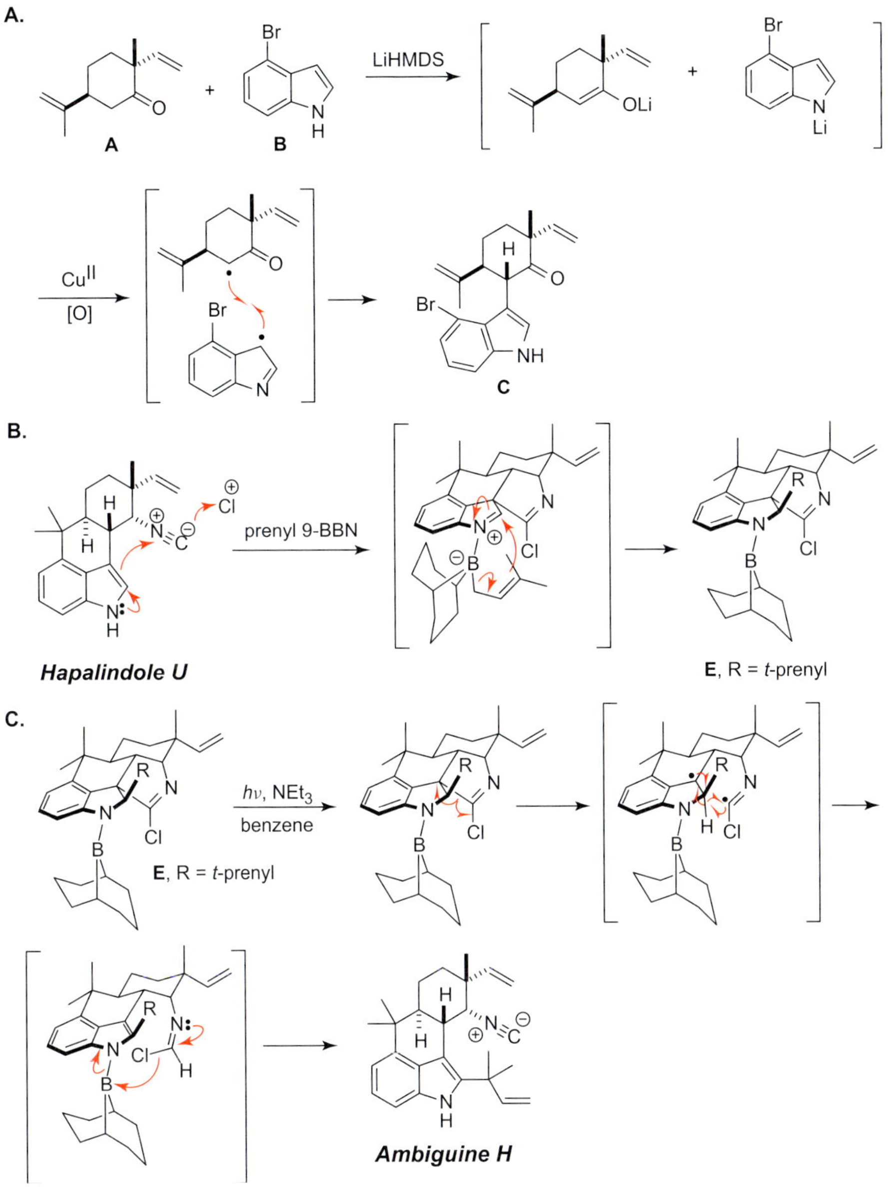
A.
Br
LiHMDS
OLi
N
Li
A
B
Cu^{II}
[O]
NH
C
B.
prenyl 9-BBN
Cl
Hapalindole U
E, R = t-prenyl
C.
hν, NEt$_3$
benzene
E, R = t-prenyl
Ambiguine H

Figure 13.7 *(continued).*

eventually achieved using Hermann's palladacycle catalyst [29] to give a 65% yield of the tetracyclic product **D**. Three additional steps provided enantiopure hapalindole U in a total of nine steps from (S)-carvone, which is a vast improvement over previously reported syntheses [28c-f].

The conversion of hapalindole U to ambiguine H was achieved with remarkable efficiency. Following a prenylation protocol reported by Danishefsky and coworkers [30], hapalindole U was exposed to *t*-BuOCl, followed by treatment with prenyl-9-BBN to give the crystalline chloroimine adduct **E**. A plausible explanation given by the authors suggested initial chlorination of the isonitrile, followed by nucleophilic attack of the electron rich indole at C-3 onto the chloro isonitrilium ion, then a β-face intramolecular attack of the appended prenyl reagent on the iminium ion (see relevant mechanism B, Figure 13.7). The stereoselective attack can be appreciated by considering the topology of the intermediate and a more facile β-face approach to the iminium ion.

The cascade of bond-breaking and bond-forming events under photochemical conditions is another remarkable feature of the last step in the synthesis. The initially produced diradical was the product of a Norrish type I cleavage, which underwent a proximity assisted H-radical transfer to the chloroimine appendage with concomitant restoration of the indole oxidation state (see relevant mechanism C, Figure 13.7). Finally formation of the isonitrile was accompanied by cleavage of the *N*-9-BBN complex to give ambiguine H in just eleven steps from (S)-carvone.

True to the title of the article, Baran and coworkers have shown that for certain classes of indole containing natural products, total synthesis can indeed be achieved with remarkable efficiency and practicality, avoiding protecting groups. Ironically, the authors used prenyl-9-BBN as a transient "protecting group" for the indole nitrogen that effectively "delivered" the prenyl group in one step, then graciously "departed" in the process of the photochemical fragmentation of **E**. Baran and coworkers have also demonstrated the power of a knowledge-based "in-house" method (radical hetero-coupling), inspired by biogenetic transformations, to synthesize other members of these complex polyfunctional indole alkaloids [28a]. The desire to obtain enantiopure ambiguine H, and related compounds in the same family, combined with the visual *relational* thinking of considering chirons derived from (R)- and (S)-carvone led to the development of highly versatile syntheses in this series.

13.7
Platensimycin

The recently discovered inhibitor of fatty acid biosynthesis, platensimycin, has an unprecedented structure consisting of a pentacyclic ketolide core motif (Figure 13.8) (see Chapter 2, section 2.1). The first total synthesis of racemic platensimycin was reported in 2006 by Nicolaou and coworkers [31], soon after the discovery of the antibiotic. Since then, approaches to the synthesis of the racemic and enantiopure core substructure have been reported by various groups [32]. Two of these approaches have relied on carvone as a starting material.

13.7.1
The Nicolaou synthesis

Nicolaou and coworkers [33] started with (*R*)-carvone, which was converted to branched derivative **A** (Figure 13.8). Thus, a 1,2-addition of a Grignard reagent in the presence of CeCl₃, followed by allylic oxidation of the resulting tertiary alcohol with transposition of the double bond gave **A**. Following an oxymercuration-carbocyclization protocol [34], treatment of **A** with Hg(OAc)₂ and NaBH₄ gave the bicyclic product **B** as a mixture of epimers. Elimination with Martin's sulfurane reagent [36] led to the exocyclic olefin in **C**, which was converted to bis-olefin **D** in a two-step process. Cleavage of the acetal afforded the aldehyde **E**, which was treated with SmI₂ to give the tricyclic product **F** as a single stereoisomer in 57% yield. Inversion of configuration by a Mitsunobu reaction, followed by treatment with base led to partial epimerization at C-9 to give **G** as a 1:1 mixture of diastereomers. Stereoselective reduction of the ketone, followed by an acidic work-up furnished the cyclic ether **H**. Oxidation with PCC gave ketone **I**, which was transformed into enone **J** and its regioisomer in a 2:1 ratio. Double alkylation led to **K**, which had previously been converted to platensimycin in a synthesis of the racemic natural product [31].

Commentary: As mentioned above, the first total synthesis of racemic platensimycin was published in 2006 by the Nicolaou group [31]. The strategy led to a tetracyclic cyclic ether intermediate (Figure 13.8, compound **J**), which has been the pivotal "relay compound" for most of the subsequent formal syntheses of platensimycin. At first glance, it may be debateable if the Nicolaou approach starting from (*R*)-carvone presents a distinct advantage, particularly in view of the absence of any direct skeletal overlap with the tetracyclic core structure. However, if one considers utilizing the available functionality in (*R*)-carvone in such ways so as to engage the isopropenyl group in an intramolecular carbocyclization, then a plausible strategy for oxacycle formation could rely on an acid-catalyzed cyloetherification. Indeed, prior reports on oxymercuration-reductive alkylation by Giese [34] provided the knowledge-based, visual *reflexive* stimulus to consider a carvone-based approach (see relevant mechanism A, Figure 13.8). To achieve this intermediate required the functionalization of (*R*)-carvone in a two-step process that introduced a usable 3-carbon branch, while transposing the enone system within the carbocyclic framework. Thus, a 1,2-addition in the presence of CeCl₃ led to the tertiary allylic alcohol, which was oxidized to enone **A** using PCC. Intramolecular carbocylization was achieved in good yield to provide a mixture of tertiary alcohols. The authors proposed a Markovnikov oxymercuration of the exocyclic alkene generating a primary radical by cleavage of the C–Hg bond under reductive conditions, followed by 1,4-addition to the enone (see relevant mechanism, A, Figure 13.8). The formation of the cyclohexanone ring was initially tried with the aldehyde **E** under Stetter conditions [37] giving directly the desired tricyclic system **F** as a 5:1 mixture of inseparable C-9 epimeric ketones.

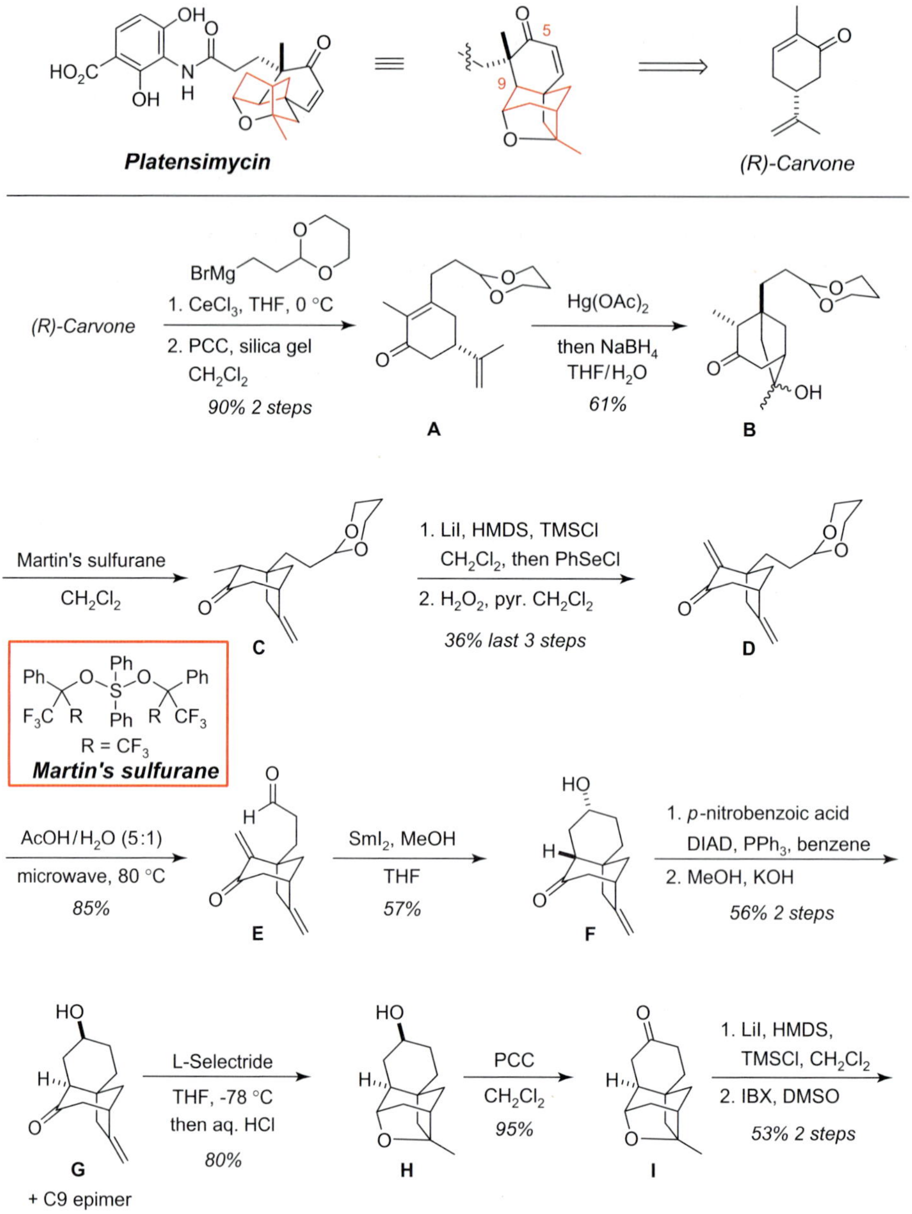

Figure 13.8 Nicolaou's approach to the synthesis of platensimycin's oxatetracyclic core motif from (R)-carvone (hidden substructure).

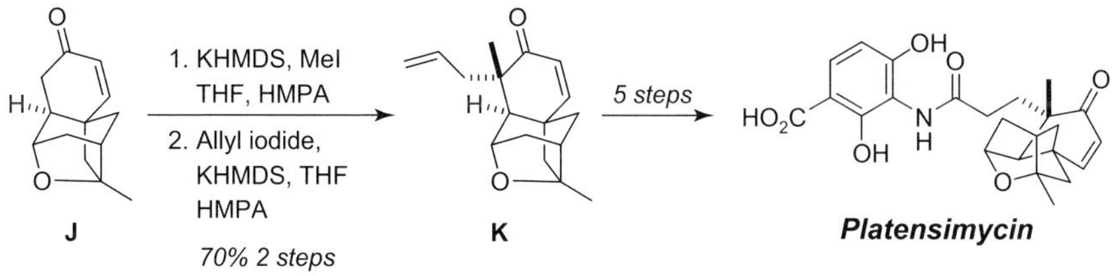

Key reactions: *Allylic oxidation with double bond transposition (**carvone** to **A**); oxymercuration-reductive radical carbocyclization (**A** to **B**); SmI₂-mediated cyclization (**E** to **F**); acid-catalyzed cycloetherification (**G** to **H**); IBX = 2-iodoxybenzoic acid*

Relevant mechanisms and transformations:

A.

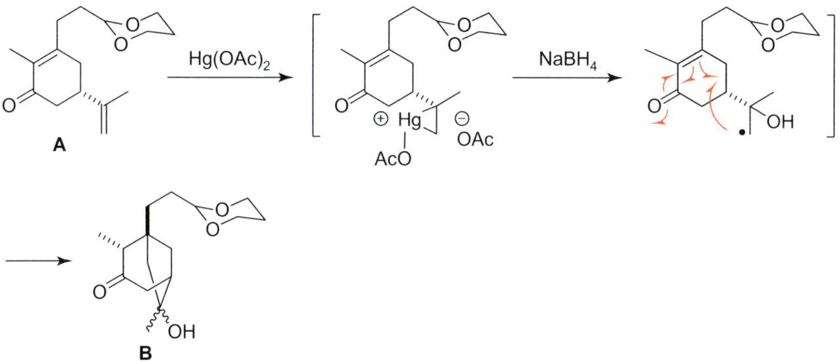

B.

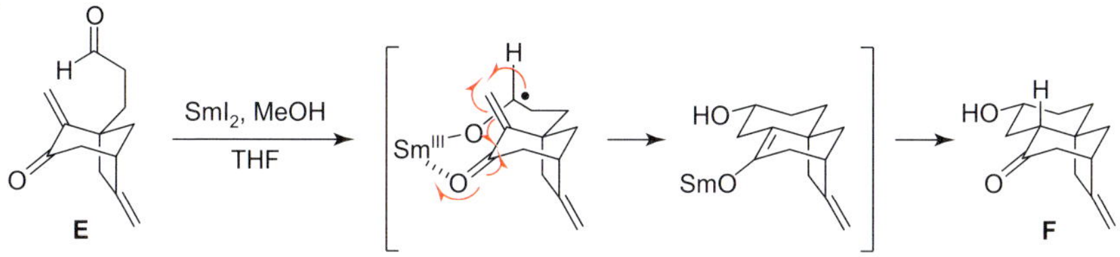

Figure 13.8 *(continued)*.

The electron transfer method using SmI₂ offered a viable solution in giving the tricyclic alcohol **F** as a single stereoisomer (see relevant mechanism B, Figure 13.8). However, epimerization at the C-9 ring junction was only possible after inversion of configuration at the C-5 position. The authors suggested a chelated Sm(III) radical intermediate leading to the equatorially disposed alcohol, which resisted equilibration, until the configuration was inverted to that of the axial alcohol. Unfortunately, this unforeseen inertness to equilibration also lengthened the synthetic sequence. All the carbon atoms of (*R*)-carvone were utilized in the elaboration of advanced ketone intermediate **J** (note color coding in Figure 13.8).

13.7.2
The Ghosh synthesis

An (S)-carvone entry to the platensimycin oxatetracyclic core motif was first reported in 2007 by Ghosh and Xi [35a]. Their strategy was based on the conversion of (S)-carvone into the known oxabicyclo[3.3.0]octane ketone **B**, which would be elaborated further to provide a suitable precursor to a critical intramolecular Diels-Alder reaction (Figure 13.9).

Thus, ketone **A**, readily available in two steps from (S)-carvone after chromatographic separation, was oxidized under Baeyer-Villiger conditions to give bicyclic lactone intermediate **B** in excellent overall yield. Further oxidation of the methyl ketone appendage gave the corresponding alcohol **C**. Protection as the TBS ether, olefination according to Petasis [38], followed by hydroboration afforded bis-protected **D** as a 2:1 mixture of isomers. Selective catalytic deprotection of the TBS ether with DDQ in aqueous THF gave the corresponding secondary alcohol, which was subsequently oxidized to the ketone and converted to **E** using an asymmetric version of the Horner-Wadsworth-Emmons reaction. Further regioselective manipulation afforded the preferentially O-protected intermediate **F**, which was oxidized to the aldehyde, and subsequently extended via a second phosphonate anion olefination reaction to give **G**. Transformation to the methoxydiene **H**, and thermal intramolecular Diels-Alder reaction afforded the oxatetracyclic core motif **I** as the O-methyl ether.

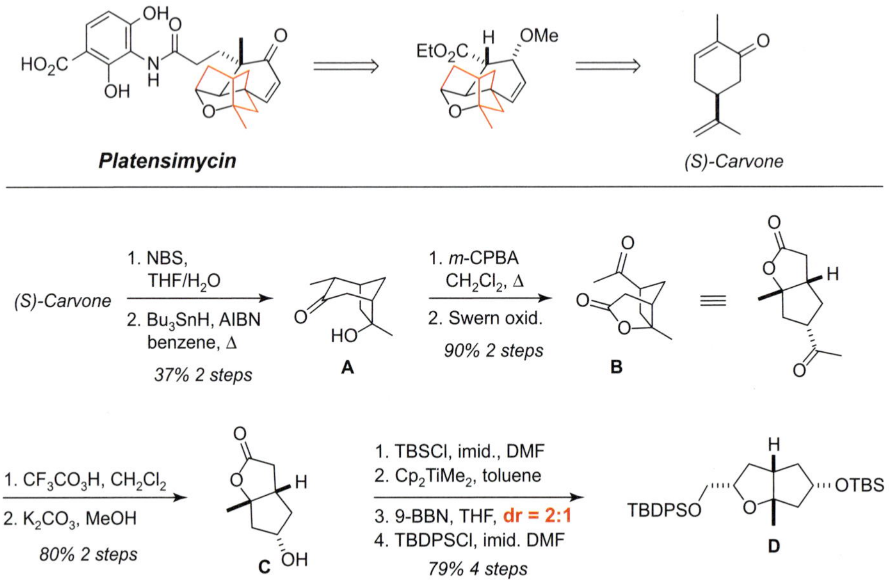

Platensimycin   (S)-Carvone

Figure 13.9 Ghosh and Xi's approach to the synthesis of platensimycin's oxatetracyclic core motif from (S)-carvone (hidden substructure).

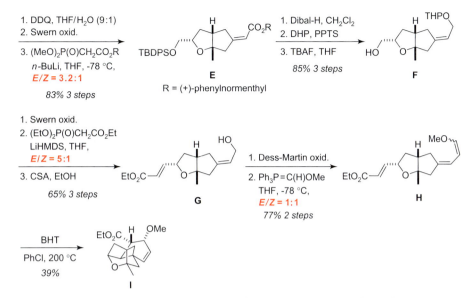

R = (+)-phenylnormenthyl

Key reactions: *Reductive radical carbocyclization (**carvone** to **A**); Baeyer-Villiger oxidation and lactone formation (**A** to **B** and **B** to **C**); Petasis olefination (**C** to **D**); asymmetric olefination (**D** to **E**); intramolecular Diels-Alder reaction (**H** to **I**); BHT = butylated hydroxytoluene*

Relevant mechanisms and transformations:

A.

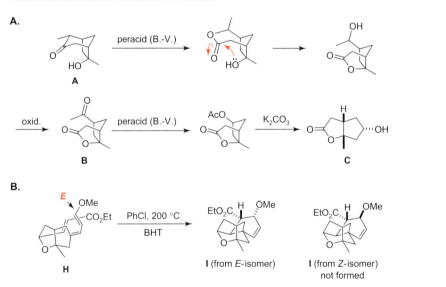

B.

Figure 13.9 *(continued)*.

Commentary: It is not uncommon to envisage the use a cyclic terpene as a starting material for the synthesis of a target molecule that contains carbocyclic subunits. Unlike the Nicolaou approach, where the (*R*)-carvone framework was used in its entirety to converge with a cyclohexane ring in platensimycin, Ghosh and Xi started with enantiomeric (*S*)-carvone, and relied on a known sequence that provided a substituted bicyclic lactone **B** as an early intermediate. This would be elaborated further to the oxabicyclic precursor **H** for a key intramolecular Diels-Alder reaction.

Analysis of the structure of platensimycin conjures a number of bond disconnections in the mind's eye. The visual connection between (*S*)-carvone and intermediate **H** is practically non-evident, unless one is familiar with the sequence shown in Figure 13.9 (see relevant transformation A). Envisaging an intramolecular Diels-Alder cyclization by engaging the seemingly remote diene and dienophile groups as depicted in **H** may require some model building to appreciate its three-dimensional structure. On the other hand, having drawn platensimycin as originally portrayed by the Merck scientists (see Chapter 2, section 2.1) [32] who determined its structure, and later by Nicolaou who reported its synthesis in racemic form [31], facilitates the visual *relational* thought process. Thus, breaking two bonds in the oxatetracyclic core motif of platensimycin (Figure 13.10, red lines) produces a Diels-Alder precursor, which is much easier to visualize in the putative three-dimensional drawing (**J**) compared to the identical, but seemingly less evident, counterpart **J**′ (Figure 13.10). It is interesting to point out the remarkable similarities (and differences) of the Nicolaou and Ghosh approaches. First, enantiomeric carvones were used by each group respectively toward the synthesis of the same platensimycin oxatetracyclic core motif. Second, central to the conception of a synthesis plan was a similar radical carbocyclization starting with (*R*)- and (*S*)-carvones. Yet, the two approaches diverge at the stage of bicyclic ketone intermediates in creatively significant ways (compare compound **C** in Figure 13.8 and compound **A** in Figure 13.9). The radical cyclization product **A** in the Ghosh and Xi synthesis (see relevant mechanism A, Figure 13.9), was separated from its epimer, and engaged in a Baeyer-Villiger oxidation *en route* to the advanced Diels-Alder precursor **H**.

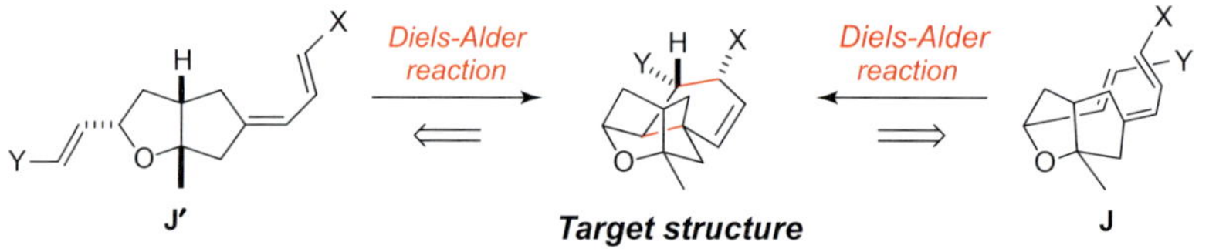

Figure 13.10 Mind's eye perspective drawing of the platensimycin oxatetracyclic core structure, and identical Diels-Alder precursors **J** and **J**′.

It should be noted that despite the requirement of high-temperature (200 °C) only the desired adduct **I** was isolated as a single stereoisomer, arising from the *E*-enol diene (see relevant transformation B, Figure 13.9). In 2009, Ghosh and coworkers completed the synthesis of the natural product using the same Diels-Alder strategy [35b].

The Ghosh and Xi approach to the oxatetracyclic core motif of platensimycin demonstrates an excellent example of combining visual *relational* and *reflexive* thought processing in synthesis planning.

13.7.3
Nicolaou's two asymmetric syntheses

Pursuant to the first total synthesis of racemic platensimycin, Nicolaou and coworkers reported two approaches to the oxatetracyclic core motif **J** (see Figure 13.8). In the first of these approaches [39], a cycloisomerization reaction inspired by Trost's Ru(II) catalyst [40] was adapted to an asymmetric version according to Zhang [41] to give the spirocyclic product **B** from **A** in 91% yield and high enantiomeric excess (Figure 13.11). It was necessary to excise the residual carboxylate function in **C**, which was accomplished using Barton's method [42]. In the process, the exomethylene radical underwent a 1,3-H shift to give an endocyclic radical, which was captured by tributyltin hydride to give **D**. Hydrolysis of the acetal group and treatment with SmI_2 in the presence of hexafluoro-2-propanol gave the tricyclic alcohol **E**, which upon exposure to TFA furnished the intended oxatetracycle (see relevant mechanism B, Figure 13.11).

In the second asymmetric synthesis [43], a chiral auxiliary approach was adopted (Figure 13.12). Thus, the Myer's enolate alkylation method [44] furnished intermediate **B** (see relevant transformation A, Figure 13.12). Conversion to the enol triflate **C**, followed by a Pd-catalyzed Kumada coupling [45] gave the allylsilane intermediate **D** in high yield. Cyclodearomatization of the phenol analogue obtained from **D**, in the presence of $PhI(OAc)_2$ in trifluoroethanol gave dienone **E** (98% *ee*) (see relevant transformation B, Figure 13.12). Treatment of the corresponding aldehyde **F** with SmI_2 as previously described by Nicolaou, gave the alcohol **G**, which upon exposure to acid, afforded the intended oxatetracycle core of platensimycin (see relevant mechanism C, Figure 13.12).

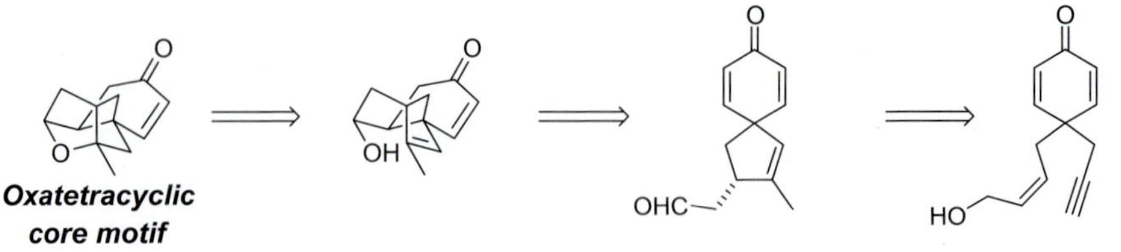

Oxatetracyclic core motif

Key reactions: Asymmetric cycloisomerization/spiroannulation (**A** to **B**); Barton decarboxylation (**C** to **D**); SmI$_2$-mediated carbocyclization (**D** to **E**); acid-mediated cycloetherification (**E** to **target**)

Relevant mechanisms and transformations

A.

B.

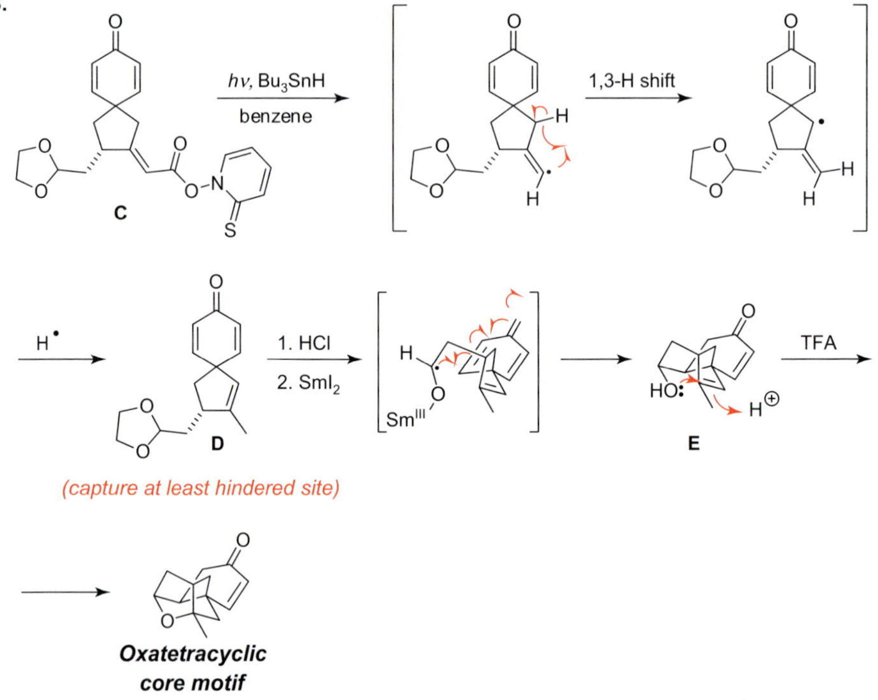

(capture at least hindered site)

Oxatetracyclic core motif

Figure 13.11 Nicolaou's catalytic asymmetric approach to the synthesis of platensimycin's oxatetracyclic core motif.

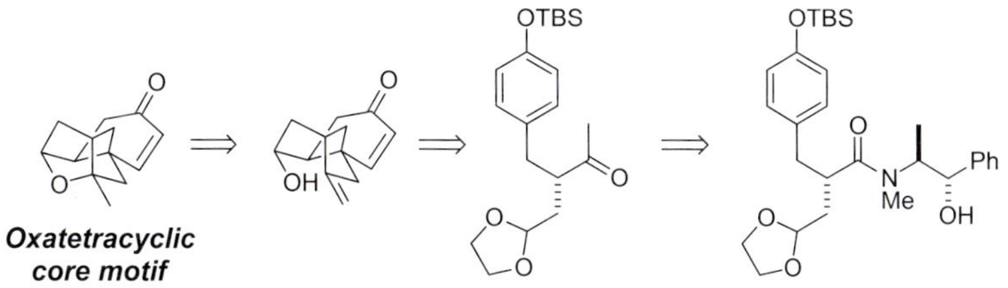

Key reactions: *Myers' asymmetric alkylation (**A** to **B**); Kumada coupling (**C** to **D**); oxidative dearomatization (**D** to **E**); SmI$_2$-mediated carbocyclization (**F** to **G**); acid-catalyzed cycloetherification (**G** to **target**)*

Relevant mechanisms and transformations

A.

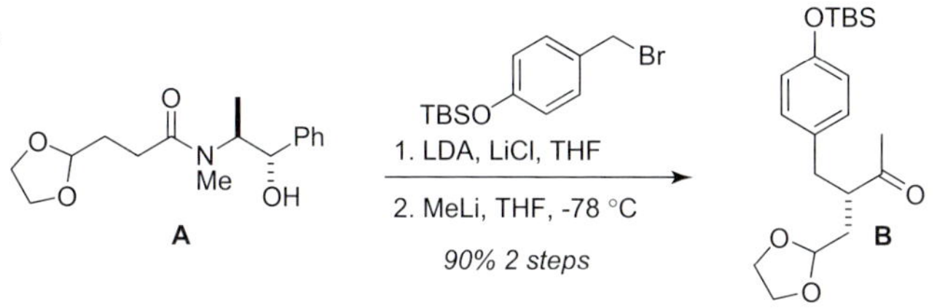

B.

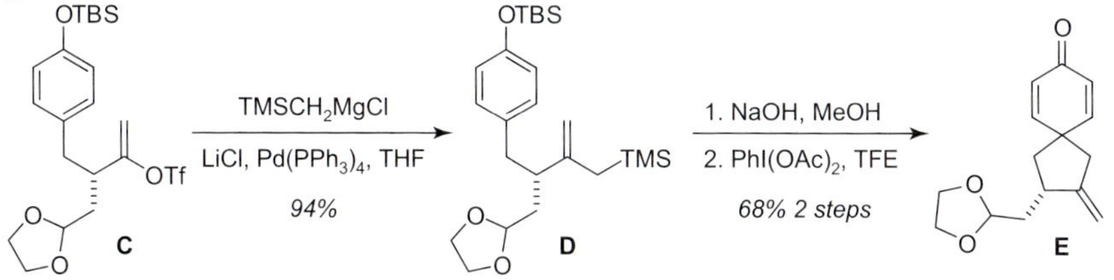

C.

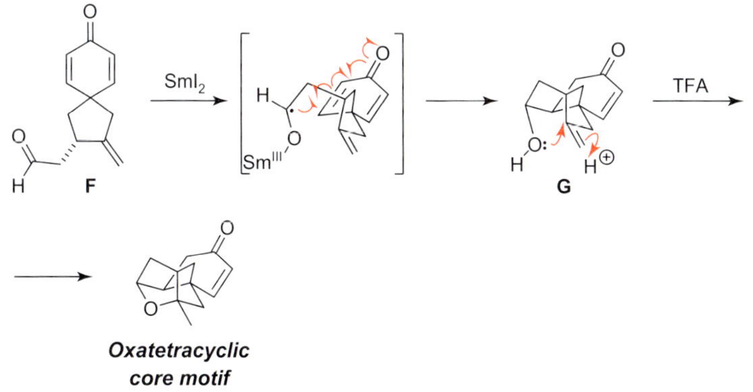

Figure 13.12 Nicolaou's chiral auxiliary-based approach to platensimycin's oxatetracyclic core motif.

13.7.4
Yamamoto's organocatalytic asymmetric synthesis

An expedient synthesis of the oxatetracyclic core motif of platensimycin was reported in 2007 by Yamamoto and coworkers [46]. A catalytic asymmetric Diels-Alder
reaction of methylcyclopentadiene and methyl acrylate (in the presence of an oxazaborolidine, Brønsted acid-assisted, chiral Lewis acid) [47] gave the norbornene
carboxylic acid derivative **A** in 92% yield and 99% *ee*. (Figure 13.13 A). An *N*-nitroso
aldol addition-decarboxylation sequence afforded enantiomerically pure ketone **B**.
A Baeyer-Villiger oxidation with H_2O_2 under basic conditions [48], followed by hydrolysis and dehydrative lactonization gave the bicyclic lactone **C**. S_N2' addition of a
vinyl cuprate to **C**, followed by lactonization of the intermediate cyclopentane in the
presence of bis-trifluoromethansulfonimide led to **D**, which was further elaborated
to **E** and **E'**. Extension to the enone **F** was followed by the critical L-proline-mediated
intramolecular Robinson annulation, which was achieved with remarkable success.
Thus, treatment of ketoaldehyde **F** with a stoichiometric amount of L-proline in
DMF effected an intramolecular Michael addition, most likely proceeding through
the iminium salt and the corresponding enamine (see relevant transformation B,
Figure 13.13). Treatment of the resulting ketoaldehyde **G** with aqueous NaOH led
to an aldol condensation and the formation of the desired oxatetracyclic core with
good diastereoselectivity (dr = 5:1). The authors rationalized the stereoselective

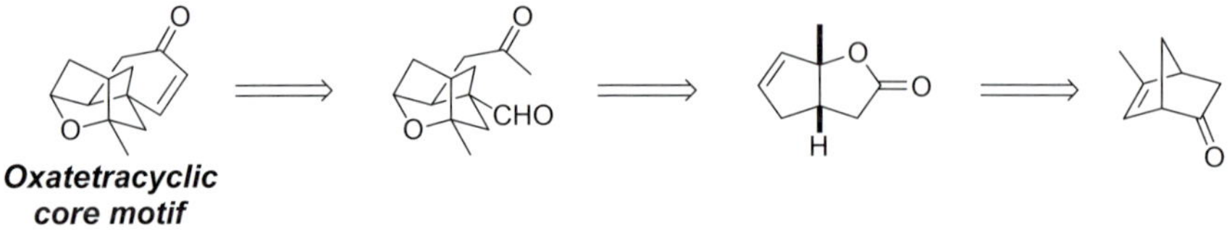

**Oxatetracyclic
core motif**

Key reactions: *Brønsted acid-assisted Diels-Alder reaction (**starting materials** to **A**); nitroso
aldol decarboxylation (**A** to **B**); Baeyer-Villiger oxidation-lactone formation (**B** to **C**); S_N2'
vinylcuprate addition (**C** to **D**); anomeric cyanation (**D** to **E**); asymmetric Robinson annulation (**F**
to **target**)*

Relevant mechanisms and transformations:

A.

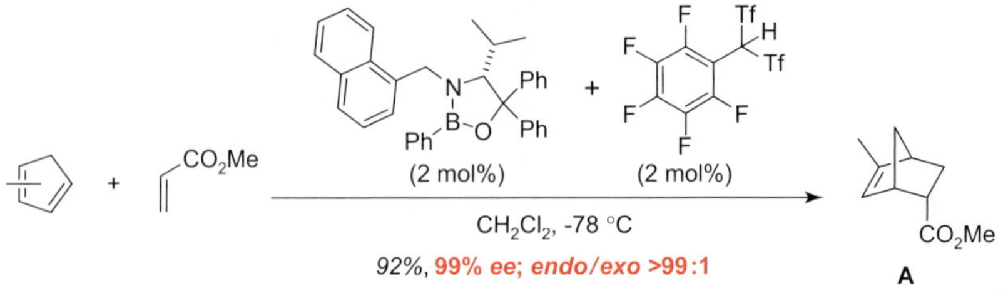

Figure 13.13 Yamamoto's approach to the platensimycin
oxatetracyclic core motif employing a catalytic Diels-Alder reaction and asymmetric Robinson annulation.

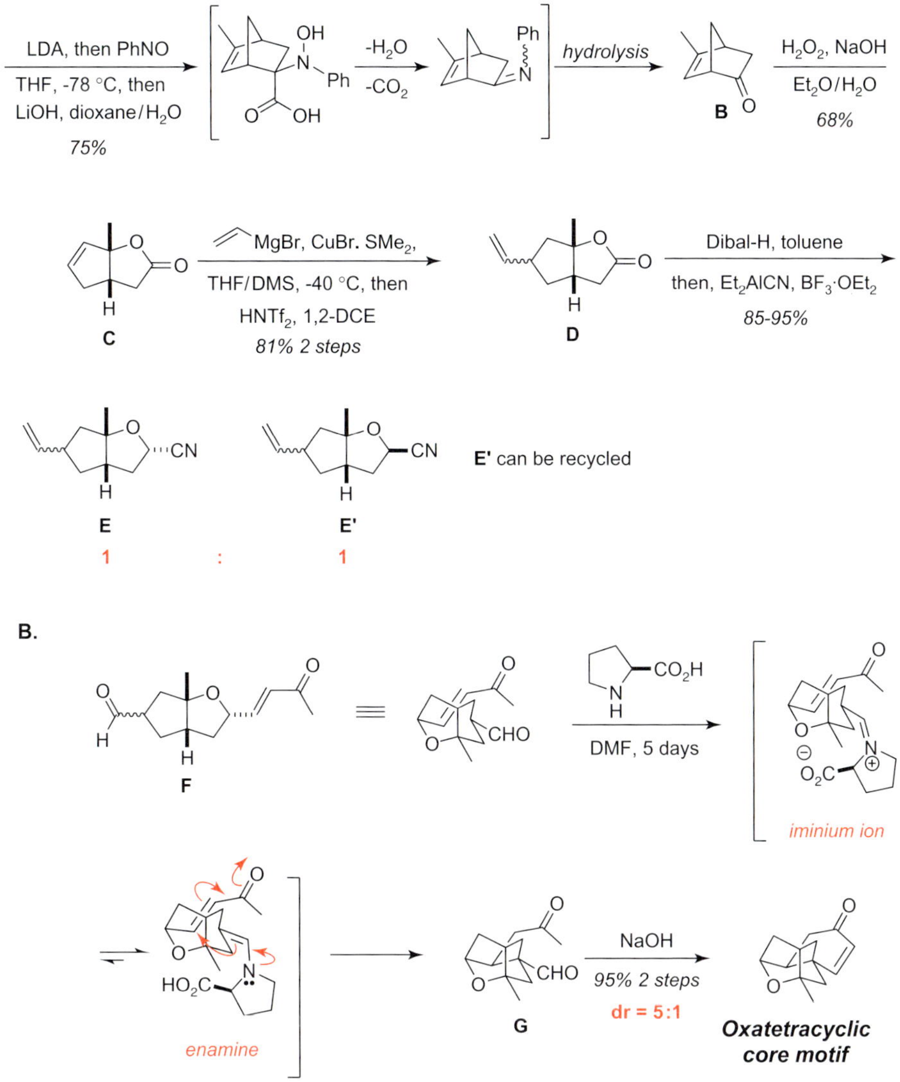

B.

Figure 13.13 (continued).

intramolecular Michael addition on the basis of a preferential *Si*-face attack due to a more favorable LUMO (π^*) energy difference between conformers.

The synthesis was completed in ten steps starting from readily available bulk chemicals (methylcyclopentadiene and methyl acrylate). It is of interest to recall that Ghosh and Xi [35] also used a Baeyer-Villiger oxidation toward a hydroxy bicyclic lactone related to Yamamoto's intermediate **C** (see Figure 13.9).

13.7.5
Corey's catalytic enantioselective synthesis

No less than four different approaches toward the synthesis of the oxatetracyclic core unit of platensimycin were reported in 2007. Two additional formal syntheses of the racemic version also appeared in 2007 [49]. A conceptually different approach was disclosed by Lalic and Corey [50] in the same year (Figure 13.14). The synthesis started with 6-methoxy-α-naphthol, which was converted to 1,4-naphthoquinone **A** in one step by oxidative ketalization with bis-trifluoroacetoxyiodobenzene [51]. Enantioselective conjugate addition of the 2-propenyl group to **A** took place using potassium 2-propenyl trifluoroborate in the presence of a Ru-BINAP catalyst and triethylamine (see relevant transformation A, Figure 13.14) [52]. Considerable experimentation delineated the critical importance of triethylamine as an essential component, and a key element, in accelerating the reaction. Five steps were needed to convert the ketone **B** into the tetrahydronaphthol **C**. Protection of the phenolic hydroxyl group, followed by treatment with bromine in CH_2Cl_2 led to tricyclic bromide **D** (see relevant transformation B, Figure 13.14). Treatment with TBAF led to quinone **E**, which was hydrogenated in a diastereoselective manner in the presence of a chiral rhodium phosphine catalyst to produce tetracyclic ketone **F**. Finally, formation of the TMS silyl enol ether, followed by oxidation with IBX (2-iodoxybenzoic acid) [53] gave the oxatetracyclic enone precursor to platensimycin. The diastereoselective bromocycloetherification reaction was rationalized on the basis of a preferential conformation, in which the 2-propenyl appendage and the OR group (R = H or MEM) are co-axial (see relevant mechanism B, Figure 13.14). As such, a simultaneous attack of the oxygen and bromine can be envisaged. The methylene "inside" rotamer was expected to be more stable by *ca.* 1.5 kcal/mol in an early transition state, compared to an alternative spatial orientation. The Corey synthesis involves less than fifteen steps with complete stereochemical control and good overall yield.

13.7.6
Platensimycin and the mind's eye

Following the first total synthesis of racemic platensimycin, two approaches to the oxatetracyclic core were based on (*R*)- and (*S*)-carvone (Nicolaou [33], in 2008 and Ghosh [35], in 2007). Three approaches (Nicolaou [39], Yamamoto [46], and Corey [50], in 2007) involved the use of catalytic asymmetric methods, while another approach made use of a chiral auxiliary (Nicolaou, 2007) [33]. An enantioselective synthesis of a reduced (de-oxo) oxatetracyclic core of platensimycin starting with the Wieland-Miescher ketone was reported by Kaliappen and Ravikumar [54a]. In 2010, Magnus and coworkers [54b] reported a very efficient and stereoselective formal synthesis of platensimycin from a tetrahydronaphthalene precursor.

It is of interest to compare the individual starting materials in each synthesis with respect to the mental/visual genesis of a synthesis plan and its relationship to one or more key reactions. To what extent the initial drawing of the target

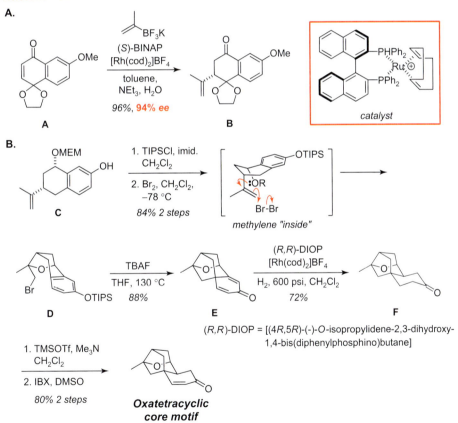

Figure 13.14 Corey's approach to the platensimycin oxate-tracyclic core employing a catalytic Michael addition and dienone homogeneous catalytic hydrogenation.

oxatetracyclic core motif influenced the disconnection of strategic bonds leading to a *reflexive* or *relational* thought process is a matter of conjecture. The diversity of the approaches is a testament to the creative ways in which the syntheses were planned. The starting materials for each of the reported approaches to the enantiopure and racemic oxatetracyclic core motif of platensimycin are shown in Figure 13.15. The reader can immediately "see" (for the most part) the convergence of the carbon framework of the starting material with a designated substructure in the target (even without color coding). Thus, the visual *relational* thought process is fully operative in most of the examples. The desire to use (*R*)- or (*S*)-carvone as a starting material sets the stage for a systematic chemical modification to advanced intermediates with predetermined deployment of usable functional groups to enable completion of the synthesis. In this regard, the Ghosh strategy differs from all others in that (*S*)-carvone is first converted to a cyclopentane derivative. Here, the visual *relational* thought process is highly dependent on a knowledge-based plan, which exploits known reactions of terpenes. All other chiral starting materials in Figure 13.15 are the result of strategic antithetic bond disconnections leading to key reactions in the forward sense. Once such reactions are part of the synthesis plan, then appropriate starting materials are chosen that permit suitable functionalization in order to provide the necessary substrates for the implementation of the key reaction(s).

The desire to use catalytic asymmetric reactions in key steps to generate enantiopure (or enantioenriched) intermediates provides the incentive to incorporate them in the synthesis plan. The Corey [50], Yamamoto [46], and Nicolaou [39] approaches shown in Figure 13.15 started with totally different achiral starting materials that were needed to implement the key catalytic step either directly (Yamamoto), or after further modification (Corey and Nicolaou).

An interesting corollary can be derived from the chosen strategies shown in Figure 13.15. The *chiron approach* starts with a designated compound such as carvone, and seeks to converge with a substructure in the target molecule (visual *relational* analysis relying on substrate control from a chiron). With a mindset focussing on *catalysis* for strategic bond forming processes, the key intermediate is generated *en route* to the target compound (visual *reflexive*, knowledge-based analysis relying on reagent control). The two philosophically and strategically different approaches can merge in cases where resident chirality can also influence transition state energies in a given catalytic reaction. Thus, the desire to combine practicality and innovation provides the incentive for creative synthesis (see Chapter 18, section 18.8).

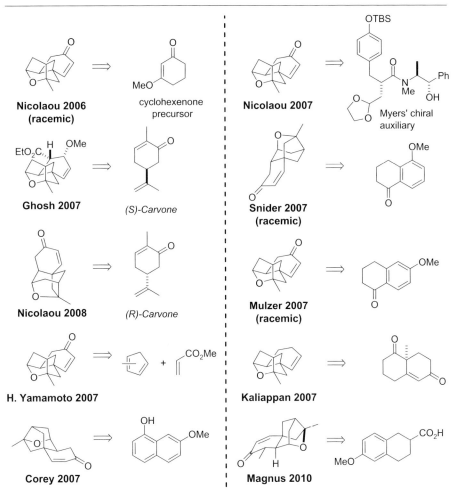

Figure 13.15 The platensimycin oxatetracyclic core as seen by different sets of "mind's eyes".

13.8
Phomactin A

The first total synthesis of the platelet activating factor antagonist phomactin A was reported in 2003 by Mohr and Halcomb [55] (Figure 13.16). Total syntheses of the racemic natural product were reported by Goldring and Pattenden in 2002 [55a] and by Hsung and coworkers in 2009 [55b]. The Halcomb strategy consisted in the use of (R)-pulegone as a precursor to the cyclohexane ring in phomactin A. Thus, C-methylation of (R)-pulegone followed by treatment with KOH effected a retro-aldol reaction to give the 2,3-dimethylcyclohexanone **A**. Base-mediated bromination via the enolate, and subsequent elimination of the α-bromoketone led to the enone **B** in good overall yield. Treatment of the lithium enolate with phenylseleno acetaldehyde, followed by mesylation [56] of the resulting alcohol led to the C-vinyl product **C** as a single diastereomer. Addition of methoxymethoxymethyl lithium generated from the corresponding tributyltin derivative [57] led to the tertiary allylic alcohol, which upon oxidation with PCC gave enone **D**. Stereoselective Luche reduction afforded the equatorial alcohol, which was epoxidized in the presence of m-CPBA to give **E** as a single diastereomer. Oxidation to the ketone, and treatment with magnesium bromide diethyl etherate in CH_2Cl_2 effected regioselective cleavage to give the bromohydrin **F**. Conversion to the trifluoroacetate was followed by elimination to the α-bromoenone and replacement of the MOM ether group with a 3,4-dimethoxybenzyl (DBM) ether, to afford **G** in good overall yield. Reduction of the ketone under Luche conditions gave the undesired secondary alcohol, which was inverted with p-nitrobenzoic acid using a Mitsunobu protocol, then protected as the TBS ether as in intermediate **H**. Halogen-metal exachange of the vinyl bromide **H** to the corresponding organolithium reagent, then treatment with epoxide **I**, prepared in twelve steps from geraniol, which included a Sharpless asymmetric epoxidation, gave an allylic alcohol that was oxidized to the ketone **J**. Treatment with TBAF, then regioselective epoxide opening under acidic conditions led to dihydro-4-pyrone **K**. Protection of the alcohol as the TES ether, followed by removal of the DMB ether group with DDQ facilitated hemiacetal formation, which was isolated as the TMS ether **L** in excellent overall yield. The key macrocyclization step was performed under Suzuki-Miyaura conditions, mediated by Pd(dppf)Cl$_2$ and AsPh$_3$ [58]. Treatment of the product with TBAF gave phomactin A.

Commentary: Analysis of the tetracyclic structure of phomactin A led Mohr and Halcomb to utilize (R)-pulegone as a starting chiron. As seen in the first steps of the synthetic sequence, an excellent level of stereochemical and functional group convergence is provided by (R)-pulegone. After C-methylation, a known retro-aldol reaction was used to arrive at **A**, effectively excising the *gem*-dimethylmethylidene appendage. Formation of the enone was followed by an indirect stereoselective α-vinylation of the enolate under the influence of the 3-methyl group (see relevant mechanism A, Figure 13.16). The hydroxymethyl group required for the dihydrofuran hemiacetal ring was then introduced, and the adduct was converted to the enone **D** setting up the necessary functionalization for a systematic and regioselective introduction of substituents culminating with vinyl bromide **H**. Coupling via

Phomactin A   (R)-Pulegone

Figure 13.16 Total synthesis of phomactin A from (R)-pulegone (partially hidden substructure).

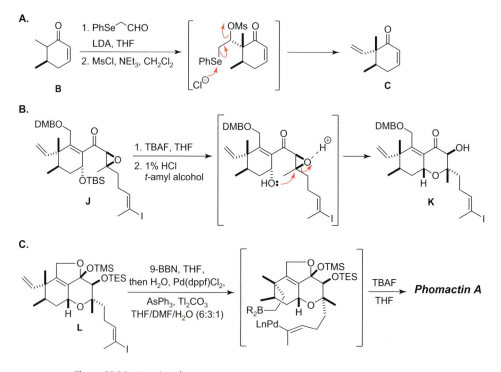

Key reactions: *Stereoselective α-vinylation (**B** to **C**); cycloetherification via intramolecular epoxide opening (**J** to **K**), B-alkyl Suzuki-Miyaura macrocyclization (**L** to **target**); DMB-ONPy = 2-(3,4-dimethoxybenzyloxy)-3-nitropyridine*

Relevant mechanisms and transformations:

Figure 13.16 *(continued).*

the vinyllithium counterpart required access to the epoxy aldehyde **I**, which was prepared from geraniol by a somewhat lengthy process. Nevertheless, elaboration to the bicyclic dihydro-4-pyrone **K** proceeded smoothly (see relevant mechanism B, Figure 13.16), setting the stage for the intramolecular *B*-alkyl Suzuki-Miyaura macrocyclization (see relevant transformation C, Figure 13.16).

A review of the individual transformations shown in Figure 13.16 demonstrates the effectiveness of the original plan. With some detours that had to be taken such as the exchange of a *O*-MOM to a *O*-DMB group, and an oxidation-reduction sequence (**F** to **G** and **G** to **H**), the original (*R*)-pulegone was systematically manipulated in a predictable way. The regioselective openings of epoxides (**E** to **F** and **J** to **K**) are also noteworthy.

13.9
Pinnaic Acid

The total synthesis of pinnaic acid from (*R*)-pulegone was described by Arimoto and coworkers [59] (Figure 13.17). Their strategy relied on the availability of chiral cyclopentenone derivative **A** in 41% overall yield from (*R*)-pulegone by a five-step degradative sequence.

Thus, a [3+2] carbocyclization utilizing Trost's trimethylenemethane reagent [60] in the presence of Pd(OAc)$_2$ gave the *cis*-fused bicyclic ketone **B** as a single stereoisomer. A Beckmann rearrangement led to lactam **C** in modest yield with recovery of starting material. Ozonolysis of the exocyclic methylene group and deoxygenation via the tosylhydrazone [61], followed by *N*-protection gave the *N*-Cbz lactam **D** in good overall yield. Reduction with LiBH$_4$, protection of the primary hydroxyl group, and treatment with ceric ammonium nitrate (CAN) afforded intermediate **E**. Oxidation to the aldehyde, and Horner-Wadsworth-Emmons olefination with ketophosphonate **F** led to *E*-enone **G** in quantitative yield. The key reductive cyclization was performed with Pd(OH)$_2$/C (Pearlman's catalyst) in the presence of acetic acid, and the resulting spirocyclic amine was protected as the *N*-TFA derivative **H**. Conversion to the terminal alkene **I** was accomplished by Grieco's method [62], proceeding through a phenylselenylation of the primary alcohol after selective removal of the *O*-TBS group. A cross-metathesis reaction with ethyl methacrylate in the presence of the Hoveyda-Grubbs second generation catalyst [63] gave the *trans*-olefin **J** in 74% yield. Removal of the second silyl ether and repeating the Grieco elimination gave the penultimate intermediate **K**. Finally, a second cross-metathesis reaction with chloro olefin **L**, prepared in a six-step sequence from but-3-yn-1-ol, followed by a global silyl ether deprotection, and hydrolysis of the trifluoroacetamide and ethyl ester groups gave the sodium salt of pinnaic acid.

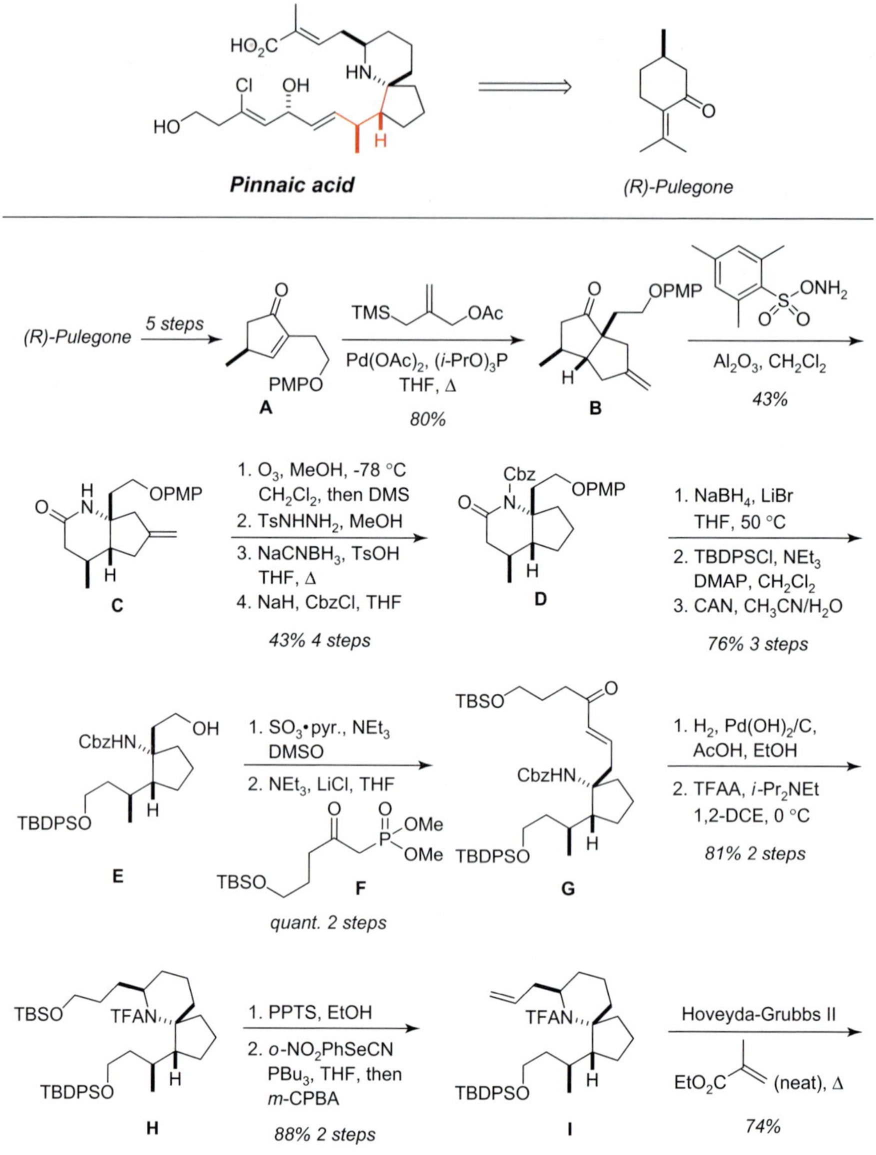

Figure 13.17 Total synthesis of pinnaic acid from (R)-pulegone (hidden substructure).

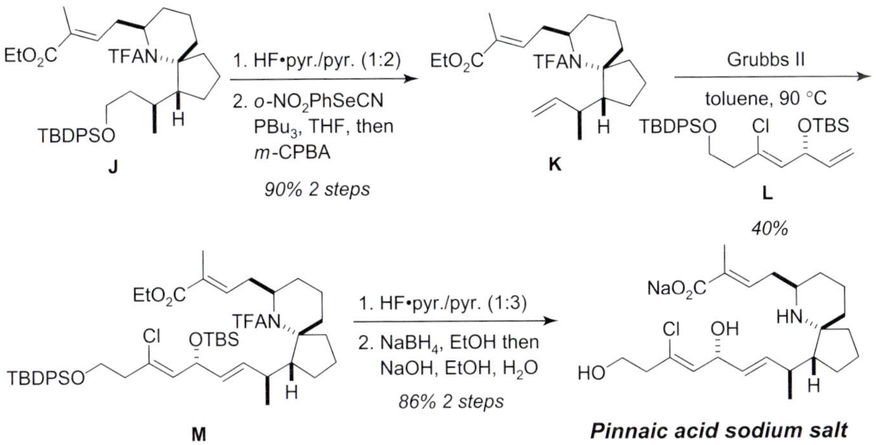

Key reactions: *Trost cycloannulation (**A** to **B**); Beckmann rearrangement (**B** to **C**); stereoselective intramolecular reductive amination (**F** to **G**); cross-metathesis with Hoveyda-Grubbs and Grubbs second generation catalysts (**I** to **J** and **K** to **M**)*

Relevant mechanisms and transformations:

A.

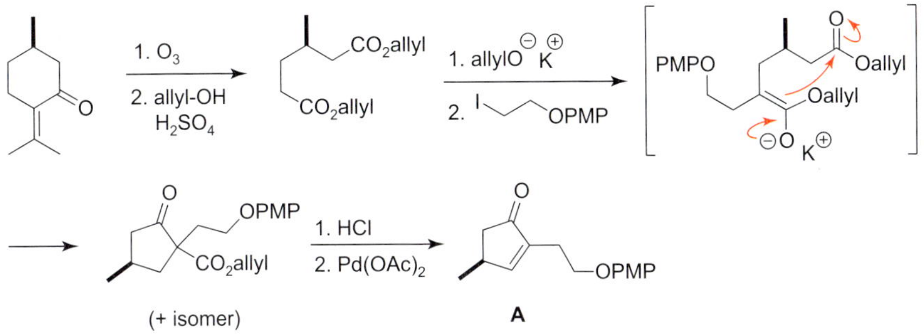

B.

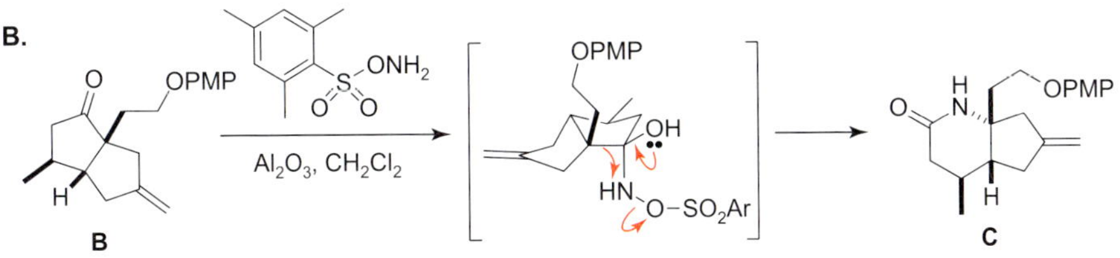

Figure 13.17 *(continued)*.

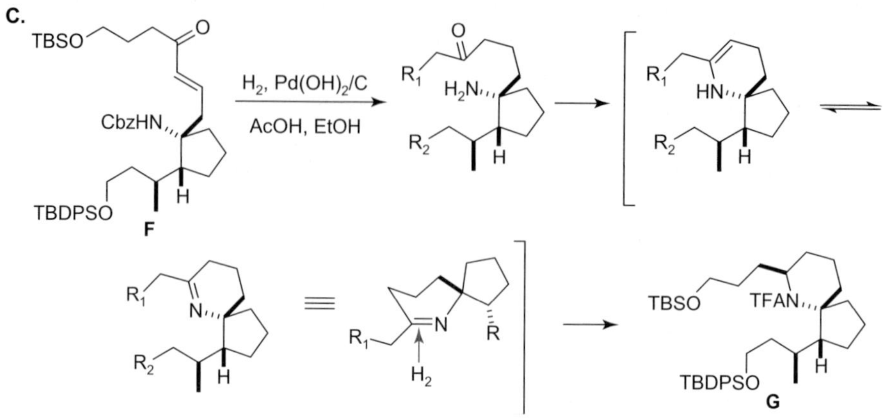

C.

Figure 13.17 (continued).

Commentary: Analysis of the 6-aza-spiro[4.5]decane core structure of pinnaic acid presents two distinct strategies. In one strategy, the piperidine ring would be the starting chiral template while in another, a cyclopentane would have to be elaborated to achieve the azaspiro system. Clearly, a major challenge resides in the stereocontrolled generation of the C-9 azaspirobicyclic tertiary carbon center. Arimoto and coworkers used (R)-pulegone as a very hidden chiron in their synthesis plan. Thus, cyclopentenone **A**, prepared from the terpene in a five-step sequence (see relevant transformations A, Figure 13.17), was used as a means to introduce a nitrogen atom through a Beckmann rearrangement. In fact, a visual *reflexive* analysis of the piperidine ring in pinnaic acid could also lead one to consider a carbocyclic oxime as a precursor provided the Beckmann rearrangement to the lactam is regioselective. With such a plan in mind, Arimoto prepared the necessary cyclopentane by a Trost [3+2] carbocyclization (**A** to **B**). Anticipating a degree of stereocontrol due to the spacial orientation of the C-methyl group in **A**, only the all-*syn* isomer **B** was formed. The Beckmann rearrangement proceeded under stereoelectronic control giving only the desired lactam (with recovery of starting material). The more highly substituted C–C bond migrated in the ring expansion as expected (see relevant mechanism B, Figure 13.17). The crucial 2,6-disubstituted piperidine ring was formed from the catalytic reduction of an appended enone **G**. The formation of the desired C-5 stereoisomer can be rationalized by a preferential attack of hydrogen on an imine or enamine intermediate (see relevant transformation C, Figure 13.17). The completion of the synthesis required two cross-metathesis reaction steps with appropriate alkenes. However, it was necessary to add extra steps in order to create the terminal alkenes (**H** to **I** and **J** to **K**).

The synthesis of pinnaic acid by Arimoto and coworkers from (R)-pulegone is another example of the implementation of a plan that relies on a knowledge-based, visual *reflexive* strategy. The deployment of the hydroxyethyl chain in **A** was a prerequisite for the stereoselectivity of the Beckmann rearrangement, thereby establishing the tertiary carbon center with the amino substituent. However, the necessary conversions of primary alcohol groups to generate alkenes (**H** to **I** and **J** to **K**) suitable for cross-metathesis was a high price to pay for an otherwise indirect, albeit interesting approach to azacyclic natural products.

13.9.1
The Danishefsky and Zhao asymmetric syntheses

The first total synthesis and proof of absolute stereochemistry of pinnaic acid was reported by Danishefsky and coworkers in 2001 [64] (Figure 13.18). The readily available Meyers lactam **A** [65] was allylated under Lewis acid-mediated conditions with allyltrimethylsilane. The enantioenriched product **B** was transformed to the C-methyl analogue **C**, and the N-Boc lactam was cleaved to the corresponding acid that was converted to the primary alcohol via reduction of the mixed anhydride. Thus, in seven steps from the racemic cyclopentanone derivative, it was possible to produce an advanced intermediate that harbored the amino group on a tertiary

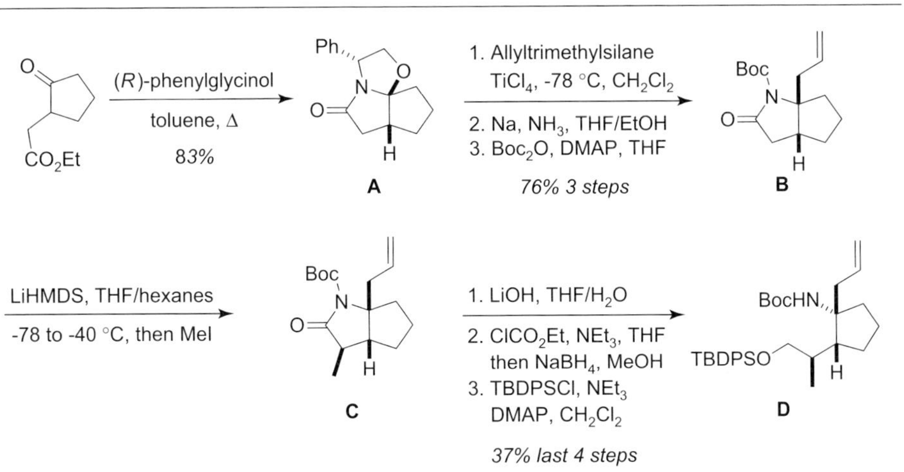

Figure 13.18 Total synthesis of pinnaic acid exploiting a Meyers lactam intermediate.

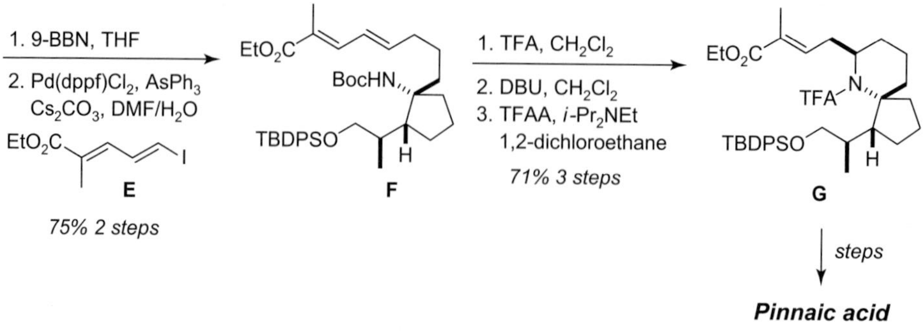

Figure 13.18 *(continued)*.

carbon atom corresponding to the azaspirocyclic core of pinnaic acid. A *B*-alkyl Suzuki-Miyaura coupling with vinyl iodide **E**, gave the extended enoate **F**. A key reaction in the synthesis of the azaspirocycle was the stereocontrolled vinylogous intramolecular Michael-type addition (**F** to **G**), in the presence of DBU as base. Subsequent chemistry was developed for the elaboration of the "southern" skipped diene appendage *en route* to pinnaic acid.

The use of resident chirality in the topologically unique lactam intermediate **A** allowed the elaboration of the cyclopentane intermediate **D** in good over-all yield starting from (*R*)-phenylglycinol and racemic α-(2-oxo-cyclopentyl)ethyl acetate. It should be noted that the nitrogen atom in **D** originates from the (*R*)-phenylglycinol chiral auxiliary (**A** to **D**). Thus, once the chiral lactam **A** had served its purpose as an excellent scaffold to deploy the *C*-allyl group, a self-immolative process was adopted in cleaving the benzylic bond under dissolving metal conditions. The highly stereocontrolled 1,6-conjugate addition of the primary amino group onto the extended enoate system (**F** to **G**) is another highlight of the synthesis.

In the Zhao synthesis of pinnaic acid [66] we also encounter a similar α-(2-oxo-cyclopentyl)carboxylic ester (methyl ester **A**, Figure 13.19) starting material as in the case of Danishefsky (ethyl ester) approach (Figure 13.18). However, asymmetry was introduced in the first step by exploiting a Noyori asymmetric hydrogenation [67] of the keto group in the presence of the $RuCl_2$-(*R*)-BINAP catalyst. The resulting bicyclic lactone was found to have an enantiomeric purity of 90%. Stereoselective *C*-methylation of the lithium enolate afforded **B**, which was reduced to a diol, and then preferentially protected as the primary *O*-benzyl ether to give **C**. Zhao's protocol to introduce nitrogen stereoselectively exploited the chemistry of nitrone anions. Thus, oxidation of **C** to the ketone, conversion to the oxime, and oxidation with *m*-CPBA gave nitrocyclopentane intermediate **D**. A key transformation involved the stereoselective Michael addition of the nitronate anion (generated with Triton B) to methyl acrylate. The resulting single diastereomer **E** was further elaborated to primary iodide **F**, and then chain-extended via dithiane anion alkylation to give the ketone **H**. Conversion of the nitro group to an amine,

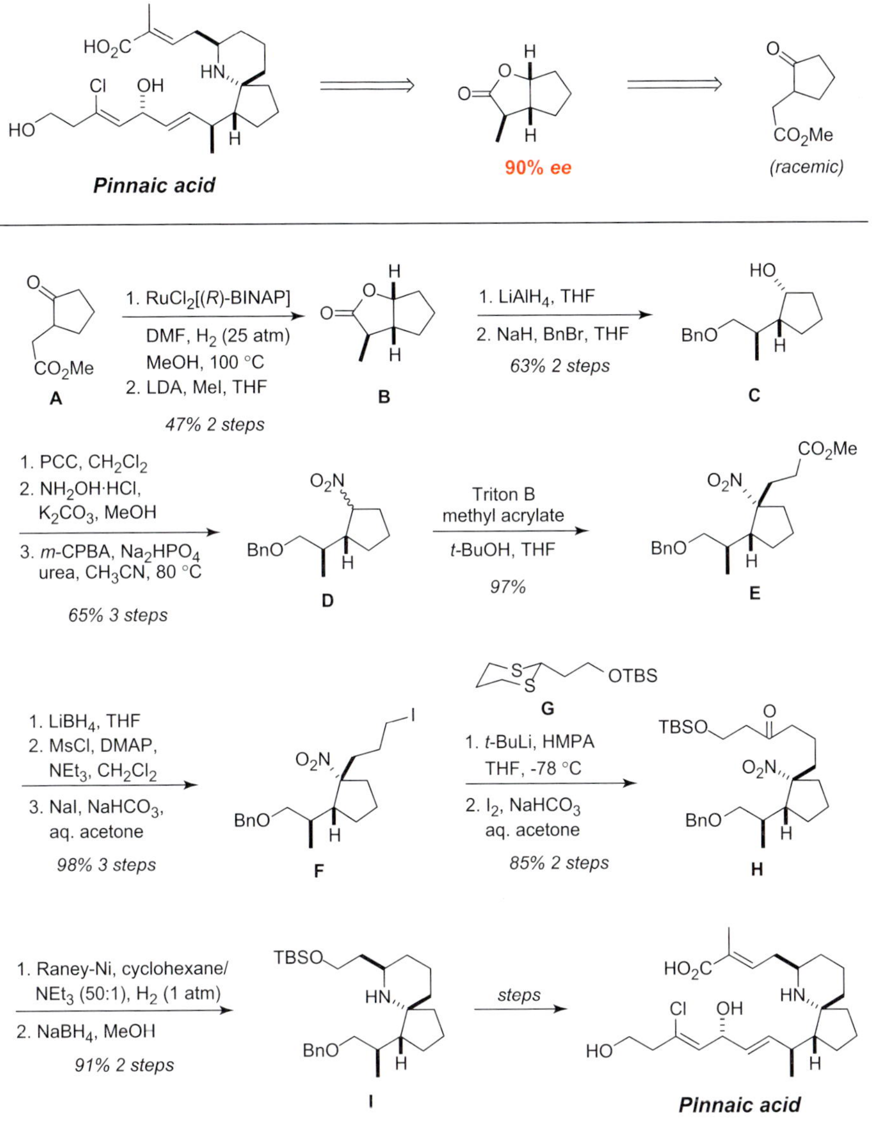

Figure 13.19 Total synthesis of pinnaic acid exploiting a Noyori catalytic asymmetric hydrogenation reaction.

and subsequent reduction of the formed imine (or enamine) with NaBH$_4$ proved to be highly selective affording the equatorially disposed appendage as in **I**. Further steps produced pinnaic acid. A total synthesis of (±)-pinnaic acid was reported in 2004 by Christie and Heathcock [59c].

The three total syntheses of enantiopure pinnaic acid by Arimoto, Danishefsky, and Zhao respectively, represent excellent examples of synthesis planning with different aspects of *reflexive* and *relational* thought processes. Pulegone (Arimoto) exemplified the *chiron approach*, while the chiral auxiliary (Danishefsky), and catalytic asymmetric reaction (Zhao) approaches showed conceptually different ways to achieve the azaspirocyclic core of pinnaic acid, setting up chirality very early in the respective sequences.

13.10
Fusicoauritone

Williams and coworkers [68] reported the first total synthesis of fusicoauritone, a representative member of a diterpenoid family featuring a rare 5-8-5 tricyclic ring system (Figure 13.20). Analysis of the structure of fusicoauritone led to the choice of readily available chiron **A** derived in six steps from (*R*)-limonene [69]. Branching at C-11 (fusicoauritone numbering) was achieved in a stereocontrolled manner through a thermal Johnson-Claisen rearrangement [70] to give **B** after reduction of the resulting ester, and protection as a MEM ether. Hydroboration of the exocyclic olefin, followed by Swern oxidation and treatment of the resulting aldehyde with (*Z*)-1-propenyllithium gave **C** as a single diastereomer in good overall yield. A second Johnson-Claisen rearrangement, followed by reduction of the ester group afforded **D**.

After exploring a number of methods to reduce the *trans*-double bond in **D**, a dissolving metal reduction using sodium metal in a concentrated solution of HMPA and *t*-BuOH afforded the saturated alcohol, which was converted to the tolylsulfone **E**. Cleavage of the MEM group, oxidation of the primary alcohol to the aldehyde, and a Wittig extension afforded the *E*-olefinic ester **F**. Partial reduction of the ester group to the aldehyde and rapid addition of the resulting product to a solution of sodium *t*-amylate in benzene gave a diastereomeric mixture of *β*-hydroxy sulfones **G** in excellent yield. Initial attempts to effect a Nazarov cyclization [71] on a modified analogue of **G** (a dolabellane-4,5-diketone), were not amenable to further modification. An alternative route involved the three-step formation of the vinyl sulfone **H**, which was first reduced to the bis-allylic alcohol, then reductively desulfonylated, and the carbonyl group restored to give **I**. Treatment with TsOH (or $BF_3 \cdot Et_2O$), gave the Nazarov cyclization product **J** in 92% yield as a mixture of C-6 diastereomers. Oxidation with *t*-butyl hypochlorite in aqueous acetone led to fusicoauritone.

Commentary: The conceptual approach to the synthesis of fusicoauritone and biogenetically related congeners harboring eight-membered core subunits was based on knowledge of biosynthetic pathways involving a dolabellyl cation (Figure 13.21). This required the synthesis of a macrocyclic dolabelladienone intermediate **I** already containing all the required stereogenic centers with their appended carbon

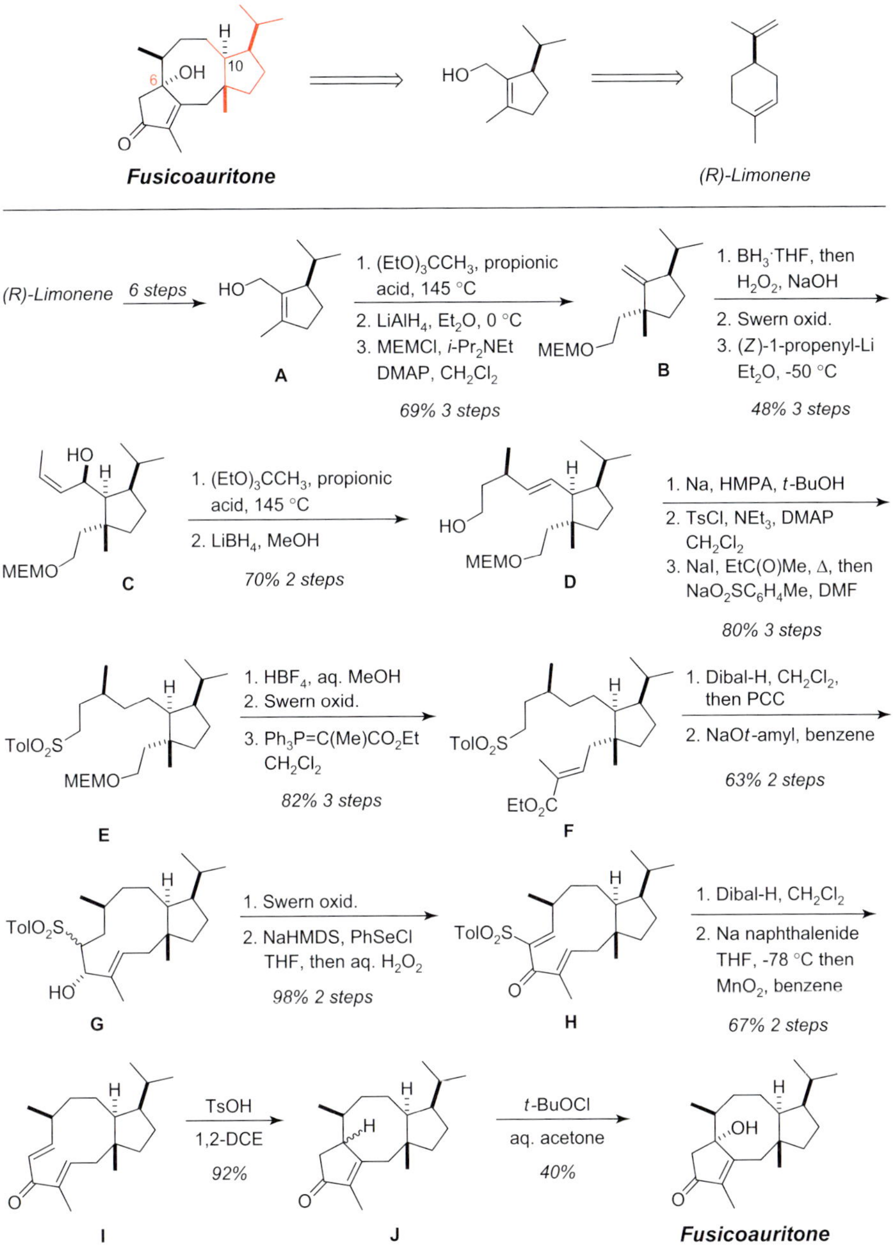

Figure 13.20 Total synthesis of fusicoauritone from (R)-limonene (partially hidden substructure).

Key reactions: *Limonene to **A**; Johnson-Claisen rearrangement (**A** to **B**; **C** to **D**); dissolving metal olefin reduction (**D** to **E**); intramolecular Julia condensation (**F** to **G**); Nazarov cyclization (**I** to **J**)*

Relevant mechanisms and transformations:

A.

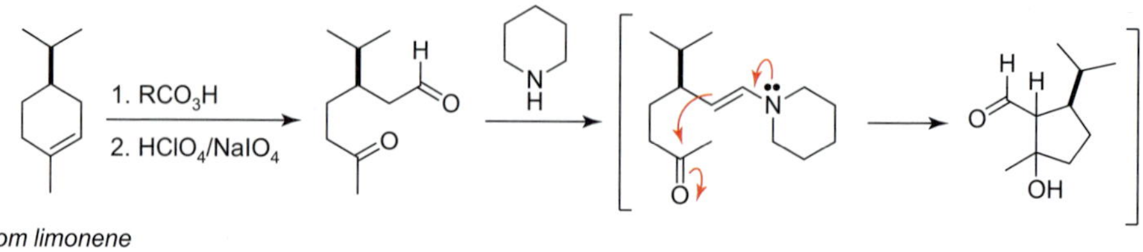

From limonene

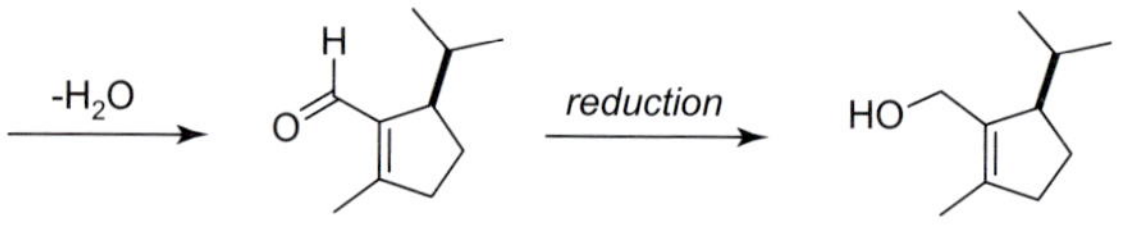

B.

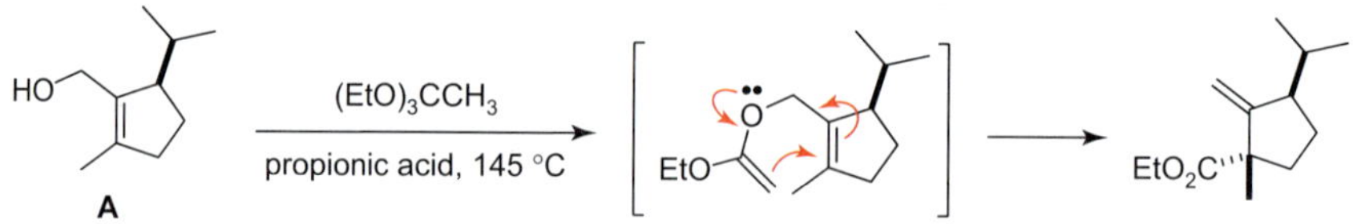

C.

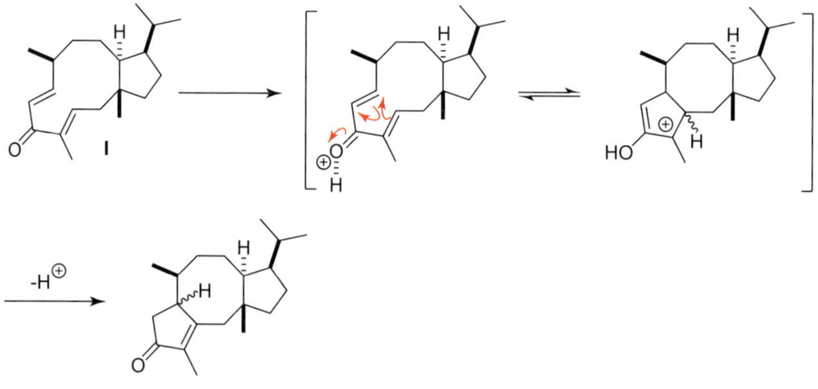

Figure 13.20 *(continued).*

substituents. Having decided on the Nazarov cyclization as the end-game strategy for the construction of the fused 5-8-5 tricycle, Williams focussed on appropriate starting materials that would provide the vicinally substituted cyclopentane ring with specific alkyl appendages. The availability of the cyclopentene **A** in six-steps from (*R*)-limonene offered the advantage of initiating the synthesis from a chiron that would secure the stereochemistry of adjacent centers in subsequent reactions (see relevant mechanism A, Figure 13.20).

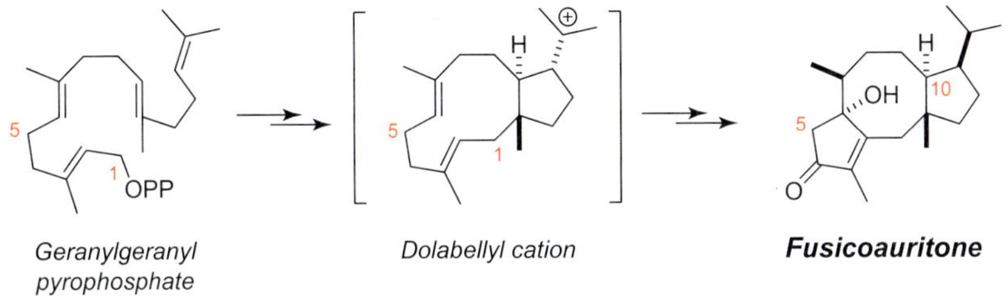

Geranylgeranyl
pyrophosphate

Dolabellyl cation

Fusicoauritone

Figure 13.21   Biogenesis of fusicoauritone and related diterpenes and sesterpenes.

Relying on the steric bias provided by the isopropyl group in **A**, a Johnson-Claisen rearrangement gave *exo*-methylene cyclopentane **B** with a fixed *C*-methyl group on a quaternary carbon (see relevant mechanism B, Figure 13.20). From this point on, the intended extensions were executed as planned *en route* to the 11-membered dolabelladienone **I**. Stereocontrolled propenylation gave **C** as a single (*Z*)-allylic alcohol, which was subjected to a second Johnson-Claisen rearrangement to set up the eventual C-7 methyl group with high stereoselectivity. The key Nazarov cyclization required the synthesis of a [9.3.0]macrocyclic precursor containing a 1,3-dienone functionality. An intramolecular Julia coupling and manipulation of the resulting β-hydroxy sulfone by a series of redox operations provided the intended Nazarov intermediate **I**. The cyclopentadienyl cation, was easily formed in the presence of TsOH, which underwent the familiar 4π electrocyclization to give tricyclic enone **J** in excellent yield (see relevant mechanism C, Figure 13.20).

The synthesis of fusicoauritone by Williams further demonstrates the interplay between visual *reflexive* (Nazarov cyclization), and *relational* (dolabellane biogenesis) thinking. With intermediate **I** already conceived "on paper," the search for a suitable carbocyclic starting chiron such as the cyclopentene **A** followed the logical visual process in the mind's eye. Knowledge of terpene transformations provided the answer.

13.11
Miscellanea

A large number of natural products and related molecules have been synthesized starting with terpenes as native chirons, or their chemically transformed variants and modified chirons [1]. The selection of molecules shown in Figure 13.22 illustrates the importance of terpenes as carbogenic building blocks for Category I, II, and III-type molecules [72] (see Chapter 9).

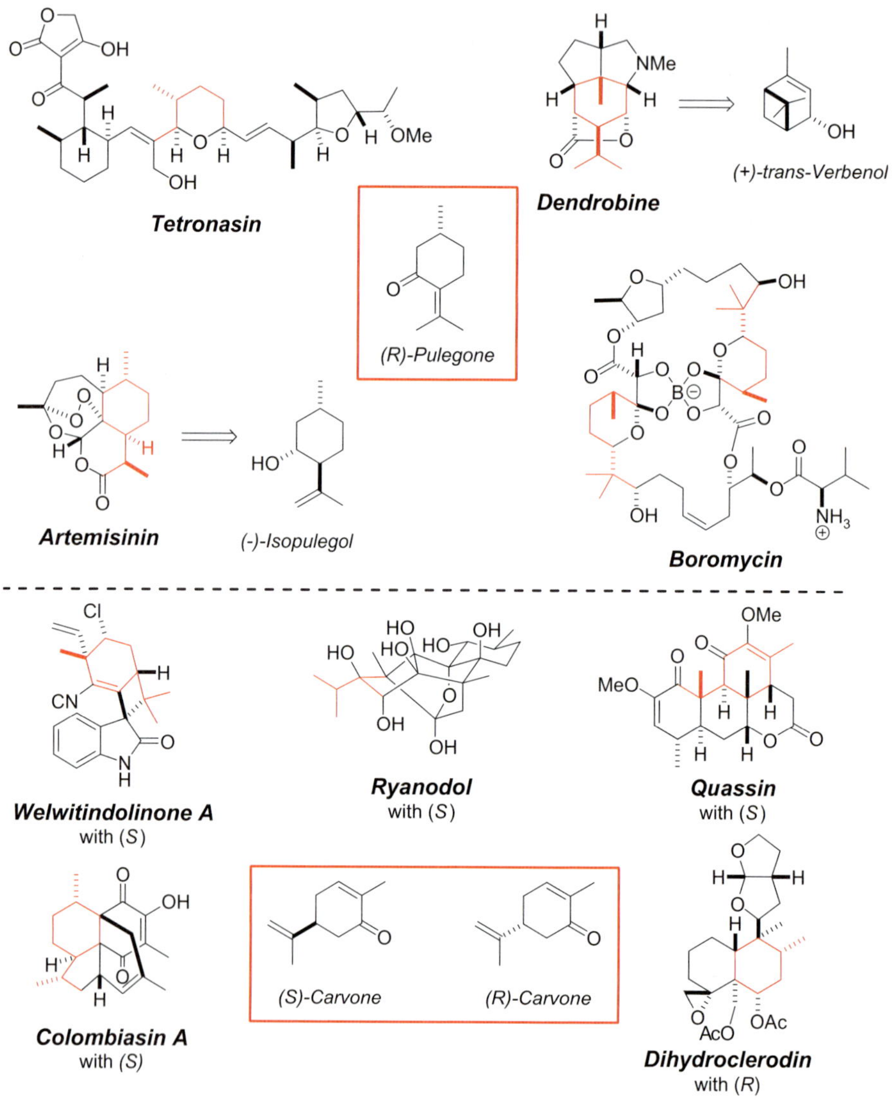

Figure 13.22 Total synthesis of selected natural products from terpenes.

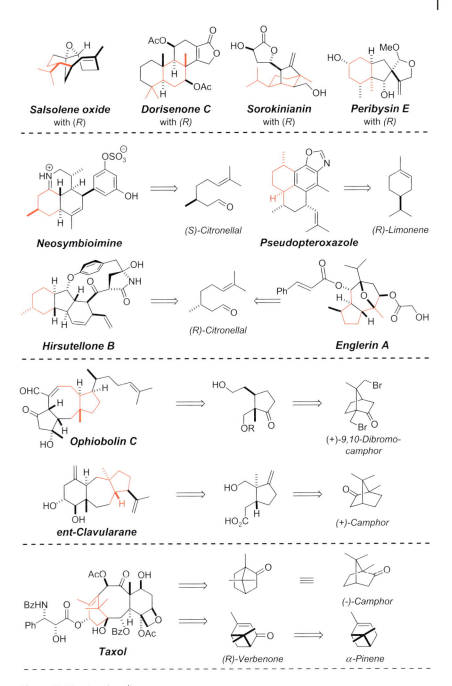

Figure 13.22 (continued).

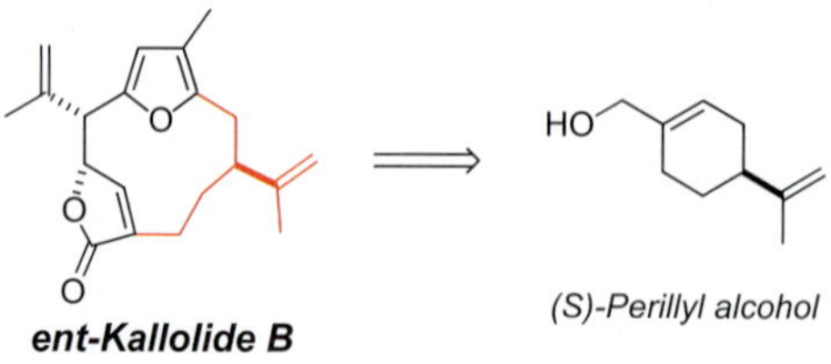

ent-Kallolide B

(S)-Perillyl alcohol

Figure 13.22 *(continued).*

References

1. For example, see: *Carbocycle Construction in Terpene Synthesis*, (a) Ho, T.-L., 1988, VCH, Weinheim; (b) Money, T. in *Studies in Natural Products Chemistry*, Atta-ur-Rahman, Ed.; 1989, vol. 4, p. 625; (c) Money, T. *Nat. Prod. Rep.* 1985, **2**, 278; *Enantioselective Synthesis*, (d) Ho, T.-L., 1992, Wiley-Interscience, NY.

2. Corey, E. J.; Pierce, H. L. *J. Am. Chem. Soc.* 1979, **101**, 5841; see also: (a) Trost, B. M.; Krische, M. J. *J. Am. Chem. Soc.* 1999, **121**, 6131; (b) Niwa, H.; Wakamatsu, K.; Hida, T.; Niiyama, K.; Kigoshi, H.; Yamada, M.; Nagase, H.; Suzuki, M.; Yamada, K. *J. Am. Chem. Soc.* 1984, **106**, 4547; (c) Miyashita, M.; Suzuki, T.; Yoshikoshi, A. *J. Am. Chem. Soc.* 1989, **111**, 3728; for reviews on sesquiterpenes, see: (d) Fraga, B. M. *Nat. Prod. Rep.* 2000, **17**, 483; *The Total Synthesis of Natural Products*, vol. 11, Part B, (e) Pirrung, M. C.; Morehead, A. T.; Young, B. G., 1999, Wiley-Interscience, NY.

3. Corey, E. J.; Enders, D. *Tetrahedron Lett.* 1976, **17**, 11.

4. Bailey, E. J.; Barton, D. H. R.; Elks, J.; Templeton, J. F. *J. Chem. Soc.* 1962, 1578.

5. Alder, K.; Schneider, S. *Justus Liebigs Ann. Chem.* 1936, **524**, 189.

6. Still, W. C.; Murata, S.; Revial, G.; Yoshihara, K. *J. Am. Chem. Soc.* 1983, **105**, 625.

7. For a review, see: (a) Paquette, L. A. *Tetrahedron* 1997, **53**, 13971; see also: (b) Still W. C. *J. Am. Chem. Soc.* 1977,

99, 4186; (c) Still W. C. *J. Am. Chem. Soc.* 1979, **101**, 2493.

8. For example, see: (a) Still, W. C.; Galynker, I. *Tetrahedron* 1981, **37**, 3981; (b) Still, W. C.; Galynker, I. *J. Am. Chem. Soc.* 1982, **104**, 1774.

9. (a) Oliver, S. F.; Högenauer, K.; Simic, O.; Antonello, A.; Smith, M. D.; Ley, S. V. *Angew. Chem. Int. Ed.* 2003, **42**, 5996; (b) Ley, S. V.; Antonello, A.; Balskus, E. P.; Booth, D. T.; Christensen, S. P.; Cleator, E.; Gold, H.; Högenauer, K.; Hünger, U.; Myers, R. M.; Oliver, S. F.; Simic, O.; Smith, M. D.; Søhel, H.; Woolford, A. J. A. *Proc. Natl. Acad. Sci. U.S.A.* 2004, **101**, 12073; see also: (c) Andrews, S. P.; Ball, M.; Wierschem, F.; Cleator, E.; Oliver, S.; Högenauer, K.; Simic, O.; Antonello, A.; Hünger, U.; Smith, M. D.; Ley, S. V. *Chem. Eur. J.* 2007, **13**, 5688.

10. Bajwa, J. S.; Anderson, R. C. *Tetrahedron Lett.* 1991, **32**, 3021.

11. Hartmann, B.; Kanazawa, A. M.; Depres, J. P.; Greene, A. E. *Tetrahedron Lett.* 1991, **32**, 5077.

12. (a) Corminboeuf, O.; Overman, L. E.; Pennington, L. D. *J. Am. Chem. Soc.* 2003, **125**, 6650; (b) Corminboeuf, O.; Overman, L. E.; Pennington, L. D. *J. Org. Chem.* 2009, **74**, 5458.

13. For the antipode, see: Fráter, G. *Helv. Chim. Acta* 1979, **62**, 641.

14. (a) Wulff, W. D.; Peterson, G. A.; Bauta, W. E.; Chan, K.-S.; Faron, K. L.; Gilbertson, S. R.; Kaesler, R. W.; Yang, D. C.; Murray, C. K. *J. Org. Chem.* 1986,

51, 277; (b) Seyferth, D. *J. Am. Chem. Soc.* 1957, **79**, 2133.

15. For a review, see: Overman, L. E.; Pennington, L. D. *J. Org. Chem.* 2003, **68**, 7143.

16. Baggiolini, E.; Hamlow, H. P.; Schaffner, K. *J. Am. Chem. Soc.* 1970, **92**, 4906.

17. Takai, K.; Oshima, K.; Nozaki, H. *Bull. Chem. Soc. Jpn.* 1983, **56**, 3791.

18. Sharma, S.; Oehlschlager, A. C. *J. Org. Chem.* 1989, **54**, 5064.

19. Orita, A.; Hamada, Y.; Nakano, T.; Toyoshima, S.; Otera, J. *Chem. Eur. J.* 2001, **7**, 3321.

20. For a review, see: Fürstner, A. *Chem. Rev.* 1999, **99**, 991.

21. Shing, T. K. M.; Yeung, Y. Y. *Angew. Chem. Int. Ed.* 2005, **44**, 7981; for (±)-samaderine B, see: Grieco, P. A.; Piñero-Nuñez, M. M. *J. Am. Chem. Soc.* 1994, **116**, 7606; for a review, see: Guo, Z.; Vangapanau, S.; Sindelar, R. W.; Walker, L. A.; Sindelar, R. D. *Curr. Med. Chem.* 2005, **12**, 173.

22. For a synthesis of (+)-quassin from (*S*)-carvone, see: (a) Shing, T. K. M.; Jiang, Q.; Mak, T. C. W. *J. Org. Chem.* 1998, **63**, 2056; (b) Kim, M.; Kawada, K.; Gross, R. S.; Watt, D. S. *J. Org. Chem.* 1990, **55**, 504; for a synthesis of (±) quassin, see: (c) Vidari, G.; Ferriño, S.; Grieco, P. A. *J. Am. Chem. Soc.* 1984, **106**, 3539.

23. Salmond, W. G.; Barta, M. A.; Havens, J. L. *J. Org. Chem.* 1978, **43**, 2057.

24. Luche, J. L.; Rodriguez-Halm, L.; Crabbé, P. *J. Chem. Soc., Chem. Commun.* 1978, 601.

25. Wilson, S. R.; Mao, D. T. *Tetrahedron Lett.* 1977, **18**, 2559.

26. Cambie, R. C.; Hay, M. P.; Larsen, L.; Rickand, C. E. F.; Rutledge, P. S.; Woodgate, P. D. *Aust. J. Chem.* 1991, **44**, 821.

27. For example, see: *Indoles*, Sundberg, R. J., 1996, p. 175, Academic Press, San Diego.

28. (a) Baran, P. S.; Maimone, T. J.; Richter, J. M. *Nature*, 2007, **446**, 404; (b) Baran, P. S.; Richter, J. M. *J. Am. Chem. Soc.* 2004, **126**, 7450; see also: (c) Vaillancourt, V.; Albizati, K. F. *J. Am. Chem. Soc.* 1993, **115**, 3499; (d)

Kinsman, A. C.; Kerr, M. A. *J. Am. Chem. Soc.* 2003, **125**, 14120; for a racemic synthesis of hapalindole Q, see: (e) Kinsman, A. C; Kerr, M. A. *Org. Lett.* 2001, **3**, 3189; for a racemic synthesis of hapalindole U, see: (f) Muratake, H.; Kumagami, H.; Natsume, M. *Tetrahedron* 1990, **46**, 6351.

29. Hermann, W. A.; Brossmer, C.; Öfele, K.; Reisinger, C.-P.; Priermeier, T.; Beller, M.; Fischer, H. *Angew. Chem. Int. Ed.* 1995, **34**, 1844.

30. Schkeryantz, J. M.; Woo, J. C. G.; Siliphaivanh, P.; Depew, K. M.; Danishefsky, S. J. *J. Am. Chem. Soc.* 1999, **121**, 11964.

31. Nicolaou, K. C.; Li, A.; Edmonds, D. J. *Angew. Chem. Int. Ed.* 2006, **45**, 7086.

32. For reviews, see: (a) Lu, X.; You, Q. *Curr. Med. Chem.* 2010, **17**, 1139; (b) Tiefenbacher, K.; Mulzer, J. *Angew. Chem. Int. Ed.* 2008, **47**, 2548; (c) Tiefenbacher, K.; Mulzer, J. *Angew. Chem. Int. Ed.* 2007, **46**, 8074.

33. (a) Nicolaou, K. C.; Pappo, D.; Tsang, K. Y.; Gibe, R.; Chen, D. Y.-K. *Angew. Chem. Int. Ed.* 2008, **47**, 944; see also: (b) Nicolaou, K. C.; Li, A.; Ellery, S. P.; Edmonds, D. J. *Angew. Chem. Int. Ed.* 2009, **48**, 6293.

34. For example, see; (a) Weinges, K.; Reichert, H. *Synlett* 1991, 785; (b) Weinges, K.; Reichert, H.; Braun, R. *Chem. Ber.* 1994, **127**, 549; see also: *Radicals in Organic Synthesis; Formation of Carbon-Carbon Bonds*, Giese, B. 1986, Pergamon Press, Oxford.

35. (a) Ghosh, A. K.; Xi, K. *Org. Lett.* 2007, **9**, 4013; for completion of the total synthesis of platensimycin, see: (b) Ghosh, A. K.; Xi, K. *J. Org. Chem.* 2009, **74**, 1163.

36. Martin, J. C.; Arhart, R. J. *J. Am. Chem. Soc.* 1971, **93**, 4327.

37. For example, see: (a) Enders, D.; Breuer, K.; Runsink, J.; Teles, J. H. *Acc. Chem. Res.* 2004, **37**, 534; (b) Johnson, J. S. *Angew. Chem. Int. Ed.* 2004, **43**, 1326.

38. (a) Petasis, N. A.; Bzowej, E. I. *J. Am. Chem. Soc.* 1990, **112**, 6392; see also: (b) Dollinger, L. M.; Ndakala, A. J.; Hashemzadeh, M.; Wang, G.; Wang, Y.; Martinez, I.; Arcari, J. T.; Galluzzo, D. J.; Howell, A. R.; Rheingold, A. L.;

Figuero, J. S. *J. Org. Chem.* 1999, **64**, 7074; (c) Singh, S. B.; Jayasuriya, H.; Ondeyka, J. G.; Herath, K. B.; Zhang, C.; Zink, D. L.; Tsou, N. N.; Ball, R. G.; Basilio, A.; Genilloud, O.; Diez, M. T.; Vicente, F.; Pelaez, F.; Young, K.; Wang, J. *J. Am. Chem. Soc.* 2006, **128**, 11916; see also: ref. 32.

39. Nicolaou, K. C.; Edmonds, D. J.; Li, A.; Tria, G. S. *Angew. Chem. Int. Ed.* 2007, **46**, 3942.

40. (a) Trost, B. M.; Toste, F. D. *J. Am. Chem. Soc.* 2000, **122**, 714; (b) Trost, B. M.; Surivet, J.-P.; Toste, F. D. *J. Am. Chem. Soc.* 2004, **126**, 15592.

41. (a) Cao, P.; Zhang, X. *Angew. Chem. Int. Ed.* 2000, **39**, 4104; (b) Lei, A.; He, M.; Zhang, X. *J. Am. Chem. Soc.* 2002, **124**, 8198.

42. Barton, D. H. R.; Crich, D.; Motherwell, W. B. *J. Chem. Soc., Chem. Commun.* 1983, 939.

43. Nicolaou, K. C.; Edmonds, D. J.; Tria, G. S.; Li, A. *Angew. Chem. Int. Ed.* 2007, **46**, 3942.

44. (a) Myers, A. G.; Gleason, J. L.; Yoon, T.; Kung, D. W. *J. Am. Chem. Soc.* 1997, **119**, 656; (b) Myers, A. G.; Gleason, J. L.; Yoon, T. *J. Am. Chem. Soc.* 1999, **117**, 8488.

45. (a) Tamao, K.; Sumitani, K.; Kumada, M. *J. Am. Chem. Soc.* 1972, **94**, 4374. (b) Kumada, M. *Pure Appl. Chem.* 1980, **52**, 669.

46. Li, P.; Payette, J. N.; Yamamoto, H. *J. Am. Chem. Soc.* 2007, **129**, 9534.

47. For example, see: (a) Futatsugi, K.; Yamamoto, H. *Angew. Chem. Int. Ed.* 2005, **44**, 1484; (b) Hasegawa, A.; Ishikawa, T.; Ishihara, K.; Yamamoto, H. *Bull. Chem. Soc. Jpn.* 2005, **78**, 1401.

48. Weinshenker, N. M.; Stephenson, R. *J. Org. Chem.* 1972, **37**, 3741.

49. (a) Zou, Y.; Chen, C.-H.; Taylor, C. D.; Foxman, B. M.; Snider, B. B. *Org. Lett.* 2007, **9**, 1825; see also: reference 32.

50. Lalic, G.; Corey, E. J. *Org. Lett.* 2007, **9**, 4921.

51. Breuning, M.; Corey, E. J. *Org. Lett.* 2001, **3**, 1559.

52. For example, see: (a) Pucheault, M.; Darses, S.; Genêt, J.-P. *Eur. J. Org. Chem.* 2002, 3552; (b) Molander, G. A.; Ellis, N. *Acc. Chem. Res.* 2007, **40**, 275;

(c) Hayashi, T.; Yamasaki, K. *Chem. Rev.* 2003, **103**, 2829.

53. Nicolaou, K. C.; Gray, D. L. F.; Montagnon, T.; Harrison, S. T. *Angew. Chem. Int. Ed.* 2002, **41**, 996.

54. (a) Kaliappan, K. D.; Ravikumar, V. *Org. Lett.* 2007, **9**, 2417; (b) Magnus, P. Rivera, H.; Lynch, V. *Org. Lett.* 2010, **12**, 5677; for a recent review on approaches to platensimycin, see: Palanichamy, K.; Kaliappan, K. *Chem. Asian J.* 2010, **5**, 668.

55. Mohr, P. J.; Halcomb, R. L. *J. Am. Chem. Soc.* 2003, **125**, 1712; for a synthesis of (±)-phomactin, see: (a) Goldring, W. P. D.; Pattenden, G. *Chem. Commun.* 2002, 1736; (b) Tang, Y.; Cole, K. P.; Buchanan, G. S.; Li, G.; Hsung, R. P. *Org. Lett.* 2009, **11**, 1591; for a synthesis of phomactin B2, see: (c) Huang, J.; Wu, C.; Wulff, W. D. *J. Am. Chem. Soc.* 2003, **125**, 1712.

56. For example, see: (a) Kowalski, C. J.; Dung, J. S. *J. Am. Chem. Soc.* 1980, **102**, 7950; (b) Clive, D. L. J.; Russell, C. G.; Suri, S. C. *J. Org. Chem.* 1982, **47**, 1632; (c) Reich, H. J.; Chow, F. *J. Chem. Soc., Chem. Commun.* 1975, 790.

57. Johnson, C. R.; Medich, J. R. *J. Org. Chem.* 1988, **53**, 4131.

58. For example, see: (a) Miyaura, N.; Ishiyama, T.; Sasaki, H.; Ishikawa, M.; Satoh, M.; Suzuki, A. *J. Am. Chem. Soc.* 1989, **111**, 314; for a review, see: (b) Cheruler, S. R.; Trauner, D.; Danishefsky, S. J. *Angew. Chem. Int. Ed.* 2001, **40**, 4544.

59. (a) Xu, S.; Arimoto, H.; Uemura, D. *Angew. Chem. Int. Ed.* 2007, **46**, 5746; (b) Hayakawa, I.; Arimoto, H.; Uemura, D. *Heterocycles* 2003, **59**, 44; for a synthesis of (±)-pinnaic acid, see: (c) Christie, H. S.; Heathcock, C. H. *Proc. Natl. Acad. Sci. U.S.A.* 2004, **101**, 12079; for a review, see: (d) Clive, D. L. J.; Yu, M.; Wang, J.; Yeh, V. S. C.; Kang, S. *Chem. Rev.* 2005, **105**, 4483.

60. Trost, B. M.; Chan, D. M. T. *J. Am. Chem. Soc.* 1979, **101**, 6429.

61. Hutchins, R. O.; Milewski, C. A.; Maryanoff, B. E. *J. Am. Chem. Soc.* 1973, **95**, 3662.

62. Grieco, P.; Masaki, Y.; Boxler, D. *J. Am. Chem. Soc.* 1975, **97**, 1597; see also:

Sharpless, K. B.; Young, M. W. *J. Org. Chem.* 1975, **40**, 947.

63. Hoveyda, A. H.; Gillingham, D. G.; van Veldhuizen, J. J.; Kataoka, O.; Garber, S. B.; Kingsbury, J. S.; Harrity, J. P. A. *Org. Biomol. Chem.* 2004, **2**, 8.

64. (a) Carson, M. W.; Kim, G.; Danishefsky, S. J. *Angew. Chem. Int. Ed.* 2001, **40**, 4453; (b) Carson, M. W.; Kim, G.; Hentemann, M. F.; Trauner, D.; Danishefsky, S. J. *Angew. Chem. Int. Ed.* 2001, **40**, 4450.

65. (a) Meyers, A. I.; Brengel, G. P. *Chem. Commun.* 1997, 1; (b) Romo, D.; Meyers, A. I. *Tetrahedron* 1991, **47**, 9503.

66. Wu, H.; Zhang, H.; Zhao, G. *Tetrahedron* 2007, **63**, 6454.

67. Kitamura, M.; Tokunaga, M.; Ohkuma, T.; Noyori, R. *Org. Synth.* 1992, **71**, 1.

68. Williams, D. R.; Robinson, L. A.; Nevill, C. R.; Reddy, J. P. *Angew. Chem. Int. Ed.* 2007, **46**, 915; for a review, see: Hiersemann, M.; Helmboldt, H. *Top. Curr. Chem.* 2005, **243**, 73.

69. White, J. D.; Ruppert, J. F.; Avery, M. A.; Torii, S.; Nokami, J. *J. Am. Chem. Soc.* 1981, **103**, 1813.

70. For example, see: Mehta, G.; Krishnamurthy, N. *Tetrahedron Lett.* 1987, **28**, 5945.

71. For reviews see: (a) Tius, M. A. *Eur. J. Org. Chem.* 2005, 2193; (b) Pelissier, H. *Tetrahedron* 2005, **61**, 6479.

72. Tetronasin: Bourque, E.; Kocienski, P. J.; Stocks, M.; Yuen, J. *Synthesis* 2005, 3219; dendrobine: Cassayre, J.; Zard, S. Z. *J. Am. Chem. Soc.* 1999, **121**, 6072; qinghaosu: Schmid, G.; Hofheinz, W. *J. Am. Chem. Soc.* 1983, **105**, 624; boromycin: White, J. D.; Avery, M. A.; Choudry, S. C.; Dhingra, O. P.; Gray, B. D.; Kang, M.-C.; Kuo, S-C. Whittle, A. J. *J. Am. Chem. Soc.* 1989, **111**, 790; welwitindolinone A: Baran, P. S.; Maimone, T. J.; Richter, J. M. *Nature* 2007, **446**, 404; ryanodol: Bélanger, A.; Berney, D.; Borschber, H.; Brousseau, R.; Doutheau, A.; Durand, R.; Katayama, H.; Lapalme, R.; Letruc, D.; Liao, C.; MacLachlan, F.; Maffrand, J.; Marazza, F.; Martino, R.; Moreau, C.; Saint-Laurent, L; Saintonge, R.; Soucy, P.; Ruest, L.; Deslongchamps, P.

Can. J. Chem. 1979, **57**, 3348; colombiasin A: Harrowven, D. C.; Pascoe, D. D.; Demurtas, D.; Bourne, H. O. *Angew. Chem. Int. Ed.* 2005, **44**, 1221; quassin: Shing, T. K. N.; Jiang, Q.; Mak, T. C. W. *J. Org. Chem.* 1998, **63**, 2056; Paquette, L. A.; Sun, L.-Q.; Watson, T. N. J.; Friedrich, D.; Freeman, B. T. *J. Am. Chem. Soc.* 1997, **119**, 2767; dehydroclerodin: Meulenars, T. M.; Stork, G. A.; Macaev, F. Z.; Jansen, B. J. M.; de Groot, A. *J. Org. Chem.* 1999, **64**, 9178; dorisenone C: Abad, A.; Agullo, C.; Cunat, A. C.; Garcia, A. B. *Tetrahedron* 2005, **61**, 1961; sorokinianin: Watanabe, H.; Onoda, T.; Kitahara, T.; Mori, K. *Tetrahedron Lett.* 1997, **38**, 6015; peribysin E: Angeles, A. R.; Waters, S. P.; Danishefsky, S. J. *J. Am. Chem. Soc.* 2008, **130**, 13765; neosymbioimine: Varseev, G. N.; Maier, M. E. *Org. Lett.* 2007, **9**, 1461; pseudopteroxazole: Davidson, J. P.; Corey, E. J. *J. Am. Chem. Soc.* 2003, **125**, 13486; hirsutellone A: Nicolaou, K. C.; Sarlah, D.; Wu, T. R.; Zhan, W. *Angew. Chem. Int. Ed.* 2009, **48**, 6870; ophiobolin C: Rowley, M.; Tsukamoto, M.; Kishi, Y. *J. Am. Chem. Soc.* 1989, **111**, 2735; *ent*-clavularane: Williams, D. R.; Coleman, P. J.; Henry, S. S. *J. Am. Chem. Soc.* 1993, **115**, 11654; taxol from camphor: (a) Holton, R. A.; Somaza, C.; Kim, H.-B.; Liang, F.; Biediger, R. J.; Boatman, P. D.; Shindo, M.; Smith, C. C.; Kim, S.; Nadizadeh, H.; Suzuki, Y.; Tao, C.; Vu, P.; Gentile, L. N.; Liu, J. H. *J. Am. Chem. Soc.* 1994, **116**, 1597; (b) Holton, R. A.; Kim, H.-B.; Somoza, C.; Liang, F.; Biediger, R. J.; Boatman, P. D.; Shindo, M.; Smith, C. C.; Kim, S.; Nadizadeh, H.; Suzuki, Y.; Tao, C.; Vu, P.; Tang, S.; Zhang, P.; Murthi, K. K.; Gentile, L. N.; Liu, J. H. *J. Am. Chem. Soc.* 1994, **116**, 1599; taxol from pinene: (a) Wender, P. A.; Badham, N. F.; Conway, S. P.; Floreancig, P. E.; Glass, T. E.; Gränicher, C.; Houze, J. B.; Jänichen, J.; Lee, D.; Marquess, D. G.; McGrane, P. L.; Meng, W.; Mucciaro, T. P.; Mühlebach, M.; Natchus, M. G.; Paulsen, H.; Rawlins, D. B.; Satkofsky, J.; Shuker, A. J.; Sutton, J. C.; Taylor,

R. E.; Tomooka, K. *J. Am. Chem. Soc.* 1997, **119**, 2755;   (b) Wender, P. A.; Badham, N. F.; Conway, S. P.; Floreancig, P. E.; Glass, T. E.; Houze, J. B.; Krauss, N. E.; Lee, D.; Marquess, D. G.; McGrane, P. L.; Meng, W.; Natchus, M. G.; Shuker, A. J.; Sutton, J. C.; Taylor, R. E. *J. Am. Chem. Soc.* 1997, **119**, 2757;   *ent*-kallolide B: Marshall, J. A.; Bartley, G. S.; Wallace, E. M. *J. Org. Chem.* 1996, **61**, 5729; englerin A: Molawi, K.; Delpont, N.; Echavarren, A. M. *Angew. Chem. Int. Ed.* 2010, **49**, 3517;   Zhou, Q. Chen, X.; Ma, D. *Angew. Chem. Int. Ed.* 2010, **49**, 3513.

14
Total Synthesis from Carbocyclic Precursors

Carbocyclic compounds possessing one or more stereogenic centers are available from natural sources, by chemical manipulations of terpenes, by microbiological transformations of aromatic compounds, and by asymmetric synthesis.

We showed examples of functionalized carbocyclic compounds derived from the above sources in Chapters 5 and 7. Here we highlight total syntheses of selected polycyclic and heterocyclic natural products starting with chiral, non-racemic carbocycles. As in the preceding four chapters, the examples involve Category I-type target molecules in an attempt to demonstrate the most useful skeletal convergence with the chosen carbocycles.

14.1
Punctatin A

The first total synthesis, and determination of absolute configuration, of the sesquiterpene punctatin A (antibiotic M95464) was reported in 1986 by Sugimura and Paquette [1] (Figure 14.1). The structural assignment of punctatin A rested on chemical transformations and X-ray analysis. The tricyclic structure harboring two quaternary stereogenic carbon atoms at one ring junction and a tertiary hydroxyl group at another, present veritable synthetic challenges. The synthesis commenced with a regioselective alkylation of a thermodynamic enolate derived from the readily available bicyclic enone **A** [2]. By controlling the reactivity of the electrophile and using 1-bromo-2-methylpropane, it was possible to obtain **B** as the exclusive product. Reduction of the ketone led to a single allylic alcohol, which was alkylated with tributyl(iodomethyl)stannane in the presence of potassium hydride to give **C**. Treatment with *n*-BuLi in hexanes led to a [2,3]-sigmatropic rearrangement [3] and installation of the hydroxymethyl group at C-7 (punctatin A numbering). Protection of the primary alcohol initially as the MEM ether, allowed exploration of subsequent steps. However, it was decided to change the MEM group to a MOM group, as shown in **D**, due to its compatibility with a hydroboration reaction. In the event, compound **D** was transformed into the alcohol, which was oxidized with PCC to afford the ketone **E**. It was necessary to epimerize the *α*-oriented axial isobutyl group, which was accomplished with NaOMe in MeOH to give **F**. Irradiation

Design and Strategy in Organic Synthesis: From the Chiron Approach to Catalysis, First Edition.
Stephen Hanessian, Simon Giroux, and Bradley L. Merner.
© 2013 Wiley-VCH Verlag GmbH & Co. KGaA. Published 2013 by Wiley-VCH Verlag GmbH & Co. KGaA.

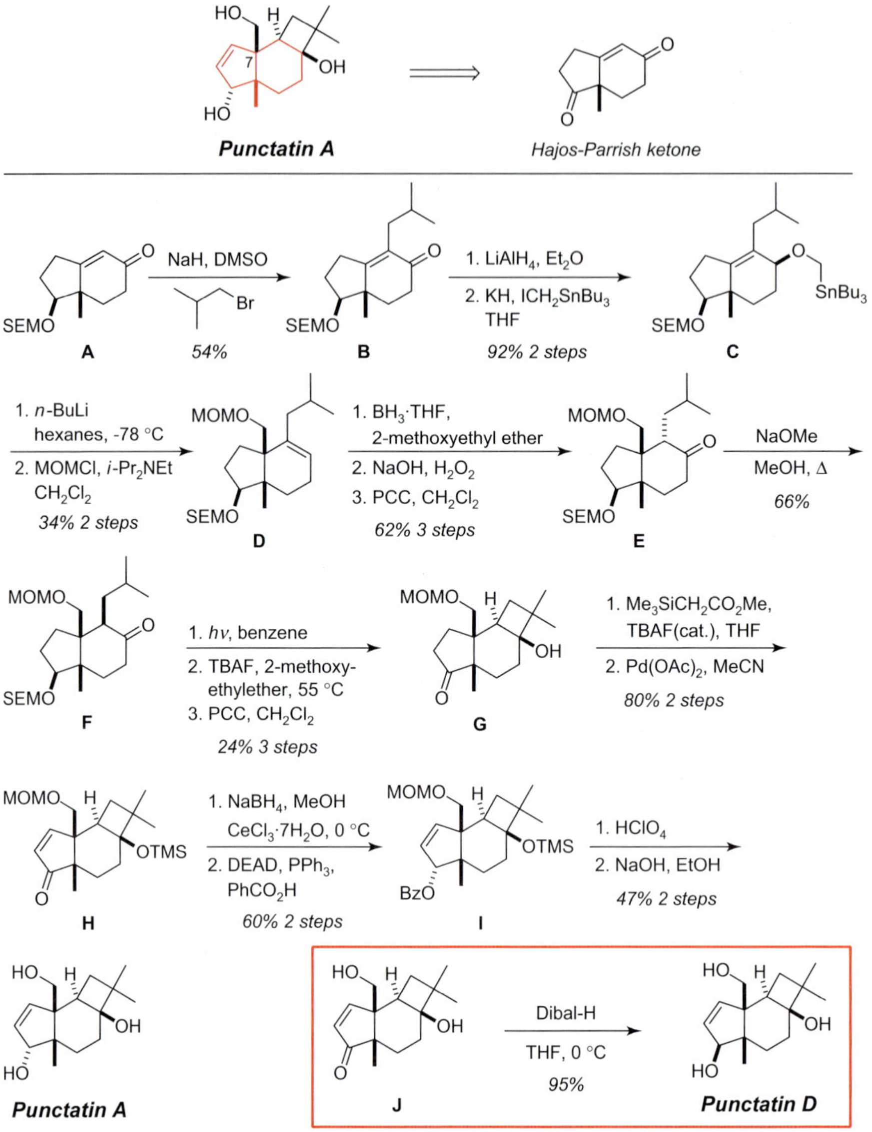

Punctatin A Hajos-Parrish ketone

Key reactions: *Thermodynamic enolate alkylation (**A** to **B**); Wittig-Still [2,3]-sigmatropic rearrangement (**C** to **D**); Norrish type II photocyclization (**F** to **G**); Pd(OAc)$_2$-mediated Saegusa oxidation of enolsilane (**G** to **H**).*

Figure 14.1 Total synthesis of punctatin A from the Hajos-Parrish ketone (apparent substructure).

Relevant mechanisms and transformations:

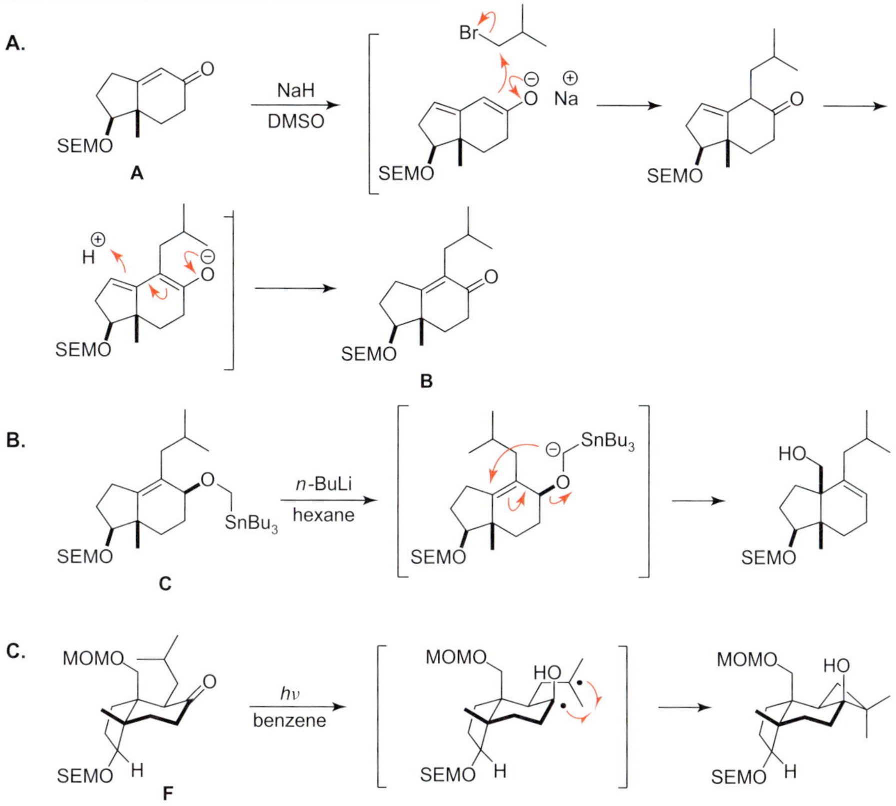

A.

B.

C.

Figure 14.1 *(continued)*.

with 254 nm light afforded the corresponding fused cyclobutanol in 49% yield. Deprotection of the silyl ether and oxidation of the alcohol led to tricyclic ketone **G**. Conversion of the cyclopentanone subunit to the enone was accomplished by first forming a silyl enol ether by treatment of **G** with methyl trimethylsilylacetate and TBAF, then by oxidation with Pd(OAc)$_2$ to give **H** [4]. Luche reduction of the ketone led to the *exo*-alcohol, which was inverted under Mitsunobu conditions to give **I** as the benzoate ester. Treatment with aqueous HClO$_4$ followed by base afforded punctatin A. Alternatively, punctatin D could be obtained by Dibal-H reduction of ketone **J**. The absolute configuration of these sesquiterpenes was secured as a result of their total synthesis.

Commentary: A visual analysis of punctatin A reveals two synthetic challenges that constitute the basis for a specific strategy to install an angular hydroxymethyl group and a fused cyclobutane. A typical case of visual *relational* analysis is evident in the choice of the starting bicyclic diketone, which is readily available from an L-proline-catalyzed intramolecular aldol reaction according to Hajos and Parrish [2] (see Chapter 5, Figure 5.12). Recognition of the excellent skeletal convergence of the Hajos-Parrish ketone with the perhydroindene segment of punctatin A also led Sugimura and Paquette to utilize the existing functionality in an advantageous way, especially with regard to the installation of the angular hydroxymethyl group and the cyclobutane ring. After unsuccessful attempts to introduce a hydroxymethyl group by conjugate addition or by photochemical methods, an alternative strategy was chosen. Alkylation of the thermodynamic enolate of **A** gave the α-2-isobutyl enone **B** (see relevant mechanism A, Figure 14.1). The implementation of the [2,3]-sigmatropic (Wittig) rearrangement according to Still [3] was carried out utilizing tributylstannylmethyl ether **C**. Smooth rearrangement took place to afford the all *cis*-configured bicyclic intermediate **D** (see relevant mechanism B, Figure 14.1). The plan called for an introduction of the cyclobutane ring based on a Norrish type II photochemical process [5]. The formation of the desired product **G** was rationalized on the basis of a favorably aligned biradical arising from the excited state of **F** (see relevant mechanism C, Figure 14.1). The well-conceived photochemical experiment also secured the *gem*-dimethyl group on the fused cyclobutane, while ensuring the desired stereochemistry of the tertiary alcohol. Clearly, the deployment of an equatorial isobutyl group adjacent to the carbonyl of bicycloalkanone **F** was done in full cognizance of the expected photochemical cyclization. The introduction of an otherwise inert hydrocarbon side-chain and its incorporation as the geminally disubstituted fused cyclobutane is a highlight of the synthesis. Introduction of unsaturation in the five-membered ring was accomplished following a Kuwajima protocol to form the silyl enol ether from ketone **G**, which was then oxidized to the corresponding enone **I** with Pd(OAc)$_2$ according to Saegusa [4]. This process also resulted in the *in situ* protection of the tertiary hydroxyl group as the TMS ether.

14.2
Acanthoic Acid

A full account of the total synthesis of the pimarane diterpene acanthoic acid was reported in 2001 by Theodorakis and coworkers [6]. A logical choice for a starting material was (*R*)-Wieland-Miescher ketone [7], which provides excellent convergence with rings *A* and *B* of the target molecule (Figure 14.2). Ketal **A** was subjected to reductive alkylation using Li/NH$_3$ to form the enolate at $-78\,^{\circ}$C. Warming the reaction to $-45\,^{\circ}$C and subsequent treatment of the enolate with methyl cyanoformate (Mander's reagent) [8] gave β-ketoester **B** according to the Coates and Shaw [9] "double reduction" protocol. Treatment of **B** with NaH and MOMCl furnished the MOM enol ether **C**. The crucial α-methylation was accomplished first by reductive cleavage of the MOM group with Li/NH$_3$,

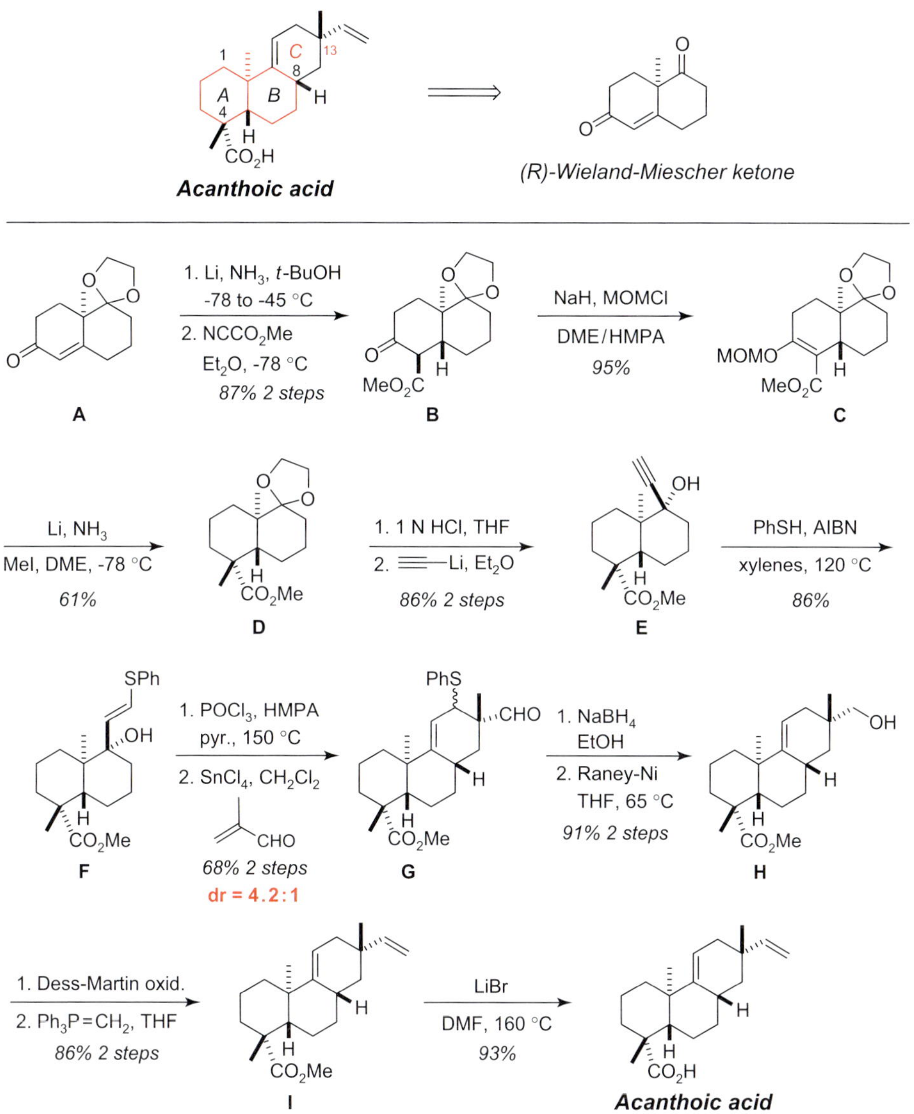

Acanthoic acid

(R)-Wieland-Miescher ketone

A

1. Li, NH₃, *t*-BuOH
   -78 to -45 °C
2. NCCO₂Me
   Et₂O, -78 °C
   87% 2 steps

B

NaH, MOMCl

DME/HMPA

95%

C

Li, NH₃

MeI, DME, -78 °C

61%

D

1. 1 N HCl, THF
2. ═══─Li, Et₂O

86% 2 steps

E

PhSH, AIBN

xylenes, 120 °C

86%

F

1. POCl₃, HMPA
   pyr., 150 °C
2. SnCl₄, CH₂Cl₂

   68% 2 steps
   dr = 4.2:1

G

1. NaBH₄
   EtOH
2. Raney-Ni
   THF, 65 °C
   91% 2 steps

H

1. Dess-Martin oxid.
2. Ph₃P=CH₂, THF
   86% 2 steps

I

LiBr

DMF, 160 °C

93%

Acanthoic acid

Key reactions: *Reductive alkylation (**A** to **B**); double reduction/enolate alkylation (**C** to **D**); POCl₃-mediated dehydration and Lewis acid-mediated Diels-Alder reaction (**F** to **G**).*

Figure 14.2 Total synthesis of acanthoic acid from (R)-Wieland-Miescher ketone (apparent substructure).

Relevant mechanisms and transformations:

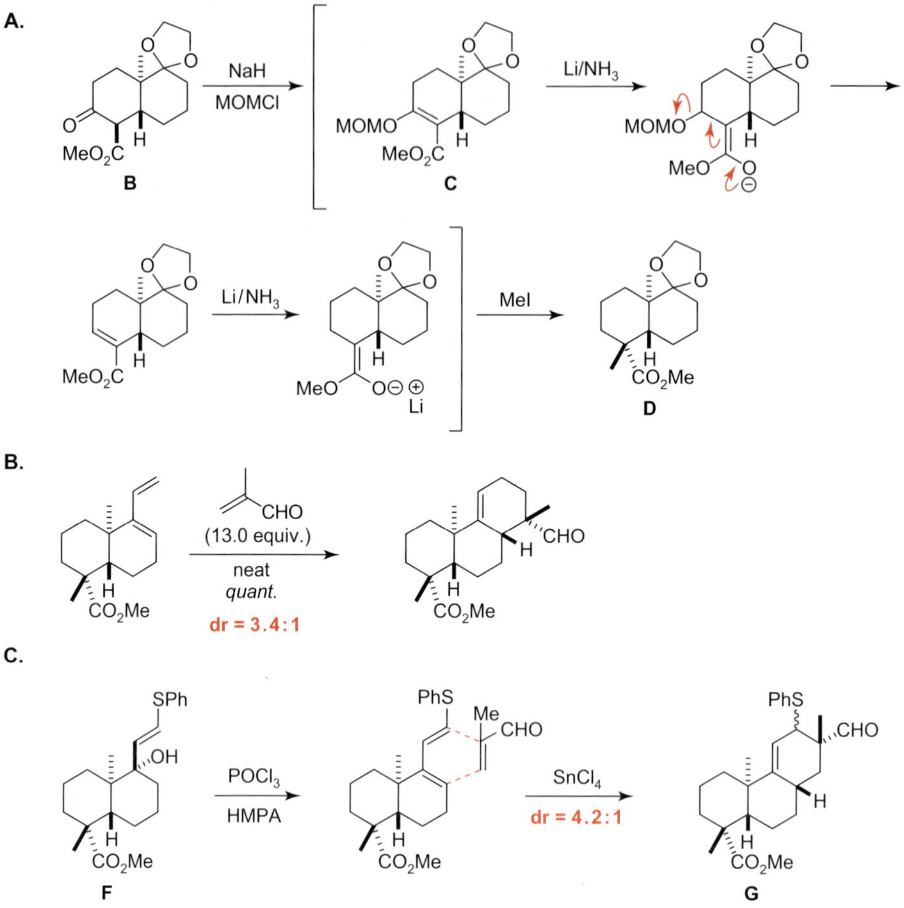

A.

B.

C.

Figure 14.2 *(continued).*

followed by treatment of the corresponding enolate with methyl iodide to give **D**. Cleavage of the ketal and stereoselective alkynylation led to **E** in excellent yield. Radical induced formation of phenylthio ether **F** was followed by dehydration with POCl₃, and a Lewis acid mediated Diels-Alder reaction to give **G** as a mixture of diastereomers (dr = 4.2:1). Reduction of the aldehyde group to the alcohol, and reductive desulfurization of the phenylthio ether led to tricyclic intermediate **H**. Separation of the major isomer followed by oxidation to the aldehyde and methylenation using a Wittig protocol led to ester **I**. Unsuccessful attempts to hydrolyze the methyl ester, under solvolytic conditions, led Theodorakis and coworkers to use LiBr in DMF at reflux to give acanthoic acid.

Commentary: Although the apparent substructure convergence of (*R*)-Wieland-Miescher ketone with rings *A* and *B* of acanthoic acid is evident, a major challenge involved the functionalization of C-4 in ring *A*. Initially, Theodorakis and coworkers had successfully investigated the introduction of methoxycarbonyl and methyl groups in a racemic form of the diketone. However, attempts to obtain an enantiomerically pure or enriched intermediate were unsuccessful. The desired result was achieved by manipulation of the readily available, enantioenriched Wieland-Miescher ketone by a series of reductive alkylations under temperature-controlled conditions (see relevant mechanism A, Figure 14.2). The initial plan was to construct ring *C* using an intermolecular Diels-Alder reaction of a bicyclic diene and methacrolein as a dienophile (see relevant mechanism B, Figure 14.2). Although good stereofacial preference was observed, resulting from an exclusive *endo*-face approach, the regiochemical outcome was not as required for the target product. It was reasoned that inverting the atomic orbital coefficients at the termini of the diene unit could result in the desired regiochemistry. Indeed, this was successfully implemented by incorporating phenylthio as a strong electron donating group at C-12. The regio- and stereoselectivity was certainly improved, although a desulfurization step was needed (see relevant transformation C, Figure 14.2).

14.3
Stachybocin Spirolactam

The first total synthesis of the spirodihydrobenzofuranlactam designated as stachybocin spirolactam was reported by Kende and coworkers [10] (Figure 14.3). (*S*)-Wieland-Miescher ketone was a convenient starting material, offering possibilities for chemical functionalization that would nicely converge with rings *A* and *B* of the target molecule. The readily available silyl enol ether of ketone **A** was alkylated in a stereoselective manner, and the product was converted into the hydrazone **B**. Treatment with iodine and DBU according to Danishefsky's protocol [11] afforded vinyl iodide **C** as the major product. Halogen-metal exchange, followed by treatment with DMF gave aldehyde **D** in excellent yield. Low-temperature lithiation of bromoester **E**, and quenching with aldehyde **D** furnished the corresponding carbinol, which was reductively deoxygenated to give **F**. Transesterification of the *t*-butyl ester in the presence of methanol and thionyl chloride, followed by sequential debenzylation with Raney-Ni and then via hydrogenolysis with Pd/C in ethanol gave triol **G** in high yield. Treatment with Amberlyst-15 in CH_2Cl_2 effected smooth spiroannulation to give **H** and an isomeric six-membered cyclic ether. Regioselective aromatic bromination with NBS, followed by chemoselective *O*-benzylation and treatment with CuCN in DMF led to the corresponding nitrile **I**. Hydrogenation of the nitrile function in **I** to give the aminomethyl group and hydrogenolysis of the benzyl ether took place in one-pot. Lactam formation in the presence of aqueous NaOH gave stachybocin spirolactam. An X-ray crystal analysis confirmed the structure of the synthetic material and led to a revision of the original proposal by Roggo [12]. An isomeric stachybotrylactam, in which the position of the lactam carbonyl was reversed compared to stachybocin spirolactam, was also synthesized by the Kende group through an *ortho*-bromination/cyanation and lactam formation protocol starting with intermediate **I**.

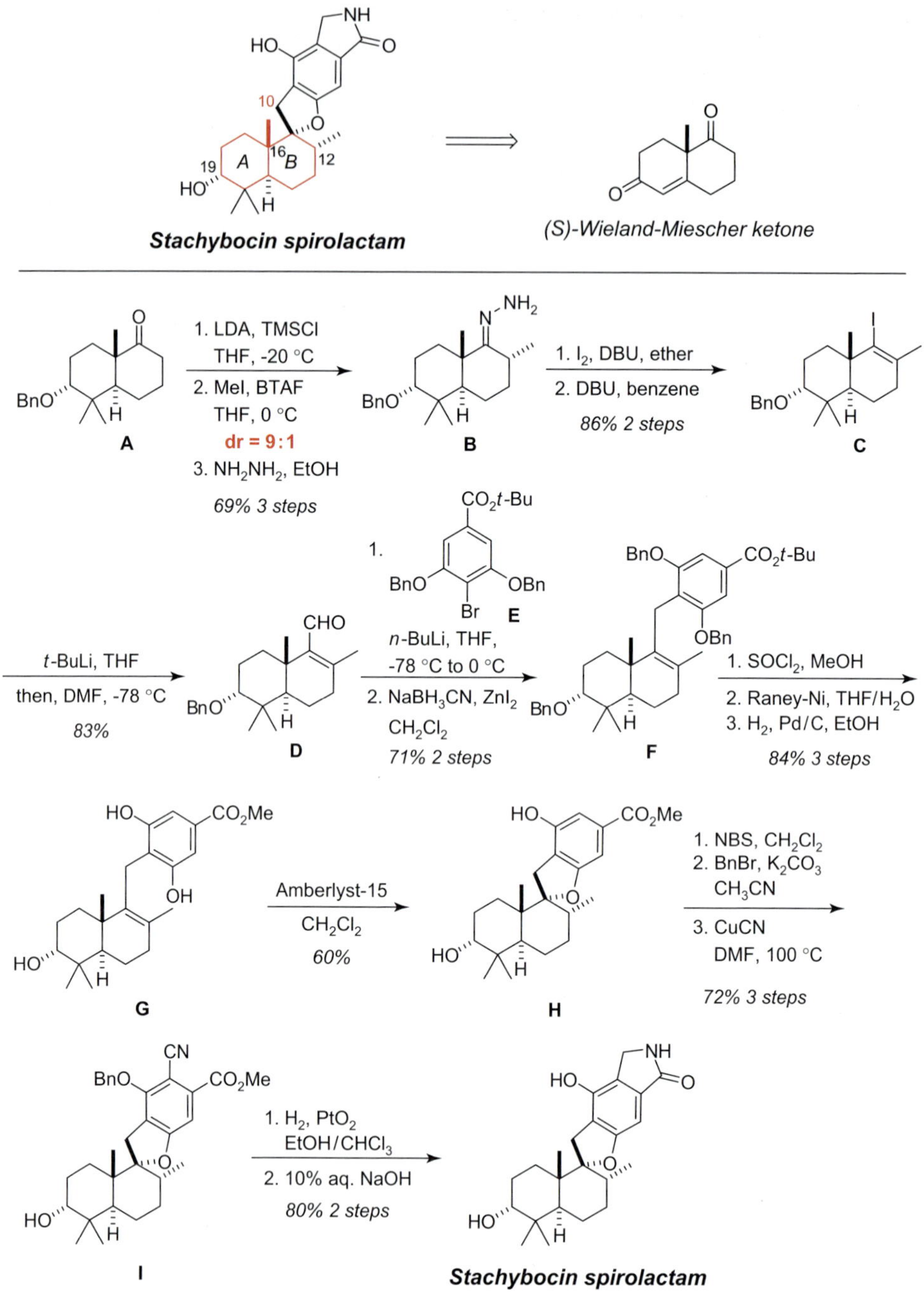

Stachybocin spirolactam

(S)-Wieland-Miescher ketone

A

1. LDA, TMSCl
   THF, -20 °C
2. MeI, BTAF
   THF, 0 °C
   dr = 9:1
3. NH₂NH₂, EtOH

69% 3 steps

B

1. I₂, DBU, ether
2. DBU, benzene

86% 2 steps

C

t-BuLi, THF

then, DMF, -78 °C

83%

D

1.
   CO₂*t*-Bu
   BnO   OBn
   Br   **E**

   n-BuLi, THF,
   -78 °C to 0 °C
2. NaBH₃CN, ZnI₂
   CH₂Cl₂

71% 2 steps

F

1. SOCl₂, MeOH
2. Raney-Ni, THF/H₂O
3. H₂, Pd/C, EtOH

84% 3 steps

G

Amberlyst-15

CH₂Cl₂

60%

H

1. NBS, CH₂Cl₂
2. BnBr, K₂CO₃
   CH₃CN
3. CuCN
   DMF, 100 °C

72% 3 steps

I

1. H₂, PtO₂
   EtOH/CHCl₃
2. 10% aq. NaOH

80% 2 steps

Stachybocin spirolactam

Figure 14.3 Total synthesis of stachybocin spirolactam from
(S)-Wieland-Miescher ketone (apparent substructure).

*Key reactions: Vinyl iodide from hydrazone (**B** to **C**); Acid-catalyzed spiroannulation (**G** to **H**); chemoselective benzylation (**H** to **I**); BTAF = benzyltrimethylammonium fluoride.*

Relevant mechanisms and transformations:

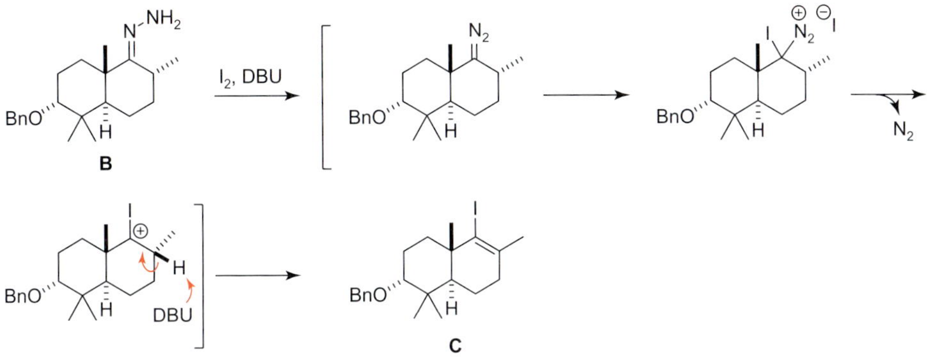

Figure 14.3 *(continued).*

Commentary: Having fully exploited the functionalization of ring *A* starting from the Wieland-Miescher ketone, there remained the challenge of introducing a *C*-methyl group at C-12 (target numbering). Methylation of the silyl enol ether prepared from **A** gave the desired α-oriented product as the major isomer. Since the plan called for the introduction of a usable functional group at the site of the carbonyl group, it was decided to utilize an oxidative olefination procedure using a hydrazone intermediate as previously shown by Barton [11b] and applied by Danishefsky [11a] in a different context. The suggested mechanism calls for the transient formation of a diazo iodide intermediate, followed by the loss of nitrogen and eventual proton abstraction from an iodo carbocation to give vinyl iodide **C** (see relevant mechanisms and transformations, Figure 14.3).

14.4
Scabronine G

The first total synthesis of scabronine G, a metabolite belonging to a broad class of angularly fused tricyclic diterpenoids was reported in 2005 by Danishefsky and coworkers [13] (Figure 14.4). The search for a suitable carbocyclic starting material that would provide good skeletal and stereochemical convergence with one or more rings in the target compound led to (*R*)-Wieland-Miescher ketone.

Readily available ketal derivative **A** was reduced under dissolving metal conditions, and the resulting enolate was treated with Mander's reagent [8] to give **B**. Conversion of the ketone to the enol triflate, and Pd-mediated hydride reduction [14] gave **C** in excellent yield. The Weinreb amide prepared from **C** was treated with vinylmagnesium bromide to give the α,β-unsaturated ketone **D**. When exposed to FeCl₃, a Nazarov-type carbocyclization [15] occurred to afford tetracyclic enone

E. Cyanide conjugate addition followed by trapping the resulting aluminum enolate with TMSCl gave the corresponding silyl enol ether. Treatment with *t*-BuOK
and *N*-(5-chloro-2-pyridyl)triflimide afforded the enol triflate **F** in excellent overall
yield. Coupling with isopropylmagnesium bromide in the presence of $ZnCl_2$ and
a Pd-catalyst according to Negishi [16] led to the isopropyl substitution product **G**.
Conversion of the cyano group to the corresponding methyl ester and hydrolysis
of the ketal gave tricyclic ketone **H**. Methoxycarbonylation followed by addition
of propanethiol led to thiopropylmethylidene ketone **I** in 90% yield over two
steps. Addition of lithium (methoxymethyl)phenyl sulfide to the ketone group gave
carbinol **J** which, when treated with $HgCl_2$, underwent a smooth ring expansion
[17] to furnish the aldehyde **K**. Thermodynamically favored isomerisation of the
cross-conjugated double bond in **K** in the presence of DBU, gave intermediate
L. However, cleavage of the methyl ester necessitated a prior protection of the
aldehyde function as the acetal, following which, scabronine G could be obtained
by treatment with base and acidification.

Commentary: The choice of (*R*)-Wieland-Miescher ketone as a starting material for
the elaboration of the 5-6-7 tricyclic skeleton of scabronine G may appear initially
to be counterintuitive. In fact, it is difficult to recognize any skeletal congruence
of the bicyclic diketone with any two of the three rings in the target molecule.
However, the location of the angular methyl group at the *BC* ring junction may
reveal the progeny of the bicyclic Wieland-Miescher ketone. Clearly then, Danishefsky's visual *relational* thinking in choosing this strategy was done with the full
knowledge that ring *A* had to be "built," and that ring *C* had to be "expanded"
to a seven-membered counterpart. Also germane to the synthesis plan was the
realization that the existing functionality in the starting (*R*)-Wieland-Miescher
ketone would be fully exploited to elaborate the fused tricyclic core structure
of scabronine G. Thus, divinyl ketone derivative **D** allowed the application of a
Lewis acid-mediated Nazarov cyclization to append a cyclopentenone motif (see
relevant mechanism A, Figure 14.4). Expansion of the cyclohexane ring in **J**
was done through a series of reactions exploiting hemithioacetal intermediates
adapted from an original precedent reported by Guerrero and coworkers [17]
(see relevant mechanism B, Figure 14.4). In the process, all the carbon atoms of
the starting Wieland-Miescher ketone were incorporated in the target structure.
More importantly, the astute chemical functionalization allowed the systematic
elaboration of rings *A* and *C* with a full awareness of the regio- and stereochemical outcome. Thus, the combination of visual *relational* and *reflexive* thinking
processes were utilized in productive ways in spite of the non-obvious overlap
of the starting Wieland-Miesher ketone as it relates to the target molecule.

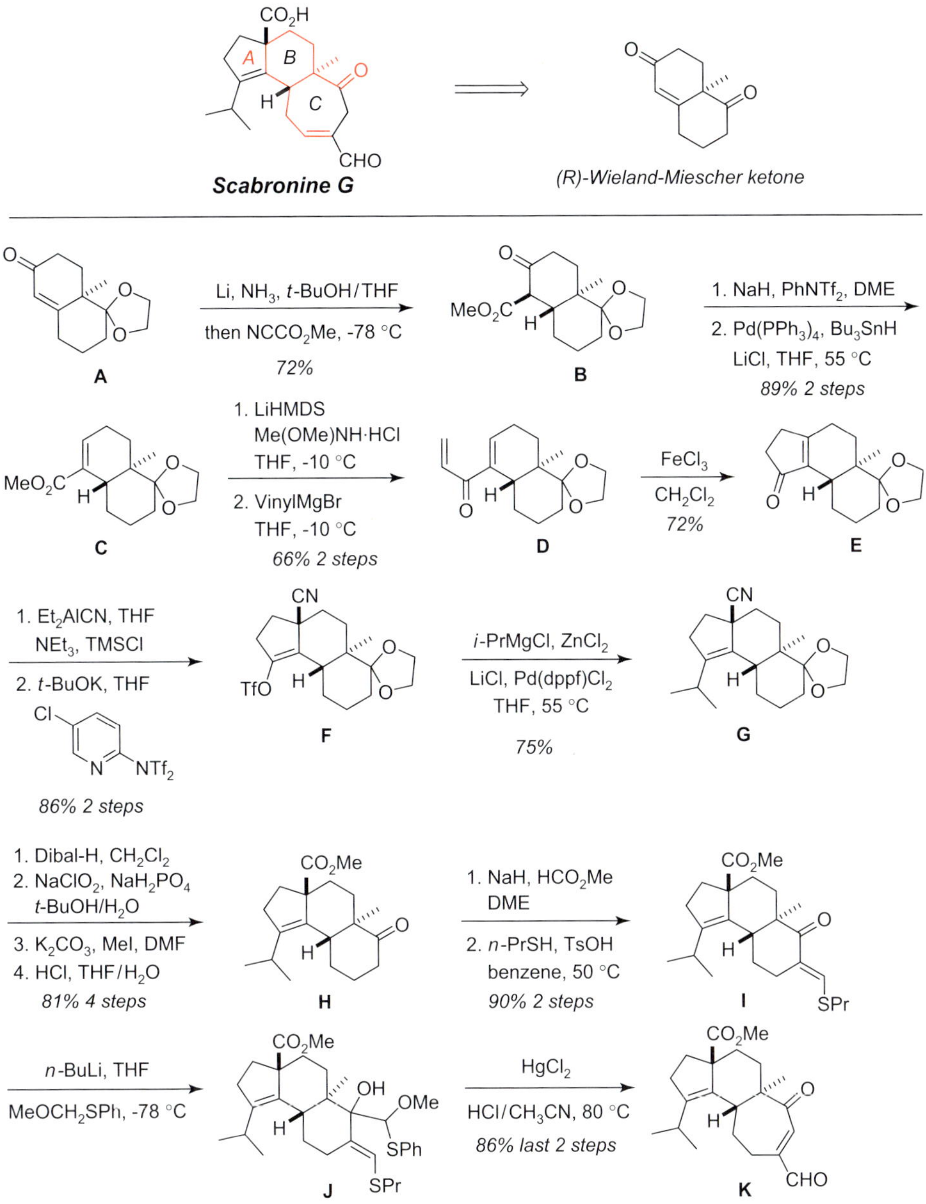

Scabronine G ⟸ (R)-Wieland-Miescher ketone

Figure 14.4 Total synthesis of scabronine G from (R)-Wieland-Miescher ketone (hidden substructure).

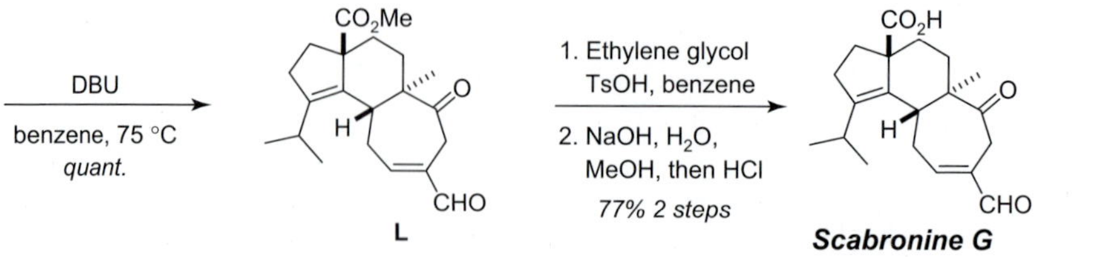

Scabronine G

Key reactions: *Kinetic enolate trapping (**A** to **B**); Lewis acid-mediated Nazarov cyclization (**D** to **E**); Negishi coupling (**F** to **G**); Lewis acid-mediated ring expansion (**J** to **K**); thermodynamic isomerization (**K** to **L**).*

Relevant mechanisms and transformations:

A.

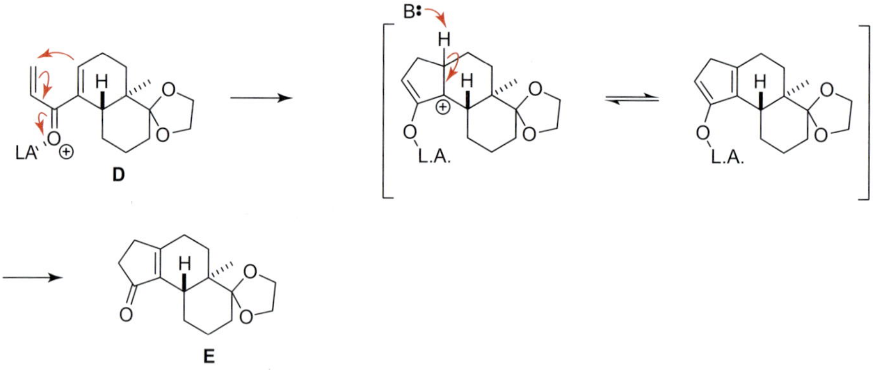

B.

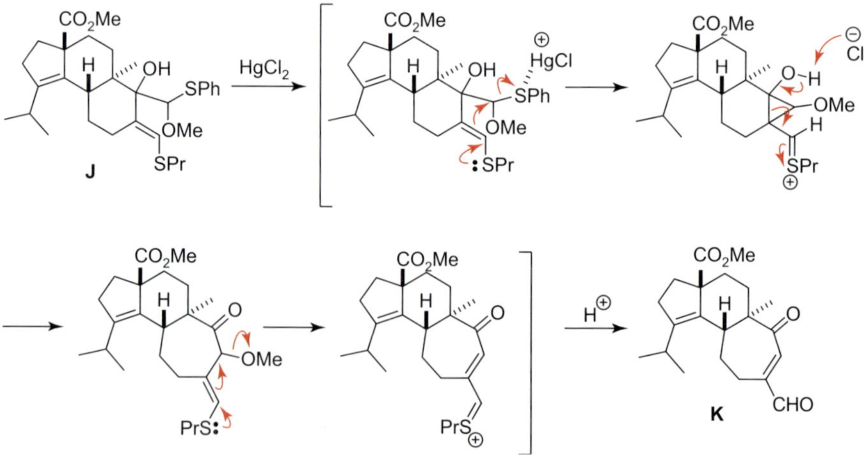

Figure 14.4 *(continued).*

14.5
Chapecoderin A

The first total synthesis of chapecoderin A, a member of the *seco*-labdane diterpenoid family was reported in 2001 by Hagiwara and coworkers [18] (Figure 14.5). The trisubstituted cyclohexane framework was related to an (*S*)-Wieland-Miescher ketone analogue, taking advantage of the chirality of the angular *C*-methyl group. Readily available diketone **A** [19] was methylated to give **B**, and the latter subjected to a Claisen reaction with the lithium enolate of *t*-butyl acetate to give carbinol **C** in excellent yield. Dehydration in the presence of $SOCl_2$, followed by reduction of the ester and acetylation afforded **D**. Ozonolytic cleavage in MeOH led to diketone **E**, which was subsequently treated with DBU and α-phenylsulfinyl-γ-butyrolactone (**F**). In the process, β-elimination of the ketoacetate was followed by a conjugate addition of the α-sulfinyl anion, then elimination of phenylsulfinic acid to give chapecoderin A, in addition to an intramolecular aldol condensation product (chapecoderin B). Dehydration of chapecoderin B in the presence of $SOCl_2$ in pyridine led to chapecoderin C. The total synthesis of these three diterpenes also allowed the assignment of their absolute configurations.

Commentary: Although there is no direct visual relationship of (*S*)-Wieland-Miescher diketone with the target molecule, Hagiwara and coworkers opted to utilize it with the intention of oxidatively modifying the cyclohexanone ring. Considering the structures of chapecoderin B and C, it can be understood that an intramolecular aldol reaction was a logical plan to convert the starting cyclic diketone to an acyclic counterpart (**D** to **E**). In the process, no carbon atoms were sacrificed from the starting diketone. The addition of a 2-carbon acetate unit starting with the Claisen reaction (**B** to **C**), eventually provided an intermediate that would generate an α,β-unsaturated ketone appendage, which would undergo conjugate addition and elimination *en route* to chapecoderin A (see relevant mechanism, Figure 14.5). This product is also a substrate for an intramolecular aldol reaction (chapecoderin B) with an option to eliminate the resulting β-hydroxy ketone (chapecoderin C).

14.6
Dragmacidin F

The first total synthesis of dragmacidin F, a member of a small family of piperazine and piperazinone-containing marine natural products, was reported in 2004 by Stoltz and coworkers [20a] (Figure 14.6). The challenge of having to functionalize a highly substituted bicyclo[3.3.1]nonane core was recognized early in the synthesis plan, and resulted in the selection of (−)-quinic acid as a suitable chiron and starting material for the synthesis of dragmacidin F. Intermediate **A**, available in three high yielding steps [21], was oxidized and the resulting ketone was subjected to a Wittig olefination to give **B**. A reductive isomerization/elimination reaction occurred in the presence of 0.5 mol% Pd/C under an atmosphere of hydrogen in MeOH to

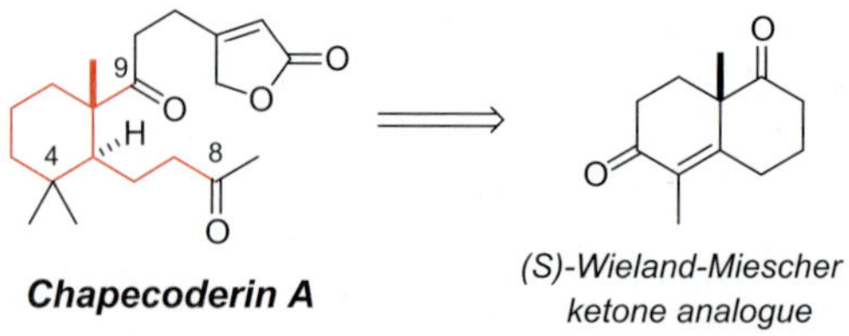

Chapecoderin A

(S)-Wieland-Miescher ketone analogue

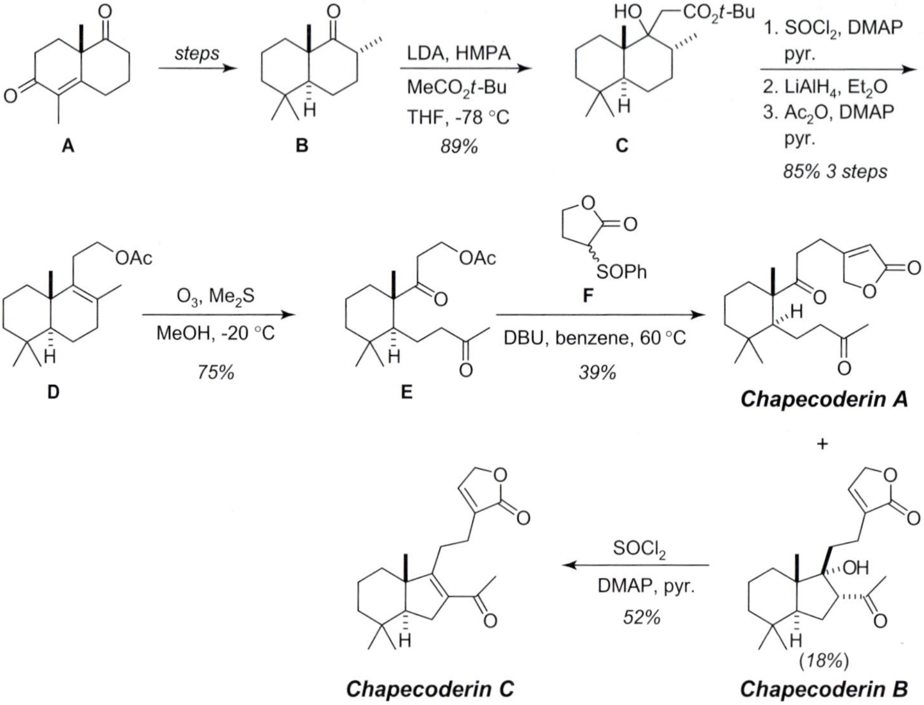

Key reactions: Sulfinyl anion conjugate addition/intramolecular aldol reaction (**E** to **Chapecoderin B**).

Relevant mechanism and transformations:

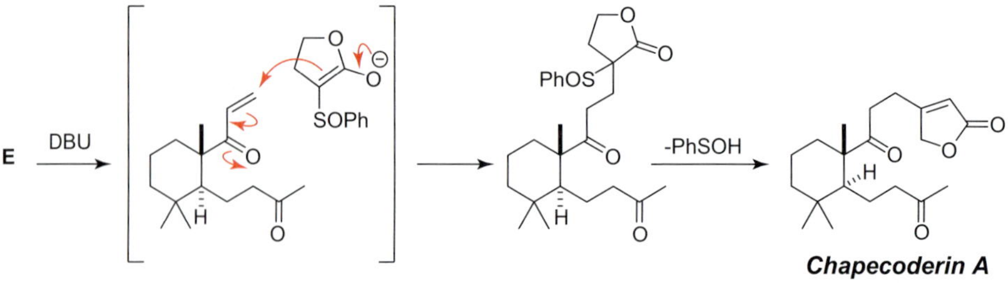

Figure 14.5 Total synthesis of chapecoderin A from (S)-Wieland-Miescher ketone (hidden substructure).

give an endocyclic cyclohexene intermediate, which was further derivatized as the Weinreb amide **C**. Displacement with 2-lithiopyrrole gave **D** in 65% yield over three steps. Carbocyclization took place in the presence of a stoichiometric amount of Pd(OAc)$_2$ and DMSO as a ligand, to give the desired bicyclic intermediate **E**. Catalytic reduction of the exocyclic methylene group, followed by *O*-methylation led to **F** in excellent yield. Application of a Suzuki-Miyaura reaction was possible with the boronic ester **G**. Thus, coupling with the dibromo indolopyrazine **H** in the presence of Pd(PPh$_3$)$_4$ under basic conditions gave **I** in 77% yield. Cleavage of the TBS ether with LiBF$_4$ in aqueous MeCN, followed by Dess-Martin oxidation afforded ketone **J**. Formation of the tosyl oxime ether **K**, then treatment with aqueous KOH, and acidification according to the conditions of the Neber rearrangement [22], eventually gave the amino ketone **L**. Finally, treatment with TMSI, followed by cyanamide in aqueous NaOH led to dragmacidin F as the TFA salt after purification by reverse phase HPLC.

Commentary: In their choice of (−)-quinic acid as a starting material, Stoltz and coworkers were acutely aware of the need to remove three hydroxyl groups, while replacing one with a nitrogen atom. Although the presence of the α-hydroxy carboxylic acid function may have established a visual relationship with a corresponding site in the target molecule, the road ahead was fraught with a number of challenges mixed with unexpected yet pleasing results. Following attempts to isomerize the exocyclic methylene group in **B** using a palladium-catalyzed π-allyl hydride addition, isomerization was eventually achieved under heterogeneous catalysis to give the endocyclic olefin as the carboxylic acid, which was then converted to the Weinreb amide.

Of equal interest was the observation that a Pd-catalyzed Heck-type cyclization with a 3-bromo-2-lithiopyrrole derivative to give **E** proceeded in low yield. Instead, a direct intramolecular Pd-mediated coupling was possible with acyl pyrrole **D** (see relevant mechanism A, Figure 14.6). Having successfully implemented the synthetic plan up to intermediate **J**, while being aided by two felicitous Pd-mediated reactions, there remained to introduce the crucial amino ketone functionality. A number of possibilities must have been planned relying on enolate amination, which, even if successful, would have required regiochemical control in view of the nature of the ketone. The choice of the Neber rearrangement was highly commendable if not audacious, in view of the lack of reports for a similar application in complex natural product synthesis (see relevant mechanism B, Figure 14.6). The total synthesis of dragmacidin D, a C-5 guanidinoethyl aromatized version of dragmacidin F was also reported by Stoltz and coworkers in 2002 [20b].

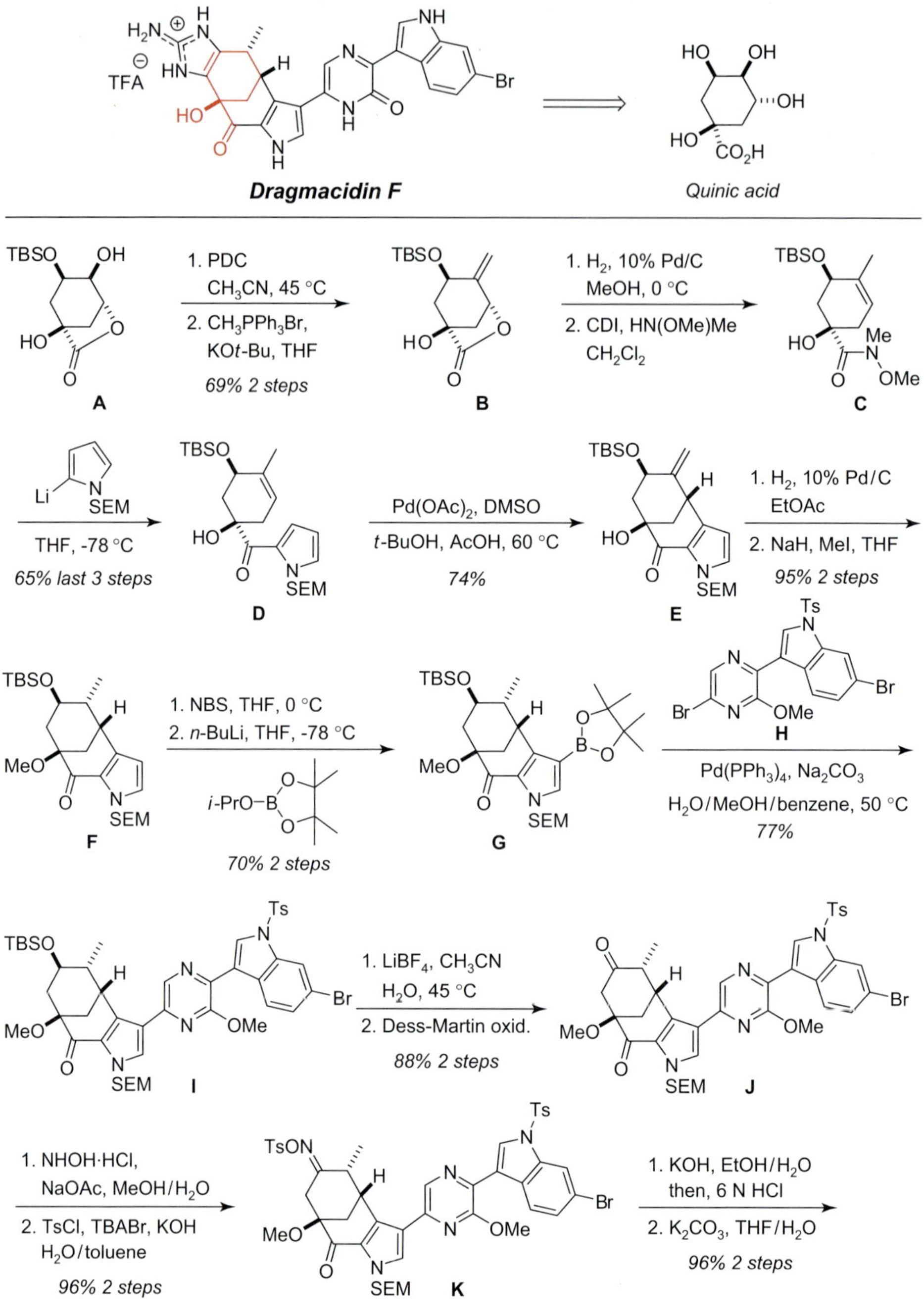

Dragmacidin F

Quinic acid

Figure 14.6 Total synthesis of dragmacidin F from quinic acid (partially hidden substructure).

Key reactions: *Pd-mediated oxidative cyclization (D to E); Suzuki coupling (G to I); Neber rearrangement (K to L).*

Relevant mechanisms and transformations:

A.

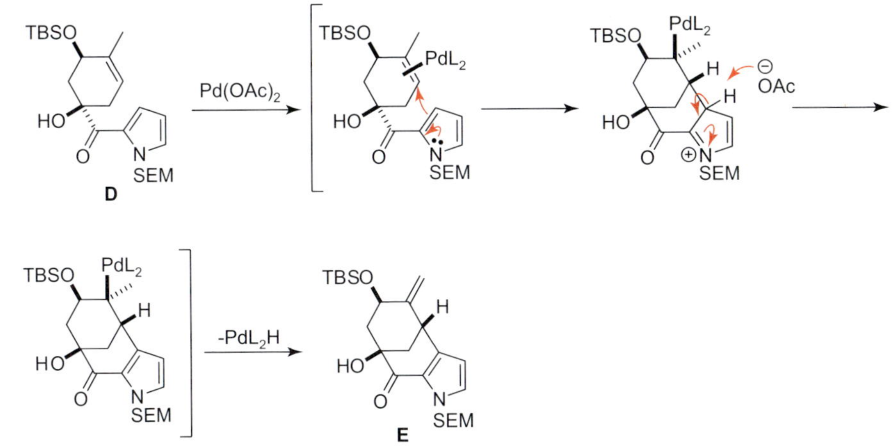

B.

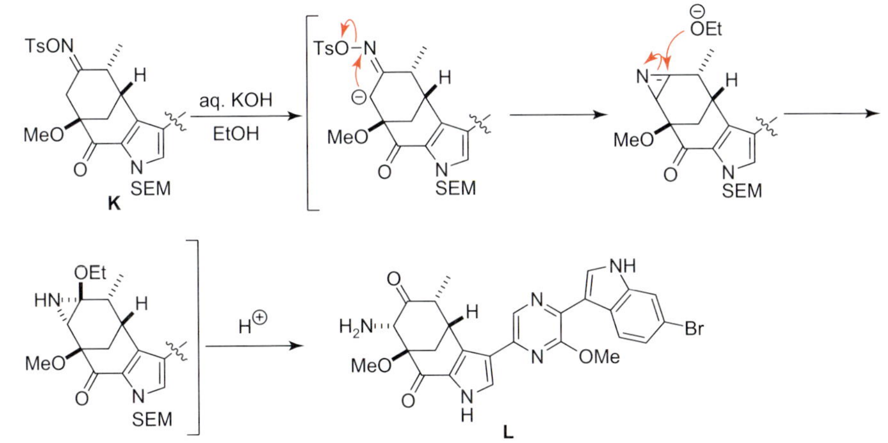

Figure 14.6 *(continued).*

14.7
Reserpine

Considering the complexity of the pentacyclic structure of reserpine, it is remarkable that the correct formula and structure first proposed in 1953 was confirmed through total synthesis by Woodward and coworkers only three years later [23] (Figure 14.7). In the interim, information about its relative and absolute stereochemistry was also forthcoming based on elegant chiroptical and chemical methods. Several total syntheses and approaches to reserpine have been reported since Woodward's landmark accomplishment [24]. As will become evident in the three syntheses discussed in this section, the major hurdle to overcome was securing the correct β-orientation of the hydrogen atom at C-3, mainly because most synthetic routes led to epimeric isoreserpine, which corresponds to the thermodynamically more favored configuration. Woodward devised an ingenious plan to overcome this hurdle by using logic, intuition, and considerable insight. Other approaches in subsequent syntheses discussed in this section relied on stereoelectronic principles to solve the same stereochemical problem. Perspective structures of reserpine and isoreserpine are depicted in Figure 14.7.

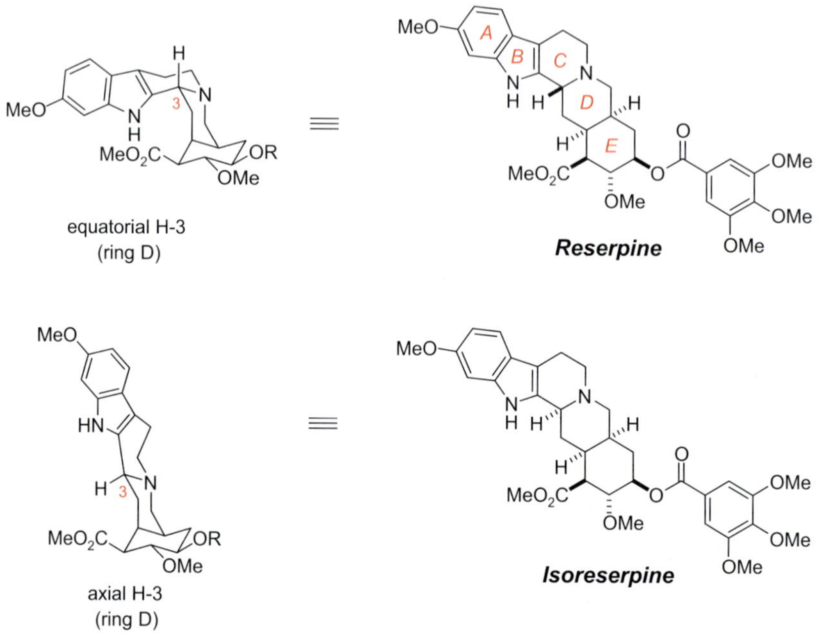

equatorial H-3
(ring D)

Reserpine

axial H-3
(ring D)

Isoreserpine

Figure 14.7 Perspective structures of reserpine and isoreserpine adapted from Stork (see ref. [27], and Figure 14.9).

14.7.1
The Woodward synthesis

The first total synthesis of reserpine was reported in 1956 by Woodward and coworkers [23], and is, to this day, considered a milestone achievement because of its tactical elegance. The enantiopure product was obtained by chemical resolution of a racemic mixture. The salient features of this synthesis are shown in Figure 14.8. Diels-Alder product **A** was reduced according to the Meerwein-Pondorff-Verley conditions to afford racemic lactone **B**. An intramolecular bromoetherification gave **C**, which upon treatment with NaOMe afforded intermediate **D**, proceeding through a β-elimination and subsequent conjugate addition of methoxide ion. Bromohydrin **E** was then oxidized to the α-bromoketone **F**, which underwent two remarkably controlled transformations in the presence of zinc in acetic acid to give **G**. Dihydroxylation to **H**, and periodate cleavage of the diol afforded aldehyde **I**. Formation of the imine with 6-methoxytryptamine (**J**), followed by treatment with POCl$_3$ and then reduction gave the C-3 epimeric isoreserpine intermediate **K**. Hydrolysis of the ester, followed by lactone formation gave **L**, which upon treatment with pivalic acid in refluxing xylene led to the desired reserpine intermediate **M** with the correct stereochemistry at C-3. Further steps afforded racemic reserpine, which was resolved as the D-camphor-10-sulfonic acid salt.

Commentary: Woodward's strategy to couple aldehyde **I** with 6-methoxytryptamine (**J**) under Bischler-Napieralski conditions was a logical way to assemble the pentacyclic structure of reserpine. The steps used to arrive at the vicinally *cis*-substituted aldehyde **I**, while intuitively unusual, deserve comment. Cognizant of the azacyclic nature of ring *D* in reserpine, it was evident that the enedione ring in the Diels-Alder product had to be extensively modified in order to eventually provide access to the precursor aldehyde **I** for the Bischler-Napieralski reaction. First, a remarkably chemoselective reduction was used to afford tricyclic lactone **B**, thereby intentionally "locking" the molecule in a conformation that would undergo a bromoetherification to **C**. The same conformation would also ensure the introduction of the methoxy group from the convex and less hindered side of the enoate produced from **C** by initial β-elimination. With the methoxy group correctly placed in **D**, Woodward devised a series of reactions leading to intermediates **E** and **F**. Reductive elimination led to **G**, thereby releasing the ether oxygen, and restoring the required substitution pattern in ring *E* (see relevant mechanism A, Figure 14.8). Oxidative cleavage of the diol in **H** led to key aldehyde **I**. The merits of locking intermediate **B** as the tricyclic lactone, thereby exploiting proximity effects in subsequent reactions, may have had a subliminal foreshadowing effect in solving an unexpected stereochemical impasse a few steps further in the synthetic scheme.

Indeed, reduction of the imine generated from the Bischler-Napieralski reaction led exclusively to the formation of the thermodynamically more stable isoreserpine (**K**). Hydride delivery occurred from the more accessible convex face (see relevant mechanism B, Figure 14.8). Woodward reasoned that changing the conformation of ring *E* by forcing it into a locked lactone as in **L** would result in an acid-catalyzed

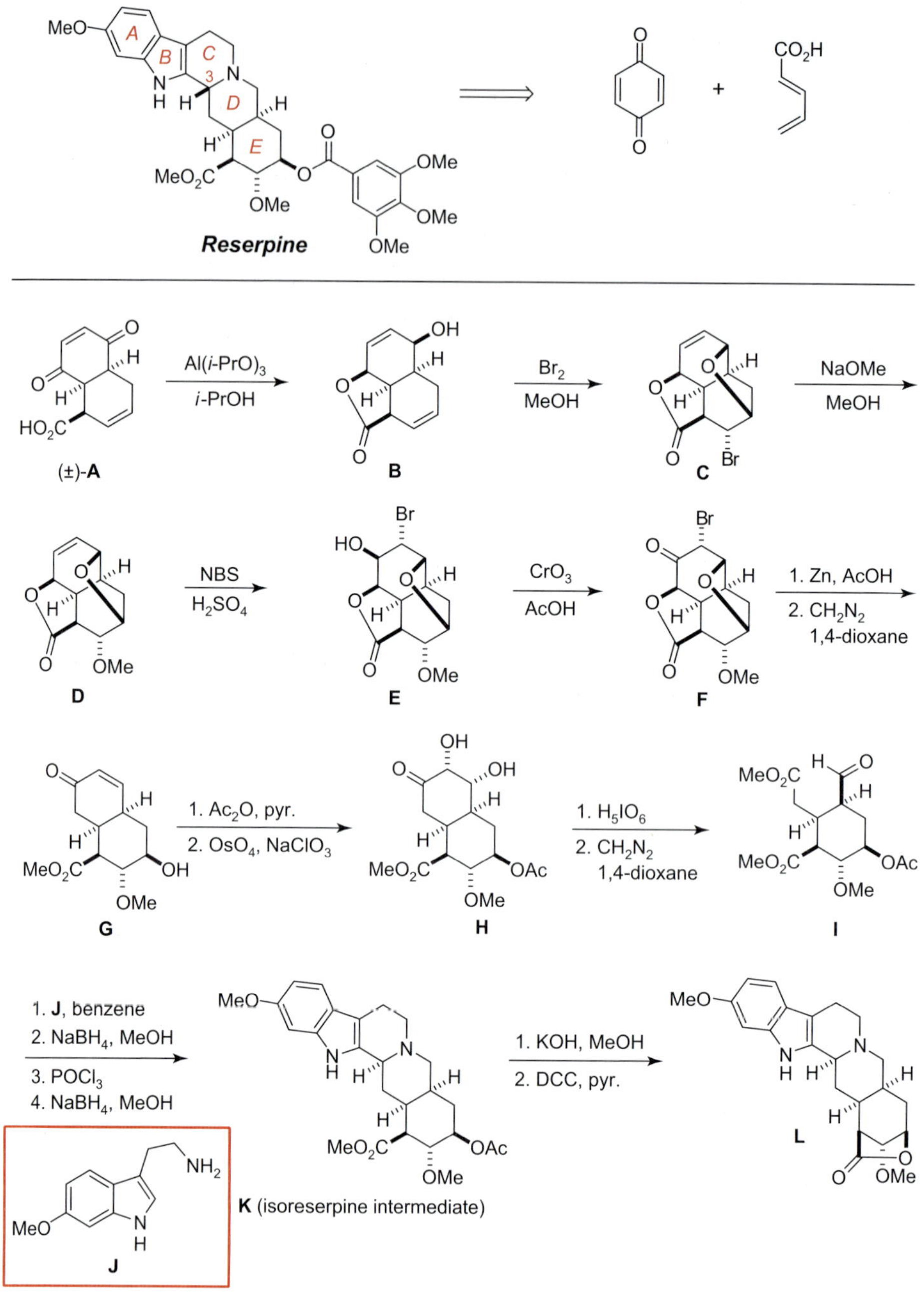

Figure 14.8 Woodward's synthesis of reserpine.

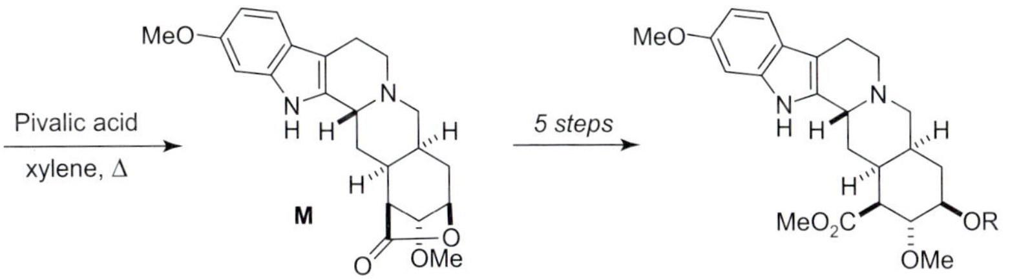

Reserpine

(R = 3,4,5-trimethoxybenzoyl)

Key reactions: *Chemoselective reduction/lactonization (**A** to **B**); bromoetherification (**B** to **C**); elimination/conjugate addition (**C** to **D**); Zn-promoted debromination and reductive lactone opening (**F** to **G**); Bischler-Napieralski cyclization (**I** + **J** to **K**); acid-catalyzed epimerization (**L** to **M**)*

Relevant mechanisms:

A.

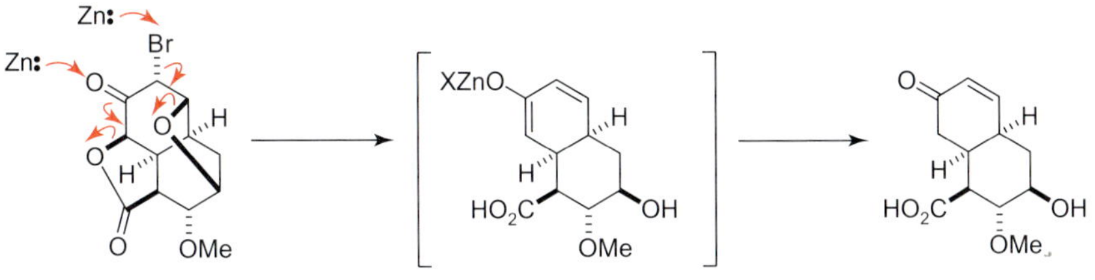

F

B.

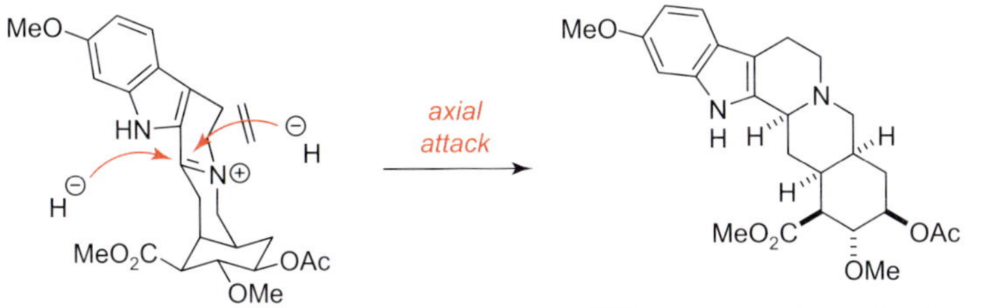

K (isoreserpine intermediate)

C.

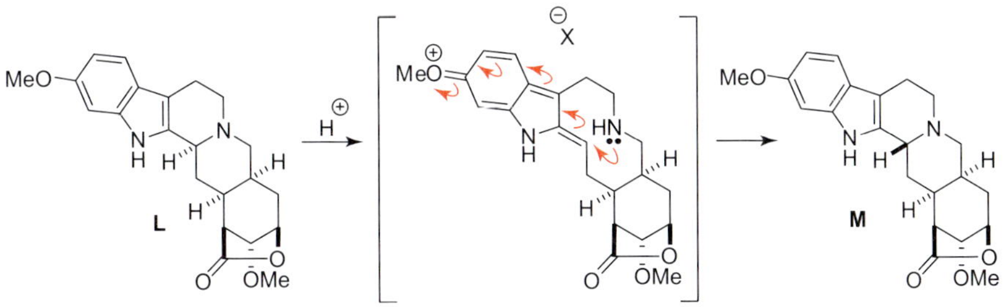

Figure 14.8 *(continued).*

ring C opening and closure, so as to favor the desired C-3 configuration (see relevant mechanism C, Figure 14.8). This imposed, conformational change proved to be highly successful and led to the reserpine precursor **M**.

Enantioselective total syntheses of reserpine have been reported by Wender [25], Martin [26], Stork [27], and Hanessian [28]. The synthesis of (±)-reserpine was reported by Pearlman [29] in 1979. Other efforts toward the synthesis of reserpine have been reviewed elsewhere [24].

14.7.2
The Stork synthesis

A regio- and stereoselective total synthesis of reserpine was reported by Stork and coworkers in 2005 [27a], following earlier studies in 1989 [27b]. In the most recent route to reserpine, the known (S)-3-cyclohexene-1-carboxylic acid (**A**) was subjected to iodolactonization according to Fukumoto and coworkers [30] to give **B**. Reduction of the lactone to the alcohol then benzylation in the presence of NaH led to epoxide **C**. Treatment with sodium phenyl selenide effected regioselective epoxide opening and oxidation to the selenoxide gave the olefin **D** after elimination of phenylselenol [30]. Transformation to ketone **E**, followed by sequential inter- and intramolecular Michael additions first onto acrylate **F** and then onto the intermediate cyclohexenone, led to the bicyclic ketone **G**. Hydrogenolysis of the benzyl ether, tosylation, and treatment with TBAF gave fluorosilane **H**, which was subjected to oxidation with *m*-CPBA, resulting in Fleming-Tamao and Baeyer-Villiger oxidations to afford **I**. Reduction with Dibal-H gave aldehyde **J**, which was treated with 6-methoxytryptamine (**K**) in the presence of KCN and AcOH to give aminonitrile **L**. Treatment with aqueous HCl gave methyl reserpate, which was esterified to give reserpine.

Commentary: As alluded to in section 14.7, one of the main challenges in the synthesis of reserpine is controlling the stereochemistry at C-3, favoring the natural isomer rather than its epimer, isoreserpine. The perspective drawings in Figures 14.7 and 14.9 show that the indole bond attached to C-3 in reserpine adopts an "axial" orientation, whereas it is "equatorial" in isoreserpine. Stork's reasoning for a stereoelectronic-based solution to this problem was a major incentive for his long-standing interest in the synthesis of reserpine. Rather than use a Bischler-Napieralski-type ring closure, as was done in the Woodward synthesis, Stork opted for a Pictet-Spengler reaction in which an iminium ion intermediate would be attacked by the tethered indole ring. He reasoned that a kinetic nucleophilic "perpendicular-chair" trajectory would lead to the desired stereochemistry at C-3.

Inspired by the prospects of overcoming the inherent thermodynamic bias, Stork devised a stepwise approach to the precursor of ring E with its full complement of five contiguous stereogenic centers. A key reaction was the double Michael addition of the dienolate generated from enone **E** onto the β-silyl enoate **F** (see relevant mechanism A, Figure 14.9). The bicyclic ketone was obtained in 88% yield as a single isomer. The decision to use cyanide ion in a Strecker-type reaction led to

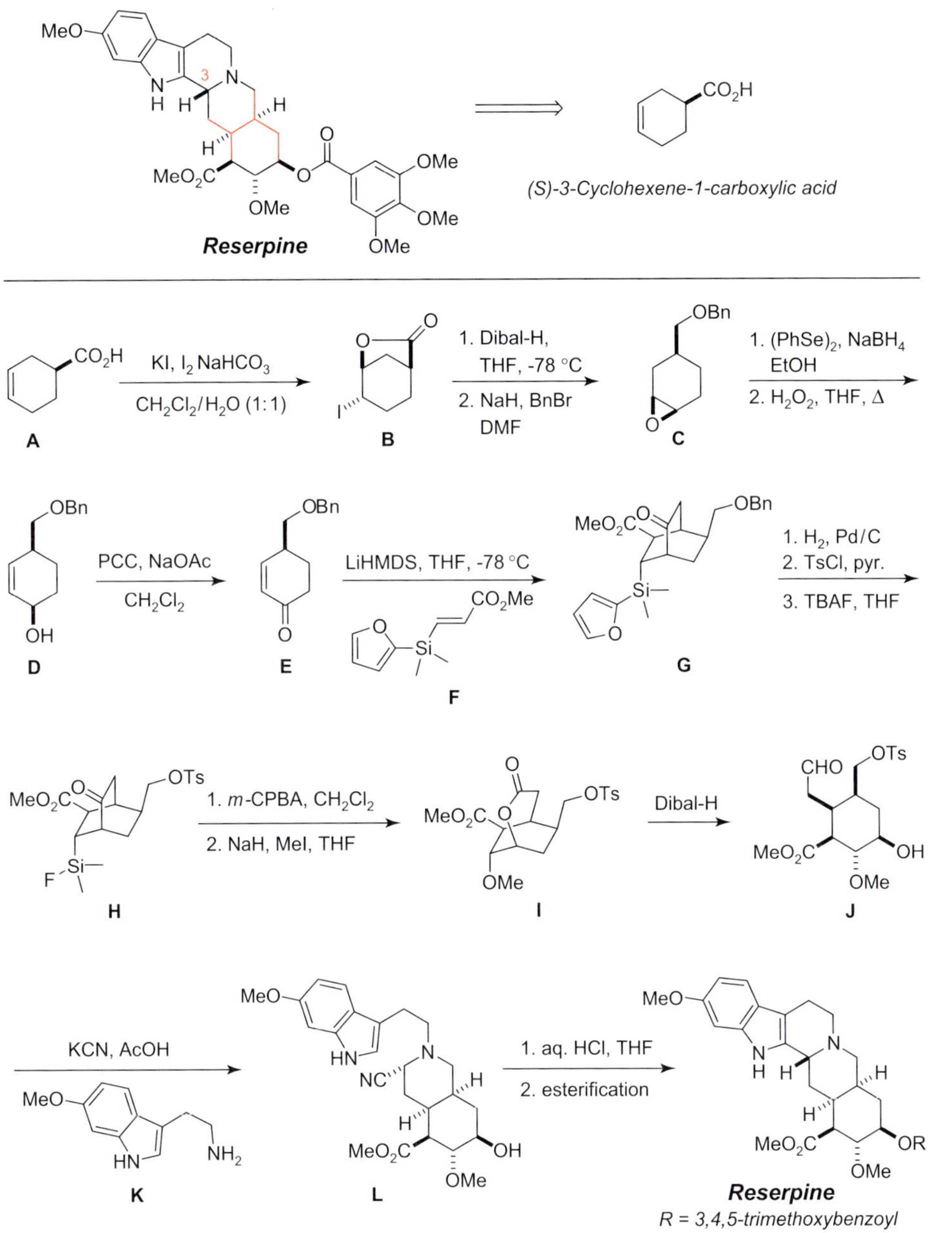

Key reactions: *Double dienolate Michael addition (**E** + **F** to **G**); α-aminonitrile iminium cyclization (**L** to methyl reserpate)*

Figure 14.9 Stork's total synthesis of reserpine from a carbocycle (hidden substructure).

Relevant mechanisms:

A.

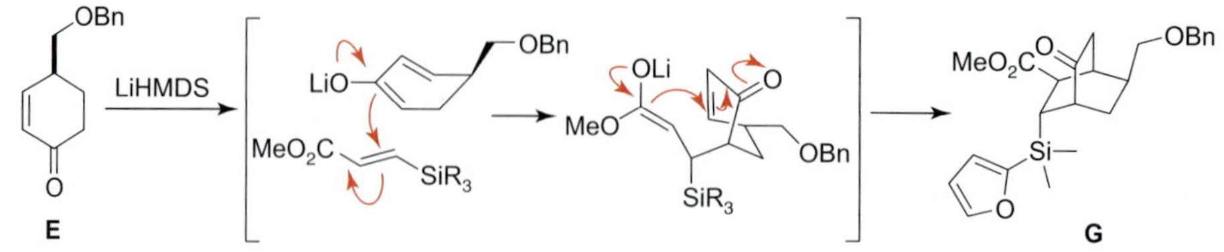

B.

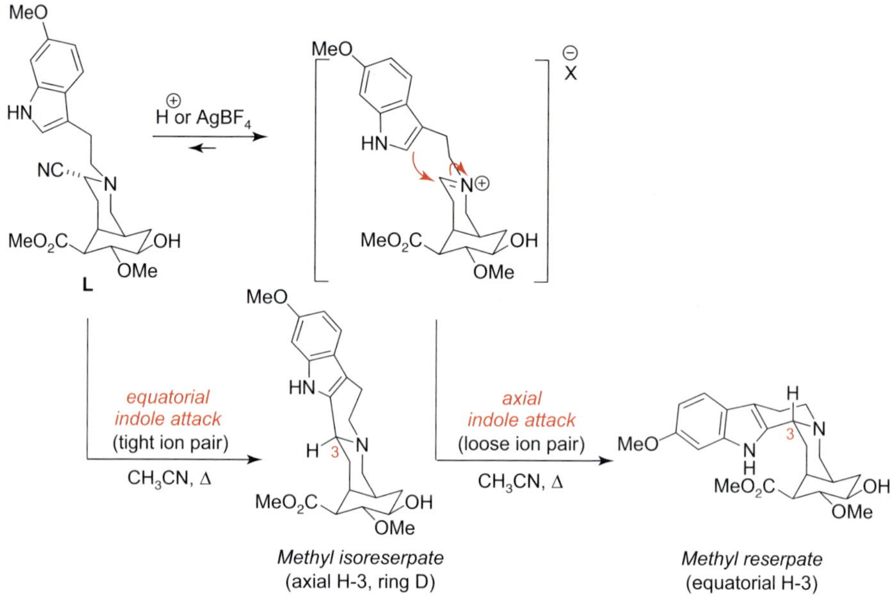

Methyl isoreserpate
(axial H-3, ring D)

Methyl reserpate
(equatorial H-3)

Figure 14.9 *(continued)*.

the isolation of aminonitrile **L** as a single isomer. Thus, in the presence of KCN, a premature Pictet-Spengler iminium ion cyclization was avoided, and resulted in displacement of the primary tosylate to give **L**. However, in refluxing acetonitrile, a 65% yield of methyl isoreserpate was obtained (see relevant mechanism B, Figure 14.9). This was rationalized on the basis of the existence of a tight ion pair arising from **L**, in which the appended indole would approach the incipient iminium ion from the β-face, giving the chair-like equatorial indole C-3 attachment. It was reasoned that allowing the cyanide counterpart to become less tightly bound, would present an opportunity for a kinetic and stereoelectronically favored axial attack of the indole moiety. This intuitive reasoning bore fruit since treatment of the aminonitrile **L** with dilute HCl (or AgBF$_4$) afforded a 90% yield of crystalline methyl reserpate.

We have seen individual solutions to the C-3 stereochemistry problem associated with the total synthesis of reserpine offered by two masters of total synthesis. Each found solutions to this problem in their own ingenious way, with Woodward relying on imposing conformational bias, and Stork resorting to stereoelectronic control.

14.7.3
The Hanessian synthesis

Focusing on ring *E* of reserpine, Hanessian and coworkers [28] chose (−)-quinic acid as a carbocyclic chiron and a suitable starting material in their total synthesis of the natural product (Figure 14.10). Although the location and stereochemistry of the hydroxyl groups at C-17 and C-18 in reserpine would be secured, there remained the challenge of introducing carbon functionality at C-15 and C-20 regio- and stereoselectively, in order to elaborate the azabicyclic precursor encompassing rings *D* and *E*. The readily available lactone **A** [21a] was methylated to **B**, then converted to ketone **C**. β-Elimination and protection of the resulting free alcohol as the TBS ether gave enone **D**. Treatment with vinylmagnesium bromide followed by esterification with chloroacetic acid, then conversion to the iodide led to iodoacetyl ester **E** as the major isomer in 72% overall yield. Radical-mediated carbocyclization [31] onto the α,β-unsaturated ester afforded the bicyclic lactone **F**. Oxidative transformation of the vinyl group to a carboxylic acid, esterification, and conversion of the lactone to the lactol with disiamylborane, led to **G** in excellent overall yield. In the presence of 6-methoxytryptamine (**H**), lactol **G** underwent a smooth acid catalyzed Pictet-Spengler reaction to give lactam **I** and its C-3 epimer in 90% yield and a ratio of 1.4:1 respectively. In order to successfully reduce the lactam carbonyl and to deoxygenate at the C-16 stereocenter, five steps were required. Protection of the tertiary alcohol as the TMS ether, followed by diborane reduction of the lactam carbonyl group, global deprotection of the silyl ethers, chemoselective reprotection of the secondary hydroxyl, and a SmI_2-mediated deoxygenation [32] of the tertiary alcohol, led to the pentacyclic methyl reserpate **K**. Deprotection of the TBS ether, followed by an esterification with 3,4,5-trimethoxybenzoyl chloride gave reserpine.

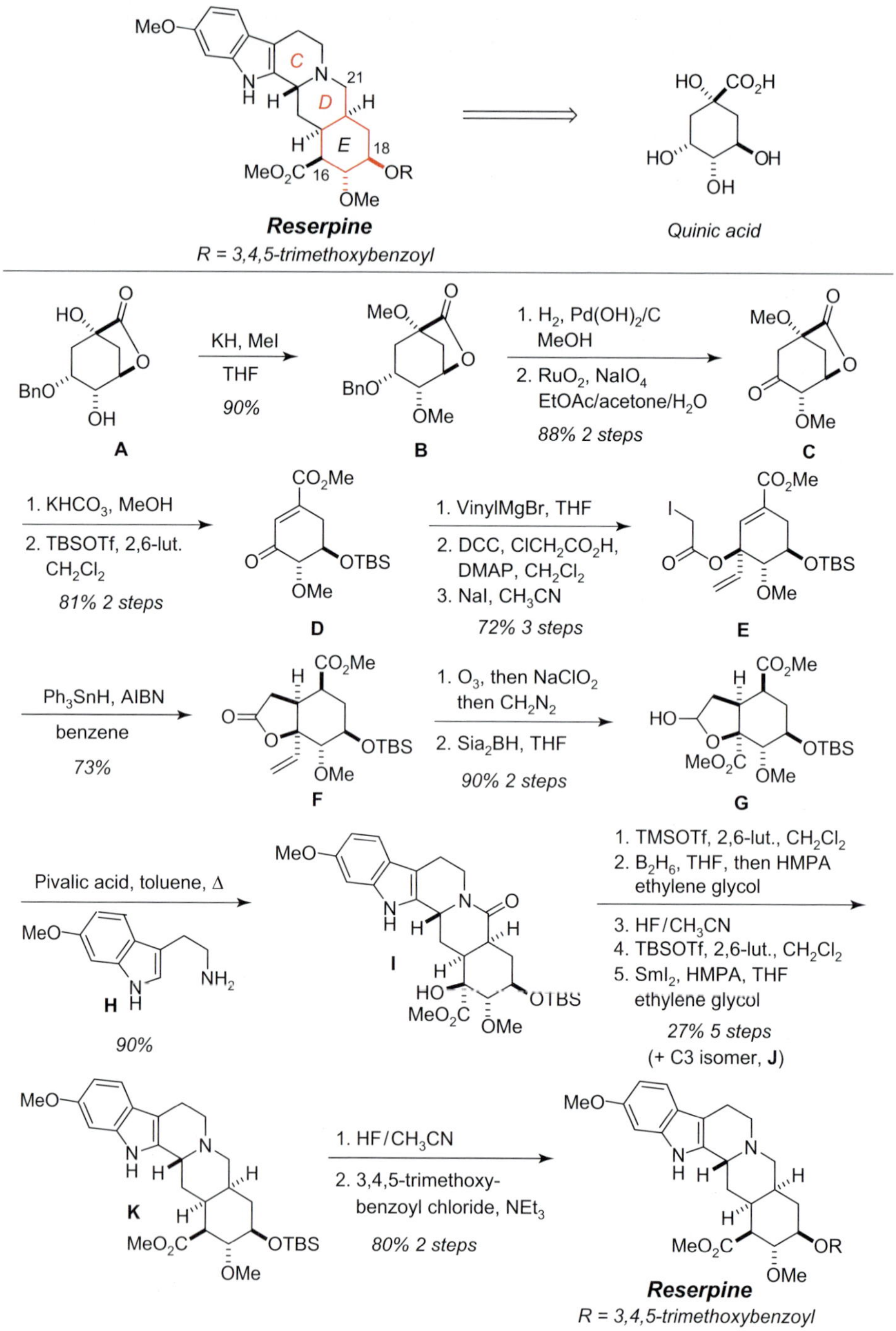

Reserpine
R = 3,4,5-trimethoxybenzoyl

Quinic acid

Figure 14.10 Hanessian's total synthesis of reserpine from quinic acid (apparent substructure).

Key reactions: *Ester-radical-induced intramolecular cyclization (**E** to **F**); Intramolecular iminium ion Pictet-Spengler cyclization (**G** + **H** to **I**); SmI$_2$-mediated deoxygenation (**K** to methyl reserpate).*

Relevant mechanisms and transformations:

A.

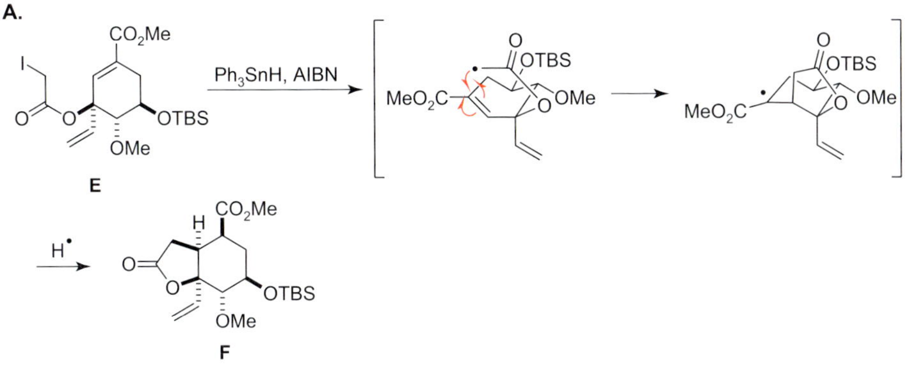

B.

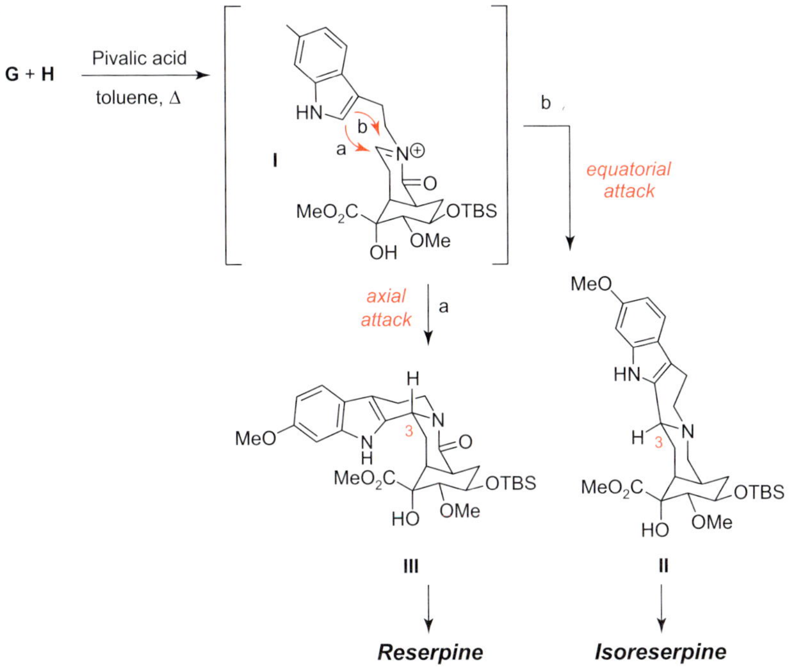

Figure 14.10 *(continued).*

Commentary: The visual relationship of quinic acid to ring *E* of reserpine as an apparent substructure is evident. In addition to the diol unit that provided a natural stereochemical convergence with C-17 and C-18 of reserpine, the position of the original carboxyl group was also seen as an asset in further elaborating

C-20. However, the lack of a useable functional group at the methylene carbon that would eventually become C-15 in reserpine presented a tactical problem. The solution relied on the stereocontrolled addition of a vinyl group onto enone **D** [33], followed by a somewhat unexploited intramolecular radical-induced Michael-type carbocyclization to give lactone **F** (see relevant mechanism A, Figure 14.10). Thus, the vinyl group was used as a masked carboxyl group, which, in turn, was conveniently obtained by ozonolysis followed by a Pinnick oxidation. In the process, a 2-carbon acetic acid appendage was successfully introduced at the site of the originally unactivated carbon atom of quinic acid. With the full complement of substituents introduced in aldehyde precursor **G**, the Pictet-Spengler reaction was attempted, cognizant that the intermediate iminium ion could be attacked by the indole moiety in two possible ways (see relevant mechanism B, Figure 14.10). Indeed, a presumed iminium ion intermediate **I** led to lactams **II** and **III** in a ratio of 1.4:1 in favor of the desired C-3 β-isomer **III**.

Although the quinic acid route to reserpine was successfully implemented in 20 steps and 2.6% overall yield, the stereochemical issue at C-3 of reserpine was not entirely resolved. Nevertheless, the precursors to the desired natural isomer accounted for 58% of the mixture of C-3 epimers in the Pictet-Spengler reaction.

14.8
Fawcettimine

An asymmetric total synthesis of fawcettimine was reported in 2007 by Toste and coworkers [34] (Figure 14.11). The first synthesis of the racemic natural product was published by Inubishi and coworkers in 1979 [35]. Substantial improvement in overall yield (10%) and number of steps (16) were reported by Heathcock and coworkers in 1986 [36].

14.8.1
Toste's synthesis of fawcettimine

In planning the synthesis of enantiopure fawcettimine, Toste chose to start with cyclohexanone **A** bearing the C-15 *C*-methyl group, and to systematically build the tetracyclic system in a stereocontrolled manner. The starting cyclohexenone **A** was obtained in 72% yield and 88% *ee* in a one-pot organocatalytic reaction from the appropriate β-ketoester and crotonaldehyde [37] (Figure 14.11). Conjugate addition of allenyltributylstannane in the presence of TBSOTf gave *trans*-silyl ether **B** with excellent diastereoselectivity. Treatment with *N*-iodosuccinimide and AgNO$_3$ led to an iodoacetylene, which was subjected to a gold-catalyzed carbocyclization [38] to afford iodohydrindenone intermediate **C** in excellent overall yield. Formation of the ethylene ketal, followed by cross-coupling under Trost's conditions [39], led to **D** in 74% yield over three steps. Conversion of the primary alcohol to the iodide, then treatment with *t*-BuOK in THF afforded tricyclic intermediate **E**. Cleavage of the ketal, followed by a regioselective hydroboration of the endocyclic double bond gave the corresponding alcohol, which was oxidized to the ketone to give a 10:1 mixture of diastereomers in favor the desired *trans*-isomer **F** in excellent overall yield. Removal of the *N*-Boc group with TFA completed the synthesis of fawcettimine in 13 steps starting from crotonaldehyde.

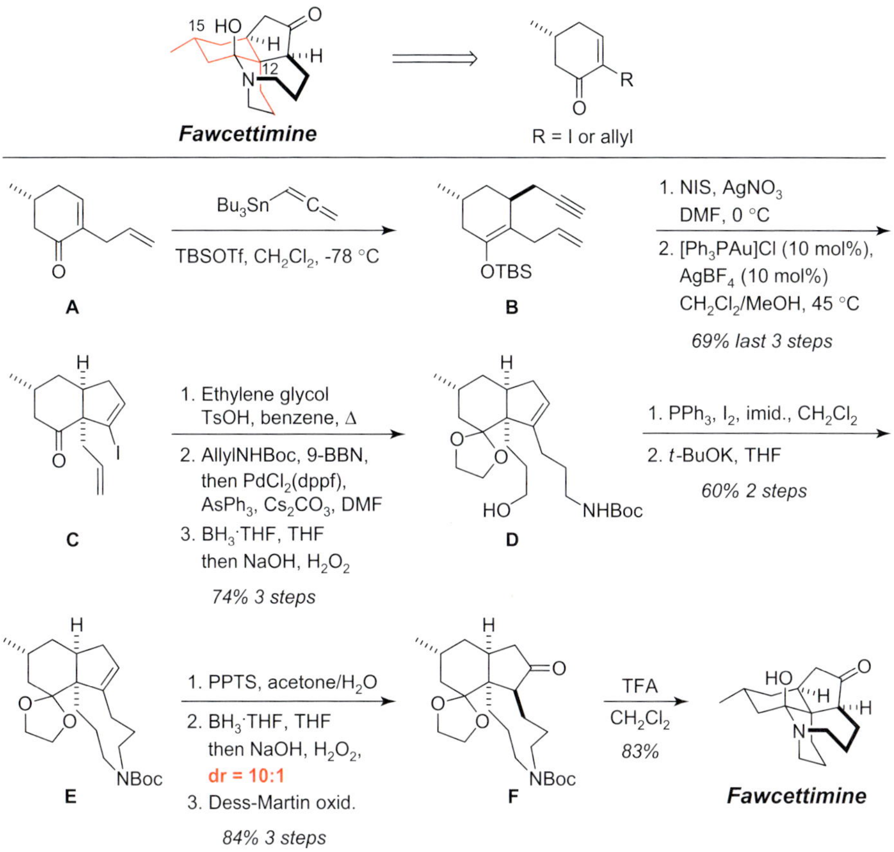

Key reactions: *Organocatalytic Robinson annulation to give **A**; allene-mediated propargylation (**A** to **B**); gold-catalyzed cyclization (**B** to **C**); Trost coupling (**C** to **D**); diastereoselective hydroboration (**E** to **F** alcohol); dppf = diphenylphosphinoferrocene.*

Figure 14.11 Toste's total synthesis of fawcettimine from a carbocycle (hidden substructure).

Relevant mechanisms and transformations:

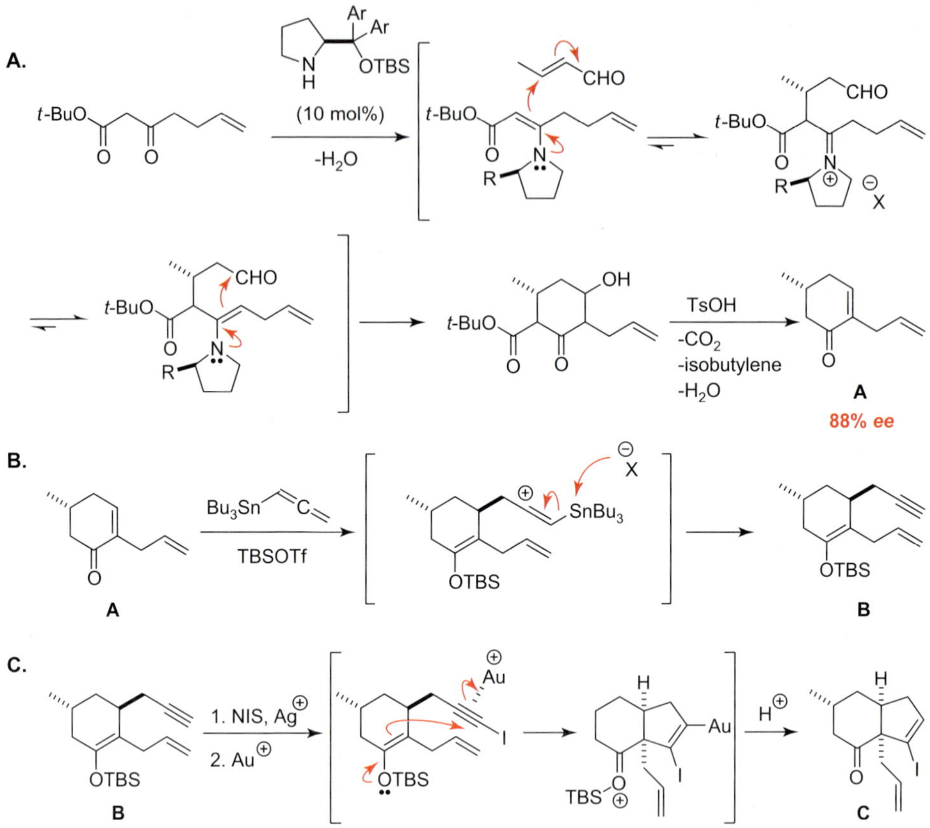

Figure 14.11 (continued).

Commentary: Visual analysis of the tetracyclic framework of fawcettimine reveals the existence of a highly congested quaternary carbon atom at C-12, and an "isolated" C-methyl group at C-15. Starting with an enantioenriched 5-methyl-2-cyclohexenone such as **A** was a logical choice, taking advantage of previous experience toward the same target molecule from Heathcock's group [36]. The challenge of establishing a quaternary stereogenic center with a vicinally deployed appendage that would be part of the seven-membered azepine ring, led Toste to consider a cyclization protocol developed in his laboratory for the construction of hydrindene systems. Although the cyclohexenone intermediate **A** could be synthesized from a known vinyl iodide precursor adopting a Suzuki-Miayura allylation, Toste resorted to a multigram scale preparation using a prolinol-catalyzed reaction (see relevant mechanism A, Figure 14.11). What followed was a sequence of steps that would lead to the application of a unique gold-catalyzed carbocyclization [39] of an iodoacetylene to form the hydrindenone core (**B** to **C**). The requisite propargyl appendage was introduced by an allene-mediated 1,4-conjugate addition (see relevant mechanism B, Figure 14.11). The iodoacetylene prepared in the presence of *N*-iodosuccinimide and AgNO$_3$ was then subjected to a *5-endo-dig* cyclization in the presence of 10 mol% Ph$_3$AuCl and AgBF$_4$ to give hydrindenone iodide **C** with the requisite allyl group at the newly created quaternary center. With this critical intermediate secured, the remaining steps involved functional group manipulations to form the azepine ring and adjustment of oxidation states. It is interesting that the hydroboration of the endocyclic double bond led to a preponderance of the desired *trans*-isomer in spite of the potential steric impediment of the azepine ring system.

In the total synthesis of fawcettimine, Toste and coworkers used one organocatalytic and two transition metal-catalyzed reactions as key steps in the elaboration of the core hydrindenone structure. A recent total synthesis of fawcettimine by Yang and coworkers [40a] made use of the same carbocyclic starting material as the Toste group. A synthesis starting from (*S,S*)-tartaric acid, as well as a formal synthesis of (±)-fawcettimine were reported by Mukai [40b] and Jung [40c] respectively.

14.8.2
Heathcock's synthesis of (±)-fawcettimine

It is of interest to outline the key steps of the Heathcock synthesis of racemic fawcettimine in the context of methodology available in the mid-eighties [36] (Figure 14.12). A key intermediate in the Heathcock synthesis plan was the hydrindanone **F** harboring the requisite vicinal *cis*-oriented side-chains. Having previously established that the configuration at C-4 in fawcettimine could be inverted to a thermodynamically more favored *trans*-isomer, the synthesis of **F** became a viable plan. Systematic functionalization of readily available racemic cyanoethyl ketone **A** was accomplished by a Lewis acid-mediated allylsilane reaction to afford adduct **B** as a mixture of isomers. Wittig extension furnished the intermediate ester **C**, which was directly subjected to an ethoxide-induced Michael addition to give the desired *5-exo-trig* cyclization product **D** in excellent yield. In order to secure the required chain length, the ester function was homologated through an Arndt-Eistert reaction of the diazoketone **E** to give **F**.

Formation of the azepine ring relied upon an intramolecular S$_N$2 displacement of the *N,O*-ditosyl intermediate resulting from LiAlH$_4$ reduction of **F**, and subsequent tosylation (**F** to **G**). Formation of the perchlorate salt enabled ozonolysis of the exocyclic olefin in **G** to give the diketone **H**. As predicted by Heathcock's NMR investigations and molecular mechanics calculations, the stereochemistry at C-4 could be inverted in the presence of a mild base (**H** to fawcettimine). Further insights into the tautomeric equilibria between the ketoamine and carbinolamine forms were contributed by Heathcock's scholarly studies in conjunction with his synthetic efforts.

Commentary: It is of interest to compare the 1986 and 2007 versions of the synthesis of fawcettimine by Heathcock [36] and Toste [34] respectively. Whereas no steps

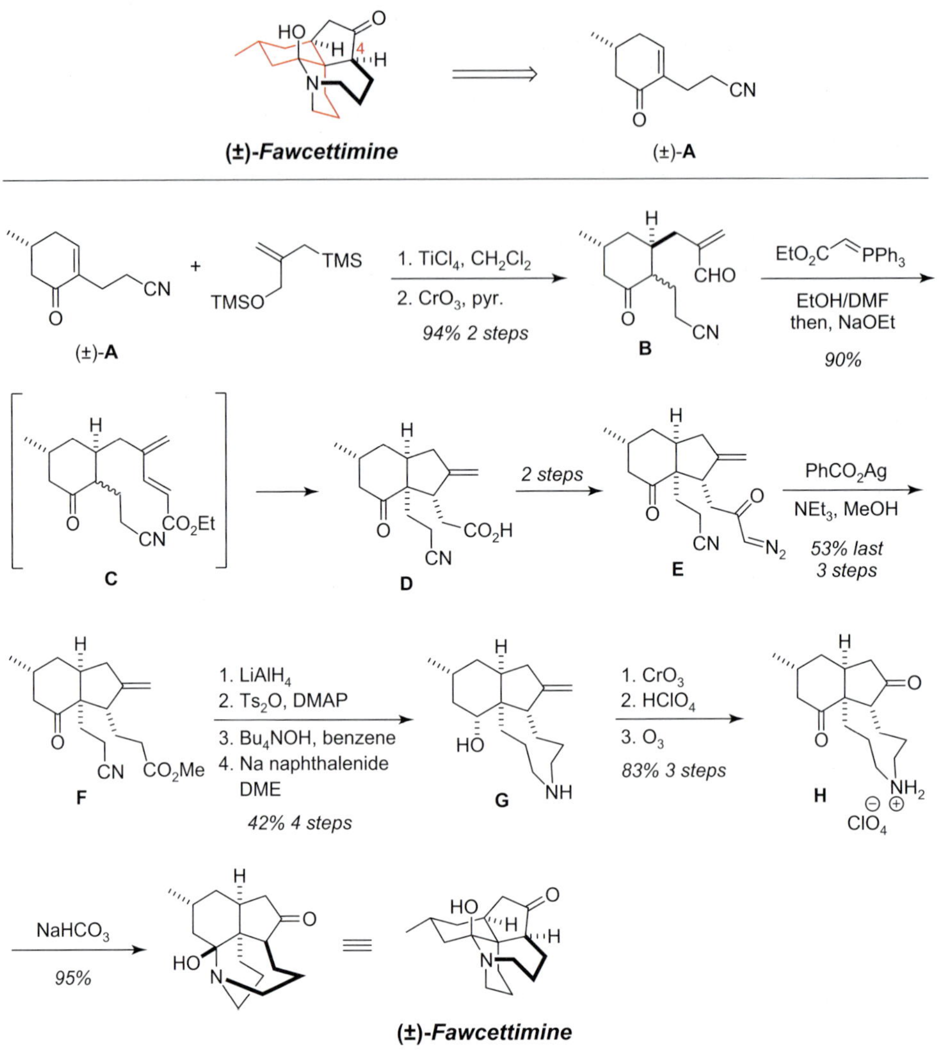

Key reactions: *Sakurai allylation (**A** to **B**); Arndt-Eistert homologation (**E** to **F**)*

Figure 14.12 Heathcock's total synthesis of (±)-fawcettimine.

in the Heathcock synthesis involved catalysis, the 16-step process was remarkably efficient with an overall yield of 10% from racemic cyanoenone **A** (Figure 14.12). Furthermore, no protecting groups were required.

The 2007 Toste synthesis of fawcettimine highlights the use of catalytic reactions in three transformations. Clearly, advances in Pd-mediated cross-coupling reactions added a new dimension to the functionalization of the intermediate vinyl iodide with the requisite number of carbon atoms (Figure 14.11,

C to D). Moreover, new methods of Au-catalyzed carbocyclizations developed by the Toste group offer alternatives to the classical Michael reactions, which were efficiently used by Heathcock.

The two syntheses of fawcettimine clearly show the evolution of powerful methodologies in C–C bond formation over the last 20 years. It is of interest to note that while the number of steps and overall yield in the synthesis of fawcettimine by Toste and coworkers remained similar to that of Heathcock, their synthesis did deliver an enantiopure natural product. However, one would not expect anything less by today's standards in natural product synthesis (see also Chapter 18, section 18.6.1).

14.9
Tamiflu

The anti-influenza compound Tamiflu has been the subject of intensive research efforts with the objective of finding an economically viable total synthesis [41]. Naturally occurring carbocyclic compounds such as quinic and shikimic acid have been used as starting materials in commercial processes [42]. Below we describe the important use of carbocyclic chirons derived from microbiological oxidations of substituted aromatic compounds, as well as from catalytic asymmetric methods towards this objective.

14.9.1
The Fang and Wong synthesis

A concise and flexible synthesis of Tamiflu was reported by Fang, Wong, and coworkers [43] starting with a 3-bromo-cis-dihydrocatechol, which can be produced on a large scale by the microbial oxidation of bromobenzene [44] (Figure 14.13). In the presence of catalytic $SnBr_4$ and N-bromoacetamide in aqueous acetonitrile [45], acetonide **A** was subjected to a smooth bromoacetamidation of the isolated double bond to give **B** in 75% overall yield for two steps. Treatment of **B** with LiHMDS led to N-acetyl aziridine **C**, which was subjected to ring-opening with 3-pentanol in the presence of $BF_3 \cdot Et_2O$ to give **D**. Cleavage of the acetonide group and treatment with α-acetoxyisobutyryl bromide according to Greenberg and Moffatt [46] gave the trans-bromo acetate **E**. Reduction of the allylic bromide with $LiEt_3BH$ proceeded with concomitant cleavage of the acetate to give **F**. Inversion of configuration with diphenylphosphoryl azide via a Mitsunobu reaction led to **G** in excellent yield. Nickel-catalyzed carbonylation [47] and formation of the ethyl ester was achieved in high yield from **G**, to give an intermediate that was subjected to catalytic Lindlar hydrogenation conditions to avoid over reduction. Tamiflu was subsequently isolated as the phosphate ester.

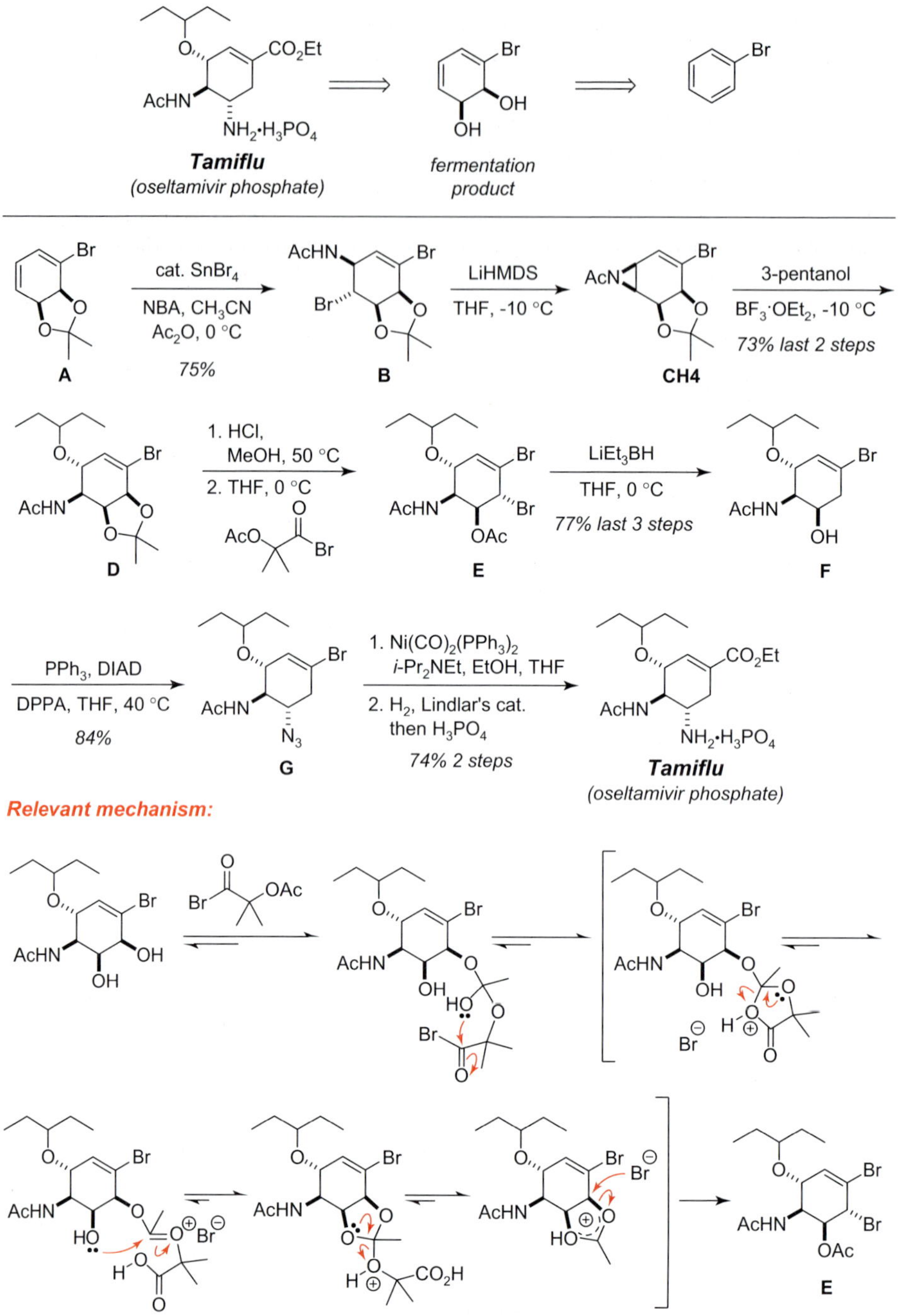

Figure 14.13 Total synthesis of Tamiflu from bromobenzene (apparent substructure).

Commentary: The carbocyclic structure of Tamiflu suggests that a number of hydroxylated carbocycles could potentially be used as starting materials for its synthesis. Indeed, a current industrial synthesis of Tamiflu starts with naturally occurring shikimic acid, and a practical 12-step synthesis from quinic acid that is amenable to kilogram-scale has been reported [42a]. Alternative syntheses have been reported where the functionalized carbocyclic core was elaborated from achiral precursors relying on innovative asymmetric processes [48]. In a previous synthesis, Fang, Wong, and their coworkers [49] utilized D-xylose as a starting material (see Chapter 11, Figure 11.20). The exploitation of a known microbiological method that converts bromobenzene to the 3-bromo-*cis*-dihydrocatechol in enantiopure form, and in large quantity, is an attractive feature of the Fang and Wong synthesis. Regio- and stereocontrolled introduction of an acetamido group without resorting to further manipulation is also a practical alternative to the use of other sources of nitrogen. Fang and coworkers relied on a number of well-precedented methods such as opening of the *N*-acetyl aziridine with 3-pentanol, and bromoacetylation using α-acetoxy-isobutyryl bromide to functionalize the requisite cyclohexene intermediate **F** (see relevant mechanism, Figure 14.13). As a consequence of the stereochemistry of the 3-bromo-*cis*-dihydrocatechol, an obligatory inversion of configuration was necessary in order to introduce the azide group as the penultimate functional group transformation. Nickel-catalyzed carbethoxylation nicely completed the planned sequence of reactions. The synthesis was executed in ten steps on gram-scale from the (1*R*-*cis*)-3-bromo-3,5-cyclohexadiene-1,2-diol. This augurs well for a large-volume production of Tamiflu in the future, although the use of the Mitsunobu reaction to introduce the azide group may present some practical issues.

14.9.2
The Hudlicky and Banwell syntheses

Total syntheses of Tamiflu have been reported that expressly avoid the use of azide (Figure 14.14). Thus, Hudlicky and coworkers [50] adopted the concept of latent symmetry as a design element in their visual analysis of the target molecule. Rather than start with readily available and enantiopure (1*R*-*cis*)-3-bromo-3,5-cyclohexadiene-1,2-diol as did Fang and coworkers [43], they used (*R,R*)-diol **A** available from the microbial oxidation of ethyl benzoate [51], thereby avoiding a Pd-catalyzed carbonylation step. Treatment of the acetonide formed from **A** with acetyl hydroxylamine led to an inverse electron demand Diels-Alder cyclization to produce oxazine **B** in 88% yield over two steps. Reductive cleavage to the allylic alcohol **C**, followed by an oxidative [3,3]-sigmatropic rearrangement [52], and formation of the oxime, led to **D**. Stereocontrolled reduction of both the oxime and alkene functional groups in the presence of Rh/Al$_2$O$_3$, and *in situ* protection of the free amine, afforded **E** in high yield. Treatment with NaOMe resulted in β-elimination and formation of the corresponding allylic alcohol, which was transformed into the aziridine **F**. Introduction of the 3-pentanol moiety using Shibasaki's method [48f] led to **G**, which was deprotected to give Tamiflu. Although the synthesis leads to known intermediates in an expeditious manner, the use of Cr and Mo reagents may be issues to consider for the prospects of an industrially viable process.

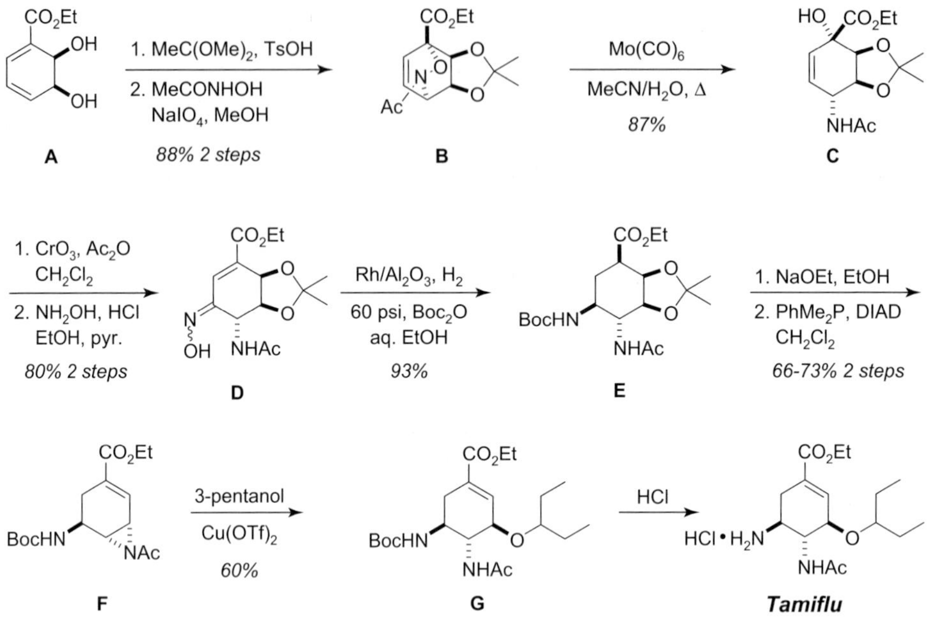

Figure 14.14 Synthesis of Tamiflu from ethyl benzoate (apparent substructure).

Banwell and coworkers [53] also commenced their synthesis with (1R-cis)-3-bromo-3,5-cyclohexadiene-1,2-diol (A), which they converted to PMB ether B in high yield by a regioselective reductive acetal cleavage (Figure 14.15). Formation of the N-tosyl carbamate was followed by an intramolecular Cu-catalyzed transformation [54] to give the aziridine D, which was stereoselectively opened with 3-pentanol to furnish E. Cleavage of the cyclic carbamate and PMB ether then led to the known diol intermediate F, which had been previously converted to Tamiflu [42]. As in the Hudlicky synthesis, the use of azide is also avoided in this approach, although the large scale manipulation of hydroxylamine in both cases may be limiting.

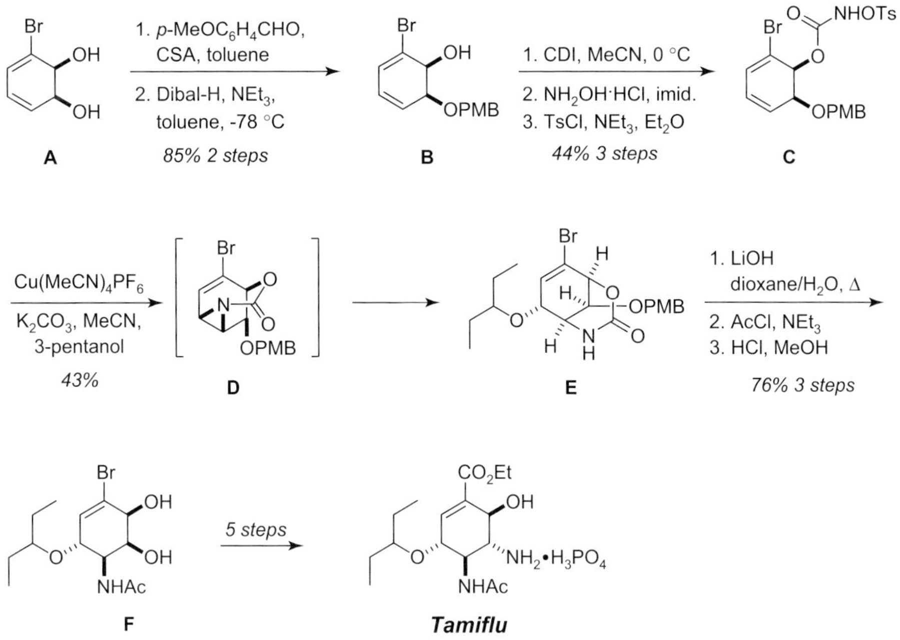

Figure 14.15 Alternative synthesis of Tamiflu from (1R-cis)-3-bromo-3,5-cyclohexadiene-1,2-diol (apparent substructure).

14.9.3
The Shibasaki catalytic asymmetric Diels-Alder synthesis

In view of the importance of Tamiflu as an anti-influenza drug, and the need to implement newer and constantly improving total syntheses, major efforts have been devoted to the development of catalytic asymmetric methods [41]. One of the most recent total syntheses reported by Shibasaki and coworkers [48a] is highlighted in Figure 14.16. A Diels-Alder reaction between TMS protected 1,3-butadiene-1-ol and dimethyl fumarate in the presence of catalyst **A**, barium isopropoxide, and CsF proceeded in excellent yield to give a 5:1 mixture of diastereomers **B** and **C** respectively. Elaboration of the enantioenriched diester **B** by effecting a double Curtius rearrangement, with careful isolation of the intermediate bis-acyl azides, and addition of *t*-BuOH led to cyclic carbamate **D** in excellent yield. Crystallization of the *N*-acetyl derivative of **D**, followed by a Pd-catalyzed addition of a malononitrile anion equivalent gave **F** in 68% yield over two steps. Face-selective epoxidation, and hydrolysis afforded the ester **G**, which was inverted through a Mitsunobu reaction. The resulting *p*-nitrobenzoate ester was saponified, and then transformed to *N*-acetyl aziridine **H**. Introduction of the 3-pentanol moiety, and deprotection was achieved as previously described by Shibasaki and others [48f] to give Tamiflu, which was isolated as the phosphate salt. Starting with cheap and readily available bulk chemicals, Shibasaki and coworkers have improved on their previous synthesis by accessing the enantioenriched carbocyclic chiron **B** in the first step of the synthesis. Although azide was not used directly in an S_N2 sense, the double Curtius rearrangement of the dicarboxylic acid derived from **B**, trapping the vicinal isocyanate with the *cis*-disposed hydroxyl group, and protection of the distal isocyanate as the *N*-Boc derivative (**D**) was conveniently performed in a "one-pot" operation with careful manipulation of the intermediate acyl azides.

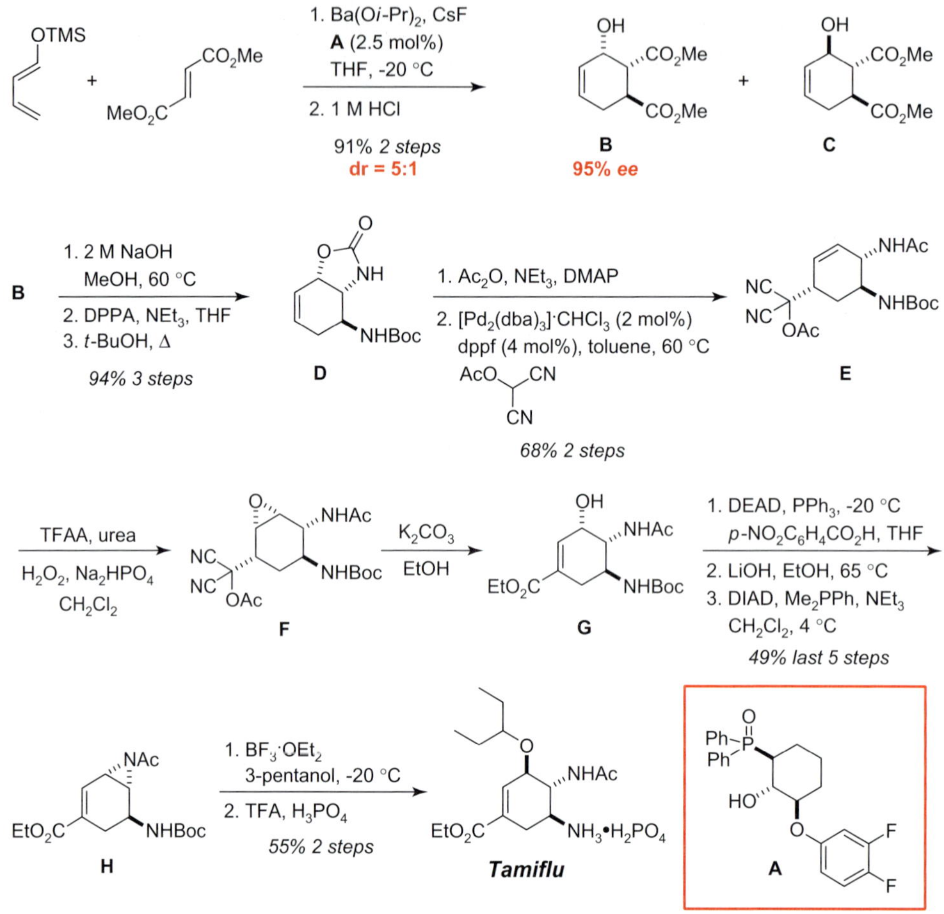

Figure 14.16 Total synthesis of Tamiflu using a catalytic asymmetric Diels-Alder reaction.

14.9.4
Tamiflu synthesis in the age of catalysis: Synopsis

The numerous total syntheses and approaches to Tamiflu have been recently reviewed [41]. In the context of this chapter dealing with carbocyclic chirons, we have highlighted three total syntheses relying on microbiologically-based methods [42, 50, 53], as well as the most recent method based on a asymmetric catalysis [48a]. Several other syntheses, primarily based on catalytic methods of preparing the starting carbocyclic chirons are referenced [48]. Although meritorious in many respects, especially for the ingenuity in design and the inventive applications of catalytic methods developed within the respective groups, none are optimized to be considered as industrially viable for large scale production. In Figure 14.17 we summarize six total syntheses of Tamiflu starting with contributions from Corey [48] and Shibasaki [48f] independently in 2006, followed by Fukuyama [48e] in 2007, Trost [48] in 2008, then Hayashi [48c] and Shibasaki [48a], both in 2009.

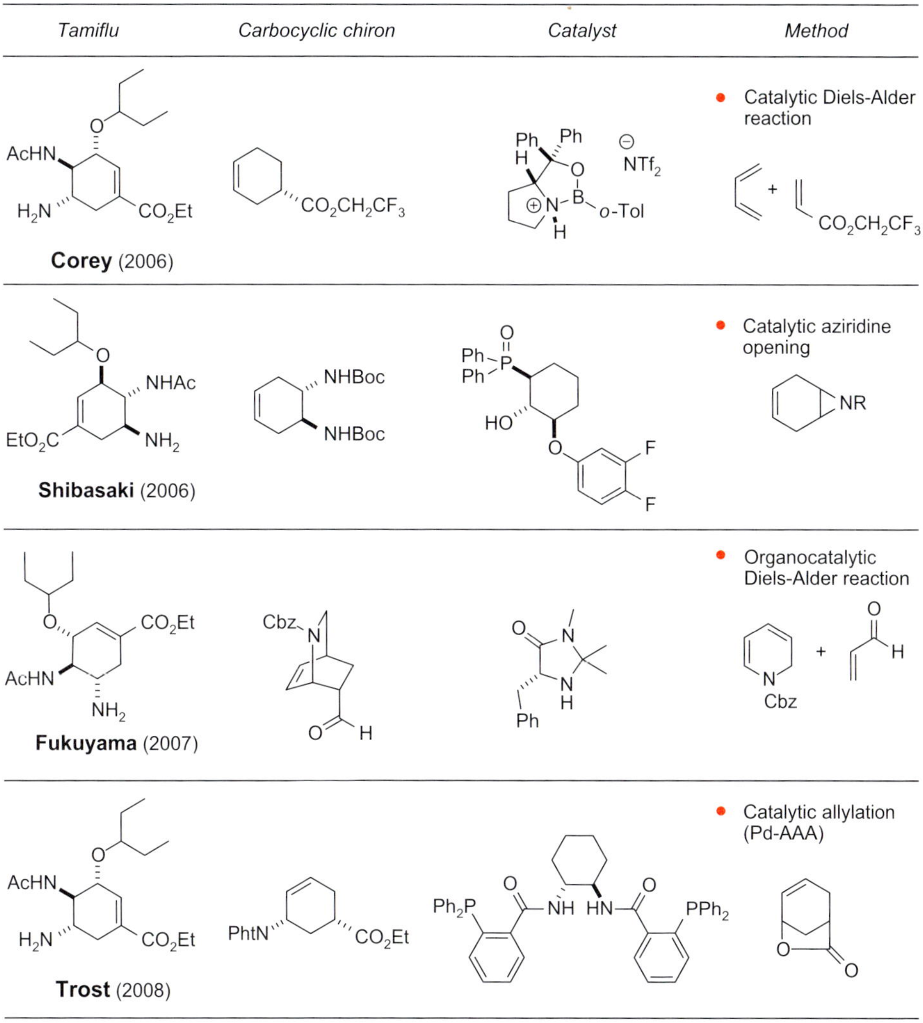

Tamiflu	Carbocyclic chiron	Catalyst	Method
Corey (2006)			• Catalytic Diels-Alder reaction
Shibasaki (2006)			• Catalytic aziridine opening
Fukuyama (2007)			• Organocatalytic Diels-Alder reaction
Trost (2008)			• Catalytic allylation (Pd-AAA)

Figure 14.17 Access to Tamiflu from carbocyclic chirons prepared by asymmetric catalysis.

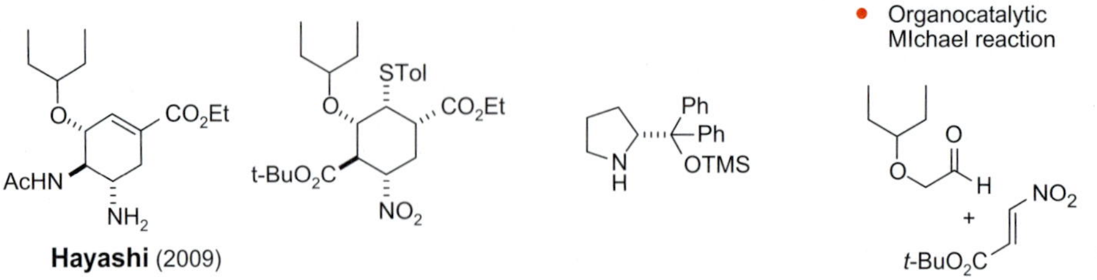

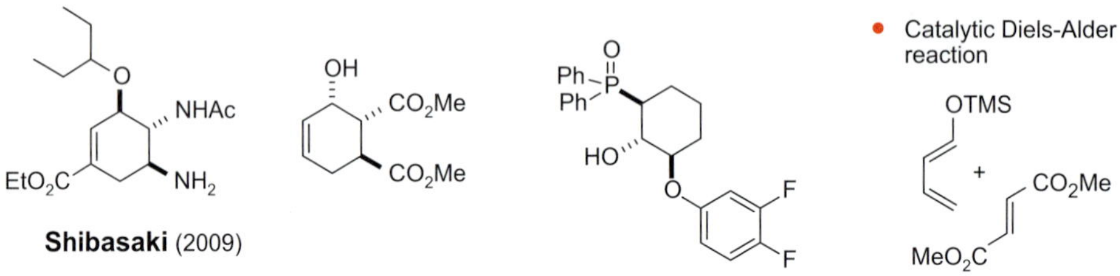

• Organocatalytic Michael reaction

Hayashi (2009)

• Catalytic Diels-Alder reaction

Shibasaki (2009)

Figure 14.17   *(continued)*.

In each case we show the catalyst, the type of reaction used, and the primary carbocyclic chiron that each group has produced. Organocatalytic and metal-based Diels-Alder reactions were used by Corey, Shibasaki, and Fukuyama. The Trost synthesis, in which a catalytic Pd-catalyzed asymmetric allylic alkylation (Pd-AAA) reaction was used, is reported to be the shortest to date (8 steps, 30% overall yield). Hayashi, relying on an organocatalytic Michael-type reaction reports a synthesis that involves three "one-pot" synthetic operations for a total of nine steps and 57% overall yield. Because of the conceptually different approaches, leading to diverse sets of carbocyclic chirons, the structures of the corresponding "Tamiflus" have been drawn in different perspectives by four of the six research groups.

14.10
Miscellanea

Many carbocyclic natural and unnatural products have been synthesized starting with appropriately functionalized chiral, non-racemic carbocycles. The Wieland-Miescher ketone and related carbocycles have been popular chirons for the synthesis of a variety of natural products [55]. In addition, a number of monocyclic and bicyclic cycloalkenones containing useful functional groups and chiral appendages have found extensive applications in total synthesis (see also Chapter 8, section 8.10.4). Selected examples are shown in Figure 14.18.

Figure 14.18 Total synthesis of natural products from carbocycles.

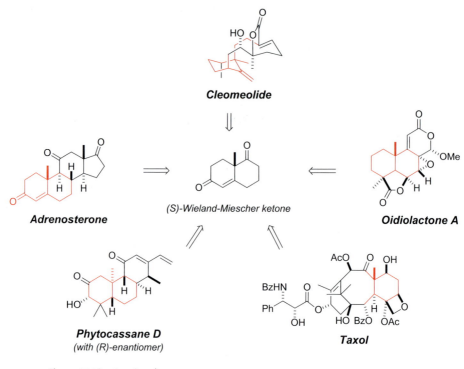

Figure 14.18 *(continued).*

Selected examples of the utilization of quinic acid in natural product synthesis are shown in Figure 14.19 [56] (see also, Chapter 7, section 7.5).

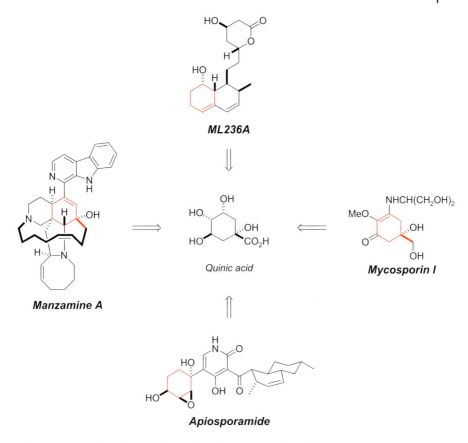

Figure 14.19 Total synthesis of natural products from quinic acid.

References

1. Sugimura, T.; Paquette, L. A. *J. Am. Chem. Soc.* 1987, **109**, 3017; Paquette, L. A.; Sugimura, T. *J. Am. Chem. Soc.* 1986, **108**, 3841.
2. Hajos, Z. G.; Parrish, D. R. *Org. Synth.* 1985, **63**, 26; Hajos, Z. G.; Parrish, D. R. *J. Org. Chem.* 1974, **39**, 1615; see also: Eder, U.; Sauer, G.; Weichert, R. *Angew. Chem. Int. Ed.* 1971, **10**, 496.
3. (a) Still, W. C. *J. Am. Chem. Soc.* 1978, **100**, 1481; (b) Still, W. C.; Mitra, A. *J. Am. Chem Soc.* 1978, **100**, 1927.
4. Nakamura, E.; Murofushi, T.; Shimizu, M.; Kuwajima, I. *J. Am. Chem. Soc.* 1976, **98**, 2346; for the oxidation of silyl enol ethers, see: Ito, Y.; Hirao, T.; Saegusa, T. *J. Org. Chem.* 1978, **43**, 1011.
5. For examples, see: (a) Fleming, I.; Kemp-Jones, A. V.; Long, W. E.; Thomas, E. J. *J. Chem. Soc., Perkin Trans. 2* 1976, 7; (b) Fleming, I.; Long, W. E. *J. Chem. Soc., Perkin Trans. 2* 1976, 14; (c) Singh, S.; Usha, G.; Tung, C.-H.; Turro, N. J.; Ramamurthy, V. *J. Org. Chem.* 1986, **51**, 941.
6. (a) Ling, T.; Chowdhury, C.; Kramer, B. A.; Vong, B. G.; Palladino, M. A.; Theodorakis, E. A. *J. Org. Chem.* 2001, **66**, 8843; (b) Ling, T.; Kramer, B. A.;

Palladino, M. A.; Theodorakis, E. A. *Org. Lett.* 2000, **2**, 2073.

7. Buchschacher, P.; Fürst, A.; Gutzwiller, J. *Org. Synth. Coll.* Vol. **VII**, 1990, 368.

8. (a) Mander, L. N.; Sethi, S. P. *Tetrahedron Lett.* 1983, **24**, 5425; see also: (b) Crabtree, S. R.; Chu, W. L. A.; Mander, L. N. *Synlett* 1990, 169.

9. Coates, R. M.; Shaw, J. E. *J. Org. Chem.* 1970, **35**, 2597; Coates, R. M.; Shaw, J. E. *J. Org. Chem.* 1970, **35**, 2601.

10. (a) Deng, W.-P.; Zhong, M.; Guo, X.-C.; Kende, A. S. *J. Org. Chem.* 2003, **68**, 7422; (b) Kende, A. S.; Deng, W.-P.; Zhong, M.; Guo, X.-C. *Org. Lett.* 2003, **5**, 1785.

11. (a) Di Grandi, M. J.; Jung, D. K.; Krol, W. J.; Danishefsky, S. J. *J. Org. Chem.* 1993, **58**, 4989; see also: (b) Barton, D. H. R.; Bashiardes, G.; Fourrey, J.-L. *Tetrahedron* 1988, **44**, 147.

12. (a) Roggo, B. E.; Petersen, F.; Sills, M.; Roesel, J. L.; Moerker, T.; Peter, H. H. *J. Antibiot.* 1996, **49**, 13; (b) Roggo, B. E.; Hug, P.; Moss, S.; Stämpfli, A.; Kriemler, H.-P.; Peter, H. H. *J. Antibiot.* 1996, **49**, 374.

13. Waters, S. P.; Tian, Y.; Li, Y.-M.; Danishefsky, S. J. *J. Am. Chem. Soc.* 2005, **127**, 13514; for a recent total synthesis of scabronine G, see: Kanoh, N.; Sakanishi, K.; Imori, E.; Nishimura, K.; Iwabuchi, Y. *Org. Lett.* 2011, **13**, 2864.

14. Scott, W. J.; Stille, J. K. *J. Am. Chem. Soc.* 1986, **108**, 3033.

15. For reviews, see: (a) Pellissier, H. *Tetrahedron* 2005, **61**, 6479; (b) Tius, M. A. *Eur. J. Org. Chem.* 2005, 2193.

16. For a review, see: Negishi, E.-i.; Liu, F. in *Metal-Catalyzed Cross-Coupling Reactions*, Diederich, F.; Stang, P. J., Eds.; 1998, Wiley-VCH, NY.

17. For example, see: Guerrero, A.; Parrilla, A.; Camps, F. *Tetrahedron Lett.* 1990, **31**, 1873.

18. Hagiwara, H.; Takeuchi, F.; Hoshi, T.; Suzuki, T.; Ando, M. *Tetrahedron Lett.* 2001, **47**, 1983; for full details and synthesis of chapecoderins A, B, and C, see: Hagiwara, H.; Takeuchi, F.; Nozawa, M.; Hoshi, T.; Suzuki, T. *Tetrahedron* 2004, **60**, 1983.

19. Hagiwara, H.; Uda, H. *J. Org. Chem.* 1988, **53**, 2308.

20. (a) Garg, N. K.; Caspi, D. D.; Stoltz, B. M. *J. Am. Chem. Soc.* 2004, **126**, 9552; see also: Garg, N. K.; Caspi, D. D.; Stoltz, B. M. *J. Am. Chem. Soc.* 2005, **127**, 5970; for a synthesis of dragmacidin D see: (b) Garg, N. K.; Sarpong, R.; Stoltz, B. M. *J. Am. Chem. Soc.* 2002, **124**, 13179; for a review, see: (c) Garg, N. K.; Stoltz, B. M. *Chem. Commun.* 2006, 3769.

21. (a) Philippe, M.; Sepulchre, A. M.; Gero, S. D.; Loibner, H.; Streicher, W.; Stutz, P. *J. Antibiot.* 1982, **35**, 1507; (b) Manthey, M. K.; González-Bello, C.; Abell, C. *J. Chem. Soc., Perkin Trans. 1* 1997, 625.

22. (a) Neber, P. W.; Friedolsheim, A. V. *Justus Liebigs Ann. Chem.* 1926, **449**, 109; (b) Ooi, T.; Takahashi, M.; Doda, K.; Maruoka, K. *J. Am. Chem. Soc.* 2002, **124**, 7640.

23. (a) Woodward, R. B.; Bader, F. E.; Bickel, H.; Frey, A. J.; Kierstead, R. W. *J. Am. Chem. Soc.* 1956, **78**, 2023; (b) Woodward, R. B.; Bader, F. E.; Bickel, H.; Frey, A. J.; Kierstead, R. W. *Tetrahedron* 1958, **2**, 11; (c) Woodward, R. B. in *Perspectives of Organic Synthesis*, Todd, A. R., Ed.; 1956, Interscience, NY.

24. For a review, see: Chen, F.-E.; Huang, J. *Chem. Rev.* 2005, **105**, 4671.

25. (a) Wender, P. A.; Schaus, J. M.; White, A. W. *J. Am. Chem. Soc.* 1980, **102**, 6157; (b) Wender, P. A.; Schaus, J. M.; Torney, D. C. *Tetrahedron Lett.* 1973, **20**, 2485; (c) Wender, P. A.; Schaus, J. M.; White, A. W. *Heterocycles* 1987, **25**, 263.

26. (a) Martin, S. F.; Rüeger, H.; Williamson, S. A.; Grzejszczak, S. *J. Am. Chem. Soc.* 1987, **109**, 6124; (b) Martin, S. F.; Grzejszczak, S.; Rüeger, H.; Williamson, S. A. *J. Am. Chem. Soc.* 1985, **107**, 4072.

27. (a) Stork, G.; Tang, P. C.; Casey, M.; Goodman, B.; Toyota, M. *J. Am. Chem. Soc.* 2005, **127**, 16255; (b) see also: Stork, G. *Pure Appl. Chem.* 1989, **61**, 439.

28. Hanessian, S.; Pan, J.; Carnell, A.; Bouchard, H.; Lesage, L. *J. Org. Chem.* 1997, **62**, 465.

29. Pearlman, B. A. *J. Am. Chem. Soc.* 1979, **101**, 6404.

30. Toyota, M.; Asoh, T.; Matsuura, M.; Fukumoto, K. *J. Org. Chem.* 1996, **61**, 8687.

31. Hanessian, S.; Di Fabio, R.; Marcoux, J.-F.; Prud'homme, M. *J. Org. Chem.* 1990, **55**, 3436 and references cited therein.

32. For example, see: (a) Kusuda, K.; Inanaga, J.; Yamaguchi, M. *Tetrahedron Lett.* 1989, **30**, 2945; (b) Hanessian, S.; Girard, C.; Chiara, J. L. *Tetrahedron Lett.* 1992, **33**, 573.

33. For early examples of ether oxygen directed attack of Grignard reagents, see: (a) Wolfrom, M. L.; Hanessian, S. *J. Org. Chem.* 1962, **27**, 1800; (b) Still, W. C.; McDonald, J. H. *Tetrahedron Lett.* 1980, **21**, 1031 and references cited therein.

34. Linghu, X.; Kennedy-Smith, J. J.; Toste, F. D. *Angew. Chem. Int. Ed.* 2007, **46**, 7671.

35. (a) Harayama, T.; Takatani, M.; Inubushi, Y. *Tetrahedron Lett.* 1979, **20**, 4307; (b) Harayama, T.; Takatani, M.; Inubushi, Y. *Chem. Pharm. Bull.* 1980, **28**, 2394.

36. (a) Heathcock, C. H.; Blumenkopf, T. A.; Smith, K. M. *J. Org. Chem.* 1989, **54**, 1548; (b) Heathcock, C. H.; Smith, K. M.; Blumenkopf, T. A. *J. Am. Chem. Soc.* 1986, **108**, 5022; for a recent total synthesis of fawcettimine, see: Otuska, Y.; Inagaki, F.; Mukai, C. *J. Org. Chem.* 2010, **75**, 3420.

37. For example, see (a) Carlone, A.; Marigo, M.; North, C.; Landa, A.; Jørgensen, K. A. *Chem. Commun.* 2006, 4928; (b) Marigo, M.; Wabnitz, T. C.; Fielenbach, D.; Jørgensen, K. A. *Angew. Chem. Int. Ed.* 2005, **44**, 794.

38. (a) Staben, S. T.; Kennedy-Smith, J. J.; Toste, F. D. *Angew. Chem. Int. Ed.* 2004, **43**, 5350; (b) Corkey, B. K.; Toste, F. D. *J. Am. Chem. Soc.* 2007, **129**, 2764; for reviews of gold(I)-catalyzed reactions, see: (c) Hashmi, A. S. K.; Buhrle, M. *Aldrichim. Acta* 2010, **43**, 27; (d) Jiménez-Núñez, E.; Echavarren, A. M. *Chem. Commun.* 2007, 333; (e) Fürstner, A.; Davies, P. W. *Angew. Chem. Int. Ed.* 2007, **46**, 3410; (f) Gorin, D. J.; Toste, F. D. *Nature* 2007, **446**, 395.

39. Staben, S. T.; Kennedy-Smith, J. J.; Huang, D.; Corkey, B. K.; LaLonde, R. L.; Toste, F. D. *Angew. Chem. Int. Ed.* 2006, **118**, 6137.

40. (a) Yang, Y.-R.; Shen, L.; Huang, J.-Z.; Xu, T.; Wei, K. *J. Org. Chem.* 2011, **76**, 3684; (b) Otsuka, Y.; Imagaki, F.; Mukai, C. *J. Org. Chem.* 2010, **75**, 3420; (c) Jung, M. E.; Cang, J. J. *Org. Lett.* 2010, **12**, 2962.

41. For selected recent reviews, see: (a) Magano, J. *Chem. Rev.* 2009, **109**, 4398; (b) Shibasaki, M.; Kanai, M. *Eur. J. Org. Chem.* 2008, 1839; (c) Abrecht, S.; Federspiel, M. C.; Estermann, H.; Fischer, R.; Karpf, M.; Mair, H.-J.; Oberhauser, T.; Rimmler, G.; Trussardi, R.; Zutter, U. *Chimia* 2007, **61**, 93; (d) Farina, V.; Brown, J. D. *Angew. Chem. Int. Ed.* 2006, **45**, 7330.

42. For example, see: (a) Rohloff, J. C.; Kent, K. M.; Postich, M. J.; Becker, M. W.; Chapman, H. H.; Kelly, D. E.; Lew, W.; Louie, M. S.; McGee, L. R.; Prisbe, E. J.; Schultze, L. M.; Yu, R. H.; Zhang, L. *J. Org. Chem.* 1998, **63**, 4545; (b) Karpf, M.; Trussardi, R. *J. Org. Chem.* 2001, **66**, 2044, and references cited therein.

43. Shie, J.-J.; Fang, J.-M.; Wong, C.-H. *Angew. Chem. Int. Ed.* 2008, **47**, 5788; see also: Nie, L.-D.; Shi, X.-X.; Ko, K. H.; Lu, W.-D. *J. Org. Chem.* 2009, **74**, 3970.

44. For example, see: (a) Hudlicky, T.; Reed. J. W. *Synlett* 2009, 685; (b) Johnson, R. A. *Org. React.* 2004, **63**, 117; (c) Banwell, M. G.; Edwards, A. J.; Harfoot, G. J.; Jolliffe, K. A.; McLeod, M. D.; McRae, K. J.; Stewart, S. G.; Vögtle, M. *Pure Appl. Chem.* 2003, **75**, 223; (d) Endoma, M. A; Bui, V. P.; Hansen, J.; Hudlicky, T. *Org. Proc. Res. Dev.* 2002, **6**, 525; (e) Boyd, D. R.; Sheldrake, G. N. *Nat. Prod. Rep.* 1998, **15**, 309.

45. Yeung, Y.-Y.; Gao, X.; Corey, E. J. *J. Am. Chem. Soc.* 2006, **128**, 9644.

46. Greenberg, S.; Moffatt, J. G. *J. Am. Chem. Soc.* 1973, **95**, 4016; see also: Boyd, D. R.; Sharma, N. D.; Llamas, N. M.; O'Dowd, C. R.; Allen, C. C. R. *Org. Biomol. Chem.* 2006, 4, 2208.

47. Blacker, A. J.; Booth, R. J.; Davies, G. M.; Sutherland, J. K. *J. Chem. Soc., Perkin Trans. 1* 1995, 2861.

48. For example, see: (a) Yamatsugu, K.; Yin, L.; Kamijo, S.; Kimura, Y.; Kanai, M.; Shibasaki, M. *Angew. Chem. Int. Ed.* 2009, **48**, 1070; (b) Nie, L.-D.; Shi, X.-X.; Ko, K. H.; Lu, W.-D. *J. Org. Chem.* 2009, **74**, 3970; (c) Ishikawa, H.; Suzuki, T.; Hayashi, Y. *Angew. Chem. Int. Ed.* 2009, **48**, 1304; (d) Trost, B. M.; Zhang, T. *Angew. Chem. Int. Ed.* 2008, **47**, 3759; for full details, see: Zhang, T.; Trost, B. M. *Chem. Eur. J.* 2011, **17**, 3630; (e) Satoh, N.; Akiba, T.; Yokoshima, S.; Fukuyama, T. *Angew. Chem. Int. Ed.* 2007, **46**, 5734; (f) Fukuta, Y.; Mita, T.; Fukuda, N.; Kanai, M.; Shibasaki, M. *J. Am. Chem. Soc.* 2006, **128**, 6312; (g) Yeung, Y.-Y.; Hong, S.; Corey, E. J. *J. Am. Chem. Soc.* 2006, **128**, 6310; see also: (h) Mandai, T.; Oshitari, T. *Synlett* 2009, 783; (i) Zutter, U.; Iding, H.; Spurr, P.; Wirz, B. *J. Org. Chem.* 2008, **73**, 4895; (j) Cong, X.; Yao, Z. J. *J. Org. Chem.* 2006, **71**, 5365; (k) Zhu, S.; Yu, S.; Wang, Y.; Ma, D. *Angew. Chem. Int. Ed.* 2010, **49**, 4656.

49. (a) Shie, J.-J.; Fang, J.-M.; Wang, S.-Y.; Tsai, K.-C.; Cheng, Y.-S. E.; Yang, A.-S.; Hsiao, S.-C.; Su, C.-Y.; Wong, C.-H. *J. Am. Chem. Soc.* 2007, **129**, 11892.

50. (a) Werner, L.; Machara, A.; Hudlicky, T. *Adv. Synth. Catal.* 2010, **352**, 195; (b) Sullivan, B.; Carrera, I.; Drouin, M.; Hudlicky, T. *Angew. Chem. Int. Ed.* 2009, **48**, 4229.

51. Zylstra, G. J.; Gibson, D. T. *J. Biol. Chem.* 1989, **264**, 14940.

52. Dauben, W. G.; Michno, D. M. *J. Org. Chem.* 1977, **42**, 682; see also: Luzzio, F. A. *Org. React.* 1998, **53**, 1.

53. Matveenko, M.; Willis, A. C.; Banwell. M. G. *Tetrahedron Lett.* 2008, **49**, 7018.

54. For example, see: Liu, R.; Herron, S. R.; Fleming, S. A. *J. Org. Chem.* 2007, **72**, 5587.

55. Pancratistatin: Tian, X.; Hudlicky, T.; Koenigsberger, K. *J. Am. Chem. Soc.* 1995, **117**, 3643; ptilocaulin: Snider, B. B.; Faith, W. C. *J. Am. Chem. Soc.* 1984, **106**, 1443; brefeldin A: Corey, E. J.; Carpino, P. *Tetrahedron Lett.* 1990, **31**, 7555; didemnenone A: Forsyth, C. J.; Clardy, J. *J. Am. Chem. Soc.* 1988, **110**, 5911; spinosin A:

Paquette, L. A.; Gao, Z.; Ni, Z.; Smith, G. F. *J. Am. Chem. Soc.* 1998, **120**, 2543; Paquette, L. A.; Collado, I.; Purdie, M. *J. Am. Chem. Soc.* 1998, **120**, 2553; hitachimycin: Smith, A. B. III.; Rano, T. A.; Chida, N.; Sulikowski, G. A.; Wood, J. L. *J. Am. Chem. Soc.* 1992, **114**, 8008; ouabain: Zhang, H.; Reddy, M. S.; Phoenix, S.; Deslongchamps, P. *Angew. Chem. Int. Ed.* 2008, **47**, 1272; cortistatin A : Nicolaou, K. C.; Peng, X.-S.; Sun, Y.-P.; Polet, D.; Zou, B.; Lim, C. S.; Chen, D. Y.-K. *J. Am. Chem. Soc.* 2009, **131**, 10587; Lee, H. M.; Nieto-Oberhuber, C.; Shair, M. D. *J. Am. Chem. Soc.* 2008, **130**, 16864.; for a synthesis from prednisone, see: Shenvi, R. A.; Guerrero, C. A.; Shi, J.; Li, C.-C.; Baran, P. S. *J. Am. Chem. Soc.* 2008, **130**, 7241; cleomeolide: Paquette, L. A.; Wang, T.-Z.; Philippo, C. M. G.; Wang, S. *J. Am. Chem. Soc.* 1994, **116**, 3367; ceroplastol I: Paquette, L. A.; Wang, T. Z.; Vo, N. H. *J. Am. Chem. Soc.* 1993, **115**, 1676; oidiolactone A: Hanessian, S.; Boyer, N.; Reddy, G. J.; Deschênes-Simard, B. *Org. Lett.* 2009, **11**, 4640; adrenosterone: Dzierba, C. D.; Zandi, K. S.; Möellers, T.; Shea, K. J. *J. Am. Chem. Soc.* 1996, **118**, 4711; phytocassane D: Yajima, A.; Mori, K. *Eur. J. Org. Chem.* 2000, 4079; taxol: Danishefsky, S. J.; Masters, J. J.; Young, W. B.; Link, J. T.; Snyder, L. B.; Magee, T. V.; Jung, D. K.; Isaacs, R. C. A.; Bornmann, W. G.; Alaimo, C. A.; Coburn, C. A.; Di Grandi, M. J. *J. Am. Chem. Soc.* 1996, **118**, 2843.

56. For a review on quinic acid, see: (a) Barco, A.; Benetti, S.; De Risi, C.; Marchetti, P.; Pollini, G. P.; Zanirato, V. *Tetrahedron: Asymmetry* 1997, **8**, 3515; (b) ML236A: Danishefsky, S. J.; Simoneau, B. *Pure Appl. Chem.* 1988, **60**, 1555; (c) manzamine: Kamenecka, T. M.; Overman, L. E. *Tetrahedron. Lett.* 1994, **35**, 4279; (d) mycosporin: White, J. D.; Cammack, J. H.; Sakuma, K. *J. Am. Chem. Soc.* 1989, **111**, 8970; (e) apiosporamide: Williams, D. R.; Kammler, D. C.; Donnell, A. F.; Goundry, W. R. F. *Angew. Chem. Int. Ed.* 2005, **44**, 6715.

15
Total Synthesis with Lactones as Precursors

With the exception of lactones derived from carbohydrates, functionally useful lactones available directly from natural sources are relatively few in number. Particularly versatile and readily available, (4R)- or (4S)-4-hydroxymethyl-γ-butyrolactones (and their 2,3-unsaturated variants), are prepared from D- and L-glutamic acids respectively in a practical two-step process [1] (see also Chapter 8, section 8.11). Here, we highlight syntheses of selected natural products starting with 5-carbon lactones as modified chirons.

15.1
Megaphone

The first total synthesis of the cytotoxic neolignan megaphone was reported in 1985 by Koga and coworkers [2] (Figure 15.1). Analysis of the structure reveals a unique cyclohexenone motif that elicits a number of synthetic approaches normally used for related carbocycles. Koga's strategy took advantage of stereocontrolled functionalization of lactone **A**, readily available from L-glutamic acid. Aldol condensation of **A** with acetaldehyde, followed by mesylation, and elimination afforded **B**, which was subjected to a conjugate addition with lithiated t-butyldimethylsilyl di(methylthio)methane, followed by addition of allyl bromide. Protodesilylation with TBAF gave lactone **C**, which was hydrolyzed to the carboxylic acid and then converted to the O-methyl acyclic ester **D**. Treatment of **D** with MeLi furnished methyl ketone **E**. Cleavage of the trityl group, and oxidation of the resulting alcohol to the aldehyde gave **F** in good overall yield. An intramolecular aldol condensation in refluxing MeOH containing NaHCO$_3$ produced cyclohexenone **G** as the major isomer, with concomitant inversion of configuration at the carbon atom bearing the methoxy group. Cleavage of the dimethyl dithioacetal, followed by condensation with 3,4,5-trimethoxyphenyllithium gave megaphone as the major isomer.

Design and Strategy in Organic Synthesis: From the Chiron Approach to Catalysis, First Edition.
Stephen Hanessian, Simon Giroux, and Bradley L. Merner.

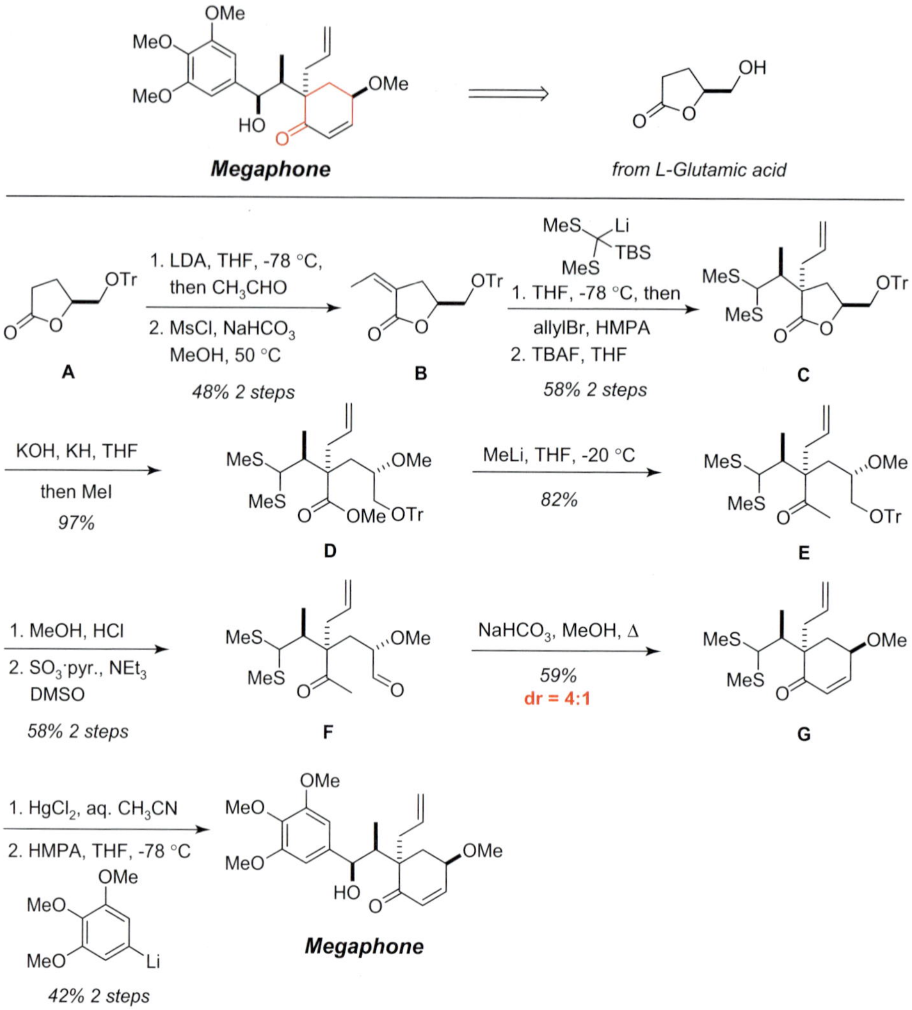

Megaphone   *from L-Glutamic acid*

Key reactions: *Stereoselective 1,4-conjugate addition (**B** to **C**); methyl ether/methyl ester formation (**C** to **D**); intramolecular aldol condesation (**F** to **G**); stereoselective arylation of aldehyde (**G** to **target**).*

Relevant mechanism:

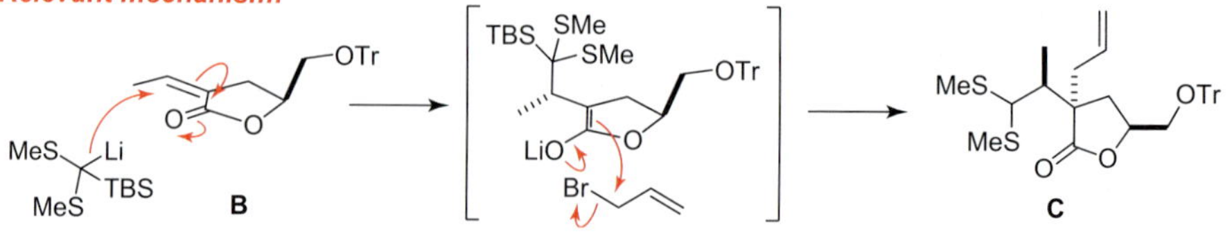

Figure 15.1   Total synthesis of megaphone from (4*S*)-4-hydroxymethyl-γ-butyrolactone (hidden substructure).

Commentary: In spite of its relatively simple structure, a number of challenges manifest themselves when considering a stereocontrolled total synthesis of mega-phone [2]. Three contiguous stereogenic carbon atoms encompass a quaternary center at the branch point of the cyclohexenone ring [3, 4]. Koga's strategy relied on two consecutive conjugate additions to the 2-ethylidene lactone **B** in generating the adjacent quaternary and tertiary centers. A bulky bis(methylthio)methyl anion reagent was added with good stereoselectivity to give an intermediate enolate, which was trapped by allyl bromide to afford lactone **C** as the major stereoisomer (see relevant mechanism, Figure 15.1). The authors attributed this highly selective sequential 1,4-addition/enolate alkylation to the effective shielding of the β-face of **B** by the trityloxymethyl group. This was based on X-ray crystallography, which showed that one of the phenyl groups of the trityl group was placed just above the double bond of the 2-ethylidene group. The sequence of reactions leading to esterification and O-methylation to furnish **D** was remarkably efficient. The mild conditions of the intramolecular aldol condensation resulted in the formation of cyclohexenone **G** as the major isomer (dr = 4:1) with concomitant epimerization of the methoxy-bearing stereocenter. It is not clear if epimerization preceded the intramolecular aldol condensation to **G** where the methoxy group could adopt a pseudoequatorial orientation.

15.2
Dihydromevinolin

A member of the mevinic acid family of fungal metabolites, dihydromevinolin exhibits potent activity as a hypocholesterolemic agent equal to other more abundant congeners. The first total synthesis of dihydromevinolin was reported by Falck and Yang [5] in 1984, and was followed by another synthesis from Hecker and Heathcock [6] in 1986.

An enantioselective total synthesis of dihydromevinolin was reported in 1990 by Hanessian and coworkers [7a] (Figure 15.2). Readily available lactone **A** [8] was converted to the 2-C-methyl analogue **B**, and then subjected to the action of lithium dimethyl phosphonoacetate to give the acyclic β-ketophosphonate **C**. Treatment with 3-phenylseleno butyraldehyde, followed by oxidative elimination furnished the diene **D** in excellent overall yield. Stereoselective reduction of the ketone, protection as the MOM ether, and desilylation gave the diol **E**, which was converted to the terminal epoxide **F**. Epoxide opening was accomplished upon addition of the lithium anion of phenylseleno acetic acid [9], followed by lactonization, and selenoxide elimination to afford butenolide **G**. Heating at reflux temperature in o-xylene gave tricyclic lactone **H** in 65% yield. A Barton-McCombie deoxygenation [10] led to **I**, which was further elaborated to give **J**. Oxidation to the aldehyde, followed by epimerization furnished **K**. A three-step sequence involving a Henry reaction [11] with nitromethane, followed by elimination and reduction gave **L**. Esterification, after deprotection of the TBS ether, then a nitronate anion-mediated conjugate addition with enone **M** in the presence of Amberlyst A-21 [11] furnished

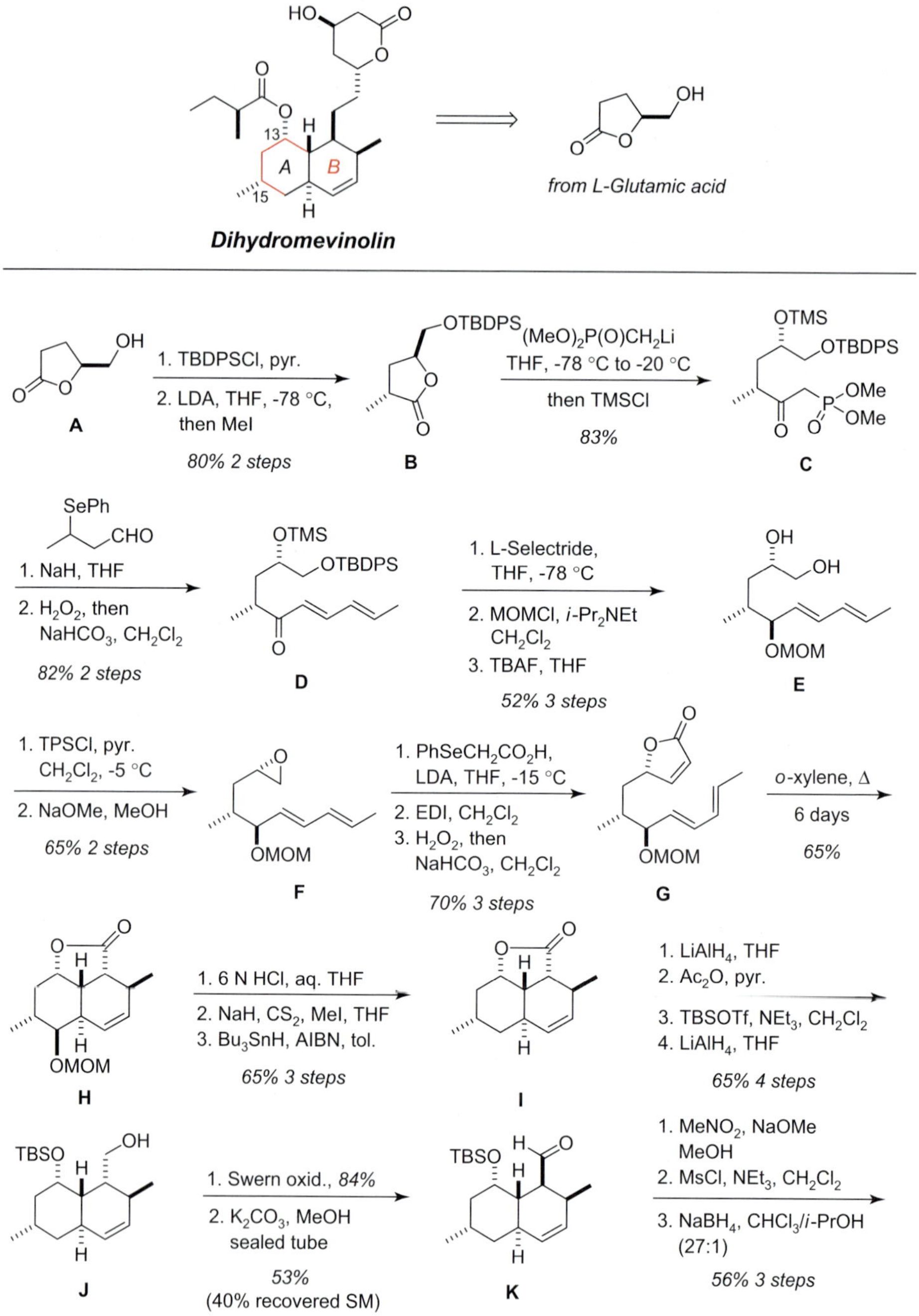

Dihydromevinolin

from L-Glutamic acid

Figure 15.2 Total synthesis of dihydromevinolin from (4S)-4-hydroxymethyl-γ-butyrolactone (hidden substructure).

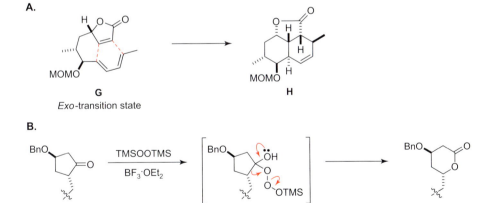

Key reactions: *Diene formation from β-ketophosphonate (**C** to **D**); epoxide to butenolide extension (**F** to **G**); intramolecular Diels-Alder reaction (**G** to **H**); nitroalkane anion conjugate addition (**L** + **M** to **N**); peroxide-mediated Baeyer-Villiger oxidation (**O-ketone** to **O-benzyl dihydromevinolin**); TPSCl = 2,4,6-triisopropylbenzenesulfonyl chloride*

Relevant mechanisms:

A.

G
Exo-transition state

H

B.

O

TMSOOTMS

BF₃·OEt₂

Figure 15.2 *(continued)*.

the adduct **N**. Reductive cleavage of the nitro group with Bu_3SnH [12], further manipulation of the ketone group, and cleavage of the TBS ether gave **O**. Swern oxidation to the ketone, followed by a bis(trimethylsilyl)peroxide-mediated [13] Baeyer-Villiger oxidation in the presence of $BF_3 \cdot Et_2O$ [14], and debenzylation, led to dihydromevinolin.

Commentary: The choice of lactone **A** as a precursor to ring *A* of dihydromevinolin is not visually obvious (see Chapter 1, section 1.8). The possible utility of this lactone as a 5-carbon chiron with matching stereogenic centers at C-13 and C-15 can be visually appreciated only at the level of intermediate **C**. Visual *reflexive* analysis of the octahydronaphthalene core structure leads one to consider an intramolecular Diels-Alder strategy to install the double bond in ring *B*. Thus, the mental jump from **B** to **C**, and from the extended diene **D** to the Diels-Alder precursor **G**, in that order, is logical. The anticipated *exo*-transition state of the Diels-Alder cyclization to give **H** was realized experimentally (see relevant mechanism A, Figure 15.2). The stereocontrolled conjugate addition of the nitronate anion generated from **L** in the presence of an anionic exchange resin proceeded with excellent selectivity setting the stage for a cyclopentanone to δ-lactone ring expansion. However, *m*-CPBA could not be used due to the presence of the double bond in ring *B*. Bis(trimethylsilyl)peroxide in the presence of $BF_3 \cdot Et_2O$ proved to be the ideal reagent to effect the Baeyer-Villiger oxidation (see relevant mechanism B, Figure 15.2). Ironically, cleavage of the *O*-benzyl ether in the presence of BCl_3 could be best achieved by close monitoring and recycling of unreacted starting material to avoid side-reactions. Nevertheless, dihydromevinolin could be isolated in crystalline form.

15.3
Mannostatin A

The stereocontrolled total synthesis of mannostatin A, an unusual aminocyclopentane triol containing a thiomethyl group, was reported in 1991 by Knapp and Dhar [15a] (Figure 15.3). D-Ribonolactone was selected as a starting material, as a six-step procedure was available to convert it into the cyclopentenol **C** from lactone **A** [16] in 47% overall yield. Treatment of **C** with *p*-methoxybenzyl isothiocyanate in the presence of NaH, followed by *S*-methylation, gave the corresponding methylthio imidoate **D**. Exposure to iodine in THF led to *trans*-iodo oxazolidinone **E** [17] in excellent yield. Removal of the *N*-PMB group gave **F**, which was treated with sodium methylthiolate in DMF to afford **G**. Cleavage of the oxazolidinone in refluxing aqueous KOH, followed by hydrolysis of the ketal group gave mannostatin A as the HCl salt in an overall yield of 32% from D-ribonolactone.

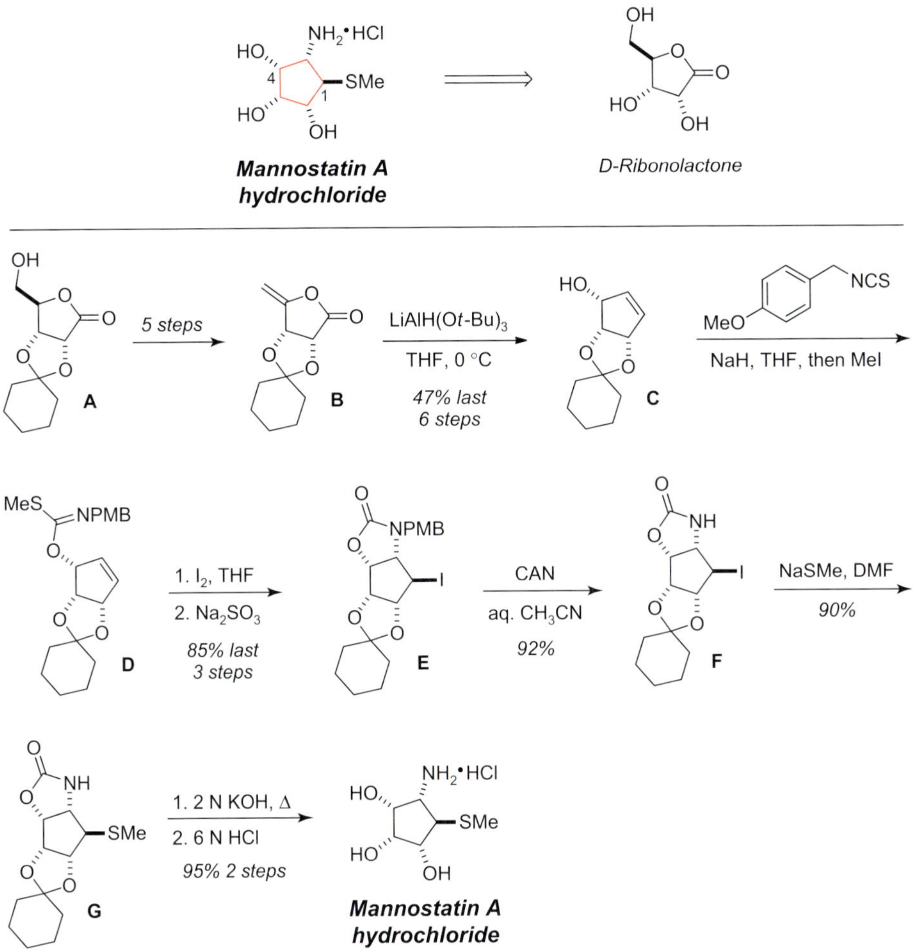

NH2•HCl
HO
4
SMe
1
HO
OH
Mannostatin A hydrochloride
OH
O
O
HO
OH
D-Ribonolactone
OH
O
O
O
O
A
5 steps
O
O
O
O
B
LiAlH(Ot-Bu)3
THF, 0 °C
47% last 6 steps
HO
O
O
C
NCS
MeO
NaH, THF, then MeI
MeS
NPMB
O
O
O
D
1. I2, THF
2. Na2SO3
85% last 3 steps
O
NPMB
O
I
O
O
E
CAN
aq. CH3CN
92%
O
NH
O
I
O
O
F
NaSMe, DMF
90%
O
NH
O
SMe
O
O
G
1. 2 N KOH, Δ
2. 6 N HCl
95% 2 steps
NH2•HCl
HO
SMe
HO
OH
Mannostatin A hydrochloride

Key reactions: *"One-pot" lactone reduction/intramolecular aldol condensation/stereoselective reduction (**B** to **C**); iodocarbamate formation (**D** to **E**); thioether formation (**F** to **G**)*

Figure 15.3 Total synthesis of mannostatin A hydrochloride from D-ribonolactone (partially hidden substructure).

Relevant mechanisms and transformations:

A.

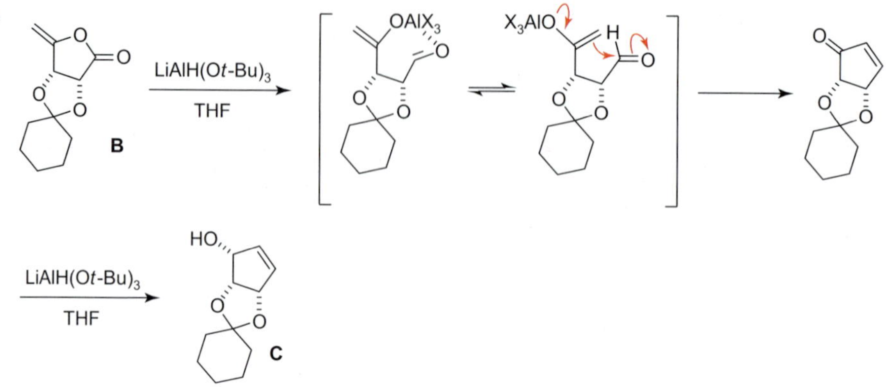

B.

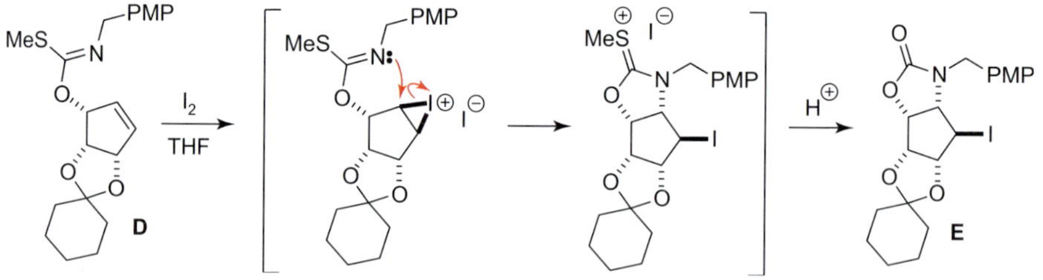

Figure 15.3 *(continued).*

Commentary: The polyhydroxy nature of mannostatin A invites the prospect of using a carbohydrate-derived starting material with a similar stereochemical pattern for the introduction of one or more of its hydroxyl groups [18]. The availability of cyclopentenol **C** in high overall yield from D-ribonolactone in six steps, encouraged Knapp and Dhar to pursue their strategy for a systematic functionalization of the endocyclic olefin. The carbocyclization of enol lactone **B** under reductive conditions, initially reported by Bélanger and Prasit [16], proceeded by reducing the lactone to a lactol, and subsequent intramolecular aldol condensation of the aluminum enolate, followed by stereoselective carbonyl reduction (see relevant mechanism A, Figure 15.3). The *trans*-iodocarbamoylation sequence (**D** to **E**) was based on previous precedents [17], securing the relative *syn*-orientation of the C-4–C-5 amino alcohol unit (see relevant mechanism B, Figure 15.3). Displacement of the iodide with sodium methylthiolate with an overall retention of configuration was rationalized by Knapp and Dhar to be the result of a possible participation of the neighboring oxazolidinone nitrogen [19]. A more plausible explanation could involve an S_N1-type reaction, where the thiolate nucleophile approaches the much less hindered *exo*-face of the tricyclic system.

In an effort to improve this synthesis of mannostatin A, Li and Fuchs [20] opted for an alternative method to introduce the nitrogen and the thiomethyl groups with the correct configuration in a single operation (Figure 15.4).

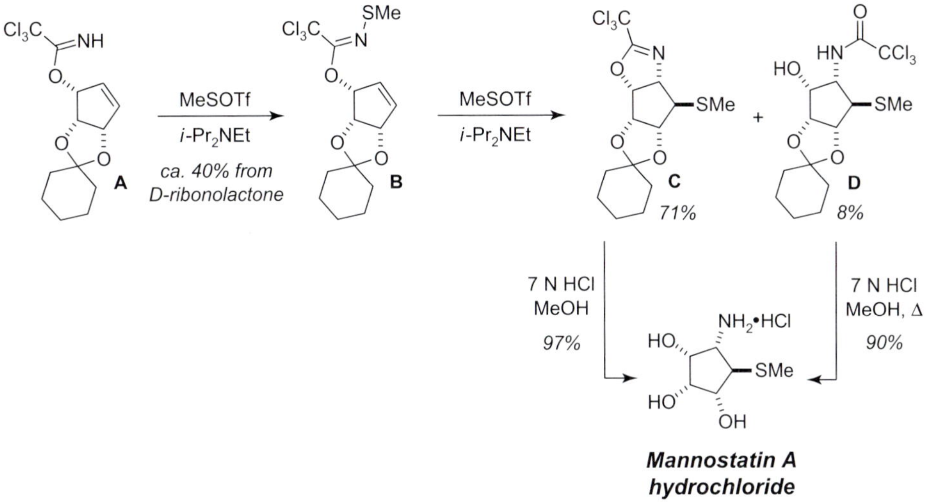

Key reactions: *Formation of N-sulfenylimidate (**A** to **B**); MeOTf-mediated intramolecular iminosulfenylation (**B** to **C**).*

Relevant mechanism:

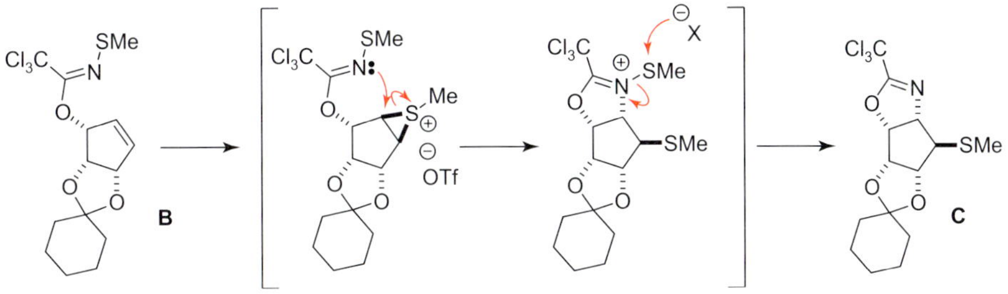

Figure 15.4 Alternative total synthesis of mannostatin A hydrochloride from D-ribonolactone.

Conversion of trichloroacetimidate **A** to the *N*-sulfenylimidate **B** was achieved with 3.3 molar equivalents of the potent thiomethylating reagent, methylsulfenyl triflate [21]. Addition of 3–4 molar equivalents of this reagent, in the presence of Hünig's base, to **B** led to the *S*-methylated 2-(trichloromethyl) oxazoline **C** in 71% yield, accompanied by the hydrolysis product **D** (8% yield). This modification led to a modest improvement in yield of mannostatin A (39% overall, 10 steps from D-ribonolactone) compared to Knapp and Dhar's synthesis [15a]. A synthesis of mannostatin A that also started with D-ribonolactone was reported in 2004 by Hu and Vasella [22]. Key steps in the enantioselective total syntheses of mannostatin A by Ganem [23], Mariano [24], and their respective groups are shown in Figures 15.5 A and B.

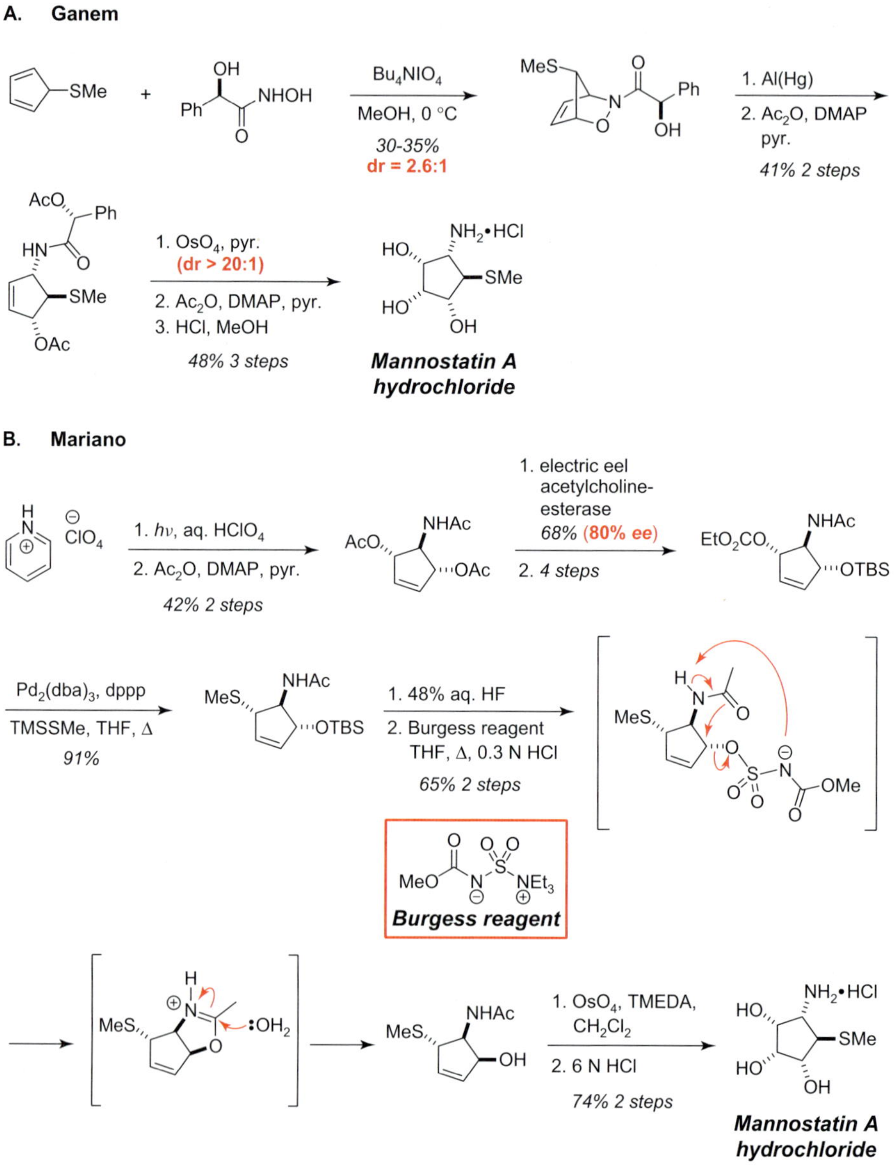

A.   Ganem

B.   Mariano

Figure 15.5 Enantioselective total syntheses of mannostatin A hydrochloride.

15.4
Furaquinocin C

The first total synthesis of furaquinocin C was reported in 1995 by Smith and collaborators [25a] (Figure 15.6). D-Ribonolactone was converted to (R)-angelicalactone, which provided an excellent starting material for a conjugate addition with the cuprate prepared from 5-iodo-2-methyl-2-pentene. The enolate of the resulting *anti*-addition product was acylated with acetyl chloride to give **A**. Introduction of the double bond was achieved by means of a selenoxide-mediated elimination to give **B**. A second cuprate addition of Me$_2$CuLi led to a 1:1.8 mixture of lactone **C** and its C-3 epimer. Formation of the TMS enol ether **D**, and the ketene acetal **E** in the presence of LDA and TMSCl, followed by treatment with quinone **F** led to the corresponding Diels-Alder product, which upon chromatography gave furaquinocin C. An identical sequence starting with the C-3 epimeric lactone led to (+)-3-*epi*-furaquinocin C. The synthesis is noteworthy for its brevity, as furaquinocin C was achieved in just six steps from (R)-angelicalactone.

Commentary: Visual analysis of furaquinocin C inspires a dual mental thought process toward a plausible synthetic approach. On the visual *relational* thinking side, the dihydrofuran harboring two vicinal alkyl substitutions including a quaternary stereogenic center, suggests a carbohydrate or related precursor, although substantial deoxygenation would be required. Smith and coworkers exploited a visual *reflexive*, reaction-based type of thinking, relying on sequential conjugate additions directed by the existing C-methyl substituent, which provided the necessary internal bias to install the quaternary stereogenic center, albeit with modest diastereoselectivity. Conjugate addition to butenolide-type precursors have ample precedents in the literature [8, 26]. The elaboration of the entire skeleton by an intermolecular regioselective Diels-Alder reaction introduces another level of disconnection in the mind's eye. Although the diastereoselectivity of the second conjugate addition reaction (**B** to **C**) was unexpectedly disappointing, the desired isomer could be separated and successfully converted to the unusual bis-TMS dienyl enol ether **E**. The total synthesis of furaquinocin C also confirmed earlier assignments of relative and absolute chemistry.

15.5
Miscellanea

Selected examples of molecules that have been synthesized starting with chiral, non-racemic lactones are shown in Figure 15.7 [27] (see also Chapter 8, section 8.11).

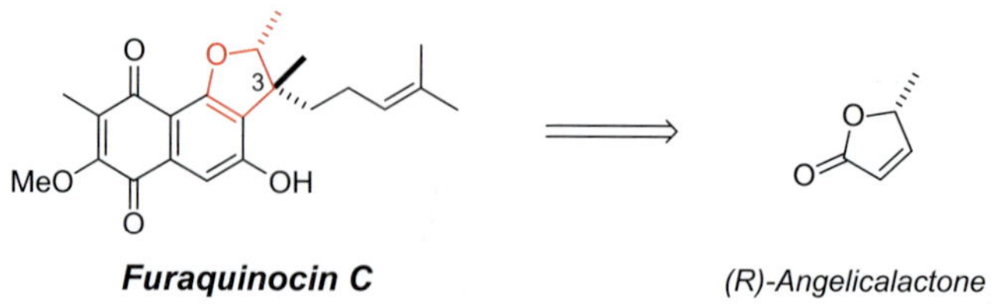

Furaquinocin C

(R)-Angelicalactone

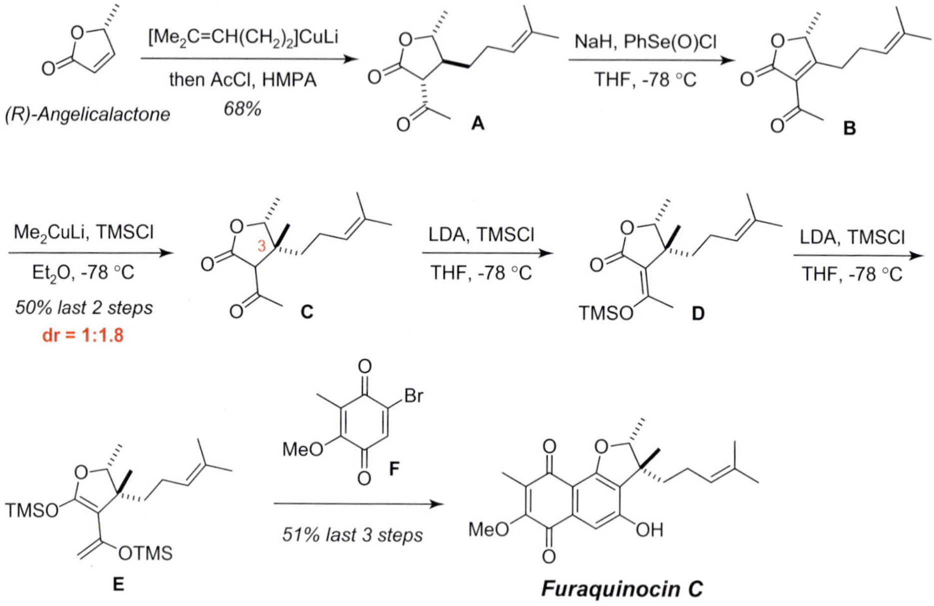

Key reactions: *Sequential cuprate addition and enolate acylation ((R)-angelicalactone to **A**); Diels-Alder reaction (**E** + **F** to **target**)*

Relevant transformation:

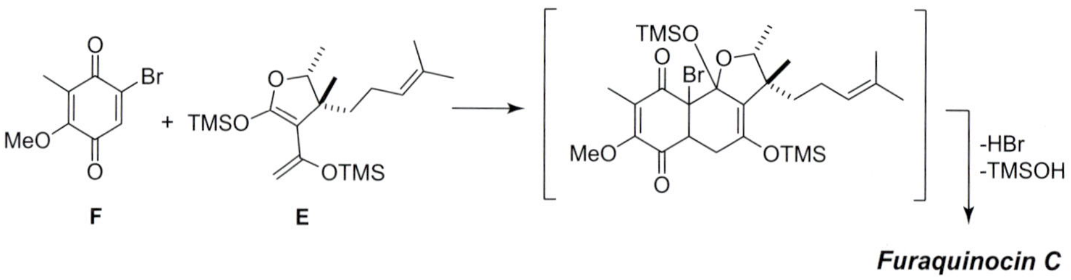

Figure 15.6 Total synthesis of furaquinocin C from (R)-angelicalactone (partially hidden substructure).

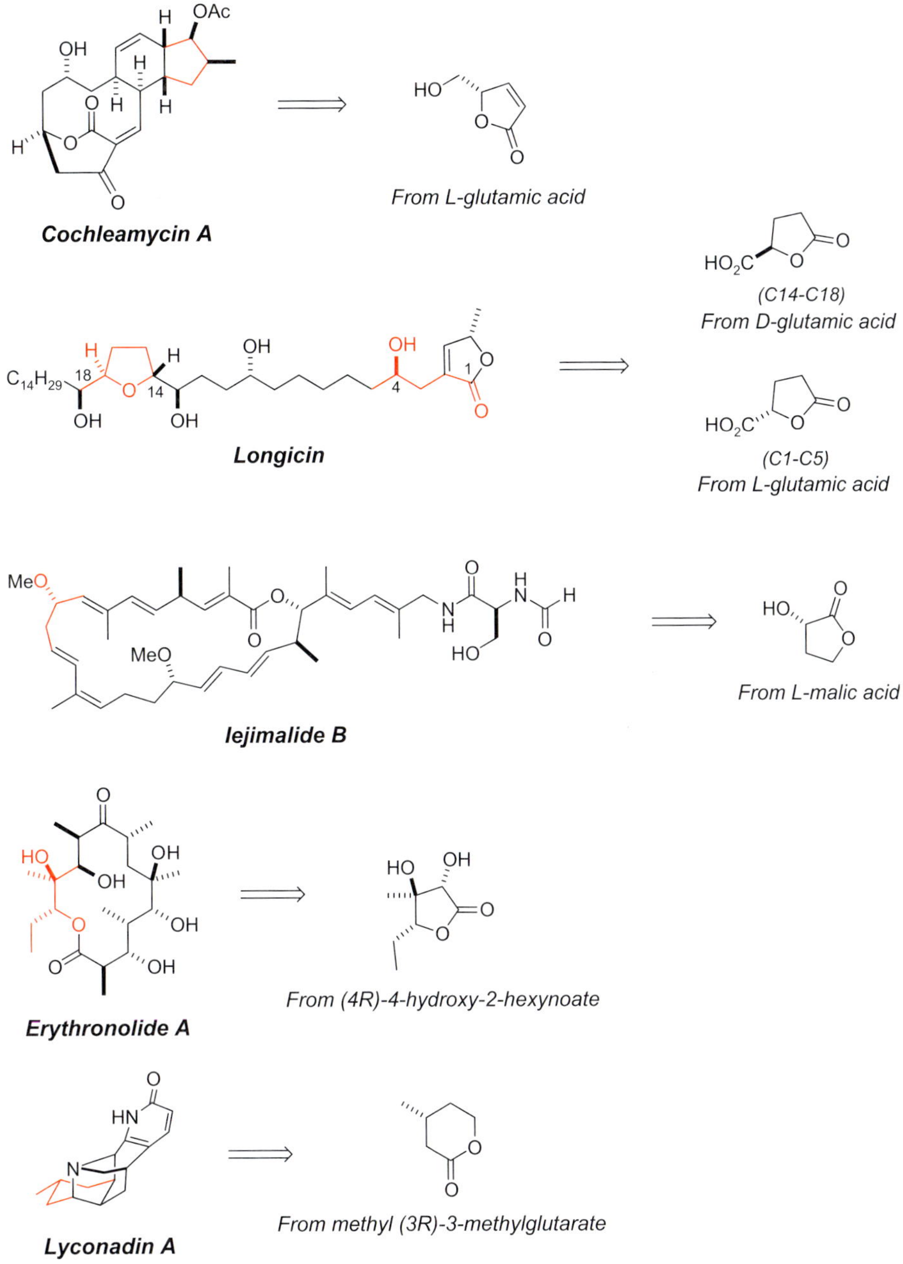

Cochleamycin A

From L-glutamic acid

Longicin

(C14-C18)
From D-glutamic acid

(C1-C5)
From L-glutamic acid

Iejimalide B

From L-malic acid

Erythronolide A

From (4R)-4-hydroxy-2-hexynoate

Lyconadin A

From methyl (3R)-3-methylglutarate

Figure 15.7 Total synthesis of selected natural products from lactone precursors.

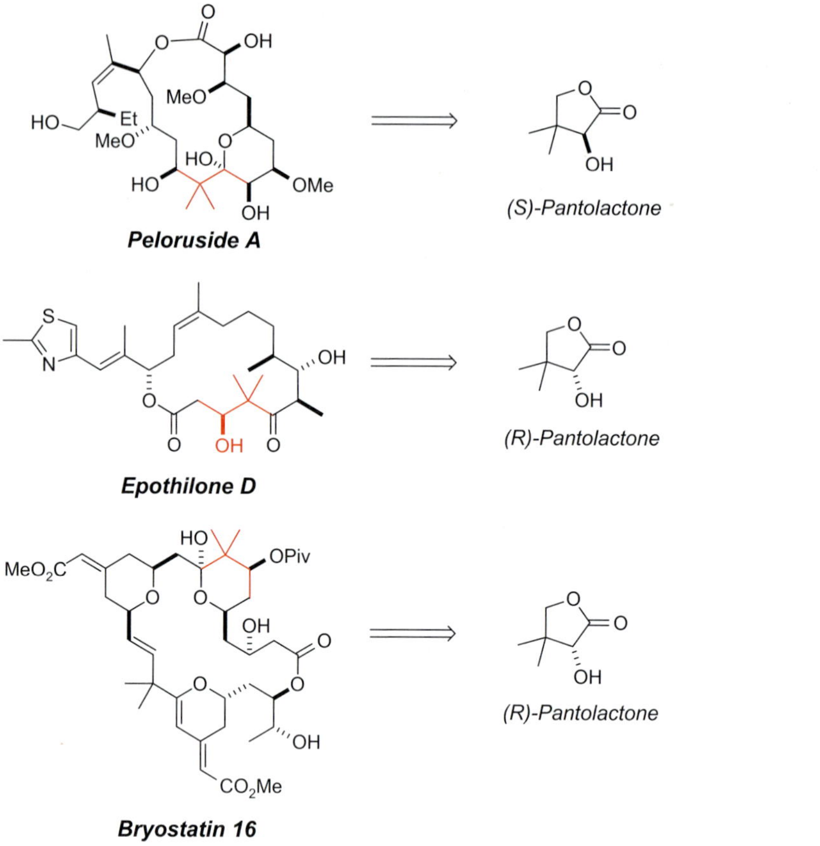

Peloruside A

(S)-Pantolactone

Epothilone D

(R)-Pantolactone

Bryostatin 16

(R)-Pantolactone

Figure 15.7 *(continued)*.

References

1. Gringore, O. H.; Rouessac, F. P. *Org. Synth.* 1984, **63**, 121.
2. (a) Tomioka, K.; Kawasaki, H.; Iitaka, Y.; Koga, K. *Tetrahedron Lett.* 1985, **26**, 903; for a synthesis of (±)-megaphone, see: (b) Büchi, G.; Chu, P.-S. *J. Am. Chem. Soc.* 1981, **103**, 2718; (c) Zoretic, P. A.; Bhaktà, C.; Khan, R. H. *Tetrahedron Lett.* 1983, **24**, 1125; (d) Hoye, T. R.; Kurth, M. J. *Tetrahedron Lett.* 1983, **24**, 4769; (e) Matsumoto, T.; Imai, S.; Usui, S.; Suetsugu, A.; Dawatsu, S.; Yamaguchi, T. *Chem. Lett.* 1984, **13**, 67.

3. For example, see: Martin, S. F. *Tetrahedron* 1980, **36**, 419 and references cited in ref. 2a.
4. For recent reviews on the construction of quaternary carbon centers by catalytic asymmetric methods, see: (a) Cozzi, P. G.; Hilgraf, R.; Zimmermann, N. *Eur. J. Org. Chem.* 2007, **72**, 5969; (b) Trost, B. M.; Jiang, C. *Synthesis* 2006, 369; (c) Christoffers, J.; Baro, A. *Adv. Synth. Catal.* 2005, **347**, 1473; (d) Douglas, C. J.; Overman, L. E. *Proc. Natl. Acad. Sci. U.S.A.* 2004, **101**, 5363;

see also: (e) Christoffers, J.; Baro, A. *Quaternary Stereocenters: Challenges and Solutions of Organic Synthesis*, Wiley-VCH, Weinheim, 2006; (f) Corey, E. J.; Guzman-Perez, A. *Angew. Chem. Int. Ed.* 1998, **37**, 388; (g) Fuji, K. *Chem. Rev.* 1993, **93**, 2037.

5. Falck, J. R.; Yang, Y.-L. *Tetrahedron Lett.* 1984, **25**, 3563.

6. Hecker, S. J.; Heathcock, C. H. *J. Am. Chem. Soc.* 1986, **108**, 4586; for a review see: Rosen, T.; Heathcock, C. H. *Tetrahedron* 1986, **42**, 4909.

7. (a) Hanessian, S.; Roy, P. J.; Petrini, M.; Hodges, P. J.; Di Fabio, R.; Carganico, G. *J. Org. Chem.* 1990, **55**, 5766; for a formal synthesis from D-glutamic acid, see: (b) Davidson, A. H.; Jones, A. J.; Floyd, C. D.; Lewis, C.; Myers, P. L. *J. Chem. Soc., Chem. Commun.* 1987, 1786.

8. Hanessian, S.; Murray, P. J.; Sahoo, S. P. *Tetrahedron Lett.* 1985, **26**, 5623.

9. Hanessian, S.; Hodges, P. J.; Murray, P. J.; Sahoo, S. P. *J. Chem. Soc., Chem. Comm.* 1986, 754.

10. Barton, D. H. R.; McCombie, S. W. *J. Chem. Soc., Perkins Trans. 1* 1975, 1574.

11. For example, see: (a) Luzzio, F. A. *Tetrahedron* 2001, **57**, 915; (b) Rosini, G.; Ballini, R.; Sorrenti, P. *Synthesis* 1983, 1014; (c) Rosini, G.; Marotta, E.; Ballini, R.; Petrini, M. *Synthesis* 1986, 237; (d) Ballini, R.; Petrini, M. *Synthesis* 1986, 1024; for a review, see: Rosini, G. in *Comprehensive Organic Synthesis*, Trost, B. M.; Fleming, I., Eds.; 1991, Vol. 2, p. 321, Pergamon, Oxford, UK.

12. For example, see: (a) Ono, N.; Fujii, M.; Kaji, A. *Synthesis* 1987, 532; (b) Otani, S.; Hashimoto, S. *Bull. Chem. Soc. Jpn.* 1987, **60**, 1825.

13. Cookson, P. G.; Davies, A. G.; Fazal, N. *J. Organomet. Chem.* 1975, **99**, C31.

14. For example, see: (a) Matsubara, S.; Takai, K.; Nozaki, H. *Bull. Chem. Soc. Jpn.* 1983, **56**, 2029; (b) Suzuki, M.; Takada, H.; Noyori, R. *J. Org. Chem.* 1982, **47**, 902.

15. (a) Knapp, S.; Murali Dhar, T. G. *J. Org. Chem.* 1991, **56**, 4096; for the synthesis of (±)-mannostatin A, see: (b) Trost, B. M.; Van Vranken, D. L. *J. Am. Chem. Soc.* 1991, **113**, 6317.

16. Bélanger, P.; Prasit, P. *Tetrahedron Lett.* 1988, **29**, 5521.

17. Knapp, S.; Patel, D. V. *J. Org. Chem.* 1984, **49**, 5072.

18. For a review on mannostatins and related natural aminopentitols, see: Berecibar, A.; Grandjean, C.; Siriwardena, A. *Chem. Rev.* 1999, **99**, 779.

19. Knapp, S.; Levorse, A. T. *J. Org. Chem.* 1988, **53**, 4006.

20. Li, C.; Fuchs, P. L. *Tetrahedron Lett.* 1994, **35**, 5121.

21. Effenberger, F.; Russ, W. *Chem. Ber.* 1982, **115**, 3719.

22. Hu, G.; Vasella, A. *Helv. Chim. Acta* 2004, **87**, 2405; for other syntheses of mannostatin A, see: Ogawa, S.; Yuming, Y. *Bioorg. Med. Chem.* 1995, **3**, 939; Trost, B. M.; Van Vranken, D. L. *J. Am. Chem. Soc.* 1991, **113**, 6317.

23. King, S. B.; Ganem, B. *J. Am. Chem. Soc.* 1991, **113**, 5089; see also: King, S. B.; Ganem, B. *J. Am. Chem. Soc.* 1994, **116**, 562.

24. Ling, R.; Mariano, P. S. *J. Org. Chem.* 1998, **63**, 6072; see also: Cho, S. J.; Ling, R.; Kim, A.; Mariano, P. S. *J. Org. Chem.* 2000, **65**, 1574.

25. (a) Smith, A. B. III; Sestelo, J. P.; Dormer, P. G. *J. Am. Chem. Soc.* 1995, **117**, 10755; for the total synthesis of furaquinocins A, B, D, and H, see: (b) Saito, T.; Morimoto, M.; Akiwama, C.; Matsumoto, I.; Suzuki, K. *J. Am. Chem. Soc.* 1998, **120**, 11633; for total syntheses of furaquinocins A, B, and E, see: (c) Trost, B. M.; Thiel, O. R.; Tsui, H.-C. *J. Am. Chem. Soc.* 2003, **1254**, 13155.

26. For example, see: (a) Takano, S.; Shimazaki, Y.; Sekiguchi, Y.; Ogasawara, K. *Chem. Lett.* 1988, **17**, 2041; (b) Hanessian, S.; Murray, P. J.; Sahoo, S. P. *Tetrahedron Lett.* 1985, **26**, 5627; (c) Seebach, D.; Knochel, P. *Helv. Chim. Acta* 1984, **67**, 261; see also: Lipshutz, B.; Sengupta, S. *Org. React.* 1992, **41**, 135.

27. Cochleamycin A: Tatsuta, K.; Narazaki, F.; Kashiki, N.; Yamamoto, J.-I.; Nakano, S. *J. Antibiot.* 2003, **56**, 584; longicin: Hanessian, S.; Giroux, S.; Buffat, M.

Org. Lett. 2005, **7**, 3989;   iejimalide B: Fürstner, A.; Nevado, C.; Tremblay, M.; Chevrier, C.; Teplý, F.; Aïssa, C.; Waser, M. *Angew. Chem. Int. Ed.* 2006, **45**, 5837; peloruside A: Evans, D. A.; Welch, D. S.; Speed, A. W. H.; Moniz, G. A.; Reichert, A.; Ho, S. *J. Am. Chem. Soc.* 2009, **131**, 3840;   epothilone D: Martin, N.; Thomas, E. J. *Tetrahedron Lett.* 2001, **42**, 8373;   erythronolide A: Stork, G.; Rychnovsky, S. D. *J. Am. Chem. Soc.* 1987, **109**, 1564;   Chamberlin, A. R.; Dezube, M.; Reich, S. H.; Sall, D. J. *J. Am. Chem. Soc.* 1989, **111**, 6247;   lyconadin A: Beshore, D. C.; Smith, A. B. III. *J. Am. Chem. Soc.* 2007, **129**, 4148; bryostatin 16: Trost, B. M.; Dong, G. *Nature* 2008, **456**, 485;   Trost, B. M.; Dong, G. *J. Am. Chem. Soc.* 2010, **132**, 16403;   see also: Trost, B. M.; Yang, H.; Thiel, O. R.; Frontier, A. J.; Brindle, C. S. *J. Am. Chem. Soc.* 2007, **129**, 2206.

16
Single Target Molecule-oriented Synthesis

As discussed in several parts of this book, natural products, varying in their structures and biological activities, have provided synthetic chemists with the incentives, and at times, the obsession to initiate and pursue exciting research programs. In many cases, these endeavours have involved years of dedicated and relentless efforts in the pursuit of specific target molecules. This time-honored practice of "total synthesis" has been one of the hallmarks of the discipline of organic chemistry over the years. Each decade has brought with it an "era" of molecular entities belonging to a biogenetically related class of natural products such as sesquiterpenes, anthracyclines, macrolides, and alkaloids to name a few. A brief survey of the annals of total synthesis in the last 50 years reveals periods during which, the "rush" toward the conquest of a given target molecule was intense. For example, there have been over a dozen individual total syntheses of molecules such as coriolin, estrone, mesembrine, vindoline, and quadrone to name a few in a period spanning 1965–1985 [1]. Although the majority of these molecules were synthesized as racemic mixtures, much innovation was introduced in devising diverse strategies of planning and execution. More recently, the challenges of total synthesis of enantiopure molecules have been admirably met in practically every class of natural product, especially with the advent of modern methodology, high field spectroscopy, and sophisticated separation techniques.

16.1
Synchronicity

The well-known expression "timing is everything" applies to many facets of our everyday lives. Arriving in "the nick of time," striking the ball "with well-timed precision," or accomplishing a task in a "timely fashion" are all too familiar events that we experience very frequently depending on the circumstances. "Timing" has its place in our lives as synthetic chemists, and it is manifested in a number of total syntheses, some of which will be briefly discussed for anecdotal reasons.

Upon completion of a total synthesis, it is customary to "write-it-up" in one of several formats depending on the nature of the work, its relative importance, presumed priority, and expected eventual impact. Following the usual editorial

Design and Strategy in Organic Synthesis: From the Chiron Approach to Catalysis, First Edition.
Stephen Hanessian, Simon Giroux, and Bradley L. Merner.
© 2013 Wiley-VCH Verlag GmbH & Co. KGaA. Published 2013 by Wiley-VCH Verlag GmbH & Co. KGaA.

procedures, the paper is published and read by those interested, admirers, and competitors alike. In academia, the objective is to publish scientific results in top-level journals which has a number of redeeming consequences. Being first to announce the total synthesis of a much publicized molecule for its therapeutic potential or for its architectural complexity for example, is the expected norm (see also Chapter 18, section 18.6.1). In industry, being first carries an overwhelmingly important weight, particularly in the case of a projected lucrative marketable drug. Priority dates on patent applications are of utmost importance and may be the difference between acquiring the rights to a product or not, with obvious mega-dollar consequences. Although the stakes are not quite as high in academia, priority in announcing the completion of a synthesis is a fiercely competitive endeavor.

Secrecy is not uncommon in academic circles when it comes to divulging the identity of a target molecule whose synthesis is in progress. At times, senior investigators become aware of each other's cachets, and they may intentionally open up the channels of communication. The friendly rivalry continues, while coworkers feel the pressure, until the results are communicated. However, under different circumstances the rivalry could turn into a race, at times shrouded with controversy. First to appear in print, while being exhilarating to one group, may be devastating to another, if only temporarily. Oftentimes, the "scoop" factor has an energizing effect that brings out the best ideas to top what was just published by another group. Rather than being a curse, time becomes the blessing of laboratory life. In actual fact, collegiality and mutual respect among peers are seldom violated. In this regard, Kipling's quote may be *à propos*:

> *"If you can meet with Triumph and Disaster, and treat those two imposters just the same …"*

16.2
Joining Forces

The term "academic freedom" has been widely known, and is usually referred to in the context of personal choices and rights in an institutional environment. It can also be loosely applied to the practice of total synthesis. In Chapters 3 and 4 we intimated the "Sinatra" (*I did it my way*), and the "Nike" (*Just do it*) approaches to synthesis [2]. The individualistic nature of the synthetic chemist compels him or her to seek new and creative ways to achieve their objectives. We are accustomed to seeing papers published by authors from one and the same institution, especially when total syntheses are concerned. A notable exception coincides with one of the most impressive accomplishments in the annals of natural products synthesis of the 20th century. The impact of this work is all the more significant if one considers the period of time in which it was achieved. Even more remarkable is the fact that two of the world's most eminent synthetic organic chemists decided to join their forces in order to conquer the total synthesis of arguably one of the most complex natural products of their era, namely vitamin B_{12} (Figure 16.1). The

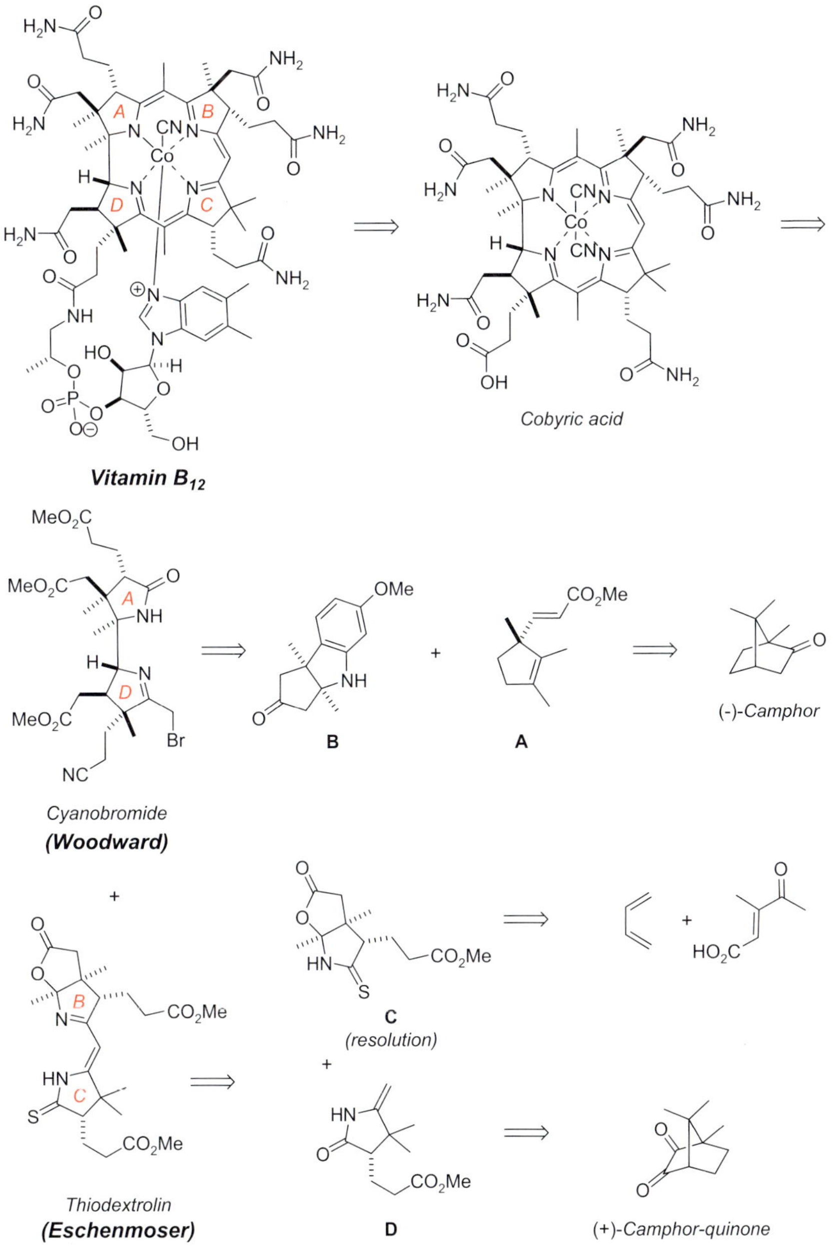

Figure 16.1 Woodward-Eschenmoser's retrosynthetic analysis of cobyric acid and vitamin B_{12}.

Woodward-Eschenmoser synthesis of this highly elaborate molecule still stands today as a milestone of ingenuity, intuitive reasoning, and especially courage. Although each group had devised and executed strategically different synthetic steps to advanced intermediates, the decision to join forces at a later stage of the assembly process, was a noble act and a testimonial to collaborative research for the benefit of science [3, 4]. With time, the alternative, "two independent camps," approach on both sides of the Atlantic Ocean would have no doubt been successful. However, the expeditious conclusion of the synthesis of cobyric acid, an immediate precursor of vitamin B_{12}, was a historical example of scholarly collaboration and mutual respect between two giants of the field. A detailed description of the Woodward-Eschenmoser synthesis is given in the primary literature, and in an exemplary monograph on the subject of total synthesis by Nicolaou and Sorensen [1]. Here, we simply highlight salient features of the synthesis of vitamin B_{12} in which, not coincidentally, terpenes were used as starting materials for segments of the target molecule. The conversion of cobyric acid to vitamin B_{12} had already been shown by Bernhauer in 1960 [5]. Once again, the impact of the achievement, announced jointly in 1973 following separate publications from each group, should be appreciated in light of the available methods of analysis and characterization of complex structures such as cobyric acid and its precursors. This monumental feat also stands as a historic example of total synthesis following structure elucidation. It should be recalled that the X-ray crystal structure of vitamin B_{12} was reported by Dorothy Crowfoot-Hodgkin [6] in 1956, yet another landmark achievement after her seminal work on the structure of penicillin [7].

Important and fundamental contributions were advanced in this joint synthetic odyssey toward vitamin B_{12}. The foundations of the Woodward-Hoffmann rules on the conservation of orbital symmetry, biogenetic hypotheses concerning the origin of life, and ingenious synthetic methods are lessons to be treasured by all [3, 4].

16.3
Back-to-back Publishing

Next to "joining forces," back-to-back publishing of papers, especially with the mutual knowledge of the authors, is the most amiable, and at times the fairest way to divulge recent results from two different laboratories. Below we present a few examples of back-to-back publications spanning the period 1967–2007, highlighting key reactions and/or delineating retrosynthetic disconnections depending on the particular synthesis.

16.3.1
Veratramine (1967)

Total syntheses of veratramine were published simultaneously in 1967 as back-to-back Communications to the Editor of the *Journal of the American Chemical Society* (Figure 16.2). The Masamune [8] strategy was to react the corresponding pyrrolidine enamine obtained from (3S)-3-methyl-5-piperidone **B** with bromide **A**, available from a member of the jervine group of steroidal alkaloids via degradation and chemical manipulation, to give **C**. Separation of isomers, and further elaboration led to 11-deoxojervine (**D**), which had been previously converted to veratramine by aromatization of the D ring. In the Johnson [9] approach, a degradation product **A'**, prepared from hecogenin was engaged in a Strecker reaction with KCN and *t*-butyl-4-amino-(3S)-3-methylbutyrate (**B'**). In both cases, advanced tetracyclic intermediates were used as starting materials. The two papers were received within 12 days of each other in the Editor's office.

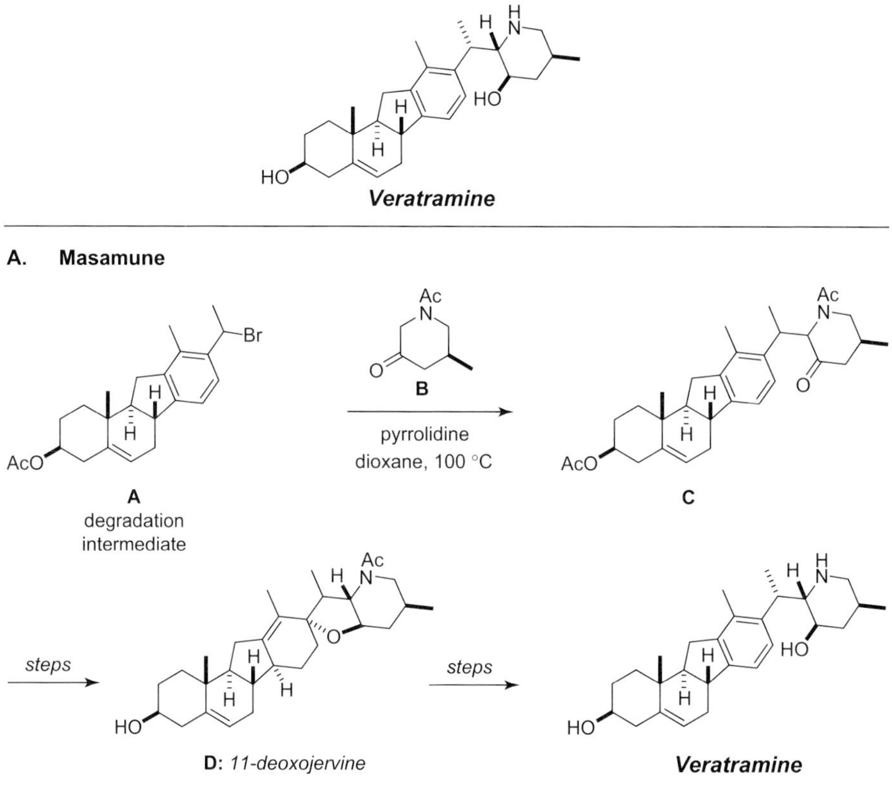

Veratramine

A. Masamune

Figure 16.2 Synthesis of veratramine by the Masamune and Johnson groups.

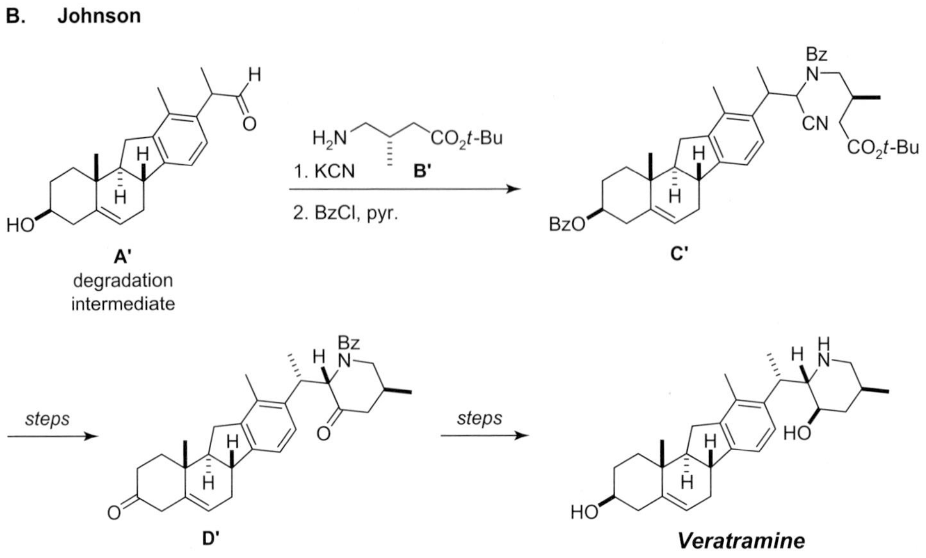

B.   Johnson

Figure 16.2   (continued).

16.3.2
(±)-Lycopodine (1968)

Back-to-back Communications with the same "received" dates describing the first syntheses of (±)-lycopodine, were published independently by Stork [10] and Ayer [11] (Figure 16.3). Key steps in the Stork synthesis relied on intermediate **B**, prepared by pyrrolidine enamine alkylation of a racemic 3-methyl cyclohexanone derivative **A**. Subsequent transannular cyclization under acidic conditions led to the tetracyclic bridged intermediate **C**, which was further manipulated by cleavage of the electron rich aromatic ring to give the keto ester **D**. Formation of the lactam, reduction of the lactam carbonyl and ketone, followed by re-oxidation of the resulting secondary alcohol gave crystalline (±)-lycopodine. The Ayer synthesis effected a Grignard addition on an iminium ion **A′** to give (±)-**B′**. Conversion to the mesylate, and subsequent cyclization of the potassium enolate led, eventually, to the natural relay intermediate **C′**, and then to lycopodine in racemic form.

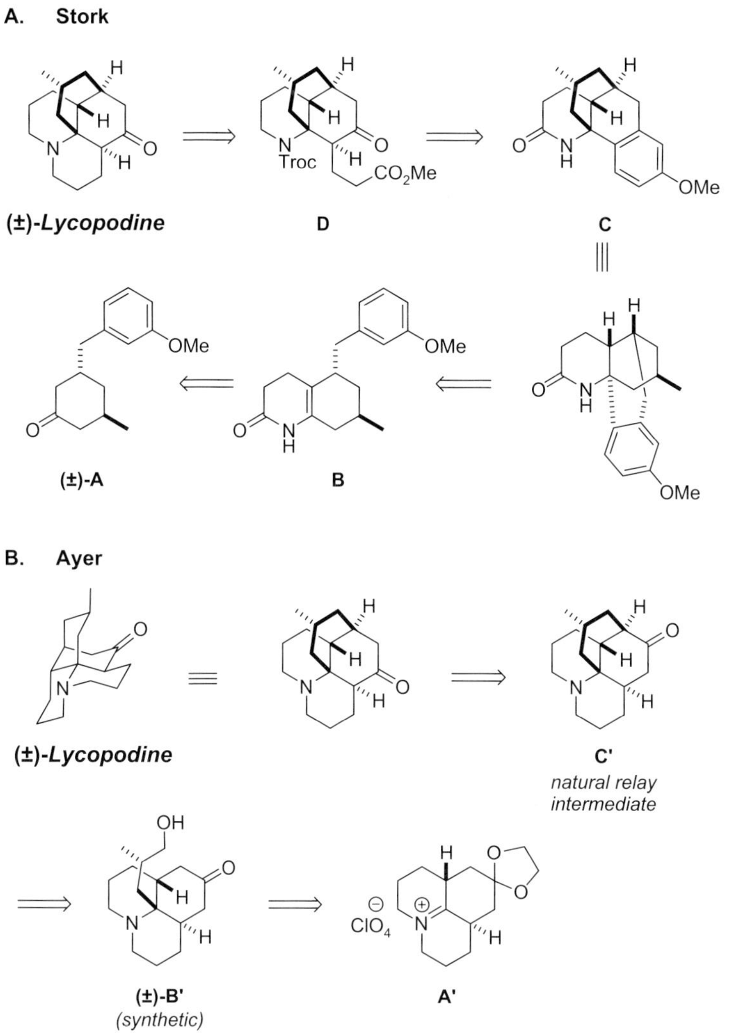

A. Stork

(±)-Lycopodine   D   C

(±)-A   B   OMe

B. Ayer

(±)-Lycopodine   C'
                                                                  natural relay
                                                                  intermediate

(±)-B'   A'
(synthetic)

Figure 16.3 Retrosynthetic analysis of (±)-lycopodine by the Stork and Ayer groups.

16.3.3
Ionomycin (1990)

The first total synthesis of ionomycin was completed by Evans and coworkers [12] during the period 1983–1985. Concurrent studies by Hanessian and coworkers [13] led to the publication of full papers, in sequence, by both groups in 1990 (Figure 16.4) [14].

Evans relied on an elegant chiral auxiliary-based approach for enolate alkylations, aldol condensations, and on directed catalytic hydrogenations to construct relevant substructures (**A-F**), harboring contiguous propionate-type triads, and alternating

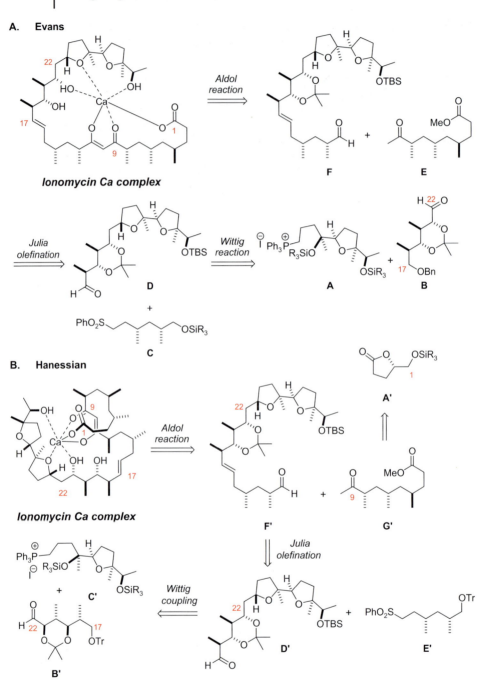

Figure 16.4 Retrosynthetic analysis of ionomycin by the Evans and Hanessian groups.

C-methyl groups (Figure 16.4). These were assembled employing Wittig (**A** and **B**) and Julia (**C** and **D**) olefinations, and an aldol reaction (**E** and **F**) *en route* to ionomycin.

In the Hanessian synthesis, a (4*S*)-4-(hydroxymethyl)-γ-butyrolactone derivative **A**′ was used as a chiral template that would eventually provide the C-1–C-22 acyclic segment of ionomycin (Figure 16.4). Stereocontrolled introduction of *C*-methyl groups was done to access **B**′, **E**′, and **G**′ utilizing a series of enolate alkylations and hydroxylations starting with **A**′, which is readily available from ʟ-glutamic acid. Geraniol was used to build the distal tetrahydrofuran motif **C**′. The final assembly relied on an aldol reaction (**F**′ and **G**′), and a Julia olefination (**D**′ and **E**′) as in the Evans synthesis. The full force of the *chiron approach*, recognizing the utility of a single amino acid as a precursor to a major portion of the C-1–C-22 acyclic chain, was a tactically viable strategy (see also Chapter 1, section 1.7).

16.3.4
Vancomycin aglycone (1998)

A classic example of back-to-back publications of complex natural products is exemplified in the five consecutive articles shared by the Evans [15] and Nicolaou [16] groups describing the total syntheses of vancomycin aglycone in 1998 (Figure 16.5 A, B). In addition to the couplings of appropriate amino acids that would provide the peptide backbone of vancomycin aglycone, a major obstacle had to be contended with in the assembly of the three macrocyclic ring systems in order to obtain the desired atropoisomers. Although both groups independently arrived at very similar penultimate precursors prior to affecting the ring *D/E* biaryl ether formation, the tactical steps for both syntheses were quite different from the outset. Cognizant of the atropoisomerism issue in the assembly step, Evans [15] first prepared the macrolactam **B** by oxidative coupling of the *A/B* rings in intermediate **A** (Figure 16.5 A). With the orientation of the *A/B* rings in place, the Evans group successfully formed the *C/D* biaryl ether linkage by a S_NAr process to give **C**. The placement of a nitro group in ring *C* was crucial for the adoption of the correct spatial orientation whereby a 5:1 ratio of the desired and undesired atropoisomers respectively was secured. Coupling with a preformed tripeptide **D** led to **E**, harboring the entire peptide sequence present in vancomycin. A second S_NAr substitution, followed by further manipulation, led ultimately to vancomycin aglycone. Inevitably, in such a complex polyfunctional system, the choices (and changes) in the protective groups were of primordial importance.

Starting with the preformed *A/B* biphenyl linked intermediate **A**′, Nicolaou and coworkers [16] chose a Cu-mediated protocol for biaryl ether formation. Unfortunately, they were only able to secure a 1:1 ratio of *C/D* biphenyl ether atropoisomers, which had to be separated. Macrolactam formation led to **B**′, which was coupled with the preformed tripeptide **C**′ to give **D**′. Cu-mediated biaryl ether formation afforded a 1:3 ratio of atropoisomers in favor of the undesired isomer. This could be thermally recycled to a 1:1 mixture of *D/E* biaryl ether atropoisomers, and separated. Deprotection, and adjustment of oxidation states gave vancomycin

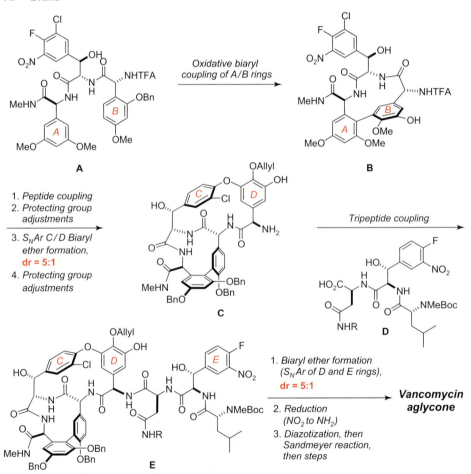

Figure 16.5 Highlights from the total syntheses of vancomycin aglycon by the Evans and Nicolaou groups.

B. Nicolaou

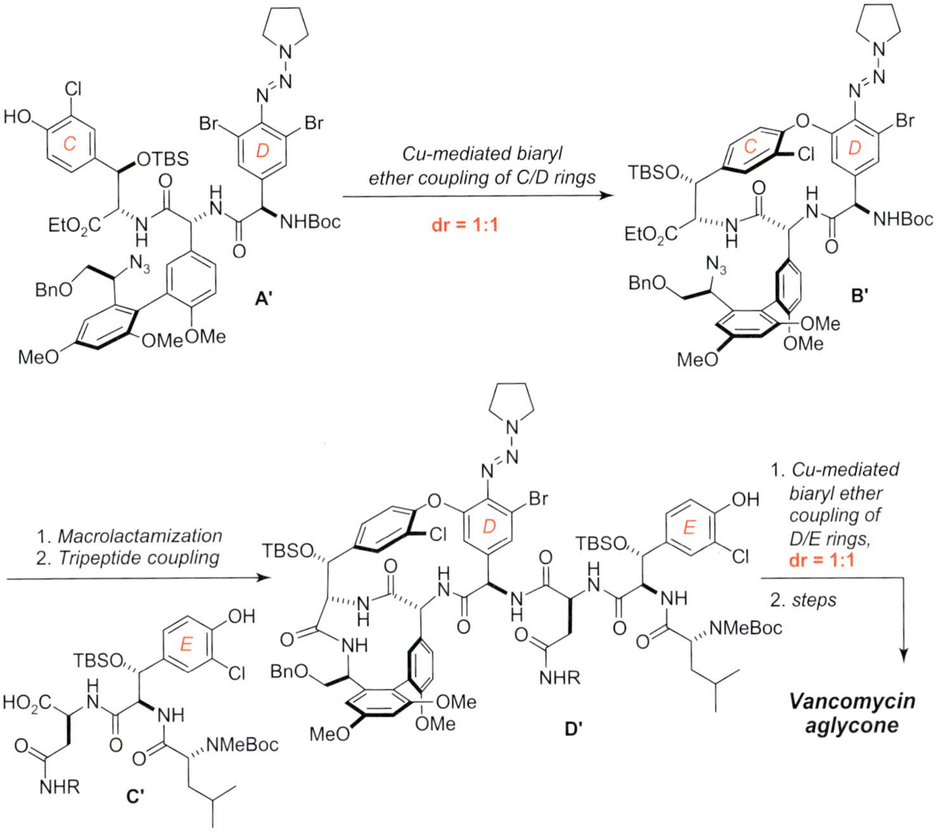

Figure 16.5 *(continued)*.

aglycone. Nicolaou used the triazene motif attached to ring *D* throughout the synthesis as a masked amine, which was eventually converted to a phenol by an indirect 4-step route. Subsequently, the aglycone was converted to vancomycin [17].

It is noteworthy that this important synthetic feat by two of the most acknowledged and masterful molecule builders was made public in one and the same issue of the Journal [18].

16.4
Same Year Publications

There are a larger number of publications announcing the total synthesis of one and the same natural product within the same year compared to back-to-back types. Occasionally, the strategies may be similar, but by and large, each synthesis may embody elements of novelty and individuality that often mirrors the preferences and biases of the respective groups. Selected examples are highlighted with an emphasis on the general strategies and particular starting materials adopting a retrosynthetic analysis format whenever appropriate.

16.4.1
(±)-Colchicine (1959)

The first total synthesis of (±)-colchicine was independently achieved by van Tamelen [19] and Eschenmoser [20] in 1959 (Figure 16.6). Each group utilized

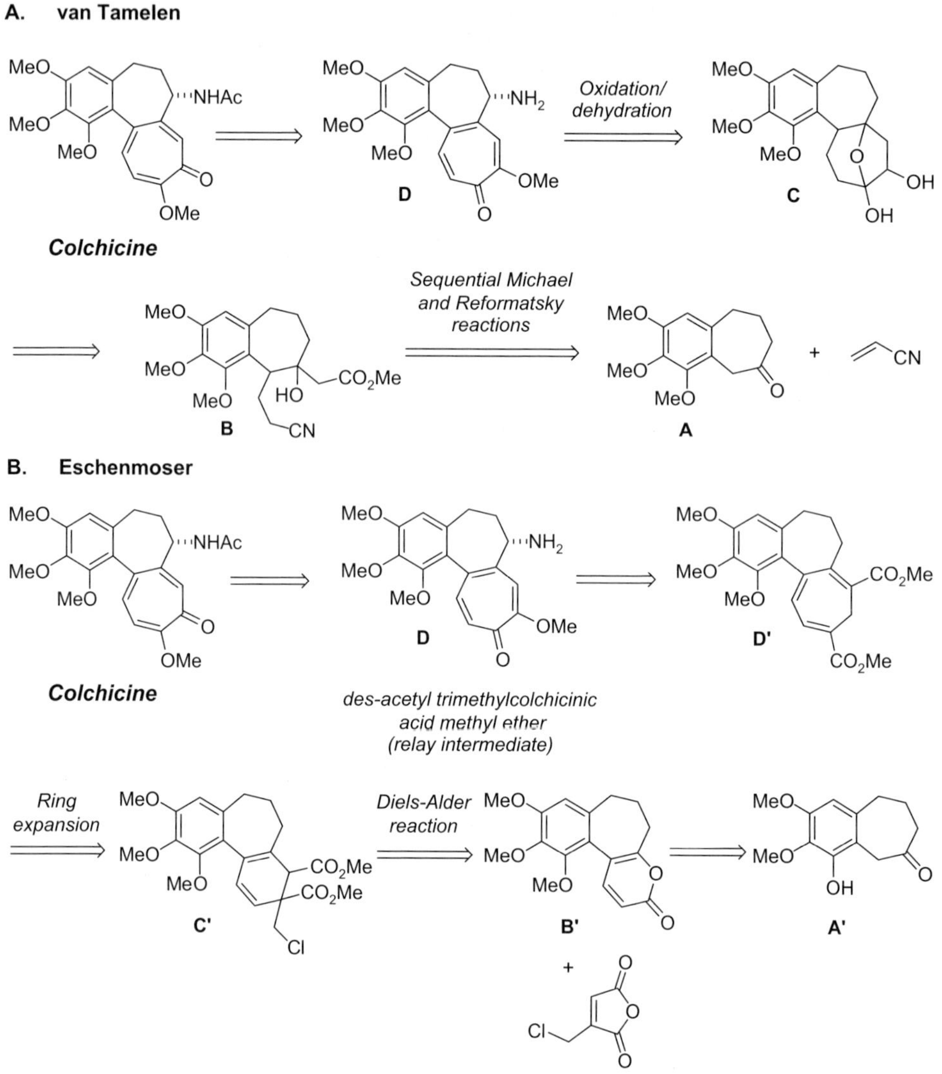

A.   van Tamelen

B.   Eschenmoser

Figure 16.6   Retrosynthetic analyses of (±)-colchicine by the van Tamelen and Eschenmoser groups.

trimethoxybenzosuberone **A** as a starting material, exploiting enolate anion conjugate addition to introduce appropriate branches that would eventually become the tropolone ring. Thus, conjugate addition to acrylonitrile, followed by a Reformatsky reaction and cyclization (**A** to **C**), allowed van Tamelen to reach des-acetyl trimethylcolchicinic acid methyl ether (**D**), which was also obtained from the natural product and used as a relay compound. Bromination of **D** with NBS, displacement with azide, reduction, N-acylation, and adjustment of oxidation states led to colchicine. The Eschenmoser strategy started from dimethyl hydroxybenzosuberone **A'**, which was converted to the pyranone **B'**. Extrusion of CO_2 in a Diels-Alder type transformation led to **C'**, which was ring-expanded to **D'**, and then converted to des-acetyl trimethylcolchicinic acid methyl ether. This intermediate was easily transformed to (±)-colchicine upon equilibration with acid, followed by N-acetylation. Both groups went on to publish details of their preliminary announcements in 1961. In 1965, Woodward reported his synthesis of colchicine [21]. A 50-year retrospective on the total synthesis of colchicine documents recent accomplishments [22].

16.4.2
(±)-Catharanthine (1970)

The first total synthesis of (±)-catharanthine was disclosed in 1970 by Büchi and coworkers [23] as a full paper. Kutney and Bylsma [24] reported a synthesis later in the same year starting with dihydrocatharanthine (Figure 16.7). Büchi utilized

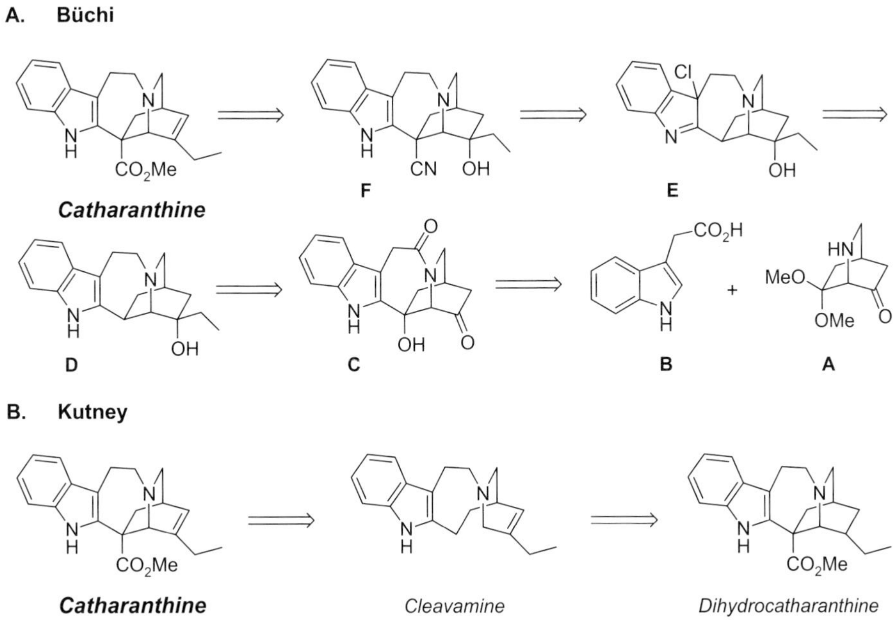

A.   Büchi

Catharanthine

B.   Kutney

Catharanthine   *Cleavamine*   *Dihydrocatharanthine*

Figure 16.7 Retrosynthetic analysis of (±)-catharanthine by the Büchi and Kutney groups.

the racemic mixture prepared from an isoquinuclidine precursor **A** as a starting material, which was converted to **C** after condensation with indole 3-acetic acid (**B**). Acid-catalyzed cyclization, followed by systematic manipulation of functional groups produced intermediates **C** then **D**, which was converted to 3-chloro indolenine **E** with *t*-butyl hypochlorite. Treatment with KCN in dimethylacetamide led to the nitrile **F**, which upon dehydration, hydrolysis and esterification, led to (±)-catharanthine.

The Kutney synthesis relied on a relay approach in which dihydrocatharanthine was elaborated into a variety of useful intermediates including cleavamine. Formation of the quaternary ammonium salt, followed by treatment with KCN and hydrolysis gave the methyl ester, which underwent a known transannular cyclization to give (±)-catharanthine.

16.4.3
(±)-Cephalotaxine (1972)

The first total syntheses of (±)-cephalotaxine were published within two months of each other in 1972 by Weinreb [25] and Semmelhack [26] independently (Figure 16.8). Weinreb started with racemic prolinol **A**, which was condensed with 3,4-methylenedioxy phenyl acetyl chloride, then cyclized to the enamide **B**. Subsequent transformations led to the diketoenamine **C**, which was converted to desmethylcephalotaxine derivative **D**. Acid-catalyzed methanolysis gave (±)-cephalotaxine.

Parallel studies in the Semmelhack group utilized pyrrolidine **A'** prepared from 2-ethoxy-pyrroline and allylmagnesium bromide. Elaboration into the diester **B'** and acyloin condensation led to **C'**, which was coupled with the mesylate derived from 3-chloropiperonyl alcohol to give **D'**. Intramolecular cyclization of the enolate via a benzyne intermediate led to (±)-cephalotaxinone, which, upon reduction with LiAlH$_4$, gave (±)-cephalotaxine.

In 1975 Weinreb and Semmelhack [27] published a joint account of their respective syntheses, as well as of other members of the Cephalotaxus family of alkaloids [28].

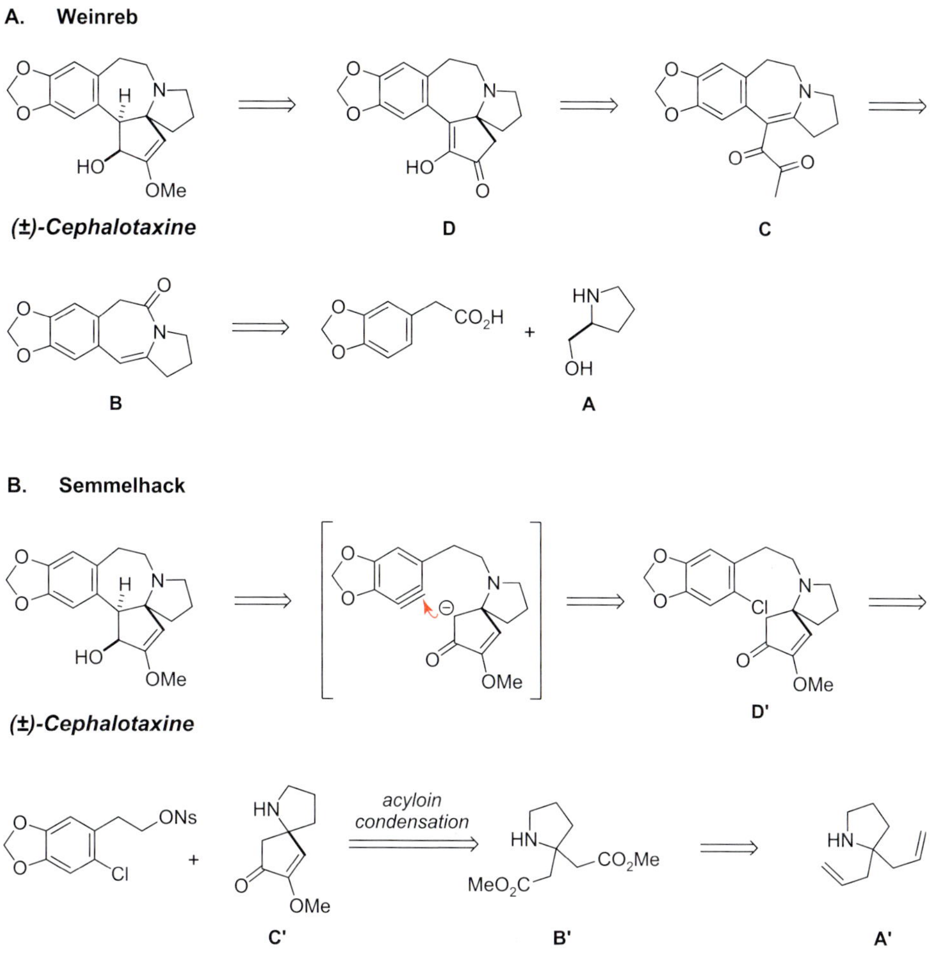

A.   Weinreb

(±)-Cephalotaxine   D   C

B   A

B.   Semmelhack

(±)-Cephalotaxine   D'

acyloin
condensation

C'   B'   A'

Figure 16.8 Retrosynthetic analysis of (±)-cephalotaxine by the Weinreb and Semmelhack groups.

16.4.4
Bleomycin A$_2$ (1982)

The total synthesis of bleomycin A$_2$, some 10 years after its isolation and subsequent structure elucidation, provided definitive confirmation of its structure, which had been revised in 1978 (Figure 16.9). Intensive studies in the groups of Umezawa [29] in Japan, and Hecht [30] in the United States spanned over several years. Each group developed methods for the synthesis of segments of bleomycin A$_2$ and its aglycone. Major challenges toward this goal involved orthogonal protecting groups, stereocontrolled glycosidations, and peptide coupling steps.

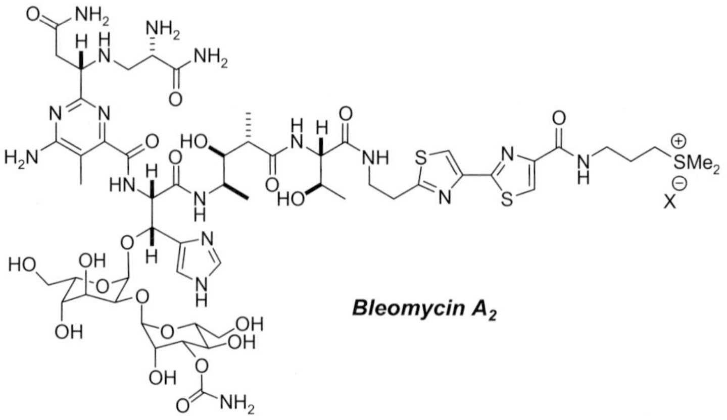

Figure 16.9 Structure of bleomycin A_2.

The synthesis of bleomycin A_2, a potent clinically used antitumor agent, was an important milestone in view of the complexity of its structure and the verification of the revised amino acid portion.

16.4.5
Kopsinine (1985)

The first total synthesis of kopsinine was reported in 1985 by Magnus and Brown [31]. Kuehne and Seaton [32] published a full account of their synthesis later in the same year (Figure 16.10). Following preliminary studies on the aspidosperma alkaloids, Magnus obtained the chiral sulfoxide **B** by an amide coupling of the enantiopure arylthiosulfinylacetic acid with the tetracyclic amine **A**. A Pummerer rearrangement, followed by intramolecular displacement and elimination, led to the diene **C**. Enolate allylation to the *endo-C*-allyl derivative **D**, and cyclization under Diels-Alder conditions led to **E** after oxidation. Further manipulation afforded the ketolactam **F**, which was converted to kopsinine after β-ketoamide cleavage, removal of the nitrogen protecting group, and reduction of the lactam carbonyl.

Kuehne and Seaton started with the racemic indoloazepine **A'**, which was transformed to the bridged azepine **B'**. Intramolecular *N*-alkylation and rearrangement afforded the pentacyclic intermediate **C'**, which was *N*-benzylated and then converted to the *N*-oxide. Oxidation with *m*-CPBA caused elimination and formation of the diene **D'**. Reduction of the *N*-oxide and heating with phenyl vinyl sulfone gave the Diels-Alder adduct **F'**, which was further elaborated to give ($\pm$)-kopsinine.

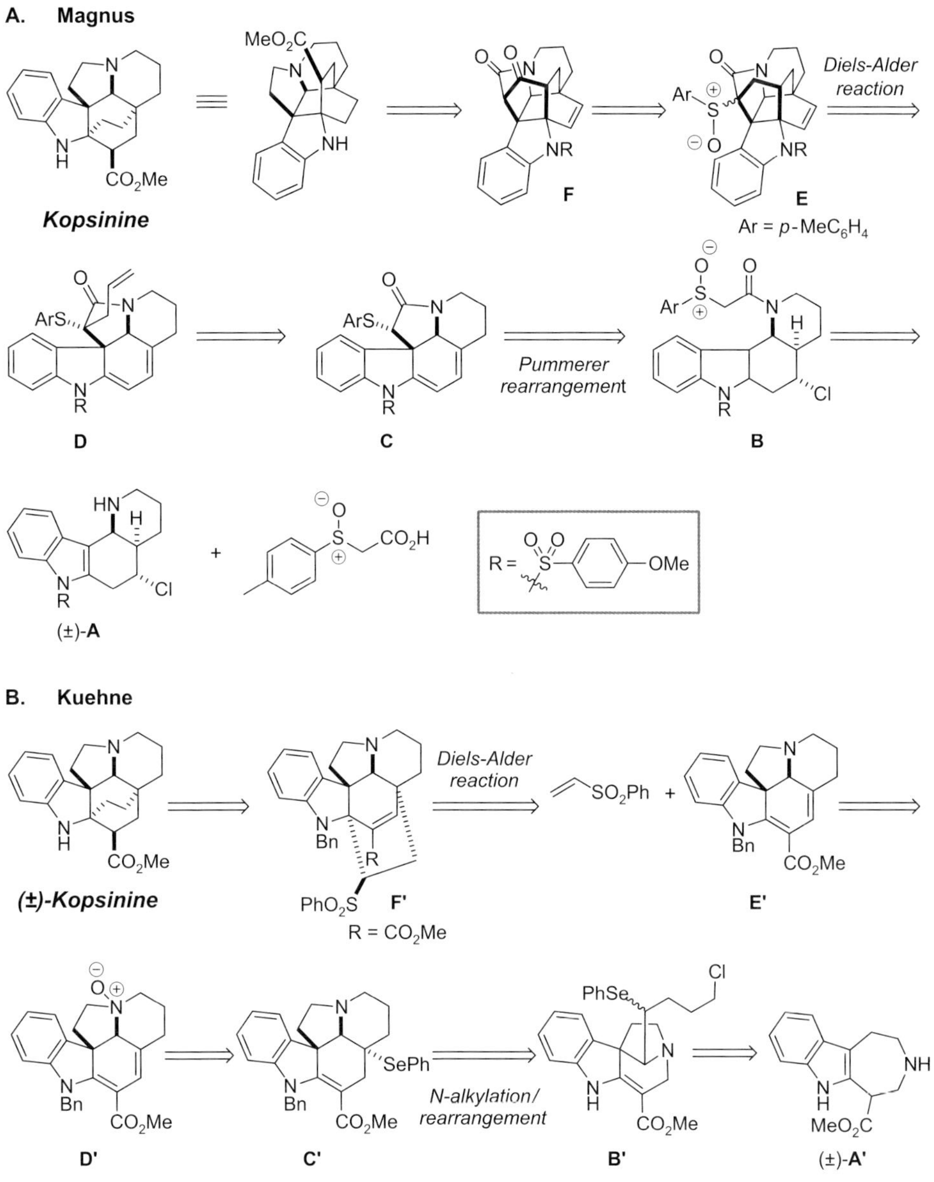

A. Magnus

Kopsinine

F

E

Ar = p-MeC₆H₄

D

C

Pummerer rearrangement

B

(±)-A

R = O_2S—OMe

B. Kuehne

(±)-Kopsinine

F'

R = CO₂Me

Diels-Alder reaction

E'

D'

C'

N-alkylation/ rearrangement

B'

(±)-A'

Figure 16.10 Retrosynthetic analyses of (-) and (±)-kopsinine by the Magnus and Kuehne groups.

16.4.6
Rapamycin (1993)

Three independent total syntheses of the potent immunosuppressive natural product rapamycin were published in 1993 (Figure 16.11). In each case, the delicate balance of functionality and the presence of several carbonyl groups required careful planning and execution in the formation of appropriate intermediates and in their eventual assembly.

In the first published synthesis, Nicolaou and coworkers [33] utilized a rather daring approach to "stich" two vinyl iodide termini of an ester-tethered advanced intermediate **C** (Figure 16.11). Previous studies in the Nicolaou group had prepared the vinyl iodides **A** and **B** in anticipation of a successful application of this intriguing macrocyclization. Amide coupling to **C**, followed by further steps led to **D**, which was successfully "stiched" with the enedistannane **E** according to the Stille protocol. Remarkably, a 28% yield of rapamycin was obtained. Considering that rapamycin encompasses a thirty one-membered macrocyclic ester, the direct bis-annulation step, especially in the absence of protective groups, was tactically innovative.

Schreiber's [34] assembly strategy of rapamycin also involved a tethered fragment coupling, albeit of a different kind (Figure 16.11). Thus, intermediate **A'**, previously obtained from a Julia-type coupling, was transformed to the C-32 ketone **B'** following a 3-step process. Upon treatment with (S)-N-Boc pipecolinal (**C'**) in the presence of SmI_2-$SmI_3(PhCHO)_2$ complex, the C-32/C-34 *anti*-diol **E'** was formed in high yield and excellent diastereomeric ratio. The high stereoselectivity in this Evans-Tishchenko [35] reaction was rationalized on the basis of a Sm-coordinated hemiacetal transition state model as depicted in expression **D'** in Figure 16.11 B. Changing O-protective groups for better compatibility, and further steps led to the aldehyde **F'**, which was extended with the phosphine oxide **G'** to give **H'**. Acetate enolate anion extension, Mukaiyama macrocyclization [36], and cleavage of the N-Boc group led to **I'**. Further manipulation gave rapamycin. Although the newly created stereogenic center at C-32 in **E'** after the Evans-Tishchenko reaction was eventually sacrificed in favor of the carbonyl as required for rapamycin, it would later serve to explore the activity of modified rapamycin analogues in conjunction with Schreiber's seminal structural and biochemical studies in this area [37].

Danishefsky's [38] synthesis of rapamycin adopted yet another innovative assembly strategy of an ester tethered intermediate (Figure 16.11). The ketoaldehyde intermediate **A''**, available from previous studies in the Danishefsky group, was coupled with the phenylsulfoxide **B''** and the product further manipulated to give the α-keto amide **C''**. Ester formation with fragment **D''** afforded ketoaldehyde **E''**, which was subjected to macroaldolization in the presence of a variety of reagents expected to produce chemically compatible metalloenolates. Ultimately, $TiCl_3(Oi\text{-}Pr)$ led to a 14% yield of a macrocyclic product, which was converted to rapamycin.

Considering the presence of highly sensitive functionalities in rapamycin, the three successful syntheses are significant achievements [39]. In fact, three conceptually different, and seldom applied methods were used to achieve stereocontrolled

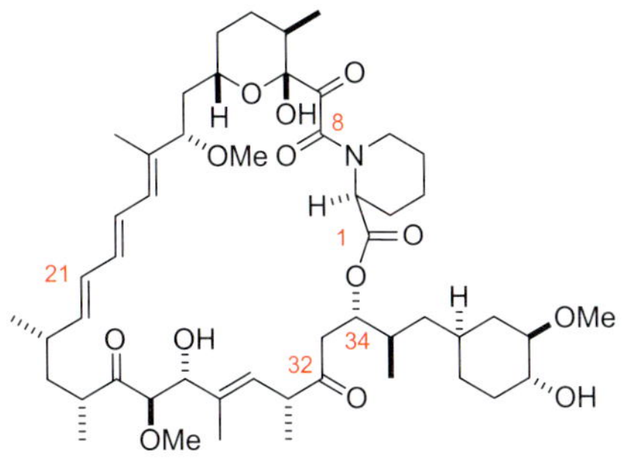

Rapamycin

A. Nicolaou

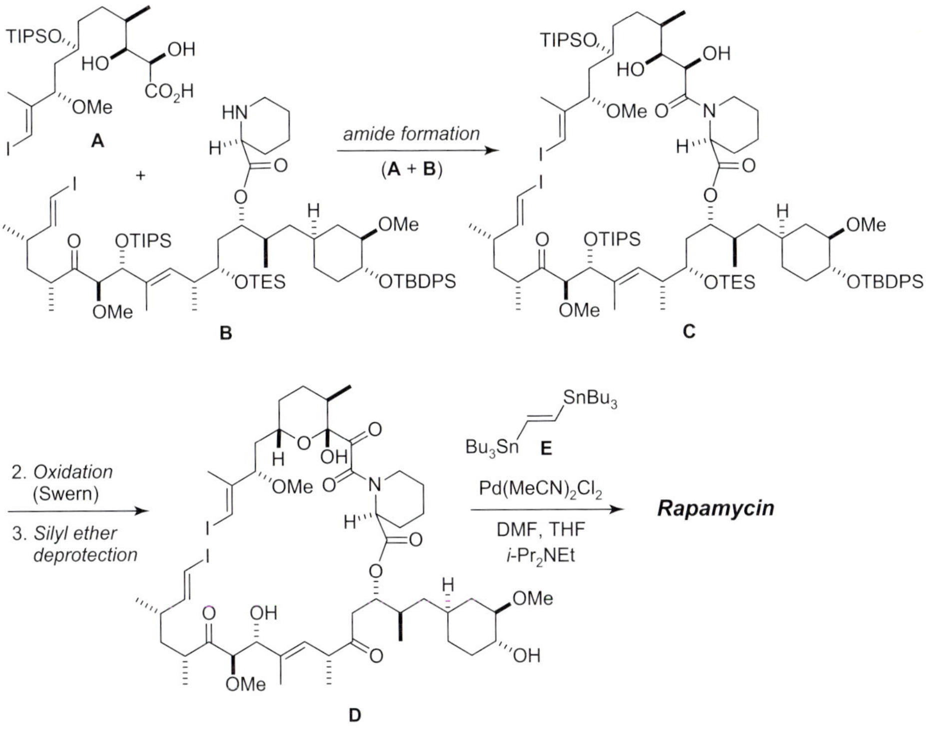

Figure 16.11 Key steps in the total syntheses of rapamycin and relevant assembly strategies by the Nicolaou, Schreiber and Danishefsky groups.

B.   Schreiber

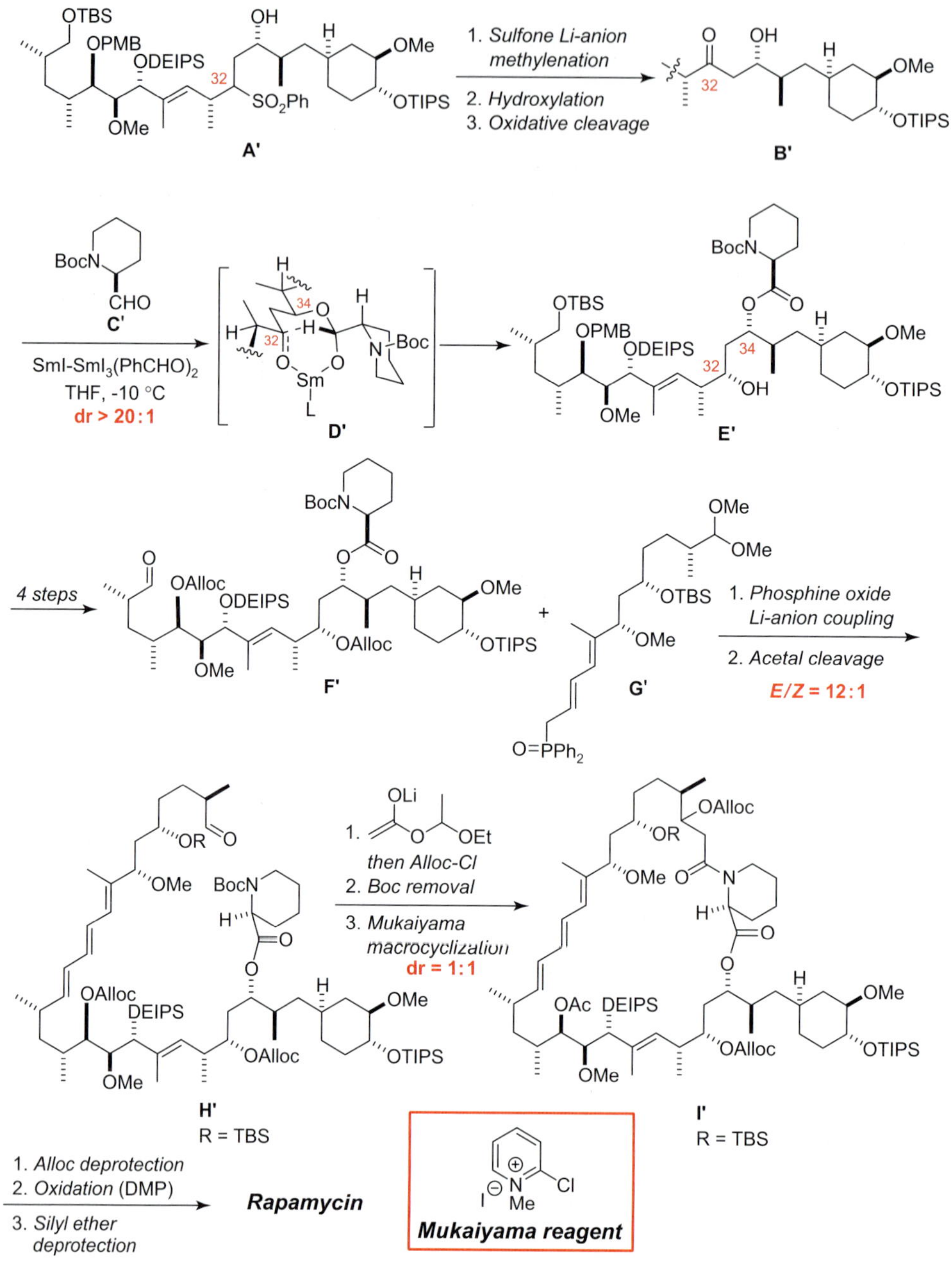

Figure 16.11   *(continued).*

C. Danishefsky

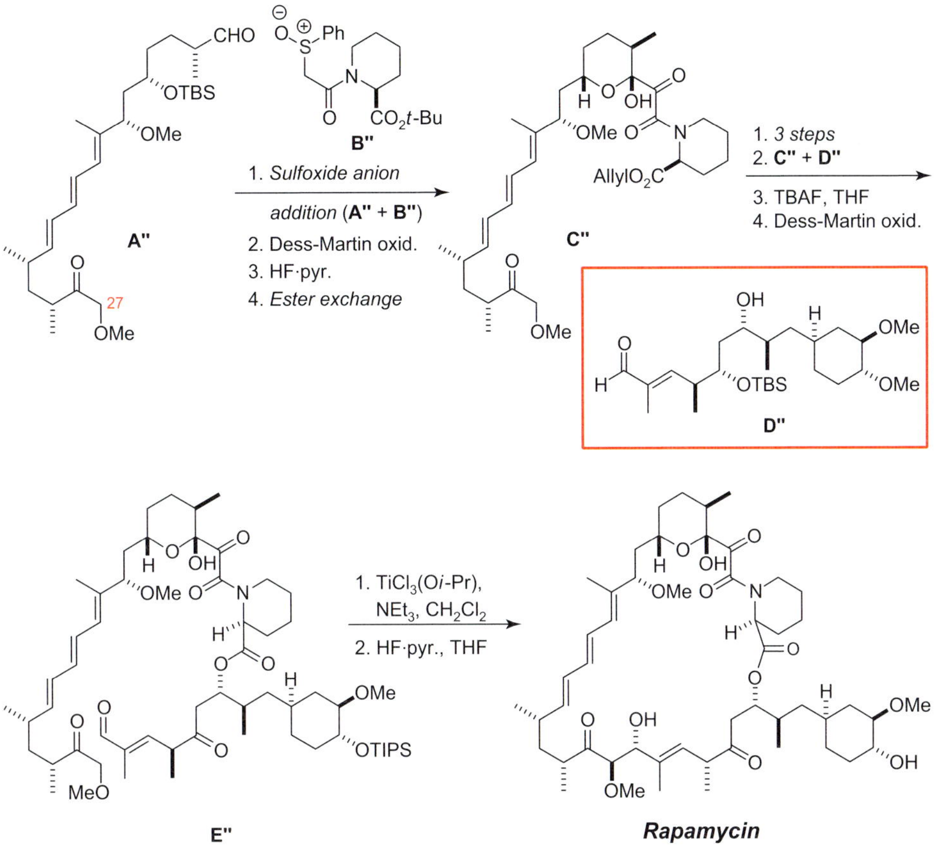

Figure 16.11 (*continued*).

or geometry-controlled bond forming reactions. It is also noteworthy that a number of chemoselective steps compatible with the presence of a tricarbonyl system (C-8–C-10), were used in all three syntheses. Thus, total synthesis was used as a forum to venture into new methodology, which adds to its value beyond simply achieving the intended goal. Rapamycin is abundantly available by fermentation and offers many opportunities for controlled degradation to valuable relay fragments as cited by Danishefsky [38].

Acknowledgements by authors for generous samples of natural products received, either for synthetic studies, or for comparison of physical data, are of common occurrence in the primary literature. In the case of rapamycin, however, acknowledgements were also made by Schreiber [37] to his colleagues Nicolaou and Danishefsky for "providing preprints that describe their total syntheses of rapamycin." Such a public statement shows collegiality at its very best, and one that can last for a professional lifetime.

16.4.7
Phomoidrides (CP molecules) (2000)

The phomoidrides are a series of architecturally and topologically fascinating natural products that exhibit inhibitory activity against squalene synthase, as well as RAS farnesyl transferase [40]. Initially, phomoidrides A and B were first isolated, and their structures determined without assigning an absolute configuration. Phomoidrides C and D were later isolated and proven to be (7R)-isomers of phomoidrides B and A respectively (Figure 16.12).

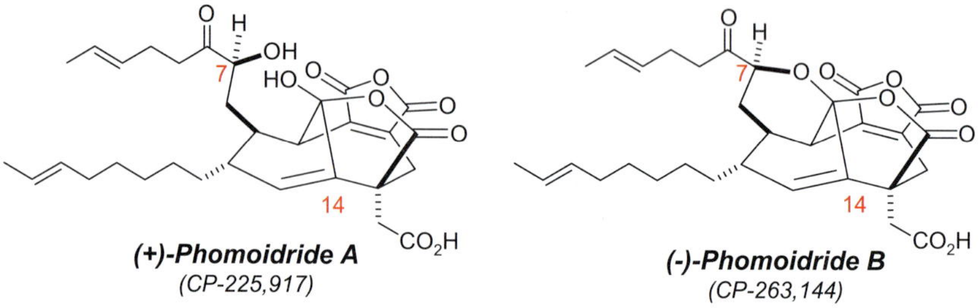

(+)-Phomoidride A
(CP-225,917)

(-)-Phomoidride B
(CP-263,144)

Figure 16.12   Structures the phomoidrides A and B.

Four independent total syntheses of phomoidride B were reported in 2000. Details and an in depth analysis of the four strategies have been reviewed by Wood and coworkers in 2003 [41].

Before embarking on a brief exposé of the syntheses, a comment on the aesthetic aspects of certain organic molecules may be *à propos*. In Figure 1.9 of Chapter 1 we showed some synthetic compounds under the subtitle "beauty is in the eye of the beholder." Phomoidride B was featured alongside the kaleidoscopic mosaic-like structures of compounds such as twistane, homopentaprismane, bullvalene, and corannulene among others.

We alluded to the "shapes and forms of molecules," and likened phomoidride B to a bird in flight. However, the serenity and grace of this imaginary metaphor assume an altogether different meaning when confronted with the reality of a synthesis plan not to mention the rigors of its execution [42]. Consider, for example, the delicate lactone acetal suspended atop the transannular bridge spanning across a cyclononadiene hydrocarbon core. Added to the challenge are the presence of a maleic anhydride appendage, and an assortment of other extended functional groups. The four syntheses will be briefly discussed in their chronological order of publication in 2000 while highlighting key steps, and at times, common starting materials. Curiously, phomoidride B is depicted in four distinctly different structural presentations by the four groups (see also section 16.4.7.5).

16.4.7.1   **The Nicolaou Synthesis** – The first total synthesis of *ent*-phomoidride B by Nicolaou and coworkers [43] also established its absolute configuration (Figure 16.13). (*R*)-Glycidol (**A**) was arbitrarily chosen as the starting chiron, which

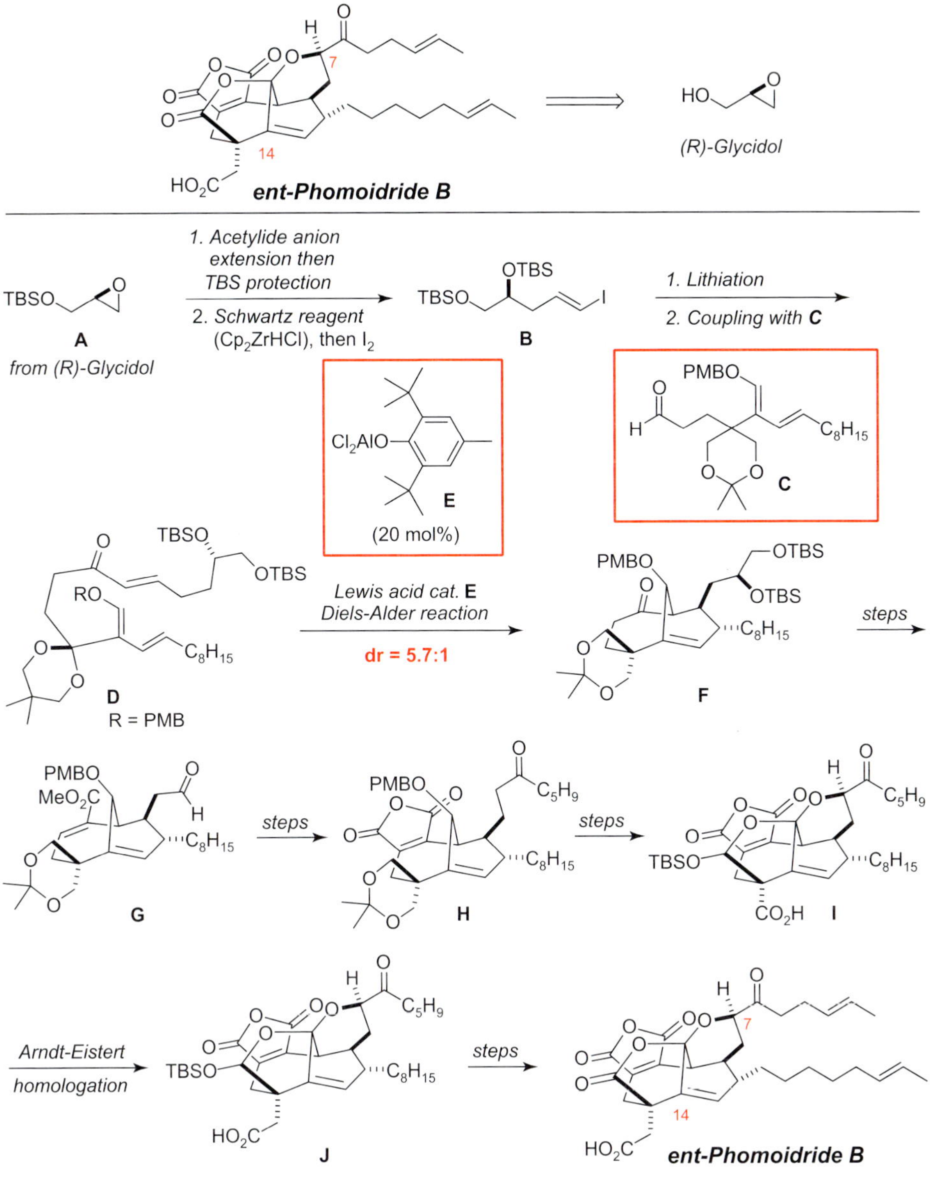

Key reactions: Intramolecular Diels-Alder (**D** to **F**); Arndt-Eistert homologation via acyl mesylate (**I** to **J**).

Figure 16.13 Total synthesis of *ent*-phomoidride B by the Nicolaou group.

was converted to vinyl iodide intermediate **B**. Conversion to the vinyllithium derivative and coupling with aldehyde **C** gave enone **D**. A critical intramolecular Diels-Alder cyclization in the presence of catalyst **E** afforded the bridged protected allylic alcohol **F** as the minor diastereomer in excellent overall yield. NMR correlation of a derivative available from an authentic racemic sample confirmed the stereochemical identity of **F**. Elaboration of available functional groups allowed safe passage to intermediates **G**, **H**, and **I**. An innovative variation of the Arndt-Eistert homologation of **I** utilized an acyl mesylate as a precursor to a diazoketone before rearrangement took place to give **J**. An indoline amide derivative of the lactone acetal formed from **J**, led to a correlation with authentic material and assignment of absolute configuration. This timely result that produced the unnatural (+)-isomer of phomoidride B, allowed other groups working in the field to consider syntheses of the natural (−)-isomer.

16.4.7.2 **The Shair synthesis** – The second total synthesis of *ent*-phomoidride B to appear in the literature was from Shair and coworkers [44] (Figure 16.14). This

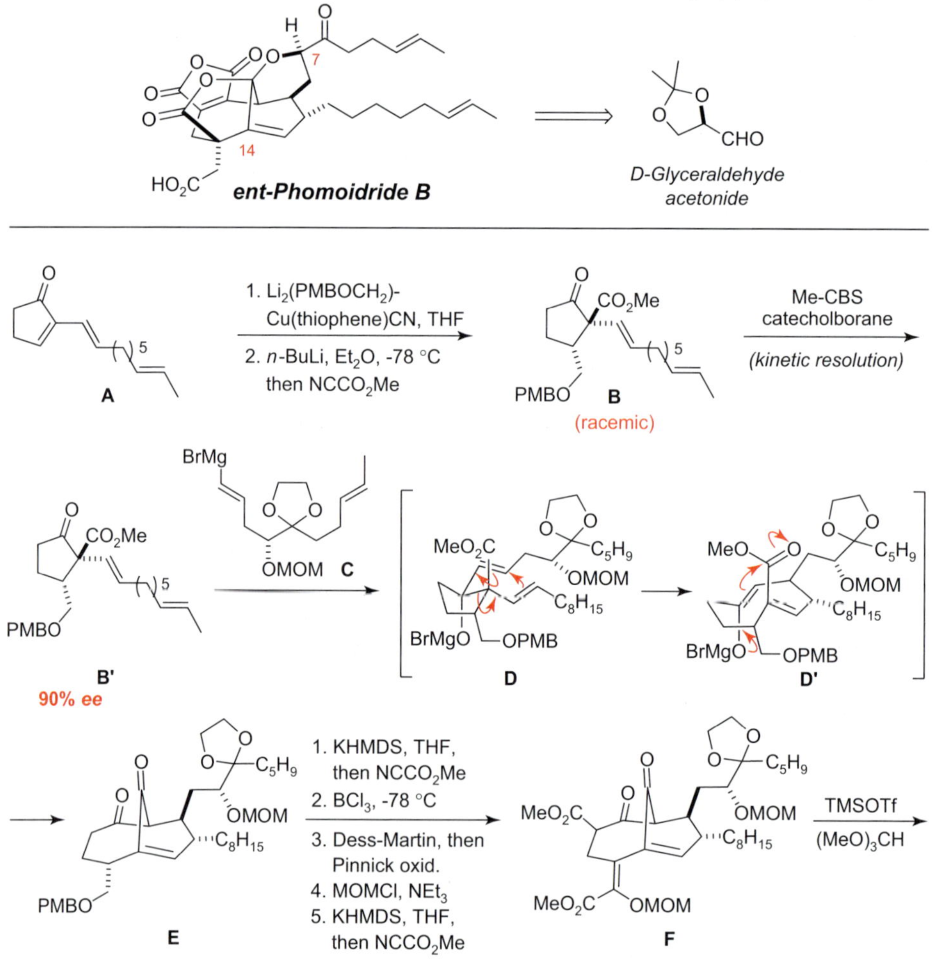

Figure 16.14 Total synthesis of *ent*-phomoidride B by the Shair group.

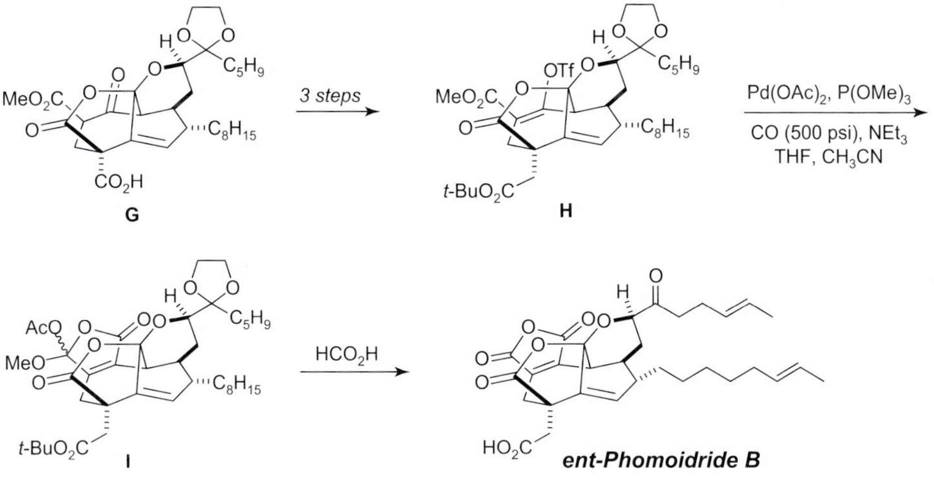

Key reactions: *CBS-mediated kinetic resolution reduction (**B** to **B'**); anionic oxy-Cope rearrangement (**B'** to **C** to **E**); transannular Dieckmann-like reaction (**D'** to **E**); TMSOTf/trimethyl orthoformate-mediated cascade reaction (**F** to **G**); Arndt-Eistert homologation (**G** to **H**).*

Figure 16.14 *(continued)*.

conceptually novel approach to the construction of the bicyclo[4.3.1]decane ring system in *ent*-phomoidride B, featured the conjugate addition and branching transformation of **A** to racemic **B**. A highly practical kinetic resolution of the ketone **B** was achieved by treatment with the Corey-Bakshi-Shibata Me-CBS/catecholborane reagent combination [45], to afford **B'** in 90% *ee* and very good yield. Addition of the Grignard reagent **C**, originally prepared from D-glyceraldehyde acetonide, led to the bridged ketone **E** in a remarkable series of reactions featuring an anion accelerated oxy-Cope rearrangement [46], followed by a transannular Dieckmann-type cyclization as shown in **D** and **D'**. Transformation to the exocyclic enol carbonate intermediate **F**, then treatment with TMSOTf and trimethyl orthoformate effected a tandem activation of the orthoformate and capture of a transient carbomethoxenium ion to give **G**. Subsequent steps involved homologation to an ester and conversion to the enol triflate **H**, which was elaborated to the carboxylic acid group, then transformed to the anhydride intermediate **I**. Hydrolysis led to *ent*-phomoidride B. The original stereogenic carbon of D-glyceraldehyde acetonide was conserved in the target molecule at C-7.

16.4.7.3 **The Fukuyama synthesis** – The first total synthesis of phomoidride B (*i.e.*, the natural isomer) to appear in print was also published in 2000 by Fukuyama and coworkers [47] (Figure 16.15). Thus, an independent confirmation of the absolute configuration of the natural (−)-isomer was provided as a result of a total synthesis. Conjugate addition of the enolate of **A** to **B** led to the ethylthio diene intermediate **C**. Auxiliary-mediated aldol condensation with aldehyde **D**, obtained from (*S*)-epichlorohydrin, followed by oxidation, gave the ketone **E**. Intramolecular Diels-Alder cyclization in the presence of ZnCl$_2$·Et$_2$O led to the bridged ketone **F**. Having served its stereocontrolling role, the chiral auxiliary was transformed to the allylthioacyl glycolate adduct **G**, which was further elaborated to the thioanhydride **H**. Chemoselective transformations, followed by an Arndt-Eistert homologation led to **I**. An oxidative lactone-acetalation via a sulfoxide intermediate gave **J**, which was oxidized to afford natural (−)-phomoidride B. Unlike the Nicolaou synthesis [43] in which (*R*)-glycidol was used in a self-immolative process, the stereochemistry of the original epoxide was maintained at C-7 in the target molecule.

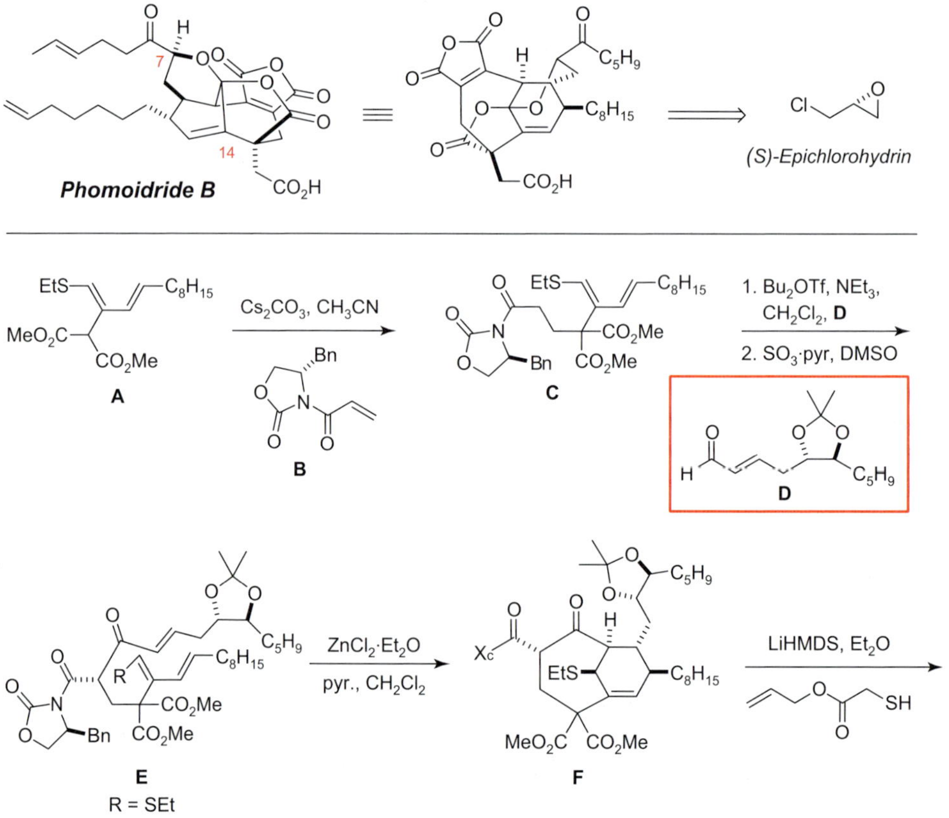

Figure 16.15 Total synthesis of phomoidride B by the Fukuyama group.

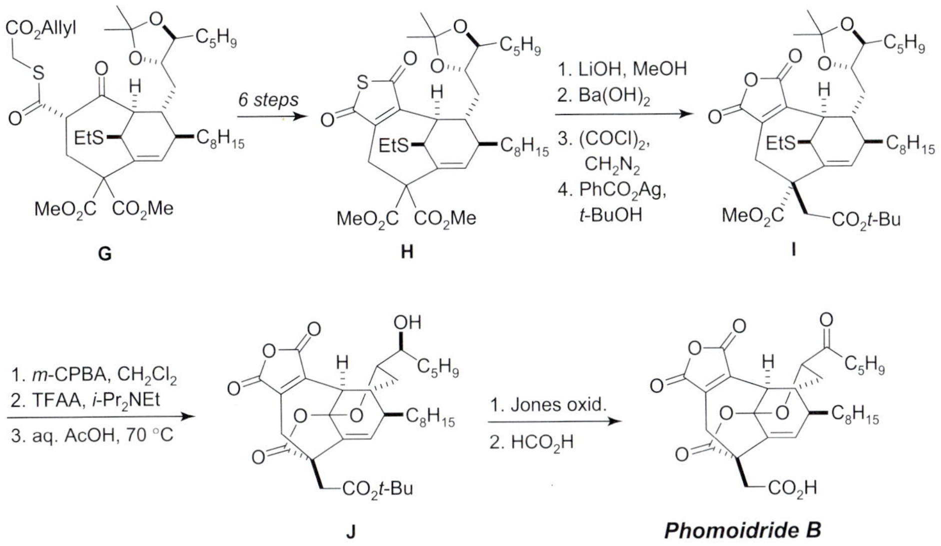

Key reactions: *Intramolecular Diels-Alder (**E** to **F**); thioanhydride formation (**G** to **H**); chemoselective hydrolysis and Arndt-Eistert homologation (**H** to **I**); thioether to ketone via sulfoxide (**I** to **J**).*

Figure 16.15 *(continued).*

16.4.7.4 The Danishefsky synthesis

The total synthesis of phomoidrides A and B by Danishefsky and coworkers in 2000 [48] also led to the discovery of phomoidrides C and D as minor constituents in fermentation broths. Difficulties encountered during the course of their synthetic efforts, with respect to securing the correct stereochemistry of the C-7 stereogenic center, led to alternative approaches to the construction of the CP molecules (Figure 16.16). As a result, total syntheses of phomoidride A, 7-*epi*-phomoidride B (phomoidride D), and phomoidride B were disclosed in the same communication [48].

A sequential aldol (via the lithium enolate of **A**) and intramolecular Heck reaction [49] between **A** and **B** led to the racemic ketone **C**. A series of steps afforded enone **D**, which was further converted to the spirocyclobutanone **E** over ten steps. Regioselective sulfenylation of the cyclobutanone ring was achieved using diphenyl disulfide. Subsequent Dess-Martin and regioselective Baeyer-Villiger oxidations led to the sulfoxide **F**. Dihydroxylation of the olefin and treatment with NaOMe gave the lactol **G**, which was directly subjected to a Swern oxidation to furnish **H**. A nine step sequence led to 7-*epi*-phomoidride B (**I**, phomoidride D). A four step sequence, which involved conversion of the maleic anhydride moiety to an ethylene diester, lactone hydrolysis, and Dess-Martin oxidation, gave **J**. Chemoselective reduction of the C-7 ketone group with LiAl(O-*t*Bu)₃H afforded the desired C-7 stereoisomer and its epimer in a 1:1 ratio. Saponification of all four ester groups and acidification led to phomoidride A, which was converted to phomoidride B upon treatment of methanesulfonic acid. Were it not for the incorrect stereochemistry at C-7 in the original synthesis, the Danishefsky approach would have led to the desired isomer in a more expedient manner. Instead a number of steps were needed to partially arrive at the correct stereochemistry from the C-7-epimer.

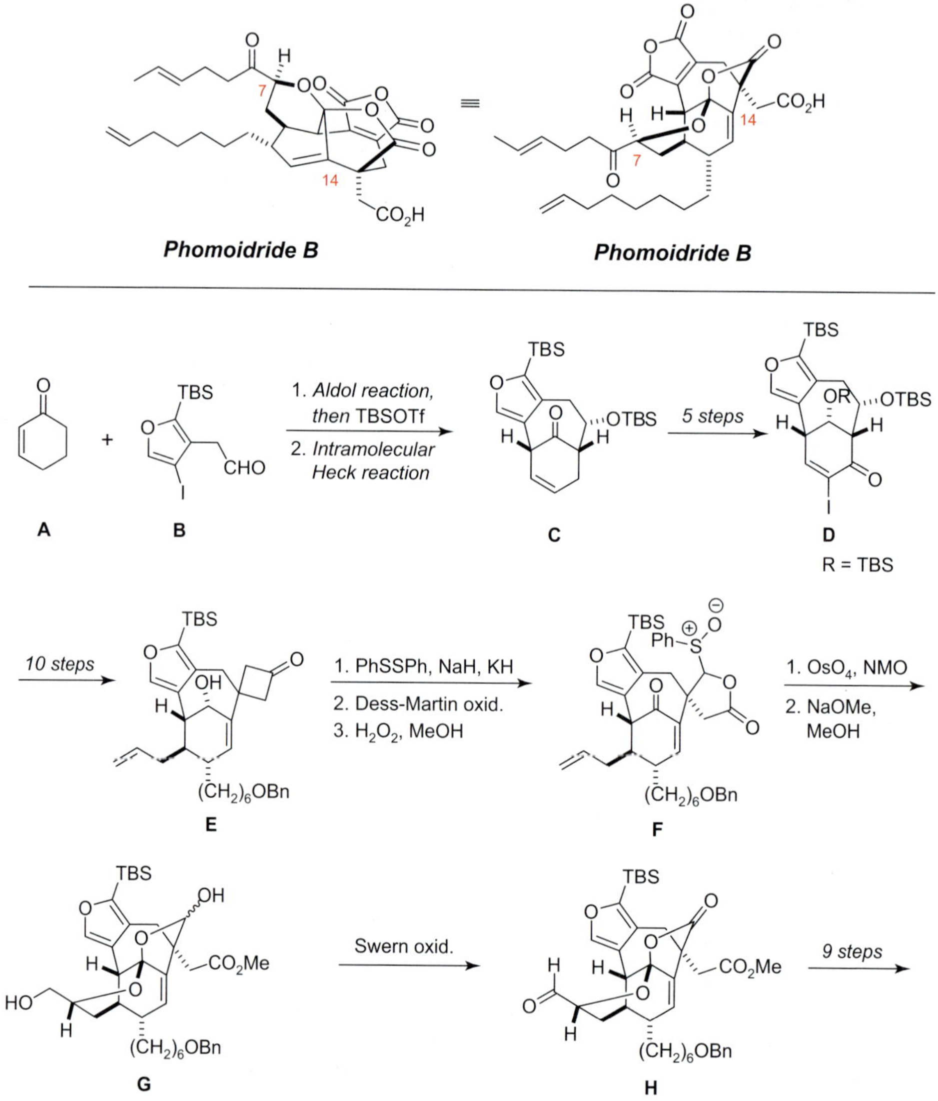

Figure 16.16 Total synthesis of phomoidride B and phomoidride A by the Danishefsky group.

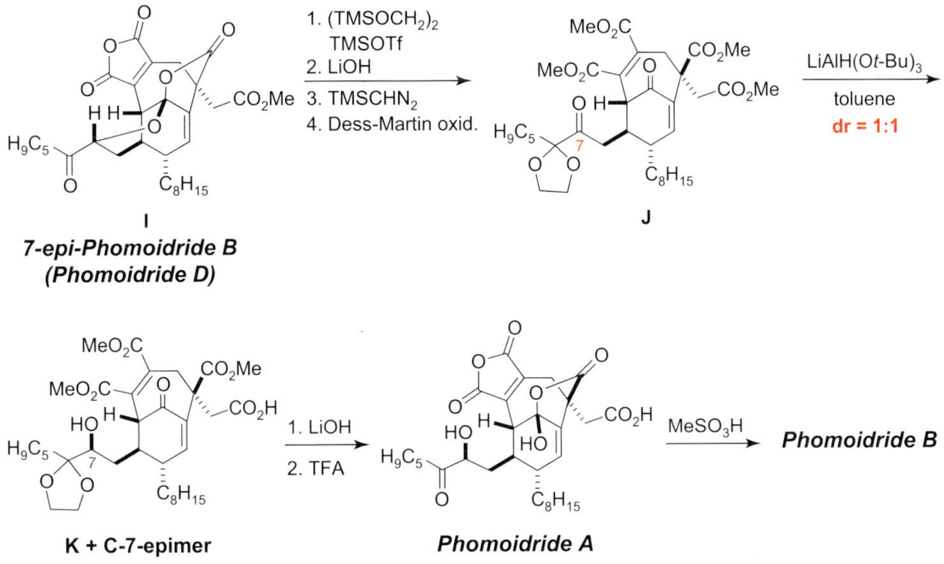

I
**7-epi-Phomoidride B
(Phomoidride D)**

J

K + C-7-epimer

Phomoidride A

Phomoidride B

Key reactions: *Sequential aldol and intramolecular Heck reaction (**A**+**B** to **C**); regioselective sulfenylation-oxidation to sulfoxide and Baeyer-Villiger oxidation (**E** to **F**); intramolecular lactol formation (**F** to **G**).*

Figure 16.16 (*continued*).

16.4.7.5 What is in a drawing? – As mentioned in section 16.4.7, the four groups chose to draw the structure of phomoidride B in different perspective representations. We have already discussed in Chapters 4 and 6 the interplay between visual *relational*, and visual *reflexive* thinking. In their respective strategies, Nicolaou [43] and Fukuyama [47] relied on an intramolecular Diels-Alder cyclization to enable the construction of the nine-membered, bridged core structure of phomoidride B. However, their original perspective portrayal of the structure may have been the visual starting point for the thought process that led to their respective disconnective analyses, and the individual choices of starting materials (Figures 16.13, 16.15). Following the successful implementation of this key reaction, and the deployment of usable functionality, the remainder of the synthesis process was concerned with peripheral and functional group modification involving chain-extensions and redox chemistry.

The same applies to the Danishefsky [48] and Shair [44] drawings, although the latter is very similar to the Nicolaou rendition. Obviously, prior studies and explorations of methods to access the core structure within the two groups have profoundly influenced the thought process that culminated with a successful synthesis. In making the "cyclohexene" core more prominently visual (Figure 16.16), Danishefsky's sequential aldol and Heck cyclization sequence can be appreciated as a key step in providing the bridged core structure, to which a furan moiety is apperded as a latent dicarboxylate.

Finally, the conceptually elegant cascade ring-closure via an anion-accelerated oxy-Cope cyclization was most probably the visual epiphany arising from the perspective portrayal of *ent*-phomoidride B in Shair's mind's eye (Figure 16.14) [44]. Curiously, the very similar perspective drawing of *ent*-phomoidride B led Nicolaou [43] to a different, subliminally directed, Diels-Alder approach (Figure 16.13).

Invariably many well-planned and executed strategies toward a given natural product synthesis may have a hidden caveat or two, that manifests itself somewhere along the path of the long journey. Sometimes, the expediency of a pivotal reaction, such as the Diels-Alder or anionic oxy-Cope cyclization, elegantly conceived and executed by Nicolaou, Fukuyama, and Shair in their individual total syntheses of *ent*-phomoidride B, may leave the intended substructure short of the required number of carbon atoms (see Chapter 18, section 18.5.3). The "price to pay" in the above three syntheses was the obligatory Arndt-Eistert homologation of a carboxylic acid to the corresponding acetic acid appendage at C-14 (Figures 16.13–16.15).

16.4.8
Borrelidin (2003–2004)

Total syntheses of borrelidin were reported in 2003 independently by the Morken [50] and Hanessian groups [51] (Figures 16.17 and 16.18). Borrelidin is a structurally distinct atypical macrolide antibiotic with no known congeners. As such, it represented a challenging target for total synthesis, especially with the seldom encountered array of deoxypropionate triads spanning C-1 to C-12. An unusual *cis-trans* cyano diene subunit presented stereochemical and functional group issues that were dealt with in different ways.

16.4.8.1 **The Morken synthesis** – The utility and power of catalysis in the creation of required critical functionalities are highlights of the Morken synthesis [50] (Figure 16.17). Thus, an iridium indane-pybox catalyzed enantioselective aldol reaction [52] with methyl acrylate and an O-differentiated glycolaldehyde gave, after further elaboration of the respective products, intermediates **A** and **B** with good diastereoselectivities (6:1) and high *ee* (>90%). The pseudoephedrine propionamide lithium anion served to install the second methyl group following a Myers alkylation protocol [53] with **A** to give **C**, which was converted to **D**. Lithiation of **B** and coupling with **D** gave the olefin **E**, which was catalytically hydrogenated and further extended to afford ynol intermediate **F**.

A Sonogashira coupling [54] with vinyl iodide **G**, prepared from (*R,R*)-1,2-cyclopentane carboxylic acid bis-(+)-menthyl ester [55], followed by regioselective desymmetrization, and Brown allylation, led to the ene-yne **H**. Palladium-catalyzed hydrostannylation, followed by iodination gave a 1:1 mixture of regioisomers. The desired isomer **I** was then converted to the cyanodiene **J**, which, was in turn, deprotected, and cyclized using the Yamaguchi lactonization [56] protocol to give borrelidin.

Borrelidin

(1:1 regioisomers)

Key reactions: Catalytic asymmetric aldol to **A** and **B**; Myers enolate alkylation (**A** to **C**); asymmetric olefin reduction (**E** to **F**); Sonogashira coupling (**F** and **G**); hydrostannylation/iodination (**H** to **I**); Pd-catalyzed cyanation (**I** to **J**). Rh[(nbd)dppb]BF$_4$ = Rhodium-norbornadiene diphenyldiphosphinobutane tetrafluoroborate

Figure 16.17 Total synthesis of borreldin by the Morken group.

16.4.8.2 The Hanessian synthesis – A conceptually different strategy that involved an iterative building of deoxypropionate subunits through conjugate addition of lithium dimethyl cuprate to α,β-unsaturated esters was the basis of the Hanessian strategy [51] (Figure 16.18). The rationale for stereochemical control in the conjugate addition reactions was predicated upon the premise that a growing acyclic chain would favor the deployment of *syn*-oriented *C*-methyl substituents to avoid 1,5-*syn*-pentane interactions [57].

Thus, conjugate addition of lithium dimethyl cuprate to intermediate **A** [58], available from D-glyceraldehyde acetonide in five steps, led to **B** as the major diastereomer. Extension to the α,β-unsaturated ester **C**, followed by two iterations

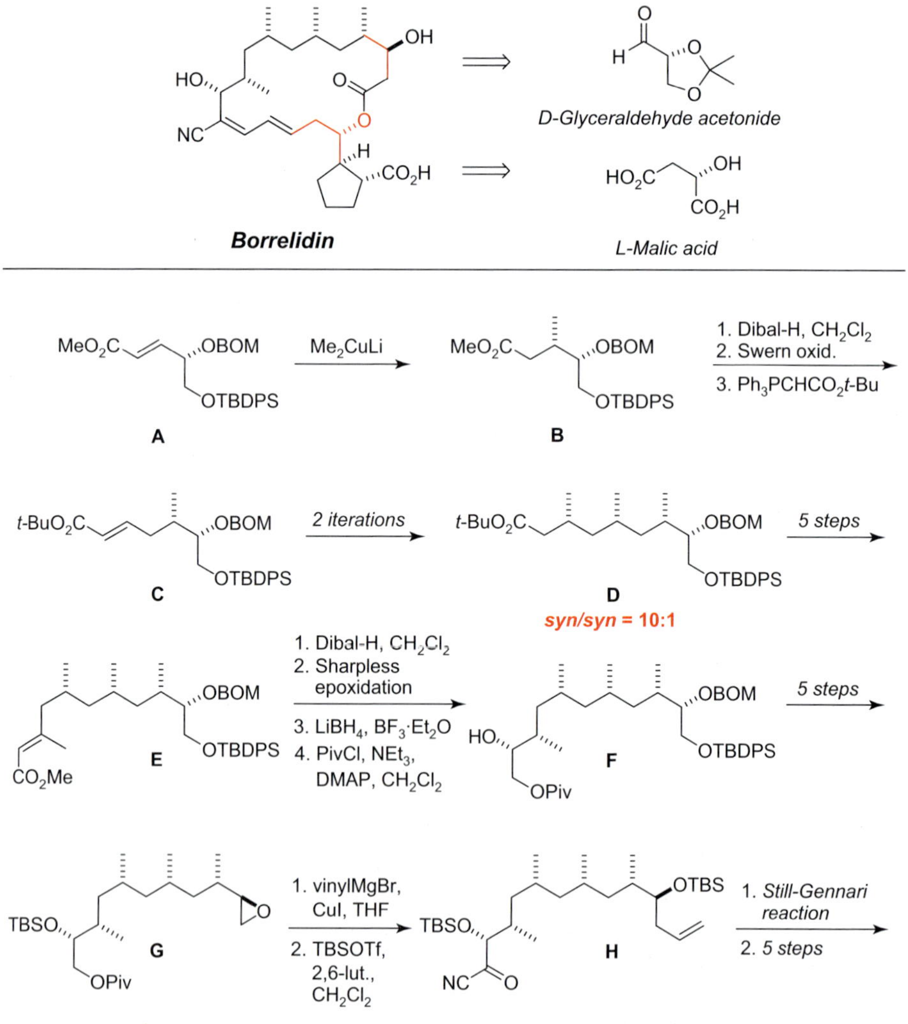

Figure 16.18 Total synthesis of borrelidin by the Hanessian group.

Key reactions: *Iterative cuprate conjugate additions (A to B; C to D, twice); Still-Gennari cis-selective olefination (H to I); Julia-Kocienski olefination (I + J to K); L-malic acid to J; TMSE = (trimethylsilylethyl)*

Figure 16.18 (*continued*).

of the conjugate addition step, gave **D** with high *syn/syn* selectivity. Further steps led to the ester **E**, which was reduced to the allylic alcohol, converted to the epoxide, and regioselectively opened to ultimately give the pivalate ester **F**. Introduction of an epoxide with inversion of configuration, opening with vinylmagnesium bromide, and further elaboration afforded an aldehyde, which was subjected to a cyanohydrin reaction followed by oxidation to give **H**. Application of the Still-Gennari [59] olefination, and further functional group manipulation gave aldehyde **I**, which was coupled with the sulfone **J** (prepared from L-malic acid) to give **K**. Deprotection and Yamaguchi lactonization led to synthetic borrelidin, which was isolated as the crystalline benzene solvate and its three-dimensional structure resolved by X-ray analysis.

16.4.8.3 **The Ōmura and Theodorakis syntheses** – Two total syntheses of borrelidin were published by Theodorakis [60], Ōmura [61], and their respective groups in 2004. Salient features highlighting the assembly of subunits are briefly shown in Figure 16.19. A noteworthy step in the Ōmura synthesis is the SmI_2-mediated intramolecular Reformatsky cyclization [62] of the vinyl bromide **H** to produce 40% of the desired *Z*-olefin (Figure 16.19 A). The deoxypropionate unit in intermediate **B** was secured by extension of **A**, obtained by an enzymatic desymmetrization of a *meso*-diacetate. Trimethylaluminum-mediated carbotitanation of **B** gave **C**, which was subjected to a chelation-controlled Rh-catalyzed hydrogenation as in the Morken synthesis [50] to give **D**. The 1,2-disubstituted cyclopentane **H** and its precursor **G** originated from the C_2-symmetrical bis-hydroxymethyl (*R,R*)-cyclopentane **F**.

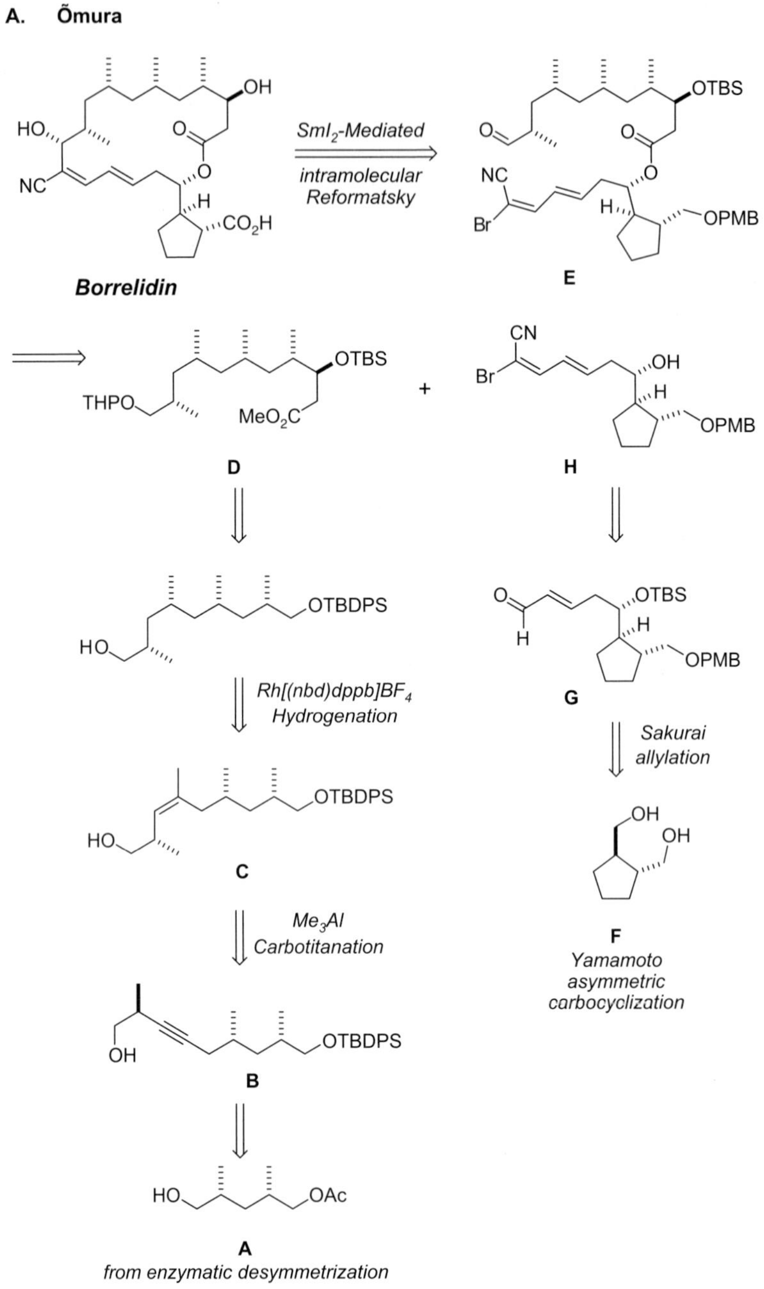

A.   Õmura

Figure 16.19   Total syntheses of borrelidin by the Õmura and Theodorakis groups.

B.   Theodorakis

Figure 16.19   *(continued)*.

A novel feature in the Theodorakis [60, 71] synthesis of borrelidin is a regiose-lective Mo-catalyzed hydrostannation of ene-yne **E'**, followed by formation of vinyl iodide **F'**. Subsequent Pd-catalyzed cyanation, as in the Morken synthesis [50], led to the desired cyanodiene. Access to the deoxypropionate subunits **C'** and **D'** was made possible by exploiting Evans [63] and Myers [53, 64] enolate alkylations respectively.

16.4.9
Amphidinolide E (2006)

Total syntheses of amphidinolide E were reported independently in 2006 by Roush [65], Lee [66], and their respective coworkers. Although both groups utilized (*R,R*)-dimethyl tartrate as a logical building block for the C-7–C-8 diol subunit, their strategies were totally different.

Roush relied on a challenging ring-closing metathesis macrocyclization of intermediate **J** (Figure 16.20). Starting with (*R,R*)-dimethyl tartrate (**A**), a series of well-precedented stereocontrolled extensions exploiting the venerable Roush allyl and crotylborations [67] led to the allyl alcohol **B**, which was subjected to a Johnson-Claisen (orthoester) rearrangement to give **C** after functional group adjustment. Crotylboration of L-glyceraldehyde 3-pentylidene (**D**) afforded **E**. Hydroboration gave **F**, which was converted to **G** and then subjected to an asymmetric silyl allylboration [68] to give **H**. A key $BF_3 \cdot Et_2O$-mediated [3+2] cyclization reaction between **C** and **H** gave the *syn*-1,5-substituted tetrahydrofuran **I** with excellent diastereoselectivity. A clever solution to a difficult esterification of **I** with diene acid **K** consisted in using a $Fe(CO)_3$ complex under Yamaguchi conditions [56]. It should be noted that the stereogenic carbon in **D** was sacrificed in a self-immolative sequence (**F** to **G**).

Lee took full advantage of (*R,R*)- and (*S,S*)-tartaric acids as sources of chirality in specific subunits in amphidinolide E (Figure 16.20). The key coupling relied on a Julia-Kocienski reaction [69] between **E** and **J**. A Suzuki coupling between **B** and **D** provided the extended diene aldehyde **E** after appropriate elaboration. The construction of the tetrahydrofuran subunit **I** was done in an intramolecular free-radical mediated cyclization of the iodide **H**. A Roush crotylboration was successfully used to extend aldehyde **G** to give **H**.

A. Roush

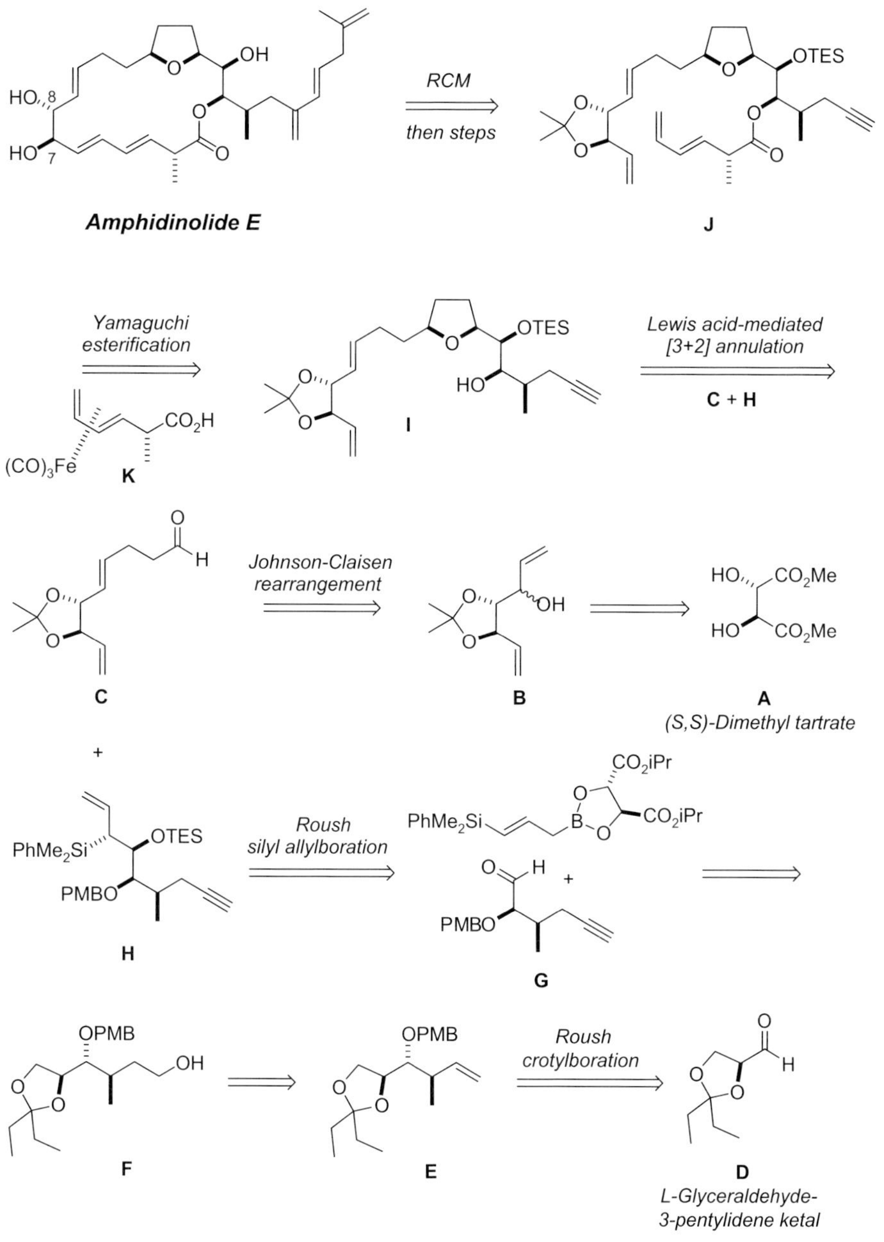

Figure 16.20 Assembly of key subunits in the total synthesis of amphidinolide E by the Roush and Lee groups.

B.   Lee

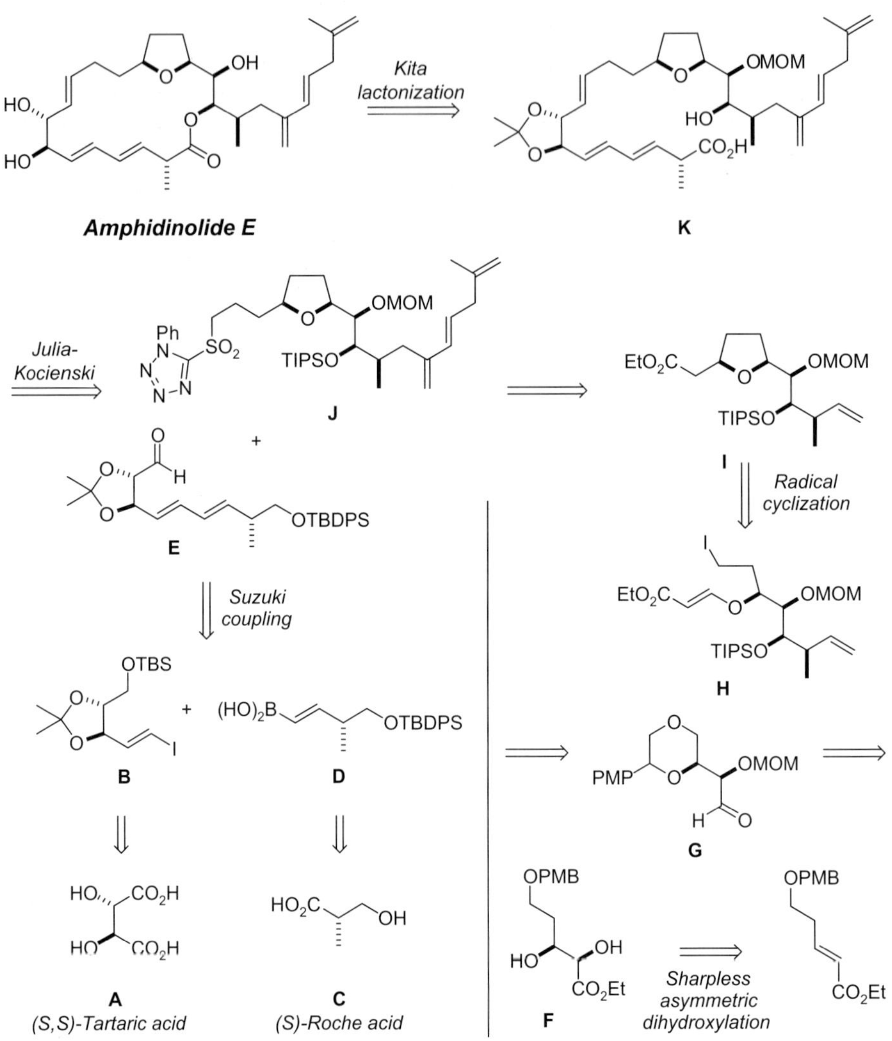

Figure 16.20 *(continued)*.

16.5
Single Target Molecules with Special Relevance

We continue this chapter with an exposé of diverse strategies directed toward the total synthesis of three molecules with special relevance, albeit in different contexts. Many examples of total syntheses of molecules with different levels of structural, functional, and stereochemical complexities were discussed in previous chapters. The motivation and incentive to pursue synthetic studies aimed primarily at natural products vary from molecule to molecule. In some historically relevant examples such as quinine, the efforts have spanned over half a century. Recently discovered natural molecules with therapeutic potential such as platensimycin (see also Chapter 2, section 2.1, Chapter 13, section 13.7), and lactacystin (see section 16.7) have ushered many innovative methods relying on catalytic asymmetric processes. Synthetic efforts in the context of life-saving drugs are best exemplified by taxol (see section 16.8).

Accomplishments in total syntheses must be judged in relation to the time period of their execution. It is to be expected that second or third-generation total syntheses of a known natural (or unnatural) product several years after it was first reported, also embody improvements on more than one front. Considering the major advances made in methodology, in separation techniques, and in spectroscopic characterization, a new synthesis of a natural product must be scrutinized for novelty, efficiency, and overall creativity in comparison to previous contributions. This places a high premium on the true value of a new disclosure relating to the total synthesis of an already known molecule.

16.6
Quinine

Quinine has been known for centuries for its medicinal properties, especially as a "cure" for malaria [70]. The annals of organic chemistry are rich with the history and properties of this unique plant alkaloid. Efforts towards its total synthesis date back to the turn of the last century [71].

16.6.1
The Stork synthesis

The first stereoselective total synthesis of (−)-quinine was only recently accomplished by Stork and coworkers (Figure 16.21) [72]. (3S)-3-Vinyl butyrolactone [73], was treated with Me₃Al and Et₂NH to give the corresponding diethylamide. Protection of the primary hydroxyl group followed by alkylation of the lithium enolate gave **A**. Cleavage of the TBS group and lactonization gave **B** as a major product (>20:1). Reduction to the lactol and treatment with a methoxymethylenephosphorane, followed by displacement of the primary hydroxyl group by azide according to Mitsunobu led to **C**. Cleavage of the enol ether to give aldehyde **D** was followed by treatment with the lithium salt of 6-methoxy-4-methylquinoline (**E**), to afford **F**. Oxidation to the ketone and Staudinger reduction of the azide led to the formation of imine **G**. Stereoselective reduction with NaBH₄, followed by removal of the TBS group gave **H** in high yield. Mesylation and intramolecular displacement led to the azabicyclo product **I**. Adopting a known benzylic oxidation reported by Uskoković and collaborators [74] completed the total synthesis of (−)-quinine.

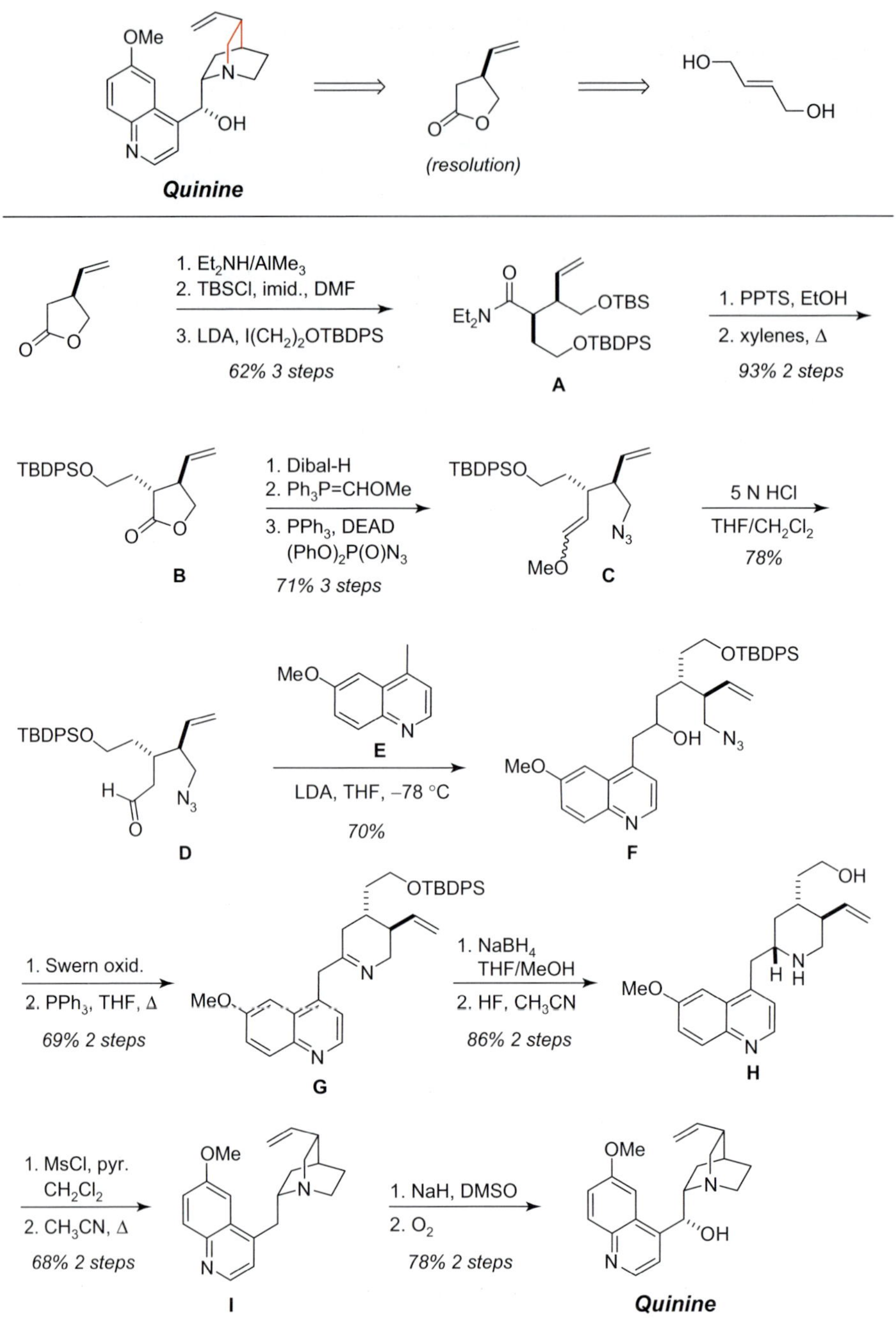

Figure 16.21 Total synthesis of quinine by the Stork group.

Key reactions: *Enolate alkylation (lactone to **A**); stereoselective cyclic imine reduction (**G** to **H**); benzylic oxidation (**I** to quinine)*

Relevant mechanisms and transformations:

A.

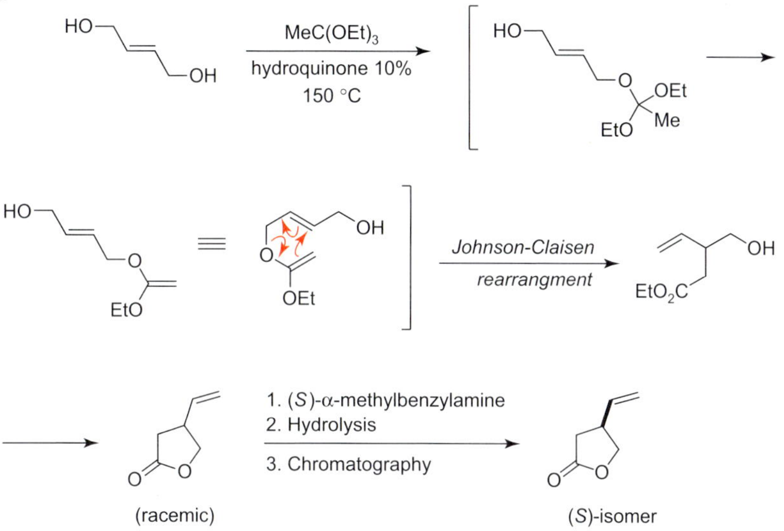

B.

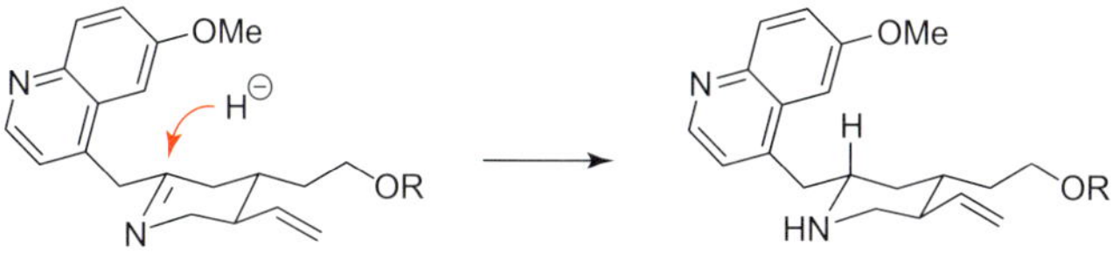

Figure 16.21 (*continued*).

Commentary: The total synthesis of quinine by Stork and coworkers starting with (3S)-3-vinyl butyrolactone is a beautiful example of visual imagery in the mind's eye. It should be noted that the vinyl group remains unscathed until quinine emerges as the victor. Key to the visualization of when and how the nitrogen atom would be introduced, is the expectation that the four carbon atoms of the starting lactone can be fully functionalized, and that the hydroxymethyl group can now accommodate an azide group. Although the starting lactone was prepared by resolution of a racemic precursor, it is of interest to recall the method of its preparation [85] (see relevant mechanism A, Figure 16.21). It should also be noted that the in principle, lactone enolate alkylation could also lead to **B**. However, the vinyl group would not provide enough stereochemical bias as observed in the acyclic amide enolate precursor to **A**. With intermediate **D** in hand, logical reactions would lead to the cyclic imine **G**. The stereochemical outcome of the hydride reduction of the cyclic imine **G** was predicated upon the premise that the vinyl and hydroxyethyl groups would adopt a diequatorial disposition in a half-chair conformation. This would result in an axial attack of hydride, placing the quinolymethyl appendage in an equatorial orientation.

16.6.2
Quinine: The Woodward and Doering formal vs total syntheses issue

In a 1944 Communication to the Editor of the *Journal of the American Chemical Society*, Woodward and Doering published a paper entitled "The Total Synthesis of Quinine" [75a]. A full paper bearing the same title appeared a year later in the same Journal [75b] (Figure 16.22). In reality, what the famous Harvard University chemists had synthesized was quinotoxin, which was resolved into its optical antipodes as the tartrate salts. The German chemists Rabe and Kindler [76] had previously converted *d*-quinotoxine, originally obtained from acid treatment of natural quinine by Pasteur in 1853 back to the natural product in a three-step process [77] (Figure 16.23). Thus, the claim for a total synthesis of quinine by Woodward and Doering was not entirely justified, even if the concluding paragraph in their 1945 paper stated: *"In view of the established conversion of quinotoxine to quinine (by Rabe and Kindler), with the synthesis of quinotoxine, the total synthesis of quinine was complete."* In today's parlance, the Woodward and Doering synthesis of quinine, would be labelled as a "formal" one. Nonetheless, much publicity was generated in the popular press and the news media when the synthesis was announced over 60 years ago, because of the crucial importance of quinine as a drug against malaria, and its limited supply during World War II [78].

Much doubt and speculation has existed over the years as to whether or not Woodward and Doering had ever produced any amount of purely synthetic quinine in their laboratory [72]. Curiously, the Rabe and Kindler experiments to convert *d*-quinotoxine to quinine remained experimentally untested in spite of the controversy. After considerable archival research [79, 80], it was concluded that Rabe and Kindler had indeed converted *d*-quinotoxine into quinine in 1918, thereby inferring that by achieving a total synthesis of *d*-quinotoxine, Woodward and Doering had also completed the formal synthesis of quinine. Having closed that chapter in the quinine saga, there remained to conclusively establish the feasibility of the Rabe and Kindler conversion of *d*-quinotoxine to quinine. In a series of carefully executed experiments, Smith and Williams [81] repeated the three-step sequence using modern methods of analysis. They also re-enacted the laboratory conditions

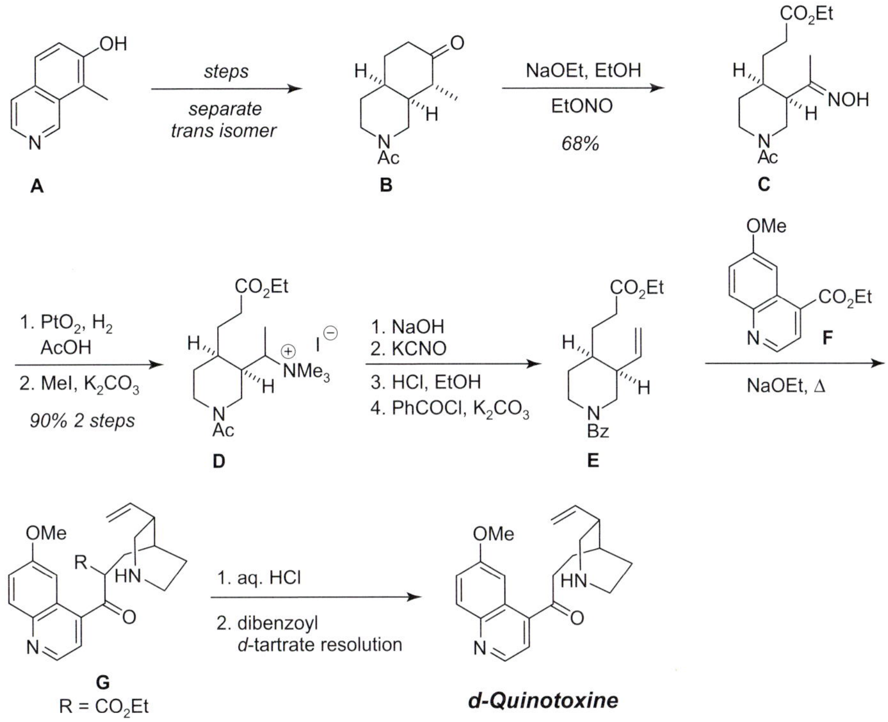

Key reactions: Oxime from α-C-methyl ketone (**B** to **C**); Hofmann elimination (**D** to **E**); enolate condensation/decarboxylation (**E** + **F** to **G**)

Relevant mechanism:

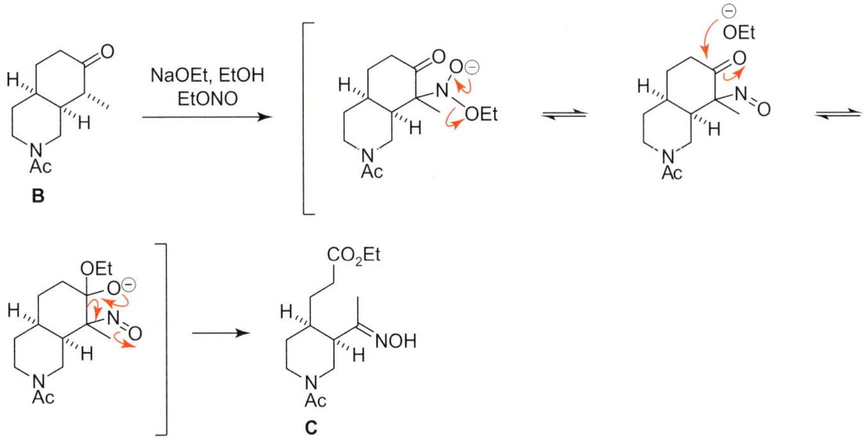

Figure 16.22 Total synthesis of *d*-quinotoxine by Woodward and Doering.

existing during the Rabe and Kindler period to the best of their abilities. Crystalline quinine, isolated as the tartrate salt, was indeed produced in the "modern" and in the "original" 1918 versions of the synthesis, although some of the intermediates could not be isolated in crystalline form. The "quinine total synthesis story" can now be put to rest with the factual knowledge provided by Smith and Williams regarding the validity of the 1918 Rabe and Kindler conversion of *d*-quinotoxine to quinine. Whether the totally synthetic (albeit resolved) *d*-quinotoxine tartrate salt obtained by Woodward and Doering in 1944 was actually also converted to quinine can now be considered to be a "formality" of anecdotal importance.

Commentary: Considering the methods and analytical techniques available to Woodward and Doering in 1944, when the completion of the total synthesis of *d*-quinotoxine (and formally quinine) was announced, the effort was of immense significance. Lacking today's methods of stereoselective and diastereoselective introduction of the vicinally *syn*-oriented acetic acid and vinyl side-chains, Woodward and Doering resorted to hydrogenation of an appropriately substituted isoquinoline **A**, to give a racemic mixture of *N*-acetyl-7-keto-8-methyl decahydroisoquinolines, from which the *cis*-isomer **B** was isolated as a crystalline solid (Figure 16.22). The incorporation of a nitrogen atom on an unactivated *C*-methyl group with concomitant cleavage of the bicyclic ketone **B** using ethyl nitrite as an electrophile is counterintuitive. The mechanism of this exceedingly mild transformation as proposed by Woodward and Doering [75], is shown in Figure 16.22.

Subsequent steps consisted of a Hofmann elimination of **D** to afford **E**, which was converted to **G** via a Dieckmann-type condensation with **F**. Decarboxylation under acidic conditions afforded racemic quinotoxine, which was resolved with dibenzoyl-*d*-tartrate to give *d*-quinotoxine. It is of interest to read a portion of the experimental section from the Woodward and Doering synthesis [75], describing, with meticulous details, the preparation of the oxime **C** and the vivid description of its crystalline characteristics.

This discourse of the Woodward and Doering formal synthesis of quinine would not be complete unless, we also show the elegant conversion of quinine to *d*-quinotoxine, and back, reported by Rabe and Kindler in 1918 (Figure 16.23) [76]. Acid-catalyzed fragmentation of the protonated quinine initially afforded the enol **A**, which is in equilibrium with *d*-quinotoxine (**B**). Treatment with HOBr led first to the *N*-bromo intermediate, which after treatment with NaOEt led, through the α-bromoketone **C**, to quinonone (**D**). Reduction with aluminum powder gave crystalline quinine in 12% yield.

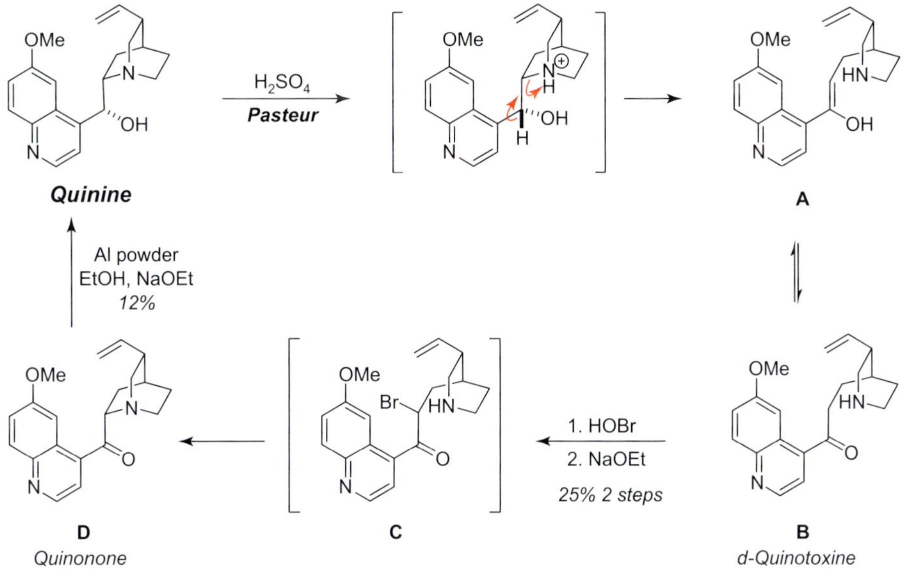

Figure 16.23 Conversion of quinine to *d*-quinotoxine (and back) by Rabe and Kindler.

16.6.3
Quinine: *Après* Woodward and Doering

During the years of the World War II, access to the cinchona bark from the Dutch East Indies, as a source of quinine was closed to North America [80]. The euphoric reaction of the press to the announcement of a "total synthesis of quinine from basic elements" by Woodward and Doering was understandable. Even today, encyclopaedias across the globe still hail the 1944 achievement as a "historic moment . . ." or "as one of the classical achievements of synthetic organic chemistry" [72, 79, 82].

Although there was a relative calm on the quinine synthesis front after such enthusiastic reports, three seminal reports were published in the period 1970–1972, describing total syntheses of quinine and quinidine.

16.6.3.1 **The Uskoković synthesis** – The first of these, came from a team of chemists at Hoffmann-La Roche Laboratories in New Jersey, headed by Uskoković [74] (Figure 16.24). Catalytic hydrogenation of 3-ethyl-4-(methoxycarbonylmethyl) pyridine (**A**) followed by classical resolution gave cincholoipon methyl ester (**B**) as the tartrate salt. Chlorination with NCS and irradiation of the intermediate *N*-chloro product **C** with a 200-W Hanovia lamp, followed by *N*-benzoylation gave the chloroethyl intermediate **D** in excellent yield. Base-induced elimination gave enantiopure *N*-benzoylmeroquinene **E**. Treatment of **E** with the lithium anion generated from the quinoline **F** afforded the ketone **G**. Conversion to an epimeric mixture of acetates (**H**), cleavage of the *N*-benzoyl group, then heating under reflux in the presence of NaOAc and AcOH in benzene resulted in the formation of a 1:1 mixture of deoxyquinine (**I**), and deoxyquinidine (**J**), epimeric at C-8. A highly stereoselective hydroxylation could be achieved with each diastereomer. Alternatively, treatment of a mixture of **I** and **J** with molecular oxygen in the presence of *t*-BuOK in DMSO and *t*-butanol gave quinine and quinidine, which could be separated by crystallization. The synthesis was completed in 17 steps to give quinine and quinidine.

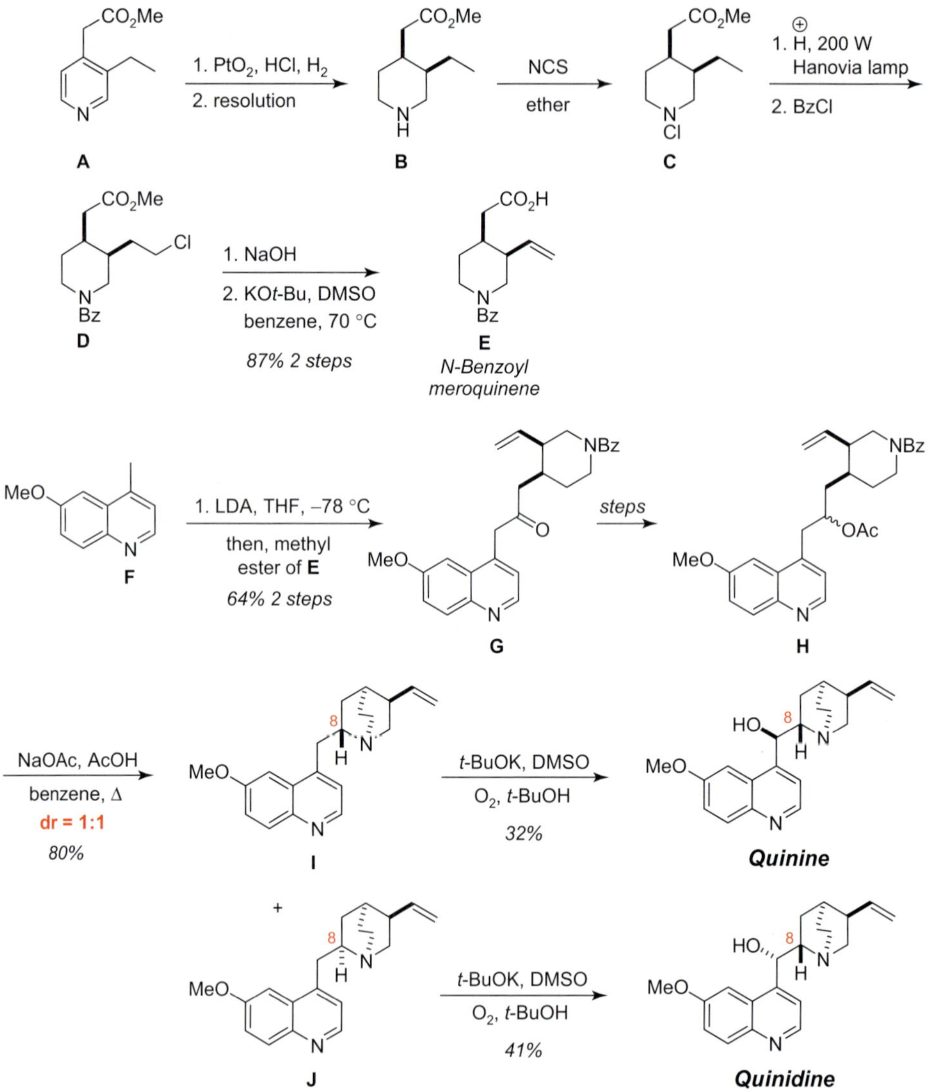

Figure 16.24 Synthesis of quinine by the Hoffmann-La Roche group.

Key reactions: *Radical N-chlorination (**B** to **C**); Löffler-Freytag rearrangement (**C** to **D**); intramolecular cyclization (**H** to **I** and **J**); benzylic oxidation (**J** to quinine)*

Relevant mechanisms and transformations:

A.

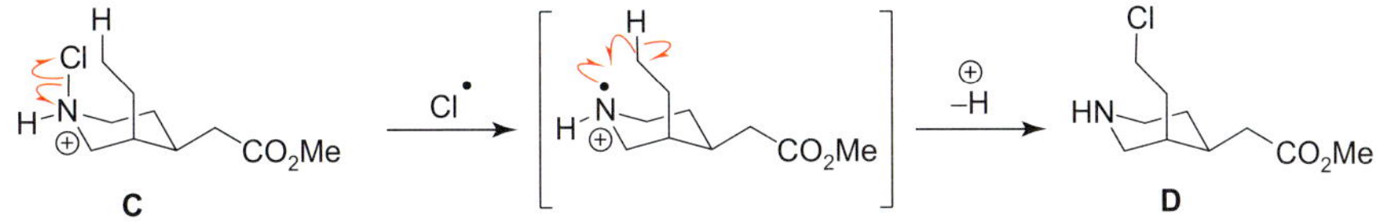

B.

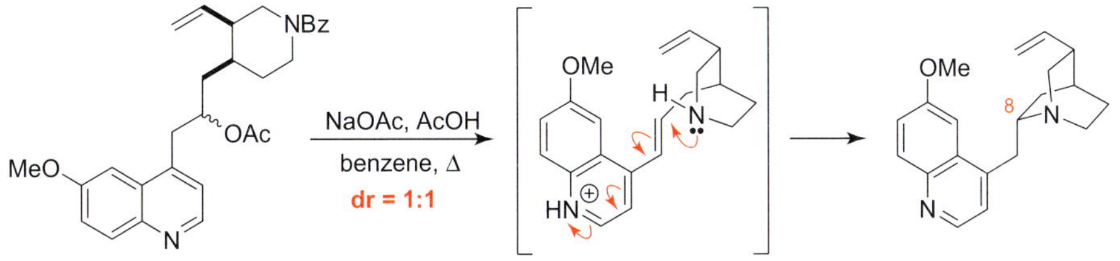

C.

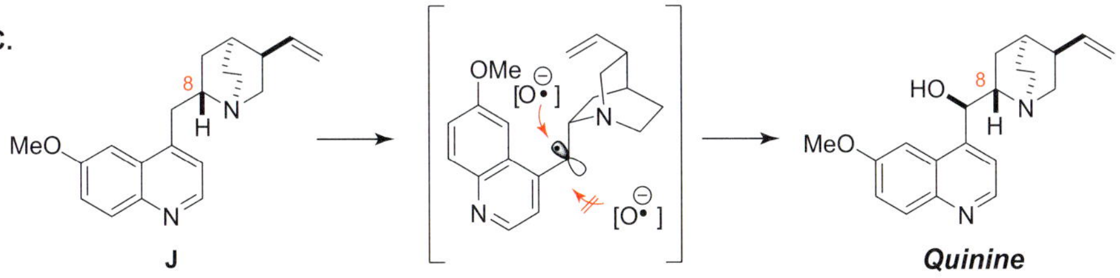

Figure 16.24 *(continued).*

Commentary: The synthesis by the Hoffmann-La Roche group embodies the elements of innovative chemistry, particularly in the application of the photochemical Löffler-Freytag rearrangement (**C** to **D**, see relevant mechanism A, Figure 16.24). The authors offered the reasonable postulate that irradiation of protonated *N*-chloro intermediate **C** generates a radical cation in which the 3-ethyl group is favorably disposed to generate a terminal radical that is captured by a chlorine radical.

The cyclization of **H** to deoxyquinine (**I**) and deoxyquinidine (**J**) (a 1:1 mixture of C-8 epimers) respectively, proceeded by an initial elimination of the acetoxy group leading to an olefin that was intramolecularly attacked by the piperidine nitrogen (see relevant mechanism B, Figure 16.24). Finally, the remarkably stereoselective hydroxylation at the benzylic (C-9) carbon atom of **I** (and/or **J**) was explained on the basis of a preferred orientation of the *N*-lone pair, and the capture of the benzylic radical from a favored pro-(*S*) face (see relevant mechanism C, Figure 16.24).

16.6.3.2 **The Gates synthesis** – A paper immediately following Uskoković's (same received dates of November 14, 1969), by Gates and coworkers [83] disclosed their total synthesis of quinine and quinidine (Figure 16.25). Condensation of the quinoline aldehyde **A** with the triphenylphosphonium salt **B** prepared from *N*-acetylmeroquinene, followed by treatment with AcOH gave *trans*-coupling product **C** as the major isomer. Treatment of **C** with NaOH, followed by refluxing in EtOH afforded deoxyquinine and deoxyquinidine cyclization products **D** and **E** in a 1:1 ratio similar to Uskoković's experience in the same series [74]. Benzylic oxygenation under slightly modified conditions afforded quinine and quinidine.

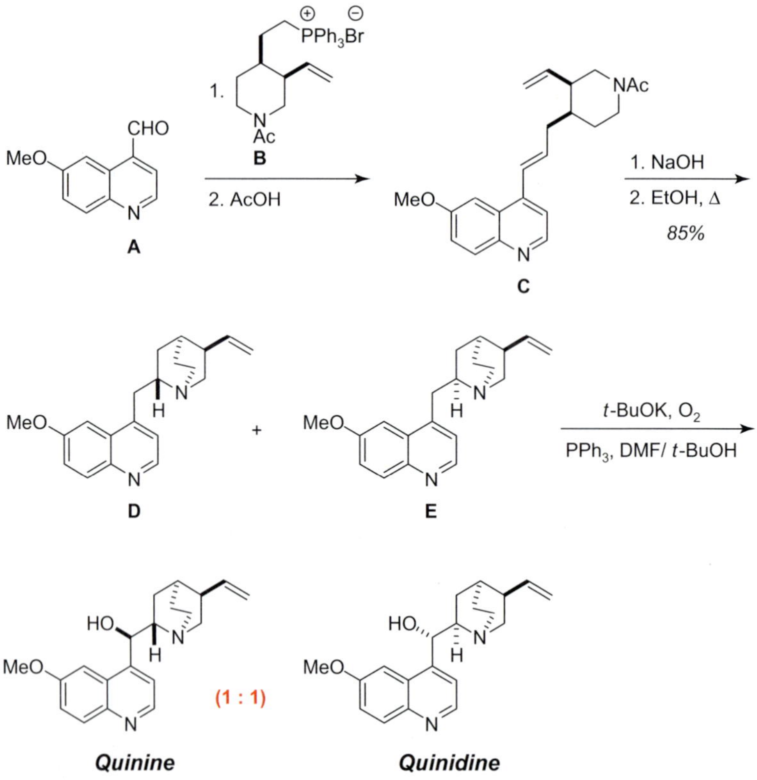

Key reactions: *Wittig olefination/isomerization (**A** + **B** to **C**); intramolecular cyclization (**C** to **D** and **E**); benzylic oxidation (**D** + **E** to quinine and quinidine)*

Figure 16.25 Synthesis of quinine and quinidine by Gates.

16.6.3.3 The Taylor and Martin synthesis – In 1972, Taylor and Martin [84] used a novel ylid-based aromatic displacement to prepare Uskoković-Gates' unsaturated intermediate **C** from chloro quinoline **A** and *N*-acetylmeroquinene aldehyde (**B**) (Figure 16.26). Treatment with base afforded a mixture of the C-8 epimeric deoxyquinine and deoxyquinidines **D**, which were oxidized according to Uskoković [74] or Gates [83] to give quinine and quinidine.

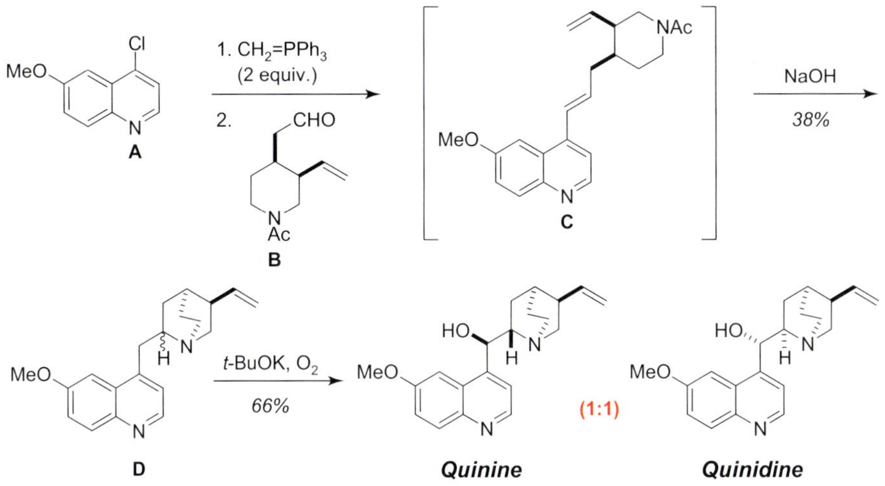

Figure 16.26 Synthesis of quinine by Taylor and Martin.

16.6.4
Quinine: Total synthesis in the modern age of catalysis

Seminal contributions to science are valued for their sustained impact from the time of their discovery to present-day practice. In the area of total synthesis, the annals of selected classic achievements will always remain as yardsticks to measure, and possibly surpass, creativity. As commented upon elsewhere [85], the impact of a given synthetic strategy toward a natural product or a molecule of interest, must be appreciated within the time frame with which the work was done. With the advent of advanced analytical and separation techniques coupled with innovative bond forming reactions involving catalytic asymmetric processes, how much more efficient and elegant should a new synthesis of quinine be than the preceding ones?

16.6.4.1 The Jacobsen synthesis – Enantioselective conjugate addition of methyl cyanoacetate in the presence of the Jacobsen aluminum-salen catalyst **B** [86], to the enamide **A** afforded **C** in 92% (Figure 16.27) [87]. Reduction of the nitrile, and cyclization led to the lactam **D** as a 1:1.7 *cis/trans* mixture. Equilibration via the lithium enolate allowed the enrichment of the mixture to the *cis*-diastereomer (3:1), after which, routine functional group manipulation and chromatographic

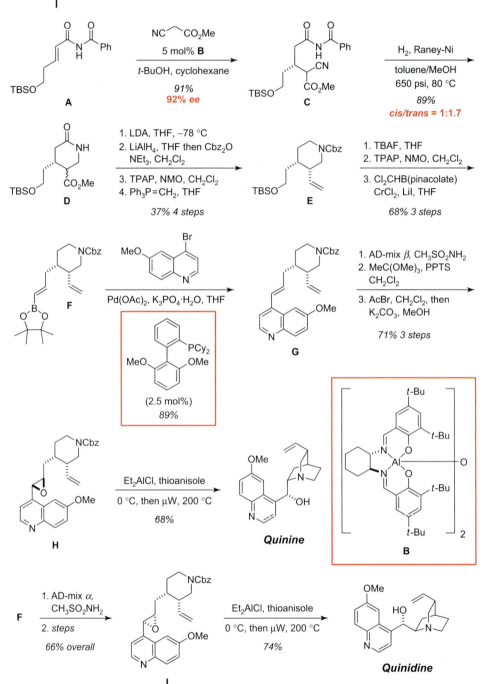

Figure 16.27 Synthesis of quinine and quinidine by Jacobsen.

separation gave meroquinene analogue **E**. Conversion to vinylboronate **F** under modified Takai conditions [88] was achieved in three steps, which was followed by a Buchwald coupling [89] affording known [83, 84] *trans*-olefin **G** in excellent yield. Sharpless asymmetric dihydroxylation (AD-mix β) [90], followed by conversion of the diol to the epoxide **H** [91], and cleavage of the N-Cbz group with Et$_2$AlCl/thioanisole [92] gave the corresponding piperidine. Heating under microwave conditions at 200 °C afforded quinine in 68% yield. Quinidine was similarly prepared using AD-mix α and proceeding with the epoxide **I**.

Commentary: The first, catalytic enantioselective synthesis of quinine and quinidine reported by Jacobsen and coworkers [87] represents a true paradigm shift in strategic planning compared to previous efforts dating back to Woodward and Doering's [75] classical approach in 1944. Absolute stereochemistry was secured early with a catalytic asymmetric conjugate addition using the Jacobsen aluminum-salen ligand (**A** to **C**). This elegant display of absolute stereochemical control (>92% *ee*) should be contrasted with the four steps that were necessary to introduce the vinylic appendage, and adjust stereochemistry (3:1 *cis/trans*). The second catalytic step led to the vinylboronate **F** (>20:1 *E/Z*), which was instrumental in the Buchwald modification of the Suzuki coupling to give the all *trans*-olefin **G**. Finally, capitalizing on a Sharpless asymmetric dihydroxylation and subsequent epoxidation, the intramolecular cyclization, already used by Uskoković [74], Gates [83], and Taylor [84] was rendered all the more efficient under microwave conditions. A 4% yield for 17 linear steps characterizes this first application of catalysis to the total synthesis of quinine and quinidine by the Jacobsen group.

16.6.4.2 The Kobayashi synthesis – An efficient and highly stereocontrolled total synthesis of quinine and quinidine was reported by Kobayashi and coworkers in 2004, followed by an improved version in 2005 [93]. Its salient features will be discussed in retrosynthetic analysis format in the context of other modern-day syntheses relying on catalysis at critical stages (Figure 16.28). The linear sequence toward the meroquinene motif started with *meso*-1,4-dihydroxy cyclopentene **A**. Enzymatic desymmetrization afforded the monoacetate **B** in >95% *ee*. Two sequential carbon functionalizations with complete transfer of chirality relied on Tsuji-Trost π-allyl Pd chemistry [94] (**B** to **C**), and a [2,3]-Claisen rearrangement using ethoxyethylene as an acetaldehyde equivalent (**C** to **D**). With the *cis*-substitution pattern secured by exploiting the chiral template provided by **B**, further systematic manipulation of functional groups resulting from the oxidative cleavage of the cyclopentene **E**, and processing via **F** and **G**, gave the meroquinene aldehyde **H** in enantiopure form. Phosphonate anion coupling between **I** and **H** afforded the *trans*-olefin **J**, which was further manipulated to give quinine. Unlike the Jacobsen synthesis [87], where two metal-based ligands were used to catalyze critical bond formation *en route* to functionalized meroquinene intermediates, Kobayashi relied on Nature's esterases as catalysts to generate a carbocyclic chiron **B**. This was subjected to systematic functionalization using a chiral template approach in which chirality was transferred intramolecularly, leading to **D**. A self-immolative

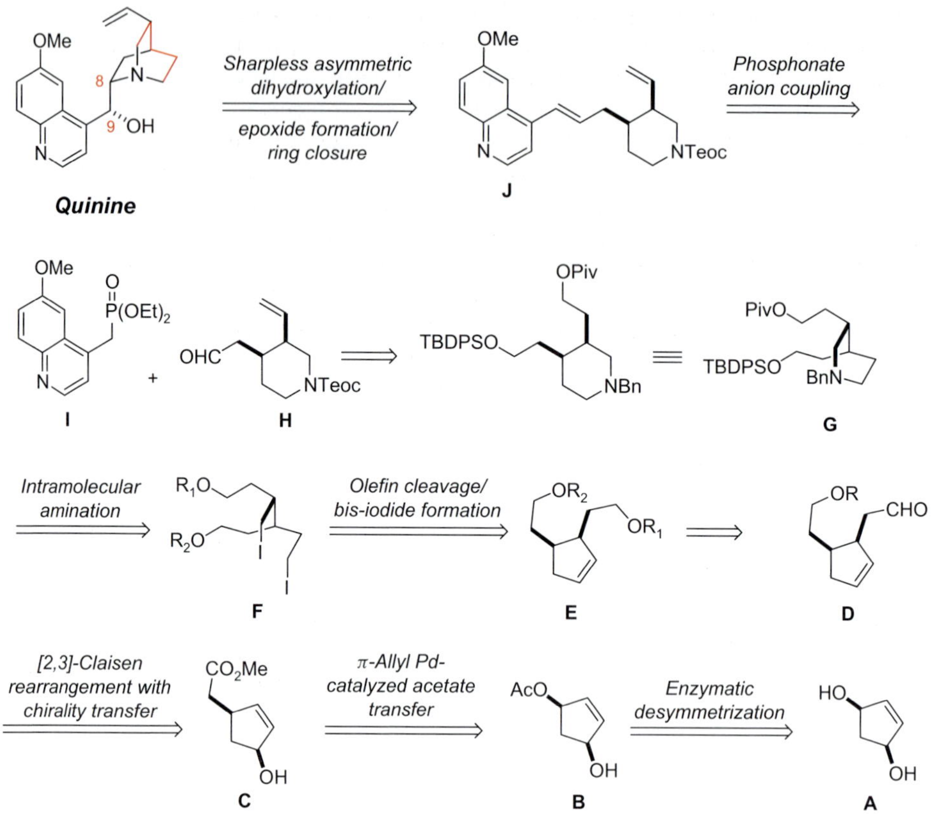

Quinine

Figure 16.28 Kobayashi's retrosynthetic analysis of quinine.

process revealed an acyclic bis-iodide **F**, which was used to construct the *cis*-oriented substitution pattern in the piperidine **G**.

It is of interest to note that except for the Woodward and Doering [75], and Stork [72] syntheses of quinine, more recent efforts by Jacobsen [87] and Kobayashi [93] relied on an intramolecular ring closure. In this regard, the Uskoković [74] and Gates [83] stereoselective benzylic C-8 hydroxylation of deoxyquinine in 1970, which was also used by Stork some 30 years later [72], must be appreciated for its operational simplicity, conceptual elegance, and mechanistic insight.

16.6.4.3 **The Williams and Krische syntheses of 7-hydroxyquinine** – In 2006, Williams and coworkers [95] reported a synthesis of (7*R*)-7-hydroxy quinine relying on a hitherto unexplored disconnection between C-3 and C-4 of the quinine structure (Figure 16.29). Stereoselective aldol condensation between the TBS-enol ether of the Williams lactone **A** [96] and aldehyde **B** led to amino ester **C** (dr > 30:1). Chain-extension and cyclization afforded piperidine **D**. Appending the OBz crotyl group on the piperidine nitrogen under basic conditions was followed by oxidation of **E** to the ketone and formation of a silyl enol ether **F**, setting the stage for an intriguing stereocontrolled ketone enolate Pd-mediated carbocyclization [97]. In the presence of TABF/Pd$_2$(dba)$_3$ and trifuryl phosphine at 85 °C in toluene, 7-oxo quinuclidine was formed stereoselectively

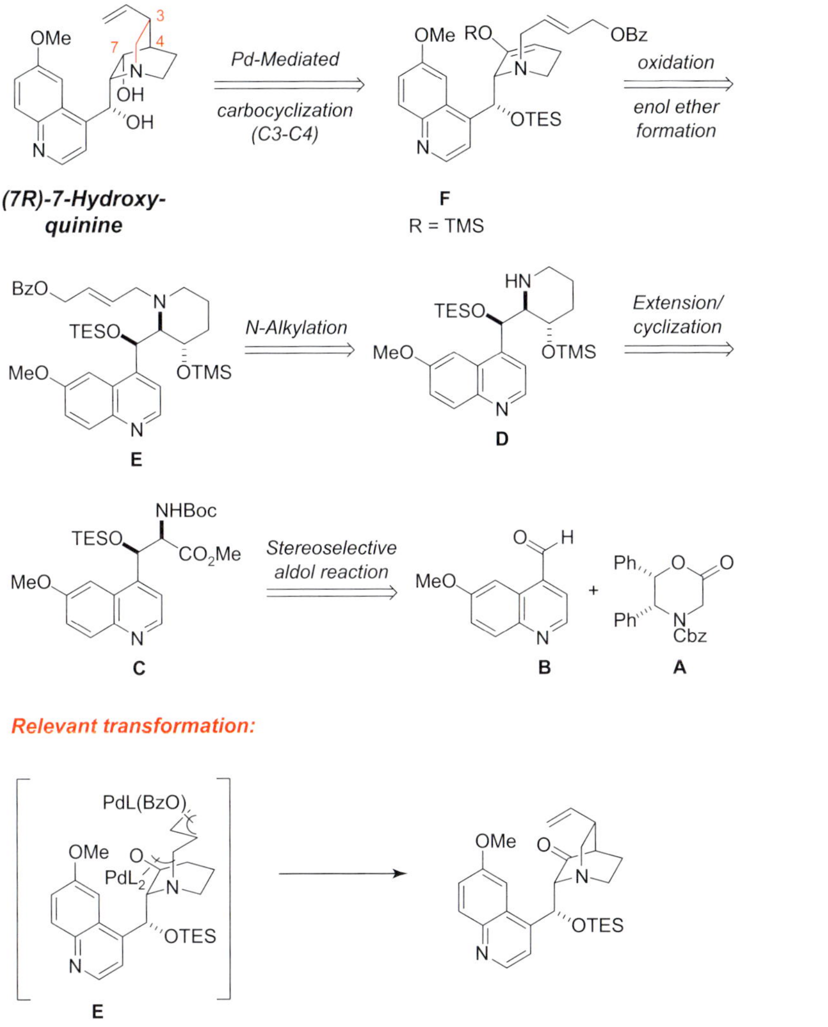

(7R)-7-Hydroxy-quinine

F
R = TMS

E

D

C

B

A

Relevant transformation:

E

Figure 16.29 Synthesis of (7*R*)-7-hydroxyquinine by Williams.

(see relevant transformation, Figure 16.29). Reduction of the carbonyl group gave predominantly (7*R*)-hydroxy quinine. A plausible mechanism was suggested by Williams as shown in Figure 16.29. It should be noted that the original glycine framework remained embedded in the target molecule as the two-carbon bridge of the aza-bicyclo[2.2.2]hexane motif.

A novel approach to meroquinene and 7-hydroxy quinine was disclosed by Krische and coworkers in 2008 [98] (Figure 16.30). A merged Morita-Baylis-Hillman and

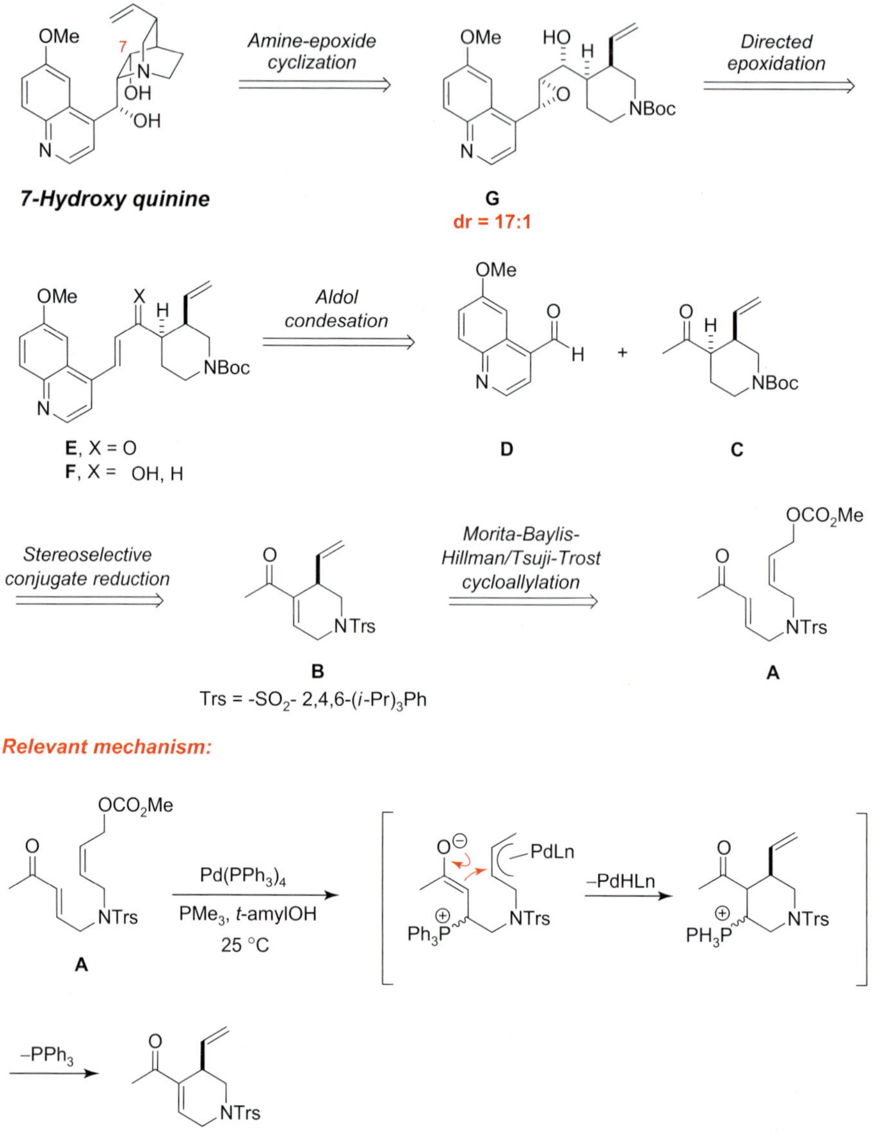

7-Hydroxy quinine

G
dr = 17:1

E, X = O
F, X = OH, H

D

C

B
Trs = -SO₂- 2,4,6-(*i*-Pr)₃Ph

A

Relevant mechanism:

A

Figure 16.30 Synthesis of (±)-7-hydroxy quinine by Krische.

Tsuji-Trost reaction was used for the intramolecular carbocyclization of the acyclic enone **A** in the presence of Pd(PPh)$_4$ and Ph$_3$P in *t*-amyl alcohol as solvent (see relevant mechanism Figure 16.30). Conjugate hydride addition to the enone **B** led to the (±)-meroquinene precursor **C** with the desired *syn*-substitution of the vinyl and methyl ketone groups. Formation of the lithium enolate, aldol condensation with 6-methoxy quinoline-4-carbaldehyde **D**, acetylation and β-elimination afforded the enone **E**. Reduction with L-Selectride (>20:1 selectivity) and hydroxyl directed epoxidation with an optimized vanadium catalyst afforded the epoxide **G** as the major diastereomer (dr = 17:1). Intramolecular cyclization was achieved in the presence of Zn(OTf)$_2$ to give (±)-7-hydroxy quinine in 70% yield. Attempts to deoxygenate the C-7 alcohol to give quinine were unsuccessful most likely due to steric hindrance. However, Krische was able to deoxygenate the C-7 alcohol derived from the ketone **E**. The resulting *trans*-olefin **F** offers an alternative, highly stereocontrolled route to the Jacobsen intermediate [87] albeit as the racemate (see compound **G** Figure 16.27).

16.6.5
The total synthesis of quinine in the mind's eye

The year 2009 marked the 65th anniversary of the total synthesis of *d*-quinotoxine by Woodward and Doering [75]. Other historic events to remember in the annals of the natural products chemistry relating to quinine are Pasteur's experiments to produce quinotoxine by acid treatment [77], Rabe's synthesis of quinine (1853) from quinotoxine (1918) [76], and Prelog's [99] establishment of the *cis*-relationship of the C-3 and C-4 substituents, hence, the absolute configuration of meroquinene (1944) by chemical degradation (Figure 16.31).

The various approaches and historical insights leading to the synthesis of quinine have been expertly reviewed in recent years [71]. It is remarkable that

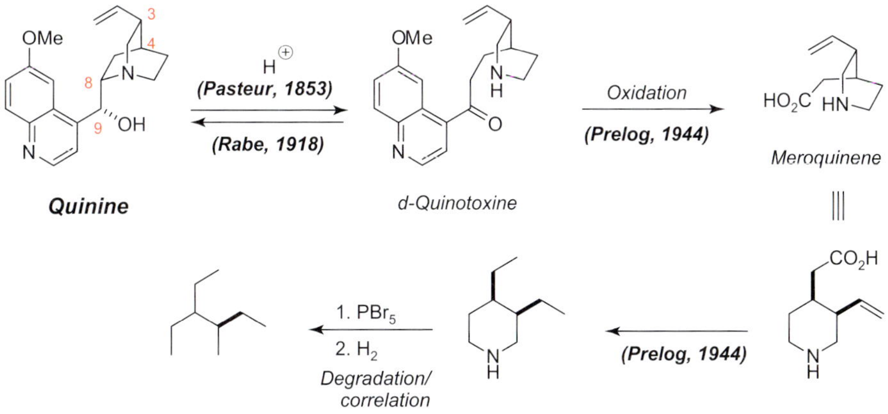

Figure 16.31 Determination of the absolute configuration of quinine and meroquinene by Prelog.

decades since the first "total synthesis of quinine" was announced by Woodward and Doering [75] there has been a renewed interest to develop innovative strategies as highlighted in the preceding sections. Although there have not been dramatic improvements in the number of steps or overall yields compared to previously reported syntheses, stereoselective bond forming processes with quasi-complete control of absolute stereochemistry have become hallmarks of the more recent contributions [72, 87, 93].

Let us review once more the synthetic approaches to quinine by recreating the visual and mental processes that led to the choice of different starting materials, especially for the construction of the quinuclidine nucleus, including the non-trivial task of securing the absolute configuration of the existing four stereogenic carbon atoms at C-3,C-4, C-8 and C-9 [100, 101] (Figure 16.31). Not until the synthesis of quinine by Stork and coworkers [72], was it possible to achieve virtually complete stereochemical control of all four stereogenic centers (Figure 16.21). This was achieved by avoiding the formation of the problematic C-8–N bond, which had previously also led to quinidine (C-8 epimer) [74, 83, 84]. Jacobsen's novel approach [87] "faltered" momentarily with the need to separate a 3:1 *cis/trans* mixture of diastereomers at the foot of the eventual vinyl group in intermediate **E** (Figure 16.27). Subsequent steps allowed him to soar again to greater stereochemical heights before the synthesis was completed. Kobayashi's decision [93] to use Nature's esterases to provide the basis for absolute control of chirality, led to total stereocontrol in conjunction with asymmetric oxidation reactions later in the synthesis, as also reported by Jacobsen [87] (Figure 16.28).

The mind's eye view of the structure of quinine was very much conditioned by the synthetic methods available to chemists during specific periods (Figure 16.32). Thus, Woodward [75] and Uskoković [74] relied on aromatic precursors and catalytic hydrogenation to create the piperidine ring with the requisite 3,4-*cis*-substitution pattern found in homomeroquinene and meroquinene respectively. Gates [83] and Taylor [84] took advantage of the methods developed by Uskoković [74]. Stork [72] utilized acyclic templates in building meroquinene-like precursors. Jacobsen [87] took it a step further by utilizing catalytic enantioselective methods. Finally, the visually hidden chiron derived from a desymmetrized cyclopentene 1,4-diol provided the chiral template to elaborate in a totally stereocontrolled approach, culminating with the last recorded total synthesis of natural quinine by Kobayashi [93].

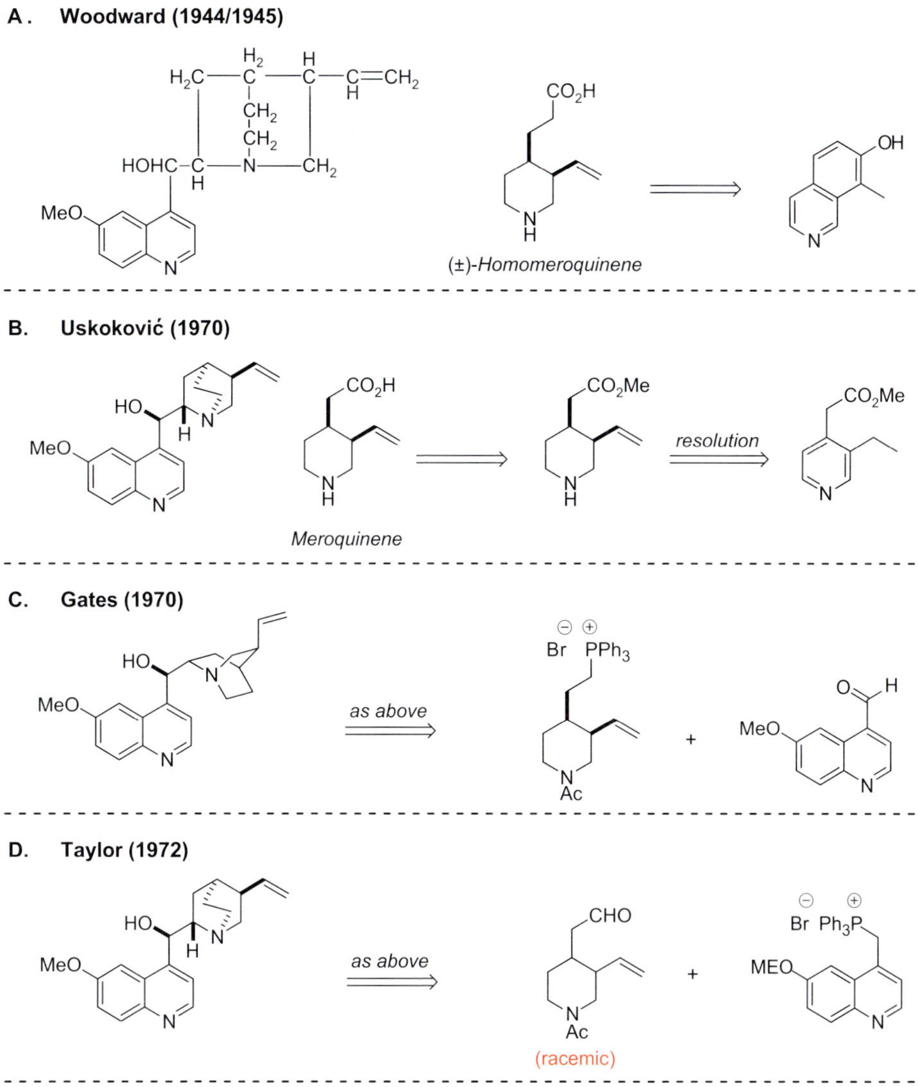

A. Woodward (1944/1945)

(±)-*Homomeroquinene*

B. Uskoković (1970)

Meroquinene

resolution

C. Gates (1970)

as above

+

D. Taylor (1972)

as above

+

(racemic)

Figure 16.32 Synthesis of quinine and the mind's eye.

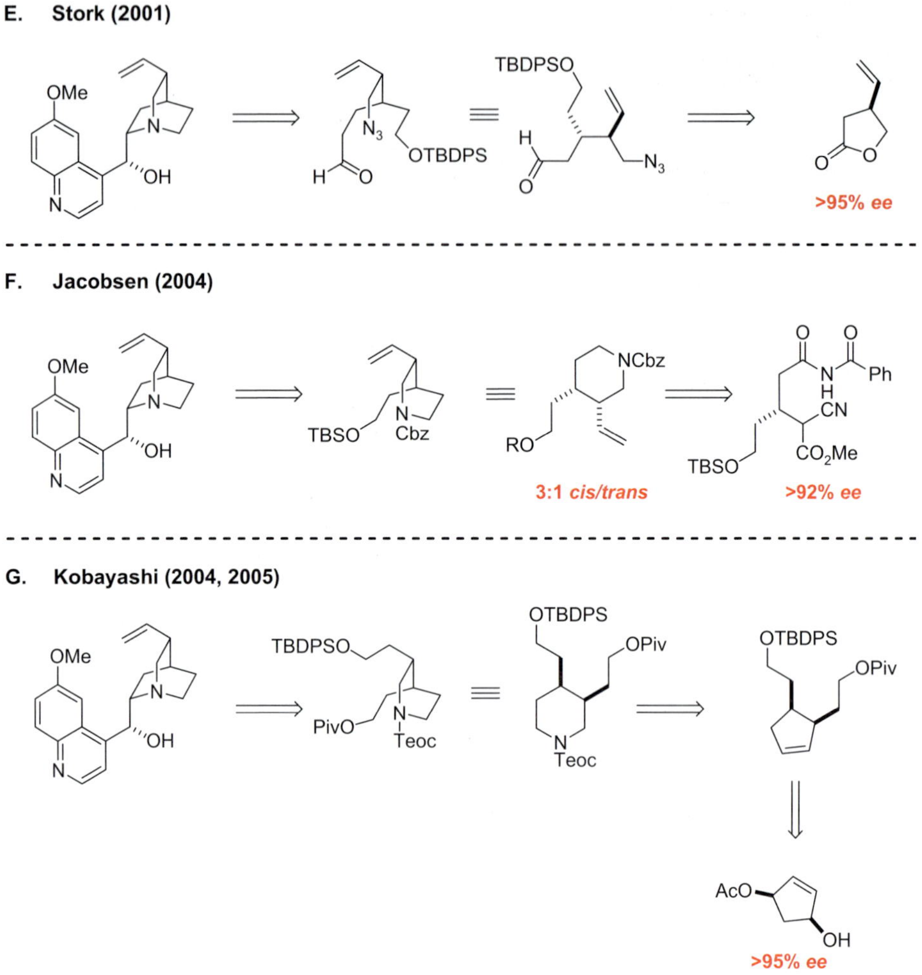

E.   Stork (2001)

>95% ee

- -

F.   Jacobsen (2004)

3:1 *cis/trans*   >92% ee

- -

G.   Kobayashi (2004, 2005)

>95% ee

Figure 16.32   (*continued*).

The different disconnections and methods to introduce chirality at C-9 in quinine are summarized in Figure 16.33. We close this section on quinine by quoting from the poignant remarks of Stork: [79]

> "*The value of quinine synthesis has essentially nothing to do with quinine . . . it is like the solution of a long-standing proof of an ancient theorem in mathematics: it advances the field.*"

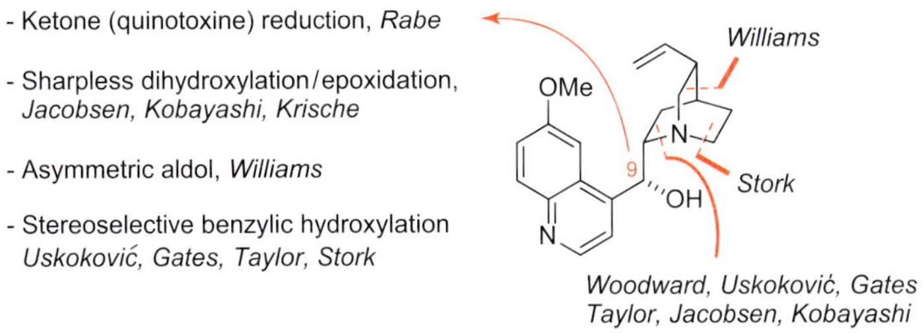

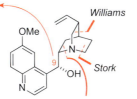

- Ketone (quinotoxine) reduction, *Rabe*

- Sharpless dihydroxylation/epoxidation, *Jacobsen, Kobayashi, Krische*

- Asymmetric aldol, *Williams*

- Stereoselective benzylic hydroxylation *Uskoković, Gates, Taylor, Stork*

Williams

Stork

Woodward, Uskoković, Gates Taylor, Jacobsen, Kobayashi

Figure 16.33 Quinine: Major bond disconnections.

16.7
Lactacystin

Lactacystin, a potent and selective inhibitor of the 20S proteasome, a proteolytic enzyme with wide implications in living cells [102], was isolated by Ōmura and coworkers in 1991 [103]. Its unique structure and absolute stereochemistry was determined by X-ray crystallography (Figure 16.34) [104]. The remarkable biological activity of lactacystin, coupled with its pH-dependent transformation to a cell-permeable active congener (−)-*clasto*-lactacystin (also known as omuralide) [105], instigated a number of studies aimed at their stereoselective syntheses [106]. In spite of the deceptively simple structure of lactacystin, a major challenge presents itself in considering the methods by which a tetrasubstituted stereogenic center at C-5 can be elaborated in a densely functionalized α-substituted pyroglutamic acid motif. An added challenge is the control of stereochemistry at the C-9 hydroxy isobutyl side-chain appendage. In a number of these syntheses, the primary target has been omuralide, which is an immediate chemical precursor of lactacystin, by opening the β-lactone ring with *N*-acetyl-ʟ-cysteine. Each of the individual syntheses have their merits with regard to the methods devised for the elaboration of the critical stereogenic centers at C-5 and C-6. While quite elegant, in some cases, the implementation of such "signature" methods have also lengthened the number of steps in the syntheses due to the necessity to introduce additional functional groups at other sites. Nevertheless, practical gram-scale syntheses of lactacystin and omuralide are now available, allowing for a better understanding of the biology of these fascinating molecules [102]. A highly potent proteasome inhibitor, salinosporamide A was also isolated from a marine actinomycete by Fenical and coworkers [107]. Its synthesis has also been completed in more than one laboratory [108].

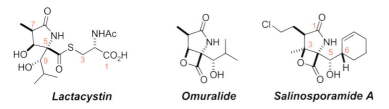

Lactacystin　　　　　*Omuralide*　　　*Salinosporamide A*

Figure 16.34 Structures of lactacystin, omuralide and salinosporamide A.

16.7.1
The first Corey synthesis

The first total synthesis of lactacystin was reported in 1992 by Corey and Reichard [109a] (Figure 16.35). An improved synthesis, amenable to large quantities was published in 1998 by Corey and coworkers [109b]. Analysis of the structure of

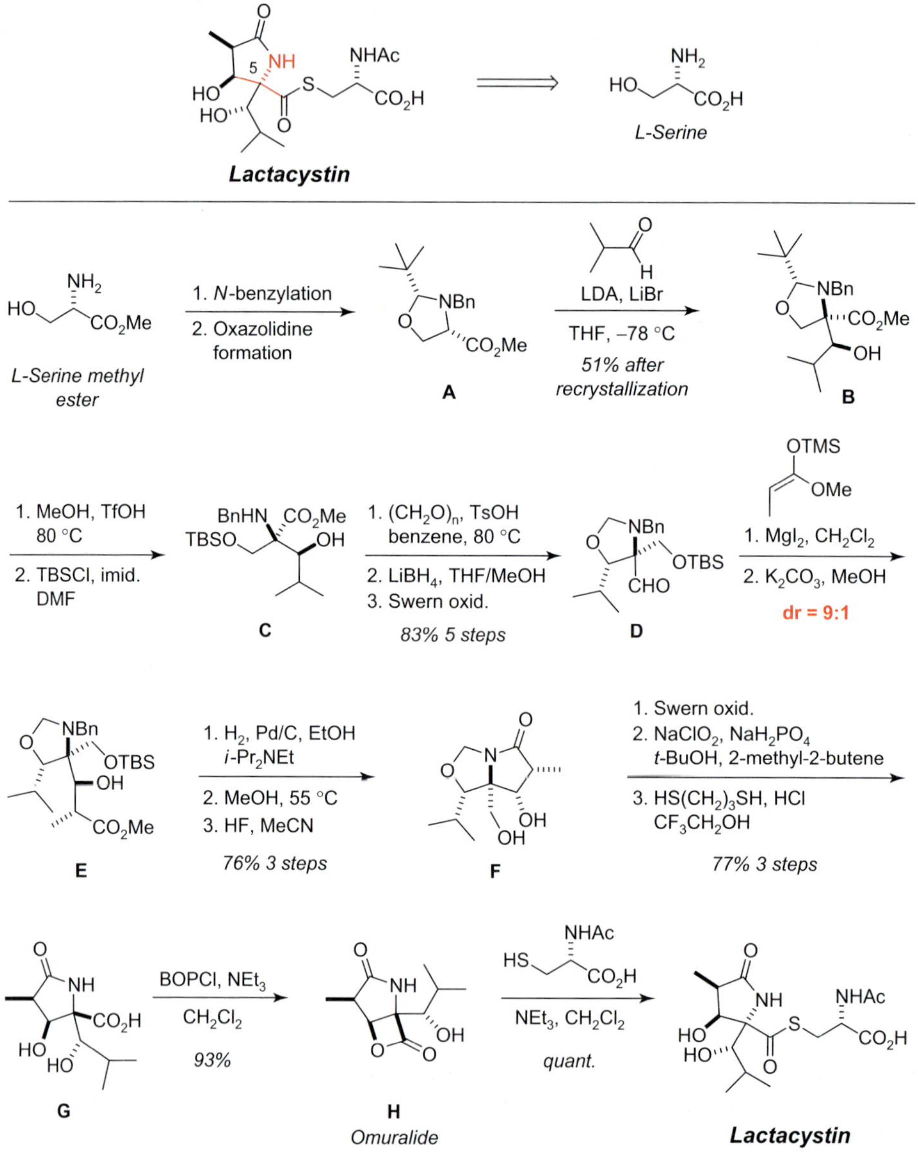

Figure 16.35 Corey's first synthesis of lactacystin.

Key reactions: *Diastereoslective aldol (**A** to **B**); Mukaiyama-like MgI$_2$-mediated aldol (**D** to **E**); intramolecular β-lactone formation (**G** to omuralide, **H**).*

Relevant mechanism:

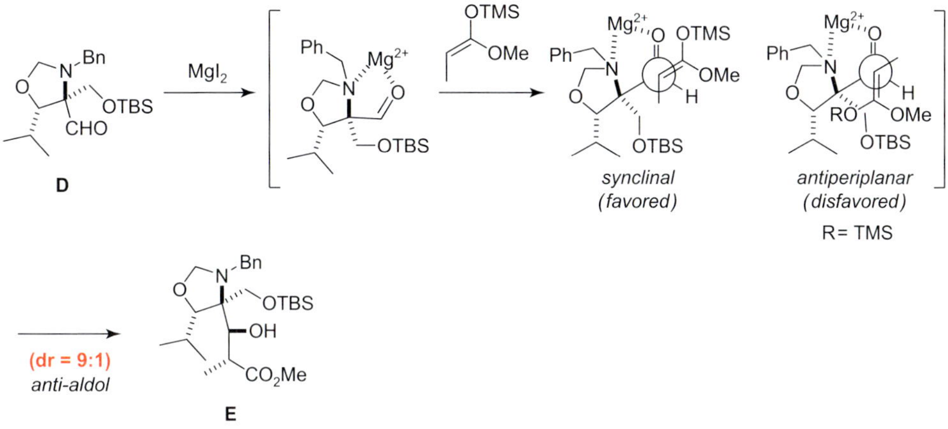

Figure 16.35 *(continued).*

lactacystin reveals L-serine as an embedded amino acid, albeit with α-substitution at C-5. Thus, the *N,O*-acetal **A**, available from L-serine in three steps, was subjected to a stereocontrolled aldol condensation with isobutyraldehyde to give **B** in 51% yield after recrystallization. Cleavage of the acetal and TBS-protection gave **C**, which was converted to the branched-aldehyde **D** in high overall yield. A Mukaiyama-type aldol reaction with the TMS ketene acetal of methyl propionate in the presence of MgI$_2$ followed by hydrolysis, led to **E** as the major isomer. Conversion to lactam **F**, and elaboration of functional groups afforded **G**, which was transformed into omuralide (**H**) via an intramolecular esterification. Thiolysis of **H** with *N*-acetyl cysteine gave lactacystin. A mechanistic rationale for the improved stereoselectivity of the Mukaiyama aldol reaction was based on the prevalence of a synclinal orientation of the achiral ketene acetal, leading to the *anti*-aldol product **E** (see relevant mechanism, Figure 16.35).

16.7.2
The second Corey synthesis

An enzymatic desymmetrization approach [110] starting with **A**, gave Corey and Nagamitsu [111] access to enantiomerically enriched (95% *ee*) product **B** (Figure 16.36). Amide formation followed by Dieckmann cyclization (see relevant mechanism A, Figure 16.36) led to **C**, which was hydroxylmethylated in a highly stereoselective manner to give **D** (dr = 9:1). The crystalline lactam was converted to **E**, which was subjected to desulfurization with Raney-Ni and then oxidized to aldehyde **F** as the major isomer (dr = 10:1). Addition of 2-propenylmagnesium

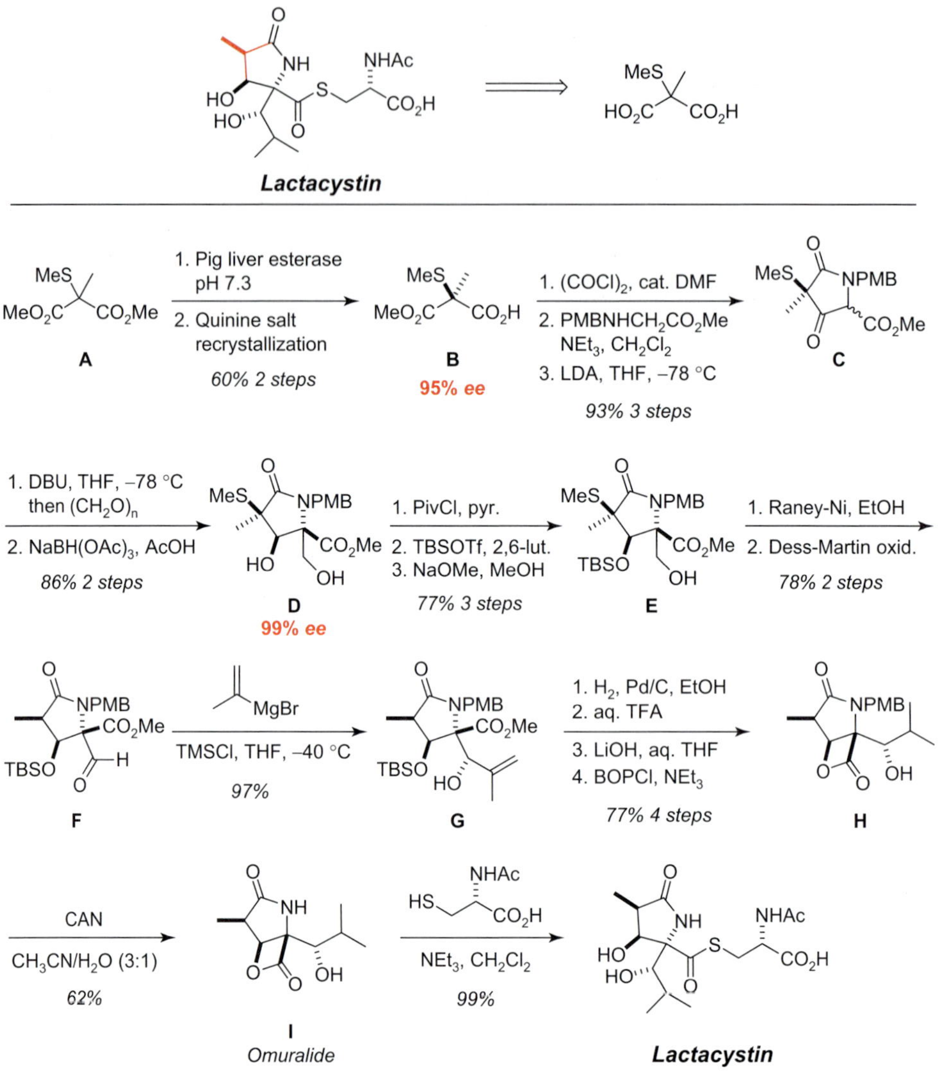

Lactacystin

Key reactions: PLE-mediated desymmetrization (**A** to **B**); Dieckmann condensation (**B** to **C**); aldol reaction and diastereoselective reduction (**C** to **D**); Grignard addition (**F** to **G**)

Figure 16.36 The second Corey synthesis of lactacystin.

bromide in the presence of TMSCl led to **G** with high stereoselectivity. Subsequent known steps led first to omuralide (**H**), then to lactacystin. This relatively short synthesis starting with an achiral malonate has the merits of high stereoselectivity and providing crystalline intermediates. In choosing the 2-methylthio-2-methyl malonate monoester, and subsequent desymmetrization to **B**, the authors were

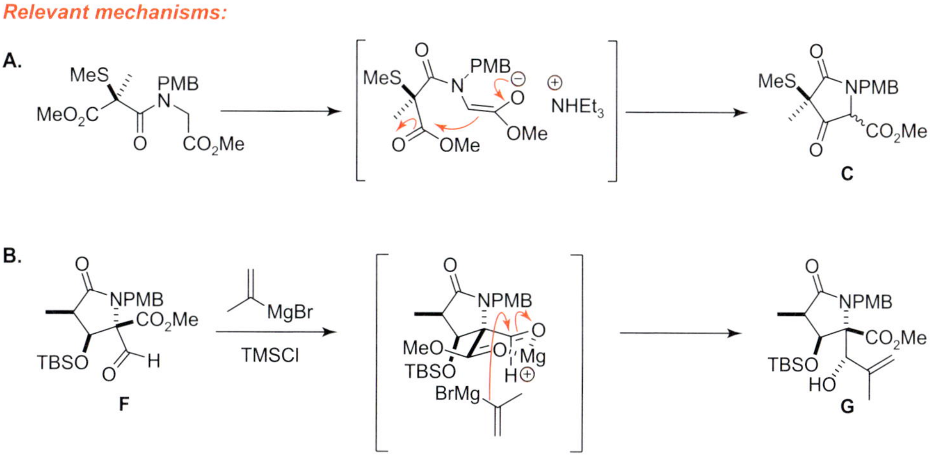

Relevant mechanisms:

A.

B.

Figure 16.36 (continued).

cognizant that Raney-Ni desulfurization (**E** to **F**) would run the risk of producing epimeric products at C-2. Not unexpectedly, the steric bias of neighboring substituents led to excellent selectivity in favor of the desired aldehyde **F**. A rationale for the stereoselective addition of the Grignard reagent is shown in Figure 16.36 (see relevant mechanism B). It is noteworthy that in the absence of TMSCl, and using isopropylmagnesium bromide, addition to the aldehyde **F** was problematic.

16.7.3
The Baldwin synthesis

A quick analysis of the structure of lactacystin will reveal an (*R*)-pyroglutamic acid motif as the heterocyclic component (Figure 16.37). Indeed, Baldwin and coworkers [112] started with *N,O*-acetal **A** [113], and proceeded to install the *C*-methyl group *en route* to unsaturated lactam **B**. The chirality at C-4 in the ring was destroyed in favor of pyrrole **C**, only to introduce it back through a vinylogous aldol reaction [114] with isobutyraldehyde. The resulting acetylated product **D** (dr = 9:1), was then converted to diol **E**, and the tertiary hydroxyl group was deoxygenated according to Barton [115] to give **F** as the major diastereomer after base-catalyzed equilibration of the *C*-methyl group. Functional group manipulation led to **G**, which was converted to lactacystin according to Corey's original protocol [109]. An overall yield of 7.5% from **A** was reported in spite of the need to separate isomers. In choosing the vinylogous aldol as a key step, it was necessary to prepare the pyrrole **C** (six steps from (*R*)-pyroglutamic acid) with a stereocontrolling (2*S*)-phenyloxazolidine appendage.

A study of Lewis acid and solvent dependence [116] revealed that only SnCl$_4$ in ether favored the formation of the desired *threo*-isomer **D** as the major product. This would imply that the tin-coordinated aldehyde approaches from the more hindered face of the oxazolidine in which the phenyl group occupies a pseudoequatorial orientation. Coordination of SnCl$_4$ to the α-lone pair on oxygen, orthogonal to the

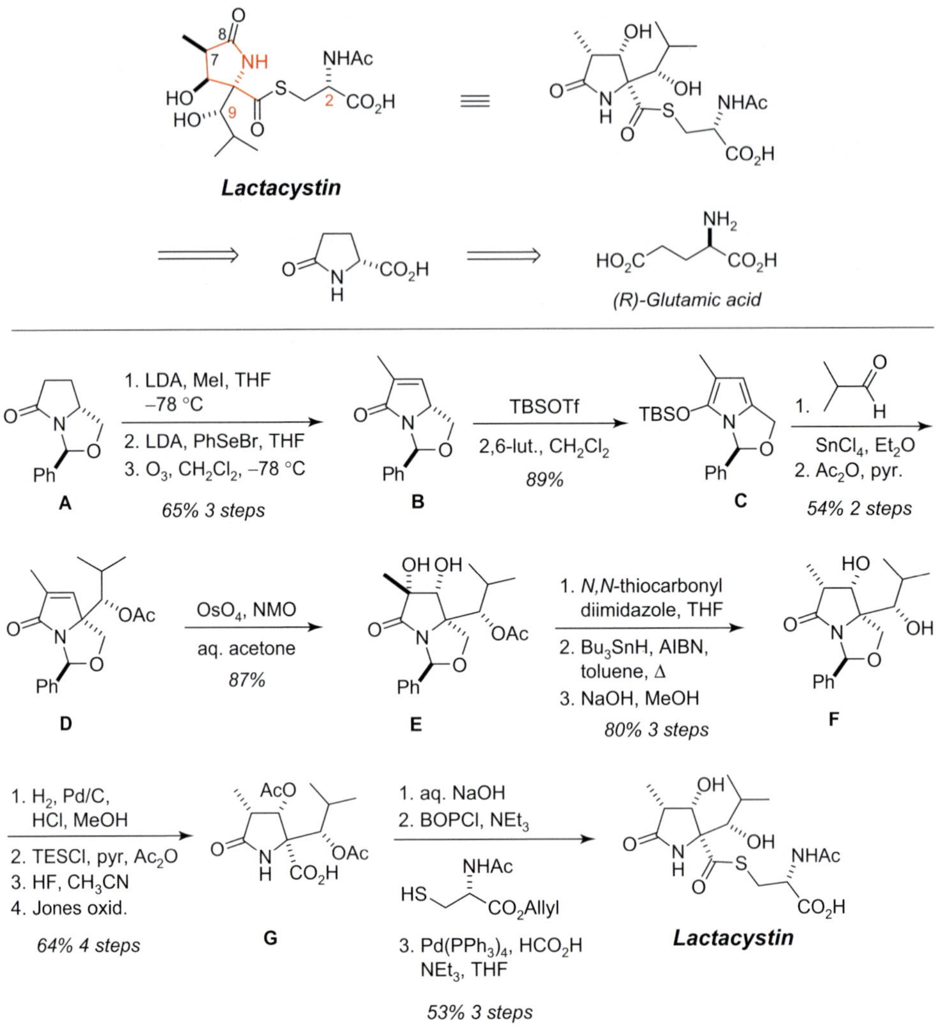

Lactacystin

(R)-Glutamic acid

Key reactions: *Vinylogous aldol (C to D); Barton–McCombie deoxygenation (E to F)*

Relevant mechanism:

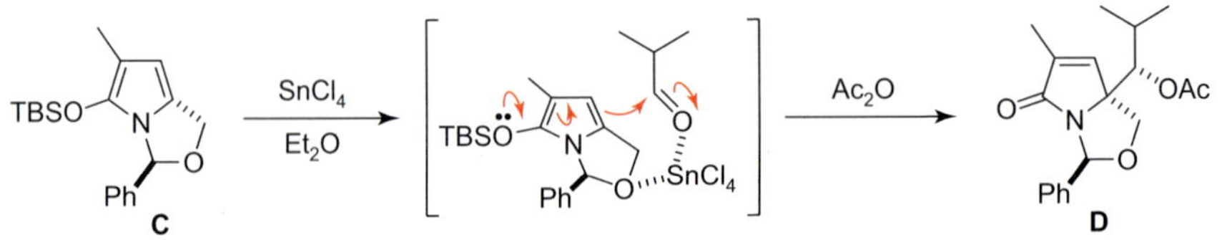

Figure 16.37 Synthesis of lactacystin by Baldwin.

plane of the pyrrole ring, would allow the proper juxtaposition of reactive sites (see relevant mechanism, Figure 16.37).

16.7.4
The Chida Synthesis

The possibility to manipulate a D-glucofuranose derivative in which a C-methyl group can be easily incorporated instigated Chida and coworkers [117] to adopt a true carbohydrate-based *chiron approach* (Figure 16.38). Thus, intermediate **A** [118], was converted to olefins **B** and **C**, then subjected to an Overman rearrangement [119] to generate a new C–N bond at C-5 of the original carbohydrate template (see relevant mechanism A, Figure 16.38). Compound **D** (dr = 4:1) so obtained,

Key reactions: Overman imidate rearrangement (**C** to **D**); Grignard extension (**F** aldehyde to **G**); stereoselective ketone reduction (ketone of **G** to **G**, recycle).

Figure 16.38 Synthesis of lactacystin by Chida.

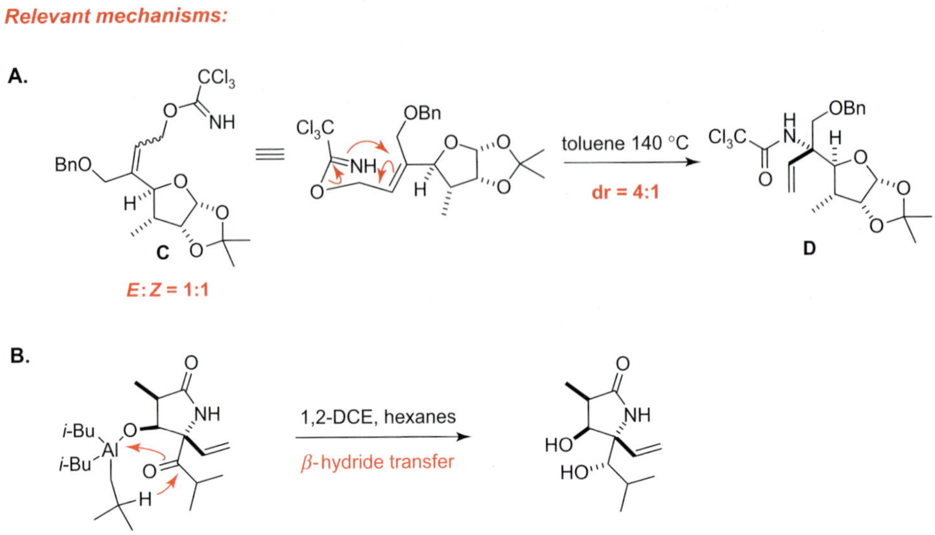

Relevant mechanisms:

A.

E:Z = 1:1

toluene 140 °C

dr = 4:1

B.

1,2-DCE, hexanes

β-hydride transfer

Figure 16.38 *(continued)*.

now harbored the crucial C-5 branching required in lactacystin. Unravelling to the aldehyde **E**, oxidation, and lactam formation afforded **F**. An unspecified method for the removal of the *O*-benzyl group in the presence of the vinylic appendage, followed by oxidation and treatment with isopropylmagnesium bromide gave a modest yield of the desired diastereomer **G**. However, oxidation of the other epimer of **G** to the ketone, and reduction with (*i*-Bu)₃Al led to, as a single isomer, **G** via a β-hydride transfer mechanism (see relevant mechanism B, Figure 16.38). Following known steps, **G** was converted to lactacystin. The merit of the D-glucose approach is in the off-template installation of the C-5 branching by the Overman rearrangement. The length of the sequence and modest stereoselectivity should be taken into account compared to more streamlined approaches.

16.7.5
The Ōmura-Smith synthesis

The first total synthesis of lactacystin in which chirality was introduced early in the sequence using a catalytic asymmetric epoxidation, was a joint contribution from the Ōmura and Smith groups [120]. Thus, a Sharpless epoxidation [121] protocol provided **A** from (*E*)-4-methyl-2-penten-1-ol (Figure 16.39). Treatment of **A** with benzyl isocyanate gave **B**, which after base-induced isomerisation led to **C** (see relevant mechanism A, figure 16.39). Jones oxidation to the acid, esterification, and hydrolysis led to a concomitant epimerization to give **D** in excellent yield. (2*R*,3*S*)-β-Hydroxy leucine **E** thus produced, was converted to the oxazoline **F**, which was subjected to a Seebach enolate hydroxymethylation [122], followed by a Moffatt oxidation [123]. Several other oxidation methods proved to be problematic. Aldehyde **G** (>98% *de*) was subjected to an asymmetric Brown crotylboration [124] to provide **H** as the major isomer (dr = 4:1) (see relevant mechanism B, Figure 16.39). Ozonolytic cleavage of the terminal double bond followed by a Pinnick oxidation [125] gave **I**. Cleavage of the oxazoline under transfer hydrogenation conditions, followed by formation of the lactam **J** and thioesterification according to the original Corey protocol [109] gave lactacystin.

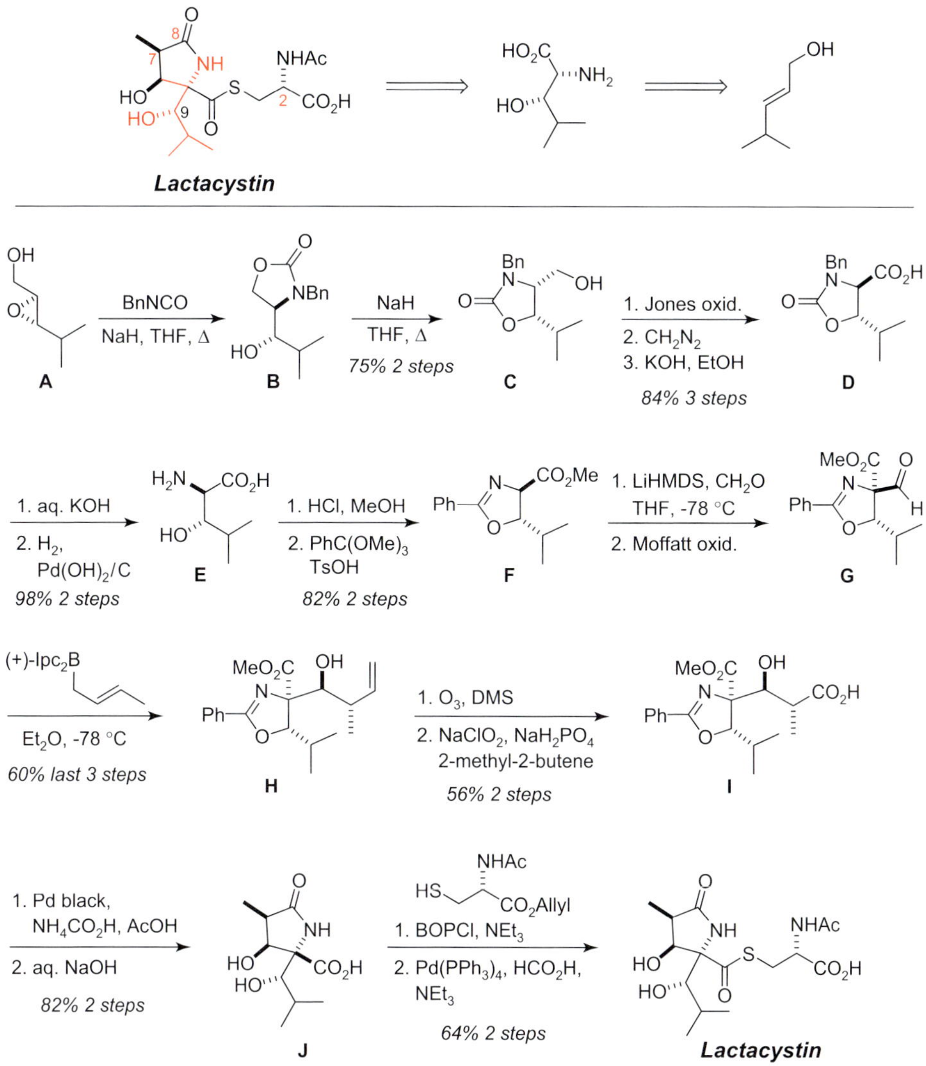

Key reactions: *Epoxide opening/cyclic carbamate isomerization (**A** to **B** to **C**); cis/trans isomerization (**C** acid to **D**); asymmetric Brown crotylboration (**G** to **H**)*

Figure 16.39 Synthesis of lactacystin by Ōmura and Smith.

Relevant mechanisms:

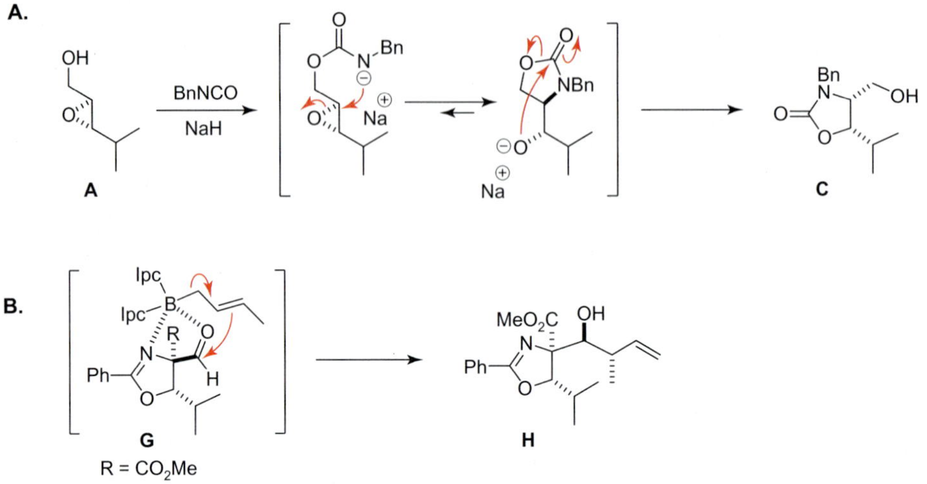

A.

B.

R = CO₂Me

Figure 16.39 *(continued).*

16.7.6
The Panek synthesis

Catalytic asymmetric aminohydroxylation was a versatile method utilized by Panek and Masse [126] to prepare (2*R*,3*S*)-3-hydroxy leucine. Adopting the Sharpless protocol [127], compound **A** was synthesized from the *p*-bromophenyl-(2*E*)-4-methyl-2-pentenoate (Figure 16.40). Methanolysis and cleavage of the *N*-Cbz group gave **B**, which was transformed to oxazoline **C** having the required *anti*-orientation. Adopting the Seebach hydroxymethylation procedure, followed by Moffatt oxidation as in the Ōmura-Smith synthesis gave **D**, which was subjected to a silylcrotylation reaction [128] in the presence of TiCl₄ to give **F** with excellent stereoselectivity. Oxidative cleavage gave **G**, which was transformed to omuralide (**H**) according to Corey [109, 111] and then further elaborated to lactacystin. The high diastereoselectivity in the silylcrotylation reaction was rationalized on the basis of a double stereodifferentiating process, whereby a chelated transition state model could be invoked (see relevant transformation, Figure 16.40).

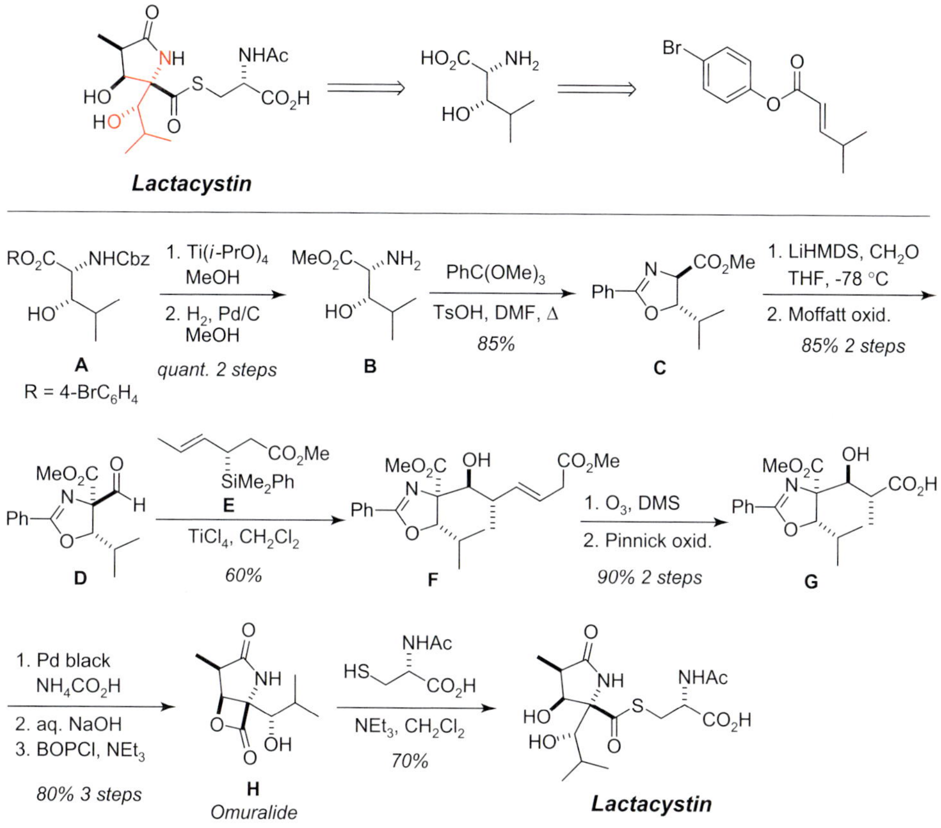

Lactacystin

A
R = 4-BrC₆H₄

Omuralide

Lactacystin

Key reactions: Sharpless asymmetric aminohydroxylation (**A**); asymmetric crotylation (**D** to **F**)

Relevant transformation:

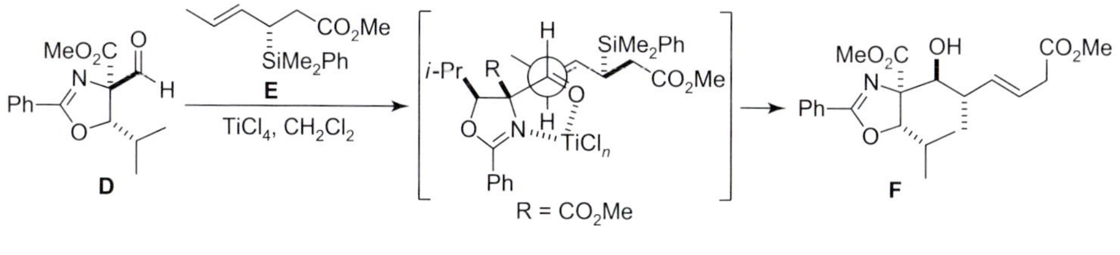

Figure 16.40 Synthesis of lactacystin by Panek.

16.7.7
The Jacobsen synthesis

The impact of modern catalytic asymmetric reactions manifests itself in the Balskus and Jacobsen [129] synthesis of lactacystin (Figure 16.41). The β-silyl imide

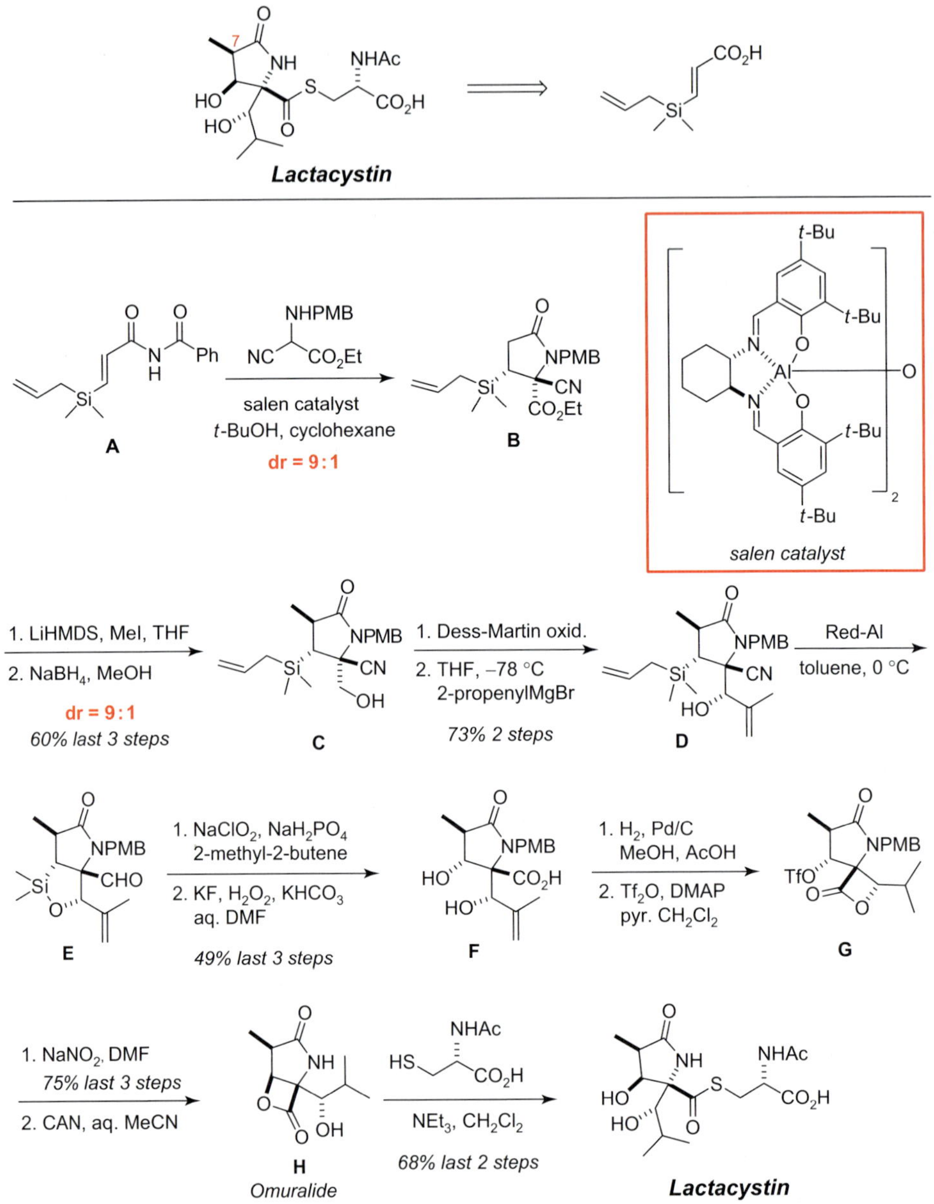

Lactacystin

salen catalyst

Omuralide

Lactacystin

Key reactions: *Catalytic asymmetric conjugate addition (A to B); Red-Al-induced cyclic siloxane formation (D to E); Tamao oxidation (E to F); β-lactone formation (F to G); nitrite-mediated inversion (G to H)*

Figure 16.41 Synthesis of lactacystin by Jacobsen.

Relevant mechanisms and transformation:

A.

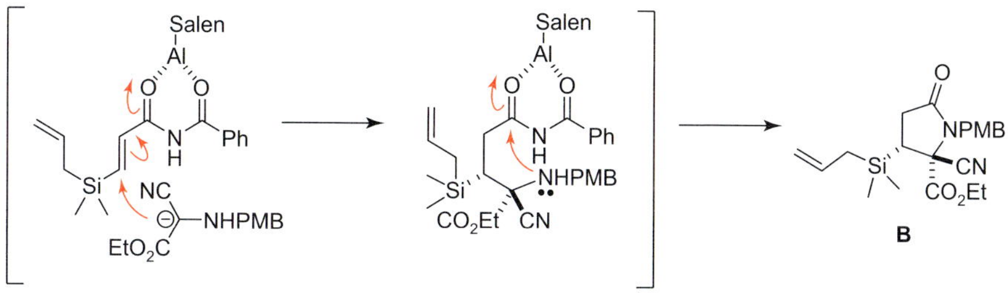

B.

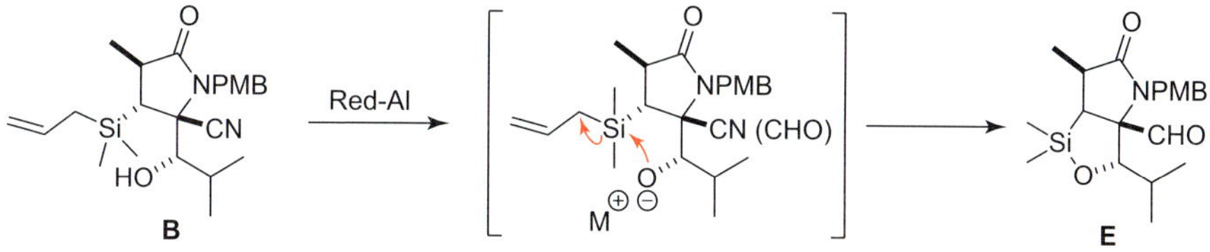

Figure 16.41 (*continued*).

A underwent conjugate addition with *N*-PMB cyanoacetate in the presence of the Jacobsen aluminum-salen complex catalyst [130] to give lactam **B** in excellent yield and good selectivity (dr = 9:1). Enolate methylation and reduction of the ester gave **C**, which was oxidized to the corresponding aldehyde, and then treated with 2-propenylmagnesium bromide to give **D** with excellent diastereoselectivity. An attempt to convert the cyano group to the aldehyde, resulted in loss of the allyl group and formation of a cyclic silyl ether **E**. Oxidation to the acid and Tamao oxidative desilylation [131] gave diol **F**, which was converted to β-lactone **G**. Configurational inversion of the triflate by $NaNO_2$, followed by cleavage of the *N*-PMB group afforded omuralide (**H**), which was converted to lactacystin according to literature precedents.

There are a number of noteworthy features of the Jacobsen synthesis, starting with the one-step formation of lactam **B**, harboring the critical C-5 substituent with the desired stereochemistry (see relevant mechanism A, Figure 16.41). Subsequent steps have relied on the topology of the lactam template and the spatial disposition of the appended substituent. The suitability of an allyldimethylsilyl group in the conjugate addition was further appreciated with its directing effect in the *C*-methylation of the lactam dianion enolate (**B** to **C**).

Other remarkable aspects in this synthesis can be found in the selectivity of the Grignard addition (aldehyde of **C** to **D**), the intramolecular alkoxide displacement of a dimethylallylsilyl group (**D** to **E**, see relevant mechanism B, Figure 16.41), the formation of a spiro β-lactone (**F** to **G**), and its compatibility with an S_N2 displacement of a neopentyl-like triflate ester at C-6, without solvolysis of the lactone.

16.7.8
The Shibasaki synthesis

Shibasaki and coworkers [132] also introduced chirality early in their synthesis of lactacystin, thereby securing the C-5 stereocenter, albeit in an indirect way (Figure 16.42). The cyclopentenone imine **A**, readily available from 3-methyl butyraldehyde was subjected to an asymmetric addition of TMSCN in the presence of the Shibasaki gadolinium-complex as catalyst [133]. The resulting nitrile **B** (98% *ee*) was converted to the *N*-Boc derivative **C**, the internal olefin was cleaved by ozonolysis, followed by adjustment of oxidation states and formation of lactam **D** in excellent overall yield. With the required functionality introduced at C-5, subsequent steps addressed the elaboration of substituents on the lactam ring and adjustment of stereochemistry. Thus, the ketone group in **D** was reduced in the presence of isopropylmagnesium bromide via β-hydride transfer mode (**D** to **E**, dr = 10:1). Conversion to α,β-unsaturated lactam **F**, and conjugate addition of a lithiosilane [134] as the zincate to **G**, was followed by a Tamao oxidation [131], to afford **H** with complete stereoselectivity. Introduction of the C-7 methyl group by enolate methylation to **I**, and functional group manipulation led to omuralide (**J**), which was converted to lactacystin as previously reported by Corey [109] and the Ōmura-Smith protocols [120].

The Shibasaki strategy to establish the stereochemistry at C-5 via a catalytic asymmetric Strecker reaction of imine **A** is reminiscent of a hidden carbocyclic chiron (see relevant mechanism A, Figure 16.42). A notable difference is that the carbocycle is used as a template to introduce the cyano group in a catalytic reaction. The inclusion of the isopropyl group in the cyclopentenone imine motif was in anticipation of an oxidative cleavage that would eventually produce the lactam **D**, in which all the carbon atoms, except for cyanide, were already present in the starting material. What was also astutely planned was the "release" of an isopropyl ketone side-chain as a result of the oxidative cleavage of the cyclopentenone **C** to give **D**. The stereoselective β-hydride transfer reduction (**D** to **E**) may have been anticipated in a chelated model, in which the ethoxycarbonyl group could adopt a pseudoaxial orientation due to $A^{1,3}$-strain (see relevant mechanism B, Figure 16.42).

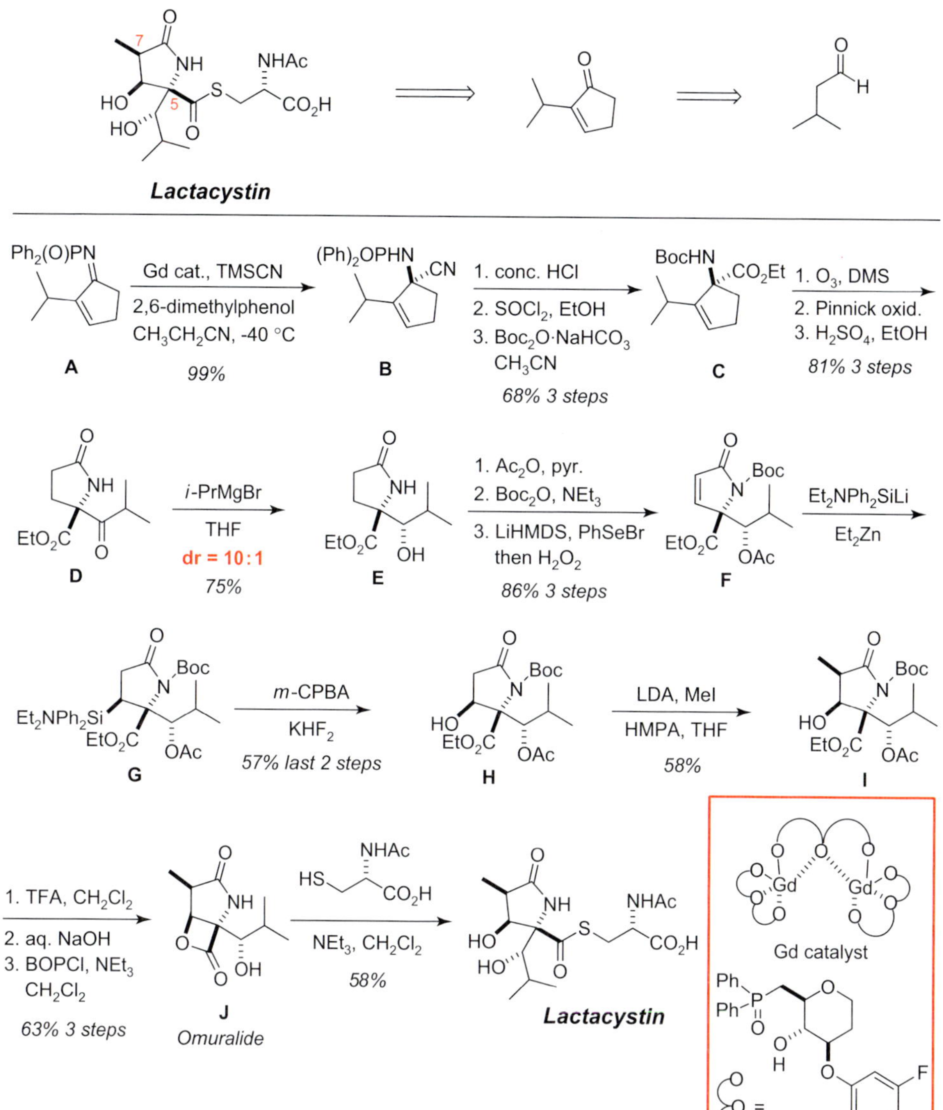

Lactacystin

Key reactions: *Catalytic asymmetric ketoimine Strecker reaction (**A** to **B**); conjugate silyl anion addition (**F** to **G**); Tamao oxidation (**G** to **H**)*

Figure 16.42 Synthesis of lactacystin by Shibasaki.

Relevant mechanisms:

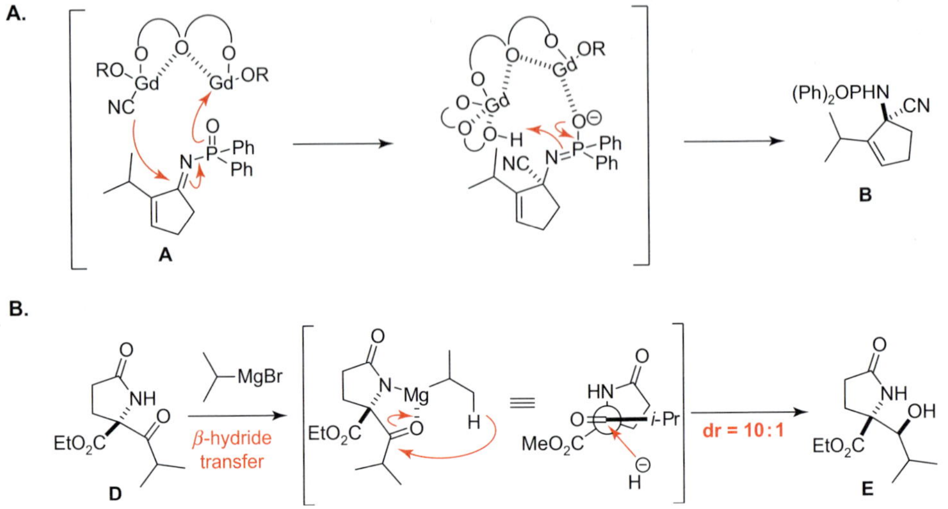

A.

B.

Figure 16.42 *(continued).*

16.7.9
Lactacystin and omuralide: Alternative methods and synthetic approaches

In addition to the eight total syntheses of lactacystin discussed in the previous section, several individually different routes to advanced intermediates have also been reported [106]. These will be briefly discussed, while highlighting key reactions and relevant mechanisms where appropriate (Figure 16.43).

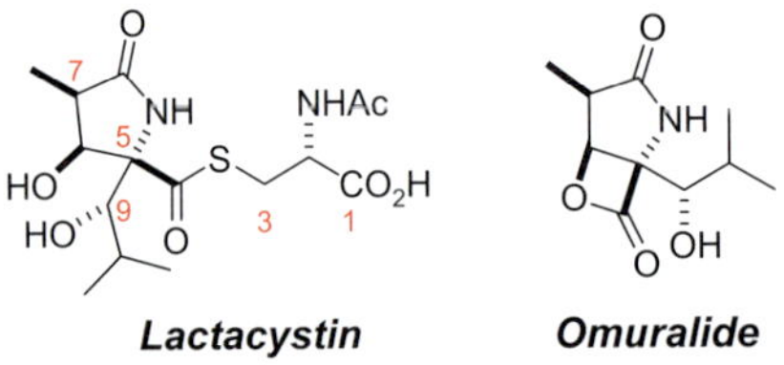

Lactacystin   *Omuralide*

Figure 16.43 Structures of lactacystin and omuralide.

16.7.10
The Kang approach

Starting with enantioenriched epoxide **A**, Kang and Jun [135] prepared allylic imidate **B**, which was subjected to a Hg(OTFA)$_2$-mediated cyclization to give **C** (Figure 16.44). Formation of lactam **D**, and extension to the ketone **E**, was followed by a stereoselective reduction mediated by a β-hydride transfer from isopropylmagnesium bromide, to ultimately give Baldwin's intermediate **F** [112].

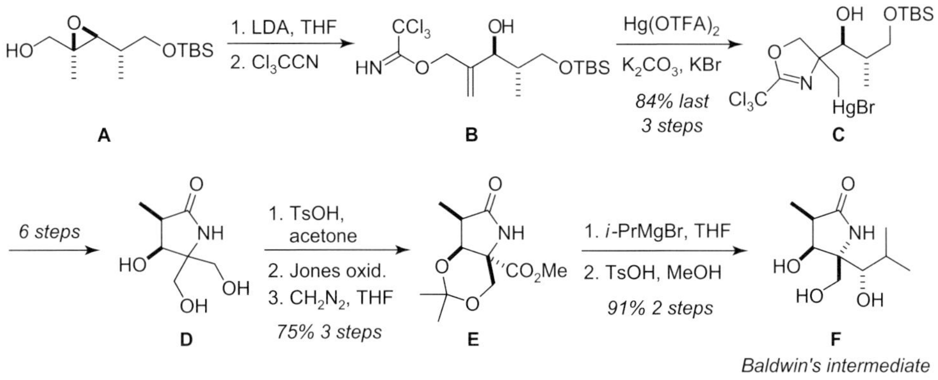

Figure 16.44 The Kang approach to Baldwin's intermediate.

16.7.11
The Adams synthesis of omuralide

Starting with the (R)-Roche ester (**A**), Adams and coworkers [136] used the aldehyde **B** in a stereoselective enolate reaction with **C**, in the presence of Me₂AlCl to give **D** (Figure 16.45). Further elaboration gave omuralide through Corey's intermediate (**E**) [109, 111]. The authors rationalized the exclusive formation of the desired diastereoisomer **D** on the basis of a *Re*-face approach of an aluminum-coordinated transition state model. It should be noted that the (S)-Roche acid derivative was previously used by Corey and Choi [137] in the total synthesis of (6R)-lactacystin.

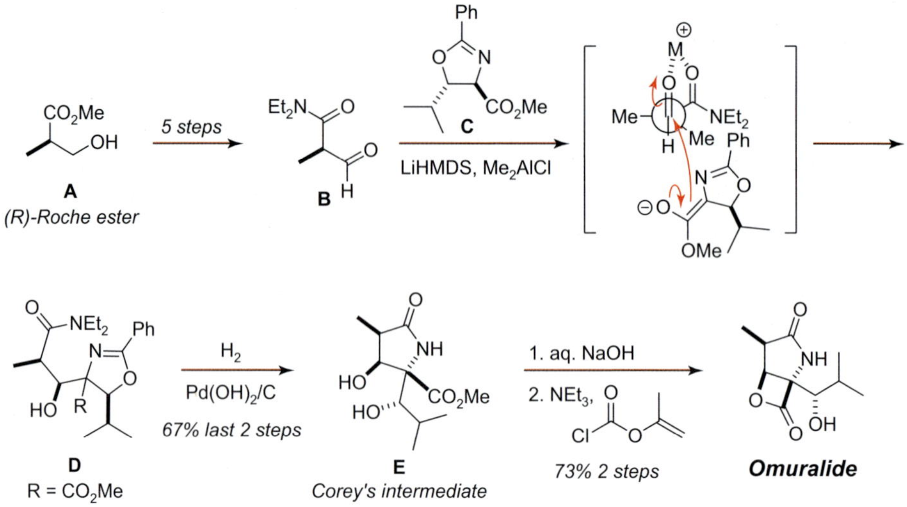

Figure 16.45 The Adams synthesis of omuralide.

16.7.12
The Ohfune approach

An asymmetric Strecker route was reported by Ohfune and coworkers [138], relying on a kinetic conformational bias in the addition of HCN to a cyclic imine (Figure 16.46). Thus, ester **A** carrying a phenylalanine moiety as a chiral inducer, was transformed to cyclic imine **B**, which would undergo a kinetically-controlled addition of cyanide ion to give α-amino nitrile **C**. Oxidative transformation to the cyclic imine **D** and hydrolysis, led to α-amino acid **E**, which was transformed to the *N,O*-methylene acetal **F**, previously utilized by Corey and Reichard [109] in their synthesis of lactacystin.

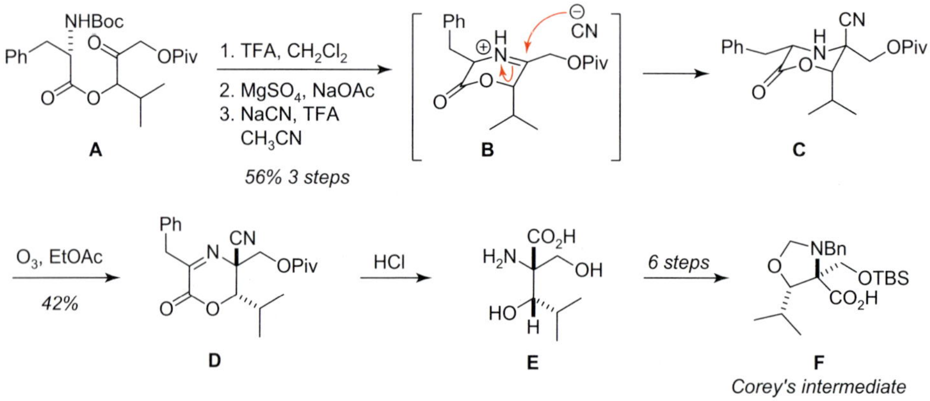

Figure 16.46 The Ohfune approach to the Corey intermediate.

16.7.13
The Pattenden approach

The Pattenden group [139] focussed on a free-radical-mediated cyclization as a key step (Figure 16.47). The enantioenriched epoxide **A** was converted to 2-trichloromethyl oxazoline **B**, then further elaborated to α-bromo amide **C** and ester **D**. Cyclization in the presence of Bu₃SnH and AIBN led to a 2:1 mixture of epimers **E**. Conversion to the ketone, and further elaboration to the 2-methylthio-2-methyl lactam analogue **F** took place with high stereoselectivity. Protection of the amide, cleavage of the silyl ethers, and reduction with Na(OAc)₃BH led to the Corey intermediate **G** [111].

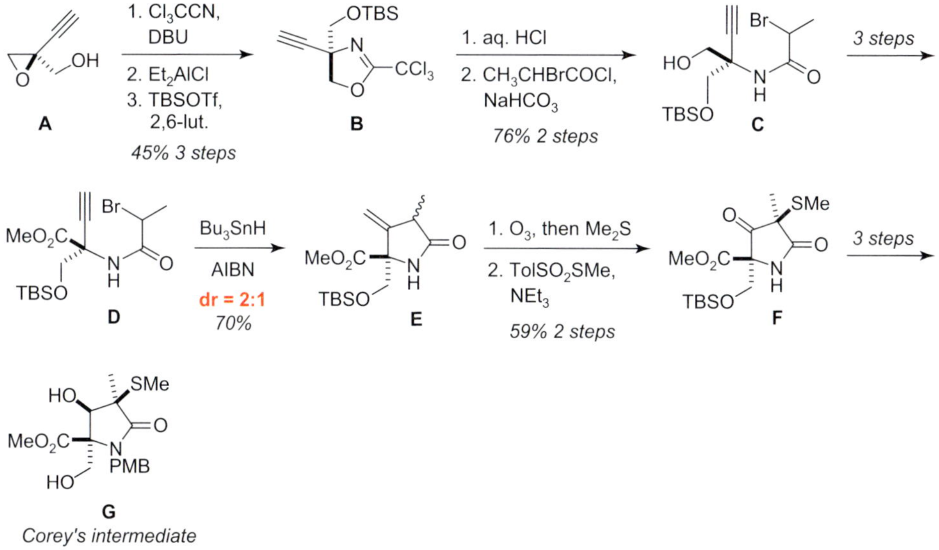

Figure 16.47 The Pattenden approach to Corey's intermediate.

16.7.14
The Hatekayama approach

Starting with the readily available synthon **A**, Hatekayama and coworkers [140] performed an asymmetric Brown crotylboration to give homoallylic alcohol **B** in excellent selectivity (99% *ee* after recrystallization) (Figure 16.48). Oxidative cleavage of the terminal double bond and lactam formation gave **C**, which was converted to acetonide **D**. Conversion to the ketone and β-hydride reduction by treatment with the isopropylmagnesium bromide, according to Kang [148], afforded alcohol **E**. Cleavage of the isopropylidene acetal furnished the Baldwin intermediate (**F**), which was converted to lactacystin in five steps. A plausible mechanism for the reduction of the ketone by β-hydride transfer was proposed.

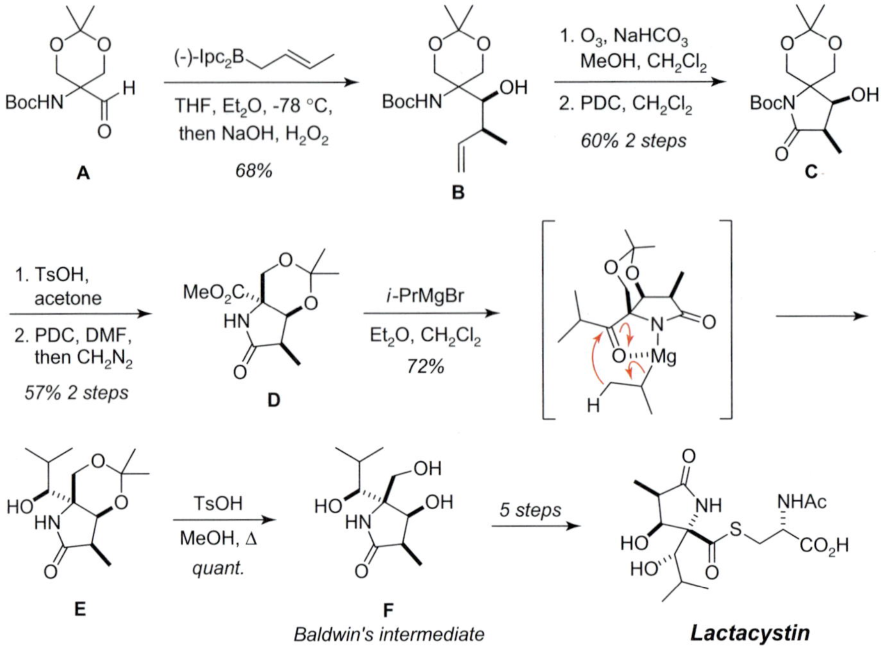

Figure 16.48 The Hatekayama approach to lactacystin.

16.7.15
The Donohue synthesis of (±)-omuralide

Using commercially available α-carbethoxypyrrole, Donohue and coworkers [141] developed a short alternative approach to racemic omuralide (Figure 16.49). Treatment of the *N*-Boc derivative with LiDBB (lithium 4,4′-di-*tert*-butyldiphenylide) and magnesium bromide in the presence of a chelating amine, generated a series of interesting metallated intermediates such as **B–D**, which led to the aldol product **E** after acetylation. Treatment with OsO₄, followed by selective Mitsunobu (S$_N$2) inversion afforded the iodide, which was reduced with catalytic InCl₃/NaBH₄ to give **F**. Functional group manipulation, then oxidation to the lactam with RuCl₃-NaIO₄ gave **G**, which was converted to the enolate, and treated with MeI. Further steps led to racemic omuralide in 14% overall yield. The exploitation of a cheap building block merits attention, although an enantioselective synthesis can then perhaps be envisaged eventually.

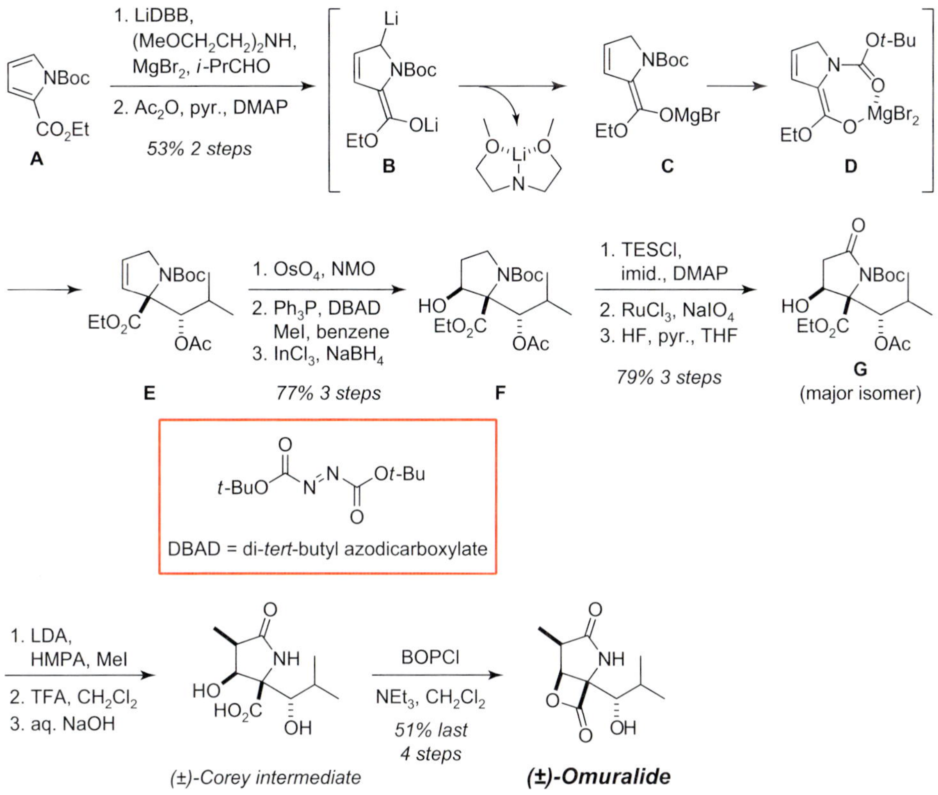

Figure 16.49 The Donohue synthesis of (±)-omuralide.

16.7.16
The Wardrop approach

A strategically different approach to the stereocontrolled formation of the trisub-stituted C-5 center in lactacystin was developed by Wardrop and Bowen [142] (Figure 16.50). The readily available epoxide **A** [121] was converted to the azido alcohol **B**, then to the benzylidene acetal **C**. Reduction of the azide, followed by alkylation with **D** gave a mixture of vinylic bromides **E**. When treated with KHMDS, the resulting alkylidene carbene **F**, formed by α-elimination, participated in an in-tramolecular 1,5-C−H insertion [143] to give **G** in 50% yield, due to the parallel formation of a propargylic amine as a result of a 1,2-migration (36%). Nevertheless, epoxidation led to **H**, and further steps afforded the known Corey intermediate **I**. Unfortunately, the control of stereochemistry at the C-7 center was only modest. A carbene insertion approach was also described by Hayes in 2002 [144].

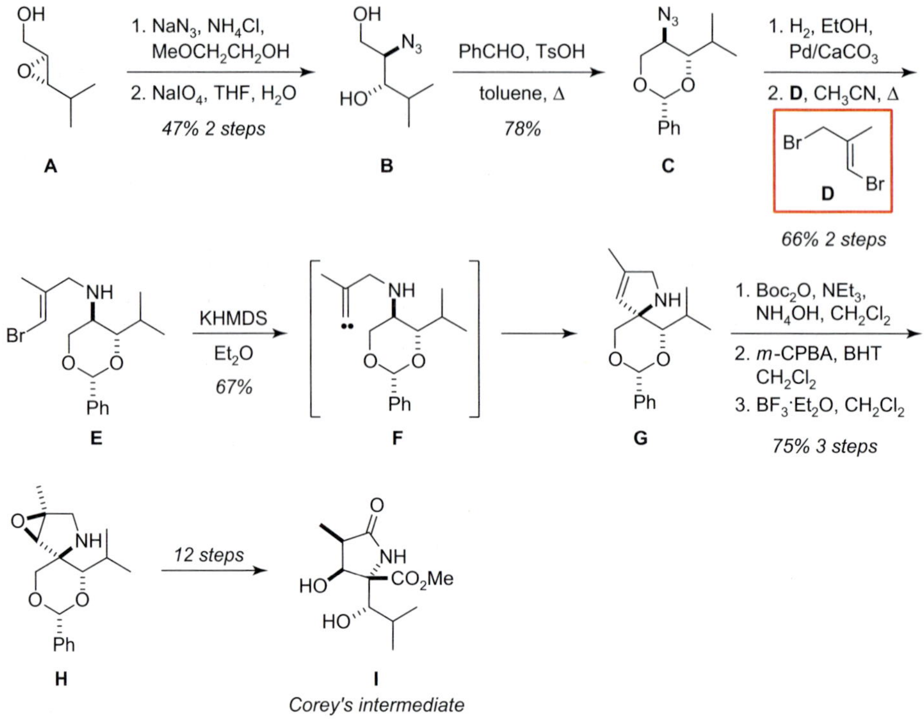

Figure 16.50 The Wardrop approach to Corey's intermediate.

16.7.17
The Hayes synthesis of lactacystin

Following studies of C–H insertion of acyclic intermediates [144], Hayes and coworkers [145] developed a variant of the Wardrop approach independently. Starting with epoxide **A** [121], known steps led to amine **B**, which was subjected to a one-pot MnO_2 oxidation of the alcohol **C**, followed by reductive amination with $NaCNBH_3$ to give **D**. Carbene insertion led to the spirocyclic amine **E**, which was oxidized to **F**. Dihydroxylation to **G**, and free-radical mediated deoxygenation of the tertiary hydroxyl group according to Barton [115] gave, after functional group manipulation, known intermediate **H**, which was converted to lactacystin in seven steps.

Figure 16.51 The Hayes synthesis of lactacycstin.

16.7.18
Total synthesis of lactacystin: Synopsis

As evidenced from the preceding section, the total synthesis of lactacystin, as well as its versatile intermediates has been an exciting endeavor for many research groups worldwide. A graphic summary of the starting materials used in the period 1992–1995, reflects upon the validity of exploiting Nature's chiral, non-racemic building blocks (chirons), such as L-serine [109], (R)-glutamic acid [112], and D-glucose [117] (Figure 16.52). The first applications of catalytic asymmetric methods was demonstrated in the synthesis of (2R,3S)-3-hydroxy leucine as a starting unnatural amino acid, by Õmura and Smith [120]. Aminohydroxylation was exploited by Panek [126]. In these and other methods, the common advanced intermediate was the fully functionalized lactam core, harboring all the requisite substituents, first reported by Corey and Reichard [109]. The use of esterases as versatile tools in the desymmetrization of achiral malonates led to a practical synthesis of lactacystin by Corey and coworkers [111]. The full force of catalytic chemical asymmetric methods is manifested in the strategies used by Jacobsen [129] and Shibasaki [132] independently. In both cases, the catalytic step led to the installation of the critical tetrasubstituted carbon center at C-5 early in the sequence. In the Jacobsen synthesis, the lactam core of lactacystin was built in a single catalytic asymmetric conjugate addition reaction, while securing the absolute stereochemistry at C-5 containing versatile functionality.

Most successful implementations of elegant strategies have to address further elaboration of "missing" functional groups, or to bring about stereochemical adjustments in order to achieve inclusion of the full complement of substituents in the intended target molecule (see Chapter 18, section 18.5.3). In the *chiron approach*, starting with amino acids for example, it was necessary to extend the carbon chain, introduce branching, then cyclize to provide the lactacystin lactam motif. Starting with D-glucose necessitated the cleavage of the anomeric carbon, thereby sacrificing the stereochemistry of C-2 in the process. The implementation of efficient asymmetric catalytic methods allowed the establishment of a critical substitution pattern at C-5, but required the introduction of peripheral groups such as *C*-methyl or hydroxyl later in the sequence in order to complete the synthesis.

Regardless of the approach, it is clear that lactacystin has provided a veritable intellectual and experimental challenge, in spite of its relatively simple structure compared to other much more ambitious synthetic objectives. Its unique biological activity in conjunction with its transformation to the corresponding β-lactone has also shown the importance of the interface between biology and chemistry.

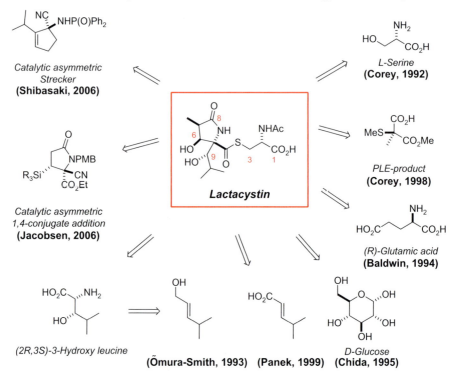

Figure 16.52 Synthesis of lactacystin: Synposis.

16.8
Taxol

Originally isolated from the Pacific yew tree [146], (−)-taxol is a potent anticancer drug which is presently marketed under the trade name of Taxol® [147]. It is produced in a semi-synthetic process starting with the more readily available baccatin III (Figure 16.53). Here was a life-saving drug in dire need of a practical synthesis on kilogram scale, reminiscent of the penicillin story from the middle of the last century. The scarcity of the natural source of taxol, incited many groups to devise methods that would produce it by synthesis. Although no less than six total syntheses of taxol have been reported to date, none are amenable to an industrial scale production [148]. The unprecedented and highly functionalized tricarbocyclic core structure harbors nine stereogenic carbon atoms, and an unusual pattern of oxygen substituents [149].

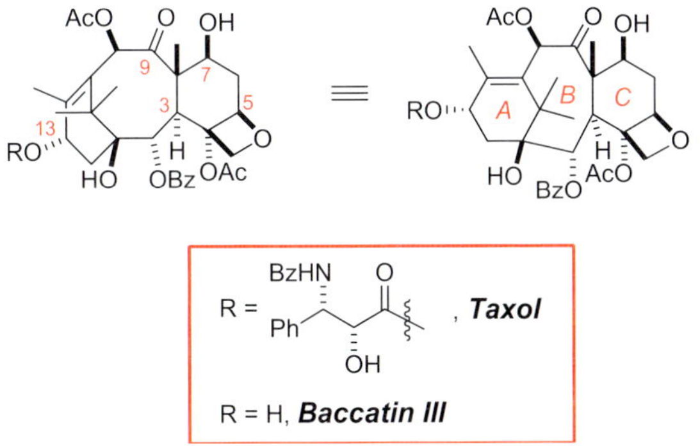

Figure 16.53 Structures of taxol and baccatin III.

The tricyclic *ABC* carbogenic core of taxol offers a number of visually obvious starting points for synthesis, especially when focussing on the six-membered rings flanking the eight-membered ring inner core (Figure 16.53). A convergent strategy of tethered cyclohexanes representing rings *A* or *B* could also be considered as a means of building the tricyclic core structure. This deceptively simple analysis presents the major challenge of introducing various levels of oxidation and branching on almost every carbon atom lining the periphery of the conformationally biased *ABC* tricyclic core structure of taxol.

We close this chapter with highlights from the key steps comprising the six reported total syntheses of taxol, starting with the two independently announced disclosures by the groups of Holton [150] and Nicolaou [151] in 1994.

16.8.1
What mad pursuit

Arguably, the total synthesis of taxol by the Holton and Nicolaou groups in 1994 was one of the milestone achievements of that decade. What started as a "magnificent obsession" turned into a fiercely competitive pursuit, culminating with individual publications that appeared in print quasi-simultaneously, albeit in two different prestigious journals [150, 151]. This monumental *tour-de-force* laid the ground work for subsequent total syntheses of taxol between 1996 and 1999.

16.8.2
The Holton synthesis of taxol

The chiral, non-racemic tricyclic core of taxol was conceived relying on a series of ingenious transformations starting with (−)-camphor [150] (see also Chapter 8, section 8.9.3.1). Previous reports from the Holton laboratory had established a sequence of transformations that involved the cycloannulation of homocamphor (**A**) to patchoulene (**B**), which upon epoxidation gave patchoulene oxide **C** [152]. Base-mediated epoxide opening gave **D**, which underwent a face-selective epoxidation to **E**, followed by a Lewis acid-promoted rearrangement to **F**. Yet another epoxidation gave **G**, which was subjected to a Grob fragmentation, followed by *O*-silylation to afford the bicyclic *AB* core system **H** (see relevant mechanism A, Figure 16.54). A stereoselective aldol condensation with 4-pentenal led to **I**. Formation of the mixed carbonate, and enolate dihydroxylation with a camphor-derived oxaziridine gave **J**. Reduction of the ketone, and transacylation afforded the cyclic carbonate **K**, which was oxidized to the ketone, then subjected to a Chan rearrangement [153] to give lactone **L** (see relevant mechanism B, Figure 16.54). Deoxygenation of the angular tertiary hydroxyl group with SmI_2 followed by an asymmetric enolate hydroxylation, and stereoselective reduction of the ketone gave **M**. Formation of a cyclic carbonate followed by oxidative cleavage of the terminal double bond to a methyl ester led to **N**. Dieckmann condensation gave the β-ketoester **O**, which was decarboxylated and *O*-protected as the BOM ether **P**. The silyl enol ether was epoxidized to give the α-hydroxy ketone **Q** as the TMS ether. The exocyclic methylene group in **R** was secured by a Grignard reaction, followed by elimination with the Burgess reagent [154]. Tosylation and face-selective dihydroxylation gave **S**, which was subjected to oxetane formation, subsequent acetylation and selective desilylation, to give **T**. Cleavage of the cyclic carbonate with PhLi, oxidation to the ketone **U**, and an acyloin-type oxidation via an intramolecular selenoxide enolate transposition followed by acetylation led to α-hydroxy ketone **V** (see relevant mechanism C, Figure 16.54). Cleavage of the TBS ether gave **W**, which was converted to taxol using the Ojima lactam (**X**) as the amino acid precursor [155].

   A total of 41 chemical steps were performed starting with patchoulene oxide (**C**) to record the first total synthesis of taxol by Holton and coworkers [150].

Figure 16.54 Holton's total synthesis of taxol.

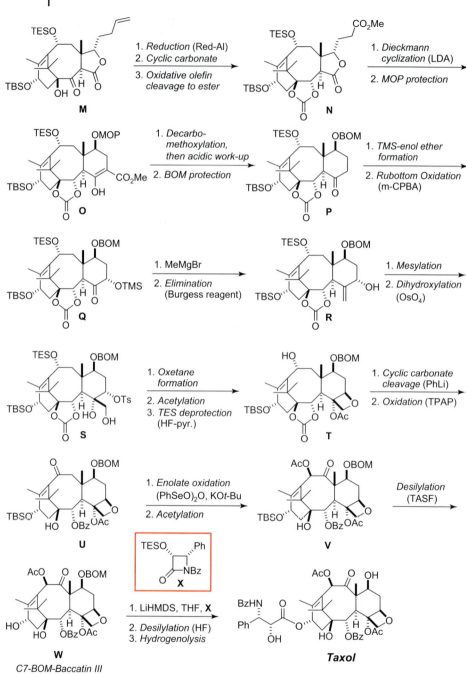

Figure 16.54 *(continued).*

Key reactions: *Epoxide opening/elimination (**C** to **D**); Lewis acid-mediated ring contraction (**E** to **F**); Grob fragmentation (**G** to **H**); Chan rearrangement (**K** to **L**); asymmetric enolate hydroxylations (**I** to **J**, **L** to **M**); Dieckmann condensation (**N** to **O**); regioselective cyclic carbonate to benzoate cleavage (**T** to **U**); acyloin-type enolate oxidation (**U** to **V**).*

Relevant mechanisms and transformations:

A.

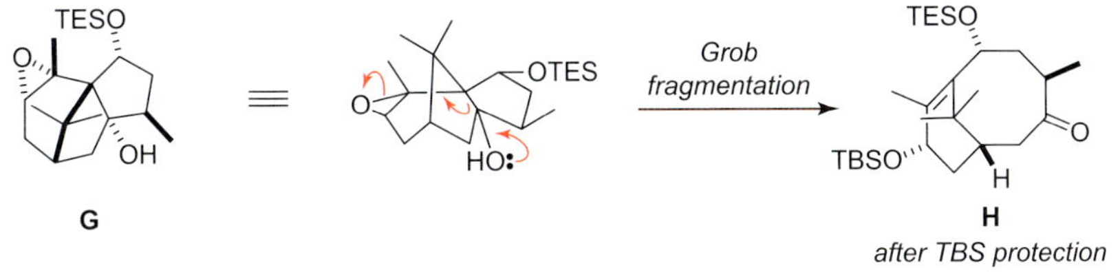

G

Grob fragmentation

H

after TBS protection

B.

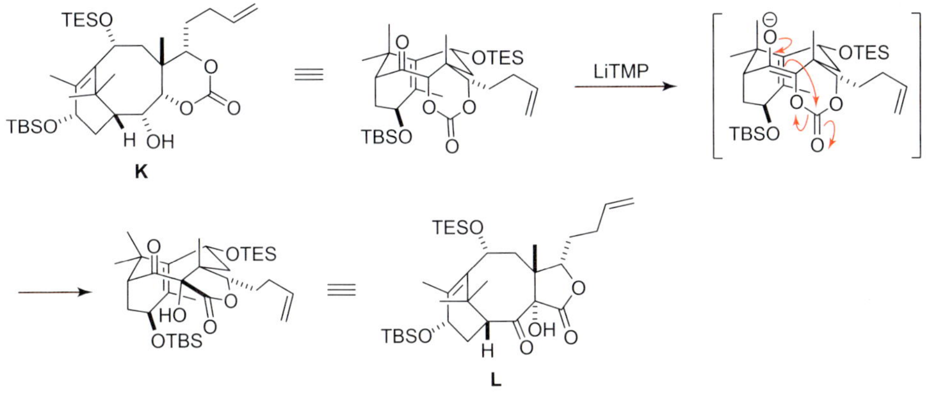

K

LiTMP

L

C.

U

1. KOt-Bu
2. (PhSeO)$_2$O

1. KOt-Bu
2. Ac$_2$O

V

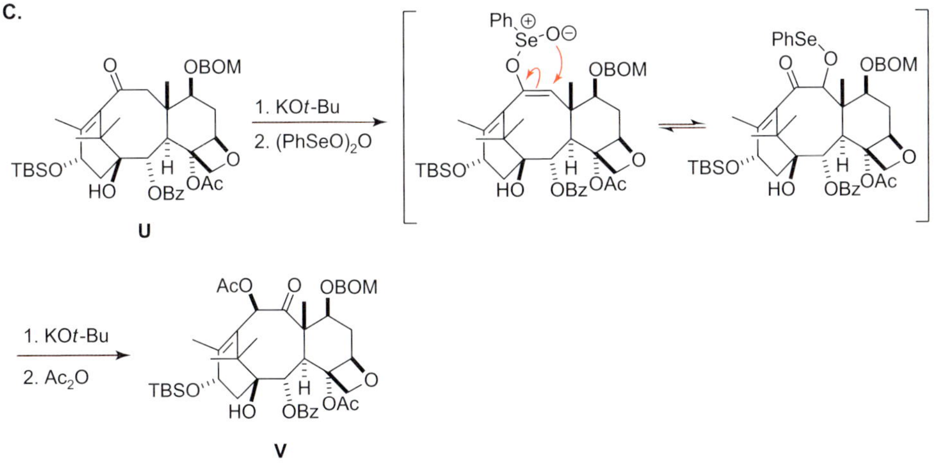

Figure 16.54 *(continued)*.

16.8.3
The Nicolaou synthesis of taxol

In the Nicolaou synthesis [151], the tricyclic core of taxol was constructed from simple laboratory chemicals consisting of a pair of dienes and dienophiles (Figure 16.55). Thus, intermediate **A**, prepared by a Diels-Alder reaction, was transformed to the phenylsulfonyl hydrazone **B**, then subjected to a Shapiro reaction [156] to give lithiated cyclohexadiene **C**, which would eventually correspond to ring *A*. A templated Diels-Alder reaction of **D** afforded bridged lactone adduct **E**, which underwent spontaneous intramolecular lactone formation to give **F** (see relevant transformation, Figure 16.55). A selective reduction of the ester with LiAlH$_4$ followed by a change in protecting groups gave the bicyclic lactone **G**, which, after protection and reduction, led to triol **H**. Formation of the spiroacetal and oxidation gave the aldehyde **I** corresponding to a functionalized precursor to ring *B*. The tricyclic core was constructed by a series of reactions that also ensured the incorporation of critical hydroxyl groups regioselectively. Thus, treatment of aldehyde **I** with lithiated intermediate **C** led to tethered adduct **J** as the major isomer. Directed epoxidation, followed by hydride opening at the tertiary carbon atom and subsequent steps furnished diol **K**.

Oxidation to the dialdehyde, and a pinacol coupling according to McMurry [157] afforded *syn*-diol **L**, which was resolved as the (1*S*)-camphanate ester, thus securing an enantioenriched advanced intermediate. Further steps to adjust oxidation states led to **M**, which was subjected to a regioselective hydroboration and hydrolysis to give the triol **N**. A series of protecting group modifications and hydroxyl group functionalizations gave **O** then **P**. The latter intermediate was subjected to an acid-induced oxetane formation, followed by acetylation of the tertiary hydroxyl group to give **Q**. Cleavage of the cyclic carbonate with PhLi, and allylic oxidation to the ketone followed by stereoselective reduction gave the protected baccatin III analogue **R**. Taxol was obtained by reaction with the Ojima lactam (**S**) [155], and subsequent deprotection of the silyl ether.

A total of 51 chemical steps were used starting with the 5-hydroxy-2-pyranone (Figure 16.55) [151].

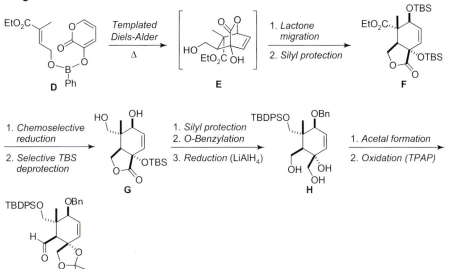

Figure 16.55 Nicolaou's total synthesis of taxol.

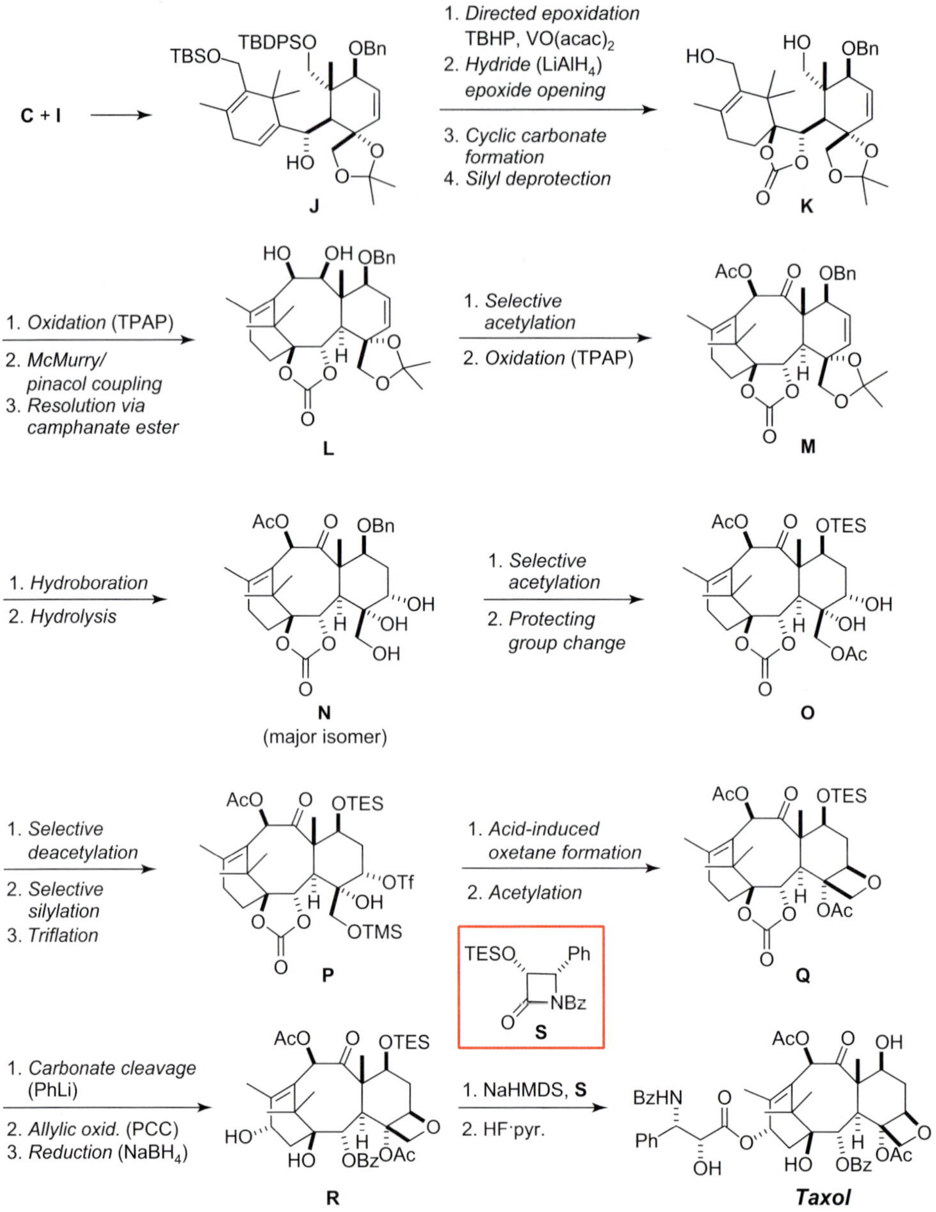

Key reactions: *Diels-Alder reaction toward **A**; templated Diels-Alder (**D** to **F**); intramolecular lactonization (**E** to **F**); chemoselective ester reduction (**F** to **G**); stereoselective lithium anion coupling (**C** + **I** to **J**); directed epoxidation and hydride opening (**J** to **K**); McMurry/pinacol coupling (**K** to **L**); regioselective cyclic carbonate cleavage to benzoate (**Q** to **R**)*

Figure 16.55 (*continued*).

Relevant transformation and mechanism:

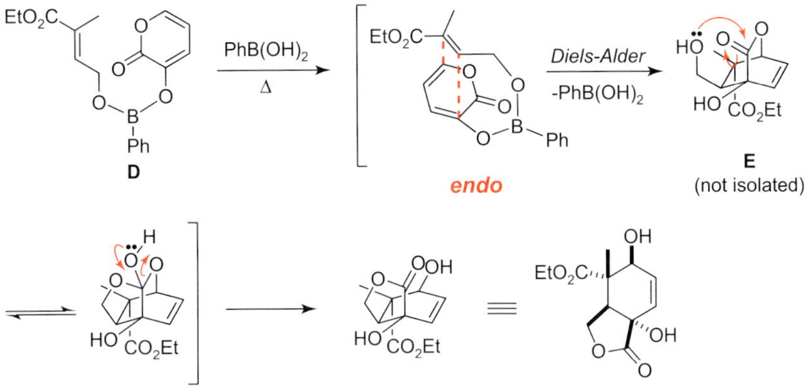

Figure 16.55 *(continued)*.

16.8.4
The Danishefsky synthesis of taxol

The choice of the Wieland-Miescher ketone by Danishefsky [158] as a precursor to taxol was no doubt to secure the angular *C*-methyl group, as well as to provide a good skeletal convergence with ring *C*, in addition to ensuring useful branch points for further elaboration to the core structure (Figure 16.56). Starting with the diketone **A**, several steps led to the iodide **B**, which was further transformed to the *O*-TMS cyanohydrin derivative **C**. Lithiation to **D** provided the cyclohexenyl moiety corresponding to ring *A*. The readily available decalone **E**, prepared from the Wieland-Miescher ketone, was transformed to spiroepoxide **F** according to Corey and Chaykovsky [159]. Regioselective Lewis acid-mediated epoxide opening led to **G**. Dihydroxylation of the double bond, selective *O*-silylation and conversion to the triflate gave **H**. Mild heating in ethylene glycol followed by *O*-benzylation led to oxetane **I** (see relevant transformation, Figure 16.56). Hydrolysis of the ketal and oxidation of the enol ether with DMDO gave α-hydroxy ketone **J**, which was further transformed to dimethyl acetal **K**. Subsequent steps led to the olefin **L** and aldehyde **M**. The tricyclic core was constructed by tethering **D** and **M** to give a preponderance of isomer **N**. Regioselective epoxidation and reductive opening led to the tertiary alcohol **O**, which was protected as the cyclic carbonate, and further elaborated to the enol triflate **P**. Extension to the terminal olefin **Q**, followed by an intramolecular Heck cyclization [160] afforded the intended tetracyclic core **R**. Regioselective epoxidation to temporarily protect the endocyclic olefin, adjustment of protecting groups, and treatment with PhLi gave the benzoate ester **S**. Conversion to the ketone, and treatment with SmI$_2$ resulted in regioselective reduction of the epoxide to give olefin **T**. Enolate oxidation and acetylation led to α-acetoxy ketone **U**. Allylic oxidation, followed by stereoselective reduction led to the *O*-protected baccatin

III derivative **V**, which was further converted to taxol according to established protocols. Starting with the Wieland-Miescher ketone, Danishefsky and coworkers required some 47 chemical steps to reach taxol.

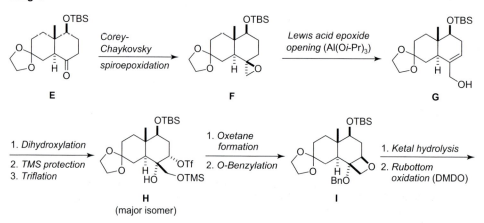

Figure 16.56 Danishefsky's total synthesis of taxol.

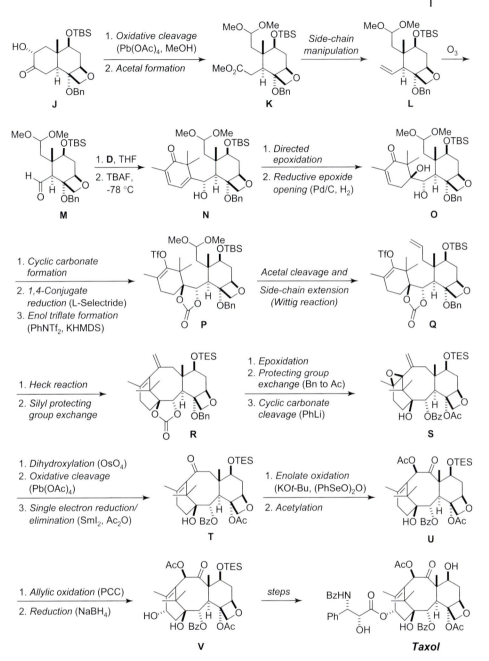

Key reactions: *Lewis acid-mediated spiroepoxide opening/elimination (**F** to **G**); oxetane formation (**H** to **I**); Rubottom oxidation (**I** to **J**); stereoselective Li-anion coupling (**D** + **M** to **N**); selective reductive epoxide opening (**N** to **O**); intramolecular Heck reaction (**Q** to **R**); acyloin-type enolate hydroxylation (**T** to **U**).*

Figure 16.56 *(continued).*

Relevant transformation:

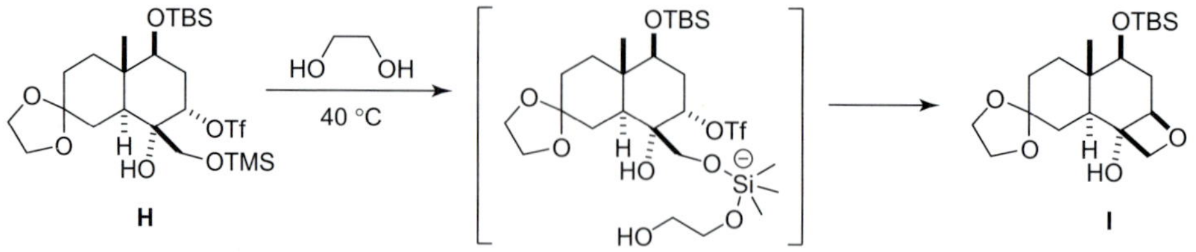

Figure 16.56 *(continued).*

16.8.5
The Wender synthesis of taxol

As discussed in Chapter 8 (section 8.9.3), there is no visual connection between taxol and α-pinene. Thus, Wender's choice of this readily available terpene as a starting chiron relied on an astute sense of knowledge-based visual *relational* thinking (Figure 16.57) [161]. A key transformation that would unfold the *AB* bicyclic core structure of taxol was a highlight of this synthesis. (*R*)-Verbenone (**A**), readily available from the air oxidation of α-pinene was transformed into the aldehyde **B**. A photoinduced 1,3-alkyl shift [162] led to **C**, which was extended to **D**. Conjugate addition with Me$_2$CuLi was followed by intramolecular cyclization to afford tricyclic intermediate **E**. Oxidation to the ketone, enolate hydroxylation, hydride reduction, formation of the isopropylidene ketal, and TBS protection gave **F**. Epoxidation with *m*-CPBA, followed by base-induced ring expansion and TIPS protection gave the *AB* bicyclic core structure **G** in excellent yield (see relevant mechanism, Figure 16.57). Enolate hydroxylation, catalytic hydrogenation of the double bond and protection of the diol as a cyclic carbonate gave **H**. Further steps involving oxidation to the aldehyde and Wittig olefination, gave **I**. Hydrolysis of the enol ether and ketal groups, selective silylation, oxidation, followed by treatment of the resulting aldehyde with the Eschenmoser reagent led **J**.

C-Allylation and BOM-protection led to the olefin **K**. An exchange of protective groups, followed by cleavage of the cyclic carbonate with PhLi afforded the benzoate ester **L**. Acetoxy transposition, and selective cleavage of the terminal olefin gave the aldehyde **M**, which was subjected to an intramolecular aldol cyclization, followed by protection as the *O*-Troc carbonate leading to **N** as the major isomer. Cleavage of the BOM group and mesylation gave **O**, which was converted to the bromide with inversion of configuration. Dihydroxylation with concomitant benzoyl migration and re-formation of a cyclic carbonate gave **P**. Debenzoylation, oxetane formation and acetylation gave **Q**, which was transformed to baccatin III (**R**), and further elaborated to taxol. The Wender synthesis starting with α-pinene is considered to be the shortest to date, comprising 37 chemical steps [161]. Were it not for the unexpected migration of the benzoyl group (**O** to **P**), which necessitated re-formation of a cyclic carbonate and cleavage with PhLi late in the synthesis, the number of steps could have been shortened. The need to invert the ring *C* hydroxyl group via a mesylate added a further two extra steps.

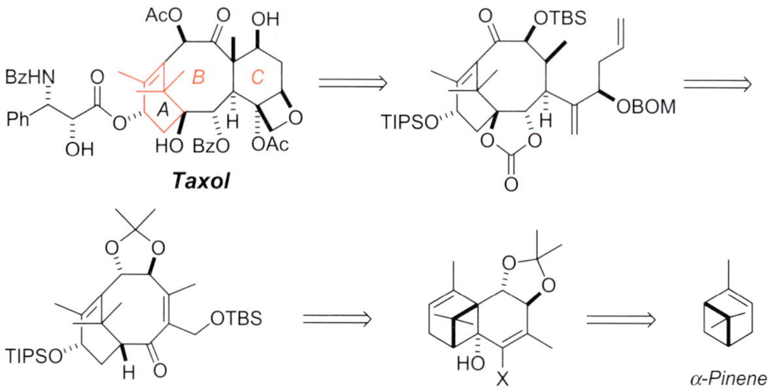

Taxol

α-Pinene

Rings A and B

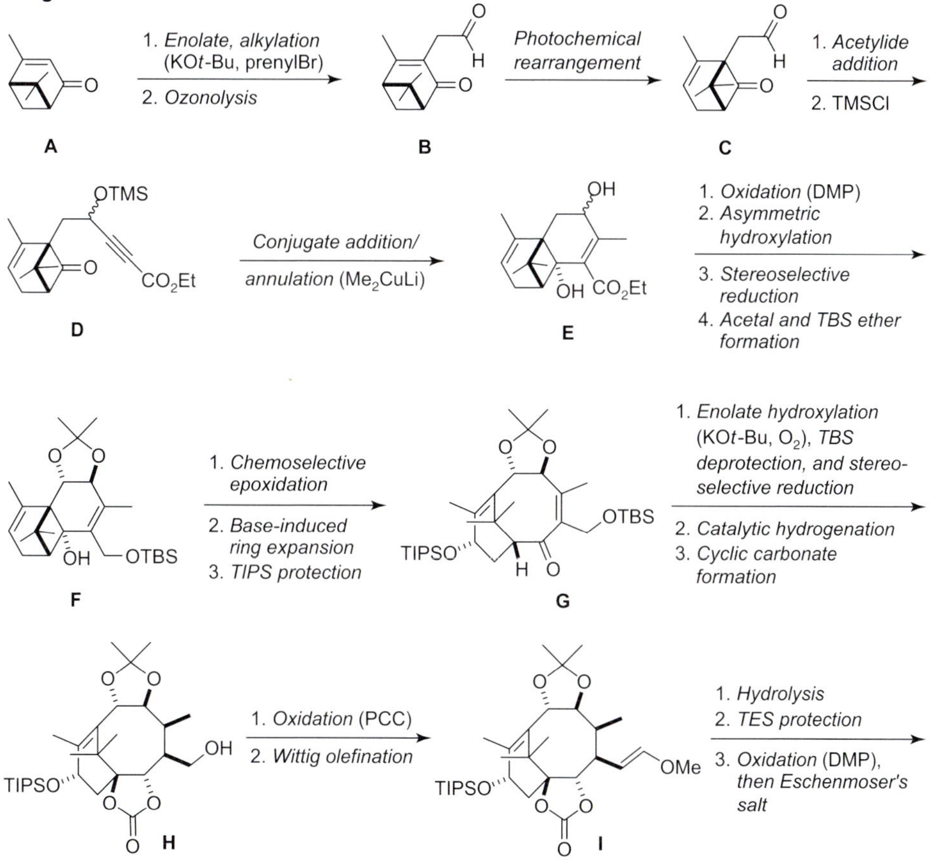

A

1. Enolate, alkylation (KOt-Bu, prenylBr)
2. Ozonolysis

B

Photochemical rearrangement

C

1. Acetylide addition
2. TMSCI

D

Conjugate addition/ annulation (Me₂CuLi)

E

1. Oxidation (DMP)
2. Asymmetric hydroxylation
3. Stereoselective reduction
4. Acetal and TBS ether formation

F

1. Chemoselective epoxidation
2. Base-induced ring expansion
3. TIPS protection

G

1. Enolate hydroxylation (KOt-Bu, O₂), TBS deprotection, and stereoselective reduction
2. Catalytic hydrogenation
3. Cyclic carbonate formation

H

1. Oxidation (PCC)
2. Wittig olefination

I

1. Hydrolysis
2. TES protection
3. Oxidation (DMP), then Eschenmoser's salt

Figure 16.57 Wender's total synthesis of taxol.

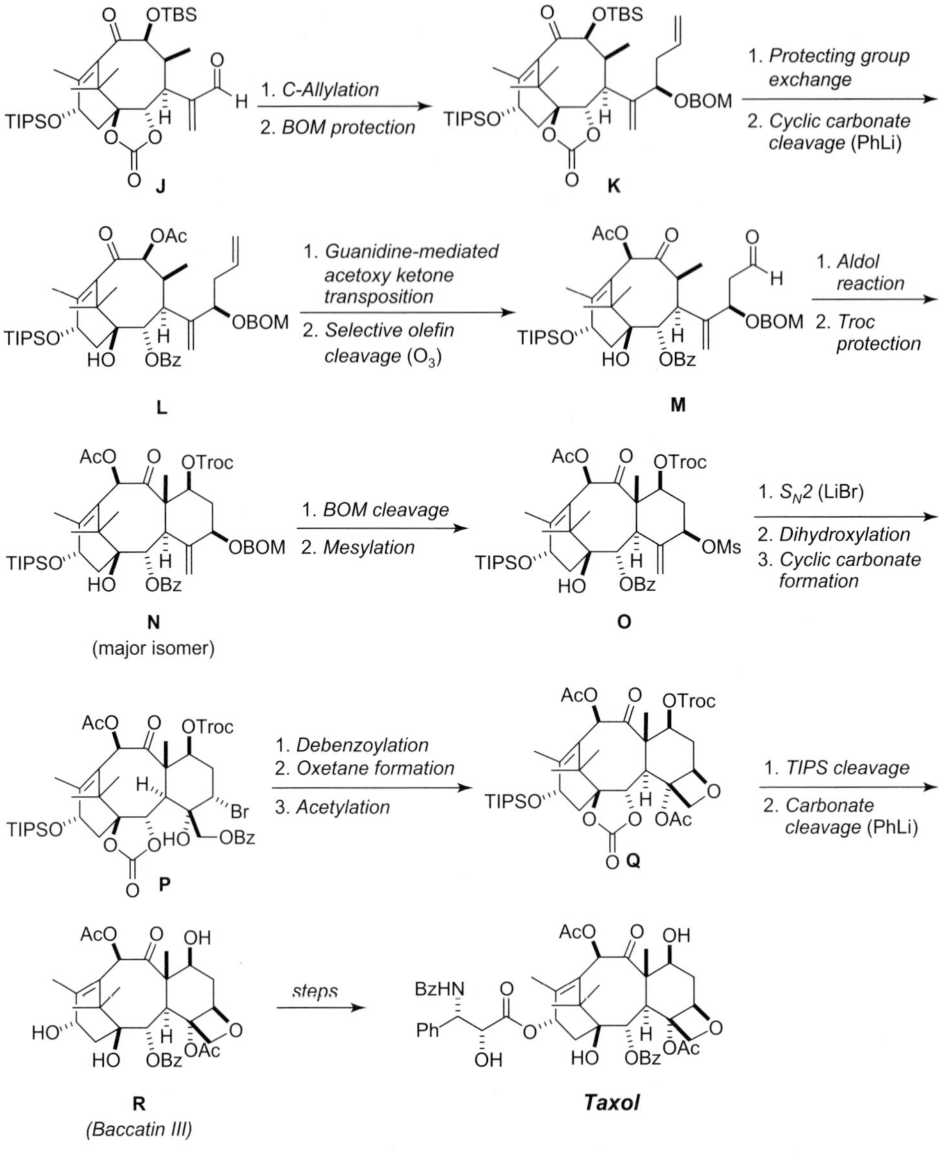

OTBS
TIPSO
J
1. C-Allylation
2. BOM protection
OBOM
K
1. Protecting group exchange
2. Cyclic carbonate cleavage (PhLi)
OAc
HO
OBz
L
1. Guanidine-mediated acetoxy ketone transposition
2. Selective olefin cleavage (O_3)
AcO
M
1. Aldol reaction
2. Troc protection
OTroc
N
(major isomer)
1. BOM cleavage
2. Mesylation
OMs
O
1. S_N2 (LiBr)
2. Dihydroxylation
3. Cyclic carbonate formation
Br
P
1. Debenzoylation
2. Oxetane formation
3. Acetylation
Q
1. TIPS cleavage
2. Carbonate cleavage (PhLi)
OH
R
(Baccatin III)
steps
BzHN
Ph
Taxol

Key reactions: *Photoinduced rearrangement (**B** to **C**); conjugate C-methylation and cyclization (**D** to **E**); Grob fragmentation (**F** to **G**); enolate hydroxylation (**G** to **H**); oxidation/ α-methylene branching (**I** to **J**); acetoxyketone transposition (**L** to **M**); intramolecular aldol (**M** to **N**)*

Figure 16.57 *(continued)*.

Relevant transformation and mechanism:

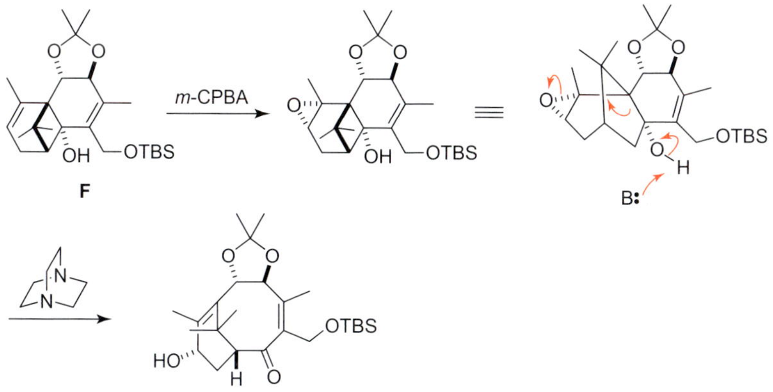

Figure 16.57 *(continued).*

16.8.6
The Kuwajima synthesis of taxol

Starting with achiral building blocks that represent rings *A* and *C*, Kuwajima [163] devised a tethered subunit strategy toward the elaboration of the tricyclic core structure of taxol (Figure 16.58). The diketone **C**, available in a few steps from **A** and **B** was converted to the enone **D**. A Sharpless asymmetric dihydroxylation of the exocyclic enol ether, followed by oxidation to the aldehyde eventually gave **E**. Further steps afforded the enantiopure aldehyde **F** representing ring *A* in taxol. An aryl lithium coupling of **G** with **F** led to the tethered intermediate **H** as a single isomer. Protection of the diol as the cyclic methylboronate, followed by a Lewis acid-mediated ring closure, and resolution of the diol led to the tricyclic intermediate **I** (see relevant mechanism, Figure 16.58). Selective reduction of the ketone, silyl protection, followed by free-radical mediated desulfurization and further chemical manipulation gave **J**. Birch reduction with K/NH_3, selective TBS deprotection, and introduction of a benzylidene acetal led to **K**. Stereoselective hydride reduction, followed by singlet oxygen-mediated dihydroxylation afforded 1,4-diol **L**, which was protected as the PMP (*p*-methoxyphenyl) acetal, then oxidized to give enone **M**. Conjugate addition of cyanide and protection as the TBS enol ether led to **N** as the major isomer. Reduction of the nitrile to the hydroxylmethyl equivalent led to **O**, which was converted to the bridged cyclopropane intermediate **P**. Cleavage with SmI_2, silyl deprotection, and protection of the diol as the phenylboronate ester led to the angular *C*-methyl group, as in **Q**. Additional protection of hydroxyl groups, oxidative cleavage of the boronate ester, followed by regioselective oxidation gave diketone **R**. Formation of the ring *C* enol triflate selectively, followed by a Pd-catalyzed cross-coupling reaction introduced the trimethylsilylmethyl group in **S**. Chlorination was followed by enolate hydroxylation and acetylation to give **U**. The wrong stereochemistry of the carbon atom bearing the acetate was corrected by a base-induced epimerization,

and the exocyclic olefin was dihydroxylated to give **V**. Subsequent oxetane formation and protecting group adjustments led to **W**. Cleavage of the cyclic carbonate
with PhLi and further elaboration of hydroxyl protecting groups led to C-7-*O*-Troc
baccatin III (**X**), which was converted to taxol. Starting from propargyl alcohol as a
precursor to intermediate **B**, the synthesis was completed in 47 chemical steps.

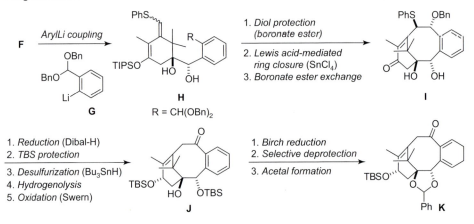

Figure 16.58 Kuwajima's total synthesis of taxol.

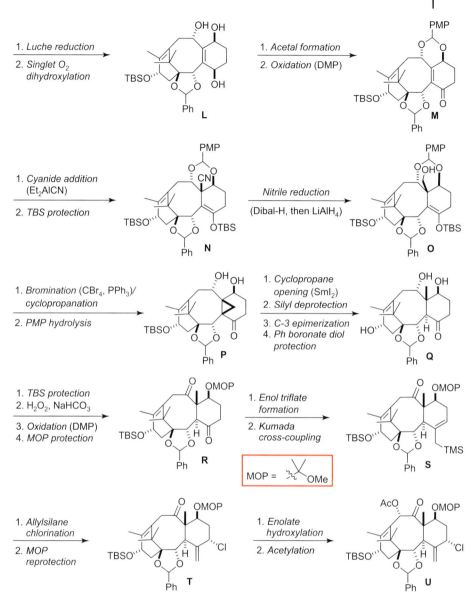

Figure 16.58 *(continued)*.

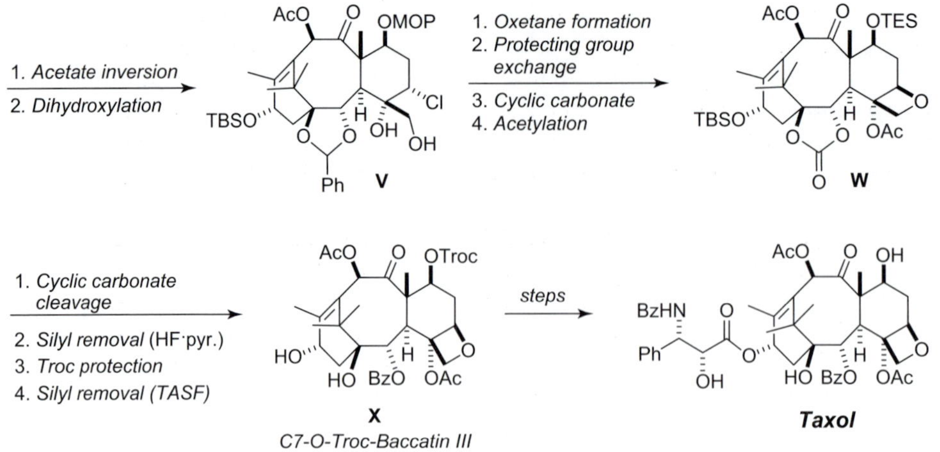

Key reactions: *Stereoselective anion coupling (**F** + **G** to **H**); Lewis acid-mediated ring closure (**H** to **I**); singlet oxygen-mediated dihydroxylation via photochemical [2+2] cycloaddition (**K** to **L**); conjugate cyanide addition (**M** to **N**); cyclopropane formation (**O** to **P**); SmI$_2$-mediated cyclopropane opening (**P** to **Q**); Kumada cross-coupling (**R** to **S**); allylsilane chlorination (**S** to **T**); enolate hydroxylation (**T** to **U**); acetate inversion (**U** to **V**).*

Relevant mechanism:

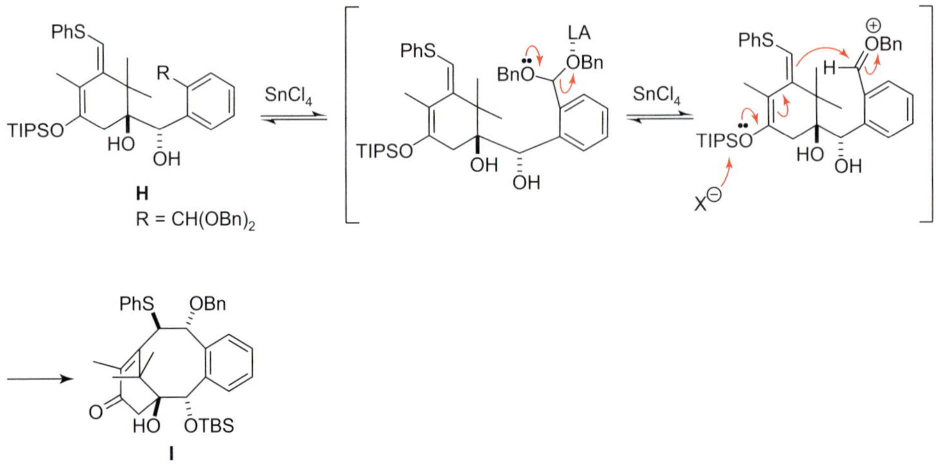

Figure 16.58 *(continued).*

16.8.7
The Mukaiyama synthesis of taxol

The choice of L-serine as a starting chiron to elaborate the eight-membered core structure of taxol represents a radically different strategy by Mukaiyama [164] (see also Chapter 8, section 8.6.3). The L-glyceraldehyde derivative **A**, prepared in a few steps from L-serine, was extended via an aldol reaction to **B**. Conversion to the aldehyde, and extension using a Mukaiyama aldol condensation [165] afforded **C** as a major diastereomer. Transformation to the methyl ketone **D**, conversion to the corresponding TMS enol ether and bromination gave the bromoketone **E**. Methylation of the lithium enolate, followed by oxidation to the aldehyde gave **F**. Treatment with SmI$_2$, followed by acetylation of the β-keto alcohol and elimination led to the cyclooctenone core intermediate **G** in excellent overall yield for the two steps. Conjugate addition of the 2-bromopentene derivative **H**, and conversion to the aldehyde gave **I**. An intramolecular aldol reaction mediated by NaOMe afforded *BC* bicycle **J**. Stereoselective reduction of the ketone, formation of an acetonide, and adjustment of functional groups gave **K**. Addition of 3-butenyl lithium, and oxidation to the ketone led to **L**. Transformation of the terminal olefin to the methyl ketone was achieved via a Wacker oxidation to give **M**. A McMurry/pinacol coupling in the presence of a low-valent titanium reagent, and *O*-benzylation led to **N**. Formation of a cyclic carbonate, selective acetylation, acetal cleavage, and oxidation gave **O**. Conversion of the diol to the cyclic thiocarbonate and dehydration, followed by allylic oxidation and stereoselective reduction introduced the hydroxyl groups in **P**. An allylic bromination in ring *C*, afforded **Q** and then **R**. Dihydroxylation to **S**, conversion to the oxetane, followed by known steps gave baccatin III (**T**), which was converted to taxol. The Mukaiyama synthesis of taxol comprised 38 steps starting with L-serine (Figure 16.59).

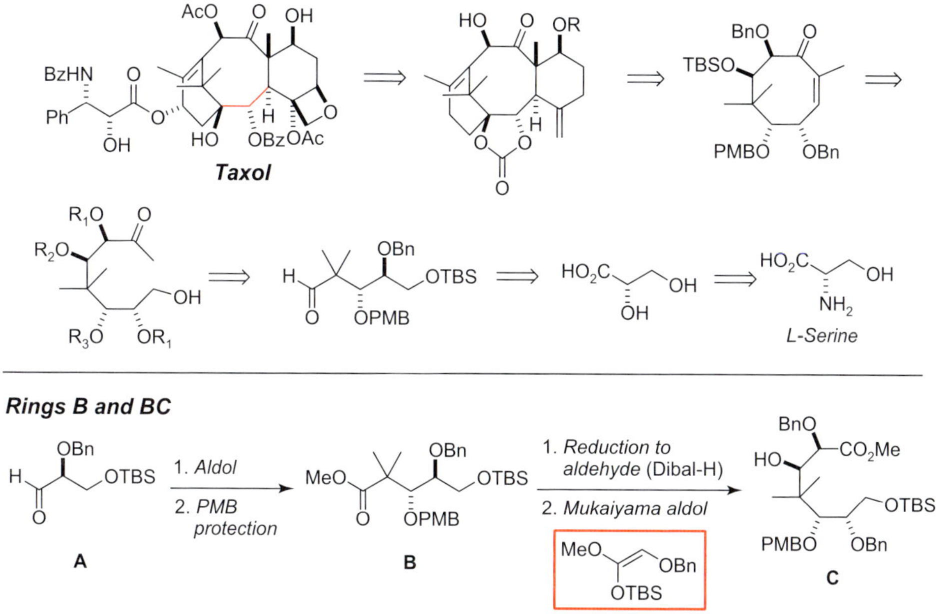

Figure 16.59 Mukaiyama's total synthesis of taxol.

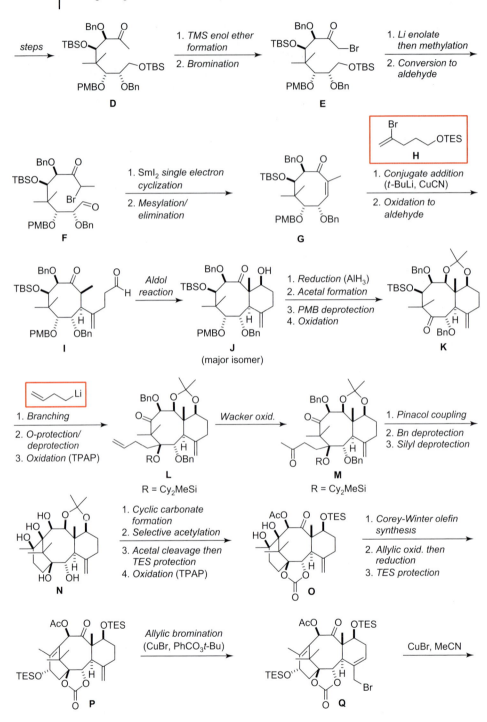

Figure 16.59 *(continued)*.

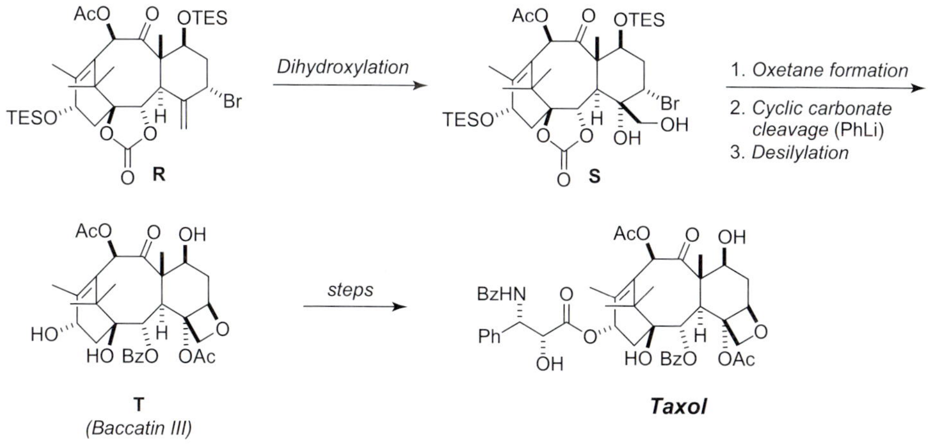

Key reactions: Stepwise aldol extensions (**A** to **B** to **C**); SmI$_2$-mediated ring closure (**F** to **G**); 1,4-conjugate addition then aldol (**I** to **J**); stereocontrolled branching then pinacol coupling (**M** to **N**); allylic bromination (**P** to **Q**)

Figure 16.59 (*continued*).

16.8.8
The six total syntheses of taxol: The calm after the storm

Some 18 years after the independent disclosures by Holton [150] and Nicolaou [151] of their heroic efforts, a new total synthesis of taxol remains, to this day, as a daunting undertaking. Seminal contributions from Danishefsky [158], Wender [161], Kuwajima [163], and Mukaiyama [164], spanning the years 1996–1999 have each demonstrated creativity and courage toward the conquest of this formidable target molecule. Curiously, no new total syntheses of taxol have been reported in the past decade in spite of its continued importance as a potent, clinically effective anticancer drug. In the absence of a new and possibly improved total synthesis, it is difficult to assess if the present-day advances in synthetic methodology and separation techniques would present an advantage over the six reported syntheses. However, in considering several of the diverse methods used to construct the *ABC* ring combinations of the taxol core structure, one comes to the realization that a high level of sophistication was already utilized by the six independent laboratories. Other than improving yields and efficiencies of certain steps, the time-honored named reactions and well-known bond construction methods remain as valid today as they were then. The difficulties of multistep syntheses based on the need to isolate and purify numerous intermediates ("stop-and-go" chemical synthesis) has been discussed by MacMillan in the context of the "taxol problem" [148]. Indeed, rare are the cases where purification of intermediates in a total synthesis is by-passed, or the need for protecting groups is avoided (see Chapter 18, section 18.5.1.1). The shortest and longest recorded synthesis of taxol starting from simple building blocks may be a matter of debate. However, it is clear that a commercially viable total synthesis of taxol must be considerably shorter than any of the reported ones, particularly in view of the quantities required [148]. Although Nature's biosynthetic pathways are sources of awe and inspiration alike, rendering them synthetically feasible by chemical means beyond the construction of the carbogenic framework of a given natural product is hard to achieve, especially for highly oxygenated molecules such as taxol. Biomimetic syntheses that utilize cascade reactions are admirably applicable to only certain families of natural products (see Chapter 4, section 4.5.4).

In reviewing the six total syntheses of taxol individually, one recognizes the many elegant ways to assemble the *ABC* core structure with a high degree of absolute stereocontrol. The Achilles heel in most of the syntheses was the need to use and interchange multiple protecting groups that would be compatible with the planned reaction conditions. Achieving the desired oxidation states of certain functional groups and the "timing" of such reactions in the context of a given sequence was also a recurring issue in all the syntheses.

As discussed in the case of many of the reported total syntheses in this book, improvements in efficiency, methodology, and even levels of creativity are the expected norm for newer disclosures of one and the same target molecules. In spite of the elegantly conceived strategies in the six total synthesis of taxol, each with its unique design and flair, none have been able to avoid the extensive functional group adjustments and modifications needed to establish the full complement of oxygen substituents (see Chapter 18, sections 18.5.2 and 18.5.3).

Methods established in the original Holton [150] and Nicolaou [151] syntheses have been instrumental in the planning and successful execution of subsequent syntheses. For example, the C-2 benzoate group was originally introduced by the nucleophilic addition of PhLi to the C-1–C-2 cyclic carbonate in advanced intermediates [150]. All subsequent total syntheses of taxol benefitted from this indirect method of introducing a benzoate ester, while taking full advantage of the cyclic carbonate as a diol protecting group over several preceding steps. In this regard, it is remarkable that such a reaction was possible in the presence of acetate esters. It is also of interest that the topology of the molecule has allowed the application of several selective *O*-protections and deprotections, mostly in favor of the desired intermediate. Also, enolate hydroxylations, either directly or proceeding through enol ethers, were successfully implemented in spite of the geometrically challenging functionalizations at ring junctions.

16.8.9
Total syntheses of taxol in the mind's eye

We end this section by returning to the conceptual origins of synthesis planning and the power of perception in the mind's of synthetic organic chemists. The fascinating interplay between visual *relational* and visual *reflexive* thinking manifests itself in the choice of staring materials, and in the strategies to construct the bi- or tricyclic core substructures of taxol (Figure 16.60).

The direct visual relationship between (−)-camphor as a starting chiron and taxol is confined to the presence of a *gem*-dimethyl groups bridging C-1 and C-11 (Figure 16.60 A). As such, only the fleeting thought (or image) of a terpene such as camphor would enter the mind's eye. However, Holton's knowledge that camphor can be ring-expanded to homocamphor, and the latter converted to patchoulene oxide according to Büchi [152], was the basis of an ingenious plan [150].

The Nicolaou [151] approach to taxol employed a pragmatic construction of rings *A* and *C* individually as racemic entities, then using an intramolecular McMurry/pinacol coupling to form the middle eight-membered ring system which, could be resolved as the (*S*)-camphanate ester (Figure 16.60 B). The visual *relational* thought process in Nicolaou's mind's eye was intimately linked, if not preceded, by the planned dialdehyde pinacol coupling reaction as a means of constructing the tricyclic *ABC* system from a tethered precursor.

The above described strategies, expertly executed by Holton [150] and Nicolaou [151], are conceptually and operationally different. They deserve recognition for having explored creatively different methods of bond construction, culminating in the first reported total syntheses of taxol in 1994.

Danishefsky [158] must have immediately recognized the relationship between the Wieland-Miescher ketone and a segment of taxol harboring the C-8 angular methyl group (Figure 16.60 C). Although the bicyclic motif in the starting chiron would not be used as such, it served to construct the "eastern" segment of the core eight-membered ring *B*, while maintaining the angular methyl group and the cyclohexanone ring intact for further elaboration. Like the Nicolaou approach [151], a tethered intermediate was generated, and engaged in an intramolecular Heck cyclization to give the intended *ABC* tricyclic core structure, already harboring the oxetane ring.

The *gem*-dimethyl C-1–C-11 bridged motif in taxol had a different visual *relational* effect in Wender's mind's eye [161] (Figure 16.60 D). In another beautiful manifestation of visual imagery and knowledge-based planning, Wender used α-pinene as a chiron to generate a bicyclic epoxide which would undergo a Grob-type fragmentation to furnish the *AB* ring system, reminiscent of Holton's studies some 2–3 years previously [150]. The elaboration of the entire *ABC* ring system would be implemented relying on an intramolecular aldol reaction (Figure 16.60 D).

The Kuwajima strategy [163] to construct the *ABC* tricyclic system capitalized on a preferred alignment of reactive functional groups in a tethered ring *A* and *C* intermediate (Figure 16.60 E). The utilization of an aromatic starting material such as *o*-bromobenzaldehyde was no doubt based on the knowledge that aromatic rings can be functionalized to 1,4-dihydroxy cyclohexenes through photochemical oxygenation of the corresponding cyclohexadienes.

Taxol

Bi- or tricyclic core strategy: **Starting materials**

A.   Holton

i.

Grob fragmentation

ii.

Dieckmann

(-)-Camphor

B.   Nicolaou

*McMurry/
pinacol coupling*

Ring C   Ring A

C.   Danishefsky

Heck reaction

Figure 16.60   Total synthesis of taxol in the mind's eye.

D. Wender

i.

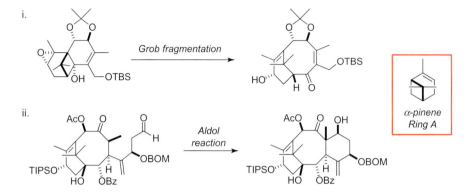

ii.

E. Kuwajima

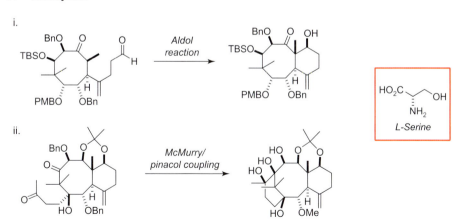

F. Mukaiyama

i.

ii.

Figure 16.60 (*continued*).

Finally, we recognize the power of a "methods and knowledge-based approach" to construct the eight-membered ring *B* of taxol with a full complement of functional groups in the Mukaiyama synthesis [164] (Figure 16.60 F). Starting with the hidden chiron L-serine, a series of planned aldol condensations was used to build the acyclic carbon backbone of ring *B*, which would undergo a SmI_2 ring closure. The elaboration of rings *BC* would be achieved via an intramolecular aldol reaction (Figure 16.60 F) [165]. The tricyclic *ABC* system was built relying on a low valent titanium-mediated McMurry/pinacol coupling reminiscent of the Nicolaou strategy (Figure 16.60 F).

It is interesting to reflect upon the six syntheses with regard to the respective starting materials, and the extent to which they provided chirality and convergence within the carbogenic landscape of taxol. Of these, the native chirons (−)-camphor, α-pinene, and L-serine have played a crucial role in securing the early chiral founding stones for the synthesis of the chemical behemoth we have come to know and respect, that is taxol.

References

1. *Classics in Total Synthesis: Targets, Strategies, Methods*, Nicolaou, K. C.; Sorensen, E. J. 1996, VCH, Weinheim
2. For example, see, Hanessian, S. *Chem. Med. Chem.* 2006, **1**, 1300.
3. For individual accounts of the collaborative effort, see Woodward, R. B. *Pure Appl. Chem.* 1973, **33**, 145; Woodward, R. B. *Pure Appl. Chem.* 1968, **25**, 519.
4. Eschenmoser, A.; Winter, C. E. 1977, *196*, 1410; see also Eschenmoser, A., *Quart. Rev.* 1970, **24**, 366; Eschenmoser, A. *Naturwissenschaften*, 1974, **61**, 513.
5. Friedrich, W.; Gross, G.; Bernhauer, K.; Zeller, P. *Helv. Chim. Acta* 1960, **43**, 704.
6. (a) Hodgkin, D. C.; Johnson, A. W.; Todd, A. R. *Chem. Soc. Spec. Publ.* 1955, **3**, p. 109; (b) Hodgkin, D. C.; Kamper, J.; Mackay, M.; Pickworth, J.; Trueblood, K. N.; White, J. G. *Nature* 1956, **178**, 64.
7. (a) Abrahamsson, S.; Hodgkin, D. C.; Maslen, E. N. *Biochem. J.* 1963, **86**, 514; (b) Hodgkin, D. C. *Adv. Sci.* 1949, **6**, 85.
8. (a) Masamune, T.; Takasugi, M.; Murai, A.; Kobayashi, K. *J. Am. Chem. Soc.* 1967, **89**, 4521; (b) Masamune, T.; Takasugi, M.; Murai, A. *Tetrahedron*

1971, **27**, 3369; see also: (c) Kutney, J. D.; Cable, J.; Gladstone, W. A. F. Hanssen, H. W.; Nair, W.; Vijay, G.; Torupka, E. J.; Warnock, W. C. *Can. J. Chem.* 1975, **53**, 1796.
9. Johnson, W. S.; de Jongh, H. A. P.; Coverdale, C. E.; Scott, J. W.; Burckhardt, U. *J. Am. Chem. Soc.* 1967, **89**, 4523.
10. Stork, G.; Kretchmer, R. A.; Schlessinger, R. H. *J. Am. Chem. Soc.* 1968, **90**, 1647; For a stereoselective total synthesis of lycopodine and previous contributions, see Yang, H.; Carter, R. G.; Zakharov, L. N. *J. Am. Chem. Soc.* 2008, **130**, 9238.
11. Ayer, W. A.; Bowman, W. R.; Joseph, T. C.; Smith, P. *J. Am. Chem. Soc.* 1968, **90**, 1649.
12. Evans, D. A.; Dow, R. L.; Shih, J. L.; Takacs, J. M.; Zahler, R. *J. Am. Chem. Soc.* 1990, **112**, 5290.
13. Hanessian, S.; Cooke, N. G.; DeHoff, B.; Sakito, Y. *J. Am. Chem. Soc.* 1990, **112**, 5276.
14. For the total synthesis of ionomycin, see: Lautens, M.; Colucci, J. T.; Hiebert, S.; Smith, N. D.; Bouchain, G. *Org. Lett.* 2002, **4**, 1879.
15. (a) Evans, D. A.; Wood, M. R.; Trotter, B. W.; Richardson, T. I.; Barrow, J. C.; Katz, J. L. *Angew. Chem. Int. Ed.* 1998,

37, 2700; (b) Evans, D. A.; Dinsmore, C. J.; Watson, P. S.; Wood, M. R.; Richardson, T. I.; Trotter, B. W.; Katz, J. L. *Angew. Chem. Int. Ed.* 1998, **37**, 2704.

16. (a) Nicolaou, K. C.; Natarajan, S.; Li, H.; Jain, N. F.; Hughes, R.; Solomon, M. E.; Ramanjulu, J. M.; Boddy, C. N. C.; Takayanagi, M. *Angew. Chem. Int. Ed.* 1998, **37**, 2708; (b) Nicolaou, K. C.; Jain, N. F.; Natarajan, S.; Hughes, R.; Solomon, M. E.; Li, H.; Ramanjulu, J. M.; Takayanagi, M.; Koumbis, A. E.; Bando, T. *Angew. Chem. Int. Ed.* 1998, **37**, 2714; (c) Nicolaou, K. C.; Takayanagi, M.; Jain, N. F.; Natarajan, S.; Koumbis, A. E.; Bando, T.; Ramanjulu, J. M. *Angew. Chem. Int. Ed.* 1998, **37**, 2717.

17. Nicolaou, K. C.; Mitchell, H. J.; Jain, N. F.; Winssinger, N.; Hughes, R.; Bando, T. *Angew. Chem. Int. Ed.* 1999, **38**, 240.

18. For the synthesis of vancomycin aglycon, see: Boger, D. L.; Miyazaki, S.; Kin, S. H.; Wu, J. H.; Castle, S. L.; Loiseleur, O.; Jiu, Q. *J. Am. Chem. Soc.* 1999, **121**, 10004; for a highlight, see Zhang, A. J.; Burgess, K. *Angew. Chem. Int. Ed.* 1999, **38**, 634.

19. (a) van Tamelen, E. E.; Spencer, T. A.Jr.; Allen, D. F.,Jr.; Orvis, R. L. *J. Am. Chem. Soc.* 1959, **81**, 6341; (b) van Tamelen, E. E.; Spencer, T. A. Jr.; Allen, D. F., Jr.; Orvis, R. L. *Tetrahedron,* 1961, **8**.

20. Schreiber, J.; Leimgruber, W.; Pesaro, M.; Schudel, P.; Eschenmoser, A. *Angew. Chem.* 1959, **71**, 637; Schreiber, J.; Leimgruber, W.; Pesaro, M.; Schudel, P.; Trelfall, T.; Eschenmoser, A. *Helv. Chim. Acta* 1961, **44**, 540.

21. Woodward, R. B. *The Harvey Lecture Series* 1963, **59**, 31.

22. Graening, T.; Schmaltz, H.-G. *Angew. Chem. Int. Ed.* 2004, **43**, 3230.

23. Büchi, G.; Kulsa, P.; Ogasawara, K.; Rosati, R. L. *J. Am. Chem. Soc.* 1970, **92**, 999.

24. Kutney, J. P.; Bylsma, F. *J. Am. Chem. Soc.* 1970, **92**, 6090.

25. Auerbach, J.; Weinreb, S. M. *J. Am. Chem. Soc.* 1972, **94**, 7172.

26. Semmelhack, M. F.; Chong, B. P.; Jones, L. D. *J. Am. Chem.. Soc.* 1972, **94**, 8629.

27. Weinreb, S. M.; Semmelhack, M. F. *Acc. Chem. Res.* 1975, **8**, 158.

28. For recent total syntheses of (−)-cephalotaxine and references to other contributions, see (a) Planas, L.; Prard-Viret, J.; Royer, J. *J. Org. Chem.* 2004, **69**, 3087; (b) Li, W.-D.; Wang, Y.-Q. *Org. Lett.* 2003, **5**, 2931; see also, Chapter 30, section 34 and refrernce 5.

29. Takita, T.; Umezawa, Y.; Saito, S.-I.; Morishima, H.; Naganawa, H.; Umezewa, H.; Tsuchiya, T.; Miyake, T.; Kageyama, S.; Umezawa, S.; Muraoka, Y.; Suzuki, M.; Otsuka, M.; Narita, M.; Kobayashi, S.; Ohno, M. *Tetrahedron Lett.* 1982, **23**, 521.

30. Aoyagi, Y.; Katano, K.; Suzuna, H.; Primeau, J.; Chang, L.-H.; Hecht, S. M. *J. Am. Chem. Soc.* 1982, **104**, 5537.

31. Magnus, P.; Brown, P. *J. Chem. Soc. Chem. Comm.* 1985, 184.

32. Kuehne, M.; Seaton, R. J. *J. Org. Chem.* 1985, **50**, 4790; for a recent synthesis of (−) -kopsinine, see: Jones, S. B.; Simmons, B.; Mastracchio, A.; MacMillan, D. W. C. *Nature* 2011, **475**, 183 and references cited therein.

33. Nicolaou, K. C.; Chakraborty, T. K.; Piscopio, A. D.; Minowa, N.; Bertinato, P. *J. Am. Chem. Soc.* 1993, **115**, 4419.

34. Romo, D.; Meyer, S. D.; Johnson, D. D.; Schreiber, S. L. *J. Am. Chem. Soc.* 1993, **115**, 7906.

35. Evans, D. A.; Hoveyda, A. H. *J. Am. Chem. Soc.* 1990, **112**, 6447.

36. Bald, E.; Saigo, K.; Mukaiyama, T. *Chem. Lett.* 1975, 1163.

37. Choi, J.; Chen, J.; Schreiber, S. L.; Clardy, J. *Science* 1996, **273**, 239.

38. Hayward, C. M.; Yohannes, D.; Danishefsky, S. J. *J. Am. Chem. Soc.* 1993, **115**, 9345.

39. For reviews on rapamycin, FK-506 and related compounds see: (a) Maddess, M. L.; Tackett, M. N.; Ley, S. V. *Proc. Med. Chem.* 2008, **66**, 5; (b) *Organic Synthesis with Carbohydrates*, Boons, G.-J.; Hale, K, J., 2000, p. 292, Sheffield Academic Press, Sheffield, UK; (c) Norley, M. C. *Contemporary Organic Synthesis* 1996, 345; (d) Sehgal, S. N.;

Molnar-Kimber, K.; Ocain, T. D.;
Weichman, B. M. *Med. Chem. Rev.*
1994, **14**, 1.

40. (a) Dabrah, T. T.; Harwood, H. J.,Jr.;
Huang, L. H.; Jankovich, N. D.;
Kaneto, T.; Li, J.-C.; Lindsey, S.;
Moshier, P. M.; Subashi, T. A.;
Therrien, M.; Watts, P. C. *J. Antibiot.*
1997, **50**, 1; (b) Dabrah, T. T.; Kaneko,
T.; Massefski, W., Jr.; Wipple, E. B. *J.
Am. Chem. Soc.* 1997, **119**, 1594.

41. Spiegel, D. A.; Njardarson, J. T.;
McDonald, I. M.; Wood, J. L. *Chem.
Rev.* 2003, **103**, 2691; see also:
Hepworth, D. *Chem. Ind.* 2000, 59.

42. The challenges involved in such an
endeavor have been likened to the
skills and bravery of Theseus who was
to sent to find his way through the
maze of the labyrinth, home of the
deadly Minotaur, and to kill him. See
Nicolaou, K. C.; Baran, P. S. *Angew.
Chem. Int. Ed.* 2002, **41**, 2678.

43. (a) Nicolaou, K. C.; Jung, J.-K.; Yoon,
W. H.; He, Y.; Zhong, Y.-L.; Baran,
P. S. *Angew. Chem. Int. Ed.* 2000, **39**,
1829; (b) Nicolaou, K. C.; Baran, P. S.;
Zhong, Y.-L.; Choi, H.-S.; Yoon, W. H.;
He, Y.; Fong, K. C. *Angew. Chem.
Int. Ed.* 1999, **38**, 1669; (c) Nicolaou,
K. C.; Baran, P. S.; Zhong, Y.-L.; Fong,
K. C.; He. Y.; Yoon, W. H.; Choi, H.-S.
Angew. Chem. Int. Ed. 1999, **38**, 1676.

44. Chen, C.; Layton, M. E.; Sheehan,
S. M.; Shair, M. D. *J. Am. Chem. Soc.*
2000, **122**, 7424; see also Chen, C.;
Layton, M. E.; Shair, M. D. *J. Am.
Chem. Soc.* 1988, **120**, 10784.

45. For a review, see Corey, E. J.; Helal,
C. J. *Angew. Chem. Int. Ed.* 1998, **37**,
1986.

46. Evans, D. A.; Golob, A. M. *J. Am.
Chem. Soc.* 1975, **97**, 4765; for a review, see: Paquette, L. A. *Tetrahedron*
1997, **53**, 13971; see also: Still, W. C.
J. Am. Chem. Soc. 1997, **99**, 4186; 1979,
101, 2493.

47. Waizumi, N.; Itoh, T.; Fukuyama, T. *J.
Am. Chem. Soc.* 2000, **122**, 7825.

48. Tan, Q.; Danishefsky, S. J. *Angew.
Chem. Int. Ed.* 2000, **39**, 4509; see
also: Kwon, Q.; Su, D.-S.; Meng,
D.; Deng, W.; D'Amico, D. C.;

Danishefsky, S. J. *Angew. Chem. Int.
Ed.* 1998, **37**, 1877.

49. For reviews, see (a) Dounay, A. B.;
Overman, L. E. *Chem. Rev.* 2003, **103**,
2945; (b) Link, J. T. *Org. React.* 2002,
60, 157; (c) De Meijere, A.; Meyer,
F. E. *Angew. Chem. Int. Ed.* 1994, **33**,
2379.

50. Duffey, M. O.; Le Tiran, A.; Morken,
J. P. *J. Am. Chem. Soc.* 2003, **125**,
1458.

51. Hanessian, S.; Yang, Y.; Giroux, S.;
Mascitti, V.; Ma, J.; Raeppel, F. *J. Am.
Chem. Soc.* 2003, **125**, 13784.

52. (a) Zhou, C.-Y.; Duffey, M. O.; Taylor,
S. J.; Morken, J. P. *Org. Lett.* 2001, **3**,
1829; (b) Taylor, S. J.; Duffey, M. O.;
Morken, J. P. *J. Am. Chem. Soc.* 2000,
122, 4528.

53. Myers, A. G.; Yang, B. H.; Chen, H.;
McKinstry, L.; Kopecky, D. J.; Gleason,
J. L. *J. Am. Chem. Soc.* 1997, **119**,
6496.

54. (a) Sonogashira, K.; Tohda, Y.;
Hagihara, N. *Tetrahedron Lett.* 1975,
16, 4467; (b) Chang, J.; Paquette, L. A.
Org. Lett. 2002, **4**, 253.

55. Misumi, A.; Iwanaga, K.; Furuta, K.;
Yamamoto, H. *J. Am. Chem. Soc.* 1985,
107, 3343.

56. Inanaga, J.; Hirata, K.; Saeki, H.;
Katsuki, T.; Yamaguchi, M. *Bull. Chem.
Soc. Jpn* 1979, **52**, 1989.

57. For a review, see Hoffmann, R. W.
Angew. Chem. Int. Ed. 2000, **39**, 2054.

58. Hanessian, S.; Sumi, K. *Synthesis* 1991,
1083.

59. Still, W. C.; Gennari, C. *Tetrahedron
Lett.* 1983, **25**, 4405.

60. Vong, B.-G.; Kim, S. H.; Abraham, S.;
Theodorakis, E. A. *Angew. Chem. Int.
Ed.* 2004, **43**, 3947; see also Vong,
B. G.; Abraham, S.; Xiang, A. X.;
Theodorakis, E. A. *Org. Lett.* 2003, **5**,
1617.

61. Nagamistu, T.; Takano, D.; Fukuda, T.;
Otoguro, K.; Kuwajima, I. S.; Harigaya,
Y.; Ōmura, S. *Org. Lett.* 2004, **6**, 1865.

62. For example, see, Fukuzawa, S.;
Matsuzawa, H.; Yoshimitsu, S. *J.
Org. Chem.* 2000, **65**, 1702.

63. For example, see, Evans, D. A.
Aldrichimica Acta 1982, **15**, 23; for
reviews, see: (a) McManus, H.; Guiry,

P. J. *Chem. Rev.* 2004, **104**, 4151; (b) Faita, G.; Jørgensen, K. A. *Chem. Rev.* 2006, **106**, 3561.

64. Myers, A. G.; Yang, B. H.; Chen, H.; Kopecky, D. *Synlett* 1997, 457.

65. Va, P.; Roush, W. R. *J. Am. Chem. Soc.* 2006, **128**, 15960.

66. Kim, C. H.; An, H. J.; Shin, W. K.; Yu, W.; Woo, S.-K.; Jung, S. K.; Lee, E. *Angew. Chem. Int. Ed.* 2006, **45**, 8019.

67. (a) Roush, W. R.; Palkowitz, A. D.; Ando, K. *J. Am.Chem. Soc.* 1990, **112**, 6348; (b) Roush, W. R.; Halterman; R. L. *J. Am. Chem. Soc.* 1986, **108**, 294.

68. (a) Roush, W. R.; Grover, P. T. *Tetrahedron Lett.* 1990, **31**, 7567; (b) Roush, W. R.; Grover, P. T. *Tetrahedron* 1992, **48**, 1981.

69. For example, see: Blackmore, P. R.; Cole, W. J.; Kocienski, P. J.; Morely, A. *Synlett* 1998, **26**.

70. For reviews on malaria, see (a) Greenwood, B.; Mutabingwa, T. *Nature*, 2002, **415**, 670; (b) Wiesner, J.; Ortmann, R.; Jomaa, H.; Schlitzer, M. *Angew. Chem. Int. Ed.* 2003, **42**, 5274; see also: *Science* 2000, **290**, 428

71. For reviews, see: Kauffman, T. S.; Rúveda, E. A. *Angew. Chem. Int. Ed.* 2005, **44**, 854; Kauffman, *Chem. Educator*, 2004, **9**, 172; see also: Nicolaou, K. C.; Snider, S. A. *Classics in Total Synthesis II: More Targets, Strategies, Methods*, 2003, Chapter 31, Wiley-VCH, Weinheim.

72. Stork, G.; Niu, D.; Fujimoto, A.; Koft, E. R.; Balkovec, J. M.; Tata, J. R.; Date, G. R. *J. Am. Chem. Soc.* 2001, **123**, 3239.

73. For example, see: (a) Kondo, K.; Mori, E. *Chem. Lett.* 1974, **741**; (b) Ishibashi, F.; Taniguchi, E. *Bull. Chem. Soc. Jpn.* 1988, **61**, 4361

74. (a) Gutzwiller, J.; Uskoković, M. *J. Am. Chem. Soc.* 1970, **92**, 204; (b) Uskoković, M.; Gutzwiller, J.; Henderson, T. *J. Am. Chem. Soc.* 1970, **92**, 203; see also Gutzwiller, J.; Uskoković, M. *J. Am. Chem. Soc.* 1978, **100**, 576.

75. Woodward, R. B.; Doering, W. E. *J. Am. Chem. Soc.* 1944, **66**, 849; Woodward, R. B.; Doering, W. E. *J. Am. Chem. Soc.* 1945, **67**, 860.

76. Rabe, P.; Kindler, K. *Ber. Dtsch. Chem. Ges.* 1918, **51**, 466; see also: Rabe, P. *Justus Liebigs Ann. Chem.* 1932, **492**, 242; Rabe, P. *Ber. Dtsch. Chem. Ges.* 1911, **44**, 2088.

77. Pasteur, L. *Compt. Rend.* 1853, **37**, 110; see als: Rabe, P. *Ann.* 1909, **363**, 366.

78. For example, see (a) *Medicinal Plants of the World*, VanWyk, B.-E.; Win, M., 2004, Timber Press, Portland, OR; (b) *Quinine: Malaria and the Quest for a Cure that Changed the World*, Rocco, F., 2004, Harper Collins.

79. Seeman, J. I. *Angew. Chem. Int. Ed.* 2007, **46**, 1378; see also: *C & E News* 2007, February 26, p. 47.

80. Smith, A. C.; Williams, R. M. *Angew. Chem. Int. Ed.* 2008, **47**, 1736.

81. Presently, the U. S. imports over 60 tons of quinine annually. It is cultivated mainly in Java and extracted from the bark of trees of Cinchona spp. Of the ca. 10% extract of alkaloids, 70% consists of quinine. Anhydrous quinine (and quinidine) are available commercially at about $1–1.5 per gram.

82. For example, see, citations in reference 14 of the Stork *et al.* paper (ref. 72 in this Chapter).

83. Gates, M.; Sugavanam, B.; Schreiber, W. L. *J. Am. Chem. Soc.* 1970, **92**, 205.

84. Taylor, E. C.; Martin, S. F. *J. Am. Chem. Soc.* 1972, **94**, 6218.

85. For example, see: Hanessian, S. in *Chemical Synthesis*, Chatgilialoglu, C.; Snieckus, V., Eds 1996, p. 61, Kluwer Publishers, the Netherlands; Hanessian, S. *Pure Appl. Chem.* 1993, **65**, 1189; Hanessian, S.; Franco, J.; Larouche, B. *Pure Appl. Chem.* 1990, **62**, 1887.

86. Taylor, M. S.; Jacobsen, E. N. *J. Am. Chem. Soc.* 2003, **125**, 11204.

87. Raheem, I. T.; Goodman, S. N.; Jacobsen, E. N. *J. Am. Chem. Soc.* 2004, **126**, 706.

88. Takai, K.; Shinomiya, N.; Kaihana, H.; Yoshida, N.; Moriwake, T. *Synlett* 1995, 963.

89. For example, see Barder, T. E.; Walker, S. D.; Martinelli, J. R.; Buchwald, S. L. *J. Am. Chem. Soc.* 2005, **124**, 4685.

90. For a review, see Kolb, H. C.; Van Nieuwenhze, M. S.; Sharpless, K. B. *Chem. Rev.* 1994, **94**, 2483.

91. Kolb, H. C.; Sharpless, K. B. *Tetrahedron* 1992, **48**, 10515.

92. Murai, A.; Tsujimoto, T. *Synlett*, 2002, 1283.

93. Igarashi, J.; Kobayashi, Y. *Tetrahedron Lett.* 2005, **46**, 6381; see also: Igarashi, J.; Katsukawa, M.; Wang, Y.-G.; Acharya, H. D.; Kobayashi, Y. *Tetrahedron Lett.* 2004, **45**, 3783.

94. For example, see Deardorff, D. R.; Linde, R. G.; Martin, A. M.; Shulman, M. J. *J. Org. Chem.* 1989, **54**, 2759.

95. Johns, D. M.; Mori, M.; Williams, R. M. *Org. Lett.* 2006, **8**, 4051.

96. Dastlik, K. A.; Sundermeier, U.; Johns, D. M.; Chen, Y.; Williams, R. M. *Synlett* 2005, 693.

97. For example, see Trost, B. M.; Sacchi, K. L.; Schroeder, G. M.; Asakawa, N. *Org. Lett.* 2002, **4**, 3427.

98. Webber, P.; Krische, M. *J. Org. Chem.* 2008, **73**, 9379.

99. Proštenik, M., Prelog, V. *Helv. Chim. Acta* 1943, **26**, 1965.

100. Prelog, V.; Zalan, E.; *Helv. Chim. Acta* 1944, **27**, 535; 545.

101. (a) Leithe, W. *Ber. Dtsch. Chem. Ges.* 1932, **65**, 660; (b) Freudenberg, K. *J. Am. Chem. Soc.* 1932, **54**, 234.

102. Ciechanover, A. *Angew. Chem. Int. Ed.* 2005, **44**, 594.

103. Ōmura, S.; Fujimoto, T.; Otoguro, K.; Matsuzaki, K.; Moriguchi, R.; Tamaka, H.; Sasaki, Y. *J. Antibiot.* 1991, **44**, 113

104. Ōmura, S.; Matsuzaki, K.; Fujimoto, T.; Kosuge, K.; Furuya, T.; Fujita, S.; Nakagawa, A. *J. Antibiot.* 1991, **44**, 117.

105. For example, see: (a) Dick, L. R.; Cruickshank, A. A.; Grenier, L.; Melandri, F. D.; Nunes, S. L.; Stein, R. L. *J. Biol. Chem.* 1996, **271**, 7273; (b) Fenteany, G.; Standaert, R. F.; Lane, W. S.; Choi, S.; Corey, E. J.; Schreiber, S. L. *Science* 1995, **268**, 726; (c) Groll, M.; Ditzel, L.; Löwe, J.; Stock, D.; Bochter, M.; Bartunik, H. D.; Huber, R. *Nature*, 1997, **386**, 463. For the first synthesis, see: Corey, E. J.; Reichard, G. A.; Kania, R. *Tetrahedron Lett.* 1993, **43**, 6977.

106. For reviews, see: (a) Shibasaki, M.; Kanai, M.; Fukuda, N. *Chem. Asian, J.* 2007, **2**, 20; (b) Corey, E. J.; Li, W.-D. Z. *Chem. Pharm. Bull.* 1999, **47**, 1.

107. Feling, R. H.; Buchanan, G. O.; Mincer, T. J.; Kauffmann, C. A.; Jensen, P. R.; Fenical, W. *Angew. Chem. Int. Ed.* 2003, **42**, 355.

108. (a) Reddy, L. R.; Saravanan, P.; Corey, E. J. *J. Am. Chem. Soc.* 2004, **126**, 6230; (b) Reddy, L. R.; Fournier, J.-F.; Reddy, B. V. S.; Corey, E. J. *J. Am. Chem. Soc.* 2005, **127**, 8974; (c) Endo, A.; Danishefsky, S. J. *J. Am. Chem. Soc.* 2005, **127**, 8298; (d) Mulholland, N. P.; Pattenden, G.; Walters, I. A. S. *Org. Biomol. Chem.* 2006, **4**, 2845; (e) Ling, T; Macherla, V. R.; Manam, R. R.; McArthur, K. A.; Potts, B. C. *Org. Lett.* 2007, **9**, 2289; (f) Ma, G; Nguyen, H.; Romo, D. *Org. Lett.* 2007, **9**, 2143.

109. (a) Corey, E. J.; Reichard, G. A. *J. Am. Chem. Soc.* 1992, **114**, 10677; (b) Corey, E. J.; Li, W.; Reichard, G. A. *J. Am. Chem. Soc.* 1998, **120**, 2330.

110. For a review, see: Ohno, M.; Otsuka, M. *Org. React.* 1989, **37**, 1

111. Corey, E. J.; Li, W.; Nagamitsu, T. *Angew. Chem. Int. Ed.* 1998, **37**, 1676.

112. Uno, H.; Baldwin, J. E.; Russell, A. T. *J. Am. Chem. Soc.* 1994, **116**, 2139.

113. (a) Thottathil, J. K.; Montiot, J. L.; Mueller, R. H.; Wong, M. K. Y.; Kissik, T. P. *J. Org. Chem.* 1986, **51**, 3140; (b) Hamada, Y.; Kawai, A.; Kohno, Y.; Hara, O.; Shiori, T. *J. Am. Chem. Soc.* 1989, **111**, 1524.

114. For a review, see: Casiraghi, G.; Zanardi, F.; Appendino, G.; Rassu, G. *Chem. Rev.* 2000, **100**, 1929.

115. Barton, D. H. R.; Subramanian, R. *J. Chem. Soc. Perkin Trans 1* 1977, 1718.

116. Uno, H.; Baldwin, J. E.; Churcher, I.; Russell, A. *Synlett* 1997, 390.

117. Chida, N.; Takeoka, J.; Ando, K.; Tsutsumi, N.; Ogawa, S. *Tetrahedron* 1997, **53**, 16287.

118. Rosenthal, A.; Sprinzl, M. *Can. J. Chem.* 1969, **47**, 3941.

119. Overman, L. E. *J. Am. Chem. Soc.* 1978, **98**, 2901.

120. (a) Nagamitsu, T.; Sunazuka, T.; Tanaka, H. Ōmura, S.; Sprengler,

P. A.; Smith, III, A. B. *J. Am. Chem. Soc.* 1996, **118**, 3584; (b) Sunazuka, T.; Nagamitsu, T.; Matsuzaki, K.; Tanaka, H.; Ōmura, S.; Smith, III, A. B. *J. Am. Chem. Soc.* 1993, **115**, 5302.

121. For reviews, see: (a) Johnson, R. A.; Sharpless, K. B.; in *Catalytic Asymmetric Synthesis*, 2nd ed., Ojima, I., Ed.; 2000, p. 103; p. 231, Wiley-VCH, Weinheim; (b) Johnson, R. A.; Sharpless, K. B. in *Comprehensive Organic Synthesis*, Trost, B. M., Ed.; 1991, vol. 7, Chapter 32, Pergamon Press, N.Y..

122. Seebach, D.; Aebi, J. D. *Tetrahedron Lett.* 1983, **24**, 3311.

123. Pfitzner, K. E.; Moffatt, J. G. *J. Am. Chem. Soc.* 1967, **87**, 5661.

124. Brown, H. C.; Bhat, K. S. *J. Am. Chem. Soc.* 1986, **108**, 293; Brown, H. C.; Bhat, K. S. *J. Am. Chem. Soc.* 1986, *108*, 5919.

125. Corey, E. J.; Myers, A. G. *J. Am. Chem. Soc.* 1985, **107**, 5574; see also Bal, B. S.; Childers, W. E., Jr.; Pinnick, H. W. *Tetrahedron* 1981, **37**, 2091.

126. Panek, J. S.; Masse, C. E. *Angew. Chem. Int. Ed.* 1999, **38**, 1093.

127. For example, see (a) Tao, B.; Schlingloff, G.; Sharpless, K. B. *Tetrahedron Lett.* 1998, **39**, 2507; (b) Li, G.; Angert, H. H.; Sharpless, K. B. *Angew. Chem. Int. Ed.* 1996, **35**, 2837; for a review, see: O'Brien, P. *Angew. Chem. Int. Ed.* 1999, **38**, 326.

128. For a review, see: Masse, C. E.: Panek, J. S. *Chem. Rev.* 1995, **95**, 1293.

129. Balskus, E. P.; Jacobsen, E. N. *J. Am. Chem. Soc.* 2006, **128**, 6810.

130. Gandelman, M.; Jacobsen, E. N. *Angew. Chem. Int. Ed.* 2005, **44**, 2393.

131. Tamao, K.; Ishida, N.; Kumada, M. *J. Org. Chem.* 1983, **48**, 2120.

132. Fukuda, N.; Sasaki, K.; Sastry, T. V. R. S.; Kanai, M.; Shibasaki, M. *J. Org. Chem.* 2006, **71**, 1220; see also ref. [101a].

133. (a) Yabu, K.; Masumoto, S.; Yamasaki, S.; Hamashima, Y.; Kanai, M.; Du, W.; Curran, D. P. Shibasaki, M. *J. Am. Chem. Soc.* 2001, **123**, 9908; (b) Masumoto, S.; Usuka, H.; Suzuki, M.; Kanai, M.; Shibasaki, M. *J. Am. Chem. Soc.* 2003, **125**, 5634.

134. Crump, R. A. N. C.; Fleming, I.; Urch, C. J. *J. Chem. Soc. Perkin Trans. I* 1994, 701.

135. Kang, S. H.; Jun, M.-S. *Chem. Commun.* 1998, 1929.

136. Soucy, F.; Grenier, L.; Behnke, M. L.; Destree, A. T.; McCormack, T. A.; Adams, J.; Plamondon, L. *J. Am. Chem. Soc.* 1999, **121**, 9967.

137. Corey, E. J.; Choi, S. *Tetrahedron Lett.* 1993, **34**, 6969.

138. Iwana, S.; Gao, W.-G.; Shinada, T.; Ohfune, Y. *Synlett* 2000, 1631.

139. Brennan, C. J.; Pattenden, G.; Rescourio, G. *Tetrahedron Lett.* 2003, **44**, 8757.

140. Ooi, H.; Ishibashi, N.; Iwabuchi, Y.; Ishihara, J.; Hatakeyama, S. *J. Org. Chem.* 2004, **69**, 7765.

141. Donohue, T. J.; Sintion, H. O.; Sisangia, L.; Harling, J. D. *Angew. Chem. Int. Ed.* 2004, **43**, 2293.

142. Wardrop, D.-J.; Bowen, E. G. *Chem. Comm.* 2005, 5106.

143. For example, see (a) Taber, D. F.; Christos, T. E.; Neubert, T. D.; Batra, D. J. *J. Org. Chem.* 1999, **64**, 9673; (b) Taber, D. F.; Neubert, J. D. *J. Org. Chem.* 2001, **66**, 143; (c) Taber, D. F.; Meagley, R. P. J. Doren, D. J. *J. Org. Chem.* 1996, **61**, 5723.

144. Green, M. P.; Prodger, J. C.; Hayes, C. J. *Tetrahedron Lett.* 2002, **43**, 6609.

145. Hayes, C. J.; Sherlock, A. E.; Selby, M. D. *Org. Biomol. Chem.* 2006, **4**, 193; see also: Hayes, C. J.; Sherlock, A. E.; Green, M. P.; Wilson, C.; Blake, A. J.; Selby, M. D.; Prodger, J. C. *J. Org. Chem.* 2008, **73**, 2041.

146. Wani, M. C.; Taylor, H. L.; Wall, M. E.; Coggon, P.; McPhail, A. T. *J. Am. Chem. Soc.* 1971, **93**, 2325; for reviews on chemistry and biology of the taxanes and taxol, see: (a) Kingston, D. G. I. *J. Org. Chem.* 2008, **73**, 3975; (b) Nicolaou, K. C.; Dai, W.-M.; Guy, R. K. *Angew. Chem. Int. Ed. Engl.* 1994, **33**, 15; (c) Guenard, D.; Gueritte-Voegelein, F.; Potier, P. *Acc. Chem. Res.* 1993, **26**, 160; (d) Rowinsky, E. K.; Onetto, N.; Canetto, R. M.; Arbuck, S. G. *Semin. Oncol.* 1992, **19**, 646; (e) Runowitz, C. D.; Wiernik, P. H.; Einzig, A. I.; Goldberg,

G. L.; Horwitz, S. B. *Cancer* 1993, **71**, 1591.

147. For example, see: (a) Gibson, D. M.; Ketchum, R. E. B.; Hirasuna, T. J.; Shuler, M. L. in *Taxol: Science and Applications*, Suffness, M. Ed., 1995, p. 71, CRC Press, Boca Raton, FL; (b) Suffness, M.; Wall, M. E. in *Taxol: Science and Applications*, 1995, p. 3, CRC Press, Boca Raton, FL; (c) *Taxane Anticancer Agents: Basic Science and Current Status*, Georg, G.; Chen, T. T.; Ojima, I.; Vyas, D. M. Eds. ACS Symposium Series 583; American Chemical Society: Washington, DC, 1995; (d) *The Story of Taxol: Nature and Politics in the Pursuit of an Anti-Cancer Drug*, Goodman, J.; Walsh, V., 2001, Cambridge University Press, U.K.

148. For comments on the taxol problem with relevance to total synthesis and scale-up, see: Walji, A. M.; MacMillan, D. W. C. *Synlett* 2007, 1477.

149. For examples of reviews, see: (a) Wender, P. A.; Natchus, M. G.; Shuker, A. J. In *Taxol: Science and Applications*, Suffness, M., Ed.; 1995; p. 123; CRC Press, Boca Raton, FL; (b) Boa, A. N.; Jenkins, P. R.; Lawrence, N. J. *Contemp. Org. Synth.* 1994, **1**, 47; (c) Kingston, D. G. I.; Molinero, A. A.; Rimoldi, J. M. In *Progress in the Chemistry of Organic Natural Products*, 1993, vol. 61, Springer-Verlag, NY; (d) Swindell, C. S. *Org. Prep. Proced. Int.* 1991, **23**, 465.

150. (a) Holton, R. A.; Somaza, C.; Kim, H.-B.; Liang, F.; Biediger, R. J.; Boatman, P. D.; Shindo, M.; Smith, C. C.; Kim, S.; Nadizadeh, H.; Suzuki, Y.; Tao, C.; Vu, P.; Gentile, L. N.; Liu, J. H. *J. Am. Chem. Soc.* 1994, **116**, 1597; (b) Holton, R. A.; Kim, H.-B.; Somoza, C.; Liang, F.; Biediger, R. J.; Boatman, P. D.; Shindo, M.; Smith, C. C.; Kim, S.; Nadizadeh, H.; Suzuki, Y.; Tao, C.; Vu, P.; Tang, S.; Zhang, P.; Murthi, K. K.; Gentile, L. N.; Liu, J. H. *J. Am. Chem. Soc.* 1994, **116**, 1599.

151. Nicolaou, K. C.; Yang, Z.; Liu, J.-J.; Ueno, H.; Nantermet, P. G.; Guy, R. K.; Clairborne, C. F.; Renaud, U.; Couladouros, E. A.; Paulvannan, K.; Sorensen, E. J. *Nature* 1994, **367**, 630;

see also: *Classics in Total Synthesis, Targets, Strategies, Methods*, Nicolaou, K. C.; Sorensen, E. J. 1996, Chapter 33, Wiley-VCH, Weinheim.

152. Büchi, G.; MacLeod, W. D.,Jr. *J. Am. Chem. Soc.* 1962, **84**, 3205.

153. Lee, S. D.; Chan, T. H.; Kwon, K.-S. *Tetrahedron Lett.* 1984, **25**, 3399.

154. Burgess, E. M.; Penton, H. R.,Jr.; Taylor, E. A. *J. Org. Chem.* 1973, **38**, 26.

155. (a) Ojima, I.; Habus, I.; Zhao, M.; Georg, G.; Jayasinghe, L. R. *J. Org. Chem.* 1991, **56**, 1681; (b) Ojima, I.; Habus, I.; Zhao, M.; Zucco, M.; Park, Y. H.; Sun, C. M.; Brigaud, T. *Tetrahedron* 1992, **48**, 6985.

156. Shapiro, R. H. *Org. React.* 1976, **23**, 405; see also Chamberlin, A. R.; Bloom, S. H. *Org. React.* 1990, **39**, 1.

157. For a review, see: McMurry, J. E. *Chem. Rev.* 1989, **89**, 1513.

158. Danishefsky, S.-J.; Masters, J. J.; Young, W. B,; Link, J. T.; Snyder, L. B.; Magee, T. V.; Jung, D. K.; Isaacs, R. C. A.; Bornmann, W. G.; Alaino, C. A.; Coburn, C. A.; DiGrandi, M. J. *J. Am. Chem. Soc.* 1996, **118**, 2843.

159. Corey, E. J.; Chaykovsky, M. *J. Am. Chem. Soc.* 1965, **87**, 1353.

160. *Palladium Reagents in Organic Synthesis*, Heck, R. F. 1985, p. 179, Academic Press, NY; for a review, see: deMeijere, A.; Meyer, F. E. *Angew. Chem. Int. Ed.* 1994, **331**, 2379.

161. (a) Wender, P. A.; Badham, N. F.; Conway, S. P.; Floreancig, P. E.; Glass, T. E.; Gränicher, C.; Houze, J. B.; Janichen, J.; Lee, D., Marquess, D. G.; McGrane, P. L.; Meng, W.; Mucciaro, T. P.; Mühlebach, M.; Natchus, M. G.; Paulsen, H.; Rawlins, D. B.; Satkofsky, J.; Shuker, A. J.; Sutton, J. C.; Taylor, R. E.; Tomooka, K. *J. Am. Chem. Soc.* 1997, **119**, 2755; (b) Wender, P. A.; Badham, N. F.; Conway, S. P.; Floreancig, P. E.; Glass, T. E.; Houze, J. B.; Krauss, N. E.; Lee, D.; Marquess, D. G.; McGrane, P. L.; Meng, W.; Natchus, M. G.; Shuker, A. J.; Sutton, J. C.; Taylor, R. E. *J. Am. Chem. Soc.* 1997, **119**, 2757.

162. (a) Hurst, J. J.; Whitham, G. H. *Proc. Chem. Soc.* 1959, 160; (b) Hurst,

J. J.; Whitham, G. H. *J. Chem. Soc.*
1960, 2864; (c) Chrétien-Bessiere, Y.;
Retermar, J.-A. *Bull. Soc. Chim. Fr.*
1963, **30**, 884; (d) Erman, W. F. *J. Am.
Chem. Soc.* 1967, **89**, 3828; Wender,
P. A.; Mucciaro, T. P. *J. Am. Chem.
Soc.* 1992, **114**, 5878.

163. (a) Kusama, H.; Hara, R.; Kawahara, S.;
Nishimori, T.; Kashima, H.; Nakamura,
N.; Morihira, K.; Kuwajima, I. *J.
Am. Chem. Soc.* 2000, **122**, 3811; (b)
Morihira, K.; Hara, R.; Kawahara, S.;
Nishimori, T.; Nakamura, N.; Kusama,
H. Kuwajima, I. *J. Am. Chem. Soc.*
1998, **120**, 12980.

164. (a) Mukaiyama, T.; Shiina, I.; Iwadare,
H.; Saitoh, M.; Nishimura, T.; Ohkawa,
N.; Sakoh, H.; Nishimura, K.; Tani,
Y.; Hasegawa, M.; Yamada, K.; Saitoh,
K. *Chem. Eur. J.* 1999, **5**, 121; (b)
Mukaiyama, T.; Shiina, I.; Iwadare,
H.; Sakoh, H.; Tani, Y.; Hasegawa, M.;
Saitoh, K. *Proc. Jpn. Acad. Ser. B.* 1997,
73, 95.

165. Mukaiyama, T.; Uchiro, H.; Shiina, I.;
Kobayashi, S. *Chem. Lett.* 1990, 1019;
see also: Mukaiyama, T. *Org. React.*
1982, **28**, 203.

17
Man, Machine, and Visual Imagery in Synthesis Planning

Synthesis planning and execution based on visual *relational* and visual *reflexive* thinking has been repeatedly demonstrated for many molecules throughout this book. In the first case, visual analysis will reveal suitable *chiral* synth*ons* (*chirons*) that can provide functional, stereochemical, and skeletal convergence with a substructure of the intended target molecule. These chirons would then be suitable starting materials to consider for a total synthesis with the added advantage that chirality can be secured at the outset, and also serve to generate new stereogenic centers by virtue of asymmetric induction.

The planning of a total synthesis based on the visual *relational* thought process is the very basis of the *chiron approach* [1]. The visual analysis of a molecule, of a medium level of complexity, will sooner or later reveal a hidden chiron in the mind's eye. However, even if it were to be present, the same chiron may be more difficult to see in the substructure of a related molecule, because of a different spatial disposition of stereogenic centers and their relative orientations around rotatable bonds.

Finding identical substructures among biogenetically related molecules that could be potentially synthesized from one and the same chiron may be challenging at times. For example, how quickly can the eye perceive the existence of one or more identical and superimposable propionate triads in erythronolide A aglycone, and monensin (Figure 17.1)? Such methyl-hydroxyl-methyl-hydroxyl motifs corresponding to two or more propionate triads are encompassed within the C-1–C-6 and C-1–C-8 substructures of erythronolide A aglycone and monensin respectively.

Design and Strategy in Organic Synthesis: From the Chiron Approach to Catalysis, First Edition.
Stephen Hanessian, Simon Giroux, and Bradley L. Merner.
© 2013 Wiley-VCH Verlag GmbH & Co. KGaA. Published 2013 by Wiley-VCH Verlag GmbH & Co. KGaA.

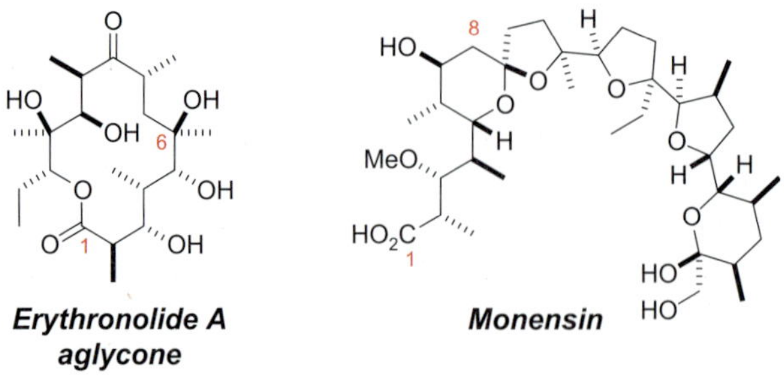

Figure 17.1 Structures of erythronolide A aglycone and monensin. Are there any superimposable subunits?.

In the absence of a means to relate these substructures in three-dimensional space, it would be difficult to establish identity, even to the stereochemically well-trained eye. The zigzag perspective drawing of the C-1–C-6 substructure within the 14-membered macrolide framework of erythronolide A aglycone will conjure a number of synthetic routes starting with chirons that match stereochemistry and functionality. However, the spatial orientation of the substituents in the C-1–C-8 substructure of monensin will present a challenge to decide which of the propionate triads is identical to the corresponding C-1–C-6 substructure in erythronolide A aglycone. A chiron available from a suitable precursor, or synthesized through a series of stereocontrolled aldol reactions, may well be an appropriate building block for the C-1–C-6 substructure of erythronolide A aglycone. Unless we can establish the absolute stereochemical relationship between the aforementioned substructures in erythronolide A aglycone and monensin, we cannot plan to use the same chiron as a common building block for both.

Relational thinking is exceedingly useful when considering the *chiron approach* as a strategy. This implies relating the target molecule (or a substructure thereof) to a chiron as two static images, representing the "end" and the "beginning" respectively, then proceeding with the process of building the molecule in a systematic manner. Being able to see the structures of compound A going to compound B in a traditional scheme, facilitates our thought process of going forward.

Leonardo da Vinci wrote: *"The eye keeps in itself the image of luminous bodies for some time"* (see Chapter 1, section 1.5). Although metaphorically speaking, da Vinci's words could also apply to molecules – it is difficult to maintain the image of a precursor and its intended product in the mind's eye without momentarily losing one or both. Thus, we are greatly dependent on the visual process for being able to relate precursors to a product in synthesis planning.

Cognizant of this need, a number of pioneering efforts have been made over the years to create computer programs that would facilitate the conception of synthetic strategies toward target molecules [2]. Below we highlight salient features of some of these programs.

17.1
The LHASA Program

The first concerted efforts toward computer-aided synthesis planning were pioneered by E. J. Corey and coworkers [3] over four decades ago. Given a target molecule, the LHASA Program (Logic and Heuristics Applied to Synthetic Analysis) has the capability of searching its knowledge base of chemically feasible transformations (transforms), thereby generating a "tree" that branches out to a variety of synthetic routes, ultimately reaching the target molecule or an intermediate. The LHASA program recognizes functional and structural disconnections based on a retrosynthetic analysis that is related to "retrons" [4]. Provisions to avoid low-percentage transformations resulting in impractical routes have been made in order to focus on the more efficient ones. Examples of "deep-perception," [5] "tactical combinations," [6] and "starting material-oriented analysis" are shown in Figure 17.2.

Searching for a "long-range"-type transform leading to aminotetralin **A**, the LHASA program suggested a retrosynthetic pathway wherein an intermolecular Diels-Alder reaction between the diene **I** and the dienophile **J** produces **H** [5]. Appropriate transformations in the forward direction would then proceed through the suggested intermediates **G** to **B**, and ultimately to **A** (Figure 17.2).

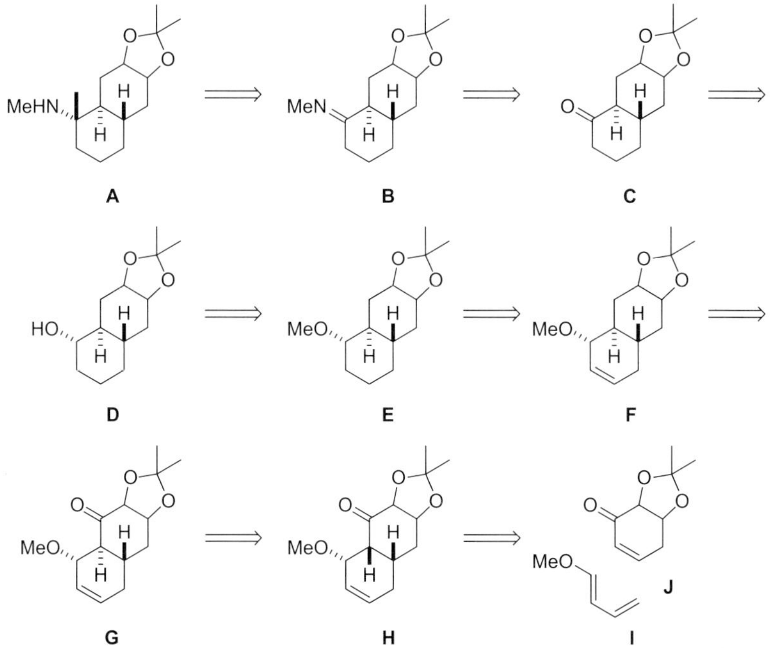

Figure 17.2 A LHASA retrosynthetic analysis involving a long-range Diels-Alder transform.

The "tactical combinations" option in the LHASA program offers a wider choice of strategies as illustrated for homogynolide B (Figure 17.3) [6]. Although the suggested routes differ from the one used in the actual synthesis, a great deal of insight can be perceived by the computer-generated pathway.

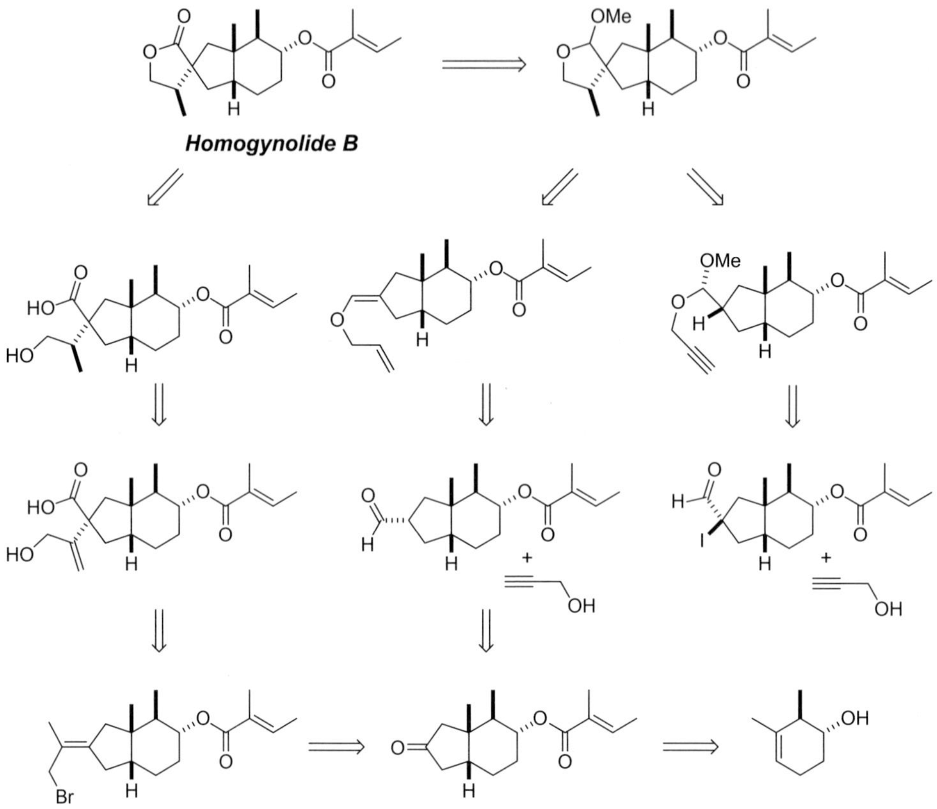

Figure 17.3 LHASA tactical combination in the retrosynthetic analysis of homogynolide B.

17.2
SYNGEN

Conceived by Hendrickson [7], the SYNGEN program relies on a set of skeletal bond disconnections (bond sets), and relating a starting material to a target molecule. Some functionalities may be present at the outset (self-consistent sequence), but others including stereochemistry could be introduced as the synthesis tree evolves. The synthetic route suggested by SYNGEN for lysergic acid is shown in Figure 17.4.

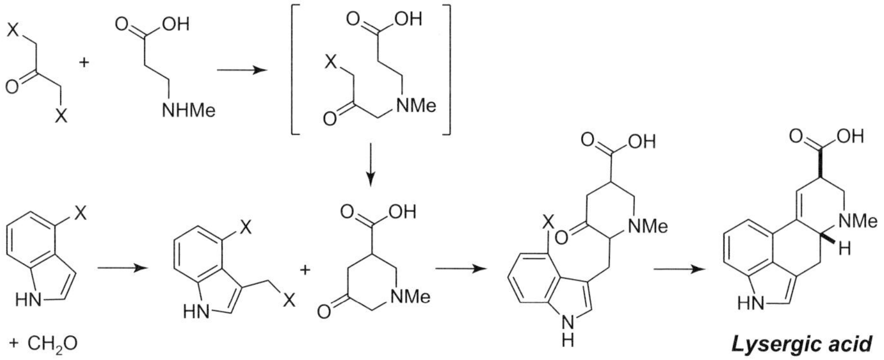

Figure 17.4 Synthetic route to lysergic acid suggested by SYNGEN.

17.3
WODCA

A more recent synthesis planning program developed by Gasteiger and cowork-ers [8] also relies on a retrosynthetic analysis approach. WODCA (Workbench for the Organization of Data for Chemical Applications) is an interactive com-puter program that combines aspects of similarity searches with physicochemical properties associated with atoms or bonds. Synthons are generated by hypothet-ical bond cleavages starting from a target molecule. A substructure-based, or a reaction-based search protocol relates the starting material to a target molecule. The data base of starting materials consists of available chemicals from the Janssen Chimica catalogue as well as starting materials from the CHIRON PROGRAM com-pound collection (see Section 17.4). An automated search for starting materials, particularly for obvious matches with target molecule substructures is available. Examples of WODCA results for lysergic acid and α-bisabolene are shown in Figures 17.5 and 17.6 respectively.

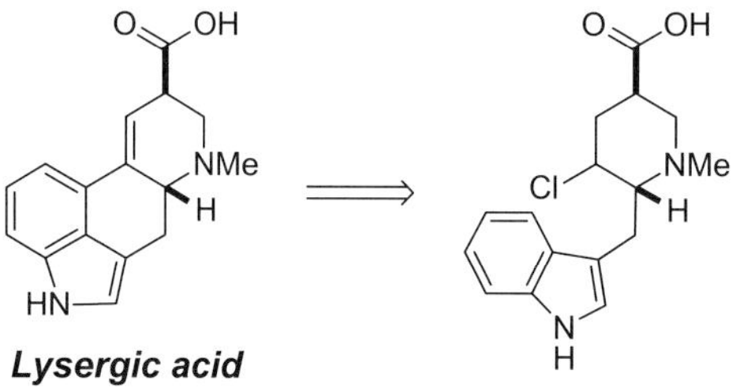

Figure 17.5 WODCA search for similar precursors for a hy-pothetical precursor to lysergic acid.

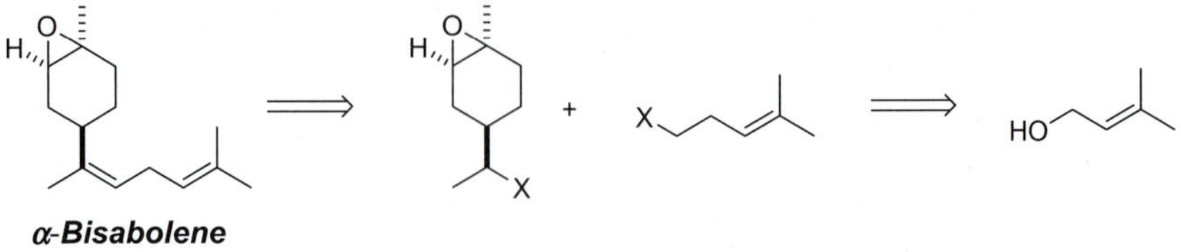

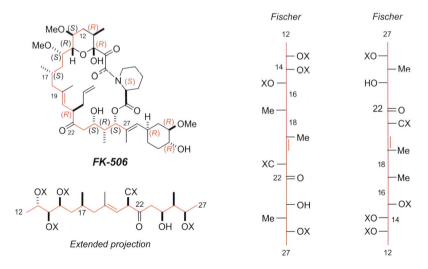

α-**Bisabolene**

Figure 17.6 WODCA disconnection of *α*-bisabolene.

17.4
The CHIRON PROGRAM

Cognizant of the difficulties in perceiving stereochemical differences between molecules and the limitations of visual decoding of non-obvious skeletal overlaps, the CHIRON PROGRAM was developed some 25 years ago [2, 9]. This program offers an interactive heuristic method to select starting materials for organic synthesis relying on the recognition of chiral substructures in target molecules. The significant features of different modules in the program are highlighted below.

17.4.1
CASA (Computer-assisted stereochemical analysis)

This option provides near instantaneous stereochemical information within a given molecule, or among several molecules drawn on the computer screen. Designation of *R*- or *S*-stereochemistry, Fischer projections, and their extended or zigzag counterparts are available by activating specific commands in a menu (Figure 17.7).

Figure 17.7 Fischer and extended (zigzag) projections of FK-506.

In Figure 17.1 we presented the case of possibly matching propionate triads in the C-1–C-6 and C-1–C-8 substructures of erythronolide A aglycone and monensin respectively. The IDENTICAL command in the CASA menu provides instant color-coded information about identical (superimposable) chiral subunits as shown in Figure 17.8. The stereochemical relationships provided by these and other capabilities of the CASA option in the CHIRON PROGRAM, greatly facilitates the choice of common precursors with matching stereochemistry and carbon skeletal convergence.

Erythronolide A aglycone   **Monensin**

Figure 17.8   The C-2–C-6 (erythronolide A aglycone) and C-4–C-8 (monensin) segments are identical.

17.4.2
CAPS (Computer-assisted precursor selection)

The main attribute of the CHIRON PROGRAM is its ability to recognize skeletal, functional, and stereochemical features in a target molecule, and to relate them to appropriate starting materials form a data base consisting of over 150,000 commercially available and literature-cited compounds. These comprise of *ca.* 5,000 carefully selected chiral, non-racemic compounds, including appropriate references and commercial sources. Compounds in the database are grouped under CLASSES, depending on their types (acyclic-3–7, carbocyclic, branched carbocyclic, heterocyclic, aromatic, etc.). After a structure is drawn on the screen, certain parameters are selected from a menu, and the search for appropriate starting materials is initiated. The program is capable of selecting those precursors in the data base that provide the best carbon skeletal, functional, and stereochemical convergence with a substructure. The results are displayed on the screen within seconds, in the form of color-coded structures, in which the overlapping bonds and atoms can be easily related to segments of the target molecule. The TRANSFORMATION OPTION displays key chemical operations, indicated by words such as; *oxidize, branch, extend, annulate,* etc. which, taken in a sequence, represent the synthesis "in a capsule." These key words are also a source of mental stimuli for the user to decide the order in which such operations can be put to practice. With each precursor comes a feasibility score, expressed as a numerical percentage, which reflects the relative ease or difficulty of each of the chemical transformations needed to achieve complete congruence of functional groups and stereochemistry.

Examples of recognition by direct overlaps of acyclic and cyclic precursors with corresponding substructures in target molecules, as well as key reactions suggested by the TRANSFORMATIONS OPTION, are shown in Figure 17.9. It is of interest that in each of these cases the target molecules have actually been independently synthesized [10] from the same precursors suggested by the CHIRON PROGRAM (see Chapters 1, 4, 10, and 14).

Multiple precursors are also found whenever possible by using the BEST COMBINA-TION option. In these cases, the program will *not* select precursors with overlapping carbon atoms. The results from kijanolide and FK-506 are shown in Figure 17.10 [2, 9]. Note the selection of a common sugar-derived precursor for two different segments of FK-506. Unlike the chirons shown in Figure 17.9, only quinic acid represents an actual precursor starting material used by White and coworkers. The others were suggested by the program from compounds in the database.

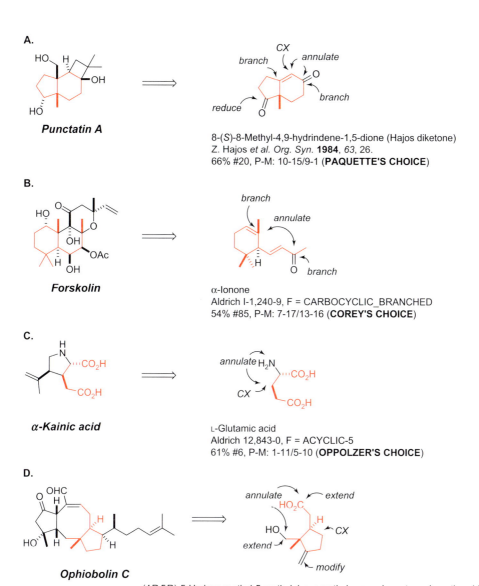

A.

Punctatin A

CX
branch   *annulate*

O

branch

reduce   O

8-(*S*)-8-Methyl-4,9-hydrindene-1,5-dione (Hajos diketone)
Z. Hajos *et al. Org. Syn.* **1984**, *63*, 26.
66% #20, P-M: 10-15/9-1 (**PAQUETTE'S CHOICE**)

B.

Forskolin

branch

annulate

branch

α-Ionone
Aldrich I-1,240-9, F = CARBOCYCLIC_BRANCHED
54% #85, P-M: 7-17/13-16 (**COREY'S CHOICE**)

C.

α-Kainic acid

annulate H₂N

CX

L-Glutamic acid
Aldrich 12,843-0, F = ACYCLIC-5
61% #6, P-M: 1-11/5-10 (**OPPOLZER'S CHOICE**)

D.

Ophiobolin C

annulate   *extend*

extend   CX

modify

(4*R*,5*R*)-5-Hydroxymethyl-5-methyl-1-exomethylene-cyclopentane-4-acetic acid
T. Money *et al. Can. J. Chem.* **1985**, *63*, 3182. F = CARBOCYCLIC_BRANCHED
49% #36, P-M: 7-8/9-24 (**KISHI'S CHOICE**)

Figure 17.9 CHIRON PROGRAM analyses of punctatin A,
forskolin, α-kainic acid, and ophiobolin C showing selected
precursors with excellent carbon framework, functional, and
stereochemical overlap.

A.

Hit # 1
Overall Score = 66%
Overlap Index = 83%
Average Precursor Score = 68%

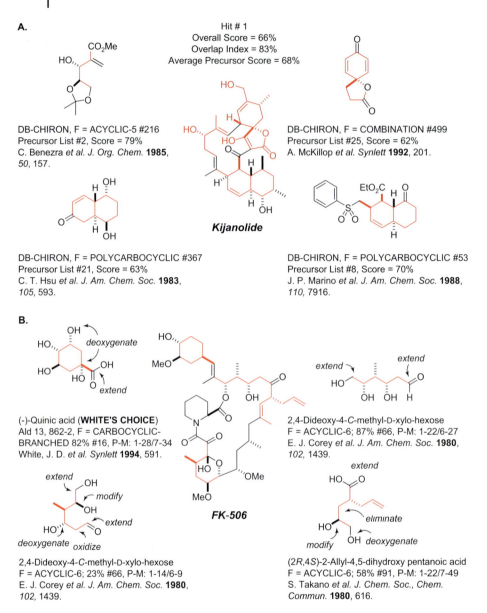

DB-CHIRON, F = ACYCLIC-5 #216
Precursor List #2, Score = 79%
C. Benezra *et al. J. Org. Chem.* **1985**, *50*, 157.

DB-CHIRON, F = COMBINATION #499
Precursor List #25, Score = 62%
A. McKillop *et al. Synlett* **1992**, 201.

Kijanolide

DB-CHIRON, F = POLYCARBOCYCLIC #367
Precursor List #21, Score = 63%
C. T. Hsu *et al. J. Am. Chem. Soc.* **1983**, *105*, 593.

DB-CHIRON, F = POLYCARBOCYCLIC #53
Precursor List #8, Score = 70%
J. P. Marino *et al. J. Am. Chem. Soc.* **1988**, *110*, 7916.

B.

(-)-Quinic acid (**WHITE'S CHOICE**)
Ald 13, 862-2, F = CARBOCYCLIC-
BRANCHED 82% #16, P-M: 1-28/7-34
White, J. D. *et al. Synlett* **1994**, 591.

FK-506

2,4-Dideoxy-4-*C*-methyl-D-xylo-hexose
F = ACYCLIC-6; 87% #66, P-M: 1-22/6-27
E. J. Corey *et al. J. Am. Chem. Soc.* **1980**, *102*, 1439.

2,4-Dideoxy-4-*C*-methyl-D-xylo-hexose
F = ACYCLIC-6; 23% #66, P-M: 1-14/6-9
E. J. Corey *et al. J. Am. Chem. Soc.* **1980**, *102*, 1439.

(2R,4S)-2-Allyl-4,5-dihydroxy pentanoic acid
F = ACYCLIC-6; 58% #91, P-M: 1-22/7-49
S. Takano *et al. J. Chem. Soc., Chem. Commun.* **1980**, 616.

Figure 17.10 BEST COMBINATION OPTION showing multiple precursors for kijanolide and FK-506.

A new dimension in heuristic analysis and perception is possible in the CLEAVE AND RESHAPE OPTION of the program. Double bonds, vicinal diols, and ketones are automatically "cleaved" and their acyclic counterparts are displayed on the screen to match target substructures [2, 9]. For example,

(1*R*,2*R*)-2-amino-4-cyclohexene-1-carboxylic acid is proposed by the program as a precursor to the β-lactam antibiotic PS-5, even though there is no visual overlap between the precursor and the target molecule. However, the TRANSFORMATION OPTION indicates an oxidative cleavage. A "cleaved and reshaped" counterpart, in which the extremities of the original endocyclic double bond have been substituted with a suitable functional group (X), will appear on the screen, providing a perfect regiochemical, stereochemical, and functional group congruence with the upper segment of PS-5 (Figure 17.11 A). Interestingly, the total synthesis of this target molecule from the same cyclic amino acid has been reported [11].

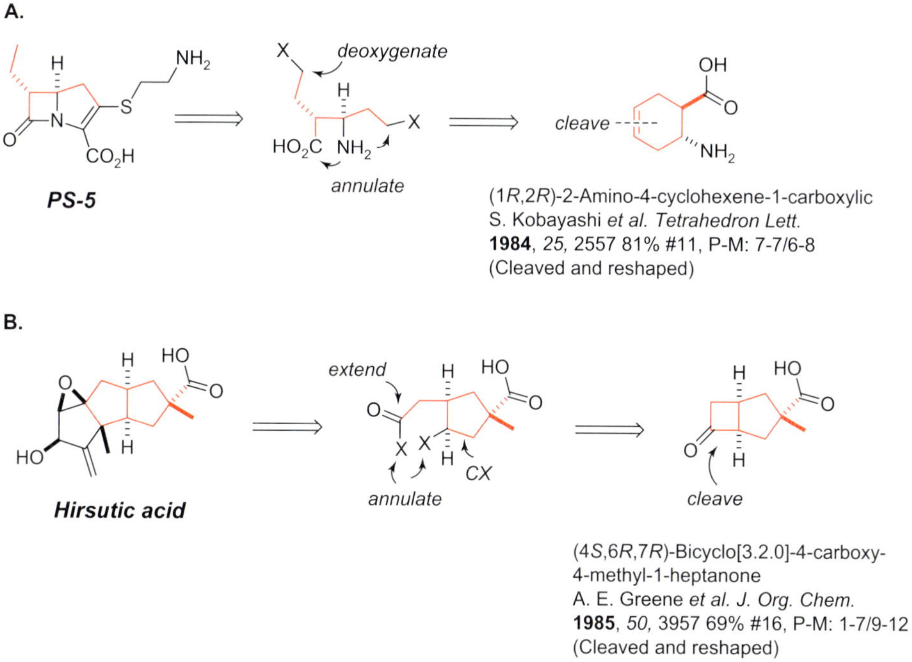

A.

PS-5

deoxygenate

annulate

cleave

(1*R*,2*R*)-2-Amino-4-cyclohexene-1-carboxylic
S. Kobayashi *et al. Tetrahedron Lett.*
1984, *25*, 2557 81% #11, P-M: 7-7/6-8
(Cleaved and reshaped)

B.

Hirsutic acid

extend

annulate

cleave

(4*S*,6*R*,7*R*)-Bicyclo[3.2.0]-4-carboxy-
4-methyl-1-heptanone
A. E. Greene *et al. J. Org. Chem.*
1985, *50*, 3957 69% #16, P-M: 1-7/9-12
(Cleaved and reshaped)

Figure 17.11 CLEAVE and RESHAPE OPTION for PS-5 and hirsutic acid.

In the case of hirsutic acid, the CYCLIC CLEAVE OPTION found (4*S*,6*R*,7*R*)-bicyclo-[3.2.0]-4-carboxy-4-methyl-1-heptanone [12] as a non-obvious precursor. Cleavage of the cyclobutanone ring by a virtual Baeyer-Villiger oxidation produces a cleaved and reshaped equivalent with the desired stereochemistry and functional groups (Figure 17.11 B).

The CHIRON Program can find applications in medicinal chemistry such as fragment-based drug discovery [13], and alternative syntheses of drug substances, including modification of existing processes [2]. It is also a powerful teaching tool showing the importance of stereochemical relationships in biosynthetic pathways [14], as well as facilitating the better visualization of perspective drawings of structures [15].

17.5
Computer-aided synthesis planning

Over the last four decades, the community of synthetic organic chemists has had the opportunity to benefit from, and contribute to, the existing computer programs whose main objective was to facilitate the task of synthesis. In spite of substantial early advances, the much-anticipated "mouse-clicking" approach to find the "best" synthetic route to a given target molecule remains somewhat arbitrary.

Ideally, a computer program that is capable of providing practical and economical routes to target molecules should do so by generating a complete synthetic scheme, inclusive of reagents, reaction conditions, and literature precedents, while providing a measure of feasibility from starting material to product. Expressing the vast knowledge base of reactions, reactivity, and strategic bond formations for a "best" synthesis scenario in machine language is a daunting, multi-task operation that is not provided by any one of the existing programs today. In fact, even with the availability of such a utopic synthesis planning super-machine, human logic and intervention is still required. No matter how useful a software program can be, the information it generates must stimulate the creative mind of the user, who usually prefers to find his or her own synthetic routes to target molecules. This divergence of cultures, which in part may be due to a measure of scepticism by an older generation of synthetic chemists, may be one reason why the development of software for modern synthesis planning has not been an active area in the past two decades.

The best heuristic, if not practical, usage of computers in synthesis planning is to take advantage of their speed in performing certain validated operations. In this context, synthetic chemists can benefit from the analogical or relational processing of skeletal similarities, the rapid stereochemical comparison between molecules, and perceived reaction pathways leading to the intended target.

Our limited powers of perception and analysis of structural, functional, and stereochemical features can be vastly enhanced without compromising our ability to think of, or to conceive of creative synthetic pathways. Thus, the fleeting existence of molecules, as Leonardo da Vinci's *luminous bodies*, in our mind's eye can be visualized at will and at length on the computer screen, until the best synthesis idea reveals itself (see Chapter 1, section 1.5).

A quote from the 1990 Nobel lecture for chemistry by E. J. Corey with regard to computers is most appropriate:

> *The field of computer-assisted synthetic analysis is fascinating in its own right, and surely one of the most interesting problems in the area of machine intelligence.*

Because of the enormous memory and speed of modern machines and the probability of continuing advances, it seems clear that computers can play an important role in synthetic design.

References

1. For example, see: Hanessian, S. *Total Synthesis of Natural Products: The Chiron Approach*, Pergamon, Oxford, 1983.
2. For a recent review, see: Hanessian, S. *Curr. Opin. Drug. Discov. Develop.* 2005, **8**, 798.
3. (a) Corey, E. J.; Wipke, W. T. *Science* 1969, **166**, 178; (b) Corey, E. J.; Long, A. K.; Rubenstein, S. D. *Science* 1985, **228**, 408.
4. Corey, E. J. *Chem. Soc. Rev.* 1988, **17**, 111.
5. Corey, E. J.; Howe, W. J.; Pensak, D. A. *J. Am. Chem. Soc.* 1974, **96**, 7724; see also: Ott, M. A.; Noordik, J. H. *J. Chem. Inf. Comput. Sci.* 1997, **37**, 98.
6. Long, A. K.; Kappos, J. C. *J. Chem. Inf. Comput. Sci.* 1994, **34**, 915.
7. (a) Hendrickson, J. B. *Angew. Chem. Int. Ed.* 1990, **29**, 1286; (b) Hendrickson, J. B.; Grier, D. L.; Toczko, G. *J. Am. Chem. Soc.* 1985, **107**, 5028; (c) Hendrickson, J. B. *Chem. Tech.* 1998, **28**, 35.
8. (a) Ihlenfeldt, W.-D.; Gasteiger, J. *Angew. Chem. Int. Ed.* 1995, **34**, 2613; see also: (b) Gasteiger, J.; Jochum, C. *Top Curr. Chem.* 1978, **74**, 93; (c) Fick, R.; Ihlenfeldt, W.-D.; Gasteiger, J. *Heterocycles* 1995, **40**, 993; (d) Gasteiger, J.; Ihlenfeldt, W.-D.; Fick, R.; Röse, J. R. *J. Chem. Inf. Comput. Sci* 1992, **32**, 700.
9. Hanessian, S.; Franco, J.; Larouche, B. *Pure Appl. Chem.* 1990, **62**, 1887; see also: Hanessian, S.; Major, F.; Leger, S. in *New Methods in Drug Research*, Makryannis, A., Ed.; p. 201, J. R. Prous Publishers, Barcelona, 1985.
10. Punctatin A: Paquette, L. A.; Sugimura, T. *J. Am. Chem. Soc.* 1986, **108**, 3841; forskolin: Corey, E. J.; Da Silva Jardine, P.; Rohloff, J. C. *J. Am. Chem. Soc.* 1988, **110**, 3672; kainic acid: Oppolzer, W.; Thirring, K. *J. Am. Chem. Soc.* 1982, **104**, 4978; ophiobolin C: Rowley, M.; Tsukamoto, M.; Kishi, Y. *J. Am. Chem. Soc.* 1989, **111**, 2735.
11. Kobayashi, S.; Kamiyama, K.; Iimori, T.; Ohno, M. *Tetrahedron Lett.* 1984, **25**, 2557.
12. Greene, A. E.; Luche, M. J.; Serra, A. A. *J. Org. Chem.* 1985, **50**, 3957.
13. For example, see: (a) Murray, C. W.; Rees, D. C. *Nature Chemistry* 2009, **1**, 187; (b) de Kloe, G. E.; Bailey, D.; Leurs, R.; de Esch, I. J. P. *Drug Discov. Today* 2009, **14**, 630; (c) Schulz, M. N.; Hubbard, R. E. *Curr. Opin. Pharmacology* 2009, **9**, 615.
14. Hanessian, S.; Botta, M.; Larouche, B.; Boyaroglu, A. *J. Chem. Inf. Comput. Sci.* 1992, **32**, 718.
15. Hanessian, S.; Franco, J.; Gagnon, G.; Laramée, D.; Larouche, B. *J. Chem. Inf. Comput. Sci.* 1990, **30**, 413.

18
The Essence of Synthesis – A Retrospective

In Chapter 1 we commented on the impact that organic synthesis has had on the well-being of humankind and society. Advances in various aspects of biology and medicine have greatly expanded our understanding of many physiological processes at the molecular level. Concurrently, the advent of efficient separation techniques, and sophisticated analytical instrument methods have enormously accelerated the enterprise of synthesis.

It is also important to point out that major developments in materials science have been made possible, because of fundamental discoveries relating to the electronic properties of organic and inorganic synthetic polymers. Cognizant of these developments in seemingly disparate fields, synthetic chemists have recognized the potential of small organic molecules as tools in newly christened subdisciplines such as chemical biology, chemogenomics, nanotechnology, and supramolecular chemistry.

In Chapter 2 (The *Why* of Synthesis) we commented on a number of reasons to "do synthesis." High on the list of priorities was the prospect of exploiting organic synthesis for the benefit of health and the well-being of humankind. Indeed, organic synthesis has played a primordial role in the development of life-saving drugs as well as life-sustaining agrochemicals. Given the necessary resources and time, practically any molecular entity of a reasonable level of complexity can be synthesized today. Molecules produced by synthesis can encompass a large diversity of functional and structural landscapes with the objective of affecting biological processes related to a variety of diseases. In this regard, Nature's molecular factories are infinitely faster in producing structurally intricate, and oftentimes therapeutically relevant products that can instil awe, admiration, and even fear in the eyes of the synthetic chemist.

Armed with an exquisite array of time-tested biochemical pathways, and a repertoire of small molecule reagents, Nature is the ultimate synthesizer, having already perfected its "system" over millennia of evolutionary selection. In contrast, only by using Nature's tools, such as genetics, can we hope to modify her pathways in order to produce "engineered" molecular entities and to test their functions in areas of interest. On the other hand, synthesis offers an advantage over Nature in achieving voluntary functional, structural, and stereochemical diversification leading, for example, to the production of both enantiomers of a

Design and Strategy in Organic Synthesis: From the Chiron Approach to Catalysis, First Edition.
Stephen Hanessian, Simon Giroux, and Bradley L. Merner.

chiral, non-racemic molecule. It is therefore opportune to once again recite words expressed by Marcellin Berthelot:

"La chimie crée son propre objet" (Chemistry creates its own object)

18.1
Lest we Forget

In Chapter 3 (The *What* of Synthesis), we briefly commented on the evolutionary trends covering a century of synthesis. Already within the early years of the past century, achievements in total synthesis were awe-inspiring considering the paucity of synthetic methods available at that time. Tropinone (Willstätter, 1901; Robinson, 1917 – see Figure 18.4), camphor (Komppa, 1903; Perkin, 1904), and α-terpineol (Perkin, 1904) were veritable masterpieces of their era [1]. Less than a generation later saw the complexity level of completed syntheses rise to include haemin (Fischer, 1929) and equilenin (Bachmann, 1939). By the middle of the 20th century, a new continent of complex structural entities was being conquered through the insightful and tactically elegant contributions of R. B. Woodward [2]. With less than twenty or so known reactions, and common laboratory reagents at his disposal, Woodward completed the total syntheses of some of the most complex natural products known at the time, including cortisone (1951), strychnine (1954), and reserpine (1958) (see Chapters 4 and 14). It is noteworthy to mention that IR, UV/Vis, and elemental analysis were the main analytical techniques used by Woodward in his remarkably successful pursuit of complex natural products. He would continue his legacy of monumental achievements with masterful syntheses, among others, of colchicine and cephalosporin C (1965), vitamin B_{12} (1973), and finally erythromycin A (1981). The last being his epitaph as a celebrated and highly revered synthetic chemist.

18.2
The Corey and Stork Schools

The annals of natural products synthesis in the second part of the last century, register a plethora of noteworthy contributions from outstanding scientists, many of whom are cited in this book. However, the period immediately following the mid-20th century, the Woodwardian era of synthesis, witnessed the emergence of two highly influential "schools" that were developed independently by E. J. Corey and G. Stork in their respective institutions. The impact of their contributions to synthetic chemistry and natural products over a period exceeding the half-century mark deserves special mention. Remarkably, seminal contributions from both of these groups continue to flourish to this day.

Corey's legendary teachings of retrosynthetic analysis are forever embedded in our minds when we think about the synthesis of most target molecules [3]. The many useful synthetic methods he has developed, including catalytic asymmetric processes, are part of our daily laboratory repertoire and reactions that we perform [4]. It is not unusual to think that at any given time, chemists are actually applying Corey's methods in different parts of the world. The variety of complex natural products synthesized in the Corey laboratories over the years, starting with longifolene (1961) and continuing with lupeol [5] (2009) (Figure 18.1 A) are, among others, a testimony of his prolific work and creative flair [3]. Perhaps a greater legacy resides in the hundreds of talented coworkers that have received their training with Corey, and

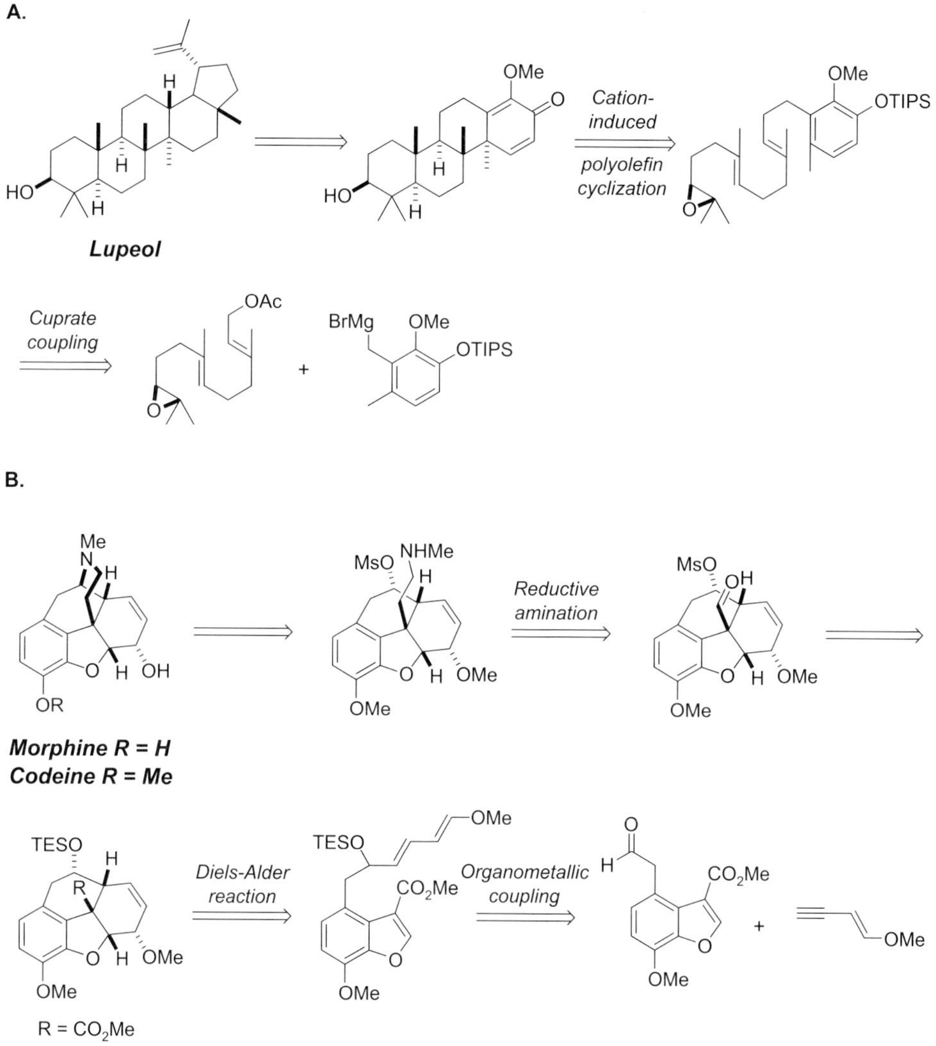

A.

B.

Figure 18.1 A. Retrosynthetic analysis of lupeol by Corey;
B. Retrosynthetic analysis of (±)-morphine and (±)-codeine by Stork.

have gone on to excel in industry as well as in academia. Noteworthy contributions from some of these scientists were also highlighted in previous chapters.

The Stork school [6] is also renowned for producing outstanding coworkers, many of whom have gone on to pursue stellar scientific careers. Examples of elegant total syntheses from some former Stork group coworkers have been featured throughout sections of this book. The rudiments of biogenetic and stereoelectronic concepts as applied to the synthesis of complex natural products, including many alkaloids, have been Stork's hallmark signature of ingenuity. The total synthesis of reserpine and quinine are exquisite examples of Stork's insightful applications of stereochemical control (see Chapter 14, section 14.7.2 and Chapter 16, section 16.6.1 respectively). The inclusion of free radical-based key steps in strategic bond-forming sequences led to a revival of these well-known, albeit seldom used, methods in natural product synthesis [8]. Many modern applications of enamine chemistry [9] have their conceptual origins in Stork's original contributions from several decades ago. The most recent total syntheses of morphine and codeine were reported by Stork and coworkers [7] in 2009 (Figure 18.1 B).

The Corey and Stork schools have produced many of the present-day brilliant practitioners of total synthesis. Together with their peers who also received superb training in different academic groups, the post Corey-Stork generation of natural product synthetic chemists have truly excelled in maintaining the highest level of creativity.

18.3
The Visual Dialogue with Molecules

The creativity of chemists in devising syntheses of complex natural products has been amply demonstrated [1]. In a large number of these cleverly crafted syntheses, the use of small naturally occurring starting materials as *chiral* synthons (*chirons*), derived from α-amino acids, carbohydrates, hydroxy acids, and terpenes was a key strategy to introduce and control absolute stereochemistry in target molecules as outlined in previous Chapters. Stoichiometric reagents, and more recently, catalytic asymmetric methods have also contributed enormously to the enterprise of total synthesis (see Chapters 2–4). No matter how simple or complex the structure of a target molecule may appear to be at first sight, the human thought process will manifest itself in different ways. Analysis of various strategies employed toward the total synthesis of a given molecule is based on the subliminal interplay between the eye and the mind's eye. Faced with the same target molecule, two synthetic chemists will generate multiple synthetically viable approaches. In spite of their similar schooling in the principles of chemical bonding, reactivity, and basic mechanistic concepts, altogether different approaches to one and the same target molecule may be proposed. This kind of creative diversity is unique to target-oriented organic synthesis, even though the basic rules of thermodynamics and chemical reactivity are closely followed in both proposals. One reason lies in the many ways in which chemical bonds within a target molecule can be broken and reformed, first virtually, and then in actual practice in the laboratory. This diversity and individuality of opinion, with regard to the design and planning of strategies toward target molecules, is one reason why the subdiscipline of organic synthesis has evolved in leaps and bounds over the last few decades.

In Chapter 4 (The *How* of synthesis), we articulated the notion of visual *reflexive* and visual *relational* thinking in synthesis planning. While still involving first eye contact with a target molecule, the visual *reflexive* thought process relies primarily on a bond construction paradigm where functional groups, and even chirality, can be generated by application of transformations such as the aldol, Diels-Alder, and other well-known reactions. The starting materials in such a strategy may be achiral or racemic, and chirality may be generated in enantioselective or diastereoselective reactions by application of stoichiometric or catalytic asymmetric methods. Retrosynthetic (antithetic/disconnective) analysis has facilitated the visual *reflexive*, or reaction-based thought process, best exemplified by the synthon approach (see Chapter 6, section 6.2.6). Thus, a knowledge and logic-based paradigm leads to specific reaction types, allowing the stepwise construction of substructures with appended functionalities, culminating with the assembly of the target molecule.

Recognizing the interrelationships of functional groups and stereochemical features in a molecule can also lead to a visual *relational* thinking modality. Thus, substructures of target molecules, in part or as a whole, may be derived from chirons such as α-amino acids, carbohydrates, hydroxy acids, terpenes, or related ones for example, either directly or indirectly after chemical modification. The modified chirons can then be integrated in the general synthesis plan at the outset, providing the desired subsets of substructures with their functional groups and stereochemical characteristics already present at strategic positions. They can also be used to generate new stereogenic centers based on their inherent resident chirality relying on substrate-controlled asymmetric induction as well as catalytic methods of bond formation. Therein lies the basic principle of the *chiron approach* in synthesis planning (see Chapter 6, section 6.2).

18.4
Total Synthesis: From whence we came . . .

The present-day level of achievements in organic synthesis, in general, and natural products in particular, is the highest ever compared to as recently as a generation ago [1] (see also Chapter 3, section 3.2). This is in part due to the innovative methodologies developed for the construction of carbogenic and heterocyclic molecules, superb advances in separation technologies, and extraordinary spectroscopic techniques.

The recent impact of powerful catalytic methods offering high levels of asymmetric induction in individual steps of a synthetic route toward a target molecule is a major advance in the science of synthesis [10]. It is also noteworthy that the implementation of catalysis in some high-volume industrial processes has been a longstanding tradition with enormous socio-economic consequences [11].

In contrast, for a good part of the 20th century, and even to this day, synthetic chemists managed to secure chirality in molecules relying primarily on chemical resolution and enzymatic desymmetrization methods (see Chapter 5). Towards the latter part of the past century, a few catalytic methods, and a handful of chiral auxiliaries relying on cleverly conceived stoichiometric ligands offered attractive methods toward the synthesis of a variety of enantiopure (or enriched) compounds [10]. However, a surge of interest in catalytic asymmetric synthesis took place some two to three decades ago, which is still steadily evolving (see also Chapter 4, section 4.6 and Chapter 5, section 5.8).

For example, the advent of the Sharpless asymmetric epoxidation reaction [12] of allylic alcohols in 1980 changed the landscape of asymmetric synthesis and paved the way for innovations in catalysis. So powerful was its impact, that entire syntheses were planned with the expectation that application of the Sharpless asymmetric epoxidation reaction at a designated stage would provide the desired epoxy alcohol with high enantioselectivity. Other asymmetric catalytic oxidation reactions soon followed [13].

The same phenomenon was experienced in asymmetric transition metal-catalyzed reductions, especially in conjunction with phosphorous containing organic ligands [14], starting with the Knowles catalytic asymmetric hydrogenation of olefins [15], and Noyori's seminal contributions to carbonyl and olefin reduction [16]. Chemical asymmetric hydride reductions of prochiral ketones using the Corey-Bakshi-Shibata oxazaborolidine ligand [17] offered a wider choice of catalytic methods to achieve high levels of stereochemically pure alcohols at various stages of a synthetic route. Examples of the above stated methods have featured prominently in many of the syntheses highlighted throughout previous chapters.

The impact of the Grubbs [18], Schrock [19], and, more recently, Hoveyda-Grubbs [20] catalysts for ring-closing metathesis reactions, as well as other extended applications, has changed the way syntheses are planned when carbocyclizations and chain-extensions (by cross-metathesis) are envisaged. Transition metal-catalyzed aryl couplings and substitutions have enormously facilitated aromatic and heteroaromatic functionalizations with important applications in the pharmaceutical industry [21]. The 2010 Nobel prize for chemistry was awarded to Richard Heck, Ei-ichi Negishi, and Akira Suzuki for their contributions to palladium-catalyzed cross coupling reactions in organic synthesis.

Many named reactions [4] relying on a metal-based catalyst for the formation of C–C bonds from functionalized alkynes, alkenes, and alkanes have become the everyday tools for assembling carbogenic molecules (see also Chapter 2, section 2.11.5). Some of these methods have become the catalytic alternatives to phosphorus and sulphur carbanion (or ylide)-based reagents.

Creative catalytic methods to functionalize unactivated C–H bonds are making steady progress with elegant examples of their use as key steps in the total synthesis of complex natural products [22] (see also section 18.6 and reference [60]).

Thus, one of the legitimate claims for having advanced the science of synthesis from whence we came only a few decades ago, can be rightfully attributed to the remarkably efficient methods for C–C, C–O, C–N, and related bond formations,

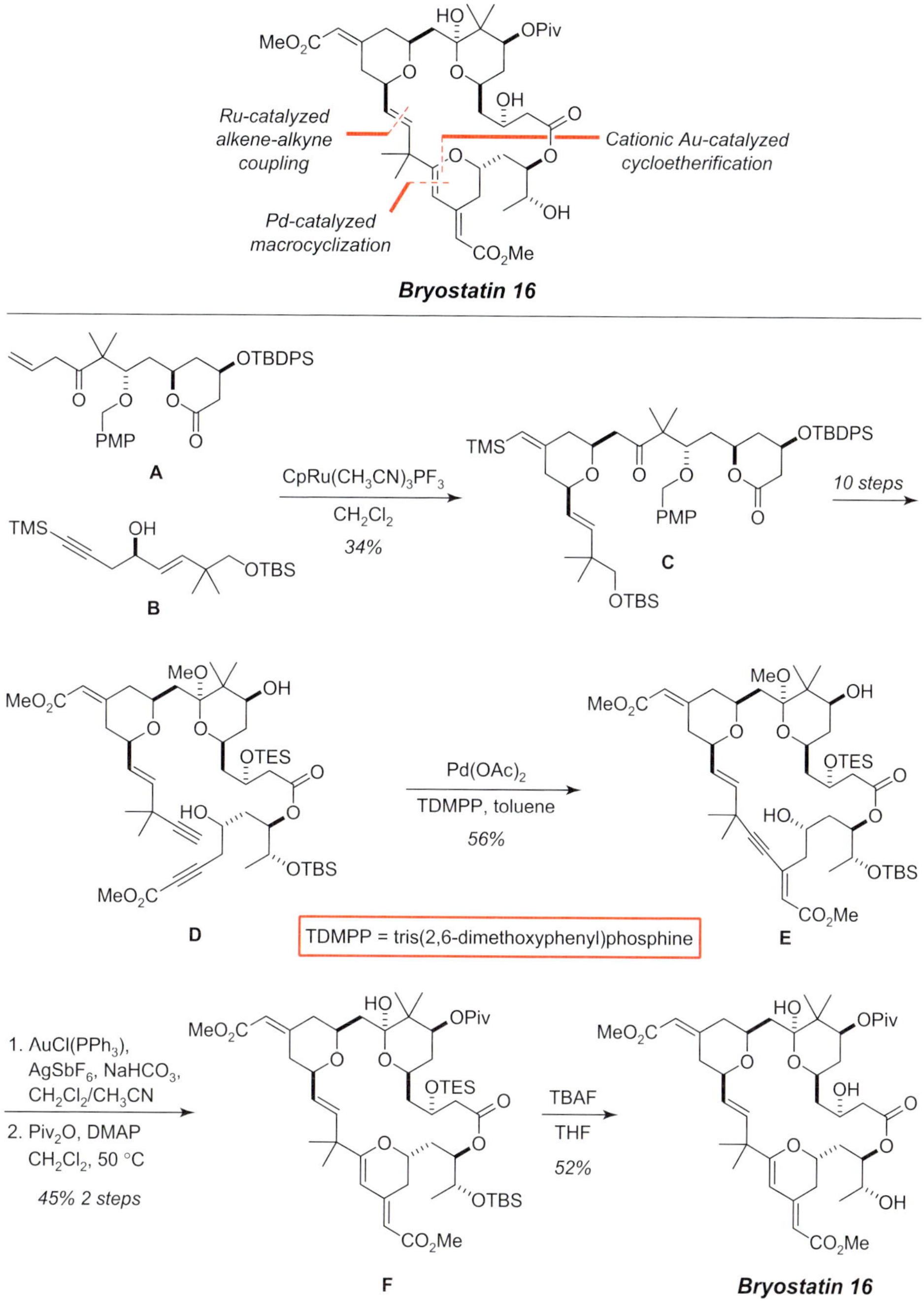

Figure 18.2 Transition metal-catalyzed key reactions in the total synthesis of bryostatin 16 by Trost and Dong.

especially mediated by transition metal catalysts [10, 23]. In fact, rare are the current syntheses in which one or more critical bond-forming steps are not done in the presence of a transition metal catalyst with an associated chiral or achiral ligand (see Chapter 4, section 4.6).

A recent total synthesis of bryostatin 16 by Trost and Dong [24] showcases, among other asymmetric reactions, the utility of Ru, Pd, and Au-catalyzed transformations (Figure 18.2).

Intermediates **A** and **B**, harboring hydroxyl and *gem*-dimethyl groups, were engaged in a "one-pot" Ru-catalyzed alkene-alkyne coupling/*O*-Michael addition sequence to give **C**. A remarkable Pd-catalyzed alkyne-alkyne coupling of **D** gave ene-yne **E**, which was subjected to a Au-catalyzed cycloetherification reaction to furnish a precursor to **F**.

Environmentally friendly and relatively benign metal-free catalysis, under the aegis of "organocatalysis," has captured the interest of synthetic chemists in the last decade [25]. The seminal papers published independently in the early part of the 1970s by scientists at Hoffmann-LaRoche in the United States of America, and at Sandoz in Basel, Switzerland lay dormant for over two decades before its enormous potential utility, and further applications were realized. The Hajos-Parish-Wiechert-Eder-Sauer reaction [26] in which L-proline was used to catalyze a an intramolecular aldol reaction of an achiral triketone to produce, after recrystallization, an enantiopure bicyclic diketone, is regarded as a milestone achievement because of the conceptual simplicity of an enamine-based mechanism [26] (see Chapter 5, section 5.8).

Impressive advances in enamine and iminium catalysis inspired by the original contributions of Barbas, Lerner, and List [27], and MacMillan [28, 29] respectively, have forged the path to explore alternative metal-free methods to traditional enolate chemistry, and to Diels-Alder, Michael, and Friedel-Crafts-type reactions, for example.

Applications of such organocatalytic and other transition metal-based asymmetric methods, in diverse areas of C–C, C–O, and C–N bond formation in natural products synthesis, while ensuring high enantio- and diastereoselectivity, are being reported with greater frequency.

Organic synthesis has also made tremendous strides in bridging the gap between structure and function as they relate to new drug entities, advanced materials, and innovative biologics. A few examples of recently marketed drugs containing one or more stereogenic centers produced by total or semi-synthesis are shown in Figure 18.3 [30].

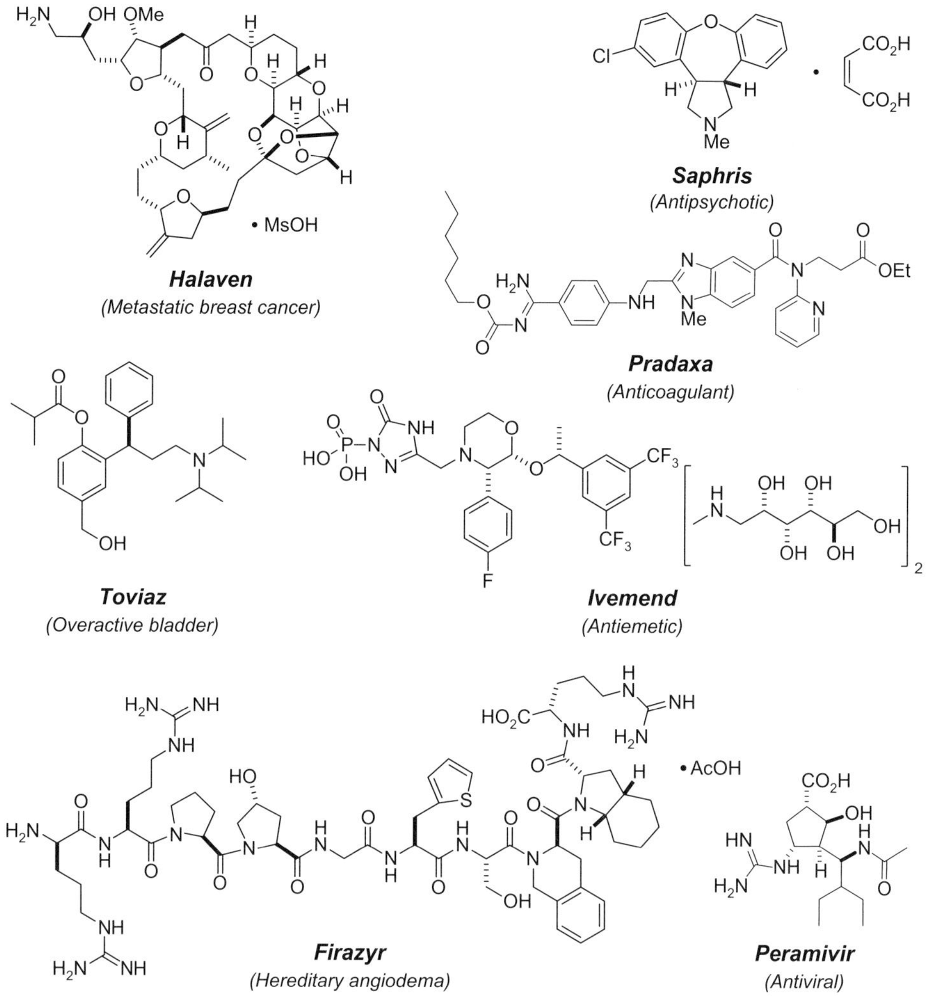
• MsOH
Halaven
(Metastatic breast cancer)
Saphris
(Antipsychotic)
Pradaxa
(Anticoagulant)
Toviaz
(Overactive bladder)
Ivemend
(Antiemetic)
• AcOH
Firazyr
(Hereditary angiodema)
Peramivir
(Antiviral)

Figure 18.3 Recently marketed synthetic and semi-synthetic drugs.

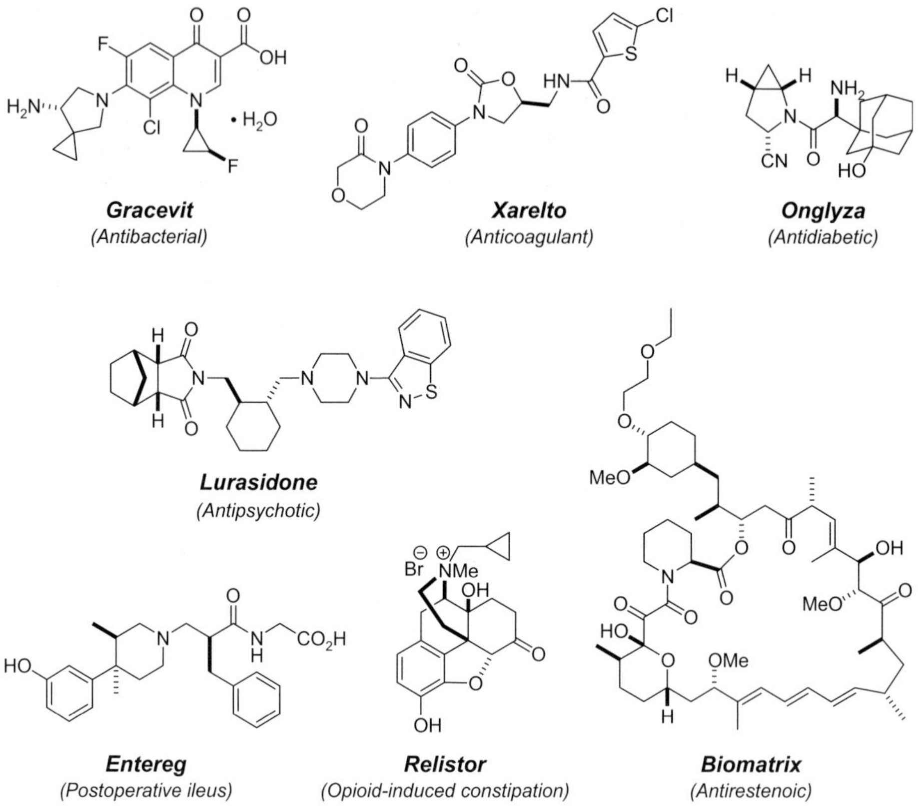

Gracevit
(Antibacterial)

Xarelto
(Anticoagulant)

Onglyza
(Antidiabetic)

Lurasidone
(Antipsychotic)

Entereg
(Postoperative ileus)

Relistor
(Opioid-induced constipation)

Biomatrix
(Antirestenoic)

Figure 18.3 (continued).

18.5
In Pursuit of the "Ideal Synthesis"

Depending on their structural complexity, the synthesis of natural products invariably involves multiple reactions, with the obligatory manipulation of diverse functional groups. Consequently, efficiency and practicality are sometimes sacrificed if only to reach the intended target molecule. In contrast, when planning the synthesis of a drug candidate, process chemists are particularly aware of the cost of goods, the potential for scale-up, and environmental problems through waste disposal. Thus, whether in an academic or an industrial environment, the Achilles heel, even of a well-planned total synthesis project, is multiple bond-forming sequences, excessive peripheral manipulation, and refunctionalization of existing or subsequently introduced functional groups on the intended carbon skeleton. Adding chirality as a pre-requisite introduces other challenges, especially in industry (see Chapter 4, section, 4.6.1). In this context, it is of interest to quote from the 1975 definition of an "ideal synthesis" by Hendrickson [31]:

The ideal synthesis creates a complex skeleton from simpler starting materials and so must link several such synthon molecules via construction reactions. Ideally, the synthesis would start from available small molecules so functionalized as to allow constructions linking them together directly, in a sequence only of successive construction reactions involving no intermediary refunctionalizations, and leading directly to the structure of the target, not only its skeleton but also its correctly placed functionality. If available, such a synthesis would be the most economical, and it would contain only construction reactions.

The 1917 synthesis of tropinone (Figure 18.4) in one synthetic operation from simple starting materials by Robinson [32] came close to an ideal synthesis, long before Hendrickson's notions of "systematic synthesis design, and numerical codification of construction reactions" [31].

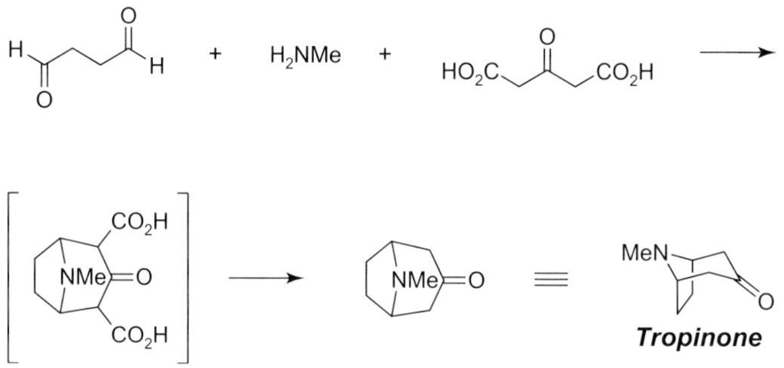

Figure 18.4 Robinson's "one-pot" synthesis of tropinone from succinaldehyde, methylamine, and acetonedicarboxylic acid.

Biomimetic cyclizations as exemplified by the elegant "one-step" synthesis of a precursor to progesterone by Johnson, of endiandric acid B by Nicolaou, and of methyl homosecodaphnyllate by Heathcock would also qualify as examples of near-ideal syntheses (see Chapter 4, section 4.5.1). However, as the level of complexity and functionality increases in an intended target molecule, so does the chasm that separates the reality of total synthesis from its "idealized" Hendricksonian paradigm.

Cognizance of such obstacles has stimulated efforts toward the development of synthetic methods that address higher efficiencies and selectivities. Terms such as "atom-economy," [33] "step-economy," [34] and "redox-economy" [35] have been suggested with good intentions. However, relevant solutions that apply to one target molecule cannot be generalized to others. For example, the vast majority of complex natural products containing polar substituents will require some measure of protecting group juggling, step-by-step segment coupling, ring formation, and obligatory functional group manipulation. Aspects of some of these issues are discussed below.

18.5.1
The problem with protecting groups – blessing or curse?

The presence of polar functional groups such as amines and alcohols will, in most cases, require the use of protecting groups in order to perform even the simplest of chemical reactions *en route* to the target molecule. In spite of the availability of a plethora of protecting groups [36], problems arising from incompatibility and lack of orthogonality can seriously compromise the successful completion of a synthesis. The annals of organic synthesis document many such cases, even necessitating a change in strategy at times. A number of these were featured in previous chapters of this book. Even when highly compatible and versatile protecting groups are used, there is a major drawback with the extra steps involved in their introduction (protection) and removal (deprotection). Unless one and the same protecting group is used, which is rarely the case in polyfunctional molecules, the number of protection-deprotection steps will inevitably rise, eventually resulting in longer sequences, multiple cycles of purifications, and overall inefficiency.

In spite of these issues, protecting groups are an integral part of many synthesis blueprints, if only reluctantly, hoping that they can withstand the rigors of the projected chemistry and gracefully depart on demand, leaving behind an unscathed intended target molecule.

It can also be argued that depending on the nature of the molecule, resorting to the use of protecting groups is a necessity rather than an option. For example, in the 1989 epic total synthesis of palytoxin, the largest man-made marine natural product of its era, Kishi and coworkers [37] synthesized a fully protected immediate precursor containing the following assortment of orthogonal protecting groups: 5 acetate esters, 4 benzoate esters, 20 *tert*-butyldimethylsilyl (TBS) ethers, 9 *p*-methoxybenzyl (PMB) ethers, 1 acetonide, 1 methyl ester, and one 2-(trimethylsilyl)ethoxycarbamate (Teoc) group (see Chapter 3, Figure 3.3). The synthesis was judiciously planned so as to assemble specific subunits already harboring different protecting groups. Inevitably, as the protecting groups were progressively removed under the mildest conditions, there was a significant loss of mass, which must have been factored in the original plan.

In any total synthesis project, considerable time and effort is dedicated to constantly "bringing up more material [38]." Except for rare cases (see below), contemplating a total synthesis of a molecule containing one or more polar groups (hydroxyl, amino, carboxylic acid, etc.), the use of protecting groups is the accepted norm, and their avoidance, however desirable, would otherwise not be realistic. The pioneering chemistry of nucleic acids, peptides, and carbohydrates done many decades ago would have been severely hampered were it not for the availability of a few robust *O*- and *N*-protecting groups that are still in use today. Since then, a steady influx of new protecting groups has been introduced, rendering the task of reaching target molecules all the more feasible [36]. Therefore, the age-old love-hate relationship with protecting groups will continue as long as we choose to synthesize molecules in the traditional ways to which we have become accustomed. However, where there is a will to minimize the use of protecting groups, there may be a way to one or more solutions.

18.5.1.1 **Protecting-group-free synthesis?** – Being aware of the pros and cons of using protecting groups, the question may be asked if their use can be avoided, or at least minimized [39]? The answer is yes, depending on the nature of the target molecule, the starting materials considered, and most importantly, the type of chemical transformations planned.

A classical example, pertaining to a complex alkaloid, is strychnine where protecting groups were hardly used in all of the 17 total and formal syntheses starting with Woodward's (see Chapter 4, section 4.4.2). Topologically compact compounds containing few polar groups are also prime candidates for consideration. For example, Heathcock's total synthesis of (±)-fawcettimine required no protecting groups (see Chapter 14, section 14.8.1). Recent syntheses of indole alkaloids, such as ambiguine H and hapalindole U by Baran and coworkers, are elegant examples showcasing the multistep construction of complex heterocyclic compounds without resorting to protecting groups (see Chapter 13, section 13.6).

There is more good news from well-documented literature precedents. For example, the temporary protection of hydroxyl groups can be circumvented when using free-radical reactions involving C–C bond formation [8]. Olefin metathesis reactions employing the Grubbs or related catalysts are tolerant of polar groups such as hydroxyl, amide, and ester groups [18, 19]. The synthesis of thienamycin utilizing a Rh(II)-mediated carbene insertion was done in the presence of an unprotected hydroxyl group already in 1980 (see Chapter 10, section 10.9). Many transition metal-catalyzed reactions as well as organometallics, especially with metals harboring low-lying d-orbitals, are tolerant of polar functional groups [23]. Stereocontrolled syntheses of O-glycosides, glycosyl-1-phosphates, and glycosyl-1-carboxylates can be achieved with anomerically activated heterocycles in the absence of protecting groups [40].

Although protecting groups will continue to be considered as an integral part of a synthesis plan, past and more recent contributions should stimulate a change in our mindset. In the process, new bond-forming reactions can be discovered as part of the design paradigm that is compatible with unprotected polar groups. Understanding the inherent reactivity of molecules in conjunction with existing functionalities, the use of certain metal-catalyzed reactions, biomimetic and related cascade reactions, are but a few exciting strategies where the practice of protecting-group-free syntheses can be contemplated.

18.5.2
The "redox economy" problem

Another aspect of synthetic operations that adds steps is the need for oxidation and reduction reactions [35]. At times, a total synthesis of a target molecule may involve multiple interconversions of hydroxyl and carbonyl groups. This may be the consequence of a particular step that generates a ketone, such as in the reaction of a Weinreb amide with an organometallic reagent, only to have it reduced to a secondary alcohol in a subsequent step. In spite of the large number of available oxidation and reduction methods, the so-called "redox" problem in total synthesis is one that affects the very essence of efficiency and practicality. Unlike the need for protecting groups however, there are a number of practical and "economical" solutions to the redox problem [35]. The principle of functional group equivalence can be considered in the original plan, where, for example, double and triple bonds can be oxidized in one step to their carbonyl derivatives, alkoxy vinyllithium reagents can be used as masked methyl ketones, functionalized organophosphorus reagents can be used to introduce unsaturation as well as polar groups. The venerable Grubbs metathesis reaction [18], and new protocols for isomerization of terminal double bonds [41], can introduce, as well as functionalize, olefinic appendages. Applications of transition metal catalysis and metal-free methods are powerful tools to avoid redox problems in synthesis (see also Chapters 3–6). Depending on the nature of the target molecule, biomimetic cascade-type reactions, biocatalytic conversions, and similar approaches inspired by Nature, such as oxidation of unactivated C–H bonds [42], are also viable solutions. Baeyer-Villiger oxidations of cyclic and acyclic ketones deliver lactones and carboxylic esters with predictable regiochemistry.

Thus, a judicious choice of carbon skeleton-forming reactions, with a conscious plan to integrate functional group equivalence and strategic redox reactions (*i.e.*, those that establish the correct functionality in the target molecule) in the overall design should be prime considerations whenever possible.

18.5.3
The "functional group adjustment" problem

As we have seen in previous chapters, truly complex molecules have been synthesized showing creativity, skill, and the courage to raise the bar to higher levels of achievement. One of the pleasures of total synthesis is in the planning phase, when we first meet the target molecule – *visually*. At first, a million or more neurons will fire instant messages between the eye-teasing molecule and the mind's eye. There will be fleeting *a-ha* moments when a critical disconnection leading to a key reverse direction reaction is conceived. Many revisions and iterations later, and after the final adrenaline rush has subsided, we invariably settle on *plan A* as we chart the path for its maiden voyage (or climb to the summit) on paper. Unbeknownst to us, a competing group may also have its own *plan A* toward the same target molecule, hopefully based on a different conceptual approach. The marathon begins with enthusiasm, high expectations, and an unbridled spirit of optimism to be shared with coworkers.

In many of the syntheses discussed in previous Chapters, the plan may have included key steps with different degrees of difficulties – some even contemplating totally unexplored chemistry. By and large, a conservative approach is usually taken when planning a total synthesis with a level of predictability that allows safe passage to the final target molecule. Whether one adopts the Nike, *Just do it*, or the Sinatra, *I did it my way* approach [43], dead-ends and detours [44] will invariably change the complexion of the synthesis. Fortunately, synthetic chemists are resourceful and able to overcome such problems aided by the plethora of existing (alternative) methods. Problem solving during a total synthesis project is an integral part of the journey and provides coworkers with the best opportunity to learn how to think and apply the thought process to overcoming impasses and road blocks along the way.

The choice of target molecule and designated route is made by the principal investigator who must also bear the burden of guiding the team to the summit. Settling on a given route implies a level of confidence in the types of reactions to be performed, the reagents to be used, and the chances of success. Indeed many ingenious syntheses have relied on one or more innovative key reactions that were successfully realized, much to the delight of all involved. However, in a number of instances, these advanced intermediates may also require *functional group adjustments* involving extension, cleavage, stereochemical inversion, and redox chemistry, which add additional steps to the synthesis.

Unlike the problems arising from unforeseen reactivity and protecting group incompatibilities, functional group adjustments are evident when the planned route is first conceived. The decision then comes down to whether or not the extra steps are worth it, if only to showcase the key reaction performed earlier in the synthesis, knowing that there will be a penalty to pay for functional group adjustments at a later stage. While in some cases this choice is made at the expense of expediency, it should not detract from the conceptual elegance of implementing the key reaction, especially if it offers new insights. Nevertheless, devising synthetic routes whereby innovative chemistry can be carried out while minimizing extra functional group adjustment steps, are important objectives to achieve better efficiency.

18.5.4
"Chiral economy"

One of the most natural and practical ways to address aspects of "chiral economy" [45] in synthesis is to start with the most appropriate chiron. This brings us back to the powers of visual perception leading to *reflexive* and *relational* thinking when planning a synthesis (see Chapter 4). The abundance of naturally occurring, enantiopure chirons provides ample opportunity to utilize carbon skeletal, functional, and stereochemical convergence with segments of the target molecule at the synthesis blueprint stage. In this respect, terpenes are ideal carbogenic chirons since they provide cyclic and acyclic variants bearing stereochemically defined alkyl groups and versatile functionalities such as carbonyls, enones, olefins, and hydroxyls. Furthermore, terpenes such as carvone, for example, can be transformed into a variety of useful congeners with minimal need for protecting groups. Enantiopure carbocyclic compounds, readily available by chemical and enzymatic asymmetric processes, are also highly versatile starting materials for the elaboration of monocyclic and polycyclic target molecules (see Chapters 6, 7, 8, 13, and 14).

The functionally richer α-amino acids, carbohydrates, and hydroxy acids will invariably require protecting groups at the outset. However, their oxidation states can be already secured and maintained in segments of the target molecules, while providing inherent resident chirality to control absolute stereochemistry in subsequent steps (see Chapters 10–12).

In conjunction with the vast number of versatile bond-forming methodologies, including catalysis, a plan that takes full advantage of the *chiron approach* can be a viable solution to the persistent aforementioned "economy" problems that we experience in synthesis today. For example, in a recent synthesis of natural (+)-bis-anthraquinone antibiotic BE-43472B, and its enantiomer, Nicolaou and coworkers [46] relied on an intermolecular Diels-Alder reaction utilizing a diene prepared from (*R*)- or (*S*)-lactaldehyde (Figure 18.5).

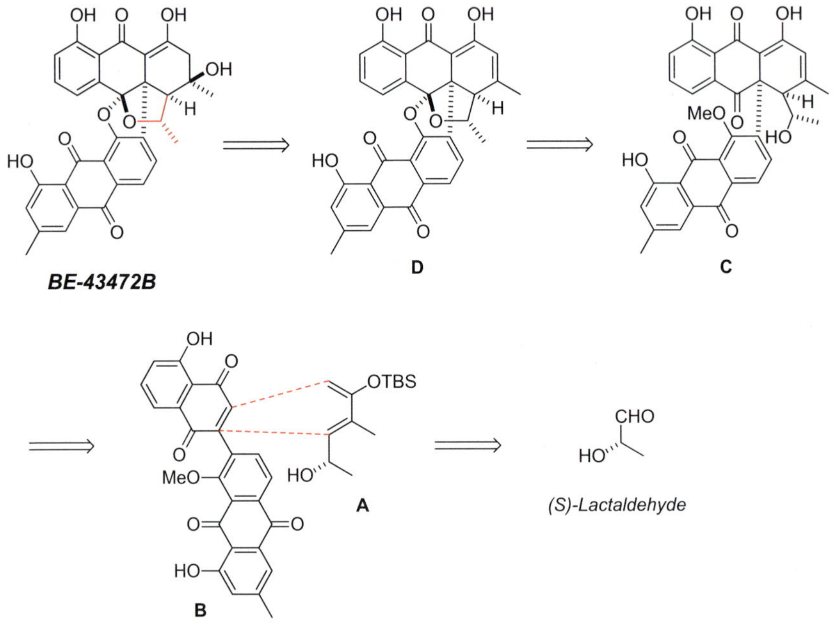

Figure 18.5 Stereocontrolled intermolecular Diels-Alder reaction involving a chiral diene (**A**) derived from (*S*)-lactaldehyde by Nicolaou.

The total synthesis of several members the crambescidin family of natural products have been extensively studied by Overman and coworkers [47] (see Chapter 3, Figure 3.2). The second synthesis of crambidine was accomplished by Gin and coworkers [48] starting with (*S*)-pyroglutamic acid as a suitable chiron (Figure 18.6). Enantiopure alcohol intermediates obtained by application of several different asymmetric catalytic reductions of the respective ketones were used to provide the chiral appendages attached to the guanidine core of crambidine. Gin and coworkers were able to assemble the complex core of the alkaloid using a unique [4+2] carbodiimide annulation (**A** and **B** to **C**) and AuCl$_3$-catalyzed intramolecular alkyne hydroamination reaction (**C** to **D**), which set the stage for a stereoelectronically controlled spirohemiaminal formation (**D** to **E**). Coupling the cesium carboxylate of **E** with iodoamide **F** led to **G**, which was deprotected to afford crambidine. Thus, the apparent pyrrolidine core could be easily prepared from (*S*)-pyroglutamic acid, which provided the necessary chirality at C-10 as well as the lactam carbonyl for activation as a thioimidate intermediate in the annulation reaction.

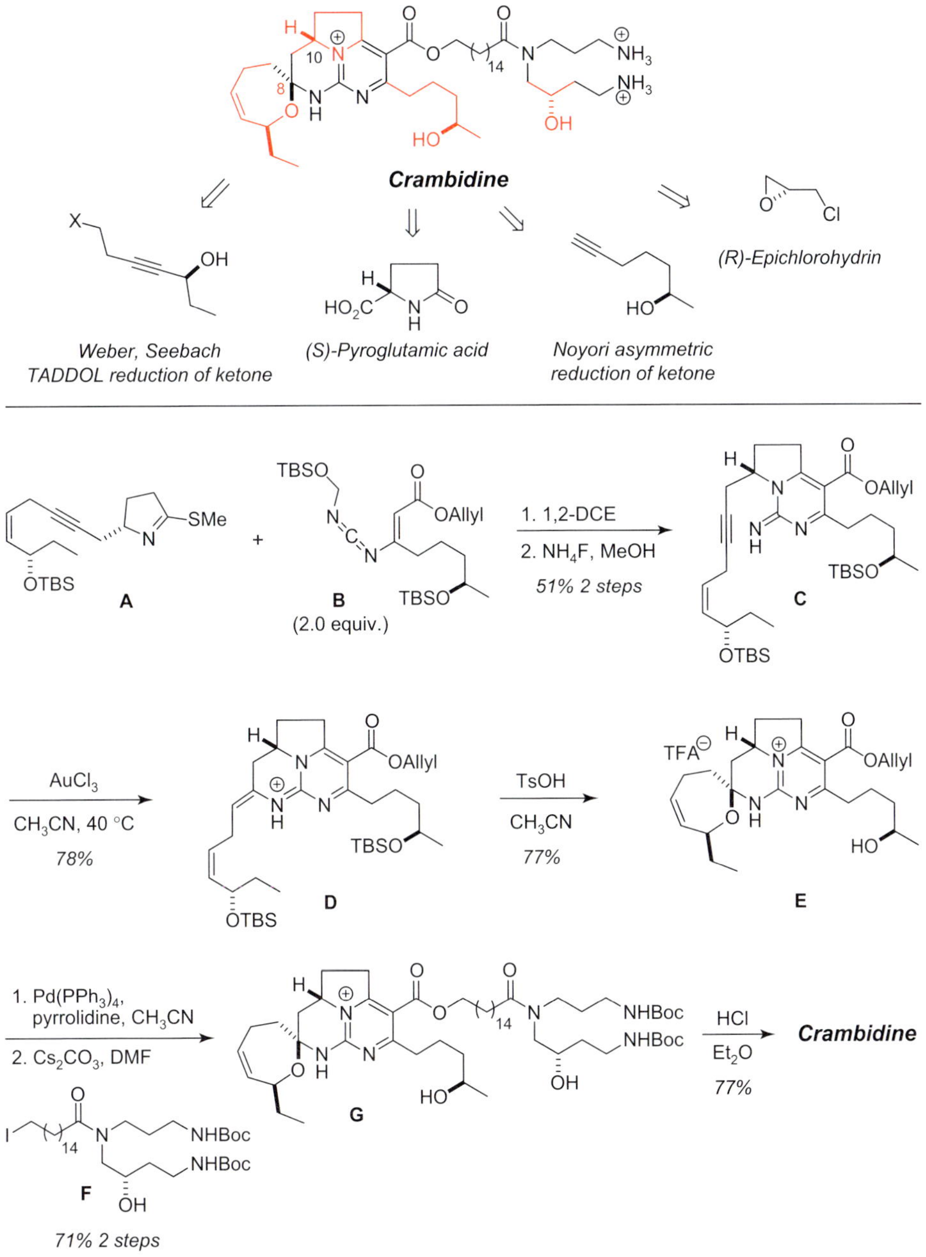

Figure 18.6 Total synthesis of crambidine using a carbodiimide annulation reaction by Gin.

Although many of the syntheses discussed in this book would fail the combined atom-step-redox, and protecting group economy test, they are nevertheless the result of heroic efforts to achieve a synthetic goal while providing excellent training to coworkers. Such periodic "tests" are also necessary to view future syntheses in a wider perspective – especially one that also takes into account our role as chemists in society as well [31c].

18.6
For the Love of Synthesis (*Synthephilia*)

Natural products varying in their structures and biological properties have provided synthetic chemists with the incentives, and at times, the obsession to initiate and pursue exciting research programs. The synthesis of many molecules that were unthinkable a few years ago can now be realized (see Chapter 3). Total synthesis continues to unravel and correct wrongly assigned original structures – a challenging research activity with very satisfying if not redeeming rewards (see Chapter 2, section 2.8).

For example, in 2008 Nicolaou and coworkers [49] revised the original structural assignment of vannusal B. In the process, known methodologies had to be adapted to construct such architecturally complex molecules harboring contiguous quaternary stereogenic centers [50] (Figure 18.7). It is noteworthy that enantiopure **A**, obtained in several steps from a lipase-mediated desymmetrization of a *meso*-1,1′-bis-cyclopentane-2,2′-diacetate, was coupled with a racemic partner **B**, to give **C** after separation of the diastereomers.

Celebrating the completion of a total synthesis project is one of the many euphoric sensations for a team that has dedicated much of its time and effort toward that objective. We previously discussed the reasons to engage in the practice of total synthesis, in spite of the arduous journey. Besides meeting the requirements of the proposed objectives, there is also the exhilaration of discovery, and a sense of sublime accomplishment that scientists know all too well. Those seemingly never ending periods when "nothing seemed to work" are soon forgotten at the sight of the final product.

The "love" of the métier of synthesis (*synthephilia*), is a common trait amongst its dedicated practitioners. Suffice it to engage a colleague in a conversation about his or her synthesis, or listen to an animated lecture describing how a recalcitrant problem was solved *en route* to the target molecule. In the final analysis, it is this inner "love" for making molecules, mixed with enthusiasm and passion, that is passed on from generation to generation of synthetic chemists.

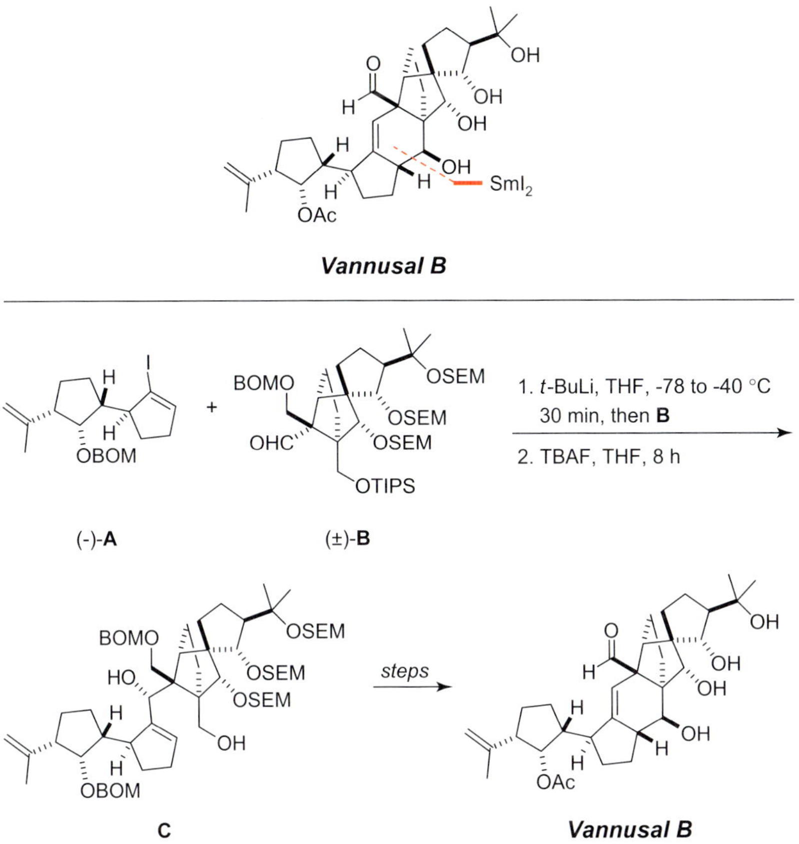

Vannusal B

(-)-**A**   (±)-**B**

1. *t*-BuLi, THF, -78 to -40 °C
   30 min, then **B**

2. TBAF, THF, 8 h

C   **Vannusal B**

Figure 18.7 Revised structure and stereochemistry of vannusal B by Nicolaou.

18.6.1
Reaching the summit

Total synthesis has been likened to mountain climbing, and further, as a "race to the summit," [51] especially for target molecules that carry a certain level of notoriety as highlighted in parts of Chapter 16 (see section 16.1). With few notable historical exceptions, the motivation for multistep syntheses of complex natural products in recent years has been to showcase creativity, to test the limits of available methodologies, and to apply newer methods developed in the senior author's laboratory. Natural product synthesis has also been considered by many as the hallowed ground for student and coworker training in preparation for research careers in industry and academia. In those instances where sufficient quantity of the final product was produced, the justification for the process may have been fulfilled by making material available for biological tests, or to gain insights with regard to specific properties. As already commented on in previous chapters, the enterprise of total synthesis, regardless of the level of complexity of the target molecule, is labor-intensive, and the path to the summit may be fraught with obstacles. However, the resolve, perseverance, and creativity of the synthetic chemist have always risen to the challenge. In general, it is only a matter of time (and especially resources), before a synthesis is completed thereby reaching the coveted summit. When more than one research group is in the race, there is the added pressure of being first to hoist the flag. However, the annals of organic synthesis have shown time and time again, that the first synthesis may not necessarily be the best. We have already commented earlier in this chapter on the utopic notion of an ideal synthesis. Until such a time comes, syntheses will be compared to each other, and will have to withstand close scrutiny with regard to efficiency, practicality, and some measure of ingenuity. Several examples of total syntheses of one and the same target molecule were shown, compared, and contrasted in earlier chapters. Ultimately, the unbiased practitioners of today, and tomorrow's new generation of synthetic chemists will be the true assessors of the best synthesis of a given target.

The lasting value of a total synthesis effort can be infinitely more useful if, as a result, new insights into fundamental concepts of reactivity, or innovative methodologies are discovered. This places a high premium on the true value of a new disclosure relating to the total synthesis of an already known molecule. Accomplishments in total synthesis must therefore be viewed not only in relation to the time period and era of their execution, but also to what they may add to existing knowledge. It is interesting that while *many* synthetic methods have been developed and enshrined as named reactions over the years [4], very few general "rules" have resulted from synthetic efforts directed towards the synthesis of natural products. The Woodward-Hoffmann rules on the conservation of orbital symmetry [52] and Baldwin's rules for ring closure [53] are the brainchildren of observations during synthetic studies related to vitamin B_{12} and the penicillins respectively. In contrast, the practitioners of organic synthesis have relied on many fundamental *effects, concepts,* and *principles* in explaining various aspects of reactivity in bond-forming processes [54].

18.7
Organic Synthesis: To where we are going

Much has been said about the future of organic synthesis by the practitioners of the art over the years [55]. Because of its enabling power to create new entities with unlimited diversity, the science of synthesis is at the center of the molecular universe. Its versatility, and utility, extends over to many other disciplines with immeasurable service to humankind. The past 100 years has already witnessed the enormous benefits provided by synthetic organic molecules starting with dyestuffs, to present-day materials for heart valves, and better medicines. The rich history and the strong foundation of organic chemistry as an exact science, inasmuch as it is governed by the rules of valence electrons and thermodynamics, will propel it to greater heights in the years to come. Traditional activities, such as the total synthesis of complex organic molecules, will no doubt continue albeit with more stringent criteria for efficiency, practicality, and purpose.

The emergence of different paradigms dealing with so-called diversity-oriented small molecule synthesis [56] is a practical and strategically clever way to explore the vast chemical space. Applied in an imaginative way, the exploitation of cascade, one-pot, multistep reactions [57] will be a powerful method to introduce functional and structural diversity while building complexity in molecules. In conjunction with automated synthesis technologies and high throughput biological screening, diversity-oriented synthesis could be enormously useful in the initial phases of a drug discovery program. A chance encounter of a synthetic small molecule within a given library of compounds with a biologically relevant enzyme or receptor may indeed lead to a new drug prototype, eventually reaching a different kind of "summit," with potential benefits to humankind.

As our understanding of biological events deepens, so will the involvement of synthetic chemistry and the molecular toolbox to probe the structure-to-function continuum. Hand in hand, synthetic chemistry and biology, in all of its fascinating forms, will combine to further unravel life's mysteries, provide miracle drugs, and aid in exploiting the full potential of the human genome. DNA-on-a-chip will have its personalized medicine component based on the most effective and safe drugs for each and every individual. Non-invasive diagnostics, implantable brain chips, neural tissue transplants, and even Paul Ehrlich's "magic bullet" approach to drug delivery and therapy [58] are realistic goals, more than a century after its prophetic prognostication.

Lessons learned from natural products will allow the synthesis of surrogates and analogues with improved activities [59]. In this context, the argument has been made that a reorientation of the traditional path followed in total synthesis of natural products be considered, at least in part by directing efforts at structure-based organic synthesis [43]. Data gleaned from X-ray and NMR studies of bioactive natural products in complex with relevant enzymes and other macromolecules can be used to design drug-like small molecules with therapeutic potential. Already, with minimal knowledge of precise molecular events, a number of highly beneficial drugs have resulted from such efforts (see Figure 18.3; also Chapters 2 and 3).

Advanced crop technology, in conjunction with safe agrochemicals, will eradicate diseases now responsible for large losses in the field resulting in famine and extreme hardship in some parts of the world. Protecting our precious environment through greener chemical methods should be a major challenge for the future.

A large number of important organic reactions already have their catalytic variants today. Even more impressive, are those catalytic *and* enantioselective reactions. Exciting recent developments in C–H bond activation [60] will expand rapidly to include a wider array of unactivated carbon atoms. The spectrum and efficacy of catalytic organic reactions including biocatalysis will continue to increase exponentially in the future, as we learn how to better harness Nature's rules, and to manipulate its genetic information (see Chapter 2, section 2.9).

Materials science and nanotechnology, based on "smart" organic molecules, will thrive even beyond their present stage of sophistication as data processing and methods of analysis improve at the atomic level [61].

Automation encompassing robotics, solid phase synthesis, and flow-chemistry will play an increasingly dominant role in laboratory practices of organic synthesis [62]. Chemical information technology will have a significant impact in integrating the many aspects of synthesis relating to drug discovery and new therapeutic agents.

Beyond these foreseeable developments, which are already on track, lie challenges yet to be defined and shaped by advances in other disciplines where organic synthesis can play an important partnering role. At the risk of prognosticating, it is safe to say that among other applications, the enterprise of synthesis will continue, for the most part, to be biologically-inspired, but chemically-driven [43].

We should also remember that the macromolecules of life such as proteins, nucleic acids, and glycolipids, are in themselves, "organic" in nature and share the same alphabet in their chemical code as do the small organic molecules that interact with them. In the words of Rustum Roy: *"Nature optimizes the system, never the components."* As synthetic chemists, we have become quite adept at optimizing the components, and are only beginning to understand the "system." In the near future, a larger effort will be dedicated to the design of small molecules to be incorporated into metabolically engineered enzymes in microorganisms, literally turning them into factories that produce novel unnatural products. The link between genes and molecules will foster the emergence of new subdisciplines and the closer integration between existing ones. Natural products will be viewed in a much broader perspective, beyond the realm of synthesis proper [63].

18.8
Synthesis at the Service of Humankind

In times of great need whole nations will rally behind a noble cause. The history of humankind registers many such heroic and united efforts to recover from natural disasters, avoid pandemics, or fight in the name of justice.

A notable example in science is the pioneering and dedicated efforts of several laboratories worldwide to develop penicillin as an antibiotic, at a time when the only viable recourse to combat infection was sulfa drugs [64]. The giants of the natural sciences such as Paul Ehrlich and Louis Pasteur among others, have laid the founding stones for the next generations to further advance the frontiers of immunology and biology that control our fate as a race.

We as a people, are the modern day beneficiaries of these discoveries, and as a consequence, we enjoy longer and healthier lives. However, many parts of the world still suffer from diseases that can be contained and cured given the means that we have at our disposal. These so-called neglected diseases still afflict large populations whose only medical resources come as aids from benevolent societies, organizations, and some pharmaceutical companies [65].

In spite of phenomenal advances in all aspects of the natural and physical sciences, life-threatening diseases continue to affect people of all ages. Major strides have been made to combat cancer and other diseases, but we are far from having a totally safe and truly effective cure for each. The transition from mouse, to monkey, to man, is enormously complicated even for the most promising drug substance.

Unravelling the genetic code holds great promise to understand human biology at the molecular and cellular levels. The discovery of key proteins that are responsible for turning on or off certain pathways leading to cell survival or death (apoptosis), must also be understood at the chemical level. This will require a greater emphasis on collaborations between scientists studying various branches of biology and those doing synthetic chemistry. In principle, this is more easily done among academic labs, rather than in competing pharmaceutical companies, and some noteworthy contributions are commendable.

In Chapter 16 (section 16.2), we commented on the notion of "Joining Forces." Two giants of organic synthesis, Robert Burns Woodward of Harvard University and Albert Eschenmoser of the ETH in Zurich, decided in the early 1970's to complete the total synthesis of vitamin B_{12} using their combined resources, rather than to continue their individual efforts. To this day, this remains as a rare major collaboration between two figureheads of our subdiscipline toward a target molecule of high relevance. Given the resources, there is no reason why major diseases cannot be tackled by combining the efforts of biologists and synthetic chemists in a sustained and fully integrated way.

Funding agencies should find creative strategies to encourage collaborations that are highly focussed, truly integrated, and dedicated to a common cause. With the best minds at work, it will only be a matter of time before dreaded diseases such as cancer, Alzheimer's, and diabetes will find safe and effective cures.

The enterprise of total synthesis is an individual and fiercely independent form of science. A particular strategy toward a target molecule also bears the imprint of the persona. Once a project is initiated, there is a great deal of excitement and passion to reach critical steps. There is euphoria and a tremendous sense of pride and accomplishment when the final target molecule is produced. In many ways, it is the individual nature of the enterprise of synthesis that brings out the best in skill and creativity. Chasing after the same target molecule could also lead to new perspectives and creates a level of competition that should benefit the science of synthesis rather than hinder it. While this time-honored tradition of synthesis should continue, there may be golden opportunities for two or more major academic groups to combine their years of experience to tackle a specific target molecule whose benefit to humankind is incontestable. Would the supply problem of life-prolonging and scarce natural products not be solved if more than one team were to collaborate from the initial blueprint up until its fruition in the laboratory, eventually leading to an industrially viable process?

Needless to say, this idealistic view of curing the ills of humankind through synthesis, presents practical and logistical problems, not the least of which is research funding, receiving due recognition, and finding ways of maintaining a measure of secrecy. Again, new initiatives can be started at the governmental as well as global funding levels to bring the best of the biological and chemical synthesis worlds together, as a unified mountain climbing team, to hoist the victory flag at the summit. Imagine the headline:

Reuters, A. P.: "Combined Chemistry/Biology Effort Produces Miracle Cancer Drug":

A team of synthetic chemists and biologists from two universities headed by professors Y and Z have succeeded in the total synthesis of "hopestatin," a clinically proven and powerful anticancer agent in a process that is suitable for large scale manufacturing. Currently isolated by a tedious extraction process from a marine organism in limited quantities, "hopestatin" will be available at reasonable cost and offer a great health benefit to the public. A spokesperson for the two teams explained that the operational strategy was inspired by elegant biological studies, and the combined expertise of the synthetic chemists was instrumental in achieving the objective of making a life-saving drug available in quantities far exceeding what nature could provide.

18.9
From the *Chiron Approach* to Catalysis

Many outstanding books, monographs, and multi-volume major works about organic synthesis, and its branches, have been published over the years (see Chapter 1, reference [7]). The community of synthetic organic chemists has benefitted enormously from the scholarly contents of these authoritative collections as reliable sources of information and references to seminal contributions spanning decades of research.

The prime objective of this book was to show the importance of the interplay between the visual and mental processes in synthesis planning and execution. Inspired by principles articulated by Sophocles and Aristotle in Chapter 1 (sections 1.5 and 1.8), we began by showing how the mind's eye perceives and analyzes through space.

This fascinating visual dialogue with a target molecule prior to formulating a synthesis plan can be perceived as a progressive epiphany of images and thoughts, which ultimately materializes in the form of a chemical entity. It is a chemical form of art and molecular architecture at its best!

As discussed in Chapter 4 (section 4.2), our visual "chemical" thinking modality about a synthesis plan toward such an entity usually follows two independent paths in the mind's eye. The visual *relational* and *reflexive* thought processes are part of the psychobiological basis of how we proceed to "do synthesis." Throughout this book, we emphasized the notion of *relational* thinking, by visually interrelating substructures in target molecules to suitable *chi*ral synth*ons* (chirons) derived from readily available small molecules. Their use as starting materials provides carbon skeletal, functional, and stereochemical convergence at the outset. The *chiron approach* to synthesis has been of general utility for virtually all classes of molecules bearing one or more stereogenic carbon atoms. In addition to its high predictive power, it combines the aesthetic features of a masterpiece painting with the space-filling ingenuity of a majestic sculpture. For example, the symbolism in many of Salvador Dali's paintings portraying images that can be viewed in different perspectives, can be likened to finding the skeletal and functional remains of a hidden chiron in the intricate architecture of a complex molecule (see Chapter 1, section 1.8).

The *reflexive* thought process is equally important in synthesis planning, because it complements the *chiron approach* in creative ways, especially with the remarkable progress made in catalytic asymmetric reactions. In tandem with existing intermediates prepared from suitable chirons, or in a *de novo* approach relying primarily on reagent controlled asymmetric induction, catalysis is a powerful method to create a variety of bonds including those carrying quaternary carbon atoms with added chirality if needed [50].

In the best of both worlds, the *chiron approach* offers Nature's small molecule chiral building blocks as starting materials, while catalysis exploits Nature's chirons as well as totally synthetic variants, as exquisite host ligands to a variety of metals, or as temporary reacting partners in catalytic reactions that take place in the absence of transition metals (organocatalysis).

In previous sections of this chapter, we discussed the need for efficient synthesis planning and execution (see section 18.5). In a recent 10-step total synthesis of actinophyllic acid, Overman and coworkers [66] demonstrate the virtues of astute visual *relational* and *reflexive* thinking (Figure 18.8).

Amino alcohol **A** was obtained in 91% *ee* from a Ru-catalyzed asymmetric hydrogenation of the corresponding ketone. Ozonolytic cleavage of the double bond in **A** led to cyclization to give the hemiaminal, followed by acetylation to afford *N*-Boc piperidine **B**. A catalytic electrophilic aromatic substitution reaction of indole **C** in the presence of 5 mol% Sc(OTf)$_3$ gave **D** in a highly efficient and diastereoselective manner (dr = 17 : 1). Chemoselective reduction of the acetate ester in **D**, followed by Swern oxidation of the resulting alcohol furnished ketone **E** in high yield. A remarkable intramolecular oxidative enolate coupling took place when **E** was subjected to LDA and an iron(III)-DMF complex to give the tetracyclic ketone **F** (see relevant mechanism A, Figure 18.8) [67]. Conversion of **F** to the allylic alcohol **G** was accomplished using a two-step sequence. Treatment of this allylic alcohol with hydrochloric acid led to a crystalline salt that was essentially enantiopure (99% *ee*). Addition of paraformaldehyde triggered a cascade of events that included an aza-Cope/Mannich reaction, followed by hemiacetal formation from which actinophyllic acid was isolated as its hydrochloride salt (see relevant mechanism B, Figure 18.8).

In reviewing Overman's synthesis plan, one should be aware of the perceptive power of tracing the origin of the starting (3*R*)-2,3-diacetoxypiperidine **B** within the polycyclic framework of actinophyllic acid. Although it is still visible after the oxidative enolate coupling, and carbonyl branching reactions (**E** to **F** to **G**), the piperidine ring can no longer be seen after the aza-Cope/Mannich reaction has been completed. It should be noted that the use of protecting groups and redox chemistry was kept to a minimum.

Actinophyllic acid

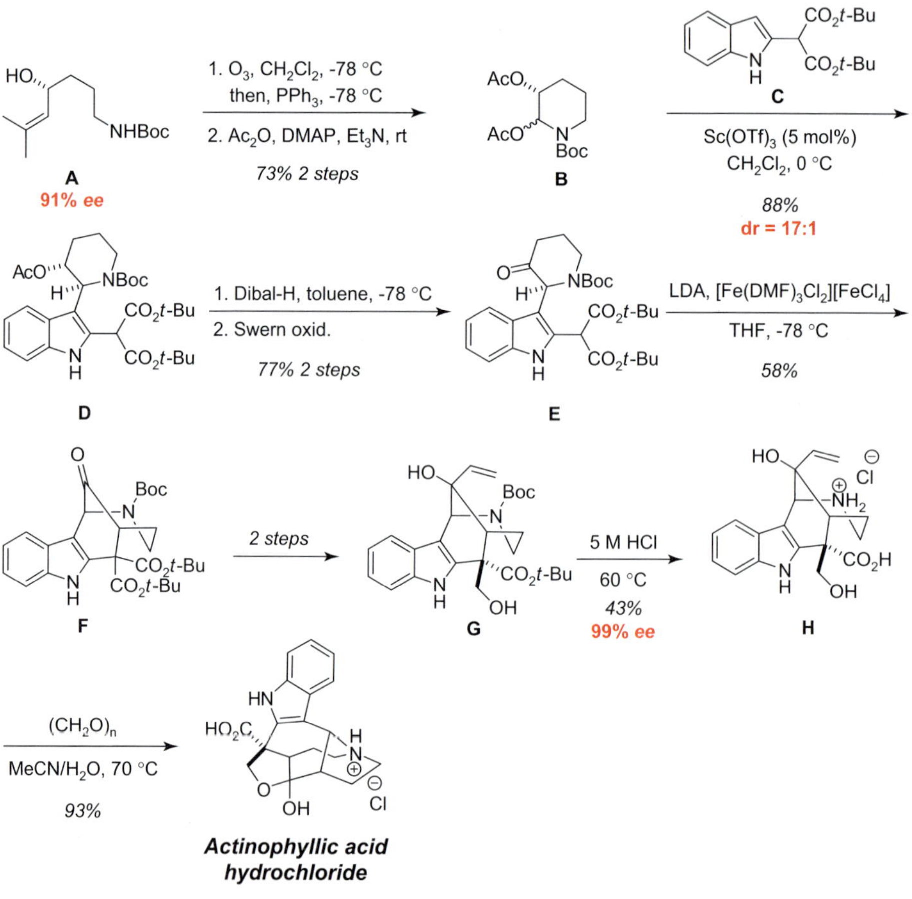

Figure 18.8 Total synthesis of actinophyllic acid by Overman.

Relevant mechanisms and transformations:

A.

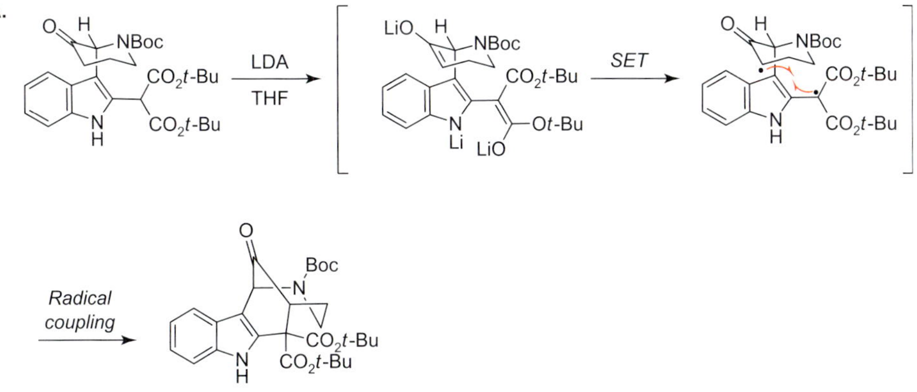

B.

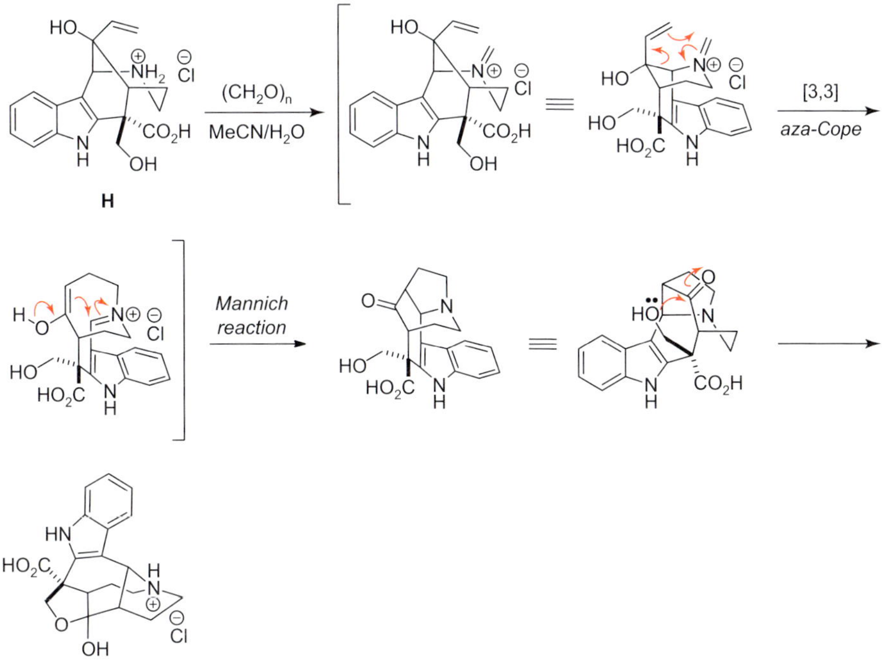

**Actinophyllic acid
hydrochloride**

Figure 18.8 *(continued)*.

18.9.1
The young, the brave, and the bold: Passing the baton

We conclude this final chapter by showcasing the prowess of a new generation of synthetic chemists in pursuit of ever-higher challenges toward the conquest of complex target molecules. Many naturally occurring alkaloids represent the ultimate in architectural and topological complexity, combining daunting concatenations of highly functionalized ring structures, with a high level of stereochemical intricacy. The total synthesis of complex alkaloids and related compounds has a rich history dating back over 50 years, which have been discussed in earlier chapters. As previously emphasized, a particularly redeeming aspect of such accomplishments is in coworker training. Indeed, over the years, many talented students have gone on to establish their own successful and independent research programs. This is quite common in academia, thus perpetuating the tradition of excellence, and assuming the roles of exemplary mentors. As with the past golden periods of synthesis, the new millennium has produced its share of young academics whose virtuosity is reflected in their daring choices of target molecules, and the innovative aspects of their approaches. Examples of such elegant total syntheses based on the *chiron approach* in conjunction with other methods of asymmetric bond formation where catalysis is featured in key steps, were amply highlighted in this book. As representatives of this new era of ingenuity in complex natural product synthesis, also containing key catalytic reactions, we show the total syntheses of himandrine, palau'amine, minfiensine, and maoecrystal Z by Movassaghi, Baran, MacMillan, and Reisman respectively [68].

18.9.1.1 **Himandrine** – The synthesis of the galbulimima alkaloid himandrine was reported in 2009 by Movassaghi and coworkers [69]. The simplification of the topological complexity of this intricate polycyclic alkaloid in the mind's eye can lead to many hypothetical substructures. Movassaghi chose to retrosynthetically disconnect the molecule in a non-obvious, yet daring way as shown in Figure 18.9. L-Alanine was used to build the cyclic imine **A**, which would be appended to the tricyclic enone **B** by means of a Michael addition. A series of highly stereocontrolled and well-documented bond-forming sequences would deliver the alkaloid in what the authors believe to be a biomimetic process in part.

The construction of the tricyclic ketone **B** commenced with the enantiomerically pure chiron **C** prepared through MacMillan's D-proline-catalyzed α-oxidation of hept-6-enal. The enantiomerically pure diol was transformed to **D** and then to **E**, and engaged in a Suzuki cross-coupling reaction to give the bromotriene **G**. The masked ketone derivative **H** was converted to **I** through a cross-metathesis reaction with acrolein. A diastereoselective Diels-Alder reaction gave the *trans*-decalin product **J** as the major diastereomer. Intramolecular Mukaiyama aldol condensation then led, after dehydration, to the desired tricyclic ketone **K**. Movassaghi chose the 2-azetidinone enamine as a masked ketone to facilitate the Diels-Alder reaction, and to exert a preference for the reactive *s*-cis conformation of the diene.

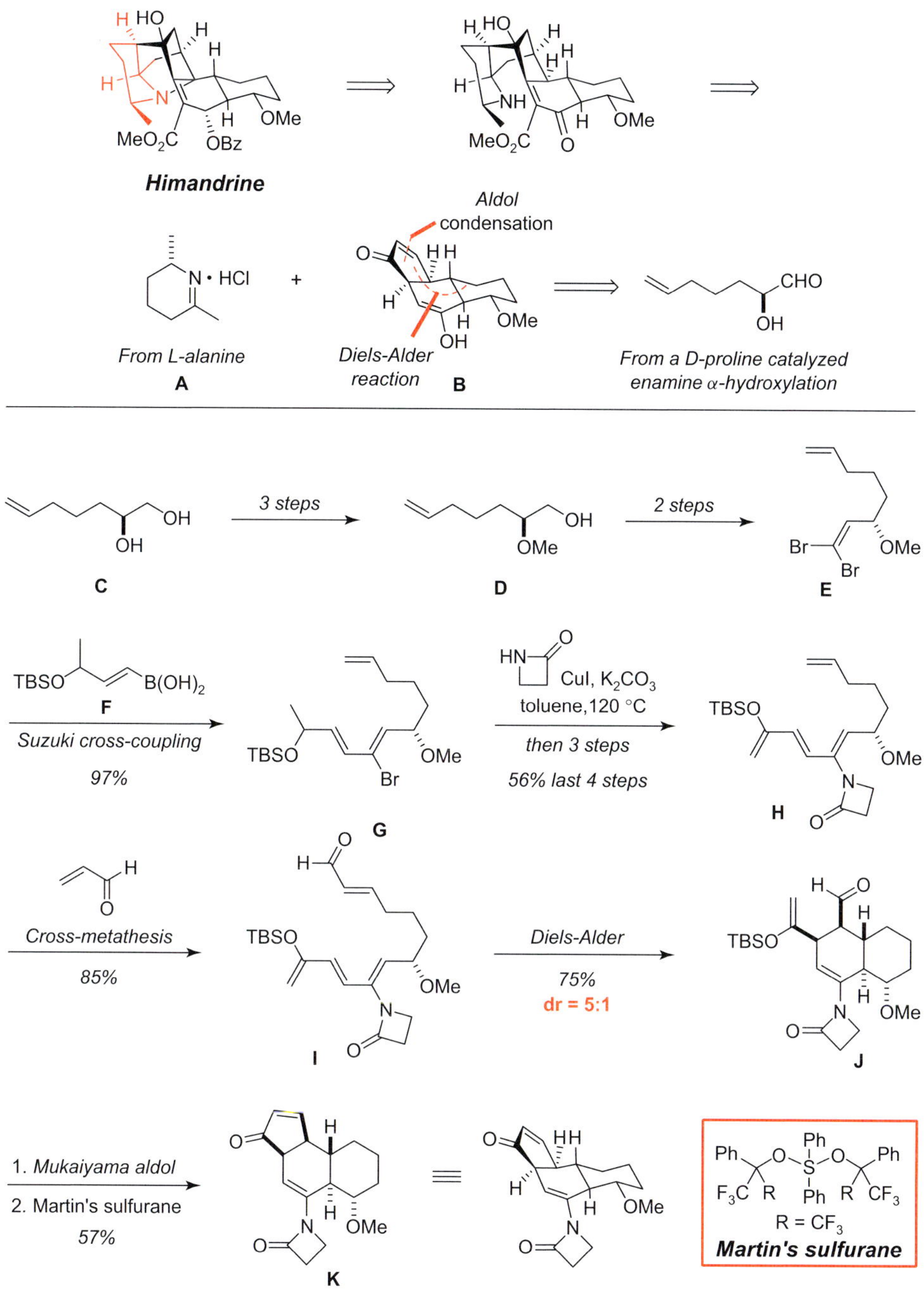

Figure 18.9 Simplification of complexity and synthesis of the tricyclic carbogenic core of himandrine (see completion of the synthesis in Figure 18.10).

Assembly of the subunits and completion of the first synthesis of himandrine is shown in Figure 18.10. Thus, conjugate addition of the anion formed from **A** to the enone **K**, proceeded with excellent stereocontrol due to blocking of the *Si*-face by the *trans*-decalin ring system to give the enamine tautomer **L**, which underwent intramolecular nucleophilic addition to the ketone affording **M**. Reduction with sodium borohydride, followed by formation of the *N*-Cbz derivative, and hydrolysis gave **N**. Treatment with the Vilsmeier reagent afforded the cyclic vinyl ether **P**. Three further steps gave **Q**, which is postulated to be a biosynthetic intermediate. Tautomerization of the intermediate **R**, followed by treatment with *N*-chlorosuccinimide led to the transient pentacyclic enol **S**, presumably via the intervening *N*-chloro intermediate. Stereoselective reduction and benzoylation furnished crystalline himandrine.

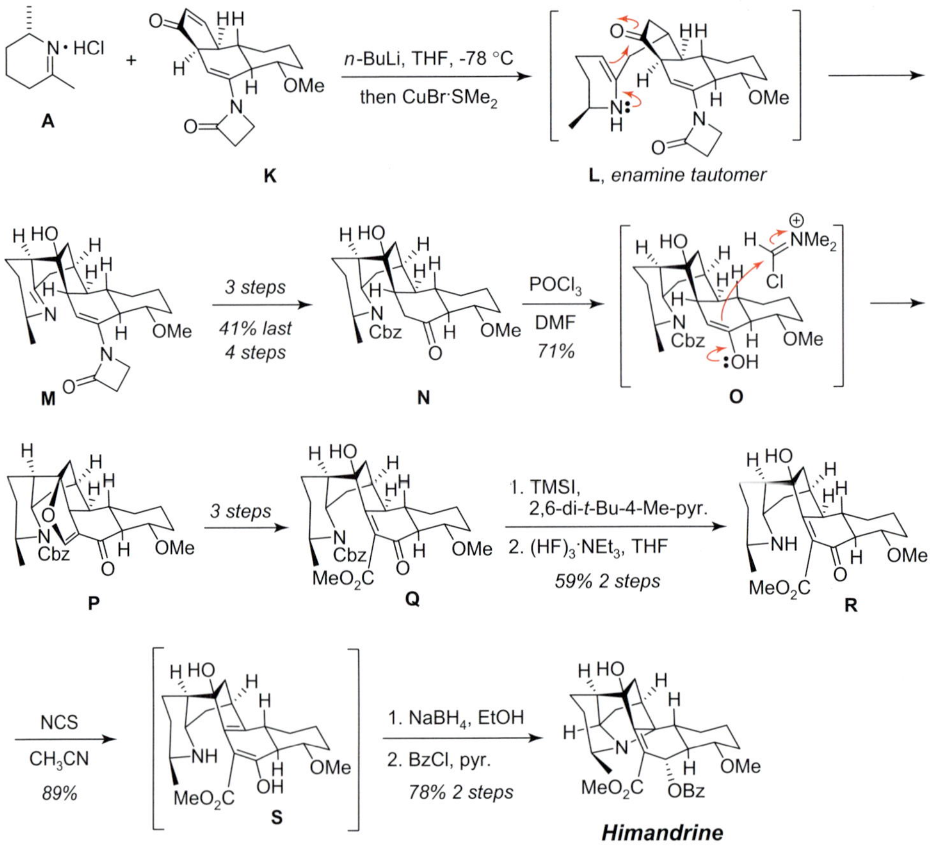

Figure 18.10 Completion of the total synthesis of himandrine by Movassaghi.

18.9.1.2 Palau'amine – The highly strained polycyclic bis-pyrrole-imidazole alkaloid palau'amine [70] has been a challenging target for total synthesis in several laboratories for some years. In spite of a structural reassignment in 2007 [71], the challenges of elaborating the highly substituted chlorocyclopentane core subunit (ring A) remained as daunting as ever [72]. In 2011, Baran and coworkers [73a] reported the first enantioselective total synthesis of palau'amine, based on a conceptually elegant strategy. The synthesis of the chlorocyclopentane core subunit *A* is shown in Figure 18.11.

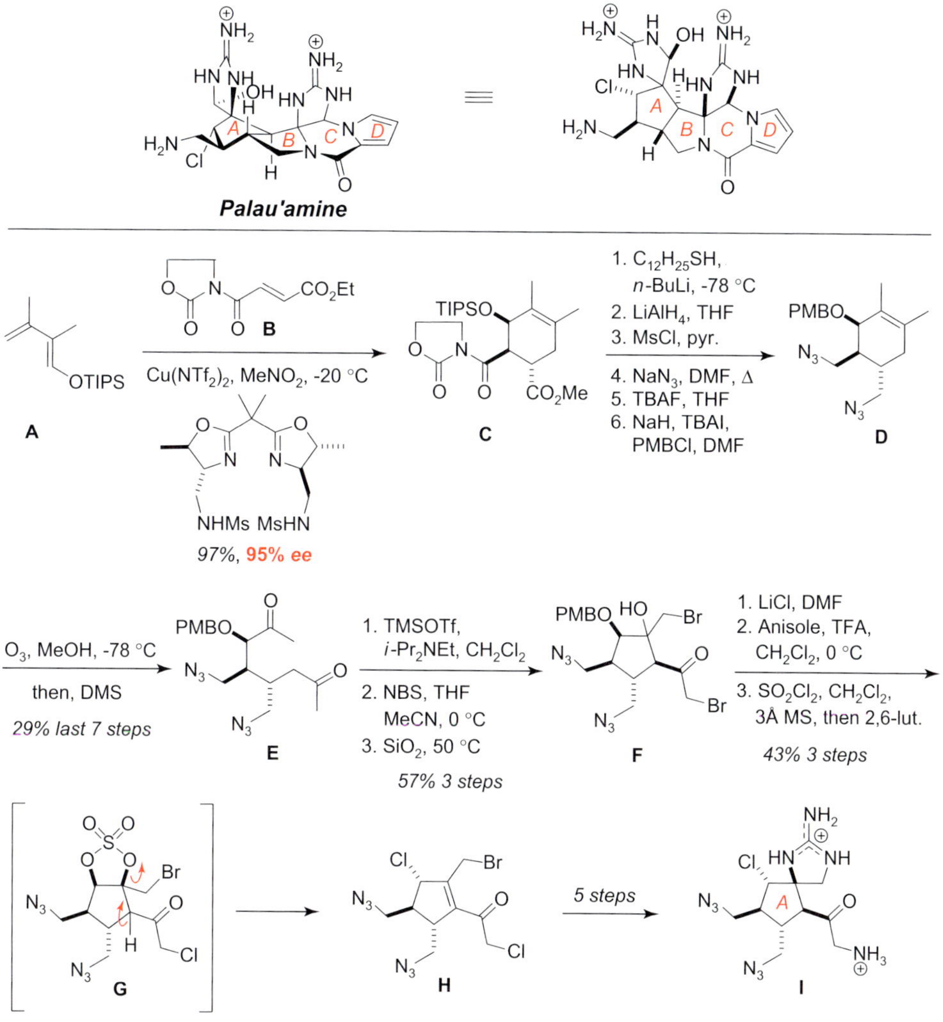

Figure 18.11 Simplification of complexity and synthesis of the pentasubstituted chlorocyclopentane core motif of palau'amine.

A Diels-Alder reaction between diene **A** and the dimethyl fumarate-derived dienophile **B**, in the presence of catalytic amounts of Cu(NTf$_2$)$_2$ and Ishihara's oxazoline ligand [73d], led to cyclohexene **C** in high yield and enantioselectivity. It is noteworthy that this reaction was operational on almost a 20 gram-scale with only small erosion of chemical yield (84%) and enantioslectivity (89% *ee*). Reduction of the ester groups, conversion to the dimesylate, and displacement with azide gave **D**, which upon ozonolytic cleavage afforded the diketone **E**. Conversion to the bis-TMS-enol ether, followed by bromination, and intramolecular aldol condensation on silica gel gave the cyclopentane **F**, which was converted to cyclic sulfate **G** prior to displacement with chloride ion to ultimately give **H**. Elaboration of the spirocyclic bis-*N*-Boc guanidine and amide formation led to the intended fully functionalized chlorocyclopentane core **I** of ring *A*. The classical stepwise construction of this subunit in a regio- and stereocontrolled manner from a carbo-cyclic precursor harboring the *trans*-bis(azidomethyl) groups differs from previous

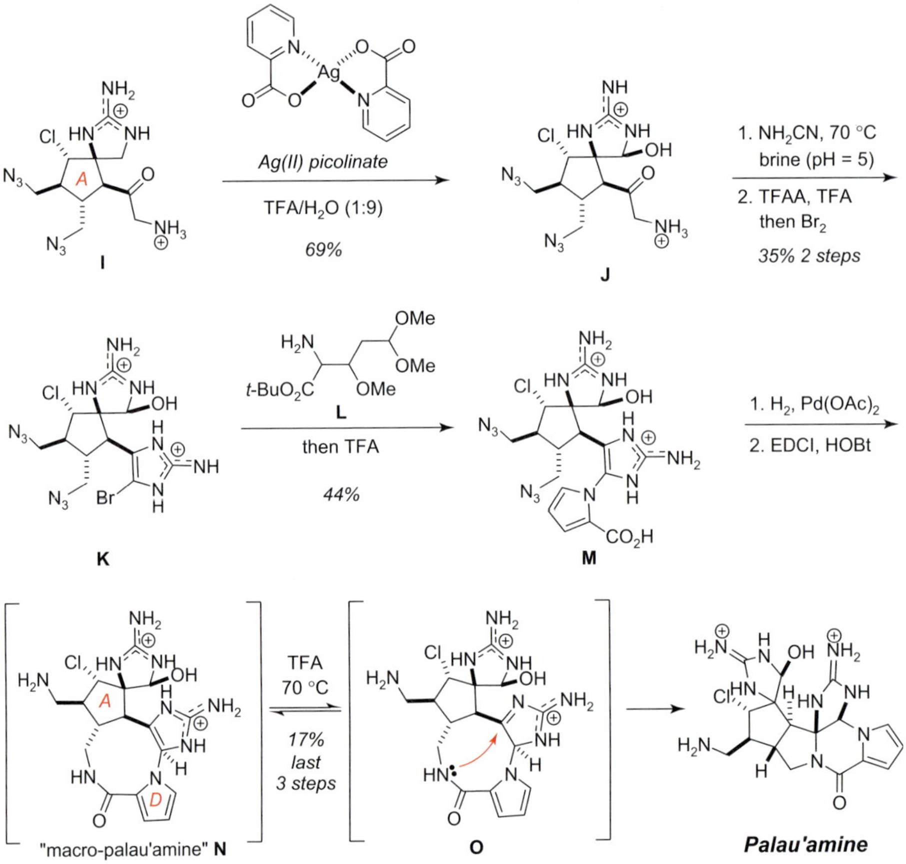

Figure 18.12 Completion of the total synthesis of palau'amine by Baran.

approaches [70]. The choice of an azide group as a latent amine was astutely planned to withstand multiple operations, before it was reduced to the desired amine in the penultimate step of the synthesis. The steps needed to complete the first total synthesis of palau'amine from advanced intermediate **I**, are shown in Figure 18.12.

Completion of the synthesis of palau'amie commenced with a stereoselective oxidation of **I** with Ag(II) picolinate to give gave the α-hydroxy guanidinium salt **J** (Figure 18.12). Elaboration of the second cyclic guanidine, followed by bromination led to **K**, which was converted to **M** by means of an acyclic pyrrole ring precursor **L**. An elegant solution to the construction of the hexacyclic ring system was found in the conversion of **M** to the macrolactam **N**, which upon exposure to acid underwent an irreversible ring contraction via **O**, to deliver palau'amine.

18.9.1.3 **Minfiensine** – A recent nine-step enantioselective total synthesis of the complex alkaloid minfiensine by MacMillan and coworkers [74] illustrates the power of an iminium-based cascade sequence as a key design element, which has become a highly successful strategy in the realm of metal-free, organocatalytic, bond-forming reactions (Figure 18.13, see also, Chapter 5, section 5.8). A comparison with previous syntheses is noteworthy [75].

Intermediate **B**, readily available from *N*-Boc tryptamine (**A**), was engaged in an intermolecular asymmetric Diels-Alder reaction catalyzed by imidazolidinone **C** to give the bridged tetracycle **D** through a series of sequential reactions (see relevant mechanism, Figure 18.13). Conversion to the allene **E** involved protection of the primary alcohol as the TES ether, reductive amination with 4-(*tert*-butylthio)but-2-ynal, followed by 6-*exo*-dig radical cyclization using t-Bu$_3$SnH. Regio- and diastereoselctive hydrogenation of the allene unit, followed by deprotection gave minfiensine.

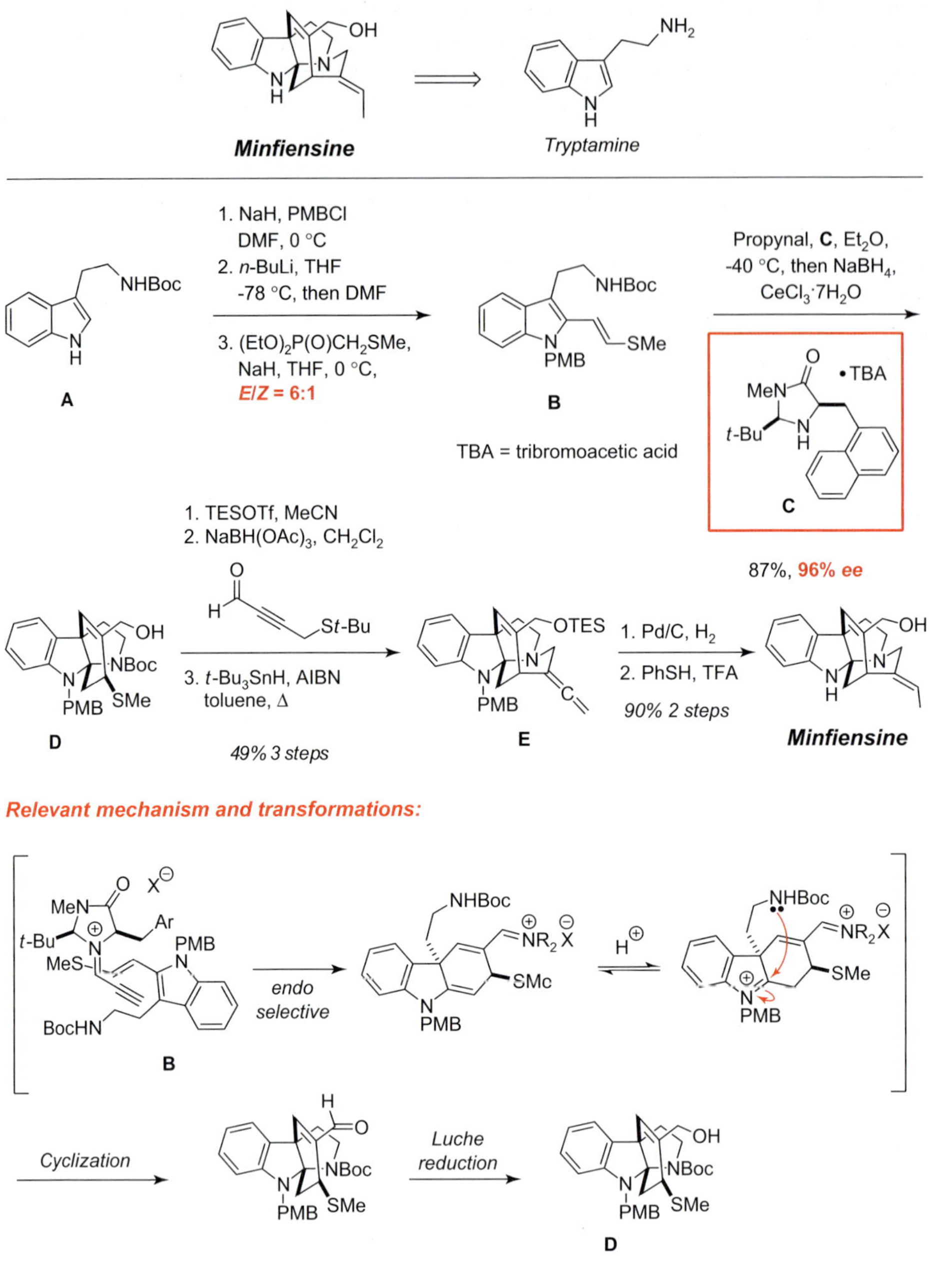

Minfiensine   **Tryptamine**

A

1. NaH, PMBCl
   DMF, 0 °C
2. *n*-BuLi, THF
   -78 °C, then DMF
3. (EtO)$_2$P(O)CH$_2$SMe,
   NaH, THF, 0 °C,
   E/Z = 6:1

B

TBA = tribromoacetic acid

Propynal, **C**, Et$_2$O,
-40 °C, then NaBH$_4$,
CeCl$_3$·7H$_2$O

C

87%, **96% ee**

D

1. TESOTf, MeCN
2. NaBH(OAc)$_3$, CH$_2$Cl$_2$

3. *t*-Bu$_3$SnH, AIBN
   toluene, Δ

49% 3 steps

E

1. Pd/C, H$_2$
2. PhSH, TFA

90% 2 steps

Minfiensine

Relevant mechanism and transformations:

B

endo selective

Cyclization

Luche reduction

D

Figure 18.13 Catalytic *endo*-[4+2] Diels-Alder cascade sequence leading to minfiensine by MacMillan.

18.9.1.4 **Maoecrystal Z** – The last synthesis entry to be discussed in this book is a 2011 contribution from Reisman and coworkers [76]. Continuing with the theme of cascade reactions for the rapid assembly of polycyclic natural products, this first enantioselective synthesis of the maoecrystal Z exemplifies the power and utility of single-electron processes for the construction of quaternary stereogenic centers and compact ring systems.

The synthesis of the tetracyclic natural product commences with the preparation of a chiral alkyl iodide. Pivaloylation of pent-4-enoic acid (**A**), followed by treatment with (*S,S*)-pseudoephedirne, and subsequent asymmetric alkylation of the Myers amide furnished **B** in good overall yield (Figure 18.14). A two-step procedure involving reductive cleavage of the auxiliary using $BH_3 \cdot NH_3$ complex and iodination gave **C** in 85% yield over two steps. From here, the readily available chiron (−)-γ-cyclogeraniol [77] (**D**) was protected as the TBS ether and then subjected to a diastereoselective epoxidation in the presence of *m*-CPBA to furnish **E** (dr = 3:1). Securing one of the two quaternary stereogenic centers present in the target molecule was achieved early on in the synthesis of maoecrystal Z by employing a radical-mediated reductive opening of the spiroepoxide function in **E**. Initial forays into this reductive coupling approach using methyl acrylate as a radical acceptor, resulted in the formation of the desired spirolactone intermediate, however, a considerable amount of an allylic alcohol byproduct was formed, presumably from a Lewis acid-mediated rearrangement of the epoxide. Using a more electrophilic radical coupling partner in the form of 2,2,2-trifluoroethylacrylate brought about a 74% yield of **F**, which was isolated as a single diastereomer. Treatment of a mixture of **F** and **C** with LiHMDS in THF gave the corresponding α-alkylated lactone as an inconsequential 1:1 mixture of diastereomers, which was directly subjected to a selenation/oxidation protocol to give **G**. Cleavage of both TBS ethers and subsequent oxidation to the dialdehyde **H**, set the stage for the pivotal radical cascade reaction. In the event, exposure of **H** to a $SmI_2/LiBr$ single-electron reductive system enabled the formation of tetracycle **I**, and four stereogenic centers in one synthetic operation. What is even more remarkable about this cascade event is that **I** was formed as a single diastereomer owing to the preference of the samarium ketyl to adopt a pseudoaxial orientation to minimize non-bonded interactions with the methylene unit of the spirolactone. Furthermore, potential and unfavorable 1,3-diaxal interactions that could arise from this orientation of the samarium ketyl are minimized by virtue of C-7 being sp^2-hybridized. Completion of the synthesis involved bis-acetylation of **I**, followed by oxidative cleavage of the olefin, and formation of the enal **J** upon treatment with Eschenmoser's salt. Chemoselective deacetylation of **J** furnished maoecrystal Z in just 12 steps for the longest linear sequence.

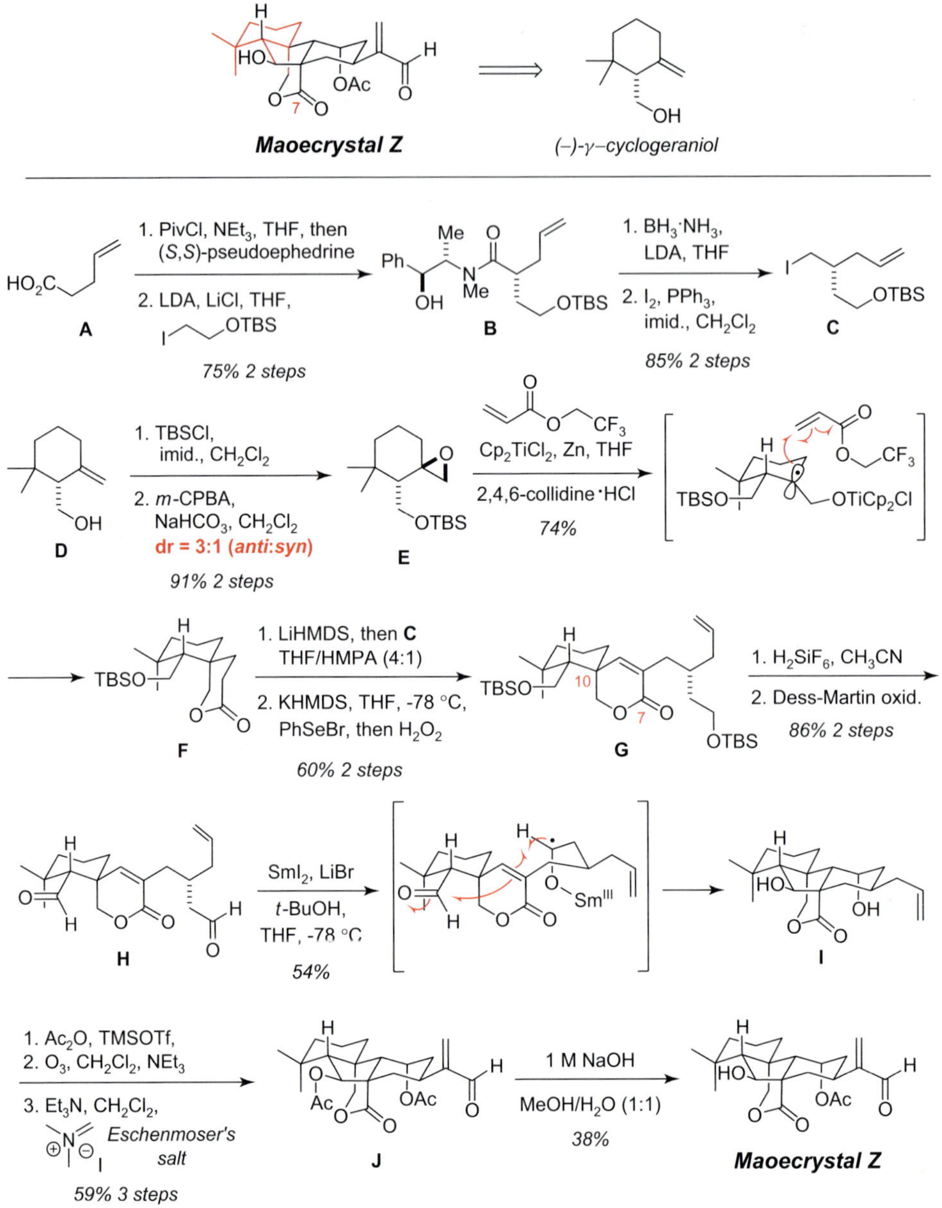

Maoecrystal Z   (–)-γ–cyclogeraniol

Figure 18.14 Synthesis of maoecrystal Z using a radical-induced cascade approach by Reisman.

18.9.2
Parting thoughts

Applying the principles of the *chiron approach* independently, and in conjunction with catalytic or stoichiometric asymmetric methods, provide powerful tools that combine the predictive power of securing substructures of intended target molecules harboring a full complement of functionality and stereochemistry, with the practicality of expedient stereocontrolled bond-forming reactions. Thus, Trost and Dong's total synthesis of bryostatin 16 (see Figure 18.2) [24], demonstrated the full impact of transition metal-based catalysis in the elaboration of the carbon framework of such a complex natural product. A single stereogenic center provided by (S)-lactaldehyde was used by Nicolaou and coworkers [46] to construct the complex carbon framework of BE-42472B by application of a highly regio- and stereoselective intermolecular Diels-Alder reaction (see Figure 18.5). Gin and coworkers [48] assembled the core structure of crambidine starting with (S)-pyroglutamic acid, in conjunction with the use of catalytic asymmetric reduction reactions of ketone intermediates to access various appendages bearing secondary hydroxyl groups (Figure 18.6). Correction of structure and a total synthesis of vannusal B was accomplished by Nicolaou and coworkers [49] (Figure 18.7). Visual *reflexive* and *relational* analysis, combined with efficiency, led to the total synthesis of actinophyllic acid by Overman and coworkers [66] (Figure 18.8). The total synthesis of the topologically and structurally intricate alkaloid himandrine by Movassaghi and coworkers [69], commenced with two enantiopure chirons that set the stage for subsequent stereocontrolled, and proximity-induced steps in the construction of the hexacyclic target molecule (see Figures 18.9 and 18.10). Being inspired by biosynthetic principles, Baran and coworkers [73] demonstrated creativity and courage in their first total synthesis of palau'amine (see Figures 18.11 and 18.12). MacMillan and coworkers [74] continue to show the virtues of enantioselective catalysis using iminium ions as reactive intermediates exemplified by the efficient and stereocontrolled total synthesis of minfiensine [75] (see Figure 18.13). The *chiron approach* coupled with the application of key radical and radical cascade reactions in the synthesis of maoecrystal Z by Reisman and coworkers [76], is exemplary of the creative flare that exists in a new generation of synthetic chemists.

18.10
A Salute to the Vanguards of Synthesis

Already a decade into the third millennium, and well into its second century of existence, organic synthesis continues to be the engine that fuels the everyday amenities we enjoy as human beings. It is safe to assume that the practice of synthesis as we presently know it, *i.e.*, a reagents and reactions-based endeavor, will continue to thrive and to be in use for the foreseeable future.

There is much exploration, learning and understanding of the basic nature of the chemistry world before we can speak of organic synthesis as a *mature* field. Nevertheless, progress made in the span of several decades has been phenomenal. What seems to be unravelling before our own eyes is a "synthesis revolution" much like the industrial revolution at the turn of the last century. The enabling aspect of synthesis creates unlimited opportunities for those who practice it, in search of new challenges to conquer, uncharted chemical space to explore, and new horizons to reach [78].

Lastly, we salute the vanguards of the field of organic chemistry who have left us with the richest of legacies. Hopefully, future generations will look upon present-day contributions with the same sense of respect and approval as we do of our mentors and predecessors. We wish them well as we close with a quote from Sir Isaac Newton [79]:

"If I have seen further, it is by standing on the shoulders of giants."

References

1. For examples of natural products of the early part of the 20th century, see: Nicolaou, K. C.; Vourloumis, D.; Winssinger, N.; Baran, P. S. *Angew. Chem. Int. Ed.* 2000, **39**, 44; see also: Chapter 1, reference 7.

2. For a brief summary, see: Kauffman, G. B. *Chem. Educator* 2004, **9**, 172; see also: *Robert Burns Woodward, Architect and Artist in the World of Molecules*, Benfey, O. T.; Morris, P. J. T. Eds.; Chemical Heritage Foundation, Philadelphia, PA, 2001.

3. For example, see: Nobel Lecture, Corey, E. J. *Angew. Chem. Int.* 1991, **30**, 455; see also: *The Logic of Chemical Synthesis*, Corey, E. J.; Cheng, X.-M., 1989, Wiley, NY.

4. For example, see: (a) *Encyclopaedia of Reagents for Organic Synthesis*, Crich, D. Ed.; 2010, Wiley-Interscience, NY; (b) *Comprehensive Organic Name Reactions and Reagents*, Vol. 1-3, Wang, Z. 2009, Wiley, NY; (c) *Name reactions: A collection of Detailed Mechanisms and Synthetic Applications*, Li. J. J. 2009 4th edition, Springer Verlag, Berlin; (d) *Strategic Applications of Named Reactions in Organic Synthesis: Background and Detailed Mechanisms*, Kürti, L.; Czakó, B. 2005, Elsevier Science, NY.

5. Surrendra, K.; Corey, E. J. *J. Am. Chem. Soc.* 2009, **131**, 13928.

6. For a retrospective, see: Hoffman, F. *Aldrichimica Acta* 1982, **15**, 3.

7. (a) Stork, G.; Yamashita, A.; Adams, J.; Schulte, G. R.; Chesworth, R.; Miyazaki, Y.; Farmer, J. J. *J. Am. Chem. Soc.* 2009, **131**, 11402; for a review, see: (b) Zezula, J.; Hudlicky, T. *Synlett* 2005, 388; for the use of catalytic reactions, see: (c) Trost, B. M.; Tang, W.; Toste, D. F. *J. Am. Chem. Soc.* 2005, **127**, 14785.

8. For example, see: (a) *Radicals in Synthesis I and II, in Topics in Current Chemistry*, Gansäuer, A., Ed., 2006, Vol. 263, 264; Springer, Berlin; (b) *Radicals in Organic Synthesis*, Renaud, P.; Sibi, M. P., 2001, Wiley-VCH, Weinheim; (c) *Advances in Free Radical Chemistry*, Zard, S. Z., Ed.; 2000, Elsevier Science, NY; (d) *An Introduction to Free Radical Chemistry*, Parson, A. F., 2000, Wiley-Blackwell, NY; (e) *Free Radical Chain Reactions in Organic Synthesis*, Motherwell, W. B.; Crich, D., 1991, Academic Press, London; (f) *Radicals in Organic Synthesis: Formation of Carbon-Carbon Bonds*, Giese, B., 1986, Pergamon, Oxford; see also: Jasperse,

C. P.; Curran, D. P.; Ferig, T. L. *Chem. Rev.* 1991, **91**, 1237.

9. For example, see: (a) *The Chemistry of Enamines*, Rapoport, Z., Ed.; 1994, Wiley, NY; (b) *Enamines*, Cook, Ed.; 1987, CRC Press, Boca Raton, FL.

10. For example, see: (a) *Encyclopaedia of Catalysis*, Horvath, I. T., Ed.; Wiley-Interscience, NY, 2003; (b) *Handbook of Organopalladium Chemistry for Organic Synthesis*, Negishi, E.-i. Wiley-Interscience, NY, 2002; (c) *Comprehensive Asymmetric Catalysis*, Jacobsen, E. N.; Pfaltz, A.; Yamamoto, H., Eds.; 2004, Springer, Heidelberg; (d) Hartwig, J. F. *Nature* 2008, **455**, 314; (e) Dick, A. R.; Sanford, M. S. *Tetrahedron* 2006, **62**, 2439; (f) Muci, A. R.; Buchwald, S. L. *Top. Curr. Chem.* 2002, **219**, 131; for selected examples in natural products synthesis, see: (g) Mohr, J. T.; Krout, M. R.; Stoltz, B. M. *Nature* 2008, **455**, 323; (h) Nicolaou, K. C.; Bulger, D. G. Jr.; Sarlah, D. *Angew. Chem. Int. Ed.* 2005, **44**, 4442; see also: Chapter 2, reference 108; Chapter 4, references [42, 50, 51]; Chapter 5, reference [42].

11. For example, see: (a) Muller, F. L.; Latimer, J. M. *Comput. Chem. Eng.* 2009, **33**, 1051; (b) Carey, J. S.; Laffan, D. Thompson, C.; Williams, M. T. *Org. Biomol. Chem.* 2006, **4**, 2337; see also: Chapter 2, references [27–29].

12. For example, see: Katsuki, T. in *Comprehensive Asymmetric Catalysis*, Jacobsen, E. N.; Pfaltz, A.; Yamamoto, H., Eds.; 1999, Vol. II, p. 621, Springer-Verlag, Berlin; see also: Chapter 2, reference 98.

13. For example, see: (a) Jacobsen, E. N.; Wu, M. H. in *Comprehensive Asymmetric Catalysis*, Jacobsen, E. N.; Pfaltz, A.; Yamamoto, H., Eds.; 1999, Vol. II, p. 649, Springer-Verlag, Berlin; (b) Wang, Z.-X.; Tu, Y.; Frohn, M.; Zhang, J.-R.; Shi, Y. *J. Am. Chem. Soc.* 1997, **119**, 11224.

14. For example, see: Brown, J. M. in *Comprehensive Asymmetric Catalysis*, Jacobsen, E. N.; Pfaltz, A.; Yamamoto, H., Eds.; 1999, Vol. I, p. 121, Springer-Verlag, Berlin.

15. (a) Vineyard, B. D.; Knowles, W. S.; Sabacky, M. J.; Bachman, G. L.; Weinkauff, D. J. *J. Am. Chem. Soc.* 1977, **99**, 5946; (b) Knowles, W. S. *Angew. Chem. Int. Ed.* 2002, **41**, 1998.

16. For example, see: (a) Noyori, R. *Acc. Chem. Res.* 1990, **23**, 345; (b) Noyori, R. in *Catalytic Asymmetric Synthesis*, Ojima, I., Ed.; 1993, p. 1, Wiley-VCH, NY.

17. Corey, E. J.; Bakshi, R. K.; Shibata, S. *J. Am. Chem. Soc.* 1987, **109**, 5551; see also: Xavier, L. C.; Mohan, J. J.; Mathre, D. J.; Thompson, A. S.; Carroll, J. D.; Corley, E. G.; Desmond, R. *Org. Synth.* 1996, **74**, 50 and Chapter 2, reference [103].

18. For example, see: *Handbook of Metathesis*, Grubbs, R. H. Ed.; 2003, Vol. 1-3, Wiley-VCH, Weinheim.

19. For example, see: Schrock, R. R. *Tetrahedron* 1999, **55**, 8141; Schrock, R. R. *J. Mol. Catal. A: Chemical* 2004, **213**, 21.

20. For example, see: Garber, S. B.; Kingsbury, J. S.; Gray, B. L.; Hoveyda, A. H. *J. Am. Chem. Soc.* 2000, **122**, 8168; (b) Chatterjee, A. K.; Choi, T.-L.; Sanders, D. P.; Grubbs, R. H. *J. Am. Chem. Soc.* 2003, **125**, 11360; for selected applications to natural products synthesis, see: (c) Hoveyda, A. H.; Malcolmson, S. J.; Meek., S. J.; Zhugralin, A. R. *Angew. Chem. Int. Ed.* 2010, **49**, 34.

21. For example, see: *Chem. Rev.* 2006, **106**, 2581. Lipton, M. F.; Barrett, A. G. Guest editors for a thematic issue; see also: Chapter 2, references [27–29]; Chapter 5, reference [41].

22. For example, see: Hinman, A.; Du Bois, J. *J. Am. Chem. Soc.* 2003, **125**, 11510.

23. For example, see: (a) Trost, B. M.; Toste, F. D.; Pinkerton, A. B. *Chem. Rev.* 2001, **101**, 2067; (b) Boudier, A.; Bromm, L. O.; Lotz, M.; Knochel, P. *Angew. Chem. Int. Ed.* 2000, **39**, 4414.

24. (a) Trost, B. M.; Dong, G. *Nature* 2008, **456**, 485; (b) Trost, B. M.; Dong, G. *J. Am. Chem. Soc.* 2010, **132**, 16403; for a discussion, see: Miller, A. K. *Angew. Chem. Int. Ed.* 2009, **48**, 3221; for a recent review of the bryostatins, see: Hale, K. J.; Manaviazar, S. *Chem. Asian J.* 2010, **5**, 704.

25. For recent relevant monographs and reviews, see: (a) *Organocatalysis*, Reetz, M.; List, B.; Jaroch, S. Weinmann, H., Eds.; 2008, Springer, Heidelberg; (b) *Enantioselective Organocatalysis*, Dalko, P. I., Ed.; 2007, Wiley-VCH, Weinheim; (c) *Asymmetric Organocatalysis: From Biomimetic Concepts to Applications in Asymmetric Synthesis*, Berkessel, A.; Groger, H. 2005, Wiley-VCH, Weinheim; (d) Dondoni, A.; Massi, A. *Angew. Chem. Int. Ed.* 2008, **47**, 4638.; (e) Melchiorre, P. Marigo, M.; Carlone, A.; Bartoli, G. *Angew. Chem. Int. Ed.* 2008, **47**, 6138; see also: Chapter 5, references [46 d-h]; Editorial by L. S. Hegedus in *J. Am. Chem. Soc.* 2009, **131**, 17995

26. (a) Hajos, Z. G.; Parrish, D. R. *J. Org. Chem.* 1974, **39**, 1615; (b) Eder, U.; Sauer, G.; Weichert, R. *Angew. Chem. Int. Ed.* 1971, **10**, 496; for mechanistic insights see: (c) Clemente, F.; Houk, K. N. *Angew. Chem. Int. Ed.* 2004, **43**, 5766; (d) Seebach, D.; Beck, A. K.; Badine, D. M.; Limbsbach, M.; Eschenmoser, A.; Treasurywala, A. M.; Hobi, R. *Helv. Chim. Acta* 2007, **90**, 425; (e) Klussmann, M.; White, A. J. P.; Iwamura, H.; Wells, D. H. Jr.; Armstrong, A.; Blackmond, D. G. *Angew. Chem. Int. Ed.* 2006, **45**, 7989.

27. List, B.; Lerner, R. A.; Barbas, C. F. III. *J. Am. Chem. Soc.* 2000, **122**, 2395; see also: Barbas, C. F. III. *Angew. Chem. Int. Ed.* 2008, **47**, 42; Mukherjee, S.; Yang, J. W.; Hoffmann, S.; List, B. *Chem. Rev.* 2007, **107**, 5471

28. Ahrendt, K. A.; Borths, C. J.; MacMillan, D. W. C. *J. Am. Chem. Soc.* 2000, **122**, 4243.

29. For a recent review, see: MacMillan, D. W. C. *Nature* 2008, **455**, 304.

30. For example, see: (a) Bronson, J.; Dhar, M.; Ewing, W.; Lonberg, N. *Ann. Rep. Med. Chem.* 2012, **47**, 499; (b) Liu, K. K.-C.; Sakya, S. M.; O'Donnell, C. J.; Flick, A. C.; Li, J. *Bioorg. Med. Chem.* 2011, **19**, 1136; (c) Munoz, B. H. *Nat. Rev. Drug Discov.*, 2009, **8**, 959; (d) Paul, S. M.; Mytelka, D. S.; Dunwiddie, C. T.; Persingh, C. C.; Munos, B. H.; Lindborg, S. R.; Schacht, A. L. *Nat. Rev. Drug Discov.*, 2010, **9**, 2035; (e)

Wild, H.; Heimbach, D.; Huwe, C. *Angew. Chem. Int. Ed.*, 2011, **50**, 7452; (f) Nadin, A.; Hattotuwagama, C.; Churcher, I. *Angew. Chem. Int. Ed.*, 2012, **51**, 1114.

31. (a) Hendrickson, J. B. *J. Am. Chem. Soc.* 1975, **97**, 5784; (b) Hendrickson, J. B. *J. Am. Chem. Soc.* 1971, **93**, 6847; see also: (c) Gaich, T.; Baran, P. S. *J. Org. Chem.* 2010, **75**, 4657.

32. Robinson, R. *J. Chem. Soc.* 1917, 762.

33. For example, see: (a) Trost, B. M. *Science* 1991, **254**, 1471; (b) Trost, B. M. *Angew. Chem. Int. Ed.* 1995, **37**, 259.

34. For example, see: (a) Wender, P. A.; Hardy, S.; Wright, D. L. *Chem. Ind.* 1997, 767; (b) Wender, P. A.; Verma, V. A.; Paxton, T. J.; Pillow, T. H. *Acc. Chem. Res.* 2008, **41**, 40.

35. Burns, N. Z.; Baran, P. S.; Hoffmann, R. W. *Angew. Chem. Int. Ed.* 2009, **48**, 2854.

36. (a) *Protective groups in Organic Synthesis* Greene, T.; Wuts, P.W.G. 4th Ed. Wiley, NY, 1999; (b) Kocienski, P. *Protecting Groups*, Thieme, Stuttgart, 1994.

37. Armstrong, R. W.; Beau, J.-M.; Cheon, S. H.; Christ, W. J.; Fujioka, H.; Ham, W.-H.; Hawkins, L. D.; Jin, H.; Kang, S. H.; Kishi, Y.; Martinelli, M. J.; McWhorter, W. W. Jr.; Mizuno, M.; Nakata, M.; Stutz, A. E.; Talamas, F. X.; Taniguchi, M.; Tino, J. A.; Ueda, K.; Uenishi, J.; White, J. B.; Yonaga, M. *J. Am. Chem. Soc.* 1989, **111**, 7525; for an essay, see: List, B. *Angew. Chem. Int. Ed.* 2010, **49**, 1730.

38. For example, see: Walji, A. M.; MacMillan, D. W. C. *Synlett* 2007, 1477.

39. For example, see: (a) Roulland, E. *Angew. Chem. Int. Ed.* 2011, **50**, 1226; (b) Young, I. S.; Baran, P. S. *Nature Chem.* 2009, **1**, 193; (c) Hoffmann, R. W. *Synthesis* 2006, 3531; (d) Koert, U. *Angew. Chem. Int. Ed.* 1995, **34**, 1370.

40. For example, see: (a) Lou, B.; Reddy, G. V.; Wang, H.; Hanessian, S. in *Preparative Carbohydrate Chemistry*, Hanessian S., Ed.; Chapter 17, p. 389, Dekker, NY, 1997; (b) Hanessian, S.; Mascitti, V.; Lu, P.-P.; Ishida, H. *Synthesis* 2002, 1959; (c) Hanessian, S.;

Lu, P.-P.; Ishida, H. *J. Am. Chem. Soc.* 1998, **120**, 13296.

41. For example, see: (a) Lim, H. J.; Smith, C. R.; Rajanbabu, T. V. *J. Org. Chem.* 2009, **74**, 4565; (b) Grotjahn, D. B.; Larsen, C. R.; Gustafsson, I. L.; Nair, R.; Sharma, A. *J. Am. Chem. Soc.* 2007, **129**, 9592; (c) Hanessian, S.; Giroux, S.; Larsson, A. *Org. Lett.* 2006, **8**, 5481.

42. For example, see: Stang, E. M. White, M. C. *Nature Chem.* 2009, **1**, 547, and references cited therein.

43. For example, see: Hanessian, S. *Chem. Med. Chem.* 2006, **1**, 1300.

44. *Dead Ends and Detours: Direct Ways to Successful Total Synthesis*, Sierra, M. A.; de la Torre, M. C. 2004, Wiley, NY.

45. The term "Chirale Ökonomie" was first used by A. Fischli. *Chimia* 1976, **30**, 4.

46. Nicolaou, K. C.; Becker, J.; Lim, Y. H.; Lemire, A.; Neubauer, T.; Montero, A. *J. Am. Chem. Soc.* 2009, **131**, 14812.

47. For example, see: Overman, L. E.; Rhee, Y. H. *J. Am. Chem. Soc.* 2005, **127**, 15652, and previous papers.

48. Perl, N. R.; Ide, N. D.; Prajapati, S.; Perfect, H. H.; Durón, S. G.; Gin, D. Y. *J. Am. Chem. Soc.* 2010, **132**, 1802.

49. (a) Nicolaou, K. C.; Ortiz, A.; Zhong, H. *Angew. Chem. Int. Ed.* 2009, **48**, 5648; Nicolaou, K. C.; Zhang, H.; Ortiz, A.; Dagneau, P. *Angew. Chem. Int. Ed.* 2008, **47**, 8605.

50. For example, see: Nie, J.; Guo, H.-C.; Cahard, D.; Ma, J.-A. *Chem. Rev.* 2011, **111**, 457; Trost, B. M.; Jiang, C. *Synthesis* 2006, 369; *Quaternary Stereocenters: Challenges and Solutions for Organic Synthesis.* Christoffers, J.; Baro, A., Eds.; 2005, Wiley-VCH, Weinheim; Douglas, C. J.; Overman, L. E. *Proc. Natl. Acad. Sci. USA* 2004, **101**, 5363; Denissova, I.; Barriault, L. *Tetrahedron* 2003, **59**, 10105; Corey, E. J.; Guzman-Perez, A. *Angew. Chem. Int. Ed.* 1998, **37**, 388; Fuji, K. *Chem. Rev.* 1993, **93**, 2037.

51. For example, see: Service, R. F. *Science* 1999, **285**, 5425.

52. (a) Woodward, R. B.; Hoffmann, R. *J. Am. Chem. Soc.* 1965, **87**, 395; for historical insights, see: Corey, E. J. *J. Org. Chem.* 2004, **69**, 2917; Hoffmann, R. *Angew. Chem. Int. Ed.* 2004, **48**, 6586.

53. Baldwin, J. E. *J. Chem. Soc., Chem. Commun.* 1976, 734. Baldwin, J. E.; Cutting, J.; Dupoint, W.; Kruse, L.; Silberman, L.; Thomas, R. C. *J. Chem. Soc., Chem. Commun.* 1976, 736.

54. Other rules have been a part of organic chemistry for generations, and can be found in general textbooks. For example, see among others: Cram, Markovnikov, Saytzeff and Fürst-Plattner rules.

55. For example, see: (a) Noyori, R. *Nature Chem.* 2009, **1**, 5; (b) Wender, P. A.; Miller, B. L. *Nature* 2009, **460**, 197; (c) Sanderson, K. *Nature* 2007, **448**, 630; (d) Kundig, P. *Science* 2006, **314**, 430; (e) Seebach, D. *Angew. Chem. Int. Ed.* 1990, **29**, 1209; (f) Whitesides, G. M. *Angew. Chem. Int. Ed.* 1990, **29**, 1209.

56. For example, see: (a) Nielsen, T. E.; Schreiber, S. L. *Angew. Chem. Int. Ed.* 2008, **47**, 48; (b) Burke, M. D.; Berger, E. M.; Schreiber, S. L. *Science* 2003, **302**, 613; (b) Burke, M. D.; Berger, E. M.; Schreiber, S. L. *Science* 2003, **302**, 613; (c) Schreiber, S. L. *Science* 2000, **287**, 1964; see also: (d) Galloway, W. R. J. D.; Diaz-Gavillon, M.; Isidro-Llobet, A.; Spring, D. R. *Angew. Chem. Int. Ed.* 2009, **48**, 2.

57. For recent reviews, see: (a) Toure, B. B.; Hall, D. G. *Chem. Rev.* 2009, **109**, 4439; (b) Shindoh, N.; Takemoto, Y.; Takatsu, K. *Chem. Eur. J.* 2009, **15**, 12168; (c) Chapman, C. J.; Frost, C. G. *Synthesis* 2007, 1; (d) Nicolaou, K. C.; Edmonds, D. J.; Bulger, P. G. *Angew. Chem. Int. Ed.* 2006, **45**, 7134; (e) Pellissier, H. *Tetrahedron* 2006, **62**, 1619; Pellissier, H. 2006, **62**, 2143; (f) *Domino Reactions in Organic Synthesis*, Tietze, L. F.; Brasche, G.; Gericke, K. M. 2006, Wiley-VCH, Weinheim; (g) Guo, H.-C.; Ma, J.-A. *Angew. Chem. Int. Ed.* 2006, **45**, 354; (h) Wasilke, J.-C.; Obrey, S. J.; Baker, R. T; Bazan, G. C. *Chem. Rev.* 2005, **105**, 1001; (i) Ramón, D. J. *Angew., Chem. Int. Ed.* 2005, **44**, 1602; (j) *Multicomponent Reactions*, Zhu, J.; Benaymé, H., Eds.; 2005, Wiley-VCH, Weinheim; (k) Tietze, L.- F. *Chem. Rev.* 1996, **96**, 115.

58. For example, see: (a) Strebhardt, K.; Ulrich, A. *Nat. Rev. Cancer* 2008, **8**, 473; (b) Tanabe, K.; Zhang, Z.; Ito, T.;

Hatta, H.; Nishimoto, S.-i. *Org. Biomol. Chem.* 2007, **5**, 3745.

59. For example, see: (a) Szpilman, A. M.; Carreira, E. M. *Angew. Chem. Int. Ed.* 2010, **50**, 9592;   (b) Bade, R.; Chan, H.-F.; Reynisson, J. *Eur. J. Med. Chem.* 2010, **45**, 5646;   (b) Kumar, K.; Waldmann, H. *Angew. Chem. Int. Ed.* 2009, **48**, 3224;   (c) Morton, D.; Leach, S.; Cordier, C.; Warriner, S.; Nelson, A. *Angew. Chem. Int. Ed.* 2009, **48**, 104; (d) Baker, D. D.; Chu, M.; Oza, U.; Rajgarhia, V. *Nat. Prod. Rep.* 2007, **24**, 1255;   (e) Chiu, Y.-W.; Balunas, M. J.; Chai, H. B.; Kinghorn, H. D. *AAPS J.* 2006, **8**, E239;   (f) Wilson, R. M.; Danishefsky, S. J. *J. Org. Chem.* 2006, **71**, 8329;   (g) Koehn, F. E.; Carter, G. T. *Nat. Rev. Drug Discov.* 2005, **4**, 206;   (h) Butler, M. S. *Nat. Prod. Rep.* 2005, **22**, 162;   (i) *Contemporary Drug Synthesis*, Li, J. J.; Johnson, D. S.; Silskovic, D. R.; Roth, B. D., 2004, Wiley-Interscience, NJ. see also: references [43] and [63].

60. *Handbook of C– H Transformations*, Dyker, G., Ed.; 2005, Wiley-VCH, NY;   For selected reviews, see: (a) Newhouse, T.; Baran, P. S. *Angew. Int. Ed.* 2011, **50**, 3362;   (b) Chen, X.; Engle, K. M.; Wang, D.-H.; Yu, J.-Q. *Angew. Chem. Int. Ed.* 2009, **48**, 5094;   (c) Thansandote, P.; Lautens, M. *Chem. Eur. J.* 2009, **15**, 5874   (d) Bergman, R. G. *Nature* 2007, **446**, 391;   (e) Brookhart, M.; Green, M. L. H.; Parkin, G. *Proc. Natl. Acad. Sci. USA.* 2007, **104**, 6908; see also: (f) Chen, M. S.; White, M. C. *Science* 2010, **327**, 566;   (g) Chen, M. S.; White, M. C. *Science* 2007, **318**, 783; (h) Godula, K.; Sames, D. *Science* 2006, **312**, 67;   (i) Hinman, A.; Du Bois, J. *J. Am. Chem. Soc.* 2003, **125**, 11510;   (j) Chen, H.; Schlecht, S.; Semple, T. C.; Hartwig, J. F. *Science* 2000, **287**, 1995; for a thematic issue, see: (k) Crabtree, R. Guest editor, *Chem. Rev.* 2010, **110**, 575.

61. For example, see: (a) *J. Am. Chem. Soc.* Editorial, 2009, **131**, 7937;   (b) Tiam, H. *Angew. Chem. Int. Ed.* 2010, **49**, 4710.

62. For example, see: (a) Baxendale, I.; Hayward, J. J.; Lanners, S.; Ley, S. V.; Smith, C. D. in *Microreactors in Organic Synthesis and Catalysis*, Wirth, T. Ed.; Wiley-VCH, Weinheim, 2008, p. 84–122; (b) Yoshida, J.-I; Nagaki, A.; Yamada, T. *Chem. Eur. J.* 2008, **14**, 1450;   (c) Mason, B. P.; Price, K. E.; Steinbacher, J. L.; Bogdan, A. R.; McQuade, T. *Chem. Rev.* 2007, **107**, 2300;   *see also: Solid Phase Synthesis*, Burgess, K., 2000, John Wiley, NY.

63. Walsh, C. T.; Fischbach, M. A. *J. Am. Chem. Soc.* 2010, **132**, 2469; see also: Walsh, C. T. *Acc. Chem. Res.* 2008, **41**, 4.

64. For historical insights, see: (a) *Penicillin: Triumph and Tragedy*, Bud, R. 2007, Oxford University Press, USA; (b) *Penicillin Man: Alexander Fleming and the Antibiotic Revolution*, Brown, K. 2005, The History Press;   (c) *Breakthrough: The True Story of Penicillin*, Jacobs, F. 2004, iUniverse Inc.;   (d) *The First Miracle Drug: How the Sulfa Drugs Transformed Medicine*, Lesch, J. E. 2006, Oxford University Press, USA; for an excellent overview of the discovery of new medicines, see: Swinney, D. C.; Anthony, J. *Nat. Rev. Drug Discov.* 2011, **10**, 507.

65. For example see: Ehmke, V.; Heindl, C.; Rottmann, M.; Freymond, C.; Schweizer, W. B.; Brun, R.; Stich, A.; Schirmeister, T.; Diederich, F. *Chem. Med. Chem.* 2011, **6**, 273;   see also: WHO websites on infectious diseases, Weekly Epidemiological Record, 2002, No. 44, p. 372; see also: Cavalli, A.; Bolognese, M. L. *J. Med. Chem.* 2009, **52**, 7339.

66. Martin, C. L.; Overman, L. E.; Rohde, J. M. *J. Am. Chem. Soc.* 2010, **132**, 4894.

67. For examples see: (a) Richter, J. M.; Whitfield, B. W.; Maimone, T. J.; Lin, D. W.; Castroviejo, M. P.; Baran, P. S. *J. Am. Chem. Soc.* 2007, **129**, 12857;   (b) for a review of earlier seminal contributions see: Csáky, A. G. Plumet, J. *Chem. Soc. Rev.* 2001, **30**, 313.

68. Academic lineage: Movassaghi, M; Ph.D., 2001 with A. G. Myers (California Institute of Technology and Harvard University); postdoctoral, 2001-2003 with E. N. Jacobsen (Harvard University); Baran, P. S; Ph.D., 2001 with K. C.

Nicolaou (The Scripps Research Institute); postdoctoral, 2001-2003 with E. J. Corey (Harvard University); MacMillan, D. W. C; Ph.D., 1996 with L. E. Overman (University of California, Irvine); postdoctoral, 1996-1998 with D. A. Evans (Harvard University); Reisman, S. E.; Ph.D., 2006 with J. L. Wood (Yale University); postdoctoral, 2006-2008 with E. N. Jacobsen (Harvard University).

69. Movassaghi, M.; Tjandra, M.; Qi, J. *J. Am. Chem. Soc.* 2009, **131**, 9648; for a synthesis of GB-13 and GB-16, see: Zi, W.; Yu, S.; Ma, D. *Angew. Chem. Int. Ed.* 2010, **49**, 5887; for a review, see: Rinner, U.; Lentsch, C.; Aichinger, C. *Synthesis* 2010, 3763.

70. (a) Kinnel, R. B.; Gehrken, H. P.; Scheuer, P. J. *J. Am. Chem. Soc.* 1993, **115**, 3376; (b) Kinnel, R. B.; Gehrken, H. P.; Swali, R.; Skoropowski, G.; Scheuer, P. J. *J. Org. Chem.* 1998, **63**, 3281.

71. (a) Köck, M.; Grube, A.; Seiple, I. B.; Baran, P. S. *Angew. Chem. Int. Ed.* 2007, **46**, 6586; (b) Grube, A.; Köck, M. *Angew. Chem. Int. Ed.* 2007, **46**, 2320; (c) Lanman, B. A.; Overman, L. E.; Paulini, R.; White, N. S. *J. Am. Chem. Soc.* 2007, **129**, 12896; (d) Buchanan, M. S.; Carroll, A. R.; Addepalli, R.; Avery, V. M.; Hooper, J. N. A.; Quinn, R. J. *J. Org. Chem.* 2007, **72**, 2309; for a related structure (carteramine A), see: Kobayashi, H.; Kitamura, K.; Nagai, Y.; Nakuo, Y.; Fusetani, N.; Van Soest, R. W. M.; Matsunaga, S. *Tetrahedron Lett.* 2007, **48**, 2127.

72. For the synthesis of the chlorocyclopentane core subunits of palau'amine, massadine and axinellamine, see: (a) Yamaguchi, J.; Seiple, I. B.; Young, I. S.; O'Malley, D. P.; Maue, M.; Baran. P. S. *Angew. Chem. Int. Ed.* 2008, **47**, 3578; (b) Breder, A.; Chinigo, G. M.; Waltman, W. W.; Carreira, E. M. *Angew. Chem. Int. Ed.* 2008, **47**, 8514; (c) Bultman, M. S.; Ma, J.; Gin, D. Y. *Angew. Chem. Int. Ed.* 2008, **47**, 6821; (d) Hudon, J.;

Cernak, T. A.; Ashenhurst, J. A.; Gleason, J. L. *Angew. Chem. Int. Ed.* 2008, **47**, 8889; for a synthesis of the pentacyclic core of palau'amine see: (e) Feldman, K. S.; Nuriye, A. Y. *Org. Lett.* 2010, **12**, 4532.

73. (a) Seiple, I. B.; Su, S.; Young, I. S.; Nakamura, A.; Yamaguchi, J.; Jørgensen, L.; Rodriguez, R. A.; O'Malley, D. P.; Gaich, T.; Köck, M.; Baran, P. S. *J. Am. Chem. Soc.* 2011, **133**, 14710; for a synthesis of (±)-palau'amine, see: (b) Seiple, I. B.; Su, S.; Young, I. S.; Lewis, C. A.; Yamaguchi, J.; Baran, P. S. *Angew. Chem. Int. Ed.* 2010, **49**, 1095; for a discussion, see: (c) Jessen, H. J.; Gademann, K. *Angew. Chem. Int. Ed.* 2010, **49**, 2972; Sakakura, A.; Kondo, R.; Matsumura, Y.; (d) Akakura, M.; Ishihara, K. *J. Am. Chem. Soc.* 2009, **131**, 17762.

74. Jones, S. B.; Simmons. B.; MacMillan, D. W. C. *J. Am. Chem. Soc.* 2009, **131**, 13606.

75. For other total syntheses of minfiensine, see: (a) Dounay, A. B.; Overman, L. E.; Wrobleski, A. D. *J. Am. Chem. Soc.* 2005, **127**, 10186; (b) Dounay, A. B.; Humphreys, P. G.; Overman, L. E.; Wrobleski, A. D. *J. Am. Chem. Soc.* 2008, **130**, 5368; (c) Shen, L.; Zhang, M. Wu, Y.; Qin, Y. *Angew. Chem. Int. Ed.* 2008, **47**, 3618 ((±)-minfiensine).

76. Cha, J. Y.; Yeoman, J. T. S.; Reisman, S. E. *J. Am. Chem. Soc.* 2011, **133**, 14964.

77. (a) Fehr, C.; Galindo, J. *Helv. Chim. Acta* 1995, **78**, 539; (b) Tanimoto, H.; Oritani, T. *Tetrahedron* 1997, **53**, 3527.

78. For example, see: Whitesides, G. M.; Deutch, J. *Nature* 2011, **469**, 21; *Chem. Eng. News* 2010, 13.

79. A quote from Sir Isaac Newton from a letter to his rival Robert Hooke, dated February 5, 1676, adapted from the original metaphor attributed to Bernard de Chartres.

Author Index [Natural Product/Target]

Abad, A.	[Dorisenone C], 515
Adams, J.	[Omuralide], 657
Al-Abed, Y.	[Trehazolin], 387
Andrade, R. B.	[Strychnine], 77
Arimoto, H.	[Pinnaic acid], 503
Armstrong, R. W.	[Calyculin], 236
Avery, M. A.	[Artemisinin], 221, 258
Ayer, W. A.	[Lycopodine], 588
Bachmann, W. E.	[Equilenin], 75
Baggiolini, E. G.	[Biotin], 339
Baldwin, J. E.	[Lactacystin], 645
Banwell, M. G.	[Tamiflu], 555
Baran, P. S.	[Ambiguine H], 481, 725
	[Hapalindole U], 481, 725
	[Palau'amine], 188, 740, 743, 749
	[Stephacidin B], 77
	[Welwitindolinone A], 514
Barden, T. C.	[Aristoteline], 215
Barner, R.	[Vitamin E], 454
Bodwell, G. J.	[Strychnine], 77
Boeckman, R. K. Jr.	[Calcimycin], 182
	[Indanomycin], 248
Boger, D. L.	[Vancomycin], 79
Bonjoch, J.	[Strychnine], 77, 114
Bosch, J.	[Strychnine], 77, 114
Brown, E.	[Isosteganone], 286
Büchi, G.	[Catharanthine], 595
	[Vindoline], 75

Design and Strategy in Organic Synthesis: From the Chiron Approach to Catalysis, First Edition.
Stephen Hanessian, Simon Giroux, and Bradley L. Merner.
© 2013 Wiley-VCH Verlag GmbH & Co. KGaA. Published 2013 by Wiley-VCH Verlag GmbH & Co. KGaA.

Burke, S. D   [Phyllanthocin], 272
Burton, J. W.   [Elatenyne], 445
Buszek, K. R.   [Octalactin A], 438

Carreira, E. M.   [Erythronolide A], 252
                           [Trehazolin], 388
                           [Zaragozic acid C], 241, 408

Cha, J. K.   [Phorbol], 77
Chackalamannil, S.   [Himbacine], 244
Chamberlin, A. R.   [Erythronolide A intermediate], 288

Chiara, J. L.   [Trehazolamine], 384
                           [Trehazolin], 384, 388

Chida, N.   [*ent*-Actinobolin], 381
                           [FR65814], 418
                           [Lactacystin], 647

Christensen, B. G.   [Thienamycin], 217, 345
Christmann, M.   [Amaminol B], 53
Ciufolini, M. A.   [FR901483], 348
Clardy, J.   [*ent*-Octalactin A], 441
Coleman, R. S.   [9a-Desmethoxy mitomycin], 394
Collum, D. B.   [Phyllanthocin], 266

Corey, E. J.   [Aplasmomycin], 266
                           [Brefeldin A], 561
                           [Ecteinascidin 743], 77, 222
                           [Erythronolide B], 75
                           [Estrone], 139
                           [Desogestrel], 139
                           [Dysidiolide], 282
                           [Forskolin], 100, 101, 707
                           [Gibberellic acid], 75
                           [Ginkgolide B], 77
                           [Lactacystin], 642, 643
                           [Longifolene], 715
                           [Lupeol], 715
                           [Picrotoxinin], 459
                           [Platensimycin], 496
                           [Pseudopteroxazole], 515
                           [Salinosporamide A], 342
                           [Tamiflu], 28, 50, 559
                           [Thromboxane B_2], 217, 237

Crimmins, M. T. [Ginkgolide B], 77
[Laulimalide], 219
[Ophirin B], 51
[Prelaureatin], 241

Danishefsky, S. J. [Calicheamicin γ_1], 77
[9a-Desmethoxy mitomycin], 394
[Eleutherobin], 257
[Epothilone A], 60
[FK-506], 309
[Gypsetin], 372
[Migrastatin], 418
[ML236A], 563
[Peribysin E], 307, 515
[*ent*-Phomoidride B], 222, 609
[Pinnaic acid], 507
[Rapamycin], 600
[Scabronine G], 529
[Taxol], 77, 562, 673
[TMC-95-B], 372
[UCS1025A], 446
[Zincophorin], 128

De Brabander, J. K. [*ent*-Palmerolide A], 238, 418
Denmark, S. E. [Brasilenyne], 435
Deslongchamps, P. [Maritimol], 51
[Ouabain], 561
[Ryanodol], 265, 514

Doering, W. E. [Quinine], 624
Donaldson, W. A. [Ambruticin S], 178
Donohue, T. J. [Omuralide], 660

Du Bois, J. [Saxitoxin], 77, 398, 400
[Tetrodotoxin], 418

Enders, D. E. [Pectinatone], 127
Eschenmoser, A. [Colchicine], 594
[Vitamin B$_{12}$], 584, 735

Evans, D. A. [Antibiotic X-206], 80
[Azaspiracid-1], 79, 253
[Calcimycin], 181, 251
[Callipeltoside A], 132
[Cytovaricin], 79
[Elaiolide], 245

[Galbulimima alkaloid 13], 231
[Ionomycin], 125, 589
[Peloruside A], 292, 580
[Phoboxazole A], 312
[Oasomycin A], 185
[Vancomycin], 79, 591
[Zaragozic acid C], 408

Falck, J. R.   [Dihydromevinolin], 569
Fang, J.-M.   [Tamiflu], 28

Forsyth, C. J.   [Didemnone A], 561
                     [Okadaic acid], 253, 310
                     [Phorboxazole A], 312

Fujiwaraa, K.   [Prelaureatin], 418
Fukumoto, K.   [Cortisone], 106

Fukuyama, T.   [Anisatin], 266
                     [Ecteinascidin 743], 77, 222
                     [Hapalindole G], 217
                     [Leinamycin], 248
                     [*ent*-Phomoidride B], 222, 608
                     [Strychnine], 77, 118
                     [Tamiflu], 28, 559
                     [Vindoline], 225
                     [Yatekamycin], 226

Fürstner, A.   [Iejemalide B], 253, 287, 579
Gallagher, P. T.   [α-Kainic acid], 334
Ganem, B.   [Mannostatin A], 576

Gates, M.   [Morphine], 75
                     [Quinine], 630

Ghosh, A. K.   [Amphidinolide W], 185
                     [Laulimalide], 219
                     [Platensimycin], 488
                     [Prezista], 138

Giese, B.   [Trehazolamine], 384
                     [Trehazolin], 384

Gin, D. Y.   [Crambidine], 728
                     [Deoxyharringtonine], 223

Gribble, G. W.   [Aristoteline], 215
Grieco, P. A.   [Calcimycin], 181
                     [Tylonolide hemiacetal], 124, 275

Grierson, D. S.   [Esperamicinone], 281
Hagiwara, H   [Chapecoderin A], 533
Halcomb, R. L.   [Phomactin A], 500

Hanessian, S.   [A-315675], 86, 372
                      [Aeruginosin 205B], 40
                      [Ajmalicine], 377
                      [Ambruticin S], 178
                      [Amphotericin B intermediate], 288
                      [Avermectin B$_{1A}$], 175
                      [Bafilomycin A$_1$], 61
                      [Borrelidin], 251, 612
                      [Chlorodysinosin A], 86, 361
                      [*ent*-Cyclizidine], 364
                      [Dihydromevinolin], 15, 569
                      [Dysinosin A], 86, 361
                      [Erythronolide A intermediate], 240, 418
                      [Ionomycin], 589
                      [Jerangolid A], 449
                      [Longicin], 579
                      [Manassantin A], 455
                      [Oidiolactone A], 562
                      [Oscillarin], 40, 86, 361
                      [Pactamycin], 367
                      [Palitantin], 279
                      [Polyoxin A], 39
                      [Reserpine], 542
                      [Thienamycin], 236
                      [Thromboxane B$_2$], 217, 237

Hart, D. J.   [Himbacine], 244
Hashimoto, S.   [Zaragozic acid C], 408

Hayashi, Y.   [Nikkomycin B], 135
                      [RK-805], 135
                      [Tamiflu], 559

Hayes, C. J.   [Lactacystin], 662
Heathcock, C. H.   [Dihydromevinolin], 569
                      [Fawcettimine], 548, 551, 725
                      [Methyl homosecodaphnyllate], 130, 723
                      [Pinnaic acid], 509
                      [Spongistatin 2], 219
                      [Tantazole B], 231
                      [Zaragozic acid A], 405

Hecht, S. M.   [Bleomycin A$_2$], 597
                        [Preussin], 230

Hiemstra, H.   [*ent*-Gelsedine], 303
Hino, T.   [Tryptoquivaline], 220
Hirama, M.   [Ciguatoxin], 418
Hofheinz, W.   [Qinghaosu], 514
Holmes, A. B.   [Octalactin], 441

Holton, R. A.   [Hemibrevetoxin B], 409
                        [Taxol], 268, 515, 666

Hudlicky, T.   [Pancratistatin], 561
                        [Tamiflu], 28, 555

Hughes, P. F.   [Balanol], 231
Hutchinson, J. H.   [California red scale pheromone], 214
Ikegami, S.   [Forskolin], 101
Inoue, M.   [Ciguatoxin], 418
Inubishi, Y.   [Fawcettimine], 548

Ireland, R. E.   [FK-506], 176
                        [Lasalosid A], 264, 416
                        [Streptolic acid], 302

Isobe, M.   [Ciguatoxin], 418
                        [Okadaic acid], 310
                        [Tetrodotoxin], 77

Jacobi, P. A.   [Saxitoxin], 77, 398
Jacobsen, E. N.   [Ambruticin S], 179
                        [Colombiasin A], 50
                        [Lactacystin], 651
                        [Quinine], 631
                        [Yohimbine], 129

Johnson, F.   [Prostaglandin F$_2\alpha$], 454
Johnson, W. S.   [Progesterone], 75, 723
                        [Veratramine], 587

Kang, S. H.   [Lactacystin intermediate], 656
Kawada, K.   [PA-48153C (Pyronetin)], 237

Kende, A. S.   [Ambruticin S], 178, 264
                        [Lankacidin C], 418
                        [Neooxazolomycin], 236
                        [Stachybocin Spirolactam], 527

Kerr, M. A.   [Hapalindole Q], 136
Kishi, Y   [Calcimycin], 182
                     [*ent*-Decarbamoyl saxitoxin], 402
                     [Halichondrin B], 79
                     [Ophiobolin C], 18, 515, 707
                     [Palytoxin], 81, 724
                     [Saxitoxin], 77, 398
                     [Tetrodotoxin], 75

Kitahara, T.   [Halicholactone], 434
Knapp, S.   [Mannostatin A], 572
                     [Trehazolin], 386

Kobayashi, Y.   [Quinine], 633
Kochetkov, N. K.   [Erythronolide B], 418

Kocieński, P. J.   [Lipstatin], 454
                     [Tetronasin], 514

Koga, K.   [Megaphone], 567
Kogen, H.   [Virantmycin], 216
Komppa, G.   [Camphor], 75
Krische, M.   [7-Hydroxy quinine], 636

Kuehne, M. E.   [Kopsinine], 598
                     [Strychnine], 77, 113

Kutney, J. P.   [Catharanthine], 595
Kuwajima, I.   [Taxol], 77, 679
LaLonde, R. T.   [Deoxynupharine], 265
Lampe, J. W.   [Balanol], 231

Lee, E.   [Ambruticin S], 178
                     [Amphidinolide E], 618
                     [(3Z)-Dactomelyne], 442

Lee, H. L.   [Biotin], 339
Ley, S. V.   [Azadirachtin], 38, 79
                     [Okadaic acid], 310
                     [Thapsivillosin F], 467
                     [Trilobolide], 467

Liu, H.-J.   [Quadrone], 267

MacMillan, D. W. C. [Callipeltoside C], 135, 253
                     [Flustramine B], 134
                     [Littoralisone], 259
                     [Minfiensine], 740, 745, 749
                     [Strychnine], 77, 121

Magnus, P.   [Kopsinine], 598
[Strychnine], 77

Maier, M. E.   [Neosymbioimine], 265, 515
Marco-Contelles, J.   [Trehazolamine], 384
[Trehazolin], 384

Mariano, P. S.   [Mannostatin A], 576
Markó, I. E.   [Ambruticin S], 179
[Jerangolid D], 450

Marshall, J. A.   [*ent*-Kallolide B], 260, 515
Martin, S. F.   [Ambruticin S], 178
[Croomine], 336
[Manzamine A], 52
[Phyllanthocin], 272
[Reserpine], 542
[Solandelactone E], 238
[Strychnine], 77

Masamune, T.   [Veratramine], 587
Matsuo, K.   [Malyngolide], 281
Maycock, C. D.   [Negamycin], 280
McWilliams, J. C.   [*ent*-Octalactin A], 441
Meyers, A. I.   [Griseoviridin], 176, 427
Michelet, V.   [Ambruticin S], 179

Miyashita, M.   [Norzoanthamine], 282
[Zincophorin], 253

Money, T.   [California red scale pheromone], 214, 255

Mori, K.   [Beetle sex pheromone], 173
[Phytocassane D], 562
[Sorokinianin], 515

Mori, M.   [Cephalotaxine], 329
[Periplanone B], 266
[Strychnine], 77, 119

Mori, Y.   [Gambierol], 80, 418
[Hemibrevetoxin B], 409

Morken, J. P.   [Borrelidin], 612
Movassaghi, M   [Ditryptophenaline], 372
[Galbulimima alkaloid B], 231
[Himandrine], 740, 749
[WIN 64821], 230

Mukaiyama, T.   [Taxol], 77, 232, 682
Mulzer, J. A.   [Elisabethin A], 252
                 [Epothilone A], 454
                 [Kendomycin], 127
                 [Laulimalide], 219

Murai, A.   [Glycinoeclepin A], 265
                 [Prelaureatin], 418

Myers, A. G.   [Neocarzinostatin], 317
                 [Stephacidin B], 77

Nagasawa, K.   [Saxitoxin], 402, 455
Nakata, T.   [Brevetoxin B], 80, 314
                 [Hemibrevetoxin B], 409
                 [Methyl sarcophytoate], 133

Nelson, S. G.   [Laulimalide], 219
Nicolaou, K. C   [Abyssomycin C], 287
                 [Amphotericin B], 418
                 [Azaspiracid-1], 79
                 [Balanol], 231
                 [BE-43472B], 728, 749
                 [Biyouyanagin A], 224
                 [Brevetoxin B], 80, 314, 418
                 [Calicheamicin γ_1], 77
                 [Colombiasin A], 51
                 [Cortistatin A], 561
                 [Endiandric acid B], 130, 723
                 [Everninomicin], 418
                 [Hemibrevetoxin B], 409
                 [Hirsutellone B], 515
                 [Maitotoxin], 41
                 [*ent*-Phomoidride B], 222, 604
                 [Platencin], 26
                 [Platensimycin], 26, 484, 491
                 [Rapamycin], 600
                 [Sarcodictyin A], 255
                 [Swinholide A], 79, 418
                 [Taxol], 77, 670
                 [Thiostrepton], 80
                 [Vancomycin], 79, 591
                 [Vannusal B], 730
                 [Zaragozic acid A], 408

Nishiyama, S.   [Breynolide sulfone], 265
Ogawa, S.   [Bengamide E], 279
                   [FR65814], 418
                   [Lycoricidine], 241, 418

Ohfune, Y.   [Domoic acid], 230
Ohno, M.   [Actinobolin], 323
                   [PS-5], 174
                   [Rhizoxin], 246

Omura, S.   [Borrelidin], 615
                   [Lactacystin], 648

Oppolzer, W.   [α-Kainic acid], 332, 707
Overman, L. A.   [Actinophyllic acid], 737
                   [Alcyonin], 40
                   [Briarellin E], 472
                   [Briarellin F], 472
                   [Gliocladin C], 372
                   [Idiospermuline], 317
                   [Manzamine A], 280, 563
                   [Ptilomycalin A], 77
                   [Pumiliotoxin 251D], 230
                   [Quadrigemine C], 133
                   [Strychnine], 77, 111

Panek, J. S.   [Lactacystin], 650
                   [Rutamycin], 252

Paquette, L. A.   [Ceroplastol I], 561
                   [Cleomeolide], 562
                   [Fomannosin], 390
                   [Punctatin A], 173, 271, 521, 707
                   [Sanglifehrin A], 252
                   [Spinosyn A], 561
                   [Taxusin], 268

Paterson, I.   [Discodermolide], 27
                   [Reidispongiolide A], 253
                   [Swinholide A], 79

Pattenden, G.   [Leinamycin], 251
                   [Phorboxazole A], 310
                   [Ulapualide A], 227

Pearlman, B. A.   [Reserpine], 542
Pearson, W. H.   [Augustamine], 239
Porco, J. A.   [Hexacyclinol], 38

Prestwich, G. D.   [Aplysistatin], 292
Rainier, J. D.   [Gambierol], 80, 418
Rapoport, H.   [Sibriosamine], 230
   [Vincamine], 305

Rawal, V   [Platencin], 26
   [Strychnine], 77

Reisman, S. E.   [Maoecrystal Z], 747
Reissig, H.-U.   [Strychnine], 77
Robichaud, J.   [Compactin], 136
Robinson, R.   [Tropinone], 723

Roush, W. R.   [Amphidinolide E], 60, 217, 618
   [Chlorotricolide], 52
   [Superstolide], 54, 455

Rychnovsky, S. D.   [Colombiasin A], 51
   [(9S)-Dihydroerythronolide A], 288
   [Hexacyclinol], 38

Saltzmann, T. N.   [Thienamycin], 217, 345
Sasaki, M.   [Gambierol], 80, 418
   [Gymnocin-A], 418

Sata, K.-i.   [Tetrodotoxin], 418
Scharf, H.-D.   [endo-Brevicomin], 220
Semmelhack, M. F.   [Cephalotaxine], 596

Shair, M. D.   [Cortistatin A], 561
   [ent-Phomoidride B], 222, 606

Shea, K. J.   [Adrenosterone], 562
Shibasaki, M.   [Lactacystin], 654
   [Strychnine], 77, 116
   [Tamiflu], 28, 138, 557
   [Xestoquinone], 60

Shiina, I.   [Octalactin], 441
Shiozaki, M.   [Trehazolin], 386

Shing, T. K. M.   [Quassin], 514
   [Samaderine Y], 477

Schmid, M.   [Vitamin E], 454
Schreiber, S. L.   [Epoxydictymene], 222
   [Motuporin], 454
   [Rapamycin], 600

Smith, A. B. III. [Discodermolide], 27, 252
[FK-506], 309
[Furaquinocin C], 286, 577
[Hitachimycin], 288, 561
[Lactacystin], 648
[Lyconadin A], 579
[Nodulisporic acid A], 281
[Penitrem D], 282
[Phorboxazole A], 312
[Spongistatin 2], 252
[Tedanolide], 455

Snapper, M. L. [Ilimaquinone], 271
Snider, B. B. [FR901483], 348
[Ptilocaulin], 561

Sorensen, E. J. [Abyssomicin C], 128
[FR182877], 53
[FR901483], 348, 350

Still, W. C. [Eucannabinolide], 463
[Monensin], 264

Stoltz, B. M [Dragmacidin F], 533
Stork, G. [Cedrol], 75
[Codeine], 715
[Cytochalasin B], 175
[Digitoxigenin], 52
[Dihydroclerodin], 514
[(9S)-Dihydroerythronolide A], 288
[Erythronolide A], 124, 275, 579
[Lycopodine], 588
[Morphine], 715
[Quinine], 621
[Reserpine], 542
[Strychnine], 77

Suzuki, T. [Laurallene], 237, 418
Takano, S. [Ajmalicine], 455
[Eburnamonine], 292
[Quebrachamine], 288

Takemoto, Y. [Halicholactone], 434
Tanaka, T. [Halicholactone], 434
Tatsuta, K. [Cochleamycin A], 579
[Nanaomycin D], 239
[Tetracycline], 417

Taylor, E. C.   [Quinine], 631
Theodorakis, E. A.   [Acanthoic acid], 524
                       [Borrelidin], 126, 615

Thomas, E. J.   [Epothilone D], 286, 580
Tomioka, K.   [Bourbonene], 288
Tomooka, K.   [Zaragozic acid A], 408
Toste, F. D.   [Fawcettimine], 548
                       [Octalactin], 441

Trost, B. M.   [Allocyathin B_2], 132
                       [Amphidinolide A], 40
                       [Aptivus], 137
                       [Aspochalasin B], 326
                       [Bryostatin 16], 292, 580, 719
                       [Hamigeran B], 61
                       [Isoquinuclidine], 279
                       [Phyllanthocin], 272
                       [Tamiflu], 28, 559

Umezawa, Y.   [Belomycin A_2], 597
Uskoković, M. R.   [Biotin], 339
                       [Quinine], 627
                       [Vitamin D intermediate], 173

Vanderwal, C. D.   [Strychnine], 77
van Tamelen, E, E.   [Colchicine], 594
Vasella, A.   [Mannostatin A], 576
Vollhardt, P. C.   [Strychnine], 77

Weinreb, S. M.   [Cephalotaxine], 596
                       [Papuamine], 274
                       [Phyllanthine], 358

Wender P. A.   [Bryostatin analogue], 86
                       [Laulimalide], 219
                       [Phorbol], 77
                       [Reserpine], 542
                       [Taxol], 77, 269, 515, 676

White, J. D.   [Boromycin], 514
                       [FK-506], 280
                       [Integerrimine], 221
                       [Integerrinecic acid], 265
                       [Latrunculin], 244
                       [Mycosporin 1], 281, 563
                       [Rhizoxin], 246

Williams, D. R.   [Apiosporamide], 563
                    [*ent*-Clavularane], 262, 515
                    [Fusicoauritone], 510
                    [Ilicicolin H], 264
                    [Milbemycin β_3], 264
                    [Myxovirescin], 418
                    [Phorboxazole A], 312
                    [Phyllanthocin], 272

Williams, R. M.   [7-Hydroxy quinine], 635
                    [Stephacidin B], 77

Wills, M.   [Halicholactone], 431
Winkler, J. D.   [Perhydrohistrionicotoxin], 232
Wipf, P.   [Tuberostemonine], 353
Wöhler, F.   [Urea], 75
Wong, C.-H.   [Tamiflu], 553

Woodward, R. B.   [Cephalosporin C], 75, 714
                    [Cortisone], 104
                    [Erythromycin A], 75, 124, 714
                    [Isoreserpine], 538
                    [Quinine], 624
                    [Reserpine], 538
                    [Strychnine], 75, 107
                    [Vitamin B$_{12}$], 75, 584, 714, 735

Yamamoto, H.   [Platensimycin], 494
Yamamoto, Y.   [Brevetoxin B], 80, 314
                    [Gambierol], 80, 418
                    [Hemibrevetoxin B], 409

Yonemitsu, O.   [Pikronolide], 418
Zakarian, A.   [Pinnatoxin A], 418
Zard, S. Z.   [Dendrobine], 514
Zhai, H.   [Absinthin], 282
Zhao, G.   [Pinnaic acid], 508
Zhu, J.   [Ecteinascidin 743], 77, 222
Ziegler, F. E.   [Calcimycin], 288
                    [9a-Desmethoxy mitomycin], 394

Chiron/Starting Material to Natural Product/Target Index

acetoveratrone
L-alanine

[strychnine], 109
[GA-13], 231
[himandrine], 741

L-allothreonine
(1R,2R)-2-amino-4-cyclohexene-1-carboxylic acid

[sibirosamine], 230
[PS-5], 174, 709

3-amino-2,3-dideoxy-D-glucose
(+)-angelicalactone

[thienamycin], 237
[furaquinocin C], 287, 577

(R)-angelicalactone
D-arabinose

[furaquinocin C], 287, 577
[hemibrevetoxin B], 409, 410
[trehazolin], 388

L-arabinose

[ambruticin S], 177, 178
[zaragozic acid A], 408

D-arabitol
D-iso-ascorbic acid

[ent-palmerolide A], 238,420
[9a-desmethoxy mitomycin A], 397
[tetrodotoxin], 419

L-asparagine
L-aspartic acid

[TMC-95-B], 372
[α-kainic acid], 334
[thienamycin], 345
[vincamine], 305

benzoic acid
1,3-butadiene
(−)-camphor

[strychnine], 118
[cortisone], 104
[taxol], 267, 515, 666, 687
[taxusin], 267
[vitamin B$_{12}$], 585

(+)-camphor
(+)-camphor-quinone

[ophiobolin C], 15, 515
[thiodextrolin], 585

Design and Strategy in Organic Synthesis: From the Chiron Approach to Catalysis, First Edition.
Stephen Hanessian, Simon Giroux, and Bradley L. Merner.
© 2013 Wiley-VCH Verlag GmbH & Co. KGaA. Published 2013 by Wiley-VCH Verlag GmbH & Co. KGaA.

camphor sulfonic acid | [quadrone], 267
(*R*)-carvone | [breynolide sulfone], 265
| [dihydroclerodin], 514
| [dorisenone C], 515
| [glycinoeclepin A], 265
| [hapalindole G], 217
| [peribysin E], 307, 515
| [picrotoxinin], 459
| [platensimycin], 485, 497
| [salsolene oxide], 515
| [sorokinianin], 515

(*S*)-carvone | [ambiguine H], 481
| [briarellin E and F], 472
| [colombiasin A], 514
| [eleutherobin], 257
| [*ent*-peribysin E], 307
| [eucannabinolide], 463
| [hapalindole U], 481
| [California red scale pheromone], 214
| [platensimycin], 488, 490, 499
| [quassin], 514
| [ryanodol], 265, 514
| [samaderine Y], 477
| [sarcodictyin A], 257
| [thapsivillosin F], 467
| [vitamin D intermediate], 173
| [welwitindolinone A], 514

L-citramalic acid | [vitamin E], 454
(*R*)-citronellal | [biyouyanagin A], 224
| [englerin A], 515
| [hirsutellone B], 515

(*S*)-citronellal | [neosymbioimine], 515
(*R*)-citronellene | [lasalosid A], 264
(*S*)-citronellene | [ambruticin S], 264
(*R*)-citronellic acid | [*ent*-octalactin A], 441
| [monensin], 264

(*R*)-citronellol | [beetle pheromone], 173, 256
| [cytochalasin B], 175
| [integerrinecic acid], 265
| [laulimalide], 220

(*S*)-citronellol

[ilicicolin H], 264
[laulimalide], 220
[littoralisone], 259
[milbemycin β₃], 264
[neosymbioimine], 265

crotonaldehyde
3-(cyanoethyl)-4-bromoanisole
(*S*)-cyclohex-3-ene-1-carboxylic acid
2-cyclohexenone

[colombiasin A], 50
[cortisone], 106
[reserpine], 542

[strychnine], 116 cyclopentadiene
[strychnine], 111

D-cysteine
L-cysteine

[griseoviridin], 176, 427
[biotin], 339
[ecteinascidin 743], 223
[latrunculin A], 244
[tantazole B], 231

2-deoxy-D-ribose
9,10-dibromocamphor

[brevetoxin B], 314, 421
[California red scale pheromone], 214
[*ent*-clavularane], 262, 515
[ophiobolin C], 15, 515

(−)-dihydrocarvone
(*R*)-epichlorohydrin

[colombiasin A], 51
[crambidine], 729
[trehazolin], 389

(*S*)-epichlorohydrin

[phomoidride B], 608
[yatekamycin], 226

D-erythronolactone

[*endo*-brevicomin], 221
[augustamine], 239, 241
[zaragozic acid C], 408

D-erythrose
D-fructose
D-galactose

[zaragozic acid C], 242
[lasalocid A], 416
[FK-506], 309
[prelaureatin], 241, 417

Garner's aldehyde
D-glucosamine

[phorboxazole A], 313
[tetracycline], 419

D-glucose

[ajmalicine], 378
[ambruticin S], 178
[FK-506], 176, 309
[fomannosin], 390
[FR65814], 417

	[ent-actinobolin], 381
	[erythronolide B], 418
	[hemibrevetoxin B], 409
	[lactacystin], 647, 664
	[lasalocid A], 417
	[laulimalide], 220
	[lycoricidine], 242, 418
	[neooxazolomycin], 236
	[pikronolide], 418
	[streptolic acid], 302
	[tetrodotoxin], 418
	[thromboxane B$_2$], 217, 237
	[trehazolamine], 385
	[trehazolin], 385, 387
	[zaragozic acid A], 406, 408

D-glutamic acid

[lactacystin], 646, 664
[longicin], 579

L-glutamic acid

[α-kainic acid], 332, 707
[cochleamycin A], 579
[dihydromevinolin], 15, 569
[longicin], 579
[megaphone], 567
[oscillarin], 361
[perhydrohistrionicotoxin], 232

D-glyceraldehyde acetonide

[borrelidin], 250, 614
[ent-decarbamoyl saxitoxin], 402
[ent-phomoidride B], 606
[indanomycin], 248
[neocarzinostatin], 317
[phorboxazole A], 312

L-glyceraldehyde acetonide

[hemibrevetoxin B], 409
[tedanolide], 455

L-glyceraldehyde diethyl ketal
D-glycerol acetonide

[amphidinolide E], 618
[saxitoxin], 398

(R)-glycidol

[ent-phomoidride B], 604
[laulimalide], 220
[spongistatin 2], 219

(S)-glycidol

[jerangolid A], 449
[migrastatin], 420
[spongistatin 2], 219

Hajos-Parrish ketone	[ceroplastol], 561
	[cortistatin], 561
	[ouabain], 561
	[punctatin A], 173, 271, 522, 739
(3*S*)-3-hydroxy butanoic acid	[griseoviridin], 428
(3*R*)-3-hydroxy butyric acid	[elaiolide], 246
(2*S*)-hydroxy butyrolactone	[iejimalide B], 287
hydroxy-β-ionone	[forskolin], 100
(2*R*,3*S*)-3-hydroxy leucine	[lactacystin], 648, 650, 663
(2*S*)-2-hydroxy-3-methyl butanedioic acid	[*ent*-octalactin A], 441
(4*R*)-4-hydroxymethyl-*γ*-butyro-lactone	[okadaic acid], 310
(4*S*)-4-hydroxymethyl-*γ*-butyrolactone	[dihydromevinolin], 569
	[ionomycin], 591
(1*R*)-1-(hydroxymethyl)-cyclohex-3-ene	[phyllanthocin], 272
(4*R*)-4-hydroxy-L-proline	[phyllanthine], 358
(*S*)-indoline-2-carboxylic acid	[virantmycin], 216
α-ionone	[forskolin], 100, 101, 706
β-ionone	[forskolin], 100, 101
L-isoleucine	[avermectin B$_{1a}$], 175
(-)-isopulegol	[artemisinin], 514
(*S*)-lactaldehyde	[BE-43472B], 728
D-lactic acid	[leinamycin], 250
	[manassantin A], 455
L-lactic acid	[ajmalicine], 455
	[himbacine], 244
	[leinamycin], 250
L-leucine	[aspochalasin B], 326
(*R*)-limonene	[fusicoauritone], 510
	[periplanone B], 266
	[pseudopteroxazole], 515
D-malic acid	[oasomycin A], 185
	[phorboxazole A], 312
	[spongistatin 2], 219
L-malic acid	[avermectinB$_{1a}$], 175
	[borrelidin], 250, 614
	[brasilenyne], 435
	[cytochalasin B], 175

[deoxyharringtonine], 223
[decarbamoyl saxitoxin], 455
[*ent*-gelsedine], 303
[epothilone A], 454
[griseoviridin], 176, 427
[halicholactone], 431
[hyperlactone C], 225
[iejimalide B], 579
[latrunculin A], 244
[lipstatin], 454
[PGF2α], 454
[phorboxazole A], 312
[saxitoxin], 455
[spongistatin 2], 219
[superstolide A], 455

D-mandelic acid   [motuporin], 454
D-mannitol   [laulimalide], 220
                         [solandelactone E], 238

D-mannose   [brevetoxin B], 314, 421
                         [everninomicin], 421
                         [hemibrevetoxin B], 409
                         [myxovirescin B], 420
                         [okadaic acid], 311
                         [trehazolin], 388

4-methoxy-toluquinone   [cortisone], 104
(*R*)-methyl adipic acid [deoxynupharidine], 265

methyl α-D-glucopyranoside   [ambruticin S], 178
                             [erythronolide A], 240

(2*S*)-2-methyl-3-hydroxy   [calcimycin], 180
propionic acid

(5*R*)-5-methyl-2-   [calcimycin], 181
cyclohexenone

(*R*)-pantolactone   [bryostatin 16], 292, 580
                             [epothilone D], 286, 580

(*S*)-pantolactone   [peloruside A], 291, 580
(*S*)-perillyl alcohol   [*ent*-kallolide B], 260, 516
(*S*)-perillaldehyde   [phyllanthocin], 266
(*R*)-α-phellandrene   [sarcodictyin A], 256
D-phenylalanine   [oscillarin], 362

L-phenylalanine

[cytochalasin B], 175
[ditryptophenaline], 372
[preussin], 230
[WIN 64821], 230

D-phenyl lactic acid
phenylhydrazine
α-pinene

[oscillarin], 362
[strychnine], 109
[aristoteline], 216
[taxol], 267, 544, 676, 687

L-pipecolic acid
(S)-N-Boc-pipecolinal
D-proline
L-proline

[FK-506], 176, 309
[rapamycin], 600
[cephalotaxine], 329
[pumiliotoxin 251D], 230

(R)-pulegone

[anisatin], 266
[aplasmomycin], 266
[artemisinin], 221, 258, 514
[boromycin], 514
[dendrobine], 514
[epoxydictymene], 222
[phomactin A], 500
[pinnaic acid], 503
[tetronasin], 514

L-pyroglutamic acid

[croomine], 336
[domoic acid], 230

(S)-pyroglutamic acid
quebrachitol

[crambidine], 728
[bengamide E], 279

(-)-quinic acid

[apiosporamide], 563
[avermectin B_{1A}], 175
[dragmacidin F], 533
[esperamicinone], 281
[FK-506], 280
[isoquinuclidine], 279
[malyngolide], 281
[manzamine A], 280, 563
[ML-236A], 563
[mycosporin I], 281, 563
[negamycin], 280
[palitantin], 279
[reserpine], 545
[Tamiflu], 27

L-rhamnose

[nanaomycin D], 239
[swinholide A], 421

D-ribonolactone — [laurallene], 237, 419
[mannostatin A], 572
[trehazolin], 387

D-ribose — [calyculin C fragment], 237
[deoxyharringtonine], 223
[9a-desmethoxy mitomycin A], 394
[pinnatoxin A], 418

(R)-Roche acid — [amphidinolide W], 185
[azaspiracid-1], 253
[calcimycin], 180
[callipeltoside C], 253
[discodermolide], 252
[erythronolide A], 252
[elisabethin A], 252
[epothilone A], 454
[FK-506], 176, 309
[iejimalide B], 253
[indanomycin], 248
[integerrimine], 221
[latrunculin A], 244
[jerangolid A], 449
[octalactin A], 438
[okadaic acid], 253, 310
[omuralide], 657
[phyllanthocin], 272
[reidispongiolide A], 253
[rhizoxin], 246
[rutamycin B], 252
[sanglifehrin A], 252
[spongistatin 2], 252
[tedanolide], 455
[zincophorin], 253

(S)-Roche acid — [amphidinolide E], 620
[calcimycin], 180
[laulimalide], 220
[octalactin A], 438
[okadaic acid], 253
[phorboxazole A], 313
[tedanolide], 455

α-santonin — [absinthin], 283
D-serine — [ent-cyclizidine], 364
[A-315675], 372

L-serine   [balanol], 231
                                            [ecteinascidin 743], 223
                                            [gliocladin C], 372
                                            [lactacystin], 642, 664
                                            [TMC-95-B], 372
                                            [taxol], 683
                                            [ulapualide A], 223

N-Boc-L-serine methyl ester   [saxitoxin], 400
(-)-shikimic acid   [Tamiflu], 27

(*R*,*R*)-tartaric acid   [FK-506], 309
                                            [idiospermulide], 318
                                            [laulimalide], 220
                                            [phyllanthocin], 272
                                            [UCS1025A], 446
                                            [zaragozic acid C], 408

(*S*,*S*)-tartaric acid   [(3*Z*)-dactomelyne], 442
                                            [amphidinolide E], 218, 619, 620

tetra-*O*-acetyl-D-glucal   [ajmalicine], 377
L-threonine   [actinobolin], 323
                                            [pactamycin], 367
                                            [salinosporamide A], 342
                                            [tantazole B], 231

tri-*O*-acetyl-D-glucal   [spongistatin 2], 219
O-trityl-(*R*)-glycidol   [phorboxazole A], 312
D-tryptophan   [tryptoquivaline], 220

L-tryptophan   [ditryptophenaline], 372
                                            [gypsetin], 372
                                            [strychnine], 113
                                            [WIN 64821], 230

L-tyrosine   [FR901483], 348
                                            [TMC-95-B], 372
                                            [tuberostemonine], 353

L-valine   [tryptoquivaline], 221
(+)-*trans*-verbenol   [dendrobine], 514

(*R*)-Wieland-Miescher ketone   [acanthoic acid], 524
                                            [penitrem D], 278
                                            [nodulisporic acid A], 278
                                            [scabronine G], 528

(*S*)-Wieland-Miescher ketone

[adrenosterone], 562
[chapecoderin A], 533
[cleomeolide], 562
[oidiolactone A], 562
[phytocassane D], 562
[stachybocin spirolactam], 527
[taxol], 562, 673

D-xylose

[amphotericin B], 421
[everninomicin], 421
[Tamiflu], 419

L-xylose

[amphotericin B], 421

Natural product/Target [Chiron]

Absinthin	[α-santonin], 282
Acanthoic acid	[(R)-Wieland-Miescher ketone], 524
Actinobolin	[L-threonine], 323
ent-Actinobolin	[D-glucose], 381
Adrenosterone	[(S)-Wieland-Miescher ketone], 562
Ajmalicine	[D-glucose], 378
	[L-lactic acid], 455
	[tetra-O-acetyl-D-glucal], 377
Ambiguine H	[(S)-carvone], 481
Ambruticin S	[(S)-citronellene], 264
	[D-glucose], 178
	[L-arabinose], 177,178
	[methyl α-D-glucopyranoside], 178
Amphidinolide E	[(S)-Roche acid], 620
	[(S,S)-tartaric acid], 218, 619, 620
	[L-glyceraldehyde diethyl ketal], 618
Amphidinolide W	[(R)-Roche acid], 185
Amphotericin B	[D-xylose], 421
Anisatin	[(R)-pulegone], 266
Apiosporamide	[(-)-quinic acid], 563
Aplasmomycin	[(R)-pulegone], 266
Aristoteline	[α-pinene], 216
Artemisinin	[(-)-isopulegol], 514
	[(R)-pulegone], 221, 258, 514
Aspochalasin B	[L-leucine], 326
Augustamine	[D-erythronolactone], 239, 241
Avermectin B$_{1A}$	[L-isoleucine], 175
	[L-malic acid], 175
	[(-)-quinic acid], 175

Design and Strategy in Organic Synthesis: From the Chiron Approach to Catalysis, First Edition.
Stephen Hanessian, Simon Giroux, and Bradley L. Merner.
© 2013 Wiley-VCH Verlag GmbH & Co. KGaA. Published 2013 by Wiley-VCH Verlag GmbH & Co. KGaA.

Azaspiracid-1   [(*R*)-Roche acid], 253
Balanol   [L-serine], 231
Biyouyanagin A   [(*R*)-citronellal], 224
BE-43472B   [(*S*)-lactaldehyde], 728
Beetle pheromone   [(*R*)-citronellol], 173, 256
Bengamide E   [quebrachitol], 279
Biotin   [L-cysteine], 339
Boromycin   [(*R*)-pulegone], 514

Borrelidin   [D-glyceraldehyde], 250, 614
                                [L-malic acid], 250, 614

Brasilenyne   [L-malic acid], 435
Brevetoxin B   [2-deoxy-D-ribose], 314, 421
                                [D-mannose], 314, 421

Breynolide sulfone   [(*R*)-carvone], 265
Briarellin E   [(*S*)-carvone], 472
Briarellin F   [(*S*)-carvone], 472
Bryostatin 16   [(*R*)-pantolactone], 292, 580

Calcimycin   [(5*R*)-5-methyl-2-cyclohexenone], 181
                                [(2*S*)-2-methyl-3-hydroxy propionic acid], 180
                                [(*S*)-Roche acid], 180
                                [(*R*)-Roche acid], 180

California red scale   [9,10-dibromocamphor], 214
pheromone   [(*S*)-carvone], 214

Callipeltoside C   [(*R*)-Roche acid], 253
ent-Clavularane   [9,10-dibromocamphor], 262, 515
Calyculin C   [D-ribose], 237
Cephalotaxine   [D-proline], 329
Ceroplastol   [Hajos-Parrish ketone], 561
Chapecoderin A   [(*S*)-Wieland-Miescher ketone], 533
Cleomeolide   [(*S*)-Wieland-Miescher ketone], 562
Cochleamycin A   [L-glutamic acid], 579

Colombiasin A   [crotonaldehyde], 50
                                [(-)-dihydrocarvone], 51
                                [(*S*)-carvone], 514

Cortisone   [1,3-butadiene], 104
                                [4-methoxy-toluquinone], 104
                                [3-(cyanoethyl)-4-bromoanisole], 106

Cortistatin A   [Hajos-Parrish ketone], 561

Crambidine   [(*R*)-epichlohydrin], 729
     [(*S*)-pyrogluatmic acid], 728

Croomine   [L-pyroglutamic acid], 336
ent-Cyclizidine   [D-serine], 364
Cytochalasin B   [(*R*)-citronellol], 175
     [L-malic acid], 175
     [L-phenylalanine], 175

(3*Z*)-Dactomelyne   [(*S,S*)-tartaric acid], 442
Decarbamoyl saxitoxin   [L-malic acid], 455
Dendrobine   [(*R*)-pulegone], 514
     [(+)-*trans*-verbenol], 514

Deoxyharringtonine   [L-malic acid], 223
     [D-ribose], 223

Deoxynupharidine   [(*R*)-methyl adipic acid], 265
9a-desmethoxy mitomycin A   [D-*iso*-ascorbic acid], 397
     [D-ribose], 494

Dihydroclerodin   [(*R*)-carvone], 514
Dihydromevinolin   [L-glutamic acid] 15, 569
     [(4*S*)-4-hydroxymethyl-γ-butyrolactone], 569

Discodermolide   [(*R*)-Roche acid], 252
Ditryptophenaline   [L-tryptophan], 372
     [L-phenylalanine], 372

Domoic acid   [L-pyroglutamic acid], 230
Dorisenone C   [(*R*)-carvone], 515
Dragmacidin F   [(-)-quinic acid], 533
Ecteinascidin 743   [L-cysteine], 223
     [L-serine], 223

Englerin A   [(*R*)-citronellal], 515
Epothilone A   [L-malic acid], 454
     [(*R*)-Roche acid], 454

Epothilone D   [(*R*)-pantolactone], 286, 580
Epoxydictymene   [(*R*)-pulegone], 222
Elaiophylin aglycone   [(3*R*)-3-hydroxy butyric acid], 246
Elisabethin A   [(*R*)-Roche acid], 252
Eleutherobin   [(*S*)-carvone], 255
Erythronolide A   [methyl α-D-glucopyranoside], 240
     [(*R*)-Roche acid], 252

Esperamicinone   [(-)-quinic acid], 281

Eucannabinolide   [(S)-carvone], 463
Everninomicin   [D-mannose], 421
    [D-xylose], 421

FK-506   [D-galactose], 309
    [D-glucose], 176, 309
    [(-)-quinic acid], 280
    [(R)-Roche acid], 176, 309
    [(R,R)-tartaric acid], 309

Fommanosin   [D-glucose], 392
Forskolin   [α-ionone], 100, 101, 706
    [β-ionone], 100, 101
    [hydroxy-α-ionone], 100

FR65814   [D-glucose], 417
Furaquinocin C   [(R)-angelicalactone], 287, 577
Fusicoauritone   [(R)-limonene], 510
Gliocladin C   [L-serine], 372
Glycinoeclepin A   [(R)-carvone], 265

Griseoviridin   [D-cysteine], 427
    [(3S)-3-hydroxybutanoic acid], 428
    [L-malic acid], 176, 427

Gypsetin   [L-tryptophan], 372
Halicholactone   [L-malic acid], 431
Hapalindole G   [(R)-carvone], 217
Hapalindole Q   [(S)-carvone], 481

Hemibrevetoxin B   [D-arabinose], 409, 410, 414
    [D-glucose], 409
    [D-mannose], 409
    [L-glyceraldehyde acetonide], 409

Himandrine   [L-alanine], 741
Himbacine   [L-lactic acid], 244
Hirsutellone B   [(R)-citronellal], 515
Hyperlactone C   [L-malic acid], 225
Idiospermulide   [(R,R)-tartaric acid], 318

Iejimalide B   [(2S)-hydroxy butyrolactone], 287
    [L-malic acid], 579
    [(R)-Roche acid], 253

Ilicicolin H   [(S)-citronellol], 264
Indanomycin   [D-glyceraldehyde acetonide], 249
    [(R)-Roche acid], 248

Integerrimine   [(R)-Roche acid], 221
Integerrinecic acid   [(R)-citronellol], 265
Ionomycin   [(4S)-4-hydroxymethyl-γ-butyrolactone], 591
Isoquinuclidine   [(-)-quinic acid], 279

Jerangolid A   [(S)-glycidol], 449
                               [(R)-Roche acid], 449

ent-Kallolide B   [(S)-perillyl alcohol], 260, 516
α-Kainic acid   [L-aspartic acid], 334
                               [L-glutamic acid], 332, 707

Lactacystin   [(2R,3S)-3-hydroxyleucine], 648, 650, 663
                               [D-glucose], 647, 664
                               [D-glutamic acid], 646, 664
                               [L-serine], 642, 664

Lasalocid A   [D-fructose], 417
                               [D-glucose], 417

Latrunculin A   [L-cysteine], 244
                               [L-malic acid], 244
                               [(R)-Roche acid], 244

Laulimalide   [(R)-citronellol], 220
                               [(S)-citronellol], 220
                               [D-glucose], 220
                               [(R)-glycidol], 220
                               [(S)-Roche acid], 220
                               [(R,R)-tartaric acid], 220

Laurallene   [D-ribonolactone], 237, 419
Leinamycin   [D-lactic acid], 249
                               [L-lactic acid], 249

Lipstatin   [L-malic acid], 454
Littoralisone   [(S)-citronellol], 259

Longicin   [D-glutamic acid], 579
                               [L-glutamic acid], 579

Lycoricidine   [D-glucose], 242, 418
Malyngolide   [(-)-quinic acid], 281
Mannostatin A   [D-ribonolactone], 572
Manzamine A   [(-)-quinic acid], 280, 563
Megaphone   [L-glutamic acid], 567
Migrastatin   [(S)-glycidol], 420
Milbemycin β₃   [(S)-citronellol], 264
ML-236A   [(-)-quinic acid], 563

Motuporin   [D-mandelic acid], 454
Mycosporin I   [(-)-quinic acid], 281, 563
Myxovirescin B   [D-mannose], 420
Nanaomycin D   [L-rhamanose], 239
Negamycin   [(-)-quinic acid], 280
Neosymbioimine   [(S)-citronellol], 265
Neocarzinostatin   [D-glyceraldehyde], 317
Neoxazolomycin   [D-glucose], 236
Oasomycin A   [D-malic acid], 185
Octalactin A   [(R)-Roche acid], 438
Oidiolactone A   [(S)-Wieland-Miescher ketone], 562
Omuralide   [(R)-Roche acid], 657

Ophiobolin C   [(+)-camphor], 15, 515
                                 [9,10-dibromocamphor], 15, 515

Okadaic acid C   [(4R)-4-hydroxymethyl-γ-butyrolactone], 312
                                 [D-mannose], 311
                                 [(R)-Roche acid], 253, 310

Oscillarin   [D-phenylalanine], 362
                                 [D-phenyllactic acid], 362
                                 [L-glutamic acid], 361

Pactamycin   [L-threonine], 367
Palitatin   [(-)-quinic acid], 279
ent-Palmerolide A   [D-arabitol], 238, 420
Peloruside A   [(S)-pantolactone], 291, 580
Penitrem D   [(R)-Wieland-Miescher ketone], 562
Perhydrohistrionicotoxin   [L-glutamic acid], 232
Peribysin E   [(R)-carvone], 307, 515
Periplanone B   [(R)-limonene], 266
PGF-2α   [L-malic acid], 454
Phomactin A   [(R)-pulegone], 500
Phomoidride B   [(S)-epichlorohydrin], 608
ent-Phomoidride B   [D-glyceraldehyde], 606

Phorboxazole A   [Garner's aldehyde], 313
                                 [D-glyceraldehyde], 312
                                 [D-malic acid], 312
                                 [L-malic acid], 312
                                 [(S)-Roche acid], 438
                                 [O-trityl-(R)-glycidol], 312

Phyllanthine   [(4R)-4-hydroxy-L-proline], 358

Phyllanthocin   [(1R)-1-(hydroxymethyl)-cyclohex-3-ene], 272
                                 [(S)-perillaldehyde], 266

	[(*R*)-Roche acid], 272
	[(*R,R*)-tartaric acid], 272
Phytocassane D	[(*S*)-Wieland-Miescher ketone], 562
Picrotoxinin	[(*R*)-carvone], 459
Pikronolide	[D-glucose], 418
Pinnaic acid	[(*R*)-pulegone], 503
Pinnatoxin A	[D-ribose], 418
Platensimycin	[(*R*)-carvone], 485, 497
	[(*S*)-carvone], 488, 490, 499
PS-5	[(1*R*,2*R*)-2-amino-4-cyclohexene-1-carboxylic acid], 174, 703
Pseudopteroxazole	[(*R*)-limonene], 515
Punctatin A	[Hajos-Parrish ketone], 173, 272, 522, 739
Quassin	[(*S*)-carvone], 514
Reidispongiolide A	[(*R*)-Roche acid], 253
Reserpine	[(*S*)-cyclohex-3-ene-1-carboxylic acid], 542
	[(-)-quinic acid], 545
Rhizoxin	[(*R*)-Roche acid], 246
Rutamycin B	[(*R*)-Roche acid], 252
Ryanodol	[(*S*)-carvone], 265, 514
Salinosporamide A	[L-threonine], 342
Salsolene oxide	[(*R*)-carvone], 515
Samaderine Y	[(*S*)-carvone], 477
Sanglifehrin A	[(*R*)-Roche acid], 252
Sarcodictyin A	[(*S*)-carvone], 255
	[(*R*)-α-phellandrene], 256
Saxitoxin	[L-malic acid], 455
	[*N*-Boc-L-serine methyl ester], 400
Scabronine G	[(*R*)-Wieland-Miescher ketone], 529
Sibirosamine	[L-allothreonine], 230
Solandelactone E	[D-mannitol], 238
Sorokinianin	[(*R*)-carvone], 515
Spongistatin 2	[tri-*O*-acetyl-D-glucal], 219
	[(*R*)-glycidol], 219
	[(*R*)-glycidol], 219
	[D-malic acid], 219
	[L-malic acid], 219
	[(*R*)-Roche acid], 252

Stachybocin spirolactam [(S)-Wieland-Miescher ketone], 527
Streptolic acid [D-glucose], 302

Strychine [acetoveratrone], 109
[benzoic acid], 118
[2-cyclohexenone], 116
[cyclopentadiene], 111
[phenylhydrazine], 109
[L-tryptophan], 113

Superstolide A [L-malic acid], 455
Tamiflu [(-)-quinic acid], 27
[(-)-shikimic acid], 27
[D-xylose], 419

Taxol [(-)-camphor], 267, 515, 666, 687
[L-serine], 683

Taxusin [(-)-camphor], 267
Tedanolide [L-glyceraldehyde], 455
[(R)-Roche acid], 455
[(S)-Roche acid], 455

Tetracycline [D-glucosamine], 417
Tetrodotoxin [D-glucose], 418
Tetronazin [(R)-pulegone], 514
Thapsivillosin F [(S)-carvone], 467

Thienamycin [L-aspartic acid], 345
[3-amino-2,3-dideoxy-D-glucose], 233

Thiodextrolin [(+)-camphorquinone], 585
Thromboxane B₂ [D-glucose], 217, 237

TMC-95-B [L-asparagine], 372
[L-serine], 372
[L-tyrosine], 372

Trehazolamine [D-glucose], 385
Trehazolin [(R)-epichlorohydrin], 389
[D-glucose], 385, 387
[D-ribonolactone], 387

Tryptoquivaline [D-tryptophan], 220
[L-valine], 221

Tuberostemonine [L-tyrosine], 353
UCS1025A [(R,R)-tartaric acid], 446
Ulapualide A [L-serine], 227

Vincamine   [L-aspartic acid], 305
Virantmycin   [(S)-indoline-2-carboxylic acid], 216
Vitamin B$_{12}$ [(-)-camphor], 585
Vitamin D intermediate   [(S)-carvone], 173
Vitamin E   [L-citramalic acid], 454
Weltwitindolinone A   [(S)-carvone], 514

WIN 64821   [L-phenylalanine], 230
                                 [L-tryptophan], 230

Yatekamycin   [(S)-epichlorohydrin], 226
Zaragozic acid A   [L-arabinose], 408
                                 [D-glucose], 406, 408
                                 [(R,R)-tartaric acid], 408

Zaragozic acid C   [D-erythronolactone], 408
                                 [D-erythrose], 242

Zincophorin   [(R)-Roche acid], 253

Key (Named) Reactions Index

Appel reaction	549
Arndt-Eistert homologation	339, 552, 605, 607, 609
azonia-Prins reaction	130, 362
oxonia-Prins-pinacol reaction	473
oxonia-Prins reaction	130, 362
Barton decarboxylation	492
Barton-McCombie deoxygenation	571, 646
Bayer-Villiger oxidation	307, 464, 488, 495, 543, 571, 610
Beckmann rearrangement	504
Birch reduction	680
Bischler-Napieralski reaction	378
Brown allyl/crotylation	311, 436, 649, 660
CBS reduction	59, 313
kinetic resolution	606
Chan rearrangement	667
Claisen condensation	233, 680
Claisen rearrangement(s)	115, 327
[2,3]-(Claisen) sigmatropic rearrangement	634
Eschenmoser-Claisen rearrangement	356
Ireland (ester enolate)-Claisen rearrangement	258, 303, 335, 417
Johnson (ortho ester)-Claisen rearrangement	257, 511, 619, 623
Overman rearrangement	647
Collins oxidation	278, 552
Cope rearrangements	
anionic oxy-Cope rearrangement	464, 466, 606
Corey-Chaykovsky reaction	674
Corey-Fuchs reaction	309
Corey-Winter olefin synthesis	684
Curtius rearrangement	403, 558

Design and Strategy in Organic Synthesis: From the Chiron Approach to Catalysis, First Edition.
Stephen Hanessian, Simon Giroux, and Bradley L. Merner.
© 2013 Wiley-VCH Verlag GmbH & Co. KGaA. Published 2013 by Wiley-VCH Verlag GmbH & Co. KGaA.

Diels-Alder reactions
intermolecular   308, 327, 525, 540, 578, 594, 599,
728, 743

intramolecular   244, 324, 478, 489, 570, 599,
605, 608, 671, 673, 715, 741

organocatalytic (intramolecular)   128, 129, 746
organocatalytic (intermolecular)   122, 136, 139, 558, 559
inverse electron demand (intramolecular)   130, 226
4π electrocyclic ring opening/   107
intramolecular Diels-Alder cascade
hetero-Diels-Alder reactions   128, 311, 313, 359, 576
asymmetric (chiral auxiliary)   179
asymmetric (catalytic)   179
Dieckmann condensation   110, 606, 625, 645, 668
Evans aldol reaction   185
Favorskii rearrangement   468
Ferrier carbocyclization   382
Ferrier glycosidation   378
Friedel-Crafts reaction   240, 330
Grignard reaction   217, 274, 365, 368, 406, 439,
466, 468, 478, 486, 546, 644, 655

Grob fragmentation   215, 269, 667, 669, 677
Heck reaction
reductive Heck cyclization   115
Heck cyclization   120, 121, 133, 242, 304, 319, 482,
610, 675

Jeffery-Heck reaction   121, 122
Henry reaction   571
Hofmann elimination   625
Horner-Wadsworth-Emmons reaction   185, 259, 311, 489, 634, 746
Jacobsen epoxidation
hydrolytic kinetic resolution (HKR)   148
Jones oxidation   277, 333, 335, 359, 439, 464,
552, 609, 646, 647, 649, 657

Julia olefination   309, 311, 511, 603
Julia-Kocieńsky olefination   615, 620
Kita lactonization   620
Kumada coupling   493, 681
Ley oxidation   468, 469, 478, 632, 672
Löffler-Freytag rearrangement   628
Luche reduction   368, 452, 478, 501, 522, 681, 746

Mannich reaction
organocatalytic   135
Mannich reaction/[3,3]-sigmatropic   114
rearrangement
aza-Cope/Mannich cyclization   112, 738
vinylogous   337
McMurry/pinacol reaction   672
Meerwein-Ponndorf-Verley oxidation   540
Michael reaction   313, 594
catalytic asymmetric   116, 259, 497, 632, 652
organocatalytic   560
vinylogous   136
Mitsunobu reaction   378, 395, 428, 452, 486, 501,
    554, 556, 558, 622, 661

double Mitsunobu amination sequence   119
Moffatt oxidation   647, 649
Morita-Baylis-Hillman reaction   343, 636
Mukaiyama aldol reaction   368, 642, 683
Mukaiyama macrolactonization   602
Nazarov cyclization   511, 531
Neber rearrangement   536, 537
Negishi coupling   412, 531
Norrish type II cleavage   389, 522
Norrish type I cleavage   482
Noyori reduction   428, 497, 509, 616
dynamic kinetic resolution (DKR)   148
Nozaki-Hiyama-Kishi reaction   313, 315, 432, 435, 440, 474
Parikh-Doering oxidation   444, 504, 568
Petasis olefination   488
Peterson olefination   412, 444, 680
Pictet-Spengler reaction   543, 546
Pinnick oxidation   259, 406, 407, 428, 606,
    642, 649, 651, 652, 663

Pummerer rearrangement   599
Reformatsky reaction   594
boron-mediated Reformatsky-type reaction   447
intramolecular SmI$_2$-mediated   616
Robinson annulation   495
organocatalytic   549
Roush allyl/crotylboration   619
Rubottom oxidation   668, 674
Saegusa oxidation   117, 277, 522
Sakurai allylation reaction   179, 507, 552, 616
Sandmeyer reaction   592
Shapiro reaction   671

Stille reaction	117, 319, 601
Still-Gennari olefination	313, 615
Strecker reaction	543, 588
catalytic	655
Suzuki reaction	307, 366, 536, 620, 632, 741
B-alkyl Suzuki-Miyaura reaction	502, 508, 549
Schmidt glycosidation reaction	317
Sharpless asymmetric aminohydroxylation	651
Sharpless asymmetric dihydroxylation (SAD)	185, 365, 446, 469, 620, 632, 634, 680
Sharpless asymmetric epoxidation (SAE)	133, 147, 303, 316, 317, 415, 614
Songashira reaction	613
Swern oxidation	261, 359, 382, 389, 392, 393, 406, 412, 413, 428, 432, 434, 451, 452, 488, 489, 511, 570, 571, 601, 610, 614, 622, 642, 667, 680, 738
Tamao-Fleming oxidation	343, 543, 652, 655
Tsuji-Trost reaction	636
Ullmann reaction	592, 593
Vilsmeyer-Haack reaction	259
Wacker oxidation	684
Wagner-Meerwein reaction	268
Weiss-Cook reaction	723
Wittig reaction	285, 315, 316, 359, 392, 412, 432, 435, 439, 468, 489, 511, 525, 552, 590, 622, 630, 631, 647, 675, 677
Wittig-Still reaction ([2.3]-rearrangement)	522
Yamaguchi esterification	619
Yamaguchi lactonization	432, 615
Yamamoto carbocyclization	616